Glossary of Meteorology

Second Edition

Todd S. Glickman
Managing Editor

American Meteorological Society
Boston, Massachusetts, U.S.A.

2000

Published by the American Meteorological Society, 45 Beacon Street, Boston, MA 02108-3693 U.S.A.

Ronald D. McPherson, Executive Director
Keith L. Seitter, Deputy Executive Director
Richard E. Hallgren, Executive Director Emeritus
Kenneth C. Spengler, Executive Director Emeritus

Funding received under Grant Number ATM-9402990 obtained through the National Science Foundation, with support from the Environmental Protection Agency, the National Oceanic and Atmospheric Administration, the U.S. Air Force, the U.S. Navy, and the Department of Energy.

Electronic composition and printing by Allen Press.

Dedicated to the memory of

Werner A. Baum

and

Charles F. Talman

Preface to the Second Edition

The first edition of the *Glossary of Meteorology* was published by the American Meteorological Society (AMS) in 1959. In the four decades between its publication and this second edition, entire discipline areas of the field of meteorology have been born and flourished. Despite this, the first edition has sold over 10,000 copies through five printings. It is a tribute to the editors of the first edition that it has withstood the tests of time and continued to be among the leading reference sources in meteorology and related sciences for four decades.

The beginning of the process to publish a second edition can be traced to two documents, "How Can We Do Another *Glossary of Meteorology*: A Discussion Paper," and "Preparatory Document for Meeting on *Glossary of Meteorology*," authored by AMS past-president Werner A. Baum in the early 1990s. These papers were discussed and reviewed in 1992 by the ad hoc Committee on the New *Glossary of Meteorology*, formed by the Council of the AMS. Over a two-day period, this committee laid the foundation for what would become the second edition. In addition to developing the overall philosophy of the second edition, the ad hoc committee developed an outline for the management structure of the enormous task that lay ahead.

In 1994, AMS submitted a proposal to the National Science Foundation (NSF) for a grant supporting the publication of an update of the *Glossary*. Later that year, the NSF awarded a grant to the AMS, with funding from five U.S. government agencies. In addition, the AMS committed to its own matching funds, the source of which included interest generated from funds on reserve. Central to the project, as in the first edition, would be drawing upon the volunteer efforts of many individuals from every facet of the Society.

The effort began in the fall of 1994 with the appointment of the *Glossary of Meteorology* Advisory Board. The immediate task of the board, under the chairmanship of Ronald Taylor, was to take as its starting point the output of the ad hoc committee and develop a detailed plan for the implementation of the project. During the early planning for the second edition, lengthy discussions were held on the intended coverage of the volume. As the AMS serves many related disciplines, how should the *Glossary* cover these? Does it make more sense to produce several independent glossaries, each for a subarea of meteorology? Should it have a different title, such as the *Glossary of Atmospheric Sciences*? While the field would be covered fully, terms from related sciences would be included if they directly impacted meteorology. This would require prudent addition of terms from areas such as physical oceanography and hydrology, fields in themselves that would not be covered in their entirety. Overall guidance was that if terms were likely to appear in AMS journals or other literature, they should be included. With meteorology becoming more integrated with other sciences, it was determined that the volume would be most useful if it were not divided into several subject areas. It was finally agreed that this second edition, as its name implies, is intended to be a glossary of the science of meteorology, as defined in the first edition:

> The study dealing with the phenomena of the atmosphere. This includes not only the physics, chemistry, and dynamics of the atmosphere, but is extended to include may of the direct effects of the atmosphere upon the earth's surface, the oceans, and life in general.

Since the publication of the first edition in 1959, the science of meteorology has evolved considerably. A number of new specialty areas have developed, and others have undergone radical changes since then (e.g. atmospheric chemistry, satellite meteorology, numerical weather prediction). The substantial growth in the number of terms for the second edition reflects these changes, rather than a change of scope or philosophy of the book.

Much of the success of the first edition has been ascribed to the fact that the science of meteorology is thoroughly covered and that terms from associated fields are also included. Yet, to ensure the focus of the book, these terms from associated fields were carefully screened prior to inclusion. The second edition encompasses the same philosophy: Terms from hydrology, oceanography, etc., are included if their inclusion is important to the understanding of meteorology as broadly defined above. The second edition follows the wise philosophy of the editor of the first edition, as stated in its preface:

> The *Glossary of Meteorology* purports to define every important meteorological term likely to be found in the literature today. It attempts to present definitions that are understandable to the generalist and yet palatable to the specialist; and it intends to be a reference book that satisfies its user in a minimum of his time.

Finally, the *Glossary* provides an excellent opportunity to collect and define terms that are either obscure or very local in nature.

The second edition maintains, in many respects, the format of the first edition. In most cases, the first sentence of the definition has "stand-alone" properties; the remainder of the definition is elaboration. Defining by pure mathematics is not permitted. A worded, physical definition must be given; it may then be elaborated by one or more equations or mathematical terms to follow-up the word-based definition. References are used sparingly within a definition, and only if they clarify the use and practice of a term. In addition, references provided should be readily accessible to most individuals. Popular slang and folk terms are included in the *Glossary*. They are colorful and add value. In addition, many are regional in nature, but show up in the literature and may be unfamiliar to those from other regions. Acronyms are included if they are part of the "scientific vocabulary." Mathematical and statistical terms are included that are widely used in the science of meteorology. As a general rule, if a mathematical or statistical term is utilized as part of another definition in the *Glossary*, it will itself be included. The *Glossary of Meteorology* consistently uses the International System of Units (Le Système International d'Unités, SI), except in special circumstances where the preponderance of uses of a term requires otherwise. Abbreviations are included as terms to be defined if the abbreviations are popularly used in the science of meteorology. They may be used as a part of the definition of a term, as long as they are not a critical part of the definition, and more clarity would be provided by spelling the abbreviated word in its entirety. Abbreviations so used conform to the standards published in the Authors' Guide (Bulletin of the American Meteorological Society, Vol. 76, No. 8, August 1995). Some terms from the first edition that were of a historical nature, or deemed to be very obscure or obsolete, have been omitted from the second edition upon the recommendation of our editors. Readers are encouraged to consult the first edition to gain a historical perspective on these terms.

Terms in this volume have been alphabetized such that those beginning with numerals or Greek letters are inserted as if they were spelled out in Roman letters. Alphabetization of compound and hyphenated words, and other style rules follow the guidance of the *Chicago Manual of Style* and the AMS Authors' Guide. Terms displayed in bold type (in the print edition) and hyperlinks (in the electronic edition) refer to other terms that are themselves defined in the *Glossary*. We have attempted to be as comprehensive as possible, though some versions of terms (plurals, modified terms, unusual combinations etc.) may not be so linked, nor are some common words that are not used in the scientific manner as they are defined in the *Glossary*.

As of the date of its first printing, the print edition of the *Glossary* is as current as our esteemed subject area editors' work permitted. Recognizing that both the science of meteorology and publishing technology is changing rapidly, the *Glossary of Meteorology* is now a "living document." It is being simultaneously published in electronic form and will be updated on a regular basis to ensure that its contents remain current.

The Council of the AMS, upon recommendation of the *Glossary* Advisory Board and concurrence of the Managing Editor, voted unanimously to dedicate this second edition to two individuals: Charles F. Talman, whose exhaustive work as librarian of the U.S. Weather Bureau compiling terms served as a basis for the U.S. Department of Commerce Weather Glossary (1946), and by extension, the AMS *Glossary of Meteorology* (1959); and to Werner A. Baum, whose leadership helped shape this project from its inception through its final stages of preparation.

Todd S. Glickman
Cambridge, Massachusetts
June, 2000

Preface to the First Edition

The *Glossary of Meteorology* purports to define every important meteorological term likely to be found in the literature today. It attempts to present definitions that are understandable to the generalist and yet palatable to the specialist; and it intends to be a reference book that satisfies its user in a minimum of his time.

The collecting of terms appearing in the *Glossary* began in 1952 with an exhaustive survey of the literature of the preceding five years. This list, plus previous collections (especially Weather Glossary, U.S. Department of Commerce, Weather Bureau, 1946; and Meteorological Glossary, His Majesty's Stationery Office, MO 225 ii, 1939 reprinted 1953), provided the platform on which the final product was assembled. In the subsequent years many of the original terms were deleted and many more added, resulting in a final count of 7247 entries. By far the majority of these terms are from the mainstream of meteorological nomenclature: synoptic, dynamic, and physical meteorology, theoretical and applied; climatology; and meteorological instrumentation. Most of the remainder are drawn from sister disciplines such as hydrology, oceanography, geomagnetism and astrophysics; some come from the basic sciences of physics and chemistry; some from applicable portions of mathematics, statistics, and electronics; and others from the folk language of weather lore through many ages. It may be argued, favorably or unfavorably, that this book is as much a glossary "for meteorologists" as "of meteorology."

Each definition represents the efforts of three or more individuals, at least two of whom are specialists in the subject area involved. Minimally, each definition passed through the chain of production and criticism consisting of subject editor, editor, reviewer, editor, technical proofreader, editor. Usually, more steps were involved. A few ground rules were set down in order to unify this production effort: mathematics would not be used as a substitute for verbal explanation; the first one or two sentences should be able to stand alone as a basic definition; the definitions should be understandable to an undergraduate in a technical college yet contain sufficient pertinent detail to satisfy the working specialist. It is a glossary of United States usage, primarily; but definitions and terms used mainly in other countries are frequently cited. One can see from the above that this *Glossary* was not merely "compiled"—it was carefully and critically assembled as the cooperative venture of many contributors.

Any reference book must provide adequate information with a minimum of effort and time required of its user. This glossary has been edited with that principle uppermost in mind. It consists of an alphabetical list, as complete as possible, of words and short phrases as they naturally appear in the literature and conversation of meteorology. Most of these terms are followed immediately by their definition. Others are followed by references to preferred synonyms. Occasionally the reader is referred to a related term whose definition contains an explanation of the term in question. With few exceptions, the "encyclopedic" approach (i.e., extensive information under a single broad heading) is avoided. However, if a reader wishes to review the related terms of a given subject, he should be able to accomplish this by making use of the internal cross-references (in boldface type) that are a feature of this volume. A few other features and conventions should be mentioned. Where synonyms are involved, the definition is found under the preferred term. (The preference is sometimes arbitrary.) Synonyms, abbreviations, and symbols appear in parentheses preceding the definition. An italicized word in the body of a definition is defined only in that context. Where a term has more than one definition, each definition is numbered and the first definition is the most common and/or meteorologically, the most important. Nearly all of the terms are singular nouns, verbs and adjectives have been kept at a minimum. Abbreviations and contractions which have become adopted as terms in themselves are included, but most common abbreviations (of units, for example) are not.

In the borderline subject areas, the lines of term inclusion/exclusion were drawn somewhat arbitrarily. If any general tendency is apparent, it is toward over-inclusion. In general, terms that are purely mathematical, physical, chemical, electronic, biological, etc., are not included. Some exceptions have been made in the case of very frequently used terms, where a brief description is given to provide, at least, orientation as to the area of reference of the term.

It is recognized that literature references are incomplete within this volume. Only those references are listed that were initially submitted by the many contributors. However, readers are referred to documentation of the rapidly swelling volume of the world's important literature in meteorology and related fields that is provided by Meteorological Abstracts and Bibliography. English abstracts of literature in 30 languages from 100 countries are presented in this monthly periodical. This journal also includes special bibliographies, author, subject and geographical indices of source material in meteorology, oceanography, hydrology, geomagnetism, cosmic rays, etc. With assistance from United States government agencies concerned with meteorological data and research, the American Meteorological Society will have prepared and published 110 special bibliographies and more than 60,000 abstracts on completion of the tenth volume in 1959.

As of this date of publication, the *Glossary of Meteorology* is respectfully up-to-the-minute. With every month, however, the growth of our science expands its vocabulary. It is hoped that this volume provides a sound basis for a continuing record of the meteorologists' language, just as it documents today's terminology and places it within the grasp of every worker in the field.

Ralph E. Huschke
Boston, Massachusetts
1959

Acknowledgments

The development of the second edition of the *Glossary of Meteorology* is the combined efforts of hundreds of individuals, the majority of whom have volunteered their time and wisdom over a multiyear period in support of this project.

The editor was extremely fortunate to be supported by a group of distinguished individuals, who composed the AMS Glossary Advisory Board. This board did more than set policy for the publication; it set the tone for the entire project. The board was chaired by Ronald C. Taylor, who provided superb leadership and guidance throughout the project. Board members William R. Bandeen, the late Werner A. Baum (who authored the original "planning documents" on behalf of the AMS Council), Russell R. Dickerson, the late Ralph E. Huschke (editor of the first edition of the *Glossary of Meteorology*), John D. Stackpole, and Ferris Webster each individually contributed in significant ways, in addition to the consensus decisions of the board.

The subject area editors compose the Editorial Board for the *Glossary of Meteorology*. These individuals were not only responsible for editing the terms in their respective areas of expertise, but for identifying, supervising, and motivating the writers and reviewers who developed the definitions. Members of the Editorial Board worked tirelessly on this project; were flooded with email, faxes, and phone calls; and successfully worked with AMS-provided software on many platforms. The editor wishes to express his thanks and admiration to these individuals for their efforts, accomplished above and beyond their other personal and professional obligations.

The nitty-gritty work involved with reviewing definitions from the first edition, and authoring and/or reviewing new definitions, fell to the large group of volunteers named below. The editor is grateful to the following writers and reviewers for their contributions: Richard Alley, Lee Anderson, Mary P. Anderson, Phil S. Anderson, Wayne M. Angevine, Phil Arkin, Dennis Atkinson, Peter G. Baines, Victor Baker, Jan Barkmeijer, Roger Bales, Peter Bannon, Roger Barry, Monika Bartelmann, Mary Barth, Randy Bassett, David Baumhefner, Kenneth Beard, William Beasley, Alfred J. Bedard Jr., William Benjey, Charles Bentley, Kenneth Bencala, Keith Beven, Niek-Jan Bink, Craig Bishop, Peter E. Black, Thomas Black, Blaine Blade, Gianluca Borzelli, Lance Bosart, Michael L. Branick, Lee Branscome, Grant Branstator, Marcia Branstetter, Raphael Bras, Bart Brashers, Stephen Brenner, David Bright, Marx Brook, John Brown, Robert A. Brown, Rodger A. Brown, Gerrit Burgers, Stephen J. Burges, Donald Burgess, Gregory Byrd, Mary Cairns, Alejandro Camerlengo, James Campbell, Kevin Carey, David A. Carter, Guy Chammas, Stanley A. Changnon Jr., Dean Churchill, Martyn Clark, Eugene Clothiaux, Samuel Colbeck, Bradley R. Colman, Martha Conklin, Dinshaw Contractor, Kerry H. Cook, Ellen J. Cooter, Steve Corfidi, Robert Davies-Jones, Donald Davis, Kenneth J. Davis, Owen Davis, Anthony DelGenio, Thomas L. Delworth, John Derber, Johanna Devereaux, Karen Devlin, Gregory A. Devoir, Leo Donner, Charles Doswell III, Thomas Dunne, Dale R. Durran, Ellsworth G. Dutton, John A. Dutton, Wesley Ebisuzaki, Brian K. Eder, David B. Enfield, Ted Engman, Dara Entekhabi, Edward S. Epstein, James Famiglietti, Graham Feingold, Brian Fiedler, Peter Finkelstein, William Finnegan, Bryce Finnerty, George Fisher, Gregory M. Flato, Robert G. Fleagle, Thomas Foken, Ralph Foster, James Franklin, Danny Fread, Paul Frenzen, Alan Fried, Michael Fritsch, Inez Fung, Frank Gallagher, Auroop Ganguly, Matthew E. Garcia, Steve Garner, Nicole Gasparini, Ron Gelaro, Marvin Geller, Harry Glahn, Anand Gnanadesikan, Joseph Golden, David C. Goodrich, Charles Tony Gordon, Stephen Graham, Jim Greenberg, Steve Griffies, Brian Gross, Richard Grotjahn, Arnold Gruber, Charles P. Guard, Alex Guenther, Dorothy K. Hall, Carl Hane, Steven R. Hanna, David Hanson, Jack Harlan, Robert Harrington, Katherine A. Harris, Giles Harrison, R. H. Pete Hawkins, Manoutcher Heidari, James

Henderson, Benjamin Herman, Douglas Hess, Peter H. Hildebrand, Richard R. Heim Jr., Karen Henry, Elisabeth Holland, Ron Holle, Frances Holt, Paul Houser, Peter Houtekamer, Robert Houze, James Hudson, Gary Hufford, Charles Hutchinson, Bisher Imam, Simon Ince, Janet M. Intrieri, Mark Iredell, Paul Janota, Iraj Javandel, Alastair D. Jenkins, Robert Johns, Gregory Johnson, Richard Johnson, Masao Kanamitsu, Tim Kane, Thomas Karl, R. Jeffrey Keeler, Edwin Kessler, Clark W. King, Peter Kitanidis, Charles Knight, Pamela N. Knox, Thomas R. Knutson, Chet Koblinsky, Chip Konrad, Philip Krider, Arun Kumar, Akhlesh Lakhtakia, Steven Lancaster, Leonard J. Lane, Rolf Langland, Christopher Landsea, Kevin Lansey, Lee Larson, Dan Law, Robert Lee, Margaret LeMone, Don Lenscho, Peter F. Lester, Dennis Lettenmaier, Gad Levy, Janice Lewis, John Lewis, Dale Linville, Robert Livezey, John Livingston, Joanne Logan, Michael Lukes, Douglas MacAyeal, Don MacGorman, Robert MacNish, Sharanya Majumdar, Frank Marks, Thorsten Markus, Cliff Mass, Larry Mays, Vlad Mazur, John A. McGinley, Ian G. McKendry, Gerald A. Meehl, A. G. C. A. Meesters, Keith Meier, Paul Menzel, Fedor Mesinger, James Metcalf, Jene Michaud, Kenneth Mielke, Ronald Miller, Chin-Hoh Moeng, Al Moller, Susan Molsberry, John B. Moncrieff, Larry Mooney, Raghuram Murtugudde, Louisa Nance, Mary Nett, Paul J. Neiman, Jeffrey Niemann, George Ohring, Jan Paegle, Robert J. Paine, David Parrish, Judith Totman Parrish, James Peak, Katherine B. Perry, Grant Petty, John Pickle, Walter Planet, Paul Polger, Evelyn Poole-Kobers, Jeffrey A. Proehl, David Pugh, James Purdom, James Purser, Michael E. Purucker, Barry Ruddick, Robert Quayle, F. Martin Ralph, Eugene M. Rasmusson, Robert M. Rauber, Ronald E. Rinehart, Robert W. Ritzi Jr., Henry W. Robinson, Peter J. Robinson, Matthew Rodell, David Rogers, Richard Rotunno, David P. Rubincam, Barry Ruddick, Chris Ruf, Catherine A. Russell, Suru Saha, Frederick Sanders, Helmuth Sandstrurn, Kenneth Sassen, Ted Scambos, John Schaake Jr., Joseph Schaefer, Hans Peter Schmid, Richard Schotland, Siegfried Schubert, Phil Schumacher, Joachim Segschneider, William Sellers, Paul Sellmann, Glenn Shaw, Joseph A. Shaw, Will Shaw, Christopher Shuman, James Shuttleworth, Vijay Singh, Mano Singham, Joe Sirutis, James E. Smith, Michael B. Smith, William D. Smyth, Brian J. Soden, William Sommers, Kevin G. Speer, R. C. Srivastava, Meric Srokosz, Jery R. Stedinger, Douw G. Steyn, Jeffrey Stith, Ronald J. Stouffer, Chris Strager, Joyce F. Strand, Troy Stribling, Bruce R. Sutherland, Edward Szoke, Richard K. Taft, Dan Tarpley, Elwynn Taylor, Karl E. Taylor, Kendrick Taylor, Eric Thaler, Dennis Thompson, Owen Thompson, Raungdet Titisuttikul, David Toll, Matthias Tomczak, Zoltan Toth, Gregory Tucker, Robert Tuleya, Martin A. Uman, Huug van den Dool, Gerbrant van Vledder, Dayton Vincent, Roger Wakimoto, John E. Walsh Jr., Dailin Wang, Jing Feng Wang, Pao Wang, Stephen Warren, Arthur Warrick, John Weaver, Tammy Weckwerth, Charles Weidman, Morris Weisman, Nigel Weiss, Jeff Whitaker, Martha Whitaker, Glenn White, C. David Whiteman, Louis Wicker, James G. Wieler, Peter Wierenga, Earle Williams, Robert G. Williams, Hugh Willoughby, James Wilson, Michael Winchell, Christian Wode, Daniel E. Wolfe, Bruce Wyman, John Wyngaard, William Yeh, Zuojun Yu, Konstantin G. Zahariev, Walter Zenk, Conrad Ziegler, and Marek Zreda.

At AMS Headquarters, the editor was extremely fortunate to work with a team of professionals dedicated to ensuring that this publication met the high standards of the first edition and the Society as a whole. Early in the project, AMS Manager of Journals Kenneth O. Wilson assisted with categorizing the terms into the subject areas, and *Glossary* Administrator Jane Dannenberg Foley had the arduous task of tracking every term and definition in the book, as well as serving as the primary liaison with the Editorial Board for administrative and database matters. Technical Editor Molly Frey, and Senior Technical Editors Richard G. Hendl and Ronald T. Podsiadlo each read through every term, posing questions for the subject area editors, incorporating modifications, and marking up the text into Standardized General Mark-up Language (SGML) required for our prepress electronic database. Copy Editor Kate O'Halloran reviewed every definition to ensure proper scientific and English usage, as well as conformance to AMS style. Carl Brooks of South End Software Corporation developed the PC-based *Glossary* database software, which enabled editors, writers, reviewers, and in-house staff to exchange, manipulate, and track the terms and their definitions. The editor also wishes to acknowledge the support AMS Executive Director Emeritus Kenneth C. Spengler has provided to the *Glossary of Meteorology*—both editions—over the past four decades. Finally, the dedicated work of AMS Executive Director Emeritus Richard E. Hallgren and Deputy Executive Director Keith L. Seitter laid the foundation for this project. It was Dr. Hallgren's vision and perseverance that brought the project to fruition, and he, with the assistance of Dr. Seitter, developed the funding proposal for this project. Throughout the project, the counsel, advice, and encouragement of Drs. Hallgren and Seitter were truly appreciated.

The second edition of the *Glossary of Meteorology* was supported in a substantial way by a grant through the National Science Foundation (NSF), with funding from the Environmental Protection Agency, the National Oceanic and Atmospheric Administration, the U.S. Air Force, the U.S. Navy, and the Department of Energy. On behalf of the AMS, and all of the past and future users of the *Glossary of Meteorology*, the editor wishes to thank these agencies for their support.

A

A/D converter—Abbreviation for analog-to-digital converter, an electronic device for quantifying a (typically) continuous-time voltage **signal** to a numeric discrete-sample sequence.

A/D converters are commonly used in imaging systems to convert **pixel** values to numeric equivalents, in **radar** systems to convert **signal intensity** to **digital** values for subsequent signal processing, and in many other areas of in situ and **remote sensing**. Important A/D converter characteristics include the number of unique quantization levels (expressed by the number of binary bits in the resulting numeric values) and the speed of quantization.

A-display—A **radar** display on which targets appear as vertical deflections from a horizontal line that represents elapsed time from the most recent transmitted **pulse**.

Distance to the **target** is indicated by the horizontal position of the deflection from the origin of the time axis; the **amplitude** of the vertical deflection is a function of the received **signal strength**. The A-display was the first type of radar display in common use. It may be produced by amplitude modulating the horizontal **sweep** of an **oscilloscope** with the received **signal**. An oscilloscope displaying this is called an **A-scope**. On an A-scope the difference between the coherent **echoes** produced by aircraft or ships and the rapidly fluctuating incoherent echoes from **precipitation** is readily apparent.

A-scope—*See* **A-display**.

AABW—Abbreviation for **Antarctic Bottom Water**.

AAMW—Abbreviation for **Australasian Mediterranean Water**.

ab-polar current—(Obsolete.) An **air current** moving away from either of the earth's poles.

abatement—The lessening or reduction of an atmospheric state that is considered detrimental to humans, animals, plants, or structures.

In **air pollution meteorology**, refers to a reduction in peak **intensity**, duration, average concentration, or **exposure** to those chemicals in the air that are considered pollutants.

abbreviated ship code—An international code assigned to a ship to identify it as the source of **meteorological observations**.

abduction—Development of a hypothesis to explain observations; frequently used in diagnostic **expert systems**; can lead to false conclusions.

For example, a particular instance of **wind** destruction by a **microburst** might be initially ascribed to a **tornado**. *Compare* **deduction**, **induction**.

aberwind—Same as **aperwind**.

abiotic—Refers to nonliving basic elements and compounds of the **environment**.

ABL—Abbreviation for **atmospheric boundary layer**.

ablation—1. All processes that remove **snow**, **ice**, or water from a **glacier**, **snowfield**, etc.; in this sense, the opposite of **accumulation**.

These processes include melting, **evaporation**, **calving**, **wind erosion**, and an **avalanche**. **Air temperature** is the dominant factor in controlling ablation, **precipitation** amounts exercising only secondary control. During the ablation season (usually summer), an ablation rate of about 2 mm h⁻¹ is typical of glaciers in a **temperate climate**.

2. The amount of **snow** or **ice** removed by the above-described processes; in this sense, the opposite of **accumulation**.

ablation area—That portion of a **glacier** surface below the **firn line** where **ablation** exceeds **accumulation**; the opposite of **accumulation area**.

abnormal—Different from **normal** in whatever sense the latter term is used.

When normal signifies typical, abnormal means unusual, lying outside the range of common occurrence. When normal signifies a mean or **median** value, abnormal implies a **deviation**, however slight, from the mean or median. *Compare* **anomalous**.

abnormal refraction—*See* **anomalous refraction**.

abnormality—In meteorology, a deviation of the weather or **climate** from the conditions normally expected at a particular time and place; a deviation from the **normal**.
Compare **anomaly**.

abroholos—Same as **Abrolhos squalls**.

Abrolhos squalls—(Also called abroholos.) **Rain** or **thundersqualls** of the frontal type experienced mainly from May through August near the Abrolhos Islands (18°S) off the coast of Brazil.

abscissa—The horizontal coordinate in a two-dimensional system of **rectangular Cartesian coordinates**; usually denoted by x; also, the horizontal axis of any graph.

absolute—1. With respect to atmospheric motions, refers to an **absolute coordinate system**; for example, **absolute vorticity** as distinguished from **relative vorticity**.

2. In **climatology**, the highest or lowest recorded value of a **meteorological element**, whether at a single **station** or over an area, during a given period.

It is most frequently applied to extremes of **temperature**: absolute maximum is the highest recorded shade temperature; absolute minimum is the lowest recorded.

3. *See* **Kelvin temperature scale**.

absolute altitude—The true vertical distance above the terrain.
Compare **true altitude**.

absolute angular momentum—The **angular momentum** as measured in an **absolute coordinate system**; hence, the **vector product** of the **position vector** of a **particle** into the **absolute momentum** of the particle.

In the **atmosphere** the absolute angular momentum M per unit mass of air is equal to the sum of the angular momentum relative to the earth and the angular momentum due to the rotation of the earth:

$$M = ur \cos \phi + \Omega r^2 \cos^2 \phi,$$

where r is the distance from the center of the earth to the particle, u the relative eastward component of **velocity**, ϕ the latitude, and Ω the angular rotation rate of the earth. Since the earth's **atmosphere** is shallow, the **variable** r is often replaced by the constant a, defined as the radius of the earth. The absolute angular momentum per unit mass is then approximated by

$$M = ua \cos \phi + \Omega a^2 \cos^2 \phi.$$

See **angular-momentum balance**, **conservation of angular momentum**.

absolute annual range of temperature—The difference between the highest and lowest **temperature** observed at a location.

absolute cavity radiometer—A combined electrical substitution and **cavity radiometer** where the electrical substitution inequivalence, efficiency of the cavity, the area of the entrance **aperture**, radiative and conductive losses, and other **energy** exchanges are accounted for such that the electrically substituted heating can be absolutely equated to the **radiant** heating of the detector.

Most currently existing absolute cavity radiometers are designed for the measurement of direct solar **irradiance**. The World Radiation Reference (WRR) **scale** for solar irradiance observations used in many meteorological and climatological applications is defined by a group of well-characterized absolute cavity radiometers maintained by the World Radiation Center (WRC) in Davos, Switzerland.

absolute coordinate system—(Or absolute reference frame.) The **inertial coordinate system** that has its origin on the axis of the earth and is fixed with respect to the stars.

Thus, any mechanical quantities in meteorology defined with respect to this frame take into account the rotation of the earth. *See* **Coriolis force**, **absolute vorticity**.

absolute drought—In British **climatology**, a **drought** period of at least 15 days during which no measurable daily **precipitation** has fallen.

In the United States similar criteria have been used to define a **dry spell** (which, in turn, has a different definition in Great Britain). Ordinarily, these criteria are applied regardless of the **season** of the year.

absolute extremes—Highest or lowest **temperature** observed over the whole period of **observation**.

absolute humidity—Same as **vapor density**.

absolute instability—1. The state of a column of air in the **atmosphere** when it has a **superadiabatic lapse rate** of **temperature** (i.e., greater than the **dry-adiabatic lapse rate**).

An **air parcel** displaced vertically would be accelerated in the direction of the displacement. The **kinetic energy** of the parcel would consequently increase with increasing distance from its level of origin. *See* **parcel method**, **conditional instability**, **absolute stability**.

2. (Also called **mechanical instability**.) The state of a column of air in the **atmosphere** when its **lapse rate** of **temperature** is greater than the **autoconvective lapse rate**.

In such a column the air density would increase with **elevation**. *See also* **autoconvection**.

absolute instrument—An instrument the **calibration** of which can be determined by means of simple physical measurements on the instrument.
Compare **secondary instrument**.

absolute isohypse—A line that has the properties of both constant **pressure** and constant height above **mean sea level**.
Therefore, it can be any **contour line** on a **constant-pressure chart**, or any **isobar** on a **constant-height chart**.

absolute linear momentum—Same as **absolute momentum**.

absolute momentum—(Also called absolute linear momentum.) The **(linear) momentum** of a **particle** as measured in an **absolute coordinate system**; hence, in meteorology, the sum of the (**vector**) momentum of the particle relative to the earth and the (vector) momentum of the particle due to the earth's rotation.

absolute monthly maximum temperature—Highest **daily maximum temperature** observed during a given calendar month over a specified period of years.

absolute monthly minimum temperature—Lowest **daily minimum temperature** observed during a given calendar month over a specified period of years.

absolute potential vorticity—Same as **potential vorticity**.

Absolute Radiation Scale—[Abbreviated ARS; also referred to as the World Radiation Reference (WRR).] A **radiation** scale for measurement of solar exitance (**irradiance**).
Prior to 1956, the Ångström Scale (ÅS) (1905) and Smithsonian Scale (SS) (1913) were used. Each **scale** was calibrated against a different **radiation** detector (i.e., the **Ångström compensation pyrheliometer** and water-stirred **pyrheliometer**, respectively), and yielded slightly different values for the irradiance, with the ÅS reading roughly 3.5% lower than the SS. The **International Pyrheliometric Scale** (IPS), defined in 1956, represented a numerical compromise between these two scales. In 1975, the IPS was replaced by the Absolute Radiation Scale (ARS). The ARS is calibrated against six **absolute cavity radiometers** maintained at the World Radiation Center in Davos, Switzerland. The **variation** among the six radiometers is about 0.3%. The IPS was found to give measured irradiance levels that were about 2%–3% percent lower than the more precise ARS.

absolute reference frame—Same as **absolute coordinate system**.

absolute refractive index—(Or absolute index of refraction.) *See* **refractive index**.

absolute stability—The state of a column of air in the **atmosphere** when its **lapse rate** of **temperature** is less than the **saturation-adiabatic lapse rate**.
An **air parcel** displaced upward by an **adiabatic process** would then be more dense than its **environment** and would tend to sink back to its level of origin. *Compare* **absolute instability**.

absolute standard barometer—**Barometer** that provides absolute measurements of **pressure** without having to be calibrated.
For example, large **sylphon** barometers with optical reading have been a preferred type of primary or absolute standard barometer.

absolute temperature extremes—Highest or lowest **temperature** observed over a specified period of **observation** used as a climatological record.
See **temperature extremes**

absolute temperature scale—(Abbreviated A.) Same as **Kelvin temperature scale**.

absolute velocity—**Velocity** as measured in an **absolute coordinate system**; hence, in meteorology, the (**vector**) sum of the velocity of a **fluid parcel** relative to the earth and the velocity of the parcel due to the earth's rotation.
The east–west component is the only one affected:

$$u_a = u + \Omega r \cos\phi,$$

where u and u_a are the relative and absolute eastward speeds, Ω the angular speed of the earth's rotation, r the radial distance of the parcel from the center of the earth, and ϕ the latitude of the parcel.

absolute vorticity—1. The **vorticity** of a fluid **particle** determined with respect to an **absolute coordinate system**.

The absolute vorticity **vector** is defined by $2\Omega + \nabla \times \mathbf{u}$, where Ω is the earth's **angular velocity** vector and **u** is the three-dimensional **relative velocity** vector.

2. The vertical component η of the absolute vorticity vector (as defined above) given by the sum of the vertical component of the **vorticity** with respect to the earth (the **relative vorticity**) ζ and the vorticity of the earth (equal to the **Coriolis parameter**) f:

$$\eta = \zeta + f.$$

absolute zero—The zero **point** of the **Kelvin temperature scale**, of fundamental significance in **thermodynamics** and statistical mechanics.

It is the linearly extrapolated **temperature** at which the volume of an **ideal gas** at constant **pressure** would vanish. All real gases become liquid or solid at sufficiently low temperatures and maintain a finite volume. Absolute zero on the Kelvin scale corresponds to $-273.15°C$.

absorbance—Used mostly by chemists, the negative **logarithm** (base 10) of the **transmittance** of an absorbing sample, often corrected for **reflection** by its container.

Despite its name, absorbance is a consequence of both **scattering** and **absorption**, although scattering is usually assumed to be negligible. To within a constant multiplier, absorbance is **absorption optical thickness** and depends on the physical thickness of the sample. *Compare* **absorptance**.

absorbed solar radiation—**Solar radiation** absorbed by the **atmosphere**'s constituent gases, suspended material, clouds, or by the earth's surface.

absorptance—One minus the sum of **reflectance** and **transmittance**.
Compare **absorbance**.

absorption—1. The process in which incident **radiant energy** is retained by a substance.

A further process always results from absorption, that is, the irreversible conversion of the absorbed **radiation** into some other form of **energy** within and according to the nature of the absorbing medium. The absorbing medium itself may emit radiation, but only after an **energy conversion** has occurred. *See* **attenuation**.

2. The taking up or assimilation of one substance by another.

The substances may combine chemically, or the absorption may just correspond to a physical **solubility**. *See* **Henry's law**.

absorption band—A connected series of closely spaced or overlapping **absorption lines**.

Absorption bands are common features in the absorption spectra of polyatomic gases. Absorption bands arise when absorbed photons are converted to simultaneous changes in more than one of the electronic, vibrational, or rotational molecular **energy** states. Changes to vibrational energy are usually accompanied by changes in rotational energy and give rise to vibrational-rotational absorption bands. Similarly, changes to molecular **electron** levels yield electronic-vibrational-rotational absorption bands.

absorption coefficient—A measure of the **extinction** due to **absorption** of **monochromatic radiation** as it traverses a medium.

Usually expressed as a **volume absorption coefficient**, k_v, with units of reciprocal length (i.e., area per unit volume), but also as a **mass absorption coefficient**, k_m, with units of area per unit mass. The two are related by $k_v = \rho k_m$, where ρ is the **density** of the absorber. *See also* **absorption optical thickness**.

absorption cross section—The area that, when multiplied by the **irradiance** of **electromagnetic waves** incident on an object, gives the **radiant flux** absorbed and dissipated by the object. Customary usage in **radar** describes the absorption cross section of an object as the area that, when multiplied by the **power density** of incident plane-wave **radiation**, gives the **power** absorbed and dissipated by the object.

The **extinction cross section** of an object is the sum of the absorption cross section and the **scattering cross section**. For a medium consisting of a **dispersion** of absorbing objects through which radiation propagates, the **volume absorption coefficient** (units: $m^2 m^{-3}$ or m^{-1}) at a given location in the medium is the sum of the **absorption** cross sections of all the objects in a unit volume centered at that location.

absorption hygrometer—(Sometimes called **chemical hygrometer**.) A type of **hygrometer** that measures the **water vapor** content of the **atmosphere** by means of the **absorption** of **vapor** by a **hygroscopic** chemical.

The amount of vapor absorbed may be determined in an absolute manner by weighing the hygroscopic material (*see* **gravimetric hygrometry**), or in a nonabsolute manner by measuring a physical property of the substance that varies with the amount of water vapor absorbed. The **humidity strip**, the **carbon-film hygrometer element**, and the thin-film capacitor are examples of the latter.

absorption line—A line of finite width in the **absorption spectrum**.

Absorption lines are characterized by their central **wavelength**, **line intensity**, and **line width**.

absorption loss—The quantity of water that is lost during the initial filling of a **reservoir** because of **absorption** by soil and rocks.

absorption optical thickness—The component of the total **optical thickness** that is due to **absorption** only.

absorption spectrum—The detailed dependence on **wavelength** of the **intensity** of **radiation** absorbed by a given medium.

Absorption spectra of gases are typically composed of discrete spectral lines and bands of overlapping lines that depend on the molecular or atomic composition of the absorbing substance and that may be used to identify it uniquely. When measuring the absorption spectrum, the medium should be considerably colder than the source of incident radiation (which should also be continuous in wavelength), in order to clearly distinguish the absorption spectrum from the analogous **emission spectrum**.

absorptivity—The fraction of incident **radiation** that is absorbed by matter.

Absorptivity may be a function of **wavelength** and/or direction, and is related to the **emissivity** of the region by **Kirchhoff's law**. The absorptivity is identically equal to unity for **blackbodies** and is independent of wavelength for **gray bodies**.

abstraction—1. The part of **precipitation** that does not become **direct runoff**.

2. The draining of water from a **stream** into another stream having a more rapid corroding action.

ACC—Abbreviation for **Antarctic Circumpolar Current**.

accelerated erosion—**Erosion** increased by human agency to beyond the **normal** geologic rate.

acceleration—The rate of change with time of the **velocity** vector of a **particle**.

If **u** is the **vector** velocity, the acceleration may be written as $D\mathbf{u}/Dt$, where D/Dt is the material (or total) derivative. For most purposes in **hydrodynamics** where **Eulerian coordinates** are employed, the acceleration is decomposed as follows:

$$\frac{D\mathbf{u}}{Dt} = \frac{\partial \mathbf{u}}{\partial t} + \mathbf{u} \cdot \nabla \mathbf{u},$$

where $\partial \mathbf{u}/\partial t$ is called the **local acceleration**, and $\mathbf{u} \cdot \nabla \mathbf{u}$ is called the **convective acceleration**.

acceleration of gravity—*See* **gravity**.

accelerometer—An instrument that measures **acceleration**.

acceptance capacity—Quantity of pollutants that a water body can accept without the **pollution** exceeding a specified level.

acceptance region—A **range** of values of a **sample** statistic used to test a hypothesis.

In the testing procedure, the sampling region is divided into an acceptance region and a **rejection region**. If the sample statistic lies within the acceptance region, the null hypothesis (a hypothesis usually based on conventional theory) is provisionally accepted. If the sample statistic lies in the rejection region, the alternative hypothesis that contradicts the null hypothesis is accepted. *See* **significance test**.

accessory cloud—A **cloud** form that is dependent, for its formation and continuation, upon the existence of one of the major **cloud genera**.

It is often an appendage of the parent cloud (as **mamma, incus, tuba, arcus**), but it also may be an immediately adjacent **cloudy** mass (as **pannus, pileus, velum**). *See* **cloud classification**.

acclimation—1. Basically, same as **acclimatization**.

2. Adaptation of living organisms to all aspects of the **environment**, not just the climatic.

Examples of acclimation would be adjustment of plants to growth at elevated atmospheric **carbon dioxide** concentrations or to soil water deficits.

acclimatization—1. (Also called acclimation.) The process by which a living organism becomes adapted to a change of climatic **environment**. There has been a growing amount of research on the acclimatization of man to extreme environments such as polar and tropical regions and high altitudes.

These studies are directed toward 1) determination of the internal physiological changes or skin changes produced by **exposure** to new climates, 2) determination of criteria for preselection (i.e., selecting the most adaptable type of man for a particular **climate**), and 3) development of external means of aiding adaptation (e.g., preconditioning, and modification of habits, diet, and clothing). As to usage, "acclimatization" has long been considered to be equivalent to "acclimation." In some quarters, however, a fine distinction is drawn by calling "acclimation" a purely natural process (or state), and "acclimatization" a process (or state) influenced by human agency. The recent trend, at least in the United States, is to use "acclimatization" as the all-inclusive technical term, and to leave "acclimation" (which never was accepted in Great Britain) to more or less loose popular usage.

2. The state or degree of adaptation to **climate**.

Castellani, Sir Aldo, 1938: *Climate and Acclimatization*, 2d ed., 146–176.
Newburgh, L. H., ed., 1949: *Physiology of Heat Regulation and the Science of Clothing*, 3–67.

accommodation coefficient—*See* **mass accommodation coefficient**.

accretion—1. (Sometimes incorrectly called **coagulation**.) In **cloud physics**, usually the growth of an **ice** hydrometeor by collision with **supercooled cloud** drops that **freeze** wholly or partially upon contact.

May also refer to the collection of smaller ice particles. This has been called a form of **agglomeration** and is analogous to **coalescence**, in which liquid drops collect other liquid drops. *See* **ice accretion**; *compare* **coagulation**.

2. In **cloud** modeling, the collection of **cloud drops** by **drizzle drops** and **raindrops**.

This nomenclature is used along with **autoconversion** and **self collection** to distinguish among three subprocesses, evident from numerical results, responsible for the growth of the **drop-size distribution** by the **collision–coalescence process**.

Kessler, E., 1969: *On the Disribution and Continuity of Water Substance in Atmospheric Circulation*, Amer. Meteor. Soc., Boston, 84 pp.

accumulated cooling—The total cooling since the time in the evening when the turbulent **heat flux** near the ground produces a net **heat** flow from the earth to the **atmosphere**; used to measure or predict evolution of the **stable boundary layer**.

It is defined as the integral of the surface kinematic heat flux $\overline{w'\theta_s'}$ over time, starting from the time t_0 just before **sunset** when the flux changes from positive to negative, and ending at any time t_e before the flux changes sign back to positive. Dimensions are **temperature** times length, such as (K·m). In the absence of **advection** and direct **radiative cooling** of the air, accumulated cooling (AC) also equals the area under the **potential temperature** profile $z(\theta)$, integrated from the surface potential temperature θ_s to the residual-layer potential temperature θ_{RL}, that is,

$$AC \equiv \int_{t_0}^{t} \overline{w'\theta_s'}\, dt \approx \int_{\theta_s}^{\theta_{RL}} z(\theta)\, d\theta.$$

accumulated temperature—(Or cumulative temperature.) In Great Britain, the accumulated excess of **temperature** above a given **standard temperature**.

It is expressed in **degree-hours** or **degree-days**. For each day, degree-hours are determined as the product of the length of time, in hours, during which temperatures are above the standard, and the amount, in degrees, by which the **mean temperature** of the period exceeds the standard. Division of the resulting degree-hour value by 24 gives a value in degree-days. Summation of either over the period of interest creates the accumulated temperature. The concept of accumulated temperature was introduced into plant geography by A. de Candolle (1855); his standard temperature was 6°C (42.8°F), below which he considered that no vegetative growth took place. It was introduced into **agricultural meteorology** in Britain in 1878, when accumulated temperatures in excess of 42°F were first published regularly in the Weekly Weather Report. These were calculated from the daily **maximum** and **minimum temperatures**. In heating calculations, a form of accumulated temperature is calculated as the number of degree-days below the standard, which is taken as 65°F in the United States and 60°F in Great Britain. Since the standard temperatures differ, degree-days are not interchangeable between the two countries.

de Candolle, A., 1855: *Géographie Botanique Raisonné*, 2 vols., Paris: V. Masson.

accumulation—1. *See* **snow accumulation**.

2. In **glaciology**, the quantity of **snow** or other solid form of water added to a **glacier** or **snowfield** by **alimentation**; the opposite of **ablation**.
Compare **snowpack**.

accumulation area—That portion of the **glacier** surface above the **firn line** where the **accumulation** exceeds **ablation**; the opposite of **ablation area**.

accumulation mode—**Aerosol** particles in the size **range** 0.5–2 μm in diameter.

The name arises from the fact that **particles** in this size range are aerodynamically stable and do not settle out, nor do they agglomerate to form larger particles; thus they tend to accumulate in the **atmosphere**. *See also* **agglomeration**.

accumulation rain gauge—(Also called accumulative rain gauge.) A class of **rain gauges** in which the **precipitation** is accumulated over time.

The depth of accumulated precipitation can be determined by the level of a float, by weighing, or by manual direct measurement of water depth. For long-term, unattended operation a known amount of liquid that prevents **evaporation** is placed in the collection container.

accumulation zone—The region of strong **radar reflectivity** within the **echo overhang** of a severe **thunderstorm**.

Within this zone large suspended **hydrometeors** are assumed to be growing.

accumulative rain gauge—*See* **accumulation rain gauge**.

accuracy—The extent to which results of a calculation or the readings of an instrument approach the true values of the calculated or measured quantities.

Compare **precision**.

acdar—Acronym for acoustic detection and ranging instruments; also used to describe an **atmospheric sounding** made with such a device.

acetaldehyde—Aldehyde, formula CH_3CHO, formed in the **oxidation** of **ethane** and other organic compounds.

It is a major precursor to the **pollutant, peroxyacetyl nitrate** (PAN).

acetic acid—Second of the series of organic (carboxylic) acids, formula CH_3COOH, systematic name ethanoic acid.

Acetic acid is thought to be formed in the **atmosphere** from the **oxidation** of **acetaldehyde**, although the exact details of the transformation are unknown. It contributes to the acidity of **rainfall** in remote (unpolluted) areas.

acetone—Simplest ketone molecule, formula CH_3COCH_3, formed in the **oxidation** of **propane** and several larger **hydrocarbons**; used widely as an organic solvent, due to its high miscibility with water.

Its **photolysis** is believed to be an important source of **odd hydrogen** radicals in the **troposphere** above the **boundary layer**.

acetonitrile—Chemical also known as methyl cyanide, formula CH_3CN.

It is emitted from incomplete combustion of vegetable matter, notably biomass, for example, cigarettes. Acetonitrile is relatively unreactive in the **troposphere** and thus reaches the **stratosphere**, where it participates in ion–molecule reactions.

acetylene—Unsaturated hydrocarbon, formula C_2H_2, systematic name ethyne; member of the alkyne family; major component of incomplete combustion.

acicular ice—"**Freshwater** ice consisting of numerous long crystals and hollow tubes having variable form, layered arrangement, and a content of air bubbles. This ice often forms at the bottom of an **ice layer** near its contact with water."[from *Glossary of Arctic and Subarctic Terms* (1955)].

Arctic, Desert, Tropic Information Center (ADTIC) Research Studies Institute, 1955: *Glossary of Arctic and Subarctic Terms*, ADTIC Pub. A-105, Maxwell AFB, AL, 90 pp.

acid deposition—The accumulation of an acidic chemical from the **atmosphere** to the surface of the earth, or to plants and structures at the surface.

Acids have high concentrations of **hydrogen** ions when dissolved in water, indicated by a **pH** less than 7. Acids can corrode metals, dissolve some types of rocks such as limestone, injure plants, and exacerbate some conditions in humans and animals. Acid deposition can occur in two forms: 1) **wet deposition** including **acid rain**, acid snow, acid hail, acid dew, acid frost, and **acid fog**; and 2) **dry deposition** including fallout of heavy **particles**, gravitational settling of lighter particles, and **interception** by and reaction with plant surfaces. Sometimes all forms of acid deposition are loosely called acid rain, although literally acid rain refers only to the liquid form. Ambient **carbon dioxide**, always present in the air, dissolves in **cloud drops** and **raindrops** creating carbonic acid with **pH** ≈ 5.6. Because this is a **normal** occurrence in the atmosphere, **rain** is defined to be acid rain only when it has pH < 5.6. However, even in remote areas, there are sufficient sulfate, **nitrate**,

ammonia, or soil cations (calcium or magnesium that are typically associated with carbonates) to cause "clean" atmospheric water to have pH in the **range** of 4.5–5.5. Polluted regions typically have pH in the range of 3–4, with values as low as 2–3. The chemicals that cause the greatest acid-deposition problems are oxides of sulfur (abbreviated as SO_x) and oxides of **nitrogen** (NO_x), which can react in the presence of atmospheric **oxidants** and water (e.g., **clouds**, **fog** and **precipitation**) to become **sulfuric acid** and **nitric acid**, respectively. These strong acids have an affinity for water, allowing droplets to grow hygroscopically in the atmosphere to produce **haze** or **smog**, even at relative humidities as low as 60% to 70%.

> Amer. Meteor. Soc., 1997: AMS Policy Statement on Acid Deposition. *Bull. Amer. Meteor. Soc*, **78**, 2263–2265.

acid dew—*See* **acid deposition**.

acid fog—Also called acid haze.

Occurrence of **fog** or **haze** in which considerable amounts of acidic material have been taken up from the gas **phase**, resulting in **pH** values less than approximately 3 in the liquid **phase**.

acid frost—*See* **acid deposition**.

acid hail—*See* **acid deposition**.

acid haze—*See* **acid fog**.

acid pollution—Chemicals not occurring naturally in the **atmosphere** that either are acidic, or that easily react or dissolve in water to become acidic.

See **acid deposition**.

acid precipitation—*See* **acid rain**.

acid rain—A popular expression for the **deposition** by **rainfall** of various airborne pollutants (especially SO_2 and NO_2) that have harmful effects on vegetation, soils, buildings and other external structures.

acid snow—*See* **acid deposition**.

acidity of water—The quantitative capacity of aqueous media to neutralize strong bases.

The acidity of a water sample is determined by titrating it with a strong base (e.g., NaOH) to a defined **pH**.

acidity profile—A record of the H^+ concentration in an **ice** core.

Unless otherwise stated, the H^+ concentration is indirectly determined by the measuring the **electrical conductivity**. Annual layering, volcanic activity, periods with abundant alkaline **dust**, and **biomass burning** can sometimes be detected.

aclinic line—(Or dip equator, magnetic equator.) The line through those points on the earth's surface at which the **magnetic inclination** is zero.

The aclinic line is a particular case of an **isoclinic line**. In South America the aclinic line lies at about 15°S latitude, while from central Africa to about Indochina it coincides approximately with the parallel of 10°N latitude. *Compare* **agonic line**, **geomagnetic equator**.

acoustic array—A sound-transmitting or receiving system with elements arranged to give desired directional characteristics.

acoustic backscattering—Scattering of **sound** or ultrasound in the direction of the source.

acoustic dispersion—The separation of a **sound wave** into its **frequency** components as it passes through a given medium.

The velocities of the **wave** components change as they pass through the medium.

acoustic Doppler current profiler—(Abbreviated ADCP.) A current-measuring instrument employing the transmission of high-frequency acoustic signals in water.

The **current** is determined by a **Doppler shift** in the backscatter echo from plankton, suspended **sediment**, and **bubbles**, all assumed to be moving with the mean speed of the water. Time **gating** circuitry is employed, which uses differences in **travel time** to divide the water column into **range** intervals called bins. The bin determinations allow development of a **profile** of **current** speed and direction over the entire water column. The ADCP can be deployed from a moving vessel, towing platform, buoy, or bottom-mounted platform. In the latter configuration, it is nonobtrusive in the water column and thus can be deployed in shipping channels.

acoustic echo sounding—Measuring the depth of the ocean by determining the time required for the echo of a **sound** impulse to return to a point near the surface (i.e., the transmitting ship).

acoustic–gravity wave—A **wave disturbance** with restoring forces that include **buoyancy** and the elastic **compressibility** of the fluid medium.

acoustic imaging—The use of acoustic **energy** to form a representation of a physical object, such as **side-scanning sonar** imaging of objects on the ocean bottom.

acoustic impedance—The total reaction of a medium to **sound** transmission through it, represented by the complex ratio of the sound pressure to the **effective flux**, that is, **particle velocity** times surface area through the medium, expressed in acoustic ohms.

acoustic intensity—The average acoustic **power** transported across a unit area, usually expressed in watts per square meter.

acoustic lens—A system of disks or other devices to spread or converge **sound waves** in a manner analogous to the way an optical lens refracts **light**.

acoustic ocean current meter—An instrument used to monitor currents by measuring the difference in **travel time** between each acoustic **pulse** transmitted in a direction opposite to the flow of a **current**, and the return pulse.

acoustic pressure—The difference between the instantaneous **total pressure** and the **static pressure**, that is, the **pressure** that would be present in the absence of the **acoustic waves**.
The **SI** unit is the **pascal**, but the **bar** is frequently used.

acoustic radar—1. Use of **sound waves** with **radar** technology for remote probing of the **lower atmosphere**, up to heights of about 1500 m, for measuring **wind speed** and direction, **humidity**, **temperature inversions**, and **turbulence**.

2. *See* **sodar**.

acoustic rain gauge—An instrument designed to determine **rainfall** over lakes and oceans.
A hydrophone is used to sense the **sound** signatures for each **drop** size as **rain** strikes a water surface. Since each sound signature is unique, it is possible to invert the underwater sound field to estimate the **drop-size distribution** within the rain. Selected moments of the drop-size distribution yield **rainfall rate**, rainfall accumulation, and other rainfall properties.
    Nystuen, J. A., et al., 1996: A comparison of automatic rain gauges. *J. Atmos. Oceanic. Technol.*, **13**, 62–73.

acoustic reflection profiling—A type of seismic profiling of the subbottom layers of the ocean floor by **acoustic waves** that reflect off the interfaces of the bottom layers.

acoustic reflectivity—The relative **reflectivity** of a specific material, that is, the tendency to deflect **sound** energy in a specific medium rather than absorb it.

acoustic refraction—The change in the direction of **sound** as it travels through a medium due to differences in physical and chemical characteristics of the medium.

acoustic resonance—A condition of an acoustic system, such as a Helmholtz **resonator**, in which the response of the system to **sound waves** becomes very large when the **frequency** of the **sound** approaches a natural vibration frequency of the air (or other material) in the system.

acoustic reverberation—The prolongation of **sound** after the original source has stopped vibrating in an enclosed space.

acoustic scattering—The irregular **reflection, refraction,** and **diffraction** of **sound** in many directions.

acoustic scintillation—Irregular fluctuations in the received **intensity** of sounds propagated through the **atmosphere** (or ocean) from a source of uniform **output**; produced by the nonhomogeneous structure of the atmosphere (or ocean) along the path of **wave propagation**.

acoustic signature—The characteristic pattern or **profile** of an object, such as a specific feature of the ocean floor, as detected by identification equipment that uses **sound** or **ultrasonic** waves.

acoustic sounding—1. The technique of **remote sensing** in which an instrument sends **acoustic waves** vertically and receives reflections from atmospheric features such as inversions or turbulent layers.

2. A representation of the **vertical profile** of one or more of the **state variables** of a fluid body (typically the ocean or **atmosphere**) deduced from variations in reflected acoustic **energy**.

3. A particular measurement of depth of water below an instrument (either at the surface or at some moored depth) computed from the **travel time** of an acoustic **pulse** emitted by the acoustic **sounder**.

acoustic thermometer—A device to measure **air temperature** based on the principle that the **speed of sound** varies as the square root of **temperature**.

The most common application is in so-called **sonic anemometers**, where the time-of-flight of **sound** pulses between a pair of **acoustic transducers** is used to determine the speed of sound. This measurement is converted to **temperature** via the formula

$$c_s^2 = 403T(1 + 0.32e/P),$$

where c_s is the speed of sound, T the **air temperature** (K), e the **partial pressure** of atmospheric **water vapor** (mb), and P the atmospheric **static pressure** (mb).

acoustic tomography—An imaging technique in which information is collected from beams of acoustic radiations that have passed through an object, generally in the form representing a two-dimensional slice through the object.

In **oceanography**, acoustic tomography consists of an inverse technique that uses acoustic signals to sample the interior of the ocean. In this way, the **temperature** structure of the interior ocean can be reconstructed from acoustic signals in multiple vertical planes.

acoustic transducer—A device, such as an underwater hydrophone, that converts acoustic **energy** into **electromagnetic energy**; or a device, such as an acoustic source, that converts **electromagnetic energy** into acoustic energy.

acoustic transponder—A device used in underwater navigation that responds with an acoustic code when interrogated by an acoustic **signal** from a ship.

acoustic velocimeter—An instrument for making direct measurements of the **speed of sound** in oceans and lakes.

acoustic wave—(Or **sound wave**.) From its Greek origin, akoustikus, meaning related to hearing, acoustic ought to be restricted to **sound** audible to humans (or animals).

Although this is often the sense in which it used, the term is also used to embrace sound of all frequencies.

acre-foot—The volume of water represented by a depth of 1 ft over an area of 1 acre.

The area of 1 acre is equal to 43 560 ft^2, 1 acre-foot is equal to 43 560 ft^3.

acrolein—Unsaturated **aldehyde**, formula CH_2CHCHO, formed as an **oxidation** product of **butadiene**, which is a common emission from automobiles.

actinic—Pertaining to **electromagnetic radiation** capable of initiating photochemical reactions, as in photography or the fading of pigments.

Because of the particularly strong action of **ultraviolet radiation** on photochemical processes, the term has come to be almost synonymous with **ultraviolet**, as in "actinic rays."

actinic balance—Same as **bolometer**.

actinic flux—The spherically integrated **radiation flux** in the earth's **atmosphere** that originates from the sun, including the direct **beam** and any **scattered** components.

This **radiation** is responsible for initiating the chemistry of the atmosphere. *Compare* **radiance, irradiance**.

actinic rays—*See* **actinic**.

actinogram—The record of an **actinograph**.

actinograph—A recording **actinometer**.
See **Robitzsch actinograph**.

actinometer—The general name for any instrument used to measure the **intensity** of **radiant energy**, particularly that of the sun.

Actinometers may be classified, accordingly to the quantities that they measure, in the following manner: 1) **pyrheliometer**, which measures the intensity of **direct solar radiation**; 2) **pyranometer**, which measures **global radiation** (the combined intensity of direct solar radiation and **diffuse sky radiation**); and 3) **pyrgeometer**, which measures the effective **terrestrial radiation**. *See* **actinometry, bolometer, dosimeter, photometer, radiometer**; *compare* **sunshine recorder**.

actinometry—A term for the measurement of **radiation**, originally more general, but now used mainly to describe photochemical techniques of measuring **ultraviolet radiation** by chemical actinometers.

actinon—A **radioactive gas**, symbol An; an inert gaseous **element** that is an **isotope** of **radon** and **thoron**.

Actinon is a member of the uranium–actinium family of radioactive elements. Its radioactive

half-life is extremely short, only 3.92 s, and its parent atoms (actinium x) are comparatively rare in the earth's crust, so actinon is considerably less important as an agency of **atmospheric ionization** than are radon and thoron.

activated complex theory—(Also called transition state theory.) Theory of chemical reaction that relates the rate of reaction to an **equilibrium** between the reactants and an activated complex or transition state, which is a maximum energy configuration of the reactants as they proceed along the reaction coordinate.

activation energy—In a chemical reaction, the **energy** that must be added to the reactants in order to allow a reaction to occur.
    See **Arrhenius equation, preexponential factor.**

active basin area—The part of the **basin** contributing to **stream flow.**

active cloud—A category of **cumulus** cloud that can distribute **air pollutants** from the **atmospheric boundary layer** to the **free atmosphere.**
    These clouds have reached their **level of free convection**, allowing **latent heat** released during water-vapor **condensation** to contribute to the positive **buoyancy** of the **cloudy** air. These clouds are usually first produced by **thermals**, but can eventually decay into passive clouds before disappearing completely. Corresponding morphological species include **cumulus mediocris, cumulus congestus,** and **cumulonimbus.**

active front—A **front**, or portion thereof, that produces appreciable **cloudiness** and, usually, **precipitation.**

active layer—(Also called frost zone, mollisol.) That part of the soil included with the **suprapermafrost layer** (i.e., existing above **permafrost**) that usually freezes in winter and thaws in summer.
    Its bottom surface is the **frost table**, beneath which may lie permafrost or **talik.** The depth of the active layer varies anywhere from a few inches to several feet. *Compare* **active permafrost.**

active network—*See* **network.**

active nitrogen—(Also known as odd nitrogen species.) Reactive forms of **nitrogen** comprising **nitric oxide**, NO, and **nitrogen dioxide**, NO_2; usually designated by NO_x.
    These oxides of are responsible for **ozone** formation in the **troposphere** and play a major part in ozone loss in the **stratosphere.** The major sources of active nitrogen in the **atmosphere** are combustion, soil emissions, **lightning**, and the reaction of **nitrous oxide** with excited **oxygen** atoms in the stratosphere.

active permafrost—Permanently **frozen ground (permafrost)** which, after thawing by artificial or unusual natural means, reverts to permafrost under **normal** climatic conditions; opposed to **passive permafrost.**
    Compare **active layer.**

active site—A site on an **ice nucleus** that forms **ice** at a lower **supersaturation** or a smaller **supercooling** than elsewhere, resulting from a local impurity or defect giving a region of local strain and deformation in the atomic arrangement of the **nucleus**, matching it more closely to the arrangement of molecules in ice, and a lower **energy** requirement for **nucleation.**

active system—A **remote sensing** system that transmits its own **electromagnetic energy**, then measures the properties of the returned **radiation.**
    Active systems require more **power** to operate than passive systems, but can be designed to have greater **sensitivity.** Examples of active systems are **radar, lidar,** and **acoustic sounders.**

activity—Measure of the effective concentration of a chemical in the gas or liquid **phase.**
    The activity is usually less than the **mass concentration**, from which it differs due to the nonideal nature of gases and solutions. In the limit of very dilute mixtures, the activity is the same as the concentration for liquid solutions, or the **partial pressure** for gases. The **activity coefficient**, usually denoted by γ or f, is the ratio of the activity to the actual physical concentration.

activity coefficient—A measure of the extent to which substances, on dissolving in water, form charged ions or associate to form multiple molecules; the amount dissolved influences colligative effects, such as **equilibrium freezing point** depression of solution drops and reaction rates in solution.

actual elevation—The vertical distance above **mean sea level** of the ground at the **meteorological station.**
    This term is denoted by the symbol H in international usage.

actual evaporation—Quantity of water evaporated from a region comprising open water or **ice** surfaces, **bare soil**, or vegetation-covered soil.

actual evapotranspiration—*See* **effective evapotranspiration**.

actual pressure—The **atmospheric pressure** at the level of the **barometer (elevation of ivory point)**, as obtained from the observed reading after applying the necessary corrections for **temperature, gravity**, and **instrument errors**.

This may or may not be the same as **station pressure**.

actual time of observation—1. The time at which the **barometric pressure** is read during a **synoptic weather observation**.

2. The time at which the **balloon** is released when an **upper-air observation** is taken.

adaptability—The ease with which an **algorithm, rule-based system**, or **neural network** developed for a given geographic location can be modified to perform equally well with input data from another location.

adaptation luminance—(Or adaptation brightness; also called adaptation level, adaptation illuminance, brightness level, field brightness, field luminance.) The average **luminance** (or **brightness**) of those objects and surfaces in the immediate vicinity of an **observer** estimating the **visual range**.

The adaptation luminance has a marked influence on an observer's estimate of the visual range because, along with the visual angle of the object under observation, it determines the observer's **threshold contrast**. High adaptation luminance tends to produce a high threshold contrast, thus reducing the estimated visual range. This effect of the adaptation luminance is to be distinguished from the influence of background luminance.

adaptation strategies—Strategies for responding to **climatic change** that reduce the consequences of the changing **climate** by adjusting the physical **environment** or the interactions between the environment and human society.

adaptive grid—A **grid** on which the number or geometric distribution of points changes in response to the characteristics of the evolving flow that is being described.

Adaptive grids are most commonly used to place higher **resolution** in regions where **error** is likely to be large, for instance, in areas where the **gradient** (or **Laplacian operator**) is large.

adaptive observational network—An **observational network** in which the location and timing of measurements of atmospheric properties are changed on a daily basis in order to minimize some measure of **forecast error**.

To achieve this aim, atmospheric measurements must adapt to the flow observed on a particular day. For example, in midlatitude regions, enhanced atmospheric measurements in regions of large horizontal **temperature** gradients can lead to a significant reduction in forecast error. As these regions of strong **baroclinicity** move from one day to the next, the locations of enhanced measurements should also move.

adaptive observations—(Also called targeted observations.) Observational data obtained specifically to improve **model** initial conditions for a numerical forecast of a selected weather feature, or to optimize a measure of forecast outcome (e.g., **error**).

Guidance for selecting adaptive observations can be obtained from model-based products, such as **singular vectors** from adjoint models or ensemble forecasts, which are used to estimate where **initial condition** error has the greatest impact on the forecast measure. Adaptive observations can be obtained from in-situ platforms such as **dropsonde** aircraft, or by direction of remote sensors including satellite or **radar** instruments. *See* **adjoint model, adjoint sensitivity**.

ADCP—Abbreviation for **acoustic Doppler current profiler**.

ADEOS—Abbreviation for **Advanced Earth Observing Satellite**.

adfreezing—The process by which one object becomes adhered to another by the binding action of **ice**.

adhesion efficiency—In **cloud physics**, the fraction of ice particles colliding with a collector ice particle that actually adheres to, or aggregates with, the collector; or, in the case of the riming process, the fraction of water drops colliding with a collector ice particle that actually adheres (freezes) to the collector.

adhesive water—Water retained by soil constituents as a result of the molecular attraction between the water and the soil.

adiabat—In most contexts, same as **dry adiabat**.

See also **saturation adiabat, pseudoadiabat**.

adiabatic—*See* **adiabatic process**.

adiabatic atmosphere—(Or dry-adiabatic atmosphere; also called convective atmosphere, homogeneous atmosphere.) A **model atmosphere** characterized by a **dry-adiabatic lapse rate** throughout its vertical extent.

Such a condition is never observed and is also rather poorly designated, since "adiabatic" represents a process, not a condition. The **pressure** in an adiabatic atmosphere decreases with height according to

$$p = p_0 \left(1 - \frac{gz}{c_{pd}T_0}\right)^{c_{pd}/R_d},$$

where p_0 and T_0 are the pressure and **temperature** (K), respectively, at **sea level**, z the geometric height, R_d the **gas constant** for dry air, c_{pd} the **specific heat** of **dry air** at constant pressure, and g the **acceleration of gravity**. *See* **homogeneous atmosphere**.

adiabatic chart—(Or adiabatic diagram.) Same as **Stüve diagram**, but often used synonymously with **thermodynamic diagram**.

adiabatic condensation point—*See* **lifting condensation level**.

adiabatic condensation pressure—Same as **condensation pressure**.

adiabatic condensation temperature—Same as **condensation temperature**.

adiabatic cooling—*See* **adiabatic process**.

adiabatic diagram—Same as **adiabatic chart**.

adiabatic equilibrium—A vertical distribution of **temperature** and **pressure** in an **atmosphere** in **hydrostatic equilibrium** such that an **air parcel** displaced **adiabatically** will continue to possess the same temperature and pressure as its surroundings, so that no restoring force acts on a **parcel** displaced vertically.

A layer in adiabatic equilibrium has uniform **equivalent potential temperature**. The state of adiabatic equilibrium is approached in a layer of air in which there is strong vertical **mixing**. *Compare* **static stability**.

adiabatic equivalent temperature—*See* **equivalent temperature**.

adiabatic expansion—*See* **adiabatic process**.

adiabatic lapse rate—In most contexts, same as **dry-adiabatic lapse rate**.
See also **saturation-adiabatic lapse rate**.

adiabatic process—A process in which a system does not interact with its surroundings by virtue of a **temperature** difference between them.

In an adiabatic process any change in **internal energy** (for a system of fixed mass) is solely a consequence of working. For an **ideal gas** and for most atmospheric systems, compression results in warming, expansion results in cooling. See **dry-adiabatic process**, **moist-adiabatic process**.

adiabatic saturation point—*See* **lifting condensation level**, **saturation level**.

adiabatic saturation pressure—(Or condensation pressure.) Pressure at the **saturation level**.

adiabatic saturation temperature—(Or condensation temperature.) Temperature at the **saturation level**.

adiabatic temperature changes—Temperature changes related to changes of **pressure** without external gain or loss of **heat**.

In a compressible fluid, such as **seawater**, **temperature** rises as the fluid is compressed and **adiabatic cooling** occurs during expansion. The latter is of practical concern when water samples taken in thermally insulated water bottles are taken at great depth and raised to the surface.

adiabatic temperature gradient—The rate of change of **temperature** due to **pressure** under adiabatic conditions.

In practice, since in the sea the pressure changes can be considered proportional to depth changes, the adiabatic temperature gradient is usually given as rate of change per unit depth, instead of per unit pressure. For practical purposes, the unit of depth is often chosen as 1000 m.

adiabatic trial—Older expression for the process of lifting a **parcel** adiabatically on a **thermodynamic diagram** to ascertain when **convective instability** will occur.

adiabatic wet-bulb temperature—*See* **wet-bulb temperature**.

adiabatically enclosed system—A thermodynamic system in which no **heat** or mass is transported across its boundaries.

adjoint assimilation—A form of variational **data assimilation** in which adjoint equations are used to obtain gradients of a **scalar** measure of a forecast (J) with respect to **model** control **variables**.
The complete **assimilation** generally involves an iterative procedure to reduce the cost-function in which a minimization **algorithm** is used to adjust initial conditions based on a **sensitivity** gradient (∇J) provided by the **adjoint model**. *See* **adjoint sensitivity**, **variational objective analysis**.

adjoint equation—An equation of the form $\mathbf{x}_0 = \mathbf{L}^T\mathbf{x}_1$, in which the **linear operator** $\mathbf{L}^T$ is the adjoint of the matrix **operator** $\mathbf{L}$ that satisfies $(\mathbf{L}^T\mathbf{x}_1,\mathbf{x}_0) = (\mathbf{x}_1,\mathbf{L}\mathbf{x}_0)$, where $\mathbf{x}_0$ and $\mathbf{x}_1$ are vectors and $(,)$ represents an **inner product**.
If $(,)$ is the standard dot product (Euclidean inner product) then $\mathbf{L}^T$ is simply the transpose of $\mathbf{L}$. *See* **adjoint sensitivity**, **adjoint model**, **tangent linear equation**.

adjoint model—A **model** composed of adjoint equations that maps a sensitivity **gradient vector**, $\nabla_x J(t_0) = \mathbf{L}^T\nabla_x J(t_1)$, from a forecast time, t_1, to an earlier time, t_0, which can be the initial time of a forecast trajectory.
J is some **scalar** measure of the forecast, $\mathbf{L}^T$ is a **linear** adjoint **operator**, and $\mathbf{x}$ is the model state vector. An adjoint model can provide a first-order (tangent linear) approximation to **sensitivity** in a **nonlinear** model. *See* **adjoint equation**, **adjoint sensitivity**, **tangent linear equation**.

adjoint sensitivity—A **gradient**, $\nabla_x J$, of some **scalar** measure of a forecast, J, with respect to the **vector** of **model** control **variables**, $\mathbf{x}$, that can include initial conditions, **boundary conditions**, and parameters.
The **inner product** $(\delta\mathbf{x}, \nabla_x J)$, where $\delta\mathbf{x}$ is a **perturbation** vector and $\nabla_x J$ is an adjoint sensitivity vector, provides δJ, a first-order (tangent **linear**) approximation to the difference, ΔJ, between an unperturbed **nonlinear** forecast and a nonlinear forecast with control **variables** perturbed by $\delta\mathbf{x}$. *See* **adjoint model**, **tangent linear equation**, **tangent linear approximation**.

adjustable cistern barometer—A **mercury barometer** that is read by first bringing the free **mercury** surface in the **cistern** to a fixed level coincident with the zero of the **scale**.
See **Fortin barometer**.

admissible concentration limit—Upper limit of concentration of a substance in water that is deemed not harmful.
The definition of harmful substance is regulatory dependant.

adret—The slope (usually equatorward, or southward in the Northern Hemisphere) of a mountain that faces into the sun.
The term is originally and most often used in referring to mountains in the Alps. Tilted toward the sun, an adret is characterized by higher temperatures, a longer **growing season**, less **snow cover** and a shorter duration of snow cover, and a higher **timber line** and **snow line** than the shaded side (the **ubac**).

adsorption—The adhesion of a thin film of liquid or gas onto a solid substance.
The solid does not chemically combine with the adsorbed substance. *See* **sorption**; *compare* **absorption**.

adsorption isotherm—A boundary on a **phase** diagram that expresses the partitioning of a compound between solid and aqueous phases.
It is the **isothermal equilibrium** relationship between the concentration of a compound sorbed to the solid phase and the concentration of the same compound in the aqueous solution in contact with the solid phase.

Advanced Earth Observing Satellite—(Abbreviated ADEOS.) A Japanese **remote sensing** satellite designed to collect worldwide environmental data from a sophisticated suite of sensors.
The core sensors on *ADEOS* are an advanced visible and **near-infrared** radiometer and an ocean color and **temperature** sensor. Additional instruments on *ADEOS* include the NASA **scatterometer**, a **total ozone** mapping **spectrometer**, an instrument to measure the **polarization** and directionality of the earth's **reflectance**, an interferometric monitor for greenhouse gases, an improved limb atmospheric spectrometer, and a **retroreflector** in space. *ADEOS* was launched on 17 August 1996, into a **sun-synchronous orbit**. The satellite failed on 29 June 1997, but will be replaced by other satellites in the ADEOS series.

Advanced Microwave Sounding Unit—(Abbreviated AMSU.) An advanced version of the **MSU** on

14

POES satellites that will replace the older MSU and **SSU** instruments, starting with *NOAA-15* launched on 13 May 1998.

Advanced TIROS-N—(Abbreviated ATN.) *See* **TIROS**.

Advanced Very High Resolution Radiometer—(Abbreviated AVHRR.) A **sensor** carried on board **NOAA** satellites used in meteorology and **oceanography** for measurements of **cloud cover** and **sea surface temperature**.

The sampling rate is 2048 pixels per scan, with a scan angle of 55.4° off **nadir** and a **pixel** resolution of 1.1 km over five spectral bands. The spectral windows of the five AVHRR **channels** are:

    Channel 1: 580–680 nm
    Channel 2: 725–1100 nm
    Channel 3: 3550–3930 nm
    Channel 4: 10.3–11.3 μm
    Channel 5: 11.4–12.4 μm.

advection—The process of **transport** of an atmospheric property solely by the mass motion (**velocity** field) of the **atmosphere**; also, the rate of change of the value of the advected property at a given point.

Advection may be expressed in **vector** notation by

$$-\mathbf{u} \cdot \nabla\phi,$$

where **u** is the **wind vector**, ϕ the atmospheric property, and $\nabla\phi$ the **gradient** of the property. In three-dimensional **Cartesian coordinates**, it is

$$-\left(u\,\frac{\partial\phi}{\partial x} + v\,\frac{\partial\phi}{\partial y} + w\,\frac{\partial\phi}{\partial z}\right),$$

where u, v, and w are the **wind** components in the eastward, northward, and vertically upward directions, respectively. The first two terms compose the **horizontal advection** and the last term is the **vertical advection**. Also, it should be noted that the property ϕ may itself be a **vector field**. Often, particularly in **synoptic meteorology**, advection refers only to the horizontal or **isobaric** components of motion, that is, the **wind field** as shown on a **synoptic chart**. Regarding the general distinction (in meteorology) between advection and **convection**, the former describes the predominantly horizontal, large-scale motions of the atmosphere, while convection describes the predominantly vertical, locally induced motions. In **oceanography**, advection refers to the horizontal or vertical flow of **seawater** as a **current**.

advection fog—1. A type of **fog** caused by the **advection** of **moist air** over a cold surface, and the consequent cooling of that air to below its **dewpoint**.

A very common advection fog is that caused by **moist air** over a cold body of water (**sea fog**).

2. Sometimes applied to **steam fog**.

advection frost—The occurrence of **frost** as a result of the horizontal **transport** (**advection**) of a cold **air mass** with **air temperature** below 0°C.

This type of **frost** is responsible for causing damage to agricultural areas of south Florida and the Rio Grande Valley of Texas during cold polar outbreaks. *Compare* **radiation fog**.

advective acceleration—*See* **acceleration**.

advective change of temperature—The contribution to local **temperature** change that is caused by the horizontal or **vertical advection** of air.

The horizontal component of change, usually the most important in the **troposphere**, is proportional to the horizontal **temperature** gradient and the magnitude of the component of the **wind** in the direction of the **gradient**. The vertical component is proportional to the **vertical velocity** and the **static stability** and depends also on whether the air is saturated.

advective-gravity flows—A type of cold-air downslope flow at the bottom of the **boundary layer** where the two dominant processes are **advection** and **buoyancy**.

Once fully developed, this idealized flow is constant in time as long as cold air is supplied or produced. Wind speed increases with the square root of downslope distance, and flow depth increases linearly with distance. Typical depths are 2–10 m. *See* **katabatic wind**, **slope flow**, **gravity flow**, **drainage wind**.

advective model—A numerical forecast model based on discrete **advection** terms only, with less or no emphasis on forcing, **dissipation**, and physics.

Advective models are most appropriately applied to nearly conserved quantities in the **atmo-**

sphere or ocean, such as **potential vorticity**. Advective models are usually for one level only, or for a vertically integrated fluid. Historically, the **barotropic model** has been remarkably successful given its simplicity. A more recent example of an advective model is the divergent **anomaly** vorticity **advection** model.

> Qin, J., and H. M. van den Dool, 1996: Simple extensions of an NWP model. *Mon. Wea. Rev.*, **124**, 277–287.

advective region—(Obsolete; also called advection region, advection layer, advective layer.) The region of the **atmosphere** just above the **tropopause**, that is, the **stratosphere**.

This term arose from the fact that this part of the atmosphere is relatively more stable than the **troposphere**, so that **temperature** changes at a point in it occur primarily by means of **advection** rather than by **convection**.

advective term—Same as **convective term**.

advective thunderstorm—A **thunderstorm** resulting from **instability** produced by **advection** of relatively colder air at high levels, or relatively warmer air at low levels, or by a combination of both conditions.

advisory—Statements issued by a **weather service** that discuss weather situations of inconvenience that do not carry the danger of **warning** criteria, but, if not observed, could lead to hazardous situations.

Some examples include **snow** advisories stating possible slick streets, or **fog** advisories for patchy fog condition causing temporary restrictions to **visibility**. *See also* **special weather statements**.

advisory area—Area over which a meteorological **advisory forecast** applies.

advisory forecast—*See* **weather advisory**.

AE—An **index** of the strength of the **auroral electrojet**, tabulated for each 2.5 minutes and measured in **nT**.

aeolian—(Also spelled eolian.) Pertaining to the action or the effect of the **wind**, as in **aeolian sounds** or aeolian deposits (of **dust**); derived from the name of the Greek god of the winds, Aeolus, whose harp was held responsible for the murmur of the gentle breezes and whose conch-shell trumpet was regarded as the source of the **gale**'s howl.

aeolian sounds—*See* **aeolian tones**.

aeolian tones—Sound, usually in the **band** of audible frequencies, associated with wake-eddy, vortex-produced **pressure** fluctuations resulting from air flow around obstacles, such as wires and twigs.

Although many such sounds are irregular noises, other familiar sounds involve fairly clear musical notes or humming sounds. The latter sounds were called aeolian tones by Rayleigh. Their **pitch** is controlled by the **frequency** with which **eddies** are formed and detached in the **wake** region on the lee side of the obstacle. The tones produced by **wind** flowing over a cylinder, including stretched wire, were shown by Strouhal in 1878 to be of frequency (pitch) f given by

$$f = 0.185u/d,$$

where u is the cross-axis wind velocity (m s^{-1}) and d is the cylinder diameter (m).

> Lord Rayleigh, 1878: *The Theory of Sound*, Vol. II, 412–413.
> Humphreys, W. J., 1940: *Physics of the Air*, 3d ed., 442–448.

aeration—In general, any process whereby a substance becomes permeated with air or other gas; can be used synonymously with "ventilation."

This term refers to the formation and renewal of **soil air**. *See* **normal aeration**.

aerial—1. Same as **antenna**.

2. Of or pertaining to the air, **atmosphere**, or aviation.

aerial perspective—Reduced **contrast** of distant objects caused by **airlight**.

In artistic practice, aerial perspective is the painting of this reduced contrast to suggest an object's distance.

AERO code—(Short for aeronautical code.) An **international synoptic code** in which selected observable meteorological elements of particular interest to aviation operations are encoded and transmitted in groups ("words") of five numerical digits in length.

aerobiology—The study of the distribution of living organisms freely suspended in the **atmosphere** and some consequences of this distribution.

It includes microorganisms and some insects, seeds, and spores. Dispersion is by the **wind**, aided in some species by special adaptations and in some by flight. Small organisms are lifted and maintained in the air by **eddy diffusion** and rising **thermal currents**, sometimes reaching heights exceeding 16 km (10 miles) in air currents and by **lateral mixing**. They may be collected by adhesive-coated slides exposed horizontally, or by culture media, but for quantitative work, volumetric sampling at the surface or by aircraft is required.

Jacobs, W. C., 1951: *Compendium of Meteorology*, 1103–1111.

aerodrome elevation—Same as **airport elevation**.

aerodrome forecast—*See* **airport forecast**.

aerodrome meteorological minima—Limiting meteorological conditions prescribed for determining the usability of an aerodrome for a particular aviation operation (e.g., **IFR** takeoff or landing)

aerodynamic—Pertaining to a body moving in a fluid, especially air.

Sometimes the term aerodynamic is applied to a dynamically stable body (e.g., aircraft or automobile) with a low drag coefficient.

aerodynamic balance—An instrument used for the measurement of the forces exerted on the surfaces of instruments exposed to flowing air.

It is frequently used in tests made on models in **wind** tunnels. *Compare* **aerostatic balance**.

aerodynamic force—The force exerted by a moving fluid (especially air) on a body immersed in it.

The component of the aerodynamic force parallel to the direction of flow is called the **drag**.

aerodynamic resistance—(Also called **drag** or aerodynamic drag.) The component of force exerted by the air on a liquid or solid object (such as a **raindrop** or airplane) that is parallel and opposite to the direction of flow relative to the object.

See **aerodynamic force**.

aerodynamic roughness length—(Also called the roughness length, z_0.) The height above the displacement plane at which the mean wind becomes zero when extrapolating the logarithmic wind-speed **profile** downward through the **surface layer**.

It is a theoretical height that must be determined from the wind-speed profile, although there has been some success at relating this height to the arrangement, spacing, and physical height of individual roughness elements such as trees or houses. The average **wind speed** $\overline{M}$ in the surface layer can be written using **Monin–Obukhov similarity theory** as

$$\overline{M} = \frac{u_*}{k} \left[\ln \left(\frac{z - d}{z_0} \right) + \psi \left(\frac{z - d - z_0}{L} \right) \right],$$

where z is height above ground, d is height of the displacement plane above ground, L is the Monin–Obukhov length **scale**, k is **von Kármán's constant**, ψ is a **stability** correction factor (= 0 for statically neutral conditions), and u_* is the **friction velocity**. To determine aerodynamic roughness, most experimentalists prefer to make **wind profile** measurements during statically neutral conditions (windy, **overcast**, negligible **temperature advection**) so that $\psi = 0$ in the equation above. *Compare* **Charnock's relation**.

aerodynamic trail—*See* **condensation trail**.

aerodynamically rough surface—A surface with irregularities sufficiently large that the **turbulent boundary layer** reaches right down to the surface.

Individual roughness elements are higher than the **laminar (viscous) sublayer**. Virtually all surfaces on the earth, except glassy-smooth water surfaces, are aerodynamically rough. *Compare* **aerodynamically smooth surface**.

aerodynamically smooth surface—A surface with irregularities sufficiently small to be entirely embedded in the **laminar sublayer**.

Thus, the smoothness of a surface will depend on the **Reynolds number**. The irregularities are taken to be sufficiently dense and similar so that only the average height need be specified. A surface is aerodynamically smooth if

$$\frac{u_* z_0}{\nu} < \mathrm{Re}_\mathrm{f},$$

where u_* is **friction velocity**, z_0 is **aerodynamic roughness length**, and ν is the **kinematic viscosity** of air. The critical **Reynolds number** is in the **range** of $2.5 \leq \mathrm{Re}_\mathrm{f} \leq 70$. *Compare* **aerodynamically rough surface**.

aerodynamics—The study of the forces exerted on and the flow around bodies, especially aircraft, moving relative to a gas, especially the **atmosphere**.

Aerodynamics is sometimes used as a synonym for the science of flight.

aerogram—*See* **thermodynamic diagram**.

aerograph—In general, any self-recording instrument carried aloft by any means to obtain meteorological data.
See **meteorograph**.

aerography—1. (Rare.) Same as **descriptive meteorology**.

2. (Rare.) The so-called nonprofessional work of meteorology: observing, **map plotting**, maintaining records, etc.

aerologation—(Also called single-grid heading.) A method utilizing **grid navigation** to simplify flight problems attendant to **pressure-pattern flight**.

A **map projection** is used in which the **great circle** course approximates a straight line, and a rectangular **grid** overlay is superimposed on the map projection and oriented along the central **meridian** of the **projection**. Then the great circle course makes equal angles with all north–south lines on the grid. When the net **drift-correction angle** is applied to correct for **wind** effect, the entire **course** may be flown on a single **heading** with respect to the rectangular grid overlay. *Compare* **single drift correction**.

aerological analysis—The study of the physical state of the **atmosphere** derived from vertical **sounding** data plotted on thermodynamic diagrams.

aerological days—Internationally agreed upon days designated for more detailed or intensive observations of the **atmosphere** over broad areas of the earth.

aerological diagram—Same as **thermodynamic diagram**.

aerological station—A surface location from which **upper-air observations** are made.

aerological table—(Obsolete.) A table used for compiling aerological observations from **upper-air soundings**.

Upper-air sounding observations are presently calculated and compiled by computer programs.

aerology—1. The study of the **free atmosphere** throughout its vertical extent, as distinguished from studies confined to the layer of the **atmosphere** adjacent to the earth's surface.

2. (Obsolete.) As officially used by the U.S. Navy until 1957, same as meteorology.

aerometeorograph—A self-recording instrument used on aircraft for the simultaneous recording of **atmospheric pressure**, **temperature**, and **humidity**.
See **meteorogram**.

aeronautical climatology—The application of the data and techniques of **climatology** to aviation meteorological problems.

aeronautical meteorological service—Service designated to provide the **meteorological information** required for air navigation.

aeronautical meteorological station—Station designated to make observations and issue meteorological reports for use in international air navigation.

aeronautical meteorology—(Or aviation meteorology.) Very generally, meteorology as applied to the effects of weather upon aviation.

aeronomy—A term denoting the physics and chemistry of the **upper atmosphere**.

It is concerned with upper-atmospheric composition (i.e., nature of constituents, **density**, **temperature**, etc.) and chemical reactions.

aerosol—A **colloidal system** in which the **dispersed phase** is composed of either solid or liquid **particles**, and in which the **dispersion medium** is some gas, usually air.

There is no clear-cut upper limit to the size of particles composing the dispersed phase in an aerosol, but as in all other colloidal systems, it is rather commonly set at 1 μm. **Haze**, most **smokes**, and some **fogs** and **clouds** may thus be regarded as aerosols. However, it is not good usage to apply the term to ordinary clouds with drops so large as to rule out the usual concept of colloidal **stability**. It is also poor usage to apply the term to the dispersed particles alone; an aerosol is a system of dispersed phase and **dispersing medium** taken together. *Compare* **airborne particulates**, **particles**, **PM-2.5**, **PM-10**.

aerosol optical depth—The **optical depth** due to **extinction** by the **aerosol** component of the **atmosphere**.

Aerosol optical depths typically decrease with increasing **wavelength** and are much smaller for **longwave radiation** than for **shortwave radiation**. Values vary widely depending on atmospheric conditions, but are typically in the **range 0.02–0.2 for visible radiation**. *See also* **turbidity**.

aerosol size distribution—The amounts of different size **particles** of solids or liquids that are suspended in air as an **aerosol**.

Particle size affects **scattering** of **sunlight** (*see* **Rayleigh scattering** and **Mie scattering**) that makes blue skies, white clouds, and hazy **smog**, and that affects **visibility**. Size affects the **nucleation** capability of **particles** to form **cloud droplets** due to both the curvature effect and the solute effect. Relative amounts of different particle sizes can be used as a **tracer** for an **air mass**, such as indicating whether it originated over continents, oceans, urban areas, or rural areas. The total abundance of particles is often proportional to the total number density of cloud droplets, which affects the size to which these droplets can grow and their resulting evolution.

aerostatic balance—(Also called air poise.) An instrument for weighing air.
Compare **aerodynamic balance**.

aestival—(Also spelled estival.) Pertaining to summer.
The corresponding adjectives for autumn, winter, and spring are **autumnal**, **hibernal**, and **vernal**.

aestivation—A state of torpidity induced in some animals by the **heat** and dryness of the summer.

Afer—(Also called Africo, Africino, Africuo, Africus ventus.) In Italy, the southwest **wind**.

afghanets—A strong, gusty, **wind** that occurs on the upper course of the Amu Darya (river) in Turkmenistan (i.e., coming from Afghanistan). It is preceded and accompanied by **duststorms**.

African jet—A low-level easterly **jet** in the summer months over the Sahara Desert of North Africa.
The **wave** disturbances drawing **energy** from this jet propagate westward to the Atlantic Ocean. They are also known as African waves. Some of them under favorable conditions become hurricanes that reach North America. *Compare* **easterly wave**.

Africo—Same as **Afer**.

afterglow—1. A broad arc occasionally seen in the solar (as opposed to antisolar) sky during the darker half of **civil twilight** and, in principle, during nautical and **astronomical twilight**.
The afterglow chiefly consists of the **purple light** and **bright segment**.
2. A **stage** of the **alpenglow** in which mountaintops are visibly colored by the **purple light**.

afterheat—(Rare.) The warm weather of **Indian summer**.

afternoon effect—The trapping of acoustic **energy** near the surface of the ocean as the result of solar heating on days with low winds.

aftersummer—(Rare.) A warm period in the fall, such as **Indian summer**.

age—In **oceanography**, the time interval between an astronomical event and the corresponding tidal phenomenon.
For example, the old oceanographic term for the **lag** in days between the occurrence of **syzygy** and the highest **spring tide** is the "age of the **tide**;" the lag between **perigee** and the highest **perigean tide** is the "age of parallax inequality."

age of the tide—*See* **age**.

ageostrophic advection—The **transport** of atmospheric properties by and in the direction of the **ageostrophic wind**.

ageostrophic wind—The **vector** difference between the real (or observed) **wind** and the **geostrophic wind**, that is, $\mathbf{u}_{ag} = \mathbf{u} - \mathbf{u}_g$.
Sometimes the magnitude of this vector difference is meant.

agglomeration—(Obsolete.) The process of growth of **hydrometeors** in **clouds** by the collection of other hydrometeors.
Another obsolete term for these processes is **coagulation**. **Accretion, coalescence,** and **aggregation** are the terms presently in use.

aggregation—1. The process of combining different surface characteristics from neighboring heterogeneous regions into an average value for the area.
It is used in **boundary layer** studies for surface fluxes, **drag**, and roughness. This process is

often necessary to define surface characteristics for numerical models that have coarse horizontal **grid** mesh and that cannot resolve the individual surface areas.

2. The process of clumping together of **snow crystals** following collision as they fall to form **snowflakes**.

This process is especially important near the **melting layer** where snow particles stick to each other more easily because of the liquid water on the surface. It also occurs at lower temperatures especially between dendritic snow crystals and occasionally rosette crystals in **cirrus**.

aggressive water—Soft, acidic water that is corrosive to metals (e.g., pipes).

agnostic chart—A **prognostic chart** that no one believes.

agonic line—The line through all points on the earth's surface at which the **magnetic declination** is zero; that is, the locus of all points at which **magnetic north** and **true north** coincide.

This line is a particular case of an **isogonic line**. The position of this line exhibits variations in time, but in 1995 was so located that it emanated from the north magnetic **pole**, trended southward through the Great Lakes region, left the American mainland near Mississippi, cut across South America to near Buenos Aires, thence through the south magnetic pole, and up in an irregular path on the other side of the earth to return to the north magnetic pole. At the present time, the North American segment of the agonic line is drifting very slowly westward. *Compare* **aclinic line**.

agricultural climatology—In general, **climatology** as applied to the effect of **climate** on crops.

It includes especially the length of the **growing season**, the relation of growth rate and crop yields to the various climatic factors and hence the optimal and limiting climates for any given crop, the value of **irrigation**, and the effect of climatic and weather conditions on the development and spread of crop diseases. This discipline is primarily concerned with the space occupied by crops, namely, the soil and the layer of air up to the tops of the plants, in which conditions are governed largely by the **microclimate**.

agricultural drought—Conditions that result in adverse crop responses, usually because plants cannot meet **potential transpiration** as a result of high atmospheric demand and/or limited **soil moisture**.

Drought severity may be defined according to the **Palmer Drought Severity Index** or functionally expressed as yield-reducing **water stress**.

agricultural meteorological station—A collection of sensors connected to a **data logger**, designed to accumulate several types of soil and **atmosphere** observations, and report weather variables related to agriculture, representing conditions for a designated area.

agricultural meteorology—In general, meteorology and **micrometeorology** as applied to specific agricultural systems and of agriculture as applied to specific atmospheric conditions.

This discipline may emphasize **atmospheric transport** of insects, pathogens, etc., that impact agriculture as well as **energy** and mass exchange of plants and animals with the atmospheric **environment**. The effect of soils and vegetation on the ratio of sensible and latent energy exchange is representative of the impact of agriculture on meteorology.

agro-met station—Abbreviation for **agricultural meteorological station**.

agroclimatic index—A measure or **indicator** of an aspect of the **climate** that has specific agricultural significance.

Examples of an agroclimatic index are average length of **growing season** (period between average last and first **freezing** temperature dates), average growing **degree-days** or **heat units** in a **growing season**, average total **chill hours** or **chill units**, and average **evapotranspiration**.

agroclimatology—*See* **agricultural climatology**.

agrometeorological forecasting—The **prediction** of certain weather conditions or patterns that may have a significant effect on agriculture.

agrometeorological station—*See* **agricultural meteorological station**.

agrometeorology—*See* **agricultural meteorology**.

Agulhas Current—(Also called Agulhas stream.) The major western semi of the subtropical **gyre** in the southern Indian Ocean and one of the swiftest **ocean currents** with mean speeds of 1.6 m s^{-1} and peak speeds exceeding 2.5 m s^{-1}.

Its total **transport** of 70 Sv (70 × 10^6 m^{3}s^{-1}) near 31°S and up to 135 Sv (135 × 10^6 m^{3}s^{-1}) near 35°S is also among the largest of all ocean currents. The Agulhas Current is fed mainly from the **East Madagascar Current** and to a smaller degree from the **Mozambique Current**. When passing the Agulhas Bank, the current produces significant **upwelling**. To the south of the Cape of Good Hope, the current flows west to southwestward first but turns sharply eastward when reaching the Agulhas Current retroflexion region near 40°S, 20°E. Eddies spawned in this region continue to move westward and turn northward to join the **Benguela Current**. The transport of water from the Indian into the Atlantic Ocean through the **eddies** is an important part of the global **ocean conveyor belt**.

Agulhas stream—*See* **Agulhas Current**.

air—Mixture of gases forming the earth's **atmosphere**, consisting of nitrogen (~78 percent), oxygen (~21 percent), **water vapor**, and other **trace gases** such as carbon dioxide, helium, argon, **ozone**, or various pollutants.

The concentration of water vapor is very variable, being a strong function of **temperature** and, hence, **altitude** in the atmosphere. **Dry air** is referred to as air from which measurable amounts of water vapor have been physically removed. Pure, dry air has a **density** of 1.293 kg m^{-3} at a temperature of 273 K and a **pressure** of 101.325 kPa. Apart from the variability of water vapor, the composition of air is essentially constant to an altitude of at least 50 km and is presently approximated as follows.

Constituent	Symbol	Volume Fraction
Nitrogen	N$_2$	.78084
Oxygen	O$_2$	.20948
Argon	Ar	9.3×10^{-3}
Carbon dioxide	CO$_2$	3.6×10^{-4}
Neon	Ne	1.82×10^{-5}
Helium	He	5.24×10^{-6}
Methane	CH$_4$	1.7×10^{-6}
Krypton	Kr	1.14×10^{-6}
Hydrogen	H$_2$	5.0×10^{-7}
Nitrous oxide	N$_2$O	3.3×10^{-7}

The concentration of ozone is variable, between 10 and 0.1 parts per million. Carbon dioxide, methane, and nitrous oxide have all been increasing since the beginning of the industrial age.

air acoustic ranging sensor—A pulsed, acoustic ranging device to determine water levels using the air column in a tube as the acoustic **sound** path.

The fundamental measurement is the time it takes for the acoustic **signal** to travel from a **transmitter** to the water surface and then back to the **receiver**. The distance from a reference point to the water surface is derived from the **travel time**. A **calibration** point is set at a fixed distance from the **acoustic transducer** and is used to correct the measured distance using the calibrated sound velocity in the tube.

air conductivity—*See* **conductivity**.

air current—Very generally, any moving **stream** of air; it has no particular technical connotation. *See* **wind**; *compare* **ocean current**.

air discharge—A form of **lightning discharge** similar to a **cloud discharge** in which the **lightning channel** propagates away from a **cloud** charge center into apparently **clear air** where it terminates.

Thus, cloud charge is moved away from its original location and **space charge** of opposite sign outside the cloud may be neutralized.

air drainage—General term for gravity-induced, downslope flow of relatively cold air.

Winds thus produced are called **gravity winds, slope winds, katabatic winds**, or **drainage winds**.

air–earth conduction current—(Also called fair-weather current.) That part of the **air–earth current** contributed by the electrical **conduction** of the **atmosphere** itself.

It is represented as a downward **current** in storm-free regions all over the world. The **conduction current** is the largest portion of the air–earth current, far outweighing the contributions made by the **precipitation current** and **convection current**, which are zero in storm-free regions. Its magnitude is approximately 3×10^{-12} amperes (A) m^{-2}, or about 1800 A for the entire earth. Such observations of the vertical **variation** of the conduction current as have been made indicate that it is approximately uniform throughout the depth of the **troposphere**, a condition that is consistent with the generally accepted view that the conduction current flows from a positively charged conducting region in the lower **ionosphere** downward to the negatively charged earth. Only in areas of temporarily disturbed weather does the conduction current become replaced by reverse flow. Accumulating evidence points to the conclusion that the conduction current continues to exist only because of the action of thunderstorms scattered at all times over the earth, which supply the positive charge to the **upper atmosphere** and negative charge to the earth. *See* **supply current**.

> Gish, O. H., 1951: *Compendium of Meteorology*, p. 113.

air–earth current—The **transfer** of **electric charge** from the positively charged **atmosphere** to the negatively charged earth.

This **current** is made up of the **air–earth conduction current**, a point-discharge current, a **precipitation current**, a **convection current**, and miscellaneous smaller contributions. Of these, the air–earth conduction current is by far the largest. This is not just true locally, but throughout the world where there are no thunderstorms occurring, which is estimated to be 80%–90% percent of the earth. The existence of this quasi-steady current in **fair** weather and the observed maintenance of the earth's net negative charge are both better established than the nature of the **supply current**, which must replenish the positive charge in the **upper atmosphere** and the negative charge on the earth.

> Gish, O. H., 1951: *Compendium of Meteorology*, p. 113.

air freezing index—*See* **freezing index**.

air–fuel ratio—The ratio of air in a fuel mixture relative to the exact amount required to convert all of a hydrocarbon fuel to water and **carbon dioxide** (known as a stoichiometric mixture).

If the air content is higher than stoichiometric, the mixture is said to be fuel-lean; if the air content is less, the mixture is fuel-rich.

air hoar—1. A deposit of **hoarfrost** on objects above the earth's surface.

Compare **surface hoar, depth hoar**.

2. (Rare.) Same as **ice fog**.

air line correction—Depth corrections applied to **sounding** lines for any portion of the line above the water, when high velocities, great depth, insufficient **sounding weight**, etc., result in large angles of inclination.

air-line sounding—A technique to determine the **water level** in a well, lake, or any water body with a **free surface**, using the principle of measuring back-pressure in a pressurized air line that extends downward below the water surface to a known **elevation**.

When air under **pressure** is introduced to the air line, it forces the water in the line out until the air bubbles from the bottom of the tube. At this time, noting the air pressure in the line, the pressure is converted mathematically into a height-of-water volume. Air-line sounding is a very effective means of measuring water levels in very deep wells.

air-line well—A well equipped with an **air-line sounding** system.

air mass—1. A widespread body of air, the properties of which can be identified as 1) having been established while that air was situated over a particular region of the earth's surface (**airmass source region**), and 2) undergoing specific modifications while in transit away from the source region.

An air mass is often defined as a widespread body of air that is approximately homogeneous in its horizontal extent, particularly with reference to **temperature** and moisture distribution; in addition, the vertical temperature and moisture variations are approximately the same over its horizontal extent. The stagnation or long-continued motion of air over a source region permits the vertical temperature and moisture distribution of the air to reach relative **equilibrium** with the underlying surface. *See* **airmass classification**.

2. In **radiation**, the ratio of the actual path length taken by the direct solar **beam** to the analogous path when the sun is overhead from the top of the **atmosphere** to the surface.

Extrapolation of surface measurements to zero air mass was the original method for estimating the value of solar **irradiance** at the top of the atmosphere.

3. *See* **optical air mass**.

air meter—A small, sensitive windmill **anemometer** having flat vanes.

It indicates the number of **linear** feet (or meters) of air that have passed the instrument during its **exposure**.

air parcel—An imaginary volume of air to which may be assigned any or all of the basic dynamic and thermodynamic properties of atmospheric air.

A **parcel** is large enough to contain a very great number of molecules, but small enough so that the properties assigned to it are approximately uniform within it and so that its motions with respect to the surrounding **atmosphere** do not induce marked compensatory movements. It cannot be given precise numerical definition, but a cubic foot of air might fit well into most contexts where air parcels are discussed, particularly those related to **static stability**. Any **fluid parcel** may be defined similarly.

air pocket—A local **downdraft** or an abrupt reduction of **headwind** or increase in **tailwind** that causes an airplane to drop suddenly.

air poise—Same as **aerostatic balance**.

air pollutants—Substances that do not occur naturally in the **atmosphere**.
Compare **designated pollutant, criteria pollutants, air toxins**.

air pollution—The presence of substances in the **atmosphere**, particularly those that do not occur naturally.
These substances are generally contaminants that substantially alter or degrade the quality of the atmosphere. The term is often used to identify undesirable substances produced by human activity, that is, **anthropogenic** air pollution. Air pollution usually designates the collection of substances that adversely affects human health, animals, and plants; deteriorates structures; interferes with commerce; or interferes with the enjoyment of life. *Compare* **airborne particulates, designated pollutant, particulates, criteria pollutants**.

air pollution alert—*See* **air pollution episode**.

air pollution control—The process of attempting to limit the amount of **air pollution** by regulating the emission of pollutants or their precursors.
Control strategies are alternative long-term policies that could reduce air pollution, as projected using air-quality modeling. For **primary pollutants** such as **sulfur dioxide**, control strategies include burning cleaner low-sulfur coal, scrubbing sulfur dioxide from the combustion products before releasing them into the **atmosphere**, or changing to alternative fuels or processes. For **secondary pollutants** such as **ozone** that are not emitted directly but are created in the **lower atmosphere** by a complex series of chemical and photochemical reactions involving NO_x and volatile **hydrocarbons**, control is achieved by changing the emissions of one or both of the primary reactants.

air pollution episode—1. An extended period of a high concentration of pollutants in the **atmosphere**.
2. A public alert or notification of unhealthful air quality.
In the U.S., the degree of alert is often based on the **pollutant standards index** (PSI):
Stage 1 episode or alert, for $200 \leq PSI < 275$. Public recommendation: avoid strenuous activities. Children and elderly people advised to stay indoors.
Stage 2 episode or alert, for $275 \leq PSI < 400$. Public recommendation: cease all physical activity, and stay indoors. Industry can be ordered to reduce emissions, and employees forced to carpool or use mass transportation for their necessary travel.
Stage 3 episode or alert, for $PSI \geq 400$. Everyone ordered to remain at home with windows closed. Minimize physical exertion and avoid driving vehicles. Automobile travel could be restricted and industries ordered to reduce emissions.

air pollution meteorology—The subdiscipline of meteorology devoted to the study of **air pollution**.
Topics include sources of pollutants, emission rates, **plume** rise, **fallout, dry** and **wet deposition**, chemistry, **precipitation scavenging, dispersion (molecular diffusion** and **turbulent transport)**, short- and long-range **transport (advection)**, trapping, venting by **cumulus** clouds, **complex terrain** and **mesoscale** circulations, receptors, impact on society, alerts and episodes, policy and regulation, modeling, **prediction**, control, and **climate change**.

air quality criteria—Quantitative and qualitative indications of the relationships between **exposure** to pollutants and effects on human health, animals, plants, and materials.
These are descriptive effects of **pollution** as a function of concentration averaged over various time durations. A time-averaged concentration is used because exposure to a high concentration of pollutants during a short time might have an impact equivalent to an exposure to a lower concentration over a longer time. *Compare* **air quality standards**.

air quality standards—Maximum legal concentration limits of **air pollutants** averaged over specified time periods.
These are prescriptive. If the actual measured concentration averaged over time exceeds the legal threshold, then the event is called an exceedence. Regulations often allow a limited number of exceedences each year, with fines or penalties imposed for too many exceedences. Regions with too many exceedences are required to develop plans to improve the air quality, for example, by changing automobile fuels and gas-station equipment or increasing the use of mass transportation. In the

United States, the standards are called **National Ambient Air Quality Standards** (NAAQS). Primary standards are designed to protect human health. Secondary standards are designed to protect crops, animals, structures, and commerce. In Canada, these legal thresholds are called National Air-Quality Objectives. Three levels of standards exist within these objectives. In ascending order of concentration limit they are Maximum Desirable Level, Maximum Acceptable Level, and Maximum Tolerable Level. *Compare* **air quality criteria, emission standard, criteria pollutants**.

air report—(Abbreviated AIREP.) A report from an aircraft in flight prepared in conformity with requirements for position and operational and/or meteorological reporting.
See **pilot report**.

air resistance—*See* **aerodynamic resistance**.

air sampler—A device used on a moving platform such as an aircraft that, when quickly opened and closed, captures a representative sample of the **atmosphere**.

air–sea interaction—The processes that occur as a consequence of the air being in contact with the sea surface, and that affect the dynamics and **thermodynamics** of the air and water **boundary layers**.
These include 1) the exchange of **momentum, heat,** mechanical **energy** (e.g., **wave** energy, **turbulence**), and mass (**water vapor**, gas species, **particulates**, sea spray, air bubbles, etc.); 2) the generation of **surface waves**; 3) the generation of turbulence; and 4) the resulting effects on the vertical profiles of **wind** and **current**.

air shower—*See* **cascade shower**.

air temperature—The **temperature** indicated by a **thermometer** exposed to the air in a place sheltered from **direct solar radiation**.

air thawing index—*See* **thawing index**.

air toxins—Hazardous **air pollutants** that are known or suspected to cause cancer or other serious health effects (such as birth or developmental defects).
The U.S. Clean Air Act Amendments of 1990 require emission reductions for 188 hazardous air pollutants from industrial factories and other sources. As of 1996, the U.S. Environmental Protection Agency has issued standards for 47 source categories, such as chemical plants, oil refineries, aerospace manufacturers, and steel mills, as well as dry cleaners, commercial sterilizers, secondary lead smelters, and chromium electroplating processors. *Compare* **designated pollutant, criteria pollutants**; *see* **downwash**.

air trap—A device incorporated in some types of **mercury barometer** to prevent air or other gaseous impurities from entering into the vacuum space.

airborne expendable bathythermograph—(Abbreviated AXBT.) An expendable instrument that is dropped from an aircraft and used to measure the **profile** of **temperature** in the water column.
The **probe** consists of a **thermistor** in a weighted, streamlined case. It falls freely at a fixed, known rate so that the elapsed time can be converted to depth. It is connected by a thin, freely unwinding wire to a small buoy with a radio **transmitter** through which the data are transmitted to the aircraft, which continues its flight.

airborne fraction—The fractional amount of **carbon dioxide**, CO_2, that remains in the gas **phase** relative to a given increase in the total amount of CO_2 (**atmosphere** and ocean combined).
The relatively high solubility of CO_2 in water causes it to partition more favorably in the oceans, and so any potential atmospheric increase is buffered.

airborne particulates—Solid **particles** suspended in the air.
Larger particles (>100 μm approximately) have terminal velocities greater than about 0.5 m s^{-1} and fall out quickly. These include **hail, snow, graupel,** insect debris, room **dust, soot** aggregates, coarse sand, gravel, and sea spray. Medium-size particles (1 to 100 μm approximately) have **sedimentation** velocities greater than 0.2 m h^{-1} and settle out slowly. These include fine **ice crystals**, pollen, hair, large bacteria, windblown dust, **fly ash**, coal dust, **silt**, fine sand, and small dust. Small particles (<1 μm, approximately) fall so slowly that they can take days to years to settle out of a quiescent **atmosphere**. For a turbulent atmosphere they may never fall out; however, they can be washed out by **rain** in a process called rainout or **washout**, leading to **wet deposition** onto the earth's surface. Examples of these particles include viruses, small bacteria, metallurgical fumes, soot, oil **smoke**, tobacco smoke, clay, and fumes. Oil and tobacco smoke are sticky, and are removed from the atmosphere when they happen to touch and stick to an object such as plant or house furnishings, in a process called **dry deposition**. *See* **criteria pollutants, precipitation**.

airborne weather radar—**Radar** equipment mounted on an aircraft for purposes of weather avoidance or **weather observation**.

Airborne weather avoidance radars are usually mounted in the nose of the aircraft, **scanning** ahead of the aircraft for weather hazards. Airborne research radars are frequently **Doppler radars** that provide measurements of **radial wind** or **turbulence** in addition to **radar reflectivity**. Such radars are typically side-pointing or **helical scanning** and may use dual beams to enable **dual-Doppler analysis** of **storm** motions. Airborne weather radars ordinarily operate at wavelengths of C-band or shorter to keep the physical size of the equipment small.

aircraft ceiling—1. After U.S. weather observing practice, the **ceiling classification** applied when the reported **ceiling** value has been determined by a pilot while in flight within one and one-half nautical miles of any runway of the airport.

Aircraft ceilings may refer to **vertical visibility** or obscuring phenomena aloft as well as to clouds, and are designated A in aviation weather observations.

2. The maximum **altitude** at which any given aircraft can be operated safely.

aircraft charge and/or charging—Same as **aircraft electrification**.

aircraft electrification—The accumulation of a net electrical charge on the surface of aircraft, or the separation of charge into two concentrations of opposite sign on distinct portions of the aircraft surface.

Net charges appear as a result of **autogenous electrification** when aircraft fly through clouds of **ice crystals** or **dust**. Charge separation occurs by **induction** when aircraft fly through regions of strong **atmospheric electric field**, as in thunderstorms. Charging may also occur by the engine exhaust carrying away a net charge leaving the aircraft charged. Development of large local charges on aircraft parts may lead to appearance of **corona discharge** or **St. Elmo's fire** and is almost always accompanied by poor radio communication due to so-called **precipitation** static.

aircraft ice accretion—Same as **aircraft icing**.

aircraft icing—(Also called airframe icing, aircraft ice accretion.) Formation of **ice**, **rime**, or **hoarfrost** on an aircraft.

aircraft meteorological station—**Meteorological station** situated aboard an aircraft.

aircraft observation—An evaluation by an **observer** aboard an aircraft in flight of one or more of the following elements: **temperature**; **humidity**; **wind speed** and **direction**; **turbulence**; **icing**; prevailing **hydrometeors** and **lithometeors**; the **state of the sky**; and the height of a specified **pressure** surface.

The **observation** includes the position and the **altitude** of the aircraft and time of observation. This term is applied to those in-flight observations that are conducted on a fairly complete and formal basis, such as in **aircraft weather reconnaissance**. *Compare* **pilot report, apob, aircraft sounding**.

aircraft report—(Abbreviated AIREP). Same as **pilot report**.

aircraft sounding—**Pressure, temperature**, and **humidity** data observed during climb or descent of an aircraft.

See **apob**; *compare* **aircraft observation**.

aircraft thermometry—The art of measuring the **temperature** of the **environment** from aircraft.

aircraft-to-satellite data relay—(Abbreviated ASDAR.) A communication method where data are sent from an aircraft to a satellite that relays the data to another location.

This method extends line-of-sight communications.

aircraft turbulence—Irregular motion of an aircraft in flight, especially when characterized by rapid up-and-down motion, caused by a rapid **variation** of atmospheric **wind** velocities.

This can occur in **cloudy** areas (particularly **towering cumulus** and **lenticular clouds**) and in **clear air**. Turbulence is the leading cause of nonfatal passenger and flight attendant injuries. The U.S. Federal Aviation Administration (FAA) classifies aircraft turbulence as follows:

Light: Causes slight, erratic changes in **altitude** and/or attitude, and rhythmic **bumpiness** as occupants feel a slight strain against seat belts.

Moderate: Similar to light, but of greater intensity, with rapid bumps or jolts, and occupants feel a slight strain against seat belts.

Severe: Turbulence that causes large, abrupt changes in altitude and attitude, and large variations in **airspeed**, with the aircraft temporarily out of control. Occupants are forced violently against their seat belts and objects are tossed about, with food service and walking impossible.

Extreme: The aircraft is tossed about so violently that it is practically impossible to control, and structural damage may occur.

See also **clear air turbulence**.

aircraft weather reconnaissance—(Or weather reconnaissance,) The making of detailed **weather observations** or investigations from aircraft in flight.

The aircraft may either fly regularly scheduled flights along a fixed route (usually over areas not covered by land or ship stations), or it may fly a special mission to survey a particular weather phenomenon such as a **thunderstorm** or **tropical cyclone**. The **RECCO code** is used to report **weather reconnaissance** data.

AIREP—Abbreviation for **air report**.

AIREP is also the name of the code used to transmit the air report information.

airfield color code—A code used to summarize, by category, actual or forecast **ceiling** and **visibility** conditions at an airfield.

Each ceiling and visibility category is assigned a specific color.

airframe icing—Same as **aircraft icing**.

airglow—(Also called light-of-the-night-sky, night-sky light, night-sky luminescence, permanent aurora.) The quasi-steady **radiant** emission from the **upper atmosphere** over middle and low latitudes, to be distinguished from the sporadic **emission** of auroras that occur over high latitudes.

Airglow is a photochemical **luminescence** (or **chemiluminescence**) arising from chemical reactions in the upper atmosphere. Many of these reactions leave molecules and atoms in excited states from which they can radiate at certain well-defined wavelengths. Emissions from molecular **oxygen** O_2, atomic oxygen O, **sodium** Na, and the **hydroxyl radical** OH are especially prominent, and measurements of airglow intensity by spectrometric techniques have provided a great deal of information about upper-atmospheric dynamics and chemistry.

airlight—**Light** scattered into quasi-horizontal viewing directions by the **atmosphere**'s molecules and small **particles** (usually excluding **fog** and **raindrops**).

On **clear** to **partly cloudy** days, airlight is dominated by singly **scattered** sunlight, but it also includes multiply scattered sunlight from clouds, the surface, and the **clear sky**. A more restrictive definition limits airlight to clear days; a less restrictive one does not limit the **elevation** angles at which airlight is seen. As **haze** concentrations increase, airlight radiances increase near the **horizon**, although not without limit. Increased airlight reduces a distant object's **contrast**, thus reducing its **visibility**. At sufficiently large horizontal optical thicknesses, airlight reduces this contrast below the **threshold contrast**, and the object is then visually indistinguishable from its surroundings. Airlight and the artistic term **aerial perspective** are related, since airlight causes the reduced contrast associated with increased distance to an object. Airlight is distinct from **airglow** that originates in the **upper atmosphere**.

airlight formula—(Also called Koschmieder's law.) A version of the **radiative transfer** equation that predicts the **luminance** due to **airlight** (L_a) in the direction of a perfectly black object near the **horizon**.

L_a can be used to calculate **daytime visual range**. For horizon sky **luminance** L_b, horizontally uniform **extinction coefficient** σ, and distance x from the observer:

$$L_a = L_b(1 - \exp(-\sigma x)).$$

airmass analysis—In general, the theory and practice of **surface synoptic chart** analysis by the so-called Norwegian methods, which involve the concepts of the **polar front** and of the broad-scale **air mass** that it separates.

Airmass analysis of surface charts may be said to consist of 1) determining the extent, the physical and **stability** properties, the movements, and the modifications of each of the air masses on the **chart**; 2) locating with some precision the **fronts** separating the air masses, and analyzing the structure and motion of the fronts; 3) analyzing **wave** perturbations on the fronts; and 4) describing and explaining the weather on the basis of the above factors.

airmass classification—A system used to identify and to characterize the different **air masses** according to a basic scheme.

A number of systems have been proposed, but the **Bergeron classification** has been the most widely accepted. In this system, air masses are designated first according to the **thermal** properties of their source regions: **tropical** (T); **polar** (P); and less frequently, **arctic** or **antarctic** (A). For characterizing the moisture distribution, air masses are distinguished as to **continental** (c) and **maritime** (m) source regions. Further classification according to whether the air is cold (k) or warm (w) relative to the surface over which it is moving indicates the low-level **stability** conditions of the air, the type of modification from below, and is also related to the weather occurring within the air mass. This outline of classification yields the following identifiers for air masses: cTk, cTw, mTk, mTw, cPk, cPw, mPk, mPw, cAk, mAk, mAw; the last of which is never used. H. C. Willett, in his classification, introduces further distinction between stable (s) and unstable (u) conditions in upper levels. Some authors may include **equatorial** (E), **monsoon** (M), or **superior air** (S) in their classifications. Others prefer to omit the arctic (A) type and describe all air masses on the basis of polar and tropical air, separated by the **polar front**.

Byers, H. R., 1944: *General Meteorology*, p. 247.
Hewson, E. W., and R. W. Longley, 1944: *Meteorology, Theoretical and Applied*, p. 249.
Willett, H. C., 1944: *Descriptive Meteorology*, 183–191.
Palmén, E., 1951: *Compendium of Meteorology*, 599–620.

airmass climatology—The representation of the **climate** of a region by the **frequency** and characteristics of the **air masses** under which it lies; basically, a type of **synoptic climatology**.

It is a development of the representation of the weather associated with winds of different directions (thermal **wind roses**, etc.) by taking account of the source and **trajectory** of the air, and it gives a more dynamic picture of the climate than do monthly averages. The first detailed study was by E. Dinies (1932). Dinies gave tables of average **temperature**, **humidity**, and **cloudiness** in Germany associated with eight classes of air in winter and summer (Durst 1951). Similar tables have since been constructed for many parts of the world.

Dinies, E., 1932: Luftkörper-Klimatologie. *Aus. d. Arch. dtsch. Seew*, **50**, No. 6.
Durst, C. S., 1951: *Compendium of Meteorology*, 967–975.

airmass modification—The change of characteristics of an **air mass** as it moves away from its region of origin.

For example, **maritime air** at midlatitudes that originally has high humidity and cool temperatures can be modified as it blows onshore over coastal mountains, where **orographic precipitation** causes the air to become drier and warmer by the time it reaches the lee side of the mountains. The rate of airmass modification depends on the differences between its original characteristics and those of the new surface over which it flows.

airmass precipitation—Any **precipitation** that can be attributed only to moisture and **temperature** distribution within an **air mass** when that air is not, at that location, being influenced by a **front** or by **orographic lifting**.

The most common form of airmass precipitation is **airmass showers**; however, a moist but **stable air mass** may produce **drizzle** independent of frontal or **orographic** influences.

airmass shower—A **shower** that is produced by local **convection** within an **unstable air mass**; the most common type of **airmass precipitation**.

Such showers are not associated with a **front** or **instability line**. They are most frequent within a **moist air** mass that is sufficiently unstable so that daytime heating at the surface can produce well-developed **cumulus** clouds. The extreme form of airmass shower is the **airmass thunderstorm**.

airmass source region—Region where **air masses** originate and acquire their horizontal homogeneous properties of **temperature** and moisture.

Horizontal homogeneity of the air mass is produced by prolonged contact (days to weeks) with the underlying surface. Main source regions are those in which the permanent or semipermanent anticyclones occur.

airmass thunderstorm—*See* **airmass shower**.

airplane observation—*See* **apob, aircraft observation**.

airport elevation—(Also called field elevation, runway elevation, aerodrome elevation, airport height.) The officially designated **elevation** of an airport above **mean sea level**, in international usage denoted by the symbol H_a.

It is the elevation of the highest point on any of the runways of the airport. *Compare* **station elevation**.

airport forecast—Aviation **meteorological forecast** for a given airport.

airport height—Same as **airport elevation**.

airshed model—A **model** designed to study and predict the air quality of a given region, usually of dimensions of the order of hundreds of kilometers.

The models typically include sources of **primary pollutants**, the meteorology of the region being modeled, and a chemical scheme that describes the production of **secondary pollutants**. Airshed models of urban regions are the most common.

airspeed—The speed on an exposed (usually airborne) object relative to the **atmosphere**.

In a **calm** atmosphere, airspeed equals **ground speed**.

airway—An officially designated air route with sectors defined as specific courses to or from directional radio stations.

airways code—*See* **U.S. airways code**

airways forecast—Same as **aviation weather forecast**.

airways observation—Same as **aviation weather observation**.

airways shelter—A small **instrument shelter** designed for use at supplementary airways weather reporting stations (SAWRS).

It is a white, louvered box mounted on a single post.

Aitken dust counter—(Or Aitken nucleus counter.) An instrument developed by John Aitken for determining the **dust** content of the **atmosphere**.

A sample of air is mixed in an expandable chamber with a larger volume of dust-free air containing **water vapor**. Upon sudden expansion, the air in the chamber cools **adiabatically** below its **dewpoint**, and **droplets** form with the dust particles as nuclei. A portion of these droplets settle on a ruled plate in the instrument and are counted with the aid of a microscope. *See* **dust counter**.
Aitken, J., 1923: *Collected Scientific Papers*, 236–246.

Aitken nucleus—Any one of the many microscopic **particles** in the **atmosphere** that serve as **condensation nuclei** for **droplet** growth in the large **supersaturations** (greater than a few hundred percent over water) produced during the rapid, near **adiabatic expansion** produced in an **Aitken dust-counter**.

These nuclei, often numbering many tens of thousands per cubic centimeter in city air, are both solid and liquid particles with diameters on the order of tenths of microns or smaller. Because of the excessive supersaturations that accompany expansions of the air sample in an Aitken dust-counter, the **nucleus** spectrum observed with this instrument does not correspond to that observed in natural cloud **condensation** processes, where supersaturations larger than one per cent over water are probably rare. On the other hand, Aitken nuclei play an important role in determining the local **electrical conductivity** of the air, because they capture small ions, becoming large ions with much lower **mobility** in the earth's fair-weather **electric field**. In air containing large numbers of Aitken nuclei, the **small ion** population is small, the **large ion** population is large, and the **air conductivity** is low. Either nucleus may also be a protoparticle for larger particles such as **cloud condensation nuclei**, the subset of Aitken nuclei responsible for the formation of **cloud droplets**.

Aitken nucleus counter—Same as **Aitken dust counter**.

alarm level—Likelihood of or confidence in a forecast or report of dangerous, threatening, or damaging weather-related phenomena.

For example, a **warning** or **watch** may be issued based on either a forecast or **observation** of severe thunderstorms, **flash flooding**, river flooding, high winds, winter storms, etc. The National Weather Service does not wait for the event to occur before warning the public; there must be some lead time associated with it, such as a **flood warning** prior to reaching **flood stage**.

Alaska Current—The eastern semi of the North Pacific subpolar **gyre**.

It is a shallow current carrying relatively warm water northward and thus has a **climate** influence similar to that exercised by the **North Atlantic** and **Norwegian Currents** on the climates of northwestern Europe, though on a smaller scale. It flows cyclonically around the Gulf of Alaska, feeding into the **Alaskan Stream**. **Freshwater** from the many rivers of Canada and Alaska reduces the water density near the coast; the result is a **pressure gradient** normal to the coast that constrains the current geostrophically to the coastal region and increases its speed to 0.3 m s^{-1}.

Alaskan Stream—The continuation of the **Alaska Current** along the southern side of the Aleutian Islands.

The distinction between the Alaskan Stream and the Alaska Current is gradual, and the two currents are sometimes regarded as one. They are, however, of different character, the Alaska Current being shallow and variable but the Alaskan Stream reaching to the ocean floor. Despite its modest speed of 0.3 m s^{-1}, it is a western **boundary current**. Most of the water of the Alaskan Stream feeds directly into the **Oyashio**. Some of its flow enters the Bering Sea between the Aleutian Islands (most of it between 168° and 172°W) and follows a **cyclonic** path before feeding into the **Kamchatka Current**, thus eventually also contributing to the Oyashio.

albedo—The ratio of reflected **flux density** to incident flux density, referenced to some surface.

Albedos commonly tend to be broadband ratios, usually referring either to the entire **spectrum** of **solar radiation**, or just to the visible portion. More precise work requires the use of spectral albedos, referenced to specific wavelengths. Visible albedos of natural surfaces **range** from low values of ~0.04 for calm, **deep water** and overhead sun, to > 0.8 for fresh snow or thick clouds. Many surfaces show an increase in albedo with increasing **solar zenith angle**. *See also* **plane albedo, planetary albedo, spherical albedo, directional-hemispherical reflectance, bihemispherical reflectance**.

albedometer—An instrument used for the measurement of the reflecting power (the **albedo**) of a surface.

A **pyranometer** adapted for the measurement of **radiation** reflected from the earth's surface is sometimes employed as an albedometer.

Fritz, S., 1948: The albedo of the ground and atmosphere. *Bull. Amer. Meteor. Soc.*, **29**, 303–312.

Alberta clipper—A low pressure system that is often fast-moving, has low moisture content, and originates in western Canada (in or near Alberta province). In the wintertime, it may be associated with a narrow but significant band of **snowfall**, and typically affects portions of the plains states, Midwest, and East Coast.

Alberta low—A **low** that originates centered on the eastern slope of the Canadian Rockies in the province of Alberta, Canada.

Formerly, it was thought that such lows actually originated (more or less independently) over this location. It is now recognized that depressions moving inland from the Pacific are the actual parent systems. Alberta lows appear as these systems enhance, or are enhanced by, the **dynamic trough** that is a typical, almost semipermanent, feature of this region. *See* **Colorado low, Alberta clipper**.

alcohol-in-glass thermometer—Same as **spirit thermometer**.

alcohol thermometer—Same as **spirit thermometer**.

aldehydes—Very reactive organic compounds that contribute to local and regional **ozone** production, and also act as the precursors of **peroxyacetyl nitrates**.

Their major atmospheric fate is reaction with **hydroxyl radicals** or **photolysis**.

Aleutian Current—(Also called the Subarctic Current.) The southern, eastward flowing **current** of the subpolar **gyre** in the North Pacific.

It is fed by the **outflow** from the **Oyashio** and lies north of the **North Pacific Current**, with which it establishes the polar front in the west and experiences much water exchange as it proceeds eastward. As it approaches the coast of North America, it divides to form the northward flowing **Alaska Current** and the southward flowing **California Current**.

Aleutian low—The low pressure center located near the Aleutian Islands on mean charts of **sea level pressure**.

It represents one of the main centers of action in the atmospheric **circulation** of the Northern Hemisphere. The Aleutian low is most intense in the winter months; in summer it is displaced toward the North Pole and is almost nonexistent. On a daily basis, the area of the Aleutian low is marked by alternating high and low pressure centers, moving generally to the eastward; it is not the scene of an intense stationary **low**. Normally the depth of **intensity** of the low pressure areas exceeds the **intensity** of the high pressure areas, so that the region is one of low pressure on the average. The travelling cyclones of subpolar latitudes usually reach maximum **intensity** in the area of the Aleutian low. The Aleutian low and its counterpart in the Atlantic Ocean, the **Icelandic low**, compose the Northern Hemisphere's **subpolar low pressure belt**.

Alexander's dark band—The dark region of the sky seen between the primary and secondary **rainbows**.

It is named after Alexander of Alphrodisias, the first person known to have commented upon it. He was the head of the Lyceum from 198 to 211 A.D.

Alfvén wave—(Also magnetohydrodynamic wave, hydromagnetic wave.) A **transverse wave** propagating along a **magnetic field** in an electrically conducting fluid (or **plasma**).

This kind of **wave** does not exist in gases or fluids that are not ionized or in a plasma free from a magnetic field. The dynamics of such waves are analogous to those in a vibrating string.

Cowling, T. G., 1976: *Magnetohydrodynamics*, Ch. 3.

Algerian Current—A narrow intense current along the Algerian coast in which the **transport** of water from the Atlantic Ocean into the **Mediterranean Sea** is concentrated.

It originates from the **Almeria–Oran front** and for the first 300 km is less than 30 km wide, with average speeds of 0.4 m s^{-1} and maximum speeds of 0.8 m s^{-1}.

algorithm—A fixed step-by-step procedure to accomplish a given result; usually a simplified procedure for solving a complex problem; also a full statement of a finite number of steps.

aliasing—1. The introduction of **error** in the **Fourier analysis** of a discrete sampling from continuous data, by which frequencies too high to be analyzed with the discrete sampling interval erroneously contribute to the **amplitude** of the lower frequencies.

Aliasing can be avoided by **filtering** out the high frequencies (using slower-response instruments or analog electronic circuits) before sampling or digitizing. *See also* **Nyquist frequency**.

2. In **radar**, **sodar**, and **lidar**, the **folding** of **target** returns from outside the **normal** unambiguous **range** interval (**range folding**) into the normal range interval, or the folding of **radial velocity** measurements outside the **unambiguous velocity interval** (**velocity folding**) into the normal velocity interval.

alidade—A stationary instrument, mounted on a stand, that measures the angle subtended at the stand by the **horizon** and an object in space.

The **clinometer** is a portable form of alidade frequently used with a **ceiling projector** to determine the height of clouds. An alidade usually measures the **elevation angle** only; a **theodolite** measures **azimuth** as well.

alignment chart—Same as **nomogram**.

alimentation—Generally, the process of providing nourishment or sustenance; thus, in **glaciology**, the combined processes that serve to increase the mass of a **glacier** or **snowfield**; the opposite of **ablation**.

The deposition of **snow** is the major form of **glacial** alimentation, but other forms of **precipitation** along with **sublimation**, refreezing of meltwater, etc. also contribute. The additional mass produced by alimentation is termed **accumulation**.

Alisov's classification of climate—A genetic scheme, based on physical causes (i.e., circulation types, airmass types, fronts, etc.), for classifying **climate** proposed by Alisov in the 1950s as distinct from an empirical method based almost entirely on observations.

Alisov, B. P., 1954: *Die Klimate der Erde*, Deutscher Verlag der Wissenschaften, Berlin, 277 pp.

alkalinity—The relative acidity of any solution expressed in a **pH** range of numbers.

The pH value is the negative common **logarithm** of the hydrogen-ion concentration in a solution, expressed in moles per liter of solution. A neutral solution, that is, one that is neither acidic nor alkaline, such as pure water, has a concentration of 10 moles per liter; its pH is thus 7. Acidic solutions have pH values ranging with decreasing acidity from 0 to nearly 7; alkaline or basic solutions have a pH ranging with increasing alkalinity from just beyond 7 to 14. In **seawater**, the alkalinity is a measure of the excess of **hydroxyl ions** over **hydrogen ions**, generally expressed as milliequivalents per liter.

alkanes—**Hydrocarbon** species (also known as paraffins) with general formula C_nH_{2n+2} where n is an integer.

Alkanes contain only saturated bonds and are relatively unreactive in the **atmosphere**, where they react by **hydrogen** abstraction to form alkyl radicals. Major sources include natural gas emissions, **biomass burning**, evaporative emissions, and fuel combustion.

alkenes—(Also called olefins.) Monoalkenes are **hydrocarbon** species with general formula C_nH_{2n}, where n is an integer, containing one unsaturated carbon–carbon bond.

Alkenes are emitted in large quantities by automobiles and by vegetation. They react moderately to very rapidly with **hydroxyl radicals** and with **ozone**, which makes them major contributors to atmospheric reactivity, but limits their concentrations in the **atmosphere**. More complex dialkenes, with more than one double bond, can also be formed.

alkylperoxy radicals—Organic free radicals formed from the addition of **oxygen** to alkyl radicals.

Their major atmospheric reaction, with NO, is an integral component of **ozone** generation in the **troposphere**.

alkynes—Highly unsaturated organic compounds containing a carbon–carbon triple bond, with general formula C_nH_{2n-2}, of which **acetylene**, or ethyne, C_2H_2 is the simplest member.

Alkynes are emitted to the **atmosphere** mainly as a result of incomplete combustion.

All-hallown summer—(Also called Allhallow summer, All Saints' summer.) In English folklore, an old name for a period, like **Indian summer**, of unseasonable warmth, supposed to occur on the eve of All Hallows day (All Saints Day, November 1). It is mentioned by Shakespeare, but its use appears to have died out. More frequently heard today are references to such as **St. Luke's summer**, **St. Martin's summer, Old Wives' summer**.

all-sky photometer—A **photometer** with a **field of view** that is wide enough to measure the amount of **light** energy emanating from the entire hemisphere or almost the entire hemisphere of sky above the instrument.

all-weather airport—An airport equipped with facilities to permit the landing of qualified aircraft and aircrewmen without regard to **operational weather limits**.

Allard's law—An equation used in predicting the **night visual range** of self-luminous targets.

Assume a **light** source at distance x from an observer and having (**monochromatic**) **luminous** power P_v in that person's direction. Then for uniform **extinction coefficient** σ, the **illuminance** E_v reaching the observer is

$$E_v = P_v x^{-2} \exp(-\sigma x).$$

When E_v equals the **threshold illuminance** for a given background **luminance** and **detection** probability, x is the **night visual range**, or maximum distance for detecting the light source.

allerheiligenwind—In the Tyrol, the **wind** that sets in after Altweibersommer ("**Old Wives' summer**").

allobaric—Of a change in **pressure**; of **pressure tendency**.

Strictly, this term could be used in many instances in place of the more frequently used term, **isallobaric**.

allohypsography—(Rare.) The pattern exhibited by a **height-change chart**.

allowed transitions—A transition between two **energy** levels in an atom or a molecule that does not violate any **selection rules**.

These transitions are characterized by large **absorption cross sections**.

alluvial—Geologic layers deposited by streams.

alluvial aquifer—Aquifer with geologic materials deposited by a **stream** and that retains a hydraulic connection with the depositing stream.

almanac—A calendar to which astronomical and other data (often weather related) are commonly added.

Almeria–Oran front—A frontal region in the western **Mediterranean Sea** between Spain and Algeria that separates salty Mediterranean **surface water** from inflowing Atlantic water.

It marks the transition from hydraulic control of the Atlantic **inflow** through the Straits of Gibraltar to **geostrophic** control of the flow. The front is associated with a strong current that feeds into the **Algerian Current**.

almwind—Local name for a **foehn** that blows from the south (Hungary) across the Tatra Mountains south of Krakow, Poland, and descends the northern valleys; similar to the Alpine **south foehn**.

It is sometimes stormy and may reach 20–25 m s^{-1} (40–50 mph) in gusts, especially in spring and fall. It raises temperatures to as much as 14°C above the **normal** for the **season**, and in winter and spring it causes avalanches. At Zakopane (in southern Poland) it sometimes blows as a **high foehn**. This **wind** occurs in front of depressions moving eastward in the Baltic.

aloegoe—One of the winds of Lake Toba, northern Sumatra.

Others include **bolon, dahatoe, loehis, nirtra, saoet, si giring giring, siroeang, tamboen**.

along-slope wind systems—Thermally forced winds that blow up or down the sidewall slopes in a mountain valley, and the accompanying countercurrents when present; the slope-wind components of the **mountain–valley wind system**.

During the daytime this system consists of upslope (**anabatic**) flow adjacent to the slope and often a compensatory return **current** directed downslope just above the upslope layer. At night it is a downslope (**katabatic**) flow layer near the slope, sometimes with a **return flow** aloft.

along-valley wind systems—Thermally forced winds blowing along the main (longitudinal) axis of a valley, sometimes accompanied by countercurrents aloft; the up- and down-valley components of the **mountain–valley wind systems**.

During the daytime, this system comprises an up-valley **wind** and its **countercurrent** if present. At night it is a down-valley wind and **return flow** if present.

along-valley winds—The component of topographically generated winds that are parallel to the valley axis and can occur during conditions of light or **calm** synoptic-scale winds such as those near the center of anticyclones.

At night, cold, dense air flowing down the river valley from the higher elevations is called a

mountain wind or **drainage wind** with typical depths of 10 to 400 m and speeds of 1 to 8 m s^{-1}, while the weaker return **circulation** aloft is called an **anti-mountain wind**. During the day, the gentle up-valley flow of warm air along the valley floor is called a **valley wind**, and the return circulation aloft is called an **anti-valley wind**. *Compare* **cross-valley winds, anabatic wind, katabatic wind**.

aloup de vent—A cold **night wind** in the vicinity of the valley of Brevenne in France.

alpach—Same as **aperwind**.

alpenglow—The occasional reappearance of **sunset** colors on a (snow-covered) mountaintop soon after **sunset** and a similar phenomenon before **sunrise**.

Alpenglow has three phases. During evening **twilight**, the first stage is the mountain peak's usual coloration seen at low sun elevations h_0 ($h_0 < 2°$). Second is the alpenglow proper that occurs a few minutes after the first color has faded (h_0 slightly less than 0°). The peaks are still in direct **sunlight**, and their colors are purer and often pinker than before. The alpenglow boundary may first occur hundreds of meters below the summit, then moves upward, and finally fades as the **atmosphere**'s **dark segment** rises. Third is the **afterglow**, which occurs nearly simultaneously with the first **purple light**. The peaks are no longer in direct sunlight; the **illumination** is more diffuse and its boundary vaguer than in the earlier stages. The third stage lasts longer than the other two ($-5° < h_0 < -9°$), and its color varies from yellow to purple. A faint second afterglow has been reported and is associated with the rare occurrence of a second purple light. The alpenglow appears to be much less common at sunrise than at sunset. The morning colors are more pink and purple, while those of evening are more orange and red.

alpha particle—(Or α-particle; symbol $_2\text{He}^4$.) Physically indistinguishable from the **nucleus** of a **helium** atom—two protons and two neutrons bound together by nuclear forces—but usually restricted to the product of nuclear reactions.

For example, alpha particles with energies of several million electron volts are emitted in the radioactive disintegration of naturally occurring isotopes of uranium and thorium. The alpha particle carries a positive charge twice the magnitude of the **electron** charge. Alpha particles are important in **atmospheric electricity** as one of the agents responsible for **atmospheric ionization**. *Compare* **beta particle, gamma ray**.

alpha ray—(Or alpha radiation.) A stream of alpha particles; can also mean a single **alpha particle**.

alpine glacier—Same as mountain **glacier**.

alpine tundra—(Also called mountain tundra.) A form of **tundra** in which the absence of trees is due to high altitude instead of high latitude.

It lies roughly between the summer isotherm of 10°C and the **snow line**.

altanus—Same as **autan**.

Alter shield—A type of **rain** or **snow gauge** shield consisting of freely hanging, spaced slats, arranged circularly around the gauge, the height of which is the same as the **gauge height**.

The purpose is to reduce undercatch due to **wind** by forcing the air toward the surface instead of accelerating over the gauge orifice. *Compare* **Nipher shield**.

Alter, J. C., 1937: Shielded storage precipitation gauges. *Mon. Wea. Rev.*, **65**, 262–265.

alternate airport—An airport (aerodrome) to which an aircraft may proceed when it becomes either impossible or inadvisable to proceed to or to land at the airport (aerodrome) of intended landing.

Alternate airports (aerodromes) include the following:

Takeoff alternate: An alternate airport (aerodrome) at which an aircraft can land should this become necessary shortly after takeoff and it is not possible to use the airport (aerodrome) of departure.

En route alternate: An airport (aerodrome) at which an aircraft would be able to land after experiencing an abnormal or emergency condition en route.

Destination alternate: An airport (aerodrome) to which an aircraft may proceed should it become impossible or inadvisable to land at the airport (aerodrome) of intended landing.

alternate forecast—Aviation forecast for another airport (aerodrome) to be used by an aircraft if it becomes inadvisable to land at the airport (aerodrome) of intended landing.

alternating unit tensor—A mathematical function with symbol ϵ_{ijk} defined to switch between the discrete values of +1, 0, and -1, depending on the values of the three indices i, j, and k:

$$\epsilon_{ijk} = \begin{cases} 1 \text{ for } i, j, k = 123, 231, 312 \\ 0 \text{ for any two indices equal} \\ -1 \text{ for } i, j, k = 321, 213, 132 \end{cases}$$

It is one of the tools used in **Einstein's summation notation** to handle operations equivalent to cross products in **vector** notation. For example: $\mathbf{A} \times \mathbf{B} = \epsilon_{ijk}A_iB_j\delta_k$, where A_i and B_j represent vectors in **summation notation**, and δ_k is the unit vector. Subscripts or indices i, j, and k must each take on the values of 1, 2, and 3, representing the Cartesian directions of x, y, and z. The alternating unit tensor is useful to describe the Coriolis term in the **equations of motion**: $\partial U_i / \partial t = \ldots + f\epsilon_{ij3}U_j + \ldots$, where f is the **Coriolis parameter**, U_i and U_j are **wind** components, and t is time. Summation notation is commonly used in **atmospheric turbulence** studies.

Stull, R. B., 1988: *An Introduction to Boundary Layer Meteorology*, 666 pp.

alti-electrograph—A balloon-borne instrument for recording the value of the **atmospheric electric field** strength within active **thunderstorms**.

A disk of specially treated paper (pole-finding paper), slowly rotated by an **aneroid** element as the **balloon** ascends, lies between two iron electrodes, the upper one being electrically connected to a **discharge** point attached to the balloon and the lower one being electrically connected to a long trailing wire. Currents passing through the paper under the influence of the **potential** difference set up in the two electrodes discolor the treated paper in a pattern with a size roughly related to the ambient **electric field strength**. A **thermal** element and a **humidity element** that record in **aerograph** fashion complete the device.

altigraph—A recording **altimeter**.

altimeter—An instrument that determines the **altitude** of an object with respect to a fixed level.

There are several types of altimeters: 1) the **pressure altimeter**, which measures **barometric pressure** and converts it to altitude above **sea level** when referenced to the existing measured **sea level barometric pressure**, or to **pressure altitude** when referenced to a **standard pressure**; 2) the radio or **radar altimeter**, which deduces altitude by measuring the time it takes a radio **signal** to travel from a **transmitter** to the underlying surface and back to the **receiver**; and 3) the **Global Positioning System** (GPS), which determines altitude by measuring time of travel of radio signals between GPS-equipped satellites and the receiver.

altimeter corrections—Corrections that must be made to the readings of a **pressure altimeter** to obtain true altitudes.

There are several sources of corrections: 1) **pressure** corrections due to changes in **sea level atmospheric pressure** readings; 2) **air temperature** corrections resulting from the differences between the actual **temperature** of the column of air beneath the **altimeter** and a **standard atmosphere** temperature **profile**; and 3) corrections for errors in the **calibration** of the mechanical parts of the altimeter.

altimeter equation—The form of the **hydrostatic equation** used to compute the **altimeter setting**, which is the **pressure** "reduced" to **sea level** using the **temperature** profile of the ICAO (International Civil Aeronautical Organization) **standard atmosphere**.

altimeter setting—Value of the **atmospheric pressure** used to adjust the subscale of a **pressure altimeter** so that it indicates the height of an aircraft above a known reference surface.

altimeter-setting indicator—A **precision aneroid barometer** calibrated to indicate directly the local **sea level** altimeter setting.

altimetry—The measurement of **altitude** or height.

In aeronautical science, altimetry is equivalent to **hypsometry**.

altithermal period—The period from about 5000–2500 B.C., proposed by Ernst V. Antevs (1952), during which the summer **temperature** of western North America was 1°–2°C warmer than today and during which the lakes of the American Great Basin dried.

Antevs, E., 1952: Cenozoic climates of the Great Basin. *Sonderdruck aus der Geologischen Rundschau*, Bd. 40, Heft 1, 94–108.

altitude—1. A measure (or condition) of height, especially of great height, as a mountain top or aircraft **flight level**.

In meteorology, altitude is used almost exclusively with respect to the height of an airborne object above the earth's surface, above a **constant-pressure surface**, or above **mean sea level**.

altocumulus

The measurement of altitude is accomplished by **altimeters** in aeronautics, and the entire study is called **altimetry**. *Compare* **elevation**.

2. In astronomy, same as **elevation angle**.

altocumulus—(Abbreviated Ac.) A principal **cloud** type (**cloud genus**), white and/or gray in color, that occurs as a layer or patch with a waved aspect, the elements of which appear as laminae, rounded masses, **rolls**, etc.

These elements usually are sharply outlined, but they may become partly fibrous or diffuse; they may or may not be merged; they generally have shadowed parts; and, by convention, when observed at an angle of more than 30° above the **horizon**, an altocumulus element subtends an angle between 1° and 5°. Small liquid water droplets invariably compose the major part of the composition of altocumulus. This results in sharpness of outline, small internal **visibility** (both common **cumuliform** characteristics), and in the occurrence of coronae and **irisation** (colored **diffraction** phenomena). With sufficiently low temperatures, **ice crystals** may appear in all forms of altocumulus, but mainly in the species **castellanus** or **floccus**, each unit of which may produce an individual **snow shower**. The **crystals** that fall from altocumulus sometimes produce **parhelia**, or a moon or **sun pillar**, any of which indicates the presence of tabular crystals. Should the composition become entirely ice crystals, the cloud would lose its characteristic sharpness of outline. Altocumulus often forms directly in **clear air**. It may be produced by an increase in size or thickening of the elements of an entire layer or patch of **cirrocumulus** (Ac cirrocumulomutatus); by subdivision of a layer of **stratocumulus** (Ac stratocumulomutatus); by transformation of **altostratus** (Ac altostratomutatus) or **nimbostratus** (Ac nimbostratomutatus); or by the spreading of **cumulus** or **cumulonimbus** (Ac cumulogenitus or Ac cumulonimbogenitus). Altocumulus frequently occurs in a given sky at different levels; also, it often is associated with clouds of other genera. **Virga** may appear with most of the species of altocumulus. This supplementary feature, however, should not be confused with the very white trails of ice crystals that frequently are formed with the dissipation of **altocumulus floccus**. When detached, the ice crystal trails are **cirrus**. Sometimes **mamma** occur with altocumulus. Cirrocumulus and stratocumulus are the clouds most easily confused with altocumulus. The elements of cirrocumulus never have shadows of their own, and nearly always are smaller. Stratocumulus elements are larger than those of altocumulus. The rolls or cells that are associated with altocumulus are thought to be a result of the **absorption** of **terrestrial radiation** and/or the presence of **wind shear**, which drives **Rayleigh–Bénard convection** or **Kelvin–Helmholtz shear instability**. *See* **cloud classification**.

altocumulus castellanus—*See* **castellanus**.

altocumulus duplicatus—*See* **duplicatus**.

altocumulus floccus—*See* **floccus**.

altocumulus lacunaris—*See* **lacunosus**.

altocumulus lacunosus—*See* **lacunosus**.

altocumulus lenticularis—*See* **lenticularis**.

altocumulus opacus—*See* **opacus**.

altocumulus perlucidus—*See* **perlucidus**.

altocumulus radiatus—*See* **radiatus**.

altocumulus stratiformis—*See* **stratiformis**.

altocumulus translucidus—*See* **translucidus**.

altocumulus undulatus—*See* **undulatus**.

altostratus— (Abbreviated As.) A principal **cloud** type (**cloud genus**) in the form of a gray or bluish (never white) sheet or layer of striated, fibrous, or uniform appearance.

Altostratus very often totally covers the sky and may, in fact, cover an area of several thousand square miles. The layer has parts thin enough to reveal the position of the sun, and if gaps and rifts appear, they are irregularly shaped and spaced. Within the rather large vertical extent of altostratus (from several hundred to thousands of feet) a very heterogeneous particulate composition may exist. In this most complete case, there may be distinguished 1) an upper part, mostly or entirely **ice crystals**; 2) a middle part, a mixture of ice crystals and/or **snowflakes** and **supercooled water** droplets; and 3) a lower part, mostly or entirely supercooled or ordinary water droplets. A number of partial combinations of these composition types may occur, but never an entire cloud like 3) above. The **particles** are widely enough dispersed so as not to obscure the sun except by

34

its thickest parts, but rather to impose a "ground-glass" effect upon the sun's **image**, and to prevent sharply outlined shadows from being cast by terrestrial objects. Halo phenomena do not occur. Altostratus is a precipitating cloud (**praecipitatio**) and therefore often is accompanied by **virga** and **mamma**. **Rain, snow, ice pellets**, etc., are present in the cloud and under its base, frequently rendering the base quite indistinct, particularly when the **precipitation** does not reach the ground. When precipitation reaches the ground, it is usually very light and of a relatively continuous nature. Altostratus may be formed by the thickening of **cirrostratus** (As cirrostratomutatus), or by the thinning of **nimbostratus** (As nimbostratomutatus). If widespread precipitation develops in **altocumulus**, altostratus may result (As altocumulogenitus). Sometimes, particularly in the **tropics**, altostratus may be produced by the spreading of the middle or upper portion of **cumulonimbus** (As cumulonimbogenitus). Cirrostratus and nimbostratus are the two other forms most easily confused with altostratus. In the first case, it should be remembered that cirrostratus does allow terrestrial shadows and frequently produces **halo** phenomena. Nimbostratus is darker colored, hides the sun, is more uniform in **optical thickness**, and always produces precipitation. At night, if precipitation does not reach the ground, it is conventional to call the doubtful layer altostratus. Any **stratiform** (layered) cloud necessarily forms because further vertical development is inhibited by the presence of a **temperature inversion**. **Rolls** and cells in altostratus are thought to be of similar origin to those in altocumulus. *See* **cloud classification**.

altostratus duplicatus—*See* **duplicatus**.

altostratus opacus—*See* **opacus**.

altostratus radiatus—*See* **radiatus**.

altostratus translucidus—*See* **translucidus**.

altostratus undulatus—*See* **undulatus**.

aluminum oxide humidity element—A **humidity**-sensing element consisting of a base of aluminium, an oxide made by anodizing the base material, and an evaporated conductive coating of metal.
  The base material acts as one electrode and the evaporated metal as the other, or outer electrode. An increase in **relative humidity** causes the impedance between the electrodes to drop. The reverse is true when the relative humidity decreases.

ambient air—1. Background, environmental, or surrounding air.
  When studying the dynamic and thermodynamic processes acting on an individual element such as an **air parcel, cloud**, smoke **plume, raindrop**, or **ice crystal**, ambient air represents the **atmosphere** outside of that element. The ambient air is often assumed to be static and of relatively large domain, within which the element resides.
  2. The air that surrounds us, within which we live.
  When **air pollutants** of high concentration from exhaust or stack gases are emitted into cleaner air, the resulting polluted mixture is called the ambient air. **National Ambient Air Quality Standards** (NAAQS) apply to this final mixture, not to the undiluted emission gases.

ambient pressure—The air pressure that is characteristic of the **atmosphere** surrounding a small-scale feature such as a **cumulus** cloud.

ambient temperature—The **temperature** that is characteristic of the **atmosphere** surrounding a small-scale feature such as a **cumulus** cloud.

ambiguity function—In **radar**, a two-dimensional function of radar range and Doppler **frequency**, determined by the transmitted **waveform**, that defines the effective **filter** response of the radar receiver. The ambiguity function is used to examine the suitability of different transmitted waveforms for achieving **accuracy, resolution**, freedom from ambiguities, and reduction of unwanted **clutter**.
  In particular, it defines 1) the **sidelobe** response to a complex radar waveform at different Doppler frequencies, and 2) the response of a **receiver** to targets at other ranges and Doppler frequencies from the desired **target**. The ambiguity function applies to both point targets and distributed targets.

Nathanson, F. E., 1991: *Radar Design Principles*, 2d ed., McGraw-Hill, 360–369.
Skolnik, M. I., 1980: *Introduction to Radar Systems*, 2d ed., McGraw-Hill, 411–420.

ammonia—A colorless gas, formula NH_3, with a sharp, irritating odor, having a **density** about six-tenths that of air at the same **temperature** and **pressure** (0.7720 g cm^{-3} at **STP**).
  A reduced **nitrogen** gas, NH_3 is emitted in large quantities from animal feedstocks, sewerage plants, etc. Ammonia is very soluble in water and is scavenged from the **lower atmosphere** by **clouds**. It is the most abundant alkaline gas in the **atmosphere** and as such plays a large role in

neutralizing acidity from sulfuric and nitric acids via formation of the **ammonium ion**. Large quantities of ammonia gas, and probably even ammonia crystals, occur in the atmospheres of the large planets Saturn and Jupiter. *See* **air**.

ammonification—The conversion of organic **nitrogen** to **ammonia** or **ammonium ion** by heterotrophic bacteria (ammonifiers) involved in the decomposition of organic matter.

ammonium chloride—Inorganic salt, formula NH_4Cl, formed from the neutralization of **hydrochloric acid** by **ammonia**.

Ammonium chloride is a low **vapor pressure** solid, which deliquesces when the **relative humidity** exceeds 80%. It is thought to be present in the urban **aerosol** present in **smog** episodes. *See* **deliquescence**.

ammonium ion—Hydrated form of **ammonia**, NH_4^+; **ammonium sulfate** and perhaps **ammonium nitrate** are common in atmospheric **aerosol**, formed by partial neutralization of **sulfuric acid** and **nitric acid**, respectively.

ammonium nitrate—Inorganic salt, formula NH_4NO_3, formed from the neutralization of **nitric acid** by **ammonia**.

Ammonium nitrate is a deliquescent solid with a low **vapor pressure**, which consequently forms in the **atmosphere** when high concentrations of ammonia and nitric acid occur, for example, in polluted urban areas. It is thought to sometimes make up a considerable fraction of the urban **aerosol** present in **smog** episodes. *See* **deliquescence**.

ammonium sulfate—Inorganic salt, formula $(NH_4)_2SO_4$.

A large constituent of tropospheric **aerosol**, due to neutralization of acidic **sulfuric acid** aerosol by **ammonia**. Much of the aerosol is partially neutralized, and the formula NH_4HSO_4, corresponding to ammonium bisulfate, is generally more appropriate. Ammonium sulfate is a solid, crystalline material that deliquesces at relative humidities greater than 75%, eventually forming liquid droplets. *See* **deliquescence**.

amorphous clouds—Clouds without any apparent structure at all, as may occur in a **whiteout** in a thick **cloud** or **fog** over a **snow** surface when one loses any sense of direction—up, down, or sideways.

amorphous frost—**Hoarfrost** that possesses no apparent simple crystalline structure; opposite of **crystalline frost**.

The lack of distinct **crystal** structure in forms of amorphous frost, however, is only a matter of **scale**. Such **frost** is built up of a multitude of units each of which has its own crystal structure, although no unit fits compatibly with its neighboring unit.

amorphous sky—A **state of the sky**, characterized by an abundance of **fractus** clouds, usually accompanied by **precipitation** falling from a higher, **overcast** cloud layer.

It is termed "amorphous" because the clouds lack any distinctive form (bands, undulations, clear-cut base, etc.).

amount of precipitation—Depth to which **precipitation** in liquid form would cover a horizontal **projection** of the earth's surface, in the absence of abstractions.

It is typically expressed in millimeters, equivalent to liters per square meter. **Snowfall** is also measured by the depth of fresh **snow** covering a horizontal surface.

amphidrome—A point in the sea where there is zero tidal **amplitude** due to canceling of **tidal waves**. **Cotidal lines** radiate from an amphidromic point and **corange lines** encircle it.

amphidromic point—*See* **amphidrome**.

amphidromic region—*See* **amphidrome**.

amplitude—Often the greatest magnitude at a given point of any spatially and temporally varying physical quantity governed by a **wave equation**; can also mean the spatial part of a time-harmonic **wave** function.

For example, in the time-harmonic (or sinusoidal) **scalar** wave function with **circular frequency** ω,

$$\psi(\mathbf{x}, t) = \phi(\mathbf{x}) \exp(-i\omega t),$$

where $\phi(\mathbf{x})$ is the (complex) amplitude of the wave, although the **modulus** of ϕ also may be called its amplitude. The (complex) amplitude of the scalar plane **harmonic** wave

$$\psi(x, t) = A \exp(ikx - i\omega t - i\theta)$$

with **wavenumber** k and initial **phase** θ is $A \exp(ikx - i\theta)$, the modulus of which, $|A|$, is also called the amplitude of the wave. In its most general sense, amplitude means extent or size. Thus the amplitude of a wave is some measure of its size.

amplitude-modulated indicator—(Also called deflection-modulated indicator.) One of two general classes of radar indicator in which the **sweep** of the **electron beam** is deflected vertically or horizontally from a **base line** to indicate the existence of an **echo** from a **target**.

The amount of deflection is usually a function of the **echo signal** strength. Common types of amplitude-modulated indicators give only the **slant range** between **radar** and target. However, since the characteristics of the **target signal** can be easily observed with this type of indicator, it is very useful in **radar meteorology**.

amplitude modulation—Changing the **amplitude** of a **carrier wave** by some means so that the resultant waveform transmits a **signal** conveying information.

For example, if the carrier wave with **angular frequency** ω is $\cos(\omega t)$, the corresponding amplitude-modulated signal is

$$f(t) = g(t)\cos(\omega t),$$

where $g(t)$ is a function that varies more slowly with time than does the carrier wave. *Compare* **frequency modulation**.

amplitude spectrum—In the Fourier or spectral **analysis** of two **time series**, a measure of the **correlation** between the two series as a function of **frequency** or **wavelength**, but ignoring any **phase** differences between the two time series.

It is defined as

$$Am(n) = Q^2(n) + Co^2(n),$$

where n is **wavenumber** (number of waves per total **period** of the time series), $Co(n)$ is the **cospectrum**, and $Q(n)$ is the **quadrature spectrum**.

Stull, R. B., 1988: *An Introduction to Boundary Layer Meteorology*, 666 pp.

AMSU—Abbreviation for **Advanced Microwave Sounding Unit**.

anabaric—Same as **anallobaric**.

anabatic—(Obsolete.) Pertaining to any upward-moving currents of air.
 See **anabatic wind**.

anabatic front—*See* **anafront**.

anabatic wind—In mountain meteorology, an **upslope wind** driven by heating (usually daytime **insolation**) at the slope surface under fair-weather conditions.

The mechanism of the anabatic wind can be described as follows. The warm surface heats a vertical column of the **atmosphere** starting at the slope surface and reaching up to a few hundred meters deep. This column is warmer than the column at the same levels over the valley or plain, resulting in hydrostatic low pressure over the slope relative to over the valley or plain. The horizontal **pressure gradient**, maximized at the slope surface, drives an **acceleration** directed toward the slope, or up the slope. Although the pressure-gradient forcing is at its maximum at the surface, **surface friction** causes the peak in the anabatic wind speeds to occur above the surface, often by several tens of meters; if the surface heating is strong, however, the **momentum** will tend to be vertically mixed. Speeds in the mountain–valley **anabatic** flow layer are often 3–5 m s⁻¹. Because heating at the surface promotes deeper **mixing** than cooling does, the heated layer, often occurring as a convective or **mixed layer**, is generally deeper than a cooled or katabatic layer. Slopes occur on many scales, and consequently anabatic flows also occur on many scales. At local scales anabatic winds are an along-slope component of **mountain–valley wind systems**. At scales ranging from the slopes of individual hills and mountains to the slopes of mountain ranges and massifs, anabatic flows represent the daytime component of **mountain–plains wind systems**. In general usage, this term does not suffer from the multiplicity of meanings that **katabatic wind** does.

Anadyr Current—A **current** through the Bering Strait that carries low-salinity water of Pacific origin into the Arctic Ocean. It is concentrated on the Siberian side with speeds near 0.3 m s⁻¹ and varies little with **season**. During winter it is augmented by additional flow from the Sphanberg Strait.

With a total **transport** of less than 0.5 Sv (0.5×10^6 m³s⁻¹), the Anadyr Current contributes little to the **mass balance** of the World Ocean but is essential to its **freshwater** budget since the **salinity** of the North Pacific is so much lower than that of the North Atlantic. In terms of freshwater transport the modest Anadyr Current is equivalent to several Amazon Rivers.

anaerobic condition—1. The absence of **oxygen**, preventing normal life for organisms that depend on oxygen.

2. Water with insufficient dissolved **oxygen** to support aerobic bacteria.

anafront—A **front** at which the warm air is ascending the **frontal surface** up to high altitudes.
With anafronts, **precipitation** may occur to the rear of the front and is sometimes associated with **cyclogenesis**. *Compare* **katafront**.

anallobar—*See* **isallobar**.

anallobaric—(Also called anabaric.) Pertaining to an increase in **atmospheric pressure**.
See **allobaric**, **isallobaric**.

anallobaric center—Same as **pressure-rise center**.

analog—1. A form of data **display** in which values are shown graphically.

2. A form of computing in which values are represented by directly measurable quantities, such as voltage or **resistance**. Analog computing methods contrast with **digital** methods in which values are treated numerically.

analog climate model—A method of predicting future **climate** by comparison with a historical situation, such as a **paleoclimate** state, in which the climatic forcing had features similar to those anticipated in the future.

analog method—A method of forecasting that involves searching historical meteorological records for previous events or flow patterns similar to the current situation, then making a **prediction** based on those past events or patterns.

analog model—An empirical forecast model based on the use of **analogs**.
The observed states subsequent to a previously observed analog of the current observed state can be used as a forecast.

analogs—Two observed states of the **atmosphere** that are very close by some measure, also applied to states of a **model**.
Formal measures of closeness include **anomaly correlation**, root-mean-square distance, and **covariance**. Usually one expects analogs to occur only during the same time of year. Atmospheric analogs that are close compared to current levels of **observational error** are unlikely to be found unless one studies a single **variable** confined to a very small area (≤1000 km radius), or otherwise reduces the **degrees of freedom** to a very small number (≤3). Analogs have a practical application in specifying the surface weather from a 3D atmospheric state produced by **NWP** and in short-term **climate prediction** for forecasts ranging from a month to several seasons. Analogs are also of interest in research related to **predictability**, short-term forecasts and **error** growth, cluster **analysis**, and estimates of dimensionality of the atmospheric **attractor**. When natural analogs are poor, improvements have been sought by combining several of them, by using anti-analogs (antilogs), and by constructing an artificial close analog by some objective weighted averaging of a set of previously observed states (**constructed analogs**). *See* **MOS**, **perfect prognosis method**.

analogue—*See* **analogs**.

analysis—1. In **synoptic meteorology**, a detailed study of the state of the **atmosphere** based on observations, usually including a separation of the entity into its component patterns and involving the drawing of families of **isopleths** for various elements.
Thus, the analysis of **synoptic** charts may consist, for example, of the drawing and the interpretation of the patterns of **wind**, **pressure**, pressure change, **temperature**, **humidity**, **clouds**, and **hydrometeors**, all based on observations taken or **forecast** simultaneously.

2. A procedure to project the state of the **atmosphere** (or any system) as known from a finite set of imperfect, irregularly distributed observations onto a regular **grid** or to represent the atmospheric state by the **amplitude** of **standard** mathematical functions.
The grid and/or functional expansion allows for subsequent forecasts by **numerical integration** or for easy diagnostic study. An analysis may be looked upon as a space–time **interpolation** system. Essential ingredients of analysis are a **background field**, usually a **short-range forecast** although **persistence** and **climatology** are options, a forecast model, knowledge of the relative **error** of the many different observational platforms employed as well as knowledge of the **error** in the **background field**, and the spatial **covariance** of the various errors. **Real time** computational considerations generally limit the complexity of methods used. Historically important is the subjective hand analysis, that is, drawing isobars on a map given **mean sea level** pressure at a set of observing stations. Most analysis systems in the past have been at best 3D spatial analyses; time–space analysis

systems have recently emerged at operational **prediction** centers. A long series of analyses is of great utility for studying the behavior of the atmospheric **climate system**. *See* **reanalysis**, **variational objective analysis**.

analysis of variance—A **statistical** technique for resolving the total **variability** of a set of data into systematic and **random** components.

The analysis of variance is fundamentally a statistical estimating and/or testing procedure. It estimates the components of **variance** due to systematic and random causes, and it leads to significance tests of these components. The statistical assumptions required for a valid test are more stringent than those for estimating the components of variance.

analyzed chart—Chart depicting the geographical distribution of meteorological conditions with the aid of fronts, isobars, etc.

analyzed map—*See* **analyzed chart**.

anaprop—*See* **anomalous propagation**.

anchor balloon—A **super-pressure balloon** suspended below a **zero-pressure balloon** to permit the zero-pressure balloon to survive night-time cooling without the expenditure of ballast and daytime warming without release of lifting gas.

The **lift** decreases during ascent and decreases during descent, providing **stability** to the zero-pressure balloon.

anchor ice—**Ice** attached to the beds of streams, lakes, and shallow seas, irrespective of its nature of formation.

On **clear**, cold nights in relatively still water, anchor ice may form directly on submerged objects. It also develops in **supercooled water** if **turbulence** is sufficient to maintain uniform **temperature** at all depths, in which case a spongy mass of **frazil** accumulates on objects exposed to rapid flow, and later **deposition** fills in the pores and creates solid ice. When the water temperature increases to above 0°C, the ice rises to the surface, often carrying with it the object on which it had accumulated. Sometimes anchor ice is erroneously called **ground ice**, a term which should be reserved for bodies of more or less **clear ice** in **frozen ground**.

anelastic approximation—An approximate system of equations for deep and shallow atmospheric **convection**.

The equations are derived under the assumptions that the percentage **range** in **potential temperature** is small and that the timescale is set by the **Brunt–Väisälä frequency**. Acoustic waves are thereby filtered—hence the term anelastic, meaning elastic **energy** is not allowed. If the vertical **scale** of motion is small compared to the depth of an **adiabatic atmosphere**, the anelastic equations reduce to the Boussinesq equations for shallow convection.

anemoclinometer—General name for an instrument that measures the **inclination of the wind** to the horizontal plane.

See **bivane**.

anemometer—The general name for instruments designed to measure either total **wind speed** or the speed of one or more **linear** components of the **wind vector**.

These instruments may be classified according to the **transducer** employed; those commonly used in meteorology include the **cup**, **propeller**, **Pitot-tube**, **hot-wire** or **hot-film**, and **sonic anemometers**.*See also* **current meter**, **wind vane**.

Middleton, W. E. K., and A. F. Spilhaus, 1953: *Meteorological Instruments*, 3d ed., rev., Univ. of Toronto Press, 135–165.

anemometer level—1. The height above the surface at which an **anemometer** is actually exposed.

2. The ideal **exposure** height of an anemometer; usually designed to be above the wakes that form behind individual **surface roughness** elements such as nearby trees, houses, or ship structures.

3. The desired or standard exposure height of an anemometer, as specified by national or international agreement.

This height is usually taken as 10 m.

anemometry—The study of measuring and recording the direction and speed (or force) of the **wind**, including its vertical component.

Systematic records of **wind direction** as indicated by wind vanes were begun in Italy in 1650 and in England in 1667. The speed of the wind was first observed by noting the rate at which light substances are carried along by the air; it is now determined by means of **anemometers**.

anemoscope—An instrument for indicating the existence and direction of the **wind**.

anemovane—A combined contact **anemometer** and **wind vane**.

aneroid—1. Literally, "not wet," containing no liquid; applied to a kind of **barometer** that contains no liquid, an aneroid barometer.

2. Same as **aneroid barometer**.

aneroid barograph—(Sometimes called aneroidograph.) An **aneroid barometer** arranged so that the deflection of the **aneroid capsule** actuates a pen that graphs a record on a rotating drum.

The magnification of the deflection of the capsule may be adjusted so that records of small fluctuations in **pressure** may be obtained. The aneroid barograph is subject to the uncertainties of the aneroid barometer and therefore must be calibrated periodically. *See also* **barograph, microbarograph**.

aneroid barometer—(Rarely called holosteric barometer, metallic barometer.) An instrument for measuring **atmospheric pressure**.

It is constructed on the following principles: An **aneroid capsule** (a thin corrugated hollow disk) is partially evacuated of gas and is restrained from collapsing by an external or internal spring; the deflection of the spring will be nearly proportional to the difference between the internal and external pressures; magnification of the spring deflection is obtained both by connecting capsules in series and by mechanical linkages. The aneroid barometer is **temperature** compensated at a given pressure level by adjustment of the **residual** gas in the aneroid or by a bimetallic link arrangement. The instrument is subject to uncertainties due to variations in the elastic properties of the spring and capsules, and due to wear in the mechanical linkages. *See* **barometer, aneroid barograph, pressure altimeter, altimeter-setting indicator**.

> U.S. Weather Bureau, 1941: *Barometers and the Measurement of Atmospheric Pressure*, Circular F, 7th ed., rev., 21–25.

aneroid capsule—(Also called pressure capsule, sylphon, expansion chamber, bellows.) *See* **aneroid barometer**.

aneroidograph—Same as **aneroid barograph**.

angel—A **radar echo** caused by a physical phenomenon not discernible by eye at the **radar** site.

Angels may appear as **coherent** or **incoherent echoes**. When diffuse and incoherent appearing, they are sometimes called **ghost** echoes. Angel echoes observed by radars with wavelengths of about 10 cm and less are usually caused by birds or insects. Radars with longer wavelengths and radar **wind profilers**, which operate in the **UHF** and **VHF** radio **frequency** bands, regularly detect echoes from the optically **clear air** that are caused by spatial fluctuations of the atmospheric **refractive index**. *See* **Bragg scattering**; *compare* **clear-air echo**.

angin—A land and sea breeze system of Malaya.

The **land breeze** is called angin-darat; the **sea breeze**, angin-laut.

angle of arrival—The angle between the propagation direction of an incident **wave** and some reference direction, which in radio engineering may be fixed relative to a **receiver** (e.g., the normal to a horizontal plane).

Compare **angle of incidence**

angle of elevation—Same as **elevation angle**.

angle of incidence—The angle between the direction of propagation of an **electromagnetic** or **acoustic wave** (or **ray**) incident on a body and the local normal to that body (although this normal may not be well defined, as that for a **cloud**, e.g.).

May also describe beams of **particles** in the broadest sense. *Compare* **angle of arrival**.

angle of minimum deviation—*See* **minimum deviation**.

angle of reflection—The angle between the direction of propagation of an **electromagnetic** or **acoustic wave** (or **ray**) reflected by a body and the local normal to that body (although this normal may not be well defined, as that for a **cloud**, e.g.).

May also describe beams of **particles** in the broadest sense. *See* **reflection**.

angle of refraction—The angle between the direction of propagation of an **electromagnetic** or **acoustic wave** (or **ray**) refracted by an **optically homogeneous** body and the local normal to that body.

Angola–Benguela front—The boundary between the southward flowing **Angola Current** and the northward flowing **Benguela Current**.

It is seen as a **temperature** front in the upper 50 m and as a **salinity** front to depths of at least 250 m.

40

Angola Current—A relatively strong southward current along the Angolan coast.

The current forms the eastern side of a cyclonic **gyre** centered on 13°S, 4°E and is driven by the extension of the **South Equatorial Countercurrent**. It reaches to depths of at least 300 m and attains its maximum speed of 0.5 m s^{-1} just below the surface.

angstrom—A unit of length equal to 0.1 nm, or 10^{-4} μm.

Ångström compensation pyrheliometer—An **absolute instrument** developed by K. Ångström for the measurement of **direct solar radiation**.

The radiation receiver station consists of two identical manganin strips, the temperatures of which are measured by attached thermocouples. One of the strips is shaded, while the other is exposed to **sunlight**. An electrical heating current is passed through the shaded strip to raise its **temperature** to that of the exposed strip. The electric **power** required to accomplish this is a measure of the solar radiation. *See* **actinometer, pyrheliometer**; *compare* **Ångström pyrgeometer**.

Ångström pyrgeometer—An instrument developed by K. Ångström for measuring the effective **terrestrial radiation**.

It consists of four manganin strips, of which two are blackened and two are polished. The blackened strips are allowed to radiate to the **atmosphere** while the polished strips are shielded. The electrical **power** required to equalize the **temperature** of the four strips is taken as a measure of the upwelling terrestrial **radiation**. *See* **actinometer, pyrheliometer**; *compare* **Ångström compensation pyrheliometer**.

Ångström turbidity coefficient—A measure of the **turbidity** of the **atmosphere**, equal to the **aerosol optical depth** at a **wavelength** of 1 μm.

angular drift—"Rock debris formed by intensive **frost action**, derived from underlying or adjacent bedrock. Contrast with **glacial drift**." [Glossary of Arctic and Subarctic Terms (1955).]

Arctic, Desert, Tropic Information Center (ADTIC) Research Studies Institute, 1955: *Glossary of Arctic and Subarctic Terms*, ADTIC Pub. A-105, Maxwell AFB, AL, 90 pp.

angular frequency—*See* **frequency**.

angular momentum—In **Newtonian mechanics** the angular momentum (or moment of momentum) **L** about a point O of a body with **linear momentum p** is the **vector** cross product

$$L = r \times p,$$

where **r** is the **position vector** of the body relative to O.

In the absence of a net **torque**, angular momentum is conserved. But angular momentum is a more fundamental quantity than that defined by this equation. For example, photons have intrinsic angular momenta (spin), which can be transferred to objects (as evidenced by **radiation torque**), and yet the **photon** has zero **rest mass**. Thus, angular momentum is best looked upon as a single entity, complete in itself, governed by the dynamical law

$$N = \frac{dL}{dt},$$

where **N** is the **torque** acting on the body with angular momentum **L**. In meteorology, it is conventional to deal with angular momentum per unit volume, which is given by the product **r** $\times \rho$**V**, where ρ is the **density** and **V** the **velocity**. *Compare* **momentum, relative angular momentum, absolute angular momentum**.

angular-momentum balance—An integral requirement for a system that is characterized by **conservation of angular momentum**.

A description of the angular-momentum balance of a physical system includes 1) a description of the partition of **absolute angular momentum** within the system, and 2) a description of the mechanism by which absolute angular momentum is transferred from one portion of the system to another and also between the system and its surroundings. *See also* **angular momentum, local angular momentum**.

Gill, A. E., 1982: *Atmosphere–Ocean Dynamics*, Academic Press, 583–587.

angular resolution—In **radar meteorology**: 1) for point **targets**, the minimum angular separation at the **antenna** for which two targets at the same **range** can be distinguished; 2) for distributed targets, the minimum angular separation at the antenna for which distinct features (such as **reflectivity** cores) of the distributed targets can be distinguished.

The angular resolution is usually taken to be the 3-dB **beamwidth**.

angular spreading—Spreading of waves in space due to differences in direction of **wave propagation**.

See also **directional distribution**.

angular-spreading factor—In early **ocean wave** forecasting methods, the fraction by which the estimated **energy** of waves leaving the generation area is multiplied to obtain the forecast wave energy at a distant point, after reduction due to **angular spreading**.

angular velocity—The rate of rotation of a **particle** about the axis of rotation, with magnitude equal to the time rate of angular displacement of any point of the body.

Angular velocity is a **vector** oriented in accordance with a **right-hand rotation** (i.e., when the fingers of the right hand are curved in the sense of rotation, the thumb points in the direction of the angular velocity vector).

angular velocity of the earth—The angular velocity of the earth (in the **absolute coordinate system**) is directed along the earth's axis toward the pole star and is equal in magnitude to 7.2921×10^{-5} rad s^{-1}.

angular wavenumber—(Also called azimuthal wavenumber, hemispheric wavenumber, zonal wavenumber.) In many meteorological contexts, the number of waves of a given **wavelength** required to encircle the earth at the latitude of the **disturbance**. If L is the wavelength, r the earth's radius, and ϕ the geographical latitude, the angular wavenumber k is given by $k = 2\pi r \cos \phi / L$.
Compare **wavenumber**.

angularity correction—Correction to be made to an observed **velocity** when the direction of the **current** is not at right angles to the **discharge section line**.

anhyetism—(Rare.) The condition of being without **rain**.

anisotropic—Not **isotropic**.

anisotropy—Refers to the nonisotropic part of the **turbulence spectrum**.
See **isotropic turbulence**.

anniversary winds—General terms for **local winds** or larger-scale **wind** systems (such as the **monsoon**, **etesians**, etc.) that recur annually.

annual cycle—That component of **variability** that is a function of the day of the year but is independent of the year.

annual exceedance series—In hydrology, a series of local maxima in a data record in which the number of local maxima is equal to the number of years of record.
See **partial-duration series**.

annual flood—The highest flow at a point on a **stream** during any particular **calendar year** or **water year**.
Flow may be expressed as maximum instantaneous **stage** or **discharge**, or the highest 24-h average. The annual flood need not exceed the established **flood stage**.

annual flood series—A **time series** of the largest **flood** that occurs during every year of the record.
See also **partial-duration series**.

annual maximum series—In hydrology, a series of data consisting of the largest values for each year.
See **annual flood**.

annual minimum series—In hydrology, a series of data consisting of the smallest values for each year.

annual range—Within a prescribed annual period, the difference between the maximum and the minimum value of a given set of numbers or of a temporally varying function.
See **range**.

annual runoff—Total volume of water that is discharged from a **drainage area** (**watershed** or catchment) and measured at a specified point on a **stream** or river during a year.

annual series—In hydrology, a series of data consisting of one value of a given **parameter** for each year.

annual storage—The storage capacity of a **reservoir** required to smooth out seasonal fluctuations in **streamflow** within a year.

anomalous—Not encompassed by rules governing the majority of cases; distinguished from **abnormal** by implying a difference of kind rather than a difference merely of degree.
This distinction between anomalous and abnormal is not strictly recognized in meteorology. As an example, a weather anomaly often implies a difference of degree.

anomalous dispersion—*See* **dispersion**.

anomalous propagation—(Sometimes abbreviated AP or anaprop.) A propagation path of **electromagnetic radiation** that deviates from the path expected from refractive conditions in a **standard atmosphere**.

In **standard propagation** conditions, radiation transmitted horizontally at the earth's surface is bent downward along a path with a radius of curvature equal to 4/3 times the radius of the earth. Subrefractive propagation causes less bending of the **ray** and superrefractive propagation causes greater downward bending than in the standard conditions. AP clutter is an extended region of ground **echoes** caused by **superrefraction**. *See* **effective earth radius**.

anomalous refraction—Literally, **refraction** that is not **normal** or typical.

Anomalous or **abnormal** is frequently used to characterize the conditions that give rise to mirages, such as in this example: "Abnormal **refraction** responsible for **mirages** is invariably associated with abnormal temperature distributions that yield abnormal spatial variations in the **refractive index**." Yet mirages and the conditions that give rise to them are very common. It would appear that, when these terms are used, normality is something defined by the **free atmosphere**, far from surfaces. Certainly, by that measure, the refractive index gradients near a surface are abnormal. It is misleading to imply that either the refractive structures or the mirages are abnormal or anomalous when they are so common near the earth's surface, merely because they are not as common farther away from that surface.

anomaly—1. The **deviation** of (usually) **temperature** or **precipitation** in a given region over a specified period from the long-term average value for the same region.

See **anomalous**, **abnormal**.

2. In **geophysics**, the local **deviation** from the long **wavelength** trend, as magnetic anomaly or **gravity** anomaly.

3. In **oceanography**, the difference between conditions actually observed at a **station** and those that would have existed had the water all been of a given arbitrary **temperature** and **salinity**.

anomaly correlation—A special case of **pattern correlation** for which the **variables** being correlated are the **departure** from some appropriately defined mean, most commonly a climatological mean.

See **pattern correlation**.

anomaly of geopotential difference—Same as **dynamic-height anomaly**.

antarctic air—A cold, dry **air mass** developed over the continent of Antarctica.

Antarctic air is generally colder at the surface in all seasons, and at all levels in austral (Southern Hemisphere) autumn and winter, than **arctic air**. See **airmass classification**.

Court, A., 1951: *Compendium of Meteorology*, p. 924.

antarctic anticyclone—(Or antarctic high.) The **glacial** anticyclone that has been said to overlie the continent of Antarctica; analogous to the **Greenland anticyclone**.

Until the International Geophysical Year there had been insufficient observational evidence to either support or contradict this theory.

Court, A., 1951: *Compendium of Meteorology*, 918–922.

Antarctic Bottom Water—(Abbreviated AABW.) A **water mass** formed by deep winter convection at the coast of Antarctica, particularly in the Weddell and Ross Seas but also at other shelf locations.

Being the densest water mass of the World Ocean, AABW is found to occupy the depth **range** below 4000 m of all ocean basins that have a connection to the Southern Ocean at that level. At the time of formation its **temperature** is close to the **freezing point** ($-1.9°C$), but to enter the oceans, AABW has to pass through and mix with the water of the **Antarctic Circumpolar Current**, which gives it its typical **salinity** of 34.7 **psu** and temperature of $+0.3°C$. Because of this it is also known as **Antarctic Circumpolar Water**, particularly in the Indian and Pacific Oceans.

Antarctic Circle—The line of latitude 66°34′S (often taken as 66½°S).

Along this line the sun does not set on the day of the **summer solstice**, about 22 December, and does not rise on the day of the **winter solstice**, about 21 June.

Antarctic Circumpolar Current—(Abbreviated ACC.) An eastward flowing current, also known as the West Wind Drift, that circles Antarctica and extends from the surface to the ocean floor.

With a volume **transport** of 130 Sv (130×10^6 m^{3}s^{-1}) it is the largest of all ocean currents. Current speed in the ACC is comparatively modest (0.1 m s^{-1}, but larger in fronts), the large transport being achieved by the current's great depth. Seventy-five percent of the transport occurs in the polar and subantarctic fronts that make up only 20% of the ACC area. Interannual **variability** is about 15% of the mean but can reach 40% on occasions. The ACC is influenced by bottom

topography, which causes deflections from its general westward path and **eddy** formation, particularly at the Scotia Ridge, the Kerguelen Plateau, and the Macquarie Ridge. The eddies are instrumental for the poleward transport of **heat** across the current, which would otherwise block meridional **heat transfer.**

Antarctic Circumpolar Water—*See* **Antarctic Bottom Water.**

antarctic climate—*See* **polar climate.**

antarctic convergence—*See* **Antarctic Polar Front.**

antarctic divergence—The region near 60°S, south of the **Antarctic Polar Front**, where high-salinity **North Atlantic Deep Water** upwells from 2500 m to just below the surface and mixes with low-salinity **Antarctic Surface Water.**

antarctic front—The semipermanent, semicontinuous **front** between the **antarctic air** of Antarctica and the **polar air** of the southern oceans; generally comparable to the **arctic front** of the Northern Hemisphere.
Compare **polar front.**

antarctic high—Same as **antarctic anticyclone.**

Antarctic Ice Sheet—The continuous **ice** mass covering most of Antarctica.

Antarctic Intermediate Water—A **water mass** identified by a **salinity** minimum found at depths between 700 and 1000 m in the Southern Hemisphere.
It is formed at various locations along the **Antarctic Polar Front** and through deep winter convection east of southern Chile and south of the Great Australian Bight. It enters all oceans with the **Antarctic Circumpolar Current** and spreads toward the **equator** between the **central water** and the **deep water.**

antarctic ozone hole—A phenomenon discovered in the mid-1980s that occurs in the winter–spring lower **stratosphere** over Antarctica.
Through a sequence involving **heterogeneous chemistry** on **polar stratospheric clouds** and (intermittent) **illumination** by **sunlight**, much or all of the **ozone** in the lower stratosphere can be photochemically destroyed. Halogen species (chlorine and bromine) contained in fairly robust molecules are transformed via heterogeneous reactions into molecules that are easily photolyzed resulting in atomic or monoxide **halogens** that lead to chemical destruction of ozone. This phenomenon also occurs over the Arctic, although to a lesser extent because of a lower incidence of polar stratospheric clouds.

Farman, J. C., Gardiner, and J. D. Shanklin, 1985: Large losses of total ozone in Antarctica reveal seasonal ClO_x/NO_x interaction. *Nature*, **315**, 208–210.
Solomon, S., R. Garcia, F. S. Rowland, and D. J. Wuebbles, 1986: On the depletion of antarctic ozone. *Nature*, **321**, 919–935.
Crutzen, P. J., and F. Arnold, 1986: Nitric acid cloud formation in the cold antarctic stratosphere: A major cause for the springtime "ozone hole". *Nature*, **324**, 651–655.
Molina, L. T., and M. J. Molina, 1987: Production of Cl_2O_2 from the self reaction of the ClO radical. *J. Phys. Chem.*, **91**, 433–436.

Antarctic Polar Front—The southern front of the **Antarctic Circumpolar Current**, also known as the antarctic convergence, that separates the **Antarctic Zone** in the south from the **polar frontal zone** in the north.
It is characterized by sea surface temperatures near 5°–6°C and a **salinity** minimum of 33.8–34.0 **psu** produced by high rainfall.

antarctic polar vortex—*See* **antarctic stratospheric vortex.**

antarctic sea smoke—*See* **sea smoke, steam fog.**

antarctic stratospheric vortex—The **vortex** in the lower **stratosphere** over the Antarctic in austral winter.

Antarctic Surface Water—The **water mass** of the **Antarctic Zone.**
It has a **temperature** of 0.0° to −1.9°C and a **salinity** below 34 **psu.**

Antarctic Zone—1. The region between the **Antarctic Polar Front** and the **Continental Water Boundary.**
2. Geographically, the region between the **Antarctic Circle** (66°32′S) and the South Pole.
Climatically, the limit of the zone may be set at about 60°S, poleward of which the prevailing westerly winds give place to easterly or variable winds. Over most of this region the average temperature does not rise above 0°C (32°F) even in summer.

antecedent—In **expert systems**, the IF part of an **IF–THEN rule**.
See also **consequent**.

antecedent precipitation index—(Abbreviated API.) A weighted summation of daily **precipitation** amounts, used as an index of **soil moisture**.
The weight given each day's precipitation is usually assumed to be an exponential or reciprocal function of time with the most recent precipitation receiving the greatest weight. The antecedent precipitation index method is commonly used to initialize some **rainfall** models.

antecedent soil moisture—The amount of moisture present in the soil at the beginning of a **storm** event, frequently expressed as an index corresponding to the **weighted average** of daily **rainfalls** for a given period of time preceding the storm event.

antenna—(Also called aerial; sometimes the more general term **radiator** is used.) A conductor or system of conductors for radiating and/or receiving radio **energy**.
As used in **radar**, the antenna is usually "directional," that is, it has the property of radiating or receiving **radio waves** in larger proportion in a given direction than in others. It includes both the basic radiating element or **feed** (such as **dipole antenna** or **horn antenna**) and its associated **reflector** for focusing the **energy**. See also **waveguide**, **loop antenna**, **beavertail antenna**.

antenna feed—The mechanism that illuminates the reflecting surface of an **antenna** or that distributes the **input** power to the radiating elements of an array antenna.

antenna gain—A measure of the effectiveness of a directional **antenna** in a given direction compared with a **standard** reference, usually an **isotropic** antenna.
Specifically, the antenna gain is the ratio of the **intensity** radiated in a given direction from the antenna to the intensity of an isotropic source emitting the same **power** as the real antenna. Sometimes the term antenna gain is used in place of axial gain, which is the value of the antenna gain in the direction of the axis of the **main lobe**. See **antenna pattern**.

antenna limit—For a particular **radar** system, the smallest measurable value of the **depolarization ratio**, a limitation imposed by imperfections in the **antenna**.

antenna pattern—(Also called radiation pattern, beam pattern, lobe pattern.) A graphical representation of the radiating properties of an **antenna** as a function of space coordinates.
Typically, the antenna pattern is a diagram showing the **radiant intensity**, the **electric field strength**, or the **radiant power** density emitted from an antenna as a function of the direction from the antenna, in the far-field region unless specified otherwise. The direction of maximum intensity defines the **main lobe** of the antenna pattern. Other local maxima, weaker than the main lobe, define the secondary lobes or sidelobes of the antenna. In the main lobe, the angular separation in a given plane between the directions at which the **intensity** falls to one-half its maximum value defines the half-power or 3-dB **beamwidth** in the plane. The directional **sensitivity** of an antenna to incoming signals is described by a function identical to the antenna pattern for transmission. Antenna patterns are also used to depict properties of the emitted **electromagnetic field** other than intensity, for example, the **polarization** as a function of position in the main lobe. See **lobe**.

antenna temperature—An **equivalent blackbody temperature** that characterizes the **noise power** at the terminals of an **antenna**.
Specifically, the antenna temperature is the **temperature** of a **blackbody** radiator that, when placed around an idealized loss-free antenna, produces the same noise power as that of the actual antenna. Antenna temperature may also refer to the **noise temperature** of the antenna as determined by not only the coupling of the antenna to environmental **noise** sources, but also noise generated by resistive losses within the antenna.

anthelic arcs—A group of **halos**, each of which passes through the **anthelion**.
Four mechanisms have been proposed for producing arcs of various shapes that nevertheless pass through the anthelion. Each of the four predicted shapes has been documented photographically. Each mechanism involves **refraction** and internal **reflection** in columnar **ice crystals**, but they differ in the particular **crystal** faces involved.

anthelion—A **luminous** white spot that appears on the **parhelic circle** 180° in **azimuth** away from the sun.
The origin of anthelion is not settled. The **brightness** of this spot may be due partly to the superposition of the parhelic circle and of the **anthelic arcs**, which intersect the parhelic circle at this point. Other candidate mechanisms giving **light** to anthelion have been suggested.

anthropoclimatology—(Obsolete.) Same as **human bioclimatology**.

anthropogenic—Human-induced or resulting from human activities; often used to refer to environmental changes, global or local in **scale**.

anthropogenic climate change—**Climate change** that occurs as a result of human activities.

anthropogenic emissions—Emissions to the **atmosphere** of gases as a result of human activity.
   The term is often used to describe, but is not restricted to, the exhaust from combustion events.

anthropogenic heat—Heat released to the **atmosphere** as a result of human activities, often involving combustion of fuels.
   Sources include industrial plants, space heating and cooling, human metabolism, and vehicle exhausts. In cities this source typically contributes 15–50 W m^{-2} to the local **heat balance**, and several hundred W m^{-2} in the center of large cities in cold climates and industrial areas. *Compare* **urban heat island**.

anti-icing—The prevention of the formation of **ice** upon any object, especially applied to **aircraft icing**.
   Compare **de-icing**.

anti-mountain wind—*See* **along-valley winds**.

anti-sea breeze—The return air flow aloft from land to sea that is part of the **sea breeze** circulation; usually observed during the day at the coast, when the land is warmer than the water.
   This **circulation** can carry **air pollutants** out to sea from coastal cities and industry.

anti-valley wind—*See* **along-valley winds**.

anticorona—Same as **glory**.

anticrepuscular arch—Same as **antitwilight arch**.

anticrepuscular rays—An extension of the **crepuscular rays** across the sky toward the **antisolar point**.
   Anticrepuscular rays seen with a **rainbow** or **fogbow** will always be radii to the bow.

anticyclogenesis—Any strengthening or **development** of **anticyclonic circulation** in the **atmosphere**; the opposite of **anticyclolysis**.
   This applies to the development of anticyclonic circulation (or the initial appearance of a **high**) where previously it was nonexistent, as well as to **intensification** of existing **anticyclonic** flow. The most common application of this term is to the formation of a new **anticyclone**. Care should be taken, however, to distinguish the increase in anticyclonic circulation from the increase in **atmospheric pressure** (**filling**), although they usually occur simultaneously. *Compare* **cyclogenesis**.

anticyclolysis—Any weakening of **anticyclonic circulation** in the **atmosphere**; the opposite of **anticyclogenesis**.
   Anticyclolysis (which refers to the **circulation**) should be distinguished from **deepening** (which refers to the **atmospheric pressure**), although the two processes usually occur together. *Compare* **cyclolysis**.

anticyclone—An atmospheric **anticyclonic circulation**, a **closed circulation**. The **wind** in an anticyclone is in the clockwise direction in the Northern Hemisphere and counterclockwise in the Southern Hemisphere.
   With respect to the relative direction of its rotation, it is the opposite of a **cyclone**. Because anticyclonic circulation and relative high atmospheric **pressure** usually coexist, the terms anticyclone and **high** are used interchangeably in common practice. *Compare* **ridge**.

anticyclonic—Having a sense of rotation about the local vertical opposite to that of the earth's rotation; that is, clockwise in the Northern Hemisphere, counterclockwise in the Southern Hemisphere, undefined at the **equator**.
   It is the opposite of **cyclonic**.

anticyclonic bora—*See* **bora**.

anticyclonic circulation—Fluid motion having a sense of rotation about the local vertical opposite to that of the earth's rotation; that is, clockwise in the Northern Hemisphere, counterclockwise in the Southern Hemisphere, undefined at the **equator**.
   It is the opposite of **cyclonic circulation**.

anticyclonic phase—*See* **foehn phase**.

anticyclonic rotation—Generally circular motion of a mass in a sense opposite to that of the earth, that is, clockwise in the Northern Hemisphere, counterclockwise in the Southern Hemisphere, undefined at the **equator**.

anticyclonic shear—Horizontal **wind shear** of such a nature that it contributes to the **anticyclonic** vorticity of the flow; that is, it tends to produce **anticyclonic rotation** of the individual air particles along the line of flow.

In the Northern Hemisphere, anticyclonic shear is present if (when one faces **downwind**) the **wind speed** decreases from left to right across the direction of flow; the opposite is true in the Southern Hemisphere. *Compare* **cyclonic shear**.

antihail rocket—A type of **cloud seeding** device that is fired into a **cloud** region where it releases a **seeding agent**.

Such rockets have been launched from the ground and from aircraft. The **seeding agent** may be generated as **smoke** from the **rocket** motor, or it may be part of the **rocket** payload. Hail suppression programs in Italy, Russia, Yugoslavia, and Switzerland have used rockets to deliver cloud seeding agents into regions of hail-producing thunderstorms where there are large amounts of **supercooled liquid water**.

Antilles Current—One of the currents of the North Atlantic subtropical **gyre** flowing along the northern side of the Greater Antilles.

Its water is supplied from the northern branch of the **North Equatorial Current**, drawn mainly from the **Gulf Stream** recirculation and therefore identical to water of the **Sargasso Sea**. East of the Straits of Florida, the Antilles Current merges with the **Florida Current**. The continuity of the Antilles Current as a steadily westward flowing current is somewhat in doubt, but westward **transport** does exist in the mean.

antilog—An artificial analog created by taking the mirror image of observed atmospheric states with respect to a climatological mean, and selecting mirror image states that are **analogs** for a given unmodified state.

antineutron—The **neutron**'s **antiparticle**, first observed in 1956.
Cork, B., G. R. Lambertson, O. Piccioni, and W. A. Wenzel, 1956: *Phys. Rev.*, **104**, 1193–1197.

antiparticle—The complementary **particle** to every **subatomic particle**, identical in mass and, if charged, identical in charge magnitude but with opposite sign.

If a particle has a magnetic (**dipole**) **moment**, its antiparticle has an equal and opposite magnetic moment. For example, the **electron** and **positron** comprise a particle–antiparticle pair. Because the electron was discovered first (and electrons are more abundant), the electron is the particle and the positron the antiparticle (antielectron), but this designation is arbitrary. The electron could just as well be called the antipositron. Collectively, antiparticles are called antimatter. When a particle and its antiparticle interact, they are annihilated, their **energy** living on in the form of **gamma rays**. The inverse process, **pair production**, is possible: A gamma ray photon interacting with an atomic **nucleus** may be annihilated, its **energy** taken up by an electron–positron pair. Some neutral particles are their own antiparticles (e.g., the **photon** and the neutral pion).

antipleion—A center of strong negative **anomaly** of a given **meteorological element**.

antiproton—The **proton**'s **antiparticle**, first observed in 1955.
Chamberlain, O., E. Segrè, C. Wiegand, and T. Ypsilantis, 1955: *Phys. Rev.*, **100**, 947–950.

antisolar point—The point on the **celestial sphere** that lies directly opposite the sun from the **observer**, that is, on the line from the sun through the observer.

antitrades—A deep layer of westerly winds in the **troposphere** above the surface trade winds of the **Tropics**.

They comprise the equatorward side of the midlatitude **westerlies**, but are found at upper levels rather than at the surface. The antitrades are best developed in the winter hemisphere and also above the eastern extremities of the subtropical highs. Farther west their base is higher and their appearance less regular. The antitrades were formerly regarded as return currents carrying away, to higher latitudes, the air that rises in the **intertropical convergence zones**, the westerly component being due to the **conservation of angular momentum** as the air moves into higher latitudes. Bjerknes (1935) showed that this explanation is incomplete: The antitrades are dynamical in origin and constitute an essential part of the **atmosphere**'s **primary circulation**.
Bjerknes, J., 1935: La circulation atmosphérique dans les latitudes sous-tropicales. *Scientia*, **57**, 114–123.

antitwilight—Same as **antitwilight arch**.

antitwilight arch—A bright wedge of pink, orange, or **purple light** that extends around the antisolar

47

horizon during **clear** twilights and that has the greatest vertical width (~3°–6°) occurring above the **antisolar point**.

The antitwilight arch rises with the antisolar point at **sunset** and sets with the antisolar point at **sunrise**. The antitwilight arch is brighter than the bluish gray **dark segment** beneath it. The boundary between these features may be distinct shortly after sunset (sun **elevation** $h_0 \sim -2°$), but it fades as evening **twilight** progresses.

anvil—Popular term for **incus**.
See **anvil cloud**.

anvil cloud—The anvil-shaped **cloud** that comprises the upper portion of mature **cumulonimbus** clouds; the popular name given to a **cumulonimbus capillatus** cloud, particularly if it embodies the supplementary feature **incus** (from the Latin for anvil).

anvil dome—See **overshooting top**.

AP—Abbreviation for **anomalous propagation**.

apaapaa—In Hawaii, a high **gale**.

aperiodic—1. Not periodic.

2. Applied to an instrument in which the indicator moves to a new position without oscillating.

aperture—1. In a unidirectional **antenna**, that portion of the plane surface that is perpendicular to the direction of maximum radiation and through which the major part of the **radiation** passes.

The physical aperture of a horn or **parabolic antenna** is identical with the actual area of the antenna face. Other types of apertures may be defined that are related to the effectiveness with which an antenna can remove **energy** from an incident **radio wave** (usually called **effective area**), and to the extent to which the intercepted energy is lost in **heat** or is reradiated.

2. In an opaque disk, the hole or window placed on either side of a lens to control the amount of **light** passing through.

aperwind—(Also called aberwind, alpach.) A warm **wind** of the Alps that thaws the **snow**.

aphelion—The point on the **orbit** of the earth (or any other body in orbit about the sun) that is farthest from the sun; the opposite of **perihelion**.

At present, aphelion occurs about July 1, when the earth is about three million miles farther from the sun than at perihelion, but the seasons in which aphelion and perihelion fall undergo a cyclic **variation** with a **period** of twenty-one thousand years. The date of aphelion passage is advancing slowly (toward dates later in the **calendar year**) at a rate of about one-half hour each year. This is a consequence of the eastward rotation of the **line of apsides** at a rate of about 11 sec of arc per year and the precession of the **equinoxes** of about 50 sec of arc per year, for a total of 61.9 sec of arc per year. See **apocenter**, **apogee**.

Apheliotes—The Greek name for the east **wind** ("blowing from the sun").

According to legend, it brings light showers. On the **Tower of the Winds** at Athens it is represented by a young man carrying fruit and grain.

API—See **antecedent precipitation index**.

apob—An **observation** of **pressure**, **temperature**, and **relative humidity** taken in the **free atmosphere** by means of an **aerometeorograph** (mounted under the wing of an aircraft).

This term was abbreviated from "airplane observation," and is rarely (if ever) used except in the above restricted sense. Compare **aircraft observation**, **pilot report**.

apocenter—The point on any **orbit** farthest from the center of attraction; the opposite of **pericenter**.
See **aphelion**, **apogee**.

apogean range—See **apogean tide**.

apogean tide—The **tide** of reduced **range** when the moon is near **apogee**, its farthest point from the earth in its elliptical **orbit**.

apogean winds—Land **winds** of Greece.

apogee—The point in an **orbit** at which any orbiting object, for example, planet, moon, artificial satellite, is farthest from the central object.
Opposite of **perigee**. Compare **perihelion**.

apparent force—(Also fictitious force, inertial force, transport force.) A force (mass times an **acceleration**) introduced on the side of an equation on which all supposedly real forces appear.

For Newton's dynamical law of motion for a body of mass m acted on by a force $\mathbf{F}$,

$$\mathbf{F} = m\mathbf{a},$$

to be valid requires that the acceleration $\mathbf{a}$ be specified relative to an **inertial reference frame**. If the acceleration in a noninertial **reference frame** (e.g., a rotating reference frame) is $\mathbf{a}^*$, then

$$\mathbf{a} = \mathbf{a}^* + \mathbf{a}_i,$$

and the previous dynamical equation may be written

$$\mathbf{F} - m\mathbf{a}_i = m\mathbf{a}^*,$$

where the quantity $- m\mathbf{a}_i$ is the inertial or apparent force and $\mathbf{a}_i$ is the inertial acceleration. Examples of apparent forces are the **centrifugal force** and the **Coriolis force**. Within classical (Newtonian) mechanics inertial forces are fictitious, merely masses times accelerations. But in general relativity, inertial forces are equivalent to real forces resulting from interactions between bodies because it is impossible to distinguish between inertial and gravitational accelerations; both are independent of the mass of the body. *See* **apparent gravity**.

> Symon, K. R., 1960: *Mechanics*, 2d ed., p. 271.
> Bergmann, P. G., 1942: *Introduction to the Theory of Relativity*, 155–156.

apparent gravity—(Also called effective gravity, virtual gravity.) For a rotating planet such as Earth the **resultant** of the force (per unit mass) due to **gravity** and the **centrifugal force**.

Apparent gravity is given by

$$\mathbf{g}^* = \mathbf{g} - \mathbf{\Omega} \times (\mathbf{\Omega} \times \mathbf{r}),$$

where $\mathbf{g}$ is the force (per unit mass) due to gravitational attraction, $\mathbf{\Omega}$ is the **angular velocity** of the planet, and $\mathbf{r}$ is the **position vector** relative to its center. Like $\mathbf{g}$, $\mathbf{g}^*$ depends only on position. Within the earth's **atmosphere** the magnitude of the centrifugal force is less than 0.03% of $\mathbf{g}$. *See* **apparent force**.

> Dutton, J. A., 1976: *The Ceaseless Wind*, p. 225.

apparent groundwater velocity—(Also called Darcy velocity.) The volume of **groundwater** passing through a unit area of **porous medium**, perpendicular to the direction of flow, in a unit of time.

apparent horizon—The line between the sky and the earth projected on the **celestial sphere**.

This line is often irregular, due to the earth's **topography**. On flat surfaces, such as the sea or a level plane, the line approximates a **great circle** if the observer's **elevation** above the surface is insignificant and a **small circle** if the elevation is appreciable.

apparent solar day—(Also called true solar day.) The interval of time between two successive transits of the sun across a **meridian**.

This interval is about four minutes longer than the sidereal day, largely because of the sun's apparent annual motion eastward along the **ecliptic** (actually, the earth's "westward" motion along its **orbit**), which motion delays the sun's return to **meridional** transit. Also, this interval is inconveniently nonuniform due to systematic variations in the earth's orbital speed around the sun and the sun's changing **declination**. The concept of the **mean solar day** has been invented to circumvent these practical difficulties.

apparent stress—A process that has the same effect as a **stress**.

An example is the **Reynolds stress**, where turbulent motions can exchange fluid from within a hypothetical cube of air with fluid of different **velocity** from outside the cube, causing the cube to deform in a manner similar to a stress acting on the face of the cube.

Applications Technology Satellite—(Abbreviated ATS.) The first experimental **geostationary satellite** series, operated by NASA, that preceded the **SMS** and **GOES** series.

ATS-1, launched in December 1966, carried the first spin scan **cloud camera**, providing full-disk **visible imagery** every 30 minutes, and introduced **WEFAX** data relay.

applied climatology—The scientific analysis of climatic data in the light of a useful application for an operational purpose.

"Operational" is interpreted as any specialized endeavor within such as industrial, manufacturing, agricultural, or technological pursuits (after Landsberg and Jacobs 1951). This is the general term for all such work and includes **agricultural climatology**, **aviation climatology**, **bioclimatology**, **industrial climatology**, and others.

> Landsberg, H. E. and Jacobs, W. C., 1951: *Compendium of Meteorology*, 976–992.

applied hydrology—Branch of **hydrology** associated with the practical use of the science to solve technical problems.

applied meteorology—A field of study where weather data, **analyses**, and forecasts are put to practical use.

Examples of applications include environmental health, **weather modification, air pollution meteorology, agricultural** and **forest meteorology**, transportation, value-added product development and display, and all aspects of **industrial meteorology**. *Compare* **applied climatology**.
> Houghton, D. D., 1984: *Handbook of Applied Meteorology*, 1328 pp.

approach channel—Channel **reach** upstream of a gauging **station** or a control structure; the reach is chosen such that flow is uniform at the station.

approach-light contact height—(Also called visual approach contact height.) The **altitude** along the instrument approach **glide path** of a landing aircraft from which the pilot will first see 500 ft of the approach light array.
> *See* **runway visual range**; *compare* **approach visibility**.

approach velocity—Mean **flow velocity**, measured a short distance **upstream** of a hydraulic structure.

approach visibility—(Sometimes called slant visibility.) The distance from which a pilot on the instrument approach **glide path** can see landing aids at the runway threshold.
> *Compare* **approach-light contact height, oblique visual range**; *see also* **runway visual range**.

APT—Abbreviation for **automatic picture transmission**.

aqueous vapor—Same as **water vapor**.

aquiclude—A geologic formation that may contain water but is incapable of transmitting water in significant quantities.

aquifer—A layer of saturated geologic materials that could yield water to springs or wells.

aquifer system—A group of two or more **aquifers** that are separated by **aquitards** or **aquicludes**.

aquifer test—A test to determine **hydrologic properties** of the **aquifer** involving the withdrawal of measured quantities of water from, or addition of water to, a well and the measurement of resulting changes in **head** in the aquifer both during and after the period of **discharge** or additions.

aquifuge—A geologic formation that has no interconnected openings and hence cannot receive or transmit water.

aquitard—A geologic formation that is not permeable enough to yield significant quantities of water to wells, but on a regional **scale** can contribute significant water to the underlying or overlaying **aquifers**.

aracaty—Northeast **wind** of Ceará, Brazil.

Arago distance—The angular distance from the **antisolar point** to the **Arago point**.

The Arago distance is sensitive to the presence of foreign **scattering** particles in the **atmosphere** since these increase the contribution of the negative (horizontal) component of **skylight** polarization and hence shift the location of the point where the negative component is just equalled by the positive component. Thus the Arago distance is a useful measure of atmospheric **turbidity**. Its value is generally close to 20°, and is a function of solar **elevation angle** and of the **wavelength** of the **light** with which **polarization** is studied.
> Sekera, Z., 1951: *Compendium of Meteorology*, 79–90.

Arago point—One of the three commonly detectable points along the vertical circle through the sun at which the **degree of polarization** of skylight goes to zero; a **neutral point**.

The Arago point, so named for its discoverer, is customarily located at about 20° above the **antisolar point**, but it lies at higher elevations in turbid air. The latter property makes the **Arago distance** a useful measure of atmospheric **turbidity**. Measurements of the location of this neutral point are typically more easily carried out than measurements of the **Babinet point** and the **Brewster point**, both of which lie so close to the sun (about 20° above and below the sun, respectively) that **glare** problems become serious.
> Neuberger, H., 1951: *Introduction to Physical Meteorology*, 196–204.

Arakawa Jacobian—A finite–difference **approximation** to the **Jacobian** operator that conserves both **kinetic energy** and **entropy**.

arc cloud—(Also known as arc cloud line.) A line of **cumuliform** clouds that forms as a result of local **convergence** along the boundary separating low-level **convective storm** outflow from the surrounding **environment**.
> *See* **gust front**.

arc cloud line—*See* **arc cloud**.

arc discharge—A **luminous**, gaseous, **electrical discharge** in which the charge **transfer** occurs continuously along a narrow channel of high **ion density**.

An arc discharge requires a continuous source of an **electric potential gradient** across the terminals of the arc. Arc discharges do not naturally occur in the **atmosphere**. *Compare* **corona discharge**, **point discharge**, **spark discharge**.

arched squall—The name applied to a **squall** in the **Tropics** when the squall cloud features a well-developed **arcus** (or **roll cloud**). It is usually a relatively **violent storm**.

See **sumatra**.

Archimedes's principle—The statement that a net upward or **buoyant force**, equal in magnitude to the weight of the displaced fluid, acts upon a body either partly or wholly submerged in a fluid at rest under the influence of **gravity**.

This force is known as the Archimedean buoyant force (or **buoyancy**), is independent of the shape of the submerged body, and does not depend upon any special properties of the fluid.

arcs of Lowitz—A **halo** in the form of arcs that pass both obliquely and vertically through the 22° parhelia.

The arcs of Lowitz are explained by **refraction** through the 60° prism of hexagonal plate **ice crystals**, oscillating as they descend. The principal axis of the plate must oscillate about the vertical by less than 30° to produce the pattern customarily called **Lowitz arcs**; larger oscillations lead to more complex arcs spreading far above and below the 22° parhelia and **halo**. See **parhelion**.

arctic air—A type of **air mass** with characteristics developed mostly in winter over arctic surfaces of **ice** and **snow**.

Arctic air is cold aloft and extends to great heights, but the surface temperatures are often higher than those of **polar air**. For two or three months in summer arctic air masses are shallow and rapidly lose their characteristics as they move southward. *See also* **antarctic air**, **airmass classification**.

arctic-alpine—Of, or pertaining to, areas above the **timber line** in mountainous regions, and to the biologic, geologic, etc., characteristics of such areas.

arctic anticyclone—*See* **arctic high**.

arctic blackout—*See* **radio blackout**.

Arctic Bottom Water—The **water mass** formed in the Arctic Ocean by a combination of **freezing** on the arctic shelf and deep winter convection in the Greenland and Norwegian Seas.

Freezing increases the **salinity** under the **ice**; the dense water sinks to the ocean floor and leaves the arctic basins to enter the Greenland and Norwegian Seas, where it mixes with water that sinks under the influence of surface cooling. The resulting water mass has a salinity of 34.95 **psu** and a **temperature** of −0.8° to −0.9°C. It fills the Arctic Ocean at all depths below 800 m, the **sill depth** to the Atlantic. It enters the Atlantic in bursts, when the passage of atmospheric depressions lifts the **thermocline** and allows Arctic Bottom Water to flow over the sill. Overflow events in the Denmark Strait and across the Iceland–Faeroe sill contribute some 5 Sv (5×10^6 m³s⁻¹) to the formation of **North Atlantic Deep Water**.

Arctic Circle—The line of latitude 66°34′N (often taken as 66½°N).

Along this line the sun does not set on the day of the **summer solstice**, about June 21, and does not rise on the day of the **winter solstice**, about December 22. From this line the number of twenty-four-hour periods of continuous day or of continuous night increases northward to about six months each at the North Pole.

arctic climate—*See* **polar climate**.

arctic desert—"Any area in the high latitudes dominated by bare rocks, **ice**, or **snow**, and having a sparse vegetation and a low annual **precipitation**." [from *Glossary of Arctic and Subarctic Terms* (1955)].

Thus stated, this includes portions of both **ice cap** and **tundra** regions of both hemispheres. The term **barrens** is sometimes used, but this has more general application.

Arctic, Desert, Tropic Information Center (ADTIC) Research Studies Institute, 1955: *Glossary of Arctic and Subarctic Terms*, ADTIC Pub. A-105, Maxwell AFB, AL, 90 pp.

arctic front—The semipermanent, semicontinuous **front** between the deep, cold **arctic air** and the shallower, basically less cold **polar air** of northern latitudes; generally comparable to the **antarctic front** of the Southern Hemisphere.

Compare **polar front**.

arctic haze—A condition of reduced horizontal and **slant visibility** (but unimpeded vertical visibility) encountered by aircraft in flight over arctic regions.

51

arctic high—(Also called arctic anticyclone, polar anticyclone, polar high.) A weak **high** that appears on mean charts of **sea level pressure** over the Arctic basin during late spring, summer, and early autumn. *Compare* **glacial anticyclone**.

Arctic Intermediate Water—A **water mass** identified by a **salinity** minimum found at a depth of about 800 m in the North Atlantic Ocean.

It is formed in two varieties in the Labrador Sea and in the Iceland Sea, from where it spreads southward but is quickly absorbed by **North Atlantic Deep Water**. The equivalent water mass in the Pacific Ocean is known as **Subarctic Intermediate Water**.

arctic mist—A **mist** of **ice crystals**; a very light **ice fog**. *Compare* **ice-crystal haze**.

arctic pack—**Pack ice** of the Arctic Ocean.

The arctic pack ice varies from about 9×10^6 km^2 to 16×10^6 km^2.

Arctic Polar Front—The **frontal zone** between the subtropical and subpolar **gyres** of the Northern Hemisphere.

In the Atlantic Ocean it is established by the meeting of the warm and saline **Gulf Stream** and the cold and fresh **Labrador Current** and extends as a **temperature** and **salinity** front, sometimes also known as the **cold wall**, from south of Newfoundland and the Grand Banks northeastward to the central North Atlantic. In the Pacific Ocean it consists of two parts, separated by the Japanese islands. The larger fresh is formed by the **confluence** of the warm and saline **Kuroshio** and the cold and fresh Oyashio and seen as a temperature and salinity front extending eastward from Japan near 35°N. The smaller fresh extends across the Sea of Japan in the west, where it separates the warm and saline **Tsushima Current** from the cold and fresh **Mid-Japan Sea Cold Current**.

arctic sea smoke—*See* **sea smoke**, **steam fog**.

arctic stratospheric vortex—The **vortex** in the lower **stratosphere** over the Arctic in **boreal** winter.

Arctic Surface Water—The **water mass** of the upper 150 m in the Arctic Ocean.

It has a **temperature** of $-1.5°$ to $-1.9°$C and a **salinity** varying between 28 and 33.5 **psu**.

arctic tree line—The northern limit of tree growth; the sinuous boundary between **tundra** and **boreal forest**; taken by many to delineate the actual southern boundary of the **Arctic Zone**.

Arctic Zone—1. (Formerly called North Frigid Zone.) Geographically, the area north of the **Arctic Circle** (66°34′N).

2. (Same as **tundra**.) Biogeographically, the area extending northward from the **arctic tree line** to the "limit of life."

It is also used for the level above the **timber line** in mountains.

arcticization—The preparation of equipment for operation in an **environment** of extremely low temperatures.

arcus—(Often called roll or shelf cloud.) A dense and horizontal **roll cloud** or wedge-shaped **accessory cloud**, sometimes appearing ragged and turbulent and other times smooth, occasionally with multiple layers.

It is seen most often in association with **cumulonimbus** clouds on the leading edge of a **gust front**. When extensive and approaching, it gives the appearance of a menacing arch. It may also occur with fronts in the absence of a cumulonimbus or in the absence of any **thermal** boundary. *See* **morning glory**.

ARDC Model Atmosphere—*See* **standard atmosphere**.

area average—The average or **mean value**, computed from observations at various locations on a horizontal plane, an **isobaric surface**, or the earth's surface; sometimes used in the study of **atmospheric turbulence** as a surrogate for the **ensemble average**.

In the real **atmosphere**, area averages can be acquired using remote sensors such as **radar**, **sodar**, and **lidar**, or they can be approximated using an appropriate flight track of a research aircraft. In numerical models, area averages can be computed using data from **grid** points at a constant **model** level. See **ergodic condition**.

area–elevation curve—Curve showing what the surface area of a **reservoir** is at an indicated **elevation** above a specific level.

area forecast—(Also called regional forecast.) A **weather forecast** for a specified geographic area; usually applied to a form of **aviation weather forecast** (to distinguish it from **terminal forecast**, **route forecast**, and **flight forecast**).

Compare **district forecast**; *see also* **ARFOR, ARFOT, ARMET.**

area of high pressure—*See* **high.**

area of influence—The area within the **cone of depression** of a discharging well or the **cone of impression** of a recharging well.

areal precipitation—Precipitation in a specific area expressed as an average depth of liquid water over the area; the average depth of **precipitation** over the specific area calculated on timescales on a **storm**, seasonal, or annual basis.

areal reduction factor—For areal **rainfall** of a given duration and **return period**, it is the ratio of the mean areal rainfall to the mean point rainfall for the same duration and return period in the same area.

ARFOR—An international code word used to indicate an **area forecast**.
See also **ARFOT, ARMET.**

ARFOT—An international code word used to indicate an **area forecast**, units in English system.
See also **ARFOR, ARMET.**

argon—Noble gas, symbol Ar, of atomic mass 40, that composes 0.93% by volume of **dry air**.
Argon is constantly produced in the earth's crust from the radioactive **decay** of ^{40}K and is subsequently released into the **atmosphere**. It is the most abundant of the inert gases in the atmosphere, and the third most abundant of all atmospheric constituents. Argon was discovered by Rayleigh and Ramsay in 1894.

Argos—A data collection and location system installed on **POES** satellites.
Argos, provided by France, allows environmental data sensed by remote platforms, fixed or moving, to be collected by POES satellites and retransmitted to ground processing stations.

arheic—(Also called areic, aretic, arhetic.) An area lacking **surface runoff** or drainage, such as deserts, in which surficial drainage is almost completely lacking, or where **rainfall** is so infrequent that all water sinks into the ground or evaporates.

arid climate—*See* **desert climate.**

arid zone—1. A region with insufficient moisture where **evaporation** exceeds **precipitation**.
2. Sometimes used synonymously with **equatorial dry zone**.

arid-zone hydrology—The study of the **hydrology** of arid or semiarid regions.

aridity—The degree to which a **climate** lacks effective, life-promoting moisture; the opposite of **humidity**, in the climate sense of the term.
The overall concept of aridity versus humidity is coming to be known as **precipitation effectiveness**. Two basic approaches have been made. The first, used by W. Köppen and modified by Bailey, does not openly define aridity, but rather assigns delimiting values of annual **precipitation** (treated with regard to distribution and **temperature**) to separate a **dry climate** from other types. The second approach actually prescribes a measure of aridity or precipitation effectiveness and uses these values as a primary parameter of classification. Of this type are Thornthwaite's **precipitation-effectiveness index** and **moisture index**, E. de Martonne's **index of aridity**, W. Gorczyński's **aridity coefficient**, Lang's **moisture factor**, and Ångström's humidity coefficient.

aridity coefficient—A function of **precipitation** and **temperature** designed by W. Gorczyński to represent the relative lack of effective moisture (the **aridity**) of a place.
It is given by (latitude factor) × (temperature range) × (precipitation ratio). The latitude factor is the cosecant of the latitude (taken as 3.0 for 0°–4°). The temperature **range** is the difference (°F) between the means of the hottest and coldest months. The precipitation ratio is the difference between the highest and lowest annual totals (adjusted to a 50-year record) divided by the average.
The value of this coefficient is about 100 in the middle of the Sahara; in the United States it ranges from 70 at Bagdad, California, to 2 at Eureka, California.

aridity index—1. As used by C. W. Thornthwaite in his 1948 **climatic classification**: an index of the degree of **water deficit** below **water need** at any given station; a measure of **aridity**.
It is calculated, independently of the opposite **humidity index**, as follows:

$$\text{aridity index} = 100d/n,$$

where d (the water deficit) is the sum of the monthly differences between precipitation and **potential evapotranspiration** for those months when the **normal** precipitation is less than the normal potential evapotranspiration; and where n is the sum of monthly values of potential evapotranspiration for the deficient months.

Thornthwaite puts the aridity index to two uses: 1) as a component of the moisture index; 2) as a basis for the more detailed classification of **moist climates** (perhumid, humid, and moist subhumid climates).

    2. *See* **index of aridity**.

arithmetic mean—1. (Also called mean, average, simple average.) One of several accepted measures of **central tendency**, physically analogous to "center of gravity."

    Pertaining to a set of numbers $x_1, x_2, \ldots, x_n$, the arithmetic mean, usually denoted by the symbol $\bar{x}$, is the sum $x_1 + x_2 + \ldots + x_n$ divided by n. Since the word "mean" is also applied to other measures of central tendency, such as weighted means, geometric means, and **harmonic** means, the adjective "arithmetic" is used for clarity. However, when used without further qualification, the term "mean" is understood as arithmetic mean. *Compare* **geometric mean**.

    2. Pertaining to a **random variable**, same as **expected value**.

ARMET—An international code word used to indicate an **area forecast**, units in metric system. *See also* **ARFOR, ARFOT**.

armouring— (Also spelled armoring.) Formation of a layer of relatively large **particles** at the surface of a deposit due to the removal of finer **sediment** particles by fluid **erosion**.

    The armoured layer generally resists further erosion.

aromatic hydrocarbons—Hydrocarbons based on the structure of **benzene**, characterized by a very stable, unsaturated ring.

    Aromatics make up a sizeable fraction of the **hydrocarbons** emitted from automobiles, and their high reactivity contributes to local **smog** formation.

Arrhenius equation—A relation used to describe the **temperature** dependence of the **rate coefficient** for a chemical reaction.

    The form of the equation is

$$k = A \exp\left(-\frac{E_a}{RT}\right),$$

where k is the rate coefficient, A is the **preexponential factor**, E_a is the **activation energy** for the reaction, R is the **gas constant**, and T is the **absolute** temperature.

Arrhenius expression—*See* **Arrhenius equation**.

arroyo—A water-carved channel or gully in arid country, usually rather small with steep banks, that is dry most of the time due to infrequent **rainfall**.

ARS—Abbreviation for **Absolute Radiation Scale**.

artesian aquifer—An **aquifer** in which the **piezometric surface** is above its upper boundary.

artesian basin—A geologic **basin** in which, due to the presence of a confining layer, **groundwater** is stored at a **pressure** greater than **hydrostatic pressure**.

artesian ground water—*See* **artesian aquifer**.

artesian well—A well in which the **water level** is higher than the top of the **aquifer**.

artificial climate—**Climate** created or modified by human activity; usually used to refer to conditions in an enclosed space, not global or local external conditions.

artificial control—A man-made structure, such as a **weir**, bridge, dam, etc., that serves as a control for a **stream gauging** station by stabilizing the **rating curve**.

artificial horizon—A planar reflecting surface that can be adjusted to coincide with the **astronomical horizon**, that is, can be made perpendicular to the **zenith**.

    This instrument is used, usually in conjunction with others, in observing celestial bodies. *See* **horizon**.

artificial ice nucleus—A man-made particulate **aerosol** used to induce the **ice** phase in cold (below 0°C) **cloud seeding** operations.

    Silver iodide aerosol is most commonly used.

artificial intelligence—(Abbreviated AI.) Multidisciplinary field encompassing computer science, neuroscience, philosophy, psychology, robotics, and linguistics, and devoted to the reproduction of the methods or results of human reasoning and brain activity.

artificial precipitation—Generally, **precipitation** that is produced as a result of human activity, such as **cloud seeding**, in contrast to natural precipitation.

artificial recharge—The process of artificially increasing the volume of water in the **saturated zone** by means of spreading water on the surface, injection wells, or inducing **infiltration** from streams or lakes.

ascendent—The negative of the **gradient**.

The ascendent of a function is a **vector** with magnitude equal to the maximum spatial rate of change of that function at a given point at a given time. It is directed toward increasing values of the function along the line of maximum change and is represented by ∇F, where F is the function and ∇ is the **del operator**.

ascending node—The northbound equatorial crossing of a **polar-orbiting satellite**, given in degrees longitude, date, and coordinated **universal time** (UTC) for any given **orbit** or pass; the point at which a satellite crosses the equatorial plane, heading north.

ASDAR—Abbreviation for **aircraft-to-satellite data relay**.

ash—1. Airborne **particulates** produced as a combustion product.

These affect air quality, and their emission and ambient concentration is often regulated. For example, coal ash can consist of oxides of silicon, aluminum, iron, calcium, magnesium, sodium, potassium, sulfur, and titanium.

2. Airborne **particulates** produced by a volcano.

These are often **fine particles** of rock with very sharp edges, which cause great abrasion and wear when ingested in the air intake of an engine, and can quickly lead to engine failure.

ash devils—Dust devils made visible by loose **ash** from the earth's surface that is picked up or entrained by the rotating column of air.

For example, after the eruption of the Mount St. Helens volcano in western Washington state during May 1980, ash devils were frequently observed in the eastern Washington **boundary layer** during the subsequent summer. *Compare* **dust devil, steam devils.**

aspect—The **compass** direction toward which a land slope faces.

The direction is taken downslope and normal to the contours of **elevation**.

aspect ratio—The ratio of height-to-length scales (D/L) characteristic of a fluid flow or of a physical **model** prototype, used for scaling of equations or for constructing physical (e.g., **wind** tunnel) models.

aspirated thermometer—A **thermometer** for which **ventilation** is provided by a suction fan.

aspiration condenser—The collecting element in the type of **ion counter** that uses the method of perpendicular velocities.

It consists of a cylindrical condenser that is electrically charged so that the radial **field** produced will collect the **ions** from the aspirated air. *See* **Weger aspirator.**

aspiration meteorograph—An instrument for the continuous recording of two or more meteorological parameters for which the **ventilation** is provided by a suction fan.

aspiration psychrometer—A **psychrometer** in which the **ventilation** is provided by a suction fan.

The **aspirator** may be driven by a spring or an electric motor. Alternatively, psychrometers with natural ventilation are also in use. *See* **ventilated psychrometer.**

aspiration thermograph—A **thermograph** for which **ventilation** is provided by a suction fan.

aspirator—A device attached to a **meteorological instrument** to provide **ventilation**.

In its usual form it consists of a suction fan.

aspre—A northeasterly **foehn** extending over the Garonne Plain in France, resulting from flow over the Massif Central.

assimilation—*See* **data assimilation.**

Assmann psychrometer—A special form of the **aspiration psychrometer**, developed by Assmann.

Two **mercury thermometers**, mounted vertically side by side in a chromium- or nickel-plated polished metal frame, are connected by ducts to an **aspirator**. Each thermometer is located inside a pair of coaxial metal tubes, highly polished inside and out, that screen the bulbs from **solar radiation**.

Middleton, W. E. K., and A. F. Spilhaus, 1953: *Meteorological Instruments*, 3rd ed., rev., Univ. of Toronto Press, 102–104.

astronomical horizon—The **great circle** that is 90° from the local vertical.

The astronomical horizon is usually more convenient to use than the **apparent horizon**.

astronomical refraction—1. (Or **atmospheric refraction**.) The difference between the angular positions of the **image** and object of a celestial object that results from **refraction** in the earth's **atmosphere**.

Astronomical refraction is greatest near the **horizon** where it normally exceeds a solar diameter. Consequently, we see the image of the sun above the horizon even though the sun itself is below, and this extends the length of the day measurably at both ends.

2. A variety of phenomena that result from the **refraction** of celestial **light** by the earth's **atmosphere**.

It is distinguished from the term **terrestrial refraction**, which is used when the source is within our atmosphere.

astronomical scintillation—*See* **scintillation**.

astronomical seeing—*See* **seeing**.

astronomical theory of climate change—*See* **Milankovitch theory**.

astronomical twilight—The **twilight** stage during which the sun's unrefracted center is at **elevation** angles $-12° > h_0 > -18°$.

During a **clear** evening's astronomical twilight, horizontal **illuminance** due to **scattered** sunlight decreases from ~0.008 **lux** to ~6×10^{-4} lux. At $h_0 = -18°$, 1) no **horizon** glow is visible at the sun's **azimuth** (the **bright segment**'s upper boundary is at the observer's **astronomical horizon**), 2) sixth-magnitude stars can be seen near the **zenith**, and 3) scattered sunlight's residual illuminance is less than that from starlight and **airglow**. *Compare* **civil twilight**.

asymmetry factor—The mean cosine of the **scattering** angle, found by integration over the complete scattering **phase function**.

The asymmetry factor is a fundamental property of a scattering medium that affects the **transfer** of **radiant energy** through the medium. Particles that are small compared to the **wavelength** of the **radiation**, such as air molecules, have asymmetry factors close to zero. Larger **particles**, such as **cloud droplets**, typically have asymmetry factors ~0.85 for **visible radiation**, consistent with strong **forward scattering**.

asymptote of convergence—Same as **convergence line**.

asymptote of divergence—Same as **divergence line**.

asymptotic expansion—A series of the form $c_0 + (c_1/x) + (c_2/x^2) + \ldots + (c_n/x^n) + \ldots$ often used for formal solution of differential equations or for function approximation when $|x|$ is large.

Though such series are divergent, terms initially decrease in size to one of minimum value at which term the series can be truncated to obtain a useful approximation.

asynchronous communication—Communication where the sender is able to send data at any time, and with any time gaps the sender desires.

asynoptic observation—A supplementary **meteorological observation** between regular **synoptic observation** periods.

Athos fall wind—A cold, northeasterly, often strong **fall wind** descending the southerly slopes of Mount Athos, on the Hagion Oros Peninsula in northeastern Greece, and extending several kilometers out over the northern Aegean Sea.

Atlantic Water—A **water mass** of Atlantic origin.

In the Arctic Ocean the term is used for the water mass that enters the region from the Norwegian Sea and, on account of its high salinity, spreads underneath the fresher **Arctic Surface Water**. It can be followed as a **salinity** maximum at 150 m near Spitzbergen and progressively deeper to 500 m in the Canada **basin**. In the scattering, Atlantic Water identifies the water mass that enters through the Straits of Gibraltar and, on account of its lower **temperature**, spreads underneath the saltier but warmer Mediterranean **surface water**. It is seen as a salinity minimum at a depth of between 20 and 50 m.

atmidometry—(Rare.) Same as **atmometry**.

atmometer—(Also called evaporimeter, evaporation gauge, atmidometer.) The general name for an instrument that measures the **evaporation rate** of water into the **atmosphere**.

Four main classes of atmometers may be distinguished: 1) large **evaporation** tanks sunk into the ground or floating in water; 2) small evaporation pans; 3) porous porcelain bodies; and 4) porous paper wick devices. The evaporation from a surface depends greatly upon the nature of the surface and the **exposure** of the surface to the atmosphere. Measured evaporation rates should be compared only between identical instruments. *See* **clay atmometer, evaporation pan, evapo-transpirometer, Livingstone sphere, Piché evaporimeter, radio atmometer**.

atmometry—(Rare; previously called atmidometry.) The science of measuring the rate and amount of **evaporation** of water.

atmoradiograph—A device for measuring the **frequency** of occurrence of **atmospherics**, the **intensity** of which is greater than a predetermined level.

atmosphere—1. A gaseous envelope gravitationally bound to a celestial body (e.g., a planet, its satellite, or a star).

Different atmospheres have very different properties. For instance, the atmosphere of Venus is very thick and **cloudy**, and is responsible for producing the very high surface temperatures on that planet by virtue of its **greenhouse effect**. On the other hand, the Martian atmosphere is very sparse. Earth's atmosphere is intermediate between these two extremes. It is distinguished from all other known atmospheres by its very active **hydrologic cycle**. One need merely examine pictures of Earth from space to appreciate the intricate **cloud** structures. Water in Earth's atmosphere plays a very important energetic role. Because of its chemical composition, most incoming **sunlight** passes through Earth's atmosphere and is absorbed at the ground. This **heat** is transported to the atmosphere through **sensible heat** and moisture fluxes. Upon **condensation**, this heat is then released into the atmosphere. The **thermodynamics** of **water vapor** is the crucial factor to the existence of **severe storms** in Earth's atmosphere. Since more **solar radiation** is absorbed in the **Tropics** than at high latitudes, the atmosphere (and the ocean) transports heat poleward. These motions are heavily altered by the effects of planetary rotation to determine the atmospheric **general circulation**. Fluid dynamical instabilities play a large role in this circulation and are crucial in determining the fluctuations in this circulation that we call "weather." The atmosphere may be conceptually divided into several **layers**, according to its **thermal** and **ionization** structure. The region where the **temperature** decreases because of the upward **heat flux** is called the **troposphere**. Above it, there is a layer in which temperature increases upward because of **ozone** absorption of **solar radiation**, the **stratosphere**. Above this, the temperature decreases in the **mesosphere**, and above this, in the **thermosphere**, the extremely energetic **radiation** causes temperature to increase with height out to the outer reaches of Earth's atmosphere, the **exosphere**. Within the mesosphere and thermosphere, solar radiation is sufficiently energetic to ionize gases. This produces the **ionosphere**.

2. As a unit of **pressure**, *see* **standard atmosphere**.

3. *See* **standard atmosphere, model atmosphere**.

atmosphere–mixed layer ocean model—A **model** of the ocean–atmosphere system that consists of a **general circulation model** of the **atmosphere** coupled with a model of the **oceanic mixed layer**.

In climatic applications, the oceanic mixed layer is that part of the ocean that stores the seasonal **heat** anomaly and is typically represented in a very simple manner. Atmosphere–mixed layer ocean models are commonly used to study **equilibrium** climatic response.

atmosphere–ocean interaction—Interactions between the **atmosphere** and the ocean, as in **air–sea interaction**, but also including large spatial scale (up to global) effects.

atmospheric attenuation—The reduction with distance from the source of the **intensity** of an acoustic or an electromagnetic **signal** propagating through the **atmosphere** caused by interaction of the signal with gaseous constituents of the atmosphere, **aerosols**, or **hydrometeors**. In general, **scattering** and **absorption** account for **attenuation**.

For **sound**, absorption is usually more important than scattering; it depends on **temperature** and **humidity**, and generally increases with increasing acoustic **frequency**. The main atmospheric constituents that absorb radar energy are **oxygen**, **water vapor**, and liquid hydrometeors. Absorption is often neglected at wavelengths of 10 cm and longer, but becomes increasingly important at shorter wavelengths. The contribution of scattering to **radar** attenuation also increases with decreasing **wavelength**. For **lidar**, scattering by molecules, aerosols, and hydrometeors dominates the attenuation, although gaseous absorption is significant at certain spectral bands and is exploited by differential absorption lidar (DIAL) to measure molecular concentrations.

atmospheric boil—*See* **scintillation**.

atmospheric boundary layer—(Abbreviated ABL; also called boundary layer, planetary boundary layer.) The bottom layer of the **troposphere** that is in contact with the surface of the earth.

It is often turbulent and is capped by a **statically stable** layer of air or **temperature inversion**. The ABL depth (i.e., the **inversion** height) is variable in time and space, ranging from tens of meters in strongly statically stable situations, to several kilometers in convective conditions over deserts. During **fair** weather over land, the ABL has a marked **diurnal** cycle. During daytime, a **mixed layer** of vigorous **turbulence** grows in depth, capped by a statically stable **entrainment zone** of intermittent turbulence. Near **sunset**, turbulence decays, leaving a **residual layer** in place of the mixed layer. During nighttime, the bottom of the residual layer is transformed into a statically **stable boundary layer** by contact with the radiatively cooled surface. **Cumulus** and **stratocumulus** clouds can form within the top portion of a humid ABL, while **fog** can form at the bottom of a stable boundary layer. The bottom 10% of the ABL is called the **surface layer**. *Compare* **Ekman layer**.

Stull, R. B., 1988: *An Introduction to Boundary Layer Meteorology*, 666 pp.

atmospheric chemistry—The study of the composition of and chemical transformations occurring in the **atmosphere**.

The discipline of atmospheric chemistry includes field measurements, computer modeling, and laboratory measurements, and requires an understanding of the interaction of the atmosphere with the **biosphere** and **anthropogenic** influences in order to be able to explain current conditions and to predict future changes.

atmospheric circulation model—A mathematical **model** for quantitatively describing, simulating, and analyzing the structure of the **circulation** in the **atmosphere** and the underlying causes.

Compare **general circulation model, hemispheric model**.

atmospheric correction—Correction made to remotely sensed **radiance** to account for effects related to the intervening **atmosphere** between the earth's surface and the satellite.

atmospheric demand—The **evapotranspiration** that would be achieved from a well-aerated soil/ plant surface at **field water-holding capacity**.

Water loss may differ from **pan evaporation** according to the index of evaporating surface to horizontal area, roughness, and other physical and biological attributes of a site.

atmospheric disturbance—1. Any interruption of a state of **equilibrium** of the **atmosphere**.

2. An area showing signs of a developing **cyclonic circulation**.

atmospheric dust—Fine **airborne particulates** of size **range** 1–100 μm.

See **dust**.

atmospheric dynamics—The study of those motions of the **atmosphere** that are associated with weather and **climate**.

In atmospheric dynamics the fluid is regarded as a continuous medium, and the fundamental laws of fluid mechanics and **thermodynamics** are expressed in terms of partial differential equations involving the fluid **velocity**, **density**, **pressure**, and **temperature**.

atmospheric electric field—A quantitative term denoting the **electric field strength** of the atmosphere at any specified point in space and time.

In areas of **fair** weather, the atmospheric electric field near the earth's surface typically is about 100 volts (V) m^{-1} and is directed vertically in such a sense as to drive positive charges downward to the earth. In areas of fair weather this field decreases in magnitude with increasing **altitude**, falling, for example, to only about 5 V m^{-1} at an altitude of about 10 km. Near **thunderstorms**, and under **clouds** of vertical development, the surface **electric field** (the electric field measured at the surface of the earth) varies widely in magnitude and direction, usually reversing its direction immediately beneath active thunderstorms. In areas of minimal local **disturbance**, a characteristic **diurnal** variation of **electric field strength** is observed. This **variation** is characterized by a maximum that occurs at about 1900 **UTC** for all points on the earth and is now believed to be produced by thunderstorms that, for geographic regions, are more numerous for the world as a whole at that **universal time** than at any other. It is now believed that thunderstorms, by replenishing the negative charge to the earth's surface, provide the **supply current** to maintain the fair-weather electric field in spite of the continued flow of the **air–earth current** that tends to neutralize that field. The **range** of the electric field in fair weather varies considerably with geographical area, from one part of the globe to another. If, however, there are no local sources of **pollution**, the surface electric field has its maximum amplitude around 1900 UTC.

atmospheric electricity—1. Electrical phenomena, regarded collectively, that occur in the earth's **atmosphere**.

These phenomena include not only such striking manifestations as **lightning** and **St. Elmo's fire**, but also less noticeable but more ubiquitous effects such as **atmospheric ionization**, the **air–earth current**, and other quiescent electrical processes. The existence of separated electric charges in the atmosphere is a consequence of many minor processes (**spray electrification, dust elec-**

trification, etc.) and a few major processes (**cosmic-ray ionization, radioactive-particle ionization**, and **thunderstorm electrification**). The details of **thunderstorm charge separation** are poorly understood. The maintenance of the prevailing **atmospheric electric field** is now widely believed to be due to **thunderstorm** effects.

2. The study of electrical processes occurring within the **atmosphere**.

atmospheric interference—Same as **atmospherics**.

atmospheric ionization—The production of ions in the **atmosphere** by the loss of an **electron** from a molecule, typically, for example, by **cosmic rays** or **cosmic radiation**.

Radioactivity at the surface can also produce **ions** in the lowest layer of the atmosphere. *See* **ionization**.

atmospheric layer—Same as **atmospheric shell**.
See also **ionosphere, troposphere**.

atmospheric model—A **model** for simulating or predicting the behavior of the **atmosphere**.

atmospheric noise—*See* **meteorological noise**.

atmospheric optics—The study of the optical characteristics of the **atmosphere** or products of atmospheric processes.

The term is usually confined to visible and near **visible radiation**. But, unlike **meteorological optics**, it routinely includes temporal and spatial resolutions beyond those discernible with the naked eye.

atmospheric oscillation—Often used synonymously with **atmospheric tide**.
See **oscillation**.

atmospheric ozone—A minor but important constituent (chemical symbol O_3) of the earth's **atmosphere**.

While it is essential for life as we know it today, it is also a toxic gas that can result in significant physiological and ecological damage if exposures exceed critical limits. In both the **stratosphere** and **troposphere, ozone** concentration levels depend on many linked chemical and meteorological mechanisms, which vary significantly with space and time. Human **pollutant** emissions often perturb these linkages, resulting in significant increases or decreases in ozone concentration. While many facets of ozone's atmospheric behavior are well understood, a large number of important uncertainties remain, the resolution of which will require substantial combined efforts by the meteorological and chemical communities. *See* **air pollution**.

Amer. Meteor. Soc., 1996: AMS Policy Statement on Atmospheric Ozone. *Bull. Amer. Meteor. Soc.*, **77**, 1247–1248.

atmospheric phenomenon—As commonly used in weather observing practice, an observable occurrence of particular physical (as opposed to dynamic or **synoptic**) significance within the **atmosphere**.

Included are all **hydrometeors** (except **clouds**, which are usually considered separately), **lithometeors, igneous meteors**, and **luminous meteors**. From the viewpoint of weather observations, **thunderstorms, tornadoes, waterspouts**, and **squalls** are also included. The above usage excludes such "phenomena" as the local or large-scale characteristics of **wind, pressure**, and **temperature**; it also excludes clouds, although it includes many products of cloud development and composition. In **aviation weather observation**, atmospheric phenomena are divided into two categories: **weather** and **obstructions to vision**.

atmospheric physics—*See* **physical meteorology**.

atmospheric polarization—Polarization in the **clear** daytime sky can be quantified by the degree of **linear** polarization P and by the vibration ellipse orientation for **skylight**'s polarized component.

For Stokes's parameters, I (the **scattered** irradiance), Q, U, and V, skylight's linear polarization P is

$$P = \frac{(Q^2 + U^2)^{1/2}}{I},$$

thus ignoring its minimal degree of **circular polarization**, $\frac{V}{I}$. P is zero at neutral points and has local maxima along a celestial **great circle** that is ~90° from the sun. At **sunrise** and **sunset**, this great circle is a sky meridian perpendicular to the **clear sky**'s **principal plane**.

atmospheric pollution—Same as **air pollution**.

atmospheric pressure—(Also called barometric pressure.) The **pressure** exerted by the **atmosphere** as a consequence of gravitational attraction exerted upon the "column" of air lying directly above the point in question.

As with any gas, the pressure exerted by the atmosphere is ultimately explainable in terms of bombardment by gas molecules; it is independent of the orientation of the surface on which it acts. Atmospheric pressure is one of the basic meteorological elements. It is measured by many varieties of **barometer** and is expressed in several unit systems. The most common unit used is the **millibar** (1 millibar equals 1000 dynes cm^{-2}). Unique to the science of meteorology is the use of inches (or millimeters) of mercury, that is, the height of a column of mercury that exactly balances the weight of the column of atmosphere the base of which coincides with that of the **mercury column**. Also employed are units of weight per area and units of force per area. A **standard atmosphere** has been defined in terms of equivalence to each of the above unit systems, and it is used as a unit itself. *See* **actual pressure, station pressure, sea level pressure.**

atmospheric radiation—1. The study of all radiative processes affecting the earth's **atmosphere.**

This discipline examines the **absorption, emission,** and **scattering** of **electromagnetic radiation** within the atmosphere; the nature and distribution of incident **solar radiation** at the top of the atmosphere; and the **reflection** and **emission** from the surface at the bottom of the atmosphere. The two main areas of importance of atmospheric radiation to meteorology are 1) the effect of **radiative heating** or cooling on **temperature,** thereby helping to define the basic structure of the earth's **climate system,** and 2) use of the spectral, angular, or **polarization** information content in measured **radiation** to deduce properties of the atmosphere or surface (i.e., **remote sensing**). *See also* **solar radiation, terrestrial radiation.**

2. **Longwave radiation** emitted by the **atmosphere.**
See also **counterradiation.**

atmospheric radiation budget—An accounting of **radiant energy** loss or gain by the **atmosphere** for some specified time and place.

The main components are the loss of longwave radiant energy to space, a net gain (usually) of **longwave radiation** from the surface, and a gain of **energy** due to the **absorption** of **shortwave radiation,** primarily by **water vapor** absorption. On a long-term average, the global atmospheric radiation budget is ≈ -100 W m^{-2}. This loss of energy is made up by the **transfer** of sensible and **latent heat** from the surface.

atmospheric refraction—**Refraction** by an **atmosphere** (usually Earth's) as a consequence of **refractive index** gradients resulting from molecular number **density** gradients arising from **pressure, temperature,** and possibly **water vapor** gradients.

Near surfaces on the earth (within a few meters or so), atmospheric refraction of visible and near-visible **light** usually is dominated by temperature gradients. Although atmospheric refraction used without qualification usually means refraction of **electromagnetic waves,** it could mean refraction of **acoustic waves.** *See* **mirage.**

atmospheric region—Same as **atmospheric shell.**
See also **ionosphere.**

atmospheric science—(Also called atmospheric sciences.) The comprehensive study of the physics, chemistry, and dynamics of the earth's **atmosphere,** from the earth's surface to several hundred kilometers; this usually includes **atmospheric chemistry, aeronomy,** magnetospheric physics, and solar influences on the entire region.

atmospheric shell—(Also called atmospheric layer, atmospheric region.) Any one of a number of strata or "layers" of the earth's **atmosphere.**

Temperature distribution is the most common criterion used for denoting the various shells. The **troposphere** (the "region of change") is the lowest 10 or 20 km of the atmosphere, characterized by decreasing **temperature** with height. The term **stratosphere** is used to denote both 1) the relatively **isothermal** region immediately above the **tropopause,** and 2) the shell extending upward from the tropopause to the **minimum temperature** level at 70–80 km; the **mesosphere** is the shell between about 20 and 70 or 80 km that has a broad **maximum temperature** at about 40 or 50 km; and the **thermosphere** is the shell above the mesosphere with a more or less steadily increasing temperature with height. The distribution of various physico-chemical processes is another criterion. The **ozonosphere,** lying roughly between 10 and 50 km, is the general region of the **upper atmosphere** in which there is an appreciable **ozone** concentration and in which ozone plays an important part in the radiative balance of the atmosphere. The **ionosphere,** starting at about 70 or 80 km, is the region in which **ionization** of one or more of the atmospheric constituents is significant. The **neutrosphere,** the shell below this, is, by contrast, relatively un-ionized. The chemosphere, with no very definite height limits, is the region in which photochemical reactions take place. Dynamic and kinetic processes are a third criterion. The **exosphere** is the region at the "top" of the atmosphere, above the **critical level of escape**, in which atmospheric **particles** can move in free orbits, subject only to the earth's **gravitation**. Composition is a fourth criterion. The **homosphere** is the shell in which there is so little photo-dissociation or gravitational separation that the mean molecular weight of the atmosphere is sensibly constant; the **heterosphere** is the region above this, where the atmospheric composition and mean molecular weight is not constant. The boundary between the two is probably at the level at which molecular **oxygen** begins to be dissociated, and this occurs in the vicinity of 80 or 90 km. For further subdivisions, *see* **ionosphere**, **troposphere**.

atmospheric sounding—Measurements, from the ground, an airborne platform, or a satellite, of atmospheric parameters (such as **temperature** or **water vapor**) at various heights or **pressure** levels.

atmospheric soundings—*See* **sounding**.

atmospheric thermodynamics—The study of **thermodynamics** as applied to atmospheres.
 Specification to the earth's **atmosphere** limits the important ranges of **temperature**, **pressure**, and gaseous components, but demands emphasis on the effects of **phase** changes of water.

atmospheric tide—(Also called atmospheric oscillation.) Defined in analogy to the oceanic **tide** as an atmospheric motion of the **scale** of the earth, in which vertical accelerations are neglected (but **compressibility** is taken into account).
 Both the sun and moon produce atmospheric tides, and there exist both **gravitational tides** and **thermal tides**. The **harmonic** component of greatest **amplitude**, the 12-hour or semidiurnal **solar atmospheric tide**, is both gravitational and thermal in origin, the fact that it is greater than the corresponding **lunar atmospheric tide** being ascribed usually to a **resonance** in the **atmosphere** with a free **period** very close to the tidal period. Other tides of 6, 8, 12, and 24 hours have been observed.
 Chapman, S., 1951: *Compendium of Meteorology*, 510–530.

atmospheric transport—The movement of chemical species through the **atmosphere** as a result of large-scale atmospheric motions.
 The motions can be localized, such as convective systems, or **synoptic** in nature.

atmospheric turbulence—*See* **turbulence**.

atmospheric water budget—The requirement that the change in the amount of water stored in the **atmosphere** moving across a region should equal the difference between **precipitation** input to and **evaporation** loss from the underlying surface.

atmospheric wave—Generally, any pattern with some roughly identifiable **periodicity** in time and/or space.
 In **atmospheric dynamics**, waves may be of **acoustic**, **gravity**, or **Rossby** type.

atmospheric window—A **range** of wavelengths over which there is relatively little **absorption** of **radiation** by atmospheric gases.
 The major windows are the visible **window**, from ~0.3 to ~0.9 μm; the **infrared** window, from ~8 to ~13 μm; and the **microwave** window, at wavelengths longer than ~1 mm. The infrared window loses much of its transparency under very humid conditions due to **continuum absorption** by **water vapor**, and can become completely opaque when clouds are present.

atmospherics—(Also called atmospheric interference, strays, sferics.) The **radio frequency** electromagnetic **radiation** originating, principally, in the irregular surges of charge in thunderstorm **lightning discharges**.
 Atmospherics are heard as a quasi-steady background of crackling **noise** (**static**) on certain radio frequencies, such as those used to broadcast AM radio signals. Since any **acceleration** of **electric charge** leads to **emission** of **electromagnetic radiation**, and since the several processes involved in propagation of **lightning** lead to very large charge accelerations, the **lightning channel** acts like a huge **transmitter**, sending out **radiation** with frequencies of the order of 10 kHz. Atmospherics may occasionally be detected at distances in excess of 3500 km (2000 mi) from their source. Advantage has been taken of this characteristic by using radio direction-finding equipment to plot **cloud-to-ground lightning** locations, and to locate active **thunderstorm** areas in remote regions and in-between weather reporting stations.

ATN—Abbreviation for **Advanced TIROS-N**.

atomic number—Total number of **protons** (and hence **electrons** for a neutral atom) in an atomic **nucleus**.

ATS—Abbreviation for **Applications Technology Satellite**.

attached thermometer—A **thermometer** that is attached to an instrument in order to determine its operating **temperature**.

It is used whenever an instrument has a significant **temperature coefficient**, as in a **mercury barometer**.

attachment—In meteorological literature, often used with particular reference to the disappearance of **free electrons** by their attachment to neutral **oxygen** atoms or molecules, thus forming negative **ions**.

The rate of the process is expressed by an attachment coefficient with **dimensions** of volume per time.

attenuation—1. (Also called **extinction**, especially in reference to optical frequencies.) A general term used to denote a decrease in **signal strength** in transmission from one point to another.

For the propagation of **electromagnetic waves** through a medium, attenuation is caused by **absorption** and **scattering**. The volume **attenuation coefficient** (m^{-1}) of such a medium is the fractional reduction of **radiance** per unit pathlength. In **radar** usage, the **specific attenuation** is the fractional reduction in **power density** per unit pathlength as plane-wave **radiation** propagates through a medium, usually expressed in decibels per kilometer. Attenuation ordinarily does not refer to the inverse-square falloff of **irradiance** or power density with **range** that is simply a consequence of **beam** divergence. *See* **extinction coefficient**.

2. The **scattering** and **absorption** of **radiant energy** by clouds or the **atmosphere** that decreases the **radiation** received by satellite sensors.

attenuation coefficient—(Also called **extinction coefficient**, especially in reference to optical frequencies.) For **radiation** propagating through a medium, the fractional depletion of **radiance** per unit pathlength, or in **radar** usage, of the **power density** of plane-wave radiation per unit pathlength.

The attenuation coefficient is defined through **Bouguer's law** as

$$\frac{dL}{L} = -\gamma \, ds,$$

where L is the **monochromatic** radiance at a given **wavelength**, γ is the attenuation coefficient, and ds is a differential increment of pathlength. So defined, the attenuation coefficient is the same as the **volume extinction coefficient** and has **dimensions** of inverse length. In radar, the usual convention is to measure the depletion of **power** on a **decibel** scale in terms of the **specific attenuation** Y, which is related to attenuation coefficient by

$$Y = 4.343\gamma,$$

where Y is in decibels per kilometer when γ is in inverse kilometers. The numerical factor in this equation is $10 \log_{10} e$.

attractor—A stable **equilibrium** state having the property that small departures from the equilibrium continually diminish.

An attractor may be represented in a **coordinate system** as a single point (the usual case) or as a bounded set of infinitely many points (as in the case of a **limit cycle**). A strange attractor is an attractor containing an infinite number of points and having the property that small changes in neighboring states give rise to large and apparently unpredictable changes in the evolution of the system. The best-known example of a strange attractor in meteorology is that discovered by E. N. Lorenz (1963) in solutions to a simplified set of equations describing the motion of air in a horizontal layer heated from below.

Lorenz, E. N., 1963: Deterministic nonperiodic flow. *J. Atmos. Sci.*, **20**, 130–141.

audibility zone—A region surrounding a **sound** source, such as an explosion, in which the **sound** can be detected, usually by a human without special aids.

The concept of an audibility zone may be applied to frequencies outside the **range** of human hearing and to detectors far more sensitive than the human ear. The existence and geometry of audibility zones depend on **temperature** and **wind** component profiles along the the path between the source and the **receiver**. Audibility zones close to a source depend on the wind and temperature profiles in the **boundary layer**. At greater distances, temperature profiles in the **stratosphere** and **mesosphere** and the location of the tropospheric **jet stream** are the primary determinants of the location and extent of audibility zones.

aufeis—(Also called icings.) The **ice** formed when water from a spring or **stream** emerges and freezes on top of previously formed ice.

aura—Same as **ora**.

aureole—A bright, faintly colored region of **light** surrounding the sun.
     The aureole is what a **corona** becomes when there is a broad **range** of **droplet** (or **particle**) sizes. Aureoles are fairly common. *See* **solar aureole**.

aurora—The sporadic **radiant** emission from the **upper atmosphere** over the middle and high latitudes.
     It is believed to be due primarily to the emission of the **nitrogen** molecule N_2, its molecular **ion** N_2^+, and atomic **oxygen** [O]. According to various theories, auroras seem definitely to be related to **magnetic storms** and the influx of charged **particles** from the sun. The exact details of the nature of the mechanisms involved are still being investigated. The aurora is most intense at times of magnetic storms (when it is also observed farthest equatorward) and shows a **periodicity** related to the sun's 27-day rotation **period** and the 11-year **sunspot cycle**. The distribution with height shows a pronounced maximum near 100 km. The lower limit is probably near 80 km. The aurora can often be clearly seen, and it assumes a variety of shapes and colors that are characteristic patterns of auroral emission. The names given to the various forms are 1) **arcs**, which are bands of **light** extending across the sky, the highest point of the arc being in the direction of the magnetic **meridian**; 2) **rays**, which may appear as single lines like a searchlight beam, or in bundles; 3) **draperies**, which have a curtainlike appearance, sharp on the bottom and tenuous in the upper parts; 4) crown or **corona**, which are seen when the rays appear to spread out from a single point in the sky; 5) **bands**, which are similar to the arcs, and may or may not have a ray structure; and 6) diffuse **luminous** surfaces, which appear as luminous clouds of indefinite shape. Sometimes the term "streamers" is used to describe the auroral forms that extend to great heights. In northern latitudes these displays are called **aurora borealis**, **aurora polaris**, or **northern lights**; in southern latitudes they are called **aurora australis**. *Compare* **airglow**.

aurora australis—(Also called southern lights.) The **aurora** of southern latitudes.

aurora borealis—(Also called aurora polaris, northern lights.) The **aurora** of northern latitudes.

aurora polaris—Same as **aurora borealis**; *see* **aurora**.

auroral arc—A special class of bandlike forms.
     The arc appears as a simple, slightly curving arch. Arcs are usually quiet and not bright. *See* **aurora**.

auroral bands—Characterized by more or less continuous lower border.
     They appear like a ribbon or sheet of luminosity, inclined in the direction of the **magnetic field**. They may be homogeneous, rayed, or striated, and can be single or multiple. *See* **aurora**.

auroral corona—In any fairly complex rayed form, viewed in the direction of the **magnetic lines of force**, rays that appear to converge, forming a fan or **corona**.
     See **aurora**.

auroral curtains—**Aurora** that give an appearance of looking at curtains or drapery as viewed from below.

auroral draperies—*See* **auroral curtains**.

auroral oval—An oval-shaped distribution of the **aurora**.
     The oval is asymmetrical, but generally around the region of the Arctic and Antarctic. The oval increases in **intensity** and size when auroral activity is more intense.

auroral rays—A form of the **aurora** consisting of rays or shafts of luminosity aligned in the direction of the geomagnetic field.
     It can consist of a single **ray**, a small bundle of rays, or many scattered rays.

auroral zone—An approximately circular region around the two geomagnetic poles within which there is a maximum of auroral activity.
     It lies about ten to fifteen degrees of geomagnetic latitudes from the geomagnetic poles. The auroral zone broadens and extends equatorward during intense auroral displays. *See* **aurora**.

austausch—Literally "exchange"; *see* **exchange coefficients**.

austausch coefficients—(Obsolete.) Same as **exchange coefficients**.

auster—Same as **ostria**.

Australasian Mediterranean Water—(Abbreviated AAMW.) A **Mediterranean Water** mass formed in the Indonesian Seas (the Australasian **Mediterranean Sea**), also known as **Banda Sea Water**.

Of Pacific origin, it undergoes strong freshening as a result of heavy **rainfall**. AAMW is characterized by uniform **salinity** of 34.7 **psu** from the surface to depths of at least 1000 m, indicating strong **mixing** in the formation region. It can be followed in the Indian Ocean as a band of low-salinity water along 10°S as far west as Madagascar.

austru—East or southeast winds in Rumania.

They are cold in winter and are drier and stronger than the **crivetz**. They may be a local name for the **foehn**. *Compare* **ostria**.

autan—(Also called **altanus**.) A strong southeast **wind** in south-central France, especially in Gascony and the upper Garonne River.

Near the Pyrenees the autan is very turbulent, growing in strength in the valleys. At Toulouse its average speed is 13 m s^{-1} (30 mph) with gusts of 20–22 m s^{-1} (45–50 mph); it tends to be strongest at midday. It increases in speed up to a height of 450 m (1500 ft), above which it weakens and veers to the south. North of Toulouse it loses its special character and becomes an ordinary southeast wind. There are two types. 1) The autan blanc brings fine dry weather, cold in winter, hot in summer, as a result of the downslope motion imposed by the Pyrenees and southern Cevennes. It occurs with an **anticyclone** centered near Denmark or moving northeastward from the Azores. It lasts for two to four days in winter, but may persist for more than a week in summer, bringing severe **drought** and desiccating the vegetation; in Catalonia (northeastern Spain) a similar wind is called the **outo**. 2) The autan noir is less frequent and rarely lasts for more than two days; it is more humid and **cloudy**, bringing **fog**, **rain**, or **snow** over high ground near the sea. As such, it is more like the **marin**, the name applied to the southeast wind out of the Cevennes where its maritime character predominates.

autobarotropic—Of, pertaining to, or characterized by a state of **autobarotropy**.

autobarotropy—The state of a fluid that is characterized by both **barotropy** and **piezotropy**, that is, **pressure** and **density** surfaces coincide.

This condition guarantees that the fluid will remain **barotropic** at all future times. Example: an ideal homentropic fluid (i.e., a fluid with homogeneous **potential temperature**) with uniform composition.

Gill, A. E., 1982: *Atmosphere–Ocean Dynamics*, Academic Press, p. 228.

autoconvection—(Also called autoconvective instability.) The supposed spontaneous initiation of **convection** in an **atmospheric layer** in which **density** increases with height and in which the **lapse rate** is greater than the **autoconvective lapse rate**.

The term is based on a false analogy with convection in an **incompressible fluid**. Convection in a gas will initiate spontaneously, through **buoyant instability**, if the lapse rate is greater than the **adiabatic lapse rate**.

autoconvective lapse rate—The **environmental lapse rate** of **temperature** in an **atmosphere** in which the **density** is constant with height (**homogeneous atmosphere**), equal to g/R, where g is the **acceleration of gravity** and R the **gas constant**.

For **dry air** the autoconvective lapse rate is approximately $+3.4 \times 10^{-4}$°C per cm. Although the term is poorly named (*see* **autoconvection**), it is useful in optics, since it defines the transition between upward and downward **refraction** of **light** in the atmosphere.

autoconversion—The initial stage of the **collision–coalescence process** whereby **cloud droplets** collide and coalesce to form **drizzle drops**.

Originally a term for the rate equation to approximate the transfer of **cloud drops** to **drizzle**. Because of the low collection efficiencies among cloud drops, autoconversion can be the limiting factor in the formation of drizzle

autocorrelation—The **simple linear correlation** of a **time series** with its own past; that is, the **correlation** of the sequence of values $x(t)$ with the sequence of values $x(t + \tau)$ occurring τ units of time later.

The time displacement τ is called the **lag**. The autocorrelation function is the autocorrelation for variable lag. The autocorrelation coefficient is the **product-moment** correlation coefficient that relates the **variables** $x(t)$ and $x(t + \tau)$. *See* **serial correlation**.

autocorrelation coefficient—*See* **autocorrelation**.

autocorrelation function—*See* **autocorrelation**.

autogenous electrification—The process by which net charge is built up on an object, such as an airplane, moving relative to air containing **dust** or **ice crystals**.

The electrification is produced by frictional effects (**triboelectrification**) accompanying contact between the object and the particulate matter. Compare **exogenous electrification**.

automated search—Examination of a set of possible solutions to find items that satisfy certain criteria.

In **expert systems**, it determines which **rules** or combinations of rules and facts can be used to solve a problem.

automatic data processing—Use of equipment, usually electronic, to perform calculations or procedures on data to enhance information content or information presentation.

automatic evaporation station—Station at which instruments make and either transmit or record **evaporation** automatically, the conversion to code form, if required, being made either directly or at an editing **station**.

automatic gain control—(Abbreviated AGC.) A process or means by which the **gain** of a circuit is automatically adjusted in a specified manner as a function of the **input** signal or other specified parameters.

Typically, AGC is used to ensure that the **output** of an amplifier remains within the design limits of a circuit.

automatic picture transmission—(Abbreviated APT.) Low-**resolution** analog transmission of **AVHRR** data from polar orbiting satellites.

Two channels of reduced-resolution (4 km) data are continuously transmitted using **analog** VHF signals (137 MHz) at reduced rates (120 lines per minute), allowing use of simple, low-cost ground **station** equipment for APT reception. APT was initiated with the **TIROS** satellite launched in December 1963.

automatic station—Station at which instruments make and record **observations** automatically.

Observations may be stored at the **station** in **analog** or **digital** form, or transmitted to a receiving station, or both.

automatic tide gauge—An automated **tide** measurement system in which water-level data and ancillary environmental data are preprocessed by a minicomputer (**data collection platform**) and transmitted to remote locations by radio, telephone, or satellite.

automatic weather station—A **weather station** at which the services of an **observer** are not required.

Most automatic weather stations are equipped with telemeter apparatus for transmitting weather information at predetermined times or when changes in value occur.

autoregressive process—A process that generates a **time series** for which representation of the current value of the measured **variable** involves a weighted sum of past values.

autoregressive series—A **time series** generated from another time series as the solution of a **linear** difference equation (usually where previous values of the output series enter into the determination of the current value).

Blackman, R. B., and J. W. Tukey, 1959: *The Measurement of Power Spectra*, 167–168.

AUTOVAP—*See* **automatic evaporation station**.

autumn—1. (Also called fall, especially in the United States.) The **season** of the year that is the transition period from summer to winter, occurring as the sun approaches the **winter solstice**.

In popular usage and for most meteorological purposes, autumn is customarily taken to include the months of September, October, and November in the Northern Hemisphere, and March, April, and May in the Southern Hemisphere.

2. Astronomically, the period extending from the **autumnal equinox** to the **winter solstice**.

autumn ice—Sea ice in early stage of formation.

It is comparatively salty and crystalline in appearance. Like **young ice**, it is not yet affected by lateral **pressure**.

autumnal—Pertaining to fall (autumn).

The corresponding adjectives for winter, spring, and summer are **hibernal**, **vernal**, and **aestival**.

autumnal equinox—The **equinox** (approximately 22 September) at which the sun crosses the **celestial equator** from the Northern Hemisphere to the Southern Hemisphere.

auvergnasse—(Also called auvergnac.) A northwest **wind** in the center of the Massif Central of France (i.e., coming from Auvergne). It is cold and generally brings **rain** or **snow**.

auxiliary ship—A **mobile ship station**, normally without certified meteorological instruments, that transmits **meteorological observations** on request in certain areas or conditions, in code or plain language.

auxiliary ship station—*See* **auxiliary ship**.

auxiliary thermometer—A **mercury-in-glass thermometer** attached to the stem of a **reversing thermometer**.

It is read at the same time as the reversing thermometer so that the **correction** to the reading of the latter, resulting from change in **temperature** since reversal, can be computed.

available head—Amount of fall in a **stream** that is available for generation of hydroelectric **power**.

available potential energy—That portion of the **total potential energy** that may be converted to **kinetic energy** in an **adiabatically enclosed system**.

If the **atmosphere** were horizontally stratified, none of its **potential energy** would be so convertible. Thus, the available potential energy for the atmosphere in a given state is the difference between the value of the total potential energy in the given state and the value it would have after an **adiabatic** redistribution of mass had produced a horizontal **stratification**.

> Lorenz, E., 1955: Available potential energy and the maintenance of the general circulation. *Tellus*, **7**, 157–167.
> Gill, A. E., 1982: *Atmosphere–Ocean Dynamics*, Academic Press, 219–225.
> Holton, J. R., 1992: *An Introduction to Dynamic Meteorology*, 3d edition, Academic Press, 243–246.

available soil moisture—The portion of water in a soil that can be readily absorbed by plant roots; generally considered to be that water held in the soil against a soil water pressure of up to approximately 1500 kPa.

available solar radiation—Total **solar radiation** intercepted by the earth, given by $2\pi rS$, where r is the earth's radius and S the **solar constant**.

available storage capacity—During operating conditions, the volume of water in a **reservoir** between the minimum and maximum water levels.

available water—The difference between the current soil moisture and the **wilting point**.

Maximum available water is the **field capacity** minus the wilting point.

avalaison—Persistent west **wind** in western France.

avalanche—1. (Also called snowslide.) A mass of **snow** (perhaps containing **ice** and rocks) moving rapidly down a steep mountain slope.

Avalanches may be characterized as loose and turbulent, or slab; either type may be dry or wet according to the nature of the snow forming it, although **dry snow** usually forms loose avalanches and **wet snow** forms slabs. A large avalanche sweeps a current of air along with and in front of it as an **avalanche wind**, which supplements its already tremendous destructive force. *See* **wind slab**; *compare* **sluff**.

2. (Also called landslide.) A mass of earth material (soil, rock, etc.) moving rapidly down a steep slope.

avalanche wind—The rush of air produced in front of an **avalanche** of **dry snow** or in front of a **landslide**.

The most destructive form, the avalanche blast, occurs when an avalanche is stopped abruptly, as in the case of an almost vertical fall into a valley floor. Such blasts may have very erratic behavior, leveling one house without damaging its neighbor.

average—1. Same as **arithmetic mean**.

2. In a very broad sense, any number lying between the extremes of a set of numbers. *Compare* **Reynolds averaging**, **ensemble average**, **area average**, **mean**, **ergodic**.

average drag coefficient—Same as **drag coefficient**, but averaged over a domain composed of different roughness elements.

The process of combining different drag coefficients from neighboring small uniform regions into an average value over a larger heterogeneous area is called **aggregation**.

average interstitial velocity—The **average velocity** of **groundwater** particles at a certain part of a saturated geologic material.

average limit of ice—A climatological term referring to the **extreme** minimum or extreme maximum extent of the **ice edge** averaged over any given month or period based on observations over a number of years.

The term should be preceded by minimum or maximum.

average net balance—The sum of **accumulation** and **ablation** over an **ice** mass (the **mass balance**) divided by the total surface area.

It is the most useful **parameter** for summarizing the change in a given **glacier** or **ice sheet**.

average over the spectrum—Net characteristics when all **eddy** sizes are considered together; used in **boundary layer** studies when considering the integral effect of all **turbulence** scales.
An example is the **mixing length**, which represents an average eddy size.

average power—In **radar**, the time-averaged **power** transmitted by an **antenna** into space.
Mathematically, average power is the product of the **peak power** transmitted, the **pulse duration**, and the **pulse repetition frequency**. *See* **transmitted power**.

average velocity—1. In hydrology, the ratio of **discharge** to the cross-sectional area perpendicular to the flow direction.
2. In hydrology, the ratio of integrated **velocity profile** to the flow depth.

AVHRR—Abbreviation for **Advanced Very High Resolution Radiometer**.

aviation climatology—Same as **aeronautical climatology**.

aviation forecast—Forecast, intended for aviation, of weather conditions at the surface and at various altitudes for a specific period of time.
See **aviation weather forecast**.

aviation meteorology—Same as **aeronautical meteorology**.

aviation observation—An **observation** of elements that pertain to aviation including some or all of the following: **station designator**; type and time of report, **sky cover**; **ceiling**; **visibility**; **weather** and **obstructions to vision**; **sea level pressure**; **temperature** and **dewpoint**; **wind direction**, **speed**, and character; **altimeter setting**; remarks; and coded data.

aviation weather forecast—(Also called airways forecast.) A forecast of weather elements of particular interest to aviation.
These elements include the **ceiling**, **visibility**, **upper winds**, **icing**, **turbulence**, and types of **precipitation** and/or **storms**. Aviation weather forecasts can be subdivided into four basic categories: **area forecasts**, **terminal forecasts**, **route forecasts**, and **flight forecasts**.

aviation weather observation—(Also called airways observation.) An evaluation, according to set procedure, of those weather elements that are most important for aircraft operations.
It always includes the **cloud height** or **vertical visibility**, **sky cover**, **visibility**, **obstructions to vision**, certain atmospheric phenomena, and **wind speed** and **direction** that prevail at the time of the observation. Complete observations include the **sea level pressure**, **temperature**, **dewpoint temperature**, and **altimeter setting**. Aviation weather observations are further classified as record, special, check, and local extra observations. The first two types are encoded and transmitted as reports on communications circuits. *Compare* **synoptic weather observation**.

Avogadro's law—Physical law stating that equal volumes of different ideal gases at the same **temperature** and **pressure** contain equal numbers of **particles** (molecules or atoms).
The interpretation of the law lies in the statistical–mechanical theory of equipartition of **energy**, and the **ideal gas** law is a consequence of this. The volume occupied by one **mole** of gas (see **Avogadro's number**) is the same for all gases (22.414 L). *See also* **Loschmidt's number**.

Avogadro's number—The number of molecules (6.02214×10^{23}) in one **mole** of gas.
That this number is a constant for permanent gases is **Avogadro's law**.

avre—In Luc-en-Diois, Department of Drôme in France, a **wind** that blows warm in winter and cool in summer.

avulsion—1. An abrupt change in the course of a **stream** or river, generally from one channel to a new one.
In this meaning, the change is considered more extensive than the cutoff of a **meander** or similar local change in channel position.
2. Any sudden cutting off or separation of land or abrupt change in the course of a **stream**, generally by breaking through the stream banks during a **flood**, including the formation of a cutoff **meander**.

AXBT—Abbreviation for **airborne expendable bathythermograph**.

axial symmetry—The symmetry that describes a three-dimensional configuration that is the same in every plane containing the axis of symmetry.
In **cylindrical coordinates** this implies independence of the azimuthal coordinate.

axis of anticyclone—The line joining the points of maximum pressure at the surface and at upper levels.

axis of contraction—With reference to a **deformation field**, that axis along which contraction (shrinking) of a fluid element is most rapid; it is normal to the **axis of dilatation**.

axis of depression—The line joining the points of minimum pressure at the surface and at upper levels.

axis of dilatation—With reference to a **deformation field**, that axis along which dilatation (stretching) of a fluid element is most rapid; it is normal to the **axis of contraction**.

axis of ridge—*See* **ridge**.

axis of trough—*See* **trough**.

axisymmetric—Having **axial symmetry**; independent of the azimuthal coordinate.

axisymmetric flow—In **hydrodynamics** and fluid mechanics, a flow in which the streamlines are symmetrically located around an axis.
    Every longitudinal plane through the axis would exhibit the same **streamline** pattern.

axisymmetric turbulence—Turbulence that is symmetrically distributed about the direction of the mean flow, as in a pipe or **wind** tunnel.

ayalas—A warm and violently squally southeast **wind** in the center of the Massif Central of France.
    In spring it causes rapid melting of the **snow**, and in autumn it brings **heavy rain**; both cause **flooding** of the rivers. *See* **marin**.

aygalas—*See* **marin**.

azimuth—The length of the arc of the **horizon** (in degrees) intercepted between a given point and a reference direction, usually north, and measured clockwise from the reference direction.
    Azimuth may be synonymous with **bearing**, a navigational term that can have different meanings. Any point in space can be located relative to an observing point by its azimuth angle, **elevation angle**, and **range**.

azimuth resolution—The angle or distance by which two targets at the same **range** must be separated in **azimuth** to be distinguished by a **radar**.
    Targets separated less than this distance appear as a single **target** on the **display**. *See* **radar resolution**.

azimuth scale—In **radar**, a circular **scale** at the outer perimeter of a PPI (**plan position indicator**) **display** on which the **azimuth** angle is indicated.

azimuthal wavenumber—Same as **angular wavenumber**.

Azores anticyclone—*See* **Azores high**.

Azores Current—One of the currents of the North Atlantic subtropical **gyre**.
    It receives some 15 Sv ($15 \times 10^6 \mathrm{m^3 s^{-1}}$) from the **North Atlantic Current** near 45°W and continues as an eastward current along 35°N to eventually feed into the **Canary Current** and close the gyre.

Azores high—The semipermanent **subtropical high** over the North Atlantic Ocean, so named especially when it is located over the eastern part of the ocean.
    The same **high**, when displaced to the western part of the Atlantic, or when it develops a separate cell there, is known as the **Bermuda high**. On mean charts of **sea level pressure**, this high is one of the principal centers of action in northern latitudes.

B

B-display—A rectangular **radar** display, with coordinates of **azimuth** and **range**, in which targets appear as intensity-modulated blips.

Because the shapes of distributed targets are distorted on these coordinates, especially at close range, this type of **display** is not popular for use in **weather radar**.

B-scope—Same as **B-display**.

B-unit—*See* **potential refractivity**.

Babinet point—One of the three commonly detectable points of zero **polarization** of **skylight**, neutral points, lying along the vertical circle through the sun; the other two are the **Arago point** and **Brewster point**.

The Babinet point typically lies only 15° to 20° above the sun, and hence is difficult to observe because of solar **glare**. The existence of this **neutral point** was discovered by Babinet in 1840.

Babinet's principle—An approximation according to which the **amplitude** of near-forward **scattering** by an opaque, planar object is the same as that of an **aperture** of the same shape and size.

Babinet's principle is sometimes combined with **Fraunhofer diffraction theory** in the development of an approximate theory of the **corona**.

back-bent occlusion—Same as **bent-back occlusion**.

back radiation—Same as **counterradiation**.

back-sheared anvil—Colloquial expression for a **cumulonimbus** anvil that spreads **upwind** into relatively strong **winds aloft**.

A back-sheared anvil implies strong divergent flow near the summit of a high-speed **convective storm** updraft. These anvils often exhibit a crisp appearance with sharp, distinctive edges.

backdoor cold front—A **cold front** that leads a cold **air mass** toward the south and southwest along the Atlantic seaboard of the United States.

This is one of the occurrences to which New Englanders give the name **sea turn**, for the cold **wind** following a backdoor cold front blows from the northeast quadrant.

background field—In **objective analysis** and **data assimilation**, an a priori estimate of the atmospheric state.

In most data assimilation systems, the background field is a forecast from the previous **analysis** time. Note that the term "first-guess field" has been used for the background field, but "background field" is currently the preferred usage.

background level—Concentration of **background pollution**.
Compare **ambient air**.

background pollution—**Air pollution** that is not produced locally.

While total concentration is the sum of locally and nonlocally produced pollution, only the locally produced pollution can be locally regulated. In such regulations, the pollutants that advect in from the outside, or which would have been present naturally, are sometimes called background pollution. *Compare* **ambient air**.

backing—1. According to general internationally accepted usage, a change in **wind direction** in a counterclockwise sense (e.g., south to southeast to east) in either hemisphere of the earth; the opposite of **veering**.

2. According to widespread usage among U.S. meteorologists, a change in **wind direction** in a counterclockwise sense in the Northern Hemisphere, clockwise in the Southern Hemisphere; the opposite of **veering**.

backing wind—In the Northern Hemisphere, a **wind** that rotates in the counterclockwise direction with increasing height.

Opposite of **veering wind**.

backlash—The play or loose motion in an instrument due to the clearance existing between mechanically contacting parts.

backlobe—In an **antenna pattern**, the **sidelobe** in the backward direction, approximately 180° from the **main lobe**.
See also **lobe**.

backpropagation—The most common method of training an artificial **neural network**.

A training set, consisting of examples of **input** data for which the **output** is known, is presented to the network, and the network weights are adjusted until the network produces results that are in agreement with the training set. This type of network is often used in **prediction** and in classification.

backscattering—The **scattering** of **radiant energy** into the hemisphere of space bounded by a plane normal to the direction of the incident **radiation** and lying on the same side as the incident **ray**; the opposite of **forward scatter**.

Atmospheric **backward scatter** depletes 6%–9% of the incident solar **beam** before it reaches the earth's surface. In **radar** usage, backscatter refers only to that radiation scattered at 180° to the direction of the incident **wave**.

backscattering cross section—1. For plane-wave **radiation** incident on a **scattering** object or a scattering medium, the ratio of the **intensity** scattered in the direction toward the source to the incident **irradiance**.

So defined, the backscattering cross section has units of area per unit solid angle, for example, square meters per **steradian**.

2. In common usage, synonymous with **radar cross section**, although this can be confusing because the radar cross section is 4π times the backscattering cross section as defined in 1) and has units of area, for example, square meters.

See **differential backscattering cross section**; *compare* **radar cross section**.

backstays of the sun—Same as **crepuscular rays**.

backtracking—A method of searching **decision trees** in stages, beginning with a **depth-first search**.

The search marks the last decision point and its alternatives. If one search path fails, the system backtracks to the most recent decision point, then follows an alternative branch. This permits searching as small a part of the tree as possible. Most instances of weather forecast **expert systems** are organized as decision trees, and use backtracking to improve their efficiency.

backward chaining—The process in **rule-based systems** that constructs a hypothesis, then tests it by working backward through the rules to see if the hypothesis is supported.

An example in weather forecasting is to assume there will be afternoon **thunderstorm** activity, then determine whether the data support this assumption. *See also* **goal-directed reasoning, forward chaining**.

backward difference—See **finite-difference approximation**.

backward scatter—Scattering into directions backward of a plane through the **scattering** center and **orthogonal** to the incident direction.

backward scattering coefficient—[Also called backscatter coefficient, backscatter(ing) cross section.] A measure of the fraction of incident **radiation** that is **scattered** directly back toward the source.

backwash—The **return flow** of water down the beach as a result of **wave** action; the opposite of **swash**.

backwater—Accumulation of water resulting from an obstruction, limited **downstream** channel capacity, **high tide**, or high stages in a connecting **stream**.

backwater curve—Profile of water surface **elevation** above a specified reference level along a flow path, usually **upstream** from an obstruction.

bad-i-sad-o-bistroz—Same as **seistan**.

Baffin Current—Southward flowing current on the eastern side of Baffin Bay with speeds of 0.2–0.4 m s^{-1}.

It is fed by low-salinity water from the Arctic Ocean, thus contributing to the **freshwater** budget between the Pacific and Atlantic, and by the **West Greenland Current**. It feeds the **Labrador Current**.

baffle—Any of several kinds of walls, blocks, or deflector vanes placed in channelized water to dissipate **energy** in order to achieve a more uniform distribution of velocities, to divert flow, or to achieve another desired engineering purpose.

baguio—(Also spelled bagio, vaguio, vario.) In the Philippine Islands, the name given to any severe **tropical cyclone**; derived from the city of Baguio, where a record 24-h **rainfall** of 46 inches occurred during the passage of a tropical cyclone in July 1911.

bai—(Also called sand mist.) In China and Japan, a "mist" that occurs in spring and fall, when loose earth is churned up by the **wind** so that clouds of **dust** rise to great heights, afterward collecting moisture and falling as colored **mist** that produces a thick coating of very fine yellow dust.

bai-u—(Also called tsuyu.) In southern Japan and in parts of China, the name of the **season** of heaviest **rainfall**.

The bai-u season (June and early July in Japan, May to July in China) is the most important period for the cultivation and transplanting of rice.The bai-u rains are also called plum rains or mold rains, with reference to the season when plums ripen and to the effects of continued dampness.

bai-u front—*See* **mei-yu front**.

balance—The sum of **accumulation** and **ablation**.

It is the change of mass per unit area relative to the previous summer surface. It is sometimes divided into a winter balance and a summer balance. *See also* **average net balance**.

balance equation—In general, an equation expressing a balance of quantities in the sense that the local or individual rates of change are zero.

More specifically, it is a **diagnostic equation** expressing a balance between the **pressure field** and the horizontal **field** of motion:

$$f \nabla_H^2 \psi + \nabla_H \psi \cdot \nabla_H f + 2(\psi_{xx}\psi_{yy} - \psi_{xy}^2) = g\nabla_H^2 z,$$

where ψ is the **streamfunction** for the nondivergent portion of the motion, f the **Coriolis parameter**, ∇_H the horizontal **del operator** in the **isobaric surface**, ∇_H^2 the corresponding **Laplacian operator**, g the **acceleration of gravity**, z the **elevation** of the isobaric surface, and subscripts denote partial differentiation. The balance equation is derived from the **divergence equation** on the assumption that the magnitude of the **horizontal divergence** is always much less than that of the (vertical) **vorticity**, and is therefore more general than the equation of **geostrophic equilibrium** to which it reduces in the absence of the **nonlinear** terms. In the case of circular contours, this nonlinear correction is the same as that introduced by the **gradient wind** assumption.

balance meter—A **net pyrradiometer** that measures the balance or difference in hemispheric (2π) **radiation** incident on both sides of a flat surface.

balance of solar radiation—*See* **net solar radiation**.

balance year—The period between minima in **net balance** on a **glacier**.

It corresponds roughly to the year between the end of one summer and the end of the following summer. Because the dates defining the balance year are meteorologically dependent, in practice glaciologists usually adhere to a "measurement year" defined by specific calendar dates.

Bali wind—The name given to a strong east **wind** at the eastern end of Java.

ball lightning—(Also called globe lightning.) A rare and randomly occurring bright ball of **light** observed floating or moving through the **atmosphere** close to the ground.

Observations have widely varying identifying characteristics for ball lightning, but the most common description is that of a sphere having a radius of 15–50 cm, orange or reddish in color, and lasting for only a few seconds before disappearing, sometimes with a loud noise. Most often ball lightning is seen in the vicinity of thunderstorms or a recent **lightning** strike, which may suggest that ball lightning is electrical in composition or origin. Considered controversial due to the lack of unambiguous physical evidence for its existence, ball lightning is becoming more accepted due to recent laboratory recreations resembling ball lightning. Despite the observations and models of these fire balls, the exact mechanism(s) for naturally occurring ball lightning is unknown.

balling—"The accumulation of **snow** in lumps under a ski runner, snow shoe, footgear, or a dog's pad. Balling occurs especially under conditions of high temperature and high humidity." [from *Glossary of Arctic and Subarctic Terms* (1955)].

> Arctic, Desert, Tropic Information Center (ADTIC) Research Studies Institute, 1955: *Glossary of Arctic and Subarctic Terms*, ADTIC Pub. A-105, Maxwell AFB, AL, 90 pp.

ballistic density—A representation of the atmospheric **density** actually encountered by a projectile in flight, expressed as a percentage of the density according to the **standard artillery atmosphere**.

Thus, if the actual density distribution produced the same effect upon a projectile as the **standard density** distribution, the ballistic density would be 100%.

ballonet—A gastight compartment of variable volume to control ascent and descent and to maintain **pressure** in a nonrigid airship.

When the ballonet is used within a **superpressure balloon**, valving of gas increases **altitude**, and pumping of air into the ballonet causes the balloon to descend. The superpressure is only marginally affected as the balloon changes altitude.

71

balloon—*See* **anchor balloon, ballonet, ceiling balloon, constant-level balloon, extensible balloon, jimsphere, kytoon, pilot balloon, racoon, robin sphere, sounding balloon, superpressure balloon, tetroon, zero-pressure balloon.**

balloon ceiling—After United States weather observing practice, the **ceiling classification** that is applied when the **ceiling** height is determined by timing the ascent and disappearance of a **ceiling balloon** or **pilot balloon.**
　　Balloon ceilings are designated B in an **aviation weather observation**, and they may pertain only to clouds or to obscuring phenomena aloft.

balloon sounding—Upper-air measurement of **temperature** and **humidity** made with a **radiosonde** attached to a free-flight **sounding balloon.**
　　Synoptic (or operational) balloon soundings are typically made twice a **day**, at 0000 and 1200 UTC, from worldwide fixed land stations and from ships.

ban-gull—Summer **sea breeze** of Scotland.

band—1. A **range** of wavelengths.
　　2. **Frequency band.**
　　3. **Absorption band.**
　　4. A **range** of **radar** frequencies, such as **X band, S band.**

band absorption—**Absorption** of **radiation** that takes place in one or more **absorption bands.** *Compare* **line absorption, continuum absorption.**

band model—One of several different treatments of **band absorption** used to approximate the transmission of **narrowband radiation** through atmospheric gases.

Banda Sea Water—*See* **Australasian Mediterranean Water.**

banded structure—In **radar**, the arrangement of **precipitation** echoes in the form of long lines or bands.

bandpass filter—A **digital**, electrical, or mechanical system designed to attenuate all signals outside a specified **frequency band**, while passing signals within this **band** relatively unchanged.

bandwidth—1. In general, a **range** of frequencies specified by the number of **hertz** contained within the **band** or by the upper and lower limiting frequencies.
　　2. The **range** of frequencies that a device is capable of generating, handling, or accommodating; usually the range in which the response is within 3 dB of the maximum response. For example, the bandwidth of a modulated **signal** or of a **bandpass filter** is commonly defined by the frequencies at which the power spectral density is 3 dB (or a factor of 2) less than that within the **band.**
　　3. The amount of **frequency** space occupied by a **signal** and required for effective transfer of information by the signal. In data transmission, the greater the bandwidth, the greater the capacity to transmit data bits.

bank—1. The sloping margin of a **stream** or river that confines flow to the natural channel during **normal** stages.
　　The top of this channel margin may be exceeded during overbank **flood** flows.
　　2. A steep slope or face, usually developed in unconsolidated material such as sand or gravel.
　　3. A shallow area in the sea or other water body, consisting of shifting **sediment**, and designated by a qualifying word, such as "gravel bank."

bank storage—Water absorbed and stored by the soil pores of the bed and banks of a **stream**, lake, or **reservoir** during higher **stage** periods and returned, fully or partially, to the water body as the water stage falls.

bankfull stage—The **stage** on a fixed **river gauge** corresponding to the top of the lowest banks within the **reach** for which the gauge is used as an **index.**
　　Compare **flood stage.**

banner cloud—(Also called cloud banner.) A **cloud** plume often observed to extend **downwind** from isolated, sharp, often pyramid-shaped mountain peaks, even on otherwise cloud-free days.
　　The Matterhorn and Mount Everest are two notable peaks where banner clouds have been frequently observed. The physics of the formation of such clouds is not completely understood. The **aerodynamics** of the flow around the peak produces flow separation and dynamically induced **pressure** reductions to the lee of the mountain peaks. The magnitude of the leeside pressure deficits increases with height to a maximum near the top of the peak, producing an upslope **pressure**

gradient and upslope flow along the lee slope of the mountain. When the air near the base of the mountain is sufficiently moist, it ascends in the upslope flow, condenses, and forms a triangular-shaped cloud, the banner cloud, to the lee of the peak. Because of its unusual shape and location, this cloud strongly resembles **snow** blowing off the peak (**snow banner**), and it is often difficult to tell the difference.

bar—A unit of **pressure** equal to 10^6 **dyne** cm^{-2} (10^6 **barye**), 1000 millibars, 29.53 inches of **mercury**.

barat—In Indonesia, a strong squally northwest or west **wind** on the north coast of Sulawesi near Menado; it occurs from December to February.

barb—(Also called feather.) A means of representing **wind speed** in the **plotting** of a **synoptic chart**; it is a short straight line drawn obliquely toward lower **pressure** from the end of a **wind-direction shaft**.

In the commonly used **five-and-ten system**, one barb represents 10 knots and a half-barb represents 5 knots. Previously, a full barb was used to represent a **Beaufort force** of 2, with a half-barb to represent a Beaufort force of 1. Customarily, the barb is often drawn obliquely or perpendicular to the **wind direction**. *See* **pennant**.

barber—1. A **severe storm** at sea during which spray and **precipitation** freeze onto the decks and rigging of boats.

2. (Also spelled berber.) In the Gulf of St. Lawrence, a local form of **blizzard** in which the wind-borne **ice** particles almost cut the skin from the face.

3. Same as **frost smoke**.

barchan—(Also spelled barchane, barkhan.) A crescent-shaped dune or **drift** of windblown sand or **snow**; the arms of the crescent point downward.

Conditions under which barchans form are a moderate supply of material (sand or snow), and winds of almost constant direction and of moderate speeds.

bare ice—**Ice** without **snow cover**.

bare soil—A soil surface devoid of any plant material.

baric analysis—*See* **isobaric analysis**.

baric topography—1. (Rare.) Same as **height pattern**.

2. Same as **pressure pattern**.

baric wind law—Same as **Buys Ballot's law**.

barih—Same as **shamal**.

barinés—Westerly **winds** in eastern Venezuela.

barocline—Same as **baroclinic**.

baroclinic—1. The **variation** with depth of motions associated with variation of **density** with depth.

The baroclinic component of the **velocity** is the total minus the **barotropic** component. In a baroclinic state, **neutral surfaces** are inclined to surfaces of constant **pressure**. The **baroclinic torque vector** is proportional to the **vector** cross product and is responsible for generating vertical shears associated with baroclinic flow.

2. Of, pertaining to, or characterized by **baroclinity**.

baroclinic atmosphere—An atmospheric state in which **density** depends upon both **temperature** and **pressure** and in which the **geostrophic wind** varies with height and is related to the horizontal temperature gradient via the **thermal wind equation**.

baroclinic disturbance—(Also called baroclinic wave.) Any **migratory** cyclone more or less associated with strong **baroclinity** of the **atmosphere**, evidenced on **synoptic charts** by temperature **gradients** in the constant-pressure surfaces, **vertical wind shear**, **tilt** of pressure **troughs** with height, and concentration of **solenoids** in the **frontal surface** near the ground.

Baroclinic disturbances play an important role in atmospheric **energy conversion** from **potential energy** to **kinetic energy**. *Compare* **barotropic disturbance**.

baroclinic forecast—A forecast of atmospheric **temperature**, winds, or **geopotential heights**, obtained by solving an initial value problem using **baroclinic model** equations.

baroclinic instability—A **hydrodynamic instability** arising from the existence of a **meridional** temperature **gradient** (and hence vertical **shear** of the mean flow and a **thermal wind**) in an **atmosphere** in **quasigeostrophic equilibrium** and possessing **static stability**.

For reasonable values of the atmospheric parameters, the **wavelength** of maximum instability corresponds to that of a **synoptic-scale** disturbance. Such a system may be interpreted as converting **potential energy** of the **basic flow** into **kinetic energy** of the unstable **perturbation**.

Holton, J. R., 1992: *An Introduction to Dynamic Meteorology*, 3d edition, Academic Press, 228–264.

baroclinic leaf—A **synoptic-scale** cloud pattern frequently observed in satellite imagery just prior to the onset of **cyclogenesis**.

The **cloud system** is an elongated pattern with well-defined borders on both sides. It is associated with midtropospheric **frontogenesis**.

baroclinic model—An **atmospheric model** that, in addition to the **advection** of the **circulation** field, includes **thermal** advection processes, an explicit representation of the **thermodynamic energy equation**, and at least two vertical data levels.

The inclusion of thermal advection processes, and in particular **differential thermal advection** between vertical levels, is essential to predict the **development** of new weather systems. Baroclinic models are the standard models used for making numerical weather forecasts by operational forecast centers.

baroclinic torque vector—*See* **baroclinic**.

baroclinic wave—1. Describes the **synoptic-scale** disturbance that grows in midlatitudes due to **baroclinic instability**.

2. Same as **baroclinic disturbance**.

baroclinicity—Same as **baroclinity**.

baroclinity—(Or baroclinicity.) The state of **stratification** in a fluid in which surfaces of constant **pressure (isobaric)** intersect surfaces of constant **density (isosteric)**.

The number, per unit area, of isobaric–isosteric solenoids intersecting a given surface is a measure of the baroclinity. If the surface is horizontal, this number N is given by

$$N = \frac{\partial \alpha}{\partial x} \frac{\partial p}{\partial y} - \frac{\partial p}{\partial x} \frac{\partial \alpha}{\partial y},$$

where α is the **specific volume** and p the pressure. Barotropy is the state of zero baroclinity. *See also* **barotropy**.

barogram—The record of a **barograph**.

barograph—A recording **barometer**.

Barographs may be classified, on the basis of their construction, into the following types: 1) **aneroid barograph** (including **microbarograph**); 2) **float barograph**; 3) **photographic barograph**; and 4) **weight barograph**. The aneroid barograph is the one most commonly used in weather stations.

U.S. Weather Bureau, 1941: *Barometers and the Measurement of Atmospheric Pressure*, Circular F, 7th ed., rev., 28–44.

barometer—An instrument for measuring **atmospheric pressure**.

Two types of barometers are commonly used in meteorology: the **mercury barometer** and the **aneroid barometer**. *See also* **barograph**.

barometer cistern—The cylindrical vessel in a **mercury barometer** into which the tube dips.

The **cistern** may be adjustable (**Fortin barometer**) or fixed (**Kew barometer**) with respect to the tube.

barometer column—Same as **mercury column**.

barometer corrections—Same as **barometric corrections**.

barometer elevation—(Or elevation of ivory point.) The vertical distance above **mean sea level** of the **ivory point** (zero point) of a **station**'s **mercurial barometer**; frequently the same as **station elevation**.

This term is denoted by the symbol H_z in international usage. The value of **atmospheric pressure** with reference to this level is termed **actual pressure**.

barometric—1. Pertaining to a **barometer** or to the results obtained by using a barometer.

2. Often used in the same sense as **pressure** (atmospheric), for example, "barometric gradient" for **pressure gradient**.

barometric altimeter—Same as **pressure altimeter**.

barometric characteristic—A distinguishing attribute or description of **atmospheric pressure**.

barometric column—Same as **mercury column**.

barometric constant—Factor relating the **pressure** and the height of a column of **mercury**, for example, 1 hPa = 0.750062 mm, 1 mm = 1.333224 hPa.

barometric correction table—Table or graph to facilitate compensation of the instrumental errors of a **mercury barometer**.

The required compensation is generally very small and is normally included in the **barometric reduction table**. *See* **compensation of instruments**.

barometric corrections—The corrections that must be applied to the reading of a **mercury barometer** in order that this observed value may be rendered accurate.

There are four kinds. 1) The **instrument correction** is the mean difference between the readings of a given mercury barometer and those of a **standard** instrument. It is a composite **correction**, including the effects of **capillarity** (*see* **capillarity correction**), **index** misalignment, imperfect vacuum, and **scale** correction, which are the **barometric errors**. 2) The **temperature correction** is applied to account for the difference between the **coefficient of expansion** of **mercury** and that of the scale. 3) The **gravity correction** is necessary because the **acceleration of gravity** varies with both **altitude** and **latitude**. 4) The **removal correction** is applied when the **barometer elevation** differs from the adopted **station elevation** and/or **climatological station** elevation. *See also* **capacity correction**.

U.S. Weather Bureau, 1941: *Barometers and the Measurement of Atmospheric Pressure*, Circular F, 7th ed., rev.

barometric errors—*See* **barometric corrections**.

barometric gradient—Same as **pressure gradient**.

barometric hypsometry—The technique of estimating **elevation** by means of **atmospheric pressure** measurements.

See **hypsometric equation, hypsometry, altimetry**.

Humphreys, W. J., 1940: *Physics of the Air*, 68–75.

barometric pressure—Same as **atmospheric pressure**.

barometric rate—(Rare.) The rate at which the **atmospheric pressure** rises or falls locally.

See **pressure tendency**.

barometric reduction table—Table for the **reduction** of station **mercury barometer** readings in conditions of **standard temperature** and **gravity** and, if required, to a **standard** level (normally **mean sea level**).

Compare **station pressure**.

barometric switch—Same as **baroswitch**.

barometric tendency—Same as **pressure tendency**.

barometric wave—Any **wave** in the **atmospheric pressure** field.

The term is usually reserved for short-period variations not associated with **cyclonic-scale** motions or with **atmospheric tides**. *See* **pressure wave**.

barometry—The study of the measurement of **atmospheric pressure**, with particular reference to ascertaining and correcting the errors of the different types of **barometer**.

baromil—The unit of length used in graduating a **mercury barometer** in the centimeter-gram-second system.

If the **barometer** is located at 45° latitude at **sea level** and its **temperature** is 0°C, a length increment of one baromil will correspond to a **pressure** increment of one **millibar**. Corrections must be applied at other locations.

baroswitch—(Also called barometric switch.) A pressure-operated switching device used in a **radiosonde**.

In operation, the expansion of an **aneroid capsule** causes an electrical contact to scan a radiosonde commutator composed of conductors separated by insulators. Each switching operation corresponds to a particular **pressure** level. The contact of an insulator or a conductor determines whether **temperature, humidity,** or reference **signal** will be transmitted.

barothermograph—An instrument that automatically records **temperature** and **pressure**.

barothermohygrograph—An instrument that automatically records **temperature, pressure,** and **humidity**.

barotropic—1. That part of the **velocity** field that is either uniform with depth or has zero horizontal component of **vorticity**.

In a purely barotropic flow, the **pressure gradients** and the **density** gradients are parallel so that their **cross product**, the **baroclinic torque vector**, is zero. Variations of fluid density are therefore not directly relevant to barotropic motion.

2. Of, pertaining to, or characterized by a condition of **barotropy**.

barotropic atmosphere—A state of the **atmosphere** in which **density** depends only upon **pressure**, that is, a state such that surfaces of constant pressure and constant density coincide, so that the **geostrophic wind** is independent of height.

barotropic disturbance—1. (Or barotropic wave.) A **wave disturbance** in a two-dimensional flow, the driving mechanism for which lies in the **variation** of **vorticity** of the basic current and/or in the variation of the vorticity of the earth about the local vertical.

Such wave disturbances are also known as **Rossby waves**. *See also* **barotropic instability**.

2. An **atmospheric wave** of **cyclonic scale** in which **troughs** and **ridges** are approximately vertical.

See **cyclone wave, short wave, barotropy**; *compare* **baroclinic disturbance**.

barotropic forecast—A forecast of the atmospheric **circulation** obtained by solving an initial value problem using a **barotropic (vorticity)** equation.

See **barotropic model**.

barotropic instability—The **hydrodynamic instability** arising from certain distributions of **vorticity** in a two-dimensional nondivergent flow.

This is an **inertial instability** in that **kinetic energy** is the only form of **energy** transferred between **current** and **perturbation**. The **variation** of vorticity in the basic current may be concentrated in discontinuities of the **horizontal wind shear** (to be distinguished from **Helmholtz instability**, where the **velocity** itself is discontinuous) or may be continuously distributed in a curved **velocity profile**. A well-known necessary condition for barotropic instability is that the basic state vorticity gradient must have both signs in the domain (i.e., counterpropagating **Rossby waves** must be possible).

barotropic model—1. Any of a number of **model atmospheres** in which some of the following conditions exist throughout the motion: coincidence of **pressure** and **temperature** surfaces; absence of **vertical wind shear**; absence of vertical motions; absence of horizontal **velocity divergence**; and conservation of the vertical component of **absolute vorticity**.

Barotropic models are usually divided into two classes: the nondivergent barotropic model and the divergent barotropic model (also called the shallow-water equations).

2. A single-parameter, single-level **atmospheric model** based solely on the **advection** of the initial **circulation** field.

The simplest form of barotropic model is based on the **barotropic** vorticity advection equation:

$$\frac{\partial \nabla^2 \psi}{\partial t} + \mathbf{V}_\psi \cdot \nabla(\nabla^2 \psi + f) = 0,$$

where ψ is the geostrophic **streamfunction**, $\mathbf{V}_\psi$ is the nondivergent **wind**, and f is the **Coriolis parameter**. The equation is derived by assuming that a vertical portion of the **atmosphere** is barotropic (i.e., **density** is constant on **pressure** surfaces) and nondivergent. Because there are no vertical variations or **thermal** advection processes in a barotropic model, it cannot predict the development of new weather systems.

barotropic pressure function—The **scalar** function in a **barotropic** fluid, with **gradient** that is equal to the **specific** pressure force:

$$\nabla \pi = \alpha \nabla p,$$

where ∇ is the **del operator**, π the barotropic pressure function, α the **specific volume**, and p the **pressure**.

barotropic torque vector—*See* **barotropic**.

barotropic vorticity equation—The **vorticity equation** in the absence of **horizontal divergence** and vertical motion, so that the **absolute vorticity** of a **parcel** is materially conserved,

$$\frac{D}{Dt}(\zeta + f) = 0,$$

where ζ is the **relative vorticity**, and f the **Coriolis parameter**.

This equation may also be interpreted as governing vertically averaged flow in which **divergence** is present but **wind direction** is constant with height. *See also* **equivalent barotropic model**.

barotropic wave—Same as **barotropic disturbance**.

barotropy—The state of a fluid in which surfaces of constant **density** (or **temperature**) are coincident with surfaces of constant **pressure**; it is the state of zero **baroclinity**.

Mathematically, the **equation of barotropy** states that the **gradients** of the density and pressure fields are proportional:

$$\nabla\rho = B\nabla p,$$

where ρ is the density, p the pressure, and B a function of thermodynamic variables, called the **coefficient of barotropy**. With the **equation of state**, this relation determines the spatial distribution of all state parameters once these are specified on any surface. For a **homogeneous atmosphere**, $B = 0$; for an **atmosphere** with homogeneous **potential temperature**,

$$B = \frac{c_v}{c_p RT},$$

where c_v and c_p are the **specific heats** at constant volume and pressure, respectively, R the **gas constant**, and T the Kelvin temperature; for an **isothermal atmosphere**, $B = 1/(RT)$. It is not necessary that a fluid that is **barotropic** at the moment will remain so, but the implication that it does often accompanies the assumption of barotropy (*see* **autobarotropic**). In this sense the assumption, or a modification thereof, is widely applied in **dynamic meteorology**. The important consequences are that **absolute vorticity** is conserved (to the extent that the motion is two dimensional), and that the **geostrophic wind** has no **shear** with height. *See* **equivalent barotropic model**, **barotropic vorticity equation**, **barotropic instability**, **barotropic disturbance**.

Holton, J. R., 1992: *An Introduction to Dynamic Meteorology*, 3d edition, Academic Press, p. 77.

barrens—Any region that is devoid of vegetation or permits only the sparse growth of very few plant species.

This term is most commonly applied to such terrain in polar regions. *See* **arctic desert**.

barrier—In polar terminology, an early term for **ice shelf**; first used by Sir James Clark Ross for the face of the Antarctic ice shelf later named for him, Ross Barrier.

barrier iceberg—Obsolescent term for **tabular iceberg**.

barrier jet—A **jet** on the **windward side** of a mountain barrier, blowing parallel to the barrier.

The jet is produced when stable **synoptic** flow at low levels approaches the barrier and is blocked (*see* **blocking**) for a significant fraction of a **day** or longer. This often occurs, for example, when a **cold front** approaches the barrier. The component of the large-scale flow perpendicular to the ridge forces the flow to ascend the barrier. Because the air column is stable, the air layer near the surface is potentially colder (by definition) than the air layer above it, and the **stratification** opposes and retards the upslope flow. As the colder air ascends, it produces higher **pressure** along the slope than at the same level over the plain, and consequently also a **pressure-gradient force** directed away from the mountains. If this pressure configuration lasts for several hours or more, Coriolis deflection accelerates the flow with a component perpendicular to the **pressure gradient**, that is, in the along-barrier direction. At timescales greater than a pendulum day—that required for **geostrophic** adjustment—these processes produce a persistent barrier jet at heights below the level of the mountain. The process of **geostrophic adjustment** also brings the flow in the jet into balance with the **thermal wind**, so an argument based on thermal wind reasoning also explains the barrier jet. Barrier jets have been documented **windward** of the Sierra Nevada in California, to the north of the Brooks Range in Alaska, and in Antarctica along the Antarctic Peninsula and the Transantarctic Mountains. Maximum speeds, which generally occur at heights just below the midway level of the mountains, reach 15–30 m s^{-1}, and the jet can extend laterally 100 km or more **upstream** of the barrier. The strong **shear** in the jet is capable of producing moderate to severe **turbulence** for low-flying aircraft.

barrier layer—The depth **range**, where it exists, between the bottom of the **oceanic surface mixed layer** and the **thermocline**, usually at a depth between 30 and 80 m.

The barrier layer is found in tropical regions where the **mixed layer** contains water of lower **salinity** than but identical in **temperature** to that of the water below. Its significance is that it acts as a barrier to the vertical penetration of **heat** into the ocean because, without a temperature **gradient** at the bottom of the mixed layer, **entrainment** of water from below does not remove any heat from the mixed layer.

barrier theory—Any notion that stationary **orographic** structures or nearly stationary atmospheric **air masses** may affect faster-moving nearby air masses to the effect of **cyclogenesis** or **anticyclogenesis**.

barye—The **pressure** unit of the centimeter-gram-second system of physical units; equal to one **dyne** per cm² (0.001 **millibar**).

basal sliding—The motion of the basal **ice** of the **glacier** relative to the material immediately beneath the glacier.

base flow—That part of the **stream** discharge that is not attributable to **direct runoff** from **precipitation** or melting **snow**; it is usually sustained by **groundwater**.

base level of erosion—1. An imaginary surface of irregular shape, inclined toward the lower end of the principal, or trunk, **stream** of a **basin**, below which the stream and its tributaries were presumed to be unable to erode.

2. More generally applied to the critical plane of **erosion**, represented approximately by **sea level** on coasts, which would be the lowest point toward which running water would usually erode.

base line—1. The reference line in a measurement by triangulation.

In meteorological observations it has several applications, for example, 1) the horizontal distance from the **observation** point to the location of a **ceiling-light** projector; 2) the horizontal distance between a **ceilometer** projector and detector; and 3) the **bearing**, distance, and slope of the line between the observational points in a **double-theodolite observation**.

2. In **dual-Doppler analysis**, the length and **bearing** of the line separating the radars.

base map—A map designed for the presentation and **analysis** of data; it usually includes only the coordinates, geographical and major political outlines, and sometimes the larger lakes and rivers.

Many modifications exist for specific uses throughout the geophysical sciences, such as the frequent inclusion of fixed reference points (or **station** positions). Mountains and contour lines are generally omitted, but high ground may be indicated by a single **contour line** and shading.

base width—On a flood **hydrograph**, a measure of the duration of direct storm **runoff**.

baseflow recession—*See* **baseflow recession curve**.

baseflow recession curve—A **recession curve** of **streamflow** so adjusted that the slope of the curve represents the **runoff** depletion rate of the **groundwater**.

A curve is formed by the observed **hydrograph** during prolonged periods of no **precipitation**.

baseflow storage—Storage within an **aquifer** that supplies water to baseflow.

baseline check—(Also known as ground check.) Calibration or quality assurance of **radiosonde** sensors before flight by comparison of radiosonde measurements with those from other instruments.

baseline monitoring—The establishment and operation of a designed surveillance system for continuous or periodic measurements and recording of existing and changing conditions that will be compared with future observations.

basic flow—The flow **field** in an **atmospheric model** used as a background state for analyzing the evolution of small **amplitude** perturbations in accordance with the **linear** dynamics of the model.

basin—1. *See* **river basin**.

2. Any body of water not having horizontal communication with the open ocean at all depths. The maximum depth at which there is horizontal communication is the **sill depth**.

basin accounting—Same as **hydrologic accounting**.

basin lag—The time difference between the centroids of **rainfall excess** and its corresponding **surface runoff** produced in the **basin**.

It is equal to the first moment of the **instantaneous unit hydrograph** of the **basin**. Conceptually, it is defined as the average length of time for which water produced by a **rainfall** remains in residence in the basin.

basin outlet—The most **downstream** point of a **drainage basin** past which all flow from an **upstream** watershed flows.

See **watershed**.

basin recharge—The difference between amounts of **precipitation** and **runoff** for a given **storm**; it is that portion of the precipitation that remains in the **basin** as **soil moisture, surface storage, groundwater**, etc.

basin response—The manner in which a **basin**'s characteristics affect the method of handling the water deposited by **precipitation** events.

basis functions—A set of functions that can be linearly combined to form a more general set of functions.

bathymetric chart—A map delineating the form of the bottom of a body of water, usually by means of depth contours (**isobaths**).

bathymetry—The study of water depths.

bathythermograph—(Abbreviated BT.) A device for obtaining, from a ship under way, a record of **temperature** against depth (strictly speaking, **pressure**) in the upper 300 m of the ocean.
    For a **thermal** element it has a xylene-filled copper coil, which actuates a stylus through a **Bourdon tube**. The pressure element is a copper **aneroid capsule** that moves a smoked glass slide at right angles to the motion of the stylus. A double **analog** record is thus obtained as the BT is lowered and recovered. This device has generally been replaced by the **expendable bathythermograph** (XBT).

bathythermograph grid—A transparent glass slide, marked with **calibration** lines of **temperature** and depth, that, when superimposed against a **bathythermograph slide**, makes it possible to read observed values of temperature and depth.

bathythermograph print—An enlarged photographic print of a **bathythermograph slide** superimposed on the appropriate **bathythermograph grid** annotated with particulars of location and date.
    The U.S. Navy Hydrographic Office maintains a master file of nearly one million **bathythermograph** prints from all the oceans.

bathythermograph slide—The smoked glass slide used in a **bathythermograph** to obtain a record of **temperature** as a function of depth.

baud—A unit of signaling speed that is the number of **signal** events per second.
    Depending on the number of bits per signal event, the **baud rate** can be different from the **bit rate**. In modern systems, a signal event represents more than one **bit**, so the baud rate is less than the bit rate (or bits per second).

baud rate—Another name for **baud**.
    Although the word "rate" in this context is redundant (since baud is the number of **signal** events per second), the combination "baud rate" is used frequently in the literature.

bay ice—1. Any recently formed **sea ice** that is sufficiently thick to impede navigation.
    2. In Labrador, one-year **ice** that forms in bays and inlets.
    3. In the Antarctic, sometimes applied to heavy **floes** recently broken away from an **ice shelf**.

Bay of Bengal Water—A **water mass** found in the upper 100 m of the eastern tropical Indian Ocean originating from the Bay of Bengal.
    It is characterized by very low **salinity** produced from monsoonal river **runoff**.

bayamo—A severe tropical **thundersquall** that occurs on the south coast of Cuba, especially near the Bight of Bayamo.

BDRF—Abbreviation for **bidirectional reflection function**.

beacon—*See* **radio beacon**.

beaded lightning—(Also called chain lightning, pearl lightning.) A particular aspect of a **normal** lightning **flash** occasionally seen when the **observer** happens to view end-on a number of segments of the irregular channel (**zigzag lightning**) and hence receives an impression of higher luminosity at a series of locations along the channel.

beam—1. A collimated source of **electromagnetic radiation** (e.g., a **laser** beam).
    2. A collimated source of atoms, molecules, or subatomic **particles** (e.g., an **electron beam**).

beam filling—In **radar**, the capacity of a **distributed target** to occupy all the space of the **pulse volume**.

beam pattern—*See* **antenna pattern**.

beam swinging—A technique used by **wind profilers** to determine the **vertical profile** of the **wind vector** above the **radar** by measuring the line-of-sight component of the **wind** sequentially along each of several different **beam** directions. The different beam directions can be obtained in various ways: 1) mechanically steering a single fixed-beam **antenna**; 2) electronically or mechanically steering the beam of a single fixed-position antenna (*see* **phased-array antenna**); or 3) switching between multiple fixed-beam antennas.

Three beam positions can be used to determine the wind vector if it is assumed that the **wind field** is the same along those beams. At least five beam positions must be used to measure both the wind and its **divergence**. Wind profilers typically use one vertical beam and oblique beams that are between 10° and 25° off vertical. Oblique angles closer to the **zenith** make separation of horizontal and vertical wind components difficult and larger zenith angles make it less likely that the wind is uniform over the horizontal distances between beams. Large zenith angles also increase the radial **range** to the **scattering** volume at a given **altitude**, thereby reducing the strength of the returned **signal**.

beam wind—Nautical term for a **crosswind**, especially a **wind** blowing 90° from a ship's **heading**.

beamwidth—A measure of the concentration of **power** in a **radar beam**. The beamwidth in a given plane is the angle subtended at the **antenna** and containing the **main lobe** within which the **intensity** in the far **field** of the transmitted **radiation** is not less than a specified fraction of the intensity in the direction of the axis of the main lobe.

Ordinarily, beamwidth is defined as the angle within which the intensity is not less than one-half the intensity on the **beam** axis (the 3-dB beamwidth). Sometimes beamwidth may refer to the **linear** extent of the beam in the cross-beam direction at some specified **range**. *See* **antenna pattern**.

bearing—The horizontal direction from one terrestrial point to another; basically synonymous with **azimuth**.

Bearing, however, may be expressed in several ways: true bearing and magnetic bearing are the angular directions in degrees measured clockwise from **true north** and **magnetic north**, respectively; **compass** bearing is expressed in terms of compass points; and relative bearing is the angular distance measured clockwise from the **heading** of a craft (in aviation, relative bearing is often referred to a clock face, that is, "3 o'clock" equals a relative bearing of 90°, etc.).

beat frequency—*See* **beating**.

beating—When two oscillating quantities with different frequencies are superposed, the **amplitude** of the combination oscillates at the **frequency** difference, the beat frequency.

The simplest example of beating is the superposition of two equal-amplitude oscillations,

$$A \cos \omega_1 t + A \cos \omega_2 t = 2A \cos\left(\frac{\omega_1 t - \omega_2 t}{2}\right) \cos\left(\frac{\omega_1 t + \omega_2 t}{2}\right),$$

which may be looked upon as an **oscillation** at the average of the two frequencies with an amplitude oscillating at their difference. Beating is often associated with waves, but nothing restricts it to waves.

Beaufort force—(Or Beaufort number.) A number denoting the speed (or "strength") of the **wind** according to the **Beaufort wind scale**.

Beaufort number—Same as **Beaufort force**.

Beaufort weather notation—A series of numbers from 0 to 12 indicating the state of the **wind** and sea.

British Meteorological Office, 1991: *Meteorological Glossary*, HMSO, London, England, 40–41.

Beaufort wind scale—A system of estimating and reporting wind speeds using a numerical **scale** ranging from 0 (**calm**) to 12 (**hurricane**).

It was invented in the early nineteenth century by Admiral Beaufort of the British Navy and was originally based on the effects of various wind speeds on the amount of canvas that a full-rigged frigate of the period could carry. It has since been modified and modernized and in its present form for international meteorological use it equates 1) **Beaufort force** (or Beaufort number), 2) **wind speed**, 3) descriptive term, and 4) visible effects upon land objects or sea surface. One land adaptation is the **NRM wind scale**.

U.S. Weather Bureau, 1955: *Circular N*, 7th ed., p. 100.
U.S. Weather Bureau, 1954: *Circular M*, 9th ed, p. 14.
Byers, H. R., 1944: *General Meteorology*, 83–85.
Donn, W. L., 1951: *Meteorology with Marine Applications*, 132–135.

List, R. J., ed., 1951: *Smithsonian Meteorological Tables*, 6th rev. ed., p. 119.

Beaumont period—Period of 48 consecutive hours in which at least 46 hourly readings have had temperatures not less than 20°C and **relative humidity** not less than 75%.

The occurrence of this period has been widely used as a criterion for issuing warnings for the onset of potato blight. *See* **operational weather limits**.

beavertail antenna—A type of **radar** antenna that forms a **beam** having a greater beam width in **azimuth** than in **elevation**, or vice versa.

In physical dimensions, its long axis lies in the plane of smaller beam width. A beam that is wide in azimuth angle and narrow in **elevation angle** is employed on height-finding radars. Search radars usually have beams narrow in azimuth but broad in elevation angle. In general, the narrower the beam, the greater the **angular resolution**.

Beer's law—*See* **Bouguer's law**.

bel—*See* **decibel**.

belat—A strong, dusty (or sandy) northwesterly **wind** that blows seaward along the south coast of Arabia.

It is most frequent in winter and early spring.

bell taper—A process of modifying a **time series** of data whereby the beginning and ending 10% of the data points are smoothly reduced in **amplitude** to approach zero at the ends.

Such conditioning of raw data is often necessary before performing a spectral or **Fourier analysis**. If this were not done, it is possible that spurious spectral signals might be introduced because the Fourier analysis assumes that the finite-duration time series is cyclic (which could cause a step **discontinuity** between the end of the series and the beginning of the repeated series). However, any time that data are modified, including the conditioning described here, the validity of the resulting **analysis** may be questioned.

Bellamy method—A method of estimating **horizontal divergence** of air, using a triangular network of **rawinsonde** stations, devised by J. C. Bellamy in the late 1940s.

At any one **altitude**, the **wind** measurements at the vertices of the rawinsonde triangle are linearly interpolated along the sides of the triangle to find the average wind component perpendicular to the side. When these normal wind components are multiplied by the length of the respective side, summed over all sides, and divided by the area of the triangle, the result is the average horizontal divergence within the triangle. Note that this method is sensitive to wind measurement errors and to the spacing between rawinsondes.

bellows—*See* **aneroid capsule**.

below minimums—Weather conditions below the **minimums** prescribed by regulation for the particular action involved, for example, landing minimums, takeoff minimums.

belt of fluctuation—(Also called zone of fluctuation, belt of phreatic fluctuation, belt of water-table fluctuation.) That part of the **lithosphere** that, because of the fluctuations of the **water table**, lies part of the time in the **zone of saturation** and part of the time in the overlying **zone of aeration**.

beltane-ree—In the Shetland and Orkney Islands, stormy weather occurring about Whitsuntide (seventh week after Easter).

Beltrami flow—A fluid motion in which the **vorticity** vector is parallel to the **velocity** vector at every point of the fluid.

Bemporad's formula—A formula for the **optical mass** m in terms of the **zenith distance** z of the sun or other celestial body;

$$m = \frac{\text{astronomical refraction (in seconds)}}{58.36'' \sin z}.$$

For values of z less than about 70°, Bemporad's formula can be replaced by the simpler approximate formula,

$$m = \sec z.$$

List, R. J., ed., 1951: *Smithsonian Meteorological Tables*, 6th rev. ed., p. 422.

Bénard cell—A form of **cellular convection**, usually hexagonal as viewed from above (when the fluid has a **free surface**), formed in thin layers of a fluid initially at rest and heated from below.

According to the usual **dynamic stability** analysis, the **Rayleigh number** must attain a critical minimum value before this type of **circulation pattern** develops. The vertical **circulation** may be upward in the core and downward on the edges, or it may be the reverse. It has been suggested that some **cloud** patterns give evidence of Bénard cells on a **large scale**.

Sutton, O. G., 1953: *Micrometeorology*, 116–125.
Brunt, D., 1951: *Compendium of Meteorology*, 1255–1262.

benchmark—A reference against which measurements may be evaluated, for instance, a runoff-measuring **station** established as representative or typical of a hydrologic region to provide a continuing series of hydrological observations that are relatively uninfluenced by past or future artificial changes.

benchmark station—A term used to designate a climatic **station** with a long record of data obtained in a single location with minimal change in the **environment** of the station.

Those stations originally designated benchmark stations are now considered part of the "Historical Climatic Network."

Benguela Current—The eastern **boundary current** of the South Atlantic subtropical **gyre**.

It transports 20–25 Sv (20–25×10^6 m^{3}s^{-1}) northward along the coast of Namibia with speeds of about 0.2 m s^{-1}, gradually leaving the coast between 30° and 25°S. As an eastern boundary current, the Benguela Current is associated with strong **coastal upwelling** that reaches as far north as 18°S. The upwelling is strongest in spring and summer (October–February) when the **trade winds** are steady; during winter (July–September) it extends northward but becomes intermittent because the **trades** are interrupted by passing low pressure systems.

bent-back occlusion—(Also called back-bent occlusion.) An **occluded front** that has reversed its direction of motion as a result of the **development** of a new **cyclone** (usually near the **point of occlusion**) or, less frequently, as the result of the displacement of the old cyclone along the **front**.

The resulting movement of the occluded front is westward and/or southward behind its associated **cold front**. It also may appear as the westward extension of the **warm front** in some cyclones. Its subsequent evolution following the development of a new cyclone can give the appearance of a cold front, with a sharp **temperature** drop and strong increase of winds.

bentu de soli—An east **wind** on the coast of Sardinia.

benzene—Organic compound of formula C_6H_6 consisting of a symmetrical six-membered ring structure.

Benzene is the parent compound of the aromatic group of compounds, consisting of toluene, the xylenes, etc., which all exhibit the benzene structure. Benzene is a major component of gasoline and is emitted into the **atmosphere** as a result of **evaporation** and incomplete fossil fuel combustion.

benzo-a-pyrene—Benzo-a-pyrene (BaP) is a polycyclic aromatic hydrocarbon (PAH), formula $C_{20}H_{12}$, consisting of five fused aromatic rings.

BaP, a very strong carcinogen, has been found in **soot** particles and other solid by-products of combustion.

berber—Same as **barber**.

berg—1. Commonly used abbreviation for **iceberg**.

2. *See* **berg wind**.

berg wind—(Literally, mountain wind.) A hot, dry, squally **wind** blowing off the interior plateau of South Africa, roughly at right angles to the coast; a type of **foehn**.

Berg winds blow mainly in winter, when a strong **anticyclone** occupies the interior plateau, producing **outflow** across the coastal regions of South Africa. They are especially frequent and noticeable on the cool west coast, where they set in suddenly during the morning, bring a **temperature** rise of 15°–20°C, giving maxima of 22°–35°C, and cause the **relative humidity** to fall from nearly 100% to 30% or less. Here, on days when the **sea breeze** sets in about noon, the **temperature** drops by 10°–16°C even more suddenly than it had risen at the onset of the berg wind. Berg wind episodes may last two to three days, causing uncomfortable weather for people and damage to crops. At Port Nolloth in southwest Africa, berg winds occur on an average of 48 days a year, June and July having nine each and the whole summer half-year (October through March) a total of only six. *See* **hot wind**; *compare* **mountain wind**.

Bergeron classification—See **airmass classification**.

Bergeron–Findeisen process—(Commonly called the ice process of precipitation, and formerly, ice-crystal theory; also Bergeron–Findeisen–Wegener process or theory, and with the names in a different order.) A theoretical explanation of the process by which **precipitation** particles may form within a **mixed cloud** (composed of both **ice crystals** and liquid water drops).

The basis of this theory is the fact that the **equilibrium vapor pressure** of **water vapor** with respect to **ice** is less than that with respect to liquid water at the same subfreezing **temperature**. Thus, within an admixture of these **particles**, and provided that the total **water content** were sufficiently high, the ice crystals would gain mass by **vapor** deposition at the expense of the liquid drops that would lose mass by **evaporation**. Upon attaining sufficient weight, the ice crystals would fall as **snow** and very likely become further modified by **accretion**, melting, and/or evaporation before reaching the ground. This theory was first proposed by T. Bergeron in 1933, and further developed by W. Findeisen. Certain of its features related to **nucleation** had been suggested by A. Wegener as early as 1911. Operation of this process requires numerous small water drops that are supercooled, which is a common feature in clouds between about 0° and −20°C or below, along with a small number of ice crystals. The crystals grow by vapor deposition at a rate (maximum at about −12°C) to give individual snow crystals in some 10 to 20 minutes. Much **cloud seeding** is based upon the introduction of artificial **ice nuclei** to supply more of the ice particles.

Wegener, A., 1911: *Thermodynamik der Atmosphäre*, Barth, Leipzig.

Findeisen, W., 1938: Die kolloidmeteorologischen Vorgänge bei der Niederschlagsbildung (Colloidal meteorological processes in the formation of precipitation). *Met. Z.*, **55**, p. 121.

Bergeron, T., 1935: On the physics of cloud and precipitation. *Proc. 5th Assembly U.G.G.I. Lisbon*, Vol. 2, p. 156.

bergy bit—A small **iceberg** or iceberg fragment; it also may be a piece of **floeberg** or **hummocked ice**; it is larger than a **growler**.

Bering Slope Current—The eastern part of the subpolar **gyre** in the deep (western) part of the Bering Sea.

Currents in this gyre are weak (0.1–0.2 m s⁻¹) but reach to the ocean floor and are associated with **eddy** formation at oceanic ridges. A southward **countercurrent** with maximum speeds of 0.25 m s⁻¹ near a depth of 150 m exists between the shelf and the northward-flowing Bering Slope Current, indicating its character as the eastern **boundary current** of a subpolar gyre.

Bermuda high—The semipermanent **subtropical high** of the North Atlantic Ocean, so named especially when it is located in the western part of the ocean.

This same **high**, when displaced toward the eastern part of the Atlantic, is known as the **Azores high**. On **mean charts** of **sea level pressure**, this high is a principal **center of action**. Warm and humid conditions prevail over the eastern United States, particularly in summer, when the Bermuda high is well developed and extends westward.

Bernal–Fowler rules—The set of rules describing the arrangement of **hydrogen** atoms in an "ideal" **ice crystal**.

These rules are 1) Each **water molecule** is oriented such that its two hydrogen atoms are directed approximately toward two of the four surrounding **oxygen** atoms (arranged almost in a tetrahedron); 2) only one hydrogen atom is present on each O-O linkage; and 3) each oxygen atom has two nearest neighboring hydrogen atoms such that the water molecule structure is preserved. Violations of these rules lead to structural defects in **ice**, responsible, among other things, for its **electrical conductivity** and its long (10^{-4} s) electrical **relaxation time**.

Bernoulli's equation—*See* **Bernoulli's theorem**.

Bernoulli's theorem—As originally formulated, a statement of the **conservation of energy** (per unit mass) for an **inviscid fluid** in **steady motion**.

The **specific energy** is composed of the **kinetic energy** $(1/2)v^2$, where v is the speed of the fluid; the **potential energy** gz, where g is the **acceleration of gravity** and z the height above an arbitrary reference level; and the **work** done by the **pressure** forces $\int \alpha dp$, where p is the pressure, α the **specific volume**, and the integration is always with respect to values of p and α on the same **parcel**. Thus, the relationship

$$\frac{1}{2}v^2 + gz + \int \alpha dp = \text{constant along a streamline}$$

is valid for **steady motion**, since the **streamline** is also the path. If the motion is also **irrotational**, the same constant holds for the entire fluid. The following special cases are important: 1) as originally formulated for a homogeneous **incompressible fluid**,

$$\int \alpha dp = \alpha p;$$

and 2) for a **perfect gas** undergoing **adiabatic** processes,

$$\int \alpha dp = c_p T,$$

where c_p is the **specific heat** at constant pressure and T the Kelvin **temperature**. If there is diabatic heating on the parcel at the rate dQ/dt per unit mass, then

$$\int \alpha dp = c_p T - \int dQ.$$

Gill, A. E., 1982: *Atmosphere–Ocean Dynamics*, Academic Press, 82–83.

Besson nephoscope—Same as the **comb nephoscope**.

Best number—See **Davies Number**.

beta drift—The **drift** of a **tropical cyclone** through the large-scale average layer-mean background **wind** in which it is embedded.

  The drift is caused by the **advection** of the background **potential vorticity** field by the **storm** circulation. In the simplest case, the background potential vorticity gradient is simply the **meridional** gradient of the **Coriolis parameter**, β, from which the term gets its name. Beta drift generally causes tropical cyclones to move poleward and westward relative to the motion they would have if the background potential vorticity field were unperturbed by the storms. This drift speed is generally around 1–2 m s^{-1}.

beta effect—Denotes how fluid motion is affected by spatial changes of the **Coriolis parameter**, for example, due to the earth's curvature.

  The term takes its name from the symbol β representing the **meridional** gradient of the Coriolis parameter at a fixed latitude. A linearly sloping lower boundary to fluid in a rotating system also experiences the beta effect.

beta gyre—**Cyclonic** or **anticyclonic** gyre (spiral motion) resulting from the **advection** of background **potential vorticity** by the **circulation** of the **tropical cyclone**.

  The influence of such gyres on the motion of the tropical cyclone is called **beta drift**.

beta particle—(Or β particle.) Physically indistinguishable from the **electron** (or **positron**) but usually restricted to products in nuclear reactions (beta decay).

  The term was coined by Ernest Rutherford, who discovered that the ionizing **radiation** emitted by uranium consisted of "at least two distinct types . . . one that is very readily absorbed . . . the α radiation, and the other of a more penetrative character . . . the β radiation." Kinetic energies of beta **particles** range from tens of thousands to millions of electron volts. Because either electrons or positrons are emitted in beta decay, the term beta particle, a relic of an era in which its identity was unknown, is falling into disuse. *See* **alpha particle**, **gamma ray**.

Boorse, H. A., and L. Motz, 1966: *The World of the Atom*, Vol. I, 437–445.

beta-plane—The derivative of the **Coriolis parameter** with respect to latitude.

  The beta-plane approximation, useful for the study of equatorial and midlatitude flow, assumes that the Coriolis parameter varies linearly with latitude. Explicitly, the Coriolis parameter is given approximately by

$$f = f_0 + \beta y,$$

in which y is the **meridional** distance from some fixed latitude where the Coriolis parameter is f_0 and β (from which the beta plane gets its name) is the meridional gradient of f at that fixed latitude.

beta plane—The **model**, introduced by C. G. Rossby, of the spherical earth as a plane with a rate of rotation f (corresponding to the **Coriolis parameter**) that varies linearly with the north–south direction y:

$$f = f_0 + \beta(y - y_0).$$

  The constant β (the **Rossby parameter**) is given its value at a central latitude on the spherical earth, $(2\Omega \cos\phi_0)/a$, where Ω is the angular speed of the earth and a its mean radius. Used chiefly in association with the **vorticity equation**, this **model** is usually accompanied by the assumption that the Coriolis parameter is constant, $f = f_0$, when it appears in undifferentiated form. Although it applies precisely to no physical situation, this model has gained wide acceptance in **dynamic meteorology**.

beta ray—(Or beta radiation.) A stream of **beta particles**; can also mean a single beta particle.

beta spiral—In **oceanography**, the turning of the current vector that arises from stratified **geo-strophic flow** when **density** is conserved following a **parcel** (as is approximately the case in the oceanic **thermocline**).

If θ is the angle of the current, U is its magnitude, f is the **Coriolis parameter**, g is the gravitational **acceleration**, and w is the **vertical velocity**, then $\partial\theta/\partial z = -(gw/fU^2)(\partial\rho/\partial z)$, so that moving upward in the water column, currents will spiral clockwise in **upwelling** zones, and counterclockwise in **downwelling** zones.

between layers—In aviation terminology, an aircraft flight conducted between layers of clouds and/or obscuring phenomena.

bhoot—Same as **bhut**.

bhut—(Or bhoot.) In India, a **dust whirl** or **dust devil**.

bidirectional reflectance distribution function—(Abbreviated BRDF.) The reflected **radiance** from a given region as a function of both incident and viewing directions.

It is equal to the reflected radiance divided by the incident **irradiance** from a single direction. *See also* **bidirectional reflectance factor, bidirectional reflection function**.

bidirectional reflectance factor—A function describing the **anisotropy** of reflected **radiance** from a surface illuminated from a single direction.

It is equal to π times the reflected radiance divided by the incident **irradiance** from a single direction. *See also* **bidirectional reflection function, bidirectional reflectance distribution function**.

bidirectional reflection function—(Abbreviated BDRF.) The relative angular dependence of the reflected radiance from a given region as function of both incident and viewing directions.

It is equal to π times the reflected **radiance** divided by the reflected **irradiance**. *See also* **bidirectional reflectance factor, bidirectional reflectance distribution function**.

bidirectional wind vane—*See* **bivane**.

biennial ice—(Also called second-year ice.) **Sea ice** that has survived one summer's melt.

biennial wind oscillation—Fluctuation of the **zonal wind** with a **period** of about two years.

It is most pronounced in the lower **stratosphere** over the equatorial region.

bifilar electrometer—An **electrometer** of the electrostatic type in which the **potential** to be measured is applied to two metal-coated quartz fibers, and the deflection due to their mutual repulsion is observed through a low-power microscope.

The bifilar electrometer is used for potential measurements in **atmospheric electricity** studies. *Compare* **unifilar electrometer**.

bihemispherical reflectance—The **albedo** of a surface under conditions of diffuse **illumination**.

billow cirrus clouds—Closely and regularly spaced wavelike **cirrus** on satellite imagery that form perpendicular to the **jet stream axis**.

These clouds are a result of strong **vertical wind shear**.

billow cloud—Popular name for **undulatus**.

Some forms are the result of **shear instability** (**Kelvin–Helmholtz**), and some result from **gravity waves**. Billow clouds are present when there is sufficient moisture present in the upward motion of the waves to make the **wave** structure visible by **condensation** of **cloud droplets**. Billows formed from gravity waves exhibit broad, nearly parallel, lines of **cloud** oriented normal to the **wind direction**, with cloud bases near an **inversion** surface. The distance between billows is on the order of 1000–2000 m.

billow wave—Atmospheric **wave motion** responsible for the formation of **billow clouds**.

bimetallic strip—A term commonly applied to a pair of metallic strips with different **thermal** expansion coefficients that have been bound together to form a single strip.

The purpose is to create a device that converts **temperature** into a mechanical **signal** either as a visible temperature reference or as part of a thermal control system (e.g., a thermostat).

bimetallic thermometer—A **thermometer** in which the sensitive element consists of a compound strip of metal formed by welding together two strips of metal having different coefficients of expansion.

The curvature of the strip is a function of its **temperature**. It is a type of **deformation thermometer**.

bimodal spectrum—In **radar**, a **Doppler spectrum** that has two modes or peaks.

There are instances in **wind profiler** applications in which one spectral mode is explained by **Bragg scattering** from **refractivity** fluctuations in the **clear air** and another by **Rayleigh scattering** by **precipitation**.

binary code—Use of binary numbers in a transmission to represent different conditions.
The key aspect of these codes is that, in **serial communications**, they can be unprintable. *See* **code forms**.

binary cyclones—Two **tropical cyclones** close enough together for their circulations to interact as the cyclones orbit around a translating point between them.

binomial distribution—A **probability distribution** that applies to experiments involving sequences of independent trials in which only two possible outcomes (e.g., success or failure) can result on each trial.
If p is the **probability** of success on each trial, and $q = 1 - p$ the probability of failure, then the probability of success occurring x times in n trials is given by the binomial distribution:

$$p(x) = \frac{n!}{x!\,(n-x)!}\,p^x q^{n-x}.$$

bio-assay—Method for testing toxic effects of water with the help of living organisms; specifically: 1) the use of a change in biological activity as an indicator of a sample's response to biological treatment; 2) determining toxicity of wastewaters by using viable organisms as test organisms.

biochemical oxygen demand—A measure of the quantity of **oxygen** used for biochemical **oxidation** of organic matter under specified conditions.

biochore—1. As used by W. Köppen in his 1931 **climatic classification**, the part of the earth's surface that is able to support plant life.
It is bounded on the one hand by the cryochore, or region of perpetual **snow**, and on the other by the xerochore, or waterless **desert**. Transition zones on either side are the bryochore, or **tundra** region, and the poëchore, or **steppe** region. The bulk of the biochore consists of the dendrochore, or treed region. *Compare* **biosphere**.
2. In **ecology**, a group or region of similar plant and/or animal life.
> Köppen, W. P., 1931: *Grundriss der Klimakunde*, 2d ed., Berlin: Walter de Gruyter.

bioclimatograph—(Rare.) A **climatic diagram** designed to show the relation between **climate** and some aspect of life.

bioclimatology—The branch of **climatology** that deals with the relations of **climate** and life, especially the effects of climate on the health and activity of human beings (**human bioclimatology**) and on animals and plants.
See **phenology, ecology**.

biogenic ice nucleus—An **ice nucleus** of biological origin, particularly bacteria (e.g., pseudomonas syringae) from plant surfaces.
These organisms have threshold temperatures as high as $-2°C$, being active at the highest **temperature** known for natural nuclei. They were first identified in leaf litter, collected worldwide. Commercial application lies in **nucleation** of water spray drops for artificial **snow** on ski slopes.

biogenic trace gases—Gases emitted to the **atmosphere** as a result of biological activity, for example, from plants, trees, and bacteria.
Examples include **isoprene, dimethyl sulfide**, and **nitrous oxide**.

biogeochemical cycle—The transformation and **transport** of substances within and among the **atmosphere, biosphere, hydrosphere**, and **lithosphere** via biological, geological, and chemical processes that are often cyclical in nature.

biomass burning—The burning of vegetation over large tracts of land, usually in tropical countries.
This biomass burning restores nutrients to the soil, but also causes plumes of **smoke** containing partially burnt gases, **secondary pollutants** such as **ozone**, and **particles** that can extend **downwind** for hundreds of kilometers.
> Crutzen, P. J., and M. O. Andreae, 1990: Biomass burning in the tropics: Impact on atmospheric chemistry and biogeochemical cycles. *Science*, **250**, 1669–1678.
> Crutzen, P. J., L. E. Heidt, J. P. Krasnec, W. H. Pollock, and W. Seiler, 1979: Biomass burning as a source of atmospheric gases CO, H_2, N_2O, CH_3Cl, and COS. *Nature*, **282**, 253–256.

biosphere—That transition zone between earth and **atmosphere** within which most forms of terrestrial life are commonly found; the outer portion of the **geosphere** and inner or lower portion of the atmosphere.

biosphere reserves—1. A protected area, or group of protected areas, set aside to support conservation of biological diversity and sustainable uses of terrestrial and coastal/marine ecosystems.

2. Areas nominated by their owner(s) and internationally designated by UNESCO as biosphere reserves under the Man and the Biosphere Program.

bipolar pattern—A **lightning** pattern revealed in **cloud-to-ground lightning** locations in which areas of predominantly negative **flash** locations are separated horizontally from positive flash locations.
This pattern has been identified in midlatitude **mesoscale convective systems** (MCSs). The convective regions of the MCSs (those with **radar** echoes greater than 35 dBZ) typically have negative lightning flashes, whereas the positive flashes are typically located in the **stratiform** region (<35 dBZ) adjacent to the **convective region**. The bipolar pattern is hypothesized to arise from 1) the **advection** of positively charged **particles** from the upper portions of the convective region to the stratiform region due to **vertical wind shear**; 2) in situ charging of **ice** particles that acquire positive charge in an **environment** of **supercooled liquid water** and **mesoscale** ascent in the **stratiform cloud** layer; or 3) some combination of the two processes.

birainy climate—(Rare.) A **climate** characterized by the regular recurrence of two distinct rainy seasons annually.
This is found in land regions near the **equator** where the heaviest rains occur shortly after the **equinoxes** (the **equinoctial rains**). It may be considered a subdivision of the **tropical rain forest climate**.

birazon—Same as **virazon**.

bird burst—Radar **echoes** caused by flocks of roosting birds that take to the air as a group and fly away in different directions.
Bird bursts typically occur near **sunrise** and have ring-shaped **reflectivity** signatures with weak echoes in the center and maximum reflectivity in circles that expand with time around the roosting site. In **Doppler radar** observations, the **velocity** signature resembles that of a **microburst** at low levels.

birefringence—(Nearly synonymous with **double refraction**.) Two different real parts of the **refractive index** for **electromagnetic waves**, identical except for their states of **(orthogonal) polarization**.
By orthogonal is meant that the waves have opposite handedness, the same ellipticity, and the major axes of their vibration ellipses are perpendicular to each other. The most general birefringence is elliptical, specific examples of which are **linear** and circular. The birefringence of a medium originates from its asymmetry. *See* **polarization**, **dichroism**; *compare* **double refraction**.

Birkeland currents—A system of electrical currents flowing along the direction of the earth's **magnetic field**, between the outer regions of the **magnetosphere** and the **ionosphere**.
Birkeland currents are confined to the auroral zones, and are intimately related to the **aurora** itself. They are named after Kristian Birkeland, an early Norwegian scientist, who first proposed the existence of such currents.

bise—(Also spelled bize.) Name given to cold northerly, northeasterly, or easterly, postfrontal winds in the Swiss Middleland (the region between the Jura and the Alps) and in various regions of eastern France.
The bise is typically driven by an **anticyclone** building to the northwest or north of the Alps. According to Wanner and Furger (1990), "the cold air flow from the north and northeast is channeled between the Jura and the Alps and leads to the formation of the bise over the Swiss Middleland and the upper Rhône Valley. Thus, the bise is normally a postfrontal phenomenon and is closely connected with the **advection** of cold and dry **continental air**. Typically the classic bise endures about one to three days." The bise is most frequent in spring, when it usually brings fine, bright weather. In winter a special case of the bise occurs when the **pressure gradient** is produced by a Mediterranean **cyclone** system to the south of the Alps, bringing **moist air** in from the Balkans. Accompanied by heavy clouds, **snow** whirlwinds in the mountains, and **rain**, snow, or **hail**, this **wind** is called a "black bise" (bise noire in Switzerland and Saône in east central France; bise nègre in Aveyron in south central France). In spring the bise can last for several days and bring damaging frosts. In the Morvan in east central France the very dry bise in March is termed hale de mars (drying wind of March). In the Drôme Valley southeast of Valence (southeast France) the name bise brume is given to a moist, mild, and sometimes foggy wind from the northwest.

Wanner, H., and M. Furger, 1990: The bise–Climatology of a regional wind north of the Alps. *Meteor. Atmos. Phys.*, **43**, p. 107.

Bishop's ring—An observation of a faint, broad, reddish-brown **corona** by the Rev. S. Bishop of Honolulu during the eruption of Krakatoa in 1883.

He described an angular radius of the inner edge of 20° and an angular width of about 10°. Subsequent observations are rare to nonexistent.

Bishop wave—A striking example of an atmospheric **lee wave**, formed in the lee of the Sierra Nevada near Bishop, California.

The phenomenon includes a **rotor cloud** and a series of **lenticular clouds** parallel to the crest of the range.

bistatic—A geometry in which the acoustic source and **receiver** are not at the same position.

bistatic radar—A **radar** system configuration with the **receiver** located at a site different from the **transmitter**.

In such a system, surfaces of constant **range** are ellipsoids with the transmitter and receiver sites as foci, and the component of **target** velocity that induces a **Doppler frequency shift** is the component normal to the ellipsoids.

bit—1. A binary unit of information.

2. A small piece of **sea ice**; a single fragment of **brash**.

bit rate—The number of bits of data sent per second.

This is frequently confused with **baud** in the literature. For bit rates above 600 bits per second, the bit rate exceeds the baud (or baud rate).

bittern—The liquid remaining after **seawater** has been concentrated by **evaporation** until salt has crystallized.

Compare **brine**.

bivane—(Or bidirectional wind vane.) A sensitive **wind vane** used in **turbulence** studies to obtain a record of the horizontal and vertical inclinations of the **wind vector**.

The instrument consists of two lightweight airfoil sections mounted orthogonally on the end of a counterbalanced rod that is free to rotate in the horizontal and vertical planes. The angular positions of the **vane** are commonly detected by **potentiometers**. Some bivanes have a propeller on the end opposite the airfoils to enable measurement of the total **wind speed**.

bize—Same as **bise**.

Bjerknes's circulation theorem—*See* **circulation theorem**.

BL—Abbreviation for **boundary layer**.

black blizzard—A colloquial term for a **duststorm** in the **Dust Bowl** of the south-central United States.

black-bulb thermometer—A **thermometer** with a sensitive element made to approximate a **blackbody** by covering it with lamp black.

The thermometer is placed in an evacuated transparent chamber that is maintained at constant **temperature**. The instrument responds to **irradiance**, modified by the transmission characteristics of its container.

black buran—Same as **karaburan**.

black fog—In the United States, colloquial name given to a dense **fog** over Cape Cod, Masachusetts, especially in the vicinity of Chatham, on the "elbow" of the Cape.

black frost—1. A **dry freeze** with respect to its effects upon vegetation, that is, the internal **freezing** of vegetation unaccompanied by the protective formation of **hoarfrost**.

A black frost is always a **killing frost**, and its name derives from the resulting blackened appearance of affected vegetation.

2. Among some fishermen, a **steam fog** that extends above the bridge level of the fishing boats. If the steam fog does not reach this height, it is a **white frost**.

black ice—1. Thin, new **ice** on freshwater or saltwater, appearing dark in color because of its transparency, which is a result of its columnar grain structure.

On lakes, black ice is commonly overlain by **white ice** formed from refrozen **snow** or **slush**.

2. A mariner's term for a dreaded form of **icing** sometimes sufficiently heavy to capsize a small ship.

3. A popular alternative for **glaze**. A thin sheet of **ice**, relatively dark in appearance, may form

when light **rain** or **drizzle** falls on a road surface that is at a **temperature** below 0°C. It may also be formed when **supercooled fog** droplets are intercepted by buildings, fences, and vegetation.

black northeaster—1. A northeasterly **gale** that occurs in southeast Australia in summer, with low pressure to the northwest and high pressure off the coast of New South Wales.

It sometimes blows for three days, with thick **overcast** weather and **heavy rain** (**visibility** as low as a quarter of a **mile**).*See* **northeaster**.

2. A northeast **gale** on the east coast of North Island, New Zealand, that lasts for several hours and is accompanied by dark clouds and **heavy rain**.

See **black squall**.

black squall—A **squall** accompanied by dark clouds and generally by **heavy rain**.

black storm—1. Same as **karaburan**.

2. Generally, any **storm** (especially a **thunderstorm**) accompanied by dense, dark clouds.

black stratus—**Stratus** clouds with tops that are warmer than the surrounding cloud-free ground.

The name is derived from satellite imagery that is often displayed as black and white imagery with warmer temperatures portrayed in darker shades of gray. In this case, **stratus** can stand out as dark (warm) objects above a lighter (cooler) background.

black wind—Same as **reshabar**.

blackbody—A hypothetical body that cannot be excited to radiate by an external source of **electromagnetic radiation** of any **frequency**, direction, or state of **polarization** except in a negligibly small set of directions around that of the source radiation.

The traditional definition of a blackbody—as one that absorbs all the **radiation** incident on it—is inadequate unless to this definition is added the requirement that the body be large compared with the **wavelength** of the incident radiation. The concept of radiation incident on a body is from geometrical (or **ray**) optics, which is never strictly valid (because all bodies are finite) and may break down completely when the body is small compared with the wavelength. This was recognized by Planck, but by almost no one who followed him. Although no strict blackbody exists, some bodies are approximately black over a limited **range** of frequencies, directions, and polarization states of the exciting radiation. *See* **blackbody radiation**, **Planck's radiation law**, **emissivity**.

Planck, M., 1959: *The Theory of Heat Radiation*, p. 2.

blackbody radiation—**Radiation**, especially its spectral distribution, from an ideal **blackbody** emitter.

blackbody temperature—The **temperature** of a body that absorbs all **radiation** incident upon it.

blackout—*See* **radio blackout**.

Blackthorn winter—In England, cold dry winds in the Thames Valley during March and April.

blad—In Scotland, a **squall** accompanied by **heavy rain**.

blash—Same as **brash**.

Blaton's formula—*See* **trajectory**.

blaze—In Scotland, a sudden blast of dry wind.

blending height—A height scale for **turbulent flow** above an inhomogeneous surface, at which the influences of individual surface patches on vertical profiles or fluxes become horizontally blended.

Below this height scale it is not permissible to treat the structure of **turbulence** as horizontally uniform. Some authors distinguish between 1) the physical blending height, where local perturbations become negligible due to turbulent **mixing**, and 2) the numerical blending height in numerical models, above which the horizontal average of locally variable profiles (e.g., **wind speed**) corresponds to a **similarity** profile. Both of these height scales are related to the length **scale** of horizontal surface variations and, typically, the physical blending height is an order of magnitude higher than the numerical blending height. *See* **effective roughness length**, **flux aggregation**.

bliffert—Same as **bluffart**.

blind drainage—Same as **closed drainage**.

blind rollers—"Long high swells that have increased in height, almost to the breaking point, as they pass over shoals or run in shallow water."

Wiegel, R. L., 1953: Waves, Tides, Currents and Beachers. *Glossary of Terms and List of Standard Symbols*, Council on Wave Research–The Engineering Foundation.

blink—A brightening of the base of a **cloud layer** caused by the **reflection** of **light** by **snow** or **ice**. See **snow blink**, **ice blink**.

blinter—In Scotland, a **gust** of **wind**.

blip—In **radar**, an **echo** as it appears on a **display**.

blirty—In Scotland, **gusts** of **wind** and **rain**; changeable, uncertain weather.

blizzard—A **severe weather** condition characterized by high winds and reduced visibilities due to falling or **blowing snow**.

The U.S. National Weather Service specifies a **wind** of 30 knots (35 miles per hour) or greater, sufficient **snow** in the air to reduce **visibility** to less than 400 m (0.25 miles). Earlier definitions also included a condition of low temperatures, on the order of $-7°C$ (20°F) or lower, or $-12°C$ (10°F) or lower (severe blizzard). The name originated in the United States but it is also used in other countries. In the Antarctic the name is given to violent **autumnal** winds off the **ice cap**. In southeastern France, the cold north wind with snow is termed blizzard (*see also* **boulbie**). Similar storms in Russian Asia are the **buran** and **purga**. In popular usage in the United States and in England, the term is often used for any heavy snowstorm accompanied by strong winds.

block—Same as **blocking high**.

blocking—1. The obstructing, on a **large scale**, of the **normal** west-to-east progress of **migratory** cyclones and anticyclones.

A blocking situation is attended by pronounced **meridional flow** in the upper levels, often comprising one or more closed **anticyclonic** circulations at high latitudes and **cyclonic** circulations at low latitudes (cut-off highs and cut-off lows). This **anomalous** circulation pattern (the "block") typically remains nearly stationary or moves slowly westward, and persists for a week or more. Prolonged blocking in the Northern Hemisphere occurs most frequently in the spring over the eastern North Atlantic and eastern North Pacific regions. *Compare* **blocking high**.

2. The retardation of stable, low-level, forced upslope flow on the **windward side** of a mountain or mountain barrier; Smith (1979) asserts that "this windward-side slowing is due to the difficulty that the heavy [cold] surface air has in running upslope."

The stable flow is characterized by having a **Froude number** much less than 1. In cases where a gapless mountain range is long enough to be a complete barrier to the flow, very **stable air** may be totally blocked or "dammed" (*see* **damming**), and the near-surface flow may be diverted back down the slope. If the blocking condition persists for more than a significant portion of a day, a **barrier jet** can form parallel to the mountain range. "In ... flow near an isolated mountain or a ridge with ends or gaps, absolute blocking of flow is not possible. The layer of dense air may pile up slightly ahead of the mountain, but this can be relieved by airflow around the mountain or through gaps in the ridge" (*see also* **gap wind**). In stable flow (Froude number less than 1) over an isolated peak, the flow in the lower portions is partially blocked and diverts around the peak, whereas in the upper portion the flow can pass upward over the peak. The boundary between the lower and upper regions has been called the **dividing streamline**.

Namias, J., and P. F. Clapp, 1951: *Compendium of Meteorology*, 560–561.
James, I. N., 1994: *Introduction to Circulating Atmosphere*, 286–291.
Smith, R. B., 1979: The influence of mountains on the atmosphere. *Advances in Geophysics*, **21**, p. 132.

blocking action—Same as **blocking**.

blocking anticyclone—*See* **blocking high**.

blocking high—(Or blocking anticyclone.) Any **high** (or **anticyclone**) that remains nearly stationary or moves slowly compared to the west-to-east motion "upstream" from its location, so that it effectively "blocks" the movement of **migratory** cyclones across its latitudes.

A blocking high may comprise a smaller-scale effect than that associated with large-scale **blocking**.

blocking oscillator—An **oscillator** that is biased in such a manner that its oscillations are periodically interrupted.

blocking situation—*See* **blocking**.

blood rain—Rain of reddish color caused by foreign matter (e.g., pollen, red **dust**) picked up by raindrops during descent.

A dust-filled **subcloud layer** is required to yield this effect, and the dust particles must contain sufficient iron oxide to be red in color. *Compare* **mud rain**, **sulfur rain**.

blout—(Also called blouter, blowther, blowthir.) In Scotland, the sudden onset of a **storm**; sudden downpour of **rain** or **hail** with **wind**.

blowby—Unburned fuel that escapes around the side of a piston in an internal combustion engine. *See also* **crankcase ventilation**.

blowing dust—**Dust** picked up locally from the surface of the earth and blown about in clouds or sheets.

It is classed as a **lithometeor** and is encoded BLDU as an **obstruction to vision** in an **aviation weather observation (METAR)**. (Encoded as BD in SAO observation format.) Blowing dust may completely obscure the sky; in its extreme form it is called a **duststorm**. A layer of **stable air** aloft tends to stop the vertical **transport** of dust by **eddies**. There is then a sharply defined upper limit to the dust layer.

blowing sand—Sand **particles** picked up from the surface of the earth by the **wind**, reducing the **horizontal visibility** to less than 11 km (about 7 statute miles).

This **lithometeor** is encoded BN as an **obstruction to vision** in a surface **aviation weather observation** and as BLSA as an obstruction to vision in a **METAR** or **SPECI** observation. In its extreme form, blowing sand constitutes a **sandstorm**.

blowing snow—**Snow** lifted from the surface of the earth by the **wind** to a height of 2 m (6 ft) or more above the surface (higher than **drifting snow**), and blown about in such quantities that **horizontal visibility** is reduced to less than 11 km (about 7 statute miles).

As an **obstruction to vision**, it is encoded BS in a surface **aviation weather observation** and as BLSN as an obstruction to vision in a **METAR** or **SPECI** observation. Blowing snow can be falling snow or snow that already accumulated but is picked up and blown about by strong winds. It is one of the classic requirements for a **blizzard**.

blowing spray—Water **droplets** displaced by the **wind** from a body of water, generally from the crests of waves, and carried up into the air in such quantities that they reduce the **horizontal visibility** to less than 11 km (about 7 statute miles).

It is encoded as BY as an **obstruction to vision** in surface **aviation weather observation** and as BLPY as an obstruction to vision in a **METAR** or **SPECI** observation.

bloxam—(Rare.) A term for the **smoothing** process used by J. C. Bloxam (1860).

He described it as follows: "The method is founded on the fact that, in meteorology, the mean of 10 or 11 consecutive days gives more correctly the **normal** constant value for the middle day of the series, than the middle day itself gives; and then the value for every day in the year having been calculated upon this principle, the whole series of amended values can in turn be subjected to the same process; and this process may be reiterated until the daily values are brought to sufficiently regular ascending and descending lines." *See* **consecutive mean**.

    Bloxam, J. C., 1860: *On the Meteorology of Newport in the Isle of Wight*, 2nd ed.

blue band—*See* **ice band**.

blue flash—Similar in most ways to the **green flash**, the wavelength-dependent **refraction** that gives a red rim on the low sun's bottom would be expected to give a blue rim on its top.

Often **scattering** diminishes the relative **spectral radiance** of the shortest wavelengths sufficiently that green is perceived, but not always.

blue ice—Pure **ice** in the form of large single crystals.

It is blue owing to the **scattering** of **light** by the ice molecules; the purer the ice, the deeper the blue.

blue-ice area—Areas where surface **ablation** has exposed **blue ice**.

These are sites, usually on large **ice** sheets, where **ice flow** has concentrated meteorites that have fallen throughout the **catchment area** of the particular blue-ice area.

blue jets—Weakly **luminous** upward propagating discharges, blue in color, emanating from the tops of thunderstorms.

Following their emergence from the top of the **thundercloud**, they typically propagate upward in narrow cones of about 15° full width at vertical speeds of roughly 100 km s^{-1} (Mach 300), **fanning** out and disappearing at heights of about 40–50 km. Their intensities are on the order of 800 kR near the base, decreasing to about 10 kR near the upper terminus. These correspond to an estimated optical **energy** of about 4 kJ, a total energy of about 30 MJ, and an energy density on the order of a few millijoules per cubic meter. Blue jets are not aligned with the local **magnetic field**.

blue moon—1. Those rare occasions when the moon takes on a markedly blue cast.

Usually, when the moon (or sun) is seen low in the sky or even high in a polluted one, it is yellowish or reddish. This is a consequence of the greater **extinction** of short **wavelength** radiation by small **particles** and molecules. Yet, there are particle sizes for which extinction is greater for long wavelengths, and the moon seen through a sky populated by such particles is bluish. It is rare that a large population of the particles of the appropriate size (and only that size) are produced, so the blue moon is very rare: 1883 (Krakatoa), 1927 (a late **monsoon** in India), 1951 (forest fires in Alberta, Canada). Because of the rarity of blue moons, the phrase, "once in a blue moon," has been used metaphorically since the midnineteenth century to denote a real, but rare event.

2. In recent times, often used incorrectly to mean the occurrence of a second full moon within a calendar month.

This strange new use of the term, popularized by a board game in the mid-1980s, arose from a mistake made by the author of a magazine article written in the 1940s. Since two full moons in one month occur fairly regularly, it should not be considered a rare event.

blue noise—Erroneous spectral **energy** added to the high-frequency (short **wavelength**) portion of the **spectrum**, such as in a **Fourier analysis** of a turbulent **time series**.

blue sun—*See* **blue moon**.

bluffart—In Scotland, a **gust** or **squall** accompanied by a sudden but short fall of **snow**.

blunk—In England, a fit of squally, tempestuous weather; a sudden **squall**.

bochorno—In Spain, a sultry **wind** (or sultry weather in general) in the Ebro valley, possibly a form of **sirocco**.

bodily tides—The deformation of the solid earth directly due to **tide**-generating forces and indirectly by the tidal load of the ocean.

The presence of bodily tides causes a change in the total tide-generating potential for the World Ocean.

body waves—**Waves** that can exist within a solid or fluid body without boundary surfaces or interfaces.

Propagation of information and **energy** in such waves is generally possible in all radial directions, but may for some types of waves be influenced by forces of, for example, **gravitation** and rotation. In the ocean, the most common body waves are **sound waves**, **light** waves penetrating through the ocean surface, and internal **gravitational waves**.

bog—1. (Also called moor or quagmire.) Area of **waterlogged**, spongy ground, generally consisting of acidic decaying vegetation that may develop into peat.

2. The vegetation characteristic of this **environment**, including various masses, sedges, and heaths.

bogus observation—A fictitious **observation** inserted into an **analysis** that is believed to improve the analysis.

bogus vortex—A representation of a tropical **vortex** based on a blend of observed surface winds and typical **tropical storm** wind profiles.

Because of limited observations and coarse **model** resolution for the production of background fields, tropical cyclones are often poorly resolved in operational analyses. To alleviate this problem, a bogus vortex can be used in the **analysis** of tropical cyclones both in global **data assimilation** systems and for initial conditions of limited area, high-resolution **hurricane** models.

bohorok—In northern Sumatra, a warm, dry local **wind** blowing on the plains of Deli, Langkat, and Lerdang.

It is caused by the east **monsoon** blowing across the Barisan Mountains and descending on the **leeward** side as a **foehn**. The bohorok occurs in May and is prevalent through September; it is feared because of its effect on the tobacco plantations on the plains of Deli. On the **windward side** of the mountains, "bohorok days" have more **rain** than usual.

boil—1. An agitated zone of water, especially at the surface of a river, spring, or the sea, caused by upward turbulent movement.

2. An upward flow of water in sand caused by **pressure** imbalances, as when overburden is removed by excavation or when water rises in an adjacent **stream** channel.

boiling point—The **temperature** at which the **equilibrium vapor pressure** between a liquid and its **vapor** is equal to the external **pressure** on the liquid.

Physically, boiling (or **ebullition**) cannot begin in a liquid until the **temperature** is raised to such a point that incipient **bubbles** forming within the liquid can grow rather than collapse. But for a bubble to grow, its internal **vapor pressure** must exceed the **hydrostatic pressure** exerted on the bubble **interface**. For liquids heated in containers that are open and fairly shallow, this hydrostatic pressure is essentially the same as the external **atmospheric pressure**, so ebullition begins when the equilibrium vapor pressure equals atmospheric pressure. In liquids perfectly free from foreign **particles** and contained in a vessel with perfectly smooth walls, boiling will not begin even at the above-described temperature, for boiling resembles **condensation** in that "nuclei" must exist to initiate the process. When a very pure liquid sample has been heated above its nominal boiling point it is said to have been superheated, a state that is very similar to the state of **supersaturation** in which a vapor may exist in a nucleus-free **environment**. Because of the **normal** decrease of **barometric pressure** with height, the nominal boiling point of water decreases 3.0°–3.5°C for each kilometer increase of **altitude** (see **hypsometer**). The boiling point is a **colligative property** of a solution; with an increase in dissolved matter, there occurs a raising of the boiling point. The boiling point of pure water at **standard pressure** is equal to 100°C (212°F) and is a **fiducial point** for **thermometer** calibration. *Compare* **ice point**.

Bolling–Allerod—A period of warm **climate** beginning abruptly approximately 14 700 years ago, following the end of the **Pleistocene**, and extending to approximately 12 700 years ago.

   This warm period ended with a return to cold conditions during the **Younger Dryas**. Traditionally, this period is divided into the Bolling (warm), Older Dryas (cold), and Allerod (warm) intervals, but recent, more detailed climatic records indicate that the entire Bolling–Allerod period was generally warm with several abrupt coolings.

   Broeker, W. S., 1992: Defining the boundaries of the late-glacial isotope episodes. *Quat. Res.*, **38, No.1**, 135–138.

bologram—The record obtained from a **bolometer**.

bolometer—(Also called actinic balance.) An instrument that measures the **intensity** of **radiant energy** by employing a thermally sensitive electrical resistor; a type of **actinometer**.

   In meteorological applications, two identical, blackened, thermally sensitive electrical resistors are used in a Wheatstone bridge circuit. Radiation is allowed to fall on one of the elements, causing a change in its **resistance**. The change is a measure of the intensity of the **radiation**.

bolon—*See* **aloegoe**.

Boltzmann's constant—(Sometimes called gas constant per molecule, Boltzmann's universal conversion factor.) The ratio of the **universal gas constant** to **Avogadro's number**; equal to 1.3804×10^{23} W K^{-1}.

bomb—An **extratropical** surface **cyclone** with a **central pressure** that falls on the average at least 1 mb h^{-1} for 24 hours.

   This predominantly maritime, cold **season** event is usually found approximately 750 km **downstream** from a mobile 500-mb **trough**, within or poleward of the maximum westerlies, and within or ahead of the planetary-scale troughs.

   Sanders, F., and J. R. Gyakum, 1980: Synoptic–dynamic climatology of the "bomb". *Mon. Wea. Rev.*, **108**, 1589–1606.

bond energy—The **energy** required to break a given chemical bond in a molecule, corresponding to the energy to separate the fragments to an infinite distance.

   For example, the oxygen–oxygen bond strength in **ozone** is defined as the difference in energy between the ozone molecule and O and O_2 fragments formed from the bond rupture.

book-end vortices—Mesoscale vortices observed at the ends of a line segment of convective cells, usually **cyclonic** on the northern end of the system and **anticyclonic** on the southern end, for an **environment** of westerly **vertical wind shear** (in the Northern Hemisphere).

   The vortices are generally strongest between 2 and 4 km above ground level, but may extend from near the surface to about 8 km above ground level. They have been observed at scales between 10 and 200 km, and often have lifetimes of several hours. In extreme cases, the larger **cyclonic** vortices may become balanced with the **Coriolis force** and last for several days. *See also* **bow echo**.

boorga—Same as **burga**.

bora—A **fall wind** with a **source** so cold that, when the air reaches the lowlands or coast, the dynamic warming is insufficient to raise the **air temperature** to the **normal** level for the region; hence it appears as a cold **wind**.

   The terms **borino** and **boraccia** denote a weak bora and a strong bora, respectively. The term was originally applied (along with **karstbora**) to the cold northeast wind on the Dalmatian coast of Croatia and Bosnia in winter when cold air from Russia crosses the mountains and descends to the relatively warm coast of the Adriatic. According to Smith (1987), the bora "has often been considered the prototype fall wind," although recent studies have revealed that some boras have **downslope windstorm** or **hydraulic jump** structure. The event often lasts a day or less, although extended events occur with enough **frequency** that "the longevity of the bora is one of its primary characteristics. A duration of four to six days is not unusual." It is very stormy and squally, the squalls sometimes reaching 50 m s^{-1} or more. F. Defant (1951) distinguishes between **cyclonic bora** (**bora scura**) with clouds and **rain**, covering the whole Adriatic and occurring with a **depression** over southern Adria, and the dry **anticyclonic bora**, with a powerful **anticyclone** over central Europe extending over Dalmatia; the latter is very violent over the land but extends only a short distance out to sea. A local bora also occurs on the east coast of the Adriatic with a **cold anticyclone** over the Balkans. The term bora is now applied to similar winds in other parts of the world. Well-known examples occur at Novorossiisk on the northern shore of the Black Sea, and in Novaya Zemlya (islands in the Russian Arctic). A squally **katabatic wind** at Alme Dagh in the Gulf of Iskenderon (eastern Mediterranean Sea) is termed rageas (also ragut, ghaziyah). The Bulgarian term is buria. In some mountainous regions of the world bora has been further generalized to represent any large mesoscale or **synoptic-scale** downslope flow of cold air, including post-arctic-frontal fall winds and cold-air downslope windstorms, which may have a hydraulic jump-like character and structure. In the case of downslope windstorms, some authors have used bora for a cold-advection flow (or one that results in cooling to the immediate lee of the mountain barrier), whereas **chinook** or **foehn** refer to a warm- or neutral-advection wind (or one that results in warming or no **temperature** change **leeward** of the barrier). Those who have attempted to classify downslope windstorms, however, have found that many cases do not fall neatly into one category or the other. *See* **Boreas, borasca**.

> Smith, R. B., 1987: Aerial observations of the Yugoslavian bora. *J. Atmos. Sci.*, **44**, 269–297.
> Defant, F., 1951: *Compendium of Meteorology*, 669–670.

bora fog—A dense "fog" caused when the **bora** lifts a spray of small drops from the surface of the sea.

It is well developed on the coast of Norway when a bora-like **fall wind** descends from the Scandinavian mountains as a sudden **storm** from the east.

bora scura—A **bora** on the Dalmatian coast of Croatia and Bosnia, with clouds and **precipitation**, the strong **pressure gradient** across the coast resulting from a **cyclone** (low pressure) system to the west.

boraccia—A strong **bora**.

borasca—(Or borasco; also called bourrasque.) Literally, "little **bora**;" a **thunderstorm** or violent **squall**, especially in the Mediterranean.

bordelais—In France, a west **wind** (i.e., blowing from Bordeaux) in Quercy (southwest France). It occurs in all **seasons**, is mild and generally rainy, but often brings violent thunderstorms in summer.

border spring—Spring formed against a subsurface barrier boundary such as a confining layer, or at a fault between a raised bedrock block and a depressed block covered with a thick **aquifer**.

bore—1. Same as **hydraulic jump**.

2. A **tidal wave** that propagates as a **solitary wave** with a steep leading edge up certain rivers.

Bore formation is favored in wedge-shaped **shoaling** estuaries at times of spring tides. Other local names include eagre (River Trent, England), pororoca (Amazon, Brazil), and mascaret (Seine, France).

Boreal—A period of more **continental climate** 5000 to 7000 years ago identified as part of the Blytt–Sernander sequence of inferred climates in northern Europe.

> Sernander, R., 1908: On the evidence of postglacial changes of climate furnished by the peat-mosses of northern Europe. *Geol. Fören. Förh.*, **30**, 456–478.

boreal forest—The forested region that adjoins the **tundra** along the **arctic tree line**.

It has two main divisions: its northern portion is a belt of **taiga** or **boreal woodland**; its southern portion is a belt of true forest, mainly conifers but with some hardwoods. On its southern boundary the boreal forest passes into "mixed forest" or "parkland," **prairie**, or **steppe**, depending on the **rainfall**.

> Hare, F. K., 1951: *Compendium of Meteorology*, 953, 957–958.

boreal region—Same as **boreal zone**.

boreal woodland—The **taiga** portion of the **boreal forest**.

boreal zone—1. Defined by W. Köppen (1931) as the zone having a definite winter with **snow**, and a short summer, generally hot. It includes a large part of North America between the **Arctic Zone** and about 40°N, extending to 35°N in the interior. In Central Europe and in Asia the boreal zone extends southward from the **tundra** to 40°–50°N.

2. A biogeographical zone or region characterized by a northern type of fauna or flora.

The term **boreal region** is used mainly by American biologists, and includes the area between the mean summer **isotherm** of 18°C or 64.4°F (roughly 45°N latitude) and the **Arctic Zone**.

> Köppen, W. P., 1931: *Grundriss der Klimakunde*, 2nd ed., Berlin: Walter de Gruyter.

Boreas—The ancient Greek name for the north **wind** (now also **borras**).

Being cold and stormy, it is represented on the **Tower of the Winds** in Athens by a warmly clad old man carrying a conch shell (probably to represent the howling of the wind). The term may originally have meant "wind from the mountains," thus the present use of **bora**.

borino—A weak **bora**.

bornan—In Switzerland, a **breeze** blowing from the valley of the Drance over the middle of Lake Geneva.

borras—Greek for "north," and by extension, "north **wind**," from the ancient Greek **Boreas**.

bottle post—Same as **drift bottle**.

bottle thermometer—A **thermoelectric thermometer** used for measuring **air temperature**.

The name is derived from the fact that the reference **thermocouple** is placed in an insulated bottle.

Bottlinger's rings—A **halo** in the form of one or more rings or **brightness** plateaus immediately surrounding the **subsun**.

The **variation** in brightness between the central subsun and the first ring is explained by the way the reflecting crystals oscillate as they fall.

bottom current—The movement of water along the bottom of reservoirs or rivers.

bottom friction—The **momentum transfer** at the lower boundary of the ocean to the solid earth by **friction** at the ocean bottom.

For the **friction velocity**, a value of $1/30$ of the **geostrophic** velocity at the upper boundary of the bottom **boundary layer** is assumed, and the **drag coefficient** is approximately 10^{-5}.

bottom ice—Same as **anchor ice**.

bottom temperature— The **temperature** of the ocean at the point where the water intersects the bottom.

bottom water—1. The **water mass** at the deepest part of the water column.

It is the densest water that is permitted to occupy that position by the regional **topography**. In the case of a **basin**, bottom water may be formed locally, or it may represent the densest water that has existed at **sill depth** in the recent past.

2. **Water masses** found at the bottom of ocean basins.

The most important bottom waters of the World Ocean are **Antarctic Bottom Water** and **Arctic Bottom Water**. Baffin Bay Bottom Water has a **salinity** of 34.49 and a **temperature** of $-0.4°C$ and is found in Baffin Bay below a depth of 1800 m; its low oxygen content of 3.6 ml l^{-1} indicates slow water renewal. Japan Sea Bottom Water has a salinity of 34.1 and a temperature of 0.04°C; it is formed by winter convection in the northern Japan Sea and occupies the Japan Sea basins at depths below 2000 m.

Bouguer–Lambert law—See **Bouguer's law**.

Bouguer's halo—*See* **Ulloa's ring**.

Bouguer's law—(Or Beer's law, Bouguer–Lambert law; sometimes called Lambert's law of absorption.) Attenuation of a **beam** of **light** by an **optically homogeneous** (transparent) medium.

First stated by Pierre Bouguer in his Essay on the Gradation of Light (1729), Bouguer also recognized that this law is independent of the **attenuation** mechanism. Although exponential attenuation with distance is attributed to Bouguer, Lambert, or Beer, the historically correct term is Bouguer's law, which is also valid for turbid media to the extent that **multiple scattering** is negligible. By a slight distortion of history we might say that Beer's law is an extension of Bouguer's law to solutions of fixed thickness but variable concentration of the absorbing solute. Mathematically, the law is written:

$$I = I_0\exp(-\tau),$$

where I_0 is the incident **radiance**, I is the transmitted radiance, and τ is the pathlength.

> Middleton, W. E. K., 1961: *Pierre Bouguer's Optical Treatise on the Gradation of Light*, Translation.
> Pfeiffer, H. G., and H. A. Liebhafsky, 1951: *J. Chem. Educ.*, **28**, 123–125.

Malinin, D. R., and J. H. Yoe, 1961: *J. Chem. Educ.*, **38**, 129–131.

boulbie—In France, a violent north **wind** in the Ariège valley south of Toulouse, especially in December and January.

It is strong enough to uproot chestnut trees; it is cold and dry, but may pick up **snow** from the ground and form drifts. *Compare* **blizzard**.

boundary conditions—A set of mathematical conditions to be satisfied, in the solution of a differential equation, at the edges or physical boundaries (including fluid boundaries) of the region in which the solution is sought.

The nature of these conditions is usually determined by the physical nature of the problem, and is a necessary part of the problem's complete formulation. Common boundary conditions for the **atmosphere** are that the **velocity** component normal to the earth's surface vanish, and that the **individual derivative** of **pressure** vanish at the upper surface. The term is also used in the context of the time evolution of an "open" **dynamical system** that interacts with other "external" systems. The state of the external systems must be specified as a boundary condition to infer the evolution of the dynamical system under consideration. For example, the evolution of the earth's atmospheric state requires the **specification** of **sea surface temperature** as a boundary condition. *See* **kinematic boundary condition, dynamic boundary condition, boundary-value problem, initial condition**.

boundary currents—Ocean currents with dynamics determined by the presence of a coastline.

They fall into two categories: 1) western boundary currents, which are narrow, deep-reaching, and fast-flowing currents, not unlike **jet streams**, associated with current instability and **eddy** shedding; and 2) eastern boundary currents, which are shallow, cover a wider region, are of moderate strength, and are often associated with **coastal upwelling** and a subsurface **countercurrent** along the **continental slope**. Both are integral parts of the **circulation** in oceanic **gyres**. The rotation of the earth causes an accumulation of **energy** on the western side, which has to be dissipated in boundary currents; this gives the western boundary currents typical widths of 100 km and typical speeds of 2 m s^{-1} and causes them to shed eddies frequently to increase the **dissipation** of energy. No similar requirement of energy dissipation exists on the eastern side, so eastern boundary currents can be broad and slow. Their special character as a boundary current results from coastal upwelling, which brings the **thermocline** to the surface and as a result produces a **temperature** front and an associated **geostrophic** maximum in the current speed, known as the **coastal jet**. Because of the **upwelling**, eastern boundary currents are atmospheric **heat** sinks. Western boundary currents are atmospheric heat sinks if they move cold water toward the **equator**, which occurs in the subpolar gyres, and atmospheric heat sources where they move tropical water into temperate regions, as in the subtropical gyres.

boundary layer—1. The layer of fluid near a boundary that is affected by **friction** against that boundary surface, and possibly by **transport** of **heat** and other variables across that surface.

In meteorology, this is the **atmospheric boundary layer**.

2. In a physical or mathematical system, a region over which some property or term in the equations varies rapidly, that is, over its full **range**; conversely, a region outside of which certain terms may be neglected.

boundary layer pumping—*See* **Ekman pumping**.

boundary layer radar—A type of **wind profiler** specially designed to study the lower part of the **troposphere**. Because clear-air reflectivities in the **boundary layer** are usually orders of magnitude higher than in the upper troposphere, these profilers can be smaller, lower powered, and less expensive than those profilers designed to cover higher regions of the **atmosphere** (see, e.g.,, **MST radars**). Boundary layer radars are generally characterized by short **pulse** lengths (less than 100 m) and the ability to make measurements starting at 100–200 m above the **radar** and extending to at least 2–4 km under typical atmospheric conditions. To satisfy these requirements, boundary layer radars usually operate in the **UHF** radar **band**.

boundary layer rolls—*See* **horizontal convective rolls**.

boundary layer separation—A condition that occurs at sufficiently high **Reynolds numbers** in which the surface **streamlines** break away from the surface.

Separation is due to the presence of a solid boundary, at which the no-slip condition—that is, the **velocity** of the fluid **particles** in contact with the surface is the velocity of that surface—is satisfied and **vorticity** is generated. Separation of a steady **boundary layer** at a plane or rounded rigid wall occurs whenever the velocity of the fluid just outside of the boundary layer decreases in the mean flow direction sufficiently rapidly and by a sufficient amount. This can be accomplished by the imposition of an opposing **pressure gradient** in the direction of flow.

Batchelor, G. K., 1967: *Fluid Dynamics*, Cambridge University Press, p. 325.

boundary mixing—Mixing occurring on sloping **topography** on the ocean margins or on seamounts primarily as the result of breaking **internal waves**.

Boundary mixing is thought to play a major role in the vertical **transport** of **heat**.

boundary of saturation—The surface between the saturated and unsaturated zones in a soil.

The boundary of saturation is the top of the **capillary fringe**.

boundary-value problem—A physical problem completely specified by a differential equation in an unknown, valid in certain information (**boundary conditions**) about the unknown given on the boundaries of that region.

The information required to determine the solution depends completely and uniquely on the particular problem. A great variety of meteorological problems are formulated as boundary-value problems. *See also* **initial-value problem**.

bounded weak echo region—(Abbreviated BWER.) A nearly vertical channel of weak radar echo, surrounded on the sides and top by significantly stronger **echo**.

The BWER, sometimes called a **vault**, is related to the strong **updraft** in a severe **convective storm** that carries newly formed **hydrometeors** to high levels before they can grow to radar-detectable sizes. BWERs are typically found at midlevels of convective storms, 3–10 km above the ground, and are a few kilometers in horizontal diameter. *See also* **weak echo region**.

Bourdon tube—A closed curved tube of elliptical cross section used in some thermometers and barometers.

The Bourdon-tube **thermometer** consists of a Bourdon tube completely filled with liquid. The expansion of the liquid due to a **temperature** change causes an increase in the radius of curvature of the tube. The curvature may then be measured by the travel of the tip of the tube. The Bourdon-tube **barometer** consists of an evacuated Bourdon tube and operates in a similar manner. In both cases the curvature is a measure of the difference between the **pressure** inside the tube and that outside.

bourrasque—Same as **borasca**.

Boussinesq approximation—An approximation to the dynamical **equations of motion** whereby **density** is assumed to be constant except in the **buoyancy** term, $-g\rho'$, of the **vertical velocity** equation, where g is the gravitational **acceleration** and ρ' is the density deviation.

The approximation is reasonable if the vertical extent of the dynamics being considered is much smaller than the density scale height—the height over which the density changes by a factor e. It is generally applicable to most oceanographic circumstances. If a system does not satisfy the Boussinesq approximation it is said to be non-Boussinesq.

Boussinesq equation—The general flow equation for two-dimensional **unconfined aquifer** flow.

Boussinesq number—The ratio of the **eddy flux** of some quantity to the molecular **flux** of that quantity.

bow echo—A bow-shaped line of convective cells that is often associated with swaths of damaging **straight-line winds** and small **tornadoes**.

 Key structural features include an intense **rear-inflow jet** impinging on the core of the bow, with **book-end** or **line-end vortices** on both sides of the rear-inflow jet, behind the ends of the bowed convective segment. Bow echoes have been observed with scales between 20 and 200 km, and often have lifetimes between 3 and 6 h. At early stages in their evolution, both **cyclonic** and **anticyclonic** book-end vortices tend to be of similar strength, but later in the evolution, the northern cyclonic vortex often dominates, giving the convective system a comma-shaped appearance.

Bowen ratio—The ratio of sensible to **latent heat** fluxes from the earth's surface up into the air.

This is equal to the **psychrometric constant** times the ratio of kinematic **temperature** flux to kinematic **moisture flux**. It can be estimated as the psychrometric constant times the ratio of **potential temperature** difference to **mixing ratio** difference, where the differences are measured between the same two heights in the atmospheric **surface layer**. Typical values are 5 over semiarid regions, 0.5 over grasslands and forests, 0.2 over irrigated orchards or grass, 0.1 over the sea, and negative in some advective situations such as over oases where **sensible heat** flux can be downward while latent heat flux is upward.

box models—Mathematical models that simulate the chemistry at a given point in space as a function of time.

Advection of chemical species in and out of the box is not considered. These models are used particularly to study the evolution of homogeneous **air masses** and for Lagrangian airshed studies.

Boyden index—A measure of the mean thermodynamic **stability** in a layer beneath 700 mb.

The Boyden index is given by $I - Z - T - 200$, where I is the Boyden index, Z is the 1000–700-mb **thickness** in dam, and T is the 700-mb **temperature** in °C.

Boyden, C. J., 1963: *Meteor. Mag.*, **92**, 198–210.

Boyle–Mariotte law—Same as **Boyle's law**.

Boyle's law—(Also called Boyle–Mariotte law, Mariotte's law.) The empirical generalization that for many so-called perfect gases, the product of **pressure** p and volume V is constant in an **isothermal process**:

$$pV = F(T),$$

where the function F of the **temperature** T cannot be specified without reference to other laws (e.g., **Charles–Gay–Lussac law**).

Boys camera—A camera used for the observation of **lightning flashes**.

The early model of this camera consists of a fixed film plate and two lenses that revolve at the opposite ends of a diameter of a circle. The **velocity** and duration of a **lightning stroke** can be computed from a comparison of the two photographs and a knowledge of the rate of rotation of the lenses. A later model consists of a fixed lens and a rotating film drum. This construction allows greater ease of interpretation.

BPI pan—A circular **evaporation pan**, six feet in diameter and two feet deep, made of unpainted galvanized iron.

The pan is buried in the ground so that about two inches of the rim extend above the surrounding ground, and the **water level** is maintained at about ground level. This installation reduces the **temperature** variations in the pan and causes its **pan coefficient** to be nearer unity than that of exposed pans. The average pan coefficient is reported to be about 0.9. The initials BPI stand for Bureau of Plant Industry (U.S. Department of Agriculture), which first introduced this instrument.

brackish water—Water containing total dissolved salts in the **range** 1000–10 000 mg^{-1}.

Bragg scattering—In **radar meteorology**, **scattering** from atmospheric **refractive index** fluctuations, components of which are resonant with the radar **wavelength**.

For **backscattering**, these components have a **scale** size of one-half the radar wavelength. *See* **clear-air echo**.

brash—1. An accumulation of **floating ice** fragments less than 2 m across, formed by breakage of other **ice** forms.

2. (Rare.) In England, a colloquial term for a sudden **gust** of **wind** or the sudden onset of a **storm**.

brave west winds—A nautical term for the strong and rather persistent westerly winds over the oceans in temperate latitudes.

They occur between latitudes 40° and 65° in the Northern Hemisphere and 35° to 65° in the Southern Hemisphere, where they are more regular and are strongest between 40° and 50°S (**roaring forties**). They are associated with the strong **pressure gradient** on the equatorial side of the frequent **depressions** passing eastward in subpolar temperate latitudes; hence they fluctuate mainly between southwest and northwest.

Brazil Current—The western **boundary current** of the South Atlantic subtropical **gyre**.

Fed by the **South Equatorial Current**, it flows as a narrow, swift current from 10°S southward along the South American coast. The Brazil Current is one of the weaker western boundary currents with a total **transport** of less than 30 Sv (30 × 10^6 m^{3}s^{-1}), half of this occurring over the shelf. Current speed is below or near 1 m s^{-1}. The current separates from the coast somewhere between 33° and 38°S, forming a front with the northward flowing **Malvinas (Falkland) Current** and continuing southward on the eastern side of the front. The separation point is more northward in summer (December–February). The southernmost extent of the Brazil Current varies between 38° and 46°S on a two-month timescale associated with **eddy** shedding. The current continues eastward as the **South Atlantic Current**.

BRDF—Abbreviation for **bidirectional reflectance distribution function**.

breadth-first search—Search strategy in which the highest layer of a **decision tree** is searched completely before proceeding to the next layer.

In this manner, no viable solution is omitted and the optimal solution is found. This strategy is often not feasible when the **search space** is large. *See also* **depth-first search**.

break—1. A sudden change in the weather; usually applied to the end of an extended period of unusually hot, cold, wet, or dry weather.

2. A hole or gap in a layer of clouds (*see* **breaks in overcast**).

3. *See* **windbreak**.

break line—*See* **tropopause break line**.

breakdown—The process by which electrically stressed air is transformed from an insulator to a conductor.

Breakdown involves the **acceleration** of electrons to **ionization potential** in the **electric field** imposed by the **thundercloud**, and the subsequent creation of new electrons that avalanche and expand the **scale** or enlarge the volume of enhanced **conductivity**. Breakdown precedes the development of **lightning**.

breakdown field—The **electric field** necessary to produce **breakdown**.

breakdown potential—Same as **dielectric strength**.

breaker—A sea **surface wave** that has become too steep to be stable and that breaks on the shore or in the open ocean.

Breakers can be classified into four categories: 1) A spilling breaker breaks gradually over a considerable distance; 2) plunging breakers tend to curl over and break with a crash; 3) surging breakers peak up, but then instead of spilling or plunging, they surge up on the beach face; 4) collapsing breakers break in the middle or near the bottom of the wave rather than at the top.

breaker depth—The depth at which a **wave** becomes unstable and breaks.

breaking-drop theory—A theory of **thunderstorm charge separation** based upon the suggested occurrence of the **Lenard effect** in thunderclouds, that is, the separation of **electric charge** due to the breakup of water drops.

This theory, advanced by Sir George C. Simpson (1927), was initially intended to account for a bipolar charge distribution within a **thundercloud** having the main positive charge center near the base of the **cloud** and the main negative charge center higher up. Simpson's theory, however, does not explain this phenomenon well because temperatures are below the **freezing point** in this region of the cloud. Evidence does remain to support a weak positive charge center that lies slightly below the lower main negative charge concentration in many, if not all, thunderclouds where the **temperature** is above the freezing point. Hence, the breaking-drop theory is best ascribed to this localized secondary positive charge center.

Simpson, G. C., 1927: The mechanism of a thunderstorm. *Proc. Roy. Soc. London A*, **114**, 376–401.

breaking-off process—Same as **cutting-off process**.

breaks in overcast—In U.S. weather observing practice, a condition wherein the **cloud cover** is more than 0.9 but less than 1.0 (to the nearest tenth).

This would appear in an **aviation weather observation** as an **overcast** sky cover, with an encoded "remark" (currently, "BINOVC") to indicate any breaks.

breakthrough curve—A plot of relative concentration of a given substance versus time where relative concentration is defined as the ratio of the actual concentration to the source concentration.

breakup—Breakup of **ice** covering a body of water at a site; depends on ice thickness, strength, **flow velocity**, and river geometry.

"Breakup connotes the end of winter to a resident of the north." (*Glossary of Arctic and Subarctic Terms*, Arctic, Desert, Tropic Information Center Pub. A–105, 1955). *See* **ice breakup**.

breakup season—The ensemble of phenomena associated with the annual disappearance of an **ice cover** on inland and coastal waters due to meteorologic (temperature, wind) and hydrologic (waves, currents, tides) factors.

breakwater—A structure, usually rock or concrete, protecting a shore area, harbor, or beach from **wave** action.

bred modes—See **bred vectors**.

bred vectors— (Also called bred modes.) Perturbations generated by the **breeding method**.

In complex physical systems the bred vectors depend on the **perturbation** amplitude, which is the only free **parameter** in the breeding technique. For example, linearly fast-growing perturbations such as convective instabilities do not amplify after reaching a certain **amplitude** since they become nonlinearly saturated (i.e., other processes prevent them from growing any further). The bred vectors are primarily used as initial perturbations in ensemble forecasting and for studying the instabilities of the **atmosphere**.

breeding method—A method of generating balanced, fast-growing perturbations to a **nonlinear** model **trajectory** for a given finite **perturbation** amplitude.

The technique consists of generating a control run of a nonlinear model for a short period of time (e.g., 6 h), then perturbing the atmospheric initial conditions and running the same **model** again for the same period of time (perturbed run). The difference between the two model runs at the final time is adjusted to the **amplitude** of the initial perturbation and is added to the new control **initial condition**. The method is then repeated. After a few days of iteration, the difference between the control and perturbed model runs represents a **sample** of fast-growing nonlinear perturbations. Multiple breeding cycles, started with different arbitrary initial perturbations, provide a broader sample of fast-growing perturbations. The technique can also be applied to a series of atmospheric **analysis** fields, where the control forecast always starts from the latest available analysis. See **bred vectors**.

breeze—1. In general, a light **wind**.

In the **Beaufort wind scale**, this is a **wind** between 4 and 10 kt (4 and 12 mph).

2. In the **Beaufort wind scale** (Beaufort **wind force** numbers 2–6), a **wind speed** ranging from 4 to 27 kt (4 to 31 mph) and categorized as follows: **light breeze**, 4–6 kt; **gentle breeze**, 7–10 kt; **moderate breeze**, 11–16 kt; **fresh breeze**, 17–21 kt; and **strong breeze**, 22–27 kt.

bremsstrahlung—Electromagnetic **radiation** resulting from the **acceleration** or deceleration of subatomic **particles** such as electrons when they interact with matter; often applied to the continuous part of the **x-ray** spectrum produced when high-energy electrons bombard a material.

Bremsstrahlung is German for "braking radiation." In cosmic ray shower production, bremsstrahlung gives rise to **emission** of **gamma rays** as electrons encounter atmospheric nuclei.

brenner—In England, a sharp **gust** of **wind** over the water.

breva—(Or breva di Como.) A strong **valley wind** blowing by day up Lake Como (in northern Italy) toward the head of the lake.

A light wind blowing off the neighboring Lake Lugano is called breva del lagio.

Brewster point—One of the three commonly detectable points of zero **polarization** of **skylight**, neutral points, along the vertical circle through the sun; the other two are the **Arago point** and **Babinet point**.

This **neutral point**, discovered by Brewster in 1840, is located about 15°–20° directly below the sun; hence it is difficult to observe because of the **glare** of the sun.

brickfielder—(Obsolete.) A term used in the early days of Sydney, Australia, when there were brickfields (presumably, areas where clay was dug and bricks were made) on the southern side of town. It refers to a **southerly burster** that raised a lot of red **dust** that covered the town.

bridled-cup anemometer—A combination **cup anemometer** and pressure-plate **anemometer**, consisting of an array of cups about a vertical axis of rotation, the free rotation of which is restricted by a suitable spring arrangement.

By proper adjustment of the force constant of the spring, it is possible to obtain an angular displacement that is proportional to **wind** velocity. The instrument was developed in 1880. The early version used four cups arranged helically, while the more modern version used a wheel with 32 cups. A bridled-cup anemometer was frequently used to measure high wind speeds.

briefing—Oral commentary on existing and expected meteorological conditions.
See **pilot briefing**.

Brier score—Useful for assessing the **accuracy** of predictions that are stated as probabilities, such as probabilities of the occurrence of **precipitation**,

$$S_B = (1/n) \sum (fp_i - v_i)^2,$$

where fp is the forecasted **probability** for an event, and the **verification** term $v = 1$ if the event did occur and $v = 0$ if it did not, for each (i) of the total number (n) forecasts.

Brier scores S_B range between 0.0 for perfect forecasting and 1.0 for the worst possible forecasting. *See* **skill**.

bright band—Radar **signature** of the **melting layer**; a narrow horizontal layer of stronger **radar reflectivity** in **precipitation** at the level in the **atmosphere** where **snow** melts to form **rain**. The bright band is most readily observed on **range–height indicator** (**RHI**) or **time–height indicator** (**THI**) displays.

As **ice crystals** fall toward warmer temperatures at lower heights, they tend to aggregate and form larger snowflakes. This growth accounts for an increase in radar reflectivity as the falling **particles** approach the **melting level**. As they cross the 0°C level, the particles begin melting from the surface inward and finally collapse into **raindrops**. The **reflectivity** maximum in the **melting layer** is explained partly by the difference in the value of the **dielectric factor**, $|K|$, of water and ice (*see* **radar reflectivity**). When a water film begins to form on a melting **snowflake**, its radar reflectivity may increase by as much as 6.5 dB because of the thermodynamic **phase change**. The reflectivity decreases below the melting level because when flakes collapse into raindrops, their fall velocities increase, causing a decrease in the number of precipitation particles per unit volume. The size of the particles also becomes smaller in the melting process, as their **density** increases from that of the snow and melting snow to that of liquid water. Both the reduction in size of the precipitation particles and the decrease in their concentration lead to a decrease in the strength of the **radar echo** at altitudes below the melting level, so that an isolated, horizontal layer of high reflectivity is established, usually centered about 100 m below the 0°C **isotherm**. The bright band is observed primarily in **stratiform** precipitation. The strong convective currents in active showers and thunderstorms tend to destroy the horizontal **stratification** essential for creating and sustaining the bright band.

bright network—*See* **network**.

bright segment—(Also called twilight arch, crepuscular arch.) During **clear** twilights, a faintly glowing band that is visible above the solar **point** when the sun's **elevation** $-7° < h_0 < -18°$ (the lower limit may occur at $h_0 > -18°$).

This band's azimuthal width is $\sim 20°–30°$, but its vertical or elevation-angle width is only a few degrees. The bright segment follows the disappearance of the **purple light** and can persist until the end of **astronomical twilight**. During clear nautical and astronomical twilights, the bright and dark segments comprise the entire sky. *See also* **afterglow**.

bright sunshine—Solar **radiation** intense enough to cast distinct shadows.
See **insolation**.

bright sunshine duration—*See* **insolation duration**.

brightness—1. A perceptual property related but not identical to **luminance**.
The brightness of an object depends not only on its luminance but also on that of its surroundings. For example, an object of uniform luminance will appear brighter against a dark background than against a bright background.

2. *See* **radiance**.

brightness level—Same as **adaptation luminance**.

brightness temperature—A descriptive measure of **radiation** in terms of the **temperature** of a hypothetical **blackbody** emitting an identical amount of radiation at the same **wavelength**.
The brightness temperature is obtained by applying the inverse of the Planck function to the measured radiation. Depending on the nature of the source of radiation and any subsequent **absorption**, the brightness temperature may be independent of, or highly dependent on, the wavelength of the radiation.

brine—**Seawater** containing a higher concentration of dissolved salt than that of the ordinary ocean.
Brine is produced by the **evaporation** or **freezing** of seawater, for, in the latter case, the **sea ice** formed is much less saline than the initial liquid, leaving the adjacent unfrozen water with increased **salinity**. The liquid remaining after seawater has been concentrated by **evaporation** until salt has crystallized is called **bittern**.

brine slush—Snow that is saturated and mixed with water on **sea ice** surfaces, or as a viscous floating mass in water after a heavy **snowfall**.

brisa—(Also spelled briza.) Spanish word for "breeze," usually applied to northeasterly winds blowing off the sea.
It has a number of specific applications: 1) On the east coast of Brazil and in Venezuela, it refers to the northeast **trade** wind; 2) at Montevideo, Uruguay, it is a **strong breeze**, possibly a **sea breeze**; 3) in the Philippines, it refers to the northeast **monsoon**; 4) in northern Puerto Rico, brisa is the northeast trade wind deflected to the east by the east–west mountain ridge; and 5) in Colombia, South America, it is a light, damp **breeze**. Brisa is also used to denote either or both components of a land and sea breeze system. *Compare* **brisote**.

brisa carabinera—Same as **carabiné**.

brisote—In Cuba, a strengthened **brisa** (over **Beaufort force** five) from the northeast.

British thermal unit—(Abbreviated B.T.U., Btu.) A unit of **energy** defined as that quantity of **heat** required to raise the **temperature** of one pound of water by one **degree** Fahrenheit; it is equal to 252.1 calories or 1055 joules.

briza—Same as **brisa**.

broad-crested weir—A **weir** extending far enough in the flow direction to cause the occurrence of **critical flow** depth over the crest.
    See **critical depth**.

broadband radiation—**Radiation** over a large enough spectral region that there is a significant change in the **Planck function**.
    The two most common broadband regions pertaining to **atmospheric radiation** are those for **solar radiation** and **terrestrial radiation**, but many other regions are also used.

Brocken bow—Same as **anticorona**.

Brocken spectre—An observer's shadow cast upon a **cloud**.
    This usually happens when the observer is on a mountain top or ridge and a low sun casts the shadow onto a **fog** or cloud in the valley below. Although the shadow is essentially the same size as the person, the observer sometimes gains the impression that it is gigantic. This is likely the result of a comparison between the nearby shadow and distant objects glimpsed through the cloud. It was named after early observations made by climbers on the Brocken, a peak in the Harz Mountains of Germany.

broeboe—A fairly strong, gusty, easterly **monsoon** wind in the north of the Spermunde Archipelago off the southwest Celebes.
    It crosses the mountains as a dry **foehn**. *See* **tenggara**, **tongara**.

broken—Descriptive of a **sky cover** of from 0.6 to 0.9 (to the nearest tenth).
    This is applied only when obscuring phenomena aloft are present, that is, not when the sky cover is composed entirely of surface-based obscuring phenomena. In **aviation weather observations**, a broken sky cover may be explicitly identified as **thin** (predominantly transparent); otherwise a predominantly opaque status is implicit. An opaque broken sky cover is the minimum requirement for a **ceiling**, and this is frequently termed broken ceiling.

bromine compounds—Any chemical compound containing bromine (Br), the 35th element in the periodic table.
    Bromine-containing compounds enter the **atmosphere** as the result of both natural and **anthropogenic** activities. The major source compounds are CH_3Br (both natural and anthropogenic in origin), the **halons** (manufactured for use as fire suppressants), and dibromomethane CH_2Br_2 (emitted from the oceans). Destruction of these compounds results in the formation of a suite of inorganic Br-containing species, including bromine atoms (Br), bromine monoxide (BrO), bromine nitrate ($BrNO_3$), hypobromous acid (HOBr), and hydrogen bromide (HBr). The chemistry that acts to interconvert these inorganic species results in depletion of **ozone** in the **stratosphere**.

brontides—Low, rumbling thunderlike sounds of short duration not originating from thunderstorms and believed to be of seismic origin.
    Different local names are used for this phenomenon in different parts of the world.
        Gold, T., and S. Soter, 1979: *Science*, **204**, 371–375.

brontometer—(Rare.) A general term to designate apparatus designed to observe the details of weather during **thunderstorms**.

brown cloud—Phenomenon particularly associated with **smog** episodes in cities such as Denver, Colorado.
    High concentrations of particulate matter suspended in the **atmosphere** lead to **light scattering**, resulting in a reduction in **visibility** and a reddish-brown sky coloration.

brown snow—**Snow** intermixed with **dust** particles; a common phenomenon in many parts of the world.
    Snows of other colors, such as **red snow** and **yellow snow**, are similarly explainable.

Brownian motion—(Also called Brownian movement.) The rapid and chaotic motion of **particles** suspended in a fluid at rest as a consequence of fluctuations in the rate at which fluid molecules collide with the particles.
    On average, the particles experience zero net force, but deviations from this average give rise to Brownian motion, so named because it reputedly was observed first in pollen grains by the botanist Robert Brown. Whether Brown actually observed what we now call Brownian motion has

been questioned on the grounds that his grains were so large that their fluctuating motion would have been imperceptible. But even if Brownian motion can ever be shown not to have been observed by Brown, his name is likely to be forever attached to this motion, especially since mathematicians have generalized this term to a broad class of **stochastic** processes. Brownian motion is more than just a scientific curiosity, having played a key role in establishing the reality of molecules and in presenting a convincing argument for the **kinetic theory** of gases.

Boorse, H. A., and L. Motz, 1966: *The World of the Atom*, Vol. I, 206–212.

Brownian rotation—Similar to and sometimes encompassed by the general term **Brownian motion**, although best reserved for the translational motion of **particles** because of a fluctuating force on them, the average of which is zero.

Brownian rotation is a consequence of a fluctuating **torque** (with zero average) on particles as a consequence of **random** molecular collisions.

brubu—A name for a **squall** in the East Indies.

Brückner cycle—An alternation of relatively cool-damp and warm-dry periods, forming an apparent **cycle** of about 35 years.

A belief in such a cycle of 35–40 years in Holland was known to Sir Francis Bacon in 1625, but it was rediscovered in 1890 by E. Brückner who regarded it as worldwide and attached great economic importance to it. Some studies have found quasiperiodic behavior on multidecadal timescales in a number of meteorological and related phenomena, but the existence of a cycle with this specific frequency remains controversial.

Brückner, E., 1890: *Klimaschwankungen seit 1700*, Nebst bemerkungen über die Klimaschwankugen der Siluvialzeit. Wien and Olmütz: Hölzel.

brughierous—The south **wind** that keeps down vegetation in the Montague-Noire region of France.

bruma—A **haze** that appears in the afternoons on the coast of Chile when sea air is transported inland.

Brunt–Väisälä frequency—1. The **frequency** N at which a displaced **air parcel** will oscillate when displaced vertically within a **statically stable** environment.

Given as

$$N = \left(\frac{g}{T_v} \frac{\partial \theta_v}{\partial z} \right)^{1/2},$$

where $g = 9.8$ m s^{-1} is gravitational **acceleration**, T_v is the average absolute **virtual temperature**, and $\partial \theta_v / \partial z$ is the vertical **gradient** of **virtual potential temperature**. Units are radians per second, although this is usually abbreviated as s^{-1}. This frequency is not defined in statically **unstable air** and is zero in statically neutral air. The frequency of internal **gravity waves** in the **atmosphere** cannot exceed the local Brunt–Väisälä frequency. This frequency is also sometimes used as a measure of the **stability** within a statically stable environment.

2. *See* **buoyancy frequency**.

Stull, R. B., 1995: *Meteorology Today for Scientists and Engineers*, 385 pp.

brüscha—The **upvalley wind** that blows from northeast by day in the Engadin, Switzerland, bringing fine weather.

It occurs as a **cold-air outbreak** when an **anticyclone** is centered over northern France.

brush discharge—Same as **corona discharge**.

bryochore—**Tundra** region.

See **biochore**

BT—Abbreviation for **bathythermograph**.

bubble bursting—Process of bursting of air **bubbles**, rising to the ocean surface after a breaking **wave**, leading to breaking of a thin film cap and, for bubbles greater than a few millimeters in diameter, subsequent ejection of a Raleigh jet.

These **particles** evaporate to leave residues that may act as **cloud condensation nuclei** in air of maritime origin.

bubble gauge—A **sensor** that determines **water level** by pumping gas into a **reservoir** and measuring the changes in **pressure**.

bubble high—(Or, simply, bubble.) A small **high**, complete with **anticyclonic circulation**, of the order of 80 to 480 km (50 to 300 miles) across, often induced by **precipitation** and vertical motions associated with thunderstorms.

These transitory small highs are relatively cold and are sometimes located behind convective **outflow** boundaries.

bubble nucleus—Center for initiation of an air bubble during **freezing** of water in a lake or **hailstone** or a **vapor** bubble in boiling.

bubbles—Air bubbles in water generated by the action of breaking waves, the impact on water of spray droplets, and by biological processes.
    They range in size from some centimeters down to microns. Small bubbles in particular can be carried down to considerable depths, as their limiting rise **velocity** is smaller than the ambient vertical water motions, and provide a significant contribution to air–sea gas **flux**.

bubbly ice—Glacier **ice** containing air bubbles.
    Air bubbles are trapped when the **ice** is formed from either water or compressed **snow**. A layer of bubbly ice is called a **white band**.

bucket temperature—The **surface temperature** of the sea as measured by a bucket-thermometer or by immersing a **surface thermometer** in a freshly drawn bucket of water.
    Compare **injection temperature**.

bucket thermometer—A water-temperature **thermometer** provided with an insulated container around the bulb.
    It is lowered into the sea on a line until it has had time to reach the **temperature** of the **surface water**, then withdrawn and read. The insulated water surrounding the bulb preserves the water reading and is also available as a **salinity** sample.

Buckingham Pi theory—A systematic method of **dimensional analysis**, whereby variables that are relevant to a particular atmospheric situation are formed into **dimensionless groups** called Pi groups.
    Because the number of dimensionless groups equals the original number of variables minus the number of **fundamental dimensions** present in all the variables, this **analysis** reduces the **degrees of freedom** for the physical situation and can be used to guide the design of measurement programs. Buckingham Pi theory is often used in **similarity theory** to identify the relevant dimensionless groups. Fundamental dimensions are length, mass, time, **temperature**, electric **current**, and **luminous** intensity. All other dimensions can be formed from combinations of these fundamental dimensions. *See* **Pi theorem**.
        Stull, R. B., 1988: *An Introduction to Boundary Layer Meteorology*, 666 pp.

budget year—In **glaciology**, the one-year period beginning with the start of the **accumulation** season at the **firn line** of a **glacier** or **ice cap** and extending through the following summer's **ablation** season.

budgets of atmospheric species—The calculation of the concentration of a gas in the **atmosphere** given by a balance between its **sources** and removal or **sink** processes.
    The sources may be direct emissions from the surface, or from reactions of other species in the atmosphere.

buffer factor—(Also called the Revelle factor.) The ratio of the fractional rise in atmospheric CO_2 content to the fractional rise in total CO_2 of a water sample at **equilibrium** with the **atmosphere**.

buildup index—1. Cumulative effects of long-term drying on current fire danger.
    2. Increase in strength of a fire management organization.
    3. Accelerated spreading of a fire with time.
    See **fire-danger meter**.

bulk average—A meteorological **variable** averaged over the vertical depth of a layer, such as the **convective mixed layer**.
    For layers that are already relatively uniform with height, the bulk average is often used as an idealized value of that variable throughout the layer. For example, the **potential temperature** is often nearly constant with height in the mixed layer, so the **vertical profile** is often idealized to be exactly constant, with value equal to the bulk average potential temperature.

bulk heat flux—1. Heat **flux** averaged over a layer of air, such as the **boundary layer**.
    2. Heat-flux **divergence** or difference between the top and bottom of a layer.

bulk method—A type of **turbulence closure** where the shape of the **vertical profile** of a **variable** is assumed to be known or idealized a priori, leaving only the parameters of that **profile** shape to be modeled or forecast.

For example, in the **convective mixed layer**, the **potential temperature** profile is idealized as uniform with height, requiring only a single value of mixed-layer potential temperature to specify the potential temperature at every height within that layer. A different example is the potential temperature in the **stable boundary layer**, which when idealized as an exponential shape requires only two variables (**temperature** decrease at the surface and *e*-folding depth) to specify the potential temperature at all heights within that layer. This type of turbulence closure is less than first-order **statistical** closure and is sometimes called **half-order closure**.

bulk mixed layer model—(Abbreviated bulk ML model.) An idealized **model** of the atmospheric **mixed layer** that assumes uniform values of variables within the ML, with sharp jumps or discontinuities across the top.

For midlatitude, fair-weather situations over land, the ML values are usually cooler in **potential temperature**, slower in **wind speed**, and higher in **humidity** than air in the **free atmosphere** just above the ML.

bulk modulus—The reciprocal of the **coefficient of compressibility**.

bulk Richardson number—1. An approximation to the **gradient Richardson number** formed by approximating local **gradients** by finite difference across layers.

The bulk Richardson number R_B is

$$R_B = \frac{(g/T_v)\Delta\theta_v \Delta z}{(\Delta U)^2 + (\Delta V)^2},$$

where g is gravitational **acceleration**, T_v is absolute **virtual temperature**, $\Delta\theta_v$ is the **virtual potential temperature** difference across a layer of **thickness** Δz, and ΔU and ΔV are the changes in horizontal **wind** components across that same layer. In the limit of layer thickness becoming small, the bulk Richardson number approaches the **gradient Richardson number**, for which a **critical Richardson number** is roughly $Ri_c = 0.25$. Gradient Richardson numbers less than this critical value are dynamically unstable and likely to become or remain turbulent. Unfortunately, a critical value is not well defined for the bulk Richardson number, leading to uncertainty in **turbulence** likelihood for values near the critical value. *See* **Richardson number**.

2. In the context of **convective storm** forecasting, a nondimensional ratio of the **convective available potential energy** (CAPE) to a measure of the **vertical wind shear**, used to characterize convective-storm types for various environments.

The vertical wind shear is one-half the square of the difference between the 6-km density-weighted mean wind speed and a 500-m mean surface layer **wind speed**. Generally, values of the bulk Richardson number less than 45 support **supercell** convection, while values greater than 45 support **multicell** or **ordinary cell** convection.

Weisman, M. L., and J. B. Klemp, 1986: Characteristics of isolated convective storms. *Mesoscale Meteorology and Forecasting*, P. Ray, Ed., Amer. Meteor. Soc., ch. 15, 504–520.

bulk stable boundary layer growth—(Abbreviated bulk SBL growth.) The increase with time of an integral measure of depth of the **stable boundary layer**.

Often the SBL has a structure with a poorly defined top. **Potential temperature** gradient and **turbulence intensity** gradually decrease to zero at the top of the SBL. Because there is no well-defined top, often an integral or *e*-folding measure of the the SBL depth is used instead. The growth of such a bulk measure of SBL depth can be used to specify the full structure of the SBL evolution.

bulk transfer coefficient—The empirical constant of proportionality in a **bulk transfer law**.

The bulk transfer coefficient for **momentum** is usually called the **drag coefficient**.

bulk transfer law—The relationship between surface **kinematic flux** of any meteorological **variable** to the product of **wind speed** times the difference of that variable between the surface and some reference height, usually taken as 10 m AGL.

For example, the bulk transfer law for kinematic **heat flux** at the surface is $\overline{w'\theta'_s} = C_H M(\theta_s - \theta_{10m})$, where C_H is the **bulk transfer coefficient** for heat, M is wind speed, θ_s is the **potential temperature** of the surface, w' is the **fluctuation** in **vertical velocity**, θ'_s is the fluctuation in potential temperature, the overbar indicates an average, and θ_{10m} is the potential temperature at height 10 m AGL.

bulk turbulence scale—An average measure of **turbulence** in the **stable boundary layer** (SBL), defined as the ratio of e-folding depth of cooling to the **temperature** decrease at the surface.

Because the SBL has no well-defined depth, and because the amount of cooling and **turbulence intensity** varies continuously with height, the bulk scale provides an overall measure of turbulence. Typical magnitudes vary from 3 m K^{-1} for light turbulence to 15 m K^{-1} for strong turbulence.

bull's eye squall—"A **squall** forming in **fair** weather characteristic of the ocean off the coast of South Africa. It is named for the peculiar appearance of the small isolated **cloud** marking the top of the invisible **vortex** of the **storm**." (H.O. Publ. 220, *Navigation Dictionary*, 1956).

bummock—From the point of view of the submariner, a downward projection from the underside of a **floating ice** canopy; the submerged counterpart of an ice **hummock**.
See also **ice keel**.

bumpiness—Rapid **variation** of the vertical component of air motion causing an aircraft to jolt alternatively upward and downward.
Bumpiness is associated generally with either **convection** currents in an unstable **atmosphere** or a flow of air across surface irregularities or both. It is more common and intense over land than over the sea. It is most marked in the lowest kilometer of the atmosphere but may extend to much higher levels, especially over mountainous terrain. Different types of aircraft may experience different types and intensities of bumpiness when flying through identical atmospheric conditions. *See also* **clear-air turbulence**.

buoyancy—1. That property of an object that enables it to float on the surface of a liquid, or ascend through and remain freely suspended in a compressible fluid such as the **atmosphere**.
Quantitatively, it may be expressed as the ratio of the **specific** weight of the fluid to the specific weight of the object; or, in another manner, by the weight of the fluid displaced minus the weight of the object.
2. (Or buoyant force, buoyancy force; also called Archimedean buoyant force.) The upward force exerted upon a **parcel** of fluid (or an object within the fluid) in a gravitational **field** by virtue of the **density** difference between the parcel (or object) and that of the surrounding fluid.
The magnitude of the **buoyancy force** F per unit mass may be determined by **Archimedes's principle** as

$$F = g\left(\frac{\rho_0}{\rho} - 1\right),$$

where g is the **acceleration of gravity**, ρ the density of the buoyed **fluid parcel** or object, and ρ_0 the density of the surrounding fluid. In the **atmosphere**, a buoyant force on an **air parcel** may be attributed directly to a local increase of **temperature** and may be written

$$F = g\left(\frac{T}{T_0} - 1\right),$$

where T and T_0 are the temperatures of the heated air and that of the **environment**, respectively. The coefficient $(T/T_0 - 1)$ is sometimes called the **buoyancy factor**. The force F is sometimes called the **reduced gravity**. *See* **free convection**.

buoyancy factor—*See* **buoyancy**.

buoyancy flux—The vertical **kinematic flux** of **virtual potential temperature** $(\overline{w'\theta_v'})$, which when multiplied by the **buoyancy** parameter (g/T_v) yields a **flux** that is proportional to buoyancy, that is,

$$Buoyancy\ Flux = \frac{g}{T_v}\ \overline{w'\theta_v'}.$$

This is also a term in the **turbulence kinetic energy** budget, describing buoyant production or **consumption** of **turbulence**. It also appears in the numerator of the **flux Richardson number**.
Stull, R. B., 1988: *An Introduction to Boundary Layer Meteorology*, 666 pp.

buoyancy force—Same as **buoyancy**.

buoyancy frequency—In a continuously **stratified fluid**, the **natural frequency** of the vertical **oscillation** of fluid parcels.
It is also called the **Brunt–Väisälä frequency**. Explicitly, the squared buoyancy frequency is

$$N^2 = -(g/\rho)d\rho/dz,$$

in which g is the **acceleration** due to **gravity** and $\rho(z)$ is **density** as a function of height z.

buoyancy length scale—A measure of the suppression of vertical turbulent motions by **statically stable** air.
Defined as:

$$l_B = \frac{\sigma_w}{N},$$

where σ_w is the **standard deviation** of **vertical velocity**, and N is the **Brunt–Väisälä frequency**. It is a useful length **scale** for the **stable boundary layer**.

buoyancy subrange—For a statically stable **atmosphere**, the portion of the **turbulence kinetic energy** spectrum at wavelengths longer than those in the **inertial subrange**, where **buoyancy** causes the spectral **energy** S to change with the -3 power of **wavenumber** k:

$$S(k) \propto N^2 k^{-3}$$

where N is the **Brunt–Väisälä frequency**.
 In conditions of strong **static stability**, turbulent motions are strongly suppressed in the vertical and have characteristics similar to **gravity waves**.

buoyancy velocity—A **vertical velocity** scale w_B for convective boundary layers, that is related to the buoyant driving force for **convection** and the depth z_i of the **mixed layer**:

$$w_B = \left[\frac{g z_i}{T_{vML}} \left(\theta_{vSfc} - \theta_{vML} \right) \right]^{1/2},$$

where g is gravitational **acceleration**, T_{vML} and θ_{vML} are the average absolute **virtual temperature** and average **virtual potential temperature** in the mixed layer, respectively, and θ_{vSfc} is the virtual potential temperature of the surface skin. It is related to the **Deardorff velocity**, w_*, by $w_B \approx 13 w_*$.

buoyancy wave—Same as **gravity wave**.
 The term "gravity wave" might be confused with waves in the **gravitation** field, rather than buoyant vertical oscillations in a **statically stable** environment, as is usually meant in meteorology.

buoyancy wavenumber—A **wavenumber** in the **turbulence kinetic energy** spectrum that separates the **buoyancy subrange** from the **inertial subrange**.
 This wavenumber k_B is given by

$$k_B = N^{3/2} \epsilon^{-1/2},$$

where N is the **Brunt–Väisälä frequency** and ϵ is the rate of **dissipation** of turbulence kinetic energy. Some measurements indicate that there is a relative minimum or gap in the **spectrum** at this wavenumber.

buoyant convection—Same as **convection**.

buoyant force—Same as **buoyancy**.

buoyant instability—That **static instability** in a system in which **buoyancy** or **reduced gravity** is the only restoring force on displacements.
 In general, a fluid is buoyantly unstable when the **environmental lapse rate** of **density** is greater than the **process lapse rate** of density. For an **incompressible fluid** this requires an increase of density with height; for the **atmosphere**, when lifting is assumed to be **adiabatic**, it requires the **lapse rate** (of **temperature**) to be greater than the **adiabatic lapse rate**.

buoyant subrange—A **range** of turbulent **eddies** in a stably stratified **atmosphere**, too small to be influenced by **shear** but large enough to be affected by **buoyancy**.
 Dimensional considerations give a power law dependence of **energy** versus **wavenumber**, or spectral slope, of $-11/5$. See **turbulence spectrum**.

buran—A strong northeast **wind** in Russia and central Asia.
 It is most frequent in winter when it resembles a **blizzard**, that is, very cold and lifting **snow** from the ground; as such it is called white buran or, on the **tundra**, purga. A similar wind in Alaska is called **burga**. The buran also occurs, but less frequently, in summer, when it raises **dust** clouds; it is then called **karaburan**.

burga—(Also spelled boorga.) A northeasterly **storm** in Alaska, bringing **sleet** or **snow**; it is similar to the winter **buran** or **purga** of Russia and Siberia.
 See **blizzard**.

Burger number—A **dimensionless number**, Bu, for atmospheric or oceanographic flow expressing the ratio between **density** stratification in the vertical and the earth's rotation in the horizontal:

$$\text{Bu} = \left(\frac{NH}{\Omega L} \right)^2 = \left(\frac{\text{Ro}}{\text{Fr}} \right)^2 = \left(\frac{R_D}{L} \right)^2,$$

where N is the **Brunt–Väisälä frequency**, Ω is the angular rotation rate of the earth, H is the **scale height** of the **atmosphere**, L is a horizontal length **scale** of typical motions, Ro is the **Rossby number**, Fr is the **Froude number**, and R_D is the Rossby **deformation radius**.

Bu is often of order one for many atmospheric phenomena, meaning that both **stratification** and rotation play nearly equal roles in governing vertical and other motions in the fluid.

Cushman–Roisin, 1994: *Introduction to Geophysical Fluid Dynamics*, 320 pp.

Burger's vortex—Exact solution of the **Navier–Stokes equations** for a steady **vortex** in which the **diffusion** of **vorticity** is balanced by **vortex stretching** in an external strain **field**.

The vortex has a Gaussian shape. For example, for a vertical **axisymmetric** vortex in an external **velocity** field given by

$$(u, v, w) = \gamma(-x/2, -y/2, z),$$

the vorticity is

$$\omega = \omega_0 \exp[(-\gamma(x^2 + y^2)/4\nu)],$$

in which ν is the **kinematic viscosity**.

buria—Bulgarian term for **bora**.

burn off—With reference to **fog** or low **stratus** cloud layers, to dissipate by heating from the sun, primarily during the early morning hours.

burning index—A relative number related to the contribution that **fire behavior** makes to the amount of effort needed to contain a fire of a specified fuel type.

The calculated burning index falls on a **scale** of 1–100: 1–11 is no fire danger; 12–35 is medium danger; 40–100 is high danger. *See* **fire-danger meter**.

burraxka silch—**Hailstorm** occurring in the Mediterranean, near Malta.

burster—(Or southerly burster; also called buster, southerly buster.) A sudden shift of **wind** to the southeast in the south and southeast parts of Australia, especially frequent on the coast of New South Wales near Sydney in summer.

It occurs in the rear of a **trough** of low pressure that is followed by the rapid advance of an **anticyclone** from west Australia. After some days of hot, dry northerly wind, heavy **cumulus** clouds approach from the south, the wind drops to **calm** and then sets in suddenly from the south, sometimes reaching **gale** force. Temperature at Sydney has fallen from 38°C to 18°C in thirty minutes. The average summer **frequency** of bursters at Sydney is 32. Similar winds are experienced in the east of South Africa, especially near Durban. *Compare* **norther**.

bush—(Rare.) Mass of spray (**dust**) swept up and around the base of a **waterspout (tornado)**.
See also **cascade**.

business as usual—One possible scenario often used to generate **climate change** predictions and evaluate potential impacts.

The "business as usual" scenario implies that no actions specifically directed at limiting greenhouse emissions will be taken by governments, companies, or individuals.

Businger–Dyer relationship—A **similarity relationship** between surface **flux** of a **variable** and the mean vertical **profile** of that variable.

These are generically of the form

$$\frac{M}{u_*} = \frac{1}{k}\left[\ln\left(\frac{z-d}{z_0}\right) + \psi\left(\frac{z-d}{L}\right)\right],$$

where M is mean wind speed, u_* is the **friction velocity** (a term representing the **momentum flux**), k is **von Kármán's constant**, z is height above ground, d is **displacement distance**, z_0 is **roughness length**, ψ is an empirical **stability** correction term, and L is the **Obukhov length**.

Stull, R. B., 1988: *An Introduction to Boundary Layer Meteorology*, 666 pp.

buster—Same as **burster**.

butadiene—Unsaturated compound consisting of two carbon–carbon double bonds, formula $CH_2CHCHCH_2$; common constituent of automobile exhaust.

Also used in the synthesis of artificial rubber for car tires.

Buys Ballot's law—(Also called baric wind law.) A law describing the relationship of the horizontal

wind direction in the **atmosphere** to the **pressure distribution**; if one stands with the back to the **wind**, the **pressure** to the left is lower than to the right in the Northern Hemisphere.

In the Southern Hemisphere, the relation is reversed. This law was formulated in 1857 by the Dutch **meteorologist** Buys Ballot and is a qualitative statement of the **geostrophic wind** relation.

BWER—Abbreviation for **bounded weak echo region**.

C

C band—*See* **radar frequency bands**.

C figure—(Obsolete.) *See* **C index**.

C index—A measure of geomagnetic activity; use discontinued in 1975 and superseded by other indices of geomagnetic activity.
See **Kp**.

C weather—(Obsolete.) Abbreviation for **contact weather**.
See **VFR weather**.

cabbeling—(Also spelled cabbaling.) Any physical process that is caused by the **nonlinear** terms in the expression of **density** as a function of S (**salinity**) and T (**temperature**) measured at constant **pressure**.
The original meaning of cabbeling is described by considering the **mixing** of two fluids of similar density but different T and S. The volume of the mixture will generally be slightly smaller than the total volume of the two original fluids (contraction upon mixing). Therefore, the mixture will have a slightly higher density than the average density of the original fluids. This densification upon mixing is thought to cause the mixed fluid to flow downward, away from the zone of mixing, and so will allow new source fluids to come in contact. In situations where vertical mixing is occurring, the effects of nonlinearity are more subtle, leading to differential **entrainment** and upward migration of the mixing interface.

cable—A nautical unit of horizontal distance defined as 0.1 **nautical mile** = 185.2 m.
Historically, a cable was defined as equal to 600 ft (100 fathoms).

cacimbo—(Also spelled caicimbo.) Local name in Angola for the wet fogs and drizzles noted with onshore winds from the **Benguela Current**.

cajú rains—In northeast Brazil, light showers occurring in October.
The **rainy season** in Brazil normally begins in December and lasts until April or May; then follows a **dry season** until the rains at the end of the year. In October of most years, however, there will be local showers known as chuvas de cajú (cashew rains), so-called because they occur at the time of blossom of the cashew tree.

cake ice—Flat pieces of **sea ice** larger than **brash**.
Cake **ice** is often tightly packed, giving it a mosaic appearance, but its surface is generally smooth in contrast to rough, hummocked **pressure ice**.

cal—Abbreviation for **calorie**.

Cal—Abbreviation for **large calorie**.

calendar year—The time interval between 1 January and 31 December, inclusive, in the Gregorian calendar.
The number of days in the calendar year varies from 365 in ordinary years to 366 in leap years.

calf—Same as **calved ice**.

calibration—The process whereby the magnitude of the **output** of a measuring instrument (e.g., the level of **mercury** in a **thermometer** or the detected backscatter **power** of a meteorological **radar**) is related to the magnitude of the **input** force (e.g., the **temperature** or **radar reflectivity**) actuating that instrument.

calibration of an instrument—The process of validating the **output** of an observing system against known reference observations or **standards**.

calibration tank—(Also called rating tank.) A tank of still water in which a **current meter** is moved at a known **velocity** for **calibration**.

California Current—The eastern **boundary current** of the North Pacific subtropical **gyre**.
It flows southward along the coast of Washington and California and is associated with **coastal upwelling** during spring and summer (April–September). The associated low **sea surface temperatures** (about 15°C) produce a coastal strip of **sea fog** along the otherwise extremely hot coastline. In autumn and winter (October–March) the **upwelling** is replaced by northward flow known as the **Davidson Current**, which reaches its peak speed of 0.2–0.3 m s^{-1} in January.

California method—In **hydrology**, a name sometimes given to a form of **frequency analysis** that employs the **return period** as the **plotting position**.

California norther—*See* **norther**.

California plotting position—For the *r*th ranked **datum** (from largest to smallest) from a **sample** of size *n*, the quotient r/n.

    The California plotting position was among the first plotting positions developed and is not recommended where probabilities of events less than the minimum in the sample are needed. *See* **plotting position**, **probability paper**.

calina—A local name given in Spain to the summer haze described by Kendrew (1937).

    He says, "In summer there is very active **evaporation** and almost complete **drought**, broken only by an occasional **thunderstorm**, and the fierce heat burns up the vegetation. Without **irrigation** the landscape is semi-desert, brown and grey are dominant colors, and **dust** is everywhere—the parched ground is thickly covered, and the air is hazy with minute dust particles which have been swept up by the strong winds. The **haze** is known as the calina, and is probably due to irregular **refraction** of the **light**, as well as the dust. The view is frequently obscured by the dismal grey calina in all the south Mediterranean lands."

    Kendrew, W. G., 1937: *The Climate of the Continents*, 3d ed., Oxford, The Clarendon Press, 478 pp.

Callao painter—Same as **painter**.

calm—1. The absence of apparent motion of the air.

    In the **Beaufort wind scale**, this condition is reported when **smoke** is observed to rise vertically, or the surface of the sea is smooth and mirrorlike. The National Weather Service reports a **wind** as calm when it is determined to have a speed of less than three knots.

    2. *See* **calm belt**.

calm belt—A latitude band in which the winds are generally light and variable (**calm**).

calms of Cancer—Along with the calms of Capricorn, the light, variable winds and **calms** that occur in the centers of the **subtropical high pressure belts** over the oceans.

    They are named after the **Tropic of Cancer** and **Tropic of Capricorn**, although their usual position is at about latitudes 30°N and S, the **horse latitudes**.

calms of Capricorn—*See* **calms of Cancer**.

calorie—(Abbreviated cal.) A unit of **energy** defined as that amount of energy required to raise the **temperature** of one **gram** of water by one **degree** Celsius (the gram-calorie or small calorie), equal to 4.1855 joules.

    The **kilogram calorie** or **large calorie** (Kcal, kg-cal, or Cal) is 1000 times as large as a calorie.

calorimeter—An instrument designed to measure quantities of **heat**; sometimes used in meteorology to measure **solar radiation**.

calved ice—(Also called calf.) A piece of **ice** floating in a body of water after **calving** from a mass of **land ice** or **iceberg**.

calving—The breaking away of a mass of **ice** from a floating **glacier**, **ice front**, or **iceberg**.

    Armstrong, T., B. Roberts, and C. Swithinbank, 1973: *Illustrated Glossary of Snow and Ice*, Scott Polar Research Institute, Special Pub. No. 4, 2d ed.; The Scholar Press Ltd., Menston Yorkshire, UK.

calvus—*See* **cumulonimbus calvus**, **cloud classification**.

camanchaca—Same as **garúa**.

Campbell–Stokes recorder—A **sunshine recorder** of the type in which the timescale is supplied by the motion of the sun.

    It consists essentially of a spherical lens that burns an **image** of the sun upon a specially prepared card. The instrument must be oriented carefully so that the timescale on the card agrees with the sun time. The depth and breadth of the **trace** may be interpreted in terms of the approximate **intensity** of the sun.

    Middleton, W. E. K., and A. F. Spilhaus, 1953: *Meteorological Instruments*, 3d ed., rev., p. 218.

campos—The **savanna** of South America.

camsin—Same as **khamsin**.

Canadian hardness-gauge—A type of **disk hardness-gauge**, especially useful in relatively soft **snow**.

Canary Current—The eastern **boundary current** of the North Atlantic subtropical **gyre**.

    It flows southward along the coast of Africa from Morocco to Senegal and is associated with **coastal upwelling**. The **upwelling** reaches its southernmost extent in winter (December–March) when the **trade winds** are strongest; it then extends past Cape Blanc (21°N), where the Canary Current separates from the African coast.

cancellation ratio—1. Broadly, a **power** ratio used in certain applications of **radar** that describes the degree to which the backscattered power from unwanted targets (**clutter**) is suppressed or canceled in one mode of measurement relative to another mode of measurement.

It is defined to yield a value greater than unity or, equivalently, a positive **decibel** quantity. For example, in some applications it is defined as the ratio of **signal intensity** backscattered from **rain** observed with **linear** polarization to the signal intensity backscattered from rain observed with **circular polarization**. Values of 25–30 dB are typical of light rain observed at wavelengths of 3–10 cm.

2. For a **dual-channel radar**, the ratio of the **power** received in the **orthogonal** channel to that received in the transmission channel for transmitted **circular polarization**. The stronger component of the **target signal** from **precipitation** is the component in the orthogonal channel because the sense of the circular polarization is reversed as its direction of propagation changes when it is **scattered** back toward the **radar**.

3. A ratio that describes the degree to which **radar** clutter is suppressed by a **clutter** filter. This usage applies particularly to **moving-target indication** (MTI) radars. Shrader and Gregers-Hansen recommend that the term "improvement factor" be used in place of "cancellation ratio" because the latter term has been used inconsistently.

Compare **circular depolarization ratio.**

Shrader, W. W., and V. Gregers-Hansen, 1990: *Radar Handbook*, 2d ed., M. I. Skolnik, ed., McGraw-Hill, MTI Radar, chapter 15, 15.1–15.72.

candela—(Abbreviated cd.) A unit of **luminous** intensity expressed in **lumen** per **steradian** (lm sr⁻¹).

The candela was first defined as 1/60 the luminous intensity, in the perpendicular direction, of a 1 cm² **blackbody** radiator at the **freezing** temperature of platinum (about 2042 K) and a **pressure** of 1 **atmosphere**. It is now defined as the luminous intensity of a **light** source producing single-frequency light at a **frequency** of 540 terahertz with a **power** of 1/683 **watt** per **steradian**. It is the **standard** unit of measure for luminous intensity adopted by the International System of Units (SI). In some texts, it is referred to as the international standard candle.

candle—The original unit of **luminous** intensity, defined as the luminous intensity of a candle of particular construction.

The candle was eventually replaced by the **international candle**, the **new candle** in 1939, and finally the **candela** in 1948.

candle ice—A form of **rotten ice**; disintegrating **sea ice** (or **lake ice**) consisting of **ice prisms** or cylinders oriented perpendicular to the original **ice** surface; these "ice fingers" may be equal in length to the thickness of the original ice before its disintegration.

Candlemas Eve winds—(Also called Candlemas crack.) Heavy winds that often occur in England in February or March. (Candlemas is 2 February.)

canigonenc—*See* **mistral.**

canopy—The vegetative covering over a surface.

The canopy is often considered to be the outer surfaces of the vegetation. Plant height and the distribution, orientation, and shape of plant leaves within a canopy influence the atmospheric **environment** and many plant processes within the canopy.

canopy temperature—The **temperature** of plants and/or the vegetative cover.

It is generally measured with **infrared** thermometers. Canopy temperature is often used to indicate vegetative water status and is used in models for estimating **transpiration** rates and **sensible heat** transport from vegetation.

Canterbury northwester—A strong northwest **foehn** wind descending the New Zealand Alps onto the Canterbury Plains of South Island, New Zealand.

The **nor'wester** (as it is called locally) is responsible for strong, warm, gusty winds that can damage crops, cause uprooting of trees in forests, damage buildings and other structures, and cause soil loss. Its drying action increases the need for **irrigation** on the Canterbury Plains. It can occur throughout the year but has highest **frequency** in spring. It generally occurs immediately prior to the passage of a **cold front** over the South Island. It is responsible for the highest recorded **air temperature** in New Zealand (42.4°C at Rangiora on the Canterbury Plains, 7 February 1973).

McKendry, I. G., 1983: Spatial and temporal aspects of the surface wind regime on the Canterbury Plains, New Zealand. *J. Climate*, **3**, 155–166.

Sturman, A., and N. Tapper, 1996: *The Weather and Climate of New Zealand*, Oxford University Press, 476 pp.

canyon wind—1. The **mountain wind** of a canyon, that is, the nighttime down-canyon flow of air caused by cooling at the canyon walls.

Because of the steepness of the slopes, canyon winds can be very strong.

2. (Or gorge wind.) Any **wind** modified by being forced to flow through a canyon or gorge, especially by a strong **pressure gradient** as is the case with **mountain-gap winds**.

The easterly **Wasatch winds** of Utah are examples. The speed of canyon winds may be increased as a **jet-effect wind** (*see also* **gap wind**), and their direction is rigidly controlled by the **topography**.

3. Same as **Wasatch winds**.

cap—(Also called **lid**.) A region of negative **buoyancy** below an existing **level of free convection** (LFC) where **energy** must be supplied to the **parcel** to maintain its ascent.

This tends to inhibit the development of **convection** until some physical mechanism can lift a parcel to its LFC. The **intensity** of the cap is measured by its **convective inhibition**. The term **capping inversion** is sometimes used, but an **inversion** is not necessary for the conditions producing convective inhibition to exist.

cap cloud—1. (Also called cloud cap.) An approximately stationary **cloud**, or **standing cloud**, on or hovering above an isolated mountain peak.

It is formed by the cooling and **condensation** in **moist air** forced up over the peak. *Compare* **crest cloud, banner cloud**; *see* **lenticularis**.

2. Same as **pileus**.

capacitance rain gauge—A **probe** consisting of a stainless steel rod covered by a Teflon sheath set inside a cylindrical rainwater **collector**.

The water surrounding the probe forms the outer plate of a coaxial-type capacitor while the metal rod forms the inner plate. The Teflon sheath serves as the **dielectric**. As the water height in the collector rises, the surface area of the capacitor increases, increasing the total capacitance. The total capacitance is measured and converted to an **analog** voltage directly proportional to the height of the water in the collector.

Nystuen, J. A., et al., 1996: A comparison of automatic rain gauges. *J. Atmos. Oceanic Technol.*, **13**, 62–73.

capacity correction—The **correction** applied to a **mercury barometer** with a nonadjustable **cistern** in order to compensate for the change in level of the cistern as the **atmospheric pressure** changes.

Thus, as the pressure falls, the height of the cistern increases, due to the exchange of **mercury** between the barometer tube and its cistern. This correction is not required if the **scale** is calibrated as in the **Kew barometer**. *See also* **barometric corrections**.

capacity of the wind—In geology, the total weight of airborne **particles** of given size, shape, and **specific gravity** that can be carried in one cubic **mile** of **wind** blowing at a given speed.

The particles considered are only those of soil and rock (detritus). A number of measurements of sand **transport** by wind in the Sahara were made by Major R. A. Bagnold, who, by theory and wind tunnel experiment, established a formula depending on **wind speed** near the ground, diameter of the sand grains, and type of surface, but not on **turbulence**. For fine sand, the lower limit of effective wind is 2.5 m s^{-1}. A similar concept with respect to waterborne material is the **sediment discharge rating**. *Compare* **competence of the wind, erosion**.

CAPE—Abbreviation for **convective available potential energy**.

Cape doctor—A strong, invigorating southeast **wind** on the south coast of South Africa.

At Cape Town it brings the "tablecloth," a sheet of hill **cloud** that covers Table Mountain. *See* **doctor, southeaster**.

capillarity—The action by which the surface of a liquid in contact with a solid (as in a capillary tube) is elevated or depressed depending on the relative attraction of the molecules of the liquid for each other and for those of the solid (e.g., the **meniscus** of a liquid column).

See **capillary action, capillary depression**.

capillarity correction—As applied to a **mercury barometer**, that part of the **instrument correction** that is required by the shape of the **meniscus** of the **mercury**.

Mercury does not wet glass and consequently the shape of the meniscus is normally convex upward, resulting in a positive **correction**. For a given **barometer**, this correction will vary slightly with the height of the meniscus. The capillarity correction can be minimized by using a tube of large bore. *See* **barometric corrections, capillary depression, capillary action**.

capillary action—The depression or elevation of the **meniscus** of a liquid contained in a tube of small diameter due to the combined effects of **gravity**, **surface tension**, and the forces of cohesion and adhesion.

When the liquid wets the walls of a container, the meniscus is shaped convex downward; if the liquid does not wet the walls of the container, the meniscus is shaped convex upward.

capillary collector—An instrument for collecting liquid water from the **atmosphere**.

The collecting head is fabricated of a porous material having a pore size of the order of 30 μm. The **pressure** difference across the water–air **interface** prevents air from entering the capillary system while allowing free flow of water.

capillary conductivity—Same as **unsaturated hydraulic conductivity**; not commonly used.

capillary depression—The depression of the **meniscus** of a liquid contained in a tube where the liquid does not wet the walls of the container (as in a **mercury barometer**).

The meniscus is shaped convex upward, and this results in a depression of the meniscus.

capillary diffusion—(Also called capillary movement.) The movement of fluids in unsaturated porous media due to **surface tension** and adhesive driving forces (**capillarity**).

capillary electrometer—An **electrometer** for measuring small **electric potential** difference based upon change of **surface tension** between **mercury** and an electrolytic solution in a capillary tube with change of **potential** difference between the liquids.

capillary forces—The mechanical forces exerted on soil water resulting from the curved **interface** between the air and water caused by the combined effect of **surface tension** and effective **contact angle**.

capillary fringe—The zone above the **water table** and below the **boundary of saturation** where the soil is saturated but at pressures less than atmospheric (i.e., under tension).

A new equivalent term gaining acceptance is **tension saturated zone**.

capillary head—(Also called capillary pressure head.) The **potential**, expressed in units of **water head**, due to the **capillarity** of a **porous medium** in the presence of a wetted liquid.

capillary hysteresis—The phenomenon that the **equilibrium** positions of the air–water interfaces in a system of pores are dependent on whether the system is increasing or decreasing in **water content** (i.e., the wetting history).

capillary interstice—A pore space in **sediment** (interstitial pore space) small enough for the occurrence of appreciable **capillary rise**.

capillary potential—For meteorological consideration, the layer of essentially **saturated soil** above the plane of **free water** (**water table**).

capillary pressure—The difference in **pressure** across the **interface** between two immiscible fluids.

The pressure difference is proportional to the **surface tension** and inversely proportional to the effective radius of the interface.

capillary ripple—Same as **capillary wave**.

capillary rise—(Also called height of capillary rise.) The height above a **free surface** to which a liquid will rise by **capillary action**.

capillary suction—Phenomenon resulting from **capillary forces** that induce a liquid to enter a **porous medium**.

capillary water—Water held in the smaller pores of a soil, generally at tensions greater than 60 cm of water.

capillary wave—(Also called ripple, capillary ripple.) A **wave** for which the primary restoring force is **surface tension**; generally taken to be one of less than 1.7-cm **wavelength**, this being the wavelength for which the theoretical **phase speed** is a minimum, and marking the transition from **gravity** to surface tension as the dominant restoring force at the sea surface. *Compare* **gravity wave**.

capillatus—*See* **cumulonimbus capillatus, cloud classification**.

capped column—A form of **ice crystal** consisting of a **hexagonal column** with plate crystals or stellar crystals at its ends and sometimes at intermediate positions.

The caps are perpendicular to the **column** and form as the **crystal** enters regions where change of **air temperature** leads to change in crystal structure from column to plate.

CAPPI—(Abbreviation for constant-altitude plan position indicator.) A composite radar **display** constructed by assembling **radar** data from many PPIs at successive **elevation** angles to obtain the pattern of the data at a specified constant **altitude**.

capping inversion—A **statically stable** layer at the top of the **atmospheric boundary layer**.

Although the word "inversion" implies that **temperature** increases with height, the word "capping inversion" is used more loosely for any stable layer (**potential temperature** increasing with height) at the top of the boundary layer. This **inversion** is a ubiquitous feature of the atmospheric boundary layer, formed because the **troposphere** is statically stable on the average, and because **turbulence** homogenizes air within the boundary layer, which by conservation of **heat** requires that a stable layer form at the top of the boundary layer. This inversion traps surface-induced turbulence and **air pollutants** below it, and causes the **free atmosphere** to not "feel" the earth's surface during **fair** weather (i.e., no **drag**, free slip, no heat or moisture from the surface, and winds are nearly **geostrophic**). *See* **lid**.

capping layer—*See* **capping inversion**.

captive balloon sounding—Sounding by means of a moored balloon carrying instruments to determine the values of one or more **upper-air** meteorological **elements**.

Especially used for studies in **micrometeorology** and **mesometeorology**.

capture—The water diverted from the boundary sources of an **aquifer** by lowering the **water table** or **piezometric surface** at a particular point in the aquifer.

carabiné—(Also called brisa carabinera, brise carabinee.) In France and Spain, a sudden and violent **wind**.

carbon—(Symbol C.) The 12th **element** in the periodic table, mass 12.000. Carbon is one of the most versatile elements and combines with itself and many other elements to form a huge variety of organic compounds, for example, **hydrocarbons** and their derivatives, some of which are found in the **atmosphere**.

Elemental carbon occurs in the atmosphere, mostly in the form of **soot** from incomplete combustion of organic matter. Smoke **particles** also have a large proportion of carbonaceous material in them. Together, soot and **smoke** account for a large part of the **reduction** in **visibility** occurring over continental and polluted regions, due to **light scattering** and light **absorption**.

carbon assimilation—Process in which **carbon** in the **atmosphere** is absorbed by a biological system and assimilated into it.

carbon-black seeding—A type of **cloud seeding** where microscopic **soot** particles are dispersed into the **atmosphere** in order to absorb **radiant energy** and thereby **heat** the surrounding air, possibly leading to **convection**.

carbon bond mechanism—A type of **lumped chemical mechanism** used in the modeling of atmospheric organic chemistry, usually in urban areas.

In the carbon bond mechanism, each **carbon** atom is treated in the **model** according to what species are bonded to it. That is, all singly bonded carbon atoms are treated together as a single entity, as are all aromatic carbon atoms, etc.

carbon budget—The change in the amount of **carbon** in a **reservoir** via fluxes of carbon into and out of the reservoir.

carbon dioxide—Colorless gas, formula CO_2, **molecular weight** 44; the fourth most abundant gas in **dry air**. The end product of the combustion or **oxidation** of organic material, including **fossil fuels**, CO_2 is a very strong **greenhouse gas** and has very important radiative effects.

Carbon dioxide is soluble in water, and the atmospheric concentration is buffered by dissolution in seawater. Carbon dioxide is taken up by the **biosphere** during **photosynthesis**, and large amounts of **carbon** can be stored in decaying organic matter.

carbon dioxide atmospheric concentrations—The amount of **carbon dioxide** in the **atmosphere**, most commonly expressed as a **mixing ratio** by volume, usually in units of parts per million by volume (ppmv).

The 1995 concentration of 360 ppmv is more than 25% greater than the preindustrial (i.e., eighteenth century) value of 280 ppmv.

carbon dioxide band—The region of the **electromagnetic spectrum** in which CO_2 plays a significant role in the **transfer** of **infrared radiation** through the **atmosphere**.

Intense **absorption** is in the **band** centered at 14.7 μm.

carbon dioxide equivalence—The contribution of various **greenhouse gases** to the enhancement of the natural **greenhouse effect**, either in terms of atmospheric concentrations of **carbon dioxide** or in terms of emission rates of carbon dioxide.

This term is most commonly used as the basis for comparison of the relative greenhouse effect of different greenhouse gases. In **climate** modeling, this term is also used to describe the practice of using carbon dioxide as a surrogate for other greenhouse gases.

carbon dioxide fertilization—Enrichment of the **atmosphere** surrounding a plant or crop, that is, the **canopy** of the crop, by increasing the concentration of **carbon dioxide**.

carbon dioxide fertilizing effect—Response of the plant or crop to **carbon dioxide fertilization** (e.g., enhancement of yield, **nitrogen fixation**, compensation for **temperature** and low light levels, and increased water use efficiency).

> Wittwer, S. H., 1995: *Food, Climate and Carbon Dioxide: The Global Environment and World Food Production,* CRC Press, Inc..

carbon disulfide—Reduced sulfur gas CS_2 formed predominantly in industrial processes, but also emitted from natural sources.

Following its **oxidation** by the **hydroxyl radical** it is transformed largely to **carbonyl sulfide** (COS).

carbon-film hygrometer element—An **electrical hygrometer** element constructed of a plastic or glass strip coated with a film of **carbon** black dispersed in a **hygroscopic** binder.

Changes in the ambient **relative humidity** lead to dimensional changes in the hygroscopic film such that the **resistance** increases proportionally with **humidity**. The successful implementation of this **sensor** type requires a well-controlled manufacturing process and very careful handling. The sensor is used in some **radiosondes** and named carbon hygristor.

carbon monoxide—A colorless, odorless, very toxic gas; formula CO, **molecular weight** 28.

It is an intermediate product in the **oxidation** or combustion of organic material. It is present in the **atmosphere** at varying levels but is found everywhere, a result of its widely distributed sources and fairly long lifetime. Carbon monoxide is oxidized to **carbon dioxide** by the **hydroxyl radical** (OH) and plays a part in local and regional air quality.

carbon pool—A **reservoir** that contains **carbon** as a principal **element** in the **biogeochemical cycle**.

carbon sink—A **reservoir** that receives **carbon** from another carbon reservoir.

Commonly used to denote a reservoir where the carbon amount increases because its total carbon received from all other reservoirs exceeds its total carbon transfer to the other reservoirs.

carbon source—A **reservoir** that provides **carbon** to another carbon reservoir.

Commonly used to denote a reservoir where the carbon inventory decreases because its total carbon emission or release to all other reservoirs exceeds its total **absorption** of carbon from all other reservoirs.

carbon tetrachloride—A significant contributor to the atmospheric chlorine budget, this compound, formula CCl_4, has been used in industrial applications as a solvent.

Its production is now banned as a result of the **Montreal Protocol** on Substances that Deplete the Ozone Layer.

carbonyl compounds—Organic compounds containing a carbon–oxygen double bond; composed of **aldehydes**, **ketones**, and organic (carboxylic) acids, plus their derivatives.

These species can be emitted directly to the **atmosphere** from natural or **anthropogenic** sources, or can be formed as intermediates in the **oxidation** of **hydrocarbons**. They are an important source of **carbon monoxide** and radicals.

carbonyl sulfide—Relatively unreactive sulfur gas (COS) that can persist in the **atmosphere** long enough for **transport** to the **stratosphere** to occur, where its **oxidation** is responsible for the maintenance of the background **stratospheric sulfate layer**.

This link between the **biosphere** and the stratosphere was recognized by Paul Crutzen in 1976. The atmospheric budget of COS is intimately linked to the biosphere, since it can be both released from, and taken up by, vegetation.

> Crutzen, P. J., 1976: The possible importance of COS for the sulfate layer of the stratosphere. *Geophys. Res. Lett.,* **3,** 73–76.

carburetor icing—The formation of **ice** in the throat of a carburetor as the result of cooling by expansion and **evaporation** of gasoline.

At one time a serious problem in aviation, carburetor heaters now render this unimportant. Carburetor icing often occurs at **ambient air** temperatures as high as 50°–60°F, the necessary cooling to the **frost point** occurring within the carburetor throat. *Compare* **aircraft icing**.

carcenet—A very cold, violent, channeled pressure-gradient or **gorge wind** in the upper Aude Valley in the eastern Pyrenees Mountains.

cardinal temperatures—Minimum and maximum temperatures that define limits of growth and development of an organism, and an optimum **temperature** at which growth proceeds with greatest rapidity.

Cardinal temperatures may vary with the stage of development.

cardinal winds—**Winds** from the four cardinal points of the **compass**; that is, north, east, south, and west winds.

Caribbean Current—A strong, swift current passing from east to west through the Caribbean Sea.

The current is the major pathway for water from the Southern into the Northern Hemisphere in the global **ocean conveyor belt**. It is also an element of the western **boundary current** system of the North Atlantic subtropical **gyre** and thus associated with high speeds and **eddy** shedding: 0.2 m s^{-1} in the Grenada Basin; 0.5 m s^{-1} in the Venezuela, Columbia, and Cayman Basins; and 0.8 m s^{-1} near Yucatan Strait. Eddies can produce occasional current reversal from westward to eastward in all basins. Most of the water from the Caribbean Current leaves through Yucatan Strait, but a small amount returns eastward from the Caribbean Sea into the Atlantic as the Caribbean Countercurrent.

Carnot cycle—An idealized reversible **work** cycle defined for any system, but usually limited, in meteorology, to a so-called **perfect gas**.

The Carnot cycle consists of four states: 1) an **isothermal** expansion of the gas at a **temperature** T_1; 2) an **adiabatic expansion** to temperature T_2; 3) an isothermal compression at temperature T_2; and 4) an **adiabatic** compression to the original state of the gas to complete the **cycle**. In a Carnot cycle, the net work done is the difference between the **heat** input Q_1 at higher temperature T_1 and the heat extracted Q_2 at the lower temperature T_2. The atmospheric **general circulation** and some storms, notably **hurricanes**, incorporate a process similar to a Carnot cycle. *See* **Carnot engine, thermodynamic efficiency**.

Carnot efficiency—Same as **thermodynamic efficiency**.

Carnot engine—An idealized reversible **heat engine** working in a **Carnot cycle**.

It is the most efficient engine that can operate between two specified temperatures; its efficiency is equivalent to the **thermodynamic efficiency**. The Carnot engine is capable of being run either as a conventional engine or as a refrigerator. *See* **reversible process**.

carrier—The unmodulated fundamental **output** of a radio or radar **transmitter**, which is capable of being modulated with information to produce a communications **signal**.

carrier frequency—The **frequency** of the unmodulated transmitted **wave** of a radio or radar **transmitter**.

carrier wave—The purely sinusoidal component of a complex **waveform** that carries a **signal** from a **transmitter** to a **receiver**.

The carrier wave itself cannot transmit information but must be changed or modulated, two common methods for doing so being **amplitude modulation** and **frequency modulation**.

carry-over—1. The portion of the **streamflow** during any month or year derived from **precipitation** in previous months or years.

2. Storage of water during a wet surplus year used for making up deficiencies in dry years.

Cartesian coordinates—A **coordinate system** in which the locations of points in space are expressed by reference to three planes, called coordinate planes, no two of which are parallel.

The three planes intersect in three straight lines, called coordinate axes. The coordinate planes and coordinate axes intersect in a common point, called the origin. From any point P in space three straight lines may be drawn, each of which is parallel to one of the three coordinate axes; each of these lines will then intersect one (and only one) of the three coordinate planes. If A, B, C denote these points of intersection, the Cartesian coordinates P are the distances PA, PB, and PC. If the coordinate axes are mutually perpendicular, the coordinate system is rectangular; otherwise, oblique. In meteorology, the most common orientation of the x, y, z **rectangular Cartesian coordinates** is such that the x axis is directed toward the east, tangent to the earth's surface; the y axis toward the north, tangent to the earth's surface; and the z axis toward the local **zenith**, perpendicular to the earth's surface. *Compare* **curvilinear coordinates**.

Cartesian tensor—A quantity specified by components that transform according to prescribed rules under rotations of (Cartesian) coordinate axes.

A Cartesian tensor of rank zero is a **scalar** and is invariant under rotations. A Cartesian tensor of rank one is a **vector**, the components of which transform under rotations according to a single 3×3 rotation matrix. Cartesian tensors of rank two have nine components that transform according to a product of two 3×3 rotation matrices. Tensors of higher rank may be defined in similar fashion. As examples related to meteorology, mass is a scalar, **velocity** is a vector, and the **stress tensor** is a Cartesian tensor of rank two. [Because of the restriction to transformation under rotation, a Cartesian tensor need not be a (general) **tensor**. The latter has components that transform in a prescribed way under arbitrary changes of coordinates.]

cascade—1. *See* **cascade shower**.

2. (Obsolete.) The spray **vortex** at the base of a **waterspout**.

cascade impactor—A low-speed impaction device consisting of a set of **impactor** plates connected in series or in parallel for use in sampling both solid and liquid **particles** suspended in the **atmosphere**.

The diameters of the nozzles or slits above each impactor plate are designed in such a manner that each sampling plate collects particles of predominantly one size **range**. This method is used to obtain different size fractions of ambient particles in the range of diameters ~ 0.5 to 30 μm.

cascade of energy—A flow of **turbulence kinetic energy** from larger **eddies** to smaller eddies.

In the **atmospheric boundary layer**, **turbulence** is usually produced at scales roughly equal to the **boundary layer** depth (order of 1 km) by **buoyancy** or **wind shear**, and is dissipated by **viscosity** into **heat** at the smallest scales (order of 1 mm). Richardson's (1922) poem eloquently describes this cascade: Big whorls have little whorls, which feed on their **velocity** / And little whorls have lesser whorls, and so on to viscosity.

Richardson, L. F., 1922: *Weather Prediction by Numerical Process.*

cascade shower—(Also air shower, cascade, extensive air shower, shower.) Multiple generations of secondary **cosmic rays** produced when primary cosmic rays interact with atoms in the **upper atmosphere**, yielding subatomic **particles** and **gamma rays**. The secondary cosmic rays in turn produce even more down through the **atmosphere**.

Billions of these particles travel downward at nearly the **speed of light** and at ground level may extend over several square kilometers (in which instance the shower may be termed an extensive air shower). The maximum flux of cosmic rays, both primary and secondary, is at an **altitude** of 20 km, and below this the **absorption** by the atmosphere reduces the **flux**, though the rays are still readily detectable at **sea level**. Intensity of cosmic ray showers has also been observed to vary with latitude, being more intense at the poles.

Friedlander, M. W., 1989: *Cosmic Rays*, pp. 13, 79.

cascading water—In reference to wells, **groundwater** that trickles or pours through cracks or perforations down the casing or uncased borehole above the **water level** in the well.

case-based reasoning—Reasoning by analogy and previous experience.

This is often the method employed by experienced **weather forecasters** and is the basis of forecasting by analogy. Successful development of **expert systems** or training of **neural networks** in meteorology often hinges on the appropriate selection of cases and their **analysis**.

case weather—As used in the Wisconsin tobacco industry, foggy weather that moistens the tobacco hanging in the sheds and makes it fit for handling and removal.

The tobacco when thus moistened is said to be "in case."

Cassegrainian mirror—An optical system for transmitting or receiving **light** in which the primary mirror has a central hole into which light is reflected from a smaller mirror with a focus coincident with that of the larger mirror.

Cassegrainian telescope—A coaxial optical system used to transmit and/or collect optical **energy**.

The concave primary mirror's surface is parabolic in shape and has a central hole into which **light** passes to or from a smaller, convex secondary mirror with a hyperbolic surface. The focus of the optical system is behind the hole in the primary mirror when collimated light illuminates the primary mirror.

castellanus—(Previously called castellatus.) A **cloud species** of which at least a fraction of its upper part presents some vertically developed **cumuliform** protuberances (some of which are taller than they are wide) that give the **cloud** a crenellated or turreted appearance.

This castellanus character is especially evident when the cloud is seen from the side. The cumuliform cloud elements generally have a common base and usually seem to be arranged in lines. The species is found only in the genera **cirrus**, **cirrocumulus**, **altocumulus**, and **strato-**

cumulus. Cirrus castellanus differs from cirrocumulus castellanus in that its vertical protuberances subtend an angle of more than 1° when observed at an angle of more than 30° above the **horizon**. When altocumulus castellanus and stratocumulus castellanus attain a considerable vertical development, they become **cumulus congestus** and often develop into **cumulonimbus**. Stratocumulus castellanus should not be confused with **stratocumulus** pierced by **cumulus**. *See* **cloud classification**.

castellatus—(Obsolete.) *See* **castellanus**.

CAT—Abbreviation for **clear-air turbulence**.

cat ice—Same as **shell ice**.

cat's nose—In England, a cool northwest **wind**.

cat's paw—A dark-looking patch on the surface of a lake or ocean, caused when a **gust** of **wind** generates **capillary waves** on the surface.

Catalina eddy—A **cyclonic** mesoscale **circulation** that forms over the coastal waters of southern California.

It is most common in the spring and early summer months but can occur at any time of the year. A Catalina eddy circulation is typically associated with a deepening of the marine layer and an associated improvement in the air quality in the Los Angeles Basin. In many cases the marine layer can deepen above 1 km and allow marine air to spill through gaps in the coastal mountains and reach interior **desert** regions. Catalina eddies are often observed to develop on the coast **downwind** of the coastal mountains during a period of offshore flow. Cyclonic **vorticity** develops over the bight of southern California in response to lowered **sea level pressure** produced by offshore, downslope flow across the Santa Ynez and San Rafael Mountains. Southeastward displacement and offshore expansion of the **cyclonic circulation** typically occur later in Catalina eddy events.

catalytic converters—Devices designed to reduce the amount of emissions from automobiles.

The current (so-called three-way) systems use a heated metal catalyst to reduce the emissions of **carbon monoxide** (CO), **hydrocarbons**, and **nitric oxide** (NO), all of which contribute to the formation of **photochemical smog**.

catalytic cycles—A set of chemical reactions in which the catalytic species involved in the reaction are constantly regenerated, allowing the reaction sequence to occur repeatedly.

For example, **ozone** destruction in the **stratosphere** occurs via a number of catalytic cycles involving nitrogen-, hydrogen-, and chlorine-containing **free radicals**.

catch—The **amount of precipitation** captured by a **rain gauge**.

catchment area—1. Same as **drainage area**.

This usage is common in British Commonwealth countries.

2. An area built specifically to collect **rainfall**.

catchment glacier—Same as **snowdrift glacier**.

cathode-ray oscillograph—(No longer commonly used.) Same as **cathode-ray oscilloscope**.

cathode-ray oscilloscope—(More commonly called oscilloscope or 'scope.) An instrument that displays visually on the face of a **cathode-ray tube** instantaneous voltages of electrical signals.

Either the **intensity** or the displacement of the **trace** may be controlled by the **signal** voltage. See **radarscope**.

cathode-ray radiogoniometer—A **radio direction finder** for **sferics** using twin **antennas** and **receivers** applied to the **CRT** deflection circuits in such a way as to yield the **source**'s direction.

cathode-ray tube—(Abbreviated CRT.) A vacuum tube consisting essentially of an **electron** gun producing a concentrated **electron beam** that impinges on a phosphorescent coating on the back of a viewing face or screen.

The excitation of the phosphor produces **light**, the **intensity** of which is controlled by regulating the flow of electrons. Deflection of the beam is achieved either electromagnetically by currents in coils around the tube or electrostatically by voltages on internal deflection plates.

cation exchange capacity—The sum total of exchangeable cations adsorbable by the **porous medium**; expressed in moles of **ion** charge per kilogram of porous media.

Cauchy number—(Also Hooke number.) In fluid mechanics, a **dimensionless group** that arises in the study of compressible flow.

The square of the **Mach number**, its physical interpretation is the ratio of inertia force (ρU^2) to **compressibility** force ($1/\kappa$), where ρ is **density**, U is a characteristic speed, and κ is compressibility.

cavaburd—(Also spelled kavaburd.) Shetland Island term for a thick fall of **snow**.

cavaliers—The local name, in the vicinity of Montpelier, France, for the days near the end of March or the beginning of April when the **mistral** is usually strongest.

caver—(Also spelled kaver.) A **gentle breeze** in the Hebrides, west of Scotland.

cavity—A region of reverse flow immediately **downwind** of an obstacle or mountain, typically found when the **static stability** of the air is near neutral (**Froude number** greater than about 1.5).

The reverse flow indicates a **circulation** or permanent **eddy** downwind of the obstacle where the **surface wind** is opposite to the **prevailing wind direction** upstream of the obstacle. The circulation continues partway up the mountain slope and then breaks away from the mountain and flows in the prevailing wind direction, finally descending to make a **closed circulation**. Such a cavity circulation causes **air pollutants** emitted near the **downwind** base of the mountain or building to be drawn back toward the obstacle and remain somewhat trapped within the circulation, leading to much larger pollutant concentration values than would be predicted by typical pollutant dispersion models.

cavity radiometer—A modern, self-calibrating **pyrheliometer** or device for measuring **direct solar radiation**.

The **receiver** (cavity) is typically conical in shape with **precision** temperature sensors and heaters. Operating at an **equilibrium** temperature, direct electrical power to the heater is decreased if **solar radiation** is added to the receiver (or added if solar radiation is removed). The electrical power necessary to maintain equilibrium is equivalent to the change in incident solar radiation.

CAVT—Abbreviation for **constant absolute vorticity trajectory**.

CAVU—[Abbreviation for **ceiling** and **visibility** unlimited (and unrestricted).] An operational term no longer formally defined in meteorology, but still commonly used in aviation, that designates a condition wherein the ceiling is more than 10 000 ft and the visibility is more than 10 miles.

CBL—Abbreviation for **convective boundary layer**.

CCL—Abbreviation for **convective condensation level**.

CCN—Abbreviation for **cloud condensation nuclei**.

CCS—Abbreviation for **co-cumulative spectrum**.

cd—Abbreviation for **candela**.

CDR—Abbreviation for **circular depolarization ratio**.

ceiling—1. After U.S. weather observing practice, the height ascribed to the lowest layer of **clouds** or obscuring phenomena when it is reported as **broken**, **overcast**, or **obscuration** and not classified "thin" or "partial." The ceiling is termed unlimited when the foregoing conditions are not satisfied. Whenever the height of a **cirriform** cloud layer is unknown, a slant / is reported in lieu of a height value. At all other times, the ceiling is expressed in feet above the surface, which is a horizontal plane with an **elevation** above **sea level** equal to the **airport elevation**. At stations where this does not apply, "surface" refers to the ground elevation at the point of **observation**.

For obscurations, the ceiling height represents **vertical visibility** into the obscuring phenomena rather than the height of the base, as in the case of clouds or obscuring phenomena aloft. In an **aviation weather observation**, the ceiling height is always preceded by a letter that designates the **ceiling classification**.

2. The greatest **altitude** to which an airborne object (aircraft, **balloon**, **rocket**, projectile, etc.) can rise, under a given set of conditions.

The ceiling of an aircraft is that altitude where the **stalling Mach number** and the buffeting **Mach number** approach identical value.

ceiling balloon—A small **balloon** used to determine the height of the **cloud base**.

The height can be computed from the ascent **velocity** of the balloon and the time required for its disappearance into the **cloud**.

ceiling classification—In **aviation weather observation**, a description or explanation of the manner in which the height of the **ceiling** is determined.

The different types of ceilings according to this classification are **aircraft ceiling**, **balloon ceiling**, **estimated ceiling**, **indefinite ceiling**, **measured ceiling**, and **precipitation ceiling**.

ceiling-height indicator—*See* **cloud-height indicator, ceiling light, ceilometer**.

ceiling light—(Also called ceiling projector) A type of **cloud-height indicator** that uses a **searchlight** to project vertically a narrow **beam** of **light** onto a **cloud base**.

The height of the cloud base is determined by using a **clinometer**, located at a known distance from the ceiling light, to measure the angle included by the illuminated spot on the **cloud**, the **observer**, and the ceiling light. *Compare* **ceilometer**.

> Middleton, W. E. K., 1969: *Invention of the Meteorological Instruments*, Johns Hopkins Press, Baltimore, p. 274.

ceiling projector—Same as **ceiling light**.

ceilometer—An automatic, active, remote-sensing instrument for detecting the presence of **clouds** overhead and measuring the height of their bases.

For optically thin clouds, such as most **cirrus**, more than one layer may be detected, but when optically thick clouds, such as liquid water **stratus**, are present, the **light** beam is unlikely to penetrate much beyond the base of the lowest liquid layer. Laser ceilometers use intense pulses of light in a very narrowly collimated, vertically directed **beam**, and have collocated **transmitter** and **receiver** systems. The **cloud base** heights may be displayed in a variety of **time-height section** images or backscatter **intensity** profile plots. Some older ceilometers use separated transmitter and receiver units. The instruments are designed to work during the day or night.

celerity—The **velocity** of a small **wave** with respect to the body of water, expressed as

$$c = (g y)^{1/2},$$

where g is the **acceleration of gravity** and y is the depth of flow.

celestial equator—The **projection** of the plane of the geographical **equator** upon the **celestial sphere**.

celestial horizon—Same as **astronomical horizon**.

celestial pole—The **projection** of the earth's axis upon the **celestial sphere**.

celestial sphere—The apparent sphere of infinite radius, having Earth as its center.

It is upon the "inner surface" of this sphere that all heavenly bodies, the **ecliptic**, and the **celestial equator** appear. Disregarding the effects of **topography** and **refraction** near the **horizon**, for practical purposes half of this sphere may be considered visible from any point on the earth's surface at any time.

cell—In **radar** usage, a local maximum in **radar reflectivity** that undergoes a life **cycle** of growth and decay.

The rising portion of the **reflectivity** maximum is indicative of **updraft**, and the later descending portion is indicative of a precipitation **downdraft**. Cells in ordinary convective storms last from 20 to 30 min, but often form longer-lasting multicell convective storms. Cells in **supercell** storms are more steady and last considerably longer. *See also* **thunderstorm cell**.

Cellini's halo—Same as **heiligenschein**.

cellular circulation—In the **general circulation** context, **circulation** in the latitude-height plane as exemplified by Hadley circulation.

See **Hadley cell, Ferrel cell**.

cellular cloud pattern—*See* **open cells, closed cells**.

cellular convection—A periodic form of **convection** separated by stream surfaces across which there is little **mixing** of the fluid.

In meteorology, the term is often used in reference to **convection** visualized by its **modulation** of **clouds**. *See* **open cells, closed cells**.

Celsius temperature scale—Same as **centigrade temperature scale**, by convention.

The Ninth General Conference on Weights and Measures in 1948 replaced the designation "degree centigrade" by "degree Celsius." Originally, Celsius took the **boiling point** of water at 1000 mb **pressure** as 0° and the **ice point** as 100°, which is inverted from the present-day **temperature scale**. *Compare* **Fahrenheit temperature scale**.

center jump—The formation of a second **pressure center** within a well-developed **pressure system**, nearly always a **low**.

The original **low pressure center** usually diminishes in magnitude as the new center deepens. The center of the **cyclone** appears to jump from the first to the second point of low pressure and can occur at a point farther along the original path of the low or may form along a new path displaced from the original low.

center of action—1. Any one of the semi-permanent **highs** and **lows** that appear on mean charts of **sea level pressure**.

As originally used by L. Teissenenc de Bort in 1881, this term was applied to maxima and minima of **pressure** on daily charts. The main centers of action in the Northern Hemisphere are the **Icelandic low**, the **Aleutian low**, the **Azores high** and/or **Bermuda high**, the **Pacific high**, the **Siberian high** (in winter), and the Asiatic low (in summer). Other less intense or less consistent mean systems may be considered. Fluctuations in the nature of these centers are intimately associated with relatively widespread and long-term weather changes.

2. As used by Sir Gilbert Walker, any region in which the **variation** of any **meteorological element** is related to weather of the following **season** in other regions.

center of falls—Same as **pressure-fall center**.

center of gravity—The point at which the resultant gravitational force acts upon an object.

The center of gravity is not necessarily inside the object. For example, the center of gravity of a ring is at the center of symmetry. If the geometry of the object does not change with time, the center of gravity will remain unchanged in relation to the object. The mathematical definition of center of gravity is

$$\mathbf{x}_c = (\int \mathbf{x}\rho dV)/(\int \rho dV),$$

where **vector x** describes the position of the object in a given **coordinate system**, $\rho = \rho(\mathbf{x})$ is the **density**, and V is the volume of the object.

center of rises—Same as **pressure-rise center**.

centered difference—See **finite differencing**.

centibar—(Abbreviated cb.) The **pressure** unit of the **meter–tonne–second system** of physical units, equal to 10 millibars or 10^4 dynes cm^{-2}.

centigrade temperature scale—(Abbreviated °C.) A **temperature scale** with the **ice point** of water at 0° and the **boiling point** at 100°.

Conversion to the **Fahrenheit temperature scale** is according to the formula °C = 5/9 (°F − 32). See **Celsius temperature scale**.

centimeter–gram–second system—(Abbreviated cgs.) A system of physical units based on the use of the centimeter, the **gram**, and the second as elementary quantities of length, mass, and time, respectively.

In this system, **density** is expressed in gm cm^{-3}, speed in cm s^{-1}, force in **dynes** (gm cm s^{-2}, **pressure** in baryes (dynes cm^{-2}), and **energy** in ergs (gm cm^{2}s^{-2}). While this is a popular system of units in nearly all fields of science and technology, some of the meter–kilogram–second system units are more convenient for certain meteorological applications.

centipoise—A convenient unit of **viscosity**, defined as one hundredth of a **poise**, or 10^{-2} gm cm^{-1}s^{-1}.

central dense overcast—The region of dense **cloud** near the core of a **tropical cyclone**.

central forecasting office—A forecast office with the responsibility for preparing **analysis**, guidance, and prognostic data, and/or for preparing forecasts and warnings over a regional or national domain to be disseminated to local **weather stations**, media, and other users of weather information.

central limit theorem—See **normal distribution**.

central moment—In **statistics**, a **moment** taken about the mean.

central pressure—At any given instant, the **atmospheric pressure** at the center of a **high** or **low**; the highest pressure in a high, the lowest pressure in a low.

Central pressure almost invariably refers to **sea level pressure** of systems on a **surface chart**.

central tendency—In **statistics**, the general level, characteristic, or typical value that is representative of the majority of cases.

Among several accepted measures of central tendency employed in data **reduction**, the most common are the **arithmetic mean** (simple average), the **median**, and the **mode**.

central water—The **water mass** of the permanent or oceanic **thermocline**, which is located at a depth of between 150 and 800 m.

Central water is formed by **subduction** in the **subtropics** between 25° and 45° latitude in both hemispheres. It therefore spans a wide **temperature** and **salinity** range, with temperature and salinity both decreasing with depth. Each ocean has its own central water with its own specific temperature–salinity relationship depending on the atmospheric conditions in the formation region. These are distinguished by appropriate names, for example, South Atlantic Central Water, Western North Pacific Central Water, etc.

centrifugal force—The **apparent force** in a rotating system, deflecting masses radially outward from the axis of rotation, with magnitude per unit mass $\omega^2 R$, where ω is the angular speed of rotation and R is the radius of curvature of the path.

This magnitude may also be written as V^2/R, in terms of the **linear** speed V. This force (per unit mass) is equal and opposite to the **centripetal acceleration**. The centrifugal force on the earth and **atmosphere** due to rotation about the earth's axis is incorporated with the **field** of **gravitation** to form the field of **gravity**.

centrifugal instability—In general, the **instability** of a rotating fluid system, usually synonymous with **rotational instability** and **inertial instability**.

centrifuge moisture equivalent—*See* **moisture equivalent**.

centripetal acceleration—The **acceleration** on a **particle** moving in a curved path, directed toward the instantaneous center of curvature of the path, with magnitude V^2/R, where V is the speed of the **particle** and R the radius of curvature of the path.

This acceleration is equal and opposite to the **centrifugal force** per unit mass.

ceraunograph—(Also spelled keraunograph.) A **radio direction-finder** for recording **sferics**.
See also **sferics receiver**.

ceraunometer—A radio **receiver** tuned to receive and count local **sferics**.

cers—A name for the **mistral** in Catalonia, Narbonne, and parts of Provence (southern France and northeastern Spain).

It is very violent and turbulent in the Aude Valley below Carcassonne with gusts often reaching 22–25 m s⁻¹ (50–55 mph). It is cold in winter, hot in summer, always dry and **clear**. A similar northerly **wind** in Spain is the **cierzo**.

certainty factor—The extent to which a statement is considered reliable.

This is a subjective factor, as opposed to **probability**, which is a mathematical likelihood of occurrence, based on previous experience. For example, a **forecaster** may state that there is a 50% probability of rain, and assign a certainty factor of 95% to that statement. There are several methods of combining certainty factors to arrive at an overall certainty. The most common of these is the Dempster–Shafer rule (Shafer 1976).

Shafer, G., 1976: *A Mathematical Theory of Evidence*, Princeton University Press.

cetane—Common name for n-hexadecane, $C_{16}H_{34}$, a large alkane with good ignition quality, which is used as a diesel fuel.

The cetane number of a fuel is a measure of the quality of a diesel fuel; it is defined as being the fraction of cetane in a blend of cetane and a less reactive compound that would have the same ignition quality as the fuel in question. *See* **alkanes, octane**.

CFCs—*See* **chlorofluorocarbons**.

CFL—Abbreviation for **Courant–Friederichs–Lewy condition**.

CFLOS—Acronym for **cloud-free line of sight**.

CFR—*See* **VFR**.

cfs—Abbreviation for cubic foot per second, a unit of **discharge** commonly used in **hydrology** and **hydraulics**.

cgs system—Abbreviation for **centimeter–gram–second system**.

chaff—Lightweight reflecting material, typically consisting of aluminum foil strips or metal-coated fibers, that is released in the **atmosphere** to produce **radar echoes**.

Initially developed as a military countermeasure, chaff is used in meteorological research to enable tracking of air motions in conditions with no natural airborne scatterers or to provide stronger echoes than those from natural scatterers.

chaff seeding—The practice of releasing radar **chaff**, which is aluminum or aluminum coated plastic strips or needles, into the **atmosphere** for the purpose of obtaining a strong **radar** reflecting **signal** in an otherwise weakly reflecting **environment**.

Chaff seeding, as a research technique, has been used to study the motion, **deformation**, **transport**, and **mixing** processes of the atmosphere on scales from a few meters to a few kilometers. Chaff seeding has also been used on the hypothesis that **corona discharge** from the needle tips could reduce **lightning** from thunderstorms. *See* **window**.

chain length—The ratio of the rate at which a **catalytic cycle** occurs to the rate of destruction of the catalytic species.

The chain length thus provides a measure of the effectiveness of a particular catalytic cycle.

chain lightning—Same as **beaded lightning**.

chain reaction—Set of reactions in which a reactive component (usually a **free radical**) is continuously regenerated, enabling the reactions to occur many times before being terminated.

A typical chain reaction consists of distinct initiation, propagation, and termination reactions. **Catalytic cycles** occurring in the **atmosphere** are entirely analogous.

chain rule—A basic **inference** rule used in most **expert systems**.

If *A* implies *B* and *B* implies *C*, then it is also true that *A* implies *C*. For example, if the fact that it is July in Boston implies that the **temperature** is above **freezing**, and having the temperature above freezing implies that no **precipitation** will reach the ground as **snow**, it follows that it will not snow in Boston in July.

challiho—In India, southerly winds, sometimes violent, causing blinding **duststorms**.

These conditions can be continuous for over two months.

chamsin—Same as **khamsin**.

chandui—Same as **chanduy**.

chanduy—(Also spelled chandui.) A cool, descending **wind** at Guayaquil, Ecuador, that blows during the **dry season** (July to November).

change chart—(Also called tendency chart.) A **chart** indicating the amount of direction of change of some **meteorological element** during a specified time interval, for example, a **height-change chart** or **pressure-change chart**.

See **differential analysis**, **tendency**.

change of phase—A process in which a substance that can exist in two or more phases is converted from one phase to another.

The most important meteorological examples are the **evaporation**, **condensation**, **freezing**, melting, **deposition**, and **sublimation** of water. *See* **phase**.

change of state—A change in the **state variables**, for example, **temperature**, **pressure**, **density**.

channel—Specific **wavelengths** of onboard satellite sensors.

channel control—A condition whereby the **stage**, and thereby the **rating curve**, of a river at a point is controlled by the **discharge** and the shape, slope, and roughness of the channel over a considerable length.

channel storage—The water volume within a specified portion of a **stream** channel at any given time.

channeling—Winds parallel to the main (longitudinal) axis of a mountain valley, resulting from the diversion of **momentum** from larger-scale winds above ridgetops into the valley.

Most commonly, direct **transfer** of momentum into the valley by **mixing** or by an impressed large-scale **pressure gradient** produces a flow along the valley axis that represents a component of the larger-scale flow, that is, in the same direction as the flow aloft. Occasionally, however, the perpendicular **geostrophic wind** relationship between the pressure gradient and the large-scale **wind** can lead to situations where the impressed pressure gradient produces a flow in the valley in the opposite direction to the along-valley component of the large-scale flow above the ridgetops.

chaos—The property describing a **dynamical system** that exhibits erratic behavior in the sense that very small changes in the initial state of the system rapidly lead to large and apparently unpredictable changes in the later state.

Chapman cycle—*See* **Chapman mechanism**.

Chapman layer—(Also called Chapman region.) The idealized height distribution of **ionization** as a function of height produced solely by **absorption** of **solar radiation**.

Named for Sydney Chapman, who first derived the shape of such a layer mathematically. Some of the basic assumptions used to develop the equation were that the ionizing **radiation** from the sun is essentially **monochromatic**, that the ionized constituent is distributed exponentially (i.e., with a constant **scale height**), and that there is an **equilibrium** between the creation of **free electrons** and their loss by **recombination**.

Chapman mechanism—A series of reactions, first proposed by Sidney Chapman in the 1930s, to explain the presence of the **ozone layer** in the earth's **stratosphere**.
The reaction sequence is as follows:

$$O_2 + h\nu \rightarrow O + O$$

$$O + O_2 + M \rightarrow O_3 + M$$

$$O_3 + h\nu \rightarrow O + O_2$$

$$O + O_3 \rightarrow O_2 + O_2$$

$$O + O + M \rightarrow O_2 + M,$$

where M is any other molecule in the termolecular reaction. Models based only on the **Chapman cycle** overpredict stratospheric O_3 levels, and it is now recognized that additional destruction of O_3 occurs through **catalytic cycles** involving oxides of **nitrogen**, chlorine, and **hydrogen**.
> Chapman, S., 1930: A theory of upper atmospheric ozone. *Mem. Roy. Meteor. Soc.*, **3**, 103–125.

Chapman region—*See* **Chapman layer**.

Chappuis bands—Spectroscopic feature of the **ozone** molecule with considerable structure related to vibrational levels in the ozone molecule occurring at a **wavelength** around 600 nm, or an **energy** of 2 eV.
The **absorption** bands lead to **photodissociation** of an ozone molecule into an O_2 molecule and an **oxygen** atom. The bands were first identified in the atmospheric **spectrum** by Chappuis (1880) and have been used to measure ozone in the **atmosphere**.
> Chappuis, J., 1880: Sur le spectre d'absorption de l'ozone. *C. R. Acad. Sci. Paris*, **91**, p. 985.

characteristic equation—1. An equation defining the **characteristics** of a set of partial differential equations.
2. A **linear** algebraic equation determining the **eigenvalues** (*see* **characteristic-value problem**) or free waves of a **boundary-value problem**.

characteristic frequency—Same as **natural frequency**.

characteristic length—Some representative length of a physical system, for example, the radius of a **vortex**, **correlation** length of atmospheric **temperature** fluctuations.

characteristic line—*See* **characteristics**.

characteristic of the pressure tendency—Shape of the curve recorded by a **barograph** during the 3-h period preceding an **observation**, represented on the **synoptic chart** by a **symbol** with a similar shape.

characteristic point—*See* **lifting condensation level**.

characteristic value—Same as **eigenvalue**.

characteristic-value problem—A problem in which an undetermined **parameter** is involved in the coefficients of a differential equation and in which the solution of the differential equation, with associated **boundary conditions**, exists only for certain discrete values of the parameter, called **eigenvalues** (or characteristic values, sometimes principal values).
An important example of a physical problem that leads to a characteristic-value problem is the determination of the modes and frequencies of a vibrating system. In this case the **dependent variable** of the differential equation represents the displacements of the system and the parameter represents the frequencies of vibration.

characteristic velocity—Some representative **velocity** in a physical system.

characteristics—Lines or surfaces associated with a partial differential equation, or with a set of such equations, which are at all points tangent to characteristic directions, determined by certain specified **linear** combinations of the equations.

The use of these lines or surfaces may facilitate the solution of the equations and is known as the **method of characteristics**. The method has been particularly successful, for example, in the problem of finite-amplitude expansion and shock waves, and the analogy between certain of these flows and a traveling **disturbance** on an **inversion** has been studied by meteorologists.

Courant, R., and K. O. Friedrichs, 1948: *Supersonic Flows and Shock Waves*, ch. 2.
Freeman, J. C., 1951: *Compendium of Meteorology*, 421–433.

charge separation—The physical process causing **cloud electrification**.

On the **particle** scale, the process can include particle collisions with selective charge **transfer** and particle capture of **small ions**. On the **cloud** scale, the process can include gravity-driven differential particle motions and **convective transport** of charged **air parcels**.

Charles–Gay–Lussac law—(Also called **Charles's law**, **Gay–Lussac's law**.) An empirical generalization that in a gaseous system at constant **pressure**, the **temperature** increase and the relative volume increase remain in approximately the same proportion for all so-called perfect gases.

Mathematically,

$$t - t_0 = \frac{1}{c} \frac{(v - v_0)}{v_0},$$

where t is temperature, v is the volume, and c is a **coefficient of thermal expansion** independent of the particular gas. If the **centigrade temperature scale** is used and v_0 is the volume at 0°C, then the value of the constant c is approximately $1/273$.

Charles's law—Same as **Charles–Gay–Lussac law**.

Charnock's relation—An empirical expression for **aerodynamic roughness length** $z_0 = \alpha_c u_*{}^2/g$ over the ocean, where u_* is the **friction velocity**, g is gravitational **acceleration**, and $\alpha_c \approx 0.015$ is the Charnock **parameter**.

It accounts for increased roughness as **wave** heights grow due to increasing surface **stress**.

chart—A map showing, principally, the **pressure pattern**, **height pattern**, or any meteorological **parameter** at a specified time.

chart datum—The **datum** to which levels on a nautical **chart** and tidal predictions are referred; usually defined in terms of a low-water tidal level, which means that a chart datum is not a horizontal surface, but it may be considered so over a limited local area.

chart plotting—The marking of meteorological observations on a **synoptic chart**.

check observation—An **aviation weather observation** taken primarily for aviation radio broadcast purposes.

It is usually abbreviated to include just those elements of a **record observation** that have an important effect on aircraft operations, such as **ceiling**, **visibility**, state of sky, **atmospheric phenomena**, **wind speed** and **direction**, **altimeter setting**, and pertinent remarks.

chemical composition of precipitation—Type and quantity of dissolved or suspended compounds in **precipitation**.

chemical energy—Energy produced or absorbed in the process of a chemical reaction.

In such a reaction, **energy** losses or gains usually involve only the outermost **electrons** of the atoms or **ions** of the system undergoing change; here a chemical bond of some type is established or broken without disrupting the original atomic or ionic identities of the constituents. Chemical changes, according to the nature of the materials entering into the change, may be induced by **heat** (thermochemical), **light** (photochemical), and electric (electrochemical) energies.

chemical hygrometer—A group of **hygrometers** based on chemical effects that depend on **humidity**.

These hygrometers are not used in practical meteorology. Examples include pictures or figures, covered with cobalt chloride, that appear blue at low and pink at high **relative humidity**. In industrial processes electrolytic methods are used, which are especially suited for extremely low humidity. *See also* **absorption hygrometer**.

chemical tracers—*See* **tracer**.

chemiluminescence—*See* **luminescence**.

chemisorption—Adsorption of a chemical onto a surface in which the strength of the interaction is somewhat stronger than pure physical **adsorption**, and more like a chemical bond.

A bond strength of around 80 kJ mol⁻¹ is often taken to be indicative of a true chemical interaction. In some cases, it is thought that chemical modification of the adsorbed species occurs.

chemosphere—The vaguely defined region of the **upper atmosphere** in which photochemical reactions take place.

It is generally considered to include the **stratosphere** (or the top thereof) and the **mesosphere**, and sometimes the lower part of the **thermosphere**. This entire region is the seat of a number of important **photochemical reactions** involving atomic **oxygen** O, molecular oxygen O_2, **ozone** O_3, hydroxyl OH, **nitrogen** N_2, **sodium** Na, and other constituents to a lesser degree. See **atmospheric shell**.

chergui—An east or southeast **desert wind** in Morocco (North Africa), especially in the north.

It is persistent, very dry and dusty, hot in summer, cold in winter. It blows with high pressure in the Mediterranean and isobars run nearly parallel with the coast. It is said to be most frequent in the 40 days following 11 or 12 July, a period known as the Smaïm. *Compare* **simoom**.

Chezy equation—An empirical equation relating the **mean velocity** of flow to channel characteristics.

It is expressed as

$$V = CR^{1/2}S_f^{1/2},$$

where C is the Chezy resistance coefficient, S_f the **gradient** of the **total head line**, and R the **hydraulic radius**. For **discharge** Q,

$$Q = AV = CAR^{1/2}S_f,$$

where A is the cross-sectional area. *See* **Manning equation**.

chi-square test—A **statistical significance test** based on **frequency** of occurrence; it is applicable both to qualitative attributes and quantitative variables.

Among its many uses, the most common are tests of hypothesized probabilities or **probability distributions (goodness of fit)**, **statistical dependence** or **independence** (association), and common **population** (homogeneity). The formula for chi square (χ^2) depends upon intended use, but is often expressible as a sum of terms of the type $(f - h)^2/h$ where f is an observed frequency and h its hypothetical value.

chibli—Same as **ghibli**.

chichili—*See* **chili**.

Chile Current—*See* **Peru/Chile Current**.

chili—A warm, dry, descending **wind** in Tunisia, resembling the **sirocco**.
In southern Algeria it is called chichili.

chill factor—*See* **wind-chill factor, wind-chill index**.

chill hour—A unit of measure for fulfillment of plant dormancy requirements prior to the start of springtime growth, especially tree fruits.

Determined by totaling the number of hours during a cold **season** when **air temperature** is below a specified value. The reference **temperature** (base temperature) for chill hour accumulation is species dependent; for example, for peach dormancy to be fulfilled, the number of hours that the air temperature is below 7°C (45°F) is accumulated. A minimum number of chill hours dependent upon species and variety must be exceeded before **normal** springtime growth can begin.

chill unit—An **index** calculated from **air temperature** to estimate fulfillment of plant dormancy requirements and the ability to start springtime growth, especially tree fruits.

Chill units are calculated in hourly time steps and summed over a time period. Cold temperatures add to chill units as the **index** ranges from 0 at temperatures less than 1°C, to 1 near 7°C, and 0 at 14°C. Temperatures above 14°C reduce chill unit accumulation as the **index** falls to -1 at 21°C. A minimum total number of chill units dependent upon species and variety must be exceeded before **normal** springtime growth can begin. Chill hour calculations frequently require hourly data from late evening to early morning.

chilled-mirror hygrometer—A device for measuring **dewpoint temperature** by chilling a mirror until **water vapor** condenses on it.

The onset of **condensation** is measured by bouncing a **light** beam off the mirror and recording when the **reflection** changes from **specular** to **diffuse**. The mirror can be cooled and heated by a variety of methods, one of which is thermoelectrically. Electronic circuits continuously monitor the light reflection and adjust the mirror **temperature** to maintain it at the **dewpoint**, even while

humidity is varying. These devices often have high absolute **accuracy** and **precision**, but somewhat slow response.

chilling—Requirement of deciduous fruit trees and some other plants for undergoing cold temperatures during the dormant period.

Sufficient chilling brings about emergence from dormancy and **normal** fruit set and growth during the following **season**. *See* **chill hour**.

chilling hour—*See* **chill hour**.

chilling injury—Physiological damage to plant parts and tissues in the **temperature range** from about 20° to 0°C.

Compare **frost damage**.

chimney cloud—A **cumulus** cloud that has much greater vertical than horizontal extent.

It frequently takes the form of a long "neck" protruding from the tops of a lower **cloud** mass where a locally strong **convection current** has penetrated the **inversion**. The shape results from the penetration into relatively **moist air** aloft which inhibits **evaporation**.

chimney plume—The **smoke** plume that is emitted from a stack (i.e., chimney) at an industrial facility.

The **plume** is usually buoyant and rises up to a final height depending on its initial **buoyancy** and **momentum** and the **stratification** of the **boundary layer**.

China Coastal Current—A **surface current** in the East China Sea and Yellow Sea flowing southward along the Chinese coast from Bohai Gulf to Taiwan.

It is driven by the northeast **monsoon** in winter and continues against the **southwest monsoon** through the summer, strengthened by river runoff from monsoonal **rainfall**. Taking in most of the waters from the Yangtze River, it contributes greatly to the increased summer transport of the **Tsushima Current**.

chinook—The name given to the **foehn** in western North America, especially on the plains to the lee or eastern side of the Rocky Mountains in the United States and Canada.

On the eastern slopes of the Rocky Mountains the chinook generally blows from the west or southwest, although the direction may be modified by **topography**. Often the chinook begins to blow at the surface as an **arctic front** retreats to the east, producing dramatic **temperature** rises. Jumps of 10°–20°C can occur in 15 minutes, and at Havre, Montana, a jump from −12° to +5°C in 3 minutes was recorded. Occasionally the arctic front is nearly stationary and oscillates back and forth over an observing **station**, causing the temperature to fluctuate wildly as the station comes alternately under the influence of warm and cold air. As in the case of any foehn, chinook winds are often strong and gusty. They can be accompanied by **mountain waves**, and they can occur in the form of damaging **downslope windstorms**. The air in the chinook originates in midtroposphere above the ridgetops, and its warmth and dryness result from **subsidence**. When moisture is present, a variety of **mountain-wave clouds** and lee-wave clouds can form, such as the **chinook arch** of the Canadian Rocky Mountains west of Calgary, Alberta. The chinook brings relief from the cold of winter, but its most important effect is to melt or sublimate **snow**: A foot of snow may disappear in a few hours. As with the foehn, researchers have attempted to classify chinooks as downslope winds with warming and **boras** as those accompanied by cooling. Again, these schemes have produced limited success because of the many ambiguous or erroneously classified cases.

chinook arch—A **mountain-wave cloud** form that occurs along the east slopes of the Rocky Mountains, particularly in the Montana–Alberta area, where the range runs roughly north–south.

The arch **cloud** has a sharp western (**upwind**) edge and a great cross-wind extent in the lee of and parallel to the Rockies. Arch clouds can stretch a few hundred km or more north–south as measured from the ground and from satellite. The arch refers to the shape of the western edge of the **wave cloud** as seen by an **observer** looking westward from the ground on the plains east (to the lee) of the mountains. The arch is especially noticeable when the blue sky to the west of it is completely **clear** of other cloud layers. Glider pilots sometimes refer to the clear area as the "window." The distinctive shape of the arch is due to the **cloud height** (usually middle or high cloud), its large cross-wind extent along the **horizon**, and the perspective of the observer, who sees the cloud and the mountain range converge with distance to the north and south. The arch cloud often presages a **chinook**. Because of the rapid **temperature** rises associated with a chinook, the arch has an important place in local **weather lore** as a **predictor**.

chinophile—Snow-loving; usually used to describe plants that have grown through a **snow cover**.

chip log—A line marked at intervals [commonly about 15 m (50 ft)], and paid out over the stern of a moving ship.

By timing the intervals at which the markers appear as the line is pulled out by a drag (the "chip"), the ship's speed can be determined. The **wave length** of **ocean waves** can be estimated by noting the position of **wave crests** relative to the markers.

chlorine compounds—Any chemical compound containing chlorine (Cl), the 17th **element** in the periodic table.

There are a number of chlorine-containing compounds found in the **atmosphere**, which are emitted from a variety of sources. Organic compounds containing chlorine are produced either via natural oceanic processes (e.g., **methyl chloride**) or for industrial purposes (e.g., **chlorofluorocarbons**). Some of these organic compounds are sufficiently long-lived to allow their **transport** to the **stratosphere**, where they are destroyed by **photolysis**. This process liberates chlorine atoms, which are active in the catalytic destruction of stratospheric **ozone**.

chlorine dioxide—A highly reactive oxide of chlorine with chemical formula OClO.

This species is of importance in the chemistry of polar stratospheric **ozone** depletion, as it is a product of the reaction of ClO with BrO. Its strong, structured **absorption spectrum** in the near-UV–visible region makes it extremely susceptible to destruction by **photolysis**, but also makes possible its **detection** via UV absorption in the **atmosphere**. Its detection in the **antarctic ozone hole** was one of the key discoveries that pointed to chlorine chemistry being the major cause of polar stratospheric **mountain-wave cloud** depletion. This compound should not be confused with the isomeric ClOO radical, which is thermally unstable (dissociating to Cl and O_2).

chlorine monoxide dimer—A chlorine compound with chemical formula Cl_2O_2.

Its chemical structure is actually ClOOCl, and it is more properly known as dichlorine peroxide. This species plays a critical role in polar **ozone** destruction. Its formation (from **recombination** of two chlorine monoxide radicals in a termolecular reaction), followed by its subsequent **photolysis** to regenerate Cl atoms, is the major ozone depletion mechanism involved in the formation of the **antarctic ozone hole**.

chlorine monoxide radical—A reactive radical species with chemical formula ClO.

Chlorine monoxide is an important inorganic chlorine compound in the **stratosphere**. This species is produced in the reaction of Cl atoms with **ozone**. Its conversion back to Cl without reforming O_3 (e.g., via reaction with O atoms) results in chlorine-catalyzed ozone loss in the stratosphere.

chlorine nitrate—A chlorine compound with chemical formula $ClNO_3$, often written as $ClONO_2$.

It is of importance as a **reservoir** species for reactive chlorine and **nitrogen** in the lower and middle **stratosphere**. It is formed in the termolecular reaction of ClO with NO_2 and is destroyed via **photolysis** by near-UV **solar radiation** or via heterogeneous reactions on **polar stratospheric clouds** and **aerosol** particles.

chlorine oxides—(Designated ClO_x.) The chemical group comprising chlorine atoms (Cl), chlorine monoxides (ClO), etc.

chlorinity—(Symbol Cl.) A measure of the chloride content, by mass, of **seawater** (grams per kilogram of seawater, or **per mille**).

Originally chlorinity was defined as the weight of chlorine in grams per kilogram of seawater after the bromides and iodides had been replaced by chlorides. To make the definition independent of atomic weights, chlorinity is now defined as 0.3285233 times the weight of silver equivalent to all the halides. The chlorinity of seawater is generally determined in order to permit the calculation of **salinity**, although other methods of determining salinity can be used. By using **normal water** as a comparison **standard**, Knudsen burettes and pipettes for the analysis, and Knudsen's tables to compute the results, determinations as accurate as those of a time-consuming gravimetric analysis can be made with a rapid titration of the seawater against silver **nitrate** solution, employing potassium chromate or other suitable indicator for the end point. It is customary to express chemical analyses of seawater in terms of chlorinity or of **chlorosity**.

chlorofluorocarbons—(Abbreviated CFCs; also called freons.) A series of compounds of entirely **anthropogenic** origin that were used in industrial applications throughout most of the twentieth century.

The major CFCs are CFC-11 ($CFCl_3$) and CFC-12 (CF_2Cl_2). These compounds have long atmospheric **lifetimes** (many decades) enabling them to be transported to the **stratosphere**. The discovery that liberation of chlorine from these species was responsible for the appearance of the **antarctic ozone hole** each spring led to the introduction of the **Montreal Protocol**, which banned the use of these species. Although they are no longer in use, their long lifetimes will lead to a very slow removal from the **atmosphere**.

chloroform—Trichloromethane, $CHCl_3$; halocarbon that was formerly in widespread use as an anesthetic.

Due to its relatively low emission rate and short lifetime, it reaches only low levels in the **atmosphere**.

chlorosity—The chloride content of one liter of **seawater**.

It is equal to the **chlorinity** of the sample times its **density** at 20°C.

chocolate gale—Same as **chocolatero**.

chocolatero—(Also called **chocolate gale**.) A moderate **norther** in the gulf region of Mexico.

See **norte**.

chocolatta north—A northwesterly **gale** of the West Indies.

chop—Short **period** surface **gravity waves** that give the sea surface a confused appearance.

choppy sea—Popularly, descriptive of short, rough, irregular **wave motion** on a sea surface.

choroisotherm—*See* **isotherm**.

chota bursát—(Hindustani for "small rain.") In India, a day or two of rainy weather that may precede the regular rains of the **monsoon**.

chou lao hu—(Chinese for "autumn tiger.") In northern China, a spell of hot weather after the beginning of the fall season.

See **Indian summer**.

chromatic scintillation—*See* **scintillation**.

chromosphere—The thin (10 000 km) layer of relatively transparent gases, predominantly **hydrogen** and **helium**, above the **photosphere** and below the **corona** of the sun.

It is best observed during total solar eclipse when its **emission spectrum** may be studied.

chronograph—A clock-driven device for recording the time of occurrence of an event.

It is often used, for example, in conjunction with a contact **anemometer** and a **wind vane** to obtain a record of **wind speed** and **direction** as a function of time. *See* **multiple register**.

chronoisotherm—*See* **isotherm**.

chronometric radiosonde—A **radiosonde** with a **carrier wave** switched on and off in such a manner that the interval of time between the transmission of signals is a function of the magnitude of the meteorological elements being measured.

chronothermometer—A **thermometer** consisting of a clock mechanism the speed of which is a function of **temperature**.

It automatically calculates the **mean temperature**.

chubasco—A severe **thunderstorm** with vivid **lightning** and violent **squalls** coming from the land on the west coast of Nicaragua and Costa Rica in Central America.

Chubascos are especially frequent in May, with a secondary maximum in October. Soon after noon, dark, heavy **clouds** form over the mountains, becoming denser and lower. At about 4 p.m. the southwesterly **sea breeze** veers to east-northeast, and the storm breaks, usually continuing until about 8 p.m.

chun fung—(Chinese for "spring wind.") In China, a spring breeze; pleasant weather at the start of the spring season.

chunyuh—In northern China, rain during the early spring that helps the winter wheat.

There is a Chinese proverb: Chunyuh is as valuable as oil.

churada—A severe rain **squall** in the Mariana Islands (western Pacific Ocean) during the northeast **monsoon**.

They occur from November to April or May, especially from January through March.

cierzo—In Spain, local term for the **mistral** in the lower valley of the Ebro.

It occurs mainly in the autumn and early winter. *Compare* **cers**.

CIN—Abbreviation for **convective inhibition**.

circle of inertia—*See* **inertial circle**.

circular cylindrical coordinates—Same as **cylindrical coordinates**.

circular depolarization ratio—(Abbreviated CDR.) The ratio of **power** received in the transmission channel to power received in the orthogonal channel of a **dual-channel radar** when a circularly polarized **signal** is transmitted.

Because the stronger component of the circularly polarized signal backscattered from **hydrometeors** has a **polarization** orthogonal to the transmitted polarization, the circular depolarization ratio yields a value less than unity or, equivalently, a negative **decibel** quantity. Note that relative to the **receiver** channels of the radar, the circular depolarization ratio is defined inversely to the **linear depolarization ratio**. *Compare* **cancellation ratio**.

circular frequency—*See* **frequency**.

circular polarization—A **polarization** state of an electromagnetic **signal** in which the **electric field** vector at a point in space describes a circle.

Relative to an observer looking in the direction of signal propagation, the electric fields of right and left circularly polarized signals rotate clockwise and counterclockwise, respectively.

circular symmetry—*See* **radial symmetry**.

circular variable—A **variable** represented by a **vector** quantity in **polar coordinates**.

Wind **velocity** is such a variable.

circular vortex—A flow in parallel planes in which **streamlines** and other isopleths are concentric circles about a common axis.

One **atmospheric model** of easterly and westerly winds is a circular vortex about the earth's **polar axis**. The **stability** properties of this **model** have been extensively investigated.

circulation—1. The flow or motion of a fluid in or through a given area or volume.

2. A precise measure of the average flow of fluid along a given closed curve.

Mathematically, circulation is the **line integral**

$$\oint \mathbf{v} \cdot d\mathbf{r}$$

about the closed curve, where $\mathbf{v}$ is the fluid **velocity** and $d\mathbf{r}$ is a **vector** element of the curve. By **Stokes's theorem**, the circulation about a plane curve is equal to the **total vorticity** of the fluid enclosed by the curve. The given curve may be fixed in space or may be defined by moving fluid parcels. *See* **circulation theorem**.

circulation index—A measure of the magnitude of one of several aspects of large-scale atmospheric **circulation** patterns.

Indices most frequently measured represent the strength of the **zonal** (east–west) or **meridional** (north–south) components of the **wind**, at the surface or at the upper levels, usually averaged spatially and often averaged in time. *See* **meridional index, zonal index**.

circulation integral—The **line integral** of an arbitrary **vector** taken around a closed curve.

Thus,

$$\oint \mathbf{a} \cdot d\mathbf{r}$$

is the circulation integral of the vector $\mathbf{a}$ around the closed curve and $d\mathbf{r}$ is an infinitesimal vector element of the curve. If the vector is the **velocity**, this integral is called the **circulation**.

circulation model—*See* **atmospheric circulation model**.

circulation pattern—The general geometric configuration of atmospheric **circulation** usually applied, in **synoptic meteorology**, to the large-scale features of **synoptic charts** and **mean charts**.

Compare **pressure pattern**.

circulation theorem—1. V. Bjerknes's circulation theorem: 1) With reference to an **absolute coordinate system**, the rate of change of the **absolute** circulation dC_a/dt of a closed individual fluid curve, that is, one that will consist always of the same fluid **particles**, is equal to the number of pressure–volume **solenoids** $N_{\alpha,-p}$ embraced by the curve

$$\frac{dC_a}{dt} = N_{\alpha,-p},$$

where the **circulation** has the same sense as the solenoids, the sense of the rotation from volume **ascendent** to **pressure gradient**.

2) With reference to a **relative coordinate system** (specifically, the rotating earth), the rate of change of **circulation** relative to the earth dC/dt of an arbitrary closed individual fluid curve is

determined by two effects: a) the **solenoid** effect that tends to change the circulation in the sense of the solenoids by an amount per unit time equal to the number of solenoids embraced by the curve; and b) the inertial effect that tends to decrease the circulation by an amount per unit time proportional to the rate at which the projected area of the curve in the equatorial plane expands:

$$\frac{dC}{dt} = N_{\alpha,-p} - 2\Omega\frac{dA}{dt},$$

where Ω is the angular speed of the earth's rotation and A is the equatorial **projection** of the curve. This is the most useful form of Bjerknes's circulation theorem. It permits the qualitative examination of many types of frictionless atmospheric motion that are too complicated for complete analytic treatment, for example, the **sea breeze**.

2. Kelvin's circulation theorem: The rate of change of the **circulation** dC/dt of a closed individual fluid curve is equal to the **circulation integral** of the **acceleration** $\dot{\mathbf{v}}$ around the curve:

$$\frac{dC}{dt} = \oint \dot{\mathbf{v}} \cdot d\mathbf{r},$$

where $d\mathbf{r}$ is a **vector** line element of the curve.

3. Höiland's circulation theorem: An arbitrary closed tubular fluid filament with constant **cross section** has a total mass **acceleration** along itself equal to the **resultant** of the force of **gravitation** along the filament:

$$\oint \rho \dot{\mathbf{v}}_a \cdot d\mathbf{r} = -\oint \rho \, d\phi_a,$$

where ρ is the fluid **density**, $\dot{\mathbf{v}}_a$ is the **absolute** vector **acceleration**, and $d\phi_a$ is the variation of the **gravitational potential** from the initial to the terminal point of the **vector** element $d\mathbf{r}$. This theorem is particularly useful in the study of the **stability** of fluid flow.

Gill, A. E., 1982: *Atmosphere–Ocean Dynamics*, Academic Press, 226–231, 237–241.

Holton, J. R., 1992: *An Introduction to Dynamic Meteorology*, 3d edition, Academic Press, 87–92.

circulation type—A characteristic **circulation** feature, typically with analogs.

An example of a circulation type is an omega **block**, characterized by a high-amplitude cut-off **ridge** surrounded to its east and west by high-amplitude cut-off **troughs**.

circumhorizontal arc—An arc parallel to the **horizon** and found at least 46° below the sun (or moon).

It is produced by the **refraction** of **light** through the 90° prisms of **ice crystals**, having entered through the vertical sides and passing out through the horizontal bases. Normally, the crystals are large, oriented, hexagonal plates. The circumhorizontal arc is the low sky counterpart of **circumzenithal arc**. The circumhorizontal arc is low in the sky when the sun is high (above 58° **elevation**); the circumzenithal arc is high in the sky when the sun is low (below 32° elevation). The most colorful circumhorizontal arcs occur when the **refraction** is close to the minimum angle of deviation, and this corresponds to a solar **elevation angle** of about 68°. In the midlatitudes, the sun only climbs to that elevation for a few hours mid day around the **summer solstice**.

circumpolar vortex—*See* **polar vortex**.

circumpolar westerlies—Same as **westerlies**.

circumpolar whirl—1. Same as **polar vortex**.

2. Same as **westerlies**.

circumscribed halo—The descriptive name given to the fully developed 22° **tangent arcs** for light-source **elevation angles** of between approximately 30° and 75°.

At these elevation angles, the upper and lower 22° tangent arcs merge to form a kidney-bean-shaped **halo** circumscribing the **halo of 22°**. Sometimes the term "circumscribed halo" is misleadingly employed as equivalent to the 22° tangent arcs.

circumsolar radiation—*See* **solar aureole**.

circumzenithal arc—An arc centered on the **zenith** and found at least 46° above the sun (or moon).

It is produced by the **refraction** of **light** through the 90° prisms of **ice crystals**, having entered through the horizontal bases and passed out through the vertical sides. Normally, the crystals are large, oriented, hexagonal plates. The circumzenithal arc is the high sky counterpart of the **circumhorizontal arc**. The circumhorizontal arc is low in the sky when the sun is high (above 58°

elevation); the circumzenithal arc is high in the sky when the sun is low (below 32° elevation). The majority of circumzenithal arcs are short-lived, short in length and faint or moderately bright. However, very colorful circumhorizontal arcs can occur when the refraction is close to the minimum angle of deviation, corresponding to a solar **elevation angle** of about 22°.

cirque glacier—Small **glacier** lying wholly within a cirque, or topographic hollow, created by **glacial** excavation located high in mountainous areas.

> Armstrong, T., B. Roberts, and C. Swithinbank, 1973: *Illustrated Glossary of Snow and Ice*, Scott Polar Research Institute, Special Pub. No. 4, 2d ed.; The Scholar Press Ltd., Menston Yorkshire, UK.

cirriform—Like **cirrus**; more generally, descriptive of clouds composed of small **particles**, mostly **ice crystals**, that are fairly widely dispersed, usually resulting in relative transparency and whiteness and often producing **halo** phenomena not observed with other **cloud** forms.

Irisation may also be observed. Cirriform clouds are high clouds (*see* **cloud classification**) existing between −25° and −85°C. As a result, when near the **horizon**, their reflected **light** traverses a sufficient thickness of air to often cause them to take on a yellow or orange tint even during the midday period. On the other hand, cirriform clouds near the **zenith** always appear whiter than any other clouds in that part of the sky. With the sun on the horizon, this type of cloud is whitish, while other clouds may be tinted with yellow or orange; when the sun sinks a little below the horizon, cirriform clouds become yellow, then pink or red; and when the sun is well under the horizon, they are gray. All species and varieties of **cirrus**, **cirrocumulus**, and **cirrostratus** clouds are cirriform in nature. *Compare* **cumuliform**, **stratiform**.

cirrocumulus—(Abbreviated Cc.) A principal **cloud** type (**cloud genus**), appearing as a thin, white patch of cloud without shadows, composed of very small elements in the form of grains, ripples, etc.

The elements may be merged or separate, and more or less regularly arranged; they subtend an angle of less than 1° when observed at an angle of more than 30° above the **horizon**. Holes or rifts often occur in a sheet of cirrocumulus. Cirrocumulus may be composed of highly **supercooled water** droplets, as well as small **ice crystals**, or a mixture of both; usually, the droplets are rapidly replaced by ice crystals. Sometime **corona** or **irisation** may be observed. **Mamma** may appear. Small **virga** may fall, particularly from **cirrocumulus castellanus** and **floccus**. (For further details, *see* **cirriform**.) Cirrocumulus, as well as **altocumulus**, often forms in a layer of **cirrus** and/or **cirrostratus** (Cc cirrogenitus or Cc cirrostratogenitus). In middle and high latitudes, cirrocumulus is usually associated in space and time with cirrus and/or cirrostratus; this association occurs less often in low latitudes. Cirrocumulus differs from these other cirriform clouds in that it is not on the whole fibrous, or both silky and smooth; rather, it is rippled and subdivided into little cloudlets. Cirrocumulus is most often confused with altocumulus. It differs primarily in that its constituent elements are very small (see above) and are without shadows. The term cirrocumulus is not used for incompletely developed small elements such as those on the margin of a sheet of altocumulus, or in separate patches at that level. See **cloud classification**.

cirrocumulus castellanus—*See* **castellanus**.

cirrocumulus floccus—*See* **floccus**.

cirrocumulus lacunosus—*See* **lacunosus**.

cirrocumulus lenticularis—*See* **lenticularis**.

cirrocumulus stratiformis—*See* **stratiformis**.

cirrocumulus undulatus—*See* **undulatus**.

cirrostratus—(Abbreviated Cs.) A principal **cloud** type (**cloud genus**), appearing as a whitish veil, usually fibrous but sometimes smooth, that may totally cover the sky, and that often produces **halo** phenomena, either partial or complete.

Sometimes a banded aspect may appear, but the intervals between the bands are filled with thinner cloud veil. The edge of a veil of cirrostratus may be straight and clear-cut, but more often it is irregular and fringed with **cirrus**. Some of the **ice crystals** that compose the cloud are large enough to fall and thereby produce a fibrous aspect. Cirrostratus occasionally may be so thin and transparent as to render it nearly indiscernible, especially through **haze** or at night. At such times, the existence of a halo may be the only revealing feature. The **angle of incidence** of **illumination** upon a cirrostratus layer is an important consideration in evaluating the identifying characteristics. When the sun is high (generally above 50° **elevation**), cirrostratus never prevents the casting of shadows by terrestrial objects; and a halo might be completely circular. At progressively lower angles of the sun, halos become fragmentary and **light** intensity noticeably decreases. Cirrostratus may be produced by the merging of elements of cirrus (Cs cirromutatus); from **cirrocumulus** (Cs cirrocumulogenitus); from the thinning of **altostratus** (Cs altostratomutatus); or from the **anvil** of **cumulonimbus** (Cs cumulonimbogenitus). Since cirrostratus and altostratus form from each other, it is frequently difficult to delineate between the two. In general, altostratus does not cause halo phenomena, is thicker than cirrostratus, appears to move more rapidly, and has a more even **optical thickness**. When near the **horizon**, cirrostratus may be impossible to distinguish from cirrus. *See* **cloud classification, cirriform**.

cirrostratus duplicatus—*See* **duplicatus**.

cirrostratus fibratus—*See* **fibratus**.

cirrostratus filosus—*See* **fibratus**.

cirrostratus nebulosus—*See* **nebulosus**.

cirrus—(Abbreviated Ci.) A principal **cloud** type (**cloud genus**) composed of detached **cirriform** elements in the form of white, delicate filaments, of white (or mostly white) patches, or of narrow bands.

These clouds have a fibrous aspect and/or a silky sheen. Many of the **ice crystal** particles of cirrus are sufficiently large to acquire an appreciable speed of fall; therefore, the cloud elements have a considerable vertical extent. Wind **shear** and variations in **particle** size usually cause these fibrous trails to be slanted or irregularly curved. For this reason, cirrus does not usually tend, as do other clouds, to appear horizontal when near the **horizon**. Because cirrus elements are too narrow, they do not produce a complete circular **halo**. Cirrus often evolves from **virga** of **cirrocumulus** or **altocumulus** (Ci cirrocumulogenitus or Ci altocumulogenitus), or from the upper part of **cumulonimbus** (Ci cumulonimbogenitus). Cirrus may also result from the transformation of **cirrostratus** of uneven **optical thickness**, the thinner parts of which dissipate (Ci cirrostratomutatus). It may be difficult at times to distinguish cirrus from cirrostratus (often impossible when near the horizon); cirrostratus has a much more continuous structure, and if subdivided, its bands are wider. Thick cirrus (usually **cirrus spissatus**) is differentiated from patches of **altostratus** by its lesser extension and white color. The term "cirrus" is frequently used for all types of **cirriform** clouds. *See* **cloud classification, cirriform**.

cirrus castellanus—*See* **castellanus**.

cirrus densus—*See* **cirrus spissatus**.

cirrus duplicatus—*See* **duplicatus**.

cirrus fibratus—*See* **fibratus**.

cirrus filosus—*See* **fibratus**.

cirrus floccus—*See* **floccus**.

cirrus intortus—**Cirrus**, the filaments of which are very irregularly curved, or more or less zigzag, and often apparently entangled.

cirrus nothus—*See* **cirrus spissatus**.

cirrus radiatus—*See* **radiatus**.

cirrus spissatus—(Previously called cirrus densus, cirrus nothus; also called thunderstorm cirrus, false cirrus.) A **cloud species** unique to the genus **cirrus**, of such **optical thickness** as to appear grayish on the side away from the sun, and to veil the sun, conceal its outline, or even hide it.

These often originate from the upper part of a **cumulonimbus**, and are often so dense that they suggest clouds of the middle level.

cirrus uncinus—A species of **cirrus** without gray parts, the elements of which are often in the form of a comma, topped with either a hook or a tuft that is not rounded.

The species **uncinus** is unique to the genus cirrus. The hook takes the shape of a reverse question mark under positive **wind shear** (wind increasing with height) conditions and the shape of a question mark under negative wind shear.

cirrus vertebratus—*See* **vertebratus**.

CISK—Abbreviation for **conditional instability of the second kind**.

cistern—The cylindrical vessel of a **mercury barometer** into which the tube dips. *See* **cistern barometer**.

cistern barometer—A **mercury barometer** in which the lower **mercury** surface is larger in area than the upper surface.

The basic construction of a cistern barometer is as follows: A glass tube 1 m in length, sealed at one end, is filled with mercury, and then inverted. The tube is mounted so that its mouth penetrates the upper surface of a **reservoir** of mercury called the **cistern** of the **barometer**. Cistern barometers are classified according to whether the cistern is fixed in volume (*see* **Kew barometer**) or variable in volume (*see* **Fortin barometer**).

civil day—A **mean solar day** reckoned from midnight to midnight, usually divided into two 12-hour divisions, although now often regarded as a single 24-hour series.

As practically applied by use of **meridional** time zones of approximately 15° width, civil time is accurate (relative to mean solar time) to within ±30 minutes, or 7½° of longitude. Minor exceptions occur when the time zones have been established according to geopolitical considerations, rather than adhering strictly to the meridians. *See* **apparent solar day, day**.

civil time—See **civil day**.

civil twilight—The period between astronomical **sunrise** or **sunset** and the time when the sun's unrefracted center is at **elevation** $h_0 = -6°$.

Local **topography** above the **astronomical horizon** will make local sunset occur before astronomical sunset. During a **clear** evening's civil twilight, horizontal **illuminance** decreases from ~585–410 **lux** to ~3.5–2 lux. As is true of nautical and **astronomical twilight**, civil twilight's length varies greatly with latitude and time of year. At $h_0 = -6°$, the ambient illuminance under clear skies historically was deemed just adequate for doing outdoor work without artificial **illumination**. However, perceptually demanding tasks, such as driving an automobile, require higher illuminances. Also at $h_0 = -6°$, luminances near the **clear sky's zenith** are low enough to make the brightest stars visible. *Compare* **astronomical twilight**.

Clapeyron–Clausius equation—Same as **Clausius–Clapeyron equation**.

Clapeyron diagram—*See* **thermodynamic diagram**.

clapotis—A French term for a **standing wave** phenomenon associated with the **reflection** of an ocean wave train from a vertical surface, such as a **breakwater** or pier.

A standing wave is a periodic vertical motion of the sea surface that does not propagate horizontally. It can be thought of as being created by the superposition of two identical waves propagating in opposite directions.

class-A evaporation pan—*See* **evaporation pan, pan evaporation**.

class-A pan—*See* **evaporation pan**.

class interval—A **range** of values of a **variable**; an interval used in dividing the **scale** of the variable for the purpose of tabulating the **frequency distribution** of a **sample**.

classical condensation theory—The theory that describes how **vapor** phase of a substance condenses into liquid and solid phases based on classical bulk thermodynamic principles.

This leads to the conclusion that **water vapor** will start condensing into liquid water or **ice** when the **environment** marginally surpasses **saturation** (100% **relative humidity** with respect to plain water or ice surface, respectively). In reality, it is almost always necessary to have foreign **particles** serving as **condensation nuclei** or **ice nuclei** for the **condensation** to occur even when the environment is somewhat supersaturated.

clathrate hydrate—A cagelike arrangement of water molecules stabilized by a central guest molecule.

Such hydrates exist at higher pressures (> 10 atmospheres) at > 100 m depth in **glaciers** where air bubbles trapped in the process of **firn** formation near the surface appear to dissolve in the **ice**.

Clausius–Clapeyron equation—(Also called Clapeyron equation, Clapeyron–Clausius equation.) The differential equation relating **pressure** of a substance to **temperature** in a system in which two phases of the substance are in **equilibrium**.

Two general expressions are

$$\frac{dp}{dT} = \frac{\delta s}{\delta v} = \frac{L}{T\delta v},$$

where p is the pressure, T the temperature, δs the difference in **specific entropy** between the phases, δv the difference in **specific volume** between the two phases, and L the **latent heat** of the **phase change**. The form most familiar in meteorology, related to the phase change between **water vapor** and liquid water, is obtained after some approximations as

$$\frac{1}{e_s}\frac{de_s}{dT} = \frac{L_v}{R_v T^2},$$

135

where e_s is the **saturation vapor pressure** of water, L_v the latent heat of **vaporization**, and R_v the **gas constant** for water vapor. A similar relation for the saturation vapor pressure in contact with an **ice** surface is obtained by replacing the latent heat of vaporization by that of **sublimation**. These equations may be integrated to obtain explicit relationships between e_s and T, given known values at some point. The most empirically accurate relationships differ slightly from results so obtained. An expression believed accurate to 0.3% for $-35°C < T < 35°C$ is given by Bolton as

$$e_s(T) = 0.6112 \exp\left(\frac{17.67T}{T + 243.5}\right),$$

where T is temperature in °C and **vapor pressure** is in kPa.

Iribarne, J. V., and W. L. Godson, 1981: *Atmospheric Thermodynamics*, D. Reidel, p. 65.

Bolton, D., 1980: The computation of equivalent potential temperature. *Mon. Wea. Rev.*, **108**, 1046–1053.

clay atmometer—An **atmometer** consisting of a porous porcelain container connected to a calibrated **reservoir** filled with distilled water.

Evaporation is determined by the depletion of water in the reservoir.

Clayden effect—When a photographic emulsion is given a very brief exposure to **light** of high **intensity**, it is desensitized toward a subsequent longer exposure to light of moderate intensity.

That is, the second exposure produces a fainter **image** than if the preexposure had not been given. This phenomenon was observed originally by Clayden when photographing **lightning** flashes, but it can be produced equally well by any type of light source, provided the intensity is sufficiently high and the duration short enough. *See* **dark lightning**.

Clayden, A. W., 1890: *Proc. Phys. Soc. (London)*, **10**, p. 180.

clean air—Air that does not contain **air pollution**.

See **NAAQS**.

clear—1. After U.S. weather observing practice, the **state of the sky** when it is cloudless or when the **sky cover** is less than 0.1 (to the nearest tenth.)

In **aviation weather observations**, a **clear sky** state is denoted by the **symbol** "O."

2. The character of the **sunrise** or **sunset** when the disk of the sun is visible at these times. *Compare* **cloudy**.

3. To change from a stormy or **cloudy** weather condition to one of no **precipitation** and decreased **cloudiness**.

4. In popular usage, the condition of the **atmosphere** when it is very transparent (as opposed to hazy, foggy, etc.) and accompanied by negligible **cloudiness**.

In **weather forecast** terminology, the maximum cloudiness considered is about 0.2. *Compare* **cloudy**, **fair**.

clear air—1. Air that is devoid of **clouds** or **fog**.

2. In some contexts, air that is devoid of any solid or liquid **particles** that would reduce **visibility**.

clear-air echo—A **radar echo** returned from a region of the **atmosphere** with no apparent meteorological scatterers such as clouds or **precipitation**. Clear-air echoes are caused by either 1) solid **target** returns from birds, insects, **dust**, or other particulate matter, or 2) spatial fluctuations of **refractivity** with **scale** sizes on the order of or smaller than the radar **wavelength**.

Two examples of the latter type of echo are **Bragg scattering** from **atmospheric turbulence** and **Fresnel reflection** or **specular reflection** from layers with sharp gradients in **refractivity**. Insects, birds, and **particulates** are the principal cause of clear-air echoes observed with radars having wavelengths of 3 cm or less, whereas high-powered radars operating at wavelengths of 10 cm or greater routinely detect backscattered signals from strong **refractive index** fluctuations. *Compare* **angel**.

clear-air turbulence—(Abbreviated CAT.) A higher **altitude** (6–15 km) **turbulence** phenomenon occurring in cloud-free regions, associated with **wind shear**, particularly between the core of a **jet stream** and the surrounding air.

It is most common near **upper air** fronts and the **tropopause**, and can often affect an aircraft without warning. Clear-air turbulence also frequently occurs close to **towering cumulus** clouds (usually within 30 km), and near mountains. Airflow disrupted by mountains and other terrain can undulate in waves of turbulence for 1000 km or more. *See also* **aircraft turbulence**.

clear ice—(Or clear icing.) Smooth compact **rime**, usually transparent, fairly amorphous, with a ragged surface, and morphologically resembling **glaze**.

This term has two different major applications. 1) Most commonly, it is used as a synonym for glaze, particularly with respect to **aircraft icing**. Factors that favor clear ice (or glaze) formation are large **drop** size, rapid **accretion** of liquid water, slight **supercooling**, and slow **dissipation** of **latent heat** of fusion. Thus, an aircraft flight through **supercooled rain** at an **air temperature** of 0° to −4°C is most conducive to clear icing. This type of **icing** does not seriously distort airfoil shape, but it does add appreciably to the weight of the craft. 2) The term may also be applied to homogeneous bodies of **glacier ice** and **lake ice**. *Compare* **rime ice**.

clear icing—Same as **clear ice**.

clear line of sight—(Abbreviated CLOS.) A **line of sight** not obstructed by **clouds** or **haze**.
Compare **cloud-free line of sight**.

clear sky—A sky free of **clouds** and other **obscurations** as observed from the point of **observation**.

clearance—In aviation terminology, authorization for an aircraft to proceed under conditions specified by an air traffic control unit.
For convenience, the term "air traffic control clearance" is frequently abbreviated to "clearance" when used in appropriate contexts. The abbreviated term "clearance" may be prefixed by the words "taxi," "takeoff," "departure," "en route," "approach," or "landing" to indicate the particular portion of flight to which the air traffic clearance relates.

clearing—A process of **clouds**, **fog**, or other **obscurations** becoming less prevalent with time.

climagram—1. Same as **climatic diagram**.
2. (Rare.) A coded numerical statement of the principal climatic elements of a location.
See **CLIMAT broadcast**.

climagraph—Same as **climatic diagram**.

CLIMAP—(Acronym for Climate: Long-range Investigation Mapping and Prediction.) A project that reconstructed **sea surface temperatures** 18 000 years before present (BP).
The **temperature** information was derived from properties of deep-ocean sediments sampled by means of corers, and provided strong support for the role of **Milankovitch** variations as a pacemaker for **climate fluctuations**. The primary attributes used were the species composition of planktonic organisms found in the sediments, and the **oxygen** isotopic composition of foraminiferan skeletons.

climat—(Or CLIMAT.) Contraction of climatological or **climatology**.

CLIMAT broadcast—A monthly broadcast of the mean values of the meteorological **elements** during the preceding month for stations of the national weather services belonging to the World Meteorological Organization (WMO).
The messages are disseminated worldwide through the Global Telecommunications System (GTS) as soon as possible after the close of a month, but not later than the fifth of the following month. This program was recommended to the predecessor of the WMO (the International Meteorological Organization) and widely implemented during 1936–39. The program, the elements broadcast, and the codes utilized have been revised several times since the program's implementation. The elements broadcast by a principal **climatological station**, according to the 1995 edition of the Manual on Codes (WMO Publication No. 306), include **atmospheric pressure**, **temperature** (means and extremes of **daily mean**, maximum, and minimum), **wind speed**, **vapor pressure**, **precipitation** [monthly total, frequency group (**quintile**) within which it falls, and **extreme** daily amount], **sunshine** duration, **standard deviation** of **temperature**, number of days with **thunderstorms** and **hail**, and number of days with precipitation, temperatures, **snow depth**, **visibility**, and **wind speed** beyond various thresholds. Provision is also made for the issuance of comparable normals, as far as possible, for the most recent climatological **standard** normal period (i.e., 1901–30, 1931–60, 1961–90, etc.).

Secretariat of the World Meteorological Organization, 1995: *Manual on Codes, WMO Publication No. 06, Annex to Technical Regulations*, Geneva, Switzerland, looseleaf.

climate—The slowly varying aspects of the atmosphere–hydrosphere–land surface system.
It is typically characterized in terms of suitable averages of the **climate system** over periods of a month or more, taking into consideration the **variability** in time of these averaged quantities. Climatic classifications include the spatial **variation** of these time-averaged variables. Beginning with the view of local climate as little more than the annual course of long-term averages of **surface temperature** and **precipitation**, the concept of climate has broadened and evolved in recent decades in response to the increased understanding of the underlying processes that determine climate and its variability. *See also* **climate system**, **climatology**, **climate change**, **climatic classification**.

climate analogs—Information about past climates used as a projection of future **climate**.

Climate analogs generally take one of the following two forms: 1) examination of climatic patterns of periods with a particular climate characteristic (e.g., warm periods) in the instrumental or paleoclimatic record; 2) studies of groups of individual years or seasons, not necessarily consecutive, selected from the instrumental record for a particular characteristic.

climate anomaly—The difference between the average climate over a period of several decades or more, and the **climate** during a particular month or **season**.
See also **climate change**.

climate change—(Also called climatic change.) Any systematic change in the long-term **statistics** of **climate** elements (such as **temperature, pressure,** or **winds**) sustained over several decades or longer.

Climate change may be due to natural **external forcings**, such as changes in solar **emission** or slow changes in the earth's **orbital elements**; natural internal processes of the **climate system**; or **anthropogenic** forcing.

climate change detection—(Also called climatic change detection.) The process of demonstrating that an observed change in **climate** is highly unusual in a **statistical** sense, although a reason for the change is not necessarily provided.

Attribution is the process of establishing the cause of a particular change in climate, including the testing of competing hypotheses. Climate change detection and attribution studies attempt to distinguish between **anthropogenic** and natural influences on climate. Any anthropogenic effect will be superimposed on the background "noise" of natural **climate variability**.

climate change projection—*See* **climate prediction**.

climate control—Term applied to schemes for artificially altering or controlling the **climate** of a region.
See **climatic control**; *compare* **weather modification**.

climate divide—*See* **climatic divide**.

climate drift—The tendency for the solution of a numerical **model** integration to move away from its initial conditions and toward its own **climate**.

Typically this new climate has some unrealistic features. Climate drift may be caused by various imbalances or errors in the model parameterizations.

climate fluctuations—(Also called climatic fluctuations.) Changes in time of one or more **variables**.

Fluctuations may be regular or irregular, but a sustained trend is not typically called a **fluctuation**.

climate forecast—*See* **climate prediction**.

climate impact assessments—Studies that attempt to evaluate or predict the impact of **climate change** on physical and ecological systems, human health, and socioeconomic sectors.

The potential sensitivity, adaptability, and vulnerability of these systems and sectors to climate change can be estimated.

climate impact studies—*See* **climate impact assessments**.

climate model—A **model** used to make forecasts or simulations of **climate**.

Most climate models are closely related to **NWP** models. In general, the equations for the **atmosphere** are those for **momentum, continuity,** the **equation of state**, the law of **thermodynamics,** and **water vapor** continuity, although simplified sets of governing equations may also be used. Depending on the climate timescale of interest, the **atmospheric model** may be coupled to similar models for the ocean (including the deep ocean for timescales greater than a year), the land surface and **biosphere**, land **snow** and **ice**, and **sea ice**. For longer timescales, variation in external forcing factors (**solar radiation**) may also be important. Because climate models are generally integrated for periods of many years, they have traditionally been run at lower **resolution** than NWP models to reduce computational costs.

climate modification—Intentional or unintentional change of **climate** caused by human activity.

climate monitoring—Long-term observations of those quantities that describe the state of the earth's surface and **atmosphere** (e.g., **temperature, pressure, precipitation**) or **climate** forcing functions (e.g., solar **output**, atmospheric **carbon dioxide** concentrations).

climate noise—(Also called climatic noise.) Variations in the state of the **climate system** that have little or no organized structure in time and/or space.

This can be a relative term, since variations with a small degree of structure may be regarded as **noise** in one context but not in another.

climate periodicity—(Also called climatic periodicity.) A **climatic cycle** in which the time interval between successive maxima and minima is constant or very nearly constant throughout the record.

climate prediction—The **prediction** of various aspects of the **climate** of a region during some future period of time.

Climate predictions are generally in the form of probabilities of anomalies of climate variables (e.g., **temperature, precipitation**), with lead times up to several seasons (*see* **climate anomaly**). The term "climate projection" rather than "climate prediction" is now commonly used for longer-range predictions that have a higher degree of uncertainty and a lesser degree of specificity. For example, this term is often used for "predictions" of **climate change** that depend on uncertain consequences of **anthropogenic** influences such as land use and the burning of **fossil fuels**.

climate projection—*See* **climate prediction**.

climate sensitivity—1. The magnitude of a climatic response to a perturbing influence.

2. In mathematical modeling of the **climate**, the difference between simulations when the magnitude of a given **parameter** is changed.

3. In the context of global **climate change**, the **equilibrium** change in global mean surface **temperature** following a unit change in **radiative forcing**.

climate signal—(Also called climatic signal.) Variations in the state of the **climate system** that have an identifiable and statistically discernible structure in time and/or space.

climate snow line—(Also called climatic snow line.) The **altitude** above which a flat surface (fully exposed to sun, **wind**, and **precipitation**) would experience a net **accumulation** of **snow** over an extended period of time. Below this altitude **ablation** would predominate.

While this concept is largely theoretical in application, it corresponds closely to the actual **firn line** of a **glacier** and to the average summer position of the **snow line** in mountainous terrain.

climate system—The system, consisting of the **atmosphere, hydrosphere, lithosphere,** and **biosphere**, determining the earth's **climate** as the result of mutual interactions and responses to external influences (forcing).

Physical, chemical, and biological processes are involved in the interactions among the components of the climate system.

climate variability—(Also called climatic variability.) The temporal variations of the atmosphere–ocean system around a mean state.

Typically, this term is used for timescales longer than those associated with **synoptic** weather events (i.e., months to millennia and longer). The term "natural climate variability" is further used to identify **climate** variations that are not attributable to or influenced by any activity related to humans.

climatic change—*See* **climate change**.

climatic classification—The division of the earth's climates into a worldwide system of contiguous regions, each one of which is defined by relative homogeneity of the **climatic elements**.

The earliest known classification of **climate**, devised by the Greeks, simply divided each hemisphere into a **mathematical climate** of three zones, the "summerless," "intermediate," and "winterless," thus accounting only for the latitudinal differences in solar effect (the Greek word *klima* means "inclination"). More recently, these zones have been labeled the **Torrid, Temperate,** and **Frigid Zones**. Apparently, the first major improvement over this classification was introduced by Alexander Supan in the nineteenth century. He based his zones on actual rather than theoretical temperatures, and named one **hot belt**, two **temperate belts**, and two **cold caps**. Supan also divided the world into 34 **climatic provinces**, with no attempt to relate similar climates of different locations. Another basic and much used approach recognizes other climatic controls as well as the sun. The resulting climates are called (with variations) **polar, temperate, tropical, continental, marine, mountain,** and probably others. Of the major climatic classifications in use today, those of W. Köppen (1918) and C. W. Thornthwaite (1931) are referred to most often. Köppen's elaborate "geographical system of climates" is based upon annual and seasonal **temperature** and **precipitation** values; his **climatic regions** are given a letter code designation. The major categories are **tropical rainy climate, dry climate, temperate rainy climate, snow forest climate, tundra climate,** and **perpetual frost climate**. Gorczyński (1948) devised a decimal number system similar to the **Köppen classification**. Thornthwaite's (1931) bioclimatological system utilizes indices of **precipitation effectiveness** to outline **humidity provinces**, and **thermal efficiency** for **temperature**

provinces; and again, a letter code designates regions. Thornthwaite (1948) introduced an approach to a "rational" classification, wherein **potential evapotranspiration** is used as a measure of **thermal efficiency**, and is compared to precipitation to form a **moisture index** and to show amounts and periods of water surplus and deficiency. Definite break points are revealed that are adaptable as climatic boundaries. Many authors have devised modified classifications to instruct a particular audience. An excellent example is that of C. E. P. Brooks (1951) in which climatic regions are defined with respect to human activity.

Supan, A., 1879: Die Temperaturzonen der Erde. *Petermanns Geog. Mitt.*, **25**, 349–358.

Brooks, C. E. P., 1951: *Climate in Everyday Life*, 17–21.

Köppen, W. P., 1918: Klassification der Klimate nach Temperatur, Niederschlag und Jahreslauf. *Petermanns Geog. Mitt.*, **64**, 193–203; 243–248.

Köppen, W. P., 1931: *Grundriss der Klimakunde*, 2d ed., Berlin: Walter de Gruyter.

Köppen, W. P., and R. Geiger, 1930–1939: *Handbuch der Klimatologie*, Berlin: Gebruder Borntraeger, 6 vols.

Thornthwaite, C. W., 1931: The climates of North America according to a new classification. *Geogr. Rev.*, **21**, 633–655.

Thornthwaite, C. W., 1948: An approach toward a rational classification of climate. *Geogr. Rev.*, **38**, 55–94.

Gorczyński, W., 1948: Decimal System of World Climates. *Przeglad Meteor. Hydrol.*, **1**, 30–43.

Trewartha, G. T., 1954: *An Introduction to Climate*, 3d ed., Appendix A, 223–238.

climatic control—Any one of the relatively permanent factors that govern the general nature of the **climate** of a portion of the earth.

These factors include 1) **solar radiation**, especially as it varies with latitude; 2) distribution of land and **water masses**; 3) **elevation** and large-scale **topography**; and 4) the atmospheric concentration of **greenhouse gases**. The types of climatic controls were used as the basis for early **climatic classifications**. All of these have been nearly constant during historical time, but most or all of them have changed during **geologic time** and have caused large-scale changes of climate. *See* **climate control**.

climatic cycle—(Also called climatic oscillation.) A slowly varying change of **climate** about a mean that recurs with some regularity but is not necessarily periodic.

The term "cycle" is used in **climatology** more loosely than in other physical sciences. It implies quasiperiodic behavior without any real **precision** in the recurrence of events. It usually indicates only that the event is more probable at the peak of the **cycle** than at the trough. The existence of numerous cycles has been postulated, ranging from those with timescales of geological length down to a few years.

climatic diagram—(Also called climagram, climagraph, climatogram, climatograph, climogram, climograph.) A graph that represents climatological information.

Usually two or more **climatic elements** are plotted to show the changes in their relationship throughout the year. Often the climatic elements are plotted on the **ordinate** axis and time (usually month) plotted on the **abscissa**. Alternatively, the 12 monthly values of one climatic element are plotted on the ordinate versus the corresponding monthly values of another climatic element plotted on the abscissa, and the resulting points are connected with a continuous line. *See* **hythergraph**.

climatic discontinuity—A **climate change** that consists of a rather abrupt and permanent change during the **period of record** from one average value (for the early part of the record) to a distinctly different average value (for the later part of the record).

climatic divide—A boundary between regions having different types of **climate**.

The most effective climatic divides are the crests of mountain ranges. The boundary between two well-defined **ocean currents** may also act as a climatic divide.

climatic element—Any one of the properties or conditions of the **atmosphere** and ocean that together define the **climate** of a place (e.g., **temperature**, **humidity**, **precipitation**).

climatic factor—1. An influence that is relevant for determining the **climate** state in a region; much the same as **climatic control**, but regarded as including more local influences, such as the effects on the climate of local **aerosol** loading or extensive paving in or near a city.

2. An influence that has a major effect on human activities or human history.

For example, it has been suggested that **climate change** may have been a factor in the disappearance of the Anasazi culture in the North American southwest.

climatic fluctuation—A climatic inconstancy that consists of any form of systematic change, whether regular or irregular, except **climatic trend** and **climatic discontinuity**.

It is characterized by at least two maxima (or minima) and one minimum (or maximum) including those at the end points of the record.

climatic index—A numerical value, often a function of the primary **climatic elements**, used to characterize the **climate** of a location or region.

climatic noise—*See* **climate noise**.

Climatic Optimum—An interval of higher summer temperatures (2°–4°C higher) that followed the retreat of the last Pleistocene glaciers in North America and Europe, varying in age with location but often between approximately 5000 and 7000 years ago.

The term is sometimes applied to warm and wet conditions that occurred during this time in some regions, although the warmth was probably accompanied by dryness in other regions.

Kutzbach, J. E., 1992: Modeling earth system changes of the past. *Modeling the Earth System*, D. Ojima, Ed., UCAR/ Office for Interdisciplinary Earth Studies, Boulder, CO.

climatic oscillation—Same as **climatic cycle**.

climatic province—A region of the earth's surface characterized by an essentially homogeneous **climate**.

A. Supan (1884) defined 34 different climatic provinces mainly by **temperature** and **rainfall**, partly by **wind** and **orography**. *Compare* **humidity province, temperature province**.

Supan, A., 1884: *Grundzüge der Physischen Erdkunde*, Leipzig, 129–134.

climatic record—The record of **climatological data** (often summarized) from a location or region.

climatic region—A region experiencing a fairly uniform **climate** according to specific criteria.

climatic risk—The **probability**, based on climatological **statistics**, that unfavorable weather will occur at a particular location or region over a certain period of time.

climatic signal—*See* **climate signal**.

climatic snow line—*See* **climate snow line**.

climatic trend—A **climate change** characterized by a reasonably smooth, monotonic increase or decrease of the average value of one or more **climatic elements** during the **period of record**.

climatic year—(Rare.) The categorizing of one year's weather at a given locality according to one of the climatic classifications.

Thus, if the weather of a given year at a certain **station** was unusually mild and rainy, falling within (for example) Köppen's definition of a **tropical climate**, that station is said to have had a "tropical year," whatever its **normal** classification.

Köppen, W. P., and R. Geiger, 1930–1939: *Handbuch der Klimatologie*, Berlin: Gebruder Borntraeger, 6 vols.

climatic zone—A belt of the earth's surface within which the **climate** is generally homogeneous in some respect; an elemental region of a simple **climatic classification**, for example, a zone defined by the latitudinal distribution of the **climatic elements**.

The expressions **polar, temperate, subtropical, tropical**, and **equatorial climate** are used to indicate the climatic zones that succeed each other from the **pole** to the **equator**.

climatization—Same as **acclimatization**.

climatograph—Same as **climatic diagram**.

climatography—A thorough, quantitative description of **climate**, particularly with reference to the tables and charts that show the characteristic values of **climatic elements** at a **station** or over an area.

The term often has a comparative geographic connotation. Like descriptive **climatology**, it is sometimes used antithetically to climatology, when the latter is narrowly used to denote only the explanation of the causes for climatic conditions. *See* **physical climate**.

Conrad, V., and L. W. Pollack, 1950: *Methods in Climatology*, 2d ed. rev. and enl., 307–331.

climatological atlas—An atlas containing maps and tables of climatic data usually including distributions of highest and lowest temperatures by month, **temperature** means by month, **diurnal** ranges of temperature by month, **frequency** of **wind direction** and **speed** for several class values, monthly average and **extreme** precipitation, number of days per month with temperatures above and/or below specified values, etc., at fixed locations over land, or within geographic areas at sea.

Marine climatic atlases often include frequencies of **sea level pressure** values, **wave** heights, gale-force winds, and other phenomena of significance to mariners.

climatological chart—A **chart** on which climatological information is presented, thus identifying the characteristic values of **climatic elements** at a **station** or over an area.

climatological data—The many types of data—instrumental, historical (such as diaries or crop records), proxy (such as tree growth rings)—that constitute the major source of information for **climate** studies.

climatological diagram—Same as **climatic diagram**.

climatological division—In the United States, regions smaller than states over which climatological **statistics** (averages and **extremes**) are computed.

The boundaries of these regions are selected such that the locations contained therein are climatologically similar.

climatological forecast—A forecast based solely upon the climatological **statistics** for a region rather than the dynamical implications of the current conditions.

Climatological forecasts are often used as a baseline for evaluating the performance of **weather** and **climate forecasts**.

climatological network—A network of **weather stations** established to observe and collect **climatological data**, usually recording 24-hour **precipitation** amounts and the daily **maximum** and **minimum temperatures** at a specified time of day.

At a limited number of stations in climatological networks, records are made of additional data, such as **evaporation, snowfall** and **snow depth**, and **soil temperatures** at several depths.

climatological observation—The type of **weather observation** taken at a **climatological substation**.

It includes recording, at least once daily, one or more of the following: **maximum temperature, minimum temperature**, and total **precipitation** since the previous **observation**; and may also include description of the weather (**cloudy, clear**, etc.) and atmospheric phenomena. Such observations compose the bulk of **climatological data** in the United States and over much of the world. In addition, **meteorological observations** are, of course, "climatological" inasmuch as they are used to derive climatological data.

climatological standard normals—Averages of **climatological data** calculated for the following consecutive 30-yr periods, established by international agreement: 1 January 1901 to 31 December 1930; 1 January 1931 to 31 December 1960; 1 January 1961 to 31 December 1990; 1 January 1991 to 31 December 2020; etc.

See **normal**.

climatological station—*See* **climatological substation, second-order climatological station, third-order climatological station**.

climatological station elevation—The **elevation** above **mean sea level** chosen as the reference **datum level** for all climatological records of **atmospheric pressure** in a given locality; not necessarily the same as **station elevation**.

This term is denoted by the symbol H_{pc} in international usage. The reduction of **pressure** to this **standard** elevation is necessary only if a **station** has changed its specific location but has stayed within the general area, so that the two (or more) sets of records remain comparable.

climatological station network—*See* **climatological network**.

climatological station pressure—The **atmospheric pressure** computed for the level of the **climatological station elevation**.

This is employed so that all **climatic records** for a given location have a common reference, and it may or may not be the same as **station pressure**. *See also* **actual pressure, sea level pressure**.

climatological substation—A weather observing **station** operated by a **cooperative observer** (unpaid volunteer) for the purpose of recording **climatological observations**.

This is a unique category of stations in the United States (there are about 6000 of them), but it compares to the **third-order climatological station** as defined by the World Meteorological Organization. *Compare* also **second-order climatological station**.

climatological summary—A table for a specific location showing means, **extremes**, and other **statistics** including the number of occasions when specific meteorological conditions or specific values of a **meteorological element** were observed or when specific values of two or more meteorological elements were observed simultaneously during a specific number of years.

climatological year—Usually, the **calendar year**; occasionally, the 12-month period beginning 1 December.

climatology—The description and scientific study of **climate**.

Descriptive climatology deals with the observed geographic or temporal distribution of **meteorological observations** over a specified period of time. Scientific climatology addresses the nature and controls of the earth's climate and the causes of **climate variability** and change on all timescales. The modern treatment of the nature and theory of climate, as opposed to a purely descriptive account, must deal with the dynamics of the entire atmosphere–ocean–land surface **climate system**, in terms of its internal interactions and its response to external factors, for example, **incoming solar radiation**. Applied climatology addresses the climate factors involved in a broad range of problems relating to the planning, design, operations, and other decision-making activities of climate sensitive sectors of modern society. *See also* **applied climatology, climate, climate system, climate change**.

climatonomy—An approach to the study of **climate** based on physics and mathematics rather than the more empirical and descriptive methods of traditional **climatology**.

This term was coined by H. Lettau (1969) to describe his one-dimensional representation of climate based on the **energy** and water balances at the surface.

Lettau, H., and K. Lettau, 1969: Shortwave radiation climatonomy. *Tellus*, **21**, 208–222.

climogram—Same as **climatic diagram**.

climograph—Same as **climatic diagram**.

clinometer—An instrument for measuring angles of **inclination**.

In meteorology, it is used in conjunction with a **ceiling light** to measure **cloud height** at night. With it the **observer** measures the angle included by the illuminated spot on the clouds, the observer, and the ceiling light.

clo—A unit of thermal insulation, usually applied to clothing or bedcovers.

It is defined as the amount of insulation necessary to maintain comfort and a mean skin temperature of 33°C (92°F) for a person who is producing **heat** at the standard metabolic rate (50 Kcal m^{-2} of body surface per hour; one met) in an indoor environment characterized by a **temperature** of 21°C (70°F), **relative humidity** of less than 50%, and air motion of 6.1 m min^{-1}. If if is assumed that 76% of the metabolic heat is lost through the clothing, the unit can be defined in physical terms as the insulation that will restrict heat loss to 1 Kcal m^{-2}h^{-1} with a temperature gradient of 0.18°C across the fabric. In the first approximation, an insulation of one clo is provided by clothing material with a total thickness of 0.64 cm and air layers (between skin and clothing and between inner and outer garments) of about 0.51 cm.

Newburgh, L. H., ed., 1949: *Physiology of Heat Regulation and the Science of Clothing*, pp. 297, 299, 445.
Winslow, C.-E. A., and L. P. Herrington, 1949: *Temperature and Human Life*, 132–140.

CLOS—Abbreviation for **clear line of sight**.

close pack ice—Floating **pack ice** in which the **ice** concentration is between 7/10 and 8/10, composed of **floes** mostly in contact.

closed basin—A land **basin** where all **surface runoff** remains within the basin, with no natural surface outlets.

closed-cell stratocumulus—A term often used to describe satellite-viewed oceanic **stratocumulus** associated with an **inversion**.

Closed-cell patterns are composed of **cloud** elements of almost hexagonal shape, bounded at their edges with a cloud-free space.

closed cells—The periodic elements comprising a **layer cloud** with a regular array of **cloud** and a connecting lattice of **clear air**.

closed circulation—Fluid flow within a closed **streamline**; a **vortex**.

closed drainage—(Also known as internal drainage, blind drainage, closed-drainage area, interior blind drainage, noncontributing area.) The surface drainage, such as the drainage toward the central part of an interior **basin**, whereby water does not reach the ocean.

closed high—A **high** that may be completely encircled by an **isobar** or **contour line**.
See **closed low**.

closed lake—A lake that does not have surface **outflow** and that loses water by **evaporation** or by seepage.

closed low—A **low** that may be completely encircled by an **isobar** or **contour line**.

(This means an isobar or contour line of any value, not necessarily restricted to those arbitrarily chosen for the **analysis** of the **chart**.) Strictly, all lows are closed. However, in weather-map analysis terminology, this designation is used commonly in two respects: 1) on surface charts, to distinguish a low from a **trough**, especially as a low develops within the trough; and 2) on upper-level charts, to accentuate the fact that the **circulation** is closed, especially at levels and over latitudes where such an occurrence is unusual. The definition of **closed high** is analogous.

closed system—1. In mathematics, a system of differential equations and supplementary conditions such that the values of all the unknowns (**dependent variables**) of the system are mathematically determined for all values of the **independent variables** (usually space and time) to which the system applies.

2. In **thermodynamics**, a system of fixed mass.

By some definitions such a system may be confined to an impermeable container. An idealized **air parcel**, not diffusing into or **mixing** with its **environment** and undergoing a **saturation-adiabatic process**, is closed, as contrasted to one undergoing **pseudoadiabatic expansion** with **precipitation** removed. *Compare* **open system, isolated system**.

3. In **synoptic meteorology**, loosely used for a **closed low** or **closed high**.

closed weather—(Obsolete.) Term for a condition of **ceiling** and **visibility** that would prohibit the pilot of an aircraft from landing or taking off from an airport, even with radio or other instrumental aids.

This term is now generally replaced by "below **minimums**."

closure assumptions—Approximations made to the **Reynolds-averaged** equations of **turbulence** to allow solutions for flow and turbulence variables.

The Reynolds-averaged equations contain **statistical** correlations such as the **variance** or **covariance** between **dependent variables** such as **velocity** or **temperature**. The equations that forecast lower-order correlations often contain unknowns of higher statistical order, a difficulty known as the **closure problem**. When the higher-order terms are approximated as empirical functions of lower-order terms and of known **independent variables**, the resulting approximate equations can then be solved. These approximations, known as closure assumptions, must satisfy **parameterization** rules.

Stull, R. B., 1988: *An Introduction to Boundary Layer Meteorology*, 666 pp.

closure problem—A difficulty in **turbulence** theory caused by more unknowns than equations.

The closure problem of turbulence is alternately described as the requirement for an infinite number of equations, which would also be impossible to solve. This problem is apparently associated with the **nonlinear** nature of turbulence, and the traditional analytical approach of **Reynolds averaging** the governing equations to eliminate **linear** terms while retaining the **nonlinear** terms as **statistical** correlations of various orders (i.e., consisting of the product of multiple **dependent variables**). The closure problem is a long-standing unsolved problem of classical (Newtonian) physics. While no exact solution has been found to date, approximations called **closure assumptions** can be made to allow approximate solution of the equations for practical applications.

cloud—1. A visible aggregate of minute water droplets and/or **ice** particles in the **atmosphere** above the earth's surface.

Cloud differs from **fog** only in that the latter is, by definition, close (a few meters) to the earth's surface. Clouds form in the **free atmosphere** as a result of **condensation** of **water vapor** in rising currents of air, or by the **evaporation** of the lowest stratum of fog. For condensation to occur at a low degree of **supersaturation**, there must be an abundance of **cloud condensation nuclei** for water clouds, or **ice nuclei** for ice-crystal clouds, at temperatures substantially above $-40°C$. The size of **cloud drops** varies from one cloud type to another, and within any given cloud there always exists a finite **range** of sizes. Generally, cloud drops (**droplets**) range from $1–100$ μm in diameter, and hence are very much smaller than **raindrops**. *See* **cloud classification**.

2. Any collection of particulate matter in the **atmosphere** dense enough to be perceptible to the eye, as a **dust** cloud or **smoke** cloud.

cloud absorption—The **absorption** of **electromagnetic radiation** within a **cloud**.

With some exceptions, clouds typically absorb $\approx10\%$ of the (broadband) **shortwave radiation** and $\approx95\%$ of the **longwave radiation** incident on their boundaries. Thin clouds, notably thin **cirrus**, may absorb considerably less. The dominant absorber tends to be the water **droplets** or **ice crystals** making up the **cloud**, with **water vapor** playing a matching role only for shortwave absorption by upper tropospheric clouds.

cloud albedo—The fraction of **solar radiation** reflected directly by **clouds** in the **atmosphere**.

This represents a major part of the **albedo** of the earth. On the average, the earth reflects 31 units of solar radiation back to the space for every 100 units received (thus, the total earth albedo is 0.31). The cloud albedo accounts for 23 units of the 31. For individual clouds, local albedo may be in excess of 0.7.

cloud amount—The amount of sky estimated to be covered by a specified **cloud** type or level (partial cloud amount) or by all cloud types and levels (total cloud amount).

The estimate is made to the nearest **octa** (eighth).

cloud attenuation—Usually, the **attenuation** of **microwave radiation** by **clouds**.

For the centimeter **wavelength** band, clouds produce **Rayleigh scattering**. The attenuation is due largely to **scattering**, rather than **absorption**, for both **ice** and water clouds. The cloud attenuation for one-way transmission is

$$\text{cloud attenuation (dB km}^{-1}) = 0.454 \frac{6\pi}{\lambda} \frac{M}{\rho} \text{ Im}(-\kappa),$$

where M (g m^{-3}) is the liquid **water content**, ρ (g cm^{-3}) the **density** of water, λ the wavelength, and $\kappa = (m^2 - 1)(m^2 + 2)^{-1}$, where m^2 is the complex **dielectric constant**. The attenuation depends only on M, and in the **range** of wavelengths from 1 to 10 cm is approximately (1 to 100) $\times$ 10^{-2}M for water clouds, and (2 to 20) $\times$ 10^{-2}M for ice clouds.

cloud band—A nearly continuous **cloud formation** with a distinct long axis, a length-to-width ratio of at least four to one, and a width greater than one **degree** of latitude.

cloud bank—Generally, a fairly well-defined mass of **cloud** observed at a distance; a cloud bank covers an appreciable portion of the **horizon** sky, but does not extend overhead.

cloud banner—Same as **banner cloud**.

cloud bar—A heavy **cloud bank** that appears on the **horizon** with the approach of an intense **tropical cyclone** (**hurricane** or **typhoon**).

It is the outer edge of the **central dense overcast** of the **storm**.

cloud base—For a given **cloud** or **cloud layer**, the lowest level in the **atmosphere** at which the air contains a perceptible quantity of cloud **particles**.

See **cloud height, ceiling**.

cloud camera—A camera system specifically designed for use in obtaining **cloud** photographs for the purposes of determining their locations, movement, or percent of sky coverage.

Video or motion picture cameras are operated in a manner that records a new **image** every second or so in order to follow cloud development (time-lapse photography). Stereographic cloud images may be obtained from cameras at separated locations. Photogrammetric analysis of images with known foreground fiducial points from cameras with known optics can yield quantitative measurements of cloud dimensions and movements. Whole-sky imaging cameras use optics or reflectors with a very wide **field of view** to record the position of nearly all clouds in the hemisphere of sky overhead to estimate the fractional sky coverage of **cloudiness**. Conversion of the photographic images to **digital** form allows automated analysis. Useful data are often limited to daylight periods.

cloud cap—Same as **cap cloud**.

cloud ceiling—The height of the lowest opaque **cloud layer** reported as **broken** (5/8 to 7/8 coverage) or **overcast**.

Expressed in above ground level (AGL) heights. *See* **ceiling**.

cloud chamber—See **Wilson cloud chamber**.

cloud chart—A cloud-observing guide showing photographs of typical **cloud** formations with corresponding cloud symbols.

cloud classification—1. A scheme of distinguishing and grouping **clouds** according to their appearance, and, where possible, to their process of formation.

The one in general use, based on a classification system introduced by Luke Howard in 1803, is that adopted by the World Meteorological Organization and published in the International Cloud Atlas (1956). This classification is based on the determination: 1) genera—the main characteristic forms of clouds; 2) species—the peculiarities in shape and differences in internal structure of clouds; 3) varieties—special characteristics of arrangement and transparency of clouds; 4) **supplementary features** and **accessory clouds**—appended and associated minor cloud forms; and 5) **mother-clouds**—the origin of clouds if formed from other clouds. The ten cloud genera are **cirrus, cirrocumulus, cirrostratus, altocumulus, altostratus, nimbostratus, stratocumulus, stratus, cumulus**, and **cumulonimbus**. The fourteen **cloud species** are **fibratus, uncinus, spissatus, castellanus, floccus, stratiform, nebulosus, lenticularis, fractus, humilis, mediocris, congestus, calvus**, and **capillatus**. The nine cloud varieties are **intortus, vertebratus, undulatus, radiatus, lacunosus, duplicatus, translucidus, perlucidus**, and **opacus**. The nine supplementary features and accessory clouds are **incus, mamma, virga, praecipitatio, arcus, tuba, pileus, velum**, and **pannus**. (Note: Although these are Latin words, it is proper convention to use only the singular endings, e.g., more than one cirrus cloud is cirrus, not cirri.)

2. A scheme of classifying clouds according to their usual altitudes.

Three classes are distinguished: high, middle, and low. High clouds include **cirrus, cirrocumulus, cirrostratus**, occasionally **altostratus**, and the tops of **cumulonimbus**. The **middle clouds** are **altocumulus, altostratus, nimbostratus** and portions of **cumulus** and **cumulonimbus**. The low clouds are **stratocumulus, stratus** and most cumulus and cumulonimbus bases, and sometimes **nimbostratus**.

3. A scheme of classifying clouds according to their particulate composition, namely, **water clouds, ice-crystal clouds**, and **mixed clouds**.

The first are composed entirely of water droplets (ordinary and/or supercooled), the second entirely of **ice crystals**, and the third a combination of the first two. Of the **cloud** genera, only **cirrostratus** and **cirrus** are always ice-crystal clouds; **cirrocumulus** can also be mixed; and only **cumulonimbus** is always mixed. **Altostratus** is nearly always mixed, but can occasionally be water. All the rest of the genera are usually water clouds, occasionally mixed; **altocumulus, cumulus, nimbostratus**, and **stratocumulus**.

> Howard, L., 1803: *On the Modifications of Clouds*, J. Taylor, London.
> World Meteorological Organization, 1956: *International Cloud Atlas*, Volumes I and II.

cloud cluster—Same as **mesoscale convective system** (MCS).
Usually refers to tropical MCSs.

cloud condensation nuclei—(Abbreviated CCN.) Hygroscopic **aerosol** particles that can serve as nuclei of atmospheric **cloud droplets**, that is, **particles** on which water condenses (activates) at supersaturations typical of atmospheric **cloud formation** (fraction of one to a few percent, depending on **cloud** type).

Concentrations of CCN need to be given in terms of a **supersaturation** spectrum covering the **range** of interest or at a specified supersaturation value. *Compare* **condensation nucleus**.

cloud condensation nuclei counter—(Abbreviated CCN counter.) Any of several devices that measure the number concentration of atmospheric **particles** upon which **water vapor** condenses at low values of **supersaturation** that are equivalent to those that occur with the formation of tropospheric liquid clouds.

The supersaturations produced in the **condensation** chambers of these devices can normally be set within a **range** from a few tenths of one percent to several percent. However, **clouds** or **haze** can begin to form on very **hygroscopic** particles at **saturation levels** well below 100%, so some devices also allow for a range of measurements at subsaturation. The concentrations of **cloud condensation nuclei** (CCN) are normally reported as the number per cubic centimeter of air activated at a specified super- or subsaturation with respect to water. *Compare* **condensation nuclei counter**.

cloud cover—(Also called cloudiness, cloudage.) That portion of the **sky cover** that is attributed to **clouds**, usually measured in tenths or eighths of sky covered.

cloud crest—Same as **crest cloud**.

cloud-detection radar—A **radar** designed for **detection** of **cloud** particles as distinguished from the larger **particles** that constitute **precipitation**.

Cloud radars typically employ short wavelengths of 1 cm and shorter (e.g., Ka **band**) and are designed for short-**range**, high-**resolution** measurements. They can also be multiwavelength or **polarimetric radars** capable of providing information on the shapes and sizes of the **cloud** particles. Radars operating at short **wavelengths** have increased **sensitivity** to all weather targets but also suffer from increased **attenuation** by clouds and precipitation.

cloud discharge—*See* **cloud flash**.

cloud drop—A spherical **particle** of liquid water, from a few micrometers to a few tens of micrometers diameter, formed by **condensation** of **water vapor** on a **hygroscopic** aerosol particle (**cloud condensation nucleus**).

Such drops, apparently suspended in the air with other drops, form a visible **cloud**. Clouds may also contain interstitial **haze** particles, smaller than a few micrometers (μm) in diameter. Activation distinguishes a cloud from a haze, which contains only or mainly unactivated droplets. **Cloud drops** differ in size from **drizzle drops** and **raindrops**. A diameter of 0.2 mm has been suggested as an upper limit to the size of drops that shall be regarded as cloud drops; larger drops fall rapidly enough so that only very strong updrafts can sustain them. Any such division is somewhat arbitrary, and active **cumulus** clouds sometimes contain cloud drops much larger than this.

cloud-drop sampler—A general term for instruments that collect **cloud droplets** to determine their size or impurities; includes devices with direct-impaction surfaces that either capture or leave an impression of the impinging droplets, such as the **cloud gun**, as well as bulk samplers that capture and collect cloud droplets for chemical analysis, such as the **rime rod**.

cloud droplet—Often used interchangeably with **cloud drop**.
When used together, a cloud droplet is considered smaller and falling more slowly than a cloud drop; a cloud drop may grow by **coalescence** with cloud droplets.

cloud echo—The **radar** signal returned from a **cloud**. A cloud echo is generally much weaker than that returned by **precipitation** and requires sensitive equipment or **cloud-detection radar** to be measurable.

cloud electrification—The process by which clouds become electrified.
Normally this will produce positively charged regions in the top of clouds and negatively charged regions in the lower part of the **cloud**.

cloud element—The smallest **cloud** or portion of cloud that can be resolved by a **remote sensing** device or other instrument.

cloud etage—The grouping of **cloud height** in the WMO classification: high etage (3–8 km); middle etage (2–4 km); and low etage (surface to 2 km).

cloud feedback—The change in the radiative effects of clouds in response to an external **climate** perturbation.
The total cloud feedback is the combined result of changes in **cloud cover**, **cloud height**, and cloud **reflectivity**. The feedback is the net effect of these changes on the amount of **sunlight** absorbed by the earth and the amount of **heat** it radiates to space. Because many different properties of clouds contribute to cloud feedback, and because different types of clouds (e.g., **cirrus** vs **stratus**) have different effects on **solar** and **terrestrial radiation**, cloud feedback is difficult to specify with confidence. Current **climate models** disagree as to whether the overall cloud feedback is positive or negative. This disagreement is a major source of uncertainty in estimates of the overall sensitivity of the global climate to **anthropogenic** perturbations.

cloud flash—(Also called intracloud flash, cloud-to-cloud flash.) A **lightning discharge** occurring between a positively charged region and a negatively charged region, both of which may lie in the same **cloud**.
The most frequent type of **cloud discharge** is one between a main positively charged region and a main negatively charged region. Cloud flashes tend to outnumber **cloud-to-ground flashes**. In general, the channel of a cloud flash will be wholly surrounded by cloud. Hence, the channel's luminosity typically produces a diffuse glow when seen from outside the cloud and this widespread glow is called **sheet lightning**.

cloud formation—1. The process by which various types of clouds are formed.
In most cases, cloud formation involves cooling by expansion of ascending **moist air**. In rare exceptions, the cooling may occur as a result of other processes such as **mixing**, as in a **contrail**. The ascent of air may result from vertical **instability**, as in most **cumulus** clouds; from undulatory motions at **inversion** surfaces, as in certain **undulatus** species; from **orographic lifting**; and at a **frontal surface**, as in many **altostratus** and other **stratiform** clouds.
2. A particular arrangement of clouds in the sky, or a striking development of a particular **cloud**.

cloud-free line of sight—(Abbreviated CFLOS.) Line of sight that is unhampered by the **clouds** present.
This is a military term and the data are used for determining the utility of various communications, surveillance, and weapons systems. Data express the **probability** that the line of sight will be unobstructed by clouds for geographic areas and **climatic regions**. *Compare* **clear line of sight**.

cloud genus—(Or cloud genera.) The WMO classification of the 10 main groups of clouds, called genera.
See **cloud classification**.

cloud gun—A device for airborne or ground-based use that exposes a soot-coated glass slide for an instant to a **cloudy** airstream.
Cloud droplets impact on and leave craters in the **soot** coating for subsequent measurement of **droplet** size spectra using a computer-driven **image** analysis system and calibrations that relate drop size to crater size.

cloud height—1. In weather observations, the height of the **cloud base** above local terrain. *Compare* **ceiling**.

2. (Rare.) The height of the **cloud top** above local terrain or above **mean sea level**.

3. (Rare.) The vertical distance from the **cloud base** to the **cloud top**; more commonly referred to as the "thickness" or "depth" of the **cloud**.

cloud-height indicator—The general term for instruments that measure the height of **cloud bases**.
Cloud-height indicators may be classified according to their principle of operation. One class of instrument is based on height determination by means of the principle of triangulation. A **beam** of **light** projected from the ground onto the base of the cloud is observed visually or electrically from a remote point. The height of the cloud is determined trigonometrically from a measurement of the angle defined by the light projector, the **observation** point, and the light spot on the cloud plus the distance between the light projector and the observation point. Examples of instruments based on this principle are the **ceilometer** and the **ceiling light**. A second class of instrument is based on **pulse** techniques. The time required for a pulse of **energy** to travel from a **radiator** located on the ground to the cloud base and back to the ground is measured electrically. The height of the cloud is computed from this transit time and a knowledge of the propagation **velocity** of the pulse. Instruments based on this principle include the **pulsed-light cloud-height indicator** and vertically directed **cloud-detection radar**.

cloud-height measurement method—Method that allows the height of a **cloud base** or top to be determined.
See **ceilometer**.

cloud image animation—The automatic sequential **display** of a series of images containing **cloud** information.
The number of images and the interimage delay are parameters that define the smoothness of the motion resulting from the animation as well as its total duration.

cloud layer—An array of **clouds**, not necessarily all of the same type, with bases at approximately the same level.
It may be either continuous or composed of detached elements.

cloud level—1. A layer in the **atmosphere** in which are found certain **cloud** genera.
Three such levels are usually defined: high, middle, and low. *See* **cloud classification**.

2. At a particular time, the layer in the **atmosphere** bounded by the limits of the bases and tops of an existing **cloud** form.

cloud line—A narrow **cloud band** in which individual elements are connected and the line is less than one **degree** of latitude in width.

cloud microphysics—Cloud processes (growth, **evaporation**, etc.) taking place on the **scale** of the individual **aerosol** or **precipitation** particle as opposed to the scale of the visual **cloud**.
See **cloud physics**.

cloud microstructure—Structure of small-scale cloud features (cells, billows, wisps, etc.) resulting from the distribution of **cloud** particles on a scale larger than the **particles** themselves but smaller than the scale of the visual cloud.
Such features may result from local motions, organized or turbulent, and be preferred sites for the development of cloud **hydrometeors**.

cloud mirror—The mirror of a **mirror nephoscope**.

cloud model—A physical or numerical framework for the **prediction** of cloud behavior.
A physical **model** might be the behavior of a lighter or denser fluid blob (**thermal**) when released into a large **environment**. A theoretical model begins with the equations of fluid flow and derives solutions in terms of the growth of **particles** in a prescribed **mixing** environment.

cloud modification—Any process by which the natural course of **development** of a **cloud** in the earth's **atmosphere** is altered, for example, by the exhaust from an aircraft engine or **smoke** and **heat** from a forest fire.

cloud motion vector—The speed and direction determined from tracking **clouds** in satellite imagery.
If the clouds tracked are passive, that is, neither growing nor decaying, then the **vector** approximates the **wind vector**.

cloud nucleus—An old term for **particles** that act as centers for **droplet** or **ice** particle formation in the **atmosphere**.

148

Current usage differentiates between **condensation**, **cloud** condensation, and **ice nucleus**.

cloud optical depth—The vertical **optical thickness** between the top and bottom of a **cloud**.

Cloud optical depths are relatively independent of **wavelength** throughout the **visible spectrum**, but rise rapidly in the **infrared** due to **absorption** by water, and many clouds approximate **blackbodies** in the **thermal infrared**. In the visible portion of the **spectrum**, the cloud optical depth is almost entirely due to **scattering** by **droplets** or **crystals**, and ranges through orders of magnitude from low values less than 0.1 for thin cirrus to over 1000 for a large **cumulonimbus**. Cloud optical depths depend directly on the cloud thickness, the liquid or **ice** water content, and the size distribution of the water droplets or ice crystals.

cloud particle—The fundamental constituents of clouds responsible for their **visibility**, consisting of water **droplets** and/or **ice crystals** having small (less than 25 cm s^{-1}) **fall velocity** through the air.

cloud particle imager—*See* **optical imaging probe**.

cloud physics—The body of knowledge concerned with physical properties of clouds in the **atmosphere** and the processes occurring therein.

Cloud physics, broadly considered, embraces not only the study of **condensation** and **precipitation** processes in **clouds**, but also **radiative transfer**, optical phenomena, electrical phenomena, and a wide variety of hydrodynamic and thermodynamic processes peculiar to natural clouds. Cloud physics is a distinct subdivision of **physical meteorology**. Early interest in this subject was stimulated by the role of clouds in aircraft safety related to **icing** and **turbulence** (The Thunderstorm Project) and the discovery of **cloud modification** techniques by **cloud seeding**. The formation of precipitation and the influence of clouds in radiative processes in the atmosphere (both solar and **thermal**) determine the key role of the subject in global **climate**.

cloud radar—A **radar** optimized to detect **clouds**.

This is usually a short **wavelength** radar often called millimeter radar, since its wavelength is usually near 8 mm (Ka band) or 3 mm (W band). Such shorter wavelengths give radars an advantage in cloud detection because of the radar **scattering cross section** for small **particles** varying as the inverse radar wavelength to the fourth power. Because **attenuation** caused by atmospheric liquid and from **water vapor** is serious at these short wavelengths, such radars are usually operated at ranges closer than about 20 km. They can be either of the **scanning** type or fixed in the vertical. Their good spatial **resolution** allows them to depict finescale cloud features and, when used in combination with other active (e.g., **lidar**) and passive (**microwave** and **IR** radiometers) sensors, can, under some conditions, be used to quantitatively map cloud ice and liquid water contents, particle sizes, and concentrations.

cloud radiative forcing—The difference between net **irradiances** measured for average atmospheric conditions and those measured in the absence of **clouds** for the same region and time period.

Cloud radiative forcing depends jointly on the amount of cloud present and the sensitivity of **radiation** to **cloud amount**. It may be partitioned into **longwave** and **shortwave** forcing terms, the combination of which typically results in a negative net forcing when referenced to the top of the **atmosphere** (i.e., satellite measurements). That is, a **clear** region typically reflects less solar, and emits more terrestrial, radiation than does the average condition, and the difference in solar **reflection** is typically greater than the difference in terrestrial **emission** to space. While the definition of cloud radiative forcing in terms of average measured values is unambiguous, the relationship between cloud radiative forcing and the **equilibrium** effects of clouds on **climate**, especially on **surface temperature**, is a complicated topic. *See also* **radiative forcing**.

cloud searchlight—Same as **ceiling light**, **ceiling projector**.

cloud sector—A differentiated zone of a **cloud system** in which the general appearance of the sky, as a whole, displays marked peculiarities.

cloud seeding—The addition of agents (**aerosol**, small **ice** particles) that will alter the **phase** and size distribution of **cloud** particles, with the intent of influencing **precipitation**.

The most frequently used agents are granulated solid **carbon dioxide** (**dry ice**), **silver iodide** aerosol for initiation of the ice phase, and salt (sodium chloride) for initiation of larger **cloud droplets**. Many other agents (e.g., organic materials, bacteria) have been tested and proposed for use. The intent of cloud seeding is to modify or alter the natural development of the cloud so as to enhance or redistribute precipitation, suppress **hail** formation, dissipate **fog** or **stratus** cloud, or suppress **lightning**. Cloud seeding may involve different techniques. Particles may be released from the ground, from aircraft, or from rockets. The goal of ice phase cloud seeding is to induce the phase transition from a **supercooled water** cloud to one composed partially or entirely of ice.

The goal of dynamic cloud seeding is to stimulate or enhance vertical air motions in the cloud through increased **buoyancy** derived from the release of **latent heat** of **freezing**. Hygroscopic **seeding** utilizes **hygroscopic** salt aerosols that readily condense water and may grow large enough to become centers for **coalescence** growth of precipitation.

cloud shield—1. In general, a broad **cloud formation** that is not more than four times as long as it is wide.

2. In **synoptic meteorology**, the principal **cloud** structure of a typical **wave cyclone**, that is, the cloud forms found on the cold-air side of the **frontal system**.

The maximum areal coverage is usually found over the region in advance of the **warm front**, and the minimum behind the **cold front**. Within the area of the cloud shield, there is an idealized but smaller **precipitation shield**.

cloud species—*See* **cloud classification**.

cloud streets—Linear **cloud** organization occurring atop the **updraft** branches of **horizontal convective rolls** when sufficient moisture is present.

Cloud streets are readily observed with satellite imagery and can extend for hundreds of kilometers, particularly where the underlying surface is uniform.

cloud symbol—One of a set of specified ideograms that represent the various **cloud** types of greatest significance or those most commonly observed.

Cloud symbols are entered on a **weather map** as part of a **station model**.

cloud system—(Or nephsystem.) An array of **clouds** and **precipitation** associated with a **cyclonic-scale** feature of atmospheric **circulation**.

Cloud systems display typical patterns and **continuity**, the **analysis** of which is termed **nephanalysis**.

cloud-to-cloud discharge—*See* **cloud flash**.

cloud-to-ground discharge—*See* **cloud-to-ground flash**.

cloud-to-ground flash—A **lightning flash** occurring between a charge center in the **cloud** and the ground.

On an annual basis, negative charge is lowered to the ground in about 95% of the flashes. The remaining flashes lower positive charge to the ground. This type of lightning flash, which can be contrasted with an **intracloud flash** or **cloud flash**, consists of one or more return strokes. The first **stroke** begins with a **stepped leader** followed by an intense **return stroke** that is the principal source of luminosity and charge **transfer**. Subsequent strokes begin with a **dart leader** followed by another return stroke. Most of the strokes use the same channel to ground. The time interval between strokes is typically 40 μs.

cloud top—For a given **cloud** or **cloud layer**, the highest level in the **atmosphere** at which the air contains a perceptible quantity of cloud **particles**.

cloud-top entrainment instability—A condition whereby **entrainment** of **dry air** into a **cloud top** causes even more entrainment, leading to the **dissipation** of the **cloud**.

When the entrained air mixes with **cloudy** air, **evaporation** of the **cloud drops** into the mixture causes the mixture to cool. As this cool mixture sinks, it generates a turbulent **circulation** that can cause more entrainment, thereby continuing the process until the cloud dissipates. While there is still debate about the requirements for such a process to occur, one of the first suggestions was that this **instability** will occur when $\Delta\theta_e < \Delta\theta_{e\,critical}$, where $\Delta\theta_e$ is the difference of **equivalent potential temperature** from just above to just below cloud top, and $\Delta\theta_{e\,critical}$ is a critical value that is near zero. While the exact value of $\Delta\theta_{e\,critical}$ has yet to be determined, it depends on the **buoyancy** of the air (i.e., on **virtual potential temperature**) and on the **temperature** change possible due to **latent heat** changes when **cloud droplets** evaporate.

cloud tracer—**Cloud** movement that is used as a measure of air motion.

cloud variety—*See* **cloud classification**.

cloud winds—**Wind** estimates based on the observed movements of identifiable **cloud elements** over short time intervals.

cloudage—Same as **cloud cover**.

cloudbow—(Also called fogbow, mistbow, white rainbow.) A large, faintly colored, circular arc formed by **light** (usually **sunlight**) falling on **cloud** or **fog**.

The apparent center of the cloudbow is the **antisolar point** (the shadow of the observer's head or possibly the shadow of the plane in which the observer is flying). The term **rainbow** is a general term for this phenomenon, a cloudbow merely being a rainbow formed in the smaller cloud or fog **droplets**. The broad whitish appearance of the cloudbow and slightly smaller angular radius certainly demonstrate that an explanation that treats light as a series of rays does not work very well for small droplets.

cloudburst—(Also called rain gush, rain gust.) In popular terminology, any sudden and heavy fall of **rain**, almost always of the **shower** type.

An unofficial criterion sometimes used specifies a rate of fall equal to or greater than 100 mm (3.94 inches) per hour. *See* **excessive precipitation**.

cloudiness—Same as **cloud cover**.

cloudy—1. The character of the **sunrise** or **sunset** when the disk of the sun is hidden at these times by **clouds** or an **obscuring phenomenon**.

Compare **clear**.

2. In popular usage, the state of the weather when **clouds** predominate at the expense of **sunlight**, or obscure the stars at night.

In **weather forecast** terminology, expected **cloud cover** of about 0.7 or more warrants the use of this term. *Compare* **clear**, **partly cloudy**.

cloudy day—A day during which the sky is more than 75% covered by opaque **clouds** for most of the day.

cloudy sky—Sky with a total cover greater than four octas (eighths).

ClO$_x$—*See* **chlorine oxides**.

cluster ions—**Ions** in which a charged molecule is surrounded by a number of loosely bound neutral molecules held by electrostatic forces.

In the **troposphere**, the associated molecules are often ambient water molecules. In the **stratosphere**, positive ions are found to cluster with water or **acetonitrile**, while negative ions are mostly associated with **sulfuric acid** or **nitric acid**.

clutter—Undesirable **echoes** on radar displays. For a **weather radar**, clutter usually refers to echoes from ground **targets** in the **main lobe** or sidelobes of the **antenna pattern** (**ground clutter**) but might also refer to apparent echoes caused by interference from other **radar** or radio sources.

For tracking or navigational radars, echoes from **precipitation** may obscure the targets of interest and are called weather clutter. *See* **cancellation ratio**, **clutter rejection**.

clutter rejection—Any of a variety of processes or techniques to eliminate the effects of unwanted signals (**clutter**) in **radar** measurements.

Clutter rejection most commonly refers to the **filtering** of the received **signal** to reject components that have specified **phase** or **amplitude** characteristics. In meteorological **Doppler radar** signal processing, for example, one approach to clutter rejection is to reject signals with Doppler velocities near zero, because stationary **scattering** objects induce no **Doppler shift**.

CME—Abbreviation for **coronal mass ejection**.

co-cumulative spectrum—(Abbreviated CCS.) In **ocean wave** studies, the integral of an **energy spectrum**.

The area under a particular **energy spectrum** from a given **frequency** value to infinity is given by the value of the CCS curve at that frequency.

co-latitude—The complement of the latitude.

See **spherical coordinates**.

coagulation—1. In **cloud physics**, an obsolete term denoting any process that converts the numerous small **cloud drops** into a smaller number of larger **precipitation** particles.

When so used, the term is employed in analogy to the coagulation of any **colloidal system**. The process can take place at temperatures both above and below 0°C for supercooled drops. *See* **coalescence**.

2. Similar to **accretion**.

3. The process whereby **aerosol** or colloidal **particles** collide with each other by **Brownian motion** and coalesce (liquid) or aggregate (solid).

coalescence—In **cloud physics**, the merging of two water drops into a single larger **drop** after collision.

Coalescence between colliding drops is affected by the impact **energy**, which tends to increase with the higher fall velocities of larger drops. Colliding drops having negligible impact energy compared to their **surface energy** behave as water spheres that collide with a **collision efficiency** (the fraction of small drops that collide with a large drop within the geometric collision **cross section**) predicted by the theory for falling spheres. The result of increasing impact energy is to flatten the colliding drops at the point of impact, impeding the drainage of the air and delaying contact between them. As the distortion relaxes, the drops rebound, reducing the **coalescence efficiency** for **cloud drops** and **drizzle drops** colliding with smaller drops. At larger impact energy, separation will occur if the **rotational** energy (fixed by **conservation of angular momentum**) is higher than the surface energy of the coalescing drops. This phenomenon, termed temporary coalescence, can result in satellite droplets considerably smaller than either of the parent drops. This phenomenon is also called partial coalescence because the large drop may gain mass as a result of the higher internal **pressure** in the small drop. At still larger impact energy, **drop breakup** occurs for the smaller drop. About 20% of the high-energy collisions between large **raindrops** (d > 3 mm) and drizzle drops (d > 0.2 mm) result in the disintegration of both drops. Other factors that affect coalescence are **electric charge** and **electric field**, both of which promote coalescence, leading to earlier onset of coalescence during an interaction so that coalescence efficiencies are increased by suppression of rebound and temporary coalescence. All of these processes are important in formation of **precipitation** in all liquid **clouds** both above and below 0°C. *See* **collision–coalescence process**.

coalescence efficiency—The fraction of all collisions between water drops of a specified size that results in actual merging of the two drops into a single larger **drop**.

In discussing the details of the growth of **raindrops** by collision and **coalescence**, it is important to distinguish clearly the terms coalescence efficiency, **collision efficiency**, and **collection efficiency**, the last being equal to the product of the first two.

coalescence process—*See* **collision–coalescence process**.

coamplitude line—*See* **corange line**.

coarse-mesh grid—A nonspecific term indicating a **grid** that has a relatively low resolution, that is, its grid points are relatively far apart.

The term is used to contrast a grid with another that has significantly higher **resolution**. *See* **fine-mesh grid**.

coarse particles—**Particles** with a diameter greater than 2 μm suspended in the **atmosphere**.

Coast Pilot—*See* **United States Coast Pilot**.

coastal climate—The **climate** in coastal regions resulting from the modification of the **macroclimate** due to the **discontinuity** in **surface roughness** at the coastline and to the different **thermal** and moisture properties of sea and land.

coastal front—A shallow (typically < 1 km deep) **mesoscale** frontal zone marked by a distinct **cyclonic** windshift in a region of enhanced **thermal** contrast (≈ 5°–10°C/10 km).

These **fronts** typically develop in coastal waters or within 100–200 km of the coast during the cooler half of the year when the land is cold relative to the ocean. In the United States coastal fronts are most frequent in New England, the Middle Atlantic states, the Carolinas, and Texas. The typical coastal front is oriented quasi-parallel to the coast and may extend for several hundred kilometers. During the winter, the coastal front may mark the boundary between frozen and nonfrozen **precipitation**. Given that coastal front development usually precedes synoptic-scale **cyclogenesis** and marks an axis of enhanced thermal contrast and a maximum in **cyclonic** vorticity and **convergence**, the coastal front often serves as a boundary along which intensifying **synoptic-scale** cyclones move poleward. Surface coastal front development typically occurs beneath the forward side of advancing **troughs** following the passage of the **ridge** axis aloft. Coastal fronts most frequently form equatorward of cold **anticyclones** where a warmer onshore flow encounters a colder **continental air** stream. Damming of cold air on coastal **orographic** barriers such as the Appalachians often appears to play an important role in coastal front development. Coastal thermal contrasts are augmented by differential **diabatic** heating where the onshore flow has passed over oceanic thermal boundaries such as the **Gulf Stream** and the adjacent continental airstream has passed over snow-covered land. Coastal fronts may form independently of cold anticyclones and associated cold air damming. In situ coastal front developments can occur near mountain barriers where upslope flow results in differential **airmass** cooling and stabilization and where offshore troughs form due to differential heating across oceanic thermal boundaries. Coastal front dissipation typically occurs with the cessation of onshore flow following **cyclone** passage.

coastal jet—*See* **boundary currents**.

coastal upwelling—The rising of water from between 200 and 400 m to the surface along coastlines where an alongshore blowing **wind** has the coast on its left in the Northern Hemisphere or on its right in the Southern Hemisphere.

Because the surface currents of the **Ekman spiral** are deflected offshore in these situations, the **surface water** is drawn away from the coast, causing the colder water from deeper layers to upwell. The associated lowering of the **sea surface temperature** results in atmospheric **heat** loss and modifies the local **climate**. The upwelled water is also rich in nutrients, and coastal upwelling regions are among the most important fishing regions of the World Ocean. The most important coastal upwelling regions are found in the eastern **boundary currents** of the subtropical **gyres**, that is, in the **Peru/Chile, California, Benguela,** and **Canary Currents**. The **Somali, East Arabian,** and **South Java Currents** develop **upwelling** on a seasonal basis.

coastal zone—A region a few kilometers wide on either side of the shoreline where local **thermal** circulations such as the **sea breeze** and **land breeze** occur.

Coastal Zone Color Scanner—(Abbreviated CZCS.) A **scanning radiometer** with six **channels** flown on *Nimbus* 7 (launched October 1978) designed to monitor ocean color and phytoplankton production in coastal areas.

Four channels are in the visible part of the **spectrum**, one in the **near-infrared**, and one in the **thermal infrared**.

coastally trapped waves—**Free waves**, having characteristics of both **Kelvin waves** and **shelf waves**, that are trapped in the vicinity of the coast.

coaxial cable—A two-conductor constant impedance transmission line consisting of one conductor centered inside and insulated from a second conductor.

cockeyed bob—A colloquial term in western Australia for a **squall**, associated with **thunder**, on the northwest coast in summer.

code forms—Use of printable characters in transmission to represent different conditions. *See* **binary code**.

coefficient of barotropy—A function of thermodynamic variables that is the coefficient of proportionality between the gradients of the **density** and **pressure** fields in a **barotropic atmosphere**. *See also* **barotropy**.

coefficient of compressibility—(Or compressibility.) The relative decrease of the volume of a system with increasing **pressure** in an **isothermal process**.

This coefficient is

$$-\frac{1}{V}\left(\frac{\partial V}{\partial p}\right)_T,$$

where V is the volume, p the pressure, and T the **temperature**. The reciprocal of this quantity is the **bulk modulus**. *Compare* **coefficient of thermal expansion, coefficient of tension**.

coefficient of consolidation—The ratio of **hydraulic conductivity**, K, to $(m_v\gamma_w)$, where γ_w is the unit weight of water and m_v is the ratio of the volume change to the increase of effective **stress**.

coefficient of continentality—*See* **continentality**.

coefficient of correlation—*See* **correlation, correlation coefficient**.

coefficient of diffusion—Same as **diffusivity**; *see* **coefficient of mutual diffusion**.

coefficient of dynamic viscosity—Same as **dynamic viscosity**.

coefficient of eddy conduction—Same as **eddy conductivity**.

coefficient of eddy diffusion—Same as **eddy diffusivity**.

coefficient of eddy viscosity—*See* **eddy viscosity, logarithmic velocity profile**.

coefficient of excess—*See* **kurtosis**.

coefficient of exchange—*See* **exchange coefficients**.

coefficient of expansion—Same as **coefficient of thermal expansion**.

coefficient of heat conduction—Same as **thermal conductivity**.

coefficient of kinematic viscosity—Same as **kinematic viscosity**.

coefficient of molecular viscosity—Same as **dynamic viscosity**.

coefficient of multiple correlation—*See* **multiple correlation**.

coefficient of mutual diffusion—A quantity in the **kinetic theory** of gases that measures the tendency of gases to diffuse into one another in nonturbulent flow.

This **diffusion coefficient** is a property of the gases in question and of the assumed nature of the molecular impacts in the **diffusion** process.

Lettau, H., 1951: *Compendium of Meteorology*, 320–334.

coefficient of piezotropy—*See* **piezotropy, equation of piezotropy**.

coefficient of polytropy—*See* **polytropic process**.

coefficient of skewness—*See* **skewness**.

coefficient of skin friction—*See* **skin-friction coefficient**.

coefficient of tension—The relative increase of **pressure** of a system with increasing **temperature** in an **isochoric** process.

In symbols this quantity is

$$\frac{1}{p}\left(\frac{\partial p}{\partial T}\right)_V,$$

where p is pressure, T temperature, and V volume. *Compare* **coefficient of compressibility, coefficient of thermal expansion**.

coefficient of thermal conduction—Same as **thermal conductivity**.

coefficient of thermal expansion—The relative increase of the volume of a system (or substance) with increasing **temperature** in an **isobaric** process.

In symbols this coefficient is

$$\frac{1}{V}\left(\frac{\partial V}{\partial T}\right)_p,$$

where V is the volume, T the temperature, and p the **pressure**. *See* **Charles–Gay–Lussac law**; *compare* **coefficient of compressibility, coefficient of tension**.

coefficient of transparency—The fraction of **direct solar radiation** that arrives at the earth's surface when the sun is at the **zenith**.

The coefficient appears in the mathematical expression of **Bouguer's law** (Beer's law, Lambert's law).

coefficient of viscosity—Same as **dynamic viscosity**.

See also **kinematic viscosity, eddy viscosity**.

coffin corner—A term used to describe the **range** of Mach numbers between the buffeting **Mach number** and the **stalling Mach number** within which an aircraft must be operated.

The buffeting and stalling Mach numbers approach each other with **altitude**; when they become the same, the **ceiling** of the aircraft is reached.

cognitive task analysis—(Abbreviated CTA.) A branch of cognitive science devoted to identifying the processes involved in performing a task requiring intelligence.

CTA is useful in the design of **expert systems**, computer interfaces, and the appropriate division of tasks between humans and computers. A recent CTA of weather forecasting in the Air Force has resulted in modifications to the forecast office and in forecaster and observer training.

coherence—1. The property of a single **wave** with a **phase** that is a continuous, **linear** function of position at a given time.

A stable local **oscillator** produces a coherent wave.

2. The property of two or more **waves** that are in **phase** both temporally and spatially. Waves are coherent if they have the same **wavelength** and a fixed phase relationship with each other. When the phase relationships are not fixed, the waves are said to be partially coherent or incoherent.

3. The **correlation coefficient** between **electromagnetic fields** at points separated in space and time, sometimes called degree of coherence. So defined, the coherence equals unity for **waves** that are perfectly coherent and is less than unity for partially coherent waves.

4. As used by Sir Gilbert Walker (1932), the **statistical** persistence exhibited by successive daily values of **atmospheric pressure** at any one location.

Walker, G. T., 1932: World weather. *Royal Meteor. Soc. Mem.*, 4.

coherence element—In **radar**, the four-dimensional volume given by the product of the **coherence time** and the **pulse volume**, within which the **target signal** is partially correlated.

To estimate the **power** of the **target signal** requires averaging over several independent samples, which means averaging signals from a **measurement cell** that is large compared with the coherence element.

coherence time—(Also called decorrelation time.) In **coherent radar** or **lidar** systems, the time interval over which the received **signal** may be regarded as approximately **monochromatic**.

For atmospheric targets this time varies from about 1 s at a **wavelength** of 1 m to 1 μs at a wavelength of 1 μm.

coherent detection—The conversion of the **intermediate frequency (IF) signal** in a **Doppler radar** or **lidar** to **I and Q channels** so that the **phase** of the **echoes** is preserved.

Typically two mixers, fed by IF reference signals in phase **quadrature**, are used to convert the **frequency** components in the IF signal, centered at the IF frequency, to two **video** signals with frequency components centered at zero frequency, without modifying the information content.

coherent echo—A **radar** echo with **phase** and **amplitude** either showing little change over successive **radar** pulses or changing in a regular and predictable way. Such an echo may arise from fixed or slowly moving point **targets** or from distributed targets in which the individual **scattering** elements do not move (or move slowly) with respect to one another.

By contrast, an **incoherent echo** is an echo with **random** phase and amplitude from **pulse** to pulse. Such echoes arise from distributed targets such as **precipitation** targets, in which the individual scatterers move with respect to one another.

coherent integration—(Also called coherent averaging or time-domain averaging.) The time-domain integration of measurements in a **coherent radar** over a sequence of pulses or over an **observation** interval, prior to estimating the **signal** properties, to improve the **signal-to-noise ratio** while minimizing signal processing.

For such integration to be effective, the integration period must be limited to the time over which the **phase** of the signal relative to a reference phase does not change substantially (i.e., the **coherence time** of the signal). The effect of the coherent integration process is to reduce the effective data sampling rate and the **Nyquist frequency**.

coherent radar—A type of **radar** that extracts additional information about a **target** through measurement of the **phase** of **echoes** from a sequence of **pulses** (or an extended **observation** interval, as in an **FM–CW radar**).

The phase information may be used to improve the **signal-to-noise ratio** (*see* **coherent integration**), to estimate the **velocity** of the target through the **Doppler effect** (*see* **Doppler radar**), or to resolve the location of the target in a **synthetic aperture radar**.

coherent scattering—In **radar**, **scattering** produced when the incident **wave** encounters a **point target** that is either fixed or moving with a constant **radial velocity**, or a **distributed target** with individual **scattering** elements fixed or slowly moving relative to one another.

Such targets give **coherent echoes**.

coherent structures—Three-dimensional regions in a **turbulent flow** with characteristic structures and lifetimes in terms of **velocity**, **temperature**, etc., that are significantly larger or longer-lived than the smallest local scales.

Much of the **turbulent transport**, conversion of mean flow **energy** into turbulent **eddies**, **nonlinear** transfer into smaller scales, and eventual **dissipation** is associated with coherent structures. Examples include gust microfronts and horseshoe vortices in the **surface layer**, **convective plumes**, and **longitudinal roll vortices**. *See* **longitudinal rolls, convective plume, Langmuir circulation.**

Etling, D., and R. A. Brown, 1993: Roll vortices in the planetary boundary layer: a review. *Bound.-Layer Meteor.*, **65**, 215–248.
Foster, R. C., 1997: Structure and energetics of optimal Ekman layer perturbations. *J. Fluid Mech.*, **333**, 97–123.
Gerz, T., J. Howell, and L. Mahrt, 1994: Vortex structures and microfronts. *Phys. Fluids*, **6**, 1242–1251.
Wilczak, J., and J. Tillman, 1980: The three-dimensional structure of convection in the atmospheric surface layer. *J. Atmos. Sci.*, p. 2424.

coherent target—A **radar** target producing a **coherent echo**.

COHMAP—(Acronym for Cooperative Holocene Mapping Project.) A project that reconstructed global **climate** at several times in the **Holocene**, using pollen, lake-level, and marine microfossils.

These reconstructions were compared with **general circulation model** results using the **boundary conditions** appropriate to each time in order to investigate the global effects of variations in

insolation resulting from the **Milankovitch variations** and the effects of the retreating **continental ice** sheets in the early Holocene.

coho—(Contraction for coherent oscillator.) An **oscillator** used in a **coherent radar** to provide a reference **phase** by which changes in the phase of the received **signal** may be recognized.
In practice, a coho usually operates at the **intermediate frequency** of the **receiver**.

col—(Also called saddle point, neutral point.) In meteorology, the point of intersection of a **trough** and a **ridge** in the **pressure pattern** of a **weather map**.
It is the point of relatively lowest **pressure** between two **highs** and the point of relatively highest pressure between two **lows**.

cold-air drop—Same as **cold pool**.

cold-air injection—A flow (15 knots or more) of cold air across an **isotherm ribbon**.
Compare **polar outbreak**.

cold air mass—*See* **airmass classification**.

cold-air outbreak—Same as **polar outbreak**.

cold-air pool—A topographic depression, such as a valley or basin, filled with cold air.
The cold air is heavy, and settles to the bottom of the depression. This air can remain stagnant, trapped by the surrounding higher terrain, resulting in long periods of poor air quality and **fog**, depending on the sources of **pollution** and amount of moisture in the air, respectively.

cold anticyclone—Same as **cold high**.

cold cap—As defined by A. Supan (1879), a region of the earth within which the **mean temperature** of the warmest month is less than 10°C. This limiting condition closely approximates the **temperature** at the **arctic tree line**, and was later adopted by W. Köppen (1918) as his boundary between the **polar climates** and **tree climates**.
Supan also defined **temperate belt** and **hot belt** in his early form of **climatic classification**.
> Supan, A., 1879: Die Temperaturzonen der Erde. *Petermanns Geog. Mitt.*, **25**, 349–358.
> Köppen, W. P., 1918: Klassification der Klimate nach Temperatur, Niederschlag und Jahreslauf. *Petermanns Geog. Mitt.*, **64**, 193–203; 243–248.

cold conveyor belt—An area in the **atmosphere** that transports cold air from one place to another.
It often refers to the low-level airflow within the relatively cold air ahead of a developing **cyclone**.

cold-core anticyclone—Same as **cold high**.

cold-core cyclone—Same as **cold low**.

cold-core high—Same as **cold high**.

cold-core low—Same as **cold low**.

cold-core rings—Large (roughly 300-km diameter) cyclonically rotating **eddies** found in the **Sargasso Sea**, containing cold **Slope Water** in their cores.
They persist for several months, occasionally interacting with the **Gulf Stream** and getting destroyed in the process. A cold-core ring is formed from a large-amplitude Gulf Stream meander that pinches off to the south, trapping relatively cold Slope Water from north of the Gulf Stream within its circumferential current. *See also* **warm-core rings**, **Gulf Stream rings**.

cold cyclone—Same as **cold low**.

cold desert—Same as **arctic desert**.

cold dome—A cold **air mass**, considered as a three-dimensional entity.
The **isentropic** surfaces bounding the cold air mass suggest the shape of a dome.

cold drop—Same as **cold pool**.

cold event—*See* **La Niña**.

cold front—Any nonoccluded **front**, or portion thereof, that moves so that the colder air replaces the warmer air; that is, the leading edge of a relatively cold **air mass**.
Compare **cold type occlusion**.

cold-front thunderstorm—A **thunderstorm** attending a **cold front**.
Formerly, the term was also applied to one of the line of thunderstorms that often appears up to a few hundred miles in advance of the **cold front**, along what is now known as an **instability line** or **squall line**.

cold-front-type occlusion—*See* **cold type occlusion**.

cold-front wave—A **disturbance** along a **cold front**, often accompanied by **clouds**, **precipitation**, and a **low pressure center**.
See **frontal wave**.

cold high—(Or cold anticyclone; also called cold-core high, cold-core anticyclone.) At a given level in the **atmosphere**, any **high** that is generally characterized by colder air near its center than around its periphery; the opposite of a **warm high**.
The **anticyclonic** intensity of a cold high decreases with height in accordance with the **thermal wind equation**. *Compare* **thermal high**.

cold low—(Or cold cyclone; also called cold-core low, cold-core cyclone.) At a given level in the **atmosphere**, any **low** that is generally characterized by colder air near its center than around its periphery; the opposite of a **warm low**.
A significant case of a cold low is that of a **cut-off low**, characterized by a completely isolated **pool of cold air** within its **vortex**. The **cyclonic** intensity of a cold low increases with height in accordance with the **thermal wind equation**.

cold occluded front—*See* **cold type occlusion**.

cold occlusion—*See* **cold type occlusion**.

cold pole—The location that has the lowest annual **mean temperature** in its hemisphere.
In the Northern Hemisphere the cold pole is usually placed at Verkhoiansk in Siberia (67°33′N, 133°24′E) with an annual mean temperature of −16°C (3°F) [January: −50°C (−59°F), July: 16°C (60°F)], but the country around Verkhoiansk is very mountainous, and lower winter temperatures are found in some of the valleys. At Oimekon, for example, the average January temperature is probably below −51°C (−60°F). In the Southern Hemisphere the cold pole is near 80°–85°S and 75°–90°E. International Geophysical Year stations located inland on Antarctica have recorded several temperatures well below −73°C (−100°F).

cold pool—1. (Also called cold drop, cold-air drop.) A region, or "pool," of relatively cold air surrounded by warmer air; the opposite of a **warm pool**.
This is usually applied to cold air of appreciable vertical extent that has been isolated in lower latitudes as part of the formation of a **cut-off low**. Cold pools are best identified as **thickness** minima on thickness charts. They are **cyclonic-scale** phenomena.
2. Any large-scale mass of cold air; a cold **air mass** or **cold dome**.

cold sector—The area within a **circulation** of a **wave cyclone** where relatively cold air can be found.
Typically, it lies behind the **cold front** associated with a **cyclone**.

cold soak—The effect of exposing equipment to low **temperatures** for an extended period of time.
Cold soak of engines necessitates preheating before their use, as lubricants have thickened, metal has become brittle, and tolerances have diminished.

> Arctic, Desert, Tropic Information Center (ADTIC) Research Studies Institute, 1955: *Glossary of Arctic and Subarctic Terms*, ADTIC Pub. A-105, Maxwell AFB, AL, 90 pp.

cold tongue—In **synoptic meteorology**, a pronounced extension or protrusion of cold air.

cold top—The small area on the **anvil** of a **convective storm** that appears colder than the majority of anvil pixels on an **infrared** satellite **image**.
The region appears colder because it is associated with an **overshooting top** that is higher than the mean anvil height. Often an overshooting top is found to be the point of origin of an **enhanced "v"** signature.

cold trough—At any given level in the **atmosphere**, any **trough** that is generally characterized by colder air near its center than its surroundings.
See **cold low**.

cold type occlusion—(Also called cold occlusion, cold occluded front, cold-front-type occlusion.) According to the Norwegian **cyclone model**, the situation where a **cold front** catches up with a portion of a **warm front** above the cold **frontal surface**.
This conceptualization of cold type occlusion development is not often observed in nature; however, it is undisputed that in strong cyclones the **low** center often retreats toward the cold air separating itself from the cold and warm fronts. A **trough** in **sea level pressure** is found between the **cyclone** center and the **wave** on the front, and this trough is the **occluded front**. Regardless of the formation processes, characteristics of a cold type occlusion are 1) a warm **temperature** or thickness ridge along the occluded front; 2) a trough in the sea level pressure field along the

occluded front; 3) relatively colder air behind the front; and 4) an increase in lower-tropospheric **static stability** behind the front.

cold wall—*See* **North Wall**.

cold wave—1. As used in the U.S. National Weather Service, a rapid fall in **temperature** within 24 hours to temperatures requiring substantially increased protection to agriculture, industry, commerce, and social activities.

Therefore, the criterion for a cold wave is twofold: the rate of temperature fall, and the minimum to which it falls. The latter depends upon region and time of year.

2. Popularly, a period of very cold weather.

colla—(Also called colla tempestada.) In the Philippines, a **fresh** or strong (less than **Beaufort force** 8) south to southwest **wind**, accompanied by **heavy rain** and severe **squalls**.

It is experienced most frequently in June and July and may persist for several days. It occurs when an extensive **trough** of low pressure runs from east to west or east-northeast to west-southwest, north of Manila, and is either stationary or moving slowly northward.

collada—A strong **wind** (16–22 m s^{-1} or 35–50 mph) in the Gulf of California, blowing from the north or northwest in the upper part of the Gulf and from the northeast in the lower part.

collar cloud—A ring of **cloud** seen occasionally at the top of a **wall cloud** (usually in wall clouds that are rotating) where the wall cloud is attached to the **updraft** base above it.

collection efficiency—In **cloud physics**, for aerodynamically interacting **cloud** and **precipitation** particles: 1) for interacting water drops, the product of **collision efficiency** and **coalescence efficiency**; 2) for interacting **ice** particles, or for water drops interacting with ice particles, the product of collision efficiency and **adhesion efficiency**.

collective—(Or sequence.) With respect to **aviation weather observations**, a group of such observations transmitted in prescribed order by stations on the same communications circuit.

collector—A class of instruments employed to determine the **electric potential** at a point in the **atmosphere**, and ultimately the **atmospheric electric field**.

All collectors consist of some device for rather rapidly bringing a conductor to the same **potential** as the air immediately surrounding it, plus some form of **electrometer** for measuring the difference in potential between the equilibrated collector and the earth itself. Collectors differ widely in their speed of response to atmospheric potential changes. When a flame is used to increase the local **ion density** and thereby facilitate potential equilibration, some advantage over a simple exposed conductor is gained; the **relaxation time** becomes of the order of one minute. Use of radioactive coatings, preferably emitters of **alpha particles**, gives a faster response, attaining relaxation times of the order of a few seconds. Certain mechanical collectors have relaxation times of the order of a few hundredths of a second. *Compare* **impactor**.

Chalmers, J. A., 1957: *Atmospheric Electricity*, 84–92.

colligative property—One of four characteristic properties of solutions, namely, the interdependent changes in **vapor pressure**, **freezing point**, **boiling point**, and osmotic pressure, with a change in dissolved matter.

If, under a given set of conditions, the value for any one property is known, the others may be computed. In general, with an increase in dissolved matter (e.g., salt in water), freezing point and vapor pressure decrease, and boiling point and osmotic pressure increase.

collision broadening—(Or pressure broadening.) The spreading of frequencies of a **spectral line** as a consequence of interactions between molecules.

Although the term "pressure broadening" is frequently encountered and is respectable, it is misleading in that **pressure** per se has nothing fundamentally to do with pressure broadening. At a fixed **temperature**, the number **density** of gas molecules is proportional to pressure. The greater the number density, the smaller the average separation between molecules, and hence the greater the **potential energy** of intermolecular interaction. Thus pressure is merely a surrogate for interaction (collision).

collision–coalescence process—In **cloud physics**, the process producing **precipitation** by collision and **coalescence** between liquid **particles** (**cloud droplets**, **drizzle drops**, and **raindrops**).

Drop breakup is a limiting factor to large **drop** growth by this process.

collision efficiency—1. The fraction of all water drops in the path of a falling larger **drop** that make contact with the larger drop.

Calculations using approximate **drag** forces between large and small **cloud drops** predict low collision efficiencies (<10%) when the large drop is less than 40 μm in diameter but collision efficiencies approaching 100% when the larger drop is greater than 80-μm diameter. Thus theory indicates that the **collision–coalescence process** for **precipitation** formation requires at least a small fraction of **cloud** drops to be larger than about 40-μm diameter. Because collision between drops does not necessarily result in **coalescence**, it is important to distinguish between collision efficiency and **coalescence efficiency**. Their product, termed the **collection efficiency**, gives the fraction of drops that collide and coalesce. *See* **warm rain process**.

  2. The fraction of drops or **aerosol** particles colliding with **precipitation** particles.
  See **accretion, washout**.

collision theory—Theory of chemical reactivity that states that only colliding molecules can undergo chemical reaction and then only if their combined **kinetic energy** exceeds a critical value.

  The maximum value for the **rate coefficient** for a simple, bimolecular, gas-phase reaction, assuming reaction occurs on every collision, is about 2×10^{-10} cm^3 per molecule per second. Actual rate coefficients are lower due to two effects—the need to overcome an **energy** barrier during collision in order for reaction to occur, and a "steric factor," which accounts for the fact that reaction can occur only when the collision occurs in a certain geometry.

colloid—*See* **colloidal system**.

colloidal dispersion—Same as **colloidal system**.

colloidal instability—A property attributed to clouds (regarded in analogy to colloidal systems or aerosols) by virtue of which the **particles** of the **cloud** tend to aggregate (through **Brownian motion**) into masses large enough to precipitate.

  The viewpoint that regards an atmospheric cloud as an **aerosol** somewhat strains the physical chemist's definition thereof, for cloud particles are much larger than the particles typically treated as colloidally dispersed materials either in a gas or in a liquid.

colloidal suspension—Same as **colloidal system**.
  See also **suspension**.

colloidal system—(Also called colloidal dispersion, colloidal suspension.) An intimate mixture of two substances, one of which, called the dispersed phase (or colloid), is uniformly distributed in a finely divided state through the second substance, called the dispersion medium (or dispersing medium).

  The dispersion medium may be a gas, a liquid, or a solid and the dispersed phase may also be any of these, with the exception of one gas in another. A system of liquid or solid **particles** colloidally dispersed in a gas is called an **aerosol**. A system of solid substance or water-insoluble liquid colloidally dispersed in liquid water is called a **hydrosol**. There is no sharp line of demarcation between true solutions and colloidal systems or between mere suspensions and colloidal systems. When the particles of the dispersed phase are smaller than about 10^{-3} μm in diameter, the system begins to assume the properties of a true solution; when the particles dispersed are much greater than 1 μm, separation of the dispersed phase from the dispersing medium becomes so rapid that the system is best regarded as a **suspension**. According to the latter criterion, natural clouds in the **atmosphere** should not be termed aerosols; however, since many **cloud** forms apparently exhibit characteristics of true colloidal suspensions, this strict physico-chemical definition is often disregarded for purposes of convenient and helpful analogy. **Condensation nuclei** and many artificial **smokes** may be regarded as aerosols.

color look-up table—(Abbreviated CLUT.) A mapping of a **pixel value** to a color value shown on a **display** device.

  Typically, a CLUT will map the **input** into a color presentation, but if all colors are of equal **intensity** for each input value, then a **gray scale** or black-and-white **image** results.

color temperature—A **temperature** describing an emitter based on the best match of the shape of its **emission spectrum** to that of a **blackbody** at that temperature.

  The match may be over only a portion of the Planck curve (*see* **Planck's law**), or even by noting the **wavelength** of **maximum** emission and using **Wien's displacement law**. Unlike the **brightness temperature**, color temperature can be used to approximate the physical temperature of objects of unknown distance (especially stars) and also of **isothermal** emitters that are optically thin.

Colorado low—In the United States, a **low** that makes its first appearance as a definite center in the vicinity of Colorado on the eastern slopes of the Rocky Mountains.

  It is, in most aspects, analogous to the **Alberta low**.

Colorado sunken pan—A type of **evaporation pan** that is about 1 m (3 ft) square and 0.5 m (18 in.) deep.

This pan is sunk into the ground to within about 5 cm (2 in.) of its rim, and the water is maintained at about ground level. It is made of unpainted galvanized iron. The **pan coefficient**, on an annual basis, is about 0.8.

colored rain—**Rain** that leaves a colored stain on the ground and on exposed objects, often red or rusty in hue.

The coloration is usually the result of rain picking up **particles** as it falls through a dust-filled **subcloud layer**. The subcloud layer, usually rich in iron oxide, may originate from an area far from the observed colored rain event. This phenomenon has been observed frequently in Italy with particles advected northward from the Sahara. *See* **blood rain**.

colorimetry—Form of **absorption** spectroscopy in which a reagent that bonds with the species of interest is added to a liquid solution, resulting in a change in color of the solution.

The method has been applied, for example, in the determination of the content of certain metals in atmospheric **aerosols**.

column—A columnar **ice crystal** with hexagonal **cross section** and having **aspect ratio** (diameter to length) as much as 10 to 1.

The column may be either solid or hollow with ends plane, pyramidal, truncated, or hollow. Pyramids and combinations of columns are included in this class. It is one of the common ice crystals found in **cirrus** clouds.

column abundance—The amount of an atmospheric **trace gas** found in a vertical column of the **atmosphere**, usually expressed in units of molecules per unit area.

Most often used in reference to **ozone**. *See* **Dobson unit**.

column model—A numerical **model** in which solutions depend only on the vertical coordinate and time.

These models are most useful near boundaries where vertical gradients dominate flow evolution. *See* **mixed-layer models**.

columnar resistance—In **atmospheric electricity**, the electrical **resistance** of a column of air 1 m square, extending from the earth's surface to some specified **altitude**.

Measurements extending to an altitude of 18 km indicate that the atmospheric columnar resistance to that height amounts to about 10^{17} ohm m^{-2}. Probably, this is only slightly less than the total columnar resistance from earth to **ionosphere**. In fact, roughly half of the total columnar resistance from the earth to 18 km is contributed by the lowest 3 km of the column where, in addition to the greater **density** of the air, the high concentration of atmospheric **particulates** leads to a relatively high population of poorly conducting **large ions** rather than the more mobile **small ions**. Total columnar resistance does not vary greatly with either time or locality. By contrast, the columnar resistance of the lowest fraction of a kilometer varies greatly, causing fluctuations in the **atmospheric electric field** at **sea level**, especially in industrial areas of highly variable atmospheric **pollution**. *See* **ion mobility**.

comb nephoscope—(Obsolete.) A **direct-vision nephoscope** constructed by L. Besson in 1897.

It consists of a comb comprising equispaced vertical rods attached to a cross piece. The comb is affixed to one end of a column 2.5–3 m (8–10 ft) long and is supported on a mounting that is free to rotate about its vertical axis. In use, the comb is turned so that the **cloud** appears to move parallel to the tips of the vertical rods.

Middleton, W. E. K., 1969: *Invention of the Meteorological Instruments*, Johns Hopkins Press, Baltimore, p. 271.

COMBAR code—(Short for combat aircraft code.) A **synoptic code** used by combat aircraft to report observable meteorological **elements** in groups of five-digit numbers.

comber—A large **wave** that rolls over or breaks on a beach, reef, etc.

combination coefficient—A measure of the specific rate of disappearance of **small ions** due to either 1) union with neutral **Aitken nuclei** to form new **large ions**; or 2) union with large ions of opposite sign to form neutral Aitken nuclei.

Dimensionally, a combination coefficient is identical with the physically similar **recombination coefficient**. Both types of combination coefficients exhibit mean values of the order of 10^{-5}cm^{3}s^{-1} at **sea level**, while the recombination coefficients are typically about one order of magnitude smaller. *See* **recombination, small-ion combination**.

combustion nucleus—A **condensation nucleus** formed as a result of industrial, **transport**, or natural combustion processes.

The chemical nature of such nuclei may vary almost as much as can the nature of reactants in combustion processes, but because of the prevalence of sulfur impurities in many fuels, the process transforming **sulfur dioxide** (SO_2) to sulfur trioxide (SO_3) and thence **sulfuric acid** is perhaps the most important in producing combustion nuclei. Sulfuric acid is very **hygroscopic** and hence can serve to nucleate atmospheric **condensation** processes. The role of the **ammonium ion** (NH_4^+), which is formed in some combustion processes, is probably of almost equal importance to that of the sulfite (SO_3^{-2}) and sulfate (SO_4^{-3}) ions. Such nuclei may be completely or (in association with **soot**) partly soluble in water.

comfort chart—As used by the American Society of Heating and Air Conditioning Engineers, a diagram showing curves of **relative humidity** and **effective temperature** superimposed upon rectangular coordinates of **wet-bulb temperature** and **dry-bulb temperature**. Upon this **chart** (one for each chosen rate of air movement) may be indicated comfort zones bounded by relative humidity and effective temperature curves; these zones may be determined for various conditions (different seasons, different nations, different races, different clothing, etc.).

comfort curve—A line drawn on a graph of **air temperature** versus some function of **humidity** (usually **wet-bulb temperature** or **relative humidity**) to show the varying conditions under which the average sedentary person feels the same degree of comfort; a curve of constant comfort.

The **effective temperature** line on a **comfort chart** (American Society of Heating and Air Conditioning) is such a curve.

comfort standard—Same as **comfort zone**.

comfort zone—(Also called comfort standard.) The ranges of indoor **temperature**, **humidity**, and air movement, under which most persons enjoy mental and physical well-being.

As represented on **comfort charts** of the American Society of Heating and Air Conditioning Engineers, comfort zones are areas bounded by curves of **effective temperature** and **relative humidity**. The limiting conditions vary somewhat according to **season** and to the native **climate** of the person or group. In the United States the comfort zone with **normal** ventilation lies between air temperatures of about 17° and 24°C (63° and 75°F) at a relative humidity of 70%, and 19°C (67°F) at a relative humidity of 30%, giving an effective temperature within a few degrees of 19°C (67°F). The limits, however, vary with the season, being higher in summer than in winter. In the United Kingdom, the comfort zone is centered on an effective temperature of about 16°C (60°F). In the Tropics the comfort zone lies between the same limits of relative humidity, but at air temperatures around 26°C (78°F). *Compare* **comfort curve**.

comma cloud system—A **cloud system** that resembles the comma punctuation mark.

The formation has a head and a tail and an **upstream** edge shaped like an "S." The comma shape results from differential rotation of the **cloud** border, and is further influenced by adjacent upward and downward vertical motions. Comma patterns vary in size from small convective complexes to large **storm** systems.

comma head—1. The rounded portion of a **comma cloud system**.

This occurs to the left of the maximum wind speed axis and contains the most rotation when viewed in motion. This region often produces most of the steady **precipitation**.

2. A **cloud** pattern seen on satellite imagery, shaped like the upper part of a comma, without the tail.

It is often associated with **cyclonic** development.

comma tail—1. The portion of a **comma cloud system** that lies to the right of, and often nearly parallel to, the axis of maximum winds.

2. A **cloud** pattern seen on satellite imagery, shaped like the lower part of a comma, without the head.

The tail is often associated with a **cold front**.

comminution—The reduction of stone to small **particles** through any natural action such as **frost action** (crevices formed through alternate **freezing** and thawing); biochemical action of plants, trees, and organisms growing in rocks; **wind corrosion** (a natural phenomenon akin to sand blasting); **ocean wave** action; **glacial** scouring; and tectonic action (the shearing and jointing of rocks caused by stresses in the earth's crust).

communication center—Center where **meteorological information** is received and relayed via **telecommunication** means.

commutator—A device for changing the direction of an electric **current**, especially for changing alternating current into direct current.

compact differencing—An "implicit" method for numerically estimating the derivatives of data on a **grid** using coefficient templates spanning a more compact **range** of grid points than the conventional "explicit" differencing template at the same order of **accuracy**.

For example, conventional fourth-order differencing on a uniform grid uses a centered template spanning five values of data; compact fourth-order differencing uses two three-point centered templates, one for the given data and one for the gridded derivatives themselves, which therefore must be solved for simultaneously. The superior accuracy of compact schemes makes them computationally advantageous, provided that the grid geometry is sufficiently regular.

compaction—Pieces of **floating ice** that are subjected to a converging motion, which increases ice **concentration** and/or produces stresses that may result in ice deformation.

compactness—*See* **concentration**.

comparative rabal—A **rabal** observation (i.e., a **radiosonde balloon** tracked by **theodolite**) taken simultaneously with the usual **rawin** observation (tracking by **radar** or **radio direction-finder**).

Its purpose is to provide a rough check on the alignment and operating **accuracy** of the electronic tracking equipment.

compass—1. An instrument for showing direction consisting of a magnetic needle or drum swinging freely on a pivot and pointing toward the earth's north magnetic pole.

The instrument contains a viewing surface marked with points of direction and/or degrees of a circle.

2. A drawing instrument used to make circles or circular arcs.

compensated scale barometer—(Also Kew pattern barometer.) **Mercury barometer** with a fixed **cistern**.

Its **scale** graduations take account of the changes in the level of the **free surface** of the **mercury** in the cistern as a function of **atmospheric pressure**.

compensation of instruments—The use of electromechanical devices to reduce (compensate for) the sensitivities of meteorological **sensors** to other parameters (e.g., the effect of **temperature** on a **pressure** sensor).

competence of the wind—In geology, the ability of the **wind** to **transport** solid **particles** either by rolling, by **suspension**, or by **saltation** (intermittent rolling and suspension).

It is usually expressed in terms of the weight of a single particle. This value varies with the **wind speed** and with the size, shape, and **specific gravity** of the particles. *Compare* **capacity of the wind**.

complete freeze-up—**Freezing** of the water in a river or shallow lake from the surface to the bed.

complex hydrograph—*See* **compound hydrograph**.

complex index of refraction—(Or complex refractive index.) See **refractive index**.

complex low—An area of low atmospheric **pressure** within which more than one **low pressure center** is found.

complex refractive index—(Or complex index of refraction.) *See* **refractive index**.

complex signal—In **radar**, a representation of the time-varying **amplitude** and **phase** of the received **signal** as the real and imaginary parts of a time-varying complex number.

These parts are called the in-phase and **quadrature** components and are measured by **coherent detection** of the received signal. The in-phase signal may be obtained by demodulating the received signal with a local **oscillator** having the same phase and **frequency** as the transmitted signal, while the quadrature signal may be obtained by demodulating the received signal with the local oscillator signal advanced or retarded in phase by 90°. *See* **I and Q channels**.

complex terrain—A region having irregular **topography**, such as mountains or coastlines.

Complex terrain can also include variations in land use, such as urban, rural, irrigated, and unirrigated. Complex terrain often generates local circulations, or modifies ambient **synoptic** weather features, to create unique local weather characteristics such as **katabatic winds**, **anabatic** clouds, and **sea breezes**. In regions of complex terrain, **weather forecast** models must have high resolution to reproduce numerically the terrain-induced weather features.

composite forecast chart—A **chart** consisting of forecasts issued from differing sources.

composite hydrograph—*See* **compound hydrograph**.

composite prognostic chart—Same as **composite forecast chart**.

composite reflectivity—A **display** or mapping of the maximum radar **reflectivity factor** at any **altitude** as a function of position on the ground.

composite vertical cross section—Graphical representation of the meteorological conditions observed at successive times in a **cross section** of the **atmosphere**.

composite water sample—Two or more samples mixed together, in known proportions, to obtain an average value of a specified characteristic.

compound centrifugal force—Same as **Coriolis force**.

compound hydrograph—(Also known as composite hydrograph, complex hydrograph.) **Hydrograph** resulting from storms with a sequence of **rainfall** events.
> The resulting hydrograph from one rainfall event continues during the hydrograph for the next rainfall event.

compressibility—The condition that the volume of a **closed system** decreases as the **pressure** on its surfaces increases. All physical substances are compressible, but the compressibility of liquids and solids is much smaller than for gases.
> The compressibility of a gas is defined by its **equation of state**, approximated adequately for many purposes by that for an **ideal gas**. *See* **coefficient of compressibility**; *compare* **incompressibility**.

compression wave—A **simple wave** or progressive **disturbance** in the one-dimensional **isentropic** flow of a compressible fluid, such that the **pressure** and **density** of a fluid **particle** increase on crossing the wave in the direction of its motion.
> A compression wave is illustrated, for example, by the compression of gas in a cylinder by means of a piston. When the gas is initially at rest in the cylinder, a compression wave may move into the undisturbed fluid at the **speed of sound** as the piston is advanced.

compressional wave—Same as **longitudinal wave**.
> *See* **compression wave**.

Compton effect—(Also Compton scattering.) Scattering of **x-rays** and **gamma radiation** by matter in which the **frequency** of the **scattered radiation** is measurably less than that of the incident **radiation** (inelastic **scattering**).
> So named because Arthur Compton was the first to explain the observed frequency shift by applying the laws of **energy** and **momentum** conservation to scattering of a **photon** by a **free electron**. Compton's experimental and theoretical investigations established the validity of the **quantum theory** of radiation, showing that photons possess momentum (and hence can exert **radiation pressure**) as well as energy.
> Boorse, H. A., and L. Motz, 1966: *The World of the Atom*, Vol. II, 902–929.

computational dispersion—*See* **numerical dispersion**.

computational instability—*See* **stability**.

computational mode—A spurious solution to a **finite-difference approximation** to a differential equation that is not related to the physical solutions of the differential equation.
> For instance, the **leapfrog differencing** scheme can introduce a computational mode.

concentration—(Also called compactness.) The fraction of sea surface area covered by **ice**, usually expressed in tenths.
> For example, 9/10 concentration implies that 90% of the local area is ice covered, with the remaining 10% being open water.

concentration basin—A semi-enclosed **basin** characterized by an excess of **evaporation** over **precipitation** plus **runoff** in which relatively fresh water enters through the strait in the surface layer, is made more saline by the net evaporation, and exits through the strait in a subsurface layer.
> The transformed saline water is more dense than the inflowing **surface water** and therefore sinks. The classic example of a concentration basin is the **Mediterranean Sea**.

concentration gradient—The change in solute concentration per unit distance in solute.
> Concentration gradients cause Fickian **diffusion** (spreading) of solutes from regions of highest to regions of lowest concentrations. In slow-moving **groundwater**, this is the dominant **mixing** process.

concentration variance—A measure of the mean-squared **variability** of concentration c of **air pollutants** or other atmospheric constituents:

$$\sigma_c^2 = \frac{1}{N} \sum_{i=1}^{N} \left(c_i - \overline{c} \right)^2 = \overline{c'^2},$$

where the overbar represents an average, N is the total number of data points, i represents the index of any one data point, and $c' = c_i - \overline{c}$ is the **deviation** of the ith data point from the mean.

concentric eyewall cycle—When a **tropical cyclone** exhibits **concentric eyewalls**, the outer eyewall often contracts and replaces the inner one.
     A **cycle** typically takes about one day to complete and is accompanied by significant fluctuations in central **surface pressure** and maximum wind speed, with maximum winds exceeding 50 m s^{-1} (96 knots). The cycle may repeat indefinitely.

concentric eyewalls—Intense **tropical cyclones** will often have two eyewalls nearly concentric about the center of the storm, the outer eyewall surrounding the inner one.
     A local **wind** maximum is generally present in each **eyewall**. Sometimes more than two eyewalls occur.

concrete minimum temperature—The reading obtained from a standard **minimum thermometer** that is exposed to the air but with its bulb in solid contact with a concrete slab.
     The slab lies horizontally in an open situation with its top almost flush with the ground. Such readings are less subject to very local influences than are **grass minimum** temperatures and are relevant to such problems as **icing** on roads and runways.

condensation—In general, the physical process by which a **vapor** becomes a liquid or solid; the opposite of **evaporation**, although on the molecular **scale**, both processes are always occurring.
     In meteorological usage, this term is applied only to the transformation from vapor to liquid; any process in which a solid forms directly from its vapor is termed **deposition**, and the reverse process **sublimation**. In meteorology, condensation is considered almost exclusively with reference to **water vapor** that changes to **dew**, **fog**, or **cloud**. Condensation in the **atmosphere** occurs by either of two processes: cooling of air to its **dewpoint**, or addition of enough water vapor to bring the mixture to the point of **saturation** (that is, the **relative humidity** is raised to 100 percent). When either of these processes occurs, condensation ensues only if **condensation nuclei** or other surfaces are present. In the complete absence of such, condensation does not occur at nominal saturation. The spontaneous formation of liquid or solid droplets from water vapor (**homogeneous nucleation**) is opposed by the surface free-energy increase that attends the creation of new surfaces of the liquid or solid **phase**. Only for extreme **supersaturation** does this free-energy balance swing in favor of **spontaneous nucleation**.

condensation level—*See* **convective condensation level**, **lifting condensation level**, **saturation level**.

condensation nuclei counter—A device that measures the number concentration of **particles** upon which a **vapor** (typically **water vapor**) condenses at high values of **supersaturation**, relative to **equilibrium vapor pressure** over the liquid **phase** of the substance.
     The value of supersaturation in these devices may be higher than that encountered in a **cloud condensation nuclei counter**, in which the supersaturation of water vapor is controlled to values that are typically found in liquid water **clouds**.

condensation nucleus—An **aerosol** particle forming a center for **condensation** under extremely high **supersaturations** (up to 400% for water, but below that required to activate **small ions**).
     High supersaturations are produced in a **condensation nuclei counter** (often using a condensate other than water) to measure the concentration of such **particles**. Supersaturations greater than a few percent are rarely produced in **clouds** in the **atmosphere** but may exist in aircraft exhaust and over hot springs. Nuclei only active at such high suersaturations may coagulate to form larger particles that become condensation nuclei. *Compare* **cloud condensation nuclei**.

condensation pressure—(Also called adiabatic saturation pressure, adiabatic condensation pressure.) The **pressure** at which a **parcel** of moist unsaturated air expanded **dry-adiabatically** reaches **saturation**.
     See **condensation temperature**, **lifting condensation level**.

condensation temperature—(Also called adiabatic saturation temperature, adiabatic condensation temperature, saturation temperature.) The **temperature** at which a **parcel** of moist unsaturated air expanded **dry-adiabatically** reaches **saturation**.
     See **condensation pressure**, **lifting condensation level**.

condensation trail—(Or contrail; also called vapor trail.) A cloudlike streamer frequently observed to form behind aircraft flying in **clear**, cold, humid air.

Condensation trails may persist and encourage the formation of a layer of **cirrus** clouds. Condensation trails may form by either of two distinct processes. First, addition of **water vapor** to the swept path of the aircraft inevitably accompanies exhaust of combustion products from the engines. If the humidifying effect of this addition overbalances the concomitant addition of the **heat** of combustion, **exhaust trails** may form depending on **mixing** with air from the **environment**. The **thermodynamics** of this process is such that the effect becomes important only for rather low **temperatures** of the order of those encountered near the **tropopause**, so this type of condensation trail is only usually observed for high-altitude flight. On occasion, exhaust provides needed **condensation nuclei**, but this effect has not been fully investigated. Second, in air that is clear, but almost fully saturated, the aerodynamic **pressure reduction** that accompanies flow of air around propeller tips and around wingtips can so cool the air as to induce **condensation** and form aerodynamic trails. The latter propeller-tip trails and wingtip trails are seldom as dense as are exhaust trails. Under some conditions the pressure reduction lowers the temperature below that for homogeneous condensation of **ice** and the trail consists of ice **particles** even at ambient temperatures as warm as −15°C. Wingtip trails only occur with aircraft of such heavy wing-loading as to yield very strong tip **vortex** circulations. Interceptor planes pulling out of dives, and hence imposing temporarily heavy wing-loading, may produce transient tip vortex trails. Faint vortex trails may appear aft of the corners of flaps during aircraft landings.

conditional distribution—The **probability distribution** of a particular **variate** (or subset of variates) when the other variates in the system considered are held fixed.

conditional instability—1. The state of a layer of unsaturated air when its **lapse rate** of **temperature** is less than the **dry-adiabatic lapse rate** but greater than the **moist-adiabatic lapse rate**.

Under such conditions a **parcel** of air at the environmental temperature is unstable to upward vertical displacements if it is saturated, unstable to downward displacements if it is saturated and contains **cloud** water, but stable to all small vertical displacements if it is unsaturated. For descending air containing only **rain** water, the **stability** depends on both the **lapse rate** and the **drop-size distribution**. This definition does not require that such a parcel be obtained by **adiabatic** displacement from any level. It also does not require that the **energy** released from latent heating (**CAPE**) be greater than the **convective inhibition** (CIN) required to bring the parcel to its **level of free convection**.

2. Similar to definition 1 except that it must be possible for a **parcel** displaced **adiabatically** from some level and with conservation of **total water mixing ratio** to attain the environmental **temperature** in a saturated state.

The choice of usage of the term "conditional instability" has been uncertain and sometimes controversial for at least 50 years. Haurwitz defined it approximately as definition 1, and this has been most frequently accepted. However, Byers used a definition similar to definition 2. Beers separated the definition into three subdefinitions, "stable type," corresponding to definition 1 when a moist **parcel** cannot be obtained, and "pseudolatent"and "real latent," corresponding to definition 2 but with the last requiring essentially that **CAPE** be greater than **CIN**. Dutton subscribes to the Haurwitz definition, while Emanuel develops a definition similar to definition 2, but with elaboration similar to that of Beers.

Haurwitz, B., 1941: *Dynamic Meteorology*, McGraw–Hill.
Byers, H., 1944: *General Meteorology*, McGraw–Hill.
Beers, 1945: Atmospheric Physics. *Handbook of Meteorology*, eds. Berry, Bollay, and Beers.
Dutton, J., 1995: *Dynamics of Atmospheric Motion*, Dover Press.
Emanuel, K., 1994: *Atmospheric Convection*, Oxford Univ. Press, 580 pp.

conditional instability of the second kind—(Abbreviated CISK.) A process whereby low-level **convergence** in the **wind field** produces **convection** and **cumulus** formation, thereby releasing **latent heat**.

This enhances the convergence and further increases convection. The atmospheric **environment** that favors CISK is found over warm, tropical oceans where there is an abundant supply of moisture, the **Coriolis force** is small, and air convergence is strong.

Charney, J. G., and A. Elliasen, 1964: On the growth of the hurricane depression. *J. Atmos. Sci.*, **21**, 69–75.

conditional mean—The **mean value** assumed by a particular **variate** when the other variates considered are held fixed.

conditional probability—The **probability** that an event A will occur, under the assumption that another event B has occurred or will occur.

The conditional probability is written $P(A|B|)$ and is expressed, "the probability of A given B."

conditional sampling—Utilizing only that portion of a **dataset** that satisfies a certain criterion.

For example, in the **boundary layer** a convective **thermal** could be defined as a flow structure that has an upward **velocity** exceeding some value for some minimum time period. If the average value of all **temperature** measurements that satisfy the above criterion is calculated, the result would give the average temperature in thermal updrafts. Such data analysis procedures allow investigators to study flow structures such as **coherent structures** that exist in the **atmosphere**.

conditional symmetric instability—*See* **slantwise convection**.

conditions of readiness—In the U.S. Navy, those preliminary measures prescribed for a given area in anticipation of hazardous and destructive weather phenomena.

The conditions are IV, III, and II for possible threat of destructive winds (of force indicated) within 72, 48, and 24 hours, respectively, and condition I for imminent destructive winds within 12 hours.

conductance—*See* **conductivity**.

conductance for moisture—The product of **bulk transfer coefficient** M_E for moisture and **wind speed** U.

This product acts like an electrical **conductance** because when used in a **bulk transfer law**, $F_E = C_E M \Delta r$, the **flux** (current) is proportional to the **conductivity** ($C_E M$) times the **potential** difference (voltage). *Compare* **eddy conductivity, conductivity, bulk transfer coefficient**.

conduction—Transport of **energy** (charge) solely as a consequence of **random** motions of individual molecules (**ions, electrons**) not moving together in coherent groups.

Conduction of energy is a consequence of **temperature** gradients; conduction of charge (electrical conduction) is a consequence of **electric potential** gradients. Conduction is distinguished from **convection** in which energy (or charge) is transported by molecules (ions, electrons) moving together in coherent groups.

conduction current—The migration of charged **particles** in a gaseous medium acted upon by an external **electric field**.

See **air–earth conduction current**.

conductive equilibrium—Same as **isothermal equilibrium**.

conductivity—1. A unit measure of electrical **conduction**; the facility with which a substance conducts electricity, as represented by the **current** density per unit electrical-potential **gradient** in the direction of flow.

Electrical conductivity is the reciprocal of electrical **resistivity** and is expressed in units such as mhos (reciprocal ohms) cm^{-1}. It is an intrinsic property of a given type of material under given physical conditions (dependent mostly upon **temperature**). Conductance, on the other hand, varies with the dimensions of the conducting system, and is the reciprocal of the electrical **resistance**.

2. *See* **thermal conductivity, thermometric conductivity**.

3. *See* **eddy conductivity**.

conductivity current—Same as **air–earth conduction current**.

conductivity–temperature–depth profiler—(Abbreviated CTD.) An electronic instrument designed to measure very accurate, nearly continuous profiles of **conductivity, temperature,** and **pressure** (depth) in the water column.

It is usually lowered on the end of a conducting cable through which the **real-time** measurements are transmitted to a recorder on the ship. Today there are also self-contained instruments (i.e., with an internal memory). Salinity is computed subsequently from the conductivity and temperature measurements. Accuracies are typically ±0.005 **psu**, 0.005 K, and 0.15% full **scale** for conductivity, temperature, and depth, respectively.

cone angle—*See* **spherical coordinates**.

cone of depression—(Also called cone of influence.) A **depression**, roughly conical in shape, formed in a **water table** or **piezometric surface** by the withdrawal of water from a well.

cone of escape—A hypothetical cone in the **exosphere**, directed vertically upward, through which an atom or molecule would theoretically be able to pass to outer space without a collision, that is, in which the **mean free path** is infinite.

Such a cone would open wider with increasing **altitude** above the **critical level of escape**, and would be nonexistent below the critical level of escape. *See* **fringe region**.

cone of impression—The roughly conical surface representing the rising surface of the **water table** or **piezometric surface** surrounding a **recharge** well.

cone of influence—Same as **cone of depression**.

cone of visibility—Term used loosely to denote the right conical space with apex at some **ground target** and within which an aircraft must be located if the pilot is to be able to discern the **target** while flying at a specified **altitude**.

Because the aircraft's altitude must be specified before this term can be given definite meaning, it might better be replaced by some such term as "circle of visibility."

cone of vision—The imaginary conical surface with the apex at a given observer's eye and the solid angle exactly filled by whatever object the observer is viewing.

confidence band—Same as **confidence interval**.

confidence coefficient—*See* **confidence interval**.

confidence interval—(Also called fiducial interval, confidence band.) A **range** of values ($a_1 < a < a_2$) determined from a **sample** by definite rules so chosen that, in repeated **random samples** from the hypothesized **population**, an arbitrarily fixed proportion ($1 - \epsilon$) of that range will include the true value α of an estimated **parameter**.

The limits (a_1 and a_2) are called confidence limits or fiducial limits, the **relative frequency** ($1 - \epsilon$) with which these limits include α is called the confidence coefficient, and the complementary **probability** ϵ is called the confidence level. As with significance levels, confidence levels are commonly chosen as 0.05 or 0.01, the corresponding confidence coefficients being 0.95 and 0.99. Confidence intervals should never be interpreted as implying that the parameter itself has a range of values; it has only one value, α. On the other hand, the confidence limits (a_1, a_2), being derived from a sample, are **random variables** the values of which on a particular sample either do or do not include the true value α of the parameter. However, in repeated samples, a certain proportion (viz., $1 - \epsilon$) of these intervals will include α, provided that the actual population satisfies the initial hypothesis.

confidence level—*See* **confidence interval**.

confidence limits—*See* **confidence interval**.

confined aquifer—**Aquifer** bounded on top and bottom by much less permeable formations.

confined groundwater— **Groundwater** in a **confined aquifer**.

confining bed—*See* **aquitard**.

confining unit—*See* **aquitard**.

confining zone—*See* **aquitard**.

confluence—The rate at which adjacent flow is converging along an axis oriented normal to the flow at the point in question.

It is the opposite of **diffluence**. In **natural coordinates** the confluence may be measured by

$$-\frac{\partial V_n}{\partial n} \text{ or } -V\frac{\partial \psi}{\partial n},$$

where V is the speed of the **wind**, the n axis is oriented 90° clockwise from the direction of the **wind vector**, V_n is the wind component in the n direction, and ψ is the **wind direction**, measured in degrees clockwise from the reference direction.

conformal map—(Also called isogonal map, orthomorphic map.) A map that preserves angles; that is, a map such that if two curves intersect at a given angle, the images of the two curves on the map also intersect at the same angle.

On such a map, at each point, the **scale** is the same in every direction. Shapes of small regions are preserved, but areas are only approximately preserved (the property of area conservation is peculiar to the **equal-area map**). The most commonly used conformal map is probably the **Lambert conic projection**, with **standard** latitudes at 30°and 60°N. On the standard latitudes, the scale is exact; between them, it is decreased by not more than about 1%; outside them, distortion increases rapidly. The Mercator and stereographic projections are also conformal maps.

Saucier, W. J., 1955: *Principles of Meteorological Analysis*, 24–38.

congelifraction—(Also called frost riving, frost splitting.) The splitting of rocks as the result of the **freezing** of the water contained in them.

The individual fragment produced by this process is called a congelifract.

congeliturbation—(Rare.) The churning and stirring of soil as a result of repeated cycles of **freezing** and thawing.

This includes such actions as **frost heaving** and surface subsidence during thaws. A body of material disturbed by frost action is called a congeliturbate.

congestus—A **cloud species** unique to the genus **cumulus**.

See **cumulus congestus, cloud classification**.

conical beam—1. In **weather radar**, the shape of the **beam** produced by a typical **parabolic antenna**.

The envelope of the **main lobe**, as defined by the half-power **beamwidth**, is a cone with the apex at the radar location.

2. In **tracking radar**, the **beam** produced by **conical scanning** methods.

conical scanning—1. In **tracking radar**, a method of angular tracking in which the direction of the **main lobe** of the **antenna pattern** is slightly offset from the axis of the **antenna**.

Rotation of the **beam** about the axis generates a cone with the antenna at the vertex and a vertex angle comparable in size to the **beamwidth**. Such an arrangement allows for accurate determination of the **bearing** and **elevation** of point targets, but is not used in **weather radar**.

2. In **weather radar**, the name sometimes applied to horizontal **scanning** because the surface swept in space by the **beam** as the **antenna** rotates in **azimuth** with fixed **elevation angle** is a cone.

conimeter—*See* **konimeter**.

coning—The nearly equal vertical and lateral **dispersion** of **air pollutants** from the centerline of a **smoke** plume, which occurs in statically neutral conditions (*see* **static stability**).

As pollutants from a **point source** are blown **downwind**, the envelope of the smoke plume (as visible by eye or as measured by the percentage decrease of concentration from that at the centerline) takes the shape of a cone. Conversely, observation of a coning smoke **plume** can be used as an indicator that the **ambient air** there is statically neutral. *Compare* **fanning, looping, fumigation**.

coniology—Same as **koniology**.

coniscope—Same as **koniscope**.

conjugate image—*See* **method of images**.

conjugate-power law—*See* **power-law profile**.

conjunction—In astronomy, the juxtaposition of the earth, sun, and one of the other planets or the moon, in which the angle subtended at the earth between the sun and the third body, in the **plane of the ecliptic**, is 0° (i.e., the third body lies either between the sun and the earth, or on the opposite side of the sun from the earth).

Compare **opposition, quadrature**.

conjunctive use—Integrated management and use of **surface water** and **groundwater**.

connate water—Water incorporated into the pores of rocks at the time when the rocks were formed; it usually has high mineral content.

The quantities of neither connate water nor **magmatic water** are appreciable in **hydrometeorology**; only **meteoric water** is considered. *See also* **fossil water**.

consecutive mean—(Also called moving average, overlapping mean, running mean.) A smoothed representation of a **time series** derived by replacing each observed value with a **mean value** computed over a selected interval.

For example, if the observations are of **daily maximum temperature** and the selected interval is five days, then the value assigned to 5 February is the mean of the daily maxima from 3 through 7 February, etc. Consecutive means are used in **smoothing** to eliminate unwanted periodicities or minimize irregular variations.

consensus average—The average value of a set of measurements determined by using the **consensus averaging** technique.

consensus averaging—A method used to estimate the true value of a quantity with measured or observed values that have **statistical** fluctuations and outliers that are presumed to be due to measurement **error**, extraneous signals, or **noise**.

The consensus averaging technique examines the set of all measured values and finds the largest subset with values within a predetermined interval of each other. If that subset has fewer than a predetermined number of values, then the entire set of measurements is rejected; otherwise, the selected subset is averaged to obtain an estimate of the true value of the measured quantity. For this technique to be valid, the true value must not change greatly during the time period over which the measurements are taken. Consensus averaging has been used in **wind** profiling to determine an estimate of the wind velocity from several measurements of the **Doppler shift** of the **radar** signal, which is often dominated by noise and contaminated by interfering radio signals, airplanes, birds, etc.

consequent—The THEN clause of an **IF–THEN rule**.
See also **antecedent**.

conservation of absolute angular momentum—*See* **conservation of angular momentum**.

conservation of absolute momentum—*See* **conservation of momentum**.

conservation of angular momentum—The principle that **absolute angular momentum** is a property that cannot be created or destroyed but can only be transferred from one physical system to another through the agency of a net **torque** on the system.
   As a consequence, the absolute angular momentum of an isolated physical system remains constant. The principle of conservation of angular momentum can be derived from Newton's second law of motion.

conservation of energy—The principle that the total **energy** of an **isolated system** remains constant.
   This principle takes into account all forms of energy in the system; it therefore provides a constraint on the conversions from one form to another. *See* **energy equation** for formulations applicable to **meteorology**.

conservation of mass—The principle (of **Newtonian mechanics**) that states that mass cannot be created or destroyed but only transferred from one volume to another.
   In meteorology, this principle is generally expressed in the form of the **equation of continuity**.

conservation of momentum—The principle that, in the absence of forces, **absolute momentum** is a property that cannot be created or destroyed.
   See **Newton's laws of motion**.

conservation of vorticity—1. The statement that in the horizontal flow of an inviscid **barotropic** fluid, the vertical component of **absolute vorticity** of each individual fluid **particle** remains constant.
   The principle was first applied to the **atmosphere** by Rossby and is the dynamical principle underlying the nondivergent **barotropic model** of the atmosphere. *See* **constant absolute vorticity trajectory**.
   2. The hypothesis that the **vorticity** of individual **eddies** is conserved during the turbulent **mixing** of a fluid.
   See **vorticity-transport hypothesis**.

conservatism—Constancy with time of a given physical property during a specific process.

conservative field—A **field** for which the **work** done in moving an isolated **particle** of unit mass around a closed path is zero:

$$\oint \mathbf{F} \cdot d\mathbf{r} = 0,$$

where **F** is the force acting on the particle and $d\mathbf{r}$ is an infinitesimal **vector** displacement along the path.
   For such a field the work done in moving a particle between any two points is independent of the path so that a **potential** Φ exists:

$$\mathbf{F} = \nabla\Phi,$$

where ∇ is the **del operator**. *See* **irrotational**.

conservative pollutants—Solids that conserve mass in natural water systems.

conservative property—A property with values that do not change in the course of a particular series of events.
   Properties can be judged conservative only when the events (processes) are specified; also, properties that are conservative for a whole system may or may not be conservative for its parts, and conversely. Applied to **airmass** properties, this term is relative.

conserved parameter diagram—Same as **conserved variable diagram**.

169

conserved variable diagram—A **thermodynamic diagram** for the **atmosphere** used for studying **cloud** processes.

One conserved **variable** representing **heat** or **temperature** is usually plotted along the **abscissa**, while the another conserved variable representing the conservation of water is plotted along the **ordinate**. The variables are chosen to be those that are conserved for both saturated and unsaturated motion. Examples of conserved variables are **equivalent potential temperature** θ_e, **liquid water potential temperature** θ_L, **moist static energy** s_e, **liquid water static energy** s_L, **total water mixing ratio** r_T, **saturation point** pressure P_{SP}, and saturation point temperature T_{SP}. These variables are not conserved for processes such as **precipitation**, **radiative cooling**, and **mixing**. When a third variable representing height z or **pressure** P is also indicated in the diagram, then there is enough information to define completely the thermodynamic state and **water content** of the air. Examples of popular sets of variables are (P, θ_e, r_T), (P, θ_L, r_T), (P, s_e, r_T), (P, s_L, r_T), and (P, P_{SP}, T_{SP}).

Bohren, C. F., and B. A. Albrecht, 1998: *Atmospheric Thermodynamics*, 402 pp.

consistent numerical scheme—A **finite-difference approximation** that approaches the corresponding differential equation as the **grid length** becomes small.

See **Lax equivalence theorem**, **convergent numerical scheme**, **stability**.

consolidated ice—An area of the sea covered by **ice** of various origins consolidated, by **wind** and currents, into a solid mass.

consolidation—1. An adjustment of soil **particles**, in response to compressive **stress**, that results in lower **porosity**.

2. Any process that increases the firmness or coherence of a loose, soft, or liquid earth material, including cementation, compaction, and **crystallization**.

3. Decrease of the volume of an **aquifer** due to pumping and lowering of the water pressure that leads to the transfer of **stress** from interstitial water to aquifer solids.

constancy—With respect to the **wind**, same as **persistence**.

constant absolute vorticity trajectory—(Abbreviated CAVT.) The path of an **air parcel** with **absolute vorticity** that remains constant in horizontal flow.

Before the advent of modern **numerical weather prediction**, these trajectories, in conjunction with the theory of **Rossby waves**, were frequently used to forecast the movement of long tropospheric waves.

Pettersen, S., 1956: *Weather Analysis and Forecasting*, 2d ed., Vol. I, 146–148, 171–176, 414–419.

constant flux layer—A layer of air tens of meters thick at the bottom of the **atmosphere** where the **variation** of vertical **turbulent flux** with **altitude** is less than 10% of its magnitude.

This layer is also called the **surface layer**, which is roughly the bottom 10%: of the **atmospheric boundary layer**. While the **flux** is not perfectly uniform with height within this layer, the idealization of a constant flux layer permits certain theoretical approaches, such as **Monin–Obukhov similarity theory** to describe the **logarithmic wind profile**.

constant-height chart—(Also called constant-level chart, fixed-level chart, isohypsic chart.) A **synoptic chart** for any surface of constant geometric **altitude** above **mean sea level** (a **constant-height surface**), usually containing plotted data and analyses of the distribution of such variables as **pressure**, **wind**, **temperature**, and **humidity** at that altitude.

A commonly analyzed constant-height chart is the **surface chart** (or "sea level chart"). All of the **upper-air** charts are considered **constant-pressure charts**.

constant-height surface—(Also called constant-level surface, isohypsic surface.) In meteorology, a surface of constant geometric or **geopotential** altitude measured with respect to **mean sea level**.

Compare **constant-pressure surface**.

constant-level balloon—(Or constant-pressure balloon.) A balloon designed to float at a constant level of **altitude**, **pressure**, **density**, or **entropy**.

See **anchor balloon**, **super-pressure balloon**, **tetroon**, **zero-pressure balloon**.

constant-level chart—Same as **constant-height chart**.

constant-level surface—Same as **constant-height surface**.

constant-pressure chart—(Also called isobaric chart, isobaric contour chart.) The **synoptic chart** for any **constant-pressure surface**, usually containing plotted data and analyses of the distribution of, for example, height of the surface, **wind**, **temperature**, and **humidity**.

Constant-pressure charts are most commonly known by their **pressure** value; for example, the

1000-mb chart (which closely corresponds to the **surface chart**), the 850-mb chart, 700-mb chart, 500-mb chart, etc. *Compare* **constant-height chart**; *see* **mandatory level**.

constant-pressure-pattern flight—A technique of **pressure-pattern flight** whereby an aircraft is navigated, in the direction of **wind** flow, along a height contour line on a **constant-pressure surface**, thereby assuring a continuous, nearly direct **tailwind**.

This is accomplished by maintaining constant **indicated altitude** on the **pressure altimeter** along with constant **absolute altitude** above a **level surface**. The latter requirement restricts the use of this method to over-ocean flight.

constant-pressure surface—(Or isobaric surface.) In meteorology, a surface along which the **atmospheric pressure** is everywhere equal at a given instant.

constant-rate dilution gauging—A method used for measurements of **surface water** discharges, by injecting a conservative **tracer** at a constant rate.

The change in concentration of samples collected **downstream**, relative to the original injected fluid, is used to calculate the flow rate along that **reach**.

constant-pressure balloon—Same as **constant-level balloon**.

constructed analogs—*See* **analogs**.

consultation—A question and answer session with a **knowledge-based system**.

It is often used in meteorological applications that require judgment on the part of the user.

consumption—The conversion of turbulent **kinetic energy** into **potential energy**.

This occurs when **turbulence** exists in a **statically stable** environment, because turbulence is moving cooler **air parcels** upward and warmer air parcels downward against the **force of gravity**. The rate of consumption is given by the **buoyancy flux**. *Compare* **dissipation**.

consumptive use—The total amount of **water** taken up by vegetation for **transpiration** or building of plant tissue, plus the unavoidable **evaporation** of **soil moisture**, **snow**, and intercepted **precipitation (interception)** associated with the vegetal growth.

Consumptive use is primarily applied to a single type of vegetation in a given area and does not include evaporation from water surfaces in or adjacent to the area; thus, it is not as general in scope as **evapotranspiration** or **duty of water**.

contact angle—The **equilibrium** angle of contact of a liquid on a rigid surface, measured within the liquid at the contact line where three phases (liquid, solid, gas) meet.

For example, water sheeting on glass has zero contact angle, but water beading on an oily surface or plastic may have a contact angle of 90° or greater.

contact flight—(Obsolete.) *See* **visual flight**.

contact nucleus—A type of **ice** nucleating **aerosol** particle characterized by its ability to initiate the ice phase in a **supercooled water** drop when it contacts the **drop** surface.

Nucleation may occur at a different **temperature** (usually higher) than would occur for a totally immersed particle. The contact can occur through a variety of processes, including **diffusion** (modified by gradients of **temperature** and **water vapor**) or collection by differential fall speed.

contact weather—Weather that permits visual air navigation.

See **VFR weather**.

contaminant—Any physical, chemical, biological, or radiological substance not normally present or found at unusually high concentrations in water or soil.

contaminate—Introduce or increase the concentration of any physical, chemical, biological, or radiological substance in the water or soil.

contamination—*See* **contaminate**.

contessa di vento—Literally, the "wind countess"; a formation of one or more **lenticular clouds** above and alee of Mount Etna in Sicily.

When more than one **cloud** is present, they are vertically stacked. *See* **la serpe**.

continental aerosol—**Aerosol** having its origin over the continents with industrial, urban, agricultural, forest, and **desert** sources, with potential for high concentrations of **hygroscopic** aerosol.

continental air—A type of **air mass** with characteristics developed over a large land area that therefore has the basic continental characteristic of relatively low moisture content.

continental anticyclone—Same as **continental high**.

continental borderland—A submarine plateau or irregular area adjacent to a continent, with depths greatly exceeding those on the **continental shelf**, but not as great as those in the deep oceans.

continental climate—The **climate** that is characteristic of the interior of a landmass of continental size.

It is marked by large annual, daily, and day-to-day ranges of **temperature**, low **relative humidity**, and (generally) by a moderate or small and irregular **rainfall**. The annual extremes of **temperature** typically occur within a month after the solstices. In its extreme form a continental climate gives rise to **deserts**. *Compare* **maritime climate**.

continental cloud—A **cloud** forming in **continental air**, characterized by relatively high concentrations of **cloud condensation nuclei** and **cloud droplets**.

These clouds are characterized by high concentrations of small drops (>500 cm^{-3}), with a somewhat narrow size distribution. *See* **maritime cloud**.

continental drift—The movement of the continents relative to each other.

Continental drift is a consequence of plate tectonics. Continents generally move with respect to each other at the rate of a few centimeters per year.

continental glacier—A continuous sheet of **land ice** that covers a very large area and moves outward in many directions.

This type of **ice** mass is so thick as to mask the land surface contours, in contrast to the smaller and thinner **highland ice**. The continental glacier of Greenland is sometimes called the Inland Ice. This term is often used to describe the great ice masses that characterized the **ice ages**.

continental high—(Or continental anticyclone.) A general area of high atmospheric **pressure** that, on mean charts of **sea level pressure**, is seen to overlie a continent during the winter.

The only really pronounced example of this is the **Siberian high**. *Compare* **glacial anticyclone**.

continental ice—Same as **continental glacier**.

continental platform—The zone that includes both the **continental shelf** or **continental borderland** and the **continental slope**.

continental shelf—The zone around the continents extending from the low-water mark seaward to where there is a marked increase in slope to greater depths.

continental shelf wave—A **free wave** that propagates along a coastal boundary in a homogenous ocean with shelf topography (i.e., a sloping bottom).

See also **coastally trapped waves**, **shelf waves**, **Kelvin wave**.

continental slope—The declivity from the outer edge of the **continental shelf** or **continental borderland** into greater depths.

Continental Water Boundary—The front between the **Antarctic Circumpolar Current** and the narrow region of westward movement close to the Antarctic coast.

It extends from the Ross Sea eastward to the Weddell Sea, where it separates the **cyclonic** movement of the Weddell Sea from the Circumpolar Current, and is sometimes also called the **Weddell Gyre Boundary**.

continentality—In **climatology**, the degree to which a point on the earth's surface is in all respects subject to the influence of a landmass; the opposite of **oceanicity** (or oceanity).

Continentality usually refers to **climate** and its immediate consequences. Usually, it is measured by the **range** of **temperature**, either the daily range or the difference between the average temperatures of the warmest and coldest months. Since the latter increases with the latitude, a convenient measure is the **annual range** of **temperature** divided by the sine of the latitude. In another form, the difference between January and July mean temperatures at a **station** is divided by the difference between the January and July means for the whole circle of latitude. An index of continentality, or coefficient of continentality, k, has been formulated by V. Conrad as follows:

$$k = \frac{1.7A}{\sin(\phi + 10°)} - 14,$$

where A is the difference between the **mean temperature** (°C) of the warmest and coldest months and ϕ is the latitude of the place in question.

Conrad, V., 1946: *Methods in Climatology*, Harvard University Press, 296–300.

continentality index—(Also called coefficient of continentality, index of continentality.) *See* **continentality**.

contingency table—A table showing the joint or compound frequencies of occurrence of two or more variables or attributes.

The simplest example is the so-called fourfold table in which two attributes A and B are each divided into two classes, say A_1, A_2, and B_1, B_2, thus giving rise to four possible combinations: (A_1, B_1), (A_1, B_2), (A_2, B_1), and (A_2, B_2). The contingency table then displays the **frequency** of each of the four combinations.

continuing current—A sustained **current** in the **lightning stroke** that flows to the ground after the **return stroke**.

Continuing currents can have durations in excess of 100 ms with magnitudes of typically 100 A. Continuing currents occur in negative and **positive cloud-to-ground lightning** flashes.

continuity—The property of a **field** such that neighboring values of a **parameter** differ only by an arbitrarily small amount if they are close enough in space and/or time.

In **synoptic meteorology**, continuity of a field is interpreted as requiring a certain smoothness of **analysis** and a similar adjustment in the time sequence of **synoptic charts**. *Compare* **discontinuity**.

continuity chart—In meteorology, a **chart**, maintained for **weather analysis** and forecasting, upon which are entered the present and historical positions of significant features of the regular **synoptic charts** at regular intervals.

Significant features include **pressure centers, fronts, instability lines, trough lines**, and **ridge lines**. *Compare* **station continuity chart**.

continuity equation—*See* **equation of continuity**.

continuous leader—Same as **dart leader**.

continuous spectrum—A **spectrum** in which **wavelengths** (and **wavenumbers** and **frequencies**) are represented by the continuum of real numbers (or a portion thereof) rather than by a discrete sequence of numbers.

A continuous function on an infinite interval, even though the function is nonzero over only a finite interval, must be represented by the **Fourier transform** rather than by **Fourier series**, and the resulting spectrum will be continuous. *See also* **discrete spectrum**.

continuous-wave radar—(Abbreviated CW radar.) A type of **radar** that transmits continuously instead of in **pulses**.

Moving targets may be recognized with such radars by observing a **Doppler shift** in the **frequency** of the received **signal**. Some forms of **modulation** of the transmitted **wave** allow measurement of **range** (e.g., **FM–CW radar**). Historically, continuous-wave radars preceded the development of the pulsed radars widely used today.

continuum absorption—A region of continuous spectral **absorption** by a gas that shows no apparent line structure.

Continuum absorption is explained semi-empirically as the overlap of absorption due to many **absorption lines** with centers far from the continuum region. A good example occurs in the 10-μm **window** region, where continuum absorption by **water vapor** can be particularly important. *Compare* **band absorption, line absorption**.

contour—Generally, an outline or configuration of a body or surface.

Often, the term is used for one of a set of lines (contour lines) drawn to represent the configuration of a surface; this is the case in meteorology, where a contour usually refers to a **contour line** of constant height on a **constant-pressure surface**. Sometimes, the term is more loosely used in the general sense of an **isopleth**. *Compare* **profile**.

contour-change line—Same as **height-change line**.

contour chart—A map depicting **contours** of a meteorological **variable**, such as height or **pressure**. *See* **constant-pressure chart**.

contour interval—The interval at which **contour lines** are drawn on a map.

contour line—(Also called contour, isohypse, isoheight.) A line of constant **elevation** above a certain reference level (usually **mean sea level**) on a previously defined surface, which may be the earth's surface, a **constant-pressure surface**, an **isentropic surface**, etc.

A contour line of a given value is the intersection of the surface in question with the **constant-height surface** of the same elevation as the value of the contour line. In meteorology, a contour line frequently refers to a line of constant height on a **constant-pressure chart**. *Compare* **isobar**.

contour map—A map that shows the configuration of a surface by means of **contour lines** drawn at regular intervals of **elevation** (contour intervals) above a reference level.

In meteorology, the **contour** analysis of a **constant-pressure chart** is a type of contour map.

contour microclimate—That portion of the **microclimate** that is directly attributable to the small-scale variations of ground level.

A great many microclimatic variations are caused by small differences in **elevation**, **steepness** of slope, and direction of slope exposure and direction of curvature. These influence the total amount of **heat** absorbed from **solar radiation**, the heat loss by outgoing **radiation**, the small-scale (**gravity**) airflow in the **boundary layer** near the soil, and the **runoff** of **precipitation**. Such microclimatic differences can be observed even on as small a **scale** as the ridges and furrows of plowed fields or mounded flower beds.

contra solem—Against the sun, hence **cyclonic**, descriptive of motion turning to the left in the Northern Hemisphere and to the right in the Southern Hemisphere; the reverse of **cum sole**.

(These terms were introduced by V. W. Ekman in 1923.) *See also* **withershins**.

contraction axis—*See* **axis of contraction**.

contraction coefficient—(Also called coefficient of contraction.) Used in fluid mechanics; the ratio of the area, measured at the **vena contracta** of a jet of liquid issuing from an opening, to the area of the opening.

contrail—Contraction for **condensation trail**.

contrail-formation graph—A graph containing the parameters **pressure**, **temperature**, and **relative humidity** for critical values at which **condensation trails** form.

The graph is used as an aid in forecasting the formation of condensation trails.

contrast—In meteorological usage, contrast C is defined by

$$C = \frac{L_0 - L_s}{L_s},$$

where L_0 is the **luminance** of a **target** object and L_s is that of its immediate visual surroundings.

Both luminances may include **airlight** or reflected **glare**, and L_s is always assumed to be greater than zero. Because **detection** outdoors often depends on luminance rather than chromaticity differences, C is usually calculated from spectrally integrated luminances. Contrast for a perfectly black target is -1 but, in principle, C has no upper limit for a self-luminous target. If $|C|$ is less than a variable threshold contrast, the target is visually indistinguishable from its surroundings.

contrast stretching—Creation of a **lookup table** that increases a **range** of **pixel** values in an **image** when mapped to the **display** device.

For example, mapping the **input** range of 10–20 to display a range of 100–200 is an example of contrast stretching. It is used to show fine detail in an image.

contrast threshold—Same as **threshold contrast**.

contraste—1. A nautical term for **winds** off the Spanish Mediterranean coast that blow from opposite directions at points not far apart.

2. (Rare.) A change of **wind direction** with height shown by the movements of **clouds** at different levels.

contributing region—(Sometimes called pulse volume, resolution volume, sampled volume.) In **radar**, the region in space that contains the targets that contribute to the received **signal** or **echo** arriving at a particular instant.

The contributing region is a volume in space determined by the **beamwidth** and **pulse length** of the radar. Specifically, the contributing region is the volume defined by one-half the pulse length in the radial direction and the beamwidth in the transverse direction. *See* **pulse volume**.

control area—A controlled airspace extending upward from a specific height above the surface of the earth in which radio contact between pilots traversing the airspace and air traffic control is required.

control day—(Also called key day.) One of several days on which the weather is supposed (according to folklore) to provide the key for the weather of a subsequent period.

The best-known of these days are in the United States, **Groundhog Day** (2 February); in England, **St. Swithin's Day** (15 July). Similar beliefs are associated with other days, mostly saints' days, in many countries. *Compare* **singularity**.

control section—(Used in **open channel flow**.) Natural or man-made physical properties at a location of a river or channel that determine the relationship between water surface **elevation** and **discharge** for some distance **upstream** and/or **downstream** of the site.

control-tower visibility—The **visibility** observed from an airport tower.

According to current U.S. weather observing practice, at civil stations the control-tower visibility becomes the official visibility for the **station** whenever the **surface visibility** becomes less than three miles.

control well—In **aquifer** testing, the pumped well or **recharge** well, as distinguished from the **observation wells**.

convection—1. In general, mass motions within a fluid resulting in **transport** and **mixing** of the properties of that fluid.

Convection, along with **conduction** and **radiation**, is a principal means of **energy transfer**. Distinction is made between **free convection** (gravitational or buoyant convection), motion caused only by **density** differences within the fluid; and **forced convection**, motion induced by mechanical forces such as deflection by a large-scale surface irregularity, **turbulent flow** caused by **friction** at the boundary of a fluid, or motion caused by any applied **pressure gradient**. Free and forced convection are not necessarily exclusive processes. On a windy day with **overcast** sky, the **heat** exchange between ground and air is an example of forced convection. On a sunny day with a little **wind** where the ground **temperature** rises, both kinds of convection take place.

2. (Or gravitational or buoyant convection.) Motions that are predominantly vertical and driven by **buoyancy** forces arising from **static instability**, with locally significant deviations from **hydrostatic equilibrium**.

Atmospheric convection is nearly always turbulent. Convection may be dry, that is, with **relative humidities** less than 100%, especially in the **boundary layer**, but is commonly moist, with visible **cumuliform** clouds. Most convective clouds are driven by positive buoyancy, with **virtual temperature** greater than the **environment**, but **clouds** with **precipitation**, **evaporation**, and/or melting can produce negatively buoyant convection. *See* **slantwise convection**.

3. As specialized in atmospheric and ocean science, a class of relatively small-**scale**, thermally (can be driven by salt concentration in the ocean) direct circulations that result from the action of **gravity** upon an unstable vertical distribution of mass. (In the case of **slantwise convection**, though, the motions are larger scale, and are driven by a combination of gravitational and centrifugal forces acting at an angle to the vertical.)

Almost all atmospheric and oceanic convection is fully turbulent and is generally composed of a collection of **convection cells**, usually having widths comparable to the depth of the convecting layer. In the **atmosphere**, convection is the dominant vertical **transport** process in **convective boundary layers**, which are common over tropical oceans and, during sunny days, over continents. In the ocean, convection is prominent in regions of high heat loss to the atmosphere and is the main mechanism of **deep water** formation. Moist convection in the atmosphere is characterized by deep, saturated **updrafts** and **downdrafts**, and unsaturated downdrafts driven largely by the **evaporation** and melting of **precipitation**. This form of convection is made visible by **cumulus** clouds and, in the case of precipitating convection, by **cumulonimbus** clouds. Moist convection and **radiation** are the dominant modes of vertical **heat** transport in the **Tropics**.

4. In **atmospheric electricity**, a process of vertical charge **transfer** by **transport** of air containing a net **space charge**, or by motion of other media (e.g., **rain**) carrying net charge.

Eddy diffusion of air containing a net charge **gradient** may also yield a **convection current**.

convection cell—1. An organized unit of **convection** within a convecting layer.

It is isolated by a stream surface, with ascending motion in the center and descending motion near the periphery, or vice versa. In laboratory convection, such cells (sometimes referred to as **Bénard cells**) are usually roughly as deep as they are wide, may take the form of squares, triangles, or hexagons, and may be laminar or turbulent and steady or oscillatory. In **atmospheric boundary layer** convection, this term refers to an organization of turbulent convection on horizontal scales at least as large as the depth of the **convective boundary layer**. In the case of cloud-topped boundary layers, convection cells may take the form of **open cells**, with broad, cloud-free areas of gentle descent surrounded by narrow **updrafts** within **cumulus** clouds, or may take the form of **closed cells** in **stratocumulus** clouds, characterized by narrow **downdrafts** at the periphery. *See also* **cell, airmass thunderstorm, ordinary cell**.

2. In the case of precipitating **moist convection**, refers to a distinct unit of **convection**, often having its own closed contours of **radar reflectivity** and a lifetime of roughly 20–30 minutes.

Such a **cell** generally begins as a cumulus **updraft**, then develops a precipitation-driven **downdraft**, and finally decays as a general, **cloudy** area containing weak downdrafts.

convection current—1. (Or convective current.) Any **current** of air involved in **convection**.

In meteorology, this is usually applied to the upward moving portion of a convection circulation, such as a **thermal** or the **updraft** in **cumulus** clouds.

2. Any net **transport** of **electric charge** effected through mass motions of some charged medium; any electric **current** induced by other than electrical forces.

In **atmospheric electricity**, the convection current is part of the **air–earth current** of charge **transfer** vertically between the earth's surface and the **upper atmosphere**. The term includes not only **eddy diffusion** currents existing in regions of net **space charge** but also currents due to fall of charged **precipitation** particles (**precipitation current**). *See* **convection**.

convection theory of cyclones—The formation of **depressions** by convective ascent of heated surface air during a sufficient interval and of sufficient magnitude for the inflowing air near the earth's surface to acquire appreciable **cyclonic rotation** in accordance with the **circulation theorem**.

Compare **wave theory of cyclones**.

convective acceleration—*See* **acceleration**.

convective activity—General term for manifestations of **convection** in the **atmosphere**, alluding particularly to the **development** of **convective clouds** and resulting weather phenomena, such as **showers, thunderstorms, squalls, hail, tornadoes**, etc.

convective adjustment—A method of representing unresolved **convection** in atmospheric models by imposing large-scale vertical profiles when convection occurs.

As originally developed, convective adjustment was applied when modeled lapse rates became **adiabatically** unstable. New temperatures were calculated for unstable layers by conserving **static energy** and imposing an **adiabatic lapse rate**. If, in addition, humidities exceeded **saturation**, they were adjusted to saturation, with excess water removed as **precipitation**. A related adjustment, (stable saturated adjustment), for stable layers with **water vapor** exceeding saturation, returned them to saturation, also conserving **energy**. More recently, convective adjustments have been developed that adjust to empirically based lapse rates, rather than **adiabatic lapse rates**, while still maintaining energy conservation. Convective adjustment is generally applied to **temperature** and **humidity** but, in principle, can also be applied to other fields affected by convection.

convective available potential energy—(Abbreviated CAPE.) The maximum energy available to an ascending **parcel**, according to parcel theory.

On a **thermodynamic diagram** this is called positive area, and can be seen as the region between the lifted parcel process curve and the environmental **sounding**, from the parcel's **level of free convection** to its **level of neutral buoyancy**. It may be defined as

$$\text{CAPE} = \int_{p_n}^{p_f} (\alpha_p - \alpha_e)\,dp,$$

where α_e is the environmental **specific volume** profile, α_p is the specific volume of a parcel moving upward moist-adiabatically from the level of free convection, p_f is the **pressure** at the level of free convection, and p_n is the pressure at the level of neutral buoyancy. The value depends on whether the **moist-adiabatic process** is considered reversible or irreversible (conventionally irreversible) and whether the **latent heat** of **freezing** is considered (conventionally not). *Compare* **convective inhibition**.

convective boundary layer—(Abbreviated CBL.) Same as **mixed layer**.

See also **atmospheric boundary layer**.

convective cell—*See* **cell**.

convective cloud—A **cloud** that owes its vertical **development**, and possibly its origin, to **convection**.

convective condensation level—(Abbreviated CCL.) On a **thermodynamic diagram**, the point of intersection of a **sounding** curve (representing the vertical distribution of **temperature** in an atmospheric column) with the **saturation mixing ratio** line corresponding to the average mixing ratio in the **surface layer** (i.e., approximately the lowest 1500 ft).

The **dry adiabat** through this point determines, approximately, the lowest temperature to which the surface air must be heated before a **parcel** can rise dry-adiabatically to its **lifting condensation level** without ever being colder than the **environment**. This temperature, the **convective temperature**, is a useful **parameter** in forecasting the onset of **convection**. *See* **conditional instability**, **level of free convection**.

convective equilibrium—Same as **adiabatic equilibrium**.

convective index—*See* stability index.

convective inhibition—(Abbreviated CIN.) The energy needed to lift an air parcel vertically and pseudoadiabatically from its originating level to its level of free convection (LFC).

For an air parcel possessing positive CAPE, the CIN represents the negative area on a thermodynamic diagram having coordinates linear in temperature and logarithmic in pressure. The negative area typically arises from the presence of a lid. Even though other factors may be favorable for development of convection, if convective inhibition is sufficiently large, deep convection will not form. The convective inhibition is expressed (analogously to CAPE) as follows:

$$\text{CIN} = -\int_{p_i}^{p_f} R_d(T_{vp} - T_{ve}) \, d\ln p,$$

where p_i is the pressure at the level at which the parcel originates, p_f is the pressure at the LFC, R_d is the specific gas constant for dry air, T_{vp} is the virtual temperature of the lifted parcel, and T_{ve} is the virtual temperature of the environment. It is assumed that the environment is in hydrostatic balance and that the pressure of the parcel is the same as that of the environment.

convective instability—1. An instability due to the buoyancy force of heavy fluid over light fluid overcoming the stabilizing influence of viscous forces.

    2. Same as potential instability.

    3. Same as thermal instability.

convective mass flux—An average vertical transport of mass for a field of cumulus clouds or thermals.

This mass flux M_c is given by

$$M_c = \rho \sigma w_{up} = \rho \sigma (1 - \sigma)(w_{up} - w_{down}),$$

where ρ is air density, σ is fraction of area covered by updrafts, w_{up} is the vertical velocity averaged across all updrafts, and w_{down} is the vertical velocity averaged over all downdrafts. With the mass flux, one can describe the average turbulent vertical flux of any conserved variable, such as potential temperature θ in cloudless thermals, by

$$\rho \overline{w'\theta'} = M_c(\theta_{up} - \theta_{down}),$$

where $\overline{w'\theta'}$ is the vertical kinematic heat flux, θ_{up} is the potential temperature averaged across all updrafts, and θ_{down} is the potential temperature averaged across all downdrafts. This method allows population-wide statistics to be described, without needing characteristics of individual updrafts and downdrafts.

convective mixed layer—Same as mixed layer.
    Compare atmospheric boundary layer.

convective overturn—1. In oceanography, same as overturn.

    2. In limnology, the process by which lake layers shift due to wind and temperature effects.

convective plume—A buoyant jet in which the buoyancy is supplied steadily from a point source; the buoyant region is continuous.

This is to be distinguished from a thermal, which is a discrete buoyant element in which the buoyancy is confined to a limited volume of fluid. A chimney plume is usually a convective plume. *See* coherent structures; *compare* coning, fanning, lofting, looping.

    Emanuel, K. A., 1984: *Atmospheric Convection*, Oxford University Press, p. 58.

convective precipitation—Precipitation particles forming in the active updraft of a cumulonimbus cloud, growing primarily by the collection of cloud droplets (i.e., by coalescence and/or riming) and falling out not far from their originating updraft.

convective rain—*See* convective precipitation.

convective region—Generally, an area that is particularly favorable for the formation of convection in the lower atmosphere, or one characterized by convective activity at a given time.

convective Richardson number—A dimensionless number that contains factors similar to the bulk Richardson number, but where the velocities, temperatures, and depths are based on mixed-layer scaling parameters.

The convective Richardson number Ri* can be written as

$$\text{Ri}^* = \frac{g\,(\Delta_{EZ}\theta_v)z_i}{\theta_v w_*^2},$$

where g is gravitational **acceleration**, $\Delta_{EZ}\theta_v$ is the change of **virtual potential temperature** across the depth of the **entrainment zone**, z_i is the depth of the mixed layer, θ_v is the average virtual potential temperature within the mixed layer, and w_* is the **Deardorff velocity**. The depth of the entrainment zone, and the **entrainment rate**, are related to this convective Richardson number.

Nelson, E., R. Stull, and E. Eloranta, 1989: A prognostic relationship for entrainment zone thickness. *J. Appl. Meteor.*, **28**, 885–903.

convective showers—Episodes of **convective precipitation** falling in an area no more than about 10 km wide and for a time period of less than about one-half hour.

convective storm—*See* **thunderstorm, cell.**

convective storm initiation mechanism—*See* **thunderstorm initiation mechanism.**

convective temperature—*See* **convective condensation level.**

convective term—(Or advective term.) Any term of the form $\mathbf{V} \cdot \nabla\zeta$ where $\mathbf{V}$ is the **velocity** field, ∇ is the del-operator, and ζ is any **field, vector,** or **scalar.**

Thus, for example, $u\dfrac{\partial u}{\partial x}$ is a **convective acceleration**. These are the nonlinear terms in Eulerian expressions for the rate of change of the **dependent variables.**

convective theory of cyclogenesis—Same as **convection theory of cyclones.**

convective transport—The component of movement of **heat** or mass induced by **thermal** gradients in fluids.

convective transport theory—A relationship between surface fluxes and state of the air in the mid **mixed layer**, which applies to situations where convective **thermals** are causing nonlocal **transport** vertically in the **atmospheric boundary layer.**

The **kinematic flux** F_S of **variable** S is

$$F_S = b_S w_B (S_{sfc\ skin} - S_{mid\text{-}ML}),$$

where w_B is the **buoyancy velocity**, $S_{sfc\ skin}$ is the value of S at the solid surface of the earth, $S_{mid\text{-}ML}$ is the value of S near the middle of the **convective mixed layer**, and b_S is an empirical **parameter** of order 10^{-3}.

convective turbulence—**Turbulence** occurring in convective storms, particularly **thunderstorms**, that is felt by aircraft.

The turbulence is caused by strong **updrafts** and **downdrafts.**

convective velocity scale—*See* **Deardorff velocity.**

convergence—1. The contraction of a **vector field**; also, a precise measure thereof.

Mathematically, convergence is negative divergence, and the latter term is used for both. (For mathematical treatment, *see* **divergence**).*Compare* **confluence.**

2. The property of a sequence or series of numbers or functions that ensures that it will approach a definite finite limit.

convergence band—Same as **convergence line.**

Convergence bands in **fair** weather can form under **thermal** updrafts, and often form a fishnet or honeycomb pattern with the **cell** size of the same order as the depth of the **mixed layer.**

convergence line—1. (Sometimes called asymptote of convergence.) Any horizontal line along which horizontal **convergence** of the airflow is occurring.

If **mass convergence** is taking place in a plane near the surface of the earth, the incoming air must be rising at the convergence line. Hence the lines are often associated with **convective clouds.**

2. Radar **signature** of horizontal **convergence** at the surface.

See **fine line.**

convergent numerical scheme—A **finite-difference approximation** with the property that the numerical solution tends to the exact solution of the differential equation as the **grid length** becomes small.

See **Lax equivalence theorem, consistent numerical scheme, stability.**

conversion factor—A numerical factor by which a quantity expressed in one system of units may be converted to another system.

conveyance—(Used in **open channel flow**.) The ratio of the **discharge**, Q, in a channel to the square root of the **energy** gradient, S_f.

In the **Manning equation** it can be expressed as

$$K = \frac{Q}{S_f^{1/2}}$$

or

$$Q = KS_f^{1/2},$$

where

$$K = \left(\frac{1}{n}\right)AR^{2/3},$$

where n is Manning's **roughness coefficient** (indicative of **resistance** to the flow) and A is the cross-sectional area of the channel.

conveyer belt—*See* **ocean conveyor belt**.

convolution—Mathematical operation that is used to describe the imperfect response or **resolution** of an instrument or a measurement.

For example, the time response of a **linear** system to an **input** function is described by

$$y(t) = \int_{-\infty}^{\infty} h(t - \alpha)x(\alpha)\, d\alpha,$$

where $x(t)$ is the input function, $y(t)$ the **output**, $h(t)$ the **weighting** function characterizing the system, and α a **variable** of integration. The output is said to be the convolution of the input with the weighting function. Many instruments, because of limited **frequency response**, have the effect of **smoothing** the input data to produce an output that is more limited than the input in its **frequency** content. This effect is more readily understood by taking the **Fourier transform** of the convolution equation, in effect transforming from the time domain to the frequency domain. The result is

$$Y(f) = H(f)X(f),$$

where Y, H, and X denote, respectively, the Fourier transforms of y, h, and x. The function $H(f)$ is called the frequency response function of the system. The magnitude of $H(f)$ determines whether frequency components that are present in the input function will also be present in the output or will be attenuated by the system. An example of a convolution process in **radar** is the smoothing of the spatial pattern of **reflectivity** as a consequence of the finite size of the **pulse volume**. Spatial irregularities with scales smaller than the pulse volume are attenuated in the measurement process by a convolution of the reflectivity field with the pulse volume.

cooking snow—Same as **water snow**.

cooling degree-day—A form of **degree-day** used to estimate the **energy** requirements for air conditioning or refrigeration; one cooling degree-day is given for each Fahrenheit **degree** that the **daily mean** temperature departs above the base of 24°C (75°F).
Compare **heating degree-day**.

cooling power—In the study of **human bioclimatology**, one of several parameters devised to measure the cooling effect of the air upon a human body. Essentially, cooling power is determined by the amount of applied **heat** required by a device to maintain it at a constant **temperature** (usually 34°C); the entire system should be made to correspond, as closely as possible, to the external **heat** exchange mechanism of the human body.
Instruments used in applying this principle include the **katathermometer**, the **frigorimeter**, and the **coolometer**. *Compare* **cooling temperature**, **operative temperature**.

Buettner, K. J. K., 1951: *Compendium of Meteorology*, p. 1115.
Gold, E., 1935: The effect of wind, temperature, humidity, and sunshine on the loss of heat of a body at temperature 98°F. *Quart. J. Roy. Meteor. Soc.*, **61**, 316–331.

cooling-power anemometer—The general term for **anemometers** operating on the principle that the **heat transfer** to air from an object at an elevated **temperature** is a function of the air speed. Examples are **hot-wire** and **hot-film anemometers**.

cooling temperature—In the study of **human bioclimatology**, one of several parameters devised to measure the cooling effect of the air upon the human body.

It is the skin **temperature** of a sphere that is in **thermal** equilibrium with the surrounding air; the sphere is supplied **heat** at a constant rate, and its heat exchange characteristics correspond to those of a blond, fair-skinned human.

coolometer—An instrument that measures the **cooling power** of the air.

It consists of a metal cylinder that is electrically heated to maintain a constant **temperature**. The **power** required is a measure of the cooling power.

cooperative observer—An unpaid **observer** who maintains a **meteorological station** for the U.S. National Weather Service.

Formerly termed a **voluntary observer** in the United States; the latter name is still used in the United Kingdom. The usual instruments furnished to such an **observer** are **maximum** and **minimum thermometers** and a nonrecording **precipitation gauge**.

coordinate axis—*See* **Cartesian coordinates**.

coordinate line—*See* **curvilinear coordinates**.

coordinate plane—*See* **Cartesian coordinates**.

coordinate surface—*See* **curvilinear coordinates**.

coordinate system—(Also called reference frame.) Any scheme for the unique identification of each point of a given continuum.

These may be points in space (**Eulerian coordinates**) or parcels of a moving fluid (**Lagrangian coordinates**). **Newton's laws of motion** do take different forms in different systems (*see* **inertial coordinate system**, **relative coordinate system**). The geometry of the system is a matter of convenience determined by the boundaries of the continuum or by other considerations (*see* **Cartesian coordinates**, **curvilinear coordinates**).

Copenhagen water—Same as **normal water**.

coplane scanning—A **radar** scanning pattern defined by a fixed plane tilted from the horizontal.

Two radars performing dual-Doppler measurements typically scan a coplane that contains both radars and is tilted a few degrees from horizontal. The **azimuth** and **elevation** angles of each radar must be varied simultaneously and in a coordinated way to scan the coplane.

copolarized signal—A received **radar** signal that has a **polarization** identical to that of the transmitted signal.

corange line—(Also called coamplitude line.) A line on a tidal **chart** joining places that have the same **tidal range** or **amplitude**; usually drawn for a particular **tidal constituent** or tidal condition (e.g., mean spring tides).

cordonazo—In Mexico, a local term for a **tropical cyclone** off the west coast.

These were formerly considered rare, but satellites reveal that they are quite common, especially in July and August.

cordonazo de San Francisco—A name given by Spanish sailors to the **autumnal equinox** because of the storms that are said to prevail about that time (St. Francis's Day; 4 October).
See **equinoctial storm**.

core sample—A sample of rock, soil, **snow**, or **ice** obtained by driving a hollow tube into the medium and withdrawing it with its contained sample or "core."

In general, the aim of core sampling is to obtain a specimen in its undisturbed natural state for subsequent analysis. The **snow sampler** is a hydrometeorological example of the type of instrument used to obtain core samples.

Coriolis acceleration—An **acceleration** of a **parcel** moving in a **relative coordinate system**.

The total acceleration of the parcel, as measured in an **inertial coordinate system**, may be expressed as the sum of the acceleration within the relative system, the acceleration of the relative system itself, and the Coriolis acceleration. In the case of the earth moving with **angular velocity** Ω, a parcel moving relative to the earth with **velocity u** has the Coriolis acceleration $2\Omega \times \mathbf{u}$. If Newton's laws are to be applied in the relative system, the Coriolis acceleration and the acceleration of the relative system must be treated as forces. *See* **apparent force**, **Coriolis force**, **inertial force**, **gravity**.

Coriolis effect—An **apparent force**, relative to the earth's surface, that causes deflection of moving

180

objects to the right in the Northern Hemisphere and to the left in the Southern Hemisphere due to the earth's rotation.

Named for Gustav Gaspard de Coriolis (1792–1843), a French mathematician who published a quantitative mathematical work on the subject in 1835.

Coriolis force—(Also called compound centrifugal force, deflecting force.) An **apparent force** on moving **particles** in a noninertial **coordinate system**, that is, the **Coriolis acceleration** as seen in this (relative) system.

Such a force is required if Newton's laws are to be applied in this system. In meteorology the Coriolis force per unit mass arises solely from the earth's rotation, and is equal to $-2\Omega \times \mathbf{u}$, where Ω is the **angular velocity of the earth** and $\mathbf{u}$ is the (relative) **velocity** of the particle. Thus the Coriolis force acts as a deflecting force, normal to the velocity, to the right of the motion in the Northern Hemisphere and to the left in the Southern Hemisphere. It cannot alter the speed of the particle. The three components toward east, north, and **zenith** are, respectively, 2Ω (v sinϕ $- w$ cosϕ), $-2\Omega u$ sinϕ, and $2\Omega u$ cosϕ, where u, v, w are the component velocities and ϕ the latitude. Since the Coriolis force is in effect proportional to the speed, its importance in any given atmospheric motion may be judged from the representative speed and duration of the motion. *See* **inertial force**.

Gill, A. E., 1982: *Atmosphere–Ocean Dynamics*, Academic Press, 72–74.

Holton, J. R., 1992: *An Introduction to Dynamic Meteorology*, 3d edition, Academic Press, 31–38.

Coriolis parameter—Twice the component of the earth's **angular velocity** about the local vertical, 2Ω sinϕ, where Ω is the angular speed of the earth and ϕ is the latitude.

Since the earth is in rigid rotation, the Coriolis parameter is equal to the component of the earth's **vorticity** about the local vertical. If the Coriolis parameter is denoted by f and the speed of a horizontally moving **fluid parcel** by V, then fV is the magnitude of the horizontal **Coriolis force** per unit mass on the parcel. *See also* **Rossby parameter**.

corn heat unit—A modification of **growing degree-days** (GDD) with both upper and lower **temperature** thresholds.

All temperatures above 86°F (30°C) are set to 86°F (30°C) and all temperatures below 50°F (10°C) are set to 50°F (10°C) before calculation of **daily mean** temperature. The reference temperature (base temperature) for corn heat units is 50°F (10°C). See **degree-day**.

corn snow—Same as **spring snow**.

corner reflector—A common type of **retroreflector**, that is, a **reflector** that returns a **ray** exactly parallel to the incident ray.

It is formed by intersecting three mutually perpendicular planes, with the center point of the reflector located at the mutual point of intersection, thus providing a **target** of eight right-angle corners. If the reflector can only be seen from one side, that one side is all that needs to be built. When the look direction has a very limited angle, only one-eighth of the structure is commonly built. The material used depends on the **wavelength** of **radiation** to be reflected.

cornice—An overhanging structure of **ice** or **snow** formed at a topographic edge (such as a mountain ridge) by wind-driven deposition.

Cornices can be extremely hazardous to mountain climbers. Cornice release can also initiate **avalanches**.

coromell—A **land breeze** from the south at La Paz, Mexico, near the south of the Gulf of California, prevailing from November to May.

It sets in at night and usually persists until 9 or 10 A.M.

corona—1. A set of one or more colored rings of small angular radii concentrically surrounding the the sun, moon, or other **light** source when veiled by a thin cloud.

The corona can be distinguished from the **halo of 22°** due to its much smaller angular radius, which is often only a few degrees, and by its color sequence, which is from bluish white on the inside to reddish on the outside, the reverse of that in the 22° halo. Further, the color sequence of the corona can be repeated. **Fraunhofer diffraction theory** is often used to provide an approximate description of the corona. This theory predicts that the center of the corona is essentially white and that the radius of a ring of a particular color is approximately inversely proportional to the **drop** radius. Consequences of this are that the rings are most discernible and have the purest colors when the droplets in a particular portion of a **cloud** are nearly uniform in size (mono disperse); the rings are most nearly circular if the droplets that produce different portions of the corona are nearly the same size (spatial homogeneity); rings are caused by droplets with a radius of less than about 15 μm (the rings from larger droplets are washed out by the angular width of the sun). When there is a broad **range** of **droplet** sizes, the rings are distinct, the colors faint, and the phenomenon is often called an **aureole**. Although it is possible for colorful coronas to be produced by **ice crystals**, usually the broad range of sizes and shapes of crystals precludes this. Similarly, coronas are rarely produced by **dust** due to the broad range of **particle** sizes normally present, but rare observations, such as **Bishop's rings**, have been reported.

2. (Also called solar corona.) The outer envelope of the sun's atmosphere consisting of ionized gases, predominantly **hydrogen** and **helium**, at temperatures that exceed one million degrees Kelvin.

The white **light** emission observed at solar eclipse or with the **coronagraph** arises from **scattering** of photospheric **radiation** from **free electrons** in the corona. The shape of the corona varies during the **sunspot cycle**. At **solar minimum** the corona has large extensions along the sun's equator, with short brush-like tufts near the poles. At **solar maximum** the equatorial extensions are much smaller and the corona is more regular in shape.

3. *See* **aurora**.

4. *See* **corona discharge**.

corona current—The electrical **current** that is equivalent to the rate of charge transferred to the air from a pointed object (or array of objects) experiencing **corona discharge**.

Ordinarily, the corona current from terrestrial objects at times of **thunderstorm** passage constitutes a **transfer** of negative charge from air to object. *Compare* **point discharge current**.

corona discharge—(Also called brush discharge, corposant.) A **luminous**, and often audible, **electric discharge** that is intermediate in nature between a **spark discharge** (with, usually, its single discharge channel) and a **point discharge** (with its diffuse, quiescent, and nonluminous character).

It occurs from objects, especially pointed ones, when the **electric field strength** near their surfaces attains a value near 1×10^5 V m^{-1}. Aircraft flying through active electrical storms often develop corona discharge streamers from antennas and propellers, and even from the entire fuselage and wing structure. So-called **precipitation** static results. It is seen also during stormy weather, emanating from the yards and masts of ships at sea. *See* **St. Elmo's fire**.

coronagraph—An instrument for photographing the **corona** and **prominences** of the sun at times other than at solar eclipse.

An occulting disk is used to block out the **image** of the body of the sun in the focal plane of the objective lens. The **light** of the corona passes the occulting disk and is focused on a photographic film. Great care must be taken to avoid light scattered from the **atmosphere** and the lenses, and from reflections in the tube of the instrument. The coronagraph is used with a narrowband polarizing **filter** or with a **spectroscope**.

coronal holes—Regions in the **solar corona** of locally lower **density** and lower **temperature**.

They are characterized by a **magnetic field** that is open to interplanetary space and are associated with high-speed **solar wind** streams. These regions appear dark in **spectroheliograms** when observed in **emission** lines from the **corona**.

coronal mass ejection—(Abbreviated CME.) Bright features that move outward through the **solar corona**.

These large expulsions of magnetized **plasma** often extend more than 90° along the solar limb and move with velocities ranging from 10 to 2000 km s^{-1}. These events disturb the **solar wind**, which in turn affects the earth's **magnetosphere**, causing **magnetic storms** and other disruptive events.

corposant—Same as **corona discharge**.

corpuscular theory of light—The hypothesis, by Sir Isaac Newton, that **light** consists of a stream of minute **particles** emitted by **luminous** bodies at very high velocities, and that the sensation of light is due to the bombardment of the retina of the eye by these particles.

corrasion—(Or wind erosion.) The abrasive action of wind-borne material, especially sand, **dust**, and **ice crystals**; a form of **weathering**.

Compare **corrosion, erosion**.

corrected airspeed—Same as **true airspeed**.

corrected altitude—The **indicated altitude** corrected for **temperature** deviation from the **standard atmosphere**.

Compare **true altitude**.

correction—Change made to a measured or observed quantity to allow for errors and thus obtain a closer approximation to the true value.

correlated-k—A technique for handling the complexity of spectral transmission in models of longwave **radiative transfer**.

Rather than integrating the **radiation** wavelength by **wavelength**, **absorption coefficients** with similar values are first grouped together, and the results for each group are summed, taking into account the relative importance of the coefficients for different atmospheric **layers**. The technique is fairly accurate and greatly reduces the amount of computation time required.

correlation—1. In general, a mutual relationship between **variables** or other entities.

In **statistical** terminology, it is a form of **statistical dependence**.

2. When used without further qualification, the **statistical** term correlation usually refers to simple, **linear correlation** between two variables x, y and is measured by the **product-moment** coefficient of correlation ρ or its **sample** estimate r defined as follows, where the respective **population** mean values of x and y are denoted by ξ and ζ, the respective **standard deviations** by $\sigma(x)$ and $\sigma(y)$, and where E is the **expected value**:

$$\rho = \frac{E[(x - \xi)(y - \zeta)]}{\sigma(x)\sigma(y)};$$

$$r = \frac{\sum(x_i - \bar{x})(y_i - \bar{y})}{[\sum(x_i - \bar{x})^2 \sum(y_i - \bar{y})^2]^{1/2}}.$$

The product moment $E[(x - \xi)(y - \zeta)]$ is usually called the **covariance** of x and y. In connection with correlation, the word "simple" is used in contradistinction to other qualifiers such as "multiple" or "partial." The word "linear" refers to a **linear** relationship between the two variables, or more precisely, to a linear approximation of the **regression function** of either **variate** with respect to the other. *See* **autocorrelation, multiple correlation, partial correlation**.

3. *See* **correlation coefficient**.

correlation coefficient—1. *See* **correlation**.

2. A measure of the **persistence** of the **eddy velocity** as a function of time and space.

Two types are distinguished: 1) In the **Eulerian correlation** coefficient, the time difference is zero,

$$R_{E^{ik}}(y_1 y_2) = \frac{\overline{u'(y_1)u'(y_2)}}{[\overline{u_i'^2(y_1)}]^{1/2}[\overline{u_k'^2(y_2)}]^{1/2}},$$

where u' is the eddy velocity. For **homogeneous** and **homologous turbulence**, this **correlation** tensor depends only on the difference $(y_2 - y_1)$; when the **turbulence** is **isotropic**, the **tensor** is spherically symmetric and $R_{E^{ik}} = R_{E^{ki}}$. 2) In the **Lagrangian correlation** coefficient, time and space are varied together in such a way that the same **fluid parcel** is being followed:

$$R_{L^{ik}}(t) = \frac{\overline{u_i'(y_1 t_1)u_k'(y_2 t_2)}}{[\overline{u_i'^2(y_1 t_1)}]^{1/2}[\overline{u_k'^2(y_2 t_2)}]^{1/2}}.$$

When the flow is one-dimensional and the **mean velocity** is much greater than the **eddy velocity**, then a fixed point experiences approximately the same sequence of fluctuations as a fluid parcel. The Lagrangian correlation coefficient can then be converted into the Eulerian by a proper scaling. These correlation coefficients have the same form and meaning when any other fluctuating quantity is used, for example, **temperature** or **pressure**.

correlation ratio—A measure of **statistical** relationship that takes into account all functional relationships between **random variables** and contrasts to the **correlation coefficient**, which measure only the **linear** relationship.

correlation triangle—The set of all possible **variances** and **covariances** between velocities in the three Cartesian directions, defined for a specified **statistical** order or **moment**.

For example, the second moment correlation triangle is

$$\overline{u'^2}$$

$$\overline{u'v'} \quad \overline{u'w'}$$

$$\overline{v'^2} \quad \overline{v'w'} \quad \overline{w'^2},$$

where squared quantities are **velocity** variances and the other terms are covariances. For first-order **turbulence closure** parameterizations, all of the terms in the correlation triangle above represent unknowns, and must be approximated. For **higher-order closure**, larger triangles of unknowns would exist.

correlative meteorology—(Rare.) The application of **statistical** correlation techniques to a series of meteorological observations.

correlogram—A graph of **autocorrelation (ordinate)** against **lag (abscissa)**.

corresponding point—A point of intersection of a **characteristic line** with a **contour line** (or **isobar**) on a **constant-pressure chart** (or **constant-height chart**), for example, on a 500-mb **chart**, the intersection of the 5400-m **contour line** with a **ridge line**.
  A corresponding point is considered as a characteristic that reappears on successive charts. *See* **singular corresponding point, principle of geometric association.**

corrosion—The gradual deterioration of material by chemical processes, such as **oxidation** or attack by acids; if caused by an atmospheric effect, a form of **weathering**.
  Of great significance is the corrosion due to the combined effects of atmospheric **temperature**, **humidity**, and suspended impurities, for example, the rusting of iron, the direct effects upon a surface wetted by acid water, or, indirectly, the rotting of wood caused by the action of fungi or bacteria in the soil and in enclosed spaces. *Compare* **corrasion, erosion.**
      Brooks, C. E. P., 1951: *Climate in Everyday Life*, 184–189.

cosine law of illumination—The incident **irradiance** on a surface illuminated by a **beam** is proportional to the cosine of the **angle of incidence**.
  The marked latitudinal **variation** in **insolation** on the earth is largely a consequence of this simple relationship. Compare **Lambert's law.**

cosmic dust—*See* **interstellar dust.**

cosmic radiation—A general term for background **radiation** from nonspecific galactic and extra-galactic sources.
  It includes both high-energy **particle** radiation in the form of **cosmic rays** and low-energy **microwave** emission with a **brightness temperature** of 2.7 K.

cosmic-ray ionization—The production of ions by **cosmic rays**.
  One or more **electrons** are removed from an atom, or molecule, or **particle**, by the cosmic ray to produce an **ion**.

cosmic-ray shower—*See* **cascade shower.**

cosmic rays—(Or **cosmic radiation**.) Without qualification, usually means the primary cosmic rays of extra-terrestrial origin that continually bombard the earth and consist mostly of high-energy **protons**, about 9% **helium** and heavier nuclei, a small percentage of **electrons**, and some **gamma rays**.
  The energies of cosmic rays are well in excess of billions of **electron volts**. Secondary cosmic rays result from interactions between primary rays and atoms in the earth's **atmosphere**. Most cosmic rays probably originate from the Milky Way galaxy, but a small fraction come from the sun as evidenced by **diurnal** variations in the cosmic ray flux.
      Boorse, H. A., and L. Motz, 1966: *The World of the Atom*, Vol. I, 669–675.
      Friedlander, M. W., 1989: *Cosmic Rays*, pp. 13, 79.

cosmogenic radioisotopes—Radioactive **isotopes** formed in the **upper atmosphere** by the action of **cosmic rays** on atmospheric constituents.
  The **carbon** isotope ^{14}C, referred to as radiocarbon, is the best-known example.

cospectrum—The real part of the **cross-spectrum** of two functions.

cotidal hour—(Also called high-water interval, lunitidal interval.) The average interval of time between the moon's passage over the **meridian** of Greenwich and the following **high water** at a specified place.

cotidal line—A line on a tidal **chart** connecting places having the same **cotidal hour** for a given lunar **tidal component**, or one connecting places that have **high water** (or **low water**) simultaneously.

cotton-belt climate—A type of warm **climate** characterized by dry winters and rainy summers; that is, a **monsoon climate**, in contrast to a **Mediterranean climate**.

cotton-region shelter—A medium-sized **instrument shelter** used extensively at **second-order stations, climatological substations**, etc.

It is a white louvered box with a flat double roof and is mounted slightly more than a meter (4 ft) above the ground on a four-legged stand.

Couette flow—The steady **laminar flow** of a **viscous fluid** between flat plates, one moving uniformly relative to the other, in which the **shearing stress** is constant within the fluid.

The **velocity** V relative to the slower-moving plate has a **linear** profile:

$$V = \frac{\tau}{\mu} \, y,$$

where τ is the **shearing stress**, μ the **dynamic viscosity**, and y the distance from the slower plate. *Compare* **Poiseuille flow**.

counter—Generic term for instruments that measure **radioactivity**.

counter-gradient—Opposite to the **gradient**.

For example, if **temperature** decreases upward, then the **temperature** gradient is positive, which would make counter-gradient negative or downward for this example.

counter-gradient flux—A **flux** of some **variable** opposite to the mean gradient of that variable.

For example, if **temperature** decreases upward, then a **counter-gradient** heat flux would be downward, from cold to hot. While this appears to violate a law of **thermodynamics** that states **heat** flows from hot to cold, those laws are found not to be violated when nonlocal motions (**air parcels** moving across finite distances) are considered. Flux is not caused by, nor related to, the local **gradient** when **coherent structures** are present.

counter sun—Same as **anthelion**.

countercurrent—A **current** flowing adjacent to another current but in the opposite direction.

counterglow—Same as **gegenschein**.

countergradient wind—Flow opposite to the expected **gradient wind**.

counterradiation—The downward **flux** of **longwave radiation** across a given surface, usually taken as the earth's surface.

Counterradiation originates in **emission** by **clouds** and **greenhouse gases** at different heights and temperatures, and is modified by subsequent **absorption** before reaching the surface in question. The long-term global average of counterradiation reaching the earth's surface is about 330 W m^{-2}, making it one of the largest terms in the **surface energy balance**.

countertrades—1. (Obsolete.) Same as **westerlies**.

2. (Obsolete.) Same as **antitrades**.

counts—(Also known as **pixel value, image level**.) The value at a particular location in an **image**; an indication of "brightness" at that location.

coupled model—A class of analytical or numerical time-dependent models in which at least two different subsystems of earth's **climate system** are allowed to interact.

These subsystems may include the **atmosphere, hydrosphere, cryosphere,** and **biosphere**. This term is most commonly used for models of the evolution and interaction of earth's atmosphere and ocean. Coupled (two way) interaction between different subsystems can be contrasted with the class of models where the evolution of a subsystem A is affected by the present state of the subsystem B, but changes in A do not have feedback on the evolution of B itself.

coupled ocean–atmosphere general circulation model—A **climate model** that includes both atmospheric and oceanic components coupled interactively.

These components are often **general circulation models** with global or regional domains. Models of this type are often used to investigate the response of **climate** to time-varying forcing, such as the gradual increase of atmospheric **greenhouse gases**.

Courant–Friederichs–Lewy condition—(Abbreviated CFL; also called Courant condition.) The property that a **finite-difference approximation** is formulated in such a way that it has access to the information that is required to determine the solution of the corresponding differential equation; violating this condition leads to a **numerical instability**.

As an example, suppose the solution to the differential equation is a **wave** traveling at speed c. If a finite-difference approximation is only able to access information on its **grid** that is traveling at speeds less than c, it violates the CFL condition and it will not be able to approximate the solution of the differential equation.

course—The direction of a line over the earth with reference to north (true or magnetic).
Because the **meridian** lines converge at the poles, the true course of a line changes continuously except when it is actually **meridional** or when it is curved so as always to cross meridians at the same angle. Magnetic course also changes but in a nonuniform manner because of the lack of complete symmetry in the **isogonic lines** of the earth's **magnetic field**. *Compare* **heading**.

covariable—Same as **covariate**.

covariance—The **expected value** of the product $(x - \xi)(y - \zeta)$, where ξ denotes the mean of x and ζ the mean of y.
See **correlation**.

covariance matrix—In **radar**, a matrix comprising the autocovariances and cross covariances of a set of **signal** amplitudes.
As applied to **polarimetric radar** signals, the diagonal terms are the respective autocovariances of the received signals of orthogonal **polarization**, that is, the **power** in the two **orthogonal** channels, and the off-diagonal terms are the cross covariances.

covariate—(Also called covariable.) An **independent variable**, or **predictor**, in a **regression equation**.
Also, a secondary **variable** that can affect the relationship between the **dependent variable** and independent variables of primary interest in a regression equation.

coverage diagram—Same as **radiation pattern**.

cow-quaker—In England, a May **storm** (after the cows have been turned out).

cow storm—In Canada, a **gale** of Ellesmere Island so strong that it "blows the horns off the cows."

cowshee—Same as **kaus**.

crab angle—Same as **drift-correction angle**.

crabbing—In aviation terminology, the motion of an aircraft in flight when a **crosswind** causes its **heading** to differ from the **course**.
See **drift, drift-correction angle**.

crachin—In the China Sea, especially in Hainan Strait and the Gulf of Tonkin north of 20°N, a period of **drizzle** or light rain with low **stratus** clouds and bad **visibility** that frequently occurs between late January and early April and may persist for several days.
It is associated with a fall of **pressure** over China and a change from a **continental** to a **maritime air** mass.

crack—A fracture in an **ice floe** that has not parted substantially.
More specifically, it is any fracture of **fast ice, consolidated ice**, or a single **floe** that may have been followed by separation ranging from a few centimeters to 1 m.

crankcase ventilation—The emission of unburned fuel to the **atmosphere** from the crankcase of an internal combustion engine.
Formerly a large source of **hydrocarbons** to the atmosphere, the rate of escape has been greatly reduced by recycling of the escaped gases to the cylinders by means of a device called the positive crankcase ventilation (PCV) valve. Crankcase ventilation contributes around 5% of unburned hydrocarbon emissions to the atmosphere.

creep—1. The movement of water under or around a structure built on permeable foundations that may lead to **erosion**.
2. The slow, downslope movement of surface soil or rock debris, usually imperceptible except when observed for long durations.

creeping—1. The motion of the **index** of an **aneroid barometer** after it has been subjected to a large and rapid change in **pressure**.
This movement is a slow adjustment of the index toward the correct pressure. The physical cause of creeping is not clearly understood.
2. *See* **soil creep**.

creeping flow—Fluid flow at very small values of the **Reynolds number**, that is, in general, a very viscous flow.

crepuscular arch—Same as **bright segment**.

crepuscular rays—(Also called shadow bands.) Literally "twilight rays," these alternating dark and light bands (shadows and **light** scattered from sunbeams, respectively) seem to diverge fanlike from the sun's position during **twilight**.

This apparent divergence of parallel **sunlight** is an artifact of **linear** perspective. Crepuscular rays may appear as 1) shadows cast across the **purple light** by high, distant **cloud** tops or 2) shadows next to light scattered from sunbeams by **haze** in the **lower atmosphere**. Sunbeams seen during the day are sometimes called crepuscular rays, even though they are observed outside twilight.

crest cloud—(Also called cloud crest.) A type of **standing cloud** that forms along a mountain range, either on the ridge or slightly above and to the lee of it.

Its process of formation and maintenance is identical to that of the **cap cloud**. *Compare* **banner cloud**; *see* **lenticularis**.

crest gauge—A device used to record the highest water stage since the last gauge reading.

crest stage—The highest **water level** attained at a measuring **station** during a **runoff** event.

crevasse—A deep rift in a **glacier**, or in any other form of **land ice**, caused by its motion.

crevasse hoar—**Ice crystals** that form and grow in **glacial** crevasses and in other cavities where a large cooled space is formed and in which **water vapor** can accumulate under **calm**, still conditions; a type of **hoarfrost**.

They have an origin similar to that of **depth hoar**; the typical crystal is a hollow cup with one side opening inward and continued in a hexagonal scroll.

criador—(Spanish term meaning "provider.") The name applied, in northern Spain, to the rain-bearing west **wind**.

Criegee biradicals—*See* **Criegee intermediate**.

Criegee intermediate—1. Proposed intermediate in the ozonolysis of **ethylene**, formula $\cdot CH_2OO\cdot$, formally a biradical species, which has two **free radical** sites.

2. Family of biradicals formed from the reaction of **ozone** with **alkenes** in the **atmosphere**. The yields of Criegee intermediates, and their subsequent reactions, are not fully characterized.

crisp variable—**Fuzzy logic** term for a **variable** that takes on a precise value as opposed to a fuzzy membership value between 0 and 1.

Crisp variables must be measurable quantities.

criteria pollutants—**Pollutants** that can injure health, harm the **environment**, and cause property damage.

In 1997, the criteria pollutants in the United States were
Carbon monoxide (CO)
Lead (Pb)
Nitrogen dioxide (NO_2)
Ozone (O_3)
Particulate matter with **aerodynamic** size less than or equal to 10 μm (**PM-10**)
Sulfur dioxide (SO_2).
Regulations govern the concentrations of criteria pollutants at receptors located in various locations within the community. *Compare* **air pollutants**.

critical area—In U.S. Navy terminology, an area in which danger is possible due to the passage of a **severe storm** (especially a **tropical cyclone**) to the extent that additional reports are necessary concerning weather conditions and the status of ships at sea.

critical depth—In a specified **stream** channel, the water depth at which the **specific energy** is the minimum for a given rate of flow.

Critical depth usually occurs at the point corresponding to an abrupt steepening of channel slopes, such as rapids.

critical depth control—A condition in a **stream** where, at a certain point, the water depth passes from above **critical depth** to below critical depth.

A change in flow conditions **downstream** of the point of critical depth is not readily transmitted above the control. Thus the relation between **stage** and **discharge** above the point of critical depth is fixed (or controlled) by the hydraulic characteristics of the channel at the section of critical depth. A critical depth control is usually found at a point where the channel slope steepens considerably, as above a rapids or waterfall.

critical discharge—For a given flow depth in an open channel, the **discharge** that will cause **critical flow** conditions at this depth.

critical drop radius—As defined by the **Köhler equation**, the size of a **hygroscopic nucleus** of given mass at which the **equilibrium** saturation ratio reaches a maximum, or the derivative of the **saturation** ratio with respect to size is zero.

This defines activation; solution droplets greater than this size are said to be activated and may grow without limit in an **environment** with lower saturation.

critical flow—The flow condition of a fluid system when one of the fundamental nondimensional parameters has a critical value, for example, flow of water in an open channel at a **Froude number** of 1, flow of gas at a **Mach number** of 1, flow in a pipe at **Reynolds number** greater than 2100, or **shear flow** in the **atmosphere** at Richardson number less than 0.25. As each **parameter** crosses its critical value, the nature of the flow changes, such as from straight to wavy, from laminar to turbulent, or from **linear** to **nonlinear**.

critical frequency—A term with several possible meanings but often that **frequency** of an incident **electromagnetic wave** at which the transmission characteristics of a medium change markedly.

As ascribed to a layer in the **ionosphere**, it is the minimum frequency of a vertically directed **radio wave** that will penetrate the layer.

critical gradient—1. The minimum channel slope that will produce **critical flow**.

2. In a **saturated soil**, the maximum **hydraulic gradient** above which **piping** or fluidization of the soil will occur.

critical height—The minimum height above aerodrome level to which an approach to landing by an aircraft can safely be continued without visual reference to the ground.

critical level—1. **Altitude** at which the relative horizontal **phase speed** of internal **gravity waves** equals the **wind speed**. As waves approach this altitude from above or below, the vertical component of **group velocity** approaches zero, causing elimination of the wave as its **energy** is absorbed and transferred to the mean wind. This causes changes in the wind-speed **profile**, which in turn can raise or lower the altitude of the critical level.

2. **Altitude** at which the **wind shear** is sufficiently strong to cause the **gradient Richardson number** to drop below its critical value, allowing **Kelvin–Helmholtz waves** to form.

These breaking waves can generate propagating internal **gravity waves** in the adjacent air above and below. *See* **buoyancy wave**.

3. The level at which the ocean **surface wave** phase speed is equal to the **wind** (or current) speed.

Considerable interaction between the **wave** field and the mean flow field can be associated with processes at this level.

4. See **critical level of escape**

critical level of escape—(Also called exobase, critical level.) The level in the **atmosphere** at which a **particle** moving rapidly upward will have a **probability** of $1/e$ of colliding with another particle on its way out of the atmosphere.

It is also the level at which the horizontal **mean free path** of an atmospheric particle equals the **scale height** of the atmosphere. The critical level of escape is the base of the **exosphere**. The height of the critical level, which can vary from 500 to 1000 km, is strongly dependent on the **temperature** of the upper **thermosphere**, and varies diurnally and with the level of **solar activity**. *See* **cone of escape, fringe region**.

critical point—The thermodynamic state in which liquid and gas phases of a substance coexist in **equilibrium** at the highest possible **temperature**. At higher temperatures than the critical no liquid **phase** can exist. For water the critical point is

$$e_s = 2.21 \times 10^5 \text{ mb}$$

$$T = 647 \text{ K}$$

$$\alpha = 3.10 \text{ gm cm}^{-3},$$

where e_s is the **saturation vapor pressure** of the **water vapor**, T is the Kelvin temperature, and α the **specific volume**.

critical region—*See* **significance test**.

critical Richardson number—The value of the **gradient Richardson number** below which air becomes dynamically unstable and turbulent.

This value is usually taken as $Ri_c = 0.25$, although suggestions in the literature range from 0.2 to 1.0. There is also some suggestion of hysteresis, where laminar air flow must drop below $Ri = 0.25$ to become turbulent, but **turbulent flow** can exist up to $Ri = 1.0$ before becoming laminar.

critical success index—*See* **skill**.

critical temperature—*See* **critical point**.

critical tidal level—A level on the shore where the emersion/submersion tidal characteristics change sharply.
　　Some biologists have suggested that zonation of plants and animals is controlled by a series of such levels, but detailed analysis of tidal **statistics** shows that the tidal transitions are seldom as sharply defined as the biological boundaries.

critical velocity—The **velocity** corresponding to the flow at **critical depth**.

crivetz—(Also called krivu, crivat.) North to east winds in Romania, especially a cold **bora**like wind characteristic of the **climate** of Romania.
　　Crivetz may occur at any **season**, but is least frequent in June and July. *See* **austru**.

Cromwell Current—*See* **Equatorial Undercurrent**.

crop calendar—List of the standard crops of a region in the form of a calendar giving the average dates of sowing and various stages of their growth to harvest.

crop coefficient—A factor K_c that is empirically determined and relates the **potential evapotranspiration** ET_0 to the crop **evapotranspiration** ET_c, that is, $ET_c = K_c \, ET_0$

crop moisture index—An **index** developed by Palmer (1968) to assess short-term crop water conditions and needs across major crop-producing regions.
　　It is based on the concept of abnormal **evapotranspiration** deficit, calculated as the difference between computed **actual evapotranspiration** (ET) and computed **potential evapotranspiration** (i.e., expected or appropriate ET). Actual evapotranspiration is based on the **temperature** and **precipitation** that occurs during the week and computed **soil moisture** in both the topsoil and subsoil layers.

> Palmer, W. C., 1968: Keeping track of moisture conditions, nationwide: The new crop moisture index. *Weatherwise*, **21**, 156–161.

cross correlation—The **correlation** between **time series** $x(t)$ and $y(t)$, where x and y may represent the same **variable** measured at different locations, or a single variable measured at one location but at different times, as for the case in which $y(t)$ represents $x(t + L)$, where L is a specified **time lag**.
　　In such cases, the two variables are usually not statistically independent, and large cross correlations between x and y can result. *See* **autocorrelation**.

cross-isobar angle—Angle between the **wind vector** and an **isobar**, at any level.
　　Such an angle is most conspicuous within the **friction layer** where the **wind** often has a component from high to low pressure.

cross polarization—*See* **cross-polarized signal**.

cross-polarized signal—A **signal** received by a **radar** that has a **polarization** orthogonal to that of the transmitted signal. Two polarizations are **orthogonal** if 1) the ellipses defined by their **electric field** vectors have the same axial ratio, 2) the major axes of the ellipses are at right angles, and 3) the electric field vectors rotate in opposite directions.

cross product—Same as **vector product**.

cross sea—A sea state of wind-generated **ocean waves** that consists of nonparallel **wave systems**.
　　A cross sea has a large amount of directional spreading.

cross section—1. In **weather analysis** and forecasting, a graphic representation of a vertical surface in the **atmosphere**, along a given horizontal line or path, and extending from the earth's surface to a given **altitude**.
　　The type of data and **analysis** presented on such a cross-sectional **chart** depends upon its purpose. In meteorology, a **synoptic** cross section is prepared from synoptic weather data. In aviation, a **flight cross section** (or route cross section) is a graphic forecast of conditions expected to be encountered along the proposed flight route; therefore, time varies along the horizontal axis of the chart. *Compare* **time section**.
　　2. Generally, a two-dimensional, representative picture of a three-dimensional entity; usually a

section or slice perpendicular to the principal axis of the entity, or passing through its center, or otherwise representative of a given aspect of the entity.

See **scattering cross section**; *compare* **profile, contour**.

cross-sectional analysis—Graphic representation of the atmospheric state in a vertical plane, usually in the form of a diagram having either the height or function of **pressure** as the vertical axis.

cross-spectrum—The **Fourier transform** of the **cross correlation** of two functions.

cross totals index—*See* **stability index**.

cross-valley wind—A **wind** that blows across the longitudinal axis of a valley from one sidewall to another.

This term is usually applied to a thermally driven (**anabatic** or **katabatic**) wind arising when unequal **insolation** on the two sidewalls causes a strong cross-valley **temperature** difference. Air then flows toward the more strongly heated sidewall. *See* **mountain–valley wind systems**.

crossover experiment—**Weather modification** experiment designed to involve two areas, only one of which is seeded at a time, with the area for **seeding** being randomly selected for each time period.

crosswind—A **wind** that has a component directed perpendicularly to the **course** (or **heading**) of an exposed, moving object; more popularly, a wind that predominantly acts in this manner.

In the broadest sense, any wind except a direct **headwind** or direct **tailwind** is a crosswind. The **drift** produced by crosswind is critical to air navigation, being especially dangerous during landing and takeoff.

crosswind gustiness—*See* **gustiness components**.

crosswise vorticity—The component of the **vorticity** vector that is perpendicular to the **flow velocity** vector.

See also **streamwise vorticity, helicity**.

CRT—Abbreviation for **cathode-ray tube**.

cryochore—A region of perpetual **snow**.

See **biochore**.

cryoconite hole—A small pit in an **ice** surface produced by the sinking of a **particle** of **dust** of cosmic origin (known as cryoconite, a fine-grained dark-colored substance).

cryogenic hygrometer—A special type of chilled mirror **hygrometer** capable of measuring very low **frost points**.

These low temperatures at the mirror are attained by evaporating low **boiling point** or cryogenic fluids. This strong cooling capability allows cryogenic hygrometers to maintain reasonable response times even at low frost points. Cryogenic hygrometers are used for airborne **humidity** measurements from aircraft or balloons, but also in semiconductor industries for monitoring very dry environmental conditions needed in manufacturing processes.

cryogenic period—An interval in **geologic time** during which geologic, geomorphic, and climatic conditions favored large-scale **glacier** formation; a period that tended to produce an **ice age** or **glacial period**.

cryology—1. The study of **ice** and **snow**.

2. In Europe, a synonym for **glaciology**.

Note: The term cryology has become almost meaningless unless it is defined in context. (*Arctic and Subarctic Terms*, Arctic, Desert Tropic Information Center Pub. A-105, 1955.)

3. The study of **sea ice**.

cryopedology—"The study of [ground movement caused by] intensive **frost action** and of **permafrost**, their causes and occurrences, and the engineering devices and practices which may be devised to overcome difficulties brought about by them." (from *Glossary of Arctic and Subarctic Terms* 1955).

Arctic, Desert, Tropic Information Center (ADTIC) Research Studies Institute, 1955: *Glossary of Arctic and Subarctic Terms*, ADTIC Pub. A-105, Maxwell AFB, AL, 90 pp.

cryoplanation—A type of **erosion** peculiar to high latitudes and/or high elevations.

Specifically, it is land reduction by the processes of intensive **frost action**, that is, **congeliturbation**, including **soil creep**, and supplemented by the erosive actions of running water, moving **ice**, and other agents.

cryoplankton—One-celled plants, usually algae, that live in **snow** and **ice** and tint their habitat red or green.

cryosphere—That portion of the earth where natural materials (water, soil, etc.) occur in frozen form.

> Generally limited to the polar latitudes and higher elevations.

cryoturbation—Same as **congeliturbation**.

cryptoclimate—(Also spelled kryptoclimate.) The **climate** of a confined space, such as the inside of a house, barn, greenhouse, or a natural or artificial cave.

> This term is often replaced by the term **microclimate**, appropriately modified, for example, house microclimate.

cryptoclimatology—The study of the **microclimate** of confined spaces.

crystal—A more or less regular periodic array of atoms, molecules, or ions, usually forming a solid.

> In everyday parlance crystal is used in a bewildering variety of ways, sometimes contradictory. Fine glassware is called crystal, although glass, an amorphous solid, is the antithesis of a crystal. A solid with facets exhibiting external symmetry may be called a crystal, although a solid without such facets may still be a crystal. A pure liquid such as water is said to be crystal clear even though transparency is not an essential property of a crystal.

crystal habit—Any characteristic external crystalline form.

> For an **ice crystal**, may refer just to the ratio of the lengths of the **crystal** parallel to its c and a axes (at 90° to or parallel to the hexagonal basal plane), but usually it simply means the crystal shape, including dendritic, skeletal, prismatic, sectors, etc. The term is not applied to the internal **crystal lattice**. Slight variations in the growth rates of different crystal planes in a given crystal structure lead to quite different crystal habits. Such growth rate variations may result from variations in **temperature** and **water vapor** supersaturation of the **environment** in which the crystal grows.

crystal lattice—An infinite regular periodic set of points in space on which the atoms, molecules, or ions of an ideal **crystal** are imagined to be located.

> No real crystal is exactly congruent with its crystal lattice.

crystalline frost—**Hoarfrost** that exhibits a relatively simple macroscopic crystalline structure; to be distinguished from **amorphous frost**.

> Crystalline frosts are classified into five forms: 1) needle; 2) featherlike; 3) plate; 4) cup; and 5) dendritic. Such **ice** forms are typically developed as a result of **deposition** at temperatures well below 0°C, the degree of **supersaturation** and **temperature** controlling the form.

crystallization—The process of formation of a **crystal** (an ordered state) from a disordered (gas) or partially ordered (liquid) state.

> Examples are the **freezing** of liquid water, the **deposition** of **water vapor** (**frost**), and crystal formation in supersaturated solutions.

crystallization nucleus—Same as **ice nucleus**, but applied mainly to the formation of **ice crystals** in a body of water.

CSI—Abbreviation for **conditional symmetric instability**.
> *See* **slantwise convection**.

CTA—Abbreviation for **cognitive task analysis**.

CTD—Abbreviation for **conductivity–temperature–depth profiler**.

cum sole—With the sun; hence **anticyclonic**; the opposite of **contra solem**.
> *See also* **deasil**.

cumulative distribution function—Same as **distribution function**.

cumulative temperature—Same as **accumulated temperature**.

cumuliform—Like **cumulus**; generally descriptive of all **clouds**, the principal characteristic of which is vertical **development** in the form of rising mounds, domes, or towers.

> This is the contrasting form to the horizontally extended **stratiform** types. Cumulus are driven by **thermal** convection and typically have vertical velocities in excess of 1 meter per second.

cumulonimbus—(Abbreviated Cb.) A principal **cloud** type (**cloud genus**), exceptionally dense and vertically developed, occurring either as isolated clouds or as a line or wall of clouds with separated upper portions.

> These clouds appear as mountains or huge towers, at least a part of the upper portions of which

is usually smooth, fibrous, or striated, and almost flattened as it approaches the **tropopause**. This part often spreads out in the form of an **anvil** (**incus**) or vast **plume**. Under the base of cumulonimbus, which is often very dark, there frequently exist **virga**, **precipitation** (**praecipitatio**), and low, ragged clouds (**pannus**), either merged with it or not. Its precipitation is often heavy and always of a showery nature. The usual occurrence of **lightning** and **thunder** within or from this cloud leads to its popular appellations: **thundercloud**, **thunderhead** (the latter usually refers only to the upper portion of the cloud), and **thunderstorm**. Cumulonimbus is composed of water droplets and **ice crystals**, the latter almost entirely in its upper portions. It also contains large water drops, **snowflakes**, **snow pellets**, and sometimes **hail**. The liquid water forms may be notably supercooled. Within a cold **air mass** in polar regions, the **fibrous ice** crystal structure may extend virtually throughout the cloud mass. Cumulonimbus always evolves from the further development of **cumulus congestus**, which, in turn, usually has resulted from the growth of **cumulus** (Cb cumulogenitus). This complete development may initiate also from **stratocumulus castellanus** (Cb stratocumulogenitus) or from **altocumulus castellanus** (Cb altocumulogenitus). In the latter case the cumulonimbus base is particularly high. It may also, but infrequently, develop from a portion of **altostratus** or **nimbostratus** (Cb altostratogenitus or Cb nimbostratogenitus). The formative process of cumulonimbus starts as a result of **convection** from the earth's surface or **instability** in the **upper air**, or both simultaneously. It therefore has a predominant **diurnal** cycle similar to that of cumulus. Cumulonimbus is rare over the polar regions, and becomes increasingly frequent with decreasing latitude, and is, in fact, an almost regular climax of the diurnal cloud **cycle** in the humid areas of the tropical regions and in humid and **unstable air** masses penetrating the temperate latitudes. Because of its great vertical size and of the magnitude and variety of forces that act within and upon it, cumulonimbus is a vertical cloud factory. In addition to the complex of accessory features it may possess, which includes **tornadoes** (**tuba**), it may also be responsible for the formation of nearly all of the other cloud genera. Cumulus congestus always preexists, and therefore is often easily confused with, cumulonimbus. A cloud is called cumulus congestus until its upper portion begins to show the diffuseness or fibrous quality indicative of ice crystal predominance. Only cumulonimbus is accompanied by lightning, thunder, or hail; only cumulus congestus can rival the **intensity** of its **shower**-type precipitation. *See* **cloud classification**, **thunderstorm**.

cumulonimbus calvus—A species of **cumulonimbus** cloud evolving from **cumulus congestus**.
    The protuberances of its upper portion have begun to lose their **cumuliform** outline; they loom and usually flatten, then transform into a whitish mass with more or less diffuse outlines and vertical **striations**. **Cirriform** cloud is not present, but the transformation into **ice crystals** often proceeds with great rapidity. Most often, this cloud is accompanied by showers. By convention, the name cumulonimbus calvus is given to any highly developed cumuliform cloud that produces **lightning**, **thunder**, or **hail**, although the top shows no obvious trace of transformation into **ice**. The species calvus is unique to the genus cumulonimbus.

cumulonimbus capillatus—A species of **cumulonimbus** cloud characterized by the presence, mostly in its upper portion, of distinct **cirriform** parts, frequently in the form of an **anvil** (**incus**), a **plume**, or a vast and more or less disorderly mass of hair.
    This **cloud** is usually accompanied by a **shower** or **thunderstorm**, often by a **squall**, and sometimes by **hail**. It generally produces very apparent **virga**. The species capillatus is unique to the genus cumulonimbus.

cumulus—(Abbreviated Cu.) A principal **cloud** type (**cloud genus**) in the form of individual, detached elements that are generally dense and posses sharp nonfibrous outlines.
    These elements develop vertically, appearing as rising mounds, domes, or towers, the upper parts of which often resemble a cauliflower. The sunlit parts of these clouds are mostly brilliant white; their bases are relatively dark and nearly horizontal. Near the **horizon** the vertical development of cumulus often causes the individual clouds to appear merged. If **precipitation** occurs, it is usually of a showery nature. Various effects of **wind**, **illumination**, etc., may modify many of the above characteristics. Strong winds may shred the clouds, often tearing away the cumulus tops to form the species **fractus**. Under certain conditions cumulus clouds may be arranged in files, **cloud streets**, oriented approximately parallel to the **wind direction**. Changes in direction of illumination and in background cause modification of color and of apparent surface relief. Cumulus is composed of a great **density** of small water droplets, frequently supercooled. Within the cloud larger water drops are formed that may, as the cloud develops, fall from the base as **rain** or **virga**. Ice crystal formation will occur within the cloud at sufficiently low temperatures, particularly in upper portions as the cloud grows vertically. Occasionally the growth of **ice crystals** at the expense of water droplets will reduce the entire cloud to diffuse trails of **snow**. Cumulus most often forms directly in **clear air** as a result of **convection** in air of sufficiently high moisture

content for a **condensation level** to be reached. As a result, a distant **diurnal** cycle of cumulus frequency is observed. Over a landmass, the cumulus maximum occurs after midday (for a horizontal extent, early afternoon; for vertical extent, somewhat later). Over a water surface, the cycle is reversed and much less obvious, with the cumulus maximum generally recognized as occurring after midnight. The vertical growth of a cumulus cell is restricted and modified by the existence and character of layers of relative **static stability** above the **cloud base**. Cumulus may also evolve from the convective transformation of **stratus** or **stratocumulus** (Cu stratomutatus or Cu stratocumulomutatus). Cumulus may be generated by **altocumulus** and, again, stratocumulus (Cu altocumulogenitus and Cu stratocumulogenitus).**Cumulonimbus** is the ultimate manifestation of the growth of cumulus; therefore, at a certain point, it is difficult to differentiate between the two. If a cloud in doubt reveals no fibrous structure, it is still cumulus; if still in doubt, cumulonimbus further differs in that it is accompanied by **lightning**, **thunder**, and sometimes **hail**. The elements of altocumulus are smaller and, along with those of stratocumulus, tend to be more merged than the separated units of cumulus. Cumulus has the unique ability to penetrate other preexisting cloud layers, sometimes partially dissipating, at other times apparently fusing with, the impaled layers. The cumulus, in this instance, retains its identity as long as it remains primarily vertically developed, is physically (although perhaps not visibly) separate from the other cloud, and has a tower- or dome-shaped summit. *See* **cloud classification, trade-wind cumulus.**

cumulus congestus—A strongly sprouting **cumulus** species with generally sharp outlines and, sometimes, with a great vertical development; it is characterized by its cauliflower or tower aspect, of large size.

Mainly in the **Tropics**, cumulus congestus may produce abundant **precipitation**. It may also occur in the form of very high towers, the tops of which are formed of kinds of **cloudy** puffs that, detaching themselves successively from the main portion of the **cloud**, are carried away by the **wind**, then disappear more or less rapidly, sometimes producing **virga**. Cumulus congestus is the result of the development of **cumulus mediocris** and, sometimes, of that of **altocumulus castellanus** or **stratocumulus castellanus**. Cumulus congestus often transforms into **cumulonimbus**; this transformation is revealed by the smooth, fibrous, or striated aspect assumed by its upper portion. The species congestus is unique to the genus cumulus.

cumulus fractus—(Previously called fractocumulus.) *See* **fractus**.

cumulus humilis—(Also called fair-weather cumulus.) A species of **cumulus** characterized by small vertical **development**, uniform flat bases and a general similarity among clouds.

Its vertical growth is usually restricted by the existence of a **temperature inversion** in the **atmosphere**; this in turn explains the unusually uniform height of the **cloud** tops of this cumulus species. A single **cloud element** that is able to penetrate the **inversion** may develop into **cumulus congestus** or even further to become **cumulonimbus**. As in all species of cumulus, **wind shear** with height may give rise to a hard appearance upshear, where cloud erosion in dry **environment** air is taking place, and a fuzzy appearance downshear . This species is unique to the genus cumulus.

cumulus mediocris—A **cloud species**, unique to the genus **cumulus**, of moderate vertical **development**, the upper protuberances or sproutings of which are not very marked; it may have a small cauliflower aspect.

This **cloud** does not give any **precipitation**, but frequently develops into **cumulus congestus** and **cumulonimbus**.

Cunnane plotting position—The Cunnane plotting position for the rth ranked (from largest to smallest) **datum** from a **sample** of size n, is the quotient

$$\frac{r - 0.4}{n + 0.2}.$$

It is used when **quantile** unbiased values are desired.
See also **plotting position, probability paper.**

Cunningham slip correction—Deviation from **Stokes's law** when **particle** size approaches the **mean free path** of air molecules.

cup anemometer—A mechanical **anemometer** with a vertical axis of rotation, usually consisting of three or four hemispherical or conical cups mounted with their diametrical planes vertical and distributed symmetrically about the axis of rotation.

The rate of rotation of the cups is a measure of the **wind speed**. In gusty **wind**, the cup anemometer overestimates the mean wind speed.

cup crystal—An **ice crystal** in the form of a hollow hexagonal cup.

Usually one side is not developed and appears to be rolled up. Cup crystals are the most common form of **depth hoar**, but are rarely observed in **snow**.

curie—A unit of **radioactivity**, defined as exactly 3.7×10^{10} disintegrations per second.

Curie temperature—(Also called Neel temperature.) The **temperature** beyond which rocks and minerals become paramagnetic, which for most geophysical purposes means they are nonmagnetic.

curl—A **vector** operation upon a **vector field** that represents the rotation of the field, related by **Stokes's theorem** to the **circulation** of the field at each point.

The curl is invariant with respect to coordinate transformations and is usually written

$$\text{curl } \mathbf{F} \text{ or } \nabla \times \mathbf{F},$$

where ∇ is the **del operator**. In **Cartesian coordinates**, if **F** has the components, F_x, F_y, F_z, the curl is

$$\left(\frac{\partial F_z}{\partial y} - \frac{\partial F_y}{\partial z}\right)\mathbf{i} + \left(\frac{\partial F_x}{\partial z} - \frac{\partial F_z}{\partial x}\right)\mathbf{j} + \left(\frac{\partial F_y}{\partial x} - \frac{\partial F_x}{\partial y}\right)\mathbf{k}.$$

Expansions in other coordinate systems may be found in any text on vector analysis. The curl of the **velocity** vector is called the **vorticity**; in a field of **solid rotation** it is equal to twice the **angular velocity**. Occasionally the vorticity is defined as one-half the curl. The curl of a two-dimensional vector field is always normal to the vectors of the field; this is not necessarily true in the three dimensions. *Compare* **divergence**.

current—1. Any movement of material in space.
See **air current, ocean current**.

2. Any movement of **electric charge** in space, by virtue of which a net **transport** of charge occurs as, for example (in **atmospheric electricity**), in a **conduction current**, **convection current**, or **precipitation current**.

current chart—A map of a water area depicting ocean-current data by current roses, vectors, or other means.

current cross section—A two-dimensional map of either the speed or of one **vector** component of **current**, generally plotted as a **contour** plot in the x–z plane, where x is horizontal distance and z is vertical distance.

current curve—A graphic representation of the flow of a **current**.
In the reversing type of **tidal current**, the curve is referred to rectangular coordinates with time represented by the **abscissa** and the speed of the current by the **ordinate**, the **flood** speeds being considered as positive and the ebb speeds as negative. In general, the current curve for a reversing tidal current approximates a cosine curve.

current ellipse—A graphic representation (a **hodograph**) of a **rotary current** (of a **tidal current**, or a **harmonic** component of a tidal current, during a complete tidal **cycle**) in which the **velocity** of the current at different hours of the tidal cycle is represented by radius vectors and **vector** angles.
A line joining the extremities of the radius vectors will form a curve roughly approximating an ellipse.

current function—Same as **Stokes's streamfunction**.

current meter—A device designed to measure speed alone or **velocity** of flowing water.
The three most common types of modern current meters are mechanical (rotor and vane), electromagnetic, and acoustic Doppler.

current pole—A pole used in measuring **surface water** current, especially from an anchored vessel such as a lightship.
The **drift** of the pole is timed as it is allowed to carry out a graduated line; the **azimuth** and speed of the line give the **current** velocity.

current profiler—In **oceanography**, an instrument that measures **current** simultaneously over a **range** of depths.
For example, an **acoustic Doppler current profiler** measures the current by the **Doppler effect** on backscattered **sound waves** over a **range** of depths.

current rose—A diagram that indicates, for a given ocean area, the average percentage of **current** setting toward each of the principal **compass** points.
The distribution of **drifts** is sometimes also indicated. *Compare* **wind rose**.

current tables—Annual tables of daily predictions of times and velocities of maximum flood currents and ebb currents and the times of **slack water** to be encountered in numerous coastal waterways.

Curtis–Godson approximation—An approximation used in longwave **radiative transfer** whereby the **transmittance** of a given path through a nonisothermal gas is related to the transmittance through an equivalent **isothermal** gas.

curve fitting—The derivation of an analytic function $f(x)$ with its graph $y = f(x)$ passing through or approximately through a finite set of data points (x_i, y_i), $i = 1, \ldots, n$.

    Curve-fitting procedures include **interpolation**, in which case $f(x_i) = y_i$ for each data point, and the **least squares** method, in which case the derived function minimizes the sum of the squares of the differences between $f(x_i)$ and y_i over all the data points. The functions used in curve fitting are usually polynomials. In least squares methods, a single **polynomial** of low degree typically is used over the entire **range** of x spanned by the data; in interpolation procedures, a separate polynomial typically is defined over each subinterval and the polynomial pieces connected through imposition of certain **continuity** conditions. *See* **spline function**.

curve of growth—For an isolated, gaseous **absorption band**, the relation between integrated **absorption** by the gas and its amount.

    That is, if

$$W = \int [1 - \exp(-\kappa x)]dv,$$

where κ is the (volumetric) **absorption coefficient** of the gas, x is the path length through it, and the **range** of integration over **frequency** v is sufficient to include essentially all absorption within the band, then W as a function of x is the curve of growth. This term, which Goody describes as "curious," is a relic from astronomical **spectroscopy**.

    Goody, R. M., 1964: *Atmospheric Radiation*, Vol. 1, p. 125.

curvilinear coordinates—Any **linear** coordinates that are not **Cartesian coordinates**.

    If u, v, w are three functions of the Cartesian coordinates x, y, z and if at least one of these functions is not a linear combination of x, y, z, then u, v, w are curvilinear coordinates of the point the Cartesian coordinates of which are x, y, z provided that the **Jacobian** $\partial(u,v,w)/\partial(x,y,z)$ is not equal to zero. Any surface along which one of the three curvilinear coordinates is constant is called a **coordinate surface**; there are three families of such surfaces. Any line along which two of the three curvilinear coordinates are constant is called a **coordinate line**; there are three sets of such lines. Three distinct coordinate lines may be drawn through each point of space. The three straight lines each of which is tangent to one of the coordinate lines at a given point in space are called the local axes. If the local axes are everywhere mutually perpendicular, the curvilinear coordinates are said to be **orthogonal** or rectangular. Examples of frequently used curvilinear coordinates are **polar coordinates** and **cylindrical coordinates**. *See also* **natural coordinates, spherical coordinates**.

custard winds—Cold easterly winds on the northeast coast of England.

cut-off high—A **warm high** that has grown out of a **ridge** and become displaced out of the basic westerly **current** and lies poleward of this current.

    Frequently, such highs are also **blocking highs**. *See* **cutting-off process, cut-off low, warm pool**.

cut-off low—A **cold low** that has grown out of a **trough** and become displaced out of the basic westerly **current** and lies equatorward of this current.

    See **cutting-off process, cut-off high, cold pool**.

cutting-off process—A sequence of events by which a **warm high** or **cold low**, originally within the **westerlies**, becomes displaced either poleward (**cut-off high**) or equatorward (**cut-off low**) out of the westerly current.

    This process is evident at very high levels in the **atmosphere**; it frequently produces, or is part of the production of, a **blocking situation**.

CW radar—Abbreviation for **continuous-wave radar**.

cyanometer—Generally, an instrument designed to measure or estimate the blueness of the sky.

    The type in most common use is the Linke-scale.

cycle—1. Any process or sequence of states in which the initial and final states of a system are the same.

2. A unit of **wave frequency**, actually one **cycle per second**.
See **kilocycle, megacycle, kilomegacycle.**

cycle per second—A unit of **frequency**, abbreviated cps, identical to **hertz**, the modern term that has supplanted it.

cyclo-geostrophic current—A **current** that is nearly in **geostrophic balance** (**pressure gradient** balanced by Coriolis term), with a correction for the centrifugal **acceleration** due to flow along curved flow lines.
  This correction term becomes important in strong circular eddying motions in which the time for a **fluid parcel** to flow around the **eddy** is comparable to a **pendulum day**.

cyclogenesis—Any **development** or strengthening of **cyclonic circulation** in the **atmosphere**; the opposite of **cyclolysis.**
  It is applied to the development of cyclonic circulation where previously it did not exist (commonly, the initial appearance of a **low** or **trough**) as well as to the **intensification** of existing **cyclonic** flow. While cyclogenesis usually occurs together with **deepening** (a decrease in atmospheric pressure), the two terms should not be used synonymously. *Compare* **anticyclogenesis.**

cyclohexane—Cyclic hydrocarbon, formula C_6H_{12}, consisting of a symmetrical six-membered ring formed from $-CH_2-$ linkages.
  Cyclohexane is widely used as a solvent and has been detected in **ambient air** samples.

cyclolysis—Any weakening of **cyclonic circulation** in the **atmosphere**; the opposite of **cyclogenesis.**
  Cyclolysis, which refers to the **circulation**, is to be distinguished from **filling**, an increase in **atmospheric pressure**, although the two processes commonly occur simultaneously. *Compare* **anticyclolysis.**

cyclone—An atmospheric **cyclonic circulation**, a **closed circulation.**
  A cyclone's direction of rotation (counterclockwise in the Northern Hemisphere) is opposite to that of an **anticyclone**. While modern meteorology restricts the use of the term cyclone to the so-called **cyclonic-scale** circulations, it is popularly still applied to the more or less violent, small-scale circulations such as **tornadoes, waterspouts, dust devils**, etc. (which may in fact exhibit **anticyclonic rotation**), and even, very loosely, to any strong **wind**. The first use of this term was in the very general sense as the generic term for all circular or highly curved wind systems. Because **cyclonic circulation** and relative low atmospheric **pressure** usually coexist, in common practice the terms cyclone and **low** are used interchangeably. Also, because cyclones are nearly always accompanied by inclement (often destructive) weather, they are frequently referred to simply as **storms.** *See* **tropical cyclone, extratropical cyclone;** *compare* **trough.**

cyclone collector—Instrument used to remove large **particles** (diameter greater than 3 μm) from **ambient air.**
  The conical geometry of the **collector** subjects the air to rotation, and the **centrifugal force** leads to **deposition** of the particles. The particles can be sampled from the walls of the device for analysis, or simply removed physically from the airflow.

cyclone family—A series of **wave cyclones** occurring in the interval between major outbreaks of **polar air.**
  The series travels along the **polar front**, usually eastward and poleward. Typically, the polar front drifts eastward and equatorward, so that each **cyclone** of the family has its origin and **trajectory** at a lower latitude than the previous cyclone.

cyclone model—A mathematical representation of **atmospheric dynamics** that attempts to simulate the atmospheric motions and conditions characteristic of **tropical, extratropical**, or **polar cyclones.**

cyclone precipitation—(Rare.) Generally, any **precipitation** associated with, or within the **circulation** of, a **cyclone**; mainly distinguished from **orographic precipitation**, but also from **airmass precipitation.**
  Most cyclone precipitation can be classed as **frontal precipitation.**

cyclone wave—1. A **disturbance** in the **troposphere** of **wave length** 1000–4000 km (**cyclonic scale**).
  They are recognized on **synoptic charts** as **migratory** high and low pressure systems. These waves have been identified with the unstable **perturbations** discussed in connection with **baroclinic instability** and **shearing instability.** *See* **short wave, barotropic disturbance.**

  2. A **frontal wave** at the crest of which there is a center of **cyclonic circulation**; therefore, the frontal wave of a **wave cyclone.**

cyclonette—(Rare.) A **tornado** or **whirlwind**.

cyclonic—Having a sense of rotation about the local vertical the same as that of the earth's rotation: that is, as viewed from above, counterclockwise in the Northern Hemisphere, clockwise in the Southern Hemisphere, undefined at the **equator**; the opposite of **anticyclonic**.

cyclonic bora—*See* **bora scura**.

cyclonic circulation—Fluid motion in the same sense as that of the earth, that is, counterclockwise in the Northern Hemisphere, clockwise in the Southern Hemisphere, undefined at the **equator**.

cyclonic phase—*See* **foehn phase**.

cyclonic precipitation—Same as **cyclone precipitation**.

cyclonic rain—*See* **cyclone precipitation**.

cyclonic rotation—Generally circular motion of a mass in the same sense as that of the earth, that is, counterclockwise in the Northern Hemisphere, clockwise in the Southern Hemisphere, undefined at the **equator**.

cyclonic scale—(Also called **synoptic scale**.) The **scale** of the **migratory** high and low pressure systems (or **cyclone waves**) of the **troposphere** with **wave lengths** of 1000–4000 km.
　　Terminology in the literature is confusing, chiefly because **cyclonic-scale** disturbances at low levels are frequently associated with large-scale disturbances in the high troposphere. *See* **barotropic instability, baroclinic instability**.

cyclonic shear—Horizontal **wind shear** of such a nature that it contributes to the **cyclonic** vorticity of the flow, that is, it tends to produce **cyclonic rotation** of the individual air particles along the line of flow.
　　In the Northern Hemisphere, cyclonic shear is present if, when one faces **downwind**, the **wind speed** increases from left to right across the direction of flow; the opposite is true in the Southern Hemisphere. *Compare* **anticyclonic shear**.

cyclostrophic flow—A form of **gradient flow** in which the **centripetal acceleration** exactly balances the **horizontal pressure force**.
　　See **cyclostrophic wind**.

cyclostrophic wind—That horizontal **wind** velocity for which the **centripetal acceleration** exactly balances the **horizontal pressure force**:

$$\frac{V_c^2}{R} = -\alpha\,\frac{\partial p}{\partial n},$$

where V_c is the cyclostrophic wind speed, R the radius of curvature of the path, α the **specific volume**, p the **pressure**, and n the direction normal to the **streamlines** toward the center of curvature of the path.
　　The cyclostrophic wind can be an approximation to the real wind in the **atmosphere** only near the **equator**, where the **Coriolis acceleration** is small; or in cases of very great **wind speed** and curvature of the path (such as a **tornado** or **hurricane**), so that the centripetal acceleration is the dominant one. *See* **Eulerian wind**.

cyclotron frequency—See **gyro-frequency**.

cylindrical coordinates—(Also called cylindrical polar coordinates, circular cylindrical coordinates.) A system of **curvilinear coordinates** in which the position of a point in space is determined by 1) its perpendicular distance from a given line, 2) its distance from a selected reference plane perpendicular to this line, and 3) its angular distance from a selected reference line when projected onto this plane.
　　The coordinates thus form the elements of a cylinder, and, in the usual notation, are written r, θ, and z, where r is the radial distance from the cylinder's z axis, and θ is the angular position from a reference line in a cylindrical **cross section** normal to the z axis. The relations between the cylindrical coordinates and the **rectangular Cartesian coordinates** (x, y, z) are $x = r\cos\theta$, $y = r\sin\theta$, $z = z$. *See also* **polar coordinates**.

cylindrical polar coordinates—Same as **cylindrical coordinates**.

cylindrical projection—A type of **map projection** in which features on a sphere are projected onto a cylinder.
　　The cylinder can be either tangent to the sphere, for which contact is along a **great circle** path, or pass through the sphere, for which contact is along two circles.

CZCS—Abbreviation for **Coastal Zone Color Scanner**.

D

d'Alembert's paradox—A hydrodynamical paradox arising from the neglect of **viscosity** in the **steady flow** of a fluid around a submerged solid body.

According to this paradox, the submerged body would offer no **resistance** to the flow of an **inviscid fluid** and the **pressure** on the surface of the body would be symmetrically distributed about the body. This paradox may be traced to the neglect of the viscous forces, which are indirectly responsible for fluid resistance by modifying the **velocity** field close to a solid body.

Birkhoff, G., 1950: *Hydrodynamics*, 10–22.

D-analysis—An **isopleth** analysis of **D-values**.

D-layer—*See* **D-region**.

D-region—The lowest region of the **ionosphere**.

The term is used somewhat loosely to describe the **ionization**, beginning about 70 km and merging with the **E-region**, that does not usually produce an echo on **normal** ionosonde recordings. The main effect of the D-region on **radio waves** is one of **absorption**, thus inhibiting long-distance propagation of **HF** and **VHF** radio waves in daytime, when D-region ionization is most intense. At low and middle latitudes, the D-region is produced mainly by the action of **solar radiation** on **nitric oxide** (NO). At high magnetic latitudes energetic **particles** of solar or auroral origin may be the principal source, in which case radio waves can be strongly absorbed at all times of day. The term **D-layer** is used occasionally by analogy with the higher E- and **F-layers**, which produce sharply defined echoes on **ionosonde** recordings.

D-value—The quantity D describing the **altitude** of a point on a **constant-pressure surface** by its **departure** from "standard" altitude:

$$D = Z - Z_p,$$

where Z is the actual altitude above **mean sea level** and Z_p is the **pressure altitude** of the same point.

dadur—In India, a **wind** blowing down the Ganges Valley from the Siwalik hills at Hardwar.

dahatoe—*See* **aloegoe**.

daily forecast—(Sometimes called short-range forecast.) A forecast for the ordinary daily range, for periods of from 12 to 48 hours in advance.

Forecasts of this type usually express in some detail, for specific geographical areas, the expected day-to-day sequence of all the aspects of weather that materially affect human activity and well-being. *See* **short-range forecast**; *compare* **medium-range forecast, long-range forecast, extended forecast**.

daily maximum temperature—Maximum **temperature** in the course of a continuous time interval of 24 hours (usually midnight to midnight **local time**).

daily mean—The long-period **mean value** of a **climatic element** on a given day of the year.

A curve of daily means throughout the year shows the annual **variation** in much greater detail than a curve based on monthly means, but unless it is based on a long period (at least 50 years) it will probably be dominated by **random** irregularities. Such long-period daily means might be referred to as "normal daily values."

daily minimum temperature—Minimum **temperature** in the course of a continuous time interval of 24 hours (usually midnight to midnight **local time**).

daily range of temperature—Arithmetic difference between **daily maximum** and **daily minimum temperature**.

daily storage—Volume of water that can be stored daily in a **reservoir** between minimum and maximum daily water levels under ordinary operating conditions.

daily variation—*See* **daily range of temperature**.

Dalton's law—The **total pressure** of a mixture of gases is the sum of the pressures each component would have if it alone occupied the volume of the mixture at its **temperature**.

Dalton's law is strictly valid only for hypothetical ideal gases but is a good approximation for atmospheric gases at **normal** terrestrial temperatures and pressures.

damming—The piling up, or complete **blocking**, of cold air approaching the slopes of a mountain barrier.

This produces high pressure along the **windward** slopes and, when it persists for more than several hours, can also produce a **barrier jet**.

damp haze—*See* **haze, smog**.

damped oscillation—*See* **oscillation**.

damped wave—Any **wave** for which the (complex) **amplitude** decreases in the direction of propagation.

damping—The more or less steady diminishing in time or space (or both) of the **amplitude** of any physical quantity.
 In the **atmosphere**, for example, acoustic, hydrodynamic, and **electromagnetic waves** are damped.

dancing dervish—Same as **dust whirl**.

dancing devil—Same as **dust whirl, dust devil**.

dangerous semicircle—The side of a **tropical cyclone** to the right of the direction of movement of the storm in the Northern Hemisphere (to the left in the Southern Hemisphere), where the winds are stronger because the cyclone's translation speed and **rotational** wind **field** are additive.
 The opposite side is termed the **navigable semicircle**. This terminology originated in the days of sailing ships. It occurred naturally since 1) the dangerous semicircle of the storm has the strongest winds and heaviest seas; 2) a sailing ship on this side tends to be carried into the path of the storm; and 3) if the storm recurves, its center is likely to cross the course of a ship running before the **wind**.

Dansgaard–Oeschger events—Warm–cold oscillations during the **last glacial** period recorded in the **oxygen** isotope record of the Greenland **ice**, and also found in biotic and isotopic indices from deep-sea sediments in the North Atlantic.
 The warm phases of these events correspond to **interstadials**, persist for a few hundred to a few thousand years, and have very rapid onset and termination (as little as a few decades). The **range** of **temperature** change inferred for the regions where the **snow** was formed that ultimately produced the Greenland ice was several degrees Celsius. Named for the ice core paleoclimatologists Willi Dansgaard and Hans Oeschger, Dansgaard–Oeschger events between 80 000 and 20 000 years ago are grouped in 10 000- to 15 000-year periods of increasing cooling. The ending of each of these groups of events is marked by a major flux of icebergs into the North Atlantic, as evidenced by associated material found in ocean floor sediments. The temperature conditions in Greenland and the North Atlantic region then return to the higher level of the beginning of the group of Dansgaard–Oeschger events. Most of the Dansgaard–Oeschger events that lasted 2000 years or more coincide with warmer conditions in East Antarctica, also revealed by analysis of oxygen isotopes in ice.
 Dansgaard, W., and H. Oeschger, 1989: Past environmental long-term records from the Arctic. *The Environmental Record in Glaciers and Ice Sheets*, H. Oeschger and C. C. Langway Jr., Eds., Wiley, 287–318.

Darcian velocity—Volume of water flow per unit area per unit time passing through a **porous medium** in the direction of interest.
 See **specific discharge**.

Darcy's law—The relationship for movement of fluids through permeable or porous media, such as soil, which states that at low **Reynolds numbers** the **flow velocity** V is proportional to **hydraulic gradient** dh/dl, where the constant of proportionality K is the **hydraulic conductivity**:

$$V = -K \frac{dh}{dl}.$$

The **hydraulic head** h is the height of fluid in a **manometer**, which is proportional to the fluid **pressure**, while l is the slant length in the medium along the flow **streamline**. The **velocity** V is really a **specific discharge** (volume flow rate of fluid per unit cross-sectional area of medium), and is sometimes called the Darcy velocity or Darcy flux. The hydraulic conductivity depends both on the **permeability** k of the medium (e.g., sand vs clay), and on the **kinematic viscosity** v of the fluid:

$$K = \frac{kg}{v},$$

where g is gravitational **acceleration**.
 Freeze, R. A., and J. A. Cherry, 1979: *Groundwater*, 15–18.

dark—1. (Obsolete.) As used in aviation observations prior to 1 September 1956, descriptive of a **cloud cover** that is composed of predominantly threatening or unusually dark-colored clouds.

It was denoted by the **symbol** "+" preceding the appropriate **sky cover** symbol. *See* **thin**.

2. For observing at night, a person needs two to three minutes to adapt to dark (**dark adaptation**) before a noninstrumental **observation**, e.g., **visibility** at night.

dark adaptation—The process by which the human eye's sensitivity increases in response to marked **luminance** decreases.

If the average luminance on the retina decreases suddenly from daytime photopic levels, a **light** source's just-detectable threshold luminance (alternatively, its **threshold illuminance** at the eye) decreases steadily for three to four minutes. This occurs because the eye's cones become much more sensitive during this period, although their sensitivity does not increase afterward. Note that cones are the only photoreceptors found in foveal (or central) vision, but that they coexist with rods outside the fovea. After about seven to ten minutes in very low-light scotopic conditions (0.1 **lux** or less), the rods become more sensitive than the cones, and visual sensitivity once again increases, reaching a maximum at about 20–30 minutes. The fully dark-adapted eye is nearly 10^5 times more sensitive than it is at daytime light levels, and the maximum rod sensitivity is about 10^3 times the maximum cone sensitivity. Although dark-adapted observers have a much lower threshold illuminance for detection, this is offset by the greatly increased **threshold contrast** in scotopic conditions. Mesopic vision occurs when both rods and cones contribute to vision at light levels between entirely photopic and scotopic conditions. Cones dominate **photopic vision** while only rods contribute to **scotopic vision**. Color vision is not possible in scotopic conditions, and the peak spectral sensitivity of scotopic vision occurs at shorter wavelengths than it does for photopic vision. Thus to the dark-adapted eye, blues and greens tend to look brighter than reds (the Purkinje shift).

dark lightning—A photographic effect in which **lightning discharges** register as dark instead of light due to multiple exposures caused by successive members of a composite **flash**.

This effect, which results from a characteristic of the exposure versus **illumination** curve of a photographic emulsion, is sometimes called the **Clayden effect** after its English discoverer.

Humphreys, W. J., 1940: *Physics of the Air*, 3d ed., p. 371.

dark segment—A blue or bluish gray wedge seen in the antisolar sky between the **astronomical horizon** and the pink, orange, or purple **antitwilight arch**.

During **clear** twilights, the dark segment is visible (assuming a flat **horizon**) at sun elevations $\sim -0.5° < h_0 < -7°$. The dark segment is sometimes called the **earth's shadow** since one of its causes is the **sunlight** shadow that the earth casts on the nearby **atmosphere**. However, atmospheric **scattering** and **extinction** make the dark segment's **luminance** and color distribution more complicated than that for simple shadows. For example, the dark segment's vertical width above the **antisolar point** is consistently greater than $|h_0|$, contrary to the behavior of a simple geometric shadow.

Darling shower—A **duststorm** caused by **cyclonic** winds in the vicinity of the River Darling in Australia.

dart leader—(Also called continuous leader.) The **leader** which, after the first **stroke**, typically initiates each succeeding stroke of a multiple-stroke **flash** lightning. (The first stroke is initiated by a **stepped leader**.)

The dart leader derives its name from its appearance on photographs taken with streak cameras. The dart leader's brightest luminosity is at its tip which is tens of meters in length, propagating downward at about 10^7 m s^{-1}. In contrast to stepped leaders, dart leaders do not typically exhibit branching because the previously established channel's low gas **density** and residual ionization provide a more favorable path for this leader than do any alternative ones.

Chalmers, J. A., 1957: *Atmospheric Electricity*, p. 239.

data acquisition system—A collection of sensors and communication links to sample or collect and then return data to a central location for further processing, display, or archiving.

data assimilation—The combining of diverse data, possibly sampled at different times and intervals and different locations, into a unified and consistent description of a physical system, such as the state of the **atmosphere**.

data collection platform—Automatic digital or analog measuring device with recording and data transmission system.

data collection system—1. Coordinated system for collection, **reduction**, and archival of observations from a hydrometeorological network.

It may include automatic transmission of the observations to a remote receiving location.

2. (Abbreviated DCS.) The **GOES** data collection system, which receives environmental data transmissions that originate at remote automatic data collection platforms and relays the **signal** to central ground stations.

data directory—A set of descriptions of a large number of datasets containing high-level information suitable for making an initial determination of the potential usefulness of a **dataset** for some application.

Information on the location of more detailed descriptions or the latest dataset itself is found in the data directory.

data logger—A device deployed in the field that records weather data from one or more external sensors.

Most modern data loggers are miniature digital computers that sample the data inputs and digitize the results at regular intervals, apply preset calibrations to convert from raw voltages or currents into meteorological variables, compute **statistics** such as mean values and variances, and record the results either internally in memory, locally on attached storage media, or transmit the results to a remote user or computer site via telephone line or radio link. Those data loggers that record the results in memory must be periodically serviced to download the data and then clear the memory to allow new data to be stored.

data window—The finite time or space interval during which measurements are obtained.

Meteorological data can be viewed as occurring along an infinite time line from the distant past into the distant future, but measurements for a limited time, for example, during a field experiment or period of routine **weather observations**, provide only a window on a portion of the time line.

dataset—A collection of data covering a particular geographic area for a particular time period.

Usually stored and accessed as a unit for processing with data models.

dataset catalog—A list of the contents of a collection of **datasets** with sufficient information to allow for identification and retrieval of datasets of interest.

dataset directory—1. An abbreviated list of the contents of a collection of **datasets** used for the retrieval of datasets.

Usually, it does not provide full information on collection and coverage of individual datasets.

2. Same as **data directory**.

datum—For marine applications, a base **elevation** used as a reference from which to reckon heights or depths.

It is called a tidal datum when defined in terms of a certain **phase** of the **tide**. Tidal datums are local datums, referenced to fixed points (bench marks), and should not be extended into areas that have differing **hydrographic** characteristics without substantiating measurements.

datum level—(Also simply datum.) Agreed-upon horizontal control surface used as a standard reference to which elevations are related.

Davidson Current—A **countercurrent** of the Pacific Ocean running north along the west coast of the United States (from northern California to Washington to at least latitude 48° N) during the winter months.

Davies number—(Also called Best number.) A **dimensionless number** (C_dRe^2; C_d = **drag coefficient**; Re = **Reynolds number**) used in computation of **terminal fall velocity** of a **precipitation** particle.

dawn—The first **light** visible in the solar sky before **sunrise** or the time of that appearance.

Dawn is synonymous with "daybreak" and the beginning of morning **twilight**. Twilight is not an antonym of "dawn," although **dusk** is used this way in nontechnical language.

day—1. A basic time increment defined by the earth's motion; specifically, a complete revolution of the earth about its own axis.

The sidereal day is defined as the time required for the earth to make one complete revolution in an **absolute coordinate system**, that is, with respect to the stars. The day in common use is the **mean solar day**, derived, by means of the **equation of time**, from the **apparent solar day**, which is determined directly from the apparent relative motion of the sun and earth. The **civil day** is a modification of the mean solar day, which renders it practical as a time measure for ordinary purposes. *See also* **year**, **lunar day**.

2. The period from midnight to midnight, local **civil time**; that is, a **civil day**.

3. The period of daylight, as opposed to that of darkness.

4. *See* **pendulum day**.

5. *See* **observational day**.

day of rain—*See* **rain day**.

daybreak—Same as **dawn**.

days with snow cover—The number of days per month with **snow cover** on the ground.
Typically expressed as the percentage of days per month that snow cover of various depths is on the ground for a location in categories from **trace** to whole inches.

daytime visual range—Same as **visual range**.

dBZ—(Sometimes written dBz.) The **logarithmic scale** for measuring **radar reflectivity factor**, referred to a value of 1 mm^6 m^{-3}, defined by

$$\zeta = 10 \log_{10}\left(\frac{Z}{Z_1}\right),$$

where Z is the **reflectivity factor** (mm^6 m^{-3}), Z_1 is 1 mm^6 m^{-3}, and ζ is the reflectivity factor in dBZ.
See **decibel**.

DCAPE—Abbreviation for **downdraft convective available potential energy**.

DCS—Abbreviation for the **data collection system** for the **GOES** satellites.

DCVZ—Abbreviation for **Denver convergence–vorticity zone**.

DDA—Abbreviation for **depth–duration–area value**.

de-briefing—In aviation terminology, the relating of factual information by a flight crew at the termination of a flight.
De-briefing information often consists of flight weather encountered, or it may deal with the condition of the aircraft, or with facilities along the airways or at the airports. *See* **pilot briefing**.

de-icing—The removal of **ice** deposited on any object, especially applied to **aircraft icing**.
Principal methods of de-icing in use today are heating, chemical treatment, and mechanical rupture of the ice deposit. *Compare* **anti-icing**.

De-Saint Venant equation—The **equation of motion** (conservation of momentum) for one-dimensional, **unsteady flow** in open channels.
The system of equations representing the **conservation of momentum** and **mass** is given by

$$\frac{\partial Q}{\partial t} + gA\,\frac{\partial y}{\partial x} + \frac{1}{A}\,\frac{\partial Q^2}{\partial x} = gA(S_0 - S_f)$$

$$\frac{\partial A}{\partial t} + \frac{\partial Q}{\partial x} - q = 0,$$

where Q is the **discharge**, A is the **cross section** area, g is the **acceleration of gravity**, y is the height of the water surface above the lowest point in the cross section, x is the distance along the channel, t is time, q is the **lateral inflow** per unit length of the channel, S_f is the **friction** slope, and S_0 is the bottom slope of the channel (positive with decline **downstream**).

Deacon wind profile parameter—A **parameter** depending on **static stability** that describes the **variation** of the **wind speed** with height.
The Deacon parameter β is defined by $du/dz = Cz^{-\beta}$, where u is the wind speed, z the height, and C a constant.

dead glacier—A **glacier** that has ceased moving due to excessive **ablation** or diminished **accumulation**; usually covered by **moraine**.

dead water—Stagnant water often associated with an **oxygen** deficit.

deaister—Same as **doister**.

dealiasing—The process of resolving ambiguous estimates of parameters of which the true magnitudes have been affected by **aliasing** or **folding**.

Deardorff velocity—A **velocity** scale w^* for the **convective mixed layer**:

$$w^* = \left[\frac{g}{T_v} z_i \overline{w'\theta_v'}_s\right]^{1/3},$$

where g is gravitational **acceleration**, T_v is **absolute** temperature, z_i is average depth of the **mixed layer**, $\overline{w'\theta_v'}_s$ is the kinematic vertical **turbulent flux** of **virtual potential temperature** near the surface.

The **velocity** scale is typically on the order of 1 m s^{-1}, which is roughly the **updraft** speed in convective thermals. This **scale** is often used in **similarity theories** for the convective mixed layer and was previously known as the **convective velocity scale**.

> Stull, R. B., 1988: *An Introduction to Boundary Layer Meteorology*, 666 pp.

deasil—An old English term meaning "with the sun," **cum sole**.
Compare **withershins**.

debacle—Breakup of river ice, usually occurring in the springtime in the Northern Hemisphere.
See also **breakup**.

decay—In ocean wave studies, the loss of **energy** from wind-generated **ocean waves** after they have ceased to be acted on by the **wind**.
This process is accompanied by an increase in **wave length** and a decrease in **wave height**.

decay area—The area into which **ocean waves** travel (as **swell**) after leaving the **generating area**.

decay distance—The distance traversed by **ocean waves** after leaving the wave **generating area**.

decay of waves—*See* **wave dissipation**.

decaying mode—*See* **neutral mode**.

decibar—A unit of **pressure** used principally in **oceanography**.
One decibar (10^5 dynes cm^{-2}) equals 0.1 **bar**. In the oceans, **hydrostatic pressure** in decibars very nearly equals the corresponding depth in meters.

decibel—(Abbreviated dB.) A logarithmic measure of the relative **power**, or of the relative values of two **flux densities**, especially of **sound** intensities and radio and radar **power densities**.
The difference n in decibels between intensities I_2 and I_1 is given by the relation

$$n = 10 \log_{10}\left(\frac{I_2}{I_1}\right).$$

Although the decibel is a measure of relative rather than **absolute** intensity, it is possible to set up an absolute scale by arbitrarily defining some particular **intensity** or power level as a reference. For **radar**, the convention is to measure the **received power** in units of milliwatts and to define 1 mW as the reference value for a decibel scale. The units are then called dBm, or decibels relative to a milliwatt. The **radar reflectivity factor**, Z, is measured logarithmically on a **scale** of dBZ, on which the reference value is $Z = 1$ mm^6 m^{-3}. In acoustic practice, an intensity of 10^{-10} μW cm^{-2} is taken as the reference intensity. This value corresponds closely to the minimal **threshold of audibility** of the human ear. The decibel is derived from the less frequently used unit the **bel**, named in honor of Alexander Graham Bell (1847–1922). Two flux densities differ by 1 bel (10 dB) when the larger is 10 times greater than the smaller. It is to be noted that the logarithmic nature of the response of sensory organs, described in the **Weber–Fechner law**, underlies the definition of the bel. *Compare* **neper**.

decile—One of a set of numbers on the random-variable axis that divides a **probability distribution** into ten equal areas.
See **quantile**.

decision tree—A graphical representation of all possible outcomes and the paths by which they may be reached; often used in classification tasks.
The top layer consists of **input** nodes (e.g., meteorological observations and data). Decision nodes determine the order of progression through the graph. The leaves of the tree are all possible outcomes or classifications, while the root is the final outcome (for example, a weather prediction or **climate** classification). Nearly all **expert systems**, and many meteorological algorithms, can most appropriately be diagrammed as a decision tree. The root of the tree represents the first test, while the leaves (nodes that do not lead to further nodes) represent the set of possible conclusions or classifications. *See also* **backward chaining**, **breadth-first search**, **depth-first search**, **forward chaining**.

declination—1. (Also called variation.) In **terrestrial magnetism**, at any given location, the angle between the geographic **meridian** and the magnetic meridian; that is, the angle between **true north** and **magnetic north**.

Declination is either "east" or "west" according as the **compass** needle points to the east or west of the geographic meridian. Lines of constant declination are called **isogonic lines** and that of zero declination is called the **agonic line**.

2. In astronomy, the angular distance between any given celestial body and the **celestial equator**, measured along a **great circle** (hour circle) passing through the **celestial poles**; thus, the astronomical analog of geographic latitude.

Declination is positive for positions north of the **equator** and negative for positions south of the equator. Declination and **right ascension** are the coordinates used in positional astronomy.

deconvolution—Any of several kinds of analyses that removes or attempts to remove the effects of **convolution** from measured data.

In **radar**, deconvolution processing may be performed on **reflectivity** data to remove the **smoothing** that arises from averaging over the **pulse volume**, on Doppler spectra to remove spreading by **turbulence**, or to improve **range resolution** by compensating for the effect of the transmitted **pulse length**. All such applications attempt to restore **resolution** that is lost in the process of measurement.

decorrelation time—*See* **coherence time**.

decoupling—A process where one layer of the **atmosphere** stops interacting with an adjacent layer.

An example is a stratocumulus-topped **turbulent boundary layer** during the night, where **infrared** radiative cooling of **cloud top** causes cold "thermals" to sink toward the ground, causing strong turbulent coupling between the **cloud** and subcloud layers. During the day, these two layers can become decoupled as the combination of solar heating and infrared cooling in the **cloud layer** combine to make the cloud layer warmer than the **subcloud layer**, with a weak stable layer in between that reduces or prevents turbulent coupling of the two layers. These turbulently decoupled layers might still interact (i.e., be slightly coupled) in other ways, such as via **radiation** or **drizzle** fallout.

deduction—Reasoning based on general truths or certainties from which inferences can be drawn for particular situations.

It may take the form of the syllogism IF A and IF B THEN C, or of **IF–THEN rules** such as those used in **expert systems**. *Compare* **induction**.

deep easterlies—Same as **equatorial easterlies**.

deep percolation—Soil water that infiltrates below the root zone toward the **water table**.

deep seepage—Same as **deep percolation**.

deep sound channel—A region in the deep ocean in which the **speed of sound** decreases to a minimum value with depth, and then increases in value as a result of **pressure**.

The **sound waves** are focused by **refraction** by the waters above and below, and can travel thousands of kilometers in this zone. *See* **sound channel**.

deep trades—Same as **equatorial easterlies**.

deep water—A **water mass** found at great depth, the most important globally being **North Atlantic Deep Water**.

See also **Indian Deep Water, Pacific Deep Water, Japan Sea Deep Water, Mediterranean Deep Water, Greenland Sea Deep Water, Norwegian Sea Deep Water**.

deep-water wave—(Also called **short wave, Stokesian wave**.) A **surface wave** the length of which is less than twice the depth of the water.

When this relationship exists the following approximation is valid:

$$c = \left(\frac{gL}{2\pi}\right)^{1/2},$$

where c is the **wave velocity**, g is the **acceleration of gravity**, and L is the **wave length**. Thus, the **velocity** of deep-water waves is independent of the depth of the water. *See* **shallow-water wave**.

deep well—A well in which water is drawn from depths exceeding 22 feet, the depth beyond which ordinary suction pumps do not operate satisfactorily.

deepening—A decrease in the **central pressure** of a **pressure system** as depicted on a **constant-**

height chart, or an analogous decrease in height on a **constant-pressure chart**; the opposite of **filling**.

The term is usually applied to low pressure rather than to high pressure, although technically it is acceptable in either sense. The deepening of a **low** is commonly accompanied by the **intensification** of its **cyclonic circulation**, and the term is frequently used to imply the process of **cyclogenesis**. Deepening can be quantitatively expressed in at least two ways: either 1) as the time rate of central-pressure decreases; or 2) as that component of the **pressure tendency** at any fixed point that is attributable neither to the motion of the pressure system relative to that point nor to the **diurnal** influence of atmospheric tides. *Compare* **cyclogenesis**.

deepening of a depression—A British term for decrease with time of the **central pressure** of a **depression**.

Defense Meteorological Satellite Program—(Abbreviated DMSP.) A series of polar orbiting sun-synchronous satellites operated by the U.S. Department of Defense.

The first in the series (DMSP Block 4) was launched in September 1966. The current DMSP **instrument payload** includes the **OLS**, **SSM/I**, and **SSM/T**.

Definitive Geomagnetic Reference Field—(Abbreviated DGRF.) A mathematical **model** of the geomagnetic field for a time in the past.

It is to be distinguished from the **International Geomagnetic Reference Field** (IGRF), which is for the present time.

deflation—In geology, the removal of loose soil and other surface material by the **wind**, leaving the rocks bare to the continuous attack of the weather.

Deflation is usually the factor responsible for the frequently stony cover of **desert** surfaces, sometimes called "desert pavement."

deflecting force—In meteorology, same as **Coriolis force**.

deflection-modulated indicator—Same as **amplitude-modulated indicator**.

deformation—The change in shape of a fluid mass by spatial variations in the **velocity** field, specifically by stretching or shearing.

A **linear** analysis of the two-dimensional velocity field can express this **field** in terms of **divergence**, **vorticity**, and deformation (or, more strictly, the rate of deformation)

$$u = u_0 + b_0 x - c_0 y + a_0 x + a_0' y + \cdots$$
$$v = v_0 + b_0 y + c_0 x - a_0 y + a_0' x + \cdots,$$

where the subscript "0" refers to a selected fixed origin, and where

$$2b_0 = \left(\frac{\partial u}{\partial x} + \frac{\partial v}{\partial y} \right)_0 = \text{divergence},$$

$$2c_0 = \left(\frac{\partial v}{\partial x} - \frac{\partial u}{\partial y} \right)_0 = \text{vorticity},$$

$$2a_0 = \left(\frac{\partial u}{\partial x} - \frac{\partial v}{\partial y} \right)_0 = \text{stretching deformation, and}$$

$$2a_0' = \left(\frac{\partial v}{\partial x} + \frac{\partial u}{\partial y} \right)_0 = \text{shearing deformation.}$$

A pure stretching deformation field ($u = a_0 x$, $v = -a_0 y$) is characterized by rectangular hyperbolic streamlines. It has two characteristic axes: an **axis of dilatation** toward which streamlines converge asymptotically, and an **axis of contraction** from which streamlines diverge asymptotically. A pure shearing deformation field ($u = a'_0 y$, $v = a'_0 x$) is also characterized by rectangular hyperbolic streamlines, but the characteristic axes are rotated 45° from the coordinate axes. The resultant deformation is a combination of the stretching and shearing deformation fields by the appropriate choice of coordinates, which define the principal characteristic axes. Deformation is a primary factor in the processes of **frontogenesis** and **frontolysis**.

Holton, J. R., 1992: *An Introduction to Dynamic Meteorology*, 3d edition, Academic Press, 266–277.

deformation axis—Usually, the major axis along which **deformation** (stretching, compression) occurs.

deformation field—*See* **deformation**.

deformation radius—*See* **Rossby radius of deformation**.

deformation thermometer—A **thermometer** using transducing elements that deform with **temperature**.

Examples of deformation thermometers are the **bimetallic thermometer** and the **Bourdon tube** type of thermometer.

deformation zone—A region of the **atmosphere** where the **stretching** or **shearing deformation** is large.

See **deformation**.

deformation zone cloud system—Clouds within a **deformation zone** in the **wind field**.

These clouds undergo elongation in one direction and contraction in the perpendicular direction. Often clouds evaporate on one side of the elongation axis producing a smooth **cloud** boundary.

deformed ice—A general term for **sea ice** that has been broken and reoriented.

It includes ridged, hummocked, and **rafted ice**. Deformed ice is distinguished by its high surface roughness.

degenerate amphidrome—A terrestrial point on a tidal **chart** from which **cotidal lines** appear to radiate.

An imaginary point where nothing happens.

deglaciation—The removal of **land ice** from an area; the opposite of **glacierization**.

degradation—The lessening of a quality of data or images because of any optical, electronic, or mechanical distortions (**noise**) in the data collection or **image** forming systems.

degree—1. A unit of **temperature**.

See **absolute temperature scale**, **Celsius temperature scale**, **centigrade temperature scale**, **Fahrenheit temperature scale**, **Kelvin temperature scale**, **Reaumur temperature scale**.

2. A unit of angular distance; 1/360 part of a circle.

See also **radian**.

degree-day—1. Generally, a measure of the **departure** of the **mean daily temperature** from a given **standard**: one degree-day for each **degree** (°C or °F) of departure above (or below) the standard during one day.

Degree-days are accumulated over a "season" at any point during which the total can be used as an **index** of past **temperature** effect upon some quantity, such as plant growth, fuel consumption, power output, etc. This concept was first used in connection with plant growth, which showed a relationship to **cumulative temperature** above a standard of 5°C (41°F). Recently, degree-days have been more frequently applied to fuel and power consumption, for example, **heating degree-day**, **cooling degree-day**. In the life sciences, the standard is often referred to as the base temperature or upper threshold, depending on whether the standard is used as a lower or upper limit, respectively. As used by the U.S. Army Corps of Engineers, Fahrenheit degree-days are computed as departures above and below 32°F, positive if above and negative if below. To avoid confusion, it might be well to call this a "freezing degree-day." The advantages and disadvantages of the latter concept are discussed well by Sakari Tuhkanen (1980). *Compare* **degree-hour**.

2. Common contraction for **heating degree-day**.

Tuhkanen, S., 1980: Climatic parameters and indices in plant geography. *Acta phytogeographica Suecica*, **67**, Svenska Vaxgeografiska Sallekapet, Uppsala, 105 pp.

degree-hour—As used by U.S. Army Corps of Engineers, the **departure** (in °F) of the hourly **temperature** from a **standard** of 32°F, positive if above and negative if below.

Degree-hours may be accumulated (summed) over any period of time, depending upon the use to which they are applied. *Compare* **degree-day**.

degree of polarization—1. The ratio of polarized to total **radiance** in some direction.

The degree of polarization ranges from 0 percent for unpolarized **radiation**, typical of **direct solar radiation**, **diffuse solar radiation** within clouds, and all terrestrial **infrared radiation**, to 100% for completely polarized radiation. The degree of polarization of **light** from a **Rayleigh atmosphere** with minimal **turbidity** is close to 100% at a **scattering** angle of 90°.

2. *See* **polarization**.

degree of saturation—The proportion of the pores in a soil filled with water. It is equal to the volume of water divided by the volume of the pores and is often expressed as a percent.

degrees of freedom—In an unconstrained dynamic or other system, the number of **independent variables** required to specify completely the state of the system at a given moment.

If the system has constraints, that is, kinematic or geometric relations between the variables, each such relation reduces by one the number of degrees of freedom of the system. In a continuous medium with given **boundary conditions**, the number of degrees of freedom is the number of **normal modes** of **oscillation**. Thus, a **particle** moving in space has three degrees of freedom; an **incompressible fluid** with a **free surface** has an infinite number of degrees of freedom.

degrees of frost—In England, the number of Fahrenheit degrees that the **temperature** falls below the **freezing point**.

Thus a day with a **minimum temperature** of 27°F may be designated as a day of five degrees of frost. *Compare* **degree-day**.

deicer—A mechanical, electrical, or chemical means for removing **ice**, commonly from an aircraft.

Mechanical deicers on aircraft deform the surface to remove accumulated ice by fracturing it. Electrical deicers remove ice by melting the bond between accumulated ice and the underlying surface. Chemical deicers lower the **freezing** temperature to prevent or remove ice from surfaces where they are applied, as on parked aircraft or runways.

del operator—The **operator** (written ∇) used to transform a **scalar** field into the **ascendent** (the negative of the **gradient**) of that **field**.

In **Cartesian coordinates** the three-dimensional del operator is

$$\nabla = \mathbf{i}\,\frac{\partial}{\partial x} + \mathbf{j}\,\frac{\partial}{\partial y} + \mathbf{k}\,\frac{\partial}{\partial z},$$

and the horizontal component is

$$\nabla_H = \mathbf{i}\,\frac{\partial}{\partial x} + \mathbf{j}\,\frac{\partial}{\partial y}.$$

Expressions for ∇ in various systems of **curvilinear coordinates** may be found in any textbook of **vector** analysis. In meteorology it is often convenient to use a **thermodynamic function of state**, such as **pressure** or **potential temperature**, as the vertical coordinate. If σ be this **parameter**, then

$$\nabla = \mathbf{i}\left(\frac{\partial}{\partial x}\right)_\sigma + \mathbf{j}\left(\frac{\partial}{\partial y}\right)_\sigma + \mathbf{k}\,\frac{\partial\sigma}{\partial z}\,\frac{\partial}{\partial\sigma},$$

where differentiation with respect to x and y is understood as carried out in surfaces of constant σ (the subscript usually being omitted). The horizontal component is now

$$\nabla_\sigma = \mathbf{i}\,\frac{\partial}{\partial x} + \mathbf{j}\,\frac{\partial}{\partial y}.$$

If the **quasi-hydrostatic approximation** is justified, as in most meteorological contexts, pressure is a useful coordinate, and

$$\nabla = \nabla_p - \mathbf{k}g\rho\,\frac{\partial}{\partial p},$$

where g is the **acceleration of gravity** and ρ the **density**. Here

$$\nabla_p = \mathbf{i}\,\frac{\partial}{\partial x} + \mathbf{j}\,\frac{\partial}{\partial y},$$

with differentiation carried out in **isobaric surfaces**.

delay—Applied to any time-varying physical quantity, usually periodic, its displacement in time relative to a similar reference quantity.

For example, the **temperature** in soil depends on depth and time. If the **surface temperature** varies sinusoidally, so does the temperature at any depth, but it is out of **phase** (is delayed) relative to the surface temperature. There is a **phase shift** or **phase delay** or phase lag of **soil temperature** relative to surface temperature. Although these three terms are most often applied to waves, nothing inherent in the concept of a delay restricts it to waves. *Compare* **lag**.

deliquescence—The process of liquefaction of a solid by dissolution in solution resulting from absorbed **water vapor**.

For atmospheric processes, the uptake of water by substances and dissolution at **relative humidity** generally present under atmospheric conditions. For example, LiCl (used to measure relative humidity by **conduction current** in older **radiosondes**) deliquesces at relative humidity in excess of 15%. The term **hygroscopic** refers to a similar process, although taking place at a higher relative humidity and a slower rate. For example, bulk sodium chloride is hygroscopic and dissolves slowly in relative humidity in excess of 76%.

Dellinger effect—Same as **fadeout**.

delta Eddington—An approximate technique for handling the complexity of shortwave **radiative transfer** in anisotropically **scattering** atmospheres.
    The approximation is fairly accurate for transmitted and reflected **irradiances** provided the **optical depth** of the medium is greater than ≈ 3 and the **absorption** is weak. It is used mainly to speed up **shortwave radiation** computations in **cloudy** atmospheres.

delta region—1. A region of **diffluence** in the **atmosphere**.
    2. A region aloft beneath which low-level **cyclogenesis** is likely to occur.
    Compare **exit region**.

delta rule—Method of training a **neural network** using **backpropagation**.
    The **error** in the **output** of the neural network (e.g., a **prediction** of minimum nighttime **temperature**) is a function of the values of the **input** parameters (**cloud cover**, 6 P.M. temperature, etc.) and the weights assigned to them. The weights are adjusted to minimize the error. Application of the delta rule results in the most rapid error reduction (learning rate). The learning rate can be adjusted as necessary to avoid being trapped in a local error minimum.

dendrite—Same as **dendritic crystal**.

dendritic crystal—(Or dendrite.) A **crystal**, particularly a planar **ice crystal**, with its macroscopic form (**crystal habit**) characterized by intricate branching structures of a treelike nature.
    Dendritic ice crystals possess hexagonal symmetry, and tend to develop when a crystal grows by **vapor** deposition at temperatures within a few degrees of $-15°C$, providing **saturation** is close to **supercooled water**. Similar forms occur by **ice** growth into supercooled liquid water at temperatures down to $-10°C$. Spatial dendrites grow in three dimensions from a central frozen **drop**.

dendrochore—A treed region.
    See **biochore**.

dendrochronology—The analysis of the annual growth rings of trees, leading to the calculation of significant indices of **climate** and general chronology of the past.
    The width of a **tree ring** was determined by the **temperature** and/or the moisture that prevailed during the year of its formation. Since stress from temperature and/or moisture variations reduces the width of the seasonal growth of a tree ring, dendrochronology has important application in the study of long-term climatic variations.

dendroclimatology—*See* **dendrochronology**.

dendrohydrology—The inference of the long-term **statistics** of various hydrological phenomena in a region using **tree rings**.

denitrification—1. The conversion of nitrite or **nitrate** to gaseous end products NO, N_2O, and N_2 by denitrifying bacteria.
    Molecular **nitrogen**, N_2, is the most abundant end product. The oxidized forms of N (nitrate, nitrite, **nitric acid**, **nitric oxide**, and **nitrous oxide**) serve as alternative **electron** acceptors in the absence of sufficient **oxygen**; thus, denitrification is primarily an anaerobic process.
    2. The removal of **active nitrogen** from the **atmosphere** through uptake into **particles**.
    The term is particularly used in context with the irreversible uptake of **nitric acid** into ice clouds (**polar stratospheric clouds**) in the **antarctic stratospheric vortex**. This reduction in the level of active nitrogen allows **ozone** depletion by **chlorine oxides** to occur unchecked.

densification on mixing—*See* **cabbeling**.

density—1. The ratio of the mass of any substance to the volume occupied by it (usually expressed in kilograms per cubic meter, but any other unit system may be used); the reciprocal of **specific volume**.
    In a continuous medium the density is defined by a limiting process and is a point function.
    2. The ratio of any quantity to the volume or area it occupies; for example, **flux density**, **power density, ion density, electron density, drainage density**.

density altitude—The **pressure altitude** corrected for **temperature** deviations from the **standard atmosphere**.

Density altitude bears the same relation to pressure altitude as **true altitude** does to **indicated altitude**.

density channel—A channel used to investigate a **density current**, for example, in experiments relating to the behavior of cold masses of air in the **atmosphere** and related frontal structures.

density correction—1. That part of the **temperature correction** of a **mercury barometer** that is necessitated by the **variation** of the **density** of **mercury** with **temperature**.

2. The **correction**, applied to the indications of a pressure-tube **anemometer** or pressure-plate anemometer, that is necessitated by the **variation** of air **density** with **temperature**.

density current—The intrusion of a denser fluid beneath a lighter fluid, due mainly to the hydrostatic forces arising from **gravity** and the **density** differences.

This term is used principally in engineering for such cases as the intrusion of saltwater below **freshwater** in an **estuary**, or for currents caused by the presence of denser water with suspended **silt** at the bottom of a lake or ocean. Many of the phenomena are quite analogous to some of those associated with cold fronts in the **atmosphere**.

density function—In **statistical** terminology, same as **probability density function**.

density of snow—The ratio of the volume of melted **snow** to the volume of the original, unmelted snow sample.

density temperature—Same as **virtual temperature**.

densus—*See* **cirrus spissatus**.

denudation—The **erosion** of the earth's surface due to **rain**, **frost**, **wind**, running water, or other actions, and the consequent uncovering of the surface that was otherwise covered.

Denver convergence–vorticity zone—(Abbreviated DCVZ.) A **mesoscale** flow feature of convergent winds, 50 to 100 km in length, usually oriented north–south, just east of the Denver, Colorado area.

The cause of the feature is an interaction of southerly low-level flow with an east–west ridge known as the Palmer Divide extending onto the eastern Colorado plains to the south of Denver. In addition to the convergent **wind field**, which has been shown to help initiate thunderstorms in the convective **season**, the flow within the convergent zone sometimes develops smaller-scale vortices, hence the name "convergence–vorticity zone."

Denver cyclone—Sometimes used synonymously with **Denver convergence–vorticity zone** or DCVZ.

More specifically, it is a subset of the DCVZ in which the flow takes on a more **cyclonic** appearance, rather than an elongated zone of convergent winds.

departure—Same as **deviation**.

depegram—(Rare). On a **tephigram**, a curve representing the distribution of the **dewpoint** for a given **sounding** of the **atmosphere**.
See **thermodynamic diagram**.

dependence—*See* **statistical independence**.

dependent meteorological office—(Abbreviated DMO.) An office that provides **meteorological service** for international air navigation in accordance with International Civil Aviation Organization specifications.

Its functions are to 1) prepare forecasts under the guidance of a **main meteorological office**; 2) supply **meteorological information** and **briefings** to aeronautical personnel; 3) supply meteorological information required by an associated **supplementary meteorological office**. *See also* **meteorological watch office**.

dependent variable—Any **variable** considered as a function of other variables, the latter being called independent.

Whether a given quantity is best treated as a dependent or **independent variable** depends upon the particular problem. *Compare* **parameter**.

depéq—Strong winds over Loet Tawar (Sumatra, Indonesia) during the **southwest monsoon**.

depergelation—The act or process of thawing **permafrost**.

depletion curve—That portion of a **hydrograph** in which **streamflow** comes primarily from **groundwater** seepage, resulting in depletion of the groundwater (or lake water, swamp water) **reservoir**.

See **recession curve**.

depolarization—The process by which a polarized **signal**, for example, a **radar** signal, loses its original **polarization** as the result of **scattering** or of propagation through an **anisotropic** medium.

The signal may experience a change of polarization, for example, from circular to elliptical as a result of **differential attenuation** or **differential phase shift**, or it may experience a loss of polarization, that is, become unpolarized, as a result of scattering in a **random** medium.

depolarization ratio—General term denoting the ratio of the **cross-polarized** to the **copolarized signal** components measured by a **polarimetric radar**.

See **circular depolarization ratio, linear depolarization ratio, elliptical depolarization ratio**.

deposit gauge—The general name for instruments used in **air pollution** studies for determining the amount of material deposited on a given area during a given time.

deposition—Processes by which traces gases or **particles** are transferred from the **atmosphere** to the surface of the earth.

Atmospheric deposition is usually divided into two categories, **wet deposition** and **dry deposition**, depending on the **phase** of the material during the deposition process. Thus, in wet deposition, the gas or particle is first incorporated into a **droplet** and is then transferred to the surface via **precipitation**. In dry deposition, the gas or particle is transported to ground level, where it is adsorbed onto a surface. The surface can be the ocean, soil, vegetation, buildings, etc. Note that the surface involved in the dry deposition may be wet or dry—the "dry" in dry deposition refers only to the phase of the material being deposited.

deposition nucleus—A solid **aerosol** particle that nucleates an **ice crystal** directly from the **vapor**, particularly (but not necessarily) at subwater **saturation** and in the absence of the bulk liquid **phase**.

The process does not preclude an intermediate adsorbed layer (**Langmuir layer**) prior to **nucleation**.

deposition velocity—In **dry deposition**, the quotient of the **flux** of a particular species to the surface (in units of concentration per unit area per unit time) and the concentration of the species at a specified reference height, typically 1 m.

Typical deposition velocities for common gas **phase** pollutants (e.g., **ozone, nitric acid**) are of order 0.01–5 cm s^{-1}.

depression—1. In general, a point or limited area of locally lower **elevation** in a particular surface.

2. In meteorology, an area of low pressure; a **low** or **trough**.

This is usually applied to a certain stage in the **development** of a **tropical cyclone**, to **migratory** lows and troughs, and to upper-level lows and troughs that are only weakly developed. This use of the term is most common in the European literature. *See* **V-shaped depression**.

3. Same as **depression angle**.

4. *See* **wet-bulb depression, dewpoint spread**.

depression angle—(Also called depression.) The angle between the **horizon** and a point below, measured along the arc that passes through the point in question and is perpendicular to the horizon.

The depression angle is the **zenith distance** of the point in question minus 90°. *Compare* **azimuth, elevation angle, zenith distance**.

depression of the dewpoint—*See* **dewpoint depression**.

depression storage—Water temporarily retained in puddles, ditches, and other depressions in the surface of the ground, and eventually evaporated or infiltrated; the small-scale counterpart of **closed drainage**.

depth–area curve—A curve showing the relation between an averaged areal **rainfall** depth and the area over which it occurs, for a specified time interval, during a specific rainfall event.

depth–area–duration analysis—Compilation of **depth–area curves** for different time intervals on a single graph.

depth–area–duration curves—**Depth–area curves** for different time intervals.

depth–duration–area value—(Commonly abbreviated DDA value.) The average depth of **precipitation** that has occurred within a specified time interval over an area of given size.

Usually, for any given **storm** or period of study, the assigned DDA values are those that represent the highest average depth for each selected **duration** and areal size.

depth–duration curve—A curve giving the relation between averaged areal **rainfall** depth and time interval for specific events.

depth–duration–frequency curve—Curves showing the time interval **frequency** for storms of specific areal **rainfall**.

depth-first search—A search **algorithm** that extends the current path as far as possible before **back-tracking** to the last choice point and trying the next alternative path.

Depth-first search generally reaches a satisfactory solution more rapidly than breadth first, an advantage when the **search space** is large. However, unlike breadth first, it does not guarantee that the optimal solution has been found. *Compare* **breadth-first search**.

depth hoar—1. **Ice crystals** (usually cup-shaped, faceted crystals) of low strength formed by **sub-limation** within **dry snow** beneath the **snow** surface; a type of **hoarfrost**.

Associated with very fast **crystal** growth under large **temperature** gradients. This is one way in which **firn** formation may begin. Depth hoar is similar in physical origin to **crevasse hoar**.

2. **Hoarfrost** composed of **crystals** that have built up a three-dimensional complex of faceted, rather than rounded, crystals.

depth-integration sediment sampling—A methodology that obtains a representative discharge-weighted water–sediment **sample** over **stream** verticals, except in an unmeasured zone near the streambed, by continuously cumulatively collecting a portion of the water–sediment mixture as the sampler traverses the vertical at an approximately constant transit rate.

depth marker—A thin board or other lightweight substance used as a means of identifying the surface of **snow** or **ice** that has been covered by a more recent **snowfall**.

depth of compensation—The depth in a body of water at which **illuminance** has diminished to the extent that **oxygen** production through **photosynthesis** and oxygen consumption through respiration by plants are equal.

It is the lower boundary of the **euphotic zone**. The illuminance at this depth is known as (or is said to have reached) the compensation point.

depth of frictional influence—The depth below the ocean surface to which the **stress** (horizontal force/unit area) of the **wind** reaches.

To measure this depth, the turbulent stress must be directly measured. This is rarely done, but is more often inferred through observations of current profiles and fitting to an **Ekman layer** profile [e.g., Eqs. (9.9) and (9.10) in Pond and Pickard (1978)]. Because of the strong effect **stratification** has on **turbulent transport**, the depth of frictional influence is also taken to be equal to the depth of the **mixed layer**, and this is in turn found by observations of **temperature** and/or **salinity** profiles.

Pond, S., and G. L. Pickard, 1978: *Introductory Dynamic Oceanography*, Pergamon Press, p. 87.

depth of runoff—**Runoff** volume per unit area (for a given time or per unit time); may be expressed as a fraction of **precipitation** volume per unit area.

depth of snow—Vertical distance between the top of a **snow** layer and the horizontal ground beneath.

The layer is assumed to be evenly spread on the surface. When the snow is not uniformly distributed, **snow depth** is measured by taking an average of multiple measurements.

depth–velocity integration method—A method for computing the **discharge** in a channel by first determining the areas of the depth–velocity curves in each vertical cross section, then plotting the area of the curve of those areas over the verticals along the water surface line.

derecho—A widespread convectively induced **straight-line windstorm**.

Specifically, the term is defined as any family of **downburst** clusters produced by an extratropical **mesoscale convective system**. Derechos may or may not be accompanied by **tornadoes**. Such events were first recognized in the Corn Belt region of the United States, but have since been observed in many other areas of the midlatitudes.

Johns, R. H., and W. O. Hirt, 1987: Derechos: Widespread convectively induced windstorms. *Wea. Forecasting,* **2**, 32–49.

DERF—Abbreviation for **dynamical extended range forecasting**.

derived gust velocity—The maximum velocity of a **sharp-edged gust** that would produce a given **acceleration** on a particular airplane flown in level flight at the design cruising speed of the aircraft and at a given air density.

The ratio of derived gust velocity to **effective gust velocity** is not constant but is of the order of 2:1.

Descartes ray—In a theory of the **rainbow** that approximates the behavior of **light** as a series of rays, the Descartes ray is the one that undergoes a minimum angle of deviation as a result of being refracted as it enters the **drop**, reflected one or more times within the drop, and then refracted again as it leaves.

The significance of the Descartes ray (in this **model**) is that the **radiance** is markedly greater than that of neighboring rays. The Descartes ray is thus useful for providing an approximate rainbow position and color order.

descendent—(Rare.) Same as **gradient**.

descending node—The southbound equatorial crossing of a **polar-orbiting satellite**, given in degrees longitude, date, and **universal time** (UTC) for any given **orbit** or pass; the point at which a satellite crosses the equatorial plane heading south.

descriptive meteorology—(Also called aerography.) A branch of meteorology that deals with the description of the **atmosphere** as a whole and its various phenomena, without going into theory.

desert—A region where **precipitation** is insufficient to support any except xerophilous (**drought** resistant) vegetation; a region of extreme aridity.
See **desert climate, trade-wind desert, equatorial dry zone**.

desert climate—(Also called arid climate.) A **climate** type that is characterized by insufficient moisture to support appreciable plant life—that is, a climate of extreme aridity.

desert devil—See **dust whirl**.

desert wind—A **wind** blowing from the **desert**.
It is very dry and usually dusty, very hot in summer but cold in winter, with a large **diurnal** range of **temperature**. Well-known examples are the **harmattan, khamsin**, and **simoom**. See also **brickfielder**.

desertification—The transformation of the **climate** of a region toward enhanced **aridity**.
Desertification can result from a decrease in **precipitation**, as well as land surface changes such as deforestation or overgrazing.

desiccation—1. In general, the process of drying up.

2. In **climatology**, a prolonged decrease or disappearance of water from a region.
This may be due to 1) a decrease of **rainfall**; 2) a failure to maintain **irrigation**; or 3) deforestation or overcropping. *Compare* **exsiccation**.

design flood—The **flood**, observed or synthetic, that is chosen as the basis for the design of a hydraulic structure.

design storm—**Rainfall** amount and distribution in space and time, used to determine a **design flood** or design **peak discharge**.

design wave—A **wave** that is chosen as a basis for the design of a structure, for example, an offshore platform or **breakwater**.
A design wave is described in terms of a **wave height, wave period**, and direction. See **extreme value distribution, extreme wave height**.

designated pollutant—1. An **air pollutant** identified as potentially harmful, deserving monitoring and careful study.

2. An **air pollutant** that is controlled by government regulation.
In the United States, there are two primary classes of designated pollutants regulated by the Environmental Protection Agency: **criteria pollutants** and **toxic pollutants**. For the 188 "toxic" air pollutants that are known or suspected to cause cancer or other serious health effects (such as birth or developmental defects), the regulations control emissions from sources such as industrial factories. See also **air toxins**.

desulfurization—Industrial process in which sulfur-containing gases are released and scrubbed from fossil fuel or from exhaust gases.
While the removal of sulfur from fuels is difficult and expensive, the SO_2 formed in combustion can often be removed easily by washing (scrubbing) with an alkaline liquid.

detection—1. The identification, at a stated level of **probability**, of the presence of a **signal** or phenomenon with certain characteristics.
Less formally, a signal may be said to be detectable if it is observable under ordinary circumstances.

2. In radio, **radar**, and **lidar**, the conversion of a **radio frequency** or **light** signal to an **IF signal**, or an IF signal to a **video signal**, accomplished in a detector or mixer.

detention storage—Water retained during a **flood** to be gradually released later at a slower rate.

deterministic—Governed by and predictable in terms of definite laws, such as dynamic equations. *Compare* **nondeterministic**.

deterministic hydrology—Analysis of hydrological processes using **deterministic** approaches, in which the hydrological parameters are based on physical relations of the various components of the **hydrologic cycle**.
   Deterministic approaches do not consider **randomness**, so that a given **input** always produces the same **output**.

detrainment—The **transfer** of air from an organized **air current** to the surrounding **atmosphere**; the opposite of **entrainment**.

detrend—To remove any background **linear** variation in a **time series** of data by first computing a linear regression (such as least squares best fit) to the data, and then subtracting it from the data.
   This aspect of conditioning data helps to remove **red noise** associated with a long-wavelength or long-period **signal** that is inadequately sampled because it is longer than the total data period of sampling.

deuterium—Isotope of **hydrogen**, having one **proton** and one **neutron** in the **nucleus**; heavy hydrogen.
   Deuterium is a nonradioactive form of hydrogen, occurring naturally with an abundance of about 0.016%. Deuterium is widely used in spectroscopic and kinetic studies, since the large relative mass difference from hydrogen can lead to very different chemical and physical properties of the compounds in which it is contained.

deuteron—**Nucleus** of heavy **hydrogen** (**deuterium**); atomic **particle** consisting of one **proton** and one **neutron**.

development—The process of **intensification** of an **atmospheric disturbance**, most commonly applied to **cyclones** and **anticyclones**.
   From the dynamical viewpoint, development usually implies the generation of **vorticity** in the **atmosphere** due to the action of the **baroclinic** structure of the flow or to the addition of **energy** to the disturbance, as distinct from the **barotropic** redistribution of **vorticity**.

deviation—(Also called departure.) In **statistics**, the difference between two numbers.
   It is commonly applied to the difference of a **variable** from its mean, or to the difference of an observed value from a theoretical value. *See* **standard deviation**, **geostrophic departure**.

devil—Same as **dust whirl**, **dust devil**.

dew—Water condensed onto grass and other objects near the ground, the temperatures of which have fallen below the **dewpoint** of the surface air due to **radiational cooling** during the night, but are still above **freezing**; **hoarfrost** may form if the dewpoint is below freezing (*see* **frost point**).
   If the **temperature** falls below freezing after dew has formed, the frozen dew is known as **white dew**. The conditions favorable to dew formation are 1) a radiating surface, well insulated from the **heat** supply of the soil, on which **vapor** may condense; 2) a **clear**, still **atmosphere** with low **specific humidity** in all but the surface layers, to permit sufficient effective **terrestrial radiation** to cool the surface; and 3) high **relative humidity** in the surface air layers, or an adjacent **source** of moisture such as a lake. Dew plays an important role in the propagation of certain plant pathogens, such as late potato blight, which require dew-covered leaves from certain stages of sporulation. Dew is responsible for the optical effect known as the **heiligenschein**.

dew cell—A type of **hygrometer** used to determine the **dewpoint**.
   The **equilibrium vapor pressure** at the surface of a saturated salt solution is less than that for a similar surface of pure water at the same **temperature**. This effect is exhibited by all salt solutions but particularly so by lithium chloride. The dew cell hygrometer works in the following manner: The lithium chloride solution is heated until a temperature is reached at which its equilibrium vapor pressure exceeds the ambient **vapor pressure**. At this point the balance will shift from **condensation** to **evaporation**, which can be detected by a characteristic decrease of the **conductivity** of the solution. By properly regulating the heating current, the **equilibrium** temperature can be transformed by means of empirical equations to the ambient **water vapor** pressure and the dewpoint. Provision is not usually made for cooling the dew cell, and consequently it will not measure vapor pressure less than the vapor pressure over saturated lithium chloride solution

at the **ambient temperature**. Dew cells are used for observational purposes, especially for **automatic weather stations**.

dew gauge—Instrument for measuring the amount of **dew** deposited per unit area.

dewbow—A **rainbow** formed in the small drops often found on grass in early morning.

While the name implies that those drops are **dew**, as a practical matter, that is probably rarely the case. Rather, the drops are usually the result of **guttation** rather than dew. When seen stretched across a meadow, the dewbow may appear in the shape of a hyperbola, but that is merely the intersection of the cone of **light** that forms the bow and the surface of the meadow. Because they are seen against a surface, dewbows sometimes are perceived as hyperbolas even though, like ordinary rainbows, they are arcs with a constant angular radius.

dewpoint—(Or **dewpoint temperature**.) The **temperature** to which a given **air parcel** must be cooled at constant **pressure** and constant water vapor content in order for **saturation** to occur.

When this temperature is below 0°C, it is sometimes called the **frost point**. The dewpoint may alternatively be defined as the temperature at which the **saturation vapor pressure** of the parcel is equal to the actual **vapor pressure** of the contained **water vapor**. Isobaric heating or cooling of an air parcel does not alter the value of that parcel's dewpoint, as long as no vapor is added or removed. Therefore, the dewpoint is a **conservative property** of air with respect to such processes. However, the dewpoint is nonconservative with respect to vertical **adiabatic** motions of air in the **atmosphere**. The dewpoint of ascending **moist air** decreases at a rate only about one-fifth as great as the **dry-adiabatic lapse rate**. The dewpoint can be measured directly by several kinds of **dewpoint hygrometers** or it can be deduced indirectly from **psychrometers** or devices that measure the water vapor density or **mixing ratio**. *See* **dewpoint formula**.

dewpoint deficit—Same as **dewpoint spread**.

dewpoint depression—(Also called spread, dewpoint spread, dewpoint deficit.) The difference in degrees between the **air temperature** and the **dewpoint**.

dewpoint formula—A formula for the calculation of the approximate height of the **lifting condensation level**.

Therefore, it is employed to estimate the height of the base of **convective clouds**, under suitable atmospheric and topographic conditions. In a practical but simplified form, it may be stated

$$H = (T - T_d)/8,$$

where H is the height above the surface in km and T and T_d are the **temperature** and **dewpoint** in degrees Celsius. This is derived from assuming that the temperature decreases 9.8°C and the dewpoint decreases 1.8°C for each km increase in **altitude**, values which are approximately correct for a well-mixed **boundary layer**. The convective-cloud height diagram is a graphic form of this relationship.

dewpoint front—*See* **dryline**.

dewpoint hygrometer—An instrument for determining the **dewpoint**.

The most widely used systems employ a small polished-metal reflecting surface, cooled and heated electrically using a Peltier-effect device, and a small **temperature** sensor embedded on the underside of the mirror. An electro-optical system is used to detect the formation of condensate and to provide the **input** to the servo-control system to regulate the temperature of the mirror. The specular **reflectance** of the bare mirror decreases with **dew** layer thickness; a preset threshold defines a specific constant dew layer thickness that is maintained by the optical servo-control. The mirror temperature equals the dewpoint.

dewpoint spread—*See* **dewpoint depression**.

dewpoint temperature—Same as **dewpoint**.

DGRF—Abbreviation for **Definitive Geomagnetic Reference Field**.

diabatic process—A thermodynamic **change of state** of a system in which the system exchanges **energy** with its surroundings by virtue of a **temperature** difference between them.
Compare **adiabatic process**.

diagnosis—*See* **analysis**.

diagnostic equation—Any equation governing a system that contains no time derivative and therefore specifies a balance of quantities in space at a moment of time (e.g., **hydrostatic equation**, **balance equation**).
Compare **prognostic equation**.

DIAL thermometer—Abbreviation for **differential absorption lidar thermometer**.

diamond dust—Small **ice crystals** falling from an apparently cloudless sky, (often, but not always, at night).
 Crystals originate from air having a higher **moisture content** above a **thermal inversion** aloft, where **mixing** leads to **nucleation** and growth of crystals at temperatures near −40°C.

diapycnal—In a direction normal to the local **isopycnal surfaces**.
 Since isopycnal surfaces are close to horizontal, the diapycnal direction is close to vertical. The **velocity** of water normal to isopycnal surfaces is called diapycnal velocity, and the fluxes of properties perpendicular to an isopycnal surface are called diapycnal fluxes.

diapycnal mixing—Mixing in a **diapycnal** direction, that is, **mixing** of fluid from one side of an **isopycnal surface** with different (**potential**) **density** fluid from the other side of the surface.

dichroism—Two different imaginary parts of the **refractive index** for **electromagnetic waves** identical except for their states of (**orthogonal**) **polarization**.
 By orthogonal is meant that the waves have opposite handedness, the same ellipticity, and the major axes of their vibration ellipses are perpendicular to each other. The most general dichroism is elliptical, specific examples of which are linear and circular. The dichroism of a medium originates from its asymmetry. *See* **birefringence**.

diel—The daily **cycle**.
 Diel is sometimes used by soil micrometeorologists and atmospheric chemists in place of the word **diurnal**.

dielectric—More or less a synonym for electrical insulator, a material with a low (compared with that of a metal) **electrical conductivity**.
 The term insulator is more precise given that at some frequencies the **relative permittivity** (or **dielectric function**) of a material usually described as an insulator may have values thought to be characteristic of metals, and at other frequencies the relative permittivity of a metal may have values thought to be characteristic of a dielectric (insulator).

dielectric constant—*See* **relative permittivity**.

dielectric factor—The factor $|K|^2$ in the **radar equation**, which depends on the **complex refractive index** of the **target**.
 See **radar reflectivity**.

dielectric function—*See* **relative permittivity**.

dielectric strength—(Sometimes called breakdown potential.) A measure of the **resistance** of a **dielectric** to **electrical breakdown** under the influence of strong electric fields; usually expressed in volts per meter.
 The dielectric strength of **dry air** at **sea level pressures** is about 3 000 000 V m⁻¹. The exact value for air depends upon geometry of the electrodes between which the **electric field** is established, upon the **humidity**, and upon whether or not water drops are present in the air.

differential (scattering) cross section—A specification of the angular distribution of the **electromagnetic energy** scattered by a **particle** or a **scattering** medium.
 The differential (scattering) cross section is defined as the ratio of the **intensity** of **radiant energy** scattered in a given direction to the incident **irradiance** and thus has **dimensions** of area per unit solid angle. The symbol σ is frequently used for **scattering cross section** and $d\sigma/d\Omega$ for the differential cross section.

differential absorption hygrometer—A type of **spectral hygrometer** using at least two closely spaced wavelengths.
 One is chosen inside an **absorption band**, the other outside and not affected by **water vapor** absorption. Subtraction of the latter measurement from the first cancels all **absorption** effects except that by the water vapor. Despite this simple principle the technical realization is difficult due to the need for adjusting and maintaining the two wavelengths very precisely. Instruments based on this technique are differential absorption lidars operating on two fixed wavelengths, or, as a recent development, tunable **diode lasers**. The **wavelength** of these lasers can be changed or tuned continuously over a small wavelength range allowing for monitoring the shape and **intensity** of a whole **absorption line** and its vicinity.

differential absorption lidar thermometer—(Abbreviated DIAL thermometer.) An active remote **sensor** that measures **temperature** using a **laser**.
 To measure temperature, a DIAL thermometer transmits three wavelengths: one on an oxygen

absorption band, one off the band, and one at the transition **wavelength** in between. With this technique, the concentrations of oxygen in various **energy** states can be measured. Many of the high rotational energy states of oxygen are temperature dependent, allowing temperature to be inferred from the DIAL measurements. This method requires very high accuracy, spectral **resolution**, and stability of the laser transmitter and detector, which are characteristics that are almost achieved at present.

differential analysis—Synoptic **analysis** of change charts or of vertical differential charts (such as **thickness charts**) obtained by the graphical or numerical subtraction of the patterns of some meteorological **variable** at two times or two levels.

Spatial differential analysis from one surface to another in the vertical is useful because of its relation to **hydrostatics**: Differential analysis of **pressure** at low levels gives the mean **isopycnic lines** for the layer; differential analysis of **temperature** and **potential temperature** at two levels approximates the mean **stability** for the layer; and differential analysis of the topographies of two different **isobaric surfaces** gives the pattern of **thickness** or mean virtual **isotherms** for the layer.
Saucier, W. J., 1955: *Principles of Meteorological Analysis*, 136–137.

differential attenuation—The difference in the loss of **power**, due to propagation, experienced by two **signals** that differ in one attribute, for example, **polarization** or **wavelength**.

As applied to **polarimetric radar** observations, the differential attenuation is defined relative to the axes of the propagation medium for which the **attenuation** is minimum and maximum. These axes are described as the principal axes of the medium.

differential backscattering cross section—The hypothetical area normal to the incident **radiation** that would geometrically intercept the total amount of radiation actually backscattered per unit solid angle (i.e., through a **scattering** angle of 180°) by a **particle**.

differential ballistic wind—In bomb ballistics, the **wind vector** that is the **weighted average** of all the differential winds from the bomb-release **altitude** down to the **target**.

The differential ballistic wind is the basis for calculation of the Q-factor in the bomb-sight setting.

differential chart—General term for a **chart** showing the amount and direction of change of a meteorological quantity in time or in space, for example, **change chart**, **vertical differential chart**.
See **differential analysis**.

differential kinematics—Method of determining from the **pressure** and **wind** fields, using kinematic equations, certain parameters relating to the movement and **development** of **synoptic** features.

differential mobility analyzer—A device used to remove from a flowing gas stream a predictable fraction of **particles** within a narrow size **range** based upon electrical **mobility**.

The electrical mobility is defined as the electrical **velocity** of the particle divided by the **field strength**. This device consists of two cylindrical electrodes, an outer ground electrode and an inner rod to which a precise negative voltage is applied. The sample flow is introduced into an annular gap between the two electrodes and flows through the device along with an inner core of particle-free sheath air without mixing. Particles with positive charges are attracted through the sheath air toward the negatively charged center electrode. Only those particles within a narrow mobility range pass through a slit near the bottom of the center electrode.

differential operator—*See* **linear operator**.

differential optical absorption—With this optical technique, one measures the difference in light **absorption** of the species of interest between a maximum and minimum in the **spectrum**.

The technique most often uses the visible or **ultraviolet** regions of the spectrum. A broadband continuous white light source, generally a high-pressure Xe lamp or incandescent quartz-iodine lamp, is used for this purpose, but light from the sun and the moon have also been used. Differential optical absorption has been used in the **atmosphere** to detect gas molecules such as **ozone**, **sulfur dioxide**, **nitrogen dioxide**, **nitrate radical** (NO_3), and **hydroxyl radical** (OH), with sensitivities ranging from the low parts-per-trillion (ppt) level to several hundred ppt.

differential phase shift—The difference in the **phase shift**, due to propagation, experienced by two **signals** that differ in one attribute, for example, **polarization** or **wavelength**.

As applied to **polarimetric radar** observations, the differential phase shift is defined relative to the axes of the propagation medium for which the phase shift is minimum and maximum. These axes are described as the principal axes of the medium.

differential reflectivity—The ratio of **radar reflectivity** measured by means of two **signals** that differ in one attribute, for example, **polarization** or **wavelength**.

As applied to **polarimetric radar** observations, the differential reflectivity is the ratio of the reflectivity observed with transmitted and received signals of horizontal polarization to that observed with signals of vertical polarization. It is commonly represented by the symbol Z_{DR}. The ratio of radar reflectivity measured with two signals of different wavelength is more commonly described as the dual-wavelength ratio.

differential thermal advection—Vertical derivative (or difference) of thermal **advection**.

differential water capacity—(Also called specific water capacity.) The rate of change of **water content** with soil water pressure.

Its value will depend on whether volumetric or gravimetric water content is used and how **pressure** is expressed.

differential wind—In bomb ballistics, the **vector** difference between the **wind** at the bomb-release **altitude** and the wind at some other specific lower altitude.

The differential winds are required for the computation of the **differential ballistic wind**.

diffluence—The rate at which adjacent flow diverges along an axis oriented normal to the flow at the point in question; the opposite of **confluence**.

The diffluence may be measured by $\partial v_n/\partial n$ or $V\partial\psi/\partial n$, where V is the speed of the **wind**, the n axis is oriented 90° clockwise from the direction of the **wind vector**, v_n is the wind component in the n direction, and ψ is the **wind direction** measured in degrees clockwise from a reference direction.

diffluent thermal ridge—A pattern of **thickness lines** that is concave toward high **thickness** and in which the thickness lines separate in the direction of the **thermal wind**.

diffluent thermal trough—A pattern of **thickness lines** that is concave toward low **thickness** and in which the thickness lines separate in the direction of the **thermal wind**.

diffraction—The process by which the direction of **radiation** is changed so that it spreads into the geometric **shadow zone** of an opaque or refractive object that lies in a radiation field.

Diffraction is an optical "edge effect," differing only in degree from **scattering**. Diffraction becomes more evident when dealing with **particles** similar to, or larger than, the **wavelength** of the radiation. In **meteorological optics**, important diffraction phenomena include the **aureole**, **Bishop's ring**, **corona**, **iridescent clouds**, etc. The principle of diffraction may also be applied to the propagation of water surface waves, as into the sheltered region formed by a barrier.

diffraction pattern—An **interference** pattern of **scattered** light centered on a **scattering** angle of 0°.

This pattern is due to the various **phase** relationships among the different paths that the **electromagnetic radiation** takes in going around, or through, a scattering particle, leading to constructive or destructive interference. In general, the larger the cross section of the **particle**, or the shorter the **wavelength** of the **radiation**, the narrower the angular spread of the diffraction pattern.

diffraction peak—The region of the **diffraction pattern** close to a **scattering** angle of 0°.

As **particles** become large compared with the **wavelength** of incident **radiation**, the diffraction peak becomes larger and tends to dominate the entire diffraction pattern, giving rise to most of the diffraction pattern effects associated with **geometric optics**.

diffraction phenomenon—A phenomenon of **meteorological optics** for which the approximate theory of **diffraction** has often been used to provide an explanation.

While no phenomenon can be correctly said to have been caused by diffraction, the theory is useful as a means of gaining insight into behavior.

diffraction zone—*See* **shadow zone**.

diffuse front—A **front** across which the characteristics of **temperature** change and **wind** shift are weakly defined.

diffuse light—**Light** that is nearly **isotropic**, or is not strongly directional.

diffuse radiation—Radiation that comes from some continuous range of directions.

This includes **radiation** that has been **scattered** at least once, and **emission** from nonpoint sources. It is to be contrasted with parallel radiation, as from a **point source**. The more **isotropic** the radiation field, the more diffuse the radiation is considered to be.

diffuse reflection—1. The change in the direction of **radiation** into all directions after encountering a **rough surface**.

Compare **specular reflection**.

2. Term frequently applied to the process by which **solar radiation** is **scattered** by **dust** and other **particles** suspended in the **atmosphere**.

3. Reflection by a **diffuse reflector**.

diffuse reflector—Any surface with irregularities so large compared to the **wavelength** of the incident **radiation** that the reflected rays are sent back in multiple directions; the opposite of a **specular reflector**, such as a mirror.

Most natural surfaces act more like diffuse reflectors than specular reflectors. Exceptions are calm water and some **ice** surfaces. To be distinguished from a **perfectly diffuse reflector**.

diffuse sky radiation—Solar **radiation** that is **scattered** at least once before it reaches the surface.

As a percentage of the **global radiation**, **diffuse radiation** is a minimum, less than 10% of the total, under **clear sky** conditions and overhead sun. The percentage rises with increasing **solar zenith angle** and reaches 100% for **twilight**, **overcast**, or highly turbid conditions. It is measured by a **shadow band pyranometer**.

diffuse skylight—Same as **diffuse sky radiation**.

diffuse solar radiation—Downward **scattered** and reflected **solar radiation** coming from the whole hemisphere of the sky with the exception of the solid angle subtended by the sun's disk.

diffusion—1. The **transport** of matter solely by the **random** motions of individual molecules not moving together in coherent groups.

Diffusion is a consequence of **concentration gradients**.

2. The process of **mixing** fluid properties by molecular and turbulent motions.

The process in **turbulent flow** is called turbulent or **eddy diffusion**. Diffusion by **turbulence** is much more rapid than diffusion by molecular motions. Turbulence with **statistics** independent of direction is called **isotropic**. In general, turbulence in the **atmosphere** and ocean is not isotropic. *See* **diffusivity**.

diffusion coefficient—*See* **diffusivity**, **coefficient of mutual diffusion**.

diffusion denuder technique—A technique for stripping a certain molecule or class of molecules from an airstream.

The molecules of interest diffuse to the walls of a flow-through reactor, and are deposited there, while particulate matter passes through the reactor unchanged. The reactors are usually cylindrical or annular tubes, to maximize the surface area for a given volume, and the inner walls can be coated with chemicals to remove the molecules of interest, for example, citric or oxalic acid for **ammonia** removal. At the end of a given time the chemicals of interest are removed from inside the denuder for quantitative chemical **analysis**.

diffusion equation—*See* **diffusivity**.

diffusion hygrometer—A **hygrometer** based upon the **diffusion** of **water vapor** through a porous membrane.

In its simplest form, it consists of a closed chamber having porous walls and containing a **hygroscopic** compound. The **absorption** of water vapor by the hygroscopic compound causes a **pressure** drop within the chamber that is measured by a **manometer**.

diffusion model—A set of mathematical equations that simulates the **diffusion** of material released in the **atmosphere** (or ocean).

Usually in the form of computer codes, diffusion models simulate diffusion in a variety of physical processes.

diffusion velocity—1. The relative mean molecular **velocity** of a selected gas undergoing **diffusion** in a gaseous **atmosphere**, commonly taken as a **nitrogen** (N_2) atmosphere.

The diffusion velocity is a molecular phenomenon and depends upon the gaseous concentration as well as upon the **pressure** and **temperature** gradients present.

2. The speed with which a **turbulent diffusion** process proceeds as evidenced by the motion of individual **eddies**.

Lettau, H., 1951: *Compendium of Meteorology*, 320–321.

diffusive convection—A form of double-diffusive **convection** that occurs when cold **freshwater** overlies warm salty water.

If a **parcel** of warm and salty water is moved upward in the water column, it loses **heat** more quickly than it loses salt, becoming colder and saltier, and thus denser, than its environs, or than the water at its initial position. It then sinks beyond its initial position into water that is saltier but warmer. It then gains heat from this water and begins to rise past its initial position. Diffusive convection transports heat more efficiently than salt and is believed to be responsible for staircase-like structures observed in the arctic **thermocline**. *See also* **double diffusive convection, Turner angle**.

diffusive equilibrium—The **steady state** resulting from the **diffusion** process, also known as **gravitational equilibrium**.

In such a state the constituent gases of the **atmosphere** would be distributed independently of each other, the heavier decreasing more rapidly with height than the lighter. However, the occurrence of turbulent **mixing** precludes establishment of diffusive equilibrium. *See* **isothermal equilibrium**.

diffusivity—(Also called coefficient of diffusion.) The ratio of the **flux** of a **conservative property** through a specified surface by **turbulence** to the **gradient** of the mean property normal to the surface.

In the special case of **isotropic turbulence** and no mean motion the Fickian diffusion equation (or Fick's equation) takes the form

$$\frac{\partial \overline{S}}{\partial t} = K_s \nabla^2 \overline{S},$$

where K_s represents the diffusivity, ∇^2 is the **Laplacian operator**, and $\overline{S}$ is the **mean value** of the property $\overline{S}$. This equation describes decreasing $\overline{S}$ where the Laplacian is negative and increasing $\overline{S}$ where the Laplacian is positive. The general case is more complex. In the statically stable **atmosphere** or ocean, the horizontal **scale of turbulence** is much greater than the vertical scale and **turbulent diffusion** in the horizontal may greatly exceed **diffusion** in the vertical. On the other hand, in the case of **buoyancy**, vertical diffusion may be greater than horizontal diffusion. *See also* **mixing, eddy flux, turbulence length scales**.

> Fleagle, R. G., and J. A. Businger, 1980: *An Introduction to Atmospheric Physics*, 2d ed., Academic Press, 64–66.
> Hinze, J. O., 1975: *Turbulence*, 2d ed., McGraw–Hill, 48–55.

digital—The presentation or recording of data or information in numerical format.

digital filter—An **algorithm** operating upon a sequence of discrete-time sampled data, designed to pass signals with selected temporal or spatial frequencies while attenuating signals with other temporal or spatial frequencies.

Such filters are often used to pass desired signals while suppressing **noise** or interfering signals. Common types are low-pass filters, which suppress high-frequency **energy**; high-pass filters, which reject low-frequency energy; and bandpass filters, which reject low- and high-frequency energy, and pass only a limited **range** of frequencies in between.

dihedral reflector—*See* **corner reflector**.

dimensional analysis—A method of determining possible relationships between meteorological variables based on their **dimensions**.

A systematic method called **Buckingham Pi theory** can be used to find such relationships. The results are often expressed as **dimensionless groups**, called Pi groups. While Buckingham Pi theory helps to identify the appropriate dimensionless groups, it cannot indicate the relationships between the groups. Such relationships must be found empirically, based on field or laboratory measurements of the dimensionless groups. When the empirical data are plotted on graphs of one dimensionless group versus another, often data from many disparate meteorological conditions will result in one common curve, yielding a **similarity relationship** that may be universal. Dimensional analysis has been used extensively and successfully in studies of the **atmospheric boundary layer**, where **turbulence** precludes other more precise descriptions of the flow because exact solutions of the **equations of motion** are impossible to find due to the **closure problem**.

> Stull, R. B., 1988: *An Introduction to Boundary Layer Meteorology*, 347–404.

dimensionless equation—*See* **nondimensional equation**.

dimensionless group—(Also nondimensional number, dimensionless number.) A dimensionless combination of several physical variables (e.g., **velocity, density, viscosity**), usually with a physical interpretation.

Dimensionless groups arise naturally in the **scale analysis** of equations. The sixth edition of the McGraw–Hill Encyclopedia of Science and Technology lists 12 pages of dimensionless groups. *See* **Boussinesq number, Cauchy number, Grashoff number, Mach number, Nusselt number, Péclet number, Prandtl number, Rayleigh number, Richardson number, Rossby number,**

Strouhal number, Taylor number; *see also* **dimensional analysis, similarity theory, Bucking-ham Pi theory**.

dimensionless number—*See* **nondimensional number**.

dimensionless parameter—*See* **nondimensional parameter**.

dimensions—*See* **fundamental dimensions**.

dimethyl disulfide—Naturally occurring sulfur gas, formula CH_3SSCH_3; minor constituent in the **natural sulfur cycle**.

dimethyl sulfide—Naturally occurring sulfur gas, formula CH_3SCH_3, emitted predominantly by phytoplankton, marine algae, seaweed, etc.
   Dimethyl sulfide (DMS) has the highest **flux** to the **atmosphere** of all biogenic sulfur gases. The **oxidation** of DMS has been linked to the formation of **aerosol** particles in the **troposphere** remote from urban areas, and thus to the potential for **climate** regulation.

dimethyl sulfoxide—Partially oxidized sulfur gas, formula $(CH_3)_2SO$, which finds large use as a solvent, but due to low volatility does not exist in the **atmosphere** in high concentrations.
   Found in marine aerosols, and thought to be a partial **oxidation** product of **dimethyl sulfide**.

dimmerfoehn—A rare and strong **south foehn** in the Alps, in which the strong **foehn** winds reach the ground only in the lower parts of the valleys and the plain or foreland.
   Under these conditions the **sea level pressure** difference between the south and north side of the Alps is greater than 12 mb. In the upper portions of the valleys, the surface winds are light and variable, as the strong winds do not penetrate to the ground. This effect has been attributed to the persistence of a cold-air layer in the upper valleys, **decoupling** the surface winds from the flow aloft and making the flow unable to follow the terrain in this region, or to the position of this region with respect to the long **mountain wave** that may accompany this foehn. The **foehn wall** and the **precipitation area** extend beyond the crest across the almost **calm** surface region in the upper valleys. This condition "is known as 'dimmer-foehn,' meaning blocked or dammed-up foehn."

   Barry, R. G., 1992: *Mountain Weather and Climate*, 2d ed., Routledge, London, p. 325.

Dines anemometer—A type of pressure-tube **anemometer**, named after the inventor, in which the **pressure head** is located on the **windward** end of the **wind vane** so that it is kept facing into the **wind**.
   The suction head is mounted concentrically with the shaft, near the bearing that supports the vane, and develops a suction that is independent of **wind direction**. The **pressure** difference between the suction head and the pressure head is proportional to the square root of the **wind speed** and is measured by a specially designed float **manometer** with a **linear** wind **scale**. The Dines anemometer was first described by William Henry Dines in 1892 and had all the essentials of the final form of the pressure-tube anemometer. With various modifications, the Dines anemometer remained a useful wind measuring device into the 1960s.

Dines compensation—Property whereby the sign of the **horizontal divergence** reverses at least once in the **troposphere** or the **stratosphere**.
   This implies that the vertically integrated **divergence** and the associated **surface pressure** tendency are small residuals of much larger contributions.

Dines radiometer—An instrument for measuring **radiant energy**.
   The Dines radiometer consists of an ether differential **thermometer** with blackened bulbs. One of the bulbs is exposed to the unknown **radiation** and the other to a **blackbody** source with a **temperature** that can be varied. Equality of radiation is indicated by the balance of the differential thermometer. *Compare* **bolometer**; *see* **actinometer, photometer, Tulipan radiometer**.

dinitrogen pentoxide—Dinitrogen pentoxide, N_2O_5, is formed from the reaction of the **nitrate radical** with **nitrogen dioxide**.
   It is, however, thermally unstable and is dissociated readily near the earth's surface to form its precursors. Higher in the **atmosphere**, it is an effective **reservoir** for **active nitrogen**. In solution it hydrolyzes to two **nitric acid** molecules. Removal of N_2O_5 on aqueous **particles** occurs very easily, causing an increase in acidity and a loss of active nitrogen ($NO + NO_2$).

diode laser—Solid-state **laser** in which lasing is achieved by passing an injection **current** through the active region of a semiconductor across the p–n junction.
   Operating in the **infrared** spectral region, 2.7–30 μm, makes diode lasers particularly useful for the measurement of certain atmospheric **trace gases** such as **nitrous oxide, carbon monoxide**, or **formaldehyde**.

Reid, J., J. Shewchun, B. K. Garside, and E. A. Ballik, 1978: High sensitivity pollution detection employing tunable diode lasers. *Appl. Optics*, **17**, 300–307.

dioxane (−1, 4)—Dioxygenated cyclic organic compound, formula $C_4H_8O_2$, formally a **cyclohexane** structure in which two diametrically opposed CH_2 groups have been replaced by **oxygen** atoms.

Dioxane is widely used as a solvent and as a stabilizer in paints, varnishes, etc; thought to be carcinogenic. Not to be confused with dioxins, formed in the combustion of chlorinated **hydrocarbons**.

dip—Same as **inclination**.

dip circle—An instrument for measuring the **inclination** of the earth's **magnetic field**.

In its simplest form it consists only of a magnetized needle mounted on a pointed bearing, and a protractor. By timing the **oscillation** of the needle, it is possible to obtain, also, relative values of the **intensity** of the earth's **magnetic field**.

dip equator—Same as **aclinic line**.

dip of the horizon—The angular difference between the horizontal and the direction to the **horizon**.

Leaving aside displacements due to **topography** (mountains and valleys), the horizon can be displaced from the horizontal by the earth's curvature, by the **elevation** of the **observer**, and by the **refraction** of **light** in the **atmosphere**.

dip pole—Same as magnetic **pole**.

dipole—1. Without qualification usually means electric dipole, a system composed of two charges of equal and opposite sign separated by a distance.

A magnetic dipole is an electric **current** loop enclosing a finite area in a plane. *See* **dipole moment**.

2. Same as **dipole antenna**.

dipole antenna—(Or, simply, dipole.) A type of **antenna** for transmitting or receiving **electromagnetic radiation**, most often in the **radio frequency band**.

The term dipole antenna has no necessary connection with **dipole** radiation in the sense usually meant in **electromagnetic theory** but arises from configuration: A dipole antenna is essentially two thin (diameter much less than length) conductors or poles separated by a gap by means of which an oscillating electric **current** is fed into (or out of) them. If the total length of the two poles is much less than the **wavelength** of their **radiation**, the dipole antenna is sometimes called a short dipole, the source of the term dipole radiation (radiation from an ideal oscillating dipole of negligible extent). Sometimes dipole antenna means an antenna with a length of half the wavelength, but a more precise term is half-wave dipole. The defining characteristic of a dipole antenna is its two identical components, not its length.

dipole moment—Without qualification usually means electric dipole moment, the product of charge and separation distance of an (electric) **dipole**.

Dipole moment is a **vector**, its direction determined by the **position vector** from the negative to the positive charge. Dipole moments (usually of molecules) are classified as permanent (the centers of positive and negative charge do not coincide even when subjected to no external **field**) and induced (**charge separation** is a consequence of an external field acting in opposite directions on positive and negative charges). Water is often given as the prime example of a molecule with a permanent dipole moment. The magnetic dipole moment of a magnetic dipole is the product of the electric **current** in the loop and the area it encloses. Magnetic dipole moment also is a vector, its direction determined by the normal to the plane of the current loop, the sense of this normal specified by the right-hand rule.

direct cell—(Also referred to as direct circulation.) A closed **thermal** circulation in a vertical plane in which the rising motion occurs at higher **potential temperature** than the sinking motion.

Such a cell converts **heat** energy to **potential energy** and then to **kinetic energy**. The importance of such cells in the **atmosphere**, in particular the subtropical and equatorial **circulation**, is much discussed in theories of the **general circulation**. It has also been suggested that the **tropical cyclone** is essentially a direct cell. *See* **indirect cell**.

direct circulation—*See* **direct cell**.

direct flow—That part of the flood **hydrograph** that represents the fastest response to **rainfall** and that is distinct from the **base flow**.

direct interaction approximation—A **turbulence closure** scheme proposed by Kraichnan that eliminates the nonlocal direct interaction between large-scale and small-scale **turbulence** components.

Monin, A. S., and A. M. Yaglom, 1975: *Statistical Fluid Mechanics*, MIT Press, p. 409.

Kraichnan, R. H., 1959: The structure of isotropic turbulence in high Reynolds numbers. *J. Fluid Mech.*, **5**, 497–542.

direct numerical simulation—Numerical simulation of fluid motions, like turbulent or transitional turbulent flows using the exact governing equations (e.g., **Navier–Stokes equations**), without any parameterizations.

See also **large eddy simulation, very large eddy simulation.**

direct product—Same as **scalar product.**

direct route—The shortest navigational distance between two points on the earth's surface.

The **great circle** is a direct route.

direct runoff—**Runoff** caused by and directly following a **rainfall** or snowmelt event; it forms the major part of the flood **hydrograph** and excludes **base flow.**

direct solar radiation—**Solar radiation** that has not been **scattered** or absorbed.

In practice, solar radiation that has been scattered through only a few degrees, characteristic of the **diffraction peak** of the **scattering function**, is unavoidably included in the operational measurement of direct solar radiation by a **pyrheliometer**. *Compare* **diffuse sky radiation, global radiation.**

direct tide—A gravitational **solar tide** or **lunar tide** in the ocean or **atmosphere** that is in phase with the apparent motions of the attracting body, and consequently has its local maxima directly under the tide-producing body and on the opposite side of the earth.

A **gravitational tide** that is in opposite phase to the apparent motions of the sun or moon is called a **reversed tide.** *See* **tide-producing force.**

direct-vision nephoscope—The class of **nephoscope** in which the **observer** notes the motion of the **cloud** by looking directly at it through the instrument.

These instruments consist of some form of rectangular gridwork supported on a column and free to rotate about the vertical axis. The observer aligns the gridwork so that the cloud appears to move parallel to its major axis. *See* **comb nephoscope, grid nephoscope.**

direction finder—An instrument consisting of two orthogonal magnetic loop antennas and associated electronics for the purpose of detecting the **azimuth** to a cloud-to-ground **lightning stroke.**

direction finder—*See* **radio direction finder.**

directional distribution—A measure of the amount of **wave** energy traveling in different directions in a **random sea.**

See also **angular spreading.**

directional-hemispherical reflectance—The **albedo** of a surface when illuminated from a single direction.

directional hydraulic conductivity—Linear proportional constant from **Darcy's law** that is specific to the direction of flow, that is, **hydraulic conductivity** that depends on direction of flow; of interest for **anisotropic** porous materials.

See **Darcian velocity.**

directivity—The ratio of the **radiation** intensity from an **antenna**, in a given direction, to the radiation intensity averaged over all directions.

The directivity is equal to the **antenna gain** plus antenna dissipation losses.

discharge—1. *See* **electric discharge.**

2. The volumetric rate of flow or volume **flux.**

discharge area—1. In subsurface **hydrology**, the area of an **aquifer** from which water is discharged by **evapotranspiration**, springs, seepage to streams, and **leakage** to other aquifers.

2. In surface **hydrology**, the area of a **stream** or a pipeline perpendicular to the **velocity** vector used for the calculation of the rate of flow.

discharge coefficient—Used in fluid mechanics; the ratio of the actual to the theoretical **discharge** through an orifice, nozzle, **weir**, or any other opening or passage.

discharge section line—Section of a river that is at a right angle with the line connecting the deepest points of the river bed.

discontinuity—The abrupt **variation** or **jump** of a **variable** at a line or surface.

Discontinuities are said to be of zero order when an undifferentiated quantity is discontinuous, or of first order when a first derivative of the quantity is discontinuous, etc. *See* **interface**, **front**, **surface of discontinuity**.

discontinuous turbulence—A situation that occurs in **statically stable** regions of the **atmospheric boundary layer** where **turbulence** is not contiguous, either vertically or horizontally.

For example, in the nighttime **stable boundary layer** one or more layers of turbulence can form that are separated by nonturbulent (laminar) layers. This situation is very difficult to **model** because the turbulence in each layer does not interact with the surface and thus can evolve separately. During daytime at the top of a **convective mixed layer**, there is usually a statically stable **capping inversion** or **entrainment zone** where turbulent **thermals** penetrating the layer are separated by regions of laminar air from the **free atmosphere**. Aircraft flying horizontally through this region would experience intermittent turbulence.

discrete spectrum—A **spectrum** in which the component **wavelengths** (and **wavenumbers** and **frequencies**) constitute a discrete sequence of values (finite or infinite in number) rather than a continuum of values.

A **Fourier analysis** of a function will yield a discrete spectrum only if the function is periodic, or is assumed to be so, or if the function is represented by a finite **sample** of its values. **Fourier series** may be used for the analysis. *See* **continuous spectrum**.

discretization—Approximating the solution of a continuous problem by representing it in terms of a discrete set of elements.

Examples include representing a continuous fluid **field** at a set of **grid** points or in terms of a sum of **basis functions**.

discriminant analysis—A variation of multiple **linear regression analysis** for **prediction** of the occurrence or nonoccurrence of an event.

To account for the nonnumerical nature of the **predictand**, a discriminant function is used as a type of **regression function** usually derived in such a way that positive values of the function correspond to "occurrence" and negative values to "nonoccurrence." In meteorology, for example, the occurrence of **precipitation** (the **predictand**) can be related to measures of **vertical velocity**, **dewpoint** temperature, **pressure** change, and other variables (the predictors) through a discriminant function. Values of the function above or below a threshold value (typically, zero) can be used to predict precipitation occurrence.

disdrometer—An instrument that measures and records the sizes of **raindrops**.

A common type of disdrometer consists of a sensitive **transponder** that measures the **momentum** of individual drops as they fall onto an exposed horizontal surface. Size is determined from momentum through **calibration**, and the **drop-size distribution** is obtained by keeping a tally of the number of drops in different size categories that fall onto the surface in a given period of time.

dish—A parabolic **reflector** type of radio or **radar** antenna.

The term is occasionally used incorrectly in referring to any type of radar antenna.

dishpan experiments—Model experiments carried out by differential heating of a fluid in a flat rotating pan.

Such experiments, by establishing similarity with the **atmosphere** through such nondimensional parameters as the **Rossby number**, **Taylor number**, etc., have reproduced many important features of the **general circulation** and of smaller scales of atmospheric motion. *See* **Rossby regime**.

Fultz, D., 1951: *Compendium of Meteorology*, 1235–1248.

disk hardness gauge—"An instrument for measuring **snow** hardness in terms of the **resistance** of snow to the **pressure** exerted by a disk attached to a spring-loaded rod; a gauge calibrated in pounds per square inch registers the amount of resistance" (Glossary of Arctic and Subarctic Terms, 1955).

Compare **Canadian hardness gauge**.

Arctic, Desert, Tropic Information Center (ADTIC) Research Studies Institute, 1955: *Glossary of Arctic and Subarctic Terms*, ADTIC Pub. A-105, Maxwell AFB, AL, 90 pp.

dispersed phase—*See* **colloidal system**.

dispersing medium—*See* **colloidal system**.

dispersion—1. The **variation** of the **complex index of refraction** with **frequency** (or, equivalently, **wavelength** in **free space**), sometimes classified as **normal** (if n increases with increasing frequency) or **anomalous** (if n decreases with increasing frequency).

But there is nothing anomalous about **anomalous dispersion**: Every material substance exhibits

223

anomalous dispersion at some frequencies. Dispersion is a consequence of the inherent frequency-dependent response of individual atoms and molecules to excitation by a time-harmonic **field**. By means of dispersion a **beam** of **light** composed of many frequencies can be spatially separated (angular dispersion) into its components as, for example, with a prism. **Rainbows** and **halos** owe their colors to angular dispersion.

2. The spreading of atmospheric constituents, such as **air pollutants**.

Dispersion can be the result of **molecular diffusion**, turbulent **mixing**, and mean **wind shear**. The displacement or **advection** of polluted air by the mean wind is usually called **transport** rather than dispersion. The amount of dispersion is usually described statistically by the **standard deviation** (σ_x, σ_y and σ_z) of **pollutant** particle locations (x, y, z) from the pollutant puff center-of-mass for isolated short releases such as explosions, or from the **plume** centerline for continuous emissions such as from a smokestack. For a plume, the local Cartesian **coordinate system** can be aligned with the x axis pointing in the mean wind direction at plume centerline height, allowing the **crosswind** and vertical dispersion to be described by σ_y and σ_z, respectively.

3. In **statistics**, the **scattering** (or degree thereof) of the values of a **frequency distribution** from their average.

dispersion coefficient—A measure of the spreading of an advecting fluid.

dispersion diagram—A graphic representation of the relationship of **frequency** to **wavelength** for **wave motion**.

The diagram is useful in determining the speed of wave motions and their **dispersion** characteristics. In ocean wave forecasting, it is a graph of the expression

$$t = 0.66\, fR,$$

showing time t in hours as a function of **wave frequency** f per second for various values of distance R in nautical miles. *See* **dispersion relationship**, **wave dispersion**.

dispersion equation—(Or frequency equation.) Same as **dispersion relation**.

dispersion medium—*See* **colloidal system**.

dispersion relation—(Also dispersion equation, frequency equation.) A functional relation between **wavenumber** and **frequency** for a plane **harmonic** wave of any kind.

That is, for a **wave** propagating in the x direction,

$$A \cos (kx - \omega t),$$

where k is wavenumber and ω is **circular frequency**, $k(\omega)$ is a dispersion relation.

dispersion relationship—A theoretical relationship between the **frequency** and **wavelength** of a **wave** that shows that waves of different frequencies travel at different (**phase**) velocities.

See **dispersion diagram**, **wave dispersion**.

dispersive flux—The mass **transfer** (mass/unit time/unit area) that occurs because of **dispersion**. It is equal to the **dispersion coefficient** multiplied by the **gradient** of the concentration with respect to distance.

dispersive medium—A medium within which **harmonic** waves move at **wave** speeds that are functions of their **wavelengths**.

See **dispersion**.

dispersivity—Equal to the **dispersion coefficient** divided by the **velocity**.

displacement distance—Same as **zero-plane displacement**.

displacement thickness—Same as **zero-plane displacement**.

display—1. In **radar**, a presentation of the **reflectivity**, **mean Doppler velocity**, or other properties of the received signals in a form to facilitate **analysis** or interpretation.

Common displays are the **PPI**, **RHI**, and **THI**.

2. Same as **indicator**.

dissipation—(Also called **viscous dissipation**.) In **thermodynamics**, the conversion of **kinetic energy** into **internal energy** by **work** done against the **viscous stresses**.

Sometimes the rate of conversion per unit volume is meant. If the **Navier–Stokes equations** of viscous flow are employed, Rayleigh's mathematical expression for the rate of viscous (or frictional) dissipation per unit volume is

$$\frac{2}{3}\mu\left[\left(\frac{\partial v}{\partial y}-\frac{\partial w}{\partial z}\right)^2+\left(\frac{\partial w}{\partial z}-\frac{\partial u}{\partial x}\right)^2+\left(\frac{\partial u}{\partial x}-\frac{\partial v}{\partial y}\right)^2\right]$$
$$+\mu\left[\left(\frac{\partial w}{\partial y}+\frac{\partial v}{\partial z}\right)^2+\left(\frac{\partial u}{\partial z}+\frac{\partial w}{\partial x}\right)^2+\left(\frac{\partial v}{\partial x}+\frac{\partial u}{\partial y}\right)^2\right],$$

where μ is the **dynamic viscosity**. The Navier–Stokes assumptions thus satisfy the primary requirement of the **second law of thermodynamics** that the rate of dissipation be positive and the process irreversible. In a turbulent fluid, which the **atmosphere** usually is, dissipation is the end result of the turbulent scale process, by which kinetic energy is transferred from its originating, or outer, scale to the dissipation scales by nonlinear dynamical interactions. Most dissipation occurs at scales near the **Kolmogorov microscale** λ_d, given by

$$\lambda_d=(v^3/\epsilon)^{1/4},$$

where v is the **kinematic viscosity** and ϵ is the rate of **energy** dissipation per unit mass. *See also* **stress tensor, energy equation.**

> Brunt, D., 1941: *Physical and Dynamical Meteorology*, 285–286.

dissipation constant—In **atmospheric electricity**, a measure of the rate at which a given electrically charged object loses its charge to the surrounding air.
If the object bears a charge q at time t, then according to a law established by Coulomb,

$$\frac{dq}{dt}=-kq,$$

where k is the object's dissipation constant expressed in reciprocal time units. It is found that k depends not only upon the geometry of the charged object but also upon the **density** of the surrounding air, its **humidity**, and its motion relative to the object.

dissipation length scale—A vertical distance related to the **intensity** and **molecular dissipation** rate of **turbulence** ϵ.
This length scale l_ϵ has been defined as

$$l_\epsilon=0.4\,\frac{\sigma_w^3}{\epsilon},$$

where σ_w is the **standard deviation** of **vertical velocity** w. It is typically of the order of millimeters.

dissipation of waves—*See* **wave dissipation.**

dissipation rate—The rate of conversion of **turbulence** into **heat** by **molecular viscosity**.
Defined as

$$\epsilon\equiv v\,\overline{\left(\frac{\partial u'_i}{\partial x_j}\right)^2}$$
$$=v\left[\overline{\left(\frac{\partial u'}{\partial x}\right)^2}+\overline{\left(\frac{\partial u'}{\partial y}\right)^2}+\overline{\left(\frac{\partial u'}{\partial z}\right)^2}+\overline{\left(\frac{\partial v'}{\partial x}\right)^2}+\overline{\left(\frac{\partial v'}{\partial y}\right)^2}\right.$$
$$\left.+\overline{\left(\frac{\partial v'}{\partial z}\right)^2}+\overline{\left(\frac{\partial w'}{\partial x}\right)^2}+\overline{\left(\frac{\partial w'}{\partial y}\right)^2}+\overline{\left(\frac{\partial w'}{\partial z}\right)^2}\right],$$

where (u', v', w') are the turbulent **perturbation** velocities (instantaneous deviations from respective mean velocities) in the (x, y, z) directions, v is the **kinematic viscosity** of air, and the overbar indicates an average. This conversion always acts to reduce **turbulence kinetic energy** and means that turbulence is not a conserved **variable**. It also causes turbulence to decay to zero unless there is continual regeneration of turbulence by other mechanisms. Turbulence **dissipation** is greatest for the smallest-size **eddies** (on the order of millimeters in diameter), but turbulence is usually produced as larger eddies roughly the size of the **atmospheric boundary layer** (on the order of hundreds of meters). The **transfer** of turbulence kinetic energy from the largest to the smallest eddies is called the inertial cascade, and the rate of this **energy transfer** is directly proportional to the dissipation rate for turbulence that is stationary (**steady state**). The medium-size eddies where turbulence is neither created nor destroyed is called the **inertial subrange**. **Similarity theory** (**dimensional analysis**) allows calculation of the dissipation rate from measurements of turbulence spectral intensity $S(\kappa)$ at **wavelength** κ, via $\epsilon=0.49S^{3/2}\kappa^{5/2}$. Typical orders of magnitude for ϵ are 10^{-2} to 10^{-3} m^2 s^{-3} during daytime **convection**, and 10^{-6} to 10^{-4} m^2 s^{-3} at night.

Stull, R. B., 1988: *An Introduction to Boundary Layer Meteorology*, 347–404.

dissipation trail—(Or distrail.) A clearly delineated limpid lane forming behind an aircraft flying in a thin cloud layer; the opposite of a **condensation trail**.

The **heat** of combustion of the aircraft fuel, released into the swept path by the exhaust stacks of the aircraft, can, under certain conditions, evaporate existing clouds (if not too dense) and yield a distrail. Clouds of low liquid **water content** and relatively high **temperature** are susceptible to distrail formation but the phenomenon is comparatively rare.

dissociation—The process in which a molecule is broken into two fragments by rupture of a chemical bond.

The dissociation may occur via a **thermal** process, or following the **absorption** of a **photon** (**photodissociation**).

distortion—Nonrealistic reproduction, or nonideal formation.

In optics, it is an imperfection in an **image** caused by an imperfection in the optical system by which the image is produced. In electronics, distortion is a change in **waveform** of a **signal** that occurs upon passage of the signal through an instrument. Crystal distortion is the expansion of a **crystal lattice** near any **crystal** boundary surface.

distortional wave—Same as **transverse wave**.

distrail—Same as **dissipation trail**.

distributed target—(Also called extended target, volume target.) A radar **target** that is large compared with the **pulse volume**, which is the cross-sectional area of the **radar beam** multiplied by one-half the length of the radar **pulse**.

Clouds and **precipitation** are examples.

distribution—Arrangement in time or space, or appointment among various classes or class intervals, especially ranges of values of a certain **variable**.

See **probability distribution, frequency distribution**.

distribution coefficient—The quantity of a solute sorbed by a solid, per unit weight of solid, divided by the quantity of the solute dissolved in the water per unit volume of water.

distribution function—(Also called probability distribution function.) A function $F(x)$ yielding the **probability** that a stated **random variable** will assume some value less than or equal to any arbitrary number x.

By definition, the distribution function is identically zero for all values of x below the least admissible value of the random variable, and identically unity for all values of x equal to or greater than the greatest admissible value of the random variable. Moreover, $F(x_2) \geq F(x_1)$ whenever $x_2 > x_1$. Sometimes, for the sake of clarity, the distribution function is called the cumulative distribution function. *Compare* **probability density function**.

distribution graph—In **hydrology**, a statistically derived **hydrograph** for a **storm** of specified **duration**, graphically representing the percent of total **direct runoff** passing a point on a **stream**, as a function of time.

This is usually presented as a **histogram** or table of percent runoff within each of successive short time intervals. In principle it is the same as the **unit hydrograph**; both are used as tools in river forecasting and for other purposes such as the comparison of **runoff** characteristics of different drainage areas.

district forecast—(Archaic.) In U.S. Weather Bureau historical records usage, a general **weather forecast** for conditions over an established geographic "forecast district."

Forecast districts are relatively large areas of the order of tens of thousands or hundreds of thousands of square miles.

disturbance—In general, any agitation or disruption of a **steady state**.

In meteorology, this has several rather loose applications: 1) used for any **low** or **cyclone**, but usually one that is relatively small in size and effect; 2) applied to an area where weather, **wind**, **pressure**, etc., show signs of the **development** of cyclonic circulation (*see* **tropical cyclone**); 3) used for any deviation in flow or pressure that is associated with a disturbed state of the weather, that is, **cloudiness** and **precipitation**; and 4) applied to any individual circulatory system within the **primary circulation** of the **atmosphere**. *See* **wave disturbance, instability**; *compare* **perturbation**.

disturbed soil sample—A sample of soil that does not preserve the pore structure and bulk **density** of the parent material.

diurnal—Daily, especially pertaining to actions that are completed within 24 hours and that recur every 24 hours; thus, most reference is made to diurnal cycles, variations, ranges, maxima, etc.

The diurnal variability of nearly all of the meteorological elements is one of the most striking and consistent features of the study of weather. The diurnal variations of important elements at the earth's surface can be summarized as follows: 1) temperature maximum occurs after local noon and minimum near **sunrise**; 2) **relative humidity** and **fog** are the reverse of **temperature**; 3) **wind** generally increases and **veers** by day and decreases and **backs** by night (*see* **heliotropic wind**, **land** and **sea breeze**, **mountain** and **valley wind**); 4) **cloudiness** and **precipitation** over a land surface increase by day and decrease at night; over **water** the reverse is true, but to a lesser extent; 5) **evaporation** is markedly greater by day; 6) **condensation** is much greater at night; 7) **atmospheric pressure** varies diurnally or semidiurnally according to the effects of **atmospheric tides**.

diurnal cooling—In the ocean, nighttime cooling due to **longwave radiation, sensible heat** flux, or **latent heat** flux that leads to the deepening of the **transient thermocline**.

diurnal heating—In the ocean, heating from **solar radiation** (primarily shortwave) that can lead to transient **thermoclines** and the **afternoon effect**.

diurnal inequality—In **oceanography**, a difference between the heights of the two **high waters** and of the two **low waters** of a **lunar day**.

diurnal tide—Once-daily tidal variations in **sea level**, which increase with lunar or solar **declination** north and south of the **equator**.

When added to semidiurnal tides they can cause a **diurnal inequality**.

diurnal vertical migration—The vertical movement in the ocean of zooplankton, which feed near the surface at night on phytoplankton in the **euphotic zone** and sink downward several hundred meters during daytime, presumably to escape larger predators.

divergence—The expansion or spreading out of a **vector field**; also, a precise measure thereof.

In mathematical discussion, divergence is taken to include **convergence**, that is, negative divergence. The mean divergence of a **field** F within a volume is equal to the net penetration of the vectors $\mathbf{F}$ through the surface bounding the volume (*see* **divergence theorem**). The divergence is invariant with respect to coordinate transformations, and may be written

$$\text{div } \mathbf{F} \text{ or } \nabla \cdot \mathbf{F},$$

where ∇ is the del operator. In **Cartesian coordinates**, if $\mathbf{F}$ has components F_x, F_y, F_z, the divergence is

$$\frac{\partial F_x}{\partial x} + \frac{\partial F_y}{\partial y} + \frac{\partial F_z}{\partial z}.$$

Expansions in other coordinate systems may be found in any text on **vector** analysis. In **hydrodynamics**, if the vector field is unspecified, the divergence usually refers to the divergence of the **velocity** field (*see also* **mass divergence**). In meteorology, because of the predominance of horizontal motions, the divergence usually refers to the two-dimensional **horizontal divergence** of the velocity field

$$\frac{\partial u}{\partial x} + \frac{\partial v}{\partial y},$$

where u and v are the x and y components of the velocity, respectively. This divergence is denoted by any of the following symbols:

$$\text{div}_H \mathbf{V}, \ \nabla_H \cdot \mathbf{V}, \ \text{div}_2 \mathbf{V}, \ \nabla_2 \cdot \mathbf{V}, \ \text{div}_p \mathbf{V}, \ \nabla_p \cdot \mathbf{V},$$

where the last two quantities involve derivatives in the **isobaric surface**. The order of magnitude of the horizontal divergence in meteorological motions is of considerable dynamic importance: The **geostrophic wind** has divergence of the order of 10^{-6}s^{-1}; the **wind field** associated with **migratory** cyclonic systems, 10^{-5}s^{-1}; motions of smaller **scale** (such as **gravity waves, frontal waves**, and cumulus **convection**) have characteristic divergence one or two orders of magnitude greater. *See* **balance equation, deformation**; *compare* **diffluence, curl, vorticity**.

divergence equation—An equation for the rate of change of **horizontal divergence** on a **parcel** (in analogy to the **vorticity equation**).

For frictionless flow this equation is

$$\frac{D}{Dt}(\nabla_p \cdot \mathbf{u}) = -(\nabla_p \cdot \mathbf{u})^2 + 2J(u, v) - \nabla\omega \cdot \frac{\partial \mathbf{u}}{\partial p} - \beta u + f\zeta - \nabla^2\phi,$$

227

where $\nabla_p \cdot \mathbf{u}$ is the horizontal divergence, $J(u, v)$ is the **Jacobian**, ω is the material rate of change of **pressure**, β is the **Rossby parameter**, ζ is the **vertical vorticity**, and ϕ is the **geopotential**. All horizontal differentiations are carried out in a **constant-pressure surface**. The divergence equation is derived by taking the **divergence** of the vector **equation of motion**. When the two terms in the divergence are neglected, this equation becomes the **balance equation**.

divergence line—(Sometimes called asymptote of divergence.) Any horizontal line along which **horizontal divergence** of the airflow is occurring.

divergence signature—A pattern in a **Doppler velocity** field representing **horizontal divergence** (or **convergence**) associated with atmospheric phenomena.

In a storm-relative **reference frame**, the idealized **signature** associated with flow diverging from a **point source** (or converging toward a **sink**) is symmetric about a line perpendicular to the **radar** viewing direction with marked radial **shear** across the core region between peak Doppler velocity values of opposite sign. The signature associated with a **divergence line** (or **convergence line**) depends on the **Doppler radar** viewing direction. When the radar viewing direction is perpendicular to the line, there is a band of radial shear across the line that corresponds to the full measure of divergence (convergence). When the viewing direction is parallel to the line, none of the divergence is sensed by the radar. At angles in between, a fraction of the divergence is sensed.

divergence theorem—(Also called Gauss's theorem.) The statement that the volume integral of the **divergence** of a **vector**, such as the **velocity** $\mathbf{V}$, over a volume V is equal to the **surface integral** of the normal component of $\mathbf{V}$ over the surface s of the volume (often called the "export" through the closed surface), provided that $\mathbf{V}$ and its derivatives are continuous and single-valued throughout V and s.

This may be written

$$\int \int \int_V \nabla \cdot \mathbf{V} \, dV = \oint \oint_s \mathbf{V} \cdot \mathbf{n} \, ds,$$

where $\mathbf{n}$ is a unit **vector** normal to the element of surface ds, and the symbol $\oint \oint_s$ indicates that the integration is to be carried out over a closed surface. This theorem is sometimes called **Green's theorem** in the plane for the case of two-dimensional flow, and Green's theorem in space for the three-dimensional case described above. The divergence theorem is used extensively in manipulating the meteorological **equations of motion**.

divergence theory of cyclones—A theory of **cyclone** development in which upper-level **divergence** in the **atmosphere**, through the principle of mass **continuity**, induces low-level **convergence** of air that develops **cyclonic circulation** through the process of **geostrophic adjustment**.

diversion of water—Transfer of water from one watercourse to another, such watercourses being either natural or man-made.

divide—1. *See* **climatic divide**.

2. *See* **drainage divide**.

dividing streamline—A **streamline** or flow boundary representing the height that separates stable flow approaching a hill or mountain into two regions: a lower region, where the flow is horizontal and splits around the obstacle (i.e., the flow is partially blocked), and an upper region, where the flow is three-dimensional and goes over the hill.

This occurs when the relationship between the flow and the obstacle height H is characterized by a **Froude number** Fr less than 1. The height of the boundary H_c has been found to be $H_c = H(1 - \text{Fr})$. Thus, it lies between the top and bottom of the hill and is lower for more stable flow. This concept has been used in **air pollution** studies to determine where pollution **plumes** are apt to impinge on a slope surface.

DMO—Abbreviation for **dependent meteorological office**.

DMSP—Abbreviation for **Defense Meteorological Satellite Program**.

Dobson spectrophotometer—A photoelectric **spectrophotometer** that is used to determine the **ozone** content of the **atmosphere**.

The instrument compares the solar energy at two wavelengths in the **absorption band** of ozone by permitting the **radiation** of each to fall alternately upon a photocell. The stronger radiation is then attenuated by an optical wedge until the photoelectric system of the **photometer** indicates that equality of radiation exists. In this manner the ratio of the radiation intensity is obtained. The ozone content of the atmosphere is computed from this value. *See* **Knudsen's tables**.

Dobson unit—A unit used in the measure of the **column abundance** of **ozone** in the **atmosphere**.

One Dobson unit is the equivalent of 2.69×10^{16} molecules of ozone per square centimeter. Alternatively, 1 Dobson unit corresponds to a layer of ozone 10 μm thick, if the ozone were held at **standard temperature and pressure** (273 K, 1 atm pressure).

doctor—1. A name for the **harmattan** on the Guinea coast of western Africa; because of its dryness, it brings relief from the hot humid onshore winds.

2. Colloquial name for the **sea breeze** in tropical and subtropical climates (such as the West Indies, South Africa, Jamaica, Australia) that is invigorating because of its relative coolness.
See **Cape doctor.**

dog days—The period of greatest heat in summer.
Although the name is popularly supposed to have been derived from the period when dogs are especially liable to go mad, it is actually taken from Sirius, the Dog Star. In ancient Greece and Rome the heliacal rising of Sirius was associated with, and believed to cause, the hot, dry, sultry season of summer. The loss of human energy and the wilting of vegetation caused by this weather led to a belief in the baleful effect of Sirius on all human affairs. No formal meteorological or climatological definition exists. In the United States dog days are considered to persist for four to six weeks between mid-July and early September; in western Europe they last from about the third of July to the eleventh of August, and also are associated with the period of greatest frequency of **thunder.**

doister—(Also spelled deaister, dyster.) In Scotland, a **severe storm** from the sea.

doldrums—(Also called equatorial calms.) A nautical term for the **equatorial trough**, with special reference to the light and variable nature of the winds.
Compare **horse latitudes**; *see* **equatorial air.**

domain—An area of human activity presumed to contain **expertise** and knowledge suitable for the basis of an expert or **knowledge-based system.**
Expert system domains should be kept as narrow as possible without becoming trivial. For example, **prediction** of **severe storms** rather than general weather prediction, or **analysis** of **cloud** type rather than the whole of **satellite meteorology.**

domain expert—A specialist who possesses knowledge and/or **expertise** about a particular **domain**, that is, a source of expertise for developing a **knowledge base** for use by an **expert system.**

domain knowledge—The facts, procedures, processes, and rules of thumb of a **domain.**

Doppler broadening—The spread of frequencies of a **spectral line** as a consequence of the **random** motions of molecules.
Because of the **Doppler effect**, the observed **frequency** of **radiation** emitted by a molecule depends on its motion relative to the observer. Even if a stationary molecule could emit radiation of only a single frequency, a gas of such molecules moving randomly in all directions with a distribution of speeds (*see* **Maxwell–Boltzmann distribution**) would be observed to emit radiation over a **range** of frequencies proportional to the **mean** molecular speed relative to the **speed of light**. This mean speed is proportional to the square root of **absolute** temperature and hence so is Doppler broadening. At **normal** temperatures and pressures Doppler broadening is dwarfed by **collision broadening**, but high in the **atmosphere** Doppler broadening may dominate and, indeed, provides a means of remotely inferring temperatures.

Doppler effect—(Also called Doppler shift.) "The change in the apparent time interval between two events which arises from the motion of an observer together with the finite **velocity** of information about the events" (Gill 1965).
Doppler effect is often used to mean frequency shifts (Doppler shift) of acoustic and **electromagnetic waves** because of relative motion between sources and observers. The relative magnitude of a Doppler shift is of order of the ratio of a characteristic speed (e.g., speed of a source) to a speed of propagation (e.g., **speed of sound, speed of light**). A shift to lower **frequency** (relative to a reference frequency) is sometimes called a red shift whereas a shift to higher frequency is sometimes called a blue shift, although no colorimetric meaning should be attached to these terms. **Acoustic waves** do not evoke sensations of color, nor do **electromagnetic waves** outside the **visible spectrum**, and even Doppler shifts of visible **light** are so small as to yield no visually perceptible color changes. According to classical theory, there is no frequency shift of **electromagnetic radiation** for motion of a **transmitter** perpendicular to the line between **receiver** and transmitter. But according to relativistic theory, even for this kind of motion there is a Doppler shift (transverse Doppler shift), although it is appreciably smaller than the longitudinal Doppler shift. *See also* **Doppler frequency shift.**
Gill, T. P., 1965: *The Doppler Effect.*

Toman, K., 1984: *EOS, Trans. Amer. Geophys. Union,* **65**, p. 1193.
Helliwell, T. M., 1966: *Introduction to Special Relativity,* 116–122.

Doppler equation—*See* **Doppler frequency shift.**

Doppler error—The **error** in measurement of the **radial velocity** of a **radar** target. The error is due to 1) **noise** in the measurement system; 2) fluctuations in the **signal** itself, which are a characteristic of echoes from distributed targets such as **precipitation**; 3) limited measurement time; and 4) atmospheric refractive effects (usually negligible) that modify the **ray** path and the **speed of light.**

Doppler frequency shift—(Also called **Doppler effect.**) In general, the change in **frequency** of a **signal** reaching a **receiver** when the receiver and the transmitting source are in motion relative to one another.

This phenomenon was first noted for **sound waves** by the Austrian physicist Christian Johann Doppler (1803–53) in 1842. In meteorology, this effect is successfully employed with remote sensors, such as **Doppler radars** and **Doppler lidars**, in which the receiver (collocated with the **transmitter**) is fixed and only the scatterers (upon which **transmitted power** impinges and is reradiated) are moving. The frequency shift, *f*, induced by a scatterer having a radial component of motion *v*, relative to the **radar** may be expressed as

$$f = -2\frac{v_r}{\lambda},$$

where λ is the **wavelength** of the transmitter, *f* is positive for motion toward the radar, and, by the usual convention, *v*, is positive for motion away from the radar. *See also* **Doppler velocity.**

Doppler lidar—A **laser radar** (**lidar**) that can determine **radial velocity** (**velocity** toward or away from the **laser**) of the air by measuring the **frequency** shift of the returning **light** that was **scattered** from atmospheric **aerosols**, compared to the original transmitted frequency.

Velocities of individual convective thermals and large-diameter turbulent **eddies** can be measured. Most lidar signals are blocked by **clouds** and **fog**, and are greatly attenuated in very polluted air. *See* **Doppler effect.**

Doppler radar—A **radar** that detects and interprets the **Doppler effect** in terms of the **radial velocity** of a **target**.

The **signal** received by a radar from a moving target differs in **frequency** from the transmitted frequency by an amount that is proportional to the radial component of the **velocity** relative to the radar. *See* **Doppler frequency shift.**

Doppler shift—*See* **Doppler effect.**

Doppler sodar—An **acoustic radar** (**sodar**) that can determine **radial velocity** (velocity toward or away from the sodar) of the air by measuring the **frequency** shift of the returning **sound waves** that were **scattered** from regions of turbulent fluctuations of **air temperature**, compared to the original transmitted frequency.

Velocities within individual surface-layer **plumes** and the bottom of convective **thermals** can be measured. Most sodar signals are blown away in strong winds, and the sodar also requires acoustic shielding from outside noise such as from roads, towns, and rustling tree leaves. *See* **Doppler effect, acoustic sounder, sodar**; *compare* **RASS.**

Doppler spectral broadening—Increase in the **Doppler spread** arising from different processes or phenomena. The **Doppler spectrum** is ordinarily regarded as a reflectivity-weighted distribution of the radial velocities of the scatterers in the **pulse volume**.

A spread of radial velocities exists because of **turbulence** and **wind shear** in the pulse volume and, for **precipitation** targets, because the **particles** have a **range** of fall velocities. The spectrum may be further broadened by processes that have little to do with the **radial velocity** distribution of the **scattering** elements. These processes include 1) the cross-beam **wind** velocity, which induces a broadening proportional to the **beamwidth** because of the radial component of the cross-beam **wind vector** pointing in directions away from the **beam** axis; 2) uncertainties in the measurement of the Doppler spectrum because of **statistical** fluctuations in the spectral estimates that arise from short sampling times; and 3) motion of the **radar beam** relative to the targets as a result of **scanning** in **azimuth** or **elevation**, which further reduces the sampling times.

Doppler spectral moments—Statistical moments of Doppler frequency or **Doppler velocity**, regarding these quantities as continuously distributed **random** variables with a **probability density function** equal to the normalized **Doppler spectrum.**

If $S(f)$ denotes the Doppler spectrum as a function of Doppler frequency f, the nth **moment** is defined by

$$\langle f^n \rangle = \frac{\displaystyle\int_{-\infty}^{\infty} S(f)f^n \, df}{\displaystyle\int_{-\infty}^{\infty} S(f) \, df},$$

where the denominator normalizes the Doppler spectrum to unit area. The nth moment in terms of Doppler velocity is related to that in terms of **frequency** by

$$\langle v^n \rangle = (-1)^n \left(\frac{\lambda}{2}\right)^n \langle f^n \rangle,$$

where λ is the radar **wavelength**. The first **moment** of **velocity** is called the **mean Doppler velocity** and the second **central moment** is called the Doppler velocity **variance**. The **Doppler spread** usually denotes the square root of the Doppler velocity variance. *See* **Doppler frequency shift**.

Doppler spectrum—In **radar**, the **power spectrum** of the **complex signal**, expressed as a function of Doppler frequency or **Doppler velocity**. Denoted by $S(v)$, it may be thought of as the reflectivity-weighted **radial velocity** distribution of the scatterers in a **distributed target**.

That is, $S(v) \, dv$ denotes the **received power** in the **velocity** interval dv. With this definition, $S(v)$ is normalized such that

$$\int_{-\infty}^{\infty} S(v) \, dv = \overline{P_r},$$

where $\overline{P_r}$ is the mean received **power** as defined by the **radar equation**.

Doppler spread—In general, the width of the **Doppler spectrum**, measured in units of either **frequency** or **velocity**.

Quantitatively, the Doppler spread is usually defined as the **standard deviation** of the **Doppler velocity** about its **mean value**, regarding the velocity as a continuously distributed **random variable** with a **probability density function** equal to the normalized Doppler spectrum. *See* **Doppler spectral moments; Doppler spectral broadening**.

Doppler velocity—The radial component of the **velocity** vector of a **scattering** object as observed by a remote **sensor**, such as a **Doppler radar** or **Doppler lidar**.

If **V** denotes the velocity vector, then the radial component, v_r, is defined by the **dot product**, **V · r**, where **r** is a unit **vector** in the pointing direction of the radar or lidar. By this definition, the Doppler velocity is positive for motion away from the radar. For distributed targets, the term Doppler velocity often refers to the **mean Doppler velocity** in the **Doppler spectrum**. Interpretation of the Doppler velocity depends on the viewing geometry and the kind of **target**. Clear-air **echoes** are assumed to move with the **wind**, so that the Doppler velocity measured at a given location in the **atmosphere** is equal to the radial component of the wind at that location. Precipitation falls relative to the air, so that the Doppler velocity of a **precipitation** target is assumed to be the sum of the **radial velocity** component of the precipitation terminal **fall velocity** and the radial component of the air motion. *See also* **Doppler frequency shift, Doppler spectral moments, Doppler spectrum**.

dosimeter—1. An instrument for measuring the **ultraviolet** in **solar** and **sky radiation**.

2. A device, worn by persons working around radioactive material, that indicates the dose of **radiation** to which they have been exposed.

dot product—Same as **scalar product**.

double correlation—1. A second-order **statistical** moment, such as a **covariance**:

$$\overline{a'b'} = \frac{1}{N} \sum_{i=1}^{N} (a_i - \overline{a})(b_i - \overline{b})$$

where N is the number of data points in the **time series**, i is the index of the data point, the overbar represents an average, the prime represents the **deviation** from the average, and a and b

231

are two variables such as **vertical velocity** and **temperature**. Used in **turbulence** equations to describe turbulent kinematic fluxes.

2. A **correlation coefficient** r defined as the **covariance** between two variables, divided by the products of the **standard deviation** σ of each **variable**:

$$r_{ab} = \frac{\overline{a'b'}}{\sigma_a \sigma_b}.$$

double diffusive convection—Fluid motion that results from the release of **potential energy** from one of two or more factors that determine the **density** of a fluid (for example, **heat** and **salinity**).

Even if the density is **statically stable**, **convection** may result if one of the factors is statically unstable. There are three major types of double diffusive convection relevant to **heat** and **mass transport** in the ocean. 1) Finger modes may occur when hot salty fluid overlays cold fresh fluid so that convection results in the form of narrow cells (salt fingers) carrying salty water downward and **freshwater** upward. 2) Diffusive modes occur when a stable salinity field is heated from below so that convection results in the form of a series of well mixed layers separated by sharp density gradients. 3) Intrusive modes occur when there are horizontal density gradients in one of the components determining the density of the fluid even if the fluid density as a whole is horizontally uniform. This **instability** develops in the form of **interleaving** intrusions.

double ebb—*See* **double tide**.

double eye structure—*See* **concentric eyewalls**.

double flood—*See* **double tide**.

double layer—1. On the surface of a substance, a layer of electric **dipoles**, the axes of which have an average orientation normal to the surface.

Double layers may appear on interfaces of solid and gas, liquid and gas, liquid and liquid, etc. They arise whenever media with different **electron** affinities (forces of attraction or **work function**) are contiguous, and dipoles are available. A net **potential** difference, the electrokinetic potential exists across a double layer.

2. *See* **electric double layer**.

double-mass analysis—In **hydrometeorology**, a test of "the consistency of the **rainfall** record at a given **station** by comparing its accumulated annual record with that of the accumulated annual, or seasonal, mean values of several other nearby stations."

Singh, V. P., 1992: *Elementary Hydrology*, p. 176.

double-mass curve—An arithmetic plot of the accumulated values of observations of two variables that are paired in time and thought to be related. As long as the relationship remains constant, the double-mass curve will appear as a straight line; a **deviation** denotes the timing of a change.

double refraction—Nearly synonymous with **birefringence** but may be applied in a restricted sense to transparent (at visible frequencies) media with sufficient birefringence that images seen through them are split.

The term originates from the Danish physician Erasmus Bartholinus, who in 1669 wrote about what he observed through the crystalline solid calcite: "objects which are looked at through the **crystal** do not show, as in the case of other transparent bodies, a single refracted **image**, but they appear double."

double register—*See* **multiple register**.

double-theodolite observation—A technique for making **winds-aloft observations** in which two **theodolites** located at either end of a **base line** follow the ascent of a **pilot balloon**.

Synchronous measurements of the **elevation** and **azimuth** angles of the balloon, taken at periodic time intervals, permit computation of the **wind speed** and **wind direction** as a function of height.

double tide—(Also known as agger, gulder.) A double-headed **tide** with a **high water** consisting of two maxima of similar height separated by a small **depression** (double high water), or a **low water** consisting of two minima separated by a small **elevation** (double low water).

Examples occur at Southampton, England, along the coast of the Netherlands, and off Cape Cod, Massachusetts.

doubly stochastic matrix—A (usually square) matrix for which the sum of each row and each column equals one.

An example is a **transilient matrix** used to describe vertically nonlocal turbulent **mixing** in the **atmosphere**, where the sum of each column must equal one to conserve a state such as **heat** or moisture, and the sum of each row must equal one to conserve **air mass**. The largest eigenvector of a doubly stochastic matrix is not larger than unity, which implies **absolute** numerical **stability** for any time step and any **grid** spacing when using the matrix to make forecasts of turbulent mixing.

down-gradient diffusion—A **random** molecular process where there is net movement of variables such as chemicals, **temperature**, moisture, or **momentum**, from regions of high concentration toward lower concentration.

A classic example is **heat** flowing from the hot end toward the cold end of a metal rod. Gradient refers to the change of mean variable across a distance, such as the temperature change across the metal rod. Down-gradient refers to movement from high to low concentration, such as from hot to cold (as opposed to **counter-gradient**, meaning from cold to hot). Many **turbulence** theories such as **K-theory** and **mixing-length theory** assume that **turbulent transport** is caused mostly by small **eddies**, resulting in a local **transport** (down-gradient **turbulent diffusion**). Such small eddy local turbulent transport is not a law, but is a **turbulence closure** assumption that works well for those atmospheric situations where small eddies dominate (such as in the atmospheric **surface layer**), but that usually fails for situations where large eddies dominate (such as in the atmospheric **mixed layer** where large convective thermals cause nonlocal turbulent transport). *Compare* **counter-gradient flux**.

down-gradient transport—The net **flux** of a **variable** from regions of high to low concentration, caused by either **molecular diffusion** or small eddy **turbulence**.
See **down-gradient diffusion**.

downburst—An area of strong, often damaging **winds** produced by a convective **downdraft** over an area from less than 1 to 10 km in horizontal dimensions.

downdraft—Small-scale downward moving **air current** in a **cumulonimbus** cloud.
See also **draft**.

downdraft convective available potential energy—(Abbreviated **DCAPE**.) The maximum energy available to a descending **parcel**, according to parcel theory.

On a **thermodynamic diagram** this is called negative area, and can be seen as the region between the descending parcel process curve and the environmental **sounding**, from the parcel's **level of free sink** to some lower level, usually the surface. Its quantitative definition is

$$\text{DCAPE} = -\int_{p_f}^{p_s} (\alpha_p - \alpha_e)\, dp,$$

where α_p and α_e are the **specific volumes** of the parcel and its **environment**, respectively, and p_s and p_f are the **surface pressure** and the **pressure** of the level of free sink, respectively.

downrush—A term sometimes applied to the strong downward-flowing **air current** that marks the dissipating stage of a **thunderstorm**.
Compare **uprush**.

downslope wind—1. A **wind** directed down a slope, often used to describe winds produced by processes larger in **scale** than the slope.

Because this flow produces **subsidence**, downslope winds experience warming, drying, increasing **stability**, and **clearing** if clouds are present.

2. Flow directed down a mountain slope and driven by cooling at the earth's surface: a component of the mountain–valley or **mountain–plains wind systems**; same as **katabatic wind**.

The many synonyms for downslope flow are sometimes used interchangeably, and this gives rise to ambiguity and confusion. Downslope can be used generically to denote any **wind** flow blowing down a slope, or it is used specifically for katabatic flows on any **scale**, such as the nocturnal slope-wind component of **mountain–valley wind systems** or mountain–plains wind systems. *See* **katabatic wind**, **gravity wind**, **drainage wind**, **fall wind**, **bora**, **foehn**, **chinook**.

downslope windstorm—A very strong, usually gusty, and occasionally violent **wind** that blows down the lee slope of a mountain range, often reaching its peak strength near the foot of the mountains and weakening rapidly farther away from the mountains.

Gust speeds in such winds may exceed 50 m s^{-1} and occasionally strong vortices capable of doing F1 to F2 damage (*see* **Fujita scale**) may occur in association with these winds. Such windstorms are most likely to the lee of elongated quasi-two-dimensional mountain ranges and can be distinguished from **gap winds**, which are confined to within or **downstream** of notable gaps or breaks in a mountain barrier and are generally weaker and less gusty. Downslope windstorms of great severity require an **upstream** mountain range having a crest at least roughly 1 km in height above terrain to its lee, and with a steep leeside slope. Meteorological conditions favoring downslope windstorms are strong **synoptic-scale** flow across the mountain barrier at the level of its crest, with the cross-range component of the flow either decreasing with height or not increasing too rapidly with height above the crest. Also favorable is high **static stability** at the level of the mountain crest in the flow approaching the mountain range, decreasing with height above. A mean-state **critical level** in the middle **troposphere**, where the flow component across the mountain drops to zero and reverses sign, is often very favorable for downslope windstorms. Downslope windstorms can be considered a **gravity wave** phenomenon in the sense that vertically propagating gravity waves launched by the passage of **stable air** over high-amplitude terrain become very steep or break, creating an internal region above the mountain that is characterized by **turbulence** and a **lapse rate** approaching the dry **adiabatic**. Such a region restricts the vertical propagation of **energy**, allowing the flow near the surface of the mountain to accelerate downslope. Downslope windstorms can also be considered **hydraulic jump** phenomena in which flow becomes supercritical above and to the lee of a mountain barrier. Downslope windstorms are often known by local names in areas where they occur throughout the world (e.g., the **bora** along the northeastern shore of the Adriatic Sea and the **Taku wind** along the Gastineau Channel in southeast Alaska).

downstream—The direction toward which a fluid is moving, usually implying the horizontal component of the mean direction or direction of the basic current; the opposite of **upstream**.

downvalley wind—(Same as **mountain breeze**.) A nocturnal, thermally forced along-valley **wind** produced as a result of **nocturnal cooling** of the valley air; a nocturnal component of the fair-weather **mountain–valley wind systems** encountered during periods of light synoptic or other larger-scale flow.

Valley cooling is accomplished by the combined effects of draining cold air off the slopes by early-evening downslope (**katabatic**) winds, and upward motion with upward cold-air **advection** from the **convergence** of katabatic flows in the valley center. Air in the valley thus becomes cooler than air at the same level over the adjacent plain (*see* **topographic amplification factor**), producing higher **pressure** in the valley. The **pressure gradient** drives a downvalley wind that begins one to four hours after **sunset**, persists for the rest of the night until after **sunrise**, and often reaches 7–10 m s^{-1} or more above the surface. The downvalley wind tends to fill the valley, that is, its depth is approximately the depth of the valley, and where mountains end and a valley empties onto the plains, the downvalley wind can become a cold-air **valley outflow jet** flowing out of the mouth of the valley. *See* **drainage wind, along-valley wind systems**.

downward terrestrial radiation—The downwelling component of **longwave radiation**.

Its value at the surface is a measure of the **greenhouse effect** of the **atmosphere**. *See also* **counterradiation**.

downward total radiation—The combination of downwelling **shortwave** and **longwave radiation**.

downwash—The downward movement of pollutants or **smoke plumes** immediately to the lee of flow obstacles, such as buildings, bluffs, or smokestacks.

This downwash is caused by **wake turbulence** or lee **cavity** circulations generated by the obstacle and brings higher-concentration pollutants down toward the ground. Stack downwash occurs when the average vertical **effluent** velocity w_0 out of the top of the smokestack is less than 1.5 times the mean wind speed M at stack top. The average height decrease of the **pollutant** plume centerline is of order $2D[(w_0/M) - 1.5]$, where D is the diameter of the top of the smokestack.

downwelling—A downward motion of surface or **subsurface water** that removes excess mass brought into an area by convergent horizontal flow near the surface.

See also **upwelling**.

downwind—The direction toward which the **wind** is blowing; with the wind.

downwind gustiness—*See* **gustiness components**.

draft—1. Generally, a small-scale current of air; usually applied to indoor air motion.

2. In aviation terminology (especially in connection with **aircraft turbulence**) a relatively small-scale current of air with marked vertical motion, that is, **updraft** or **downdraft**.

drag—(Or resistance.) The frictional impedance offered by air to the motion of bodies passing through it. More precisely, the component of **aerodynamic force** parallel to the direction of mean flow.

At very low speeds, most of the drag exerted by the air on a body moving through it is due to **viscous drag** (or **skin friction**) acting through a fairly thin boundary layer. In the case of spheres, the low-speed **air resistance** is given by **Stokes's law**. For higher speeds, so-called **form drag** or **pressure drag** arises as a result of separation of the **laminar boundary layer** creating a **wake** region of chaotic flow in which the **pressure** is reduced. In general, for large **Reynolds numbers**, form drag is far more significant than viscous drag. The **velocity** dependence of air resistance changes more or less continuously from **linear** dependence in the viscous range to velocity-squared dependence at high speeds. The latter dependence is given by **Rayleigh's formula**,

$$\text{drag} = \frac{1}{2} C_D \rho L^2 U^2,$$

where ρ is the **density** of the medium, L a characteristic linear dimension of the body, and U the speed of the body relative to the fluid. The **drag coefficient** C_D is proportional to an experimentally determined power of the Reynolds number. At speeds approaching that of **sound**, an additional source of drag arises out of the formation of **shock waves**.

drag anemometer—An **anemometer** that measures the **wind** velocity by sensing the **drag** force on an object, commonly a cylinder or sphere, placed in the flow.

   The force is proportional to the square of the **wind speed**.

drag coefficient—A dimensionless ratio of the component of force parallel to the direction of flow (**drag**) exerted on a body by a fluid to the **kinetic energy** of the fluid multiplied by a characteristic surface area of the body.

   In symbols, the drag coefficient C_D is

$$C_D = \frac{\text{force of resistance}}{\frac{1}{2}\rho U^2 L^2},$$

where ρ is the fluid **density**, U the speed, and L a **characteristic length** of the body. **Rayleigh's formula** expresses the drag coefficient as proportional to a power of the **Reynolds number** (Re),

$$C_D = \text{const.}(\text{Re})^n.$$

For **Couette flow**, $n = -1$ and the constant is 2. In most **aerodynamic** experiments n lies between $-1/4$ and $-1/2$. The drag coefficient for the force exerted by the **atmosphere** on the earth is called the **skin-friction coefficient**.

drag law—The relationship between **wind speed** and force caused by the **wind** against objects or along surfaces.

   Force per unit area of the wind parallel to the ground is called surface **stress**, τ, resulting in a drag law of the form

$$\tau = C_D \rho M^2 \text{ or } u_*^2 = C_D M^2,$$

where C_D is a **drag coefficient**, M is wind speed, ρ is air **density**, and u_* is **friction velocity**, which when squared equals the kinematic surface **stress** τ/ρ. Most atmospheric flows near the surface are turbulent, resulting in drag coefficients that must be determined empirically rather than by using molecular properties such as **viscosity** or **Reynolds number**. *See* **drag**.

drag sphere anemometer—A device that measures **wind speed** by the **drag** force the **wind** produces on a solid sphere.

   Wind speed is proportional to the square root of the drag force. The sphere is often mounted on the end of a rod or pole, with strain gauges attached to the rod to measure the force indirectly.

dragon—(Rare.) Nautical term for a **waterspout**.

drain—1. To provide channels, such as open ditches or drain tile, so that excess water can be removed by surface or by internal flow.

   2. To lose water from the soil by **percolation**. (Glossary of Terms in Soil Science, Agriculture Department of Canada, Publication 1459, revised 1976)

drainage area—Any portion of the earth's surface within a physical boundary defined by topographic slopes that diverts all **runoff** to the same drainage outlet.

drainage basin—The total area **upstream** of a specific point, drained by the **stream** and its tributaries.

drainage density—One of the means of numerically characterizing a **stream** for study and comparative purposes, computed by dividing the length of the stream channel by its **drainage area**; usually expressed in miles per square **mile**.

drainage divide—The line following the ridges or summits that form the exterior boundary of a **watershed**, **river basin**, or catchment.

   A drainage divide separates the area **tributary** to a particular **stream** from the area's tributaries to adjacent streams.

drainage network—Same as **stream network**.

drainage well—A well designed for the purpose of removing the gravity-drained water from a saturated, **porous medium**.
It is used in dams to reduce the amount of seepage entering the **downstream** face of the dam.

drainage wind—Cold-air-runoff winds that are produced when air in contact with terrain surfaces is cooled and flows downslope and/or downvalley.
This generic term is often used to indicate aggregate downslope (**katabatic**) and downvalley flows, when it is difficult to distinguish between the two. This happens frequently in basins, at the upper end of valleys, in complicated **topography** where the downslope and downvalley directions are not perpendicular, and in simple valleys when the weaker and shallower downslope flows are masked or overwhelmed by the stronger downvalley flow. Over even gently sloping topography, drainage winds also refer to **gravity winds** that drain cold air into **frost hollows**, river valleys, and other lower-lying terrain. See **downslope wind**.

drawdown—The decline in the water table level or piezometric level versus distance from a pumping well, or versus time at a given distance from a pumping well, resulting from the continuous pumping from a well discharging at a known rate.

drawdown curve—A plot of the decline of **water table** or piezometric level versus distance from a pumping well, or versus time at a given distance from a pumping well, resulting from the continuous pumping from a well discharging at a known rate.

drawoff—Water withdrawn from a surface **reservoir**.

dreikanter—A **ventifact** with three edges and three facets formed by **wind** abrasion.
Conditions of formation seem to require regions having scanty vegetation, strong winds, and sand and pebbles as a common surface cover. They occur in many parts of the Arctic, particularly in **periglacial** regions. As such they are used as tentative indicators of past conditions, both climatic and geological.

dried ice—**Sea ice** from the surface of which meltwater has disappeared after the formation of cracks and **thaw** holes.
During the period of drying, the surface whitens.

drift—1. The effect of the **velocity** of fluid flow upon the velocity (relative to a fixed external point) of an object moving within the fluid; the **vector** difference between the velocity of the object relative to the fluid and its velocity relative to the fixed reference.
In air navigation, drift is often couched in terms of angular difference between **heading** and **course**, and thus can be produced only by a **crosswind**; when the **wind** velocity is parallel to the heading of the aircraft (direct **headwind** or **tailwind**), the drift is considered to be zero. The calculation of the drift (**leeway**) effects upon an ocean-going vessel is complicated by having to consider the combined effects of two fluids in motion.
2. "In geology, materials in **transport** by **ice**; deposits made by **glacial** ice on land, in the sea, and in bodies of meltwater." (*Glossary of Arctic and Subarctic Terms*, 1955.)
3. The speed of an **ocean current**.
In publications for the mariner, drifts are usually given in miles per day or in knots.
4. *See* **snowdrift**.
5. The horizontal track of an object, for example, **clouds**, caused by the **wind** or fluid motion. Also refers to a shift in the **calibration** of a satellite sensor or change in the orbital track of a satellite.

Arctic, Desert, Tropic Information Center (ADTIC) Research Studies Institute, 1955: *Glossary of Arctic and Subarctic Terms*, ADTIC Pub. A-105, Maxwell AFB, AL.

drift bottle—(Also called satellite.) A bottle, of one of various designs, that is released into the sea for use in studying currents.
It contains a card, identifying the date and place of release, to be returned by the finder with the date and place of recovery. The bottle should be so designed and ballasted as to minimize direct **wind** effects. Since the path of a bottle can only be estimated between release point and recovery, and generally only a few percent are returned, this is an inefficient, although inexpensive, technique. *Compare* **drift card**.

drift card—A card, such as is used in a **drift bottle**, encased in a buoyant, waterproof envelope and released in the same manner as a drift bottle, for the purpose of estimating the **surface current** in a body of water.
Cheaper and lighter than bottles, drift cards are especially suited to dropping in large quantities

from aircraft, and it is assumed that the card, having less freeboard than a bottle, is less affected by **wind**.

drift-correction angle—(Or drift correction; also called crab angle.) In air navigation, the angle between an aircraft's **course** and its **heading** required to maintain that course against the **wind**.
See **single drift correction, multiple drift correction**.

drift current—The current computed by dividing the distance moved by a real object, or hypothetical **particle** moving with the fluid, by the time during which the movement takes place.

drift ice—Any **sea ice** that has drifted from its place of origin.
The term is used in a wide sense to include any areas of sea ice, other than **fast ice**, no matter what form it takes or how disposed.
U.S.N. Hydrographic Office, 1952: *A Functional Glossary of Ice Terminology*, Circular M.

drift station—A research **station** deployed on a drifting **ice floe**.

drifter—Any object, for example, a **Swallow float**, that is passively drifting with the prevailing currents and can be used to obtain approximately Lagrangian current measurements.

drifting buoy—A **drifter** that is fixed to a buoy that can be tracked by satellite, radio, or other means.
Surface and subsurface drifting buoys can obtain measurements of **hydrographic** (e.g., tem-**perature**) and chemical (e.g., fluorometry) parameters in addition to providing position information that is used to calculate currents. Surface and subsurface drogues can provide information on the currents at any chosen depth or **density** horizon.

drifting dust—Surface **dust** (very fine solid **particles** of order 10–100 μm) that is moved by the **wind**.
See **saltation, drifting sand, airborne particulates**.

drifting sand—Surface sand that is moved by the **wind**, such as that which forms sand dunes.
See **saltation, drifting dust, airborne particulates**.

drifting snow—Snow raised from the surface of the earth by the **wind** to a height of less than 6 ft above the surface.
In **aviation weather observations**, drifting snow is not regarded as an **obstruction to vision** because it does not restrict **horizontal visibility** at 6 ft or more above the surface. When **snow** is raised 6 ft or more above the surface, it is classified as **blowing snow**.

driller's log—A record of the geologic materials encountered during drilling.
The data for a driller's log are collected by obtaining samples of the formations during the drilling process, and noting where and at what depth(s) the samples were collected and how and where the samples change with depth. The thickness of geologic formations and the depth to water are also noted. An ideal driller's log will also have a detailed and accurate description of the surface location of the well, borehole, or **piezometer** that is drilled.

drip—Condensed or otherwise collected moisture falling from leaves, twigs, etc.

driven snow—Snow that has been moved by **wind** and collected into **snowdrifts**.
A **wind speed** of about 13 mph will move light surface snow. In weather reports, snow lifted less than 6 ft above the surface is **drifting snow**, more than 6 ft is **blowing snow**.

driving-rain index—Quantity that gives a measure of the amount of wind-driven **rainfall** on a vertical surface.
It is the product of the average rainfall and the average **wind** over a specified period.

drizzle—(Sometimes popularly called mist.) Very small, numerous, and uniformly distributed water drops that may appear to float while following **air currents**.
Unlike **fog** droplets, drizzle falls to the ground. It usually falls from low **stratus** clouds and is frequently accompanied by low **visibility** and fog. In **weather observations**, drizzle is classified as 1) light, the rate of fall being from a **trace** to 0.3 mm (0.01 in.) per hour; 2) moderate, the rate of fall being from 0.3 to 0.5 mm (0.01–0.02 in.) per hour, and 3) heavy, the rate of fall being greater than 0.5 mm (0.02 in.) per hour. When **precipitation** equals or exceeds 1 mm (0.04 in.) per hour, all or part of the precipitation is usually **rain**. However, true drizzle falling as heavily as 1.3 mm (0.05 in.) per hour has been observed. By convention, **drizzle drops** are taken to be less than 0.5 mm (0.02 in.) in diameter. Larger drops are considered **raindrops**. *Compare* **mist**.

drizzle drop—A **drop** of water of diameter 0.2–0.5 mm (0.008–0.02 in.) falling usually (but not always) from low **stratus** or **stratocumulus** cloud.
Although this is the correct term for this size **range**, all water drops of diameter greater than

0.2 mm are frequently termed **raindrops**, as opposed to **cloud drops**. Should such drops reach the ground, they can be felt on the upturned face.

drogue—A sea anchor or other parachute-shaped device for use in water.

Drogues suspended at desired depths from buoys are used to determine the **set** and **drift** of currents at those depths, by tracking the motions that they give to the buoys at the surface.

drop—Liquid **particle**, with shape maintained as a balance between **surface tension** and air **drag** when falling at terminal **velocity** under **gravity** in the **atmosphere**; drops less than 1 mm (0.04 in.) are approximately spherical.

The shape may also be influenced by ambient **electric field** or **thunderstorm** strength.

drop breakup—The disruption of **raindrops** caused primarily by collisions with other raindrops.

The distribution of **drop** fragments resulting from collisions is a complicated function of drop sizes, but generally the number of fragments increases with drop size. Thus, larger raindrops are more susceptible to breakup. Drop breakup is a limiting factor to the growth of raindrops by the **collision–coalescence process**.

drop-size distribution—The **frequency distribution** of **drop** sizes (diameters, volumes) that is characteristic of a given **cloud** or of a given fall of **rain**.

Most natural clouds have unimodal (single maximum) distributions, but occasionally bimodal distributions are observed. In convective clouds, the drop-size distribution is found to change with time and to vary systematically with height, the modal size increasing and the number decreasing with height. For many purposes a useful single **parameter** representing a given distribution is the volume median diameter, that is, that diameter for which the total volume of all drops having greater diameters is just equal to the total volume of all drops having smaller diameters. The drop-size distribution is one of the primary factors involved in determining the **radar reflectivity** of any fall of **precipitation**, or of a **cloud** mass.

drop-size spectrometer—(Formerly called drop-size meter; also called **disdrometer**.) Any device that measures the size distribution of **cloud droplets** or **raindrops**, for example, certain optical **particle** probes.

droplet—A small spherical **particle** of any liquid; in meteorology, particularly a water droplet.

There is no defined size limit separating droplets from drops of water, but it is sometimes convenient to denote two disparate size ranges, such as the oft-used distinction of liquid **cloud** particles (droplets) from liquid **precipitation** (drops), thereby implying that a maximum diameter of 0.2 mm (0.008 in.) is the limit for droplets. *See* **cloud droplet, drizzle drop, drop, raindrop.**

droplet spectrum—Same as **drop-size distribution**.

dropsonde—(Also called parachute radiosonde.) A **radiosonde** with a parachute dropped from an airplane carrying receiving equipment for the purpose of obtaining an **upper-air sounding** during descent.

dropsonde observation—An evaluation of meteorological data received from a descending **dropsonde**.

The dropsonde is a small expendable instrument package that is released from an aircraft. As it descends, it radio transmits **pressure**, **temperature**, and **relative humidity** data back to the aircraft. By tracking the position of the dropsonde using radio-navigation techniques [e.g., **Global Positioning System** (GPS)], **wind speed** and direction data can be obtained. The processed data are usually presented in terms of height, **temperature**, **dewpoint**, and **winds** at mandatory and significant pressure levels. A dropsonde observation is comparable to a **rawinsonde observation**. Data from the dropsonde are usually received and processed in the aircraft that dropped the instrument.

drosometer—An instrument used to measure the amount of **dew** formed on a given surface.

One type consists of a hemispheric glass vacuum cup exposed to the **atmosphere**. Dew forming on the glass surface automatically collects in the bottom of the cup and is weighed at the end of the **exposure** period. Another form of drosometer, the **Duvdevani dew gauge**, consists of a block of wood with its surface treated in such a manner that dew forms in characteristic patterns. Photographs are supplied with each instrument to enable the **observer** to match the dew formation with a set of standards corresponding to a dew "fall" of from 0.01 to 0.45 mm (0.0004–0.018 in.).

drought—(Sometimes spelled drouth.) A period of abnormally dry weather sufficiently long enough to cause a serious hydrological imbalance.

Drought is a relative term, therefore any discussion in terms of **precipitation** deficit must refer to the particular precipitation-related activity that is under discussion. For example, there may be a shortage of precipitation during the **growing season** resulting in crop damage (**agricultural drought**), or during the winter **runoff** and **percolation** season affecting water supplies (**hydrological drought**). *Compare* **dry spell**; *see* **absolute drought, partial drought**.

drought index—Computed value related to some of the cumulative effects of a prolonged and **abnormal** moisture deficiency; an index of **hydrological drought** corresponding to levels below the mean in streams, lakes, reservoirs, and the like.

　　However, an index of the **agricultural drought** must relate to the cumulative effects of either an **absolute** or **abnormal** transpiration deficit (World Meteorological Organization). *See* **Palmer Drought Severity Index, standard precipitation index**.

drouth—Same as **drought**.

droxtal—A tiny **ice** particle, about 10–20 μm in diameter, formed by direct **freezing** of **supercooled water** droplets with little growth directly from the **vapor**.

　　The term combines the words **drop** and **crystal**. Droxtals cause most of the **visibility** reduction in **ice fog**.

dry adiabat—A line of constant **potential temperature** on a **thermodynamic diagram**.

　　In terms of pressure p, and **specific volume** α, the equation for a dry adiabat may be written

$$p\alpha^{c_p/c_v} = \text{constant},$$

where c_p and c_v are the **specific heats** of **dry air** at constant pressure and volume, respectively. Meteorologically, the dry adiabat is intended to represent the lifting of dry air in a **dry-adiabatic process**. Since this is also an **isentropic** process, a dry adiabat is an **isentrope**. *Compare* **saturation adiabat**; *see* **Poisson equation**.

dry-adiabatic atmosphere—Same as **adiabatic atmosphere**.

dry-adiabatic lapse rate—1. A **process lapse rate** of **temperature**, the rate of decrease of temperature with height of a **parcel** of **dry air** lifted by a reversible **adiabatic process** through an **atmosphere** in **hydrostatic equilibrium**.

　　This **lapse rate** is g/c_{pd}, where g is gravitational **acceleration** and c_{pd} is the **specific heat** of dry air at constant **pressure**, approximately 9.8°C km^{-1}. Potential temperature is constant with height in an **atmospheric layer** with this lapse rate.

　　2. The **adiabatic lapse rate** of unsaturated air containing **water vapor**.

　　This differs from definition 1 by the factor

$$\frac{1 + r_v}{1 + r_v c_{pv}/c_{pd}} \approx 1 - 0.85 r_v,$$

where r_v is the **mixing ratio** of water vapor and c_{pv} is the **specific heat** of water vapor.

dry-adiabatic process—1. An **adiabatic process** in a hypothetical **atmosphere** in which no moisture is present.

　　2. An **adiabatic process** in which no **condensation** of its **water vapor** occurs and no liquid water is present.

　　See **dry-adiabatic lapse rate**.

dry air—1. In **atmospheric thermodynamics** and chemistry, air that is assumed to contain no **water vapor**.

　　Compare **moist air**.

　　2. Generally, air with low **relative humidity**.

dry-bulb temperature—Technically, the **temperature** registered by the **dry-bulb thermometer** of a **psychrometer**.

　　However, it is identical to the temperature of the air and may also be used in that sense. *Compare* **wet-bulb temperature**.

dry-bulb thermometer—In a **psychrometer**, the **thermometer** that has a dry bulb and therefore directly measures **air temperature**.

　　See **wet-bulb thermometer**.

dry climate—1. In W. Köppen's 1918 **climatic classification**, the major category (B climates), which includes **steppe climate** and **desert climate**.

　　These climates, unlike the others in his work, are defined strictly by the amount of annual

precipitation as a function of seasonal distribution and of annual **temperature** (see formulas under **steppe climate**). In contrast are the **rainy climates**.

2. In C. W. Thornthwaite's 1948 **climatic classification**, any **climate** type in which the seasonal **water surplus** does not counteract seasonal water deficiency; thus it has a **moisture index** of less than zero.

These types include the **dry subhumid, semiarid**, and **arid climates**. In contrast are the **moist climates**. The dry climates are subdivided further according to values of **humidity index** into the following: little or no water surplus; moderate winter water surplus; moderate summer water surplus; large winter water surplus; large summer water surplus.

Köppen, W. P., 1918: Klassifikation der Klimate nach Temperatur, Niederschlag und Jahreslauf. *Petermanns Geog. Mitt.*, **64**, 193–203; 243–248.

Thornthwaite, C. W., 1948: An approach toward a rational classification of climate. *Geogr. Rev.*, **38**, 55–94.

dry convection—A vertical exchange of air without **precipitation** at the ground.

dry deposition—The process by which atmospheric gases and **particles** are transferred to the surface as a result of **random** turbulent air motions.
See also **deposition**; *compare* **wet deposition**.

dry enthalpy—*See* **enthalpy**.

dry fog—1. A **fog** that does not moisten exposed surfaces.

2. (Rare.) A condition of reduced **visibility** due to the presence of **dust, smoke**, or **haze** in the air.

It is not a true **fog**.

dry freeze—The **freezing** of the soil and terrestrial objects caused by a reduction of **temperature** when the adjacent air does not contain sufficient moisture for the formation of **hoarfrost** on exposed surfaces.

With respect to vegetation alone, this is termed a **black frost**. A dry freeze is usually considered to be a more local and short-period (probably radiative) phenomenon than a **freeze**.

dry growth—The condition in the **accretion** of **supercooled water** droplets onto **ice** in which the **temperature** of the ice remains below the **freezing point** and the surface is mostly dry, except locally immediately after the **drop** collisions.

The process of dry growth produces ice with many air cavities and bubbles, called **rime**. *Compare* **wet growth**; *see* **hailstone**.

dry haze—**Haze** occurring with conditions of low **relative humidity**.

dry ice—Solidified **carbon dioxide** that, at −78.5°C and **ambient pressure**, changes directly to a gas as it absorbs **heat**.

It is used as a coolant to induce the **ice** phase for **supercooled cloud** and **fog** modification procedures.

dry-ice seeding—*See* **cloud seeding**.

dry permafrost—**Permafrost** that contains little or no **ice**; it is loose and crumbly.

dry season—In certain types of **climate**, an annually recurring period of one or more months during which **precipitation** is at a minimum for that region; the opposite of **rainy season**.

dry snow—**Snow** from which a snowball cannot readily be made.

dry spell—A period of **precipitation** below a specified amount.

The specific period and **amount of precipitation** vary depending on the particular activity under discussion.

dry stage—Part of an obsolete conceptual **model** of **air parcel** ascent referring to that portion of the ascent during which the parcel is unsaturated.

Other portions of the ascent were described as the **rain stage**, the **snow stage**, and the **hail stage**.

dry static energy—(Also called the Montgomery streamfunction.) A **thermodynamic variable** similar to **potential temperature**, except that the concept of **static energy** assumes that any **kinetic energy** is locally dissipated into **heat**.

The amount of this dissipative heating is often negligible. When dry static energy, s, is expressed in units of kJ kg^{-1}, the resulting values are of order 300 kJ kg^{-1}, which reinforces the analogy with potential temperatures in units of Kelvin. Dry static energy is conserved during unsaturated vertical and horizontal motion, and is defined as

$$s = c_p T + gz,$$

where c_p is the **specific heat capacity** of air at constant **pressure**, T is **absolute** temperature, g is gravitational **acceleration**, and z is height. The reference height can be arbitrary; it is sometimes taken as $z = 0$ at $P = 100$ kPa to be consistent with potential temperature, or it can be defined relative to the local ground or **sea level**. *Compare* **moist static energy, liquid water static energy, saturation static energy**.

dry subhumid climate—*See* **subhumid climate**.

dry tongue—In **synoptic meteorology**, a pronounced protrusion of relatively **dry air** into a region of higher **moisture content**, sometimes associated with the **jet stream** or **jet streaks** at upper levels.

This term is more frequently used than its opposite, **moist tongue**.

dry year—Years in which **streamflow** records show **runoff** significantly less than the mean annual runoff.

dryline—A low-level **mesoscale** boundary or transition zone hundreds of kilometers in length and up to tens of kilometers in width separating **dry air** from **moist air**.

The length of the dryline is related to large-scale terrain or large-scale weather system features, whereas its width is related to mesoscale processes. In its quiescent state, the dryline may be considered the intersection of the top of a low-level moist layer with large-scale features of sloping terrain. In this state the shallow layer of moisture near the higher terrain is eroded by turbulent **mixing** with daytime heating. Moisture gradients are additionally strengthened by horizontal **convergence** resulting from downward **transport** of horizontal **momentum** in the **dry air**. In a more dynamically active state the dryline often advances away from the higher terrain as an integral component of an **extratropical cyclone** or **frontal wave**. In such cases it extends equatorward from the cyclone or wave. In this state moisture gradients and boundary motion are largely influenced by downward transport of horizontal momentum resulting from larger-scale sinking in the dry air. The dryline is found all over the world. In the United States the dryline, which marks the boundary between moist air from the Gulf of Mexico and dry **continental air** from the west, is found in the Plains region. It is most often present during the spring, where it is often the site of **thunderstorm** development. Typically the dryline in the United States advances eastward during the day and retreats westward at night.

dryth—In England, a dry northerly or easterly **wind**.

dual-channel radar—A **radar** capable of simultaneously receiving signals with polarizations identical and **orthogonal** to that of the transmitted **signal**.

Compare **dual-polarization radar, polarimetric radar**, and **polarization-diversity radar**.

dual-Doppler analysis—An analysis technique that makes use of **radial velocity** measurements from two or more **Doppler radars** or **lidars** to determine the **wind field** within an area of **precipitation** or any other region of space in which there are adequate radar **targets**. Observations are made from two or more separate vantage points of the radial velocities of **echoes** from a given region of the **atmosphere**.

Typically, the radial components are combined to deduce the **field** of the horizontal **wind vector** on one or more surfaces in the atmosphere. The surfaces may be those that are actually scanned (often by **coplane scanning**) or a series of horizontal planes. The three-dimensional wind vector may then be estimated by applying the **continuity equation**, which requires computing the field of **horizontal divergence** at each of the altitudes within a layer and integrating the divergence over **altitude** using assumed values of the **vertical velocity** at the top or bottom of the layer as **boundary conditions**.

dual-frequency radar—Same as **dual-wavelength radar**.

dual-polarization radar—A **radar** capable of transmitting and receiving two **orthogonal** polarizations.

The transmitted **polarization** must be switchable at a rate that is fast compared with the timescale of changes in the **scattering** properties of the **target** and the propagation medium. *Compare* **dual-channel radar, polarimetric radar, polarization-diversity radar**.

dual-wavelength radar—A **radar** capable of transmitting signals having two wavelengths and measuring separately the **echoes** at the two **wavelengths**. Properties of the **scattering** medium (**cloud, precipitation**, or the **clear air**) may be deduced from the difference in **reflectivity** or the difference in **attenuation** at the two wavelengths.

See **differential reflectivity, differential attenuation**.

duct—Applied to the **atmosphere** and ocean, any region with vertically varying properties such that waves of any kind (e.g., **electromagnetic** and **acoustic**) launched in certain directions are guided by or trapped within the region rather than propagating radially from their source.

For a duct to exist, **attenuation** must be negligible over distances comparable to the characteristic linear dimensions of the duct.

duff—The vegetative matter, such as leaves, twigs, dead logs, etc., that covers the ground in the forest; unconsolidated decomposing and partially decomposed organic material immediately under a layer of leaf litter.

Duff forms a layer about 5 cm (2 in.) thick that overlies the soil of a forest floor. Its **thermal** insulation is an important factor in the formation of **permafrost**, and the quality and **moisture content** of duff is significant in considerations of forest fire hazard (**fire weather**).

duplexer—A **radar** device that allows a single **antenna** to serve as both the **transmitter** and the **receiver**.

Modern radars use a duplexer consisting of a ferrite circulator and a passive, radioactive gas-primed TR tube to protect the receiver.

duplicatus—A **cloud variety** composed of superimposed layers, sheets, or patches.

These stratified parts, at slightly different levels, are sometimes partly merged. This variety modifies the species **fibratus**, **uncinus**, **stratiformis**, and **lenticularis** in the genera **cirrus**, **cirrostratus**, **altocumulus**, and **altostratus**. *See* **cloud classification**.

Dupuit–Forchheimer assumptions—Simplifying approximation that water flow is horizontal and evenly distributed with depth below a **water table**.

The approximation is valid for mildly sloping water tables and for flow that is constrained to shallow depths.

Forchheimer, P., 1930: *Hydraulik*, 3d ed., Teubner, Leipzig, Berlin.

duration—In connection with **wind waves**, the duration is the length of time that winds generating **surface waves** have been present.

This, in addition to the **fetch** (which is the distance over which these winds act) and **wind direction**, determines the **amplitude** and direction of the wind-generated waves.

duration curve—A curve indicating the total time during which event data exceed specified values.

duration statistics—Statistics of the durations of **calm** periods and storms in connection with **wind waves**.

The expected durations are useful when planning activities that may require a certain number of hours, days or weeks to be completed. For example, the expected **duration** of wave heights below a specific value are useful when planning critical construction operations at offshore locations.

düsenwind—(Literally "jet wind" or "blast wind.") The **mountain-gap wind** of the Dardenelles; a strong east-northeast **wind** that blows out of the Dardanelles into the Aegean Sea, penetrating as far as the island of Lemnos.

It is caused by a **ridge of high pressure** over the Black Sea. *See* **jet-effect wind**.

dusk—The part of morning or evening **twilight** between complete darkness and **civil twilight** (i.e., the combined nautical and astronomical twilights).

In nontechnical usage, dusk is the evening counterpart of **dawn**. In technical usage, morning dusk is the first part of dawn.

dust—Solid materials suspended in the **atmosphere** in the form of small irregular **particles**, many of which are microscopic in size.

It imparts a tannish or grayish hue to distant objects. The sun's disk is pale or colorless or has a yellowish tinge at all periods of the day. Dust cannot be a stable component of the atmosphere because it must eventually fall back to the earth's surface when winds and **turbulence** become too weak to bear it aloft. Dust is due to many natural and artificial sources, for example, volcanic eruptions, salt spray from the seas, blowing solid particles, plant pollen, bacteria, and **smoke** and ashes from forest fires and industrial combustion processes. It was once thought that dust particles were a main source of **condensation nuclei**; this is no longer regarded as probable as most dusts are not sufficiently **hygroscopic**. *Compare* **smoke**, **haze**; *see* **duststorm**, **dust devil**.

dust avalanche—An **avalanche** of dry loose **snow**.

Dust Bowl—A name given, early in 1935, to the region in the south-central United States affected by **drought** and **duststorms** at that time.

It included parts of five states: Colorado, Kansas, New Mexico, Texas, and Oklahoma. It resulted from a long period of deficient **rainfall** combined with loosening of the soil by destruction of the natural vegetation. The name has since been extended to similar regions in other parts of the world.

dust counter—(Also called Kern counter, nucleus counter.) General term for an instrument that measures the size and number of **dust** particles in a known volume of air.

See **Aitken dust counter, cascade impactor, Owens dust recorder, nucleus counter, konimeter.**

dust devil—A well-developed **dust whirl**; a small but vigorous **whirlwind**, usually of short **duration**, rendered visible by **dust**, sand, and debris picked up from the ground.

Dust devils occasionally are strong enough to cause minor damage (up to F1 on the **Fujita scale**). Diameters **range** from about 3 m to greater than 30 m; their average height is about 200 m, but a few have been observed as high as 1 km or more. They have been observed to rotate anticyclonically as well as cyclonically. Although the **vertical velocity** is predominantly upward, the flow along the axis of large dust devils may be downward. Large dust devils may also contain secondary vortices. Dust devils are best developed on a hot, **calm** afternoon with **clear** skies, in a dry region when intense surface heating causes a very steep **lapse rate** of **temperature** in the lowest 100 m of the **atmosphere**.

dust-devil effect—In **atmospheric electricity**, a rather sudden and short-lived change of the vertical component of the **atmospheric electric field** that accompanies the passage of a **dust devil** near an instrument sensitive to the vertical **gradient**.

Such changes may be either positive or negative and the charge is probably produced by **triboelectrification**.

dust electrification—The electrification of **dust** by frictional effects produced by the movement of the dust over surfaces.

dust haze—*See* **haze.**

dust horizon—The top of a **dust** layer confined by a low-level **temperature inversion** with the appearance of the **horizon** when viewed from above against the sky.

In such instances the **true horizon** is usually obscured by the dust layer. Similarly defined are **fog horizon** and **smoke horizon**.

dust wall—The leading edge of a **duststorm**, which looks like a knobby vertical or convex wall when viewed from the **clear air** ahead of the storm.

Sharply defined dust walls are caused by **gust fronts** at the leading edge of **thunderstorm** straight-line **outflow** winds along the surface. Sometimes **surface friction** will slow the air immediately touching the surface resulting in a dust wall that curves back at the bottom. *See also* **haboob.**

dust well—A pit in an **ice** surface produced by small, dark particles on the ice.

This phenomenon occurs primarily during sunny conditions when the **dust** absorbs more solar **shortwave radiation** than the regional ice causing the dust to melt down into the ice. *Compare* **cryoconite hole.**

dust whirl—(Also called dancing dervish, dancing devil, devil, satan, shaitan; and, over **desert** areas, desert devil, sand auger, sand devil.) A rapidly rotating column of air (**whirlwind**) over a dry and dusty or shady area, carrying **dust**, leaves, and other light material picked up from the ground.

When well developed it is known as a **dust devil**. Dust whirls typically form as the result of strong **convection** during sunny, hot, **calm** summer afternoons. This type is generally several meters in diameter at the base, narrowing for a short distance upward and then expanding again, like two cones apex to apex. Their height varies; normally it is only 30–100 m, but in hot **desert** country it may be as high as 1 km. Rotation may be either clockwise or counterclockwise. Dust whirls move erratically, from one patch of heated air to another, and generally slowly. In desert country it is not unusual for three or more desert devils to be visible at the same time. Another type of vigorous dust whirl occurs under the bases of **cumulonimbus** or **cumulus** clouds, almost always on or near a **wind-shift line**. These vortices often inflict little or no damage and are short-lived, but occasionally represent the first visible sign of a developing **tornado**. Another form of dust whirl, often seen at street corners, is merely an **eddy** caused by the meeting of winds blowing along two intersecting streets. Such whirls are small and very short-lived. *Compare* **duststorm.**

duster—Same as **duststorm.**

dustfall jars—Devices for the removal of relatively large **particles** from **ambient air** by means of gravitational **sedimentation**.

duststorm—(Or dust storm; also called duster, **black blizzard**.) An unusual, frequently **severe weather** condition characterized by strong winds and dust-filled air over an extensive area.

Prerequisite to a duststorm is a period of **drought** over an area of normally arable land, thus providing the very **fine particles** of **dust** that distinguish it from the more common **sandstorm** of **desert** regions. A duststorm usually arises suddenly in the form of an advancing **dust wall** that may be many kilometers long and and a kilometer or so deep, ahead of which there may be some **dust whirls**, either detached or merging with the main mass. Ahead of the dust wall the air is very hot and the **wind** is light. In U.S. weather observing practice, if **blowing dust** reduces **visibility** to between 5/8 and 5/16 **statute mile**, a "duststorm" is reported; if the **visibility** is reduced to below 5/16 statute mile, it is reported as a "severe duststorm." Duststorm winds can also be associated with **thunderstorm** outflows and **gust fronts**. While these are often shorter-lived than synoptically forced duststorms, they can be quite intense, with an impressive leading edge of the dusty gust front, sometimes called a dust wall. *Compare* **haboob**.

duty of water—The total volume of **irrigation** water required for a particular type of crop to mature.

It includes **consumptive use**, **evaporation** and seepage from ditches and canals, and the water eventually returned to streams by **percolation** and **surface runoff**.

Duvdevani dew gauge—*See* **drosometer**.

Dvorak technique—An **analysis** procedure, named after Vern Dvorak, for determining **hurricane** intensity from **cloud** patterns in satellite images.

> Dvorak, V. F., 1975: Tropical cyclone intensity analysis and forecasting from satellite imagery. *Mon. Wea. Rev.*, **103**, 420–430.
> Dvorak, V. F., 1984: Tropical cyclone intensity analysis using satellite data. NOAA Technical Report NESDIS 11, U.S. Dept. of Commerce, Washington, D.C., 47 pp.

dwigh—(Also called dwey, dwoy.) In Newfoundland, a sudden **shower** or **snowstorm**.

dye laser—Laser in which the **radiation** from a fixed-wavelength **laser** is focused into an organic dye, which then emits at a longer **wavelength**.

The resulting radiation, usually in the visible **range**, is tunable, so a much wider range of molecules can be detected. The lasers can be either pulsed or continuous. Uses include **lidar**, sensing of atmospheric **trace gases**, particularly **free radicals**, and the **detection** of atmospheric gases in laboratory kinetics experiments.

dynamic boundary condition—The condition that the **pressure** must be continuous across an **internal boundary** or **free surface** in a fluid.

This condition is applied in meteorology, for example, at **frontal surfaces** and at the **tropopause**. See also **boundary conditions**.

dynamic climatology—The **statistical** collation and study of observed elements (or derived parameters) of the **atmosphere**, particularly in relation to the physical and dynamical explanation or interpretation either of the contemporary **climate** patterns with their **anomalous** fluctuations or of the long-term climate changes or trends.

dynamic flux—Transport of a quantity, expressed in units of that quantity crossing a unit area per unit time.

Examples are **heat flux** in units of $J\ m^{-2}\ s^{-1}$ (equivalent to $W\ m^{-2}$), **momentum flux** in units of $kg\ (m\ s^{-1})(m^2\ s)^{-1}$ (equivalent to $kg\ m^{-1}\ s^{-2}$), or mass **flux** in units of $kg\ m^{-2}\ s^{-1}$. When a flux in dynamic units is divided by air density (or for heat flux by air **density** times **specific heat** of air at constant **pressure**), the result is the flux in kinematic units, which is often easier to measure by conventional weather instruments such as **anemometers** and **thermometers**.

dynamic height—1. (Also called geodynamic height.) The height of a point in the **atmosphere** expressed in a unit proportional to the **geopotential** at that point.

Since the geopotential at **altitude** z is numerically equal to the **work** done when a **particle** of unit mass is lifted from **sea level** up to this height, the **dimensions** of dynamic height are those of **potential energy** per unit mass. The **standard** unit of dynamic height is the **dynamic meter** (or **geodynamic meter**). One of the practical advantages of the dynamic height over the geometric height is that when the former is introduced into the **hydrostatic equation** the height acceleration of **gravity** is eliminated. In meteorological height calculations **geopotential height** is more often used than dynamic height. In **oceanography**, dynamic computations are also based upon units of dynamic height (or dynamic depth).

2. In **oceanography**, represents the ability of a column of water to do **work** due to differences in **geopotential** (the potential for **gravity** to do work because of height of the water relative to some reference level).

The dynamic height is computed from the measured **density** distribution. **Geopotential height** differences, expressed by changes in dynamic **topography**, are a measure of the horizontal **pressure gradient** force.

dynamic-height anomaly—(Also called anomaly of geopotential differences.) In **oceanography**, the excess of the actual **geopotential** difference, between two given **isobaric surfaces**, over the geopotential difference in a homogeneous water column of **salinity** of 35 psu and **temperature** 0°C.

dynamic initialization—A procedure that adjusts the initial conditions of an **NWP** model so that unwanted high-frequency oscillations are removed.
   The procedure requires that the NWP model be integrated forward and backward around the initial time using a heavily damped **finite-difference equation**.
   Daley, R., 1991: *Atmospheric Data Analysis*, 322–332.

dynamic instability—1. *See* **instability**.
   2. Same as **inertial instability**.

dynamic meteorology—The study of atmospheric motions as solutions of the **fundamental equations of hydrodynamics** or other systems of equations appropriate to special situations, as in the **statistical** theory of **turbulence**.
   The restrictions of this definition suffice to distinguish dynamic meteorology from other fields, for example, **physical meteorology** or **synoptic meteorology**, such distinctions being a function of the state of the science rather than of the subject matter itself.

dynamic meter—(Also called geodynamic meter.) The **standard** unit of **dynamic height**, defined as 10 m² s⁻²; it is related to the **geopotential** φ, the geometric height z in meters, and the **geopotential height** Z in geopotential meters by

$$d\phi = 10 \ d\psi = 9.8 \ dZ = gdz,$$

where g is the **acceleration of gravity** in meters per second squared. (Some sources prefer to give the constants 10 and 9.8 the units of meters per second squared so that the units of φ and Z would be the same as those of the geometric height.)
   The dynamic meter is about 2% longer than the geometric meter and the **geopotential meter**.

dynamic oceanography—Application of the concepts and methods of dynamics (a branch of physics) to the ocean; the study of the effects of dynamical forces on ocean behavior, motion, and water properties.
   The term "dynamic method" is used to describe a method whereby the **baroclinic** component of the **geostrophic current** is estimated through measurements of **density** structure.

dynamic pressure—(Also called velocity pressure, stagnation pressure.) In engineering fluid mechanics, the **kinetic energy**, $(1/2)\rho V^2$, of the fluid, where ρ is the **density** and V the speed.
   This applies in cases where this quantity may be conveniently considered as adding to the **static pressure**; that is, the dynamic pressure at a given point is the difference between the static pressure at that point and the **total pressure** at the stagnation point of the same **streamline**. This concept must be distinguished from the **hydrodynamic pressure**, and the terminology is confusing in meteorological contexts.

dynamic range—A measure of the ability of an amplifier, **transducer**, **receiver**, or other kind of **sensor** to measure both weak and strong signals.
   Specifically, the **range** of **input** levels, ordinarily expressed in **decibels**, over which the system can operate within some specified range of performance; for example, the range over which the response of the system is **linear** or approximately linear.

dynamic similarity—Airflow or **wind** characteristics that behave alike in different situations.
   For example, a vertical **wind profile** in one location might have the same shape as a wind profile in a different location, except for differences in depth or **wind speed**. When such vertical profiles are expressed in terms of dimensionless height and dimensionless wind profile, then the profiles often become a common curve that can be described by a single **dimensionless equation**. *See* **similarity theory**.

dynamic stability—A measure of the ability of a fluid to resist or recover from finite perturbations of a **steady state**.
   A negative value of dynamic stability is equivalent to **dynamic instability**.

dynamic trough—Same as **lee trough**.

dynamic viscosity—(Also called coefficient of molecular viscosity, coefficient of viscosity.) A fluid property defined as the ratio of the **shearing stress** to the **shear** of the motion.

It is independent of the **velocity distribution**, the dimensions of the system, etc., and for a gas it is independent of **pressure** except at very low pressures. For the dynamic viscosity μ of a **perfect gas**, the **kinetic theory** of gases gives

$$\mu = \frac{1}{3}\rho cL,$$

where ρ is the gas **density**, c is the average speed of the **random** heat motion of the gas molecules and is proportional to the square root of the **temperature**, and L is the **mean free path**. For **dry air** at 0°C, the dynamic viscosity is about 1.7×10^{-4} g cm^{-1}s^{-1}. While the dynamic viscosity of most gases increases with increasing temperature, that of most liquids, including water, decreases rapidly with increasing temperature. *See* **kinematic viscosity, eddy viscosity, Newtonian friction law**.

dynamical extended range forecasting—(Abbreviated DERF.) Forecasting the state of the **atmosphere** (using a **numerical weather prediction** model with observed initial conditions) beyond the range for which individual **cyclonic-scale** disturbances are predictable.

Although, in general, the useful information regarding cyclonic-scale disturbances has been lost in the extended range, the behavior of large-scale, lower-frequency phenomena may still be predictable.

dynamical forecasting—*See* **numerical forecasting**.

dynamical sublayer—A layer of order 1–10 m in thickness adjacent to the surface, where influences of **stratification** and **Coriolis force** are negligible, but far enough away from the surface that viscous and individual roughness-element effects are negligible.

Namely, it is where $z << |L|$, where z is height above the surface and L is **Obukhov length**, the depth of the sublayer.

Brutsaert, W., 1982: *Evaporation into the Atmosphere*, Kluwer, 57–64.

dynamical system—In the most general sense, any process or set of processes that evolves in time and in which the evolution is governed by some set of physical laws.

The term is also used to refer to mathematical models that evolve in time. These systems can range from relatively simple, for instance, systems of a few **variables** governed by a few equations, to extremely complex, like the complete **climate system**.

dynamo theory—The hypothesis, first proposed by Balfour Stewart, that explains the regular daily variations in the earth's **magnetic field** in terms of electrical currents in the lower **ionosphere**, generated by tidal motions of the ionized air across the earth's magnetic field.

Whitten, R. C., and I. G. Poppoff, 1971: *Fundamentals of Aeronomy*, Wiley & Sons, New York, p. 219.

dyne—The unit of force in the **centimeter–gram–second system** of physical units, that is, one gm cm s^{-2}, equal to 7.233×10^{-5} poundal.

dyster—Same as **doister**.

E

E-ε closure—A **turbulence closure** of one-and-a-half **statistical** order, where forecast equations are retained for mean wind and **temperature**, for **wind** and temperature **variance**, and for **molecular dissipation** of wind and temperature.

Other unknowns such as fluxes or variables of higher statistical order are approximated or parameterized as a function of the mean quantities, variances E, and **dissipation rates** ϵ.

E-layer—*See* **E-region**.

E-region—(Also called Heaviside layer, Kennelly–Heaviside layer.) The region of the **ionosphere** usually found at an **altitude** between 100 and 120 km.

It exhibits one or more distinct maxima and sharp gradients of free **electron density**. It is most pronounced in the daytime but does not entirely disappear at night. Ionosonde recordings show that the E-region is often subdivided into two or more "E-layers," while localized and intermittent regions of fairly high ionization, known as sporadic E-layers, are also frequently observed. The E-region is produced by **absorption** of **solar radiation** at a variety of extreme ultraviolet (EUV) and **x-ray** wavelengths.

eagre—Same as **bore**.

EAPE—Abbreviation for **evaporative available potential energy**.

earth—The solid, liquid, and gaseous parts of the planet taken as a whole.

Near-earth space (such as the **magnetosphere**) is often included.

earth–air current—*See* **air–earth conduction current**.

earth–atmosphere radiation budget—The combination of the **atmospheric radiation budget** and the **surface radiation budget**.

This is also the net **radiant flux density** (solar plus terrestrial) at the top of the **atmosphere**, averaged over a specified time interval.

earth current—A large-scale surge of **electric charge** within the conductive earth, associated with a disturbance of the **ionosphere**.

Current patterns of quasi-circular form and extending over areas the size of whole continents have been identified and are known to be closely related to solar-induced variations in the extreme **upper atmosphere**.

earth-current storm—(Rarely, electrical storm.) Irregular fluctuations in an **earth current**, often associated with **electric field** strengths as large as several volts per kilometer, in the earth's crust, superimposed on the **normal** diurnal **variation** of the earth currents.

Such storms are closely related to **magnetic storms**.

earth hummock—(Also called earth mound, knoll.) A low, round, natural, or man-made hill.

earth inductor—A coil of wire that is rotated in the earth's **magnetic field**.

The electrical **current** induced in this coil is a measure of the **magnetic field intensity**.

earth–ionosphere waveguide—The physical structure formed by the conductive earth and the conductive lower **ionosphere** (D region) that together sandwich the highly resistive **atmosphere** between them.

This **waveguide** sustains global **electromagnetic wave** propagation from the lower end of the **Schumann resonance** band (5 Hz) to the waveguide cut-off **frequency** near 1500 Hz, beyond which the **radiation** will not be retained within the waveguide.

earth mound—Same as **earth hummock**.

Earth Observing System—(Abbreviated EOS.) A major NASA initiative to develop state-of-the-art **remote sensing** instruments for global studies of the land surface, **biosphere**, solid earth, **atmosphere**, and oceans.

Current plans call for three major satellite platforms, two in polar and one in a low inclination **orbit**. One polar orbiter (EOS-AM) will have a midmorning **descending node** crossing time, while the other (EOS-PM) will have an **equator** crossing time in the afternoon. The third (EOS-CHEM) will be in a low inclination orbit. Each EOS satellite is designed to make maximum use of simultaneous, complementary views of the earth from a wide variety of instruments. EOS-AM will be the first satellite launched in this series, followed by EOS-PM, and the EOS Chemistry Mission (EOS-CHEM). The AM mission will be based on a suite of instruments including an advanced spaceborne **thermal** emission and **reflection** radiometer, a **cloud** and earth **radiant energy** system, a multiangle imaging spectroradiometer, a moderate **resolution** spectroradiometer, and a **sensor** to measure **pollution** in the **troposphere**. Instruments carried on the PM platform will include the advanced **microwave** scanning radiometer-EOS, the moderate resolution spectroradiometer, an advanced microwave sounding unit, an atmospheric **infrared** sounder, a **humidity** sounder for Brazil, and the cloud and earth radiant energy system. The EOS-CHEM will monitor **atmospheric ozone, aerosols,** and **pollution** using the high-resolution dynamics limb **sounder,** the microwave limb sounder, an **ozone** monitoring instrument, and the tropospheric **emission** spectrometer.

Earth Radiation Budget Experiment—(Abbreviated ERBE.) An instrument package flown on **ERBS** (launched October 1984), and also on the *NOAA-9* and *-10* polar-orbiting satellites (launched December 1984 and September 1986, respectively).

ERBE consists of separate **scanning** and nonscanning instruments that were designed to determine the monthly average **radiation** budget on regional, zonal, and global scales. An earlier earth radiation budget instrument was flown on *Nimbus-6* and *-7* (launched June 1975 and October 1978).

Earth Radiation Budget Satellite—(Abbreviated ERBS.) *See* **Earth Radiation Budget Experiment**.

Earth Resources Technology Satellite—(Abbreviated ERTS.) A sun-synchronous **polar-orbiting satellite** designed for **remote sensing** and mapping of land areas.

Launched in July 1972, the satellite was renamed *Landsat-1*.

earth's shadow—Closely related to, but not identical to, the **dark segment** seen in **clear** twilight skies.

earth stripe—*See* **patterned ground**.

earth structure—The spatial distribution of various components and properties of the earth, such as the core, mantle, elastic properties, etc.

earth-synchronous orbit—See **geosynchronous satellite**.

earth temperature—The **temperature** inside the earth.

May also refer to the **surface temperature**.

earth thermometer—Same as **soil thermometer**.

earth tide—Tidal movements of the solid earth due to direct gravitational forcing and loading of the adjacent seabed by marine **tides**.

earthlight—(Also called earthshine.) Light reflected by the earth; analogous to moonlight.

Earthlight is what enables one to see the portions of the moon that are not sunlit.

earthquake—The sudden, cyclic movement of the earth caused by the release of strain inside the earth.

This movement causes faulting. Earthquakes often occur along tectonic plate boundaries.

earthquake flood—*See* **tsunami**.

earthshine—Same as **earthlight**.

East African Coastal Current—*See* **Zanzibar Current**.

East Antarctic Ice Sheet—That portion of the **Antarctic Ice Sheet** lying predominantly in the Eastern Hemisphere.

A line following the Transantarctic Mountains to the Antarctic Peninsula serves as the boundary between the East and **West Antarctic Ice Sheets**.

East Arabian Current—The continuation of the **Somali Current** along the eastern Arabian peninsula during the **southwest monsoon** (June–October) with peak speeds in May of 0.5–0.8 m s^{-1}.

It is associated with **coastal upwelling** during May–September, when sea surface temperatures along the coast are lowered by about 5°C.

East Auckland Current—The continuation of the **East Australian Current** along the eastern coast of New Zealand's North Island; the current is thus part of the western **boundary current** system of the South Pacific subtropical **gyre**.

During summer (December–March) it continues southward as the **East Cape Current** to reach the Chatham Rise. During winter (June–August) some of it separates from the shelf and flows eastward into the Pacific, forming a **temperature** front along 29°S.

East Australian Current—The western **boundary current** of the South Pacific subtropical **gyre**.

It originates in the Coral Sea near 1°S from the **South Equatorial Current** and flows southward along the east Australian coast. Although it is the weakest of all western boundary currents with a mean transport of little more than 15 Sv (15×10^6 m³s⁻¹), its speed is rarely less than 1.5 m s⁻¹. The current is stronger in summer (December–March). It separates from the Australian coast between 31° and 34°S to flow toward the northern tip of North Island, New Zealand, shedding about three **eddies** per year in the process. Its eastward passage from Australia to New Zealand is known as the Tasman Front, which separates warm tropical water in the Coral Sea from the subtropical water of the Tasman Sea.

East Cape Current—*See* **East Auckland Current**.

East Greenland Current—A southward flowing current along Greenland's east coast that forms part of the North Atlantic subpolar **gyre** and at the same time constitutes the major **outflow** route of Arctic water into the Atlantic.

This water has a **salinity** of 30–33 **psu** and a **temperature** below −1°C. Some of it is diverted just north of Denmark Strait and northeast of Iceland into the East Iceland Current, which carries it toward the Norwegian Sea as part of the formation process of **Arctic Bottom Water**. The remainder is joined south of Denmark Strait and southwest of Iceland by the northwestward flowing **Irminger Current**, which brings the water of the subpolar gyre. Transport estimates are 5 Sv (5×10^6 m³ s⁻¹) for the East Greenland Current and 8–11 Sv ($8–11 \times 10^6$ m³ s⁻¹) for the Irminger Current. The combined flow continues around the southern tip of Greenland into the **West Greenland Current**.

East Iceland Current—*See* **East Greenland Current**.

East Indian Current—A northward flowing western **boundary current** found along the east coast of southern India from February to August.

Its speed is consistently above 0.5 m s⁻¹, reaching a maximum of 0.7–1.0 m s⁻¹ in May and June. The current is remarkable for its independence of the **monsoon** cycle; during February–April (the late Northeast Monsoon **season**) it runs into the **wind**. *See also* **East Indian winter jet**.

East Indian winter jet—A southward flowing powerful western **boundary current** that replaces the northward flowing **East Indian Current** along the east coast of southern India during October–December.

Its speed is consistently above 1 m s⁻¹. It continues westward south of Sri Lanka, where it establishes a strong current shear but little water exchange with the eastward flowing **Indian equatorial jet** to the south. It feeds its water into the Arabian Sea.

East Kamchatka Current—*See* **Kamchatka Current**.

East Korea Warm Current—A current branching off from the **Tsushima Current** as it enters the Japan Sea through the Korea Strait.

It follows the Korean coast northward to 36°–38°N where it meets the **North Korea Cold Current** to establish the polar front of the Japan Sea. It continues northeastward along the southern side of the polar front, shifting its path every few months and shedding **eddies** along the way. It rejoins the Tsushima Current before reaching 40°N.

East Madagascar Current—One of the western **boundary currents** of the subtropical **gyre** in the southern Indian Ocean.

It originates near 20°S where the **South Equatorial Current** splits east of Madagascar into a northward and a southward flowing branch. The current carries the 20 Sv (20×10^6 m³ s⁻¹) of the southern branch southward along Madagascar as a swift, deep, and narrow boundary current. South of Madagascar some of this water moves westward to join the **Agulhas Current**, but most of it flows northward along the west coast of Madagascar to at least 15°S before also turning westward to join the **Mozambique Current**.

easterlies—Any **winds** with components from the east, usually applied to broad currents or patterns of persistent easterly winds, the "easterly belts," such as the **equatorial easterlies**, the **tropical easterlies**, and the **polar easterlies**.

See also **trade winds**.

easterly belt—*See* **easterlies**.

easterly trough—*See* **easterly wave**.

easterly wave—A **migratory** wavelike **disturbance** of the **tropical easterlies**.

It is a **wave** within the broad easterly current and moves from east to west, generally more slowly than the current in which it is embedded. Although best described in terms of its wavelike characteristics in the **wind field**, it also consists of a weak **trough** of low pressure. Easterly waves do not extend across the **equatorial trough**. To the west of the **trough line** in an easterly wave over the ocean, there is generally found **divergence**, a shallow moist layer, and exceptionally fine weather. The moist layer rises rapidly near the trough line; in and to the east of the trough line intense **convergence**, much **cloudiness**, and **heavy rain** showers prevail. This asymmetric weather pattern may be greatly distorted by **orographic** and **diurnal** influences if the wave passes over land areas. Easterly waves occasionally intensify into **tropical cyclones**. *Compare* **equatorial wave**.

ebb current—The movement of a **tidal current** away from the shore or down a **tidal river** or **estuary**.

ebb interval—The interval in time between consecutive cycles of tidal ebbs.
   See also **ebb current**.

ebb strength—The magnitude (speed) of the ebb tide.
   See also **ebb current**.

ebb tide—*See* **ebb current**.

Ebert ion counter—An **ion counter** in which the free airstream flows between polarized condenser plates that attract and capture **ions** of opposite sign.
   It is used for the measurement of the concentration and **mobility** of **small ions** in the **atmosphere**.

ebullition—The boiling of a liquid; specifically, the formation within the liquid of bubbles of that liquid's **vapor** and the vigorous ascent of the bubbles to the liquid's surface, a process that usually begins at that liquid's nominal **boiling point**.
   Ebullition produces a much greater rate of escape of liquid molecules into the vapor phase than does **evaporation**, since the **effective area** for **phase change** is much greater in a boiling liquid than in a liquid evaporating from the same container, in that each bubble affords surface area for local evaporation.

ECD—Abbreviation for **electron capture detection**.

echo—In **radar**, a general term for the appearance, on a radar **display**, of the radio **signal** scattered or reflected from a **target**. The characteristics of a **radar echo** are determined by 1) the **waveform**, **frequency**, and **power** of the incident **wave**; 2) the **range** and **velocity** of the target with respect to the radar; and 3) the size, shape, and composition of the target.
   See also **target signal**, **blip**.

echo amplitude—1. The **amplitude** of the vertical deflection of an **echo** on an **A-display**.
   2. Less commonly, the **amplitude** of the **target signal**.

echo box—A type of electronic instrument used to test and adjust **radar** equipment.
   It operates on the principle of a cavity **resonator**. A small amount of **electromagnetic energy** from the transmitting **antenna** is fed into a small cavity (or box), the volume of which can be adjusted to resonate electrically, or "ring," to signals of this **frequency**, like a tuning fork responding to a musical note of proper pitch. This **resonance** is detected by the radar receiver. The amount of resonance appearing at the **receiver** output (the **oscilloscope**) is a function of the **power** transmitted, the tuning of the cavity, and the tuning and amplification of the receiver. Accordingly, the echo box provides a test of the overall efficiency of the radar system, eliminating atmospheric variables.

echo contour—A trace through points of equal **signal intensity** on a radar **display**.
   The properties of **echo** contours are not standardized. For example, the **contour** intervals may be fixed or variable, may be corrected for range or for various sources of **attenuation**, and may be averaged in time. *See* **isoecho**.

echo-free vault—Same as **bounded weak-echo region**.

echo frequency—Ambiguous term that could denote 1) the **fluctuation** rate of the **echo amplitude**, or 2) the Doppler **frequency** of the **target signal**.

echo intensity—1. The **brightness** or brilliance of an **echo** as it appears on an intensity-modulated **display**.
   2. Less commonly, the **intensity** of the **target signal**.

echo overhang—In the **radar echo** associated with a severe **thunderstorm**, the portion of the echo that is located above the **weak-echo region** on the low-altitude **inflow** side of the **storm**.
   The overhang consists of **precipitation** particles diverging from the storm's summit that descend as they are carried **downwind**. If the storm echo develops a **bounded weak-echo region**, it is found within the echo overhang.

echo power—The electrical strength, or **power**, of the **echo** received by a **radar**, normally measured in decibels relative to a milliwatt (dBm).
See **radar equation**.

echo pulse—A **pulse** of radio **energy** received at a **radar** after **reflection** from a **target**; that is, the **target signal** of a **pulsed radar**.

echo signal—Same as **target signal**.

echo sounder—A device that uses **sound waves** to measure the depth of **surface water** bodies.

ecliptic—The **great circle** in the **celestial sphere** that is the apparent annual path of the sun around the earth.
The plane in which this circle lies is the **plane of the ecliptic**. The angle that this plane makes with the plane passing through the earth's **equator** is the **obliquity of the ecliptic**. With reference to the actual motion within the solar system, the plane of the earth's orbit is equivalent to the plane of the ecliptic. That is, the intersection of the plane of the earth's orbit with the earth's surface is the same as the **projection** of the ecliptic upon the earth's surface. The most northern and southern points on this line of intersection define the latitudes of the **Tropic of Cancer** and **Tropic of Capricorn**, respectively. *See* **solstice**, **equinox**; *compare* **zodiac**.

ecnephias—A **squall** or **thunderstorm** in the Mediterranean

ecoclimatology—Same as **ecological climatology**.

ecological climatology—(Also called ecoclimatology.) A branch of **bioclimatology** that studies the relations between organisms and their climatic **environment**.
It includes the physiological adaptation of plants and animals to their **climate**, and the geographical distribution of plants and animals in relation to climate.

ecology—The study of the mutual relations between organisms and their **environment**.

economic yield of aquifer—Maximum pumping rate from an **aquifer** that can be economically sustained without adversely affecting the water supply or the water's quality.

ecosphere—The **atmosphere**, oceans, **biosphere** (plant and animal life), and top portion of the earth's crust, which are involved in complex interactions defining **ecology**.
Namely, this is the mutual relation between organisms and their **environment** over the spherical surface of the earth.

ecosystem—Organisms and the **environment** in which they interact.
Sometimes the definition is extended to include the processes of interaction. *See* **ecology**.

ecotone—In **ecology**, a zone of transition from one major plant community to another.
For example, the forest–tundra ecotone in high northern latitudes is a zone of patchy and often stunted tree growth intermixed with areas of **tundra**.

edaphic—1. Of or relating to the soil.
2. Resulting from or influenced by the soil rather than the **climate**.

eddy—1. By analogy with a molecule, a "glob" of fluid within the fluid mass that has a certain structure and life history of its own, the activities of the bulk fluid being the net result of the motion of the eddies.
The concept is applied with varying results to phenomena ranging from the momentary spasms of the **wind** to **storms** and **anticyclones**.
2. Any **circulation** drawing its **energy** from a flow of much larger **scale**, and brought about by **pressure** irregularities, as in the lee of a solid obstacle.
3. In studies of the **general circulation**, departures of a **field** (e.g., **temperature** or **relative vorticity**) from the **zonal** mean of that field.
4. A **closed circulation** system produced as an offshoot from an **ocean current**.
Eddies are the result of the **turbulence** of the oceanic **circulation** and are common throughout the World Ocean. The corresponding features in the **atmosphere** are the **wind** currents around high and low pressure disturbances. Oceanic **cyclonic** eddies have a shallow **thermocline** at the center and are therefore also known as cold-core eddies; **anticyclonic** eddies are associated with a depressed thermocline in the center and are also known as warm-core eddies. The most prominent eddies are those shed by western **boundary currents**, also known as rings; they are about 200 km in diameter and reach beyond a depth of 1500 m. Another class of eddies is produced by **shear** between currents flowing in opposing directions. These eddies tend to be smaller (10–50 km in diameter) and shallower.

eddy accumulation—A technique for estimating vertical turbulent **trace gas** fluxes that does not require fast response sensors.

Air is collected in two reservoirs, one when the **vertical velocity** is positive and the other when the vertical velocity is negative. The volume rate at which the sample is collected is proportional to the magnitude of the vertical velocity. The **flux** is estimated from the difference in mass collected in each reservoir divided by the collection time.

eddy advection—Transport of a fluid property by **eddies** too large to be modeled by **eddy diffusivity** in a **turbulent flow**.

Examples are the **transport** by **organized large eddies** in the **boundary layers** of the **atmosphere** and the ocean. These large eddies have dimensions comparable to the height of the boundary layers.

eddy coefficients—Same as **exchange coefficients**.

eddy conduction—*See* **eddy flux**.

eddy conduction coefficient—*See* **eddy conductivity**.

eddy conductivity—(Also called coefficient of eddy conduction, eddy conduction coefficient.) The **exchange coefficient** for **eddy heat conduction**.

eddy continuum—In general, the continuum hypothesis used to generalize Newton's laws for point masses to continuous media such as fluids or gases.

At any point in the continuous media it must be possible to identify volumes small enough that derivatives exist, yet large enough to contain sufficient **particles** that the average macroscopic property (such as **momentum**, mass, or **temperature**) may be defined. The eddy continuum hypothesis is used to extend the concept of a fluid continuum to a **turbulent flow** in which small-scale **eddies** are substituted for fluid particles. To apply the eddy continuum hypothesis, a **scale analysis** must be performed to verify that the averaging periods are sufficiently long that enough small-scale turbulent eddies are sampled that their dynamical effect on the mean flow is captured. If the eddies' length scales are comparable to the mean flow length scales, the eddy continuum most likely can be defined. The eddy continuum hypothesis is often applied in the study of large-scale organization (of the smaller-scale turbulent eddies) in the **planetary boundary layer** turbulence.

Brown, R. A., 1990: *Fluid Mechanics of the Atmosphere*, Academic Press, 490 pp.

eddy correlation—1. The **covariance** between two **variables**, associated with turbulent motions.

For example, if the overbar represents a **mean value**, and the prime denotes a **deviation** from the mean, then

$$\overline{w'\theta'} = \frac{1}{N} \sum_{i=1}^{N} (w_i - \overline{w})(\theta_i - \overline{\theta})$$

is the eddy correlation of **vertical velocity** w and **potential temperature** θ, where i is a data index and N is the total number of data points. If one of the two variables is a **velocity**, such as in the example here, then the eddy correlation represents a **kinematic flux** associated with **turbulence**. In this example, the **correlation** is a vertical kinematic **heat flux** (units of K m s^{-1}), which can be transformed into a **dynamic flux** (W m^{-2}) by multiplying by air density times **specific heat** at constant **pressure**.

2. A method of measuring the **flux** densities of mass, **heat**, and **momentum** across a plane at a point in **turbulent flow**.

For vertical fluxes, the fluxes are found as the **covariance** of the fluctuations in the **vertical wind velocity** with local variations in concentration, heat content, or horizontal **wind** velocity, respectively. For conservative quantities in the atmospheric **surface layer**, such vertical fluxes are virtually equal to the fluxes at the surface of the earth. Measurements from towers require sensors with a speed of response that is typically no larger than about 1 s, but the required response varies with height, **wind speed**, and amount of buoyancy-induced **mixing**. Observations from eddy correlation systems on fixed-wing aircraft in the **planetary boundary layer** typically require a response that is about ten times faster than stationary systems.

eddy covariance—The most direct approach for determining a vertical **turbulent flux**.

The flux is calculated as the average of the instantaneous product of **vertical velocity** and a **scalar** quantity (such as a concentration).

eddy diffusion—1. Same as **turbulent diffusion**.

2. **Mixing** that is caused by **eddies** that can vary in size from the smallest scales (**Kolmogorov scale**) to subtropical **gyres**.

eddy diffusion coefficient—Same as **eddy diffusivity**.

eddy diffusivity—(Also called coefficient of eddy diffusion, eddy diffusion coefficient.) The **exchange coefficient** for the **diffusion** of a **conservative property** by eddies in a **turbulent flow**.
See also **diffusivity**.

eddy flux—(Or turbulent flux.) The rate of **transfer** of a conservative fluid property through a surface by turbulent **eddies**.

In the case of an unchanging mean state, the eddy flux of a property S in the z direction is expressed by

$$F_s = \overline{\rho}\ \overline{ws},$$

where lower case represents **turbulence** values (fluctuations from the mean), w represents **vertical velocity**, ρ **density**, and the bar represents the mean over a chosen period. By analogy with **molecular diffusion**, the eddy flux in the **planetary boundary layer** is often expressed by

$$F_s = -\rho K_s \frac{\partial S}{\partial z},$$

where K_s is the turbulent transfer coefficient (also **turbulent exchange** coefficient, or **eddy diffusivity**). This can be generalized to include **eddy diffusion** in the horizontal (x, y) plane. If S is replaced by $c_p T$, where c_p represents **specific heat** at constant **pressure** and T represents **temperature**, the vertical turbulent **heat flux** (or eddy heat flux) can be represented by

$$F_h = c_p \overline{\rho}\ \overline{w\theta} = -\rho K_h \left(\frac{\partial \overline{T}}{\partial z} + \Gamma \right),$$

where θ is the temperature fluctuation, K_h the eddy thermal **diffusion coefficient** (or **eddy conductivity**), and Γ represents the **adiabatic lapse rate**. Similarly, eddy **momentum flux** is represented by

$$F_m = -\rho K_m \frac{\partial U}{\partial z},$$

where U represents the mean speed in the direction of the mean wind, and K_m represents the eddy viscosity coefficient (or **eddy viscosity**). Also, the turbulent transfer of **water vapor** may be expressed in **energy** units by

$$F_e = -\rho L K_e \frac{\partial \overline{q}}{\partial z},$$

where L represents the **latent heat** of vaporization, K_e is the **eddy diffusion coefficient**, and $\overline{q}$ is the **specific humidity**. *See also* **diffusion, turbulence**.

Fleagle, R. G., and J. A. Businger, 1980: *An Introduction to Atmospheric Physics*, 2d ed., Academic Press, 266–272.

eddy heat conduction—*See* **eddy flux**.

eddy heat flux—*See* **eddy flux**.

eddy kinetic energy—*See* **turbulence kinetic energy**.

eddy shearing stress—The **shearing stress** produced by turbulent **eddies** in analogy to the shearing stress related to **viscosity** in a **laminar flow**.

eddy spectrum—*See* **turbulence spectrum**.

eddy-stress tensor—*See* **Reynolds stresses, stress tensor**.

eddy stresses—Same as **Reynolds stresses**.

eddy transfer coefficients—Same as **exchange coefficients**.

eddy transport—*See* **eddy flux**.

eddy velocity—(Also called fluctuation velocity.) The difference between the **mean velocity** of fluid flow and the instantaneous **velocity** at a point.

For example,

$$u = \breve{U} - U,$$

where u is the **eddy** velocity, $\breve{U}$ is instantaneous velocity, and U is mean velocity. Over the same interval that defines the mean velocity, the average value of the eddy velocity is necessarily zero.

eddy viscosity—The turbulent **transfer** of **momentum** by **eddies** giving rise to an internal fluid **friction**, in a manner analogous to the action of **molecular viscosity** in **laminar flow**, but taking place on a much larger **scale**.

The value of the **coefficient of eddy viscosity** (an **exchange coefficient**) is of the order of 1 $m^2 s^{-1}$, or one hundred thousand times the molecular **kinematic viscosity**. Eddy viscosity is often represented by the symbol K, and the **turbulence** parameterization that uses eddy viscosity is called **K-theory**. In this theory, the **eddy flux** in kinematic units is related to the mean vertical **gradient**, such as in this example for vertical **flux** of horizontal **momentum**:

$$\overline{u'w'} = -K \frac{\partial \overline{U}}{\partial z},$$

where w is **vertical velocity**, U is horizontal **wind** in the x direction, the overbar represents an average, and the prime denotes the **deviation** or **perturbation** from an average. Eddy viscosity is a function of the flow, not of the fluid. It is greater for flows with more turbulence. The eddy viscosity or K-theory approach is a **parameterization** for the eddy **momentum flux** (Reynolds stress) that works reasonably well when only small eddies are present in the flow, but that behaves poorly when large-eddy **coherent structures**, such as thermals in the **convective mixed layer**, are present. *See* **Reynolds stresses, eddy correlation**; *compare* **transilient turbulence theory**.

edge wave—A specific wave form that propagates along a boundary.

The presence of the rigid boundary alters the **wave** dynamics. A simple example of an edge wave is the rotary wave that can easily be generated in a bathtub.

8D technique—(Also called frost point technique.) A technique for using the **radiosonde observation** to determine the presence of liquid water droplets in supercooled clouds in saturated or nearly saturated layers of air.

For each reported level in the **sounding**, the negative value of eight times the **dewpoint spread** (−8D) is plotted on the **pseudoadiabatic chart** (or equivalent chart). Where the **temperature** sounding lies to the left of the −8D curve, liquid **droplet** clouds are considered to be present, and **icing** is possible on aircraft flying in the **cloud layer**.

effective aperture—Same as **effective area**.

effective area—1. (Also called effective aperture.) In **antenna** design, the ratio of the **received power** available at the terminals of an antenna to the **power** per unit area in the incident **wave**.

For all antennas, effective area A is related to **gain** G at a given **wavelength** λ by the equation

$$\frac{A}{G} = \frac{\lambda^2}{4\pi}.$$

The effective area of an ideal antenna is equal to its physical area S. In practice, A/S for **microwave** antennas is always less than 1, a representative value for paraboloids being 0.6. *See* **aperture**.

2. Same as **scattering cross section**.

effective atmosphere—Same as optically effective atmosphere.

effective earth radius—In **radar** and radio propagation studies, a value for the radius of the earth that may be used in place of the actual radius to correct for **refraction** by the **atmosphere**.

The effective earth radius is a convenient fiction that makes straight the actual curved path of a radio **ray** in the atmosphere by presenting it relative to an imaginary earth with a radius larger than the radius of the real earth, thus maintaining the relative curvature between earth and radio ray. The effective earth radius R_e is approximately

$$\frac{1}{R_e} = \frac{1}{R} + \frac{dn}{dz},$$

where R is the radius of the earth and dn/dz is the **refractive index** gradient of its atmosphere. The effective earth radius is well defined only to the extent that dn/dz is constant. Moreover, because n varies with **frequency**, so does the effective earth radius. The **index of refraction** of the **atmosphere** ordinarily decreases with **altitude** so that radar or **radio waves** that are transmitted horizontally are refracted downward. The radius of curvature of the propagating rays depends on the vertical **gradient** of the refractive index. For **standard refraction** conditions, the effective earth radius is 4/3 times the geometric radius.

effective evapotranspiration—Actual amount of water lost to **evapotranspiration** from the soil–plant continuum by an actively growing plant or crop.

Water loss through evapotranspiration depends upon plant and soil characteristics, and upon the amount of **available water** in the soil. *See also* **potential evapotranspiration**.

effective flux—The sum of molecular and **turbulent fluxes** within the bottom few millimeters of the earth's surface.

Immediately at the surface of the solid earth, turbulent fluxes must be zero. At this **interface**, the **flux** of sensible or **latent heat** must be totally transported by molecular motions (**conduction** and **diffusion**). Within a few millimeters above the surface, **turbulence** takes over to **transport** the fluxes, while the molecular effects become negligible. For simplicity, the sum of the molecular and turbulent fluxes is sometimes called the **effective turbulent flux**; frequently, "effective" is left out entirely. Thus, the expression "surface turbulent flux" really means "surface effective flux."

effective gust velocity—The vertical component of the **velocity** of a **sharp-edged gust** that would produce a given **acceleration** on a particular airplane flown in level flight at the design cruising speed of the aircraft and at a given air density.

See **derived gust velocity**.

effective permeability—The **permeability** of a soil not fully saturated with the fluid of interest.

It is equal to the **intrinsic permeability** multiplied by the **relative permeability**.

effective porosity—1. The volume of the voids that conduct water divided by the volume of the sample; equal to or less than the **porosity**.

2. The portion of the **porosity** that can be drained by **gravity** forces; equal to the porosity minus the residual water content.

effective precipitation—1. That part of **precipitation** that reaches **stream** channels as **direct runoff**.

2. In **irrigation**, the portion of the **precipitation** that remains in the soil and is available for **consumptive use**.

effective rainfall—Same as **effective precipitation**.

effective roughness length—The value of the **aerodynamic roughness length** over a heterogeneous surface that relates the horizontally averaged **wind velocity profile** to the spatially averaged or aggregated **shear** stress or **momentum flux**.

The effective roughness length is defined only for heights above the numerical **blending height** and is dependent on the magnitude, distribution, and length **scale** of **surface roughness** variations. The effective roughness length is typically larger than linearly or geometrically averaged roughness lengths. *See* **flux aggregation**.

effective snowmelt—That part of snowmelt that reaches **stream** channels as **runoff**.

effective stack height—The final **equilibrium** height of the centerline of an **effluent** (smoke) **plume** from a smokestack or other source.

The effective stack height is the sum of the actual physical height of the top of the stack, plus any plume rise due to **buoyancy** or initial **momentum** (inertia) of the rising effluent, minus any **downwash** such as stack downwash, building downwash, or terrain downwash. The downwashes are associated with **wake turbulence** behind objects on the ground.

effective temperature—1. The **temperature** at which motionless saturated air would induce, in a sedentary worker wearing ordinary indoor clothing, the same sensation of comfort as that induced by the actual conditions of temperature, **humidity**, and air movement. Effective temperature is used as a guide in air-conditioning practice, and, on the **comfort chart** (American Society of Heating and Air Conditioning Engineers), it appears as a family of curves that serves as one coordinate in defining **comfort zones**.

Compare **sensible temperature, cooling temperature, operative temperature**.

2. In **phenology**, the **temperature** at which plants, especially cereals, begin to grow; generally about 0°C (32°F), for example, winter wheat, to 7°–10°C (45°–50°F), for example, maize.

See **degree-day**.

3. With respect to **radiation**, the **blackbody temperature** that would yield the same amount of radiation as that emitted.

The effective temperature of the earth as a whole is about 254 K, corresponding to an average **emittance** of 236 W m^{-2}.

effective turbulent flux—Same as **effective flux**.

effective velocity—The **flux** of **groundwater** divided by the **effective porosity**.

effluent—Term commonly used in water quality for liquid, partially or completely treated, that flows out of a **reservoir, basin**, sewage treatment plant, or industrial treatment plant.

effluent limitations—Limitations or specifications placed by regulatory agencies on the quantity and quality of the **effluent** leaving a **reservoir**, waterway, or sewage (or industrial) treatment plant.

effluent seepage—A **groundwater hydrology** term describing the slow and diffused **discharge** of **groundwater** to the surface or to **streams**.

effluent stream—(Also called gaining stream.) A **stream** with a flow maintained by **base flow** during long rainless periods.

efflux velocity—The average flow rate of material emitted into the **atmosphere** from a **source** such as a smokestack.
This is the average speed of gas out of the top of a smokestack.

Egyptian wind—In Egypt and the Gulf of Suez, a westerly **wind**, frequent during winter and accompanied by **fog** and **dust**.

EHF—Abbreviation for extremely high frequency.
See **radio frequency band**.

eigenvalue—A **scalar** value λ that permits nonzero solutions y of equations of the form

$$Ly = \lambda y,$$

where L is a **linear operator** and where y can represent a **vector** or a function that is subject to certain **boundary conditions**. When y is a vector, L represents a matrix and y is termed an eigenvector. When y is a function, L can represent a differential or integral form, in which case y is called an eigenfunction.

eight-inch rain gauge—A nonrecording cylindrical **rain gauge** with a **collector** diameter of 20.3 cm (8 in.).
The **rain** drains into a collector tube that amplifies the depth by a factor of 10. The depth in the collector tube is measured by a wooden ruler. Formerly in common use in the United States.
NOAA/National Weather Service, 1972: *National Weather Service Observing Handbook No. 2, Substation Observations*, NOAA, US GPO.

eighteen-degree water—*See* **mode water**.

Einstein's summation notation—A special notation for writing **vector** equations using the **scalar** components of vectors rather than the vectors themselves.
This notation has found widespread use in the study of **atmospheric turbulence**, where the many vector and **tensor** components of a governing equation can be concisely represented by a much smaller number of terms. The key to this notation is the use of indices (subscripts) i, j, and k, which can each take on the values of 1, 2, or 3 to represent the three Cartesian directions (x, y, z). *See also* **alternating unit tensor**.
Stull, R. B., 1988: *An Introduction to Boundary Layer Meteorology*, 57–63.

ejection chamber—The chamber that houses **dropsonde** equipment just prior to and during release from an aircraft.
A similar piece of equipment is the dropsonde dispenser.

Ekman boundary conditions—These are the four mathematical/physical conditions used when integrating the fourth-order differential equations governing the **Ekman layer**.
These conditions are

$$U = 0 \text{ and } V = 0 \text{ at } z = 0$$

$$U \to G \text{ and } V \to 0 \text{ as } z \to \infty,$$

where (U, V) are the horizontal winds to the east and north, respectively; z is height above the surface; and G is the **geostrophic wind** speed. The first two conditions state that **friction** at the ground causes the **wind speed** to be zero there, while the last two conditions state that the winds become **geostrophic** where friction is negligible at large distances above the surface.

Ekman layer—(Sometimes called spiral layer.) A hypothetical layer of air at the bottom of a statically **neutral atmosphere** surrounding a rotating planet, where **surface friction** and small **eddies** (local **turbulence**) cause ageostrophic, cross-isobaric winds.
The spiral shape of a **hodograph** of this flow is called an **Ekman spiral**. The Ekman layer is not observed in the earth's **atmosphere** because of the existence of large eddies (nonlocal turbulence) creating a **mixed layer**, and because the earth's **troposphere** is **statically stable** on average, causing a thermodynamic **cap** to the **boundary layer** rather than a dynamic cap at a theoretical Ekman

layer depth of 20 u_*/f, where u_* is the **friction velocity** and f is the **Coriolis parameter**. *Compare* **atmospheric boundary layer**.

Ekman number—A **dimensionless number** relating the ratio of eddy **viscous forces** to **Coriolis forces**.

Explicitly, the vertical Ekman number is

$$E_v = 2K_v/(fD^2),$$

in which K_v is the vertical **eddy viscosity**, f the **Coriolis parameter**, and D the characteristic vertical length **scale**. The horizontal Ekman number is

$$E_b = 2K_b/(fL^2),$$

in which K_b is the horizontal eddy viscosity and L is the characteristic horizontal length scale. The Ekman number gives a measure of the rate at which stresses at a boundary (e.g., **wind**-induced stresses at the ocean surface) are communicated to the fluid interior.

Ekman pumping—The mechanism by which the effects of **boundary layer** momentum fluxes are communicated directly to the neighboring (essentially inviscid) fluid.

The mechanism involves a forced **secondary circulation** referred to as Ekman pumping or Ekman suction, depending on its sign. The effect is important in both the **atmosphere** and the ocean.

Holton, J. R., 1992: *An Introduction to Dynamic Meteorology*, 3d edition, Academic Press, 133–139.

Ekman spiral—1. As used in meteorology, an idealized mathematical description of the **wind** distribution in the **atmospheric boundary layer**, within which the earth's surface has an appreciable effect on the air motion.

The **model** is simplified by assuming that within this layer the **eddy viscosity** and **density** are constant, the motion is horizontal and steady, the isobars are straight and parallel, and the **geostrophic wind** is constant with height. The x direction is taken along the **pressure gradient**; the resulting approximate equations for the component wind speeds U and V in the x and y directions, respectively, at any level z are

$$U = -Ge^{-\beta}\sin\beta, \quad V = G(1 - e^{-\beta}\cos\beta),$$

where G is the geostrophic wind speed, $\beta = z(f/2K_M)^{1/2}$, f is the **Coriolis parameter**, and K_M is the **eddy viscosity**. The lowest level H where $U = 0$, so that the true wind and the geostrophic wind have the same direction, is called the **geostrophic wind level** (or **gradient wind level**). It is given by

$$H = (2K_M/f)^{1/2}(\tfrac{3}{4}\pi + \alpha_0),$$

where α_0 is the angle between the **surface wind** and the surface isobars. At this height the magnitude of the true wind will exceed that of the geostrophic wind by a small amount, depending on the value of β. The Ekman spiral is an equiangular spiral having the geostrophic wind as its limit point. Below the geostrophic wind level the wind blows across the isobars toward low pressure, at an angle that is a maximum at the surface and does not exceed 45°. The **deviation of the wind vector** from the geostrophic wind vector diminishes upward at an exponential rate. The theory of this spiral was developed by Ekman in 1902 for motion in the upper layers of the ocean under the influence of a steady **wind**. It was applied to the **atmosphere** by Åkerblom in 1908.

2. As originally applied by Ekman to **ocean currents**, a graphic representation of the way in which the theoretical wind-driven currents in the surface layers of the sea vary with depth.

In an ocean that is assumed to be homogeneous, infinitely deep, unbounded, and having a constant **eddy viscosity**, over which a uniform steady **wind** blows, Ekman has computed that the current induced in the surface layers by the **wind** will have the following characteristics: 1) At the very surface the water will move at an angle of 45° **cum sole** from the **wind direction**; 2) in successively deeper layers the movement will be deflected farther and farther cum sole from the wind direction, and the speed will decrease; and 3) a **hodograph** of the **velocity** vectors would form a spiral descending into the water and decreasing in **amplitude** exponentially with depth. The depth at which the **vector** first points 180° from the **wind vector** is called the **depth of frictional influence** (or depth of frictional resistance). At this depth the speed is $e^{-\pi}$ times that at the surface. The layer from the surface to the depth of frictional influence is called the layer of frictional influence. If the velocity vectors from the surface to the depth of frictional influence are integrated, the resultant vertically integrated motion is 90° cum sole from the wind direction.

Ekman transport—The total **transport** resulting from a balance between the **Coriolis force** and frictional **stress** at the bottom.

The Ekman transport is

$$M_{ek} = \tau/\rho f,$$

where τ is the **stress**, ρ is the **density**, and f is the **Coriolis parameter**. The **divergence** and **convergence** of the Ekman transport driven by the **wind stress** is responsible for the wind-driven **gyres**.

El Niño—A significant increase in **sea surface temperature** over the eastern and central equatorial Pacific that occurs at irregular intervals, generally ranging between two and seven years.

El Niño conditions, which are often characterized by "warm events," most often develop during the early months of the year and decay during the following year. The term was originally applied by fishermen of northern Peru to a warm annual southward coastal current that develops shortly after the Christmas season; hence the Spanish name referring to "the Christ Child." The name subsequently became more commonly used in reference to the occasional very strong coastal warmings that are associated with torrential rains in the **desert** coastal regions of southern Ecuador–northern Peru. The current definition of El Niño developed following the discovery that the coastal warmings are simply part of a larger-scale phenomenon arising from coupled ocean–atmosphere interactions across a broad expanse of the equatorial Pacific. *See also* **La Niña, Southern Oscillation, ENSO**.

El Niño–Southern Oscillation—*See* **ENSO**.

electric charge—A fundamental property of matter. This property exhibits two states, positive and negative, that result in the action of electric forces in the presence of an **electric field**.

These two states were identified and named by Benjamin Franklin. The positive charge on the **proton** and the negative charge on the **electron** represent the fundamental charge, 1.6×10^{-19} coulomb (C).

electric currents in the atmosphere—*See* **air–earth current**.

electric discharge—(Or, simply, discharge; also gaseous electric discharge, gaseous discharge.) The flow of electricity through a gas, resulting in the **emission** of **radiation** that is characteristic of the gas and of the **intensity** of the **current**.

electric double layer—An interfacial region, near the boundary between two different phases of a substance, in which physical properties vary markedly (in contrast with those in the bulk phases).

For electrically conducting phases, charge distribution occurs in this interfacial region, which may be approximated as two parallel sheets of charge of opposite sign, hence the term **double layer**. This name is retained even if the interfacial region is more complex. Double layers arise from an excess of charge, which may be **electrons**, **ions**, or oriented **dipoles**, in the interfacial region.

Hampel, C. A., Ed., 1964: *The Encyclopedia of Electrochemistry*, p. 464.
Kortya, J., and J. Dvo, 1987: *Principles of Electrochemistry*, pp. 148, 207.

electric field—1. A **vector field**, usually denoted by **E**, defined as follows: at a given time and at each point in space the force experienced by a positive charge (sometimes called a test charge) at that point divided by the magnitude of the charge, taken to be sufficiently small that it does not affect the positions and velocities of all other charges.

The set of all vectors thus obtained is the electric field, although this term is often used for its value at any given point. The magnitude of the **vector** is the **electric field intensity** and the direction of the vector is parallel to the **lines of force**.

2. Same as **electric field strength**.
See **atmospheric electric field**.

electric field intensity—The magnitude of the **electric field** at any point.

Strictly speaking, the electric field is the set of all values of the **electric field strength**, but electric field and electric field intensity (as well as electric field strength and electric **vector**) are used more or less interchangeably. The trend is to use electric field both for the **field** taken as a whole and for its value at any point, context being sufficient to determine precise meaning.

electric field strength—(Also called **electric field**, electric intensity, **electric field intensity**, **electric potential gradient**, **field strength**, and probably others.) The electrical force exerted on a unit positive charge placed at a given point in space.

The electric field strength is expressed, in the **mks system** of electrical units, in terms of volts per meter and is a **vector** quantity. The electric field strength of the **atmosphere** is commonly referred to as the **atmospheric electric field**.

electric intensity—Same as **electric field strength**.

electric lines of force—(Also called electric field lines.) Imaginary curves in space tangent to the **electric field** at each point; analogous to **streamlines** in fluid mechanics.

electric potential—That function of position Φ the negative **gradient** of which is the **(static) electric field**:

$$\mathbf{E} = -\nabla\Phi.$$

Physically, the difference between the electric potential at any two points is the amount of **work** done by an external force when moving a unit charge at constant **velocity** along any path connecting these two points.

electric potential gradient—Same as **electric field strength** except of opposite sign. *See also* **potential gradient**.

electric potential gradient—Same as **electric field strength**.

electric storm—Same as **electrical storm**.

electric thermometer—*See* **electrical thermometer**.

electrical breakdown—The sudden decrease of **resistivity** of a substance when the applied **electric field strength** rises above a certain threshold value (the substance's **dielectric strength**).

For air at normal **pressures** and **temperatures**, experiment has shown that the **breakdown** process occurs at a **field strength** of about 3×10^6 V m^{-1}. This value decreases approximately linearly with pressure, and is dependent upon **humidity** and traces of foreign gases. In the region of high field strength just ahead of an actively growing **leader** in a **lightning stroke**, breakdown occurs in the form of a rapidly moving **wave** of sudden **ionization** (**electron avalanche**). The **dielectric strength** in a **cloud** of water drops is less than that in cloud-free humid air.

electrical conductivity—Measure of the ability of material to conduct an electrical **current**.

For water samples, it depends on the concentration and type of ionic constituents in the water and the **temperature** of the water; it is expressed in millisiemens per meter, or micromhos per centimeter.

electrical discharge—*See* **electric discharge**.

electrical hygrometer—A **hygrometer** that uses a transducing **element** with electrical properties that are a function of atmospheric **water vapor** content.

Examples of such a **transducer** are the **humidity strip**, the **carbon-film hygrometer element**, thin-film capacitors, aluminium oxide humidity elements, and **goldbeater's skin hygrometers**.

electrical log—A record obtained in a well, by means of a traveling electrode, that gives a detailed picture of the characteristics and thickness of the various geologic strata, and an indication of the water quality, by measuring the apparent **resistivity** of the materials surrounding the well bore.

electrical mobility analyzer—*See* **differential mobility analyzer**.

electrical storm—1. Popular term for **thunderstorm**.

2. Sometimes applied to a relatively rare condition of disturbed **atmospheric electric field** in the **lower atmosphere** that arises when strong winds are blowing and much **dust** is in the air, but there is no **thunderstorm** activity.

Triboelectrification due to the **blowing dust** may charge fences and other metallic objects to such an extent that slight shocks are felt upon touch.

3. Same as **earth-current storm**.

electrical substitution radiometer—A **radiometer** for which the **output** of the **thermal** detector is measured as the detector is alternately exposed to a **radiant energy** source and then to a known internal electrical heating.

Radiant heating per unit area of an entrance **aperture** is then equated to the electrical **energy** used to equivalently **heat** the detector.

electrical thermometer—A **thermometer** that uses a transducing element with electrical properties that are a function of its **thermal** state.

Common meteorological examples of such thermometers are the **resistance thermometer** and the thermoelectric or **thermocouple thermometer**.

electricity of precipitation—*See* **precipitation current**.

electro-osmosis—In subsurface **hydrology**, the movement of liquid in a **porous medium** due to differences in **electric potential**.

electrode effect—The accumulation of an excess of **ions** of positive sign in the neighborhood of a negative electrode, and vice versa, when ions are continuously produced in the space above the electrode and move under the influence of the electrode's **field**.

> Chalmers, J. A., 1957: *Atmospheric Electricity*, pp. 25, 148.

electrojets—Strong concentrated electric currents flowing in the lower **ionosphere**.

The **equatorial electrojet** flows along the earth's **magnetic dip** equator and is present at all times, while the auroral electrojet is a more sporadic phenomenon occurring in association with auroral displays at high magnetic latitudes.

electrolytic strip—Same as **humidity strip**.

electromagnetic acoustic probe—(Abbreviated EMAC probe.) A device tested in the late 1950s and 1960s that used acoustic waves to create clear-air targets for a **continuous-wave radar**.

The acoustic **source** produces **sound waves** that perturb the atmospheric **refractive index**. The perturbations propagate at the local **speed of sound** and serve as **targets** for the **radar**. When the acoustic **wavelength** is matched to half the radar wavelength, a **resonance** condition develops that leads to the production of a detectable **radar echo** (*see* **Bragg scattering**). The **Doppler shift** of this radar echo is a function of the **virtual temperature** and the **wind** component along the direction of the **radar beam**. **Radio acoustic sounding systems** (RASS) also use this principle, but with a vertically pointed **beam** so that the Doppler shift depends on virtual temperature and the vertical wind component. The vertical wind component in RASS is removed either by time averaging or by making independent measurements of the wind. The EMAC probe was designed to make wind measurements with an operating range of 400 m. The virtual temperature (and corresponding speed of sound) was assumed, based on an independent measurement at the radar site.

electromagnetic energy—The **energy** of **electromagnetic radiation**.

This may be thought of in a quantum mechanical sense as the sum of all the **photon** energies composing the **radiation**, or in a classical sense as the combined energy contained in the oscillating **electric** and **magnetic fields**.

electromagnetic field—A **field** entity comprising the **electric field E** and the **magnetic field** (or magnetic induction) **B**.

All electric and magnetic fields are electromagnetic fields because the decomposition of these fields is not unique. This is evident from the Lorentz force (*see* **magnetic induction**) on a charge, which depends on its **velocity** in a given **inertial coordinate system**. In another inertial coordinate system velocity, and hence the Lorentz force, is different. But ultimately it is only by means of the Lorentz force that fields are observed. Thus, two observers moving with a constant velocity relative to each other will not agree on whether a particular field is solely electric or solely magnetic.

electromagnetic radiation—**Energy** propagated in the form of an advancing **electric** and **magnetic field** disturbance.

The term **radiation**, alone, is commonly used for this type of energy, although it actually has a broader meaning. In the classical **wave theory of light** (or **electromagnetic theory**) the propagation is thought of as a continuous wavelike **disturbance** of the electric and magnetic fields, which oscillate in planes **orthogonal** to each other and to the direction of propagation. The **quantum theory** of electromagnetic radiation adds the perspective that these disturbances also have particle-like attributes, being quantized into **photons** of minimum energy that have finite **momentum**. The observable properties and physical effects of various portions of the **electromagnetic spectrum** are of considerable importance in meteorology and are discussed under their respective names. *See* **cosmic rays, gamma rays, x-rays, ultraviolet radiation, visible radiation, infrared radiation, microwave radiation, radio waves**.

electromagnetic spectrum—The ordered sequence of all known **electromagnetic radiations**, extending from the shortest **cosmic rays** through **gamma rays, x-rays, ultraviolet radiation, visible radiation, infrared radiation**, and including **microwave** and all other radio **wavelengths**.

The division of this continuum of wavelengths (or frequencies) into a number of named subportions is rather arbitrary and, with one or two exceptions, the boundaries of the several subportions are only vaguely defined. Nevertheless, to each of the commonly identified subportions there correspond characteristic types of physical systems capable of emitting **radiation** at those wavelengths. Thus, gamma rays are emitted from the nuclei of atoms as they undergo any of several types of nuclear rearrangements; visible **light** is emitted, for the most part, by atoms with planetary electrons undergoing transitions to lower **energy** states; infrared radiation is associated with characteristic molecular vibrations and rotations; and **radio waves**, broadly speaking, are emitted by virtue of the accelerations of **free electrons** in metals as, for example, the moving electrons in a radio **antenna** wire.

electromagnetic theory—The branch of physics primarily concerned with the group of forces associated with **electric charges**.

electromagnetic wave—An **oscillation** of the **electric** or **magnetic field** associated with the propagation of **electromagnetic radiation**.

Electromagnetic waves are characterized by their **wavelength** or **wavenumber, amplitude**, and polarization characteristics. They propagate at the **speed of light**.

electrometeor—A visible or audible manifestation of **atmospheric electricity**.

This includes, therefore, not only **igneous meteors**, but also the sounds produced by them, principally **thunder**.

electrometer—An instrument for measuring differences of **electric potential**.

See **unifilar electrometer, bifilar electrometer, capillary electrometer, quadrant electrometer, vacuum-tube electrometer;** *compare* **electroscope**.

electron—The negatively charged **subatomic particle** with charge 1.60218 x 10^{-19} coulombs and **rest mass** 9.10939 x 10^{-31} kg.

Each **element** in the periodic table is characterized by a fixed number of electrons orbiting a much heavier **nucleus**. The mass of the electron is about 1/1836 that of the **proton**. *See* **free electron, beta particle**.

Boorse, H. A., and L. Motz, 1966: *The World of the Atom*, Vol. I, 408–426.

electron avalanche—The process in which a relatively small number of **free electrons** in a gas that is subjected to a strong **electric field** accelerate, ionize gas atoms by collision, and thus form new electrons to undergo the same process in cumulative fashion.

All streamers in a **lightning discharge** propagate by formation of electron avalanches in the regions of high **electric field strength** that move ahead of their advancing tips. Particularly in the case of the intense return **streamer**, avalanche processes are enhanced by formation of photoelectrons as a result of **ultraviolet radiation** emitted by the excited molecules in the region just behind the tip. An avalanche cannot possibly begin until the local electric field strength is high enough to accelerate a free electron to the minimum ionizing speed in the space and time interval corresponding to one **mean free path** of the electron, for upon collision, the electron usually loses its forward motion in the direction of the field. Maintenance of an avalanche requires a large reservoir of charge, such as accumulates more or less periodically in active **thunderstorms**.

electron beam—A stream of **electrons** confined and focused by a **magnetic field**.

In a **cathode-ray tube** (CRT), the electron beam is scanned across the tube's phosphorescent surface, as it is modulated in **intensity**, to produce a visual **image**. The **beam** emanates from the tube's cathode and is the cathode ray.

electron capture detection—(Abbreviated ECD.) A detector used in a gas chromatograph that operates by sensing a decrease in the number of free electrons in the detector cavity due to their capture by electrophilic compounds eluting from the column of the chromatograph.

The **free electrons** are produced by beta decay of a radioactive **isotope** (e.g., ^{63}Ni, **half-life** 92 years) in the detector cavity.

electron density—The number of **electrons** per unit volume of a substance.

Often applied in a restricted sense to **free electrons**.

electron precipitation—Term used to denote the injection of energetic **electrons** usually along **magnetic field** lines into the **upper atmosphere** from the **magnetosphere** or directly from the sun.

electron volt—(Abbreviated eV.) The **kinetic energy** acquired by an **electron** accelerated from rest through a **potential** difference of one volt (1.6 x 10^{-19} J).

electronic excitation—The process by which a molecule is excited from a low-lying electronic state to a higher **energy** electronic state; this can occur by **absorption** of a **photon**, by the action of an **electric discharge**, or as a result of a chemical reaction.

electronic theodolite—*See* **radar theodolite, radio direction finder**.

electronic thermometer—A **thermometer** that uses electronic circuitry to detect temperature-induced changes in a **thermal** element and display the corresponding **temperature** in a **digital** form.

The most common **sensor** of this type is a **thermistor**.

electronic transitions—A change in the electronic configuration of a chemical species that occurs following the **absorption** of a **photon** (usually in the visible or **ultraviolet** region of the **spectrum**).

electroscope—A general name for instruments that detect the presence of (but do not necessarily measure) small electrical charges by electrostatic means.
Compare **electrometer**.

electrosphere—A layer of the **atmosphere**, beginning a few tens of kilometers above the surface of the earth and extending to the **ionosphere**, in which the **electrical conductivity** is so high that the layer is essentially at a constant **electric potential**.

electrostatic electrometer—*See* **unifilar electrometer**, **bifilar electrometer**, **quadrant electrometer**.

electrostatic precipitator—A device for removing particulate matter from smokestack exhaust gas by imparting an **electric charge** to the **particles** and then attracting them to a metal plate or screen of opposite charge before the gas is exhausted out of the top of the stack.

Elektro—*See* **Geostationary Operational Meteorological Satellite**.

element—1. Smallest entity that is capable of possessing chemical characteristics and that cannot be changed into smaller neutral units by chemical reaction.
Elements in their pure form may exist in atomic forms, for example, He, Ne, or may be associated into molecular units such as H_2, O_2, graphite, or diamond.

2. Any one of the properties or conditions of the **atmosphere** that together specify the physical state of weather or **climate** at a given place for any particular moment or period of time; the climatic **elements**, meteorological elements.

3. The smallest definable object of interest in a scene.
It is a single item in a collection, **population**, or **sample**.

elephanta—(Also called elephant, elephanter.) A strong southeasterly **wind** on the Malabar Coast of India in September and October, at the end of the **southwest monsoon**.
It brings **thundersqualls** and **heavy rain**.

elerwind—A **wind** of Sun Valley, north of Kufstein, in the Tyrol.

elevated convection—Convection that originates from an **atmospheric layer** above the **boundary layer**.

elevation—1. A measure (or condition) of height, especially with respect to the height of a point on the earth's surface above a reference plane (usually **mean sea level**), as "station elevation."
The term **altitude** (e.g., "high-altitude station") and the general term "height" are also used in this sense.

2. Same as **elevation angle**.

elevation angle—(Also called elevation.) The angle between the **horizon** and a point above the horizon, measured along the arc that passes through the **zenith** and the point in question.
In astronomy this is termed **altitude**. *Compare* **azimuth**, **depression angle**, **zenith distance**.

elevation capacity curve—Measures the volume (capacity) of a **reservoir** below a certain **elevation**.

elevation of ivory point—Same as **barometer elevation**.

ELF—Abbreviation for **extremely low frequency**.
See **radio frequency band**.

Eliassen–Palm flux—A **vector** quantity with nonzero components in the latitude–height plane, the direction and magnitude of which determine the relative importance of **eddy heat flux** and **momentum flux**.
When the Eliassen–Palm flux (EPF) vector points upward, the **meridional** heat flux dominates; when the EPF vector points in the meridional direction, the meridional flux of **zonal** momentum dominates. The **divergence** of the Eliassen–Palm flux is more frequently used as a diagnostic tool, as it is proportional to the **eddy** potential **vorticity** flux. In the special case when the EPF divergence is zero, as for steady, frictionless, **linear** waves, **thermal wind** balance is maintained in an idealized zonally symmetric **atmosphere** through the induction of a mean **meridional circulation**, even though the waves produce significant eddy heat and momentum fluxes.

elliptical depolarization ratio—The ratio of **power** received in the transmission channel to power received in the **orthogonal** channel of a **dual-channel radar**, when an elliptically polarized **signal** is transmitted. For a weather target, this ratio depends, in general, on the ratio of the major to the minor axis of the **polarization** ellipse, the orientation of the ellipse, and the depolarizing characteristics of the **hydrometeors** that constitute the **target**.

Typically, the ratio will be less than unity (or a negative **decibel** quantity) when the major and minor axes are comparable in size and the polarization ellipse is close to circular, but will be greater than unity (or a positive decibel quantity) as the polarization ellipse becomes sufficiently elongated on either of its axes. A greater **depolarization ratio** is achievable with elliptical than with **circular polarization** by matching the axial ratio and the orientation of the transmitted signal to the depolarizing characteristics of the target. *Compare* **cancellation ratio, circular depolarization ratio.**

elliptical polarization—A **polarization** state of an **electromagnetic wave** in which the **electric field** vector at a point in space describes an ellipse. Definition of the polarization state includes the direction of rotation of the electric field (right or left), the ellipticity, and the orientation of the major axis of the ellipse.

Compare **circular polarization.**

Elmo's fire—Same as **Saint Elmo's fire.**

Elsasser's radiation chart—A **radiation chart** developed by W. M. Elsasser for the graphical solution of the **radiative transfer** problems of importance in meteorology.

Given a **radiosonde** record of the vertical **variation** of **temperature** and **water vapor** content, one can find with this chart such quantities as the effective **terrestrial radiation**, net **flux** of **infrared radiation** at a **cloud base** or a **cloud top**, and **radiative cooling** rates.

elve—Transient laterally extensive **illumination** of the **airglow** layer, at about 90 km, over **thunderstorms**, and associated with the electromagnetic **pulse** from the **return stroke** of a **lightning flash** to ground.

elvegust—(Also called sno.) A cold descending **squall** in the upper parts of Norwegian fjords.

EMAC probe—Abbreviation for **electromagnetic acoustic probe.**

emagram—*See* **thermodynamic diagram.**

emanometer—An instrument for the measurement of the **radon** content of the **atmosphere**.

Radon is removed from a sample of air by **condensation** or **adsorption** on a surface, and is then placed in an **ionization chamber** and its activity determined.

embata—A local onshore southwest **wind** caused by the reversal of the northeast **trade winds** in the lee of the Canary Islands.

embryo—An incipient **nucleus** of many molecular dimensions for the initiation of one **phase** in another, as **ice** in **supercooled water**.

emissary sky—A sky of **cirrus** clouds that are either isolated or in small, separated groups; so called because this formation is often one of the first indications of the approach of a **cyclonic** storm.

emission—With respect to **radiation**, the generation and sending out of **radiant energy**.

The emission of radiation by natural emitters is accompanied by a loss of **energy** and is considered separately from the processes of **absorption** or **scattering**.

emission factor—For **air pollution** produced during an industrial process, the ratio of output (**air pollutants** or other by-product such as waste **heat**) to some measure of productivity, such as input quantity of fuel or raw materials or amount of final product.

As written by the U.S. Environmental Protection Agency, an emission factor is a representative value that attempts to relate the quantity of a pollutant released to the **atmosphere** with an activity associated with the release of that pollutant. These factors are usually expressed as the weight of pollutant divided by a unit weight, volume, distance, or duration of the activity emitting the pollutant (e.g., kilograms of particulate emitted per megagram of coal burned).

emission inventory—A catalog or list of each emission source within a region, along with the location and emission height, type of **air pollutants** emitted, rate of emissions, initial emission characteristics such as **buoyancy** and initial **momentum**, and emission schedule if not continuous.

Such a catalog often includes **anthropogenic** sources such as individual point sources from industry including some farming activities, line sources along roads, and area sources associated with quasi-uniform distributions of sources such as houses in a residential neighborhood. Biogenic area sources are also included, such as lawns, parks, farmland, pasture, and forests.

emission line—A line of finite width in the **emission spectrum**, characteristic of gaseous emission. Emission lines are characterized by their central **wavelength, line intensity**, and **line width**.

emission rate—The mass of **air pollutants** emitted from a **source** every second, for example, 1 kg s⁻¹ of **sulfur dioxide** from the top of a smokestack.

emission spectrum—The detailed dependence on **wavelength** of the **intensity** of **radiation** emitted by a given **radiator**.

Emission spectra are typically composed of discrete spectral lines and bands of overlapping lines, which depend on the molecular or atomic composition of the emitting substance and which may be used to identify it uniquely.

emission standard—A legal restriction on the maximum amount of **air pollutants**, especially airborne toxic chemicals, that are allowed to be emitted from any single source such as a smokestack.

The standard can also be expressed as the maximum concentration allowed in the exhaust gas at stack top or within the **ventilation** air.

emissive power—Same as **emittance**.

emissivity—The ratio of the **power** emitted by a body at a **temperature** T to the power emitted if the body obeyed **Planck's radiation law**.

Strictly, emissivity should be qualified by the **frequency**, direction, and even **polarization** state of the emitted **radiation**. This is recognized in qualifiers such as **monochromatic** (at a given frequency) as opposed to total (over a broad **range** of frequencies), and directional (the ratio for a particular direction) as opposed to hemispherical (the ratio for a hemisphere of directions). Contrary to a widespread misconception, the upper limit of emissivity is not 1. This upper limit is valid (approximately) only for bodies large compared with all relevant wavelengths. *See also* **blackbody, emissive power, emittance**.

emittance—(Or emissive power.) Emitted **irradiance**.

empirical flood formula—Empirical formula expressing peak **flood** discharge as a function of **catchment area** and other factors.

encroachment method—A technique used in **boundary layer** meteorology to denote a type of **mixed layer** growth rate that is proportional to the rate of warming of the mixed layer divided by the ambient **lapse rate** immediately above the top of the mixed layer.

In other words, heating from the surface causes the depth of the mixed layer to increase. Also known as the thermodynamic method, the **mixed-layer top** can be found by finding the area under the early morning **sounding** (between the sounding and an **adiabat** for warmer air) that is equal to the area under a plot of the surface **heat flux** versus time up to the time of interest. *Compare* **energetics method**.

end of storm oscillation—(Abbreviated EOSO.) A characteristic in the **electric field** signature at the ground near the end of a **thunderstorm**.

The electric field changes from upward-directed (foul-weather **polarity**) to downward-directed (fair-weather polarity) and often retains the latter polarity for several tens of minutes before returning momentarily to foul-weather polarity. Positive ground flashes and **spider lightning** are occasionally observed during the period of reversed **field** polarity. The origin of the EOSO is only partly understood.

endorheic lake—A lake without surface or subsurface **outflow**, where the **inflow** water is lost by **evaporation**.

endothermic reaction—A chemical reaction that absorbs **heat** from its surroundings.

energetics—The study of **energy conversion** and **transfer** in a physical system.

energetics method—A technique used in **boundary layer** meteorology to describe a type of mixed-layer growth that is governed by the amount of **turbulence kinetic energy** that is available for conversion into **potential energy** during **entrainment** of warm free-atmosphere air downward into the **mixed layer**.

Compare **encroachment method**.

energy—A measurable physical quantity, with **dimensions** mass times **velocity** squared, that is conserved for an **isolated system**.

Energy of motion is **kinetic energy**; energy of position is **potential energy**. *See* **energy conversion, internal energy, enthalpy**.

Feynman, R. P., R. B. Leighton, and M. Sands, 1963: *The Feynman Lectures in Physics*, Vol. I, p. 4-2.

energy balance—The balance between the net warming or cooling of a volume and all possible **sources** and **sinks** of **energy**.

The main sources and sinks of energy typically include the net fluxes of **sensible heat**, latent

heat, and **radiant energy**. Conservation of energy requires that the energy received by a surface must equal that lost from the surface plus that stored. For water and land surfaces, the main source of energy is **net radiation**, which equals the sum of short and long waveband **radiation** downward minus radiation reflected or emitted upward. This energy is normally transferred into the soil (soil **heat flux**), into the air (sensible heat flux), or into latent heat flux (**evapotranspiration** or ET). Small amounts of the incoming energy can change the **heat** content of water or crops at the surface or are converted to other forms of energy (e.g., **photosynthesis**). Energy balance is often used to estimate evapotranspiration by 1) measuring net radiation, soil heat flux, and sensible heat flux; 2) entering those values into an energy balance equation; and 3) solving for the latent heat flux (ET). Under hot, dry, windy (**advection**) conditions, heat from the air in addition to net radiation is sometimes available at an underlying cool surface. Advection can potentially increase **evaporation** rates to higher than the energy available from net radiation alone. *See also* **surface energy balance**.

energy cascade—The **transfer of energy** from larger **eddies** to smaller eddies in the **inertial subrange** of the **spectrum** of turbulence kinetic energy.

energy-containing eddies—A **range** of **eddy** sizes containing much of the **energy** in the **turbulence spectrum**.
    Approximately centered on **integral length scales**, these eddies are larger than those in the **inertial subrange** and smaller than the semipermanent eddies associated with the spectrum's formation. *See* **turbulence length scales**.

energy conversion—(Also energy transformation, **energy transfer**.) A process in which **energy** changes from one form to another.
    Energy is conserved for a system that does not interact with its surroundings, and the total energy of such a system may often be expressed as the sum of energies of different kinds:

$$E = E_1 + E_2 + E_3 + \cdots.$$

Thus if E_1 decreases in any process, E_2, etc., must increase correspondingly for E to remain constant, and we may say that energy of type 1 has been converted into energies of type 2, 3, etc.

energy density spectrum—(Sometimes called energy spectrum.) The square of the **amplitude** of the (complex) **Fourier transform** of an **aperiodic** function.
    Thus, if $f(t)$ is the given function, its Fourier transform is

$$F(\omega) = \frac{1}{2\pi} \int_{-\infty}^{\infty} f(t)e^{-i\omega t}dt,$$

and the energy density spectrum is $|F(\omega)|^2$. It is assumed that the total **energy** $\int_{-\infty}^{\infty} |f|^2 dt$ is finite.

energy equation—1. Thermodynamic energy equation; same as the **first law of thermodynamics**.

    2. Mechanical energy equation (or kinetic energy equation): an expression for the rate of change of **kinetic energy**, which is obtained by **scalar** multiplication of the three-dimensional **vector** equation of motion by the vector velocity **u**; it may be written in the form

$$\frac{\partial}{\partial t}\left(\frac{1}{2}\rho\mathbf{u}^2\right) = -\nabla \cdot \left(\frac{1}{2}\rho\mathbf{u}^2 + \rho\phi + p\right)\mathbf{u} - \frac{\partial(\rho\phi)}{\partial t} + p\nabla \cdot \mathbf{u} - \mathbf{u} \cdot \mathbf{F},$$

where $\phi = gz$ is the **geopotential** energy, ρ is **density**, p is **pressure**, **F** is the vector frictional force per unit volume, and ∇ is the **del operator**.

    3. Total energy equation: An expression relating all forms of **energy** obtained by combining the thermodynamic energy equation with the **mechanical energy equation**. When integrated over a fixed volume of the **atmosphere**, this equation takes the form

$$\frac{\partial}{\partial t}\int \rho(c_v T + \frac{1}{2}\mathbf{u}^2 + \phi)dV = \int (\rho\, c_v T + \frac{1}{2}\rho\mathbf{u}^2 + \rho\phi)V_n ds$$

$$+ \int pV_n ds + \int \rho Q dV - \int \mathbf{u} \cdot \mathbf{F}dV,$$

where dV is the volume element, ds is the element of the surface of the volume, and V_n is the inwardly directed **velocity** normal to the surface of the volume.
    This equation expresses the fact that the combined internal, kinetic, and **potential energy** in a given volume can vary only as a result of 1) the **transport** of these forms of energy across the boundaries of the volume; 2) the **work** done by **pressure** forces on the boundary; 3) the addition or removal of **heat**; and 4) the dissipational effect of **friction**.

Gill, A. E., 1982: *Atmosphere–Ocean Dynamics*, Academic Press, 76–82.

energy grade line—The line above a **datum** used for expressing the total **energy** of a flow.

energy level—Any of the possible discrete energies of an atom, molecule, or **nucleus**.
According to **Newtonian mechanics**, the **energy** of a body is a continuous **variable**, but according to the **quantum theory**, measured energies of bound states are discrete, some said to be allowed, others forbidden. We are not aware of this discreteness for macroscopic objects because of the extreme (relative) smallness of the separation between their energy levels. During transitions from one energy level to another, quanta of **radiant energy** are emitted or absorbed, their **frequency** depending on the difference between energy levels. Emissions of this type are responsible for the **aurora borealis**, for example.

energy spectrum—Same as **energy density spectrum**.

energy transfer—Any process in which a system interacts with its surroundings in such a way that the **energy** of the system increases (or decreases) while that of the surroundings decreases (or increases) by the same amount.
See also **energy conversion**.

energy transformation—Same as **energy conversion**.

engineering hydrology—Branch of **applied hydrology** that deals with engineering applications such as planning, designing, operating, and maintaining **water resources** projects.

enhanced greenhouse effect—The additional **greenhouse effect** due to an increase in the concentration of **greenhouse gases**.
Because of the complexity of the **feedback** processes within the **climate system**, the enhanced greenhouse effect is usually referred to in terms of the **radiative forcing** that results at the **tropopause**, after the **stratosphere** has come into a new **radiative equilibrium**. For example, thus defined, the enhanced greenhouse effect due to an effective doubling of **carbon dioxide** concentration from its preindustrial baseline is about 4 W m^{-2}. Note the difference in meaning from the greenhouse effect, which refers to the entire natural process and which results in a climatological average counterradiation of about 330 W m^{-2} to the surface. A comparable value for the enhanced greenhouse effect at the surface will only be obtained once the new **equilibrium** temperature of the surface is known, since they are strongly interrelated.

enhanced image—(Also called enhanced picture.) An **image** that has had processing (e.g., **contrast stretching**, **pseudocoloring**, **unsharp masking**) applied to improve information presentation.

enhanced network—*See* **network**.

enhanced "v"—A **signature** on **infrared** satellite imagery that depicts a warm, wedge-shaped region stretching from the upshear edge of a **thunderstorm** anvil, downshear along its long axis.
This so-called warm **wake** is surrounded by long, narrow regions of colder pixels along either side, forming an apparent V shape. The warmer pixels may be stratospheric **cirrus** "blow-off" from **overshooting tops**, meaning that they are higher than the mean **anvil** height, even though they are warmer. The **signature** typically forms on convective storms possessing extremely strong updrafts.

enhancement curve—A mathematical function or set of tabulated values that can be used to process **digital** imagery to emphasize specific features or areas of interest.

ENP—In air navigation, abbreviation for "estimated time over next reporting position."

ensemble average—An average taken over many different flow realizations that have the same **initial** and **boundary conditions**.
In the limit of the **sample** size going to infinity, the ensemble average approaches the **ensemble mean** and may be a function of both time and position. When the flow is steady and homogeneous, the ensemble, space, and time means are equal. *See* **ergodicity**.
Hinze, J. O., 1975: *Turbulence*, 2d ed., McGraw–Hill, p. 5.

ensemble average—1. The value of a meteorological **variable** found by averaging over many independent descriptions or realizations of that variable.
While this type of average often forms the basis of theories and simulations (numerical or physical models) for **turbulence**, it is usually impossible to compute in real life because we cannot control the **atmosphere** in order to reproduce multiple ensemble members. In real life, an **ergodic** hypothesis is often used, where time or space averages are assumed to be reasonable approximations to ensemble averages.
2. In **numerical weather prediction**, the average found by averaging over many different

forecasts for the same domain and time period, but starting from slightly different **initial conditions** or using different numerical models or parameterizations.

This ensemble average is usually more accurate than any single **model** run because it partially counteracts the sensitive dependence to initial conditions associated with the **nonlinear** equations that govern the **atmosphere**.

ensemble forecast—A set of different forecasts all valid at the same forecast time(s).

The differences between the forecasts can provide information on the **probability distribution** of the predicted variables. The forecasts in the ensemble may have different **initial conditions**, **boundary conditions**, **parameter** settings, or may even be from entirely independent **NWP** models.

ensemble mean—The average of a predicted **variable** or **field** over an ensemble of forecasts.

In taking the mean, one filters those aspects of the forecasts that are not predictable.

> Leith, C. E., 1974: Theoretical skill of Monte Carlo forecasts. *Mon. Wea. Rev.*, **102**, 409–418.

ensemble spread—The average difference between the individual **ensemble forecasts** of a quantity and the **ensemble mean** forecast of the quantity.

The ensemble spread is often used to predict the magnitude of the **forecast error**. *See* **spread skill correlation**.

ENSO—Acronym for El Niño–Southern Oscillation, coined in the early 1980s in recognition of the intimate linkage between **El Niño** events and the **Southern Oscillation**, which prior to the late 1960s had been viewed as two unrelated phenomena.

The global ocean–atmosphere phenomenon to which this term applies is sometimes referred to as the "ENSO **cycle**." *See also* **El Niño, La Niña, Southern Oscillation**.

> Bjerknes, J., 1969: Atmospheric teleconnections from the equatorial Pacific. *Mon. Wea. Rev.*, **97**, 163–172.
> Philander, S. George, 1990: *El Niño, La Niña, and the Southern Oscillation*, Academic Press, International Geophysics Series, Vol. 46.

enstrophy—One-half the square of the **relative vorticity**.

The term was popularized by C. Leith and is based on the modern Greek στρωφη, meaning "act of turning." Enstrophy is a conservative quantity in two-dimensional inviscid flow. However, when **viscosity** is finite, enstrophy tends to be selectively decayed relative to more rugged integrals such as **energy** and **angular momentum**.

> Leith, C. E., 1968: Diffusion approximation for two-dimensional turbulence. *Phys. Fluids*, **11**, 671–673.

enthalpy—A thermodynamic state function H defined as

$$H = U + pV,$$

where U is the **internal energy**, p is **pressure**, and V is volume.

Specific enthalpy of a homogeneous system, h, is its enthalpy divided by its mass, m, defined by

$$h = H/m = u + pv,$$

where u is **specific** internal energy and v is **specific volume**. With aid of the **gas laws**, the specific enthalpy of an **ideal gas** may also be written as

$$h = c_p T + \text{constant},$$

where T is **temperature** and c_p is the **specific heat** at constant pressure. The specific enthalpy of a liquid, h_l, is

$$h_l = c_l T + \text{constant},$$

where c_l is the liquid's specific heat, which is nearly independent of pressure and specific volume. For a system consisting of a mixture of components, the total enthalpy is the mass-weighted sum of the enthalpies of each component. Thus, the total enthalpy of a system consisting of a mixture of **dry air**, **water vapor**, and liquid water is

$$H = (m_d c_{pd} + m_v c_{pv} + m_w c_w)T + \text{constant},$$

where m_d, m_v, and m_w are the masses of dry air, water vapor, and liquid water, respectively; c_{pd} and c_{pv} the specific heats of dry air and water vapor; and c_w is the specific heat of liquid water. This quantity is commonly called **moist enthalpy**, with specific moist enthalpy given by $h = H/(m_d + m_v + m_w)$. With the aid of the definition of the latent heat of vaporization (*see* **latent heat**), moist enthalpy may also be written as

$$H = (m_d c_{pd} + m_t c_w)T + m_v L_v + \text{constant},$$

where m_t is the mass of **vapor** plus liquid and L_v is the latent heat of vaporization. Similar relations can be written to include the effects of **ice**. In an **adiabatic, reversible process**, enthalpy and specific enthalpy are conserved, although the component specific enthalpies may not be, due to the exchange of enthalpy between components in **phase** changes.

entity-type entrainment mixing—An idealization of **cloud** mixing where parcels of **cloudy** and noncloudy air move with no dilution to their respective levels of neutral **buoyancy** within a cloud, after which they mix with the **environment**.

entrainment—1. In meteorology, the **mixing** of environmental air into a preexisting organized **air current** so that the environmental air becomes part of the current; the opposite of **detrainment**.
　　Entrainment of air into **clouds**, especially **cumulus**, is said to be inhomogeneous when the timescale for mixing of environmental air is very much greater than the timescale for **drop** evaporation. Under these conditions, which are often found when environmental air is first entrained into cumulus, regions of cloud and entrained air are intertwined, with **evaporation** occurring only on the edges of the **interface** between the cloudy and entrained environmental air.
　　2. The process by which turbulent fluid within a **mixed layer** incorporates adjacent fluid that is nonturbulent, or much less turbulent; thus entrainment always proceeds toward the nonturbulent layer.
　　In the absence of **advection** effects, this tends to deepen the mixed layer.

entrainment coefficient—The ratio of lateral **entrainment velocity** to plume-rise **velocity** of a smoke plume.
　　Smoke plumes that rise through the **environment** due to **buoyancy** or **momentum** become diluted with environmental air, where the rate of dilution is proportional to the rise rate of the **plume**. The entrainment coefficient is this constant of proportionality. This concept of lateral **entrainment** does not work well for convective **thermals** in the **atmospheric boundary layer**. One reason is that **boundary layer** thermals are part of a large overturning **circulation** involving the whole boundary layer during statically unstable conditions of **free convection**, rather than being thin currents of rising smokestack air.

entrainment rate—The rate at which **entrainment** proceeds.
　　In the absence of **advection** effects, the entrainment rate is equal to the rate of deepening of the **mixed layer**.

entrainment velocity—The volume of entrained air per unit area per unit time, which has units of **velocity**.
　　It is a measure of dilution of an entity, such as a rising **smoke plume**, or of a whole layer, such as the **atmospheric boundary layer**. For a growing atmospheric **mixed layer**, the rate of rise of the top of the mixed layer z_i equals the entrainment velocity w_e minus any large-scale **subsidence** w_s that is imposed at the top of the **boundary layer**. During sunny days over land, the entrainment velocity is proportional to the **heat flux** from the surface divided by the **potential temperature** change across the **entrainment zone** (i.e., the strength of the **capping inversion**). On windy days, the entrainment velocity is proportional to the **turbulence kinetic energy** at the top of the boundary layer and is inversely proportional to the **potential temperature** change across the entrainment zone. *See* **entrainment rate**.

entrainment zone—(Also called entrainment layer.) A layer of intermittent **turbulence** and overshooting **thermals** at the top of the **convective mixed layer** where the **free atmosphere** is entrained into the top of the **boundary layer**.
　　The entrainment zone is thinner when a stronger **temperature inversion** caps the boundary layer and thicker when turbulence and thermals are more vigorous.

entrance region—The region of **confluence** at the **upwind** extremity of a **jet stream**; the opposite of **exit region**.

entropy—A thermodynamic **state variable** denoted by S (s denotes **specific entropy**, entropy per unit mass).
　　The rate of change of entropy of a thermodynamic system is defined as

$$dS/dt = Q/T,$$

where Q is the heating rate in a **reversible process** and T is **absolute** temperature. Integration of this equation yields the entropy difference between two states. The entropy of an **isolated system** cannot decrease in any real physical process, which is one statement of the **second law of thermodynamics**. The **specific entropy** of an **ideal gas**, s_g, may be expressed as

$$s_g = c_{pg} \ln T - R_g \ln p_g + \text{constant},$$

where c_{pg} is the **specific heat** at constant pressure of that gas, R_g is its **gas constant**, and T and p_g are its **temperature** and **pressure**. The entropy of a liquid, s_l, is

$$s_l = c_l \ln T + \text{constant},$$

where c_l is the specific heat of the liquid.

envelope soliton—A group of **waves** with an envelope (the curve touching the surface near the **wave crests**) that propagates without change of form, as a **solitary wave**.

Under certain circumstances, **ocean wave** groups have approximately this behavior.

environment—External conditions and surroundings, especially those that affect the quality of life of plants, animals, and human beings.

In agriculture the environment includes the air, soil, and water conditions.

environmental acoustics—The general term applied to the study of a variety of atmospheric **sound** propagation phenomena, including frequency-dependent molecular **absorption**, **refraction** by **wind** and **temperature** gradients, **diffraction** by **topography**, buildings, trees, etc., and **scattering** by **turbulence**.

Also of interest are sounds, some of atmospheric origin, ranging from **infrasonic** emissions by **thunderstorms** and large explosions to **ultrasonic** emissions by high-power turbines and transducers used for materials testing.

environmental chambers—Large reactors, usually made of glass or polytetrafluoroethylene (PTFE) film, in which gas-phase chemical reactions are carried out to simulate changes occurring in the **atmosphere**.

Either natural **sunlight** or artificial **ultraviolet** lighting can be used to initiate the chemistry. The buildup of products such as **nitrogen dioxide** and **ozone** is monitored using techniques such as **gas chromatography** or **infrared spectroscopy**.

environmental lapse rate—The rate of decrease of **temperature** with **elevation**, $-\partial T/\partial z$, or occasionally $\partial T/\partial p$, where p is **pressure**.

The concept may be applied to other atmospheric variables (e.g., **lapse rate** of **density**) if these are specified. The environmental lapse rate is determined by the distribution of temperature in the vertical at a given time and place and should be carefully distinguished from the **process lapse rate**, which applies to an individual **air parcel**. *See* **autoconvective lapse rate**, **superadiabatic lapse rate**.

environmental tracer—A substance that occurs naturally in the environmental system that can be used as a **tracer**.

eolation—The **erosion** of land surfaces by wind-driven **dust** or sand.

See also **corrasion**.

Eole—A French satellite, launched in August 1971, that collected meteorological data from 500 free-flying constant-level balloons to study hemispheric **circulation** patterns.

eolian—Same as **aeolian**.

EOS—Abbreviation for **Earth Observing System**.

EOSO—Abbreviation for **end of storm oscillation**.

ephemeral data—Data that help to characterize the conditions under which the **remote sensing** data were collected, that may be used to calibrate the **sensor** prior to **analysis**, or that include such information as the positioning and spectral stability of sensors, sun angle, or platform attitude.

ephemeral lake—Lake that becomes dry during a **dry season** or in dry years.

ephemeral stream—A **stream** channel that carries water only during and immediately after periods of **rainfall** or **snowmelt**.

Compare **intermittent stream**, **perennial stream**.

ephemeris—The position of a satellite in space as a function of time.

Ephemeris data are used for **gridding** satellite imagery. Since ephemeris data are based solely on the predicted position of the satellite, an ephemeris is susceptible to errors from vehicle pitch, orbital eccentricity, and the oblateness of the earth.

epilimnion—The layer of water above the **thermocline** in a **freshwater** body.

epipycnal—In a direction that lies within the plane of an **isopycnal surface**.

The epipycnal direction is close to horizontal.

epipycnal mixing—Mixing of fluid in an **epipycnal** direction, that is, **mixing** of fluid with adjacent fluid from the same **density** surface.

episode—In the context of **air pollution**, a time period during which **pollution** standards are exceeded or, more loosely, during which **pollutant** levels are significantly greater than **normal** or background levels.

episode criteria—Levels of **criteria pollutants** that, when exceeded, constitute a **smog** alert as set by the state of California.
  Such levels are regarded as being dangerous to certain portions of the population if experienced for a short time.

epitaxis—(Also called oriented overgrowth.) The state or process wherein one crystalline material builds up a layer (usually thin) of its own **crystal lattice** upon the surface of some other **crystal** with a lattice geometry similar in crystal symmetry and lattice spacing to its own.
  A close match of lattice geometry gives rise to **ice** nucleation, at only a few degrees below 0°C, whereas a small water drop may normally supercool some tens of degrees K. The effect has been demonstrated for **water vapor** growth to **ice crystals** and **freezing** of supercooled droplets on large single crystals, oriented with respect to the base common occurrence crystal such as **silver iodide**. Molecule size and structure also influence the effect.

epoch—1. The formal geochronologic unit, longer than a **geologic age** and shorter than a **geologic period**, during which the rocks of the corresponding series were formed (from Glossary of Geology 1997).
  2. A term used informally to designate a length (usually short) of geological time, for example, **glacial epoch**.
  3. In **paleomagnetism**, a date to which measurements of a time-varying quantity are referred, for example, "a **chart** of **magnetic declination** for epoch 1965.0," or informally as a magnetic **polarity** epoch.
  Magnetic polarity refers to whether the geomagnetic field was like it is today (normal polarity) or whether the north and south poles are reversed from their present configuration (reversed polarity).

American Geological Institute, 1997: *Glossary of Geology*, 4th ed., J. A. Jackson, Ed., p. 213.

epoxide—Cyclic compound formed by adding a single **oxygen** atom across the double bond of an **alkene**.
  These reactive compounds are used in industrial-scale organic synthesis.

equal-area map—A flat map so drawn that equal units of actual (or represented) area in any two portions of the map have identical map areas.
  For maps representing portions of the earth's surface this is obtained by continuously changing the scales of the meridians and parallels. Equal-area maps covering the whole globe, such as the Sanson–Flamsteed sinusoidal **projection**, are approximately elliptical. In Lambert's azimuthal equal-area projection, the parallels of latitude come closer together as the **equator** is approached. Any map of the whole world or of a hemisphere necessarily distorts the shape of a region far from the center of the map, but such maps are useful for some climatological studies in which the correct representation of area is important. For limited parts of the globe equal-area projections are quite practical. *Compare* **conformal map**.

equation of barotropy—*See* **barotropy**.

equation of continuity—(Or continuity equation.) A hydrodynamical equation that expresses the principle of the **conservation of mass** in a fluid.
  It equates the increase in mass in a hypothetical fluid volume to the net flow of mass into the volume. The equation of continuity is usually written in either of two forms:

$$\frac{\partial \rho}{\partial t} + \nabla \cdot (\rho \mathbf{u}) = 0;$$

or

$$\frac{D\rho}{Dt} + \rho \nabla \cdot \mathbf{u} = 0,$$

where ρ is the fluid **density** and $\mathbf{u}$ the **velocity** vector. Under the quasi-static approximation with **pressure** as vertical coordinate, the equation takes the form

$$\nabla_p \cdot \mathbf{v} + \frac{\partial \omega}{\partial p} = 0,$$

where p is the pressure, $\mathbf{v}$ is the horizontal velocity, $\omega = Dp/Dt$ the vertical p velocity, and ∇_p the **del operator** in the **isobaric surface**.

equation of motion—*See* **equations of motion**.

equation of piezotropy—(Also called **physical equation**.) An equation relating the **thermodynamic variables** in processes of a piezotropic fluid.
    In its general form it expresses the **density** ρ as a function of the **pressure** p:

$$\rho = \rho(p).$$

The derivative $d\rho/dp$ is called the **coefficient of piezotropy**. The most familiar such equation is that for polytropic changes of state in an **ideal gas**:

$$p\rho^{-\lambda} = \text{constant},$$

where λ is the **modulus** of the **polytropic process**. Prior to the discovery of the **first law of thermodynamics** in the midnineteenth century, the equation of piezotropy was used to complete the hydrodynamic equation set consisting of the **equations of motion** and the **conservation of mass**.

equation of state—(Also known, for an **ideal gas**, as the **Charles–Gay–Lussac law**, or Charles law, or Gay–Lussac's law.) An equation relating **temperature**, **pressure**, and volume of a system in thermodynamic **equilibrium**.
    The equation of state for an ideal gas of N molecules in a volume V is

$$pV = NkT,$$

where p is pressure, T is **absolute** temperature, and k is **Boltzmann's constant**. After division by the mass of the volume, this may also be written as

$$p\alpha = RT,$$

where α is the **specific volume** and R is an individual **gas constant**. For a mixture of gases, a similar equation may be written with the aid of **Dalton's law**, where R becomes a **weighted average** of the individual gas constants.

equation of static equilibrium—*See* **hydrostatic equation**.

equation of time—The difference at any instant between the apparent solar time and the mean solar time as measured at a specified place; it is the difference between the hour angles of the apparent sun and the mean sun.
    Tables of the equation of time are available in astronomical almanacs. The equation of time is zero at four times during a year, on about 15 April, 15 June, 31 August, and 24 December, at present. The algebraic sign associated with the equation of time varies from one source to another, a fact that must be kept in mind in abstracting values from almanacs. *See* **apparent solar day**, **mean solar day**.

equations of motion—A set of hydrodynamical equations representing the application of Newton's second law of motion to a fluid system.
    The total **acceleration** on an individual fluid **particle** is equated to the sum of the forces acting on the particle within the fluid. Written for a unit mass of fluid in motion in a **coordinate system** fixed with respect to the earth, the **vector** equation of motion for the **atmosphere** is

$$\frac{D\mathbf{u}}{Dt} = -2\Omega \times \mathbf{u} - g\mathbf{k} - \frac{1}{\rho}\nabla p + \mathbf{F},$$

where $\mathbf{u}$ is the three-dimensional **velocity** vector, Ω the **angular velocity of the earth**, $\mathbf{k}$ a unit vector directed upward, ρ the **density**, p the **pressure**, g the **acceleration of gravity**, and $\mathbf{F}$ the **frictional force** per unit mass. The usual form for the **scalar** equations of motion in **spherical coordinates** (λ, ϕ, r), with λ the longitude, ϕ the latitude, and r the radius from the center of the earth, is as follows:

$$\frac{\partial u}{\partial t} + u\frac{\partial u}{r\cos\phi\,\partial\lambda} + v\frac{\partial u}{r\,\partial\phi} + w\frac{\partial u}{\partial r} - \left(2\Omega + \frac{u}{r\cos\phi}\right)(v\sin\phi - w\cos\phi) + \frac{1}{\rho}\frac{\partial p}{r\cos\phi\,\partial\lambda} = F_\lambda,$$

271

$$\frac{\partial v}{\partial t} + u\,\frac{\partial v}{r\cos\phi\partial\lambda} + v\,\frac{\partial v}{r\partial\phi} + w\,\frac{\partial v}{\partial r} + \left(2\Omega + \frac{u}{r\cos\phi}\right)u\sin\phi + \frac{vw}{r} + \frac{1}{\rho}\,\frac{\partial p}{r\partial\phi} = F_\phi,$$

$$\frac{\partial w}{\partial t} + u\,\frac{\partial w}{r\cos\phi\partial\lambda} + v\,\frac{\partial w}{r\partial\phi} + w\,\frac{\partial w}{\partial r} - \left(2\Omega + \frac{u}{r\cos\phi}\right)u\cos\phi - \frac{v^2}{r} + \frac{1}{\rho}\,\frac{\partial p}{\partial r} + g = F_r.$$

Most global **numerical weather prediction** models and **general circulation** models use an approximate version of the above nonhydrostatic **primitive equations**. This version involves the approximation of the vertical equation of motion by the **hydrostatic equation** and the selective approximation of $r = a + z$ by $r \approx a$, where a is the constant radius to **mean sea level** and z is the height above mean sea level. These approximations result in the quasi-static primitive equations

$$\frac{\partial u}{\partial t} + u\,\frac{\partial u}{a\cos\phi\partial\lambda} + v\,\frac{\partial u}{a\partial\phi} + w\,\frac{\partial u}{\partial z} - \left(2\Omega + \frac{u}{a\cos\phi}\right)v\sin\phi + \frac{1}{\rho}\,\frac{\partial p}{a\cos\phi\partial\lambda} = F_\lambda,$$

$$\frac{\partial v}{\partial t} + u\,\frac{\partial v}{a\cos\phi\partial\lambda} + v\,\frac{\partial v}{a\partial\phi} + w\,\frac{\partial v}{\partial z} + \left(2\Omega + \frac{u}{a\cos\phi}\right)u\sin\phi + \frac{1}{\rho}\,\frac{\partial p}{a\partial\phi} = F_\phi,$$

$$\frac{1}{\rho}\,\frac{\partial p}{\partial r} + g = 0.$$

See **Newton's laws of motion, vorticity equation.**

equator—Geographically, on the earth's surface, the imaginary **great circle** of latitude 0°, which is equidistant from the poles, and which separates the Northern Hemisphere from the Southern Hemisphere.
   See **meteorological equator, heat equator, thermal equator, celestial equator, aclinic line, geomagnetic equator.**

equatorial acceleration—A state in which the equatorial **atmosphere** of a celestial body has a larger **absolute** angular **velocity** than the more poleward portions of the atmosphere.
   An equatorial acceleration is exhibited by the Sun, Jupiter, and Saturn. The term is a misnomer since it involves a lateral angular **shear**, not an **acceleration**.

equatorial air—According to some authors, the air of the **doldrums** or the **equatorial trough**, to be distinguished somewhat vaguely from the **tropical air** of the **trade-wind** zones.
   Tropical air "becomes" equatorial air when the former enters the equatorial zone and stagnates. There is no significant distinction between the physical properties of these two types of air in the lower **troposphere**.
      Grimes, A., 1951: *Compendium of Meteorology*, p. 881.

equatorial bulge—The gross deviation of the earth from a spherical shape due to its rotation.
   The approximate shape of the earth is an ellipsoid. The equatorial radius is about 7 km greater and the polar radius about 14 km less than the corresponding radius of a sphere with the same volume as the earth.

equatorial calms—Same as **doldrums.**

equatorial climate—A climatic zone with a **climate** typical of regions along the **equator.**
   See **tropical climate.**

equatorial convergence zone—Same as **intertropical convergence zone.**

Equatorial Countercurrent—*See* **North Equatorial Countercurrent, South Equatorial Countercurrent.**

equatorial crossing time—The local solar time of a satellite's passage over the **equator.**
   Orbital numbers are assigned at the ascending crossing point and are advanced by one when the polar-orbiting spacecraft completes a full revolution about the poles and again reaches the equator.

equatorial current system—The system of **ocean currents** found in the upper Atlantic and Pacific Oceans between 20°S and 20°N.
   Its major components are the westward flowing North and South Equatorial Currents (**NEC**

and **SEC**), which occupy most of the region. The **Equatorial Intermediate Current** (EIC) is a subsurface band of intensified westward movement between 2°S and 2°N at depths between 300 and 1000 m. All other components of the system are narrow (200–400 km) bands of eastward flow: the North and South Equatorial Countercurrents (**NECC** and **SECC**), found at depths between 0 and 200 m and located between 5° and 10°N and 5° and 10°S, respectively; the **Equatorial Undercurrent** (EUC), found on the **equator** at a depth of 200 m; and the North and South Subsurface Countercurrents (NSCC and SSCC), found on either side of the EIC at depths between 400 and 700 m. The major elements of the equatorial current system, particularly the NEC, SEC, NECC, and EUC, are also seen in the Indian Ocean during the northeast **monsoon** season (December–April) but are significantly modified during the **southwest monsoon** season. In addition, the **Indian equatorial jet** is a feature not seen in the other oceans.

equatorial deep jets—A vertical stack of alternately eastward and westward **ocean currents** below the **thermocline** within a degree of the **equator**.

First discovered in the Indian Ocean, equatorial deep jets have been most extensively observed and described in the Pacific. There they have a dominant vertical wavelength of 300–400 m, and are most clearly seen at depths between 500 and 2000 m. The eastward and westward **relative current** extreme values may be superimposed on a larger vertical scale flow that may be either eastward or westward. For example, an eastward jet may appear in a given measurement as a relative minimum in a larger-scale westward flow. On a given longitude the depths of the jets vary interannually, but there is no clear evidence for steady vertical propagation.

equatorial dry zone—An arid region existing in the **equatorial trough** resulting from low-level wind **divergence** and **subsidence**.

The most famous dry zone is situated a little south of the **equator** in the central part of the equatorial Pacific.

equatorial easterlies—(Also called deep trades, deep easterlies.) As used by some authors, the **trade winds** in the summer hemisphere when they are very deep, extending to at least 8–10 km **altitude**, and generally not topped by upper **westerlies**.

If upper westerlies are present, they are too weak and shallow to influence the weather. In the winter hemisphere, these **easterlies** are restricted to a narrow belt along the **equator**. *Compare* **tropical easterlies**.

equatorial electrojet—The large eastward flow of electrical **current** in the **ionosphere** that occurs near noon within 5° of the **magnetic equator**.

It causes an abnormally large maximum value of the northern component of the geomagnetic **field** near local noon for stations in this region.

equatorial forest—Same as **tropical rain forest**.

equatorial front—Same as **intertropical front**.

Equatorial Intermediate Current—*See* **equatorial current system**.

equatorial Kelvin wave—An eastward propagating **wave** centered about the **equator** with **zonal** but no **meridional** velocity.

The zonal velocity has a Gaussian meridional structure centered about the equator with **standard deviation** equal to the equatorial **Rossby radius of deformation**. In the **shallow water approximation** the waves are nondispersive with **frequency**

$$\omega = \pm \, ck$$

in which k is the along-boundary **wavenumber**, and the **phase speed**

$$c = (gH)^{1/2},$$

with g the **acceleration of gravity** and H the mean fluid depth. The equatorial Rossby radius of deformation in this case is

$$R = (c/\beta)^{1/2},$$

in which β is the meridional gradient of the **Coriolis parameter** at the equator. A **Kelvin wave** propagating equatorward along the western boundary of an ocean basin can generate an equatorial Kelvin wave when reaching the equator. Upon reaching the eastern boundary of an ocean basin, an equatorial Kelvin wave can generate a coastal Kelvin wave propagating poleward along the eastern boundary of the ocean basin.

equatorial meteorology—Same as **tropical meteorology**.

equatorial orbiting satellite—A satellite with its orbital plane coincident with the earth's equatorial plane.

This **orbit** will have zero **inclination** angle.

equatorial Poincare wave—By analogy with a **Poincare wave** in a channel, an **inertio-gravity wave** confined to a region about the **equator**.

The **meridional** velocity of the nth mode of these waves has meridional structure of the form

$$\exp(-y^2/2R^2)\ H_n(y/R),$$

in which y is the meridional distance from the equator, R is the equatorial **Rossby radius of deformation**, and H_n is the nth Hermite **polynomial**. In the **shallow water approximation**

$$R^2 = (gH)^{1/2}/\beta,$$

in which β is the meridional gradient of the **Coriolis parameter** at the equator, g is the **acceleration of gravity**, and H is the mean fluid depth.

equatorial radius of deformation—*See* **Rossby radius of deformation**.

equatorial rain forest—Same as **tropical rain forest**.

equatorial stratospheric waves—Waves that propagate from the tropical **troposphere** upward into the **stratosphere**, namely, **Kelvin** and **Yanai** (mixed Rossby–gravity) waves triggered by **cumulus** convection.

These waves play an integral role in the stratospheric **quasi-biennial oscillation**.

equatorial tide—**Tide** occurring when the moon is near the **equator**; **diurnal inequality** is at a minimum.

Compare **tropic tide**, **spring tide**, **neap tide**.

equatorial trough—1. The quasi-continuous belt of low pressure lying between the **subtropical high pressure belts** of the Northern and Southern Hemispheres.

This entire region is one of very homogeneous air, probably the most ideally **barotropic** region of the **atmosphere**. Yet **humidity** is so high that slight variations in **stability** cause major variations in weather. The position of the equatorial trough is fairly constant in the eastern portions of the Atlantic and Pacific, but it varies greatly with **season** in the western portions of those oceans and in southern Asia and the Indian Ocean. It moves into or toward the summer hemisphere. It has been suggested that this name be adopted as the one general term for this region of the atmosphere. Thus, the equatorial trough would be said to contain regions of **doldrums**; portions of it could be described as **intertropical convergence zones**; and within it there might be detected **intertropical fronts**. However, one weakness of this nomenclature is that it alludes specifically and only to the existence of a **trough** of low pressure. Perhaps an even more general term might be preferable, for example, atmospheric equator.

2. Same as **meteorological equator**.

Riehl, H., 1954: *Tropical Meteorology*, p. 238.
Berry, F. A., E. Bollay and N. R. Beers, Eds., 1945: *Handbook of Meteorology*, 776–777.

Equatorial Undercurrent—(Abbreviated EUC.) A **subsurface current** flowing eastward along the **equator**.

A narrow, swift-flowing ribbon with a thickness of 200 m and a width of at most 400 km, it displays the largest current speeds of the **equatorial current system**. In the Pacific, where it is also known as the **Cromwell Current**, it flows with a speed of 1.5 m s^{-1} at a depth of 200 m in the west, rising to a depth of 40 m in the east. In the Atlantic its core is at a depth of 100 m and its speed exceeds 1.2 m s^{-1}. In the Indian Ocean it exists as a flow ribbon centered on a depth of 200 m during the northeast **monsoon** season (December–April); during the remainder of the year this flow gets incorporated into the eastward flowing **southwest monsoon** current. In all oceans the EUC swings back and forth between two extreme positions 90–150 km either side of the **equator** with a two- to three-week **period**.

equatorial upwelling—The rising of water along the **equator** from about 200 m to the surface.

It occurs in the Atlantic and Pacific Oceans where the Southern Hemisphere **trade winds** reach into the Northern Hemisphere, giving uniform **wind direction** on either side of the equator. Because the surface currents of the **Ekman spiral** are deflected to different sides of the **wind** in both hemispheres, the **surface water** is drawn away from the equator, causing the colder water from deeper layers to upwell.

equatorial vortex—A closed **cyclonic circulation** (**closed low**) within the **equatorial trough**; it develops from an **equatorial wave**.
 Some equatorial vortices intensify to become **tropical cyclones**.

equatorial water—The **water mass** of the permanent or oceanic **thermocline** (100–600 m) in the tropical Pacific from 20°S to 15°N.
 South Pacific Equatorial Water (SPEW) occupies the region south of the **equator**; it is derived from **central water** and an injection of high-salinity water formed by **evaporation** near Polynesia that sinks to a depth of 200 m. North Pacific Equatorial Water (NPEW) is found north of the equator; it is formed by subsurface **mixing** between SPEW and central water. Indian Central Water is a historical term used for **Australasian Mediterranean Water** in the western Indian Ocean before the Indonesian Seas were recognized as its **source region**.

equatorial wave—A wavelike **disturbance** of the **equatorial easterlies** that extends across the **equatorial trough**.
 Equatorial waves are especially frequent over the western Pacific Ocean, and many develop into equatorial vortices. *Compare* **easterly wave**.

equatorial waveguide— The equatorial zone that acts as a **waveguide** in that **disturbances** are trapped in the vicinity of the **equator**.
 See also **equatorially trapped waves**.

equatorial waves—A general description of **wave motion** supported and confined by the change in sign of the **Coriolis parameter** at the **equator**.
 A decomposition of the linearized **equations of motion** in **vertical modes** centered on the equator leads to a **meridional** structure that is a product of a Gaussian function and Hermite polynomials, with plane waves in the **zonal** direction. These solutions include trapped **gravity waves**, **Rossby waves**, and mixed **Rossby–gravity waves**, and together with **Kelvin waves** form a complete set.

equatorial westerlies—The westerly winds occasionally found in the **equatorial trough** and separated from the midlatitude **westerlies** by the broad belt of easterly **trade winds**.
 As the air flow in the **lower atmosphere** is mostly easterly in and about the equatorial trough, the existence of westerlies on mean charts in some areas has been a subject of much interest and speculation. In some regions, this **abnormality** can be explained as the result of limited areas of west winds on the equatorward side of frequent westward moving cyclones in the equatorial trough. Elsewhere (notably over the Indian Ocean during the Northern Hemisphere summer), the equatorial westerlies may result from the deflection of Southern Hemisphere air as it flows northward across the geographical **equator** as part of the **monsoon**. Equatorial westerlies can also be induced on the western side of a **large scale** localized heating near the equator, such as in the maritime continent.
> Palmer, C. E., 1951: *Compendium of Meteorology*, p. 876.
> Riehl, H., 1954: *Tropical Meteorology*, 3–4.

equatorially trapped waves—Planetary waves propagating parallel to the **equator**, the **meridional** structures of which are trapped in the vicinity of the equator.
 Included are equatorial **Kelvin waves**, **Rossby waves**, mixed **Rossby–gravity waves**, and **gravity waves**. The equatorial trapping is due to the **variation** of the **Coriolis parameter** and the fact that the parameter itself is zero at the equator. *See also* **equatorial waveguide**.

equideparture—(Rare.) Equality of **deviation** (or departure) from a **normal** value.
 Lines or curves of equideparture are called isametrics. They may be drawn on a map to show, for example, areas where the **atmospheric pressure** is above or below normal by the same amount.

equigeopotential surface—Same as **geopotential surface**.

equilibrium—1. In mechanics, a state in which the **vector** sum of all forces, that is, the **acceleration** vector, is zero.
 In **hydrodynamics**, it is usually further required that a **steady state** exist throughout the atmospheric or fluid **model**. The equilibrium may be stable or unstable with respect to displacements therefrom. *See also* **hydrostatic equilibrium, geostrophic equilibrium, instability**.
 2. In **thermodynamics**, any state of a system that would not undergo change if the system were to be isolated.
 Processes in an **isolated system** not in equilibrium are irreversible and always in the direction of equilibrium.

equilibrium climate—A statistically stationary state of **climate**; in mathematical modeling of climate, the nondrifting state eventually reached under some specified external "boundary" conditions (such as the trace **greenhouse gas** concentrations in the **atmosphere**, the **solar constant**, and, perhaps, the geographical distributions of vegetation and **glacial** ice).

equilibrium climate response—In mathematical modeling of **climate**, the **equilibrium climate** achieved after allowing the **climate system** sufficient time to completely adjust to a change in external forcing.

equilibrium drawdown—**Drawdown** of the **water table** or **piezometric head** when a steady-state condition has been reached.

equilibrium line—The boundary on a **glacier** between the **ablation area** and **accumulation area**.
No net mass is gained or lost at this location. In the absence of **superimposed ice**, this line is equal to the **snow line** at the end of the **mass balance** year.

equilibrium-line altitude—The **elevation** of the **equilibrium line**.

equilibrium line of glacier—*See* equilibrium line.

equilibrium paraboloid—*See* paraboloid.

equilibrium range—1. (Also known as the universal equilibrium range.) The **range** of higher-frequency **eddies** in the **turbulence kinetic energy** spectrum within which the **energy cascade** takes place.
The equilibrium range includes the **inertial subrange** and lies between the range of **energy-containing eddies** and the **viscous subrange**.
2. That part of the ocean wave **spectrum** where a balance exists among the **input** of **energy** from the **wind**, the **transfer** of energy to different **wave** components (**nonlinear** wave–wave interactions) and the **dissipation** of energy due to breaking.

equilibrium solar tide—A theoretical concept in analogy to Laplace's oceanic **equilibrium tide**; roughly, the form of the **atmosphere** determined solely by gravitational forces in the absence of any rotation of the earth relative to the sun.

equilibrium spheroid—The shape that the earth would attain if it were entirely covered by a tideless ocean of constant depth.
Compare **geoid**.

equilibrium theory—In general, any theory derived from the assumption of the existence of time invariants.

equilibrium tide—The hypothetical **tide** that would be produced by the lunar and solar tidal forces in the absence of ocean constraints and dynamics.

equilibrium vapor pressure—The **pressure** of a **vapor** in **equilibrium** with its condensed **phase** (liquid or solid).
This equilibrium is dynamic in that **evaporation** and **condensation** are both occurring but are in equilibrium. It is a function of **temperature** only and is nearly independent of the existence or **density** of other gases in a system. It may, however, be dependent on the shape of the surface of the condensed phase. More generally, the equilibrium vapor pressure with respect to small droplets containing soluble impurities may be quite different from the **saturation vapor pressure**.

equilibrium well discharge—Discharge due to pumping under steady-state conditions; that is, with unchanging **drawdown**.

equinoctial gales—A series of **equinoctial storms**.

equinoctial rains—Rainy seasons that occur regularly in many places within a few degrees of the **equator** at or shortly after the **equinoxes**.
This characteristic of two distinct annual **precipitation** maxima is found chiefly within a **tropical rain forest climate**. The main regions are the Congo Valley, the greater Amazon Valley, and the East Indies. In most places, the spring maximum is the greater. *Compare* **zenithal rains**.

equinoctial storm—(Also called line storm, line gale.) Severe storms in the United Kingdom and North America that are popularly believed to accompany the vernal and autumnal equinoxes.
The belief may have originated in the mid-1700s when sailors observed the West Indian hurricanes, which are most frequent at about the **autumnal equinox**. Statistics of **gale** frequency do not show that storms are especially frequent in temperate latitudes about 22 September, though it does sometimes happen that the first **severe storm** of the winter half-year occurs toward the end of September. The belief has been extended by analogy to the **vernal equinox**, about 21 March.

equinoctial tide—**Tide** occurring when the sun is near **equinox**.

During this period, spring-tide ranges are greater than average.

equinox—1. Either of the two points of intersection of the sun's apparent annual path and the plane of the earth's **equator**, that is, a point of intersection of the **ecliptic** and the **celestial equator**.

2. Popularly, the time at which the sun passes directly above the **equator**; the "time of the equinox."

In northern latitudes the **vernal equinox** falls on or about 21 March, and the **autumnal equinox** on or about 22 September. These dates are reversed in the Southern Hemisphere. *Compare* **solstice**.

equiparte—(Also called equipatos.) In Mexico, during October to January, heavy, cold **rains** that last for several days.

equipatos—Same as **equiparte**.

equipluve—*See* **isomer**.

equipotential line—A line along which hydraulic **potential** remains constant.

equipotential surface—A surface defined by the function

$$\Phi(x, y, z) = \text{constant},$$

where Φ is the **potential** function associated with any **field** that can be written

$$\mathbf{F} = -\nabla\Phi.$$

Examples are gravitational, electrostatic, and magnetostatic equipotential surfaces.

equiscalar surface—In meteorology, a surface along which a specified **scalar** quantity, that is, **pressure** or **temperature**, is constant.

equivalent altitude of aerodrome—**Altitude** that, in a **standard atmosphere**, corresponds to an air density equal to the mean atmospheric **density** for the **season** at an aerodrome.

equivalent barotropic level—The level in an **equivalent barotropic model** at which motions are equivalent to the vertical average of motions in the corresponding column.

equivalent barotropic model—An enhanced form of **barotropic model** in which the **variation** of **wind** with height is vertically averaged assuming that the **thermal wind** is in the same direction as the **geostrophic wind** at all heights.

The resulting single-parameter **model** for **geostrophic** streamfunction is

$$\frac{\partial \nabla^2 \psi^*}{\partial t} + \mathbf{V}^* \cdot \nabla \left(\nabla^2 \psi^* + f + \frac{A(p_0)f_0 g}{RT_0} H \right) = 0,$$

which resembles the **barotropic vorticity equation** except for the third term inside the brackets. ψ^* and $\mathbf{V}^*$ are the vertically averaged geostrophic **streamfunction** and geostrophic wind, respectively. The coefficients p_0, f_0, T_0, g, and R are customary parameters and are defined at a reference latitude at the lower boundary. A is a vertical averaging coefficient also defined at the lower boundary. H is the terrain height, allowing the equivalent barotropic model to represent **orographic** forcing in a crude way. This makes the equivalent barotropic model the most popular barotropic model for real data numerical forecasts.

equivalent blackbody temperature—1. For a radiating body, the **temperature** of a **blackbody** that emits the same **radiant flux density**.

So defined, the equivalent blackbody temperature T_E is given by

$$T_E = \left(\frac{E}{\sigma} \right)^{1/4},$$

where E is the radiant flux density emitted by the body, with **SI** units of watts per square meter, and σ (5.670×10^{-8} W m^{-2}K^{-4}) is the **Stefan–Boltzmann constant**. This definition is not restricted to solid bodies but applies also to gases and liquids.

2. (Also known as **brightness temperature**.) The apparent **temperature** of a nonblackbody determined by measurement with a **radiometer**.

equivalent footcandle—Same as **footlambert**.

equivalent head wind—The **equivalent longitudinal wind** that opposes the flight of an aircraft. *See* **equivalent tail wind**.

equivalent longitudinal wind—For an aerial route, that fictitious uniform **wind** everywhere parallel to the air route that would produce the same mean aircraft speed with respect to the earth's surface as the actual wind.

This is the component of the wind parallel to the line of flight integrated along the entire route. *See* **equivalent head wind, equivalent tail wind.**

equivalent potential temperature—(Also called **wet-equivalent potential temperature**.) A thermodynamic quantity, with its natural **logarithm** proportional to the **entropy** of **moist air**, that is conserved in a **reversible moist adiabatic process.**

It is given most accurately as

$$\theta_e = T \left(\frac{p_0}{p_d}\right)^{R_d/(c_{pd} + r_t c)} \mathrm{H}^{-r_v R_v/(c_{pd} + r_t c)} \exp\left[\frac{L_v r_v}{(c_{pd} + r_t c)T}\right],$$

where θ_e is the equivalent potential temperature, c_{pd} is the **heat capacity** at constant **pressure** of **dry air**, r_t is the **total water mixing ratio**, c is the heat capacity of liquid water, T is the **temperature**, R_d is the **gas constant** for dry air, p_d is the **partial pressure** of dry air, p_0 is a reference pressure (usually 100 kPa), L_v is the **latent heat** of **vaporization**, r_v is the vapor **mixing ratio**, R_v is the **gas constant** for **water vapor**, and H is the **relative humidity**. Neglect of the quantity $r_t c$, where it appears, yields a simpler expression with good **accuracy.**

Paluch, I., 1979: *J. Atmos. Sci.*, **36**, 2467–2478.
Emanuel, K. A., 1994: *Atmospheric Convection*, Oxford University Press, 580 pp.

equivalent reflectivity factor— The **radar reflectivity factor** of a **target** consisting of water drops small compared with the **radar** wavelength, which would produce the same **reflectivity** as that of a **target** with unknown properties.

Mathematically, the equivalent **radar reflectivity factor**, Z_e, of a given target is defined by

$$Z_e = \frac{\eta \lambda^4}{0.93\pi^5},$$

where η is the reflectivity of the target, λ is the radar **wavelength**, and 0.93 is the **dielectric factor** appropriate for water. This definition is based on the **Rayleigh scattering** approximation, which requires that the water drops have diameters no larger than about one-tenth the radar wavelength. For targets consisting of sufficiently small water drops, it follows from the definition that the reflectivity factor, Z, and Z_e are equivalent. But if the composition and sizes of the **particles** are unknown, it is appropriate to regard the reflectivity factor that is determined from the measured reflectivity as Z_e. *See* **radar equation.**

equivalent tail wind—The **equivalent longitudinal wind** that aids the flight of an aircraft.
See **equivalent head wind.**

equivalent temperature—1. Isobaric equivalent temperature: The **temperature** that an **air parcel** would have if all **water vapor** were condensed at constant **pressure** and the **enthalpy** released from the **vapor** used to heat the air:

$$T_{ie} = T\left(1 + \frac{Lw}{c_p T}\right),$$

where T_{ie} is the isobaric equivalent temperature, T the **temperature**, w the **mixing ratio**, L the **latent heat**, and c_p the **specific heat** of air at constant pressure. This process is physically impossible in the **atmosphere.**

2. Adiabatic equivalent temperature (also known as pseudoequivalent temperature): The **temperature** that an **air parcel** would have after undergoing the following process: **dry-adiabatic** expansion until saturated; **pseudoadiabatic expansion** until all moisture is precipitated out; dry-adiabatic compression to the initial **pressure.**

This is the equivalent temperature as read from a thermodynamic **chart** and is always greater than the isobaric equivalent temperature:

$$T_{ae} = T \exp\frac{Lw}{c_p T},$$

where T_{ae} is the adiabatic equivalent temperature.

equivalent width—A spectral measure of the total **absorption** of **radiant energy** by an **absorption line** or **absorption band.**

It is the **bandwidth** of a hypothetical perfect absorber that would absorb the same amount of **energy** as the absorption line or absorption band.

ERBE—Abbreviation for **Earth Radiation Budget Experiment**.

ERBS—Abbreviation for **Earth Radiation Budget Satellite**.

erf—Abbreviation for for error function. Same as **probability integral**.

erg—The unit of **energy** in the **centimeter–gram–second system** of physical units, that is, one dyne-centimeter.

One erg is equal to 10^{-7} **joule** or to 2.389×10^{-8} cal.

ergodic—A system in which the time mean of every measurable function of the system equals its space mean.

For example, **isotropic** homogeneous **turbulence** is often assumed to be ergodic, that is, the time mean at any given moment (e.g., **velocity** or velocity **variance**) coincides with the spatial mean.

ergodic condition—An idealized atmospheric state where the **ensemble average** equals both the time and spatial averages.

While most **turbulence** theories are based on the ensemble average, most calculations of averages in the real **atmosphere** are averages over time of a fixed point, or averages over a line or a volume at an instant in time (i.e., a snapshot). By making the assumption **ergodic**, values measured in the real atmosphere can be utilized within the **Reynolds averaged** equations for **turbulent flow**.

ergodicity—(Also called statistical stationarity.) For **turbulent flow**, the property of having the spatial, temporal and **ensemble averages** all converge to the same mean.

This can only be true if the flow is stationary and homogenous. As a consequence, the **auto-correlation** of an **ergodic** flow **variable** is zero as either the averaging length, time, or number of realizations goes to infinity. *See* **ensemble average**.

Hinze, J. O., 1975: *Turbulence*, 2d ed., McGraw–Hill, p. 5.

erosion—The movement of soil or rock from one point to another by the action of the sea, running water, moving **ice**, **precipitation**, or **wind**.

Erosion is distinct from **weathering**, for the latter does not necessarily imply **transport** of material. Where human agency has increased erosion beyond the **normal** geologic rate, it is termed **accelerated erosion**. Wind erosion is a very important factor in the continued redistribution of earth surface material. Two measures of this effect are used by geologists, **capacity of the wind** and **competence of the wind**. *Compare* **corrasion, corrosion, eolation**.

erosion ridge—"One of a series of small ridges on a **snow** surface formed by the corrasive action of wind-blown snow.

"Erosion ridges may be formed either at right angles to the **wind** or aligned with the wind." (Glossary of Arctic and Subarctic Terms 1955.) *Compare* **sastruga**.

Arctic, Desert, Tropic Information Center (ADTIC) Research Studies Institute, 1955: *Glossary of Arctic and Subarctic Terms*, ADTIC Pub. A-105, Maxwell AFB, AL.

error—1. *See* **instrument error, observational error, random error, systematic error, standard error**.

2. *See* **forecast error**.

error distribution—The **probability distribution** of **random errors**, typically a **normal distribution** with a zero mean; that is,

$$f(v) = \frac{1}{\sigma\sqrt{2\pi}} \, e^{-v^2/2\sigma^2} \; (-\infty < v < \infty),$$

where v is the random error, and σ is the **standard deviation** of v (in this connection, σ is commonly called the **standard error**); $f(v)$ denotes the error distribution (or **normal curve of error**).

error function—(Sometimes abbreviated erf.) Same as **probability integral**.

error of estimate—Same as **residual**.

ERS—Abbreviation for **European Remote Sensing Satellites**.

Ertel potential vorticity—*See* **potential vorticity**.

ertor—The effective (radiational) **temperature** of the **ozone layer** (region).

ERTS—Abbreviation for **Earth Resources Technology Satellite**

escape speed—Same as **escape velocity**.

escape velocity—(Or escape speed.) The minimum speed that an object directed radially outward from an astronomical body must have in order to escape its gravitational attraction.

The escape velocity for an object at a distance R from the center of a spherically symmetric body is

$$(2gR)^{1/2},$$

where g is the corresponding **acceleration** due to **gravity**. For R equal to the radius of the earth and g its surface acceleration due to gravity, the escape velocity is about 11 200 m s^{-1}. The earth retains its **atmosphere** only because this escape velocity is appreciably larger than the mean speed of air molecules.

ESSA—The first series of operational meteorological satellites launched by the United States.

Based on the earlier **TIROS** series of satellites, the nine satellites launched between 1966 and 1969 are often referred to as the **TIROS Operational System**.

estimated ceiling—After U.S. weather observing practice, the **ceiling classification** applied to a **ceiling** height that is determined in any of the following ways: 1) by means of a convective-cloud height diagram or **dewpoint formula**; 2) from the known heights of unobscured portions of natural landmarks, or objects more than one and one-half nautical miles from any runway of the airport; 3) on the basis of observational experience, provided the sky is not obscured by surface-based **hydrometeors** or **lithometeors**, and other guides are lacking or considered unreliable; or 4) determined by **ceilometer** or **ceiling light** when the penetration of the light beam is in excess of **normal** for the particular height and type of layer, or when the **elevation angle** of the **clinometer** or ceilometer-detector scanner exceeds 84°.

Such a ceiling is denoted E in aviation weather reports.

estival—Same as **aestival**.

estuary—1. The portion of a river that is affected by **tides**.

2. A semi-enclosed body of water where the **salinity** of ocean water is measurably reduced by **freshwater** input.

Estuaries are very important nursery regions for many coastal ocean species of fish and invertebrates.

ETA—Abbreviation for estimated time of arrival.

eta vertical coordinate—A vertical coordinate for atmospheric models defined with a steplike representation of **topography**, having mountains formed of **grid** boxes of the **model**.

The vertical coordinate surfaces are quasi-horizontal, intersecting model mountains or forming their nearly horizontal upper sides. It is defined by

$$\eta = \frac{(p - p_T)}{(p_S - p_T)} \frac{(p_{ref}(z_S) - p_T)}{(p_{ref}(0) - p_T)},$$

where p is **pressure**; the subscripts T and S stand for the top and the ground surface values of the **model atmosphere**, respectively; z is geometric height; and $p_{ref}(z)$ is a suitably defined reference pressure as a function of z. The ground surface heights z_S are permitted to take only a discrete set of values, those of elevations at which the interfaces of the chosen model layers are located when pressure is equal to the reference pressure $p_{ref}(z)$. The "Eta Model" using the eta coordinate became the primary short-range operational forecasting model of the U.S. National Weather Service in 1993. *See* **sigma vertical coordinate**, **isentropic vertical coordinate**, **hybrid vertical coordinate**, **vertical coordinate system**.

ETD—Abbreviation for estimated time of departure.

ETE—In air navigation, abbreviation for estimated time en route.

etesian climate—Same as **Mediterranean climate**.

etesians—The prevailing northerly winds in summer in the eastern Mediterranean and especially the Aegean Sea; basically similar to **monsoon** and equivalent to the **maestro** of the Adriatic Sea.

According to the ancient Greeks, the etesians blow for 40 days, beginning with the heliacal rising of Sirius. They are associated (along with the **seistan** and **shamal**) with the deep low pressure area that forms in summer over northwest India. They bring **clear** skies and dry, relatively cool weather. In Greece the etesian **wind** is locally named the sleeper. In Turkey it is the **meltém**. The Romans used the word also for the **southwest monsoon** of the Arabian Sea.

ethane—Second member of the **alkane** family, formula C_2H_6, ethane is a naturally occurring hydrocarbon produced from natural gas, **biomass burning**, ocean, and terrestrial vegetation.

It is fairly unreactive and some can be transported to the **stratosphere**.

ethene—Simplest **alkene**, formula C_2H_4; a major emission of fossil fuel combustion, particularly automobiles.

It is also emitted from oceans, vegetation, and as a product of **biomass burning**.

ethylene—Same as **ethene**.

EUC—Abbreviation for **Equatorial Undercurrent**.

Eulerian coordinates—Any system of coordinates in which properties of a fluid are assigned to points in space at each given time, without attempt to identify individual fluid parcels from one time to the next.

Since most observations in meteorology are made locally at specified time intervals, an Eulerian system is usually, though by no means always, more convenient. A sequence of **synoptic charts** is an Eulerian representation of the data. Eulerian coordinates are to be distinguished from **Lagrangian coordinates**. The particular **coordinate system** used to identify points in space (Cartesian, cylindrical, spherical, etc.) is quite independent of whether the representation is Eulerian or Lagrangian. *See* **equations of motion**.

Eulerian correlation—(Sometimes called synoptic correlation.) The **correlation** between the properties of a flow at various points in space at a single instant in time.

Compare **Lagrangian correlation**; *see* **correlation coefficient**.

Eulerian current measurement—The measurement of **current** by an instrument fixed in space, for example, a moored **current meter** or a bottom-mounted **acoustic Doppler current profiler**.

The Eulerian current is computed by averaging the measured fluid **velocity** over a suitable interval (usually a few minutes).

Eulerian equations—Any of the **fundamental equations of hydrodynamics** expressed in **Eulerian coordinates**.

These are so commonly used that the designation "Eulerian" is often omitted. *See*, for example, **equations of motion**.

Eulerian wind—In the classification of Jeffreys, a **wind** motion only in response to the **pressure force**.

In symbols,

$$\frac{D\mathbf{v}}{Dt} = -\alpha \nabla_H p,$$

where $\mathbf{v}$ is the horizontal **velocity**, α the **specific volume**, ∇_H the horizontal **del operator**, and p the **pressure**. The **cyclostrophic wind** is a special case of the Eulerian wind, which is limited in its meteorological applicability to those situations in which the **Coriolis effect** is negligible.

eupatheoscope—An instrument designed by A. F. Dufton in 1929 to record the warmth of a room from the point of view of comfort. It consists of a blackened hollow copper cylinder heated by a **carbon** filament and a metal filament lamp, which are controlled by a thermostat to maintain a **temperature** of 75°F. The **power** used is interpreted in terms of **equivalent temperatures**.

euphotic zone—The layer of a water body that receives sufficient **sunlight** to support effective **photosynthesis**.

euraquilo—(Also called euroaquilo, euroclydon.) A stormy **wind** from northeast or north-northeast in Arabia and the Near East.

See **gregale**.

eurithermic—Adaptable to a wide **range** of **temperatures**.

It is used in reference to organisms.

European Remote Sensing Satellites—(Abbreviated ERS.) A series of satellites designed to study the earth's land, **atmosphere**, and oceans.

The major instruments on ERS are an active **microwave** instrument, a **radar altimeter**, and an along-track **scanning radiometer**. The active microwave instrument operates in three modes: **SAR** image mode, SAR wave mode, and wind **scatterometer** mode. The scanning radiometer is made up of two instruments, an **infrared** radiometer and a microwave **sounder**. *ERS-2* also has a global **ozone** monitoring experiment. *ERS-1* was launched in 1991 and *ERS-2* in 1995. Both satellites operate in sun-synchronous, near-polar orbits.

Euros—The Greek name for the stormy, rainy southeast **wind**.

On the **Tower of the Winds** at Athens it is represented by an old man, warmly clothed and wrapped in his mantle.

eustatic—*See* **eustatic sea level changes**.

eustatic sea level changes—Worldwide changes of **sea level** due to the increase in the volume of water in the ocean basins.

Volume changes are due to mass increases from melting of grounded **ice** and **thermal** expansion or contraction of the oceans as they warm or cool. Over geological time the shape and volume of the ocean basins themselves also evolve.

evanescent level—A theoretical boundary between a region in the fluid where waves of some **frequency** are propagating and a region in the fluid where waves of the same frequency do not propagate (where they are evanescent).

This is also called a **critical level** but is distinct from the critical level where the background flow has the same speed as the **phase speed** of the waves.

evaporation—(Also called vaporization.) The physical process by which a liquid or solid is transformed to the gaseous state; the opposite of **condensation**.

Evaporation is usually restricted in use to the change of water from liquid to gas, while **sublimation** is used for the change from solid to gas. According to the **kinetic theory** of gases, evaporation occurs when liquid molecules escape into the **vapor** phase as a result of the chance acquisition of above-average, outward-directed, translational velocities at a time when they happen to lie within about one **mean free path** below the effective liquid surface. It is conventionally stated that evaporation into a gas ceases when the gas reaches **saturation**. In reality, net evaporation does cease, but only because the numbers of molecules escaping from and returning to the liquid are equal, that is, evaporation is counteracted by condensation. Because the molecules that escape the condensed **phase** have above-average energies, those left behind have below-average energies, which is manifested by a decrease in **temperature** of the condensed phase (unless compensated for by **energy transfer** from the surroundings). *See also* **evapotranspiration**.

evaporation capacity—*See* **potential evaporation**.

evaporation fog—Fog formed as a result of **evaporation** of water that is warmer than the air.

If formed over a water surface, it is sometimes termed **steam fog**. If formed from evaporation of **rain** falling through the air, it is sometimes termed rain fog, precipitation fog, or **frontal fog**.

evaporation frost—1. **Frost** caused by evaporative cooling due to dry winds over a moist surface, where the **wind** temperature is above **freezing**, but cooling below freezing occurs due to **latent heat** loss.

2. Agricultural term, where **evaporation** freezes the crop.

evaporation gauge—A general name for devices such as **atmometers**, **evaporation pans**, and **lysimeters** that measure **evaporation** in some fashion.

evaporation opportunity—*See* **relative evaporation**.

evaporation pan—A type of **evaporation gauge** or **evaporimeter**; it is a pan used in the measurement of the **evaporation** of water into the **atmosphere**.

The U.S. Weather Bureau evaporation pan (Class-A pan) is a cylindrical container fabricated of galvanized iron or other rust-resistant metal with a depth of 25.4 cm (10 in.) and a diameter of 121.9 cm (48 in.). The pan is accurately leveled at a site that is nearly flat, well sodded, and free from obstructions. The **water level** is maintained at between 5 and 7.5 cm (2 and 3 in.) below the top of the rim, and periodic measurements are made of the changes of the water level with the aid of a **hook gauge** set in the **still well**. When the water level drops to 17.8 cm (7 in.), the pan is refilled. Its average pan coefficient is about 0.7. *See also* **BPI pan**, **Colorado sunken pan**, **floating pan**, **screened pan**.

evaporation power—*See* **potential evaporation**.

evaporation rate—Quantity of **actual evaporation** per unit area per unit time.

evaporative available potential energy—(Abbreviated EAPE.) Analogous to **convective available potential energy**, except that it is related to the negative **buoyancy** associated with evaporative cooling of liquid water within a sinking cloudy **air parcel**.

On a **thermodynamic diagram**, if a line is drawn corresponding to a sinking cloudy air parcel (follow a **moist adiabat** downward while the parcel is cloudy, and then after all liquid water has evaporated, continue below **cloud base** by following a **dry adiabat**), the EAPE is the area between this air-parcel line and a line corresponding to the environmental **sounding**. The resulting EAPE can be related to the **kinetic energy** associated with a negatively buoyant, sinking air parcel, and is useful for determining attributes of **downdrafts** and **downbursts** from **thunderstorms**.

evaporative capacity—*See* **potential evaporation**.

evaporative power—*See* **potential evaporation**.

evaporativity—*See* **potential evaporation**.

evaporimeter—A class of **evaporation gauges**, which includes **evaporation pans** and **atmometers**, that measure **evaporation** from free-standing water or a thoroughly wetted surface.

evapotranspiration—1. The combined processes through which water is transferred to the **atmosphere** from open water and **ice** surfaces, **bare soil**, and vegetation that make up the earth's surface.

2. (Also called flyoff, water loss, total evaporation.) The total amount of water transferred from the earth to the **atmosphere**.

This is the most general term for the result of this composite process; **duty of water** and **consumptive use** have more specific applications.

evapotranspirometer—A type of **lysimeter** that measures the rate of **evapotranspiration**.

It consists of a vegetation soil tank so designed that all water added to the tank and all water left after evapotranspiration can be measured, usually by weighing.

evapotron—A device that uses **eddy correlation** to measure the turbulent vertical **flux density** of **water vapor** within a few meters of the surface of the earth.

Evapotrons were originally designed with **analog** electronics to make the necessary **covariance** calculations in **real time**. This term is not in common use for systems that make the covariance calculations with modern digital electronics.

exceedance interval—The average number of years between the occurrence of an event and that of a "greater" event.

exceedance probability—For any threshold, x, the **probability** that during the year the **random variable** in question, X, will exceed some x; exceedance probability = $P[X > x]$.

exceptional visibility—**Visibility** such that objects are readily visible at great distance.

excess—*See* **kurtosis**.

excess rain—Volume resulting from **rainfall** at a rate greater than the **infiltration rate**.

This term is used to describe **effective precipitation**.

excessive precipitation—**Precipitation** (generally in the form of **rain**) of an unusually high rate.

The expression is usually used qualitatively, but several meteorological services have adopted quantitative limits. In general, formulas for determining criteria for excessive precipitation take the following form:

$$R = \frac{at}{b + t},$$

where R is the amount of **rainfall** in t minutes, and a and b are constants. For longer **storm** periods, the expressions used are often of the type

$$R = a + b \log T,$$

where T is the time in hours.

exchange coefficients— [Also called turbulent transfer coefficient, eddy diffusivity, austausch coefficients (obsolete).] The ratio of the **turbulent flux** of a **conservative property** through a surface to the **gradient** of the mean of the property normal to the surface.

See **diffusion, eddy flux, turbulence**.

excited state—Configurations of an atom or molecule that contain more **energy** than the **ground state**.

Excited electronic or vibrational states usually occur following the **absorption** of **radiation** of the correct **frequency**, and are lost by processes such as **emission** of the radiation (**fluorescence**) or by reaction. Many electronically excited states are more reactive than the ground state, for example, the first excited state of atomic **oxygen**, O(^{1}D).

exhalation—1. In soil science, the process by which radioactive gases escape from the surface layers of soil or loose rock where they are formed by decay of radioactive salts.

The exhalation of radioactive gases, notably **radon** and **thoron**, increases with **soil temperature** and so normally exhibits a single daily maximum around midday. Decreases of **atmospheric pressure** normally increase the exhalation, and **freezing** of the surface soil layers usually greatly reduces it.

2. The streaming forth of volcanic gases; also the escape of gases from a **magnetic field**.

> Israël, 1951: *Compendium of Meteorology*, 155–158.
> American Geological Institute, 1997: *Glossary of Geology*, 4th ed., J. A. Jackson, Ed., p. 221.

exhaust trail—(Or engine exhaust trail.) A **condensation trail** that forms when the **water vapor** from fuel combustion of the aircraft engine is mixed with and saturates (or supersaturates) the air in the **wake** of the aircraft.

Exhaust trails are of more common occurrence and of longer duration than **aerodynamic trails**.

exit jet—*See* **outflow jet**.

exit region—1. The region of **diffluence** at the **downwind** extremity of a **jet stream**; the opposite of **entrance region**.

2. Same as **delta region**.

exogenic influences—Influences on the earth's **climate** that originate outside the **atmosphere** or **hydrosphere**, for example, **solar** or **cosmic radiation**, variations in the earth's orbit, etc.

exogenous electrification—The separation of **electric charge** in a conductor placed in a preexisting **electric field**.

This is especially applied to the **charge separation** observed on metal-covered aircraft. It is the result of an **induction charging mechanism**, and does not by itself create any net total charge on the conductor. It is to be distinguished, therefore, from **autogenous electrification**.

exorheic—A **basin** or region characterized by external drainage or draining to the ocean.

exosphere—The outermost, or topmost, portion of the **atmosphere**.

Its lower boundary is the **critical level of escape**, usually located at 500–1000 km above the earth's surface. In the exosphere, the air density is so low that the **mean free path** of individual **particles** depends upon their direction with respect to the local vertical, being greatest for upward moving particles. It is only from the exosphere that atmospheric gases can, to any appreciable extent, escape into outer space. *See* **cone of escape, fringe region**.

expansion wave—(Also called rarefaction wave.) A **simple wave** or progressive **disturbance** in the **isentropic** flow of a compressible fluid, such that the **pressure** and **density** of a fluid **particle** decrease on crossing the wave in the direction of its motion.

This may be illustrated, for example, by the withdrawal of a piston from a gas-filled cylinder. When the gas is initially at rest in the cylinder, an expansion wave may move into the undisturbed fluid at the **speed of sound** as the piston is withdrawn. The expansion wave is a finite-amplitude disturbance changing shape as it propagates, but it may or may not be accompanied by a **shock wave** front, with the associated discontinuities. *See* **compression wave**.

expectation—Same as **expected value**.

expected value—(Also called expectation, mathematical expectation.) The **arithmetic mean** of a **random variable**, conceptually similar to the **simple average** but broader in scope.

If $g(x)$ is a continuous function of x, then the expected value of $g(x)$, denoted by $E[g(x)]$, is the integral (or sum, if x is discrete) of $g(x)$ times the **probability** element of x. Thus, if x is continuous with **probability density function** $f(x)$ defined in the **range** $a < x < b$, then

$$E[g(x)] \equiv \int_{a}^{b} g(x) f(x) dx;$$

or, if x is discrete with possible values $x_1, x_2, \ldots, x_n$ and probability function $f(x_i)$, then

$$E[g(x)] \equiv \sum_{i=1}^{n} g(x_i) f(x_i).$$

This reduces to the mean of x itself in the case $g(x) = x$.

expendable bathythermograph—(Abbreviated XBT.) A device for obtaining a record of **temperature** as a function of depth to 1800 m from a ship with a speed as high as 15 m s^{-1}.

Temperature is measured with a **thermistor** within an expendable casing. Depth is determined by a priori knowledge of the rate at which the casing sinks and the time of each recorded data value. A fine wire provides a data transfer line to the ship for shipboard recording. Airborne versions

(**AXBT**) are also used; these use radio frequencies to transmit the data to the aircraft during deployment. *See* **conductivity–temperature–depth profiler**.

experimental basin—A **watershed** instrumented with hydrologic and hydrometeorologic equipment to measure components of the **hydrologic cycle**, for the purpose of understanding (and thereby modeling) hydrologic processes.

expert system—Problem-solving and decision-making system based on knowledge of its task and logical rules or procedures for using the knowledge.

Both the knowledge and the logic are codified from the experience of human specialists in the field (the **domain** experts). Codification can take many forms, including **rules**, frames, or scripts. Examples in meteorology include systems used for the interpretation of satellite imagery and for **fog** and **turbulence** forecasting. *See also* **knowledge-based system**.

expertise—A capability of a person to perform an operation in a limited **domain** with exceptional results when compared to others capable of performing the same operation.

Expertise may depend on abstractions, such as individual mental models, rather than on knowledge alone.

explained variance—*See* **regression**.

explanation—**Expert system** feature that reveals the motivation, justification, or rationalization of its conclusions by presenting goals, facts, and **heuristic** rules that affected or determined the choice of conclusions.

explicit time difference—*See* **implicit time difference**.

explosive warming—Same as **stratospheric warming**.

exposure—1. A qualitative term referring to the surroundings of a **site**, especially to how well its openness will allow in situ measurements made at that site to represent the general area. Exposure is of particular concern for **wind**, **temperature**, **precipitation**, and **radiation** measurements. *See* **instrument exposure**.

2. The amount of time an individual is affected by a particular phenomenon, such as exposure to **sunlight** as affects UV-induced sunburn, **air pollutants**, or nuclear **radiation**.

3. *See* **instrument exposure**.

exposure of instruments—The appropriate placement of an instrument so that its indications represent the real state of the **atmosphere** unmodified by nearby influences.

exsiccation—Drying up by the removal of moisture.

In **climatology** it implies the loss of moisture by draining or by increased **evaporation**, rather than by a decreased supply of water from **precipitation**. *Compare* **desiccation**.

extended forecast—A forecast of weather conditions for a period extending beyond three or more days from the day of issuance.

The U.S. National Weather Service issues extended forecasts for the three- to five-day period ahead. *Compare* **long-range forecast**, **medium-range forecast**.

extended-range forecast—Same as **extended forecast**.

extended source—A producer of **pollutants** with an emission orifice generally more than about 10 m in size.

Examples include two-dimensional "area" sources such as agricultural and forest burning, volatile organic emissions from lagoons or other water treatment areas, and other open, expansive emission sources, or three-dimensional "volume" sources such as open buildings, tanks with several orifices, and other such structures.

extended target—*See* **distributed target**.

extensible balloon—A balloon made of a material of low modulus of elasticity such as natural or synthetic rubber.

Sounding balloons made of rubber expand to five or six times their flaccid diameter before bursting.

extensive air shower—*See* **cascade shower**.

extensive property—Same as **extensive quantity**.

extensive quantity—(Or extensive property.) Any property of a **parcel** or system that is a function of the mass of the system, for example, **kinetic** or **internal energy**.

See **intensive quantity**.

external forcings—**Boundary conditions** (such as surface **drag** against the surface) and body forcings (such as heating caused by **infrared radiation** divergence within the air) that are imposed on the **atmosphere** from outside the domain of interest.

The exact definition depends on the **scale** context of the problem. For example, boundary layer meteorologists might consider **geostrophic wind** to be imposed as an external forcing at a single **weather station**, while general circulation dynamicists might calculate geostrophic wind as an internal **variable** based on the global differential heating imposed on the earth as an external forcing by the sun.

external Rossby radius of deformation—*See* **Rossby radius of deformation**.

external water circulation—That part of the **hydrologic cycle** pertaining to water that evaporates from the sea surface and subsequently falls as **precipitation** on the continent.

external wave—A **wave** in fluid motion having its maximum amplitude at an external boundary such as a **free surface**.

Any **surface wave** on the free surface of a homogeneous **incompressible fluid** is an external wave. *Compare* **internal wave**.

extinction—(Sometimes called **attenuation**.) The removal of **radiant energy** from an incident **beam** by the processes of **absorption** and/or **scattering**.

extinction coefficient—(Also called **attenuation coefficient**, especially in reference to **radar** frequencies.) For **radiation** propagating through a medium, the fractional depletion of **radiance** per unit path length.

The **volume extinction coefficient** is defined through **Beer's law** as

$$\frac{dL}{L} = -\gamma \, ds,$$

where L is the **monochromatic** radiance at a given **wavelength**, γ is the volume extinction coefficient, and ds is an increment of path length. The mass extinction coefficient equals the volume extinction coefficient divided by the **density** of the medium. Thus, in **SI** units, the volume extinction coefficient has units of inverse meters and the mass extinction coefficient has units of square meters per kilogram. In general, **extinction** of **radiant energy** is caused by **absorption** and **scattering**. The extinction coefficient is the sum of the **absorption coefficient** and the **scattering coefficient**, and generally depends on wavelength and **temperature**. *Compare* **specific attenuation**.

extinction cross section—(Also called attenuation cross section, especially for **radar** and **radio wave** propagation.) The area that, when multiplied by the **irradiance** of **electromagnetic waves** incident on an object, gives the total **radiant flux** scattered and absorbed by the object.

Customary usage in radar describes the **attenuation** cross section as the area that, when multiplied by the **power density** of incident plane-wave **radiation**, gives the **power** removed from the **beam** by **absorption** and **scattering**. For a propagation medium consisting of a **dispersion** of scattering and absorbing objects, the **volume extinction coefficient** ($m^2 m^{-3}$ or m^{-1}) at a given location in the medium is the sum of the extinction cross sections of all the objects in a unit volume centered at the location. *Compare* **scattering cross section**, **absorption cross section**.

extraordinary ray—(Or extraordinary wave.) *See* **ordinary ray**.

extrapolation—The extension of a relationship between two or more **variables** beyond the range covered by knowledge, or the calculation of a value outside that range.

In **synoptic meteorology**, extrapolation commonly refers to the forecasting of the position of a weather-pattern feature based solely upon recent past motion of that feature. *Compare* **interpolation**.

extraterrestrial radiation—**Electromagnetic energy** in a specified **frequency band** originating from sources other than those related to the earth, for example, **cosmic** or **solar radiation**.

extratropical—In meteorology, typical of occurrences poleward of the belt of **tropical easterlies**.

extratropical cyclone—(Sometimes called extratropical low, extratropical storm.) Any **cyclonic-scale** storm that is not a **tropical cyclone**, usually referring only to the **migratory** frontal cyclones of middle and high latitudes.

Compare **subtropical cyclone**.

extratropical low—Same as **extratropical cyclone**.

extratropical storm—Same as **extratropical cyclone**.

extreme—In **climatology**, the highest and, in some cases, the lowest value of a **climatic element** observed during a given time interval or during a given month or **season** of that period.

If this value were the greatest extreme for the whole **period of record** for which observations are available, the value would be the **absolute** extreme.

extreme temperature—Same as **temperature extremes**.
See also **absolute temperature extremes**.

extreme value analysis—Analysis of **stochastic** processes for the purpose of estimating the probabilities of rare events.

Such **analysis** is often made difficult by uncertainties in the **statistics** due to an inherent scarcity of data. In meteorology, extreme value analyses have been performed for the **prediction** of damaging **rain**, maximum **frost** penetration, and extreme **winds**.

extreme value distribution—1. A **probability distribution** for a **random variable** obtained from another random variable x, where each **sample** value for the derived distribution is obtained by selecting the maximum amplitude from a sequence of **random** values of x.

The most common extreme value distribution is the Gumbel distribution, defined by the cumulative **probability** function

$$F(x) = e^{-e^{-\lambda(x-x_0)}},$$

where x is assumed to be exponentially distributed with mean and **standard deviation** equal to λ and where $e^{\lambda x_0}$ is equal to the sample size N. The double exponential form is valid when N is large. (Keeping 1962)

2. A **probability distribution** indicating the **probability** that a measured quantity will exceed a prescribed value during the course of a given period (say, one year); used in the design of offshore and coastal structures.

In common parlance, the "100-year wave" is the **wave height** that has a probability of 0.01 of being exceeded within a 12-month period. *See* **design wave, extreme wave height**.

Keeping, E. S., 1962: *Introduction to Statistical Inference*, 206–207.

extreme wave height—The **wave height** that is exceeded on average only once in a specified period, usually 50 or 100 years.

It is often used as a basis for the **design wave**. *See* **extreme value distribution**.

extremely high frequency—(Abbreviated EHF.) *See* **radio frequency band**.

extremely low frequency—(Abbreviated ELF.) *See* **radio frequency band**.

eye—In meteorology, usually the "eye of the storm" (**hurricane, typhoon**), that is, the roughly circular area of comparatively light winds found at the center of a severe **tropical cyclone** and surrounded by the **eyewall**.

The winds increase gradually outward from the center but can remain very light up to the inner edge of the eyewall. No **rain** occurs and in intense tropical cyclones the eye is **clear** with blue sky overhead. Most, but not all, tropical cyclones with maximum winds in excess of 40 m s^{-1}(78 knots) have eyes visible on satellite imagery. Eye diameters vary from 10 to more than 100 km.

eye of the wind—*See* **teeth of the gale**.

eyewall—A ring of **cumulonimbus** that encircles the **eye** of a **tropical cyclone**.

In **radar** depictions, the clouds must occupy at least 180° of arc to be called an eyewall.

F

ƒ-plane—A local approximation of the spherical earth as a plane **normal** to the zenithal component of the earth's rotation.

The rotation rate ƒ is assumed to be constant on the plane. This approximation is valid in describing atmospheric or oceanic motions with time scales smaller than or comparable to $1/f$.

ƒ-plane approximation—Used in **atmospheric dynamics**; an approximation in which the **Coriolis parameter** ƒ is assumed to be invariant with latitude.

F-ratio—(Also called variance ratio.) Ordinarily, the ratio of two independent estimates of a common **variance**.

It is used as a **significance test** in **analysis of variance** and is known in this regard as the F-test.

F-region—The general region of the **ionosphere** containing the F_1-**layer** and F_2-**layer**.

F-scale—Abbreviation for **Fujita scale**.

F-test—*See* **F-ratio**.

facsimile chart—(Popularly called fax chart, fax map.) In meteorology, any display form of weather information, usually a type of **synoptic chart**, that has been reproduced by **facsimile equipment**.

Master charts are plotted and analyzed at central **weather stations** and are transmitted to individual weather stations. Other fields of data are developed by local software from an array of **digital** data supplied by regional centers. Transmission of weather information by facsimile is rapidly becoming obsolete.

facsimile equipment—(Popularly called fax.) Apparatus used for the electrical transmission of a graphic or printed record, for example, a **weather map** or a letter, usually over telephone wires but occasionally by radio.

The received **image** is built up from dots or lines that may be of constant **intensity** or of varying intensity, depending on the type of system employed. *See* **pixel**.

faculae—Irregular patches of bright material present in the solar **photosphere** forming a veined **network** in the vicinity of sunspots.

These regions are a few hundred degrees Kelvin hotter than their surroundings. They have low contrast on the solar disk and are best observed near the solar limb. These features relate directly to **plage** or brightenings in the overlying **chromosphere**.

fadeout—*See* **fading**.

fading—1. (Also called fadeout.) Applied to an **electromagnetic wave**, often a **radio wave**, a drop in its **power** or **field strength** below a specified level.

2. Weakening of a received radio or **radar** signal caused by changes in the propagation medium.

The term was originally applied to long-period changes in **signal strength** over **bistatic radar** links caused by **interference** between the **signal** reflected by the **ionosphere** and the direct, unreflected signal or **ground wave**. It now refers to interference phenomena caused by any type of **multipath transmission** either of tropospheric or ionospheric origin.

Fahrenheit temperature scale—(Abbreviated °F.) A **temperature scale** with the **ice point** at 32° and the **boiling point** of water at 212°.

Conversion with the **centigrade temperature scale** (°C) is accomplished by the formula: °F = (9/5)°C + 32. *Compare* **Celsius temperature scale**.

fair—With respect to weather, generally descriptive of pleasant weather conditions, with due regard for location and time of year.

It is subject to popular misinterpretation, for it is a purely subjective description. When this term is used in weather forecasts (National Weather Service), it is meant to imply 1) no **precipitation**; 2) less than 0.4 **sky cover** of low clouds; 3) no other extreme conditions of **cloudiness**, **visibility**, or **wind**; 4) **unrestricted visibility**; and 5) light winds of generally less than 10 knots (5 m s⁻¹). *Compare* **cloudy**.

fair-weather cumulus—Same as **cumulus humilis**.

fair-weather current—Same as **air–earth conduction current**.

fair-weather electricity—The distribution of ions and currents in the **atmosphere** and at the surface of the earth that occurs during fair weather or in areas where there is no **thunderstorm** activity.

This distribution sets up a downward-directed **electric field** referred to as a fair-weather **field**.

Falkland Current—*See* **Malvinas Current**.

fall—Same as **autumn**.

fall equinox—Same as **autumnal equinox**.

fall velocity—*See* **terminal fall velocity**.

fall wind—A **wind** that accelerates as it moves downslope because of its low **temperature** and greater **density**.

A fall wind is a larger-scale phenomenon than the individual-slope **scale** and is produced by accumulated cold air spilling down a slope or over a mountain range. The cold air often either accumulates on a plateau or other elevated terrain, or is part of an extensive cold **air mass** approaching a mountain range as a **cold front**. Fall winds may have a hydraulic character similar to water flowing over a dam, and one of the details of this flow is that the **acceleration** of the cold air begins to occur before the crest of the mountain range and therefore before the down-sloping portion of the **topography**. Fall winds are especially well developed as strong easterly winds on the coast of Norway, and for some distance inland; here they give a narrow strip of fine weather along the shore. They are also well developed on the northern coast of the Aegean Sea. At the southeastern tip of rocky Hagion Oros Peninsula in Greece, Mt. Athos rises to 2033 m (6670 ft) and descends steeply to the sea; northerly winds are disturbed by this great mass and descend as the cold northeasterly **Athos fall wind**, often of **gale** force, extending several kilometers out to sea. On the coast of Peru the name is given to sudden heavy **gusts** that often come down from the high land after onset of the **sea breeze**. At Rio de Janeiro, Brazil, descending **squalls** from the northwest are termed *terre altos*. In the Antarctic fall winds off the **inland ice** form violent **blizzards**. Other examples of fall winds are the **mistral**, **papagayo**, and **vardar**. Some authors have generalized this term to refer to downslope winds forced by large meso- and synoptic-scale processes (i.e., scales larger than that of an individual slope), even if they do not represent flows of colder air. Thus, under this nonstandard definition, the **foehn** and **chinook** could be considered fall winds.

Smith, R. B., 1987: Aerial observations of the Yugoslavian bora. *J. Atmos. Sci.*, **44**, 269–297.

falling limb—Same as **recession**.

falling sphere—A technique for the measurement of atmospheric **density** and **temperature** in the **middle atmosphere**.

Radar tracking of the descent rate of small spheres released from **rockets** gives a measure of atmospheric **drag**, from which densities and temperatures can be inferred.

falling tide—*See* **ebb current**.

fallout—The descent to the ground of **dust** and other debris raised to great heights in the atmosphere by a violent explosion, especially applied to **radioactive fallout** from an atomic or thermonuclear explosion.

fallout winds—Those tropospheric **winds** that carry **radioactive fallout** materials (primarily materials produced by atmospheric tests of nuclear weapons).

These winds are observed by standard **winds-aloft observation** techniques. Fallout winds are used to construct radioactive fallout plots and ground fallout plots.

fallstreaks—Same as **virga**.

Fallstreifen—Same as **virga**.

false alarm rate—*See* **skill**.

false cirrus—Same as **cirrus spissatus**.

false warm sector—The sector, in a horizontal plane, between the **occluded front** and a **secondary cold-front** of an **occluded cyclone**.

family of chemical species—A group of chemical species that are rapidly interconverted by chemical reactions.

For example, **hydrogen** atoms (H), **hydroxyl radical** (OH), and **hydroperoxyl radical** (HO_2) are referred to as the "odd hydrogen" family. Other families are the NO_x family (NO and NO_2), and ClO_x family (predominantly Cl and ClO). The concept of chemical families is often encountered in the modeling of **atmospheric chemistry**, where the treatment of groups of interconverting species as families can reduce the complexity of the **model** calculation.

family of tornadoes—A sequence of long-lived tornadoes produced by a "cyclic" **supercell** storm.

Tornadoes touch down at quasi-regular intervals (typically 45 min). Usually a new **tornado** develops in a new **mesocyclone** just after an old tornado has decayed in an old, occluded neighboring mesocyclone. Sometimes, two successive tornadoes may overlap in time for a few minutes. The two mesocyclones may rotate partially around each other. If the damage tracks of the tornadoes appear to form a wavy broken line, the family is classified as a series mode. In the more common parallel-mode family, the damage tracks are parallel arcs with each new tornado forming on the right side of its predecessor. The parallel mode is subcategorized into left turn and right turn, according to the direction in which the paths curve.

> Davies–Jones, R., 1986: Tornado dynamics. *Thunderstorm Morphology and Dynamics*, E. Kessler, Ed., Univ. of Okla. Press, p. 223.

fanning—A pattern of smokestack **plume** dispersion in a statically stable **atmosphere**, in which the plume spreads out in the horizontal like an oriental fan and meanders about at a fixed height with little vertical spread.

> *Compare* **looping, coning, fumigation**

far infrared—The portion of the **electromagnetic spectrum** lying between the **middle infrared** and **microwaves**.

> This covers the **wavelength** range approximately from 15 μm to 1 mm, but usage varies.

farmer's year—In Great Britain, the 12-month period starting with the Sunday nearest 1 March.

> *Compare* **grower's year**.

fast Fourier transform—(Abbreviated FFT.) An **algorithm** to compute rapidly the **digital** form of the discrete **Fourier transform**.

> The procedure requires that the length of the data series undergoing transformation be an integral power of two, which is often achieved by truncating the series or extending it with zeros. More flexible mixed-radix algorithms have been developed that allow data series of any length. They gain efficiency as the prime factors of the length become small, and they are equivalent to the FFT when the data series length is a power of two.

fast ice—(Also called landfast ice.) **Sea ice** that is immobile due to its attachment to a coast, usually extending offshore to about the 20-m **isobath**.

> In protected bays and inlets, fast ice is smooth and level, typically reaching a thickness of between 2 and 2.5 m. Along exposed coastlines, fast ice may be greatly deformed.

fast ion—Same as **small ion**.

fast-response sensors—In situ instruments that rapidly respond to turbulent fluctuations in atmospheric variables (e.g., **velocity**, **temperature**, and **humidity**).

> This capability typically requires a **sensor** time constant of < 1 s for tower-based sensors and in the **range** of 0.01–0.1 s for aircraft sensors.

fastest mile—Over a specified period (usually the 24-hour **observational day**), the fastest speed, in miles per hour, of any "mile" of **wind**.

> The accompanying direction is specified also. The fastest mile is the reciprocal of the shortest interval (in 24 hours) that it takes one **mile** of air to pass a given point. This record is maintained only at weather stations that have a **multiple register** and thus have a time-record of the passing of each mile of wind. *Compare* **peak gust**.

fastest mile wind—*See* **fastest mile**.

fata bromosa—An apparent **fog** in the distance where no fog actually exists.

> The appearance is sufficiently compelling that one can observe ships, say, vanishing into or emerging from the seeming fog. The fata bromosa seems to be an infrequent partner of the **superior mirage**. The name translates as "fairy fog." *Compare* **fata morgana**.

fata morgana—A **mirage**, but the specific physical circumstances under which fata morgana should be applied to a particular sighting are ill-defined.

> The best that can be said is that the mirage interacts with features in the landscape to present a scene of sufficient ambiguity that viewers often arrive at quite fanciful interpretations. Cases reported include those of the apparent sighting of cities, mountains, forests, or islands (sometimes metamorphosing one into the other) in places where it was known that no such things existed. Indeed, the mirage's eponym is the fairy Morgan, the legendary half-sister of King Arthur, whose magical powers enabled her to create castles in the air ("fata" is Italian for "fairy"). It is not that any particular occurrence is the least bit mysterious or that either the meteorological or optical behavior is difficult to understand. Indeed, most cases of the fata morgana are undoubtedly multiple-image **superior mirages**, sometimes with marked magnification, sometimes complicated by periodic inhomogeneities, sometimes augmented by an **inferior mirage**. *Compare* **fata bromosa**.

fathom—The common unit of depth in the ocean, equal to six feet.

It is also sometimes used in expressing horizontal distances, in which case 100 fathoms is equivalent to one **cable** or very nearly to one-tenth of a **nautical mile**.

fathom curve—Same as **isobath**.

favogn—A **foehn** of the Swiss Alps.

fax—Frequent abbreviation for **facsimile equipment**.

fax chart—Same as **facsimile chart**.

fax map—Same as **facsimile chart**.

feather—1. *See* **ice feathers**.

2. Same as **barb**.

FEAWP—Abbreviation for **field excursion associated with precipitation**.

feed—(Or antenna feed.) The **source** of **illumination** for an **antenna** reflector.

feedback—A sequence of interactions that determines the response of a system to an initial **perturbation**.

Feedbacks may either amplify (**positive feedback**) or reduce (**negative feedback**) the ultimate state of the system.

feeding—(Rare.) Descriptive of a windstorm of increasing **intensity**.

feeling bottom—As waves approach the coast and move into shallow water, the sea bed affects their propagation, leading to phenomena such as **wave** refraction and **wave breaking**.

The latter is easily observed on any beach; as the waves approach the beach they steepen and break. As a **rule-of-thumb**, waves feel the bottom when the water depth is less than approximately one-quarter of their **wavelength**.

Fengyun 2—(Chinese for "wind and cloud.") A geostationary **meteorological satellite** launched by China and stationed over the Indian Ocean at 105°E longitude.

After a successful launch on 10 June 1997 the satellite began regular operations late in 1997. Multi-spectral imagery is provided by a modified version of the **VISSR** instrument, very similar to those used on the **GMS** satellites. In early 1998 the satellite encountered difficulties with its data link. A separate series of Chinese polar orbiting satellites have also used the Fengyun name (*Fengyun 1A* and *1B*).

Fermat's principle—In its more or less original form, **light** travels between two points in the least possible time.

More correctly, the path taken by a light ray between two points in an **optically homogeneous** media is such that the **line integral** along the path,

$$\int_1^2 n \, ds,$$

where n is the **refractive index**, is an extremum (a minimum or a maximum or even stationary) relative to all possible paths. Because n is inversely related to **phase velocity**, this integral is proportional to a time, the constant of proportionality being the free-space **speed of light**. Fermat was guided by a "metaphysical principle that nature performed its actions in the simplest and most economical ways."

Sabra, A. I., 1967: *Theories of Light: From Descartes to Newton*, p. 136.

Ferrel cell—A zonally symmetric **circulation** that appears to be **thermally indirect** (when viewed using height or **pressure** as the vertical coordinate) first proposed by William Ferrel in 1856 as the middle of three **meridional** cells in each hemisphere.

A similar type of cell was described by Matthew Maury in 1855 and James Thomson in 1857 at about the same time. The Ferrel cell has sinking motion in the same latitudes as the **Hadley cell**, but has rising motion in higher latitudes (approximately near 60°). The Ferrel cell is maintained by **heat** and **momentum** fluxes due to large-scale **eddies** and by **diabatic processes**; these processes are illustrated by the **Kuo–Eliassen equation**.

fetch—1. The distance **upstream** of a measurement **site**, receptor site, or region of meteorological interest, that is relatively uniform.

If a measurement site is located in the middle of a farm field with homogeneous land use, and if there are no changes to the land use and no obstructions such as trees or buildings immediately upstream of the site, then the site is said to have "large fetch". Large fetch is usually considered good if the measurements are to be representative of the **atmosphere** over the farm field. Similarly, measurements over a homogeneous forest could also have large fetch if there are no clearcuts or changes in the tree characteristics upstream of the measurement site.

    2. (Also called **generating area**.) An ocean area where waves are generated by a **wind** having a constant direction and speed.

    3. The length of the fetch area, measured in the direction of the **wind** in which ocean waves are generated. In many cases, the fetch is limited by the **upwind** distance to the coast.

46° lateral arcs—A **halo** in the form of arcs in the vicinity of the **halo of 46°**, the form of which changes markedly with solar **elevation**.

    Sometimes given the names supralateral and infralateral, depending on whether the arcs are mainly above or below the sun. The supralateral arc appears to the sides and above the halo of 46° when the sun is low; it is concave toward the sun but vanishes for solar elevations above 32°. The infralateral arc can be either convex or concave. For some elevations, either arc may or may not be tangent to the halo of 46°. These arcs are explained by **refraction** through the 90° prism ends of columnar **ice crystals** oriented with their long axis horizontal.

4D chart—A **chart** showing the field of **D-values** (deviations of the actual altitudes along a **constant-pressure surface** from the **standard atmosphere** altitude of that surface) in terms of the three **dimensions** of space and one of time.

    It is a form of a four-dimensional **display** of **pressure altitude**. The space dimensions are represented by D-value contours, and the time dimension is provided by **tau-value** lines. 4D charts are used primarily for relatively long-distance preflight planning over ocean areas. *See also* **S-values**.

FFT—Abbreviation for **fast Fourier transform**.

fiard—*See* **fjord**.

fibratus—(Formerly called filosus.) A **cloud species** characterized by a fine hairlike or striated composition, the filaments of which which are usually distinctly separated from each other.

    The extremities of these filaments are always thin and never terminated by tufts or hooks. This species is found mainly in the genera **cirrus** and **cirrostratus**. Cirrostratus fibratus may develop from **cirrus fibratus** or **cirrus spissatus**. *See* **cloud classification**.

fibrous ice—Same as **acicular ice**.

Fick's equation—*See* **diffusivity**.

Fickian diffusion equation—*See* **diffusivity**.

fido—(Or FIDO.) A system for dissipating **fog** artificially, in which gasoline or other fuel is burned at intervals along an airstrip that is to be cleared of fog.

    The name was formed from the four initial letters of the project name: Fog Investigation Dispersal Operations. Fido systems were used at many bomber bases in England during World War II, and commercial transports have been brought into fog-covered airports in the United States using similar methods.

fiducial interval—Same as **confidence interval**.

fiducial limits—*See* **confidence interval**.

fiducial point—1. A **point** (or line) on a **scale** used for reference or comparison purposes.

    In the **calibration** of meteorological thermometers, for example, the fiducial points are 100°C (212°F) and 0°C (32°F), which correspond to the **boiling point** and **ice point** of water at **standard pressure** (760 mm of **mercury**).

    2. A reference point of known location on a map, graph, or **image** used to set the position for overlays of similar maps, graphs, or images.

fiducial temperature—That **temperature** at which, in a specified latitude, the reading of a particular **barometer** requires no temperature or latitude correction.
    See **barometric corrections**.

field—In its restricted physical sense, any physical quantity that varies in three-dimensional space (and possibly time), usually continuously except possibly on surfaces or curves.

    Field quantities often satisfy partial differential equations. An example of a **scalar** field is the **temperature** $T(x, y, z, t)$ at time t at each point (x, y, z) of a solid body; an example of a **vector field** is the (local) **velocity** field $\mathbf{v}(x, y, z, t)$ in a fluid, the separate parts of which are in motion relative to each other. The **continuity** of these fields is a mathematical fiction, obtained by averaging over volumes containing many atoms or molecules but still small on a macroscopic **scale**.

field capacity—The amount of water left in the soil after it has been saturated and allowed to drain by **gravity** for 24 hours.
See also **detention storage**, **gravitational water**, **retention storage**.

field changes—The rapid variations of the **electric field** at the earth's surface and beneath, within, and above **thunderclouds**.
Used to determine quantitative estimates of the charge transferred during a **lightning discharge**, heights of the charge centers (regions or volumes of net charge in the **atmosphere**), and many other features of thunderclouds.

field elevation—Same as **airport elevation**.

field excursion associated with precipitation—(Abbreviated FEAWP.) A characteristic feature in **electric field** records directly beneath a **thundercloud** in which the electric field momentarily reverses **polarity** during a burst of intense **precipitation**.
The timescale for these events is of the order of minutes. The physical origins of the FEAWP are still unresolved.

field mill—An instrument that obtains a continuous measurement of the sign and magnitude of the local **electric potential gradient** by alternately shielding and exposing a conductor that is grounded through a **resistance** to develop an alternating **potential** that is proportional to the **field**.

field of view—(Abbreviated FOV.) *See* **instantaneous field of view**.

field permeability—**Permeability** measured under field conditions.

field strength—In **radar**, the strength of the **electric field** produced at a point by the transmitted **radio waves**, measured in microvolts or millivolts per meter.

field water-holding capacity—Amount of water held in a soil after drainage of **gravitational water**.
Gravitational water typically drains from a soil within a few days of wetting.

FIFOR—An international code word used to indicate a **flight forecast**.
A FIFOR is a trip forecast, that is, a forecast that progresses along with the aircraft.

filament—Dark, sinuous ribbon observed on the solar disk.
Filaments consist of higher **density**, and lower **temperature** material supported by solar magnetic fields that absorb the relatively intense disk **emission** (for example, as seen in Hα emission of neutral **hydrogen**). As these features rotate across the sun's limb they appear as quiescent **prominences** and are seen in **elevation** above the edge of the sun.

filling—An increase in the **central pressure** of a **pressure system** on a **constant-height chart**, or an analogous increase in height on a **constant-pressure chart**; the opposite of **deepening**.
The term is commonly applied to a **low** rather than to a **high**. Since filling is almost always accompanied by a decrease in the **intensity** of **cyclonic circulation**, it is frequently used to imply the process of **cyclolysis**. Filling can be defined quantitatively in at least two ways: 1) as the time rate of central-pressure increase; or 2) as that component of the **pressure tendency** that cannot be attributed to either the motion of the pressure system relative to that point or to the influence of **atmospheric tides**.

filling of a depression—*See* **filling**.

film crust—A type of **snow crust**; a very thin layer of **ice** upon a **snow** surface, formed by the **freezing** of meltwater or **rain** into a continuous film; not as thick as an **ice crust**.

film renewal model—Same as **surface renewal model**.

filter—1. Any material that, by **absorption** or **reflection**, selectively modifies the **radiation** transmitted through an optical system.
2. To remove a certain component or components of **radiation**, usually by means of a filter, although other devices may be used.
3. A mathematical tool used to remove information within certain **frequency** (or **wavelength**) limits from a time (or space) series of data.
Filters are commonly used to remove high (low) frequency **oscillations** from a dataset when oscillations of low (high) frequency are of interest.

filtered equations—A system of approximate hydrodynamical equations from which certain families of solutions of the original equations, normally corresponding to timescales shorter than those of interest, are automatically excluded.

See **filtering approximations**.

filtered model—A **model** composed of **filtered equations**.
Compare **unfiltered model**.

filtering—1. Any process by which the Fourier components of a temporally or spatially varying **output** are different in **amplitude** or **phase** from the corresponding components of an **input**.

Cellophane is a (temporal) **filter**: The amplitudes of the spectral components of white **light** (input) incident on the cellophane are different from those of the light it transmits (output). A **thermometer** is a low-pass **filter**; all components of **air temperature** are fed into the thermometer but it passes only the low-frequency components and suppresses the high-frequency components. Optical devices of all kinds are also spatial filters: They may transmit each spatial component of a pattern differently.

2. The separation of a wanted component of a **time series** from any unwanted residue (**noise**). *See* **noise filtering**.

filtering approximations—A set of mathematical approximations introduced into a system of hydrodynamical partial differential equations to **filter** out or exclude solutions corresponding to those physical disturbances that are believed to contribute only negligibly to the problem at hand.

In the theory of cyclonic-scale atmospheric flow, for example, the introduction of the **quasi-hydrostatic approximation** effectively excludes solutions of the **equations of motion** corresponding to atmospheric **sound waves**, while the selective use of the **quasigeostrophic approximation**, as in the **vorticity equation** with the **divergence** eliminated, effectively filters out solutions corresponding to high-speed atmospheric **gravity waves**. Through the use of such filtering approximations, the equations of motion can be focused upon the desired components of the flow.
Charney, J. G., 1948: On the Scale of Atmospheric Motion. *Geofys. Publ.*, **17**, 3–17.

filtration—The process of forcing a fluid (liquid or gas) through a **filtering** medium (e.g., granular material such as sand, **silt**, diatomaceous earth, fine cloth, unglazed porcelain, or specially prepared paper) resulting in the removal of suspended or colloidal matter.

Findeisen–Bergeron nucleation process—*See* **Bergeron–Findeisen process**.

fine line—A narrow **radar echo** indicating a boundary (such as a frontal boundary, **gust front**, or **dryline**) across which a **density** or moisture discontinuity exists.

The **reflectivity** is apparently explained by **scattering** from the **refractive index** gradients across the **discontinuity** and from insects and insect-eating birds that are concentrated along the line. *See* **clear-air echo**.

fine-mesh grid—A nonspecific term indicating that a **grid** has relatively high resolution, that is, that its grid points are relatively near to each other.

In the past, a grid length of less that 200 km was generally considered to be fine mesh, but with the rapid growth of computer power permitting the use of extremely high-resolution grids, the term is now used in a more contextual manner. *Compare* **coarse-mesh grid**.

fine-mesh model—A numerical **model** that carries out its computations on a **fine-mesh grid**.

fine particles—Particles occurring in the **atmosphere** with diameters less than 2 μm.

This size **range** includes **Aitken nuclei** and **particles** in the **accumulation mode**.

Fineman nephoscope—One adaptation of the **mirror nephoscope**.

finestructure—Small-scale structure in the fields of **temperature**, **salinity**, **density**, and/or **velocity**.

The term finestructure refers in the ocean to structures of vertical **scale** smaller than 100 m and larger than 1 m, usually associated with small-scale physical factors that may cause **mixing**, but not with the mixing events themselves. Examples are **interleaving** motions, **inertial** and **internal waves**, and **vertical modes**. Note the overlap in the vertical scale with that of **microstructure**.

finger rafting—A particular form of **rafted ice**, typical of thin ice, whereby overlapping occurs in alternating, interlocking segments (like the interlaced fingers of clasped hands).

fingering—The phenomenon describing preferential fluid pathways through porous media.

fingerprint method—The identification of an observed **climate signal** with a structure characteristic of a predicted **climate change**.

The fingerprint method requires the analysis of a **multivariate signal**, involving changes in a single **climate** variable at many places or levels in the **atmosphere**, or changes in two or more different variables. This method is used in attempts to attribute observed climate change to a given cause, such as increased **greenhouse gas** concentrations.

finite depth—An ocean wave is said to be in finite depth if the presence of the sea bed affects it. Generally this depth is taken to be less than one-quarter of the **wavelength** of the **wave**.

finite-difference approximation—The difference between the values of a function at two discrete points, used to approximate the derivative of the function.
   The derivative $f'(x)$ of a function $f(x)$ at an arbitrary point x is usually approximated by finite differences in one of three ways:

$$f'(x) \doteq \frac{f(x + a) - f(x)}{a} \equiv \frac{\Delta f}{a},$$

where Δf is called a forward difference;

$$f'(x) \doteq \frac{f(x + a/2) - f(x - a/2)}{a} \equiv \frac{\delta f}{a},$$

where δf is called a centered difference;

$$f'(x) \doteq \frac{f(x) - f(x - a)}{a} \equiv \frac{\nabla f}{a},$$

where ∇f is called a backward difference (not to be confused with the **gradient**). Of these approximations, the centered difference is the most accurate; whether it is the most convenient or accurate for the problem as a whole depends on the character of the equations involved. Higher derivatives are approximated by iteration of these formulas.

finite-difference equation—An equation resulting from the use of **finite-difference approximations** to the derivatives in an ordinary or partial differential equation.
   A finite-difference equation may have a solution (or solutions) the behavior of which is quite unlike that (or those) of the differential equation approximated, and care must be taken to exclude or minimize any such solutions in the computation. Finite-difference equations are extensively used in meteorology, a realistic **atmosphere** usually being described by equations without tabulated solutions. *See* **computational mode**.

finite differencing—A numerical method for solving differential equations in which the original equations are approximated by a set of algebraic equations that are solved on a network of discrete **grid** points.

finite-element model—A **model** with **dependent variables** represented as a finite series of piecewise-developed **polynomial** basis functions (finite elements), as opposed to globally defined Fourier or **spherical harmonic** functions.
   Each basis function is nonzero only over a very small part of the domain. Examples are hat functions and piecewise-defined cubic splines.

fire behavior—The manner in which fire reacts to the variables fuel, weather, and **topography**.

fire-danger meter—1. A device for combining ratings of several variable factors into numerical classes of fire danger.
   2. A graphical aid used in fire-weather forecasting to calculate the degree of forest-fire danger (or **burning index**).
   Commonly in the form of a circular slide rule, the fire-danger meter relates numerical indices of 1) the seasonal stage of foliage, 2) the cumulative effect of past **precipitation** or lack thereof (**buildup index**), 3) the measured **fuel moisture**, and 4) the speed of the **wind** in the woods. The fuel moisture is determined by weighing a special type of wooden stick that has been exposed in the woods, its weight being proportional to its contained water.

fire weather—Weather variables, especially **wind**, **temperature**, **relative humidity**, and **precipitation**, that influence fire starts, **fire behavior**, or fire suppression.

firm yield—Estimated maximum amount of water that can be supplied by a **reservoir** under specified conditions.

firn—Old **snow** that has become granular and compacted (dense) as the result of various surface metamorphoses, mainly melting and refreezing but also including **sublimation**.
   The resulting **particles** are generally spherical and rather uniform. Firnification, the process of

firn formation, is the first step in the transformation of snow into **land ice** (usually **glacier ice**). Some authorities restrict the use of firn to snow that has lasted through one summer, thereby distinguishing it from **spring snow**. Originally, the French term, "névé," was equivalent to the German term, "firn," but there is a growing tendency, especially among British glaciologists, to use "névés" for an area of firn, that is, generally for the **accumulation area** above or at the head of a **glacier**

firn field—1. An area of **firn** that is not part of a **glacier**.

    2. Same as **accumulation area**.

firn ice—Same as **iced firn**.

firn limit—Same as **firn line**.

firn line—The boundary of the area of **snow** on a **glacier** surviving one year's **ablation**, thus becoming **firn**.

    In the absence of **superimposed ice**, this limit is equivalent to the **equilibrium line**. *See* **climate snow line**.

firn snow—Same as **firn**.

firnification—The process of **firn** formation.

firnspiegel—A thin, highly reflective sheet of **clear ice** formed at the **snow** surface; formed in spring when subfreezing air temperatures combine with subsurface melting (due to penetration of solar **energy** into the snow).

first-guess field—*See* **background field**.

first gust—The sharp increase in **wind speed** often associated with the early mature stage of a **thunderstorm cell**.

    It occurs with the passage of the **discontinuity** zone that is the boundary of the cold-air **downdraft**. The first gust can reach destructive speeds.

first-hop wave—The first **reflection** of a **lightning stroke** from the **ionosphere** that arrives after the original **ground wave**.

first law of thermodynamics—The total **internal energy** U of an **isolated system** is constant.

    If a thermodynamic system is not isolated, its internal energy may change because of two distinguishable macroscopic processes: working (a force exerted through a distance) and heating (**energy** exchange by virtue of a **temperature** difference between the system and its surroundings). The first law may be written

$$dU/dt = Q + W,$$

where Q is the rate of heating and W is the rate of working on the system. For a simple system in which working is solely a consequence of volume changes, the rate of working is given by

$$W = -p\,dV/dt,$$

where p is **pressure** and V is volume, provided volume changes at a sufficiently slow rate (quasi-static process) that the pressure is approximately uniform.

first-order climatological station—A **meteorological station** at which autographic records or hourly readings of **atmospheric pressure**, **temperature**, **humidity**, **wind**, **sunshine**, and **precipitation** are made, together with observations at fixed hours of the amount and form of **clouds** and notes on the weather.

    Compare **first-order station**

first-order closure—An approximation made to solve **turbulence** equations that assumes turbulent fluxes of a quantity such as **moisture** flow down the **mean** gradient of moisture, where the rate of flow is proportional to an **eddy diffusivity**.

    The symbol for eddy diffusivity is often K, hence this theory is also known as **K-theory**. By first order, it is implied that any turbulence **statistics** of second or higher order (**variances**, **covariances**, etc.) that appear in the governing equations are replaced by approximations that depend only on first-order statistics (i.e., mean values of the **dependent variables** and of **independent variables**). Such an approximation reduces the number of unknowns in the governing equations, allowing them to be mathematically closed, thereby allowing them to be solved for the approximate flow state. First-order closure can be applied locally (as in K-theory) or nonlocally (as in **transilient turbulence theory**). *See also* **closure assumptions**, **closure problem**.

first-order reactions—A chemical reaction involving only one chemical species, in which the rate of decrease of the concentration of the reactant is directly proportional to its concentration.

The constant of proportionality is the first-order **rate coefficient** k, which has units of inverse time: $d[A]/dt = -k[A]$, where $[A]$ is the concentration of species A. The first-order rate equation is often expressed in its integrated form, $[A] = [A]_0 \exp(-kt)$, where $[A]_0$ is the initial concentration of species A. A common example of a first-order reaction is a radioactive decay process. In the **atmosphere**, **photolysis** reactions obey first-order kinetics. Also, unimolecular decomposition reactions at their high-pressure limit empirically follow first-order kinetics.

first-order station—After U.S. National Weather Service practice, any **meteorological station** that is staffed in whole or in part by National Weather Service (Civil Service) personnel, regardless of the type or extent of work required of that station.

Compare **first-order climatological station**; *see also* **second-order station**.

first-year ice—**Sea ice** that has not yet experienced summer melt.

First-year ice is distinguished from older ice primarily by having a higher **salinity**. Undeformed first-year ice differs from older ice in that it is smoother and lacks refrozen melt ponds. First-year ridges are distinguished by being larger, more angular, and more porous than multiyear ridges.

Fisher & Porter rain gauge—A weighing-recording **rain gauge** that mechanically converts the depth of accumulated **precipitation** during a given period of time to a code disk position.

The code disk position is recorded on punched paper tape at selected intervals. The time interval of recording is 15 minutes and the accumulated precipitation is read to the nearest 2.5 mm (0.10 in.). The gauge can be modified for remote transmission.

> NOAA/National Weather Service, 1972: *National Weather Service Observing Handbook No. 2, Substation Observations*, NOAA, US GPO..

fitness figure—(Also called fitness number.) In the United Kingdom, a measure of the fitness of the weather at an airport for the safe landing of aircraft.

The figure F is computed on the basis of corrected values of **visibility** and **cloud height**. Observed visibility is adjusted according to **intensity** of **precipitation**; and cloud height is corrected for height of nearby obstructions and **cloud amount**. Further corrections are applied for the cross-runway component of the **wind**. The most comparable device widely used in the United States is the **sliding scale** of **ceiling** and visibility.

five-and-ten system—The most common system for representing **wind speed**, to the nearest 5 knots, in symbolic form on **synoptic charts**.

It consists of drawing the appropriate number of half-barbs, barbs, and pennants from the end of the **wind-direction shaft**. In this system, a half-barb represents 5 knots, a **barb** 10 knots, and a **pennant** 50 knots.

five-day forecast—A **weather forecast** for the five-day period ahead.

This term is most often used by the media.

five-point method—Method used to determine the **average velocity** in the vertical in a deep **stream** (depths greater than 20 ft or 6 m) of depth D, on the basis of a **weighted average** of **velocity** observations at depths 0, $0.2D$, $0.6D$, $0.8D$, D below the water surface.

fix—A geographical position determined by visual reference to the surface, by reference to one or more radio NAVAIDs, by celestial plotting, or by another navigational device.

fixed-beam ceilometer—*See* **ceilometer**.

fixed-level chart—Same as **constant-height chart**.

fixed time broadcast—Broadcast made according to a schedule of transmissions agreed internationally or nationally.

fixing of atmospheric gases—The incorporation of gaseous molecules into biological systems or by industrial processes.

The term is most commonly applied to **nitrogen fixation**, which provides a mechanism for biota to acquire nutrient **nitrogen** from the relatively inert nitrogen in the **atmosphere**.

fjord—(Sometimes spelled fiord, fiard.) A deep-water inlet, usually surrounded by mountains; specifically a submerged U-shaped valley carved out by **glacial** action.

The fjord is characteristic of the coastal regions of Norway, western Scotland and Ireland, Greenland, Labrador, Alaska, British Columbia, southern Chile, the Antarctic peninsula, southwest New Zealand, and other high-latitude oceanic islands (Iceland, Spitzbergen, Kerguelen, etc.).

flame collector—A device used in atmospheric electric measurements for the removal of **induction** charge on apparatus.

It is based upon the principle that products of combustion are ionized and will consequently conduct electricity from charged bodies. *See* **collector**.

flame ionization detection—A type of gas detector used in **gas chromatography**.

Individual components of a gas mixture elute from the column and are burned in a **hydrogen** flame, where a large quantity of **ions** are produced. These ions are vented through a **collector** tube, which is maintained at some electrical **potential** so that a **current** passes through the ions produced in the flame to the collector. The current measured is proportional to the quantity of eluted compound. The technique has found the most use for the measurement of organic compounds, particularly **hydrocarbons**, in the **atmosphere**.

flan—In Scotland, a sudden **gust** or **squall** of **wind** from land.

Flanders storm—In England, a heavy fall of **snow** coming with the south **wind**.

flanking line—An organized lifting zone of **cumulus** and **towering cumulus** clouds, connected to and extending outward from the mature **updraft** tower of a **supercell** or strong **multicell convective storm**.

The flanking line often has a stair-step appearance, with the tallest clouds adjacent to the mature updraft tower.

flare—A bright, **transient** event within the sun's **chromosphere** and **corona**.

Flares produce enhanced **emission** at radio frequencies and in the **ultraviolet** and **x-ray** spectral regions as well. They may also produce increased **particle** emission, often with **ions** of **cosmic ray** energies. Flares usually appear within minutes and fade within an hour. They are localized to small areas (typically $<10^{-3}$ of the solar disk) and usually occur within solar active regions.

flare echo—A weak **radar echo** in the form of a tapered radial extension beyond the far edge of a severe **thunderstorm**, associated with the presence of large **hydrometeors**, especially **hail**, in the storm.

The phenomenon is explained by what is referred to as "three-body **scattering**." Some of the **radar** energy is **scattered** from the hail region to the ground, reflected back to the hail region, and then scattered a third time, eventually returning to the radar as a weak **signal**. From the perspective of the radar, the signal travels out and back with a timing indicating that the **scattering** region is beyond the back edge of the storm. Flare echoes are observed more commonly with radars operating at shorter wavelengths, evidently because of deviations from **Rayleigh scattering**.

flash—1. *See* **lightning flash**.

2. *See* **green flash**.

flash flood—A **flood** that rises and falls quite rapidly with little or no advance **warning**, usually as the result of intense **rainfall** over a relatively small area.

Some possible causes are **ice jams**, dam failure, and **topography**.

flash photolysis systems—Technique for measuring the rate of a chemical reaction in which a short, intense flash of **light** initiates a chemical reaction and some spectroscopic **probe** is used to measure directly the change in concentration of one of the reactants, most often a **free radical**.

The light pulse is provided by a flash lamp or a **laser**, while the **detection** technique is usually either **absorption** or **fluorescence** spectroscopy. Reactions occurring on timescales of less than 1 μs can be measured using this technique.

flashy stream—**Stream** that rapidly collects flows from the steep slopes of its catchment (**watershed**, **basin**) and produces **flood** peaks soon after the **rain**.

Its flow quickly subsides after the cessation of **rainfall**.

flat low—An area of low pressure characterized by uniform **barometric pressure** without a well-defined horizontal **gradient**.

flaw—A British nautical term for a sudden **gust** or **squall** of **wind**.

Fleet Broadcast—That **high frequency** radio broadcast (in addition to the **General Broadcast**) to all navy ships and merchant ships in which warnings are given.

flight briefing—Same as **pilot briefing**.

flight cross section—Graphic representation of the meteorological conditions observed simultaneously in a **vertical section** of the **atmosphere** taken along the predicted horizontal path of the aircraft.

Compare **route cross section.**

flight documentation—Written or printed documents, including charts or forms containing **meteorological information** for a flight.

flight forecast—Statement of the meteorological conditions expected to be encountered successively in flight.
See **FIFOR.**

flight information center—A center established to provide flight information and alerting services.

flight information region—Airspace of defined dimensions within which flight information and alerting services are provided.

flight level— A surface of constant **atmospheric pressure** that is related to a specific pressure **datum** (1013.25 hPa) and is separated from other such surfaces by specific pressure intervals.
It is conventionally given a numerical value to the nearest 1000 ft in units of 100 ft in accordance with the structure of the **ICAO Standard Atmosphere.** For example, the 500-hPa level is written as FL 180, the **ICAO** standard height being 18 289 ft.

flight plan—A detailed statement of the intended route, **altitude,** and time of a proposed aircraft flight.
Most flight plans also include notation of existing and forecast weather conditions at destination (and **alternate airport,** if required) as well as **airspeed** and **ground speed** estimates. *Compare* **clearance, pilot briefing.**

flight train—Equipment for launching a **radiosonde** and maintaining it aloft, typically consisting of a **radiosonde balloon,** a parachute, a de-reeler, and a string to mate them together.

flight-weather briefing—Same as **pilot briefing.**

flist—In Scotland, a sudden **shower** accompanied by a **squall.**

float barograph—A type of recording **siphon barometer.**
The mechanically magnified motion of a float resting on the lower **mercury** surface is used to record **atmospheric pressure** on a rotating drum.

float-type rain gauge—A class of **rain gauge** in which the level of the collected rainwater is measured by the position of a float resting on the surface of the water.
This instrument is frequently used as a **recording rain gauge** by connecting the float through a linkage to a pen that records on a clock-driven chart.
Middleton, W. E. K., and A. F. Spilhaus, 1953: *Meteorological Instruments,* 3d ed., rev., Univ. of Toronto Press, 121–122.

floating ice—Any form of **ice** found floating in water.
The principal kinds of floating ice are **lake ice,** river ice, **sea ice** that forms by the **freezing** of water at the surface, and **glacier ice** (ice of land origin) formed on land or in an **ice shelf.** The concept includes ice that is stranded or grounded.

floating pan—An **evaporation pan** in which the **evaporation** is measured from water in a pan floating in a larger body of water.

floccus—Cloud species in which each **cloud element** is a small tuft with a **cumuliform** or rounded appearance, the lower part of which is more or less ragged and often accompanied by **virga.**
This species is found in the genera **cirrus, cirrocumulus, altocumulus,** and sometimes also in **stratocumulus.** Cirrocumulus floccus sometimes evolves as the result of the dissipation of the common base of **cirrocumulus castellanus;** in like manner, altocumulus floccus may evolve from **altocumulus castellanus.** Cirrus floccus differs from cirrocumulus floccus in that its elements subtend an angle of greater than 1° when observed at an angle of more than 30° above the **horizon.**
See **cloud classification.**

floe—Any relatively flat piece of **ice** 20 m or more across.
It may be composed of several fragments bonded together.

floeberg—A mass of **hummocked ice,** formed by the piling up of many ice **floes** by lateral pressure; an extreme form of **pressure ice.**
It may be more than 50 ft high and resemble an **iceberg.**

flood—The overflowing of the **normal** confines of a **stream** or other body of water, or the accumulation of water over areas that are not normally submerged.

flood channel—An enlarged **stream** channel carrying water during **floods.**

flood current—The movement of a **tidal current** toward the shore or up a **tidal river** or **estuary**. *See* **flood tide**.

flood forecast—Forecasts used to provide warnings for people to evacuate areas threatened by **floods**, and to help in the operation of flood control structures.

flood forecasting—The use of real-time **precipitation** and **streamflow** data in rainfall-runoff and **streamflow routing** models to forecast flow rates and water levels for periods ranging from a few hours to days ahead, depending on the size of the **watershed** or **river basin**.

flood frequency distribution—The **probability distribution** that describes the likelihood of different annual maximum **floods**.

The cumulative flood frequency distribution, $F(x)$, specifies the **probability** that the annual maximum flood, X, is less than any given value.

flood icing—Same as **icing**.

flood interval—The interval in time between consecutive cycles of tidal **floods**.
See also **flood current**.

flood irrigation—(Also called level basin system.) Irrigation by quickly inundating a flat field to a predetermined depth.

flood marks—Natural marks left on a shoreline, trees, or other objects indicating the maximum stage of **floods**.

flood mitigation—Methods of reducing the effects of **floods**.
These methods may be structural solutions (e.g., reservoirs, levees) or nonstructural (e.g., land-use planning, early warning systems).

flood plain—Typically, a low area adjacent to a river or other body of water that is subject to **flooding**.

flood probability—*See* **flood frequency distribution**.

flood proofing—Methods for reducing **flood** damage in flood-prone areas, particularly in the **flood plain**.
Flood proofing involves techniques such as raising the foundations of new structures above a designated flood elevation, or, for existing buildings, altering exterior walls, windows, and entrances to reduce the **frequency** of damage to the building.

flood routing—(Also known as storage routing, **streamflow routing**.) A mathematical procedure for predicting the changing magnitude, speed, and shape of a **flood wave** as a function of time at one or more points along a waterway or channel.

flood stage—An arbitrarily fixed and generally accepted **gauge height** or **elevation** above which a rise in the water surface elevation is termed a **flood**.
It is commonly fixed as the **stage** at which **overflow** of the **normal** banks or damage to property would begin.

flood strength—The magnitude (speed) of the **flood tide**.
See also **flood current**.

flood tide—The **tidal current** associated with the increase in the height of a **tide**; the opposite of ebb current. *See* **flood current**.

flood warning—Advance notice that a **flood** may occur in the near future at specified locations.

flood watch—A product issued by an NWSO (National Weather Service Office) or NWSFO (National Weather Service Forecast Office) to advise of the possibility that **flooding** will occur in a specified area over a period longer than six hours.

flood wave—The rise in **stage**, culminating in a crest and followed by **recession** to lower stages, associated with a **flood**.

flooding—1. Water overflowing the **bankfull stage** of a natural or artificial waterway.
2. Accumulation of water caused by **surface runoff** in low-lying areas not usually submerged.

flooding ice—Same as **icing**.

floods-above-base series—*See* **partial-duration series**.

Florida Current—One of the western **boundary currents** of the North Atlantic subtropical **gyre**.
A deep, narrow, and swift current, it originates from the **Loop Current** of the Gulf of Mexico.

It enters the Atlantic Ocean through the Florida Straits with a **transport** of 30 Sv (30 × 10⁶ m³ s⁻¹) and speeds of 1.8 m s⁻¹ and follows the coast northward for 1200 km to Cape Hatteras (35°N). Input from the **Antilles Current** and **entrainment** from the **Gulf Stream** recirculation in the **Sargasso Sea** increase its transport to 70–100 Sv (70–100 × 10⁶ m³ s⁻¹) before it leaves the coast at Cape Hatteras to continue as the Gulf Stream.

flow law—For **ice**, the physical relation between **stress** in a material and the strain rate or **deformation** rate of the material.

It represents a macroscopic, continuum approximation of the processes occurring within the material and typically contains parameters (often called constants) that must be determined from experiments. Most flow "laws" are simplifications of some more general relation and are valid only within some **range** of **temperature**, stress, etc., for which a given deformational process is dominant.

flow line—The path of a fluid **particle** in a two-dimensional steady-state regime.

flow net—A set of **flow lines** and corresponding intersecting **equipotential lines** arranged in a prescribed way.

flow pattern—Distribution of the **troughs** and **ridges** as depicted on **upper-air charts**.

flow velocity—The volume of fluid passing through a unit area perpendicular to the direction of flow in a unit of time.

flowmeter—A device for measuring the rate of fluid flow, usually in pipes.

fluctuation—**Variation**, especially back and forth between successive values in a series of observations; or, variations of data points about a smooth curve passing among them.

fluctuation velocity—Same as **eddy velocity**.

flue gas—The gases from combustion or **oxidation** processes.

fluid parcel—A continuous mass of fluid.

flume—1. In **open channel flow**, a manufactured conduit (usually rectangular in shape) with a portion having a constricted **cross section**, which may be used to control and measure **streamflow**.

2. In **geomorphology**, a narrow, steep-sided valley.

3. In **hydraulics**, a channel used for studying the flow of fluids under **gravity**. The fluid is pumped from a sump to a stilling tank that acts as a **reservoir** from which the fluid is discharged down the flume at varying speeds. The motion is viewed from above or through glass sides.

fluorescence—The **emission** of **radiation** associated with the relaxation of an atom or molecule from an excited **energy level** to a lower (usually **ground state**) level.

The emission can be in the visible or **ultraviolet** if an electronic transition is involved, or in the **infrared** if it is a vibrational transition. *See also* **resonance fluorescence, laser-induced fluorescence, luminescence.**

flurry—1. (United States.) **Snow flurry**.

2. (Archaic.) A sudden and brief wind **squall**.

fluvial dynamics—A branch of **potamology** dealing with the action of forces on riverbed materials and with water flow in watercourses.

fluvial morphology—The science of the formation of riverbeds, **flood** plains, and **stream** forms by the action of water.

flux—1. The rate of flow of some quantity, often used in reference to the flow of some form of **energy**.
See also **power**.

2. In the field of **atmospheric turbulence** and **boundary layers**, often used as a contraction for **flux density**; namely, the flow of a quantity per unit area per unit time. These fluxes can be defined in two forms: dynamic and kinematic. The **dynamic flux** of a quantity is the flow of that quantity per unit area per unit time, where often the word dynamic is assumed if it is not explicitly stated. The advantage of a **kinematic flux** is that it has units that are more easily measured by a conventional **meteorological instrument**. The units are usually a **velocity** (m s⁻¹) times a **temperature** (K), **specific humidity** (kg_{water}/kg_{air}), or **wind speed** (m s⁻¹).

flux adjustment—In **climate** modeling, the practice of modifying the fluxes (of **heat** and water) between the **atmosphere** and ocean in coupled atmosphere–ocean models.

This modification is designed to minimize the **climate drift** that occurs during **model** integration. These **flux** adjustments are typically a function of location and **season**.

flux aggregation—The process by which an effective horizontal average or aggregate of turbulent fluxes is formed over inhomogeneous surfaces.

The aggregated **flux** differs from the **spatial average** of **equilibrium** fluxes in an area, due to nonlinear advective enhancement associated with local **advection** across surface transitions. Aggregated fluxes can be related to vertical profiles only above the **blending height**.

flux-corrected transport—A general methodology for constructing a smooth monotonic flux-form **transport** scheme.

The **flux** computed by a monotonic, but diffusive, low-order scheme is combined (nonlinearly) with the flux computed by an accurate, but dispersive, high-order scheme to obtain the net transport flux. The high-order scheme is locally given the greatest possible weight without introducing oscillations (ripples) to the solution.

Zalesak, S. T., 1979: Fully multidimensional flux-corrected transport algorithms for fluid. *J. Comput. Phys.*, **31**, 335–362.

flux correction—Same as **flux adjustment**.

flux density—(Sometimes, **flux**.) The rate at which any physical quantity is transported (usually in a hemisphere of directions) across a unit area.

The term "flux" indicates flow (rate of **transport**), and "density" here is an areal (as opposed to volumetric) density. Flux densities often may be written in the form pnv, where n is the number density (volumetric) of carriers of a property p (e.g., mass, charge, **momentum**, **energy**) and v is their speed.

flux density—In **radiation**, the **radiant energy** per unit time crossing a plane surface of unit area. Units are W m^{-2}. *See also* **irradiance**, **emittance**.

flux-density threshold—Same as **threshold illuminance**.

flux emissivity—The ratio of emitted **flux density** from a given surface to that from a **blackbody** at the same **temperature**.

The adjective **flux** is typically applied only when the **emissivity** is not constant with direction, as from surfaces that do not emit according to **Lambert's law**.

flux of radiation—*See* **radiant flux**.

flux-profile relationships—1. An equation that relates the vertical **turbulent flux** of a quantity to the shape or slope of the **vertical profile** of the **mean value** of that quantity, usually applied in the **surface layer** of the **atmospheric boundary layer**.

For example, in the surface layer the **wind speed** increases approximately logarithmically with height, at a rate that depends on the **friction velocity** u_* and the **Obukhov length** L. Because both these last two variables depend on surface fluxes, it is possible to find these fluxes from measurements of the mean wind **profile**, a technique that can be easier and less costly than measuring the fluxes directly.

2. Same as **universal functions**.

flux-ratio method—A procedure originally proposed by Ball in 1960 to estimate the **entrainment velocity** w_e into the top of the **convective boundary layer** by requiring a fixed value for the ratio A of entrained **heat flux** to surface heat flux ($\overline{w'\theta'}$, in kinematic form):

$$w_e = A\overline{w'\theta'}_s/\Delta\theta,$$

where $\Delta\theta$ is the **potential temperature** change across the stable layer at the top of the **mixed layer**.

The **flux** ratio, sometimes called the Ball ratio, is empirically estimated to be $A = 0.2 \pm 0.05$. This method works quite well for pure **free convection** but is inadequate when a significant portion of **turbulence** is generated by **wind shear**. *See* **eddy correlation**.

Ball, F. K., 1960: Control of inversion height by surface heating. *Quart. J. Roy. Meteor. Soc.*, **86**, 483–494.

flux Richardson number—A **dimensionless number** defined as the ratio of the **buoyancy** term of the **turbulence kinetic energy** budget equation and the negative of the **shear** terms of that same equation:

$$R_f = \frac{(g/T_v)\ \overline{w'\theta_v'}}{\overline{u'w'}\ (\partial\overline{U}/\partial z) + \overline{v'w'}\ (\partial\overline{V}/\partial z)},$$

where g is gravitational **acceleration**, z is height, θ_v is **virtual potential temperature**, T_v is the **virtual temperature**, (U, V) are the horizontal Cartesian **wind** components from the west and the south, primes denote deviations from mean values, and overbars denote the mean or average. It is a measure of **dynamic stability**, which describes the capability of the flow to remain turbulent by **wind shear** overpowering any **static stability**. Normally, the denominator is negative. Thus, when the numerator is positive, the flux Richardson number R_f is negative for statically **unstable air**. When the flux Richardson number is less than 1, the flow is dynamically unstable and is turbulent. When the flux Richardson number becomes greater than 1, the flow becomes dynamically stable, **turbulence** would tend to decay, and the flux Richardson number would become undefined. *Compare* **gradient Richardson number, bulk Richardson number, critical Richardson number.**

flux transmittance—The ratio of transmitted **irradiance** to the incident irradiance.
Used to distinguish from the **transmittance** of **radiance** in a specific direction only.

fluxplate—An instrument that records **heat flux** electrically.

fly ash—Fine **airborne particulates** carried out of a boiler by waste gases after the combustion of solid fuel, for example, coal, and expelled as noncombustible airborne emissions or recovered as by-products for commercial use.
See **aerosol, air pollution.**

flyoff—(Obsolete.) Same as **evapotranspiration.**

FM–CW radar—(Frequency-modulated, **continuous-wave radar**.) A **radar** that transmits a **waveform** that is continuously and regularly modulated in **frequency**, typically using a **linear** sweep or **sawtooth wave** for the **frequency modulation**.
In such a radar the **range** to a **target** is determined by the frequency of the **signal** it returns. Extremely high range **resolution** can be achieved by measuring the returned signal in narrow frequency bands.

foam crust—A **snow** surface feature that looks like small overlapping waves, like sea foam on a beach.
Foam crust occurs during the **ablation** of the snow surface and may further develop into a more pronounced wedge-shaped form known as **plowshares.**

foehn—(Or föhn.) A warm, dry, **downslope wind** descending the lee side of the Alps as a result of **synoptic-scale**, cross-barrier flow over the mountain range.
The winds are often strong and gusty, sometimes forming downslope windstorms as a result of **mountain wave** activity. The air in the near-surface flow originates at or above the main crest height of the Alpine barrier, and achieves its warmth and dryness as a result of **adiabatic** descent. The foehn often replaces a retreating cold **air mass** from a **polar** or **arctic front**, producing dramatic **temperature** rises that reach 10°C and occasionally even 20°C or more, sometimes in a matter of minutes. This is especially true of the **south foehn**, which blows from northern Italy, where the air is warm, to the north of the Alps (Austria, Germany, Switzerland), where the air is cooler and could be cold **arctic air** as just described. The **north foehn**, blowing from a cooler to a warmer region, produces less dramatic temperature changes. The air in the foehn, originating from the mid **troposphere**, is characteristically clean. Its warm temperatures rapidly melt (or sublimate) **snow**, sometimes producing **flooding**, and the extreme dryness can lead to dangerous **fire weather** conditions. The Alpine foehn has been extensively studied by European scientists, and it is recognized as the type wind for similar downslope winds, resulting from cross-barrier flow, in other parts of the world. In other mountain ranges the foehn has a variety of local names, including **chinook** in the Rocky Mountains in North America; **zonda** for a westerly foehn from the Argentine Andes; ljuka in Carthinia (northwestern Croatia); **halny wiatr** in Poland; **austru** in Romania; and **favogn** in Switzerland. A northeasterly foehn descending the Massif Central in France and extending over the Garonne Plain is locally called **aspre**. A dry wind from the northwest descending the coastal hills in Majorca is named the **sky sweeper**. In New Zealand a foehn blowing from the New Zealand Alps onto the Canterbury Plains is the **Canterbury northwester**. A cross-barrier flow that produces strong winds and cooling is called a **bora** in many parts of the world. Many authors have attempted to classify strong **wind** events as foehn (or chinook) or bora, for example, for climatologies. These studies have had mixed success: Many wind events are easy to classify, but a number of events are difficult, depending on the data available (most studies attempt to use surface data) and the method used to differentiate between the two types of events. *See* **foehn phase, high foehn.**

Defant, F., 1951: *Compendium of Meteorology*, 667–669.

foehn air—The air associated with **foehn** winds, very warm and dry.

foehn bank—A **bank** of **orographic** stratiform **crest cloud** occurring over a mountain barrier during **foehn** conditions; same as **foehn wall.**

foehn break—Same as **foehn gap**.

foehn cloud—Any cloudform associated with the **foehn**, usually referring to **standing clouds** of two types, **orographic clouds** and **mountain wave clouds**.
   Orographic clouds may include **crest clouds** and the **foehn wall**. Wave clouds may consist of **lenticular** (including altocumulus standing lenticular, or ACSL) clouds, lee-wave clouds and cloud bands, and **rotors**. *See also* **Bishop wave, chinook arch, contessa di vento, Moazagotl**.

foehn cyclone—A **cyclone** formed (or at least enhanced) as a result of the **foehn** process on the lee side of a mountain range.
   A mid-latitude, north–south mountain range provides the best inducement for this type of **development**. The **Alberta low** and the **Colorado low** of the eastern slopes of the Rockies are good examples. *See* **dynamic trough**.

foehn gap—A break in an extensive **cloud** deck or **cloud shield**, usually a band parallel to and **downwind** of the mountain ridge line.
   Especially visible in satellite pictures, this cloud-free zone results from the strong **sinking** motion on the lee side of a mountain barrier during a **foehn** or **chinook**.

foehn island—Isolated penetrations of the **foehn** through a cold-air layer remaining in the lowlands to reach the ground or floor of the valley.

foehn nose—A characteristic **deformation** of isobars on a surface **synoptic** weather **chart** during a well-developed **foehn**.
   The flow produces a **pressure ridge** over and just **windward** of the mountain range and a **lee trough** on the **downwind** side. The isobars bulge correspondingly, giving a noselike **profile** in the pressure ridge over the mountains. This configuration is prominent and characteristic of the foehn over the Alps, but over the Rocky Mountains the **ridge** pattern is less pronounced (perhaps because of the more complicated **upstream** orography or the presence of cold-air layers, which obscure the surface-pressure pattern, in the valleys of the western United States).

foehn pause—1. (Also called sturmpause.) A temporary cessation of the **foehn** at the ground, due to the formation or intrusion of a cold air layer that lifts the foehn above the valley floor.
   This often happens for a few hours in the early morning, just before **sunrise**, and brings about a rapid fall of **temperature** and increase of **relative humidity** that persists for an hour or more. Pause may also result from the advance or retreat of the **mountain wave** pattern that often accompanies the foehn.
   2. The boundary between **foehn air** and its surroundings.

foehn period—The duration of continuous **foehn** conditions at a given location.

foehn phase—One of the three stages to describe the development of the **foehn** in the Alps.
   They are 1) the **preliminary phase**, when cold air at the surface is separated from warm **dry air** aloft by a **subsidence inversion**; 2) the **anticyclonic phase**, when the warm air reaches valley stations as the result of the cold air retreating out toward the plain; and 3) the **stationary phase** or **cyclonic phase**, when the **foehn wall** forms and the **downslope wind** becomes appreciable.

foehn storm—A type of destructive **storm** that frequently occurs in October in the Bavarian Alps.
   After a **clear** dawn a **high foehn** develops. At about 8:00 A.M. **local time** clouds form on the mountains, **temperature** rises rapidly in the valleys, and at about 9:30 A.M. the **foehn** begins to blow from the south-southwest, bringing dirty, reddish-yellow **rain**. The storm ceases about 5:00 P.M. local time. During the storm the **atmospheric pressure** fluctuates rapidly.

foehn trough—The lee (or dynamic) **trough** formed in connection with the **foehn**.

foehn wall—The **leeward** edge of the **orographic** stratiform **cap cloud** as seen from the lee side of a mountain barrier, preceding or during a **foehn** or **chinook** event.
   The edge is generally abrupt and resembles a wall of **cloud** ("foehnwand" in German). This cloud often signifies the occurrence of **orographic precipitation**, especially **snowfall** in the cold **season**, over the peaks. *See* **foehn cloud**.

foehn wave—**Mountain waves, lee waves,** or trapped lee waves in the air stream flowing over the mountain barrier that occur in association with **foehn** conditions.
   The **Moazagotl** is one example of a foehn wave made visible by lee-wave clouds.

fog—Water droplets suspended in the **atmosphere** in the vicinity the earth's surface that affect **visibility**.
   According to international definition, fog reduces visibility below 1 km (0.62 miles). Fog differs from **cloud** only in that the base of fog is at the earth's surface while clouds are above the surface.

When composed of **ice crystals**, it is termed **ice fog**. Visibility reduction in fog depends on concentration of **cloud condensation nuclei** and the resulting distribution of **droplet** sizes. Patchy fog may also occur, particularly where air of different **temperature and moisture content** is interacting, which sometimes make these definitions difficult to apply in practice. Fogs of all types originate when the temperature and **dewpoint** of the air become identical (or nearly so). This may occur through cooling of the air to a little beyond its dewpoint (producing **advection fog**, **radiation fog** or **upslope fog**), or by adding moisture and thereby elevating the dewpoint (producing **steam fog** or **frontal fog**). Fog seldom forms when the dewpoint spread is greater than 4°F. According to U.S. weather observing practice, fog that hides less than 0.6 of the sky is called **ground fog**. If fog is so shallow that it is not an **obstruction to vision** at a height of 6 ft above the surface, it is called simply **shallow fog**. In **aviation weather observations** fog is encoded F, and ground fog GF. Fog is easily distinguished from **haze** by its higher **relative humidity** (near 100%, having physiologically appreciable dampness) and gray color. Haze does not contain activated droplets larger than the critical size according to Köhler theory. **Mist** may be considered an intermediate between fog and haze; its **particles** are smaller (a few μm maximum) in size, it has lower relative humidity than fog, and does not obstruct visibility to the same extent. There is no distinct line, however, between any of these categories. Near industrial areas, fog is often mixed with **smoke**, and this combination has been known as **smog**. However, fog droplets are usually absent in **photochemical smog**, which only contains unactivated haze droplets.

fog bank—Generally, a fairly well-defined mass of **fog** observed in the distance, most commonly at sea.

    This is not applied to patches of **shallow fog**.

fog chamber—Experimental apparatus for investigating the interaction between liquids and solid surfaces.

    The method involves the introduction of fine solid **particles** into a humidified reactor, followed by rapid cooling, to condense a layer of liquid onto the particles. Samples of the **droplets** can then be analyzed after a given time to quantify the extent of reaction that has occurred.

fog day—A day when **fog** is observed at an **observation** point.

fog deposit—The formation of an **ice** coating when **fog** contacts exposed surfaces that are at temperatures below **freezing**.

fog dispersal—The process whereby natural **fog** may be dissipated, as by artificial local surface heating or stirring, water spray, or **ice** seeding when the fog is supercooled.

fog dissipation—The natural process of **fog** disappearance, as by solar heating on **sunrise** or increase of **winds aloft** and vertical **mixing** with drier, warmer air.

fog drip—Water dripping to the ground from trees or other objects that have collected the moisture from drifting **fog**.

    In some instances the dripping is as heavy as light **rain**, as sometimes occurs among the redwood trees along the coast of northern California. During the foggy but almost rainless summers of California, fog drip prevents an excessive aridity in the coastal forests. As much as 0.05 in. of water, equivalent to a moderate **shower**, has been collected from a California fog in a single night. From the **aerodynamic** principles underlying the theory of **collection efficiency** it is clear that the needlelike leaves of conifers are better adapted to the removal of droplets from drifting fog than are the broad leaves of most deciduous trees. This characteristic may have been crucial in the evolutionary development of the redwoods in the limited areas wherein they thrive. Man-made fog-water collectors have been constructed in western Chile that utilize fog drip to provide water to the local population.

fog drop—An elementary **particle** of **fog**, physically the same as a **cloud drop**.

fog forest—The dense, rich, forest growth that is found at high or medium-high altitudes on tropical mountains.

    This occurs when the tropical rainforest penetrates altitudes of **cloud formation**; the **climate** found there is excessively moist and not too cold to prevent plant growth.

fog horizon—The top of a **fog** layer that is confined by a low-level **temperature inversion** in such a way as to give the appearance of the **horizon** when viewed from above against the sky.

    The **true horizon** is usually obscured by the fog in such instances. *Compare* **dust horizon**, **haze horizon**, **smoke horizon**.

fog patches—Small areas of **fog** of little vertical extent [normally greater than 2 but less than 6 m (6–20 ft)] that reduce **horizontal visibility**.

Stars and the sun may be seen through the patches.

fog point—Same as **dewpoint**.

fog scale—A classification of **fog** intensity based on its effectiveness in decreasing **horizontal visibility**. Such practice is not current in U.S. weather observing procedures.

fog wind—(Also called nebelwind.) The humid east **wind** that crosses the divide of the Andes east of Lake Titicaca and descends on the west in violent squalls; probably the same as **puelche**.

Owing to the **adiabatic** heating of the descending air, the clouds rapidly disintegrate, leaving a sharply defined **cloud** wall, similar to the **foehn wall**.

fogbow—Same as **cloudbow**.

föhn—*See* **foehn**.

folding—A term used to denote the **aliasing** of radar **echoes** in **range** or **velocity** as a result of sampling limitations.

folding frequency—(Also called Nyquist frequency.) The highest **frequency** that can be measured using discretely sampled data.

It is given by $n_f (\text{rad s}^{-1}) = \pi / \Delta t$, where n_f is the Nyquist frequency and t is the time increment between observations. Stated another way, a minimum of two data points is needed to define a **wave**, hence the **period** of the smallest wave (i.e., the period corresponding to the Nyquist frequency) that can be measured is $2\Delta t$. The word "folding" comes about because any frequencies that are higher than the Nyquist frequency in a continuous **signal** will be aliased or folded into lower frequencies when the signal is discretely sampled. To avoid this severe problem, the original signal must be filtered by **analog** or physical methods to remove all frequencies higher than the Nyquist frequency before the signal is sampled or digitized.

following wind—Generally, same as **tailwind**; specifically, a **wind** blowing abaft of the beam of a vessel; the opposite of **opposing wind**.

F_1-layer—The layer of the **ionosphere** that exists as an appendage on the lower part of the **F_2-layer** during the day.

It exhibits a distinct maximum of **free-electron density**, except at high latitudes in winter, when it is not detectable. Its **virtual height** ranges from 200 to 300 km, being lowest at around noon. The F_1-layer is formed by **absorption** of **solar radiation** in the extreme ultraviolet (EUV) range of wavelengths and is often roughly similar in shape to an idealized **Chapman layer**.

foot-candle—A unit of **illuminance** or **illumination** equal to one **lumen** per square foot.

This is the illuminance provided by a **light** source of **luminous** intensity one **candela** (formerly **candle**) at a distance of 1 ft, hence the name. Full **sunlight** with **zenith** sun produces an illuminance of about 10 000 foot-candles on a horizontal surface. Full moonlight provides an illuminance of about 0.02 foot-candles. Adequate illumination for reading is taken to be about 10 foot-candles; that for machine shop work is about 40 foot-candles. *Compare* **lux**.

foot-pound—A unit of **energy** equal to 1.356 joules.

footlambert—A unit of **luminance** (or **brightness**) equal to $1/\pi$ candles ft^{-2}.

In the United Kingdom this is also called the equivalent **foot-candle**.

footprint—1. In **micrometeorology**, the region of ground that affects a **turbulent flux** measurement above the surface.

Also known as the source-weight **distribution function**, the footprint can figuratively be described as the **ensemble average** field-of-view of a turbulent flux measurement. The footprint function is derived from a suitable **model** of **turbulent transport**. Alternately, an analogous footprint can be defined for **scalar** concentrations or for **radiative fluxes**.

2. In **remote sensing**, the instantaneous field-of-view of an airborne remote sensing instrument.

footprint modeling—A modeling approach to determine the relative importance of **upwind** source areas on the value of an atmospheric **variable** at a given height.

The **footprint** is dependent on atmospheric **stability** and the surface type and is different for scalars than for higher-order **turbulence** statistics (such as turbulent fluxes). Footprint models use either Eulerian or Lagrangian **dispersion** theory. The former is usually based on idealized assumptions (e.g., horizontal homogeneity, restriction of the vertical range), whereas the latter (**stochastic particle modeling**) may take into account more realistic situations but is computationally much more expensive.

forano—A **sea breeze** of Naples in Italy.

forbidden transitions—Spectroscopic transitions that have very low probability due to the violation of certain **selection rules**.

Such transitions are of low intensity and are not important in the **atmosphere** unless the molecule involved has a high abundance, for example, O_2. If the transition to the **ground state** is forbidden, such as $O(^1D)$ to $O(^3P)$, then the excited atom can have a lifetime long enough to engage in important reactions.

force of gravity—*See* **gravity**.

force-restore method—A method to approximate the **surface energy** budget that accounts for storage of **heat** in the top layer of soil.

In this **model** the ground is conceptually split into two layers, a relatively thin layer near the top with uniform **temperature**, and a deep soil layer also of uniform but different temperature. The thickness of the top layer is carefully chosen based on the region where a periodic temperature surface forcing has appreciable effect. The net result is that **flux** from the deep soil layer tends to restore the top layer, opposing any radiative forcing from the **atmosphere**. This method is an alternative to a multilayer soil model, which is computationally more expensive.

forced cloud—Small **cumulus** clouds that are everywhere negatively buoyant compared to the surrounding **environment**, but that exist because the inertia of the rising **thermals** feeding them from underneath is sufficient to penetrate the **lifting condensation level** (LCL).

Morphologically, these are often **cumulus humilis** clouds and are typically found at the top of the **convective boundary layer** during daytime over land, when a strong **temperature inversion** aloft prevents the clouds from growing deeper. *Compare* **active cloud, passive cloud**.

forced convection—*See* **convection**.

forced oscillation—*See* **oscillation**.

forced wave—Any **wave** that is required to fit irregularities at the boundary of a system or satisfy some impressed force within the system.

The forced wave will not in general be a characteristic mode of **oscillation** of the system. It cannot be exhibited independently unless the system admits no **free waves**. A homogeneous **incompressible fluid** bounded by two rigid surfaces is an example of such a system. The gravity **lee wave** in the **atmosphere** is an example of mixed forced and free waves. *See* **resonance, oscillation**.

forcing function—An inhomogeneous term appearing in a differential equation, that is, a term that can be a function of the **independent variable** but not of the **dependent variable** in the equation.

forecast—*See* **weather forecast**.

forecast amendment—An update to a forecast based on predetermined amendment criteria.

The criteria are designed to fit a particular forecast, that is, **wind speed, temperature**, or **precipitation**.

forecast area—A specific area over which the conditions stated in a **weather forecast** are valid.

forecast chart—A forecast of one or more specific meteorological **elements** for a specified time period and a specified surface or airspace, depicted graphically on a **chart**.

forecast district—*See* **district forecast**.

forecast error—A measure of the difference between a forecast and the corresponding **verification** from **analysis** or observations.

forecast evaluation—The process and practice of determining the quality and value of forecasts. *Compare* **forecast verification**.

forecast lead time—The length of time between the issuance of a forecast and the occurrence of the phenomena that were predicted.

forecast period—The time interval for which a forecast is made.

forecast-reversal test—A test used to evaluate the adequacy of a given method of **forecast verification**.

The same **verification** method is applied, simultaneously, to a given forecast and to a fabricated forecast of opposite conditions. Comparison of the verification scores gives an indication of the value of the verification system.

forecast updating—Revising a previous estimate of a forecast (i.e., **flood, streamflow, precipitation**, etc.) by using information acquired between the time of the estimate and the present time.

forecast verification—Any process for determining the **accuracy** of a **weather forecast** by comparing the predicted weather with the observed weather of the **forecast period**.

Principal purposes of forecast verification are to test forecasting skills and methods. *Compare* **forecast evaluation**.

forecaster—In meteorology, same as **weather forecaster**.

forecasting error—Difference between a forecast value and the observed value.

Forel scale—A **scale** of yellows, greens, and blues for recording the color of **seawater**, as seen against the white background of a **Secchi disk**.

forerunner—The **low-frequency** ocean **swell** that commonly arrives at the coast before the primary swell generated by a distant **storm**.

forest climate—*See* **humid climate**.

forest meteorology—Study and applications of weather and **climate** data to the reproduction, growth, and harvesting of forests.

forestry—The science of the development and care of forests; the management of growing timber.

Climatology plays an important part, including the study of the general livability of a forest to unfavorable weather conditions such as heavy **snow**, **frost**, **glaze**, strong **winds**, and **drought** (**fire weather**), and also the **analysis** of the **microclimate** of the forest itself and the effect of this on the reproduction, spread, and control of pests.

forked lightning—The common form of **cloud-to-ground discharge** always visually present to a greater or lesser degree that exhibits downward-directed branches from the main **lightning channel**.

In general, of the many branches of the **stepped leader**, only one is connected to the ground, defining the primary, bright **return stroke** path; the other incomplete channels decay after the ascent of the first return stroke. *Compare* **streak lightning**, **zigzag lightning**.

form drag—1. In flow over or past a solid obstacle, that part of the **momentum** or **kinetic energy** loss associated with the flow stagnation and related **pressure distribution** on the obstacle.

The form drag scales as the square of the **relative velocity**, but also depends sensitively on the form of the obstacle. *See* **gravity wave drag**.

2. *See* **drag**.

formaldehyde—Simplest member of aldehyde family, formula CH_2O.

Formaldehyde is very soluble in water; solutions of formaldehyde (formalin) are used in biological labs to preserve samples. Formaldehyde is fairly toxic; health concerns are associated with its emission from foam, cavity insulation, or new plastic materials such as upholstery, carpet, etc. Formaldehyde is present in the **atmosphere** as an intermediate in the **oxidation** of **methane** and many other **hydrocarbons**; its **photolysis** is a major source of **free radicals** in the atmosphere.

formation fluid—Fluid naturally present in a rock unit (formation).

formic acid—Simplest of the organic (carboxylic) acids, formula HCOOH, systematic name methanoic acid.

Formic acid is found in both the gas and aqueous phases in the **atmosphere**, and is thought to be formed as a product of **formaldehyde** oxidation.

Fortin barometer—An **adjustable cistern barometer**.

Forty Saints' storm—A southerly **gale** in Greece occurring a little before the **equinox** in March.

forward chaining—Processing method in a rule-based **expert system** that begins by identifying a chain of true **rules** that lead to a conclusion.

Compare **backward chaining**.

forward difference—*See* **finite-difference approximation**.

forward scatter—Scattering into directions forward of a plane through the **scattering** center and **orthogonal** to the incident direction.

Forward scatter makes up half of the total **Rayleigh scatter**. This fraction increases with increasing **particle** size at a given **wavelength**, so that large particle scatter is predominantly forward scatter. *See also* **asymmetry factor**.

forward scattering spectrometer probe—(Often abbreviated FSSP.) *See* **optical scattering probe**.

forward visibility—In aviation terminology, the **visibility** from an aircraft along its line of flight.

Forward visibility is important to aircraft crew members because it determines the maximum forward distance at which other aircraft can be seen. The companion terms for other aspects of visibility from aircraft are **vertical visibility** and **slant visibility**.

fossil fuels—Combustible materials derived from the long-term decomposition of organic matter.
Fossil fuels comprise coal, oil, and natural gas. The fuels are rich in **carbon** and high in **energy** content. Other minor constituents are sulfur and **nitrogen**.

fossil ice—**Ground ice** found in regions of **permafrost**, or in other regions where present-day temperatures are not low enough to have formed it; **ice** that was formed in the geologic past.

fossil permafrost—Same as **passive permafrost**.

fossil water—Water that was present when the rock was formed.
Compare **connate water, meteoric water**.

Foucault pendulum—A pendulum with precession, relative to an observer fixed to the earth, produced by the effect of **Coriolis acceleration**.
If set swinging at the North Pole in a given plane in space, its **linear momentum** perpendicular to the plane is zero, and it will continue to swing in this invariable plane while the earth rotates beneath it with a **period** of one day. At a latitude L, the **frequency** of precession is $1/T \sin L$ where T is one day.

four-dimensional data assimilation—The process of incorporating (assimilating) observations into a forecast model over a period of time to create an estimate of the atmospheric state.
See **data assimilation**.

four-dimensional variational assimilation—A **variational objective analysis** of the atmospheric state performed in the three space **dimensions** and the time dimension with a forecast model usually used to exactly constrain the solution in time.

Fourier analysis—The representation of physical or mathematical data by the use of the **Fourier series** or **Fourier integral**.

Fourier coefficients—*See* **Fourier series**.

Fourier integral—The representation of a function $f(x)$ for all values of x in terms of infinite integrals in the form

$$f(x) = \frac{1}{2\pi} \int_{-\infty}^{\infty} \int_{-\infty}^{\infty} f(t) \cos [u(t - x)] \, dt du.$$

See also **Fourier transform, Fourier series**.

Fourier series—The representation of a function $f(x)$ in an interval $(-L, L)$ by a series consisting of sines and cosines with a common **period** $2L$, in the form

$$f(x) = A_0 + \sum_{n=1}^{\infty} \left(A_n \cos \frac{n\pi x}{L} + B_n \sin \frac{n\pi x}{L} \right), \quad -L < x < L,$$

where the **Fourier coefficients** are defined as

$$A_0 = \frac{1}{2L} \int_{-L}^{L} f(x) dx,$$

$$A_n = \frac{1}{L} \int_{-L}^{L} f(x) \cos \frac{n\pi x}{L} dx,$$

and

$$B_n = \frac{1}{L} \int_{-L}^{L} f(x) \sin \frac{n\pi x}{L} dx.$$

When $f(x)$ is an even function, only the cosine terms appear; when $f(x)$ is odd, only the sine terms appear. The conditions on $f(x)$ guaranteeing **convergence** of the series are quite general, and the series may serve as a root-mean-square approximation even when it does not converge. If the function is defined on an infinite interval and is not periodic, it is represented by the **Fourier integral**. By either representation, the function is decomposed into periodic components the frequencies of which constitute the **spectrum** of the function. The Fourier series employs a **discrete spectrum** of **wavelengths** $2L/n$ ($n = 1, 2, \ldots$); the Fourier integral requires a **continuous spectrum**. *See also* **Fourier transform**.

309

Fourier transform—An analytical transformation of a function $f(x)$ obtained (if it exists) by multiplying the function by e^{-iux} and integrating over all x,

$$F(u) = \int_{-\infty}^{\infty} e^{-iux} f(x) dx \; (-\infty < u < \infty),$$

where u is the new **variable** of the transform $F(u)$ and $i^2 = -1$.

If the Fourier transform of a function is known, the function itself may be recovered by use of the inversion formula:

$$f(x) = \frac{1}{2\pi} \int_{-\infty}^{\infty} e^{iux} F(u) du \; (-\infty < x < \infty).$$

The Fourier transform has the same uses as the **Fourier series**: For example, the integrand $F(u)$ exp (iux) is a solution of a given **linear** equation, so that the integral sum of these solutions is the most general solution of the equation. When the variable u is complex, the Fourier transform is equivalent to the **Laplace transform**. *See also* **Fourier integral**, **spectral function**.

FOV—Abbreviation for **field of view**.

foveal vision—(Also called photopic vision.) That part of an **image** focused on the fovea, a region of the retina near the optical axis of the eye.

The fovea has a high density of the color-sensitive receptors known as cones, but it has no rods. Foveal vision permits much higher **resolution** than does **parafoveal vision** and is the **normal** mode of vision under daytime conditions. The vision in the fovea is exclusively photopic while in the parafoveal region it can be either photopic or **scotopic**. *See* **dark adaptation**.

fowan—A dry, scorching **wind** of Great Britain and the Isle of Man.

foyer—The place of origin of a group of **atmospherics**.

FPP scale—Abbreviation for **Fujita–Pearson scale**.

fps system—A system of measurement in which the fundamental units are foot, pound, and second.

fractal—An irregular geometric object that is self-similar to its substructure at any level of refinement.

The fractal dimension is a measure of the irregularity of the boundary of the object. Clouds have been shown to exhibit fractal characteristics with **cloud** perimeters having fractal dimension of about 1.35. The result follows from the observation that over a large size **range**, cloud area is proportional to cloud perimeter raised to a power slightly less than 1.5 (rather than proportional to the square of the perimeter, as is the case in Euclidean geometry); fractal dimension of the perimeter then is calculated as the ratio of 2 (the exponent in the Euclidean power law) to the exponent in the observed power relation.

Lovejoy, S., 1982: Area-perimeter relation for rain and cloud areas. *Science*, **216**, 185–187.

fractocumulus—(Obsolete.) *See* **fractus, cumulus fractus**.

fractostratus—(Obsolete.) *See* **fractus, stratus fractus**.

fractus—A **cloud species** in which the **cloud** elements are irregular but generally small in size, and that presents a ragged, shredded appearance, as if torn.

All of these characteristics change ceaselessly and often rapidly. Stratus fractus is distinguishable from cumulus fractus by its smaller vertical extent, darker color, and by the greater **dispersion** of its **particles**. Cumulus fractus actually looks like a ragged **cumulus** cloud. The species fractus occurs only in the genera mentioned above. *See* **cloud classification, scud**.

frank—Nautical term for a steady **wind**.

fraternal twin integrations—Two forecast integrations started from slightly different (or possibly identical) initial conditions, using two different **NWP** models, aimed at studying **forecast error** growth.

While **identical twin integrations** differ only due to initial value differences, fraternal twins also diverge due to differences in **model** formulation.

Fraunhofer diffraction theory—An approximate theory of near-forward **scattering** of electromagnetic waves by objects large compared with the **wavelength**.

This theory is independent of the composition of the objects and the state of **polarization** of the **illumination**.

Fraunhofer lines—Dark **absorption** lines in the **solar spectrum** seen against the bright continuum **spectrum** of the **photosphere**.

Associated with specific atomic **energy level** transitions, these absorption features occur as a consequence of **wavelength** variations in the atomic **opacity** and the outward decrease in photospheric **temperature** (after the German physicist Joseph Fraunhofer, 1787–1826).

frazil—(Or frazil crystals; also called needle ice.) **Ice crystals** that form in **supercooled water** that is too turbulent to permit **coagulation** into **sheet ice**.

This is most common in swiftly flowing streams, but is also found in a turbulent sea (where it is called **lolly ice**). It may accumulate as **anchor ice** on submerged objects obstructing the water flow.

frazil ice—1. An accumulation of **frazil** in a body of water.

2. The initial stage of **ice** formation in turbulent water.

Frazil ice consists of platelets or discs roughly 1 mm in diameter. These small platelets clump together to form **shuga**, and eventually (if sufficient open water area exists) form **pancake ice**. Frazil ice may form in open water leads and around the ice margins in the Arctic, but it is most common in the Antarctic. It may also form in turbulent rivers in winter, particularly in rapids.

freak wave—(Also called rogue wave.) An unexpectedly occurring **wave** of great height (and also **steepness**), often found to occur where waves meet opposing currents, such as in the **Agulhas Current** off South Africa.

Freak waves are known to cause serious damage to ships.

free acidity—The titratable acidity in the aqueous phase of a soil.

It may be expressed in milliequivalents per unit mass of soil or in other suitable units.

free air—Very generally, air that is not modified by local influence to an appreciable extent.
Compare **free atmosphere**.

free atmosphere—(Sometimes called **free air**.) That portion of the earth's **atmosphere**, above the **planetary boundary layer**, in which the effect of the earth's **surface friction** on the air motion is negligible, and in which the air is usually treated (dynamically) as an **ideal fluid**.

The base of the free atmosphere is usually taken as the **geostrophic wind level**.

free-board—The vertical distance between **normal** maximum operating level at a **reservoir** and the top of the dam, or between the **design flood** stage and the top of a levee.

Adequate free-board is employed as a safety interval and provides protection against overtopping by **wave** action, debris, etc.

free convection—*See* **convection**.

free convection scaling—Velocity and length scales that apply to the **atmospheric boundary layer** during conditions of a statically unstable **atmosphere** with vigorous **thermal** circulations and negligible shear-generation of **turbulence**, such as during a sunny day over land with **calm** winds.

The length **scale** is z_i, the **mixed-layer** depth, and the **velocity** scale is w^* the **Deardorff velocity**. A **free convection** timescale can be created as $t^* = z_i/w^*$. A free convection temperature scale $\theta^* = \overline{w'\theta}_s/z_i$ can be found using the surface kinematic **heat flux**, $\overline{w'\theta'}$. Similar scales can be generated for other variables such as **humidity**.

Stull, R. B., 1988: *An Introduction to Boundary Layer Meteorology*, 666 pp.

free convection scaling velocity—*See* **Deardorff velocity**.

free electron—An **electron** not bound to an atom or molecule.

In this sense a (negative) **beta particle** is a free electron.

free foehn—Same as **high foehn**.

free nappe—Term used in **hydraulics**; **nappe**, the underside of which is not in contact with the **overflow** structure and is at ambient **atmospheric pressure**.

free radical—A chemical species that possesses an unpaired **electron**.

These compounds, such as chlorine monoxide (ClO), **hydroxyl radical** (OH), and **nitric oxide** (NO), are characterized by a high chemical reactivity. Free radicals are often generated in the **atmosphere** from the **photodissociation** of more stable species. Their subsequent chemistry is central to such environmental issues as stratospheric **ozone** depletion and tropospheric **oxidant** production.

free space—A hypothetical medium devoid of all matter (i.e., a perfect vacuum).

No such medium exists except as an idealization. Even interstellar space is sparsely populated with atoms (mostly **hydrogen**), molecules, and **particles**.

free-stream Mach number—The **Mach number** of the total airframe (entire aircraft) as contrasted with the **local Mach number** of a section of the airframe.

free streamline—A **streamline** separating fluid in motion from fluid at rest.
Both **pressure** and speed are constant along a free streamline.

free surface—The upper surface of a layer of liquid at which the **pressure** on the liquid is equal to the external **atmospheric pressure**, assumed constant.
The existence of a free surface is expressed, in **hydrodynamics**, by the relation $Dp/Dt = 0$ on the free surface, where p is the total fluid pressure; this relation is often referred to as the free-surface condition and is a special case of a **dynamic boundary condition**.

free-surface conditions—Mathematical conditions on the **velocity** and **stress** at the material boundary of a fluid.
The first condition is that the tangential component of velocity is continuous across a material boundary separating a fluid and another medium. The second condition is that the difference between the values of the stress on two surface elements parallel to the boundary and immediately on either side of it is a normal force due wholly to **surface tension**.
Batchelor, G. K., 1967: *An Introduction to Fluid Dynamics*, Cambridge University Press, 148–151.

free turbulence—**Turbulence** in a free stream, not directly affected by the presence of a boundary.

free water—Liquid water within a mass of deposited **snow**.

free-water content—*See* **water content**.

free wave—Any **wave** not acted upon by any external force except for the initial force that created it; a wave solution satisfying a homogeneous **equation of motion** and homogeneous **boundary conditions**.
In a system with no impressed forces, a free wave has zero **amplitude** at the boundaries of the system. The **phase speed**, **wavelength**, etc., of the free wave or waves are characteristics of the system. A simple example of such a wave in meteorology is a **billow cloud** layer over level ground. In a steady-state solution, free waves have arbitrary amplitude. These may be specified by initial conditions to determine the solution completely. A free wave on a water surface is one created by a sudden impulse, thereafter influenced only by **friction**, the dimensions of the basin, and the dispersive character of the water medium it moves in. Most ocean **surface waves**, except **tidal waves**, are free waves. *Compare* **forced wave**; *see* **oscillation**.

freeze—1. *See* **freezing**.
2. The condition that exists when, over a widespread area, the **surface temperature** of the air remains below **freezing** (0°C) for a sufficient time to constitute the characteristic feature of the weather.
This is a general term, and the time period necessary is usually considered to be two or more days; only the hardiest herbaceous crops survive. It differs from a **dry freeze** or **black frost**, for these terms are usually used to describe purely local freezing due to **chilling** of the surface air by rapid **radiation** from a restricted portion of the earth. *Compare* **light freeze**, **hard freeze**, **killing freeze**.

freeze-free period—The period, usually expressed in days, between the last occurrence of **freezing** temperatures (0°C) in the spring and the first occurrence in the autumn.

freeze probability curve—Statistical curve relating **probability** of first and last **freeze** (0°C) occurrence in the fall and spring to specific dates; a tool to assess freeze risk.
May be applied to other critical threshold temperatures such as −1°C, −2°C, −4°C, etc.

freeze–thaw pattern—Pattern of alternating temperatures from above to below **freezing** causing freezing and thawing of upper soil layer, often resulting in damage to plant root systems, particularly small winter grains, most commonly observed with soils near **saturation** in surface layers.

freeze-up—The seasonal formation of continuous **ice cover** on a body of water.

freezing—1. The **phase** transition of a substance passing from the liquid to the solid state; solidification; the opposite of **fusion**.
In meteorology, this almost invariably applies to the freezing of water. The **phase change** from the gaseous to the solid state is **deposition**. Like **condensation**, the freezing of water involves the process of **nucleation**. *See* **ice point**, **freezing point**, **true freezing point**, **melting point**.
2. Said of an **environment** when its **temperature** is equal to or less than 0°C (32°F).
See **freeze**.

freezing degree-day—*See* **degree-day**.

freezing drizzle—Drizzle that falls in liquid form but freezes upon impact to form a coating of **glaze**.
In U.S. **aviation weather observations**, this **hydrometeor** is encoded ZL. The physical cause of this phenomenon is the same as that for **freezing rain**.

freezing fog—A **fog** the droplets of which **freeze** upon contact with exposed objects and form a coating of **rime** and/or **glaze**.

freezing index—(Also called "coldness sun.") As used by the U. S. Army Corps of Engineers, the number of Fahrenheit **degree-days** (above and below 32°F) between the highest and lowest points on the cumulative degree-days **time curve** for one **freezing season**.
For a critical review of the topic see Sakari Tuhkanen (1980). "The freezing index is used as a measure of the combined duration and magnitude of below **freezing** temperatures occurring during any given freezing season. The index determined for air temperatures at 4.5 feet above the ground is commonly designated as the **air freezing index**, while that determined for temperatures immediately below a surface is known as the surface freezing-index." (from Glossary of Arctic and Subarctic Terms 1955). *Compare* **thawing index**.

> Tuhkanen, S., 1980: Climatic parameters and indices in plant geography. *Acta phytogeographica Suecica*, **67**, Svenska Vaxgeografiska Sallekapet, Uppsala, 105 pp.
> Arctic, Desert, Tropic Information Center (ADTIC) Research Studies Institute, 1955: *Glossary of Arctic and Subarctic Terms*, ADTIC Pub. A-105, Maxwell AFB, AL, 90 pp.

freezing level—Commonly, and in aviation terminology, the lowest **altitude** in the **atmosphere**, over a given location, at which the **air temperature** is 0°C; the height of the 0°C constant-temperature surface.
This simple concept may become slightly complicated by the existence of one or more "above-freezing layers" formed by **temperature** inversions at altitudes higher than the above-defined freezing level. In **cloud physics** terminology, this is more accurately termed the **melting level**, for melting of **ice** always occurs very near 0°C, but liquid **cloud drops** may remain supercooled to much colder temperatures. *See* **icing level**.

freezing-level chart—A **synoptic chart** showing the height of the 0°C constant-temperature surface by means of **contour lines**.

freezing nucleus—Any particle immersed within **supercooled water**, initiating the growth of an **ice crystal** to be compared with **particles** nucleating directly from the **vapor** phase (**deposition nucleus**).
Similar particles may nucleate at somewhat different temperatures (a few degrees) depending on the process. Observations of natural freezing nuclei indicate that there is normally present in the **atmosphere** a large variety of such particles with varying activation temperatures (temperatures at which they become effective nucleators). Certain bacteria from vegetation (pseudomonas syringae) nucleate **ice** at temperatures as high as −2°C; mineral particles (e.g., clays: kaolinite and montmorillonite) at −10° to −20°C; artificial nuclei (e.g., **silver iodide**, lead iodide, and metaldehyde), as **smoke**, can be found to nucleate at intermediate temperatures, i.e., −5° to −10°C. The origin, distribution, and composition of these particles is highly variable; some are composed of a mixture with a **hygroscopic** component that dilutes prior to **nucleation** of the water by the freezing nucleus.

freezing point—(Also called apparent freezing point.) The **temperature** at which a liquid solidifies under any given set of conditions.
It may or may not be the same as the **melting point** or the more rigidly defined **true freezing point** or (for water) **ice point**. It is not an equilibrium property of a substance; it applies to the liquid **phase** only. The freezing point is somewhat dependent upon the purity of the liquid; the volume and shape of the liquid mass; the availability of **freezing nuclei**; and the **pressure** acting upon the liquid. The freezing point is a **colligative property** of a solution and becomes proportionately lower with an increasing amount of dissolved matter. Therefore, since natural water almost invariably contains some solutes, its freezing point is found to be slightly below 0°C. For example, bulk samples of normal **seawater** freeze at about −1.9°C (28.6°F).

freezing precipitation—Any form of liquid **precipitation** that freezes upon impact with the ground or exposed objects, that is, **freezing rain**, **freezing drizzle**, or **freezing fog**.
Compare **frozen precipitation**.

freezing rain—Rain that falls in liquid form but freezes upon impact to form a coating of **glaze** upon the ground and on exposed objects.
In **aviation weather observations**, this **hydrometeor** is encoded ZR. While the **temperature** of the ground surface and glazed objects is typically near or below **freezing** (0°C or 32°F), it is necessary that the water drops be supercooled before striking. Freezing rain can sometimes occur

on surfaces exposed to the air (such as tree limbs) with air temperatures slightly above freezing in strong winds. Local evaporational cooling may result in freezing. Freezing rain frequently occurs, therefore, as a transient condition between the occurrence of **rain** and **ice pellets** (**sleet**). When encountered by an aircraft in flight, freezing rain can cause a dangerous **accretion** of **clear icing**.

freezing season—As used by the U.S. Army Corps of Engineers, the period of time between the highest point and the succeeding lowest point on the **time curve** of cumulative **degree-days** above and below 32°F; the opposite of **thawing season**.

A less rigorous, but more commonly used, definition is the number of days or months between the first day of fall or winter and the last day of the same winter or the following spring on which the **air temperature** is below 0°C.

freezing spray—Sea spray transported through the air at temperatures below 0°C.

Fremantle Doctor—Local name for the strong **sea breeze** that brings significant lowering of summertime temperatures to the shores of southwestern Australia.

freon—Registered trade name for the **chlorofluorocarbons** (**CFCs**).

frequency—1. The rate of recurrence of any periodic phenomenon, often associated with waves of all kinds.

Without qualification frequency often means temporal frequency, the rate of recurrence of a time-varying function, but could mean spatial frequency, the rate of recurrence of a space-varying function. Spatial frequency is the reciprocal of the repeat distance (sometimes the **wavelength**). The **dimensions** of (temporal) frequency are inverse time. A common unit for frequency is **cycle per second**, formerly abbreviated cps, but superseded by **hertz**, abbreviated as Hz. The symbol ν is often used for frequency but f is common in engineering. Period is inverse frequency. Related to frequency, and applied especially to sinusoidally varying quantities, is angular or **circular frequency**, often denoted by $\omega = 2\pi\nu$, with units radians per unit time (e.g., radians per second).

2. In **statistics**, the number of times a specified event occurs in a given series of observations; for example, the number of rainy days observed at a particular **station** during a certain period of time.

In many types of studies (hydrometeorological, especially) the reciprocal of frequency, the **recurrence interval**, is used.

frequency agility—The use of multiple **transmitter** frequencies on successive pulses in a **radar** system to reduce the **variance** of a measurement by taking advantage of the greater **statistical independence** of measurements made at different frequencies.

frequency analysis—Based upon the available record, frequency analysis involves the choice of a **frequency distribution** to describe the phenomena of interest and the estimation of the parameters of that distribution, so as to obtain a description of the relationship between different values of a **variable** and their **exceedance probability**.

frequency band—A set of frequencies lying within a definite **range**.
See **radar frequency bands**, **radio frequency band**.

frequency coding—A type of **modulation** applied to a **radar** in which the **carrier frequency** is systematically altered within a transmitted **pulse** to achieve higher **range resolution**.
See **pulse compression**.

frequency distribution—1. A curve the coordinates of which are the values of the **variable** and the **frequency** of occurrence. It is often presented as a **histogram**.
Compare **probability distribution**.

2. (Also called frequency curve.) The relationship between the **probability** that a threshold is exceeded and the threshold.

In a graphical representation, the curve is often plotted with **frequency** or probability on the vertical y axis and the corresponding **rainfall** or **flood** flow values on the horizontal x axis.

frequency-domain averaging—A technique for **smoothing** the **statistical** fluctuations in power spectra, thus improving the detectability of signals in **noise**.

Two essentially equivalent techniques can be used on power spectra derived from MN data points: 1) M spectra each containing N spectral **power density** points are averaged together in each **frequency** bin, producing a single averaged N-point **spectrum**. 2) The spectral points in a single MN-point spectrum are averaged together in groups of M points, also producing a single averaged N-point power. Such techniques are widely used for **Doppler spectrum** measurements in **wind profilers** and **MST radars**.

frequency equation—*See* **dispersion relation**.

frequency function—Same as **probability density function**.

frequency meter—An instrument for measuring the **frequency** of **oscillation** of a periodic **signal**.

frequency modulation—Changing the **frequency** of a **carrier wave** by some means so that the **resultant** waveform transmits a **signal** conveying information.

For example, if the carrier wave with **angular frequency** ω is $\cos(\omega t)$, the corresponding frequency-modulated signal is

$$f(t) = a \cos[\omega t + f(t)],$$

where a is a constant and $|df/dt| \ll \omega$. Frequency modulation is used in some **continuous-wave radars** to permit range determination, but it is not generally used in **weather radars**. *Compare* **amplitude modulation**.

frequency response—A rating that indicates the **frequency** range over which an instrument will respond uniformly or within specified limits.

The frequency response is an important **parameter** in evaluating the dynamic response of an instrument.

fresh—Descriptive of air that is stimulating and refreshing.

See **fresh breeze**, **fresh gale**.

fresh breeze—In the **Beaufort wind scale**, a **wind** with a speed from 17 to 21 knots (19–24 mph) or Beaufort Number 5 (force 5).

fresh gale—In the **Beaufort wind scale**, a **wind** with a speed from 34 to 40 knots (39–46 mph) or Beaufort Number 8 (force 8).

freshet—1. The annual spring rise of **streams** in cold climates as a result of melting **snow**.

2. A **flood** resulting from either **rain** or melting **snow**.

In this sense it is usually applied only to small **streams** and to **floods** of minor severity.

3. A small **freshwater** stream.

freshwater—Water that contains less than 1000 mg l^{-1} of dissolved solids.

Fresnel reflection—In **radar**, a **scattering** mechanism proposed to explain certain kinds of **clear-air echoes** observed by **UHF** and **VHF** radars.

Such echoes are observed by vertically pointing **radars** operating at wavelengths of about 1 m and longer. They are in the form of thin, horizontal layers that exhibit strong aspect sensitivity, in the sense that the **reflectivity** for a vertical **beam** is greater than that for off-vertical beams. They are thought to be explained by partial reflections from thin layers containing sharp vertical gradients of **refractivity**. The layers have vertical extents that are comparable to or less than a **wavelength** and horizontal extents that are as large as the width of the first **Fresnel zone**, namely, $(z\lambda)^{1/2}$, where z is the **altitude** and λ is the radar wavelength. Echoes explained by Fresnel reflection have longer **coherence** times than those explained by **Bragg scattering** from beam-filling echoes and are more in the nature of specular reflections. Sometimes a distinction is made between Fresnel reflection and Fresnel **scattering**. The term scattering is used if there are several or many thin reflective layers in the **pulse volume**; **reflection** is reserved for the situation of only one layer in the pulse volume.

Röttger, J., and M. F. Larsen, 1990: UHF/VHF radar techniques for atmospheric research and wind profiler applications. *Radar in Meteorology*, D. Atlas, Ed., American Meteorological Society, 241–242.

Fresnel zone—A circular zone centered about the direct path between a **transmitter** and a **receiver** (or between **radar** antenna and **target**), so defined that the distance along a path from transmitter to receiver through a point within the zone has a path length equal to some value between $[L + n\lambda/2]$ and $[L + (n + 1)\lambda/2]$, where L is the length of the direct path, λ is the **wavelength**, and n is a positive integer or zero.

The first Fresnel zone is the zone defined by $n = 0$ and containing the minimum path length. Fresnel zones are a useful concept for analyzing the **interference** between the direct **signal** on a propagation path and signals reflected by objects that are displaced from the direct path. Thus, for a given path, reflected radio **energy** arriving at the receiver from any point will have a **phase** determined by the particular Fresnel zone in which the point is located. Reflected signals from zones defined by even values of n will interfere constructively with the direct signal; those from zones defined by odd values of n will interfere destructively.

friagem—(Also spelled vriajem.) A period of cold weather in the middle and upper parts of the Amazon Valley and in eastern Bolivia.

Such periods occur during the **dry season** in the Southern Hemisphere winter.

friction—(Or frictional force.) The mechanical resistive force offered by one medium or body to the relative motion of another medium or body in contact with the first.

Solid bodies in relative motion display sliding and rolling friction that depend upon the forces pressing the bodies together, but that are nearly independent of the shapes or relative speeds of the bodies. The **resistance** of fluids to the relative motion of a solid body is, however, dependent upon the relative speed and the shape of the body, as well as upon the character of the flow itself (*see* **drag**). Turbulent resistance of the earth on the **atmosphere** (**surface friction**) has been represented as proportional to both the first and second power of the speed of the low-level winds. The first representation leads to the cross-isobar frictional **wind** and to an estimate of the low-level **frictional convergence**, while the second representation has been employed in studies of **momentum** and **energy** abstraction from the atmosphere at the earth's surface. The word "friction" is often inappropriately used in place of turbulent drag when describing the differences between the **atmospheric boundary layer** and the rest of the **troposphere**.

friction coefficient—Same as **skin-friction coefficient**.

friction head—The **head** that is lost by fluid flowing in a **stream** or conduit due to **friction** per unit weight of fluid.

friction layer—An alternative, somewhat inappropriate, name for the **atmospheric boundary layer**.

In the real **atmosphere**, turbulent **drag**, rather than molecular **friction**, is responsible for reducing **wind speeds** in the **boundary layer**. Another inappropriate name for the friction layer is the **Ekman layer**, where **turbulence** is modeled similar to molecular friction except that an **eddy viscosity** is used in place of a **molecular viscosity**.

friction loss—The irrecoverable conversion of flow **energy** into **thermal energy** due to **shear** stress and the generation of turbulent vortices.

friction velocity—A reference **wind** velocity defined by the relation

$$u_* = (|\tau/\rho|)^{1/2},$$

where τ is the **Reynolds stress**, ρ the **density**, and u_* friction velocity.

Using the surface kinematic **momentum** fluxes in the x and y directions $(\overline{u'w'}_s, \overline{v'w'}_s)$ to represent surface **stress**, the friction velocity can be written as

$$u_* = [(\overline{u'w'}_s)^2 + (\overline{u'w'}_s)^2]^{1/4}.$$

It is usually applied to motion near the ground where the **shearing stress** is often assumed to be independent of height and approximately proportional to the square of the **mean velocity**. The friction velocity is, therefore, exactly the **velocity** for which this square law would be valid.

Sutton, O. G., 1953: *Micrometeorology*, p. 76.

frictional convergence—The coming together (usually horizontally) of air due to **drag** against the surface being different at different locations.

A classic example is the **wind** blowing toward shore from over the ocean. The ocean is relatively smooth with little **frictional drag**, while the land, with trees and buildings, is rougher and has more drag. When the wind from the ocean reaches land, it will slow down due to the increased drag. Thus, air will be flowing toward the shore from the ocean faster than it will leave the shore over land, causing horizontal **convergence**. Mass continuity requires updrafts in these regions, thereby leading to enhanced **cloudiness** and possibly **precipitation**. *See* **aerodynamic roughness length**.

frictional dissipation—*See* **dissipation**.

frictional divergence—Opposite of **frictional convergence**; namely, the spreading apart of air horizontally caused by a change in air **drag** against the surface.

frictional drag—1. The molecular (viscous) retarding force on an object, such as air, as it moves across the earth's surface.

In the **atmospheric boundary layer**, turbulent **drag** dominates by many orders of magnitude over molecular frictional drag.

2. The retarding force that an object such as a **raindrop** or airplane encounters as it moves through the **atmosphere**.

frictional force—Same as **friction**; *see also* **drag**, **drag coefficient**, **skin-friction coefficient**.

frictional secondary flow—Same as **secondary flow**.

frictional skin drag—A **viscous force** that tends to retard the **velocity** of air flowing along a surface.
The total **drag** against the air is the sum of **form drag**, **skin drag**, and **wave drag**.

frictional torque—The **torque** exerted by the force of **friction**.
In meteorology, this term usually refers to the effect of **surface friction**. The eastward component of the frictional torque is of primary interest because, together with the eastward component of the **mountain torque**, it constitutes the only significant mechanism by which the **atmosphere** can gain or lose **absolute angular momentum** about the earth's axis. This torque may be expressed by the integral

$$-a \int_s \tau_x \cos \phi\, ds,$$

where a is the earth's radius, ϕ is the latitude, τ_x is the eastward component of the **frictional force** per unit area, and ds is an areal element of the earth's surface. The frictional torque serves to impart the earth's **angular momentum** to the air, thus decreasing or increasing the (relative) **wind speed**.

friendly ice—From the point of view of the submariner, an **ice** canopy containing many large skylights or other features that permit a submarine to surface.
There must be more than 10 such features per 56 km (30 nautical miles) along the submarine's track.

Frigid Zone—(Obsolete.) Based on an ancient **climatic classification** scheme using solar illumination geometry, the region poleward of the Arctic or Antarctic Circles (66°33′N and 66°33′S, respectively).
Compare **Temperate Zone, Torrid Zone.**

frigorimeter—An instrument designed to measure the physiological **cooling power** in milligram-calories. It consists of a blackened copper sphere, diameter 7.5 cm, the surface of which is maintained electrically at 36.5°C against the **heat** losses due to all meteorological conditions of the **ambient air**.

fringe region—(Also called spray region.) The upper portion of the **exosphere**, where the **cone of escape** equals or exceeds 180°.
In this region the individual atoms have so little chance of collision that they essentially travel in free orbits, subject to the earth's **gravitation**, at speeds imparted by the last collision. *See* **critical level of escape.**

frog storm—(Rare; also called whippoorwill storm.) The first stormy weather in spring after a warm period.

front—1. In meteorology, generally, the **interface** or transition zone between two **air masses** of different **density**.
Since the **temperature** distribution is the most important regulator of atmospheric density, a front almost invariably separates air masses of different temperature. Along with the basic density criterion and the common temperature criterion, many other features may distinguish a front, such as a pressure **trough**, a change in **wind direction**, a moisture discontinuity, and certain characteristic **cloud** and **precipitation** forms. The term front is used ambiguously for 1) frontal zone, the three-dimensional zone or layer of large horizontal density gradient, bounded by 2) frontal surfaces across which the horizontal density gradient is discontinuous (frontal surface usually refers specifically to the warmer side of the frontal zone); and 3) **surface front**, the line of intersection of a frontal surface or frontal zone with the earth's surface or, less frequently, with a specified **constant-pressure surface**. Types of front include **polar front, arctic front, cold front, warm front,** and **occluded front.** *See also* **anafront, katafront, intertropical front, secondary front, upper front.**

2. *See* **wave front.**

front-passage fog—*See* **frontal fog.**

frontal analysis—Analysis of the structure and development of a region in the **atmosphere** in terms of **fronts** and **air masses**.

frontal contour—The line of intersection of a **front** (frontal surface) with a specified surface in the **atmosphere**, usually a **constant-pressure surface**.
With respect to only one surface, this line is usually called, simply, the front. A **frontal-contour chart** is a single **synoptic chart** on which are drawn all available frontal contours, thereby clearly showing the three-dimensional configuration of all fronts within the area of the chart. *Compare* **frontal profile.**

frontal-contour chart—*See* **frontal contour.**

frontal cyclone—In general, any **cyclone** associated with a **front**; often used synonymously with **wave cyclone** or with **extratropical cyclone** (as opposed to **tropical cyclone**, which is nonfrontal).

frontal fog—Fog associated with frontal zones and frontal passages.
   It is usually divided into three types: warm-front prefrontal fog; cold-front **post-frontal fog**; and frontal-passage fog. The first two types are a result of **rain** falling into cold **stable air** and raising the **dewpoint temperature**. Frontal-passage fog can result from the "mixing of warm and cold **air masses** in the frontal zone" or by "sudden cooling of air over moist ground."
   Byers, H. R., 1944: *General Meteorology*, 518–519.
   George, J. J., 1951: *Compendium of Meteorology*, 1183–1184, 1187.

frontal inversion—A **temperature inversion** (**temperature** rising with height) in the **atmosphere**, encountered upon vertical ascent through a sloping **front** (or frontal zone).

frontal lifting—The forced ascent of the warmer, less dense air at and near a **front**, occurring whenever the relative velocities of the two **air masses** are such that they converge at the front.
   See **convection**.

frontal model—A simplified representation of the characteristic distribution of **temperature**, **winds**, **relative humidity**, **cloudiness**, and **rainfall** associated with typical cold and warm **fronts** in the **atmosphere**.

frontal occlusion—*See* **occluded front**.

frontal passage—(Acronym fropa.) The passage of a **front** over a point on the earth's surface; or, the transit of an aircraft through a **frontal zone**.

frontal-passage fog—*See* **frontal fog**.

frontal precipitation—Any **precipitation** attributable to the action of a **front**; used mainly to distinguish this type from **airmass precipitation** and **orographic precipitation**.

frontal profile—The outline of a **front** as seen on a vertical **cross section** oriented normal to the **frontal surface**.
   See **profile**; *compare* **frontal contour**.

frontal strip—(Rare.) The presentation of a **front** on a **synoptic chart** as a **frontal zone**; that is, two lines, rather than a single line, are drawn to represent the boundaries of the zone.

frontal surface—*See* **front**.

frontal system—Simply, a system of fronts as they appear on a **synoptic chart**.
   This is used for 1) a continuous **front** and its characteristics along its entire extent, including its warm, cold, stationary, and occluded sectors, its variations of **intensity**, and any frontal cyclones along it; and 2) the orientation and nature of the fronts within the **circulation** of a **frontal cyclone**.

frontal theory—Theory of the formation and **development** of **air masses**, **cold fronts**, and **warm fronts** in the **atmosphere**, and of the formation and **development** of **extratropical** depressions in relation to air masses and **fronts**.

frontal thunderstorm—An individual **thunderstorm** the initiation of which resulted from rising motion associated with a **front**, or a thunderstorm within a convective system generated and organized by frontal rising motion.

frontal wave—A horizontal wavelike **deformation** of a **front** in the lower levels, commonly associated with a maximum of **cyclonic circulation** in the adjacent flow.
   It may develop into a **wave cyclone**.

frontal zone—*See* **front**.

frontogenesis—1. The initial formation of a **front** or frontal zone.
   2. In general, an increase in the horizontal **gradient** of an **airmass** property, principally **density**, and the development of the accompanying features of the **wind field** that typify a **front**.

frontogenetical function—A kinematic measure of the tendency of the flow in an **air mass** to increase the horizontal **gradient** of a **conservative property** θ, defined by the equation

$$F = \frac{D}{Dt}|\nabla_H \theta|,$$

where ∇_H is the horizontal **del operator**.

When θ is identified with the **potential temperature**, this function measures the **frontogenesis** in an air mass.

Pettersen, S., 1956: *Weather Analysis and Forecasting*, 2d ed., Vol. I, 200–201.

frontolysis—1. The **dissipation** of a **front** or frontal zone.

2. In general, a decrease in the horizontal **gradient** of an **air mass** property, principally **density**, and the **dissipation** of the accompanying features of the **wind field**.

fropa—An acronym, and formerly the aviation weather communications code word, for **frontal passage**.

frost—1. The fuzzy layer of **ice crystals** on a cold object, such as a window or bridge, that forms by direct **deposition** of **water vapor** to solid **ice**.

2. The condition that exists when the **temperature** of the earth's surface and earthbound objects fall below **freezing**.

Depending upon the actual values of ambient-air temperature, **dewpoint**, and the **temperature** attained by surface objects, frost may occur in a variety of forms. These include a general **freeze**, **hoarfrost** (or **white frost**), and **dry freeze** (or **black frost**). If a frost period is sufficiently severe to end the **growing season** (or delay its beginning), it is commonly referred to as a **killing frost**. *See* **frost day**, **ground frost**.

3. *See* **frozen ground**.

4. Same as **hoarfrost**.
Compare **rime**.

frost action—In general, cycles of **freezing** and thawing of water contained in natural or man-made materials.

This is especially applied to the disruptive effects of this action. In geology, two basic types of frost action are described: 1) **congelifraction**, the shattering or splitting of rock material; and 2) **congeliturbation**, the churning, **heaving**, and thrusting of soil material.

frost belt—(Or frost dam.) A ditch constructed to assist the early and rapid **freezing** of the soil in order to block the seepage of **subsurface flow** from entering a critical area.

frost blister—A seasonal **frost mound** produced through doming of seasonally **frozen ground** by a subsurface accumulation of water under high hydraulic potential during progressive **freezing** of the **active layer**.

(Glossary of Permafrost and Related Ground-Ice Terms, National Research Council of Canada, NRCC 27952, Technical Memorandum No. 142, 1988.)

frost boil—1. An accumulation of water and mud released from **ground ice** by spring thawing.

2. A small mound of fresh soil material, formed by **frost action**.

(Glossary of Permafrost and Related Ground-Ice Terms, National Research Council of Canada, NRCC 27952, Technical Memorandum No. 142, 1988.)

frost churning—Same as **congeliturbation**.

frost climate—The coldest **temperature province** in C. W. Thornthwaite's 1931 **climatic classification**. It is the **climate** of the **ice cap** regions of the earth, that is, those regions perennially covered with **snow** and **ice**.

It is equivalent to the more commonly used term **perpetual frost climate**, to the colder of Köppen's (1918) **polar climates**, and to Nordenskjöld's (1928) **high arctic climate**.

Thornthwaite, C. W., 1931: The climates of North America according to a new classification. *Geogr. Rev.*, **21**, 633–655.

Köppen, W. P., 1918: Klassifikation der Klimate nach Temperatur, Niederschlag und Jahreslauf. *Petermanns Geog. Mitt.*, **64**, 193–203; 243–248.

Nordenskjöld, O., and L. Meeking, 1928: *The Geography of the Polar Regions*, American Geographical Society, Special Pub. No. 8, New York.

frost dam—Same as **frost belt**.

frost damage—Physiological damage to plants and plant tissues during occurrences of **frost** with temperatures generally below 3°C.

frost day—An **observational day** on which **frost** occurs; one of a family of climatic indicators (e.g., **thunderstorm day**, **rain day**).

The definition is somewhat arbitrary, depending upon the accepted criteria for a frost observation. Thus, it may be 1) a day on which the minimum **air temperature** in the **thermometer shelter** falls below 0°C (32°F); 2) a day on which a deposit of **white frost** is observed on the ground; 3) in British usage, a day on which the **minimum temperature** at the level of the ground or on the tops of low, close-growing vegetation falls to −0.9°C (30.4°F) or below (also called a "day with ground frost"); and perhaps others. The present trend is to drop such terms in favor of something less ambiguous, such as "day with minimum temperature below 0°C (32°F)."

frost depth—Depth into soil that **frost** has penetrated at any given time.
See **frost line**.

frost fan—*See* **frost protection**.

frost feathers—Same as **ice feathers**.

frost flakes—Same as **ice fog**.

frost flowers—Same as **ice flowers**.

frost fog—Same as **ice fog**.

frost-free season—The period, usually expressed in days, between the last observed occurrence of **frost** in the spring and the first observed occurrence of frost in the autumn.

frost hazard—The risk of damage by **frost**.
It may be expressed as the **probability** or **frequency** of **killing frost** on different dates during the **growing season**, or as the distribution of dates of the last killing frost of spring or the first of autumn. A strict application of the concept would take into account the actual species or group of plants that might potentially be killed, as different plants sustain **frost damage** at different temperatures. **Wind-chill factors** should also be considered.

frost haze—Same as ice-crystal **haze**.

frost heaving—The upward or outward movement of the ground surface (or objects on, or in, the ground) caused by the formation of **ice** in the soil.
(Glossary of Permafrost and Related Ground-Ice Terms, National Research Council of Canada, NRCC 27952, Technical Memorandum No. 142, 1988.)

frost hollow—A local bowl-shaped region or depression in the surface in which, in suitable conditions, cold air accumulates during the night as the result of cold **air drainage** called **katabatic wind**.
Such regions are subject to a greater incidence of **frost**, and to more severe frosts, than are the surrounding areas of nonconcave shape. *See* **frost pocket**.

frost in the air—Same as **ice crystals**.

frost line—Maximum depth of **frozen ground** during the winter.
The term may refer to an individual winter, to an average over a number of years, or to the greatest depth since observations began. The frost line varies with the nature of soil and the protection afforded by vegetal ground cover and **snow cover**, as well as with the amount of seasonal cooling. *Compare* **frost table**, **permafrost table**.

frost mist—1. A very light **ice fog**, that is, same as **ice-crystal haze**, **arctic mist**.
2. See **ice crystal**.

frost mound—(Also called ice mound, ground ice mound.) A conical mound on a land surface, caused by the **freezing** of water in the ground.
It is a product of **frost heaving**, but is unusual in that it requires a great concentration of water in a relatively small subsurface volume. Usually a frost mound is of seasonal duration. *Compare* **pingo**, **earth hummock**, **ground ice**; *see* **frost blister**.

frost plants—Same as **cryoplankton**.

frost pocket—Common agricultural/gardening term for **frost hollow**.

frost point—*See* **dewpoint**.

frost point hygrometer—An instrument similar to the **dewpoint hygrometer**, but especially suited for low **frost points**.
The strong cooling necessary for these measurements is provided either by multistage Peltier devices or by evaporating low-boiling-point (cryogenic) fluids like liquid **nitrogen**. Using this technique, frost points down to −100°C can be measured. Electrical **resistance** wires are used to provide heating.

frost-point technique—Same as **8D technique**.

frost protection—Methods of a passive or active nature that serve to reduce the damage to plants during a **frost** period.

Passive methods include variety selection, site selection, and planting date. Active methods include heaters, fans, **flooding**, sprinkling, **windbreaks**, mulching, etc.

frost ring—A false annual growth ring in the trunk of a tree, due to out-of-season defoliation by **frost** and subsequent regrowth of foliage.

A frost ring is identified by a thin light ring within the annual growth ring.

frost riving—Same as **congelifraction**.

frost smoke—1. (Sometimes called barber.) A rare type of **fog** formed in the same manner as a **steam fog**, but at colder temperatures so that it is composed of **ice** particles instead of water droplets.

Thus, it is a type of **ice fog**.

2. Same as **steam fog**.
Compare **ice crystals**.

frost snow—Same as **ice crystals**.

frost splitting—Same as **congelifraction**.

frost table—An irregular surface in the ground that, at any given time, represents the penetration of thawing into seasonally **frozen ground**.

In regions of **permafrost**, when the **active layer** is thawed completely, this coincides with the **permafrost table**. *Compare* **frost line**.

frost zone—The layer of ground subject to seasonal **freezing**.

In regions of **permafrost**, this corresponds to the **active layer**.

frostburn—Damage to skin tissue resulting from contact of bare skin with metal surfaces at below-freezing temperatures.

frostless zone—(Also called thermal belt, thermal zone, green belt, verdant zone.) That warmest part of a slope above a valley floor lying between the layer of cold air that forms over the valley floor on **calm**, clear nights and the cold hilltops or plateaus.

The air flowing down the slopes is warmed by **mixing** with the air above ground level and to some extent also by **adiabatic** compression. The frostless zone is not a fixed belt but varies in level from night to night and **season** to season according to the initial **temperature**, the length of the night, and the clearness of the sky. Its lower limit is sometimes clearly marked by the upper limit of **frost damage** to crops, following the hillsides at a small angle to the horizontal. *See* **thermal belt**.

Froude number—1. The nondimensional ratio of the **inertial force** to the **force of gravity** for a given fluid flow; the reciprocal of the **Reech number**.

It may be given as

$$\text{Fr} = V^2/Lg,$$

where V is a **characteristic velocity**, L a **characteristic length**, and g the **acceleration of gravity**; or as the square root of this number.

2. For atmospheric flows over hills or other obstacles, a more useful form of the Froude number is

$$\text{Fr} = \frac{V}{N_{BV}L_w},$$

where N_{BV} is the **Brunt–Väisälä frequency** of the ambient **upstream** environment, V is the **wind speed** component across the mountain, and L_w is the width of the mountain.

Fr can be interpreted as the ratio of natural **wavelength** of the air to wavelength of the mountain. Sometimes π will appear in the numerator, and other times the ratio will be squared. When Fr = 1, the natural wavelength of the air is in **resonance** with the size of the mountain and creates the most intense **mountain waves**, which can sometimes contain **lenticular clouds** and **rotors** of reverse flow at the surface. For Fr < 1, some of the low-altitude upstream air is blocked by the hill, short-wavelength waves separate from the top of the hill, and the remaining air at lower altitudes flows laterally around the hill. For Fr > 1, very long wavelengths form **downwind** of the hill, and can include a **cavity** of reverse flow just to the lee of the hill near the surface. Another form of the Froude number, using $(z_i - z_{hill})$ in place of L_w, is useful for diagnosing downslope windstorms and **hydraulic jump**, where z_i is the depth of the **mixed layer** above the base of the mountain, and z_{hill} is the height of the mountain.

frozen fog—Same as **ice fog**.

See **freezing fog**.

frozen ground—Soil within which the moisture has predominantly changed to **ice**, the unfrozen portion being in **vapor** phase.

Ice within the soil bonds (adfreezes) adjacent soil particles and renders frozen ground very hard. "Permanently" frozen ground is called **permafrost**. "Dry" frozen ground is relatively loose and crumbly because of the lack of bonding ice. Frozen ground is sometimes inadvisedly called **frost** or **ground frost**.

frozen-in-field—A **magnetic field** in a (zero **resistivity**) electrically conducting fluid (or **plasma**), so named because when the plasma moves, the **magnetic lines of force** may be said to move with it.

The field lines are fixed within the plasma. Because a plasma is an electrical conductor, when it moves in a magnetic field an electric **current** results, which in turn produces a magnetic field that adds to the existing field. The net field (assuming negligible resistivity) is as if the existing field lines had moved with the plasma.

frozen precipitation—Any form of **precipitation** that reaches the ground in frozen form, that is, **snow**, **snow pellets**, **snow grains**, **ice crystals**, **ice pellets**, and **hail**.

frozen turbulence—**Turbulence** that advects with the mean wind and is not statistically changed during the **advection** process.

In measurements, frozen turbulence is also referred to as **Taylor's hypothesis**, which allows **time series** measured at a single point to be interpreted as spatial variations.

frozen-turbulence approximation—*See* **Taylor's hypothesis**.

F_2-layer—The highest permanently observable layer of the **ionosphere**.

It exhibits a distinct maximum of **free-electron density** occurring at a height that ranges from about 225 km in the polar winter to over 400 km in daytime near the **magnetic equator**. Like the other ionospheric **layers**, the F_2-layer is formed by **absorption** of short-wavelength **solar radiation**, but its behavior and properties are more complex. Unlike the other ionospheric layers, the F_2-layer tends to rise during the middle of the day, except at middle to high latitudes in winter. Its maximum electron density occurs during the day, its minimum usually just before **sunrise**. It is the layer that is most useful for long-range radio transmission. *See* **F_1-layer**.

fuel moisture—*See* **fire-danger meter**.

Fujita–Pearson scale—(Abbreviated FPP scale.) Characterizes a **tornado's intensity** (F) by its path length (PL) and path width (PW).

The six-point scale for PL and PW is as follows.

Scale (F)	PL	PW
0	< 1.6 km	< 16 m
1	1.6–5.0 km	16–50 m
2	5.1–15.9 km	51–160 m
3	16–50 km	161–508 m
4	51–159 km	509–1448 m
5	160–507 km	1449–4989 m

The six-point scale for damage intensity is presented under the **Fujita scale**.

Fujita scale—(Also known as the F-scale.) Relates **tornado** intensity indirectly to structural and/or vegetative damage.

The estimated **wind speed** is calculated using the following formula: $V = 6.30 \ (F+2)^{1.5} \ \text{m s}^{-1}$. A six-point scale has been developed that corresponds to the following wind-speed estimates:

F0 (light damage): 18–32 m s^{-1}
F1 (moderate damage): 33–49 m s^{-1}
F2 (considerable damage): 50–69 m s^{-1}
F3 (severe damage): 70–92 m s^{-1}
F4 (devastating damage): 93–116 m s^{-1}
F5 (incredible damage):117–142 m s^{-1}.

Although extremely dependent on the design of a structure and the tree type, the following visual characteristics of the damage have been assigned to the F-scale.

F0 - Some damage to chimneys; branches broken; shallow-rooted trees knocked over.
F1 - Surface of roofs peeled off; mobile homes pushed off foundations or overturned; moving autos pushed off road.

F2 - Roofs torn off frame houses; mobile homes demolished; boxcars pushed over; large trees snapped or uprooted.

F3 - Roofs and some walls torn off well-constructed houses; trains overturned; most trees in forest uprooted; heavy cars lifted off ground and thrown.

F4 - Well-constructed houses leveled; structures with weak foundations blown off; large missiles generated.

F5 - Strong frame houses lifted off foundations and carried considerable distances; automobile-sized missiles flying through the air for distances in excess of 100 m; trees debarked.

Fujita, T., 1981: Tornadoes and downbursts in the context of generalized planetary scales. *J. Atmos. Sci.*, **38**, 1511–1534.

Fujiwhara effect—The tendency of two nearby **tropical cyclones** to rotate cyclonically about each other as a result of their circulations' mutual **advection**.

This occurs with some frequency in the northwestern Pacific **basin**, where it presents a significant forecast challenge, but happens more rarely in other ocean basins. *See* **binary cyclones**.

fulchronograph—A device for the measurement of lightning-stroke **current** as a function of time.

It consists of a number of magnetic links mounted on the perimeter of a rapidly rotating wheel that is located close to the conductor bearing the **lightning** surge. The **resolution** of the instrument is sufficient to determine current peaks as a function of time, but not the current wave shape.

fulgurite—A glassy, rootlike tube formed when a **lightning stroke** terminates in dry sandy soil.

The intense heating of the **current** passing down into the soil along an irregular path fuses the sands. Concurrently, **vaporization** of **soil moisture** and possibly even vaporization of the sandy materials causes the molten material to be expanded into a tube with a diameter that may be over 2.5 cm (1 in), but with a very thin wall. Fulgurites have been recovered in lengths of over 1.5 m (5 ft).

fully arisen sea—Same as **fully developed sea**.

fully developed sea—(Also called fully arisen sea.) A sea for which the **input** of **energy** to the waves from the local **wind** is in balance with the **transfer** of energy among the different **wave** components, and with the **dissipation** of energy by **wave breaking**.

Usually this is taken to mean that the **sea state** is independent of the distance (**fetch**) over which the wind blows and the time (duration) for which it has been blowing. Some of the wave components of the **energy spectrum** may not have their maximum amount of spectral energy in a fully developed sea, as an "overshoot" tends to occur during the process of wave growth.

fully penetrating well—A well that is open to the entire thickness of an **aquifer**.

fully rough flow—Turbulent flow for which

$$\frac{u_*\epsilon}{\nu} \geq 100,$$

where u_* is the **friction velocity**, ν is the **kinematic viscosity**, and ϵ is the average height of the surface irregularities.

When a flow is fully rough, it becomes almost independent of the **viscosity**.

fumigation—**Mixing** downward of an elevated **plume** of **air pollution** (often embedded in a layer of statically **stable air** such as a **temperature inversion**) into a turbulent **mixed layer** that has grown into and entrained the plume.

Compare **looping, coning, fanning, entrainment zone**.

fumulus—A contraction of the words fume and **cumulus**, indicating water-droplet clouds that form within the top of rising **plumes** from smokestacks, cooling towers, or open fires.

fundamental dimensions—There are seven basic dimensions in science, from which all other dimensions are derived (Table 1); two supplementary dimensions are listed in Table 2.

Historically, a variety of units have been used as measures of each dimension (e.g., length units: meters, inches, nautical miles, light years). One set has been adopted as the international system (**SI**) of units. The first letters of the first three units of this international system are "m, k, s"; hence, this system of units is sometimes called the MKS system. Derived units are formed from combinations of basic units. Examples of some derived units that are used in meteorology are listed in Table 3. A prefix (Table 4) can be added to these units to indicate larger or smaller values, such as kilometer (km), which is 1000 m.

Table 1. Fundamental dimensions.

Fundamental dimension	Unit	Abbreviation
length	meter	m
mass	kilogram	kg
time	second	s
electrical **current**	ampere	A
temperature	Kelvin	K
amount	**mole**	mol
luminous intensity	**candela**	cd

Table 2. Supplementary dimensions.

Dimension	Unit	Abbreviation
plane angle	**radian**	rad
solid angle	**steradian**	sr

Table 3. Examples of derived dimensions.

Dimension	Unit (Abbrev.)	Composition
force	**newton** (N)	$kg\ m\ s^{-2}$
energy	**joule** (J)	$kg\ m\ ^2s^{-2}$
power	**watt** (W)	$kg\ m\ ^2s^{-3}$
pressure	**pascal** (Pa)	$kg\ m\ ^{-1}s^{-2}$

Table 4. Prefixes. (*U.S. size designations.)

Multiplier	Size*	Name	Abbrev.
10^{18}	quintillion	exa	E
10^{15}	quadrillion	peta	P
10^{12}	trillion	tera	T
10^{9}	billion	giga	G
10^{6}	million	mega	M
10^{3}	thousand	kilo	k
10^{-3}	thousandth	milli	m
10^{-6}	millionth	micro	μ
10^{-9}	billionth	nano	n
10^{-12}	trillionth	pico	p
10^{-15}	quadrillionth	femto	f
10^{-18}	quintillionth	atto	a

fundamental equations of hydrodynamics—The **equations of motion**, the **equation of continuity**, the **energy equation**, the **equation of state**, and the **equation of continuity** for water substance, considered together as a **closed system** of equations.

A simplified physical **model** can dispense with certain of these equations without sacrificing completeness; for example, in two-dimensional homogeneous incompressible flow, **kinetic energy** is the only form of **energy**, and the equations of motion and continuity form a closed system.

fundamental frequency—The lowest **frequency** or longest **wavelength** defined by the length of the data set.

It has exactly one **cycle** per total **length of record**.

fundamental unit—A unit measure of a basic physical quantity such as mass, length, time; for example, one gram, one centimeter, one second, respectively. Other quantities, such as temperature, may be considered fundamental and assigned a fundamental unit.

funnel—*See* **funnel cloud**.

funnel cloud—A funnel-shaped **cloud** of **condensation**, usually extending from a deep **convective cloud**, and associated with a violently rotating column of air that is not in contact with the ground (hence not a **tornado**).

Funnel clouds can occur through a variety of processes in association with **convection**; not all funnel clouds go on to become tornadoes. For example, small funnel clouds are infrequently seen extending from small, dissipating **cumulus** clouds in environments with significant **vertical wind shear** in the cloud-bearing layer.

Bluestein, H., 1994: High-based funnel clouds in the Southern Plains. *Mon. Wea. Rev.*, **122**, 2631–2638.

funneling—A situation in which the **wind** is flowing through a channel that becomes narrower at the far end causing the flow to be directed along the channel and to accelerate as the narrow end is approached.

An example is wind is trapped in a valley that becomes narrower in the **downwind** direction. **Air pollution** emitted into valley air can be funneled in this manner.

furiani—In Italy, a southwest **wind**, vehement and short-lived, followed by a **gale** from the south or southeast, which blows in the vicinity of the Po River.

fusion—The **phase** transition of a substance passing from the solid to the liquid state; melting.

In meteorology, fusion is almost always understood to refer to the melting of **ice**, which, if the ice is pure and subjected to one **standard atmosphere** of **pressure**, takes place at the **ice point** of 0°C (*see* **melting point**). Additional heating at the melting point is required to fuse any substance. The **specific enthalpy** that must be added to fuse ice is called the **latent heat** of fusion and is approximately 3.35×10^5 J kg^{-1}.

fuzzy knowledge—Information characterized as being not quantifiable, for example, the concept of being "foggy."

It is often confused with **uncertainty**, which pertains more to whether a quantity is known or not. For example, one can report present conditions as more or less "foggy," depending on the current visibility. Uncertainty, on the other hand, occurs when one tries to predict the **probability** that the **visibility** will decrease within the next few hours.

fuzzy logic—A system of logic dealing with the concept of partial truth with values ranging between "completely true" and "completely false."

It is often confused with **probability**, which represents the degree of possibility of an occurrence. Fuzzy logic sets need not sum to 1 as do probabilities.

fuzzy set—A set where every element's degree of membership is specified by a value between 0 and 1.

The elements have different degrees of belonging to the set. For example, membership of a **temperature** of 0°C in the set "cold" might be 0.2, but membership of −20°C might be 0.9.

fuzzy variable—In **fuzzy logic**, a quantity that can take on linguistic rather than precise numerical values.

For example, a fuzzy variable, **temperature**, might have values such as "high," "medium," and "low."

G

G-layer—*See* **G-region**.

G-region—(Obsolete.) A region of **free electrons** in the **ionosphere** occasionally observed above the **F$_2$-layer**.
  The existence of this region has not been definitely established.

G system—Same as **grid navigation**.

gaign—A cross-mountain **wind** that causes clouds to form on the crests of mountains in Italy.

gain—An increase or amplification.
  There are two general uses of the term in **radar meteorology**: 1) **antenna gain** (or **gain factor**), which is the ratio of the **power** transmitted along the **beam** axis to that of an **isotropic radiator** transmitting the same total power; and 2) **receiver gain** (or **video gain**), which is the amplification given a **signal** by the **receiver**.

gain factor—Same as **gain**.

gaining stream—*See* **effluent stream**.

gale—1. In general, and in popular use, an unusually strong **wind**.
  2. In storm-warning terminology, a **wind** of 28–47 knots (32–54 mph).
  In the **Beaufort wind scale**, a **wind** with a speed from 28–55 knots (32–63 mph) and categorized as follows: **moderate gale**, 28–33 knots (Force 7); **fresh gale**, 34–40 knots (Force 8); **strong gale**, 41–47 knots (Force 9); and **whole gale**, 48–55 knots (Force 10).

gale warning—A **storm warning**, for marine interests, of sustained winds of 28–47 knots inclusive (32–54 mph) either predicted or occurring and not associated with tropical cyclones.
  The storm-warning signals for this condition are 1) two triangular red pennants by day and 2) a white lantern over a red lantern by night.

Galerkin approximation—An approximation to a differential equation that uses a **linear** combination of **basis functions** with weights determined by minimizing a globally integrated measure of the **error** in the solution.
  The resulting error is **orthogonal** to each basis function in an integral sense. Finite-element approximations result when piecewise linear basis functions are selected. Spectral approximations are produced when **spherical harmonic** basis functions are selected over spherical domains. *See* **finite-element model**.

Galerkin methods—Methods used for the solution of integral equations or for the solution of boundary-value problems that can be transformed to integral-equation form.
  As an example of the Galerkin method, in the integral equation

$$y(x) = (Fx) + \lambda \int_a^b K(x, s) y(s) ds,$$

the **dependent variable** $y(x)$ is approximated by a **linear** combination of the **polynomial** functions, $1, x, x^2, \ldots, x^n$ as

$$y(x) \approx \sum_{j=0}^{n} A_j x^j.$$

The approximating polynomial is substituted in the integral equation and the assumption made that exact equality holds when the resulting relation is multiplied by x^i, $i = 0, \ldots, n$, and integrated from a to b. There results a system of $n + 1$ linear equations in the A_j, the solution for which yields the coefficients in the polynomial approximation of $y(x)$.

galerne—(Also called galerna, galerno, giboulée.) A squally northwesterly **wind**, cold, humid, and showery, that occurs in the rear of a low pressure area over the English Channel and off the Atlantic coast of France and northern Spain.

Galilei—(Abbreviated Gal, gal.) The unit of **acceleration** in the **centimeter–gram–second system** of units, equal to 1 cm s^{-2}; commonly used in gravimetry.

gallego—A cold, piercing, northerly **wind** in Spain and Portugal.

gallium—(Symbol Ga.) A metallic **element**, atomic number 31 and atomic weight 69.72, which is soft enough to cut with a knife.

Its **melting point** is very low, 29.74°C (85.46°F) and its **boiling point** is 1700°C (3092°F). Gallium is used as a substitute for **mercury** in high-temperature thermometers.

gamma distribution—A continuous **random variable** x is said to have a gamma distribution if the **range** of x consists of all positive real numbers and if x possesses a **density function** $f(x)$ that satisfies the equation

$$f(x) = \frac{1}{\beta\Gamma(p)} (x/\beta)^{p-1} e^{-x/\beta} \ (x > 0),$$

where β and p are positive numbers and where Γ is the gamma function defined by the integral

$$\Gamma(p) = \int_0^\infty t^{p-1} e^{-t} dt$$

for $p > 0$.

The mean of the density function is βp, and the **standard deviation** is $\beta(p)^{1/2}$. The gamma distribution, which is a special case of the **Pearson distribution**, is skewed to the right. In meteorology, **scalar** multiples of the gamma distribution are used to represent size distributions of **cloud drops**. *See* **skewness**.

gamma ray—(Or γ-ray, γ ray, γ radiation, gamma radiation.) **Electromagnetic radiation** originating from transitions between **energy** levels of atomic nuclei.

A **nucleus** formed as a consequence of beta or alpha emission sometimes exists briefly in an excited **energy level** and makes a transition to a lower energy level accompanied by emission of a gamma ray photon with energy equal to the difference between the energies of the initial and final levels. Gamma ray energies from radioactive decay lie in the approximate **range** 10 keV–6 MeV. Gamma rays are also emitted in nuclear reactions. The boundary between x-rays and gamma rays is fuzzy, the latter term being most often used for electromagnetic radiation of nuclear origin.

Boorse, H. A., and L. Motz, 1966: *The World of the Atom*, Vol. I, 446–448.

gamma-ray snow gauge—An instrument designed to determine the **water content** of **snow** by measuring the amount of **gamma radiation** absorbed by the snow overlying the **transmitter** that is placed at the surface.

Because the earth naturally emits gamma radiation, by the same concept, aircraft measurements of gamma radiation can be used to estimate the water content of snow on the ground along the aircraft's flight path.

gap in spectrum—*See* **spectral gap**.

gap wind—A strong, low-level **wind** through either a relatively level channel between two mountain ranges or a gap in a mountain barrier; originally applied to strong (10–20 m s⁻¹) easterly winds through the Strait of Juan de Fuca between the Olympic Mountains of western Washington State and the mountains of Vancouver Island, British Columbia, Canada.

There they have been defined as "a flow of air in a **sea level** channel that accelerates under the influence of a **pressure gradient** parallel to the axis of the channel." As in the case of **mountain-gap winds**, this term has also been applied to pressure-gradient winds accelerating through a gap in a mountain barrier. The pressure gradient often results from a stable, post-cold-frontal **anticyclone** approaching the barrier and being partially blocked (*see* **blocking**) as it ascends the barrier, except for the flow through the gap or channel. The **tehuantepecer** of Central America is a well-known gap wind by this definition. These flows have sometimes been referred to as **jet-effect wind** and **canyon wind**.

Overland, J. E., and B. A. Walter, 1981: Gap winds in the Strait of Juan de Fuca. *Mon. Wea. Rev.*, **109**, 2221.

garbi—*See* **garbin**.

garbin—(Or garbis, gherbine, gherbino.) A **sea breeze**.

In southwest France it refers to a southwesterly sea breeze that sets in about 9 A.M., reaches it maximum towards 2 p.m. and ceases about 5 P.M. In Catalonia (northeast Spain) the sea breeze is called garbi but is easterly. The Spanish equivalent is garbino.

gargal—*See* **gregale**.

Garrett–Munk spectrum—An analytic approximation to the **internal gravity wave** spectrum (for **ocean waves** with large horizontal extent compared with the vertical extent) that is observed to have the same structure throughout the deep ocean.

The **energy density spectrum** is proportional to

$$Ef/[\omega(\omega^2 - f^2)^{1/2}] \approx Ef/\omega^2 \ (\text{for } \omega >> f),$$

for waves of **frequency** ω in which f is the **Coriolis parameter** and E is a nondimensional **parameter** empirically determined to be $E \approx 6 \times 10^{-5}$. For a wide range of observations E has been found to vary by no more than a factor of 2.

garúa—(Or camanchaca.) A dense **fog** and/or **drizzle** from low stratus on the west coast of South America.

It creates a **raw**, cold **atmosphere** that may last for weeks in winter and supplies a limited amount of moisture to the area.

gas chromatography—Analytical separation technique where the minor components in a mixture of gases are separated and resolved into individual components.

The technique requires the transmission of the gas sample through a column in the chromatograph using a mobile phase or carrier gas. The column is either packed or coated with a material for which the gases to be separated have an affinity and the strength of this affinity largely determines the time any individual component is retained in the column. Various detectors are employed in gas chromatography, from very specific compound-responsive detectors (flame photometric detector, **electron** capture detector, **photoionization** detector, etc.) to some very generally sensitive detectors (flame **ionization** detector, **thermal conductivity** detector, atomic emission detector, etc.) Gas chromatography is commonly used for the quantification of **halocarbons** and **hydrocarbons** in the **atmosphere**.

gas constant—The constant factor in the **equation of state** for ideal gases.

The **universal gas constant** is

$$R^* = 8.316963 \text{ J mol}^{-1}\text{K}^{-1}.$$

The gas constant for a particular gas is

$$R = R^*/m,$$

where m is the **molecular weight** of the gas. For a mixture, the "molecular weight" is a weighted mean of the molecular weights of the components:

$$m = \left(\frac{f_1}{m_1} + \cdots + \frac{f_n}{m_n} \right)^{-1},$$

where $m_1, \cdots, m_n$ are the molecular weights of the n gases, and $f_1, \cdots, f_n$ are their masses relative to the total mass of the mixture. The gas constant for **dry air** is

$$R_d = 2.870 \times 10^2 \text{ W kg}^{-1}\text{K}^{-1}.$$

The gas constant for **water vapor** is

$$R_v = 4.615 \times 10^2 \text{ W kg}^{-1}\text{K}^{-1}.$$

For **moist air**, the variable percentage of water vapor is taken into account by retaining the gas constant for dry air while using the **virtual temperature** in place of the **temperature**. *See also* **Boltzmann's constant**.

gas constant per molecule—Same as **Boltzmann's constant**.

gas filter correlation spectroscopy—A form of **nondispersive infrared spectrometry** that takes advantage of the banded nature of gas-phase **infrared** spectra.

The **background level** of **radiation** is measured by inserting a concentrated sample of the gas to be analyzed, thus **filtering** out wavelengths specific to that molecule. The technique has found application for the **remote sensing** of atmospheric species from satellites, for example, HNO_3 (Gille and Russell 1984). It has also been used in ground-based applications such as the measurement of emissions from automobiles (Bishop et al. 1989).

Gille, J. C., and J. M. Russell III, 1984: The Limb Infrared Monitor of the Stratosphere: Experimental description, performance, and results. *J. Geophys. Res.*, **89**, 5125–5140.

Bishop, G. A., J. R. Starkey, A. Ihlenfeldt, W. J. Williams, and D. H. Stedman, 1989: IR long-path photometry: A remote sensing tool for automobile emissions. *Analytical Chemistry*, **61**, 671A–677A.

gas laws—(Also called ideal-gas laws.) Usually, the laws associated with an **ideal gas**.

Modifications to these, such as **Van der Waal's equation**, are necessary for real gases, but modifications are small for atmospheric gases at **normal** environmental conditions. *See* **Boyle's law, Charles–Gay-Lussac law, Dalton's law**.

gas-particle distribution factor—The fraction of the total atmospheric burden of a given **element** that is found in the particulate **phase**.

These factors are most often given for sulfur, **carbon**, and **nitrogen**.

gas-phase kinetics—The branch of physical chemistry that deals with the determination of the rates and mechanisms of chemical reactions occurring in the gas **phase**.

gas–solid reactions—A reaction that occurs at the surface of a solid **particle**, between a gas-phase species and a species in the condensed **phase**.
　　An example is the **hydrolysis** of N_2O_5 on a polar stratospheric **cloud**, in which a gas-phase N_2O_5 molecule is converted to two HNO_3 molecules by reaction with condensed water. *See also* **heterogeneous chemistry**.

gas thermometer—A **thermometer** that utilizes the **thermal** properties of gas.
　　There are two forms of this instrument: 1) a type in which the gas is kept at constant volume, and **pressure** is the thermometric property; and 2) a type in which the gas is kept at constant pressure, and volume is the thermometric property.

gas transfer—(Also called gas exchange.) The movement of gas species from the air into the water column, or vice versa.
　　Physical mechanisms by which this can occur include **molecular diffusion, turbulent diffusion** and **eddy** motion in both the air and the water column, the **evaporation** of spray, and the dissolution of air bubbles in water.

gaseous discharge—Same as **electric discharge**.

gaseous electric discharge—Same as **electric discharge**.

gate-to-gate azimuthal shear—Marked difference in **Doppler velocity** values at azimuthally adjacent **range gates** indicating the presence of the radial component of **vorticity**. Vorticity may be due to the presence of a tornado-sized **vortex** (**tornadic vortex signature**) or it may simply be caused by **wind shear** across a **discontinuity** boundary.

gating—(Or range gating.) The use of **digital** or electrical methods in **radar** to eliminate or reject the **target** signals from all targets that are outside certain **range** limits.
　　Such methods make it possible to measure properties of the echoes from particular targets without interference from the signals returned from closer or more distant targets.

gauge height—Height of water surface above a gauge **datum**.
　　It is used interchangeably with the terms **stage** and **water level**; gauge height is more appropriate when used with a reading on a gauge.

gauge relation—An empirical curve relating **stream** discharge or **stage** at a point on a stream to **discharge** or stage at one or more **upstream** points and, possibly, to other parameters.

gauge zero—The **elevation** at which a water-level gauge zero is set relative to a general **datum**.

gauging section—Cross section of an open channel or **stream**, where measurements of flow depth and **velocity** are made in order to determine **rating curves**.

gauging site—The geographic location of a **stream gauge**.
　　See **gauging section**.

gauss—A cgs unit of **magnetic induction** (or magnetic **flux** per unit area across an area at right angles to the **magnetic field**), equal to one maxwell per square centimeter.
　　Cgs and mks units of magnetic induction are related by 10^4 gauss = 1 weber m^{-2} = 1 Tesla (T), where 1 weber = 1 (**newton** meter)/amp = 1 volt s. The **induction** of the earth's magnetic field in the United States is of order 0.5 gauss, with the magnetic field oriented about 20° from **zenith**. Magnetic induction inside superconducting magnets can be as high as 20 T (2×10^5 gauss), while magnetic induction produced by the human spine is of order 15×10^{-15} T (1.5×10^{-10} gauss).

Gauss curve—The normal **probability density function**.
　　See **normal distribution**.

Gauss elimination—A systematic procedure for solving systems of **linear** algebraic equations of the form

$$a_{1,1}x_1 + a_{1,2}x_2 + \cdots + a_{1,n}x_n = b_1$$

$$a_{2,1}x_1 + a_{2,2}x_2 + \cdots + a_{2,n}x_n = b_2$$

$$\cdots \cdots \cdots \cdots \cdots \cdots \cdots \cdots$$

$$a_{n,1}x_1 + a_{n,2}x_2 + \cdots + a_{n,n}x_n = b_n,$$

where the x_j represent unknown quantities and where the $a_{i,j}$ and b_i are prescribed constants. Multiples of one equation are combined with another to eliminate variables successively until the system has the upper triangular form

$$a_{1,1}x_1 + a_{1,2}x_2 + \cdots + a_{1,n}x_n = b_1$$

$$a_{2,2}x_2 + \cdots + a_{2,n}x_n = b_2$$

$$\cdots \cdots \quad \cdots \cdots \quad \cdots \cdots \quad \cdots \cdots$$

$$a_{n,n}x_n = b_n,$$

where the constants in all but the first equation have been altered by the elimination process. The last equation is solved for x_n and the resulting value substituted in the preceding equation which then can be solved for x_{n-1}. The back-substitution process is continued until all the remaining unknowns are determined.

Gauss's theorem—Same as **divergence theorem**.

Gaussian distribution—Same as **normal distribution**.

Gaussian grid—In global atmospheric models using the **spectral method**, the physical **grid** required for alias-free evaluation of the quadratic terms in the **equations of motion** (e.g., **horizontal advection**).

The **resolution** of the grid is determined by the spectral truncation and the requirements for the transformations of the **spherical harmonics** (Fourier transformation in longitude and Gaussian **quadrature** of the Legendre polynomials in latitude).

Gaussian plume models—Models for the **dispersion** of gases or **particles**, usually from a **point source**.

The concentration inside the **plume** is predicted using Gaussian **statistics**, with the center line of the plume (the point source advected by the mean flow) at the maximum of the **Gaussian distribution**, and with the **standard deviation** of the Gaussian distribution an increasing function of time or **downwind** distance. *See* **air pollution**.

Gaussian process—(Also called stationary Gaussian process, stationary Gaussian time series.) A **stationary time series** in which the joint **probability distribution** of any sequence of values, $x(t_1)$, $x(t_2)$, . . ., $x(t_n)$, is a multivariate **normal distribution**, or a stationary **random process** that is completely determined by its **spectrum** or **autocorrelation function**.

Gaussian wave packet—A collection of **waves**, the amplitudes of which have a Gaussian **dependence** on **wavenumber** about some central wavenumber.

Gay–Lussac's law—Same as **Charles–Gay–Lussac law**.

GCA—Commonly used abbreviation for **ground-controlled approach**.

GCA minimums—The minimum **ceiling** and/or **visibility** under which an aircraft may be landed with the use of **ground-controlled approach** (GCA).

The pilot must be qualified to use GCA (and the aircraft must be equipped with the necessary radio equipment) before GCA minimums are applicable to the particular flight. At most airports GCA minimums are less than other minimums because of the double check on the "glide path" position of the aircraft.

GCM—Abbreviation for **general circulation model**.

GDD—Abbreviation for **growing degree-day**.

GDH—Abbreviation for **growing degree-hour**.

GDPS—Abbreviation for **Global Data Processing System**.

gebli—Same as **ghibli**.

geg—A desert **dust whirl** of China and Tibet.

gegenschein—A round or slightly elongated area of **light** seen in the night sky along the **zodiac**.
     Like the **zodiacal light**, the gegenschein is mainly **sunlight** scattered by **particles** in interplanetary space along the **ecliptic**. The zodiacal light is **forward scattered** while the gegenschein is **backscattered** light.

gending—A local dry wind in the northern plains of Java, resembling the **foehn**.
     It is caused by a **wind** crossing the mountains near the south coast and pushing between the volcanoes.

General Broadcast—That radio broadcast, to all navy ships and merchant ships, on which are given: marine warnings; area forecasts; map analyses and forecasts; surface weather reports; and **upper air**, **aircraft report**, and satellite information.
     See **Fleet Broadcast, MERCAST**.

general circulation—1. (Also called planetary circulation.) In its broadest sense, the complete **statistical** description of **large scale** atmospheric motions.
     These **statistics** are generated from the ensemble of daily data and include not only the temporal and spatial mean flows (e.g. **zonal westerlies** and **easterlies**) but also all other mean properties of the **atmosphere** that are linked to these flows (e.g., semipermanent **waves** and **meridional cells**) that together form the general circulation. The general circulation also includes higher-order statistics that measure the spatial and temporal **variability** of the atmosphere necessary to understand the large-scale temporal and spatial mean state of the atmosphere (e.g., seasonal changes and the effects of transient **cyclones**). *Compare* **planetary circulation**.

    2. *See* **primary circulation**.
        Grotjahn, R., 1993: *Global Atmospheric Circulations: Observations and Theories*, 3–4, 160–161.
        Holton, J. R., 1992: *An Introduction to Dynamic Meteorology*, 3d ed., 310–312.
        Peixoto, J. P., and A. H. Oort, 1992: *Physics of Climate*, 8–26.

general circulation model—(Abbreviated GCM.) A time-dependent numerical **model** of the **atmosphere**.
     The governing equations are the conservation laws of physics expressed in finite-difference form, spectral form, or finite-element form. Evolution of the model circulation is computed by time integration of those equations starting from an **initial condition**. The GCM can be used for weather prediction or for **climate** studies. *Compare* **atmospheric circulation model, hemispheric model**.

general circulation of the atmosphere—*See* **general circulation**.

general forecast—*See* **weather forecast**.

generalized absorption coefficient—A single, effective value for the **absorption coefficient** that mimics the integrated behavior of all absorption coefficients within the **band**; used in conjunction with a **generalized transmission function** to yield the integrated transmission through an **absorption band**.

generalized coordinates—(Also called **Lagrangian coordinates**.) Any set of coordinates specifying the state of the system under consideration.
     Usually employed in problems involving a finite number of **degrees of freedom**, the generalized coordinates are chosen so as to take advantage of the constraints of the system in reducing the total number of coordinates.

generalized hydrostatic equation—The vertical component of the **vector** equation of motion in **natural coordinates** when the **acceleration of gravity** is replaced by the **virtual gravity**.
     For most purposes it is identical to the **hydrostatic equation**.

generalized transmission function—The **transmission function** that results when specific band models are adapted to more general situations.
     They provide a wavelength-integrated **transmission function** over part or all of the **band**.

generate and test—A method of problem solving that consists of generating possible scenarios and testing each until a satisfactory solution is found.
     Not all possible scenarios are tested, and the solution found may not be optimal.

generating area—An area of the ocean where the action of the **wind** causes waves to be generated. Once generated, the waves generally propagate away from the area. *See* **fetch, swell**.

generating cell—In **radar**, a small region of locally high **reflectivity** from which a trail of **hydrometeors** originates.

It is postulated that **snow crystals** are formed and grow in the generating cells and that the cells are maintained by **convection** induced by the release of **latent heat** accompanying the **crystal** growth. The shape of the **snow** trail below a generating cell depends on the fall speed of the snow and the **vertical profile** of the horizontal **wind**.

generic rule—An **IF–THEN rule** that applies equally well to all geographic locations and is independent of the data source.

genitus—Used when a **cloud** forms from a part of another cloud.

It is then given the name of the appropriate genus, followed by the name of the genus of the **mother-cloud** with the addition of the suffix "genitus" (e.g., **stratocumulus** cumulogenitus). *Compare* **mutatus**.

World Meteorological Organization, 1956: *International Cloud Atlas*, Volume I, p. 6.

Genoa cyclone—(Or Genoa low.) A **cyclone**, or **low**, that appears to have formed or developed in the vicinity of the Gulf of Genoa (Ligurian Sea).

GENOT—Aviation communications code word for "general notice."

GENOTs are originated by the U.S. Federal Aviation Administration and appear on all teletype weather services. They carry general information of interest to airmen, forecasters, and others connected with aircraft operation and **weather service**. *Compare* **NOTAM**.

gentle breeze—In the **Beaufort wind scale**, a **wind** with a speed from 7 to 10 knots (8–12 mph) or Beaufort Number 3 (Force 3).

geocentric reference system—A **coordinate system** with its origin at the earth's center, as distinct from one with an origin on the earth's surface.

geodesic grid—A **grid** formed by a collection of geodesics, or, on the spherical earth, arcs of great circles.

A typical example is the icosahedral, or icosahedral-hexagonal, grid, consisting of twenty equilateral triangular faces expanded onto a sphere, and further subdivided into smaller triangles. Other possibilities exist, such as grids obtained by the expansion of a cube with each of its sides subdivided into smaller elements by arcs of great circles.

geodesy—The study of the shape, size, and **gravity** of the earth.

Geodetic Satellite—(Acronym *GEOSAT*.) A U.S. Navy oceanographic satellite that used a **radar altimeter** to obtain closely spaced, precise mapping of the earth's **geoid** over the ocean.

GEOSAT was launched on 12 March 1985 and produced maps of seafloor **topography** with unprecedented **resolution**.

geodynamic height—Same as **dynamic height**.

geodynamic meter—Same as **dynamic meter**.

geographic coordinates—Same as **spherical coordinates**.

geographic effect on winds—The relationship between **topography** and winds.

There are two main mechanisms: 1) topographically modified flow, where existing synoptic-scale winds are modulated or redirected by the presence of mountains or terrain features; and 2) geographically generated flow, where solar heating and **nocturnal cooling** generate **anabatic** and **katabatic** flows and mountain and valley circulations during conditions of weak **synoptic** winds.

geographic horizon—*See* **horizon**.

geographically possible sunshine duration—(Or topographically possible sunshine duration.) *See* **insolation duration**.

geoid—The particular **geopotential surface** that most nearly coincides with the mean level of the oceans of the earth.

For mapping purposes it is customary to use an ellipsoid of revolution as an adequate and convenient approximation to the geoid. The **dimensions** and orientation of the assumed ellipsoid may represent an attempt to find the ellipsoid that most nearly fits the geoid as a whole, or they may represent an attempt to fit only a particular part of the geoid without regard to the remainder of it. When mention is made of the dimensions of the earth, reference is usually made to the dimensions of the ellipsoid most nearly representing the geoid as a whole. The actual geoid can depart from a best-fitting sphere in places by as much as 100 m.

geoisotherm—Same as **isogeotherm**.

geologic age—The fourth-order division of **geologic time**, delimited by very minor changes, usually in **sea level** and/or **climate** or in the biota and usually local, that is, limited to a single continent or portion thereof.
   In general, two or more geologic ages constitute a **geologic epoch**, and geologic ages may be defined differently in different parts of the world.

geologic epoch—The third-order division of **geologic time**, delimited by partial withdrawal of the sea from land masses and by gentle crustal disturbances in localized areas.
   Two or more epochs are required to make up a **geologic period**, and, in turn, two or more periods are needed to constitute a **geologic era**.

geologic era—The primary and largest division of **geologic time**.
   Limits correspond with major, global crustal events, changes in **sea level** and/or **climate**, or biotic changes. Five geologic eras have been established: Archeozoic [before 2500 million years ago (Ma)], Proterozoic (2500 to 570 Ma), Paleozoic (570 Ma to ca. 250 Ma), Mesozoic (ca. 250 to ca. 70 Ma), and Cenozoic (since ca. 70 Ma). All eras are divided into at least two **geologic periods**.

geologic period—The secondary division of **geologic time**, delimited by moderate but usually global crustal events, changes in **sea level** and/or **climate**, or biotic changes, sometimes in a relatively localized area.
   Two or more periods are required to make up a **geologic era**, and each period comprises two or more **geologic epochs**.

geologic thermometer—Same as **geothermometer**.

geologic time—Time as considered in terms of the history of the earth.
   It is divided into **geologic eras, periods**, and **epochs**. Depending on the part of the geologic time scale, increments are as long as tens of millions of years or as short as hundreds of years. In general, geologic time is more finely divided closer to the present.

geological climate—Same as **paleoclimate**.

geological era—Same as **geologic era**.

geomagnetic coordinates—A system of **spherical coordinates** based on the best fit of a centered **dipole** to the actual **magnetic field** of the earth.
   The **field** due to an earth-centered **magnetic dipole** is given by the first three terms of the **spherical harmonic** expression of the **IGRF** (or **DGRF**) **model**.

geomagnetic equator—A **great circle** on the earth's surface that is everywhere equidistant from the geomagnetic poles, that is, the **equator** in the system of **geomagnetic coordinates**.
   See **aclinic line**.

geomagnetic latitude—A coordinate used in **geomagnetism** bearing the same relation to the geomagnetic **dipole** equator as geographic latitude does to the geographic **equator**.
   When **meridional** variations in phenomena closely related to the earth's **magnetic field** are plotted according to geomagnetic rather than geographic latitude, more clear-cut relations are usually depicted. **Cosmic ray** intensities and auroral frequencies are examples of quantities best studied as functions of this type of latitude.
   Campbell, W. H., 1997: *Introduction to Geomagnetic Fields*, Cambridge Univ. Press, p. 32.

geomagnetic meridian—A **great circle** of the earth through the **geomagnetic poles**.

geomagnetic pole—A **pole** in the geomagnetic system of coordinates.
   Compare **magnetic dipole**.

geomagnetic storm—*See* **magnetic storm**.

geomagnetism—1. (Also called terrestrial magnetism, geomagnetic field.) The earth's **magnetic field** or the geophysical phenomena caused or affected by this field.
   2. The scientific study of the earth's **magnetic field**, including its **variation** in space and time, and its relation to other geophysical phenomena (e.g., **aurora**).
   Geomagnetism belongs to the same family of earth sciences as **geodesy** and **geomorphology**.

geometric mean—For n positive numbers, the positive nth root of their product; that is, for the set of positive numbers $x_1, x_2, \ldots, x_n$, the geometric mean is the quantity $(x_1 x_2 \cdots x_n)^{1/n}$.

Compare **arithmetic mean**.

geometric optics—The application of **ray tracing** to explain **scattering** and refractive effects by **particles** that are very large compared with the **wavelength** of the **radiation**.

Geometric optics provides useful explanations for atmospheric features such as **rainbows, halos**, and many other **ice crystal** optical displays.

geometrical horizon—*See* **horizon**.

geomorphology—(Formerly physiography.) The science dealing with the form and surface configuration of the solid earth.

It is primarily an attempt to reveal the complex interrelationships between the origin (and therefore material composition) of surface features on the one hand, and the causes of the surface alteration (**erosion**, weather, crustal upheaval, etc.) on the other hand.

geophysical day— An internationally agreed-upon day designated for more detailed or intensive observations of the **atmosphere** over broad regions of the earth.

See **aerological days**.

geophysics—The physics of the earth and its **environment**, that is, earth, air, and (by extension) space.

Classically, geophysics is concerned with the nature of the physical occurrences at and below the surface of the earth including, therefore, geology, **oceanography**, **geodesy**, seismology, **hydrology**, etc. The trend is to extend the scope of geophysics to include meteorology, **geomagnetism**, astrophysics, and other sciences concerned with the physical nature of the **universe**. Geophysics uses analytical and mathematical, rather than purely descriptive, techniques.

geopotential—The **potential energy** of a unit mass relative to **sea level**, numerically equal to the **work** that would be done in lifting the unit mass from sea level to the height at which the mass is located; commonly expressed in terms of **dynamic height** or **geopotential height**.

The geopotential Φ at height z is given mathematically by the expression

$$\Phi = \int_0^z g\,dz',$$

where g is the **acceleration of gravity**.

geopotential field—Distribution of the **geopotential** altitude of an **isobaric surface**.

geopotential height—The height of a given point in the **atmosphere** in units proportional to the **potential energy** of unit mass (**geopotential**) at this height relative to **sea level**.

The relation, in **SI** units, between the geopotential height Z and the geometric height z is

$$Z = \frac{1}{980} \int_0^z g\,dz',$$

where g is the **acceleration of gravity**, so that the two heights are numerically interchangeable for most meteorological purposes. Also, one **geopotential meter** is equal to 0.98 **dynamic meter**. *See* **dynamic height**.

geopotential meter—A distance of 1 m in **geopotential height**.

geopotential surface—(Also called **equipotential surface**, level surface.) A surface of constant **geopotential**, that is, a surface along which a **parcel** of air could move without undergoing any changes in its **potential energy**.

Geopotential surfaces almost coincide with surfaces of constant geometric height. Because of the poleward increase of the **acceleration of gravity** along a surface of constant geometric height, a given geopotential surface has a smaller geometric height over the poles than over the **equator**. *See* **potential, geopotential height, dynamic height**.

geopotential thickness—The difference in the **geopotential height** of two **constant-pressure surfaces** in the **atmosphere**, proportional to the appropriately defined mean air **temperature** between the two surfaces:

$$\Delta\phi = -R \int_{p_1}^{p_2} T\,\frac{dp}{p} = RT_m \ln\left(\frac{p_1}{p_2}\right),$$

where $\Delta\phi$ is the geopotential thickness; R the **gas constant** for air; p_1 and p_2 the **pressure** at the lower and upper **isobaric** surfaces, respectively; T the Kelvin temperature; and T_m the **mean temperature**.

geopotential topography—(Also called absolute geopotential topography.) The **topography** of any surface as represented by lines of equal **geopotential**.

These lines are the contours of intersection between the actual surface and the level surfaces (which everywhere are normal to the direction of the **force of gravity**), and are spaced at equal intervals of **dynamic height**.

George's index—Same as **K index**.

GEOSAT—Acronym for **Geodetic Satellite**.

geosphere—The "solid" portion of the earth, including **water masses**; the **lithosphere** plus the **hydrosphere**.

Above the geosphere lies the **atmosphere** and at the **interface** between these two regions is found almost all of the **biosphere**, or zone of life. *Compare* **geoid, equilibrium spheroid**.

Geostationary Meteorological Satellite—(Abbreviated GMS.) A series of satellites operated by the Japanese Meteorological Agency.

The GMS program was initiated with the launch of *GMS-1* in July 1977. Multi-spectral imagery is provided by a slightly modified version of the **VISSR** instrument used on **GOES** satellites. The GMS is positioned at 140°E longitude.

Geostationary Operational Environmental Satellite—(Abbreviated GOES.) Applies to both the satellites themselves and to the overall system of geostationary observations used by the United States.

The current operational series of GOES satellites were preceded by the **ATS** and **SMS** satellites, with *GOES-1* being launched on 16 October 1975. The early GOES (1 through 7) were spin-stabilized spacecraft, while the latest GOES are **three-axis stabilized**. Two GOES satellites are normally in operation, one at 75° W longitude and the other at 135° W longitude. Before launch, GOES satellites are given a letter designation (e.g., GOES-J) that is changed to a number designation (e.g., *GOES-9*) when the satellite becomes operational. The current generation of GOES satellites supports separate imager and **sounder** systems, **SEM** and **DCS**. The imager is a five-channel **scanning radiometer** with a 1-km **resolution** visible **channel**, along with slightly lower resolution images in the midinfrared, **water vapor**, and **thermal** IR bands. The sounder has 18 **thermal infrared** bands and a low-resolution visible **band**.

Geostationary Operational Meteorological Satellite—(Abbreviated GOMS; also referred to as Elektro or Electro.) A Russian satellite system designed to monitor central Asia and the Indian Ocean at 76°E longitude.

The satellite reached geostationary **orbit** in November 1994, but has not produced reliable operational imagery.

geostationary satellite—A satellite in a west-to-east **orbit** at an **altitude** of 35 786 km (19 600 n mi) above the **equator**.

At this altitude, it circles the axis of the earth once every 24 hours, making its speed in orbit synchronous with the earth's rotation. A geostationary orbit is geosynchronous, but it is also required to have zero **inclination** angle (orbital plane coincides with the earth's equatorial plane) and zero eccentricity (a perfectly circular orbit). Geostationary satellites (such as **GOES, Meteosat**, and **GMS**) remain essentially stationary over a given geographical point above the equator. *See* **geosynchronous satellite**.

geostrophic—Referring to the balance, in the **atmosphere**, between the horizontal **Coriolis forces** and the horizontal **pressure forces**.

See **geostrophic wind, geostrophic equilibrium, geostrophic balance**.

geostrophic adjustment—The process by which an unbalanced atmospheric flow field is modified to **geostrophic equilibrium**, generally by a mutual adjustment of the atmospheric **wind** and **pressure** fields depending on the initial horizontal **scale** of the **disturbance**.

geostrophic advection—The **transport** of atmospheric properties by and in the direction of the **geostrophic wind**.

geostrophic approximation—1. The assumption that the horizontal **wind** may be represented by the **geostrophic wind**.

2. Same as **quasigeostrophic approximation**.

geostrophic balance—Describes a balance between **Coriolis** and horizontal **pressure-gradient forces**.

Explicitly,

$$fv = (1/\rho)\partial p/\partial x,$$

and

$$fu = -(1/\rho)\ \partial p/\partial y,$$

in which f is the **Coriolis parameter**, u and v the **zonal** and **meridional** components of **velocity**, x and y the zonal and meridional coordinates, p the **pressure**, and ρ the **density**.

geostrophic coordinates—The **coordinate system** used in the **semigeostrophic equations**.
Geostrophic coordinates (X, Y, Z, T) are related to physical space coordinates (x, y, z, t) by

$$X = x + \frac{v_g}{f}, Y = y - \frac{u_g}{f}, Z = z, T = t,$$

where u_g and v_g are the eastward and northward **geostrophic winds** and f is the **Coriolis parameter**. Through this coordinate transform the semigeostrophic equations arise. The **geostrophic approximation** is not used as frequently in the semigeostrophic equations as in the **quasigeostrophic equations**; thus the semigeostrophic equations produce a more accurate representation of certain phenomena, such as **fronts** and **jets**.

geostrophic current—A **current** in which the balance in the horizontal components of the **equations of motion** is between the horizontal **pressure gradient** and the **Coriolis force**.
The vertical component is in **hydrostatic balance** and the **pressure** increases with depth in proportion to the mass of water above. If pressure is mapped on a **level surface** (**geopotential**), then **geostrophic flow** is parallel to the isobars, with high pressure to the right (left) of the flow in the Northern (Southern) Hemisphere. For the **geostrophic balance** to hold, the flow must be steady, very weak, large-scale, and friction-free.

geostrophic departure—(Also called geostrophic deviation, **ageostrophic wind**.) The **vector** (or sometimes only the magnitude) difference between the real (or observed) **wind** and the **geostrophic wind**.
In the **atmospheric boundary layer**, winds can be subgeostrophic due to turbulent **drag** against the surface. This causes a steady-state geostrophic departure, written here as separate Cartesian components $(V - V_g)$ and $(U - U_g)$, or

$$(V - V_g) = C_D MU/f_c z_i$$

$$(U - U_g) = -C_D MV/f z_i$$

where (U, V) are horizontal wind components, C_D is a **drag coefficient**, M is total **wind speed**, (U_g, V_g) are geostrophic winds components, f_c is the **Coriolis parameter**, and z_i is depth of the boundary layer. The equations above are only approximate (they assume a slab boundary layer with no **entrainment**), but illustrate the effect.

geostrophic deviation—Same as **geostrophic departure**, **ageostrophic wind**.

geostrophic dividers—An instrument, based on the principle of the **geostrophic wind scale**, for determining the **geostrophic wind** from data on a **weather map**.

geostrophic drag coefficient—The ratio of the **Reynolds stress** at the surface to the square of the **geostrophic wind** velocity.

geostrophic equilibrium—A state of motion of an **inviscid fluid** in which the horizontal **Coriolis force** exactly balances the **horizontal pressure force** at all points of the **field** so described:

$$2\Omega \times \mathbf{v}_g = -\alpha \nabla_H p,$$

where Ω is the **vector** angular **velocity** of the earth, $\mathbf{v}_g$ the **geostrophic wind** velocity, α the **specific volume**, p the **pressure**, and ∇_H the horizontal **del operator**.
With respect to cyclone-scale motions in **extratropical** latitudes, the **free atmosphere** frequently approaches a state of geostrophic equilibrium.

geostrophic flow—A form of **gradient flow** where the **Coriolis force** exactly balances the **horizontal pressure force**.
See **geostrophic wind**.

geostrophic flux—The **transport** (**flux**) of an atmospheric property by means of the **geostrophic wind**.

336

When applied to the horizontal average flux by an **eddy**, the geostrophic flux of **geostrophic** momentum will be zero unless the eddy has horizontal axis tilts.

Grotjahn, R., 1993: *Global Atmospheric Circulations: Observations and Theories*, 107–108.

geostrophic momentum approximation—The selective approximation of true **momentum** by the corresponding **geostrophic** momentum.

The approximation is usually made in conjunction with the assumption that the **Coriolis parameter** f is a constant, in which case it takes the form

$$\frac{\partial u_g}{\partial t} + u\,\frac{\partial u_g}{\partial x} + v\,\frac{\partial u_g}{\partial y} + w\,\frac{\partial u_g}{\partial z} - fv + \frac{1}{\rho}\frac{\partial p}{\partial x} = 0,$$

$$\frac{\partial v_g}{\partial t} + u\,\frac{\partial v_g}{\partial x} + v\,\frac{\partial v_g}{\partial y} + w\,\frac{\partial v_g}{\partial z} + fu + \frac{1}{\rho}\frac{\partial p}{\partial y} = 0,$$

where u, v, and w are the eastward, northward, and upward components of the total **wind**, ρ is the **density**, p the **pressure**, and

$$u_g = -\frac{1}{f\rho}\frac{\partial p}{\partial y} \text{ and } v_g = \frac{1}{f\rho}\frac{\partial p}{\partial x}$$

the eastward and northward components of the **geostrophic wind**. In the geostrophic momentum approximation the geostrophic wind is the advected quantity but the **advection** is by the total (geostrophic plus **ageostrophic**) wind. When the geostrophic momentum approximation is combined with **geostrophic coordinates**, the resulting equations are termed the **semigeostrophic equations**.

geostrophic vorticity—The **vorticity** of the **geostrophic wind**.

If **pressure** is taken as the vertical coordinate and the **variability** of the **Coriolis parameter** is neglected, the geostrophic relative vorticity ζ_g is given by

$$\zeta_g = \frac{g}{f}\,\nabla_p^2 z,$$

where g is the **acceleration of gravity**, f the Coriolis parameter, z the height of the **constant-pressure surface**, and ∇_p^2 is the isobaric **Laplacian operator**. *See* **thermal vorticity**.

geostrophic wind—That horizontal **wind** velocity for which the **Coriolis acceleration** exactly balances the **horizontal pressure force**:

$$f\mathbf{k} \times \mathbf{v}_g = -g\nabla_p z,$$

where $\mathbf{v}_g$ is the geostrophic wind, f the **Coriolis parameter**, $\mathbf{k}$ the vertical unit **vector**, g the **acceleration of gravity**, ∇_p the horizontal **del operator** with **pressure** as the vertical coordinate, and z the height of the **constant-pressure surface**.

The geostrophic wind is thus directed along the **contour** lines on a constant-pressure surface (or along the isobars in a **geopotential surface**) with low elevations (or low pressure) to the left in the Northern Hemisphere and to the right in the Southern Hemisphere. The speed of the geostrophic wind V_g is given by

$$V_g = -\frac{g}{f}\frac{\partial z}{\partial n},$$

where $\partial z/\partial n$ is the slope of the **isobaric surface** normal to the contour lines to the left of the direction of motion in the Northern Hemisphere and to the right in the Southern Hemisphere. The geostrophic wind is defined at every point except along the **equator**. The validity of this approximation in dynamic theory depends upon the particular context. *See* **geostrophic equilibrium**, **geostrophic approximation**, **quasigeostrophic approximation**; *compare* **gradient wind**.

geostrophic wind level—(Also called gradient wind level.) The lowest level at which the **wind** becomes **geostrophic** in the theory of the **Ekman spiral**, proportional to $(\nu/\sin\phi)^{1/2}$, where ν is the kinetic **eddy viscosity** and ϕ the latitude.

In practice it is observed that the geostrophic wind level is often between 1 and 2 km, and it is assumed that this marks the upper limit of frictional influence of the earth's surface. The geostrophic wind level may be considered to be the top of the **Ekman layer** and **planetary boundary layer**, that is, the base of the **free atmosphere**.

geostrophic wind scale—A graphical device used for the determination of the speed of the **geostrophic wind** from the **isobar** or contour-line spacing on a **synoptic chart**.

It is a **nomogram** representing solutions of the geostrophic wind equation:

$$V_g = \frac{1}{f\rho} \frac{\partial p}{\partial n},$$

where **geopotential height** is the vertical coordinate; or

$$V_g = \frac{g}{f} \frac{\partial z}{\partial n},$$

where **atmospheric pressure** is the vertical coordinate. In the above equations, V_g is the speed of the **geostrophic wind**, ρ the **density** of the air, f the **Coriolis parameter**, p the **pressure** at a fixed geopotential height, z the height of a **constant-pressure surface**, and n horizontal distance measured normal to the flow. The n axis is directed to the right of the flow in the Northern Hemisphere and to the left of the flow in the Southern Hemisphere. In the **nomogram**, standard values of ρ or g are usually adopted. The **gradient** of pressure or height is approximated by the finite difference ratio, $\Delta p/\Delta n$ or $\Delta z/\Delta n$, in which a standard difference in pressure or height is adopted; Δn then represents the normal distance between **isobars** or **contour lines** drawn. The nomogram often utilizes Δn as the **abscissa** and the latitude as **ordinate**, so that the speed of the geostrophic wind may be read from a family of lines of the graph.

geosynchronous satellite—A satellite in an equatorial or near-equatorial **orbit** that orbits with the same **angular velocity** as the earth, making one revolution in 24 hours.
Note that a geosynchronous orbit is not necessarily also in a geostationary orbit. *See* **geostationary satellite**.

geothermal gradient—Same as **thermal gradient**.

geothermometer—(Also called geologic thermometer.) A **thermometer** designed to measure temperatures in deep-sea deposits or in bore holes deep below the surface of the earth.

Gerdien aspirator—An instrument used for the determination of the **electrical conductivity** of the atmosphere.

Gerstner wave—A finite-amplitude rotational **gravity wave**.
The Gerstner wave formulas are exact solutions of the **equations of motion** for a homogeneous **incompressible fluid** with a **free surface**.
Lamb, H., 1953: *Hydrodynamics*, 6th ed., 421–423.

gestalt—(German: shape or form.) In **synoptic meteorology**, a complex of weather elements occurring in a familiar form, although this term is not commonly used.
The complex, though not necessarily referring to basic hydrodynamical or thermodynamical quantities, may persist for an appreciable length of time and is often considered to be an entity in itself. An example of a gestalt is a **warm front** and its associated pattern of **cloudiness** and **precipitation**.

geyser—A hot spring that ejects intermittent jets of water and **steam**.
The action results from heating of **groundwater** circulating through hot rock under conditions that prevent continuous circulation.

gharbi—A fresh westerly **wind** of oceanic origin in Morocco.
Compare **garbin**.

gharra—Hard **squalls** from the northeast in Libya and Africa.
They are sudden and frequent and are accompanied by **heavy rain** and **thunder**.

ghaziyah—*See* **bora**.

gherbine—*See* **garbin**.

ghibli—(Also called chibli, gebli, gibleh, gibli, kibli.) A hot dust-bearing **desert wind** in Tripolitania (northwestern Libya), similar to the **foehn**.
In Morocco, the analogous gibla is a hot dry wind from between southeast and south. It means "the direction in which one turns," that is, the traditional direction of Mecca. *See* **chili, khamsin, sirocco, suahili**.

GHIS—Abbreviation for **GOES High-Resolution Interferometer Sounder**.

ghost—In **radar**, a diffuse **echo** in the apparently **clear air** that is caused by a **cloud** of point targets such as insects or by **Bragg scattering** from spatial variations of the **refractivity** of truly clear air.
Ghost echoes are incoherent echoes with characteristics similar to those of weather echoes. *Compare* **angel**; *see* **clear-air echo**.

Gibbs free energy—Same as **Gibbs function.**

Gibbs function—(Also called Gibbs free energy, thermodynamic potential.)
A mathematically defined **thermodynamic function of state**, which is constant during a reversible isobaric–isothermal process. The most important such process in meteorology is the change in **phase** of water substance. In symbols the Gibbs function g is

$$g = h - Ts,$$

where h is **specific enthalpy**, T is Kelvin **temperature**, and s is **specific entropy.** *Compare* **Helmholtz function.**

gibla—See **ghibli.**

gibleh—Same as **ghibli.**

gibli—Same as **ghibli.**

giboulée—Same as **galerne.**

Gibraltar outflow water—**Mediterranean Water** that leaves the Alboran Sea through the Strait of Gibraltar below a depth of 150 m, penetrating the North Atlantic Central Water in the Gulf of Cadiz.
Initially, Gibraltar outflow water is advected by a turbulent contour current. It transforms into a neutrally balanced middepth flow at between 700 and 1200 m south of the Algarve coastline (south Portugal). Also called a Mediterranean Water tongue, its high salt content is caused by the **arid climate** of the European **Mediterranean Sea.**

gigahertz—(Abbreviated GHz; formerly called kilomegacycle.) A unit of **frequency** equal to one billion (10^9) cycles per second.

Gill anemometer—A lightweight version of a **propeller anemometer** sometimes used instead of the more expensive **sonic anemometer** for the measurement of turbulent **velocity** fluctuations.
The "traditional" Gill propeller anemometer is a three-axis arrangement, which is particularly suited for the measurement of the vertical **wind.** The so-called K-Gill propeller vane has two propellers, mounted at an angle of 90°, that are aligned with the mean wind by a **vane.** One particular type of sonic anemometer is sometimes referred to as "Gill sonic anemometer" in the literature of the early 1990s.

glacial—1. Pertaining to **ice**, especially in great masses, such as sheets of **land ice** or glaciers.
2. Pertaining to an interval of **geologic time** that was marked by an equatorward advance of **ice** during an **ice age**; the opposite of **interglacial phase.**
These intervals are variously called **glacial periods, glacial epochs**, glacial "stages," etc.

glacial anticyclone—(Or glacial high.) A type of **semipermanent anticyclone** that has been said to overlie the **ice caps** of Greenland and Antarctica.
As presented by W. H. Hobbs, it was thought that these anticyclones played a dominant part in the world atmospheric **circulation**, but modern theory backed by limited observation has tended to diminish their importance and even to question their reality. *See* **thermal high.**
Hare, F. K., 1951: *Compendium of Meteorology*, 961–963.

glacial drift—All rock material in **transport** by **glacial** ice, and all deposits predominantly of glacial origin made in the sea or in bodies of glacial meltwater, including rocks rafted by icebergs.
"Glacial drift occurs as scattered rock fragments, as till [rocks mixed with finer material], and as outwash [fine material with no rocks]. Contrast with **angular drift.**"(from *Glossary of Arctic and Subarctic Terms* 1955).
Arctic, Desert, Tropic Information Center (ADTIC) Research Studies Institute, 1955: *Glossary of Arctic and Subarctic Terms*, ADTIC Pub. A-105, Maxwell AFB, AL, 90 pp.

glacial epoch—1. Any of the **geologic epochs** characterized by an **ice age.**
Thus, the **Pleistocene epoch** may be termed a "glacial epoch."
2. Generally, an interval of **geologic time** that was marked by a major equatorward advance of **ice.**
This has been applied to an entire **ice age** or (rarely) to the individual **glacial** "stages" that make up an ice age. The term "epoch" here is not used in the most technical sense of a **geologic epoch.**

glacial geology—The study of land features resulting from **glaciation.**

glacial high—*See* **glacial anticyclone.**

glacial maximum—The time of greatest **ice** volume and/or position of greatest areal extent of any **glacierization**.

glacial period—1. Any of the **geologic periods** that embraced an **ice age**.

For example, the **Quaternary period** may be called a "glacial period."

2. Generally, an interval of **geologic time** that was marked by a major equatorward advance of **ice**.

This may be applied to an entire **ice age** or (rarely) to the individual **glacier** "stages" that make up an ice age. The term "period" here is not used in the most technical sense of a **geologic period**.

glacial phase—Same as **glacial period**.

glacial rebound—Vertical raising of a portion of the earth's crust following the removal of an **ice** mass.

Glacial rebound is the reaction to **deglaciation**.

glacial trough—Large area, such as those in the Alps, carved out to depths of hundreds to thousands of meters by the erosive action of **glaciers**.

glaciation—1. Alteration of any part of the earth's surface by passage of a **glacier**, chiefly by **glacial** erosion or deposition; distinguish from **glacierization**.

2. As used in many texts, particularly with respect to the **ice ages**, same as **glacierization**, for example, "**Pleistocene glaciation**."

3. The transformation of **cloud** particles from **supercooled water** drops to **ice crystals**.

Thus, a **cumulonimbus** cloud is said to have a glaciated upper portion.

glaciation limit—1. For a given locality, the lowest **altitude** at which **glaciers** can develop.

2. Same as **glacial maximum**.

glacier—A mass of **land ice**, formed by the further recrystallization of **firn**, flowing continuously from higher to lower elevations.

This term covers all such **ice** accumulations from the extensive **continental glacier** to tiny **snowdrift glaciers**. Nearly all glaciers are classified according to the topographical features with which they are associated, for example, highland glacier, **plateau glacier**, **piedmont glacier**, **valley glacier**, **cirque glacier**. They are also classified according to their seasonal temperatures, or melting characteristics, as **temperate glaciers** or **polar glaciers**. If a glacier is flowing, it is active or living; but an active glacier may be advancing or retreating depending upon the rate of flow compared to the rate of **ablation** at the terminus. A glacier that has ceased to flow is termed stagnant or dead.

glacier berg—*See* **iceberg**.

glacier flood—A sudden release of substantial amounts of meltwater from a **glacier**.

glacier flow—The movement of **ice** caused by gravitational force.

See **flow law**.

glacier front—(Also called glacier terminus.) The leading edge of a **glacier**.

glacier ice—Any **ice** that is or was once a part of a **glacier**.

It has been consolidated from **firn** by further melting and refreezing and by **static pressure**. Firn becomes glacier ice once the pockets of air between individual ice grains are no longer interconnected. Glacier ice may be found in the sea as an **iceberg**.

glacier tongue—An extension of a **glacier** or **ice stream** projecting seaward, usually afloat.

Armstrong, T., B. Roberts, and C. Swithinbank, 1973: *Illustrated Glossary of Snow and Ice*, Scott Polar Research Institute, Special Pub. No. 4, 2d ed.; The Scholar Press Ltd., Menston Yorkshire, UK..

glacier wind—(Or glacier breeze.) A shallow **gravity wind**, along the icy surface of a **glacier**, caused by the **temperature** difference between the air in contact with the glacier and **free air** at the same **altitude**.

The glacier wind does not reverse itself diurnally as do mountain and valley winds, but it reaches its maximum intensity in the early afternoon. The glacier wind is characterized by strongly **turbulent flow**. *See* **katabatic wind**.

Defant, F., 1951: *Compendium of Meteorology*, p. 663.

glacierization—The covering of a land area by **glacier ice**.

This term was coined by G. Taylor in the Antarctic and introduced by Wright and Priestly (1922) to distinguish the act of **glacial** inundation from its geologic consequences (**glaciation**).

It is growing in use in Great Britain but still is considered unnecessary by some American geologists, who use "glacier covering."

Wright, Sir C. E., and Priestley, Sir R. E., 1922: *Glaciology*, British (Terra Nova) Antarctic Expedition 1910–1913; Harrison and Sons Ltd, London, 581 pp.

glaciology—The study of **snow** and **ice** on the earth's surface.
It includes the study of **glaciers**, snow, and **sea ice**.

glaçon—Very generally, a piece of **sea ice** that is smaller than a medium-sized **floe**.

glare—Light sufficiently intense to cause physical discomfort or to reduce **contrast** below the level at which detail is discernible.

glare ice—Any highly reflective sheet of **ice** on water, land, or **glacier**.

glass—In nautical terminology, a contraction for "weather glass" (a **mercury barometer**).

glaves—(Also called glave, glavis.) A **foehn**like **wind** of the Faroe Islands.

glaze—(Also called glaze ice, glazed frost, verglas.) A coating of **ice**, generally clear and smooth, formed on exposed objects by the **freezing** of a film of **supercooled water** deposited by **rain**, **drizzle**, **fog**, or possibly condensed from supercooled **water vapor**.
Glaze is denser, harder, and more transparent than either **rime** or **hoarfrost**. Its **density** may be as high as 0.8 or 0.9 g cm^{-3}. Factors that favor glaze formation are large **drop** size, rapid **accretion**, slight **supercooling**, and slow **dissipation** of **heat of fusion**. The opposite effects favor rime formation. The accretion of glaze on terrestrial objects constitutes an **ice storm**; as a type of **aircraft icing** it is called **clear ice**. Glaze, as well as rime, may form on ice particles in the **atmosphere**. Ordinary **hail** is composed entirely (or nearly so) of glaze; the alternating clear and opaque layers of some hailstones represent glaze and rime, deposited under varying conditions around the growing **hailstone**. *Compare* **rime**, **rime ice**, **hard rime**, **soft rime**.

glaze ice—Same as **glaze**.

glaze storm— A **precipitation** event during which a coating of **ice** forms on exposed objects by the **freezing** of **supercooled water** deposited by **rain**, **drizzle**, and/or **fog**.
See **ice storm**.

glazed frost—British term for **glaze**.

glide path—A descent profile determined for vertical guidance during a final approach or landing of an aircraft.
See **instrument landing system**.

glime—An **ice** coating with a consistency intermediate between **glaze** and **rime**.

glitter—The spots of **light** reflected from a **point source** by the surface of the sea.

global area coverage—One full **orbit** of **AVHRR** data at reduced resolution (3.3 km x 4.0 km) stored on **digital** tape recorders for subsequent playback to **NOAA** Command and Data Acquisition ground stations.

global baseline datasets—Datasets designed to provide a complete historical record of the **climate**.
Such datasets include parameters such as **surface air temperatures**, **sea surface temperatures**, tropospheric and stratospheric temperatures, atmospheric **water vapor**, and **radiative fluxes** at the top of the **atmosphere**. For maximum value, such datasets should be designed to minimize the inhomogeneities associated with changes in instrumentation, **instrument exposure**, or observing techniques.

global circuit—The structure and combination of processes set up by the conductive earth, the conductive **ionosphere**, and all agents of electrification within the resistive **troposphere**.
The so-called dc global circuit is characterized by the ionosphere potential and the ac global circuit refers to the **Schumann resonances** within the **earth–ionosphere waveguide**.

global circulation—*See* **general circulation**.

global circulation model—*See* **general circulation model**.

global climate system—Same as **climate system**.

Global Data Processing System—(Abbreviated GDPS.) The coordinated global system of meteorological centers and arrangements for the processing, storage, and retrieval of **meteorological information** within the framework of the World Weather Watch.
It includes World Meteorological Centers, **Regional Meteorological Centers**, and National Meteorological Centers.

Global Level of the Sea Surface—(Abbreviated GLOSS.) A worldwide network of **sea level** gauges, defined and developed under the auspices of the Intergovernmental Oceanographic Commission.
Its purpose is to monitor long-term variations in the level of the sea surface globally by reporting the observations to the **Permanent Service for Mean Sea Level**.

global model—*See* **general circulation model**.

Global Observing System—The coordinated global system of methods, techniques, and facilities for making observations on a world-wide scale within the framework of the World Weather Watch.

Global Positioning System—(Abbreviated GPS.) A navigation system based on a constellation of 24 low earth-orbiting satellites having highly accurate clocks and the computational capacity to triangulate positions near the earth's surface.
Developed by the U.S. Department of Defense, the system has the capability of determining position to an **accuracy** of 30–100 m. If systems at two locations are used with long integration times, positions may be determined within millimeters of a known reference position.

global radiation—The total of **direct solar radiation** and **diffuse sky radiation** received by a unit horizontal surface.
Global radiation is measured with a **pyranometer**.

global radiation—**Solar radiation**, direct and diffuse, received from a solid angle of 2π steradians on a horizontal surface.

Global Telecommunications System—(Abbreviated GTS.) The globally integrated communications that provides for the collection and distribution of meteorological data within the World Weather Watch.

glory—Small, faintly colored rings of **light** surrounding the **antisolar point**, seen when looking down at a **water cloud**.
Having a radius of only a few degrees, the glory often surrounds an airplane's shadow cast on a **cloud** or a mountain climber's shadow cast on **fog** in a valley. (The shadow of the observer plays no role in the phenomenon other than as easy way of quickly finding the **antisolar point**.)The glory is not as easily described by simple theory as is the **corona**. Nevertheless, some similarities hold: the angular size of a particular ring is approximately inversely proportional to **drop** size. The result is that glories are formed by droplets with radii smaller than about 25 μm (the rings from larger droplets are washed out by the angular width of the sun). Similarly, a broad **droplet** distribution will destroy the glory.

GLOSS—Abbreviation for **Global Level of the Sea Surface**.

glow discharge—Generic term for any gaseous **electrical discharge** that produces luminosity.
Thus **corona discharge** is an example of glow discharge, but **point discharge** is not. Relatively high electric field strengths are required for glow discharges, for the **density** of radiatively recombining gas atoms and molecules must be high. While **spark discharge** and **lightning** fall under this definition, this term is most commonly applied to the more continuous, quiescent, and less brilliant discharges.

GMS—Abbreviation for **Geostationary Meteorological Satellite**.

goal-directed reasoning—A problem-solving procedure in which candidate solutions are generated from each possible goal of a **knowledge-based system** and evidence is then gathered to determine whether each is acceptable for the current situation.
See also **backward chaining**.

GOES—Abbreviation for **Geostationary Operational Environmental Satellite**.

GOES High-Resolution Interferometer Sounder—(Abbreviated GHIS.) A Michelson **interferometer** being considered for geostationary **sounding** applications.

GOES Variable—(Abbreviated GVAR) The format of **GOES** I-M imager and **sounder** data as retransmitted to ground stations.

Gold slide—A slide rule used on British ships to compute **barometric corrections** and reduction of **pressure** to **sea level**.
It includes the effects of **temperature**, latitude, **index correction**, and **barometric** height above sea level.

goldbeater's skin hygrometer—A **hygrometer** using goldbeater's skin as the sensitive element.
Variations of the physical dimensions of the skin caused by its **hygroscopic** character indicate atmospheric **relative humidity**. The length of a piece of goldbeater's skin changes between 5%

and 7% for a change in **humidity** from 0% to 100%. The **time constant** of response becomes extremely long both at low ambient temperatures and at very high and very low relative humidity. (Note: Goldbeater's skin is the prepared outside membrane of the large intestine of an ox; it is used in goldbeating to separate the leaves of the metal.)

golfada—A heavy **gale** of the Mediterranean.

GOMS—Abbreviation for **Geostationary Operational Meteorological Satellite**.

goniometer—An instrument used for measuring geometric angles.
> *See* **radio direction finder**.

goodness of fit—In general terms, a quantitative measure of the ability of an assumed functional form to fit a given set of data.
> In meteorology, "goodness of fit" usually refers to the size of the **residual variance** for a **regression function** used to fit a set of data. *See* **regression**.

gorge wind—A **gap wind** or **canyon wind** through a gorge.

gosling blast—(Also called gosling storm.) A sudden **squall** of **rain** or **sleet** in England.

gowk storm—(Also called gowh's storm.) In England, a **storm** or **gale** occurring at about the end of April or the beginning of May.

GPS—Abbreviation for **Global Positioning System**.

grab sampling—Rapid collection of whole-air samples into a suitable container, such as an evacuated canister or a polytetrafluoroethylene (PTFE) bag.

gradex method—The substitution of the frequency distributions of **rainfall** depths for the **frequency distribution** of **flood** flows.
> This implicitly assumes that critical storms are large enough to overwhelm variations in **soil moisture**, rainfall timing, **rainfall intensity** variations, and other parameters.

gradient—1. The space rate of decrease of a function.
> The gradient of a function in three space dimensions is the **vector** normal to surfaces of constant value of the function and directed toward decreasing values, with magnitude equal to the rate of decrease of the function in this direction. The gradient of a function f is denoted by $-\nabla f$ (without the minus sign in the older literature) and is itself a function of both space and time. The **ascendent** is the negative of the gradient. In **Cartesian coordinates**, the expression for the gradient is

$$-\nabla f = \frac{\partial f}{\partial x}\,\mathbf{i} + \frac{\partial f}{\partial y}\,\mathbf{j} + \frac{\partial f}{\partial z}\,\mathbf{k}.$$

For expressions in other coordinate systems, see Berry et al. (1945).
> 2. Often loosely used to denote the magnitude of the gradient or ascendent (i.e., without regard to sign) of a horizontal **pressure field**.
> > Berry, F. A., E. Bollay, and N. Beers, 1945: *Handbook of Meteorology*, 224–225.

gradient current—In **oceanography**, a current determined by the condition that the horizontal **pressure gradient** due to the (hydrostatic) distribution of mass balances the **Coriolis force** due to the earth's rotation.
> The gradient current corresponds to the **geostrophic wind** in meteorology. In practice, the distribution of **density** is determined by measurements of **salinity** and **temperature** at a series of depths in a number of positions. From this the **geopotential topography** of any **isobaric surface** relative to any other isobaric surface may be computed and the horizontal pressure gradient may be expressed by the **geopotential** slope of the isobaric surface. In this way relative isobaric surface currents are obtained, corresponding to **thermal wind** in meteorology. If one isobaric surface is known to be level, the **absolute** geopotential topography of any other surface may be computed by reference to this, and hence absolute gradient currents are obtained. Where no isobaric surface is known to be level, the total gradient current will consist of the **relative gradient current**, due to the distribution of density, and the **slope current**, due to that portion of the **inclination** of the isobaric surfaces that is not the result of the distribution of density. *See also* **geostrophic current**.

gradient flow—Horizontal frictionless flow in which **isobars** and **streamlines** coincide, or equivalently, in which the **tangential acceleration** is everywhere zero.
> The balance of normal forces (**pressure force**, **Coriolis force**, **centrifugal force**) is then given by the **gradient wind** equation. Important special cases of gradient flow, in which two of the

normal forces predominate over the third, are **geostrophic flow**, **inertial flow**, and **cyclostrophic flow**.

gradient Richardson number—A dimensionless ratio, Ri, related to the buoyant production or **consumption** of **turbulence** divided by the **shear production** of turbulence.

It is used to indicate **dynamic stability** and the formation of turbulence:

$$\text{Ri} = \frac{\dfrac{g}{T_v}\dfrac{\partial \theta_v}{\partial z}}{\left(\dfrac{\partial U}{\partial z}\right)^2 + \left(\dfrac{\partial V}{\partial z}\right)^2},$$

where θ_v is **virtual potential temperature**, T_v is virtual absolute temperature, z is height, g is gravitational **acceleration**, and (U, V) are the **wind** components toward the east and north. The **critical Richardson number**, Ri_c, is about 0.25 (although reported values have ranged from roughly 0.2 to 1.0), and flow is dynamically unstable and turbulent when $\text{Ri} < \text{Ri}_c$. Such turbulence happens either when the **wind shear** is great enough to overpower any stabilizing buoyant forces (numerator is positive), or when there is **static instability** (numerator is negative). *Compare* **bulk Richardson number**, **flux Richardson number**, **Froude number**.

gradient transport theory—A first-order **turbulence closure** approximation that assumes that turbulent fluxes of any variable flow down the local **gradient** of that mean variable; analogous to molecular **transport**.

This local turbulence closure approach assumes that **turbulence** consists of only small **eddies**, causing diffusion-like **transport**. An example is

$$\overline{w'c'} = -K\frac{\partial \overline{c}}{\partial z},$$

where the vertical **kinematic flux** $\overline{w'c'}$ of a **pollutant** is modeled as being equal to an eddy thermal diffusivity K times the vertical gradient of mean concentration $\overline{c}$. This theory is also called **K-theory** or eddy-viscosity theory. *Compare* **higher-order closure**, **nonlocal closure**, **transilient turbulence theory**.

gradient wind—Any horizontal **wind** velocity tangent to the **contour line** of a **constant-pressure surface** (or to the **isobar** of a **geopotential surface**) at the point in question.

At such points where the wind is **gradient**, the **Coriolis acceleration** and the **centripetal acceleration** together exactly balance the **horizontal pressure force**:

$$\frac{V_{gr}}{R} + fV_{gr} = -g\frac{\partial z}{\partial n},$$

where V_{gr} is the gradient wind speed, R the radius of curvature of the path, f the **Coriolis parameter**, g the **acceleration of gravity**, z the **elevation** of the **constant-pressure surface**, and n the direction normal to the **streamline** and **contour** toward decreasing elevation, that is, to the left of the direction of flow in the Northern Hemisphere and to the right in the Southern Hemisphere. Since $R > 0$ for **cyclonic** flow and $R < 0$ for **anticyclonic** flow, it follows that the cyclonic gradient speed is less than (and the anticyclonic gradient speed is greater than) the **geostrophic wind** speed for the same latitude and **pressure force**.

Holton, J. R., 1992: *An Introduction to Dynamic Meteorology*, 3d edition, Academic Press, 67–69.

gradient wind level—Same as **geostrophic wind level**.

grading—The degree of **mixing** of **particle** size classes in a **sediment**.

Well-graded sediments are those with a more or less uniform distribution of sizes; poorly graded implies uniformity in size or lack of a continuous distribution.

graduation—In **statistics**, same as **curve fitting**.

grains of ice—A British term for **ice pellets (sleet)**.

Sleet, in turn, has a different meaning in Great Britain than it does in the United States. In British terminology, **sleet** is **precipitation** in the form of a mixture of **rain** and **snow**.

gram—A cgs unit of mass; originally defined as the mass of 1 cubic centimeter of water at a **temperature** of 4.5°C but now taken as the one-thousandth part of the **standard** kilogram, a mass preserved by the International Bureau of Weights and Measures at Sevrés, France.

gram calorie—(Abbreviated gm cal, cal.) *See* **calorie**.

gram-mole—(Often called gram-molecular weight.) A mass of a substance in grams numerically equal to its **molecular weight**.

Example: A gram-mole of salt (NaCl) is 58.44 grams.

granular snow—1. Same as **snow grains**.

2. Same as **spring snow**.

granules—Convective cells that penetrate from below into the solar **photosphere**.

These cells have an irregular shape with a typical size of 1 to 2 arc seconds (1000 km). They are relatively bright, separated by narrow dark lanes.

graphing board—A board that holds graph paper on which is plotted information obtained from a **pilot-balloon observation**.

The vertical **scale** of the board, engraved on a metal plate, is an **altitude** scale indicating the assumed height of the balloon at the end of each minute of flight. The horizontal scale is engraved on a movable T-square and is calibrated in units of direction and speed. The result is a curve of **wind speed** and **direction** against height.

Grashoff number—(Abbreviated Gr.) A **nondimensional parameter** used in the theory of **heat transfer**, defined by

$$\text{Gr} = L^3 g \frac{T_1 - T_0}{\nu^2 T_0},$$

where L is a representative length, T_1 and T_0 representative temperatures, g the **acceleration of gravity**, and ν the **kinematic viscosity**.

The Grashoff number is associated with the **Reynolds number** and the **Prandtl number** in the study of **convection**.

grass—(Obsolete.) Background **noise** on a radar **A-display**.

grass minimum—The **minimum temperature** shown by a **minimum thermometer** exposed in an open situation with its bulb at the level of the tops of the grass blades in short turf.

grass temperature—The **temperature** registered by a **thermometer** with its bulb at the level of the tops of the grass blades in short turf.

grassland climate—*See* **subhumid climate**.

graupel—Heavily rimed **snow** particles, often called **snow pellets**; often indistinguishable from very small **soft hail** except for the size convention that **hail** must have a diameter greater than 5 mm.

Sometimes distinguished by shape into conical, hexagonal, and lump (irregular) graupel.

gravimetric hygrometry—A method using **absorption** of **water vapor** by a desiccant from a known volume of air.

The mass of water vapor is determined through the weighing of the drying agent before and after absorbing the vapor. This method is used mostly in national **calibration** laboratories for providing an **absolute** reference **standard**.

gravitation—(Also called **gravity**, force of gravitation.) The mutual attraction between bodies possessing mass.

Within **Newtonian mechanics** the gravitational force between two point masses (bodies small in extent compared with r, the distance between them) m_1 and m_2 is

$$F = \frac{G m_1 m_2}{r^2},$$

where the Newtonian constant of gravitation G is 6.672×10^{-11} m³ kg s⁻². This force is directed along the line between the two point masses. Forces between extended bodies can be determined by conceptually dividing them into sufficiently small parts, calculating the gravitational forces between them according to the previous equation, and adding these forces vectorially. The gravitational **field** intensity (or simply gravitational field) of a system of masses at any point is the gravitational force the system would exert on a unit mass at that point. Thus gravitational, electric, and magnetic fields are defined in a similar way. A gravitational field is an **acceleration** field, the acceleration due to (not of) gravity. *See* **apparent gravity**.

gravitational convection—*See* **convection**.

gravitational equilibrium—Same as **diffusive equilibrium**.

gravitational head—The **potential energy** per unit weight of fluid due to its **elevation** above an arbitrary **datum**.

gravitational instability—Same as **buoyant instability**.

gravitational potential—The negative of the **gravitational potential energy** per unit mass.
The **gradient** of the gravitational potential is the gravitational **field**. *See* **gravitation**; *compare* **geopotential**.

gravitational potential energy—The **potential energy** a body has by virtue of its position in the gravitational **field** of other bodies.
See **gravitation, potential energy**.

gravitational stability—*See* **buoyant instability**.

gravitational tide—An **atmospheric tide** due to gravitational attraction of the sun or moon.
The semidiurnal **solar atmospheric tide** is partly gravitational; the semidiurnal **lunar atmospheric tide** is totally gravitational. *See* **thermal tide, tide-producing force**.

gravitational water—Water in the larger, noncapillary pores of the soil that is free to drain because there is insufficient tension to hold the water against the force of gravity; it generally flows from the soil in the first 24 hours following its appearance in the soil profile.
See also **detention storage, field capacity, retention storage**.

gravitational wave—Same as **gravity wave**.

gravity—*See* **gravitation**.

gravity—(Or force of gravity.) The force imparted by the earth to a mass that is at rest relative to the earth.
Since the earth is rotating, the force observed as gravity is the **resultant** of the force of **gravitation** and the **centrifugal force** arising from this rotation. It is directed normal to **sea level** and to its **geopotential surfaces**. The magnitude of the force of gravity at sea level decreases from the poles, where the centrifugal force is zero, to the **equator**, where the centrifugal force is a maximum but directed opposite to the force of gravitation. This difference is accentuated by the shape of the earth, which is nearly that of an oblate spheroid of revolution slightly depressed at the poles. Also, because of the asymmetric distribution of the mass of the earth, the force of gravity is not directed precisely toward the earth's center. The magnitude of the force of gravity per unit mass (**acceleration of gravity**) g may be determined at any latitude ϕ and at any geometric height z (meters) above sea level in the **free air** from the following empirical formula:

$$g = g_\phi - (3.085462 \times 10^{-4} + 2.27 \times 10^{-7}\cos 2\phi)z$$

$$+ (7.254 \times 10^{-11} + 1.0 \times 10^{-13}\cos 2\phi)z^2$$

$$- (1.517 \times 10^{-17} + 6 \times 10^{-20}\cos 2\phi)z^3 \ (cm \ s^{-2}),$$

where $g_\phi = 980.6160 (1 - 0.0026373 \cos 2\phi + 0.0000059 \cos^2 2\phi)$ is the sea level value of gravity (cm s^{-2}) at latitude ϕ. This formula as applied near the earth indicates that gravity changes very little with height or latitude, so that for rough calculations a constant value of 980 cm s^{-2} may be used. Besides these variations in the magnitude of the force of gravity, there are more localized variations controlled by the **topography** of the earth's surface, and the distribution of mass beneath. The magnitude of the force of gravity is usually called either gravity, acceleration of gravity, or **apparent gravity**. *See* **virtual gravity, geopotential height, standard gravity**.
    List, R. J., Ed., 1951: *Smithsonian Meteorological Tables*, 6th rev. ed., 488–494.

gravity correction—*See* **barometric corrections**.

gravity current—A flow driven by horizontal **pressure gradients**, due typically to the flow of a fluid of one **density** intruding into an ambient fluid of different density.
In nonrotating fluid, the head of the current travels with uniform horizontal speed

$$C(g'H)^{1/2},$$

in which g' is the **reduced gravity**, H is the current depth, and C is a constant of order unity. **Sea breezes**, landslides, and some instances of cold water formation are common manifestations of gravity currents in geophysical circumstances.

gravity flow—1. In general, water flow in which gravitational forces predominate.
    2. In **glaciology, glacial** movement in which **ice flow** results from the downslope component of gravitational force.

gravity wave—(Also called gravitational wave.) A **wave disturbance** in which **buoyancy** (or **reduced gravity**) acts as the restoring force on parcels displaced from **hydrostatic equilibrium**.

There is a direct oscillatory conversion between **potential** and **kinetic energy** in the **wave motion**. Pure gravity waves are stable for fluid systems that have **static stability**. This static stability may be 1) concentrated in an **interface** or 2) continuously distributed along the axis of **gravity**. The following remarks apply to the two types, respectively. 1) A **wave** generated at an **interface** is similar to a **surface wave**, having maximum amplitude at the interface. A plane gravity wave is characteristically composed of a pair of waves, the two moving in opposite directions with equal speed relative to the fluid itself. In the case where the upper fluid has zero **density**, the interface is a **free surface** and the two gravity waves move with speeds

$$c = U \pm \left[\frac{gL}{2\pi} \tanh\left(\frac{2\pi H}{L} \right) \right]^{1/2},$$

where U is the current speed of fluid, g the **acceleration of gravity**, L the **wavelength**, and H the depth of the fluid. For **deep-water waves** (or Stokesian waves or short waves), $H \gg L$ and the **wave speed** reduces to

$$c = U \pm \left(\frac{gL}{2\pi} \right)^{1/2}.$$

For **shallow-water waves** (or Lagrangian waves or long waves), $H \ll L$, and

$$c = U \pm (gH)^{1/2}.$$

All waves of consequence on the ocean surface or interfaces are gravity waves, for the **surface tension** of the water becomes negligible at wavelengths of greater than a few centimeters (*see* **capillary wave**). 2) Heterogeneous fluids, such as the **atmosphere**, have static stability arising from a **stratification** in which the **environmental lapse rate** is less than the **process lapse rate**. The atmosphere can support short internal gravity waves and long external gravity waves. The short waves (of the order of 10 km) have been associated, for example, with **lee waves** and **billow waves**. Such waves have vertical accelerations that cannot be neglected in the vertical equation of **perturbation motion**. The long gravity waves, moving relative to the atmosphere with speed $\pm (gH)^{1/2}$, where H is the height of the corresponding **homogeneous atmosphere**, have small vertical accelerations and are therefore consistent with the **quasi-hydrostatic approximation**. In neither type of gravity wave, however, is the **horizontal divergence** negligible. For meteorological purposes in which neither type is desired as a solution, for example, **numerical forecasting**, they may be eliminated by some restriction on the magnitude of the horizontal divergence. The above discussion is based upon the **method of small perturbations**. In certain special cases of water waves, for example, the **Gerstner wave** or the **solitary wave**, a theory of finite-amplitude disturbances exists. *See* **shear-gravity wave**.

Gill, A. E., 1982: *Atmosphere–Ocean Dynamics*, Academic Press, 95–188.

gravity wave drag—A generally **zonal** acceleration produced by upward propagating **gravity waves** at levels where the waves break.

Gravity waves grow in **amplitude** as they propagate upward from their surface **source** regions into lower-density regions. When large amplitudes are attained, the waves break, leading to **turbulence** and **dissipation**. This produces a zonal force per unit mass on the zonal mean flow. Gravity wave drag plays an important role in explaining the zonal mean flow and **thermal** structure at higher atmospheric levels, particularly in the **mesosphere**.

gravity wave drag parameterization—A **parameterization** designed to approximate the effect on the resolved flow of the **gravity wave drag** that would be generated by unresolved subgrid-scale **topography** in an **atmospheric model**.

gravity wind—(Also called drainage wind.) A **wind** resulting from cold air running or flowing down a slope, caused by greater air density near the slope than at the same **altitude** some distance horizontally from the slope.

Generally used when cold air is locally generated by a chilled slope surface during periods of weak **synoptic** or other larger-scale winds, as with **katabatic** or **drainage winds**. Slopes can be gentle as found over rolling **topography** or into river valleys, or steep as on mountains or mountain ranges. Although usually applied to smaller (individual slope) scale flows, the term is occasionally used to include **fall winds**, or air advected from a cold **source region** elsewhere, then spilling over and accelerating down a slope, as with a **bora** or a **cold front** passage over **orography**. *See* **downslope wind**.

gray absorber—An absorber acting like a **graybody**.

In practice, natural surfaces can be approximated as gray absorbers over only limited spectral ranges. The cloud-free **atmosphere**, however, is a selective absorber, and should not be approximated as a gray absorber.

gray ice—Young **ice** 10–15 cm thick, less elastic than **nilas**.

graybody—A hypothetical "body" that absorbs some constant fraction, between 0 and 1, of all **electromagnetic radiation** incident upon it.

This fraction is the **absorptivity** and is independent of **wavelength**. As such, a graybody represents a surface of absorptive characteristics intermediate between those of a white body and a **blackbody**. No such substances are known in nature.

grayscale—A black-and-white **image** (with zero or more shades of gray between pure black and pure white).

Also, the mapping relating an **image level** to a **display** level on a **monochromatic** display.

grease ice—(Also called ice fat, lard ice.) A thin skin of **frazil** crystals coagulated on the sea surface having a dark, greasy appearance.

It precedes the development of **shuga**.

Great Basin high—A **high-pressure system** centered over the Great Basin of the western United States.

It is a frequent feature of the **surface chart** in the winter season.

great circle—A line formed by the intersection of the surface of a sphere and a plane that passes through the center of the sphere.

great-circle course—A **course**, route, or track along a **great circle** over the earth's surface.

The great-circle course is the least distance between two points on a sphere. The angle between the great circle and **true north** changes along the course except along a **meridian** or the **equator**. *See* **rhumb-line course, grid navigation**.

Great Whirl—*See* **Somali Current**.

grécale—*See* **gregale**.

greco—An Italian name for the northeast **wind**.

It was given by Roman sailors to the northeast wind in the Gulf of Lions because it came from the direction of the Greek colony of Marsala (Marseilles). Wind names of similar origin are common in the western Mediterranean, for example, **gregale**.

green belt—Same as **frostless zone**.

green flash—A flash of green **light** seen on or (seemingly) adjacent to the upper rim of the low sun (at either **sunrise** or **sunset**).

The green flash is a **mirage**, but the **image** formed in this case is of a portion of the sun rather than of an earthbound object. In addition to the displacement and **distortion** that is characteristic of mirages, there is also significant **dispersion**. The upper edge of the low sun normally has a thin green rim (occasionally blue) that is too narrow to be seen by the naked eye unless the rest of the sun is obstructed, say, by the **horizon**. It is often asserted that the green flash is seen in this way: a mere **transient** view of the **green rim** between **obscuration** by the rest of the sun and obstruction by the horizon. Yet such a sequence produces a singularly poor flash. Rather, the remarkable flashes always seem to involve multiple and magnified images of the green rim. Indeed, the presence of such multiple images of a small portion of the sun is a good **indicator** of a forthcoming flash. The optical **signature** of multiple images is a serrated edge to the sun. The **refraction** that displaces the image of the low sun up from the position it would occupy in the absence of an **atmosphere** does so more strongly for shorter wavelengths. This leads to a red rim on the bottom of the sun and a blue or green rim on the top.

green iceberg—Desalinated seawater frozen to the base of some Antarctic **ice shelves** that becomes exposed to view when an **iceberg** separates from the shelf and capsizes.

The green color of the **ice**, and of the seawater from which it froze, is due to dissolved organic matter.

green moon—*See* **blue moon**.

green rim—(Also called green segment.) The upper rim of the low sun usually displays a green rim.

The **transient** sighting of this rim, between its **obscuration** by the rest of the sun and its obstruction by the **horizon**, is sometimes credited with being the origin of the **green flash**. But, while the color of the flash is provided by the green rim, the size and transient nature has more to do with multiple-image **mirages**.

Green's function—(Also called influence function.) A function that is the known solution of a homogenous differential equation in a specified region and that may be generalized (if the equation is **linear**) to satisfy given boundary or initial conditions, or a nonhomogeneous differential equation.

It is thus an alternative method to the **Fourier transform** or **Laplace transform**, applicable to many of the same problems. The Green's function method takes a fundamental solution and assigns it a weight at each point, say, of the boundary, according to the value of the given boundary condition there; the Fourier method analyzes the entire boundary condition into **wave** components thus assigning each of these a weight or **amplitude**. Both methods then obtain the final solution by summation or integration.

Green's theorem—A form of the **divergence theorem** applied to a **vector field** so chosen as to yield a formula useful in applying the **Green's function** method of solution of a **boundary-value problem**.

The most common form of the theorem is

$$\int_V (\phi \nabla^2 \psi - \psi \nabla^2 \phi)dV = \oint_S \left(\phi \, \frac{\partial \psi}{\partial n} - \psi \frac{\partial \phi}{\partial n} \right)dS,$$

where dV and dS are elements of the volume V and closed bounding surface S, respectively; ϕ and ψ are any twice differential functions with continuous second partial derivatives in V; n is the outer normal to S; and ∇^2 is the **Laplacian operator**.

green segment—Same as **green rim**.

green snow—A **snow** surface that has attained a greenish tint as a result of the growth within it of certain microscopic algae (**cryoplankton**).

green sun—*See* **blue moon**.

green thunderstorm—Any **thunderstorm** that is perceived by observers to be green.

The perceptually dominant **wavelength** of **light** from green thunderstorms ranges from blue-green to yellow-green. The purity of the color is generally low and the physical mechanism that causes the green appearance is not understood. Although green clouds often occur in conjunction with **severe weather**, there is no evidence to support anecdotal attributions of the cause of this green to specific characteristics of severe storms, such as **hail** or **tornadoes**.

Bohren, C. F., and A. B. Fraser, 1993: Green thunderstorms. *Bull. Amer. Meteor. Soc.*, **74**, 2185–2193.

greenhouse effect—The heating effect exerted by the **atmosphere** upon the earth because certain **trace gases** in the atmosphere (**water vapor**, **carbon dioxide**, etc.) absorb and reemit **infrared radiation**.

Most of the **sunlight** incident on the earth is transmitted through the atmosphere and absorbed at the earth's surface. The surface tries to maintain **energy balance** in part by emitting its own **radiation**, which is primarily at the **infrared** wavelengths characteristic of the earth's **temperature**. Most of the **heat** radiated by the surface is absorbed by trace gases in the overlying atmosphere and reemitted in all directions. The component that is radiated downward warms the earth's surface more than would occur if only the direct sunlight were absorbed. The magnitude of this enhanced warming is the greenhouse effect. Earth's annual mean surface temperature of 15°C is 33°C higher as a result of the greenhouse effect than the **mean temperature** resulting from **radiative equilibrium** of a **blackbody** at the earth's mean distance from the sun. The term "greenhouse effect" is something of a misnomer. It is an analogy to the trapping of heat by the glass panes of a greenhouse, which let sunlight in. In the atmosphere, however, heat is trapped radiatively, while in an actual greenhouse, heat is mechanically prevented from escaping (via **convection**) by the glass enclosure.

greenhouse gas stabilization—Alteration of the **anthropogenic** emission rates of **greenhouse gases** in order to maintain atmospheric concentrations of greenhouse gases at a particular level.

The amount by which **anthropogenic emissions** of greenhouse gases must be reduced to stabilize at present-day concentrations is substantial for most of the greenhouse gases.

greenhouse gases—Those gases, such as **water vapor**, **carbon dioxide**, **ozone**, **methane**, **nitrous oxide**, and **chlorofluorocarbons**, that are fairly transparent to the short wavelengths of **solar radiation** but efficient at absorbing the longer wavelengths of the **infrared radiation** emitted by the earth and **atmosphere**.

The trapping of **heat** by these gases controls the earth's **surface temperature** despite their presence in only **trace** concentrations in the atmosphere. **Anthropogenic emissions** are important additional sources for all except water vapor. Water vapor, the most important greenhouse gas, is thought to increase in concentration in response to increased concentrations of the other greenhouse gases as a result of feedbacks in the **climate system**.

349

Greenland anticyclone—The **glacial** anticyclone that is supposed to overlie Greenland; analogous to the **antarctic anticyclone**.

Greenland high—*See* **Greenland anticyclone**.

Greenland Ice Sheet—The contiguous **ice sheet** covering most of the Greenland subcontinent.
Strictly speaking, it does not also refer to the adjacent small **ice caps** and **glaciers** that are physically separated from the main ice mass.

Greenland Sea Deep Water—The coldest (below −1.1°C) and densest component of the **water masses** that mix to produce the **Arctic Bottom Water**.

grégal—*See* **gregale**.

gregale—The Maltese and best-known variant of a term for a strong northeast **wind** in the central and western Mediterranean and adjacent European land areas (stronger than the **levante**).
It occurs either with high pressure over central Europe or the Balkans and low pressure over Libya, when it may continue for up to five days, or with the passage of a **depression** to south or southeast, when it lasts only a day or two. It is most frequent in winter. The weather varies with the type of **pressure distribution** and the onshore or offshore direction of the wind. In Malta the gregale raises dangerous seas in the harbor. The principal variant is the grégal of the Côte d'Azur (French Mediterranean coast), humid and rainy because it is a wind off the Mediterranean deflected by the Alpine Massif (gargal of the Roussillon region). This is mainly a summer wind and dangerous for vines as it favors mildew. In Provençe and Languedoc (southern France) it is the gregau, in Spain and Minorca guergal, and in Corsica grécale; these winds are cold and dry and bring spring frosts. *Compare* **greco, levantera, lombarde**.

gregau—*See* **gregale**.

gregori wind—In the Tirol, an east **wind** during March and April.

grenade sounding— A method of **atmospheric sounding** in which **wind** velocities and temperatures are inferred, as functions of height to about 90 km, from analysis of the speed of travel of **sound waves** from a series of grenades fired at height intervals of a few kilometers from an ascending **meteorological rocket**.
The sounds of detonation are received at an array of microphones on the ground. The position and time of the grenade bursts are determined by optical or radio means.

grid—A set of points arranged in an orderly fashion on which specified variables are analyzed or predicted.
Various forms of horizontal and vertical grids, each with particular characteristics, have been devised for use in **numerical weather prediction**.

grid heading—In air navigation, a **heading** in terms of **grid meridians**.

grid length—The distance between adjacent points on a **grid**.
In cases where all variables are not described at all grid points, the grid length is generally considered to be the distance between a grid point and its nearest neighbor at which the same variables are located. *See* **staggered grid**.

grid meridians—An arbitrary set of straight lines constructed parallel to the 180° and 0° meridians.
They are used for **grid** navigations; **grid north** replaces **true north**.

grid navigation—(Sometimes called grid system, G system.) A system of navigation by use of an arbitrary set of **grid** lines, as constructed to navigation by use of latitude and longitude.
A grid navigation chart contains a set of straight lines (grid navigations) parallel to the 180° and 0° meridians. The lines are used like true meridians on a Mercator map. With respect to the **grid meridians**, great circles and rhumb lines are identical. Thus, some navigational problems are greatly simplified when **grid north** is used instead of **true north**. *See* **aerologation**.

grid nephoscope—A **direct-vision nephoscope** that consists of a grid work of bars mounted horizontally on the end of a vertical column and made free to rotate about the vertical axis.
The **observer** rotates the **grid** and adjusts his position until some feature of the **cloud** appears to move along the major axis of the grid. The **azimuth** angle at which the grid is set is taken as the direction of cloud motion.
Middleton, W. E. K., 1969: *Invention of the Meteorological Instruments*, Johns Hopkins Press, Baltimore, 270–271.

grid north—*See* **grid meridians**.

grid system—Same as **grid navigation**.

grid telescoping—The placement of sequentially finer grids within each other in order to obtain the highest **resolution** possible over a limited region with decreasing resolution outside of that area.

This permits a given numerical **model** to run over a large area while incurring the expense of high resolution only over a small region of particular interest. *See* **variable resolution model, nested grids**.

grid turbulence—**Turbulence** generated by a grid or screen, usually in a **wind** tunnel or water channel.

The turbulence is approximately homogeneous and the **turbulence kinetic energy** decays as $1/x$, where x is the distance **downstream**.

gridding—The procedure of locating environmental satellite imagery on a certain geographic **reference frame**, or placing latitude and longitude information on a satellite image.

Gringorten plotting position—The Gringorten plotting position for the rth ranked (from largest to smallest) **datum** from a **sample** of size n is the quotient:

$$\frac{r - 0.44}{n + 0.12}.$$

See **plotting position, probability paper**.

Gringorten, I. I., 1963: A plotting rule for extreme probability paper. *J. Geophys. Res.*, **68**, No. 3, 813–814.
Gumble, E. J., 1941: Probability interpretation of the observed return periods of floods. *Trans., Amer. Geophys. Union*, **21**, 836–850.

gross-austausch—The exchange of airmass properties and the associated **momentum** and **energy** transports produced on a worldwide scale by the **migratory** large-scale disturbances of middle latitudes.

When the atmospheric **circulation** is regarded as a large-scale **turbulence** process, the **cyclones** and **anticyclones** are considered to be **eddies** superposed on the average **zonal wind** currents. The **mixing length**, that is, the average distance over which these traveling eddies maintain the characteristics of their original **environment**, has a value near 10^8 cm. The coefficient of turbulent mass exchange or the exchange coefficient, which is a measure of the intensity of the large-scale exchange processes and hence of the intensity of the **general circulation**, has a magnitude of between 10^8 and 10^6 gm cm^{-1}s^{-1} (compared with a value of about 10^2 gm cm^{-1}s^{-1} for small-scale turbulence). On the basis of results obtained so far, there is considerable doubt whether turbulence concepts can be applied meaningfully to the large-scale features of the **atmosphere**.

Haurwitz, B., 1941: *Dynamic Meteorology*, 265–267.

gross interception loss—Total **rainfall** intercepted before reaching ground surface or open water body.

gross primary production—The amount of **energy** fixed by **photosynthesis** over a defined time period.

Grosswetterlage—"The mean pressure distribution (at **sea level**) for a time interval during which the positions of the stationary (**steering**) **cyclones** and **anticyclones** and the steering within a special **circulation** region remain essentially unchanged."

A Grosswetterlage is thus a certain large-scale **circulation pattern** that is mostly synonymous with a particular **long wave** regime.

Baur, F., 1951: *Compendium of Meteorology*, p. 825.

ground check—1. (Also called baseline check.) A procedure followed prior to the release of a **radiosonde** in order to obtain the **temperature** and **humidity** corrections for the **radiosonde system**.

2. Any instrumental check prior to the ground launch of an airborne experiment.

ground-check chamber—A chamber used to check the sensing elements of **radiosonde** equipment.

The chamber houses sources of **heat** and **water vapor** plus instruments for measuring **temperature, humidity**, and **pressure**. A motor-driven fan maintains air circulation in the chamber.

ground clutter—Radar **echoes** from trees, buildings, or other objects on the ground.

Such echoes may be caused by the **reflection** of **energy** back to the **radar** in the **main lobe** or sidelobes of the **antenna pattern** and, in **weather radar** applications, interfere with the meteorological echoes at the same **range**.

ground-controlled approach—(Commonly abbreviated GCA.) A system for the precise control of the approach and landing of aircraft by an operator on the ground.

Precision approach **radar** is employed to view the aircraft, and the radar operator "talks" the pilot down by giving him his position and suggested corrections for maintaining the "perfect approach." Currently, GCA is a required facility at an **all-weather airport**. If perfectly operated, this system would ensure safe landing under zero **ceiling** and **visibility** conditions; however, to provide a safety factor, **GCA minimums** are prescribed at nearly all installations.

ground data—Same as **ground truth**.

ground discharge—Same as **cloud-to-ground discharge**.

ground flash—Same as **cloud-to-ground flash** or **cloud-to-ground discharge**.

ground flux—The **transport** of an atmospheric quantity into or out of the ground.
　　The most common example is the transport of **sensible heat** into the ground due to **radiative heating** of the ground surface.

ground fog—1. According to U.S. weather observing practice, a **fog** that hides less than 0.6 of the sky and does not extend to the base of any clouds that may lie above it.
　　As an **obstruction to vision** in an **aviation weather observation**, ground fog is encoded GF.
　　2. *See* **radiation fog**.

ground frost—1. In British usage, a **freezing** condition injurious to vegetation, which is considered to have occurred when a **minimum thermometer** exposed to the sky at a point just above a grass surface records a **temperature (grass temperature)** of $-0.9°C$ ($30.4°F$) or below.
　　Since 1961 in Britain the **statistics** refer to the "number of days with **grass minimum** temperature below $0°C$" rather than to ground frost. A fuller discussion is given in McIntosh (1963). *See* **frost**.
　　2. *See* **frozen ground**.
　　　　McIntosh, D. H., 1963: *The Meteorological Glossary*, Her Majesty's Stationery Office, London, p. 12.

ground ice—(Also called subsoil ice, subterranean ice, underground ice, **stone ice**.) A body of **clear ice** in **frozen ground**.
　　Ice of this nature is most commonly found in more or less **permanently frozen ground (permafrost)** and may be of sufficient age to be termed **fossil ice**. *Compare* **frost mound, icing**.
　　　　Muller, S. W., 1947: *Permafrost, or Permanently Frozen Ground, and Related Engineering Problems*, p. 217.

ground ice mound—Same as **frost mound**.

ground inversion—An air layer with its base at the ground surface and in which **temperature** increases with height.
　　These often form at night over land under **clear** skies and are **statically stable**. *See* **inversion, lapse rate**.

ground layer—1. An air layer next to the ground surface with its vertical extent controlled by **atmospheric turbulence** and **surface roughness**.
　　See **surface boundary layer**.
　　2. The top layer of soil that interacts with the **atmosphere**.

ground speed—The speed of an airborne object relative to the earth's surface.
　　It is the magnitude of the **vector** sum of the object's **velocity** with respect to the air and the **wind** velocity, or, expressed in a different manner, the algebraic sum of the aircraft's **airspeed** and the **wind factor**.

ground state—The lowest **energy level** of an atom or molecule, corresponding to the most stable configuration of the atoms and electrons.
　　At the relatively cold temperatures encountered in the **atmosphere**, most species normally exist in their ground states. Excitation to higher energy levels usually occurs via the **absorption** of **radiation**.

ground streamer—An upward advancing column of high ionization (a **streamer** or arc) that typically ascends from a point on the earth's surface toward a descending **stepped leader**.
　　The ground streamer usually joins the stepped leader about 50 m above the ground, after which the upward propagating **light** and **current** of the **return stroke** begin. Ground streamers occur because of the very high electric field intensities that build up directly below the descending, charged stepped leader. Often, more than one ground streamer starts up from the general area under a descending leader, but usually only one makes contact with the leader.

ground swell—**Swell** as it passes through shallow water; it is characterized by a marked increase in height in water shallower than one-tenth **wave length**.
　　To the seaman, ground swell is an indication of **shoal** water; to the shore-dweller, it is often an indication of approaching bad weather.

ground target—Any radar **target** on the ground.

Such targets include buildings, mountains, hills, trees or shrubs, and automobiles, which may be detected in the **main lobe** or sidelobes of the **antenna** and may produce **coherent** or **incoherent echoes** that can interfere with the **detection** or quantitative measurement of weather echoes. *See* **ground clutter**.

ground-to-cloud discharge—A **lightning discharge** in which the original **leader** process starts upward from some object on the ground; the opposite of the more common **cloud-to-ground discharge**.

Ground-to-cloud discharges most frequently emanate from very tall structures that, being at the same **potential** as the earth, can exhibit the strong **field** intensities near their upper extremities necessary to initiate leaders.

ground track—Term used in **remote sensing** to represent the vertical **projection** of the actual flight path of an **aerial** or space vehicle onto the surface of the earth or other celestial body.

ground truth—Any measurement of an observed quantity that can be used to validate or verify a new (often **remote sensing**) measurement or technique.

ground visibility—In aviation terminology, the **horizontal visibility** observed at the ground, that is, **surface visibility** or **control-tower visibility**.

ground wave—A **radio wave** that propagates by means of interaction with the earth's surface, as opposed to free-space or **sky wave** propagation.

At low frequencies, ground waves can propagate thousands of kilometers following the earth's curvature.

Barton, D. K., and S. A. Leonov, eds., 1997: *Radar Technology Encyclopedia*, p. 470.

Groundhog Day—2 February; in American folklore, a day that is popularly supposed to provide the key to the weather for the remainder of the winter.

Specifically, if the groundhog (woodchuck, *Marmota monax*), upon emerging from its hole, casts a shadow, it will return underground, thereby foreboding more wintry weather. There is no convincing **statistical** evidence to support this belief. This date, 2 February, is actually Candlemas Day, which is associated with similar beliefs in Europe, but in the United States, the popularity of this legend has come to overshadow the original significance of the day. One of the most popular Groundhog Day celebrations occurs in Punxsutawney, Pennsylvania, when the media gather at **sunrise** to see whether the local groundhog (named Phil) casts a shadow. *See* **control day**.

grounding line—The junction between **land ice** (such as an **ice sheet** or **glacier**) and **floating ice** (such as an **ice shelf**, **fast ice**, or **sea ice**).

groundwater—(Also called **phreatic water**.) Subsurface water that occupies the **zone of saturation**; thus, only the water below the **water table**, as distinguished from **interflow** and **soil moisture**.

groundwater basin—A **groundwater flow** system bounded by **groundwater divides**.

groundwater dam—A geological feature (e.g., a fault) or rock unit of low permeability that impedes the horizontal movement of **groundwater flow**, thereby causing high horizontal hydraulic gradients evident in the **water table** or other **potentiometric surface**.

groundwater dating—Use of radioactive isotopes present in a **groundwater** sample to estimate the length of time the groundwater has been in the subsurface.
See also **radioactive dating**.

groundwater depletion curve—(Also called **groundwater recession**.) *See* **baseflow recession curve**.

groundwater divide—The boundary between **groundwater** basins; defined by a line connecting the high points on the **water table** or other **potentiometric surface**.
Groundwater flows away from a groundwater divide.

groundwater flow—The rate of **groundwater** movement through the subsurface.

groundwater hydrology—(Sometimes referred to as **hydrogeology**.) That branch of **hydrology** devoted to the study of the movement and occurrence of water in the **saturated zone**.

groundwater level—The level of **groundwater** in an **unconfined aquifer** below which the **porous medium** is saturated.

groundwater mining—(Also known as groundwater overdraft, groundwater overexploitation.) Extracting **groundwater** at a rate that exceeds its **recharge**, so that there is a net loss of groundwater.

groundwater mound—The surface representing the rise in **water table** or **potentiometric surface** caused by the injection of water.

groundwater recession—Same as **baseflow recession**.

groundwater runoff—See **base flow**.

groundwater station—A **station** in a study area designated and designed for collection of **groundwater** information such as **water level, temperature**, or water quality.
See also **observation well, monitor well**.

groundwater storage—The estimated amount of water stored in the **saturated zone** and the estimated amount of recoverable water stored in the **unsaturated zone** of an **aquifer**.

groundwater table—Same as **water table**.

group velocity—1. The **velocity** of the envelope of a group of waves of nearly equal frequencies.
   From the **dispersion relation** $k(\omega)$ the group velocity is defined as $d\omega/dk$, as distinguished from the **phase velocity** (or **phase speed**) ω/k. The origin of this term is made clearer by considering the superposition of two equal-amplitude plane **harmonic** waves with **wavenumbers** $k \pm \Delta k$ and frequencies $\omega \pm \Delta\omega$:

$$2A\cos(kx - \omega t)\cos(\Delta kx - \Delta\omega t).$$

Because $|\Delta\omega/\omega| \ll 1$, this composite **wave** may be looked upon as a high-frequency wave, with phase velocity ω/k, modulated by a wave of much lower **frequency** $\Delta\omega$. The envelope of the high-frequency wave is a low-frequency wave propagating with the group velocity $\Delta\omega/\Delta k$. Note the similarity with **beating**. Indeed, a group of waves may be looked upon as a moving beat.
   2. The **velocity** at which a group of waves, and the **wave** energy, travels.
   In **deep water**, on the basis of **linear water wave theory**, it can be shown to be equal to one half the **phase velocity**.

grower's year—In Great Britain, the 12-month period starting 6 November.
   Compare **farmer's year**.

growing degree-day—(Abbreviated GDD.) A **heat index** that relates the development of plants, insects, and disease organisms to environmental **air temperature**.
   GDD is calculated by subtracting a base **temperature** from the **daily mean** temperature and GDD values less than zero are set to zero. The summation over time is related to development of plants, insects, and disease organisms. The reference temperature (base temperature) below which development either slows or stops is species dependent. For example, cool **season** plants (canning pea, spring wheat, etc.): base temperature is 40°F (5°C); warm season plants (sweet corn, green bean, etc.): base temperature is 50°F (10°C); and very warm season plants (cotton, okra, etc.): base temperature is 60°F (15°C). *See* **degree day, heat unit**

growing degree-hour—(Abbreviated GDH.) A **heat index** that relates the development of plants, insects, and disease organisms to environmental **air temperature**.
   Analogous to **growing degree-day** (GDD), GDH is calculated by subtracting a base **temperature** from an hourly environmental temperature measurement and values less than zero are set to zero. A summation of hourly GDH is related to development. The temperature threshold (base temperature) below which development either slows or stops is species dependent.

growing mode—*See* **neutral mode**.

growing season—Generally, the period of the year during which the **temperature** of cultivated vegetation (i.e., the temperature of the vegetal **microclimate**) remains sufficiently high to allow plant growth.
   This is an important concept in **agricultural climatology**, but it suffers greatly from vagueness and complexity. The growing season is highly variable due to plant varieties as related to temperature sensitivity. Currently, the most common measure of this period, "the average length of growing season," is defined as the number of days between the average dates of the last **killing frost** (*see* **frost**) in **spring** and the first killing frost of **autumn**. The lack of a positive, practical definition for (and means of determining) a "killing" frost seriously limits the scientific usefulness of this measure. To provide some economic significance, the effective growing season is defined as the length of growing season that prevails in 80% of the years. Another measure, the **frost-free season**, is defined as the interval between the last and first occurrences of 32°F temperatures in spring and fall. This may be observed exactly, but its relationship to the local **microclimate** is variable and nonspecific, and it does not consider differences in types of vegetation. Still a fourth measure, the vegetative period or vegetation **season**, attempts to allow for the greater microclimatic **temperature**

range and for the general growth retardation by cold temperatures, and is defined as the summer period confined between occurrences of 42°F (or 41°F or 43°F) temperatures. At best, any of the above is an index of growing season length, rather than a direct measure of it. Basically, the growing season (and "killing frost") should be defined biologically rather than meteorologically and should consider the detailed microclimate, plant resistance to frost, growth rate versus temperature, and probably other factors.

growler—A small piece of floating **sea ice**, usually a fragment of an **iceberg** or **floeberg**.

It floats low in the water, and its surface often is heavily pitted. It is smaller than a **bergy bit** and often appears greenish in color.

GTS—Abbreviation for **Global Telecommunications System**.

guba—In New Guinea, a rain **squall** on the sea.

guergal—*See* **gregale**.

Guiana Current—An **ocean current** flowing northwestward along the northern coast of South America (the Guianas).

The Guiana Current is an extension of the **South Equatorial Current** (flowing west across the ocean between the **equator** and 20°S), which crosses the equator and approaches the coast of South America. Eventually, it is joined by part of the **North Equatorial Current** and becomes, successively, the **Caribbean Current** and the **Florida Current**.

Guinea Current—An **ocean current** flowing along the south coast of northwest Africa into the Gulf of Guinea.

Gulf of Papagayo—A gulf along the northern Pacific coast of Costa Rica that is frequently affected by strong offshore winds blowing from the Caribbean across Lake Nicaragua.

See **Papagayo, Tehuano**.

Gulf of Tehuantepec—A gulf along the southern Pacific coast of Mexico that in winter is affected by frequent strong winds blowing from the Gulf of Mexico southward across the isthmus of Tehuantepec into the Pacific.

See **Tehuano, Papagayo**.

Gulf Stream—One of the western **boundary currents** of the North Atlantic subtropical **gyre** and one of the swiftest **ocean currents** with one of the largest transports.

A deep, narrow, and swift current, it continues from the **Florida Current** for 2500 km in a northeastward direction, penetrating into the Atlantic as a free jet. It reaches its maximum transport of 90–150 Sv (90–150 × 10^6 m³s⁻¹) near 65°W before beginning to lose water to the **Sargasso Sea**; this water rejoins the Florida Current as the Gulf Stream recirculation. Some 50–90 Sv (50–90 × 10^6 m³s⁻¹) continue northeastward past the Grand Banks (50°W), where the current is also known as the **Gulf Stream Extension** and the **North Atlantic Current** (or North Atlantic Drift). It forms a marked **temperature** and **salinity** front with the **Labrador Current**, which meets the Gulf Stream Extension from the north and then flows parallel to it. As a free jet, the Gulf Stream develops instabilities in the form of meanders that eventually break off as **eddies**, also known as rings. **Cyclonic (cold core) rings** contain cold Labrador Current water and drift slowly southwestward into the Sargasso Sea. **Anticyclonic (warm core) rings** contain warm Sargasso Sea water and drift southwestward in the **Slope Water** found between the Gulf Stream and the **continental shelf**.

Gulf Stream Extension—*See* **Gulf Stream**.

Gulf Stream rings—**Eddies** formed by pinching off of meanders in the **Gulf Stream**.

Eddies forming to the north (coastward side) of the Gulf Stream are called **warm-core rings** (**anticyclonic**) because they encapsulate relatively warm **Sargasso Sea** water in the pinching off process, and those forming to the south (oceanward side) are called **cold-core rings** because they encapsulate relatively cold **Slope Water** in their centers.

Gulf Stream System—The western **boundary current** of the North Atlantic subtropical **gyre** and its eastward continuation, consisting of the **Florida Current**, the **Gulf Stream**, the **Gulf Stream Extension**, and its continuation in the **North Atlantic** and **Azores Currents**.

gully-squall—A nautical term for a violent **squall** of **wind** from mountain ravines on the Pacific side of Central America.

gush—Same as **cloudburst**.

gust—1. A sudden, brief increase in the speed of the **wind**.

It is of a more **transient** character than a **squall** and is followed by a **lull** or slackening in the

wind speed. Generally, winds are least gusty over large water surfaces and most gusty over rough land and near high buildings. According to U.S. weather observing practice, gusts are reported when the peak wind speed reaches at least 16 knots and the **variation** in wind speed between the peaks and lulls is at least 9 knots. The duration of a gust is usually less than 20 s.

2. With respect to **aircraft turbulence**, a sharp change in **wind speed** relative to the aircraft; a sudden increase in **airspeed** due to fluctuations in the airflow, resulting in increased structural stresses upon the aircraft.

3. (Rare.) Same as **cloudburst**.

gust amplitude—Maximum value of a **gust** during a given period.

gust decay time—Time interval between the moment of attainment of a **gust peak speed** and the end of a **gust**.

gust duration— Time period between the beginning and end of a **gust** of **wind**, usually less than 20 s.

gust formation time—Time interval between the beginning of a **wind** gust and the **gust** peak **wind speed**.

gust frequency—Number of wind **gusts** occurring during the **gust frequency interval**.

gust frequency interval—Specific time interval over which the **gust frequency** is determined.

gust front—The leading edge of a **mesoscale** pressure dome separating the **outflow** air in a **convective storm** from the environmental air.
    This boundary, which is marked by upward motion along it and downward motion behind it, is followed by a **surge** of gusty winds on or near the ground. A gust front is often associated with a **pressure jump**, **wind** shift, **temperature** drop, and sometimes with heavy **precipitation**. Gust fronts are often marked by **arcus** clouds.

gust-gradient distance—The horizontal distance along an aircraft flight path from the "edge" of a **gust** to the point at which the gust reaches its maximum speed.

gust load—The combined effect, on an aircraft structure, of a **gust** and the pilot's effort to counteract the gust.
    American commercial airplanes are designed for a gust load equivalent to that which would be produced by a gust having an **effective gust velocity** of 30 ft s^{-1} at an **airspeed** equal to the design cruising speed of the aircraft.

gust peak speed—Instantaneous **wind speed** at the moment the **gust amplitude** is attained.

gust probe—An air velocity sensing instrument usually mounted on the front of an aircraft that resolves turbulent fluctuations in all three components relative to the aircraft.
    Instruments used for this purpose include differential pressure probes, vanes, and heated wires (**hot-wire anemometers**). To measure the air velocity relative to the earth, the gust probe measurements must be corrected for the aircraft attitude and **velocity**.

gust vector—The transitory **vector** departure, lasting for a fairly short time, of the **wind** velocity from its **mean value**.

gustiness—In general, a quality of air flow characterized by **gusts**.

gustiness components—1. Turbulent fluctuations of the three **orthogonal** wind components (u', v', w') in the **atmosphere**.
    If the measurements are tower-based, the longitudinal (x) component is commonly along the mean wind direction, the lateral (y) component is perpendicular to the longitudinal component on the horizontal plane to the left when facing the **downwind** direction, and the vertical (z) component is upward, perpendicular to the longitudinal and lateral components. For aircraft measurements, the gustiness components may also be in the airplane **coordinate system**, with the longitudinal component along the longitudinal axis of the aircraft, the lateral perpendicular to the longitudinal in the horizontal plane of the aircraft, and the vertical upward, perpendicular to the horizontal plane of the aircraft.

2. The ratios, to the mean wind speed, of the average magnitudes of the component fluctuations of the **wind** along three mutually perpendicular axes.
    For a mean wind speed U, the gustiness components along the x axis (*longitudinal* or *downwind gustiness*), the y axis (*lateral* or *crosswind gustiness*), and the z axis (*vertical gustiness*) are given by

$$g_x = \frac{\overline{|u'|}}{U}, g_y = \frac{\overline{|v'|}}{U}, g_z = \frac{\overline{|w'|}}{U}.$$

Here $\overline{|u'|}$, $\overline{|v'|}$, $\overline{|w'|}$ are the average magnitudes of the **eddy** velocities along the x, y, and z axes, respectively.

3. (Or intensity of turbulence.) The ratios of the root-mean-squares of the **eddy** velocities to the mean wind speed:

$$g_x = \frac{\overline{u'^2}^{1/2}}{U}$$

for the longitudinal component, and similarly for the lateral and vertical components.

The gustiness components are sometimes defined as the square of the above expressions.

gustiness factor—A measure of the intensity of **gusts** given by the ratio of the total **range** of **wind speed** between gusts and the intermediate periods of lighter **wind** to the mean wind speed, averaged over both gusts and lulls.

gustnado—Colloquial expression for a short-lived, shallow, generally weak **tornado** found along a **gust front**. Gustnadoes are usually visualized by a rotating **dust** or debris cloud.

See **nonsupercell tornado**.

gustsonde—An instrument, dropped from high altitude and carried by a stable parachute, used to measure the vertical component of **turbulence** aloft.

It consists of an **accelerometer** and radio telemetering equipment.

Guti weather—In Zimbabwe (formerly Rhodesia), a dense **stratocumulus** overcast, frequently with **drizzle**, occurring mainly in early summer.

The condition is associated with easterly winds that invade the interior, bringing in cool and stable **maritime air** when an **anticyclone** moves eastward south of Africa.

guttation—The water exuded from leaves as a result of root pressure.

Sometimes guttation is confused with **dew**, although the origin and appearance of the resulting drops are different. On grass, dew appears as many drops that cover the surface of the blade; dew forms by **condensation** from **water vapor** in the **atmosphere**. Guttation appears as a single large pendant drop at the tip of the blade; guttation forms by the extrusion of liquid water from the moist ground. Sometimes one is seen, sometimes the other, sometimes both. When a wet **heiligenschein** is seen on the grass it is mainly caused by dewdrops; when a **rainbow** is seen on the grass it is mainly caused by guttation drops. See also **dewbow**.

guttra—In Iran, sudden **squalls** in May.

guxen—Cold **wind** of the Alps in Switzerland.

Guyana Current—One of the western **boundary currents** of the Atlantic Ocean and part of the pathway for water from the southern into the Northern Hemisphere in the global **ocean conveyor belt**.

Flowing northward along the northern coast of South America, it originates from the **North Brazil Current**, receives a contribution from the **South Equatorial Current**, and continues as the **Caribbean Current**. The continuity of the Guyana Current as a permanent northward current is in doubt, but northward flow does exist in the mean. **Eddies** related to flow **instability** typical for western boundary currents have been reported.

guzzle—In the Shetland Islands, an angry blast of **wind**, dry and parching.

GVAR—Abbreviation for **GOES Variable**.

gyres—Oceanic current systems of planetary scale driven by the global **wind** system.

The subtropical gyres are driven by the **trade winds** and by the **westerlies** of the temperate regions, the subpolar gyres by the westerlies and the **polar easterlies**. Gyres consist of a narrow, swift-flowing western **boundary current**, an eastward-flowing **zonal** current, a broad and slow-moving eastern boundary current, and a westward flowing zonal current. Eight gyres are distinguished in the World Ocean: In the Atlantic, the **Brazil**, **South Atlantic**, **Benguela**, and **South Equatorial Currents** form the subtropical gyre of the Southern Hemisphere; the **Gulf Stream**, **Azores**, **Canary**, and **North Equatorial Currents** form the subtropical gyre in the Northern Hemisphere; the **Labrador**, **North Atlantic**, **Irminger**, and **East Greenland Currents** form the subpolar gyre. In the Pacific, the **East Australian**, **South Pacific**, **Peru/Chile**, and South Equatorial Currents form the subtropical gyre of the Southern Hemisphere; the **Kuroshio**, **North Pacific**, **California**, and North Equatorial Currents form the subtropical gyre of the Northern Hemisphere; the **Oyashio**, **Aleutian**, California, and **Alaskan Currents** and the **Alaskan Stream** form the subpolar gyre; a second subpolar gyre exists in the Bering Sea. In the Indian Ocean, the **Agulhas**, **South Indian**, **West Australian**, and South Equatorial Currents form the only subtropical gyre.

gyro-frequency—(Also cyclotron frequency, Larmor frequency.) The **frequency** of a moving charged **particle** or **ion** in a **magnetic field**.

The **trajectory** of a charged particle acted on solely by a magnetic force is a helix with its axis parallel to the **magnetic induction**. In any plane perpendicular to the magnetic induction the particle describes a circular orbit with gyro-frequency qB/m, where q is charge, m is mass, and B is the magnitude of magnetic induction.

H

haar—A name applied to a wet **sea fog** or very fine **drizzle** that drifts in from the sea in coastal districts of eastern Scotland and northeastern England.
It occurs most frequently in summer.

haboob—(Many variant spellings, including habbub, habub, haboub, hubbob, hubbub.) A strong **wind** and **sandstorm** or **duststorm** in northern and central Sudan, especially around Khartoum, where the average number is about 24 a year.
The name comes from the Arabic word habb, meaning "wind." Haboobs are most frequent from May through September, especially in June, but they have occurred in every month except November. Their average duration is three hours; they are most severe in April and May when the soil is driest. They may approach from any direction, but most commonly from the north in winter and from the south, southeast, or east in summer. The average maximum wind velocity is over 13 m s^{-1} (30 mph) and a speed of 28 m s^{-1} (62 mph) has been recorded. The sand and **dust** form a dense whirling wall that may be 1000 m (3000 ft) high; it is often preceded by isolated **dust whirls**. During these storms, enormous quantities of sand are deposited. Haboobs usually occur after a few days of rising **temperature** and falling **pressure**.
<blockquote>Sutton, L. J., 1925: Haboobs. Quart. J. Roy. Meteor. Soc., 51, 25–30.</blockquote>

Hadley cell—A direct thermally driven and zonally symmetric **circulation** under the strong influence of the earth's rotation, first proposed by George Hadley in 1735 as an explanation for the **trade winds**.
It consists of the equatorward movement of the trade winds between about latitude 30° and the **equator** in each hemisphere, with rising **wind** components near the equator, poleward flow aloft, and, finally, descending components at about latitude 30° again. In a dishpan experiment, a Hadley cell is any direct thermally driven vertical cell of the approximate **scale** of the dishpan.

Hadley regime—In a dishpan experiment, a flow dominated by a single large **Hadley cell**.

Hagen–Poiseuille flow—Same as **Poiseuille flow**.

hail—Precipitation in the form of balls or irregular lumps of **ice**, always produced by convective clouds, nearly always **cumulonimbus**.
An individual unit of hail is called a **hailstone**. By convention, hail has a diameter of 5 mm or more, while smaller **particles** of similar origin, formerly called **small hail**, may be classed as either **ice pellets** or **snow pellets**. **Thunderstorms** that are characterized by strong updrafts, large liquid water contents, large cloud-drop sizes, and great vertical height are favorable to hail formation. The destructive effects of hailstorms upon plant and animal life, buildings and property, and aircraft in flight render them a prime object of **weather modification** studies. In **aviation weather observations**, hail is encoded A.

hail stage—Part of an obsolete conceptual **model** of **air parcel** ascent referring to a portion of the ascent during which the **parcel** temperature remains at the **freezing point** until all the **rain** produced previously has frozen.
Other portions of the ascent were described as the **dry stage**, **snow stage**, and **rain stage**.

hailpad—A device used to obtain data on the size distribution and mass of hailstones.
A hailpad usually consists of a plastic foam panel covered by aluminum foil or white latex paint and set in a frame that is hammered into the ground. Hail that impinges on the pad leaves dents in it. The dimensions of the dents are analyzed to obtain the **hailstone** size and mass data.
<blockquote>Long, A. B., et al., 1980: The hailpad: materials, data reduction and calibration. J. Appl. Meteor., 19, 1300–1313.</blockquote>

hailstone—A single unit of **hail**, ranging in size from that of a pea to that of a grapefruit (i.e., from 5 mm to more than 15 cm in diameter).
Hailstones may be spheroidal, conical, or generally irregular in shape. The spheroidal stones often exhibit a layered internal structure, with layers of **ice** containing many air bubbles alternating with layers of relatively **clear ice**. These probably correspond to **dry growth** and **wet growth** and are called **rime** and **glaze**, respectively. The conical stones fall with their bases downward without much tumbling and are often smaller and not as layered. Irregular hailstones often have a lobate structure and are not composed of smaller hailstones frozen together. Hailstones grow by **accretion** of **supercooled water** drops and sometimes also by **accretion** of minor amounts of small ice particles. Large hail may contain liquid water and be spongy (an intimate mixture of ice and water) in some regions; it is usually solid ice with **density** greater than 0.8 g cm^{-3}. Small hail may be indistinguishable from large **graupel** (**snow pellets**) except for the convention that hail must be larger than 5 mm in diameter. The **density** of **small hail** can be much less than 0.8 g cm^{-3} if

they are dry; if partly melted such hailstones become spongy. The largest recorded hailstone in the United States fell in a **hailstorm** in Coffeyville, Kansas on 3 September 1970. It weighed 766 g, had a longest dimension about 15 cm, and had protrusions (lobes) several centimeters long on one side that formed as it grew.

hailstorm—Any **storm** that produces **hailstones** that fall to the ground; usually used when the amount or size of the **hail** is considered significant.

hair hygrometer—A **hygrometer** that measures **relative humidity** by means of the **variation** in length of a strand of human hair.

The length variation of a properly treated hair is 2%–2.5% when the **humidity** changes from 0%–100%. The hair hygrometer is considered to be a satisfactory instrument in situations where extreme and very low humidities are seldom or never found. The rate of response is very dependent on **air temperature**; the **lag time** increases with decreasing **temperature**. For air temperatures between 0° and 30°C and relative humidities between 20% and 80%, a good hair hygrometer should indicate 90% of a sudden change in humidity within about three minutes.

halcyon days—A period of fine weather.
The term originated in Greek mythology.

hale de mars—*See* **bise**.

half-arc angle—The **elevation angle** of that point that a given observer regards as the bisector of the arc from his **zenith** to his **horizon**; a measure of the apparent degree of flattening of the dome of the sky.

Because to almost all persons the sky appears not as a hemispheric dome but rather as a more oblate surface, the half-arc angle designated by most observers is substantially less than 45°, usually falling within the interval 20°–35°. The more remote the landmarks on an observer 's horizon, the smaller in general will be his designated value of the half-arc angle. Also, presence of a **cloud** deck serves to produce an apparent flattening effect. No fully satisfactory explanation exists to account for the apparent flattening of the sky.

half-life—The time over which the number of radioactive nuclei of a given species decays exponentially to half its initial value.

Half-lives range from small fractions of a second to billions of years. Half-life is also applied to any exponential decay. That is, if a physical quantity N at any time t is related to its initial value N_0 by

$$N = N_0 e^{-t/\tau},$$

the half-life for this process is $\tau \ln 2$, where τ is the **mean life**.

half-order closure—A method to approximate the effects of **turbulence** by retaining a **prognostic equation** for mean variables (such as **wind** or **temperature**), but where a **profile** shape of those mean variables is assumed a priori.

For example, in the **boundary layer**, if the daytime profile of **potential temperature** is assumed to be uniform with height, then only one temperature forecast equation is needed for the whole layer. Similarly at night, if an exponential shape is assumed for the potential temperature profile in the **stable boundary layer**, then only one temperature forecast equation is needed for the cooling at the surface. This approach is less computationally expensive than solving forecast equations at every height within the **mixed layer** or stable boundary layer. *See* **closure assumptions, first-order closure, higher-order closure, nonlocal closure**.

half-period zone—Same as **Fresnel zone**.

half-power points—The points on the **radiation pattern** of an **antenna** where the **transmitted power** is one-half that of the maximum, both measured at the same **range**.

The maximum usually lies along the axis of the **beam**, and therefore the angle subtended at the antenna by the half-power points is used to define a beam width. The beam width need not be the same for all planes through the axis of the beam.

half-tide level—*See* **mean tide level**.

half-width—*See* **line width**.

Hall effect—The **electric field** produced by an electric **current** in a conductor in a **magnetic field**.

The electric field, which is perpendicular to both the current and the magnetic field, is the reaction **field** (in a bounded conductor) that balances the magnetic force on moving charges. It is named after Edwin H. Hall, who in 1879 first discovered this effect in laboratory investigations of gold. The Hall effect is a determinant in the behavior of the electrical currents generated by winds in the lower **ionosphere**, since these winds advect the ionized layers across the earth's magnetic field and produce a complex electrical current system in the ionosphere. The current system in turn produces small changes in the earth's magnetic field as measured at the surface. *See also* **magnetic induction**.

Hallett–Mossop process—(Also called rime splintering.) One of the mechanisms thought to be responsible for secondary **ice** production, when **ice crystal** concentrations in clouds well in excess (x 10 000) of the **ice nucleus** concentration are found.

Ice **particles** are produced in the **range** of **temperature** $-3°$ to $-8°C$ (with a maximum at $-4°C$) as **graupel** grows by **accretion** provided the **cloud droplet** spectrum contains appropriate numbers of droplets smaller than 12 μm and greater than 25 μm. About 50 **splinters** are produced per milligram of accreted ice.

Halmahera Eddy—A **recirculation** system between the **South Equatorial Current** and the **North Equatorial Countercurrent** in the extreme west of the Pacific Ocean, east of Halmahera.

The **circulation**, about 1000 km in diameter, is clockwise from April–November but reverses during December–March when the Philippines experience **monsoon** winds blowing from the northeast in the Northern Hemisphere and from the northwest south of the **equator**.

halny wiatr—Local name for a **foehn** wind in mountainous regions of Poland.

halo—Any of a family of colored or whitish rings, arcs, pillars or spots of **light** that appear in the sky and are explained by the **reflection** or **refraction** of light by **ice crystals**.

They are usually found in the vicinity of the light source, the most important of which are the sun and moon, but may arise from artificial lights if seen, say, through an **ice fog**. Halos exhibiting some prismatic coloration are explained, at least in part, by the refraction of light by the ice crystals. However, the color is usually fairly pale, being best on a red edge next to the light. The exceptions, in having very good colors, are the circumhorizontal and circumzenithal arcs, the positions of which are not determined by the minimum angle of refraction. Halos that are white, or show the same color as the light source itself, are explained by the reflection of light off the **crystal** faces. Whether explained by reflection or refraction, the pattern that emerges depends upon the crystal type, crystal orientation (actually the **probability** of various orientations within a **population** of crystals), and the **elevation angle** of the light (sun). With such a rich range of possibilities, a large variety of halos are theoretically possible and over 50 different halo phenomena have been documented photographically. Some halos predicted theoretically have yet to be reported, others that have been reported have yet to be explained. The most common halo is the **halo of 22°**. Other frequently seen halos are the **parhelia** of the halo of 22°, the **sun pillar**, the 22° **tangent arcs**, the **circumzenithal arc**, the **halo of 46°**, and the **parhelic circle**. On very rare occasions an observer's sky will be filled with a display with 10, 20, or more different halos, usually persisting for only a few minutes. Much supernatural lore has been prompted by observations of old of such events. Halos must be distinguished from optical phenomena arising from water drops, such as the **rainbow**, **corona**, and **glory**.

halo of 46°—A **halo** in the form of a circle, or portion of a circle, with an angular radius of about 46° about a **light** source, such as the sun or moon.

The coloration is reddish on the inner edge to bluish on the outer edge. This halo is much less common that the **halo of 22°**. The 46° halo is explained by the **refraction** of light passing through the 90° prism formed between the side and basal faces of a hexagonal **ice crystal**. The minimum angle of **deviation** for this ice prism is about 46°. Closely associated with this halo are the 46° infralateral arcs and the 46° supralateral arcs. In particular, the shape of the 46° supralateral arc often follows the uppermost parts of the 46° halo so closely that the two are almost impossible to distinguish. In fact, upon examination, a large fraction of the halos commonly interpreted as being 46° halos turn out to be pieces of 46° supralateral arcs. Which of the two halos (46° halo or 46° supralateral arc) is more frequent has not been settled.

halo of 22°—A **halo** in the form of a circle, or portion of a circle, with an angular radius of about 22° about a **light** source, such as the sun or moon.

This is the most common of all halos. The sky is darker just to the inside of the halo than it is to the outside. The halo exhibits a pale coloration from a reddish tint on the inside fading to a bluish white on the outside. The 22° halo is explained by the **refraction** of light that enters one prism face and leaves by the second prism face beyond, thus being refracted by a prism with an effective angle of 60°. The **angle of minimum deviation** for an **ice** prism of this prism angle is about 22°, so such light does not appear inside the halo, accounting for the darker region there. The minimum angle of **deviation** varies slightly with **wavelength**, with the longer wavelengths being deviated least. This causes the reddish inner edge, outside which the additional contributions from light of increasingly shorter wavelengths decrease the color purity. The orientation that a **crystal** must have to contribute light to the halo depends on both the **elevation angle** of the light source and the portion of the halo in question, but the **probability** that a crystal will have a particular orientation depends upon its type and size. Consequently, it is frequently the case that for a given **population** of crystals and sun height, only a portion of the halo will be seen, while a change in one or the other might enable the full circle to form. It is not the case, despite being frequently asserted, that a view of the full circle requires crystals to have **random** orientations.

halocarbons—A collective term for the group of partially halogenated organic species, including the **chlorofluorocarbons**, **hydrochlorofluorocarbons**, **hydrofluorocarbons**, **halons**, **methyl chloride**, **methyl bromide**, etc.

These compounds, some of which are naturally occurring (largely through production in the oceans) and some of which are of **anthropogenic** origin (use as solvents, foam blowing agents, refrigerants, etc.), are the major source of the **halogens** (F, Cl, Br, I) to the **stratosphere**.

halocline—A vertical **salinity** gradient in some layer of a body of water that is appreciably greater than the gradients above and below it; also a layer in which such a **gradient** occurs.

The principal haloclines in the ocean are either seasonal, due to **freshwater** inputs, or permanent.

halogens—The elements fluorine (F), chlorine (Cl), bromine (Br), iodine (I), and astatine (At), which make up the seventh period in the periodic table of the elements.

The first four members of the period are present in **trace** levels in the **atmosphere** as the result of both natural and **anthropogenic** activity. The chemistry of **chlorine compounds** and **bromine compounds** is particularly important in the catalytic destruction of **ozone** in the **stratosphere** and is responsible for the occurrence of the **antarctic ozone hole** each spring. Because of this role in ozone depletion, the production of chlorinated and brominated organic compounds is being phased out.

halons—A group of brominated organic compounds used as fire retardants, including CF_3Br (halon 1301), CF_2Br_2 (halon 1202), CF_2BrCl (halon 1211), and CF_2BrCF_2Br (halon 2402).

The **transport** of these compounds to the **stratosphere**, followed by their **photolysis** to release Br atoms, gives this group a very high ozone depletion potential. The production of these compounds is now banned as a result of the **Montreal Protocol** and its subsequent amendments.

hanging dam—Accumulation of **frazil** under an **ice cover** on a river that reduces the **cross section** of the **stream**.

hanging glacier—A **glacier** lying above a cliff or steep mountainside.

As the glacier advances, its **calving** can cause **ice** avalanches.

hanning—A procedure for **smoothing** spectra in order to reduce uncertainties due to truncation of the original data.

harbor oscillations—*See* **seiche** and **tsunami**.

hard freeze—A **freeze** in which seasonal vegetation is destroyed, the ground surface is frozen solid underfoot, and heavy **ice** is formed on small water surfaces such as puddles and water containers.

It is to be distinguished from a hard frost (**black frost**).

hard frost—Same as **black frost**.

hard radiation—Quantum **radiation** (**particles** or **photons**) of high penetrating power, typically of **high frequency** and short **wavelength**.

A 10-cm thickness of lead is usually used as the criterion upon which the relative penetrating power of various types of **radiation** is based. Hard radiation will penetrate such a shield; **soft radiation** will not.

hard rime—Opaque, granular masses of **rime** deposited chiefly on vertical surfaces by a dense **supercooled fog**.

Hard rime is more compact and amorphous than **soft rime** and may build out into the **wind** as glazed cones or **ice feathers**. The **icing** of ships and shoreline structures by supercooled spray from the sea usually has the characteristics of hard rime.

hardpan—A dense, hard soil layer that restricts the penetration of water and roots.

It is formed by cementation of particles with sesquioxides, organic matter, silica, or calcium carbonate. The hardness does not change appreciably with changes in **moisture content**, and the soil does not **slake** in water.

harmattan—(Also spelled harmatan, harmetan, hermitan.) A dry, dust-bearing **wind** from the northeast or east that blows in West Africa especially from late November until mid-March.

It originates in the Sahara as a **desert wind** and extends southward to about 5°N in January and 18°N in July. It is associated with the high pressure area that lies over the northwest Sahara in winter and the adjoining part of the Atlantic in other seasons. In summer the cooler onshore **southwest monsoon** undercuts it, but the harmattan continues to blow at a height of about 2 km (3000–6000 ft) and sometimes deposits **dust** on ships at sea. This conflict of winds causes the so-called West African tornadoes. *See* **doctor.**

harmonic—1. Any integral multiple of the lowest (or fundamental) **frequency** of a physical system. For example, the motion of a taut, gently plucked violin string is the superposition of sinusoidal motions with frequencies ω_0, ω_1 $2\omega_0$, ω_2, $3\omega_0$, . . . where ω_0 is the **fundamental frequency** and ω_1, ω_2, . . . are the harmonics (or overtones), resulting in a harmonious composite **sound.**

2. A sine or cosine component of the **Fourier series** representation of an empirical or theoretical function.

harmonic analysis—1. A **statistical** method for determining the **amplitude** and **period** of certain **harmonic** or **wave** components in a set of data with the aid of **Fourier series.**

Harmonic analysis has been used in meteorology, for example, to determine periodicities in climatic data (Conrad 1950); to determine the wavelengths most strongly represented in **general circulation** flow patterns; and to determine the **spectrum** of turbulent **eddies** (Sutton 1953).

2. The representation of tidal variations as the sum of several **harmonics**, each of different **period, amplitude,** and **phase.**

The periods fall into three tidal species: long period, **diurnal**, and semidiurnal. Each tidal species contains groups of harmonics that can be separated by analysis of a month of observations. In turn, each group contains constituents that can be separated by analysis of a year of observations. In shallow water, harmonics are also generated in the third-diurnal, fourth-diurnal, and higher species. These constituents can be used for **harmonic prediction** of tides.

> Conrad, V., 1950: *Methods in Climatology*, 119–154.
> Sutton, O. G., 1953: *Micrometeorology*, 96–103.

harmonic constant—*See* **harmonic analysis.**

harmonic function—Any solution of the **Laplace equation.**

harmonic mean—For a set of numbers, the reciprocal of the **arithmetic mean** of their reciprocals.

harmonic prediction—*See* **harmonic analysis.**

Hartley bands—*See* **ozone.**

haster—In England, a **violent storm** of **rain.**

haud—In Scotland, a **squall.**

haugull—(Also called havgull, havgula.) Cold, damp **wind** blowing from the sea during summer in Scotland and Norway.

haze—Particles suspended in air, reducing **visibility** by **scattering** light; often a mixture of aerosols and **photochemical smog.**

Many aerosols increase in size with increasing **relative humidity** due to **deliquescence**, drastically decreasing visibility. On Köhler curve plots of **saturation** relative **humidity** versus **aerosol** particle radius, **equilibrium** haze **particles** are to the left of the peak, while growing **cloud droplets** are to the right. Many haze formations are caused by the presence of an abundance of **condensation nuclei** which may grow in size, due to a variety of causes, and become **mist, fog,** or **cloud.** Distinction is sometimes drawn between **dry haze** and **damp haze**, largely on the basis of differences in optical effects produced by the smaller particles (dry haze) and larger particles (damp haze), which develop from slow **condensation** upon the **hygroscopic** haze particles. Dry haze particles, with diameters of the order of 0.1 μm, are small enough to **scatter** shorter wavelengths of **light** preferentially though not according to the inverse fourth-power law of **Rayleigh scattering.** Such haze particles produce a bluish color when the haze is viewed against a dark background, for **dispersion** allows only the slightly bluish **scattered** light to reach the eye. The same type of haze, when viewed against a light background, appears as a yellowish veil, for here the principal effect is the removal of the bluer components from the light originating in the distant light-colored background. Haze may be distinguished by this same effect from mist, which yields only a gray **obscuration**, since the particle sizes are too large to yield appreciable differential scattering of various wavelengths. *See* **smaze, arctic haze;** *compare* **particulates.**

haze aloft—**Haze** layer with a base not in contact with the earth's surface.

haze droplet—Any small liquid **droplet** (less than a μm in diameter) contributing to an atmospheric **haze** condition.

In certain industrial areas, such droplets may be entirely nonaqueous (largely **hydrocarbons**). Most haze droplets are salts dissolved in water in an **environment** below water **saturation** (<100% **relative humidity**) with **equilibrium** size given by the **Köhler equation**. Near seacoasts, droplets of **seawater** are responsible for haze; even far inland, some haze conditions have been shown to be due to such salt-solution droplets. Many inland haze particles are composed of ammonium sulphate (bisulphate), probably produced from industrial **air pollution** or locally in some areas with volcanic activity as **sulfuric acid**, reacting with **ammonia** from agricultural activities.

haze horizon—The top of a **haze layer** that is confined by a low-level **temperature inversion** and has the appearance of the **horizon** when viewed from above against the sky.

In such instances the **true horizon** is usually obscured by the haze layer. *See also* **haze line, dust horizon, fog horizon, smoke horizon.**

haze layer—A layer of **haze** in the **atmosphere**, usually bounded at the top (**haze line**) by a **temperature inversion** and frequently extending downward to the ground.

Necessary for the existence of this phenomenon are, of course, a source of haze particles and a relatively stable **stratification** of **temperature** in the **atmosphere** either within or immediately above the haze layer. A haze layer may vary in extent from the type developed locally over an urban area at night to a layer covering thousands of square miles as within an old and **stable air mass**.

haze level—Same as **haze line**.

haze line—(Or haze level.) The boundary surface in the **atmosphere** between a **haze layer** and the relatively clean, transparent air above; the top of a haze layer.

This "line" usually exists near the base of a **temperature inversion** that prevents the upward spreading of the **haze** particles. It is frequently a very distinct **discontinuity** and is particularly striking when viewed from an aircraft that is either just passing through it or is flying high aloft with the sun above the **horizon**. In the latter case, **glare** due to the upward **scattering** of **sunlight** from the haze can produce a very limited **cone of visibility**. *See also* **haze horizon.**

hazemeter—Name sometimes applied to a **transmissometer**.

Hazen method—A method of fitting a frequency curve to an observed series of **floods** on the assumption that the logarithms of the **variate** are normally distributed.

Hazen plotting position—For the rth ranked (from largest to smallest) **datum** from a **sample** of size n, the quotient

$$\frac{2r-1}{2n}.$$

The Hazen plotting position is the original **plotting position** and has been recommended for use with lognormal distributions; however, it leads to excessively high return periods for large flows.

While it has many desirable mathematical properties, the Hazen plotting position produces biased estimates of pertinent **statistics**. *See* **plotting position, probability paper.**

HCFCs—Abbreviation for **hydrochlorofluorocarbons**.

HCMM Satellite—Abbreviation for **Heat Capacity Mapping Mission Satellite**.

head—The **energy** of fluid per weight units; dimensionally expressed as length unit, for example, Newton × meters/Newton = meters.

heading—The direction toward which an aircraft or ocean vessel is oriented.

A heading may be with reference to **true north** or **magnetic north**. The heading and **course** may be different, especially in air navigation, because of **drift**.

headwaters—The upper reaches of a **stream** near its **source**.

headwind—(Also called opposing wind.) A **wind** that opposes the intended progress of an exposed, moving object, for example, rendering an airborne object's **airspeed** greater than its groundspeed; the opposite of a **tailwind**.

This effect is particularly critical in air navigation; it is of interest to note that a **wind direction** may have no component opposing the intended **course** of an aircraft, but because of **drift** effects, may have a component opposing the aircraft's **heading**.

heap clouds—A popular name for **cumulus**.

heat—1. (Or heat content.) A form of **energy** transferred between systems, existing only in the process of **transfer**.

2. Same as **enthalpy**.

Heat, used as a noun, is confusing and controversial in its scientific meaning. The differential of heat is considered imperfect in that its value depends on the process applied. In the thermodynamic definitions in this glossary, heat is avoided as a noun or adjective except where required by established use. The process of heating is, however, defined as the net **absorption** of **internal energy** by a system.

heat balance—*See* **energy balance**.

heat budget—*See* **energy balance**.

heat burst—(Also **heat thunderstorm**.) Localized, sudden increase in **surface temperature** associated with a **thunderstorm**, **shower**, or **mesoscale convective system**, often accompanied by extreme drying.

The **temperature** jump can be so extreme that it is at times referred to as a "hot blast of air." Occurs in association with precipitation-driven downdrafts penetrating a shallow surface stable layer and reaching the ground.

heat capacity—(Also called thermal capacity.) The ratio of the **energy** or **enthalpy** absorbed (or released) by a system to the corresponding **temperature** rise (or fall).

Heat capacities are defined for particular processes. For a constant volume process,

$$C_v = \frac{\partial U}{\partial T},$$

where U is the **internal energy** of a system and T is its temperature. For a constant **pressure** process,

$$C_p = \frac{\partial H}{\partial T},$$

where H is the system enthalpy. The heating rate, Q, for a constant volume process is

$$Q = C_v dT/dt \ (V = \text{constant}),$$

whereas in a constant pressure process,

$$Q = C_p dT/dt \ (p = \text{constant}).$$

See **specific heat capacity**.

Heat Capacity Mapping Mission Satellite—(Abbreviated *HCMM Satellite*.) An experimental NASA satellite (launched April 1978 as an Applications Explorer Mission) that carried a two-channel imaging **radiometer** that measured reflected **solar radiation** and **thermal radiation** emitted by the earth's surface.

heat capacity method—Technique for indirectly determining **soil moisture content** by measuring the **heat capacity** of the soil, which varies approximately linearly with **water content**.

heat conductivity—1. Same as **thermal conductivity**.

2. Less frequently, same as **thermometric conductivity**.

heat engine—(Same as **heat transfer**.) A device for producing **work** by virtue of a **temperature** difference.

The **atmosphere** may be considered such a device. *See* **Carnot engine**.

heat equator—1. (Also called **thermal equator**.) The line that circumscribes the earth and connects all points of highest mean annual **temperature** for their longitudes.

The course of the heat equator varies with the arrangement of continents and **ocean currents**. It does not even approximately parallel the geographic **equator**, but ranges from about 20°N in Mexico to about 14°S latitude in Brazil. From West Africa to the East Indies, the heat equator lies north of the geographic equator; from New Guinea to 120°W longitude, it lies south of the geographic equator.

2. The approximate latitude of highest mean annual **surface temperature** (about 10°N).

heat flux—(Or moisture flux.) A quantity measured according to the formula

$$B = \lambda \frac{dt}{dz},$$

where λ is the **thermal conductivity** of the medium (i.e., soil, air) that the **heat** (or moisture) is moving through.

This may be expressed as **flux** per unit area for heat or moisture.

Mann, J., et al., 1979: *Agrometeorology*, Springer–Verlag, 35–45.

heat function—Same as **enthalpy**.

heat index—As used by C. W. Thornthwaite in his 1948 **climatic classification**, a function of **temperature** designed to have low magnitude under cold conditions, increasing exponentially with increasing temperature. For a given **station**, it is numerically equal to the sum of the 12 monthly values of the expression

$$\left(\frac{t}{5}\right)^{1.514},$$

where t is the **normal** monthly temperature in °C.

The heat index is used in the computation of **potential evapotranspiration**, which, in turn, is one of the basic parameters of the 1948 classification.

Thornthwaite, C. W., 1948: An approach toward a rational classification of climate. *Geogr. Rev.*, **38**, 55–94.

heat island—*See* **urban heat island**.

heat lightning—Nontechnically, the luminosity observed from ordinary **lightning** too far away for its **thunder** to be heard.

Since such observations have often been made with **clear** skies overhead, and since hot summer evenings particularly favor this type of **observation**, there has arisen a popular misconception that the presence of diffuse flashes in the apparent absence of thunderclouds implies that lightning is somehow occurring in the **atmosphere** merely as a result of excessive **heat**.

heat low—Same as **thermal low**.

heat of condensation—(Or latent heat of condensation.) *See* **latent heat**.

heat of fusion—(Or latent heat of fusion.) *See* **latent heat**.

heat of sublimation—(Or latent heat of sublimation.) *See* **latent heat**.

heat of vaporization—(Or latent heat of vaporization.) *See* **latent heat**.

heat thunderstorm—In popular terminology, a **thunderstorm** of the **air mass** type that develops near the end of a hot, humid summer day; this term has no precise technical meaning.

heat transfer—Energy **transfer** as a consequence of **temperature** differences.
See also **convection, conduction, radiation**.

heat transfer coefficient—The rate of **heat transfer** per unit area per unit **temperature** difference.
See also **eddy heat conduction**.

heat unit—Term used to describe what has been traditionally called a **growing degree-day**.

Also used to describe the result of using different formulas for calculating growing degree-days. This term is often used to avoid the confusion of the term "growing degree-day" wherein more than one "day" can be accumulated within a 24-h period due to differences in plant type.

heat wave—(Also called hot wave, warm wave.) A period of abnormally and uncomfortably hot and usually humid weather.

To be a heat wave such a period should last at least one day, but conventionally it lasts from several days to several weeks. In 1900, A. T. Burrows more rigidly defined a "hot wave" as a spell of three or more days on each of which the maximum shade **temperature** reaches or exceeds 90°F. More realistically, the comfort criteria for any one region are dependent upon the **normal** conditions of that region. In the eastern United States, heat waves generally build up with southerly winds on the western flank of an **anticyclone** centered over the southeastern states, the air being warmed by passage over a land surface heated by the sun. *See also* **hot wind**.

Ward, R. de C., 1925: *The Climates of the United States*, 383–395.

heating degree-day—A form of **degree-day** used as an indication of fuel consumption; in U.S. usage, one heating degree-day is given for each **degree** that the **daily mean** temperature departs below the base of 65°F (where degrees Celsius are used, the base is usually 19°C).

In accumulating degree-days over a "heating season," days on which the **mean temperature** exceeds 65°F are ignored.
See **cooling degree-day**.

heave—1. The motion imparted to a floating body by **wave** action.

It includes both the vertical rise and fall, and the horizontal **transport**.

2. The up-and-down motion of the **center of gravity** of a ship.

See **surge, sway, ship motion.**

heaving—Soil heaving, including **frost heaving.**

heavy ion—Same as **large ion.**

heavy rain—**Rain** with a rate of accumulation exceeding a specific value that is geographically dependent.

heavy water—(Also known as deuterium oxide.) Water composed of **oxygen** and **deuterium.**

hectopascal—(Abbreviated hPa.) A hectopascal is equal to 100 **pascals** or 1 **millibar.**

Hector—Term used by World War II pilots to refer to **thunderstorms** that form regularly over the Tiwi Islands (Melville Island and Bathurst Island) just to the north of Darwin, Australia.

The pilots used these storms as navigation aids to return to Darwin after flying sorties.

height—*See* **altitude, elevation.**

height analysis method—Procedure used to analyze and monitor atmospheric conditions by studying the heights of fixed **pressure** levels.

height-change chart—A **chart** indicating the change in height of a **constant-pressure surface** over a specified previous time interval; comparable to a **pressure-change chart.**

See **height-change line.**

height-change line—(Also called contour-change line, isallohypse.) A line of equal change in height of a **constant-pressure surface** over a specified previous interval of time; the lines drawn on a **height-change chart.**

height-finding radar—A **radar** designed for accurate determination of **target** altitude.

The **beamwidth** of such radars is generally much narrower in the vertical than in the horizontal.

height pattern—(Also called isobaric topography, baric topography, pressure topography.) In meteorology, the general geometric characteristics of the distribution of height of a **constant-pressure surface** as shown by **contour lines** on a **constant-pressure chart.**

Compare **pressure pattern, circulation pattern.**

height vertical coordinate—A **vertical coordinate system** based on the geometric or **geopotential height.**

heiligenschein—A diffuse bright region surrounding the shadow an observer's head casts on a irregular surface.

It is most apparent when the sun is low in the sky and when the surface is **dew**-covered. The explanation of the heiligenschein varies depending upon whether it is seen over a dry or a dew-covered surface. When an observer's shadow is cast on a dry, irregular surface (such as gravel or vegetation), each irregularity near the **antisolar point** covers its own shadow. In other directions, the average brightness results from a mixture of sunlit and shaded surfaces. The lower the sun in the sky, the longer the shadows and so the greater the **contrast** with the brighter region near the antisolar point. While evident over virtually any irregular surface, the ready appearance of the heiligenschein in sunlit wooded areas when seen from an airplane has spawned the epithet, the "hot spot in the forest." Observations of the "hot spot in the forest." are undoubtedly all the more striking when the plane is high enough that its own shadow (the umbra portion) has vanished. The presence of dew on some species of grass greatly enhances the heiligenschein. Dewdrops held off the surface of the leaf by small hairs focus **sunlight** on the leaf where it is diffusely reflected. The **drop**, acting in a manner similar to the lens in a lighthouse, then collects a large fraction of this diffusely reflected **light** that would have otherwise gone in other directions and sends it back toward the source and the observer. The heiligenschein is occasionally called **Cellini's halo**, after Benvenuto Cellini who described its behavior in his memoirs of 1562. He even pointed out that "it appears to the greatest advantage when the grass is moist with dew," but felt its appearance bespoke "the wondrous ways of [God's] providence toward me." Shades of this interpretation are also found in the name heiligenschein: It is German for "the light of the holy one."

helical scanning—A **scanning** technique in airborne **Doppler radar** that makes use of both the aircraft motion and scanning to the side of the aircraft to survey the volume of **atmosphere** surrounding the flight track.

Typically, the radar **antenna** is located in the tail of the airplane and rotates continuously about an axis aligned with the airplane. The pointing direction may be normal to the flight track or, as common for dual-beam airborne radars, may include pointing angles both forward and rearward of the direction normal to the flight track. The fore and aft directions provide measurements for **dual-Doppler analysis** that can be used to determine the three-dimensional **velocity** vector of the **echoes**.

helicity—One-half the **scalar product** of the **velocity** and **vorticity** vectors.

It is a conserved quantity if the flow is inviscid and homogeneous in **density**, but is not conserved in more general viscous flows with **buoyancy** effects. The concept is useful in understanding severe **convective storms** and **tornadoes**, since in strong updrafts the velocity and vorticity vectors tend to be aligned, yielding high helicity. Three-dimensional **turbulence** containing a nonzero **mean value** of helicity may develop an inertial decay range, but the development is slowed by helicity. The reluctance of helical turbulence to **cascade** into an **inertial range** means that small-scale atmospheric flows with high helicity are less unstable and more predictable than small-scale flows with low helicity. *See also* **storm-relative environmental helicity, streamwise vorticity**.

heliograph—An instrument that records the duration of **sunshine** and gives a qualitative measure of the amount of sunshine by the action of the sun's rays upon blueprint paper; a type of **sunshine recorder**.

helion—(Plural: helia.) Suffix meaning sun or solar in terms related to **meteorological optics**. *Compare* **selene**.

helioseismology—Study of the the internal structure (i.e., **density, temperature,** abundance, etc.) and dynamics (i.e., mixing, rotation rate, etc.) of the sun from measurements of its surface oscillations.

heliostat—A clock-driven instrument mount that automatically and continuously points in the direction of the sun.

It is used with a **pyrheliometer** when continuous **direct solar radiation** measurements are required.

heliotropic wind—A subtle **diurnal** component of the **wind** velocity leading to a diurnal shift of the wind or turning of the wind with the sun, produced by the east-to-west progression of daytime surface heating.

helium—(Symbol He.) Lightest of the **noble gases**, atomic number 2, atomic weight 4.003.

It has an atomic mass of 4 (2 protons and 2 neutrons) and is a colorless, monatomic **element**. It is the sixth most abundant gas in **dry air**. Helium is very light, having a **specific gravity** referred to air of 0.138. This element is unique in that its existence on the sun was recognized prior to its discovery on the earth. Spectroscopic discovery in 1868 by Janssen in the **solar spectrum** was followed by terrestrial detection in 1889 and chemical isolation in 1894 by Ramsay. Because helium is not flammable and has a lifting power 92% of that of **hydrogen**, it is widely used as the inflation gas for lighter-than-air craft and for meteorological balloons. Helium nuclei are identical to radioactive alpha **particles** and helium is formed as a product of radioactive decay of certain elements, particularly the uranium group. Consequently, helium is often associated with deep deposits of fossil oil or gas and is released on mining.

helm—*See* **helm wind**.

helm bar—*See* **helm wind**.

helm wind—A strong cold northeasterly **wind** blowing down into the Eden valley from the western slope of the Crossfell Range (893 m or 2930 ft) in northern England.

Beyond the stormy northeasterly flow, there is a belt of **calm** on either side of which may be a light westerly wind. The helm wind occurs when the general direction of the wind is between north-northeast and east. A line of **cloud** (the helm) forms along the crest of the **ridge**, and above the **calm belt** a narrow, nearly stationary roll of cloud (the helm bar) rotates about a horizontal axis parallel to the helm bar. It may occur in any month but is most frequent in winter and spring.

Helmholtz equation—A **linear** second-order partial differential equation of the form

$$\nabla^2 \psi + k\psi = 0,$$

where ∇^2 is the **Laplacian operator** and k is a constant.

This equation occurs frequently in **dynamic meteorology**, for example, in **numerical forecasting** or in the study of **lee waves**. In classical physics it is the equation of the vibrating membrane. The case $k = 0$ reduces to a **Laplace equation**.

Helmholtz free energy—(Also called Helmholtz function, work function.) A **thermodynamic function of state** that, in a reversible **isothermal process**, increases with **work** done on the system.

In typical notation the Helmholtz free energy is

$$F = U - TS,$$

where F is the Helmholtz free energy (sometimes designated as A), U is the **internal energy**, T is **temperature**, and S is the **entropy**. By use of the **first law of thermodynamics** for reversible processes, the rate of change of the Helmholtz free energy is given by

$$\frac{dF}{dt} = -S\frac{dT}{dt} - \frac{dW}{dt},$$

where W is the work done by the system. *Compare* **Gibbs function**.

Helmholtz function—*See* **Helmholtz free energy**.

Helmholtz instability—(Also called shearing instability.) The **hydrodynamic instability** arising from a **shear**, or **discontinuity**, in **current** speed at the **interface** between two fluids in two-dimensional motion.

The **perturbation** gains **kinetic energy** at the expense of that of the basic currents. According to the theory of **small perturbations**, waves of all wavelengths on such an interface are unstable, their rate of growth being $e^{\mu t}$ with μ given by

$$\mu = \frac{\pi}{\lambda}\,|U - U'|,$$

where λ is the **wavelength** and U and U' the **current** speeds of the two fluids. Such waves are called Helmholtz waves or shear waves, and move with a **phase speed** c equal to the mean of the current speeds,

$$c = \frac{1}{2}\,(U + U').$$

With an assumed **density** difference in the fluids, **gravity waves** may also be generated. The combination of these effects yields a critical wavelength λ_c,

$$\lambda_c = \frac{2\pi}{g}\left(\frac{\rho\rho'}{\rho^2 - \rho'^2}\right)(U - U')^2,$$

where ρ and ρ' are the densities of the lower and upper fluids, respectively. Waves shorter than the critical are unstable, longer waves, stable. This analysis has been applied to **billow clouds**; however, the critical wavelength is considered too small (of the order of a few kilometers) for this sort of **instability** to be the explanation for the growth of **cyclonic** disturbances on atmospheric **fronts**. *See* **shearing instability**.

Haurwitz, B., 1941: *Dynamic Meteorology*, 282–292, 307–309.

Helmholtz's theorem—The statement that if F is a **vector field** satisfying certain quite general mathematical conditions, then F is the sum of two vectors, one of which is **irrotational** (has no **vorticity**), the other **solenoidal** (has no **divergence**).

Thus, the horizontal **velocity** field, for example, may be expressed by

$$\mathbf{v} = \nabla_H\chi + \mathbf{k}\times\nabla_H\psi = \mathbf{v}_\chi + \mathbf{v}_\psi,$$

where $\mathbf{v}_\chi$ is **irrotational**, that is, $\nabla_H\times\mathbf{v}_\chi = 0$; $\mathbf{v}_\psi$ is solenoidal, that is, $\nabla_H\cdot\mathbf{v}_\psi = 0$; $\mathbf{k}$ is a unit **vector** directed vertically; and χ and ψ are **scalar** functions that may be computed from the given **wind field**.

Charney, J. G., 1955: *J. Mar. Res.*, **14**, 477–498.
Salby, M. L., 1996: *Fundamentals of Atmospheric Physics*, 375–377.

Helmholtz wave—An **unstable wave** in a system of two homogeneous fluids with a **velocity** discontinuity at the **interface**.

See **Helmholtz instability**.

hemispheric flux—*See* **radiant flux**.

hemispheric model—A numerical **model** constructed assuming a boundary condition along the **equator**.

In many applications, the boundary condition is based upon vanishing **meridional** velocity along the equator. In some spectral formulations, the boundary condition may be formulated using symmetric or antisymmetric expansion (basis) functions. *Compare* **atmospheric circulation model**, **circulation model**, **general circulation model**.

hemispheric wavenumber—Same as **angular wavenumber**.

Henry's law—The concentration of a chemical species in a liquid solution is proportional to its gas-phase **partial pressure**.

The constant of proportionality is called the Henry's law constant and provides a measure of the **solubility** of a particular compound. For molecules that form ions in solution, Henry's law is modified to take account of the increased solubility as a function of **pH**.

Henyey–Greenstein phase function—A single-parameter analytic form for an **anisotropic, scattering** phase function.

The Henyey–Greenstein phase function is written as

$$p(\theta) = \frac{1 - g^2}{(1 + g^2 - 2g\cos\theta)^{3/2}},$$

where θ is the scattering angle. The **parameter** g is conveniently equal to the **asymmetry factor**.

Henyey, L. G., and J. L. Greenstein, 1941: *Astrophys. J.*, **93**, p.70.

hermitan—Same as **harmattan**.

hertz—The **standard** unit of measurement for **frequency** in cycles/second.

Prefixes include kilo (10^3), mega (10^6), and giga (10^9), abbreviated kHz, MHz, and GHz; replaces cycles per second in modern usage.

Herzberg bands—*See* **oxygen**.

Herzberg continuum—The portion of the **UV** spectrum of molecular **oxygen** (O_2) between 200 and 240 nm.

The **absorption** results from forbidden **electronic transitions** from the molecular **ground state** to a number of electronically **excited states**. **Photolysis** of O_2 is dominated by the Herzberg continuum throughout the **stratosphere**.

heterodyne—Pertaining to the process of **mixing** two radio signals of different frequencies to produce a third **signal** that is the difference of the two, that is, to produce **beating** between the two frequencies.

heterogeneous chemistry—A wide-ranging subject that consists of chemistry involving two phases, usually one or more gaseous reactants, and a condensed **phase** substrate where the reaction occurs, either liquid or solid.

Generally the substrate facilitates the reaction and many heterogeneous reactions do not proceed in the gas phase. Reactants include free radicals as well as closed-shell molecules. Often, a reactant will hydrolyze (react with H_2O) heterogeneously; in this case a major constituent of the substrate is a reactant. Common substrates are **fog** and **rain** droplets as well as **aerosol** in the **lower atmosphere** and **sulfuric acid** aerosol (*see* **stratospheric sulfate layer**) and **water** ice **particles** in the **upper atmosphere**. *See* **polar stratospheric clouds**.

heterogeneous fluid—A fluid in which the **density** varies from point to point.

For most purposes, the **atmosphere** must be treated as heterogeneous, particularly with regard to the decrease of density with height. *Compare* **homogeneous fluid**.

heterogeneous nucleation—*See* **nucleation**.

heterosphere—The upper portion of a two-part division of the **atmosphere** according to the general homogeneity of atmospheric composition; the layer above the **homosphere**.

The heterosphere is characterized by **variation** in composition and mean molecular weight of constituent gases. This region starts at 80–100 km above the earth and therefore closely coincides with the **ionosphere** and the **thermosphere**. *See* **atmospheric shell**.

heuristic—In **artificial intelligence**, a rule of thumb, generally based on expert experience or common sense rather than an underlying theory or mathematical **model**, that can be incorporated in a **knowledge base** and used to guide a problem-solving process.

Most procedures used by human weather forecasters are heuristic, as are many pattern-recognition techniques in **radar** and **satellite meteorology**.

Hevel's halo—A historical **halo** candidate, of which no convincing photographic documentation is known.

The possibility of this halo is based on early drawings of what seems to be a faint white halo of 90° radius centered on the sun. But good theoretical support is lacking, and subsequent reports seem to be misidentifications of the subhelic arcs.

hexagonal column—One of the many forms in which **ice crystals** are found in the **atmosphere**.

This particular **crystal habit** of **ice** is characterized by hexagonal cross section in a plane perpendicular to the long direction (principal axis, optic axis, or c axis) of the columns. It differs from that found in hexagonal platelets only in that environmental conditions have favored growth along the principal axis rather than perpendicular to that axis. Growth by **vapor** deposition at temperatures of from $-3°$ to about $-8°C$ and also at lower **temperature** below $-25°C$ leads to growth of columnar crystals, though other **crystal** features (needles, scrolls) also appear in this temperature interval, depending on the degree of **water vapor** supersaturation and the crystal **fall velocity**. *See* **column**.

hexagonal platelet—A small **ice crystal** of hexagonal tabular form.
   The distance across the hexagonal facet of the **crystal** may be as large as 1–2 mm and the ratio of thickness to diameter as small as 1/100. This crystal form usually grows by **vapor** deposition at temperatures between $-8°$ and $-25°C$. At temperatures between $-12°$ and $-16°C$, as crystals grow and attain a higher **fall velocity**, provided that the **saturation** ratio approaches water saturation, the corners sprout and form dendritic (treelike) side arms. *See* **dendrite**.

HF—Abbreviation for high frequency.
   See **radio frequency band**.

HFCs—Abbreviation for **hydrofluorocarbons**.

hi-reference signal—The audio-frequency **signal** transmitted by the Diamond–Hinman **radiosonde** when the **baroswitch** pen passes each fifteenth contact of the **commutator**, up to a number determined by the design of the commutator, and each fifth contact thereafter.
   This signal is transmitted so that the **pressure**, **temperature**, and **humidity** may be more readily distinguished.

hibernal—Pertaining to winter.
   The corresponding adjectives for spring, summer, and fall are **vernal**, **aestival**, and **autumnal**.

hidden layer—In artificial neural networks, a layer of nodes between the **input** and **output** layers that contains the weights and processes data.
   The values within the hidden layer are constantly adjusted during the training of the **neural network**, until the desired output is reached.

HIFOR—An international code word used to abbreviate "high-level forecast."

hig—(Also called ig.) In England, a sharp, short-lived **storm** of **rain** or **wind**.

high—In meteorology, an area of high pressure, referring to a maximum of **atmospheric pressure** in two dimensions (closed isobars) in the synoptic surface chart, or a maximum of height (closed contours) in the **constant-pressure chart**.
   Since a high is, on the **synoptic chart**, always associated with **anticyclonic circulation**, the term is used interchangeably with **anticyclone**. *Compare* **low**.

high aloft—Same as **upper-level anticyclone**.

high-altitude station—A weather observing **station** at a sufficiently high elevation to be nonrepresentative of conditions near **sea level**; 2000 m (about 6500 ft) has been given as a reasonable lower limit.
   High-altitude stations may be divided into 1) mountain stations, which are freely exposed on or near the summits of peaks and at which observations nearly approximate the free-air conditions at their level, and 2) plateau stations, which have an extreme kind of **continental climate**.

high arctic climate—In Nordenskjöld's **climatic classification** (1928), the **climate** of those parts of the world where the average temperature of the warmest month is below 0°C.
   It is equivalent, therefore, to the **perpetual frost climate** of W. Köppen (1918), and to C. W. Thornthwaite's **frost climate** (1931). *See* **polar climate**.

Nordenskjöld, O., and L. Meeking, 1928: *The Geography of the Polar Regions*, American Geographical Society, Special Pub. No. 8, New York.
Köppen, W. P., 1918: Klassification der Klimate nach Temperatur, Niederschlag und Jahreslauf. *Petermanns Geog. Mitt.*, **64**, 193–203; 243–248.
Thornthwaite, C. W., 1931: The climates of North America according to a new classification. *Geogr. Rev.*, **21**, 633–655.

high-based thunderstorm—*See* **high-level thunderstorm**.

high cloud—*See* **cloud classification**.

high foehn—(Also called free foehn.) The presence of south foehnlike conditions at higher elevations of the Alps while the lower elevations and the plains or "foreland" to the north are under a cold **air mass**.

Under these conditions, the mountains are often warmer than the lowlands. The warming is attributed to subsiding air in a **synoptic** anticyclone above the cold surface air. Although the winds in the warm air are apt to have a southerly component, the high foehn does not necessarily have strong, gusty **foehn** winds. However, this situation often precedes the foehn at the surface.

high fog—In the United States, the frequent **fog** on the slopes of the coastal mountains of California, especially applied when the fog overtops the range and extends as **stratus** over the **leeward** valleys.

high frequency—(Abbreviated HF.) *See* **radio frequency band**.

high-frequency radar—Radar operating at wavelengths of 10–100 m with the received **signal** being **scattered** by **ocean waves** of half the **radar** wavelength.

The received signal is Doppler shifted by an amount corresponding to the **phase speed** of this ocean wave component. The **surface current** in the radial direction can be computed from the difference between the observed **Doppler shift** and that which is expected from the surface-wave **dispersion relation**. *See also* **Doppler radar**, **marine radar**, **microwave radar**, **synthetic aperture radar**.

high index—A relatively high value of the **zonal index** which, in middle latitudes, indicates a relatively strong westerly component of **wind** flow and the characteristic weather features attending such motion.

A **synoptic** circulation pattern of this type is commonly called a "high-index situation."

high-level anticyclone—Same as **upper-level anticyclone**.

high-level cyclone—Same as **upper-level cyclone**.

high-level ridge—Same as **upper-level ridge**.

high-level thunderstorm—Generally, a **thunderstorm** based at a comparatively high altitude in the **atmosphere**, roughly 2400 m or higher.

These storms form most strikingly over arid regions, and frequently their **precipitation** is evaporated before reaching the earth's surface.

high-level trough—Same as **upper-level trough**.

high-performance liquid chromatography—(Abbreviated HPLC; also called high-pressure liquid chromatography.) Chromatographic technique similar to **gas chromatography**, except that various individual solvents, combinations of solvents, and changing concentrations of solvent are used in combination with stationary column phases to separate components of a liquid matrix.

Various detectors are used in HPLC, including **UV** (fixed and **variable** wavelength), **refractive index**, and **conductivity** detectors.

high polar glacier—In Ahlmann's (1935) **glacier** classification, a **polar glacier** with **firn** in the **accumulation area** that is 100 m or more thick and that does not melt appreciably in summer.

Ahlmann, H. W., 1935: Contribution to the physics of glaciers. *Geographic Journal*, **86**, 97–113.

high pressure center—*See* **pressure center**.

high pressure system—*See* **high**.

high-reference signal—*See* **hi-reference signal**.

High Resolution Infrared Radiation Sounder—(Abbreviated HIRS.) One of three **sounders** that comprise the **TIROS** Operational Vertical Sounder package on **NOAA** polar orbiter satellites.

HIRS-2, derived from the experimental HIRS-1, which flew on *Nimbus 6* launched in June 1975, measures incident **radiation** in 1 visible and 19 **infrared** spectral regions, with a **nadir** resolution of 17 km.

High Resolution Picture Transmission—(Abbreviated HRPT.) Full **resolution** digital **AVHRR** data (all **channels**) transmitted at **S-band** by **POES** satellites.

The HRPT data stream is equivalent to **local area coverage** data, but is transmitted as a continuous broadcast by the satellite as it orbits the earth.

high tide—Same as **high water**.

high-volume filter—Instrument for measuring total **particulate matter** in the **atmosphere**.

Air is drawn into the instrument at a high flow rate of 1–2 $m^3 min^{-1}$ and passes through a filter bed that traps **particles**. The high-volume filter is the current standard recommended by the U.S. Environmental Protection Agency for measurement of particulate matter in the atmosphere.

high water—(Popularly called high tide.) The maximum water level reached in a tidal **cycle**.

high-water interval—Same as **cotidal hour**.

higher-order closure—An approximation to **turbulence** that retains prognostic equations for mean variables (such as **potential temperature** and **wind**) as well as for some of the higher-order **statistics** including **variance** (such as **turbulence kinetic energy** or **temperature** variance) or **covariance** (kinematic fluxes such as for **heat** and **momentum**).

Regardless of the **statistical** order of the forecast equations, other high-order statistical terms appear in those equations, the solutions of which require approximations known as **turbulence closure** assumptions. While usually considered more accurate than first-order closures (**K-theory**), higher-order closure solutions are computationally more expensive. Turbulence closures are often classified according to two attributes: the order of statistical closure, and the degree of nonlocalness. Common types of higher-order closure include (in increasing statistical order): one-and-a-half order closure (also known as k-ϵ closure), **second-order closure**, and **third-order closure**. All turbulence closures are designed to reduce an infinite set of equations that cannot be solved to a finite set of equations that can be solved approximately to help make weather forecasts and describe physical processes. *See* **Reynolds averaging**; *compare* **zero-order closure**, **half-order closure**.

highest astronomical tide—The highest **water level** that can be predicted to occur under any combination of astronomical conditions.

highland climate—Same as **mountain climate**.

highland ice—An unbroken, but limited, sheet of **land ice** lying on relatively flat plateau country.

It is thin enough to show the main contours of the underlying land, in contrast to a **continental glacier**.

hill fog—**Fog** formed by **orographic lifting** of **moist air** to **condensation** over a hill.

See **orographic fog**.

Himawari—("Sunflower" in Japanese.) The name sometimes given to the Japanese **GMS** satellites, especially *GMS-1* launched on 14 July 1977.

hippy—A heave–pitch–roll **sensor** consisting of vertical **acceleration** and tilt sensors, used to measure the height and directionality of ocean surface **gravity waves**.

HIRS—Abbreviation for **High Resolution Infrared Radiation Sounder**.

histogram—A graphical representation of a **frequency distribution**.

The **range** of the **variable** is divided into class intervals for which the **frequency** of occurrence is represented by a rectangular column; the height of the column is proportional to the frequency of observations within the interval.

historical climate—The **climate** of the historical period, which may be taken as the past 7000 years; generally distinguished from the period of time during which regular daily observations from meteorological instruments are available, which begins generally in the seventeenth to eighteenth centuries A.D.

Thus, historical climate is the climate of a period of history for which no instrumental observations exist, but its main characteristics may be reconstructed from written descriptions.

hoar—Same as **hoarfrost**.

hoar crystal—An individual **ice crystal** in a deposit of **hoarfrost**.

Such crystals always grow by **deposition**.

hoarfrost—A deposit of interlocking **ice crystals** (**hoar crystals**) formed by direct **deposition** on objects, usually those of small diameter freely exposed to the air, such as tree branches, plant stems and leaf edges, wires, poles, etc.

Also, **frost** may form on the skin of an aircraft when a cold aircraft flies into air that is warm and moist or when it passes through air that is supersaturated with **water vapor**. The deposition of hoarfrost is similar to the process by which **dew** is formed, except that the **temperature** of the befrosted object must be below **freezing**. It forms when air with a **dewpoint** below freezing is brought to **saturation** by cooling. In addition to its formation on freely exposed objects (**air hoar**), hoarfrost also forms inside unheated buildings and vehicles, in caves, in crevasses (**crevasse hoar**), on **snow** surfaces (**surface hoar**), and in air spaces within snow, especially below a **snow crust** (**depth hoar**). Hoarfrost is more fluffy and feathery than **rime**, which in turn is lighter than **glaze**. Observationally, hoarfrost is designated light or heavy (frost) depending upon the amount and uniformity of deposition.

Hobbs's theory—*See* **glacial anticyclone**.

hodogram—*See* **hodograph**.

hodograph—In general (mathematics), the locus of one end of a **variable** vector as the other end remains fixed.

A common hodograph (or hodogram) in meteorology represents the vertical distribution of the horizontal **wind**.

hodograph analysis—A method of **analysis** of a **wind** sounding at a **station**.

Contrary to mathematical convention, individual wind vectors at selected levels are plotted, typically on a polar coordinate diagram, from the region of the diagram toward the direction from which the wind blows. The lengths of the vectors are proportional to the corresponding wind speeds. The sense of each **vector** is toward the origin. The lines joining the starting points of airflow at successive levels form a **hodograph** (or hodogram). Each such line corresponds to the **wind shear** vector in the layer concerned.

hohlraum—An experimental source of **blackbody radiation** in the form of an **isothermal** cavity with a small opening.

The walls of the cavity are partially reflective and are kept at constant **temperature**, maintaining **radiative equilibrium** within the cavity. A hohlraum is also a perfect absorber since it traps all **radiation** that enters the small **aperture**.

Höiland's circulation theorem—*See* **circulation theorem**.

Holmboe instability—An **instability** of an unbounded stratified parallel **shear flow** to the development of cusplike waves that propagate with **phase speed** along the flow direction significantly different from the speed of the inflection point of the **shear**.

The propagating phase speed of Holmboe instability distinguishes it from **Kelvin–Helmholtz** instability. The instability occurs only in **stratified fluid** due to a resonant coupling between an **internal gravity wave** and a wavelike **disturbance** where the background shear varies vertically.

Holocene epoch—The last 10 000 years of **geologic time**.

holosteric barometer—(Rare.) Same as **aneroid barometer**.

("Holosteric" means "wholly made of solids," while "**aneroid**" means "devoid of liquid.")

homobront—(Rare.) Same as **isobront**.

homoclime—**Stations** with similar **climatic diagrams**.

homogeneous atmosphere—1. A hypothetical **atmosphere** in which the **density** is constant with height.

The **lapse rate** of **temperature** in such an atmosphere is known as the **autoconvective lapse rate** and is equal to g/R (or approximately 3.4°C/100 m), where g is the **acceleration of gravity** and R is the **gas constant** for air. A homogeneous atmosphere has a finite total thickness that is given by $R_d T_v/g$, where R_d is the **gas constant** for **dry air** and T_v is the **virtual temperature** (K) at the surface. For a **surface temperature** of 273 K, the vertical extent of the homogeneous atmosphere is approximately 8000 m. At the top of such an **atmosphere** both the **pressure** and **absolute** temperature vanish.

2. Same as **adiabatic atmosphere**.

homogeneous fluid—Strictly, a fluid with uniform properties throughout, but meteorologists sometimes designate as homogeneous a fluid with constant **density**.

No fluid is homogeneous in an **absolute** sense. Homogeneity must be specified relative to a **characteristic length** (e.g., the size of the **probe** used to measure the properties of the fluid). *Compare* **heterogeneous fluid**.

homogeneous nucleation—*See* **nucleation**.

homogeneous turbulence—**Turbulence** in which spatial derivatives of all mean turbulent quantities are negligible.

Sufficiently far **downstream**, **grid turbulence** is considered to be approximately homogeneous.

homogeneous wave—A **wave** for which the **surfaces of constant phase** and of constant **amplitude** coincide.

A wave that is not homogeneous is inhomogeneous. An example of an inhomogeneous wave is a **radio wave** transmitted in water illuminated at oblique incidence.

homologous turbulence—**Turbulence** with a constant average shear **stress** throughout the **field**, for example, **Couette flow**.

See also **isotropic turbulence**.

homopause—The top of the **homosphere**, or the level of transition between it and the **heterosphere**. It probably lies between 80 and 90 km, where molecular **oxygen** begins to dissociate into atomic oxygen. The homopause is somewhat lower in the daytime than at night.

homosphere—The lower portion of a two-part division of the **atmosphere** according to the general homogeneity of atmospheric composition; opposed to the **heterosphere**.
The homosphere is the region in which there is no gross change in atmospheric composition, that is, all of the atmosphere from the earth's surface to about 80 or 100 km. *See* **atmospheric shell**.

homothermy—Condition that occurs when water has an **isothermal** temperature **profile**.

hook echo—A pendant, curve-shaped region of **reflectivity** caused when **precipitation** is drawn into the **cyclonic** spiral of a **mesocyclone**.
The hook echo is a fairly shallow feature, typically extending only up to 3–4 km in height before becoming part of a **bounded weak echo region** (BWER).
Fujita, T., 1958: Mesoanalysis of the Illinois tornadoes of 9 April 1953. *J. Meteor.*, **15**, 288–296.

hook gauge—An instrument used to measure changes in the level of the water in an **evaporation pan**.
It consists of a pointed metal hook, mounted in the vertical, the position of which with respect to its supporting member may be adjusted by means of a micrometer arrangement. The gauge is placed on the **still well** and a measurement taken when the point of the hook just breaks above the surface of the water.

Hooke number—*See* **Cauchy number**.

Hopfield bands—*See* **oxygen**.

horizon—One of several lines or planes used as reference for observation and measurement relative to a given location on the surface of the earth, and referred generally to a horizontal direction (i.e., at right angles to the **zenith**).
Considerable contradiction exists between the nomenclatures for the several concepts of horizon. Aside from the distinctly different geological horizons (strata of earth material), it may be said that there are two types of horizons: earth–sky horizons (1, 2, and 3 below) and celestial horizons (4 and 5 below). Meteorology is primarily concerned with the former, astronomy with the latter. Specifically, the following constitute the major variant usages, with suggested nomenclature along with other names that have been applied. 1) **Local horizon**: the actual lower boundary of the observed sky or the upper outline of terrestrial objects including nearby natural obstructions. 2) **Geographic horizon** (also called apparent horizon, local horizon, visible horizon): the distant line along which earth and sky appear to meet. In both popular usage and weather observing, this is the usually conception of horizon. Nearby prominences are said to obscure the horizon and are not considered to be a part of it. The minimum desirable **horizon distance** should be of the order of three miles. 3) **Sea level horizon** (also called ideal horizon, sensible horizon, sea horizon, visible horizon, apparent horizon): the apparent junction of the sky and the **sea level** surface of the earth; the horizon as actually observed at sea. This type of horizon is used as the reference for establishing times of **sunrise** and **sunset**. 4) **Astronomical horizon** (also called sensible horizon, real horizon): the plane that passes through the observer's eye and is perpendicular to the zenith at that point; or, the intersection of that plane with the **celestial sphere** (i.e., a **great circle** on the celestial sphere equidistant from the observer's zenith and **nadir**). It is the **projection** of a horizontal plane in every direction from the point of orientation. 5) **Celestial horizon** (also called rational horizon, geometrical horizon, true horizon): the plane, through the center of the earth, that is perpendicular to a radius of the earth that passes through the point of observation on the earth's surface; or, the intersection of that plane with the celestial sphere. *See also* **artificial horizon, fog horizon, haze horizon, smoke horizon**.

horizon distance—The distance, at any given **azimuth**, to the point on the earth's surface constituting the **horizon** for some specified observer.
In the particularly simple case of an observer at sea, the horizon distance s in miles corresponding to an observer whose eye is at a height h feet above **sea level**, is given by the relation

$$s = 1.317h^{1/2},$$

which holds to a good degree of approximation as long as both s and h are small compared to the earth's radius.

horizontal advection—*See* **advection**.

horizontal convective rolls—(Also known as horizontal roll vortices, boundary layer rolls.) Counterrotating horizontal vortices that commonly occur within the **convective boundary layer**; their major axes are aligned with the mean boundary layer wind-shear **vector**.

The depth of the **roll** circulations is consistent with the depth of the **boundary layer**; the **wavelength**, measured from **updraft** to updraft in the cross-roll direction, is about three times the boundary layer depth.

horizontal divergence—A measure of the local spreading or **divergence** of a **vector field** in a horizontal plane, commonly written $\nabla_H \cdot \mathbf{V}$.

In **Cartesian coordinates**,

$$\frac{\partial u}{\partial x} + \frac{\partial v}{\partial y} = \nabla_H \cdot \mathbf{V},$$

where u and v are the components of the vector field $\mathbf{V}$ along the horizontal axes x and y, respectively. The horizontal divergence of the **velocity** field is related to the vertical motion and local **pressure** variations through the **equation of continuity** and **equations of motion**. Horizontal divergence is often referred to simply as divergence.

horizontal pressure force—(Often called, simply, pressure force.) The horizontal **pressure gradient** per unit mass, $-\alpha\nabla_H p$, where α is the **specific volume**, p the **pressure**, and ∇_H the horizontal component of the **del operator**.

This force acts normal to the horizontal isobars toward lower pressure. It is one of the three important forces appearing in the horizontal **equations of motion**, the others being the **Coriolis force** and **friction**. *See* **pressure force**.

horizontal roll vortices—*See* **horizontal convective rolls**.

horizontal time section—*See* **time section**.

horizontal visibility—Maximum distance at which an **observer** can see and identify an object lying close to the horizontal plane on which he or she is standing.

horizontal wind shear—A horizontal **variation** in **current** speed within a flow.

horn antenna—(Also called horn radiator.) A type of radio **antenna** often used because of its directional properties and simple structure.

It consists essentially of flared extensions of a metallic **waveguide**. Large horn antennas are used primarily in research applications to produce very narrow vertical or horizontal beams.

horn radiator—Same as **horn antenna**.

horse latitudes—The belts of latitude over the oceans at approximately 30°–35°N and S where winds are predominantly **calm** or very light and weather is hot and dry.

These latitudes mark the **normal** axis of the **subtropical highs**, and move north and south by about 5° following the sun. The two calm belts are known as the **calms of Cancer** and **calms of Capricorn** in the Northern and Southern Hemispheres, respectively; in the North Atlantic Ocean, these are the latitudes of the **Sargasso Sea**. The name is believed to have originated in the days of sailing ships, when the voyage across the Atlantic in those latitudes was often prolonged by calms or baffling winds so that water ran short, and ships carrying horses to the West Indies found it necessary to throw the horses overboard. *Compare* **doldrums**.

hot belt—As presented by A. Supan (1879), the belt around the earth within which the annual **mean temperature** exceeds 20°C.

He also defined a **temperate belt** and **cold cap**. *See* **climatic classification**.

Supan, A., 1879: Die Temperaturzonen der Erde. *Petermanns Geog. Mitt.*, **25**, 349–358.

hot-film anemometer—Similar to a **hot-wire anemometer**, with the exception that the **sensing element** is a thin film of conductive material on the surface of a nonconductive rod.

hot spot in the forest—*See* **heiligenschein**.

hot wave—*See* **heat wave**.

hot wind—General term for **winds** characterized by intense **heat** and low **relative humidity**, such as summertime **desert** winds or an extreme **foehn**.

hot-wire anemometer—An **anemometer** that utilizes the principle that the **convection** of **heat** from a body is a function of its **ventilation**.

In its usual form it consists of a thin platinum wire heated to approximately 1000°C so that its **temperature** is relatively independent of **ambient temperature** variation. **Wind speed** is determined by measuring either the **current** required to maintain the hot wire at a constant temperature or the **resistance** variation of the hot wire while a constant current through the wire is maintained. Wind speeds as low as a few centimeters per second can be measured in this manner. The **response time** constant of the wire can be made very small. *See* **hot-film anemometer**.

hour-out line—Same as **time curve**.

hourly distance scale—*See* **time-distance graph**.

hourly observation—Commonly used term for **record observation**.

hourly time-front computer—A computer that shows the hourly travel distance of a vehicle under the influence of either a single or a complex set of **velocity** vectors.

One of the primary functions of the computer is to facilitate flight planning by the **wave-front method**. A special type of computer has been designed to be used on 4D charts.

Hovmüller diagram—A diagram that shows **isopleths** of atmospheric **variation**, such as **pressure** or **thickness**, usually averaged over a band of latitude.

Time is usually on one axis, longitude the other. This product demonstrates the progression of large-scale atmospheric features over a long period of time.

HO$_x$—Odd **nitrogen**; generally taken to mean the sum of hydroxyl (OH) and hydroperoxyl (HO$_2$) radicals.

HO$_x$ species play a central role in atmospheric **oxidation** processes.

hPa—Abbreviation for **hectopascal**.

HPLC—Abbreviation for **high-performance liquid chromatography**.

HRPT—Abbreviation for **High Resolution Picture Transmission**.

hubbob—Same as **haboob**.

Huggins band—A relatively weak **absorption band** of **ozone** lying approximately between 310 and 340 nm.

Absorption of **direct solar radiation** in this band led to the first positive identification of ozone in the **atmosphere**.

human bioclimatology—A major branch of **bioclimatology** that deals with effects of **climate** upon man. Currently, its major emphasis is on 1) the **heat balance** of the human body under different conditions of **air temperature**, **humidity**, and **wind**; 2) the effects of **radiation**, especially nuclear and **ultraviolet**, on genetics and general health; 3) the effects of atmospheric conditions and of types of changes of weather and climate on human health, vigor, and disease; and recently, 4) the effects of electrical conditions, including the atmospheric **potential gradient** and **longwave** radiations.

Buettner, K. J. K., 1951: *Compendium of Meteorology*, 1112–1125.

human climatology—Same as **human bioclimatology**.

Humboldt Current—*See* **Peru/Chile Current**.

humid climate—(Also called forest climate.) The **climate** of a region where the typical vegetation is forest.

The humid climate is one of Thornthwaite's **humidity provinces**.

Thornthwaite, C. W., 1931: The climates of North America according to a new classification. *Geogr. Rev.*, **21**, 633–655.

humid mesothermal climate—Following C. W. Thornthwaite's 1931 terminology, a name sometimes given to **temperate rainy climates**.

See **humid climate, mesothermal climate**.

Thornthwaite, C. W., 1931: The climates of North America according to a new classification. *Geogr. Rev.*, **21**, 633–655.

humid microthermal climate—Following C. W. Thornthwaite's 1931 terminology, a name sometimes given to a **snow forest climate**.

See **humid climate, microthermal climate**.

Thornthwaite, C. W., 1931: The climates of North America according to a new classification. *Geogr. Rev.*, **21**, 633–655.

humidity—1. Generally, some measure of the **water vapor** content of air.

The multiplicity of humidity measures is partly due to different methods of measurement and partly because the conservative measures (**mixing ratio**, **specific humidity**) cover an extremely wide **dynamic range**, as a result of the rapid **variation** of **saturation vapor pressure** with **temperature**.

2. Popularly, same as **relative humidity**.

humidity element—The **transducer** of any **hygrometer**, that is, that part of a hygrometer that quantitatively "senses" atmospheric **water vapor**.

humidity index—As used by C. W. Thornthwaite in his 1948 **climatic classification**, an index of the degree of **water surplus** over **water need** at any given **station**. It is calculated, independent of the opposing **aridity index**, as follows:

$$\text{humidity index} = 100s/n,$$

where s (the water surplus) is the sum of the monthly differences between **precipitation** and **potential evapotranspiration** for those months when the **normal** precipitation exceeds the latter, and n (the water need) is the sum of monthly potential evapotranspiration for those months of surplus.

The humidity index has two uses in Thornthwaite's classification: 1) as a component of the **moisture index**; and 2) as a basis for detailed classification of **dry climates**.

Thornthwaite, C. W., 1948: An approach toward a rational classification of climate. *Geogr. Rev.*, **38**, 55–94.

humidity mixing ratio—*See* **mixing ratio**.

humidity province—In C. W. Thornthwaite's 1931 **climatic classification**, a region in which the **precipitation effectiveness** of its **climate** produces a definite type of biological consequence, in particular the climatic climax formations of vegetation (**rain forest, tundra**, etc.).

Five main classes of humidity province are distinguished, bounded by values of precipitation effectiveness index (P–E index): 1) **wet** or rain forest; 2) **humid** or forest; 3) **subhumid** or grassland; 4) **semiarid** or steppe; and 5) **arid** or desert. Thornthwaite (1948) used values of **moisture index** to limit similar but purely climatic (nonbiological) zones as follows: 1) **perhumid**; 2) humid; 3) subhumid; 4) semiarid; and 5) arid. The **moist climates** are those with a positive moisture index; **dry climates** have negative values. *Compare* **hyetal region, climatic province, temperature province**.

Thornthwaite, C. W., 1931: The climates of North America according to a new classification. *Geogr. Rev.*, **21**, 633–655.
Thornthwaite, C. W., 1948: An approach toward a rational classification of climate. *Geogr. Rev.*, **38**, 55–94.

humidity strip—(Also called electrolytic strip.) A flat plastic strip bounded by electrodes on two sides and coated with a **hygroscopic** chemical compound such as lithium chloride.

The electrical **resistance** of this coating is a function of the amount of moisture absorbed from the **atmosphere** and the **temperature** of the strip. Humidity strips have been used in **radiosondes**, but are replaced today by thin-film capacitors or carbon-film **hygrometer** elements. *See* **dew cell**.

humilis—A **cloud species** unique to the genus **cumulus**.
See **cumulus humilis, cloud classification**.

hummock—A mound of broken ice projecting upward, formed by **ice** deformation.
The submerged counterpart of a hummock is termed a **bummock**. *See also* **ridge**.

hummocked ice—**Pressure ice** characterized by haphazardly arranged mounds or hillocks ("**hummocks**").

This has less definite form than **rafted ice** or **tented ice**, but in fact may develop from either of those as melting, **sublimation**, or drifting changes the sharper ice edges into more rounded shapes.

hurly-burly—A **thunderstorm** in England.

hurricane—(Many regional names.) A **tropical cyclone** with 1-min average surface (10 m) winds in excess of 32 m s⁻¹ (64 knots) in the Western Hemisphere (North Atlantic Ocean, Caribbean Sea, Gulf of Mexico, and in the eastern and central North Pacific east of the date line).

The name is derived from "huracan," a Taino and Carib god, or "hunraken," the Mayan storm god. For a more complete discussion, *see* **tropical cyclone**.

hurricane band—(Also called spiral band, **hurricane radar band**.) A hurricane radar band of circular or spiral shape associated with a **tropical cyclone** (**hurricane** or **typhoon**).

Made evident by **radar** observations, hurricane bands typically curve cyclonically inward toward the center of the **storm**. The bands may be classified as primary if they merge into the **eyewall** encircling the **eye** of the storm, or secondary if they are disconnected from the eyewall. Hurricane bands generally move slowly around the center of the storm in the direction of the hurricane circulation. *See* **banded structure**.

hurricane bar—*See* **cloud bar**.

hurricane beacon—An air-launched balloon designed to be released in the **eye** of a **tropical cyclone**, float within the eye at predetermined levels, and transmit radio signals for **RDF** positioning.

hurricane core—The innermost region of a **tropical cyclone**, encompassing the **eye** and the **eyewall**.
In the core, the local **Rossby number**, the ratio of the swirling **wind** to the **Coriolis parameter** times the distance to the **circulation** center, significantly exceeds unity, so the flow is approximately **cyclostrophic**.

hurricane eye—*See* **eye**.

hurricane-force wind—In the **Beaufort wind scale**, a **wind** with a speed equal to or greater than 64 knots (73 mph) or Beaufort Number 12 (Force 12).
Hurricane-force winds are not exclusive to hurricanes; they occur quite often in strong nontropical storms such as the **northeaster**, or even in severe **thunderstorms**.

hurricane hunters—Popular term for aircraft and/or personnel engaged in **tropical cyclone** reconnaissance.

hurricane radar band—*See* **hurricane band**.

hurricane surge—Same as **hurricane wave**; *see also* **storm surge**.

hurricane tide—Same as **storm surge**.

hurricane warning—A **warning** of impending sustained surface winds of **hurricane** force (64 knots/ 73 mph or greater) within 24 hours or less.
A hurricane warning can remain in effect when dangerously **high water** or a combination of dangerously high water and exceptionally high waves continue even though the winds may be less than hurricane force. For maritime interests, the storm-warning signals are 1) two square red flags with black centers by day and 2) a white lantern between two red lanterns by night. *See* **storm warning**.

hurricane watch—An announcement for a specific area that **hurricane** conditions (sustained winds of 64 knots/73 mph or higher) pose a threat within 36 hours.
Residents are cautioned to take stock of their preparedness needs, but, otherwise, are advised to continue normal activities.

hurricane wave—*See* **storm surge**.

hurricano—1. *See* **hurricane**.
   2. (Rare.) A **waterspout**.
   (Shakespeare, W., King Lear, Act III, Scene 2; in the First Folio, spelled hyrricano.)

Huygens's construction—(Also Huygens's principle.) An approximate geometrical procedure for determining the propagation of electromagnetic (and other) waves.
According to this construction, every point of a **wave front** in a medium at any instant is the source of secondary spherical wavelets propagating with the **phase velocity** of the medium. The envelope of all these wavelets then determines the wave front at a later instant. This construction is approximate if for no other reason than that often the complicated electromagnetic fields in matter are not simple waves with well-defined fronts. Moreover, the Huygens construction in its original form requires an ad hoc obliquity factor in order to obliterate that part of the complete envelope that is not observed. This construction deflects attention from physical explanations of the propagation of electromagnetic waves: Only (charged) matter is the source of such waves, not wave fronts. Huygens's construction is sometimes useful in obtaining approximate mathematical solutions to some problems in **wave propagation**, but for many problems (e.g., **scattering** of a plane **electromagnetic wave** by an arbitrary sphere) this construction is useless.

hybrid—A four-port **waveguide** device used in **microwave** circuits having the property that a **signal** incident at one port is divided equally between two other ports, provided that the fourth port is terminated by a matched load.
A particular type of hybrid, known as a 90° hybrid, yields outputs that differ in **phase** by 90°.

hybrid vertical coordinate—A vertical coordinate for atmospheric models that changes from one **standard** coordinate (for instance, the terrain-following sigma **coordinate system**) to another coordinate system as a function of height.
For instance, **hybrid** coordinates allow a **model** to accrue the benefits of **pressure** or **isentropic** coordinates in the upper model layers where the use of a terrain-following coordinate is disadvantageous while avoiding technical difficulties associated with the use of a nonterrain following coordinate near the ground. *See* **isentropic vertical coordinate**, **height vertical coordinate**, **vertical coordinate system**.

hydraulic analogy—The analogy between the flow of a shallow liquid and the flow of a compressible gas.

Various phenomena such as shock waves (**hydraulic jumps**) occur in both systems. The analogy requires neglect of vertical accelerations in the liquid, and restrictions on the ratio of **specific heats** for the gas.

hydraulic barrier—A general term that refers to a **groundwater flow** boundary, usually induced by **groundwater** pumping, that significantly impedes the movement of dissolved contaminants.

hydraulic conductivity—The proportionality constant between the volumetric **flux** and the **hydraulic gradient**, as in **Darcy's law**.

It includes the effects of the pore structure, and the **viscosity** and **density** of the water.

hydraulic dispersion—Same as **mechanical dispersion**.

hydraulic grade line—The line of the **piezometric head**, or the line of the **potential energy** of flow.

hydraulic gradient—The slope of the **hydraulic grade line**.

hydraulic head—The sum of the **elevation** head and the **pressure head**.

Same as **piezometric head**.

hydraulic jump—The sudden and usually turbulent passage of water in an open channel from low **stage**, below **critical depth**, to high stage, above critical depth.

During this passage, the **velocity** changes from supercritical to subcritical. There is considerable loss of **energy** during the jump. For meteorological applications, *see* **pressure jump**. *See also* **bore**, **critical velocity**.

> Lamb, H., 1953: *Hydrodynamics*, 6th ed., sec. 187.

hydraulic mean depth—Ratio of the cross-sectional area of the **contour** of the river bottom at a point in an open channel to the surface width, that is, $D = A/B$.

hydraulic radius—Ratio of the cross-sectional area of the flow at a point in an open channel or closed conduit to the **wetted perimeter**, that is, $R = A/P$.

hydraulic resistivity—The ratio of one over the average **hydraulic conductivity** of an **aquifer** $(1/K)$.

hydraulic routing—Methods of **flood routing** that are based on the **equation of continuity** and various forms (extent of approximation) of the **momentum** equation of the flow.

hydraulic similarity—A geometric and **dynamic similarity** that exists between two flow phenomena.

In other words, in these two hydraulically similar situations, all homologous **dimensions** and homologous forces are in the same ratios.

hydraulic structures—Structures installed in natural or man-made waterways to impound, direct, control, or measure the flow of water.

Examples are dams and **spillways**, sluice gates, **weirs**, and **flumes**.

hydraulics—Branch of fluid mechanics that deals with the flow of water in open channels, closed conduits, **hydraulic structures**, and hydraulic machines (pumps, turbines, presses, etc.).

hydrocarbons—Strictly speaking, organic molecules consisting of just **carbon** and **hydrogen**; often loosely applied also to derivatives of hydrocarbons containing **oxygen**, **halogens**, etc.

The atmospheric burden of hydrocarbons is provided from both natural and **anthropogenic emissions**.

hydrochemical facies—Distinct zones that have cation and anion concentrations of diagnostic chemical character of water solutions in hydrologic systems, which is describable within defined composition categories.

hydrochloric acid—*See* **hydrogen chloride**.

hydrochlorofluorocarbons—(Abbreviated HCFCs) A collection of partially chlorinated and fluorinated **hydrocarbons** (mostly methanes and ethanes), used as refrigerants, foam-blowing agents, and solvents. These species have been developed as replacements for the now-banned **chlorofluorocarbons**.

Examples include HCFC-141b (CH_3CFCl_2) and HCFC-142b (CH_3CF_2Cl). These compounds are less harmful to the **ozone layer** than the chlorofluorocarbons, due to their shorter atmospheric **lifetimes**. However, because their **ozone** depletion potentials are nonzero, these compounds will only be in use temporarily and will ultimately be phased out of production.

hydrodynamic dispersion—Longitudinal and lateral spreading (at a macroscopic level) of a solute being advected through porous media due to **mechanical dispersion** and **molecular diffusion**.

hydrodynamic equations of motion—The partial differential equations that determine the motion of a fluid.

As a minimum, these equations determine the fluid **velocity** vector and **pressure**, and express the conservation of fluid **momentum** and mass.

hydrodynamic instability—*See* **instability**.

hydrodynamic pressure—The difference between the **pressure** and the **hydrostatic pressure**.

This concept is useful chiefly in problems of the **steady flow** of an **incompressible fluid** in which the hydrostatic pressure is constant for a given **elevation** (as when the fluid is bounded above by a rigid plate), so that the external force **field** (**gravity**) may be eliminated from the problem. If p^* is the hydrodynamic pressure, ρ the **density**, and V the speed, **Bernoulli's equation** gives

$$p^* + \frac{1}{2}\rho V^2 = \text{constant along a streamline.}$$

See also **static pressure**.

hydrodynamic stability—*See* **stability**.

hydrodynamically rough surface—Any surface that by **surface friction** extracts **momentum** from the fluid flowing over it.

There are a number of measures of the roughness of a surface that relate to the degree of **mixing** in the fluid caused by this frictional overturning of the fluid. *See* **roughness length**, **friction velocity**, **zero-plane displacement**; *compare* **aerodynamically rough surface**, **aerodynamically smooth surface**.

hydrodynamics—The study of fluid motion.

"Fluid" here refers ambiguously to liquids and gases. Although "classical" hydrodynamics was primarily concerned with incompressible fluids, the term **aerodynamics** has been reserved for such a specialized aspect of compressible fluid flow that most of meteorological dynamics is best included under the general heading of "hydrodynamics." W. and J. Bjerknes refer to the hydrodynamics of compressible fluids as **physical hydrodynamics**.

hydrofluorocarbons—(Abbreviated HFCs.) A collection of partially fluorinated **hydrocarbons** in use or under development as replacement compounds for the **chlorofluorocarbons** (CFCs).

The lack of chlorine in these compounds eliminates their potential to destroy stratospheric **ozone**. Also, the presence of H atoms makes them susceptible to removal from the **atmosphere** via reaction with hydroxyl radicals, reducing their effectiveness as greenhouse warming gases. The most widely used of this class of compounds is HFC-134a (CF_3CFH_2), which is currently being used in automobile air conditioners and domestic refrigeration applications.

hydrogen—(Symbol H.) A colorless and odorless gaseous **element**, atomic number 1, atomic weight 1.008; the lightest and apparently the most abundant chemical element in the **universe**.

However, it is found only in **trace** quantities in the observable portion of our **atmosphere**, in the amount of only about 0.00005% by volume of **dry air**. Hydrogen H_2 has a **molecular weight** of 2.0160 and **specific gravity** referred to air of 0.0695. The low escape **velocity** of hydrogen, combined with absence of active generating mechanisms on the earth, holds its concentration to negligible amounts. Its concentration increases above altitudes of several hundred kilometers as a consequence of diffusive separation. At one time, hydrogen was commonly used for inflating meteorological balloons, but because of its dangerous combustibility, it has been replaced by **helium**.

hydrogen chloride—A toxic, colorless, strongly acidic gas (HCl) that dissolves readily in water to form **hydrochloric acid**.

In the **troposphere**, **hydrochloric acid** can be produced from the reaction of **nitric acid** or **sulfuric acid** with sea-salt **particles** (NaCl). It is also involved in the formation of **ammonium chloride** aerosol. HCl is also the most abundant form of chlorine in an inorganic compound found in the **stratosphere**. Produced there predominantly from the reaction of Cl atoms with **methane** and molecular **hydrogen**, it acts as a temporary, relatively unreactive reservoir species for chlorine.

hydrogen ion—A positively charged species of chemical, symbol H^+, the ionized form of the **hydrogen** atom.

The hydrogen ion is hydrated in aqueous solutions and is usually written as H_3O^+. Neutral water (**pH** 7) contains hydrogen ions at a concentration of 10^{-7} mol L^{-1}. Dissolution of acids in water leads to an increase in the hydrogen ion concentration, and thus a decrease in the PH.

hydrogen peroxide—Strong oxidizing agent, formula H_2O_2, that is formed in the **atmosphere** from the reaction between two **hydroperoxyl radicals**

$$HO_2 + HO_2 \rightarrow H_2O_2 + O_2.$$

Hydrogen peroxide is very soluble in water droplets, leading to a high rate of liquid-phase **oxidation** for soluble sulfur compounds, particularly **sulfur dioxide**.

hydrogen sulfide—Reduced sulfur gas, formula H_2S, emitted by rotting vegetation; also found in emissions from terrestrial **wetlands**, marshes, etc.

H_2S is a common reduced sulfur gas in terrestrial systems.

hydrogeochemistry—The study of the chemical composition of natural waters.

hydrogeological boundary—The approximate outer limits of a **porous medium** with significant differences in hydrogeological properties from that of the adjacent media.

hydrogeology—The science of the occurrence, distribution, and movement of water below the surface of the earth with emphasis on geologic aspects.

hydrograph—A record and graphical representation of **discharge** as a function of time at a specific location, for example, the discharge at a point in a **stream** or the discharge from a pumping well.

hydrographic—Referring to study of water features (oceans, lakes, rivers), including physical characterisics (**oceanography**, limnology) and elements affecting safe navigation.

hydrographic station—Same as **serial station**.

hydrographical network—Aggregate of rivers and other permanent or temporary watercourses, lakes, and **reservoirs** over any given area.
    See also **stream network**.

hydrography—The measurement and study of depths and currents in open seas, lakes, **estuaries**, and rivers.

hydrologic accounting—(Also called basin accounting, water budget.) *See* **water budget**.

hydrologic balance—*See* **water budget**.

hydrologic budget—*See* **water budget**.

hydrologic cycle—(Also called hydrological cycle.) The **cycle** in which water evaporates from the oceans and the land surface, is carried over the earth in atmospheric **circulation** as **water vapor**, precipitates again as **rain** or **snow**, is intercepted by trees and vegetation, provides **runoff** on the land surface, infiltrates into soils, recharges **groundwater**, discharges into streams, and ultimately, flows out into the oceans, from which it will eventually evaporate again.

hydrologic equation—An equation evaluating the amounts of water flowing in any part of the **hydrologic cycle**. It is expressed as

$$I - O = \Delta S,$$

where I is **inflow** into system during a defined period, O is **outflow** from a system during a defined period, and ΔS is change in storage in the system during the period.

hydrologic properties—A general term that refers to the collective hydraulic characteristics of a soil or rock such as the **conductivity**, **permeability**, **water content**, and **pressure head**, and the interrelationships among those properties.

hydrologic regime—Spatial and temporal variations of the components in a **water budget**.

hydrologic year—(Also known as hydrological year.) *See* **water year**.

hydrological basin—*See* **watershed**.

hydrological cycle—Same as **hydrologic cycle**.

hydrological drought—(Also hydrologic drought.) Prolonged period of below-normal **precipitation**, causing deficiencies in water supply, as measured by below-normal **streamflow**, lake and **reservoir** levels, **groundwater levels**, and depleted **soil moisture content**.

hydrological element—1. Aspect of hydrologic processes and phenomena, such as **precipitation**, **streamflow** (**stage**), etc.

    2. In hydrologic modeling, an area with similar **hydrologic properties** (soil, vegetation, etc.).

hydrological forecast—A statement of expected hydrological conditions for a specified period and for a specified locality.

hydrological forecasting—The process of generating a hydrologic forecast for a specific point and time.

hydrological network—Set of stations designed to measure the spatial and temporal distribution of **hydrologic properties**, such as **rainfall**, **streamflow**, etc.

hydrological routing—Methods of **flood routing** that are based on the combination of the **continuity equation** and a storage relationship.

hydrological warning—Emergency announcement of information on an anticipated, potentially dangerous, hydrologic event.
An example is a **flood warning** of an impending high flow in a channel.

hydrology—The scientific study of the waters of the earth, especially with relation to the effects of **precipitation** and **evaporation** upon the occurrence and character of water in streams, lakes, and on or below the land surface.
In terms of the **hydrologic cycle**, the scope of hydrology may be defined as that portion of the **cycle** from precipitation to evaporation or return of the water to the seas. Applied hydrology utilizes scientific findings to predict rates and amounts of **runoff** (river forecasting), estimate required **spillway** and **reservoir** capacities, study soil–water–plant relationships in agriculture, estimate **available water** supply, and for other applications necessary to the management of **water resources**. *Compare* **hydrography**.

hydrolysis—The process by which a substrate is split to form two end products by the intervention of a molecule of water, or the reaction of a substrate with a **water molecule**.

hydromagnetics—Same as **magnetohydrodynamics**.

hydrometeor—Any product of **condensation** or **deposition** of atmospheric **water vapor**, whether formed in the **free atmosphere** or at the earth's surface; also, any water particle blown by the **wind** from the earth's surface.
Hydrometeors may be classified in a number of different ways, of which the following is one example: 1) liquid or solid water particles formed and remaining suspended in the air, for example, damp (high **relative humidity**) haze, **cloud**, **fog**, **ice fog**, and **mist**; 2) liquid **precipitation**, for example, **drizzle** and **rain**; 3) **freezing precipitation**, for example, **freezing drizzle** and **freezing rain**; 4) solid (frozen) precipitation, for example, **snow, hail, ice pellets, snow pellets (soft hail, graupel), snow grains**, and **ice crystals**; 5) falling **particles** that evaporate before reaching the ground, for example, **virga**; 6) liquid or solid water particles lifted by the wind from the earth's surface, for example, **drifting snow, blowing snow**, and **blowing spray**.

hydrometeor charge—The net charge, for example, on a **raindrop** or a water drop.

hydrometeorology—1. Study of the atmospheric and terrestrial phases of the **hydrological cycle** with emphasis on the interrelationship between them.
2. Meteorology plus **hydrology**.
Many countries use the word in this sense to name the official service charged with the dual responsibility of weather and hydrologic functions.
3. (Rare.) That branch of meteorology that deals with the **hydrometeors**.

hydrometer—An instrument used for measuring the **specific gravity** of a liquid.

hydrometric station—A **station** on a river, lake, **estuary**, or **reservoir** where water quantity and quality data are collected and recorded.
Such data may include **stage** (water surface elevation), **discharge, sediment concentration**, water **temperature**, chemical and biological properties of water, **ice** formations, and other characteristics.

hydrometry—1. Measurement and computation of **streamflow**.
2. Techniques and instrumentation for the measurement and analysis of water.
3. The science of the determination of the **specific gravity** of a liquid.

hydroperoxides—Oxygenated compounds that are organic derivatives of **hydrogen peroxide**.
The simplest is methyl hydroperoxide, formula CH_3OOH. These compounds are formed in the **oxidation** of **hydrocarbons** in relatively **clean air**, where oxides of **nitrogen** are not abundant. The presence of the alkyl group renders them much less soluble than hydrogen peroxide and therefore more susceptible to long-range **transport** in the **atmosphere**.

hydroperoxyl radical—**Free radical** of formula HO_2, formed by the **transfer** of a **hydrogen** atom to molecular **oxygen**.

The hydroperoxyl radical is a chain carrier in the **atmosphere** and in combustion systems. In the **stratosphere**, it is involved in **ozone**-destroying chain reactions. In the **troposphere**, it is formed as an intermediate in the **oxidation** of most **hydrocarbons**.

Bate, D. R., and M. Nicolet, 1950: The photochemistry of atmospheric water vapor. *J. Geophys. Res.*, **55**, p. 301.

hydrophotometer—An instrument for measuring the **extinction coefficient** in water.
See **photometer**.

hydrophyte—A vascular plant that requires an abundance of water for growth; a plant growing in water or soils too **waterlogged** for most other plants to survive.

hydrophytic plant—A plant requiring abundance of water to grow, common in marshes and **wetlands**.

hydrosol—A **colloidal system** in which the **dispersion medium** is water.
The **dispersed phase** may be a solid, a gas, or another liquid. *Compare* **acrosol**.

hydrosphere—The water portion of the earth as distinguished from the solid part, called the **lithosphere**, and from the gaseous outer envelope, called the **atmosphere**.
The hydrosphere includes **snow, ice,** and **glaciers**. Sometimes the water in the atmosphere, which includes **water vapor, clouds,** and all forms of **precipitation** while still in the atmosphere, is included in the term hydrosphere. *See also* **biosphere, geosphere**.

hydrostatic approximation—1. The assumption that the **atmosphere** is in **hydrostatic equilibrium**.
2. Same as **quasi-hydrostatic approximation**.
3. An approximation in geophysical fluid dynamics that is based on the assumption that the horizontal **scale** is large compared to the vertical scale, such that the vertical **pressure gradient** may be given as the product of **density** times the gravitational **acceleration**.
See **hydrostatic balance**.

hydrostatic assumption—The assumption that the **force of gravity** is balanced by the vertical component of the atmospheric **pressure gradient** force, as must occur in the absence of atmospheric motions.

hydrostatic balance—Describes a balance between vertical **pressure gradient** and **buoyancy** forces. Explicitly,

$$dp/dz = -\rho g.$$

When this balance does not hold, the fluid is nonhydrostatic.

hydrostatic equation—The form assumed by the vertical component of the **vector** equation of motion when all **Coriolis**, earth curvature, frictional, and vertical **acceleration** terms are considered negligible compared with those involving the vertical **pressure force** and the **force of gravity**.
Thus

$$\frac{\partial p}{\partial z} = -\rho g,$$

where p is the **pressure**, ρ the **density**, g the **acceleration of gravity**, and z the geometric height. For cyclonic-scale motions the **error** committed in applying the hydrostatic equation to the **atmosphere** is less than 0.01%. Strong vertical accelerations in thunderstorms and mountain waves may be 1% of **gravity** or more in extreme situations.

hydrostatic equilibrium—The state of a fluid with surfaces of constant **pressure** and constant mass (or **density**) coincident and horizontal throughout.
Complete balance exists between the **force of gravity** and the **pressure force**. The relation between the pressure and the geometric height is given by the **hydrostatic equation**. The analysis of atmospheric **stability** has been developed most completely for an **atmosphere** in hydrostatic equilibrium. *See* **parcel method, slice method, quasi-hydrostatic approximation**.

hydrostatic instability—Same as **static instability**.

hydrostatic model—An atmospheric model in which the **hydrostatic approximation** replaces the vertical **momentum** equation.
This implies that vertical **acceleration** is negligible compared to vertical **pressure gradients** and vertical **buoyancy** forces, a good approximation for **synoptic** and subsynoptic scales of motion. Hydrostatic models have been successfully applied with horizontal resolutions as small as about 10 km, resolving even some **mesoscale** circulations. Global and regional weather prediction models have traditionally been hydrostatic models.

hydrostatic pressure—The **pressure** in a fluid in **hydrostatic equilibrium**.
  Compare **static pressure**; *see also* **hydrodynamic pressure**.

hydrostatic stability—Same as **static stability**.

hydrostatics—That part of fluid mechanics restricted to fluids in which the **velocity** (**linear** or angular) of mass motion does not vary from point to point.
  Although the combining form hydro comes from a Greek word meaning water, the term hydrostatic is used for gases as well as liquids. When applied to the **atmosphere, hydrostatic equilibrium** and **hydrostatic pressure** logically ought to become aerostatic equilibrium and aerostatic pressure, but these terms are rarely used.

hydroxyl emission—(Also known as Meinel bands.) Spectroscopic feature seen in the **upper atmosphere**; a form of **chemiluminescence** due to the formation of vibrationally excited **hydroxyl radicals** in the reaction of **hydrogen** atoms with **ozone**, $H + O_3 \rightarrow OH + O_2$.
  The origin of the emission was correctly identified by A. Meinel in the 1950s. The emission is a component of the **airglow**.
  Meinel, A. B., 1950: OH emission bands in the spectrum of the night sky. *Astrophys. J.*, **111**, 555–564.

hydroxyl ion—Negatively charged species of chemical formula, OH^-.
  Neutral water (**pH** 7) contains 10^{-7} mol L^{-1} hydroxyl ions. Dissolution of bases in water leads to an increase in the hydroxyl ion concentration, and thus an increase in the pH.

hydroxyl radical—The hydroxyl radical, OH, is responsible for the **oxidation** of most of the compounds that are released into, or formed in, the **atmosphere**.
  It is ubiquitous in the atmosphere, although its mean concentration is about $10^6 cm^{-3}$ and the **mixing ratio** seldom exceeds 1 ppt (1 part in 10^{12}). In the **lower atmosphere** it is formed by the reaction of excited **oxygen** atoms $O(^1D)$ with water or by the reaction of **hydroperoxyl radicals** with **nitric oxide**. In the **stratosphere** it is involved in ozone-destroying **catalytic cycles**. The capacity of the atmosphere to clean itself via the occurrence of oxidation reactions (oxidizing capacity) is usually related to the level of OH present.
  Levy, H., 1972: Photochemistry of the lower troposphere. *Planet. Space Sci.*, **20**, 919–935.

hyetal—Of or pertaining to **rain**.
  Compare **pluvial**.

hyetal coefficient—(Rare.) Same as **pluviometric coefficient**.

hyetal equator—A line (or transition zone) that encircles the earth and lies between two belts that typify the annual time distribution of **rainfall** in the lower latitudes of each hemisphere; a form of **meteorological equator**.
  It lies slightly north of the geographic **equator**, reaching its most northerly position at about 10°N latitude near the mouth of the Orinoco River in South America. The hyetal equator is more or less centrally situated in the belt of tropical rainfall, which has two rainy seasons and generally one main **dry season**, the latter occurring in the winter of the corresponding hemisphere.

hyetal region—A region in which the amount and seasonal **variation** of **rainfall** are of a given type.
  According to Köppen in his **climatic classification**, the main types of **hyetal** regions are 1) **desert**, with rainfall rare and irregular; 2) winter-dry, with main **rainy season** in summer, characteristic of a **monsoon climate**; 3) summer-dry, with rainy season in winter, such as a **Mediterranean climate**; 4) **rain** at all seasons, but not evenly distributed through the year; 5) rain on more than half the days in every month of the year. Minor types may be distinguished in which the **season** of rainfall maximum is either broken by a short **dry season** or is displaced toward spring or autumn. *Compare* **humidity province**.
  Köppen, W. P., and R. Geiger, 1930–1939: *Handbuch der Klimatologie*, Berlin: Gebruder Borntraeger, 6 vols.

hyetograph—A **chart** or function describing the temporal distribution of **precipitation** during a **storm** event, at a point, or over an area.

hyetography—The study of the annual **variation** and geographic distribution of **precipitation**.

hyetology—(Rare.) The science that treats the origin, structure, and various other features of all the forms of **precipitation**.

hygristor—An electric **humidity** sensor element often used in **radiosonde** equipment that relies on changes in the **resistance** of a humidity-sensitive component.
  A **carbon** humidity element is used in the United States. It comprises a polystyrene slide or strip with two metal electrodes along the long edges sprayed with a mixture of carbon particles and a cellulose binder. The binder changes its volume with **relative humidity** in such a way that

it separates the carbon particles from each other as the humidity increases, thus increasing the resistance between the electrodes. The relative humidity is calibrated as a function of resistance.

hygrodeik—A form of **psychrometer** with **wet-** and **dry-bulb thermometers** mounted on opposite edges of a specially designed graph of the **psychrometric tables**.

It is so arranged that the intersections of two curves determined by the wet- and dry-bulb readings yield the **relative humidity, dewpoint,** and **absolute humidity**.

hygrogram—The record made by a **hygrograph**.

hygrograph—A recording **hygrometer**.

hygrokinematics—The descriptive study of the motion of water substance in the **atmosphere**.

Vapor motion and **cloud** imagery from satellite observations are displayed on television weathercasts.

hygrology—The study that deals with the **water vapor** content (**humidity**) of the **atmosphere**.

hygrometer—Any instrument that measures the **water vapor** content of the **atmosphere**.

There are six basically different means of transduction used in measuring this quantity and hence an equal number of types of hygrometers. These are 1) the **psychrometer**, which utilizes the thermodynamic method; 2) the class of instruments that depends upon a change of physical **dimensions** due to the **absorption** of moisture (*see* **hair hygrometer, torsion hygrometer, goldbeater's-skin hygrometer**); 3) those that depend upon **condensation** of **moisture** (*see* **dewpoint hygrometer, frost point hygrometer**); 4) the class of instruments that depend upon the change of chemical or electrical properties due to the absorption of moisture (*see* **absorption hygrometer, electrical hygrometer, carbon-film hygrometer element, dew cell**); 5) the class of instruments that depend upon the **diffusion** of water vapor through a porous membrane (*see* **diffusion hygrometer**); and 6) the class of instruments that depend upon measurements of the absorption spectra of water vapor (*see* **spectral hygrometer**).

Middleton, W. E. K., and A. F. Spilhaus, 1953: *Meteorological Instruments*, 3d ed., rev., Univ. of Toronto Press, 105–116.

hygrometric formula—Same as **psychrometric formula**.

hygrometric tables—Tables that are set up to easily convert the measured **dry-** and **wet-bulb temperatures** to common **humidity** measures like **dewpoint temperature** or **relative humidity**.

See also **psychrometric tables**.

hygrometry—The study that treats the measurement of the **humidity** of the **atmosphere** and other gases.

hygroscope—An instrument showing qualitatively whether the air is dry or damp, usually by the change in appearance or **dimensions** of some substance (*see* **chemical hygrometer**).

Hygroscopes are frequently used as home weather predictors, for example, "weather houses" in which the appearance of the "old man" or the "old woman" is determined by the twisting and untwisting of a piece of catgut in response to changes of **humidity**.

hygroscopic—1. Pertaining to a marked ability to accelerate the **condensation** of **water vapor**; in general usage, the ability of a crystalline solid, (salt, brown sugar) to absorb **vapor** but at such a low rate under most conditions that it does not dissolve completely.

In meteorology, this term is applied principally to those **condensation nuclei** composed of salts that yield aqueous solutions of a very low equilibrium **vapor pressure** compared with that of pure water at the same **temperature**. Condensation on **hygroscopic nuclei** may begin at a **relative humidity** much lower than 100% (about 76% for sodium chloride); below this value **particles** remain dry. There is often a hysteresis such that particles remain liquid as the relative humidity falls and are present as a supersaturated solution. On so-called nonhygroscopic nuclei, which merely furnish sufficiently large (by molecular standards) wettable surfaces, relative humidity of nearly 100% is required to cause condensation. "Damp **haze**" is formed of hygroscopic particles in the process of slow growth in relatively **dry air** as it cools.

2. Descriptive of a substance, the physical characteristics of which are appreciably altered by effects of **water vapor**.

The **hygroscopicity** of certain materials has been advantageously utilized in **humidity** measurement and control devices, for example, the hair element of a **hair hygrometer**.

hygroscopic coefficient—The **moisture weight percentage** of a soil after equilibration with a **water vapor** saturated **atmosphere** at a particular **temperature**.

hygroscopic moisture—Moisture held in a soil that is in **equilibrium** with atmospheric **water vapor**.

hygroscopic nucleus—A **particle** that acts as center for water condensation at **relative humidity** below, at, and above **saturation** (100%).

hygroscopic water—1. Tightly held water on soil **particle** surfaces as a result of adhesion; this water is essentially unavailable to vegetation.

2. Water held by soil under specific **relative humidity** and **temperature** conditions (usually 98% relative humidity and 25°C).

hygroscopicity—The relative ability of a substance (as an **aerosol**) to adsorb **water vapor** from its surroundings and ultimately dissolve.

hygrothermogram—The record of a **hygrothermograph**.

hygrothermograph—A **recording instrument** combining, on one record, the **variation** of atmospheric **temperature** and **humidity** content as a function of time.

The most common hygrothermograph is a hair **hygrograph** combined with a **thermograph**.

hygrothermometer—An instrument that measures both atmospheric **water vapor** content and **temperature**.

See **hygrothermograph**.

hyperbolic point—(Sometimes called neutral point.) A **singular point** in a **streamline** field that constitutes the intersection of a **convergence line** and **divergence line**.

It is analogous to a **col** in the **field** of a single-valued **scalar** quantity.

hypersonic—Referring usually to an object moving more than five times the **speed of sound** (Mach 5) in the gas or liquid surrounding it.

Compare **supersonic**.

hypochlorous acid—A weak, unstable acid, formula HOCl.

Hypochlorous acid is also a minor component of the gas-phase inorganic chlorine budget in the **stratosphere**. It is formed there largely from reaction of **hydroperoxyl radicals** (HO_2) with chlorine monoxide (ClO) radicals, and from the **hydrolysis** of **chlorine nitrate** on stratospheric aerosols. It is subject to quite rapid **photolysis** in the sunlit **atmosphere**.

hypolimnion—The layer of water below the **thermocline** in a **freshwater** body.

hypothermia—A fall in the **temperature** of an animal body below the usual level.

This state is brought about when the homeostatic mechanisms fail to maintain adequate production of **heat** under conditions of extreme cold.

hypsithermal period—The period from about 7000 to 500 B.C., proposed by E. S. Deevey and R. F. Flint (1957), during which global **climate** was thought to be warmer than today.

Spanning the **Boreal** through **SubBoreal** periods of the Blytt–Sernander sequence of inferred climates in northern Europe, it includes both wet and dry periods. It is followed by the general expansion of **glaciers** known as the **Neoglacial**.

Deevey, E. S., and R. F. Flint, 1957: Postglacial hypsithermal interval. *Science*, **125**, 183–184.

Sernander, R., 1908: On the evidence of postglacial changes of climate furnished by the peat-mosses of northern Europe. *Geol. Fören. Förh.*, **30**, 456–478.

hypsography—The **height pattern** of a physically defined surface, as revealed by **contour** lines.

In meteorology, this may refer to a **constant-pressure surface** or to an **isentropic surface** in the **atmosphere**. *Compare* **pressure pattern**.

hypsometer—Literally, an instrument for measuring height; specifically, an instrument for measuring **atmospheric pressure** by determining the **boiling point** of a liquid at the **station**.

The relationship between the boiling point of a liquid and atmospheric pressure is given by the **Clapeyron–Clausius equation**. The sensitivity of the hypsometer increases with decreasing **pressure**, making it more useful for high-altitude work. Consequently, hypsometers are frequently used for height estimation.

hypsometric equation—An equation relating the **thickness**, h, between two **isobaric** surfaces to the **mean temperature** of the layer:

$$h = z_2 - z_1 = \frac{R\overline{T}}{g} \ln \left(\frac{p_1}{p_2} \right)$$

where z_1 and z_2 are geometric heights at **pressure** levels p_1 and p_2, respectively; R is the **gas constant** for **dry air**; $\overline{T}$ is the mean temperature of the layer; and g is **gravity**.

The hypsometric equation is derived from the **hydrostatic equation** and the **ideal gas law**.

hypsometry—The science of height measurement.

hythergraph—A type of **climatic diagram** where the coordinates are some form of **temperature** versus a form of **humidity** or **precipitation**.

A common, specific use is to show the annual "march" of mean monthly values of temperature and precipitation at a given **station**. Also, a **comfort chart** may be considered a hythergraph. There are many other possibilities.

I

I and Q channels—(Abbreviation for in-phase and quadrature channels.) In **Doppler radar** and **lidar** receivers, the **quadrature** video channels produced by **coherent detection** of the **IF signal**.

The **energy** in the two channels is in **phase** quadrature, and they may be considered together to be a single **complex signal** with **frequency** components from $-F$ to $+F$, where F is the passband limit of the low-pass **filter** used in the demodulator.

IAC—Abbreviation for **international analysis code**.

IAS—Abbreviation for **indicated airspeed**.

Ibe wind—A local strong **wind** that blows through the Dzungarian Gate (in western China), a gap in the mountain ridge separating the **depression** of Lakes Balkash and Ala Kul from that of Lake Ebi Nor.

The wind resembles the **foehn** and brings a sudden rise of **temperature**, in winter from about $-26°$ to about $-1°C$.

IBL—Abbreviation for **internal boundary layer**.

ICAO—(Abbreviation for International Civil Aviation Organization.) The international civil authority that sets the standards and practices for global air traffic operations.

ICAO Standard Atmosphere—*See* **standard atmosphere**.

ice—The solid, crystalline form of water substance; it is found in the **atmosphere** as **snow crystals**, **hail**, **ice pellets**, etc., and on the earth's surface in forms such as **hoarfrost, rime, glaze, sea ice, glacier ice, ground ice, frazil, anchor ice**, etc.

This form of water is, strictly speaking, called ice I, the only one of the several known modifications of solid water substance that is stable at commonly occurring temperatures and pressures. (Some of the other forms have very unusual properties, ice VII, for example, being stable only at pressures above 22 400 kg cm^{-2}, but then existing at temperatures up to about 100°C.) Ice has an open structure because the water molecules bond to their neighbors covalently only in four directions; it therefore floats on higher **density** water, where broken molecular bonds permit closer packing. All commonly occurring forms of ice are crystalline, although large single crystals are relatively rare except in **glaciers**. The **ice crystal** lattice possesses hexagonal symmetry that manifests itself in the gross forms of such single crystals as are sometimes found in **snow**. At an air pressure of one atmosphere, ice melts at 0°C by definition of the **Celsius temperature scale**. (Strictly at **equilibrium** among water, ice, and **vapor** occurs at +0.01°C, the **triple point**.) On the other hand, ice does not invariably form in liquid water cooled below this **temperature**; it has a tendency to supercool, more so in the absence of ice **nuclei**. *See* **Bernal–Fowler rules, ice crystal**.

ice accretion—The process by which a layer of **ice** (**icing**) builds up on solid objects that are exposed to **freezing precipitation** or to **supercooled fog** or **cloud droplets**.

At the earth's surface this usually refers to **glaze** formation, and the amount of ice can be roughly measured by an **ice-accretion indicator**. For airborne objects, ice accretion refers to any type of **aircraft icing**. *See* **accretion**.

ice-accretion indicator—An instrument used to detect the occurrence of **freezing precipitation**.

It usually consists of a strip of sheet aluminum about 4 cm (1.5 in.) wide and is exposed horizontally, faceup, in the **free air** about 1 m above the ground. *Compare* **icing-rate meter**.

ice age—A major interval of **geologic time** during which extensive ice sheets (**continental glaciers**) formed over many parts of the world.

The best known **ice ages** are 1) the Huronian in Canada, occurring very early in the Proterozoic era (2700–1800 million years ago); 2) the pre-Cambrian and early Cambrian, which occurred in the early Paleozoic era (about 540 million years ago) and left traces widely scattered over the world; and 3) the Permo-Carboniferous, occurring during the late Paleozoic era (from 290 million years ago), which was extensively developed on Gondwana, a large continent comprising what is now India, South America, Australia, Antarctica, Africa, and portions of Asia and North America. The term ice age is also applied to advances and retreats of **glaciers** during the Quaternary era.

ice and snow albedo—*See* **albedo**.

ice and snow albedo–temperature feedback—A **positive feedback** mechanism involving ice and **snow cover, surface albedo**, and **temperature**.

For example, given an initial warming, a decrease in snow and **ice cover** occurs, lowering the surface albedo. This causes an increase in the **absorption** of **solar radiation**, which amplifies the initial increase in temperature.

ice apron—A thin mass of **snow** and **ice** adhering to a mountainside.

> Armstrong, T., B. Roberts, and C. Swithinbank, 1973: *Illustrated Glossary of Snow and Ice*, Scott Polar Research Institute, Special Pub. No. 4, 2d ed.; The Scholar Press Ltd., Menston Yorkshire, UK.

ice band—A long strip of **ice floes**, parallel to but separated from the edge of the main **ice pack**.

ice bands—**Ice layers** formed by separate **accumulation** events and exposed at a **glacier front** or a **crevasse** wall.

ice bay—(Also called ice bight.) A baylike recess in the edge of a large **ice floe** or **ice shelf**.

ice belt—*See* **ice band**.

ice bight—Same as **ice bay**.

ice blink—A relatively bright region on the underside of clouds produced by the **reflection** of **light** by **ice**.

This term is used in polar regions where it contributes to the **sky map**; ice blink is not as bright as **snow blink**, but is brighter than the reflection of light by land or water.

ice blister—*See* **icing**.

ice breakup—Disintegration of an **ice cover** on land, river, or coastal waters as a result of **thermal** and mechanical processes.

See **breakup**.

ice-bulb temperature—Same as the **wet-bulb temperature** when the latter is below 0°C and the water on the **thermometer** is frozen.

Note that in dry conditions the wet bulb of a **psychrometer** may **freeze** when the **air temperature** is appreciably above the **freezing point**.

ice cap—A dome-shaped perennial cover of **ice** and **snow** over an extensive portion of the earth's surface.

Ice caps are considerably smaller than **ice sheets**, and it is felt improper to refer to the **Greenland** and **Antarctic Ice Sheets** as ice caps. Most ice caps are most probably remnants of the **Quaternary Ice Age**. The term was first used for the supposedly perennial **ice cover** at both poles of the earth. However, since it has been found that the ice of arctic waters is largely seasonal, the use of this term to denote arctic **polar ice** is now considered improper.

ice-cap climate—Same as **perpetual frost climate**.

ice clearing—Same as **polyn'ya**.

ice cover—A layer of **ice** on top of some other feature.

Usually used in reference to an **ice layer** at the surface of a lake or pond.

ice crust—1. A type of **snow crust**; a layer of **ice**, thicker than a **film crust**, upon a **snow** surface.

It is formed by the **freezing** of meltwater or rainwater that has flowed into the snow surface.

2. Same as **ice rind**.

ice crystal—Any one of a number of macroscopic, crystalline forms in which **ice** appears, including **hexagonal columns**, **hexagonal platelets**, **dendritic crystals**, **ice needles**, and combinations of these forms.

The **crystal lattice** of ice is hexagonal in its symmetry under most atmospheric conditions. Varying conditions of **temperature** and **vapor pressure** can lead to growth of crystalline forms in which the simple hexagonal pattern is present in widely different habits (a thin hexagonal plate or a long thin hexagonal column). In many ice crystals, trigonal symmetry can be observed, suggesting an influence of a cubic symmetry. The principal axis (c axis) of a single **crystal** of ice is perpendicular to the axis of hexagonal symmetry. Planes perpendicular to this axis are called basal planes (a axes related to the prism facets) and present a hexagonal cross section. Ice is **anisotropic** in both its optical and electrical properties and has a high dielectric constant (even higher than water) resulting from its water dipole structure. The electrical **relaxation time** for water is much shorter than for ice (10^9 Hz compared with 10^4 Hz), resulting from a **chain reaction** requirement for molecules to relax through defects in the ice lattice. In the **free air**, ice crystals compose cirrus-type clouds, and near the ground they form the **hydrometeor** called, remarkably enough, "ice crystals" (or **ice prisms**). They are one constituent of **ice fog**, the other constituent being **droxtals**. On terrestrial objects the ice crystal is the elemental unit of **hoarfrost** in all of its various forms. Ice crystals that form in slightly **supercooled water** are termed **frazil**. Ice originating as frozen water (e.g., **hail**, **graupel**, and **lake ice**) still has hexagonal symmetry but lacks any external hexagonal form. Analysis of their sections (0.5 mm) in polarized **light** reveals different crystal shapes and orientations, depending on the **freezing** and any annealing and subsequent recrystallization process.

ice-crystal cloud—A **cloud** consisting entirely of **ice crystals** (such as **cirrus**); to be distinguished in this sense from **water clouds** and **mixed clouds**.

Ice-crystal clouds have a diffuse and fibrous appearance, quite different from that typical of water droplet clouds, resulting from growth in much weaker updrafts and different fall speeds of a wider size **particle** spectrum.

ice-crystal fog—Same as **ice fog**.

ice-crystal haze—A type of very light **ice fog** composed only of **ice crystals** (no **droxtals**), and at times observable to altitudes as great as 7000 m.

It is usually associated with **precipitation** of ice crystals. Observed from the ground, ice-crystal haze may be dense enough to hinder observation of celestial bodies, sometimes even the sun. Looking down from the air, however, the ground is usually visible and the **horizon** only blurred. For very sparse ice-crystal haze during daytime, **sunlight** reflecting from **crystal** faces produces sparkling in the air; hence the name **diamond dust** for these crystals. *Compare* **arctic mist**.

ice day—In **climatology**, a day on which the maximum air **temperature** in a **thermometer shelter** does not rise above 0°C (32°F), and **ice** on the surface of water does not **thaw**.

This term is not used in the United States, but is used in the United Kingdom, throughout most of Europe, and probably in many other parts of the world.

ice desert—Any polar area permanently covered by **ice** and **snow**, with no vegetation other than occasional **red snow** or **green snow**.

ice drift—Movement of **ice fields** or **floes** in water bodies caused by **wind** or currents.

ice edge—The demarcation between open water and **sea ice**.

ice fat—Same as **grease ice**.

ice feathers—(Also called frost feathers.) A type of **hoarfrost** that is formed on the **windward side** of terrestrial objects and on aircraft flying from cold to warm air layers.

Ice feathers are made up of single, columnar **ice crystals**, some of which grow out from others at large angles and thus build up a delicate spatial array of tiny crystals. See **air hoar**.

ice field—A large, level area of **ice**, either of **sea ice** ("more than five miles across") or an **ice cap** or **highland ice**.

ice floe—*See* **floe**.

ice flow—The motion of **ice** driven by gravitational forces (*see* **glacier flow**) or, for **sea ice**, **wind** and water currents.

ice flowers—1. Delicate tufts of **hoarfrost** that occasionally form in great abundance on an **ice** or **snow** surface (**surface hoar**); it also forms as a type of **crevasse hoar** or window frost.

2. Formations of **ice crystals** on the surface of a quiet, slowly **freezing** body of water.

ice fog—(Also called ice-crystal fog, frozen fog, frost fog, frost flakes, **air hoar, rime fog, pogonip**.) A type of **fog**, composed of suspended **particles** of ice, partly **ice crystals** 20 to 100 μm in diameter, but chiefly, especially when dense, **droxtals** 12–20 μm in diameter.

It occurs at very low temperatures, and usually in **clear, calm** weather in high latitudes. The sun is usually visible and may cause **halo** phenomena. Ice fog is rare at temperatures warmer than −30°C, and increases in **frequency** with decreasing **temperature** until it is almost always present at **air temperatures** of −45°C in the vicinity of a source of **water vapor**. Such sources are the open water of fast-flowing streams or of the sea, herds of animals, volcanoes, and especially products of combustion for heating or propulsion. At temperatures warmer than −30°C, these sources can cause **steam fog** of liquid water droplets, which may turn into ice fog when cooled (*see* **frost smoke**). See **ice-crystal haze, arctic mist**.

ice foot—Sea ice firmly frozen to the shore at the high-tide line and unaffected by **tide**.

This type of **fast ice** is formed by the **freezing** of **seawater** during **ebb tide** and of spray. It is separated from the floating sea ice by a **tide crack**; in many areas it offers a fairly level, continuous route for surface travel.

ice forecast—Describes the predicted position of **ice** boundaries and expected ice phenomena (ice concentration, distribution, stage of development, thickness and direction of drift, number and size of **icebergs**) for a specified period and for a specified locality, based on forecast meteorological and oceanographic conditions and the regional ice climatology.

An ice forecast is often issued to cover the period between the current ice analysis and the next scheduled ice analysis.

ice formation on aircraft—Ice formation can occur on aircraft either on the ground or in flight.

Ice **accretion** in flight may constitute a danger by affecting the **aerodynamic** characteristics, engine performance, or in other ways. There are four types of **airframe icing**: 1) **rime**: a light, white opaque deposit that forms generally at temperatures well below 0°C in clouds of low water content, consisting of small **supercooled water** droplets; 2) **clear ice** or **glaze**: a coating of clear ice that forms in clouds of high water content consisting of large (greater than 40 μm in diameter) supercooled water droplets in the form of **drizzle** or **rainfall** on aircraft with a **temperature** near or below 0°C; 3) mixed **ice** or cloudy ice: a rough, cloudy deposit that occurs in clouds containing a large **range** of **drop** sizes or a mix of **ice crystals**, **cloud droplets**, and **snowflakes**; and 4) **hoarfrost**: a white crystalline coating of ice that forms in **clear air** by **deposition** of **water vapor** when an aircraft surface is colder than the **frost point** of the air; this can occur when an aircraft moves rapidly (usually in descent) from very cold air into a region with warm and relatively **moist air**.

ice front—The seaward facing, clifflike edge of an **ice shelf** (so called by the British Antarctic Place–Names Committee).

> Armstrong, T., B. Roberts, and C. Swithinbank, 1973: *Illustrated Glossary of Snow and Ice*, Scott Polar Research Institute, Special Pub. No. 4, 2d ed., The Scholar Press Ltd., Menston Yorkshire, UK.

ice island—A form of **tabular iceberg** found in the Arctic Ocean, with a thickness of 30–50 m and from a few thousand square meters to 500 km² in area.

Ice islands often have an undulated surface, which gives them a ribbed appearance from the air.

> Armstrong, T., B. Roberts, and C. Swithinbank, 1973: *Illustrated Glossary of Snow and Ice*, Scott Polar Research Institute, Special Pub. No. 4, 2d ed., The Scholar Press Ltd., Menston Yorkshire, UK.

ice jam—1. An accumulation of broken river **ice** caught in a narrow channel.

Ice jams during **freeze-up** are quite porous, whereas **breakup** jams may comprise solid flows, frequently producing local floods during a spring breakup.

2. Fields of lake or **sea ice** thawed loose from the shores in early spring and blown against the shore, sometimes exerting great pressures.

ice keel—The submerged counterpart of an ice ridge.

ice layer—An **ice crust** that has been covered by **new snow**.

When exposed at a **glacier front** or in crevasses, the ice layers viewed in **cross section** are termed **ice bands**.

ice limit—The climatological position of the extreme minimum or maximum of **sea ice** extent.

ice mixing ratio—(Also called ice water mixing ratio.) The ratio of the mass of **ice** per unit mass of **dry air** in a sample containing cloud ice and/or **frozen precipitation**.

ice mound—1. *See* **frost mound**.

2. *See* **icing**.

ice multiplication—A process from which more **ice** particles are produced from existing **ice crystals** in clouds.

Sometimes known as ice enhancement. The process is inferred from the observation that ice particle concentration often exceeds that of **ice nuclei**, sometimes by several orders of magnitude. Currently the following mechanisms are thought to be responsible for the ice multiplication phenomenon: 1) mechanical fracturing of ice crystals during **evaporation**; 2) shattering or partial fragmentation of large drops during **freezing**; and 3) ice splinter formation during the riming of ice particles (**Hallett–Mossop process**).

ice needle—A long, thin **ice crystal**, axis coincident with the c axis of **ice** and with the cross section perpendicular to its long dimension being, at least in part, hexagonal.

Ice needles grow in a narrow **range** of **temperature** near −4°C and also at much lower temperatures, below −25° to −50°C depending on ice **supersaturation**.

ice nuclei counter—(Abbreviated IN counter.) Any of several devices for counting atmospheric **particles** that serve as heterogeneous **ice nuclei** that are suited by composition to catalyze the formation of **ice crystals** in the **atmosphere**.

The devices operate on varied principles, include means to cool and moisturize the air within a chamber or over a nucleus-collection filter, and are intended to measure the number concentration of ice nuclei that form ice crystals as a function of temperatures that would occur in subzero tropospheric clouds (thus, 0° to near −40°C). The product is the **nucleation** activity temperature spectrum. Some, but by no means all, types of ice nuclei counters are designed to attempt to replicate **supersaturations** that occur where ice crystals form in natural tropospheric clouds.

ice nucleus—Any **particle** that serves as a **nucleus** leading to the formation of **ice crystals** without regard to the particular physical processes involved in the **nucleation**.

The process is referred to as **heterogeneous nucleation**, as opposed to **homogeneous nucleation**, which depends on the formation of an **ice** particle large enough to grow by **random** motion of water molecules alone. Four processes are generally distinguished: 1) **deposition** (**sorption**; previously called **sublimation**), where the ice phase forms directly from **water vapor**; 2) **condensation** freezing, where the ice **phase** forms in a supercooled solution following growth and dilution of a **cloud condensation nucleus**; 3) contact **freezing**, where a supercooled **droplet** nucleates following contact of an ice nucleating **aerosol**; 4) immersion freezing, where the nucleating particle is completely immersed in the supercooled liquid, which nucleates with sufficient cooling. Because of this multiplicity of nucleation mechanisms it is often difficult to deduce the processes active in a given cloud. Artificially generated aerosols such as **silver iodide** show activity by all four mechanisms, but at different rates. For natural ice nucleating aerosols, activities in all modes do not generally occur. Observations strongly suggest that, whatever their physico-chemical nature, most natural nuclei act through a freezing process rather than by deposition.

ice pack—*See* **pack ice**.

ice pellets—A type of **precipitation** consisting of transparent or translucent pellets of **ice**, 5 mm or less in diameter.

They may be spherical, irregular, or (rarely) conical in shape. Ice pellets usually bounce when hitting hard ground and make a sound upon impact. Now internationally recognized, ice pellets includes two basically different types of precipitation, known in the United States as 1) **sleet** and 2) **small hail**. Thus a two-part definition is given: 1) sleet or **grains of ice**, generally transparent, globular, solid grains of ice that have formed from the **freezing** of **raindrops** or the refreezing of largely melted **snowflakes** when falling through a below-freezing layer of air near the earth's surface; 2) small hail, generally translucent **particles**, consisting of **snow pellets** encased in a thin layer of ice. The ice layer may form either by the **accretion** of droplets upon the snow pellet or by the melting and refreezing of the surface of the snow pellet. *Compare* **hail**, **graupel**.

ice plain—An area of slightly grounded **ice** in the mouth of some **ice streams**.

Typically an area of very low basal **stress**.

ice point—The **temperature** at which a mixture of air-saturated pure water and pure **ice** may exist in **equilibrium** at a **pressure** of one **standard atmosphere**.

The ice point is often used as one **fiducial point** (0°C or 32°F) in establishing a **thermometric scale** because it is reproduced relatively easily under laboratory conditions. The ice point is frequently called the **freezing point**, but the latter term should be reserved for the much broader reference to the solidification of any kind of liquid under various conditions. *See* **melting point**; *compare* **boiling point**.

ice pole—(Also called pole of inaccessibility.) The approximate center of the most consolidated portion of the arctic **pack ice**, near 83° or 84°N and 160°W.

This term was falling into disuse until its reintroduction with reference to antarctic International Geophysical Year (IGY) activity.

ice prisms—**Ice crystals** having well defined crystalline facets, usually on both basal planes and on prism planes.

ice raft—A discrete block of slow-moving **land ice** that has been incorporated en masse into either an **ice stream** or an **ice shelf**.

ice regime phase—The state or condition of the **ice** on a river, lake, or other body of water, caused by the **thermodynamics** of weather conditions.

ice rind—(Also called ice crust.) A thin but hard layer of **sea ice**, river ice, or lake ice.

Apparently this term is used in at least two ways: 1) for a new encrustation upon old **ice**; and 2) for a single layer of ice usually found in bays and fjords where **freshwater** freezes on top of slightly colder **seawater**. *See* **ice crust**.

ice rise—A usually dome-shaped mass of **ice** resting on rock and surrounded either by an **ice shelf**, or partly by an ice shelf and partly by the sea. No rock is exposed.

Armstrong, T., B. Roberts, and C. Swithinbank, 1973: *Illustrated Glossary of Snow and Ice*, Scott Polar Research Institute, Special Pub. No. 4, 2d ed., The Scholar Press Ltd., Menston Yorkshire, UK.

ice run—Movement of **ice** or **slush** ice with the current of a **stream**, particularly in the initial stage in the spring or summer **breakup** of river ice.

Ice Saints—St. Mamertus, St. Pancras, and St. Savertius or St. Gervais, whose feast-days fall on 11, 12, and 13 May, respectively. These days are associated with May frosts in the folklore of a large part of Europe.

ice sheet—*See* **continental glacier**.

ice shelf—(Also called shelf ice; formerly **barrier**.) A thick **ice** formation with a fairly level surface, formed along a polar coast and in shallow bays and inlets, where it is fastened to the shore and often reaches bottom.

An ice shelf may grow hundreds of miles out to sea. It is usually an extension of **land ice**, and the seaward edge floats freely in **deep water**. The **calving** of an ice shelf forms **tabular icebergs** and **ice islands**.

ice spicule—*See* **spicule**.

ice storm—(Also called silver storm.) A **storm** characterized by a fall of **freezing** liquid **precipitation**. The attendant formation of **glaze** on terrestrial objects creates many hazards.

ice stream—A fast-moving section of an **ice sheet** contained within the ice sheet.

Motion of an ice stream is dominated by **basal sliding**.

ice strip—Same as **ice belt**.

ice structure—The arrangement of **water molecules** in an **ice crystal**.

Under **normal** atmospheric temperatures and pressures between 0° and −100°C, water molecules arrange themselves into an hexagonal crystalline structure called ice-Ih. When viewed along the principal c axis these molecules form spatial hexagonal rings lying above each other, each water molecule surrounded by four others, in a near tetrahedral arrangement.

ice tongue—Any narrow extension of a **glacier** or **ice shelf**, such as a projection floating in the sea or an **outlet glacier** of an **ice cap**.

ice water mixing ratio—Same as **ice mixing ratio**.

iceberg—A large mass of floating or stranded **ice** that has broken away from a **glacier**; usually more than 5 m above **sea level**.

The unmodified term "iceberg" usually refers to the irregular masses of ice formed by the **calving** of glaciers along an orographically rough coast, whereas tabular icebergs and ice islands are calved from an **ice shelf**, and floebergs are formed from **sea ice**. In decreasing size, they are classified as: **ice island** (few thousand square meters to 500 km² in area); **tabular iceberg**; iceberg; **bergy bit** (less than 5 m above sea level, between 1 and 200 m² in area); and **growler** (less than 1 m above sea level, about 20 m² in area).

iced firn—Firn that has become permeated by meltwater and then refrozen; a late stage in the formation of **land ice** from **snow**.

icefall—That portion of a **glacier** where a sudden steepening of descent causes a chaotic breaking up of the **ice**.

Icelandic low—1. The low pressure center located near Iceland (mainly between Iceland and southern Greenland) on mean charts of **sea level pressure**.

It is a principal **center of action** in the atmospheric **circulation** of the Northern Hemisphere. It is most intense during winter, having a January central pressure below 996 mb. In summer, it not only weakens but also tends to split into two centers, one near Davis Strait and the other west of Iceland. Like its Pacific counterpart, the **Aleutian low**, its daily position and **intensity** vary greatly so that it is best regarded as a region where **migratory** lows tend to slow up and deepen.

2. Any **low**, on a **synoptic chart**, centered near Iceland.

icicle—Ice in the shape of a narrow cone, hanging point downward from a roof, fence, cliffside, etc.

An icicle is formed when above-freezing water, for example, snowmelt or groundwater, runs or drips into subfreezing air. The water freezes as it drips or runs, forming a narrow cone pointed downward and growing in both length and width, widest at its top. Most icicles are found hanging from the edges of heated, snow-topped roofs, with any water that has not frozen in its downward traverse forming ice on the surfaces below.

icing—1. In general, any deposit or coating of **ice** on an object, caused by the impingement and **freezing** of liquid (usually supercooled) **hydrometeors**; to be distinguished from **hoarfrost** in that the latter results from the **deposition** of **water vapor**.

The two basic types of icing are **rime** and **glaze**. *See* **aircraft icing, carburetor icing**.

2. [Also known as flood icing, flooding ice, **aufeis** (German), naled (Russian).] A mass or sheet of **ice** formed during the winter by successive **freezing** of sheets of water that may seep from the ground, from a river, or from a spring.

icing intensity—Rate at which **ice accretion** occurs, expressed in units of depth per unit time.

icing level—Lowest height above **sea level** at which an aircraft in flight may encounter **icing**.
See **freezing level**.

icing mound—*See* **ground ice mound**.

icing-rate meter—An instrument for the measurement of the rate of **ice accretion** on an unheated body, for example, rotating cylinders or discs, stationary airfoils, vibrating rods, and electrical-impedance devices.

ideal fluid—1. Sometimes used for an inviscid, **incompressible fluid**.

2. Same as **inviscid fluid**.

ideal gas—(Also called **perfect gas**.) A gas for which the **potential energy** of interaction between molecules is independent of their separation and hence is independent of gas volume.

Thus, the **internal energy** of an ideal gas depends only on its **temperature**. To a very good approximation, atmospheric gases at **normal** terrestrial temperatures and pressures are ideal.

ideal-gas laws—Same as **gas laws**.

ideal horizon—*See* **horizon**.

identical twin integrations—Two forecast integrations with the same **NWP** model, started from slightly different initial conditions, aimed at studying atmospheric **error** growth.
Compare **fraternal twin integrations**.

IF—Abbreviation for **intermediate frequency**.

IF signal—(Abbreviation for intermediate frequency signal.) The **signal** at an intermediate stage of **radar** and **lidar** receivers, chosen at a **standard** frequency (often 30 or 60 MHz) where amplifiers and filters are commonly available.

The radar or lidar **echoes** are converted to IF signals by a mixer, which shifts the **frequency** of the signals through the use of a local **oscillator**. Information is extracted from IF signals by **detection** or **coherent detection**.

IF–THEN rule—(Also called **production rule** or, simply, **rule**.) The basic structure of most **expert systems**.

The necessary conditions are introduced by an "IF" keyword, and the consequences, if those conditions are met, by a "THEN" keyword.

IFF—(Abbreviation for identification: friend or foe.) A system of transponders carried by an aircraft that makes it possible to locate and identify the aircraft by **radar**.

Although the system was originally developed for military uses, it has also been used in non-military applications for tracking aircraft on radar displays. When a radar signal is detected by the **transponder** on the aircraft, the transponder sends back a coded message that is detected by the radar. The radar then plots on its **display** (usually a **PPI**) an indication of both the location and the identity (coded) of the aircraft.

IFR—The commonly used abbreviation for **instrument flight rules**; in popular aviation terminology, descriptive of the conditions of reduced **visibility** to which instrument flight rules apply.
Compare **VFR**.

IFR flight—Same as **instrument flight**.

IFR terminal minimums—The **operational weather limits** concerned with minimum conditions of **ceiling** and **visibility** at an airport under which aircraft may legally approach and land under **instrument flight rules**.

These minimum values frequently are in the form of a **sliding scale**, and also vary with aircraft **type**, pilot experience, and from airport to airport.

IFR weather—Same as **instrument weather**; *see* **IFR**.

ig—Same as **hig**.

igneous meteor—In U.S. weather observing practice, a visible **electrical discharge** in the atmosphere.

Lightning is the most common and important type, but types of **corona discharge** are also included. *Compare* **electrometeor**.

IGRF—Abbreviation for **International Geomagnetic Reference Field**.

illuminance—The photometric equivalent of **irradiance**.

Illuminance is obtained by integrating **spectral irradiance** weighted by **luminous efficiency** over the **visible spectrum**. *Compare* **luminance**.

illumination—In **radar**, a term sometimes used to describe 1) the **irradiance** or **power density** incident on a **target**, or 2) the irradiance or power density supplied by a **feed** to an **antenna**.

illuminometer—Same as **photometer**.

ILS—Abbreviation for **instrument landing system**.

image—A two-dimensional array of data with values at each element of the array related to an **intensity** or a color.

An image is typically defined as the result of some type of image collection system; however, it could be the representation in two **dimensions** of any data by intensity or color.

image enhancement—The process of changing the **display** levels in an **image** to highlight particular information in the image.

This includes, but is not limited to, **contrast** improvement, edge enhancement, spatial **filtering**, **noise** suppression, **image** smoothing, and image sharpening. The result of this process is an **enhanced image**.

image level—*See* **pixel value**.

image processing—The use of automated or manual techniques to provide the means of assessing, preprocessing, extracting features, classifying, identifying, and displaying the original or processed imagery for subjective evaluation, interpretation, and further interaction with the data.

imbibition—Absorption of fluid by a solid or **colloid**, as in a seed swelling in the act of water absorption prior to germination.

immiscible displacement—In porous mediums, the simultaneous movement of two or more immiscible fluids (i.e., fluids that cannot be mixed together, such as water and oil).

The fluids retain distinct phases separated by a **phase** interface surface. *See* **multiple-phase flow**; *compare* **miscible displacement**.

impactometer—Same as **impactor**.

impactor—(Or impactometer.) A general term for instruments that sample particles suspended in the **atmosphere** by impaction.

This instrument is based on the principle that **particles** in an airstream will continue in a straight line due to their inertia when the flow of the air bends sharply; if a surface to which they can adhere is present, they will strike it and may stick. *See* **cascade impactor**, **rotating multicylinder**; *compare* **collector**.

impinger—Type of **impactor** in which **particles** are collected in a liquid rather than on a solid surface.

A drawback is that soluble gases can also be collected.

implicit time difference—A **finite-difference approximation** in which the terms producing time change are specified at the predicted time level.

The approximation

$$(f^{n+1} - f^{n-1})/2\Delta t = g(f^{n+1})$$

(where superscript n denotes a **point** in time, separated by step Δt from the prior $[n - 1]$ and subsequent $[n + 1]$ discrete time levels) is an implicit time difference approximation to the differential equation $df/dt = g(f)$. Implicit approximations may be more difficult to implement than explicit time differences. For explicit time differences, $g(f)$ would be specified as $g(f^n)$ or $g(f^{n-1})$. Implicit time differences are relatively more stable and allow larger time steps than explicit time differences. *Compare* **leapfrog differencing**.

Improved TIROS Operational System—(Abbreviated ITOS.) The name given to *ITOS-1* and *NOAA-1 to -5*.

ITOS satellites were launched between 1970 and 1976. *See* **TIROS**.

IN counter—Abbreviation for **ice nuclei counter**.

inaccuracy—The difference between the **input** quantity applied to a measuring instrument and the **output** quantity indicated by that instrument.

The inaccuracy of an instrument is equal to the sum of its **instrument error** and its **uncertainty**.

inactive front—(Or passive front.) A **front**, or portion thereof, that produces little **cloudiness** and **precipitation**, as opposed to an **active front**.

inadvertent climate modification—Unintentional changes in **climate** caused by **anthropogenic** activities, whether local, regional, or global.

inadvertent cloud modification—Unintentional changes of **clouds** due to **anthropogenic** activities.

It has been argued that the release of **cloud condensation nuclei** worldwide by aircraft, industrial, and agricultural activities may lead to changes of **cloud albedo** both at **cirrus** levels and in boundary layer **stratus/stratocumulus**, thus influencing the earth's **radiation balance** and overall **climate**.

inadvertent weather modification—Unintentional change of weather due to **anthropogenic** activities.

inch of mercury—A common unit used in the measurement of **atmospheric pressure**.

1) One inch of mercury (in Hg) is defined as that **pressure** exerted by a 1-in. column of **mercury** at **standard gravity** and a **temperature** of 0°C:

$$1 \text{ in Hg} = 25.4 \text{ mm Hg} = 33.864 \text{ mb} = 1.00005 \text{ in Hg} (45°)$$

This is a unit recommended for meteorological use.

2) One 45° inch of mercury [in Hg (45°)] is defined as that pressure exerted by a 1-in. column of mercury at 45° latitude at **sea level** and a temperature of 0°C.

It is evident that for most purposes these two units are interchangeable. When this is not the case, the unit should be carefully specified. Metric, rather than the English, units of length are used in many branches of science, and in other parts of the world. The early development and continued widespread use of the **mercury barometer** has fostered this manner of expressing atmospheric pressure. Although it has largely been replaced by the **hectopascal** (hPa) in most meteorological work, inches of mercury is still used in **altimetry**, and it remains the most common form of **barometer** scale **calibration**.

incident solar flux—*See* **insolation**.

inclination—1. (Also called dip.) In **terrestrial magnetism**, the angle through which a freely suspended magnet would dip below the **horizon** in the magnetic north–south **meridional** plane; one of the magnetic elements.

At the **aclinic line** (**dip equator**) the inclination is zero; at either magnetic **pole** (**dip pole**) the inclination is 90°.

2. The angle between the plane of the satellite **orbit** and the earth's equatorial plane.

An inclination angle of less than 90° is referred to as a **prograde orbit**, while an inclination angle greater than 90° is called a **retrograde orbit**.

inclination of the axis of a cyclone—A measure of the angle from horizontal of a line that extends through the geographic centers of **circulation** of a **cyclone** on multiple levels in the **atmosphere**.

inclination of the axis of an anticyclone—A measure of the angle from horizontal of a line that extends through the geographic centers of **circulation** of an **anticyclone** on multiple levels in the **atmosphere**.

inclination of the wind—Departure from horizontal of a **streamline**.

incoherent echo—A radar echo with **phase** and **amplitude** that varies randomly and unpredictably from **pulse** to pulse.

Such echoes are caused by **incoherent scattering**. *Compare* **coherent echo**.

incoherent scatter radar—A **radar** for measuring many of the properties of the **ionosphere** and the neutral **upper atmosphere**.

The radar uses a technique that is much more sensitive and has greater spatial **resolution** than more conventional **ionosondes**, but requires powerful radar transmitters and large antennas.

incoherent scattering—Scattering produced when an incident **wave** encounters randomly moving **scattering** elements causing the **scattered** field to exhibit **random** variations in **phase** and **amplitude**.

In **radar**, incoherent scattering accounts for **incoherent echoes**. *Compare* **coherent scattering**.

incoming solar radiation—*See* **insolation**.

incompressibility—An idealized condition in which **compressibility** vanishes.

Most liquids are nearly incompressible. For shallow **layers** of the **atmosphere** with **velocity** fluctuations much less than the **speed of sound**, incompressibility is a good approximation and is often made for mathematical convenience. For a fluid with a **density** depending on **temperature**, incompressibility does not imply non-**divergence**. *See* **anelastic approximation, Boussinesq approximation**.

incompressible fluid—A fluid in which the **density** remains constant for **isothermal** pressure changes, that is, for which the **coefficient of compressibility** is zero.

Expansion and contraction of an incompressible fluid under diabatic heating or cooling is thus allowed for. In the more usual problem of isothermal processes, the fluid may or may not be stratified (have density differences within it), but motion of a **parcel** from higher to lower **pressure** or vice versa will not change the density of that parcel. Stated mathematically, the density gradient $\nabla \rho$ and the **local derivative** $\partial \rho / \partial t$ may not be zero, but the **individual derivative** $D\rho / Dt$ vanishes. By the **equation of continuity**, it follows that the total **divergence** vanishes:

$$\nabla \cdot \mathbf{u} = \frac{\partial u}{\partial x} + \frac{\partial v}{\partial y} + \frac{\partial w}{\partial z} = 0,$$

where $\mathbf{u}$ is the **velocity** with components u, v, and w. For many purposes in meteorology, the **atmosphere** is treated as a **heterogeneous fluid** in which only vertical motions show **compressibility**. Together with the assumption of **hydrostatic equilibrium**, this has the effect of eliminating compression waves (including **sound waves**).

incus—(Also called **anvil, anvil cloud, thunderhead**.) A supplementary **cloud** feature peculiar to **cumulonimbus capillatus**; the spreading of the upper portion of **cumulonimbus** when this part takes the form of an anvil with a fibrous or smooth aspect.

See **cloud classification**.

indefinite ceiling—(Formerly called ragged ceiling.) After U.S. weather observing practice, the **ceiling classification** applied when the reported **ceiling** value represents the **vertical visibility** upward into surface-based atmospheric phenomena (except **precipitation**).

Such phenomena include **fog, blowing snow**, and all of the **lithometeors**. All indefinite ceilings are estimations, but one of the following must be used as a guide: 1) the distance an **observer** can see vertically into the obstruction; 2) the height corresponding to the top of a ceiling-light **beam**; 3) the height at which a **ceiling balloon** completely disappears; 4) the height determined by the **sensor** algorithm at automated stations. The letters "VV" (vertical visibility) are used to designate an indefinite ceiling.

independence—*See* **statistical independence, independent variable**.

independent pixel approximation—In **radiation** modeling, the assumption that the radiative properties of a given horizontal region (or **pixel** in a satellite image) may be considered in isolation from neighboring pixels.

Its purpose is to allow the application of one-dimensional **radiative transfer** theory. The **accuracy** of this approximation depends on the size of the pixel and the degree of heterogeneity within the pixel and between neighboring pixels.

independent variable—Any of those variables of a problem, chosen according to convenience, that may arbitrarily be specified, and that then determine the other or **dependent variables** of the problem.

The independent variables are often called the coordinates, particularly in problems involving motion in space. Dependent and independent variables can be interchanged, for example, height and **pressure**.

independently distributed—Having the property of **statistical independence**.

index—1. The indicating part of an instrument; for example, the hand of a watch or the **meniscus** of a **mercury column**.

2. *See* **circulation index**.

3. *See* **zonal index**.

index correction—The **correction** applied to a **mercury barometer** to compensate for **temperature** and **gravity**.

index cycle—A roughly cyclic **variation** in the **zonal index**.

The average length of the index cycle is six weeks, varying from about three to eight weeks. It is most pronounced in the winter months.

Namias, J., and P. F. Clapp, 1951: *Compendium of Meteorology*, p. 561.

index error—The **error** in a **mercury barometer**'s **index** revealed by comparison with a **standard** instrument; normally a constant.

The index is graduated on the assumption that the reading represents the actual difference in level between the upper and lower **mercury** surfaces.

index of aridity—A measure of the **precipitation effectiveness** or **aridity** of a region, proposed by De Martonne (1925), given by the following relationship:

$$\text{index of aridity} = \frac{P}{T + 10},$$

where P (cm) is the annual **precipitation** and T (°C) the annual **mean temperature**.

De Martonne, E., 1925: *Traité de Géographie Physique*, Paris.

index of continentality—(Also called coefficient of continentality, continentality index.) *See* **continentality**.

index of evaporating surface of horizontal area—Ratio of **water loss** for area of vegetative cover to **evaporation** from free water surface, that is, **class-A evaporation pan**.

index of refraction—*See* **refractive index**.

index of wetness—A numerical index, often expressed as a percentage, corresponding to the ratio of the **runoff** from a **basin** in a given year to the annual average.

Could be restricted to a particular **season** of interest.

Indian Deep Water—Occupies the depth **range** between 1500 and 3800 m in the Indian Ocean.

It is characterized by a **salinity** maximum of 34.75–34.80 **psu**, indicating that its origin is **North Atlantic Deep Water** imported eastward past the Cape of Good Hope. *See* **deep water**.

Indian equatorial jet—An intense eastward flow of about four-week duration found at the **equator** in the Indian Ocean during the transition periods between the **monsoon** seasons (April–June and October–December).

Speeds at the equator can exceed 1 m s⁻¹ but fall off to less than 0.2 m s⁻¹ at 2°S and 2°N. The jet is the result of **sea level** adjustment to the change of **wind direction** from one monsoon season to the next.

Indian National Satellite System—(Acronym INSAT.) A series of multipurpose geostationary satellites designed to support communications, television broadcasting, and meteorological observations.

INSAT-1 (launched August 1983) was positioned at 74°E longitude, and was the first **three-axis stabilized** geostationary satellite. Not all satellites in the INSAT series have carried meteorological imagers.

Indian spring low water—A tidal **low water** datum, designed for regions of mixed tides, that is depressed below **mean sea level** by the sum of the amplitudes of the principal semidiurnal **lunar** and **solar tides** and the principal **diurnal tides** (M2 + S2 + K1 + O1); originally developed for parts of the Indian Ocean.

Indian summer—A period, in mid- or late autumn, of abnormally warm weather, generally **clear** skies, sunny but hazy days, and cool nights.

In New England, at least one **killing frost** and preferably a substantial period of normally cool weather must precede this warm spell in order for it to be considered a true "Indian summer." It does not occur every year, and in some years there may be two or three Indian summers. The term is most often heard in the northeastern United States, but its usage extends throughout English-speaking countries. It dates back at least to 1778, but its origin is not certain; the most probable suggestions relate it to the way that the American Indians availed themselves of this extra opportunity to increase their winter stores. The comparable period in Europe is termed the **Old Wives' summer**, and, poetically, may be referred to as **halcyon days**. In England, dependent upon dates of occurrence, such a period may be called **St. Martin's summer**, **St. Luke's summer**, and formerly **All-hallow summer**.

indicated airspeed—(Abbreviated IAS.) The **airspeed** read or recorded directly from an airspeed indicator.

Indicated airspeed is usually lower than the actual airspeed and must be corrected for both **temperature** and **density** to yield **true airspeed**. The **correction** is quickly accomplished on an ordinary navigational computer.

indicated altitude—The **altitude** read directly from a **pressure altimeter** when set to the prescribed **altimeter setting**.

This value differs from the **corrected altitude** as a function of the difference between the actual **density** of the underlying air and that of the **standard atmosphere**. The vertical separation of aircraft on airways is based on indicated altitude, and in general, standard aircraft operating procedure calls for the use of indicated altitude.

indicator—(Sometimes called **display**.) An instrument used to reveal but not necessarily measure the presence of an electrical quantity.
　　It is used to display the **output** of a **sensing element** after suitable amplification and modification. In **radar** the term is used to refer to the cathode-ray oscilloscopes, or other recording devices, where the **echoes** returned from **targets** are presented visually or graphically. *See* **radarscope**.

indicator function—A **signal** that is used to decide which subsets of data to include in an analysis.
　　For example, an aircraft flying through a **convective boundary layer** will fly sometimes within **thermal** updrafts and sometimes between thermals. Positive vertical velocities that exceed some threshold could be used to indicate when a measurement is being made in a thermal. By averaging only those temperatures that were obtained within thermals as defined by the indicator function, an average temperature for the thermals can be found. This method of using indicator functions to select portions of a larger dataset is called **conditional sampling**.

indifferent equilibrium—Same as **neutral equilibrium**.

indifferent stability—Same as **neutral stability**.

indirect cell—A **closed circulation** in a vertical plane in which the rising motion occurs at lower **potential temperature** than the descending motion, thus forming an **energy** sink.
　　See **direct cell**.

indirect circulation—A term usually used in the context of a "thermally indirect circulation" that refers to a **circulation** that has ascending motion in a region of relatively low **temperature** and descending motion in a region of relatively high temperature.
　　It therefore is sustained by dynamical processes rather than by **thermal** processes.

individual derivative—(Also called material derivative, particle derivative, substantial derivative.) The rate of change of a quantity with respect to time, following a **fluid parcel**.
　　For example, if $\phi(x, y, z, t)$ is a property of the fluid and $x = x(t)$, $y = y(t)$, $z = z(t)$ are the **equations of motion** of a certain **particle** of this fluid, then the **total derivative**,

$$\frac{D\phi}{Dt} = \frac{\partial \phi}{\partial t} + \frac{\partial \phi}{\partial x}\frac{dx}{dt} + \frac{\partial \phi}{\partial y}\frac{dy}{dt} + \frac{\partial \phi}{\partial z}\frac{dz}{dt} = \frac{\partial \phi}{\partial t} + \mathbf{u} \cdot \nabla\phi$$

(where $\mathbf{u}$ is the **velocity** of the fluid and ∇ is the **del operator**), is an individual derivative. It gives the rate of change of the property of a given **parcel** of the fluid as opposed to the rate of change at a fixed geometrical point, which is usually called the **local derivative**. The term $\mathbf{u} \cdot \nabla\phi$ is called the **advective term**, expressing the **variation** of ϕ in a parcel moving into regions of different ϕ.

Indonesian Throughflow—The **transport** of upper ocean water from the Pacific to the Indian Ocean through the Indonesian Seas.
　　The throughflow is an important link in the **ocean conveyor belt**. It is driven by the **sea level** difference between the Pacific and Indian Oceans and is at its maximum, with 12–20 Sv (12–20 $\times$ 10^6 m^3 s^{-1}), during May–September, when it opposes the **wind**. The minimum, 2–5 Sv, occurs during November–March, when it follows the wind.

induced recharge—The water entering into an **aquifer** from a **stream** or body of water as a result of lowering the **water table** or potentiometric **head** in an aquifer.

induction—1. Reasoning from particular instances to general conclusions; not logically valid.
　　Compare **abduction**, **deduction**.
　　2. *See* **magnetic induction**.

induction charging mechanism—A physical process for **particle** charging involving the collision of pairs of particles in an ambient **electric field**.
　　Electric charge induced on the particle surfaces by the ambient electric field is made available for **transfer** when the two particles come into contact. Subsequent differential particle motions under **gravity** are postulated to result in large-scale **charge separation**. The specific role of specific charging in the electrification of thunderclouds has not been resolved.

induction icing—The formation of **ice** in the air intake channels of jet engines.

See **aircraft icing**.

industrial climatology—A type of **applied climatology** that studies the effect of **climate** and weather on industry's operations.

The goal of industrial climatology is to provide industry with a sound **statistical** basis for all administrative and operational decisions that involve a weather factor.

industrial meteorology—Generally, the application of meteorological data and techniques to industrial, business, or commercial problems.

Fleming, J. R., 1996: *Historical Essays on Meteorology, 1919–1995,* 417–510.

inert gases—Same as **noble gases**.

inertia wave—1. Any **wave motion** in which no form of **energy** other than **kinetic energy** is present.

In this general sense, **Helmholtz waves**, **barotropic disturbances**, **Rossby waves**, etc., are inertia waves.

2. More restrictedly, a **wave motion** in which the source of **kinetic energy** of the **disturbance** is the rotation of the fluid about some given axis.

In the **atmosphere** a westerly **wind** system is such a source, the inertia waves here being, in general, stable. A similar analysis has been applied to smaller vortices, such as the **hurricane**. *See* **inertial instability**

inertial circle—(Or circle of inertia.) A loop in the path of a **parcel** in **inertial flow**, which is approximately circular if the latitudinal displacement is small.

inertial-convective subrange—A range of **eddies** in the **scalar** spectrum in which the **diffusivity** is unimportant and that are therefore simply advected by the mean flow; so called as an analogy with the **inertial subrange** of the **turbulence kinetic energy** spectrum.

inertial coordinate system—A system in which the **(vector) momentum** of a **particle** is conserved in the absence of external forces.

Thus, only in an inertial system can **Newton's laws of motion** be appropriately applied. For all purposes in meteorology, a system with origin on the axis of the earth and fixed with respect to the stars (**absolute coordinate system**) can be considered an inertial system. When relative coordinate systems are used, moving with respect to the inertial system, apparent forces arise in Newton's laws, such as the **Coriolis force**.

Pedlosky, J., 1987: *Geophysical Fluid Dynamics,* 2d ed., 14–21.

inertial current—A current in which the dominant balance is between the inertial and the Coriolis terms in the **equation of motion**, causing **streamlines** to be curved to the right (left) in the Northern (Southern) Hemisphere.

If we think of the streamlines in an inertial current as being locally circular with radius of curvature R, and speed V along the streamlines, the balance of forces in the radial direction is

$$V^2/R = fV,$$

where f is the **Coriolis parameter**. Thus, the radius of curvature is V/f, which is about 10 km for a 1 m s^{-1} current.

inertial-diffusive subrange—The range of **wavenumbers** within the **inertial subrange** where **diffusivity** becomes important for reducing fluctuations of **temperature** or other **scalar** quantities.

inertial flow—Flow in the absence of external forces; in meteorology, frictionless flow in a **geopotential surface** in which there is no **pressure gradient**.

The centrifugal and **Coriolis accelerations** must therefore be equal and opposite, and the constant inertial **wind speed** V_i is given by

$$V_i = fR,$$

where f is the **Coriolis parameter** and R the radius of curvature of the path. The inertial path is **anticyclonic** in both hemispheres, more strongly curved near the poles than near the **equator**, resembling a series of similar loops called inertial circles. The **inertial frequency** with which these loops are described by the **parcel** is approximately equal to $f/2\pi = 2\sin\phi$ per sidereal day, where ϕ is the latitude. All the loops are bounded on the north and south by the same parallels of latitude, but the parcel has a net longitudinal movement as it describes them. The **inertial period** (the reciprocal of inertial frequency) is just one-half **pendulum day**.

inertial force—1. (Or inertia force.) A force in a given **coordinate system** arising from the inertia of a **parcel** moving with respect to another coordinate system.

For example, the **Coriolis acceleration** on a parcel moving with respect to a coordinate system fixed in space becomes an inertial force, the **Coriolis force**, in a coordinate system rotating with the earth.

2. *See* **apparent force**.

inertial forecast—A forecast based on the supposition that the initial conditions of weather and its elements will persist throughout the entire period of the forecast.

inertial frequency—*See* **inertial flow**.

inertial instability—1. (Also called dynamic instability.) Generally, **instability** in which the only form of **energy** transferred between the **steady state** and the **disturbance** is **kinetic energy**.
See **Helmholtz instability, barotropic instability**.

2. The **hydrodynamic instability** arising in a rotating fluid mass when the **velocity distribution** is such that the **kinetic energy** of a **disturbance** grows at the expense of kinetic energy of the rotation.

For a small plane-symmetric displacement (**wavenumber** zero) using the **parcel method**, this criterion for **instability** is that the **centrifugal force** on the displaced parcels is larger than the centrifugal force acting on the **environment**. On the assumption that **absolute angular momentum** is conserved, this states that the fluid is unstable if absolute angular momentum decreases outward from the axis;

$$R \frac{\partial \omega_a}{\partial R} + 2\omega_a < 0,$$

where ω_a is the **absolute** angular speed and R the distance from the axis. If this criterion is applied to rotation of the **westerlies** about the earth's axis, the angular speed of the earth is so large that the inequality fails and the **disturbance** is stable. If applied to a system rotating about a local vertical, the criterion might be satisfied in low latitudes where the component of the earth's rotation about the local vertical is small. Inertial instability has been suggested in connection with the genesis of hurricanes. *See* **rotational instability**.

Holland, J. R., 1992: *An Introduction to Dynamic Meteorology*, 3d edition, Academic Press, 207–208.
Eliassen, A., and E. Kleinschmidt, 1957: Dynamic Meteorology. *Handbuch der Geophysik*, Vol. XLVIII, 64–72.

inertial lag—A **delay** in the response of a flow to the forces acting upon it.

inertial oscillation—A periodic motion in which the fluid inertia is balanced purely by the **Coriolis force**.

Fluid **parcel** motions are typically horizontal and circular, with constant speed and a **velocity** vector that constantly veers to the right in the Northern Hemisphere and to the left in the Southern Hemisphere. The inertial period, the **period** of an inertial oscillation, is $2\pi/f$, where f is the **Coriolis parameter**.

inertial period—*See* **inertial flow, inertial oscillation**.

inertial range—(Also called **inertial subrange**.) The range of length scales over which **energy** is transferred and **dissipation** due to **molecular viscosity** is negligible.

The **power spectrum** has power law behavior over the inertial range. In **two-dimensional turbulence** the power spectrum is theoretically proportional to k^{-3} in which k is the **wavenumber**, and in three-dimensional **turbulence** the power spectrum is theoretically proportional to $k^{-5/3}$. The latter is known as the Kolmogorov minus 5/3 law.

inertial reference frame—Within **Newtonian mechanics**, a **reference frame** relative to which every point mass not subjected to a net force is unaccelerated.

Within relativistic mechanics, a reference frame is inertial in a (local) region of space and time if every point mass in this region remains in uniform motion. According to the principle of relativity, all the laws of physics have the same form (and contain the same numerical constants) when expressed relative to any inertial reference frame.

Taylor, E. F. and J. A. Wheeler, 1966: *Spacetime Physics*, 9–12.

inertial stability—A state of flow that transfers no **kinetic energy** from the **steady state** to a flow **disturbance**.

inertial sublayer—A sublayer in wall-bounded **shear** flows characterized by a sufficiently large **Reynolds number** and a **logarithmic velocity profile** (e.g., the atmospheric **surface layer**); so called as an analogy with the **inertial subrange** of the **turbulence kinetic energy** (TKE) **spectrum**, where **viscosity** provides a **sink** for **momentum** as **dissipation** provides a sink for TKE in the inertial subrange.

inertial subrange—An intermediate range of turbulent scales or wavelengths that is smaller than the **energy-containing eddies** but larger than viscous **eddies**.

In the inertial subrange, the net **energy** coming from the energy-containing eddies is in **equilibrium** with the net energy cascading to smaller eddies where it is dissipated. Thus the slope of the **energy spectrum** in this range remains constant. Kolmogorov showed that the slope is $-5/3$ based on dimensional arguments, namely, $S \propto \epsilon^{2/3} k^{-5/3}$, for ϵ representing **viscous dissipation** rate of **turbulence kinetic energy**, k is **wavenumber** (inversely proportional to the **wavelength**), and S is spectral energy in a Fourier decomposition of a turbulent **signal**. *Compare* **energy spectrum**, **spectral gap**; *see also* **Kolmogorov's similarity hypotheses**.

inertial wave—A **wave** that is caused by a breakdown in the **geostrophic equilibrium** and describes the motion of a water parcel under the influence (balance) of the **Coriolis** and **inertial forces**.
See **inertial oscillation**.

inertio-gravity wave—An **internal gravity wave** propagating under the influence of both **buoyancy** and **Coriolis forces**.

The **dispersion relation** is given by **frequency**

$$\omega = \pm(N k_h + f k_v)/|\mathbf{k}|,$$

in which N is the **buoyancy frequency**, f is the **Coriolis parameter**, and k_h and k_v are the horizontal and vertical components, respectively, of the **wavenumber** vector $\mathbf{k}$. For all wavenumbers, inertio-gravity waves have frequency smaller than N and greater than f. Their **group velocity** is perpendicular to the **phase velocity** such that the vertical component of the group velocity is opposite in sign to the vertical component of the phase velocity. For an upward propagating inertio-gravity wave in the Northern Hemisphere, the **perturbation** wind **vector** turns anticyclonically with height.

inference—A logical process of drawing conclusions from a collection of data and relationships between data and potential conclusions.

Examples are **chain rule**, **modus ponens**, **modus tollens**, and **resolution**.

inference engine—The part of a rule-based **expert system** that makes logical **inferences** or decisions.

inferior mirage—A **mirage** in which the **image** or images are displaced downward from the position of the object.

If only a single image of distant objects is seen, then the term **sinking** is often applied: A horizontal surface appears to curve downwards with increasing distance and terminate in a relatively nearby optical **horizon**. The inferior mirage is most striking when it exhibits two images; the second, lower image is always inverted and of reduced magnification. Sometimes textbooks suggest that there is but a single image: the lower, inverted one. The upper erect image is claimed to be the object. However, both are images, and have positions and magnifications that differ from that of the object. Also, the lower inverted image is sometimes misinterpreted as having resulted from a **reflection** and when this is seen over land, it leads to the assumption that there must be water in the distance causing the reflection. This is the origin of the long association of the mirage and illusory water, and this leads to the assumption that water is present on a dry surface. The mirage owes its name (from se mirer, to look in a mirror) to this impression of arising from a reflection, having been named by French mariners for images seen at sea. For vertical objects seen beyond the optical horizon, typically the lower portion of the object cannot be seen, and an upper portion of the object is seen twice: erect, and inverted. The farther away the object, the more of the lower portion of it will have vanished so that, for example, the upper decks of a distant ship might appear erect and inverted and apparently floating above and disconnected from the optical horizon while the lower decks will not be seen at all. Sometimes a scene such as this is misinterpreted as resulting from a **superior mirage** by a person who thinks the ship's images have been lifted up from the horizon. Actually, in this case, everything is displaced, but the horizon has merely been displaced more. Inferior mirages occur over a surface when the **temperature** decreases with height. The formation of a two-image inferior mirage also requires that the temperature gradient decrease with height. These conditions are met when the surface is relatively warm, resulting in an upward **heat flux**, such as over sun-warmed ground or a lake at night. *See* **sinking**, **stooping**; *compare* **superior mirage**, **towering**, **looming**.

infiltration—The passage of water through the soil surface into the soil.

infiltration capacity—Maximum volumetric rate at which water can be absorbed by a porous material, per unit area, under given conditions.

infiltration coefficient—The ratio of **infiltration rate** to **rainfall** or sprinkler **irrigation** rates.

infiltration gallery—An unlined or partially lined horizontal conduit constructed horizontally or with low gradient into a water-bearing **porous medium** for the purpose of collecting the water from the medium by **gravity**.

infiltration index—Rate of **infiltration** calculated from records of **rainfall** and **runoff**.
There are alternative values depending on the method of calculation.

infiltration rate—The rate at which a liquid enters a porous material, expressed as volumetric rate per unit area.

infiltration routing—Procedure of computing the downward movement of water through an unsaturated bed by taking into account stepwise movement of the **wetting front** and the changes of water stored in each soil layer.

infiltration well—A well drilled into the **unsaturated zone** for the purpose of increasing **recharge** by **gravity**.

infiltrometer—A device designed to measure the rate of **infiltration** of water into soil.

inflow—Flow of water into a **stream**, lake, **reservoir**, container, **basin**, **aquifer system**, etc.

influence function—Same as **Green's function**.

influent seepage—Movement of water from a **free water** source at the land surface toward the **water table**; commonly used to describe flow from a **stream**.

infralateral tangent arcs—Two oblique, colored arcs, convex toward the sun and tangent to the **halo of 46°** at points below the **altitude** of the sun.
These arcs are produced by **refraction** (90° effective prism angle) in hexagonal columnar **ice crystals** the principal axes of which are horizontal but randomly directed in **azimuth**. If the sun's **elevation** exceeds about 68° the arcs cannot appear. A complementary pair of arcs, the supralateral **tangent arcs**, occasionally may be observed above the solar **altitude**.
> Humphreys, W. J., 1940: *Physics of the Air*, 3d ed., 532–533.

infrared—(Abbreviated IR.) Pertaining to or same as **infrared radiation**.

infrared image—Satellite imagery sensed in the 3–13-μm **wavelength** region of the **electromagnetic spectrum**.
Usually refers to the **thermal infrared**, particularly the 10–12.5-μm **window**.

Infrared Interferometer Spectrometer—(Abbreviated IRIS.) A Michelson **interferometer** that flew on *Nimbus* 3 and 4, launched in April 1969 and April 1970, respectively.
IRIS measured a broad **spectrum** from 5 to 20 μm, with a **resolution** of 150 km.

infrared picture—*See* **infrared image**.

infrared radiation—That portion of the **electromagnetic spectrum** lying between visible **light** and **microwaves**.
The **wavelength** range is approximately between 720 and 1 mm. In meteorology, this **range** is often further divided into the **solar infrared** and **terrestrial radiation**, with the division occurring around 4 μm. Dominant absorbers of infrared radiation include the earth's surface, **clouds**, **water vapor**, and **carbon dioxide**. By **Kirchhoff's law**, these are also good emitters of infrared radiation.

infrared radiometry—The technique of measuring **radiant energy**, especially radiant energy in that portion of the total **electromagnetic spectrum** lying within the **infrared** region (wavelengths between 3 and 13 μm).

infrared spectrometer—An instrument often flown on a satellite that measures intensities of **electromagnetic radiation** at wavelengths within the **infrared** portion of the **spectrum**.

infrared spectroscopy—The study of the interaction of substances with **infrared** electromagnetic **radiation** in the 3–33-μm spectral region.
In this spectral region the **infrared** (IR) radiation interacts with vibrational–rotational **energy** levels of the substance under study. IR spectroscopy can be used to determine the concentration of the sample under study or to study the spectral characteristics of the sample. In the former case, the **absorption** of IR radiation is related to the sample concentration through the Beer–Lambert law. In studying spectral features, one obtains information about functional groups, interatomic distances, bond-force constants, and molecular charge distributions. *See* **tunable laser spectroscopy**.

infrasonic—Referring to **sound** frequencies lower than those at the lower limit of average, unimpaired human hearing, about 16–30 Hz.
Compare **ultrasonic**.

infrasonic observatory—A system for measuring the small **pressure** variations associated with **sound waves** below the **range** of human hearing (< 20 Hz).
Such systems are currently being evaluated for meteorological applications (e.g., **tornado** detection).

infrasound—Sound at **infrasonic** frequencies.

ingesting—The receipt of data by a computer system.
The source of the data is often another computer or data distribution system.

initial condition—The state of a time-dependent **dynamical system**, for instance, an **NWP** model, at a given time used to start a forecast of the future state of the system.
See **analysis**.

initial detention—Same as **surface storage**.

initial rainfall—**Rainfall**, measured at the onset of a **storm**, that includes **interception** and **depression storage**, prior to reaching the capacity of the depression storage.

initial-value problem—A class of problems that have solutions determined by time integration of a set of differential equations from some initial state.
For example, the **equilibrium** state of a damped **linear** system can be obtained by time-integrating the governing set of equations until a **steady state** is reached. A short-term **NWP** forecast is another example of an initial-value problem. *See* **transient problem**; *compare* **boundary-value problem**.

initialization—Any method that modifies observed atmospheric **initial conditions** so that high-frequency oscillations are removed from a subsequent **forecast** with an **NWP** model.
See **dynamic initialization**, **normal mode initialization**.

injection temperature—The **temperature** of the **seawater** as measured at the seawater intakes in the engine room of a vessel.

inland ice—1. An older reference to the **Greenland Ice Sheet**.
2. Same as **land ice**, particularly the more interior portions of an **ice sheet**.

inland sea breeze—A **circulation** similar to a **sea breeze**, except not at a shore.
The inland sea breeze is a very weak **thermal** circulation caused by **temperature** contrast between different land surfaces and is sometimes observed between cool irrigated farm land and neighboring dry **desert** land. This phenomenon is observed only when the **synoptic-scale** winds are very light.

inner eyewall—*See* **concentric eyewalls**.

inner layer—For flow over a hill, the bottom layer of air in the **boundary layer** that slows down relative to winds at the same height upstream of the hill in response to **drag** against the surface.

inner product—Same as **scalar product**.

input—(Or input signal.) The quantity to be measured (or modulated, or detected, or operated upon) that is received by an instrument.
Thus, for a **thermometer**, **temperature** is the input quantity.

INS—(Abbreviation for inertial navigation system.) A type of dead-reckoning navigational system, used on aircraft and other vehicles, which is based on the measurement of accelerations.
Accelerations are measured by devices such as gyroscopes, stabilized with respect to inertial space. Navigational information such as vehicle **velocity**, orientation, and position is determined from these measurements by computers or other instrumentation.

INSAT—Acronym for **Indian National Satellite System**.

insolation—1. (Contracted from incoming solar radiation.) In general, **solar radiation** received at the earth's surface.
See **terrestrial radiation**, **direct solar radiation**, **global radiation**, **diffuse sky radiation**, **atmospheric radiation**.
2. The amount of **direct solar radiation** incident upon a unit horizontal surface at a specific level on or above the surface of the earth.
Compare **solar constant**, **total solar irradiance**.

insolation duration—1. Bright **sunshine** duration: interval of time during which **solar radiation** is intense enough to cast distinct shadows.

2. Geographically or topographically possible **sunshine** duration: maximum interval of time during which **solar radiation** can reach a given surface.

3. Maximum possible **sunshine** duration: interval of time between the rising and setting of the upper limb of the sun.

inspectional analysis—The **reduction** of the mathematical equations of a problem to nondimensional units of space, time, and mass; or testing such equations for invariance under any group of transformations.

The procedure stands in close relation to **dimensional analysis** and usually gives rise to a set of **nondimensional numbers** appearing as coefficients in the governing equations. These can always be arranged to be the same as the nondimensional numbers of **parameters** of a corresponding **dimensional analysis**.

instability—A property of the **steady state** of a system such that certain disturbances or **perturbations** introduced into the steady state will increase in magnitude, the maximum perturbation **amplitude** always remaining larger than the initial amplitude.

The **method of small perturbations**, assuming **permanent waves**, is the usual method of testing for instability; unstable perturbations then usually increase exponentially with time. An unstable **nonlinear** system may or may not approach another steady state; the method of small perturbations is incapable of making this **prediction**. The **small perturbation** may be a **wave** or a **parcel** displacement. The **parcel method** assumes that the **environment** is unaffected by the displacement of the parcel. The **slice method** has occasionally been used as a modification of the parcel method to gain a little information about the interaction of parcel and environment. Stability as defined above is an asymptotic concept; other definitions are possible. Precision is required of the user, and caution of the reader. The concept of instability is employed in many sciences. In meteorology the reference is usually to one of the following.

1. Static instability (or hydrostatic instability) of vertical displacements of a parcel in a fluid in **hydrostatic equilibrium**. (*See* **conditional instability**, **absolute instability**, **convective instability**, **buoyant instability**.)

2. Hydrodynamic instability (or dynamic instability) of parcel displacements or, more usually, of waves in a moving fluid system governed by the **fundamental equations of hydrodynamics**, to which the **quasi-hydrostatic approximation** may or may not apply. (*See* **Helmholtz instability**, **inertial instability**, **shearing instability**, **baroclinic instability**, **barotropic instability**, **rotational instability**.)

The space **scale** of unstable waves is important in meteorology: Thus Helmholtz, baroclinic, and barotropic instability give, in general, unstable waves of increasing **wavelength**. The timescale is also important: A perturbation that grows for two days before dying out is effectively unstable for many meteorological purposes, but this is an **initial-value problem** and one cannot assume the existence of permanent waves. These meteorological types of hydrodynamic instability must not be confused with the phenomenon often referred to by mathematicians and physicists by the same term. A great deal of study has been devoted to the problem of the onset of **turbulence** in simple flows under laboratory conditions, and here **viscosity** is a source of instability. *See* **computational instability**.

instability chart—1. *See* **stability chart**.

2. *See* **isentropic thickness chart**.

instability index—*See* **stability index**.

instability line—(No longer frequently used.) Any nonfrontal line or band of **convective activity** in the **atmosphere**.

It is a general term and includes the developing, mature, and dissipating stages. However, when the mature stage consists of a line of active thunderstorms, it is properly called a **squall line**; therefore, in practice, instability line often refers to the less active phases. *See also* **pseudo front**, **prefrontal squall line**

Fulks, J. R., 1951: *Compendium of Meteorology*, 647–652.

instability shower—*See* **convective showers**.

instantaneous field of view—The scan spot size or instantaneous geographic coverage of a satellite sensor.

instantaneous unit hydrograph—**Unit hydrograph** resulting from a unit amount of **effective precipitation** applied to a **drainage basin** in an infinitesimally short time.

instrument correction—The mean difference between the readings of a given instrument and those of a **standard** instrument.

See **barometric corrections**.

instrument error—The correctable part of the **inaccuracy** of an instrument.

instrument exposure—The physical location of an instrument.

The effect of immediate **environment** upon the **representativeness** of the measurements obtained by meteorological instruments is considerable and is not always correctable. The purpose of the **instrument shelter** is to provide as good an **exposure** as possible.

instrument flight—(Also called IFR flight, on instruments.) An aircraft flight conducted by use of navigational instruments that permit the flight crew to proceed without reference to visual landmarks or celestial fixes.

instrument flight rules—(Abbreviated IFR.) A set of regulations set down by the U.S. Civil Aeronautics Board (in Civil Air Regulations) to govern the operational control of aircraft on **instrument flight**.

IFR is seldom used to denote the rules themselves, but is in popular use to describe the weather and/or flight conditions to which these rules apply.

instrument landing system—(Abbreviated ILS.) A navigational aid used to facilitate the landing of an aircraft in **instrument weather** at an airport.

The instrument landing system consists of two parts: 1) a directional guide to bring the plane to the correct runway; and 2) a **glide path** to bring the plane down at the correct glide angle or slope to touch the runway at the correct point. Current safety requirements demand that visual contact with the ground be possible within the final several hundred feet of descent.

instrument payload—*See* **payload**.

instrument shelter—(Or thermometer shelter; also called thermoscreen, thermometer screen) A boxlike structure designed to protect certain meteorological instruments from **exposure** to direct **sunshine**, **precipitation**, and **condensation**, while at the same time providing adequate **ventilation**.

Instrument shelters are usually painted white, have louvered sides, usually a double roof, and are mounted on a stand a meter or so above the ground with the door side facing poleward. Instrument shelters are meant to house thermometric instruments, such as **psychrometers**, **maximum** and **minimum thermometers**, **hygrothermographs**, etc. *See* **airways shelter**, **cotton-region shelter**, **Stevenson screen**.

instrument weather—(Also called IFR weather.) In aviation terminology, route or terminal weather conditions of sufficiently low **visibility** to require the operation of aircraft under **instrument flight rules**.

insulated stream—A **stream** or stream **reach** that is hydrologically separated from the **saturated zone** of the underlying formation by an **aquitard**.

Compare **perched stream**.

insulation—A nonconducting material designed to reduce the **transfer** of **energy** or electric **current** between two materials.

integral depth scale—An average height that is weighted by some other characteristic of the **vertical profile**.

For example, the **potential temperature** profile in the **stable boundary layer** at night often has an exponential shape because the greatest cooling has occurred nearest the ground and the **temperature** change decreases with height. By finding the area under the potential temperature change curve and dividing by the temperature change at the surface, a height scale is obtained that is an integral measure of the depth of the stable boundary layer. For this particular example of an exponential **profile**, the integral depth corresponds to the e-folding depth of the profile.

integral length scales—Of the three standard **turbulence length scales**, the ones that are measures of the largest separation distance over which components of the **eddy** velocities at two distinct points are correlated.

They characterize the energy-containing **range** of eddy length scales. In the most general form, the integral scales (expressed here as a **tensor**) are functions of position and are defined in terms of the normalized two-point **velocity** correlations. *Compare* **Taylor microscale**, **Kolmogorov microscale**.

integration method—A method of estimating a quantity, such as the mean streamflow **velocity** or mean sediment concentration along a **vertical profile**, whereby a measurement instrument is raised and lowered across the profile at a constant rate to achieve a uniform sampling of the profile.

intensification—*See* **intensity**.

intensity—1. In general, expresses the rate of **transfer** per unit area of some condition or physical quantity, such as **rainfall, electromagnetic energy, sound**, etc.

2. (Or **radiant intensity**.) Radiant **power** per unit solid angle; in **SI** units, W sr^{-1}.

3. In **synoptic meteorology**, the general strength of flow around an individual **cyclone** or **anticyclone** (most often applied to the former).

This concept is commonly used in terms of a process, "intensification," or descriptively, as an "intense **low**."

Palmer, J. M., 1993: *Metrologia*, **30**, 371–372.

intensity–duration–frequency curve—Rainfall intensity–duration–frequency curves describe the **rainfall intensity** for a given **exceedance probability** over a **range** of durations.

They are important for the design of many facilities when it is not clear what the critical duration will be. Often a family of curves is provided where each curve corresponds to the duration–intensity relationship for a specific exceedance probability.

intensity-modulated indicator—A general class of radar **display** or **indicator** in which echoes from targets are presented as areas of **light** with **intensity**, brilliance, or color normally a function of the **power** of the **target signal**.

These indicators usually show the position of a **target** in terms of spatial coordinates. **B-scopes**, **PPI** scopes, and **RHI** scopes are intensity modulated. *See* **intensity modulation**.

intensity modulation—Change in a **carrier** signal **intensity** for the purpose of transmitting information.

See **modulation**.

intensity of turbulence—Same as **gustiness components**.

intensive property—Same as **intensive quantity**.

intensive quantity—(Or intensive property.) Any property of a **parcel** or system that is not altered by removal of mass from the system, for example, **temperature**.

Any **specific** quantities are, by definition, intensive. *Compare* **extensive quantity**.

interactive image processing—Application of **image enhancement** techniques in such a way as to present the results of the enhancement within seconds after allowing a user to change a **parameter** or technique.

interception—1. The **amount of precipitation** caught on vegetation or structures that is subsequently evaporated without reaching the ground.

2. The process by which **precipitation** is caught and retained on vegetation or structures, which afterward either reaches the ground as **throughfall** or is evaporated.

As a general rule, this loss to **runoff** or stream **discharge** only occurs at the beginning of a **storm**.

3. The loss of **sunshine**, a part of which may be intercepted by hills, trees, or tall buildings. This loss must be accounted for when evaluating instrumental records of sunshine.

4. The loss of a portion of the **solar spectrum** due to **absorption** and **scattering** by atmospheric gases and **aerosols**; commonly refers to the absorption of **ultraviolet radiation** by **ozone** and aerosols.

interception of precipitation—*See* **interception**.

interceptometer—A **rain gauge** that is placed under trees or in foliage to determine the **rainfall** in that location.

By comparing this **catch** with that from a **rain gauge** set in the open, the amount of rainfall that has been intercepted by foliage is determined.

interchange coefficient—Same as **exchange coefficient**.

intercloud discharge—Same as **cloud-to-cloud discharge**.

interdiurnal variation—Arithmetic difference between the mean daily values of a **meteorological element** on two consecutive days.

interface—(Also called internal boundary.) A surface separating two fluids, across which there is a

discontinuity of some fluid property, such as **density, velocity,** etc., or of some derivative of one of these properties in a direction normal to the interface.

Therefore, the **equations of motion** do not apply at the interface but are replaced by the kinematic and dynamic **boundary conditions.** *See* **surface of discontinuity.**

interfacial layer—*See* **interface.**

interfacial tension—The tangential force at the surface between two liquids, or a liquid and a solid, caused by the difference in attraction between the molecules of each **phase.**

Expressed as a force per unit length or as an **energy** per unit area.

interference—The superposition of two or more waves resulting in an **amplitude** of the composite **wave** not necessarily the algebraic sum of the amplitudes of each of its components.

The simplest example is the superposition of two one-dimensional **scalar** waves with equal amplitude A:

$$A \cos (kx - \omega t + \phi_1) + A \cos (kx - \omega t + \phi_2) = B \cos (kx - \omega t + \psi),$$

where

$$B = A(2 + 2 \cos (\phi_2 - \phi_1)),$$

$$\tan \psi = \frac{\sin \phi_1 + \sin \phi_2}{\cos \phi_1 + \cos \phi_2}.$$

The **amplitude** B of the sum of these waves lies between 0 (destructive interference) and $2A$ (constructive interference) depending on the phase difference $\Delta\phi = \phi_2 - \phi_1$:

$$B = 0 \text{ when } \Delta\phi = \pm\pi, \pm3\pi, \ldots,$$

$$B = 2A \text{ when } \Delta\phi = 0, \pm2\pi, \ldots$$

Destructive and constructive interference are the two extremes of interference not, as is often implied, the only two possibilities. Interference requires **coherence** of the waves, by which is meant a definite and fixed **phase** difference between them.

interference lobe—*See* **lobe.**

interference region—That region in space in which **interference** between **wave trains** occurs.

In **microwave** propagation, it refers to the region bounded by the **ray** path and the surface of the earth that is above the **radio horizon.** Interference lobes and height-gain patterns are formed in this region by the addition of the direct and the surface-reflected **wave.** By contrast, the **diffraction zone** lies below the radio horizon.

interferometer—An instrument that compares or combines **radiation** from a source that has propagated on two or more paths.

interferometry—The use of **interference** phenomena for purposes of measurement.

In **radar,** one use of interferometric techniques is to determine the **angle of arrival** of a **wave** by comparing the **phases** of the **signals** received at separate antennas or at separate points on the same **antenna.** Another interferometric application is to shape and steer the beams of **phased-array antennas** by adjusting the phases of the different elements of the array.

interflow—The water, derived from **precipitation,** that infiltrates the soil surface and then moves laterally through the upper layers of soil above the **water table** until it reaches a **stream** channel or returns to the surface at some point downslope from its point of **infiltration.**

Although readily defined, interflow is difficult to measure and quantify. *Compare* **subsurface flow.**

intergelisol—Same as **pereletok.**

interglacial—A period of warm **climate** during the **Pleistocene** (and earlier **glacial epochs**) during which **continental glaciers** retreated to minimum extent.

Interglacials have been of approximately 10 000 years duration, spaced at approximately 100 000-year intervals over the last 1 000 000 years. The last 10 000 years, or postglacial, is generally considered to be an interglacial.

interglacial phase—Generally, an interval of **geologic time** that was marked by a major poleward retreat of **ice.**

This may be applied to an entire interval between **ice ages** or (rarely) to the individual "stages" that make up an **interglacial** period.

interleaving—A type of quasi-horizontal **mixing** in which **layers** of fluid ("intrusions") flow laterally in alternating directions.

Their **signature** is **temperature** (T) and/or **salinity** (S) inversions in a **profile** of T and S versus depth, indicating that layers of water have moved laterally from opposing directions. The term "interleaving layer" connotes an intrusive motion of large extent in the direction along the front.

intermediate frequency—1. (Abbreviated IF.) The **beat frequency** used in **heterodyne** reception resulting from the combination of the received radio-frequency **signal** and a locally generated signal.

The intermediate-frequency signal is usually the difference between the above signals and is employed to avoid the difficulty of the direct amplification of radio-frequency signals, which is technically more difficult to accomplish.

2. *See* **IF signal**.

intermediate ion—An atmospheric **ion** of size and **mobility** intermediate between the **small ion** and the **large ion**.

Whereas small and large ions are readily detected in all localities, intermediate ions have been reported only in a limited number of cases. The mobility of this class of ions lies generally in the interval from 0.01 to 0.1 cm s^{-1} (V cm^{-1})$^{-1}$. In general, the present understanding of the nature and origin of intermediate ions is limited.

intermediate standard times—Times at which surface **synoptic** observations are made, that is, 0300, 0900, 1500 and 2100 **UTC**.

intermediate water—As a general term, any **water mass** found at intermediate depth in the ocean.

Antarctic Intermediate Water is the most important of these, followed by **Subarctic Intermediate Water** and **Arctic Intermediate Water**. Other water masses identified as intermediate water are Atlantic Intermediate Water in Baffin Bay, also called **Polar Atlantic Water**, identified by a **temperature** maximum at a depth of about 500 m resulting from **inflow** from the **West Greenland Current**; Arctic Intermediate Water in Baffin Bay, identified by a temperature minimum at a depth between 50 and 200 m resulting from inflow of arctic water from the north; and Levantine Intermediate Water in the Eurafrican **Mediterranean Sea**, identified by a **salinity** maximum at a depth between 150 and 400 m and formed when cold winter winds, descending on the region between Rhodes and Cyprus and on the northern and central Adriatic Sea, result in the cooling and sinking of **surface water**.

intermittency—The property of **turbulence** within one **air mass** that occurs at some times and some places and does not occur at intervening times or places.

Whereas the classical theory of **homogeneous turbulence** relies on the assumption that the **turbulence energy** dissipation rate ϵ is constant in space, in reality ϵ is not always constant. Those inhomogeneities may lead to intermittent turbulence. To predict a turbulent or nonturbulent (laminar) behavior within an air mass properly, the complete vertical profiles of either the **virtual potential temperature** or the **buoyancy** must be known. Turbulence is often intermittent in the **stable boundary layer** (e.g., nocturnal) and in the **entrainment zone** capping the **convective mixed layer** (e.g., daytime).

intermittent stream—A **stream** that carries water a considerable portion of the time, but that ceases to flow occasionally or seasonally because bed seepage and **evapotranspiration** exceed the **available water** supply.

Compare **ephemeral stream, perennial stream**.

intermonthly variation—Arithmetic difference between the mean monthly values of a **meteorological element** on two consecutive months.

internal boundary—Same as **interface**.

internal boundary layer—(Abbreviated IBL.) A layer within the **atmosphere** bounded below by the surface, and above by a more or less sharp **discontinuity** in some atmospheric property.

Internal boundary layers are associated with the **horizontal advection** of air across a **discontinuity** in some property of the surface (e.g., **aerodynamic roughness length** or surface **heat flux**) and can be viewed as layers in which the atmosphere is adjusting to new surface properties. *See* **thermal internal boundary layer, mechanical internal boundary layer**.

internal energy—The **energy** of a system exclusive of its **kinetic energy** of mass motion and its **potential energy** arising from external forces.

The internal energy of a system of molecules is the sum of their translational kinetic energies, their vibrational (kinetic and potential) and rotational (kinetic) energies, and the **total potential energy** arising from forces between molecules. An **ideal gas** is defined as one for which the intermolecular **potential energy** is zero. The internal energy of such a gas depends only on its **temperature**.

internal friction—Same as **viscosity**.

internal gravity wave—(Also called internal waves, **gravity waves**.) A **wave** that propagates in **density**-stratified fluid under the influence of **buoyancy** forces.
   The **dispersion relation** is given by **frequency**

$$\omega = \pm(Nk_h)/|\mathbf{k}|,$$

in which N is the **buoyancy frequency** and k_h is the horizontal component of the **wavenumber** vector $\mathbf{k}$. For all wavenumbers, internal gravity waves have frequency smaller than N. Their **group velocity** is perpendicular to the **phase velocity** such that the vertical component of the group velocity is opposite in sign to the vertical component of the phase velocity.

internal Rossby radius of deformation—*See* **Rossby radius of deformation**.

internal tides—**Tidal waves** that propagate at **density** differences within the ocean.
   They travel slowly compared with surface **gravity waves** and have wavelengths of only a few tens of kilometers, but they can have amplitudes of tens of meters. The associated internal currents are termed **baroclinic** motions.

internal water circulation—The conceptual **hydrological cycle** for a specified continental surface that comprises water evaporating from the surface, condensing in the overlying **atmosphere**, and falling back to the surface.
   In reality, some of the evaporated water leaves the region in the **wind** and is replaced with **water vapor** brought into the region by wind.

internal wave—A **wave** in fluid motion having its maximum amplitude within the fluid or at an internal boundary (**interface**).
   The concepts of internal and **external waves** originated in the study of **gravity waves** in homogeneous incompressible fluids, and it makes no difference in the dynamics of the wave whether the **static stability** of the fluid is concentrated in a **free surface** or in an interface. However, internal waves in a fluid with continually varying **density** have maximum amplitudes and nodal surfaces within the fluid itself, so that these are properly distinguished from external waves. *Compare* **surface wave**.

international analysis code—(Abbreviated IAC.) An internationally recognized code for communicating details of **synoptic chart** analyses.

international candle—(Obsolete.) A unit of **luminous** intensity, defined as a fixed fraction of the luminous intensity of a standardized electric lamp operated under prescribed conditions.
   The international candle was adopted as a more precise measure of the luminous intensity than the **candle**. It was replaced by the "new candle" in 1939, and finally the **candela** in 1948.

International Geomagnetic Reference Field—(Abbreviated IGRF.) A geomagnetic **field** defined by a mathematical formula and believed to be the best analytical description of the complete geomagnetic field.
   This formula includes a series of **spherical harmonic** terms, each of which has a coefficient that determines the strength of that term. Since the geomagnetic field changes with time, these coefficients are upgraded every five years by a committee of the International Association of Geomagnetism and Aeronomy. The IGRF is thought to represent the geomagnetic field due to deep earth processes. It does not include wavelengths shorter than a few thousand kilometers since these are thought to be due to crustal magnetization rather than deep earth processes. Accordingly the value of the geomagnetic field measured at a location may differ somewhat from that given by the IGRF.

international index numbers—A system of designating **meteorological observing stations** by number, established and administered by the World Meteorological Organization.
   Under this scheme, specified areas of the world are divided into blocks, each bearing a two-number designator; stations within each block have an additional unique three-number designator, the numbers generally increasing from east to west and from south to north. The international language of this system facilitates quick identification of the source of any **meteorological report**.

International Pyrheliometric Scale—(Abbreviated IPS.) A **radiation** scale for measurement of solar exitance (**irradiance**).

411

Defined in 1956, it replaced the Ångström and Smithsonian scales, and was itself replaced in 1975 by the **Absolute Radiation Scale**.

International Satellite Cloud Climatology Project—(Abbreviated ISCCP.) A World Meteorological Organization project, which began in July 1983, to collect visible and 11-μm **infrared** satellite data from polar and geostationary platforms and process them into **cloud** climatology.
Satellite data are sampled at intervals of 30 km and 3 h.

international standard atmosphere—Same as **ICAO Standard Atmosphere**; *see* **standard atmosphere**.

international synoptic code—(Or synoptic code.) A code approved by the World Meteorological Organization in which the observable meteorological elements are encoded and transmitted in "words" of five numerical digits length.

international synoptic surface observation code—A code to report surface observations.
References: 1) International **synoptic** codes FM 12-XI SYNOP Report of surface observations from a fixed land **station**; 2) FM 13-XI SHIP Report of surface observations from a mobile land station; and 3) FM 14-XI SYNOP MOBIL Report of surface observations from a mobile land station.
World Meteorological Organization, 1995: *Manual on Codes, VI.1 Part A*, WMO-No. 306, p. 1.9-A-1.

International System of Units—(Abbreviated SI in all languages.) A system of physical units in which the fundamental quantities are length, time, mass, electric **current**, **temperature**, **luminous** intensity, and amount of substance, and the corresponding units are the meter, second, kilogram, ampere, kelvin, **candela**, and **mole**.
It has been given official status and recommended for universal use by the General Conference on Weights and Measures. It is also known (in French) as Système International d'Unités.

International Table calorie—(Abbreviated ITcal.) Same as **calorie**; sometimes called the international calorie or international steam table calorie.

International Units—The original definitions of electrical units (including those of **power** and **energy**) given in terms of experimentally determined quantities.
They were originally designated practical units, and later international units, and the unit symbols given by the subscript "int." The relationships between the international and **absolute** units of electromotive force (the volt) and **resistance** (the ohm) were formally defined at the CIPM of 1946, and the relationships between other international and absolute units may be calculated from them.

interplanetary dust—**Interstellar dust** in the region between planets in Earth's solar system.

interplanetary magnetic field—(Abbreviated IMP.) The **magnetic field** found far from any bodies of the solar system.

interpluvial—Periods of **dry climate** within the **Pleistocene epoch** during which lakes dried.
Generally corresponding to **glacials** in regions with monsoonal **climate** and to **interglacials** in regions with **Mediterranean climate**.

interpolation—The estimation of unknown intermediate values from known discrete values of a **dependent variable**.
Various methods are available in one dimension for fitting polynomials or other functions to the known points, the elaborateness of the technique used depending on (among other things) the number and **accuracy** of the known values. The **analysis** of a **weather chart** is an interpolation and **smoothing** in two dimensions. *Compare* **extrapolation**.

interpulse period—(Abbreviated IPP; sometimes called pulse period.) In **radar**, the time between successive; the reciprocal of the **pulse repetition frequency**.

interquartile range—*See* **quartile**.

interrupted stream—A **stream** that contains stretches of both perennial and intermittent flow.

interstadial—A brief (1000 years), relatively warm period within a **glacial**, during which **climate**, **glaciers**, and **sea levels** are between glacial and **interglacial** extremes.
A period of glacier retreat as opposed to a "stadial" period of glacier advance.

interstellar dust—(Also called cosmic dust.) Small, solid **particles** in the supposedly empty space between stars.
Dark blotches in the Milky Way are not due to the absence of stars but rather to **attenuation** of starlight by interstellar dust. *Compare* **interplanetary dust**.

interstice—The nonsolid space in a **porous medium** (i.e., a pore space or void) occupied by air, water, or other fluids.

intertropical convergence zone—1. (Also called ITCZ, equatorial convergence zone.) The axis, or a portion thereof, of the broad trade-wind current of the **Tropics**.

This axis is the dividing line between the **southeast trades** and the **northeast trades** (of the Southern and Northern Hemispheres, respectively). It is collocated with the ascending branch of the **Hadley cell**. At one time it was held that this was a **convergence line** along its entire extent. It is now recognized that actual **convergence** occurs only along portions of this line. For further discussion, *see* **equatorial trough**; *see also* **intertropical front, doldrums**.

2. Same as **meteorological equator**.

intertropical front—(Or equatorial front; also called tropical front.) A **front** presumed to exist within the **equatorial trough** separating the air of the Northern and Southern Hemispheres.

It has been generally agreed that this front, if one exists, cannot be explained in the same terms as the fronts of higher latitudes. However, the extent to which **frontal theory** is to be modified and the nature of the modifications are as yet very controversial questions. *See also* **intertropical convergence zone, doldrums**.

intortus—A **cloud** unique to the genus **cirrus**.
See **cirrus intortus, cloud classification**.

intracloud discharge—Same as **cloud discharge**.

intracloud flash—A **lightning discharge** occurring between a positive charge center and a negative charge center, both of which lie in the same **cloud**; starts most frequently in the region of the strong **electric field** between the upper positive and lower negative **space charge** regions.

In summer **thunderstorms**, intracloud flashes precede the occurrence of **cloud-to-ground flashes**; they also outnumber cloud-to-ground flashes. Intracloud **lightning** develops bidirectionally like a two-ended tree: one end of the tree is a branching negative **leader**, the other is a branching positive leader. Later in the **flash**, fast negative leaders similar to **dart leaders** (also called **K changes**) appear in the positive end region and propagate toward the flash origin. In weather observing, this type of **discharge** is often mistaken for a cloud-to-cloud flash, but the latter term should be restricted to true intercloud discharges, which are far less common than intracloud discharges. Cloud discharges tend to outnumber cloud-to-ground discharges in semiarid regions where the bases of thunderclouds may be several kilometers above the earth's surface. In general, the channel of a **cloud flash** will be wholly surrounded by cloud. Hence the channel's luminosity typically produces a diffuse glow when seen from outside the cloud, and this widespread glow is called **sheet lightning**.

intrinsic permeability—Permeability at **saturation**.
See **effective permeability, relative permeability**.

intrinsic wave frequency—The **frequency** of **oscillation** measured by a **sensor** that moves with the mean wind.

If a **wave** of **wavelength** λ has intrinsic frequency f_i and is imbedded in a mean wind speed of U, then the local **wave frequency** f in Hz measured by a **sensor** fixed to the ground is

$$f = f_i + U/\lambda.$$

This relationship applies to internal **gravity waves** in the **atmosphere**. A special case is for standing **mountain waves** (in which occur **lenticular clouds**) where $f = 0$, because the intrinsic frequency exactly counteracts the **wind** effect.

intromission—The **lateral mixing** of ambient environmental air into the edges of **mixed-layer** convective thermals or **cumulus** clouds, leaving a large undiluted core in the middle of the **thermal**.

While this term is closely related to lateral **entrainment**, the phrase "lateral entrainment" is often associated with idealizations where the **ambient air** is mixed quickly across the whole diameter of the **plume**, leaving no undiluted core. While the lateral entrainment model has been applied successfully to smokestack plumes of meters to tens of meters in diameter, it is not appropriate for mixed-layer thermals of order 1 km in diameter.

intromission zone—The outside part of convective thermals that experience **mixing** with the ambient **environment**.

These thermals are somewhat cylindrically shaped in the **convective boundary layer** rather than bubble-shaped as was proposed by some classical theories. The inside of a **thermal** is called a protected core and is usually not contaminated with entrained environmental air.

intrusions—*See* **interleaving**.

invasion of air—An injection of air from one area or level to another.

The air is typically the opposite of the surrounding air, that is, a cold air invasion into an area of warm air.

inventories of anthropogenic pollutants—Databases on the rates of emission of **anthropogenic** compounds to the **atmosphere**.

To calculate the effects of these emissions, inventories are made available by various agencies, based on **statistics** regarding amounts of fuel combusted, percentages of various gases typically found in the emissions, etc.

inverna—A southeast **wind** of Lake Maggiore, Italy.

inverse-square law—A relation between physical quantities of the form: x is proportional to $1/y^2$, where y is most often a distance, and x is often a force or **flux**.

An example of the inverse square law is the decrease of radiative flux with distance from a **point source**, as is often used to approximate **radiation** reaching the earth from the sun.

inversion—In meteorology, a **departure** from the usual decrease or increase with **altitude** of the value of an atmospheric property; also, the layer through which this departure occurs (the "inversion layer"), or the lowest altitude at which the departure is found (the "base of the inversion").

inversion layer—*See* **inversion**.

inversion of rainfall—*See* **precipitation inversion, zone of maximum precipitation**.

inverted barometer effect—Adjustment of **sea level** to changes in **barometric pressure**; in the case of full adjustment, an increase in barometric pressure of 1 mb corresponds to a fall in sea level of 0.01 m.

If there is this full adjustment, the observed pressures at the sea bed are unchanged.

inviscid fluid—(Also called **ideal fluid**, perfect fluid.) A **nonviscous fluid**, that is, a fluid for which all surface forces exerted on the boundaries of each small element of the fluid act normal to these boundaries.

By definition, therefore, the **stress tensor** reduces to the **pressure**, a point-function **scalar** in the fluid. Thus, in the dynamics of an inviscid fluid, as opposed to a real **viscous fluid**, 1) no restraints are placed on the tangential component of the flow at a solid bounding surface, and 2) there is no **dissipation** of kinetic into **thermal energy** within the fluid. In the **free atmosphere** the flow is often treated as inviscid, and the viscous forces may be neglected for many purposes. Where an inviscid fluid flows along a surface, that surface is said to be a free slip surface.

Sutton, O. G., 1953: *Micrometeorology*, chap. 2.

ion—1. In chemistry, atoms or specific groupings of atoms that have gained or lost one or more electrons, as the "chloride ion" or "ammonium ion."

Ions are most familiar in aqueous solutions and in **crystal** structures, but they also exist in the gas **phase** at all altitudes in the **atmosphere**. They are most abundant, and have the greatest importance, in the **ionosphere**, between about 70 and 300 km in **altitude**.

2. In **atmospheric electricity**, any of several types of electrically charged submicroscopic **particles** normally found in the **atmosphere**.

Atmospheric ions are of two principal types, **small ions** and **large ions**, although a class of **intermediate ions** has occasionally been reported. The **ionization** process that forms small ions depends upon two distinct agencies, **cosmic rays** and radioactive emanations. Each of these consists of very energetic particles that ionize neutral air molecules by knocking out one or more planetary electrons. The resulting **free electron** and positively charged molecule (or atom) very quickly attach themselves to one or, at most, a small number of neutral air molecules, thereby forming new small ions. In the presence of **Aitken nuclei**, some of the small ions will in turn attach themselves to these nuclei, thereby creating new large ions. The two main classes of ions differ widely in **mobility**. Only the highly mobile small ions contribute significantly to the **electrical conductivity** of the air under most conditions. The intermediate ions and large ions are important in certain **space charge** effects, but are too sluggish to contribute much to **conductivity**. The processes of formation of ions are offset by certain processes of destruction of the ions. *See* **recombination, ion mobility**.

Wait, C. R., and W. D. Parkinson, 1951: *Compendium of Meteorology*, 120–127.

Chalmers, J. A., 1957: *Atmospheric Electricity*, 55–80.

ion-capture theory—A theory of **thunderstorm charge separation** advanced by C. T. R. Wilson (1916).

According to this theory, the lower negative charge of a **thundercloud** is generated by the accumulation there of raindrops that have captured predominantly negative **ions** in their descent through the **cloud**. The preferential **capture** of negative ions by such drops is said to be due to the **polarization** of the drops in the **normal** atmospheric **electric field** existing between the negatively charged earth and positively charged **ionosphere**. The lower halves of the falling drops therefore would attract and capture negative charges while their upper halves would be unable to draw in positive charges with comparable efficiency; thus a net negative charge would build up on the drops. This theory is generally regarded today as incapable of accounting for any important portion of thunderstorm charge separation, for it is quantitatively inadequate in view of typical ion densities. *See* **precipitation current**.

Chalmers, J. A., 1957: *Atmospheric Electricity*, p. 262.
Wilson, C. T. R., 1916: On some determinations of the sign and magnitude of electric discharges in lightning flashes. *Proc. Roy. Soc. A*, **92**, 555–574.

ion chromatography—A form of high-pressure liquid chromatography using a **conductivity** detector where a combination of weak ionic solvents are used to separate anions and cations of a solution, with the contribution of the solvent to **conductivity** suppressed just prior to **detection**. The technique is useful for measuring anions such as sulfate, **nitrate**, and chloride in **hydrometeors**.

ion concentration—Same as **ion density**.

ion counter—An apparatus that counts the number of unit charges of electricity that are contained in a sampled volume of the **atmosphere**.
  The general procedure is to pass a sample of the atmosphere through a charged cylindrical condenser. The change in the **potential** across the condenser is a measure of the ionic charge contained in the sample volume. The change in potential depends upon such factors as the polarizing potential of the condenser, the **mobility** and charge of the **ions**, volume and length of the condenser, and sample flow rate. An ion counter is an **ionization chamber** in which there is no internal amplification by gas multiplication.

ion density—(Or ion concentration.) In **atmospheric electricity**, the number of **ions** per unit volume of a given **sample** of air; more particularly, the number of ions of given type (positive **small ion**, negative small ion, positive **large ion**, etc.) per unit volume of air.

Wait, C. R., and W. D. Parkinson, 1951: *Compendium of Meteorology*, 124–125.

ion mean life—The average time interval under specified atmospheric conditions between the formation and destruction of an **ion** of any given type.
  The **mean life** of **small ions** in **clean air**, for example, over the sea, is four to five minutes, but in polluted air it is generally less than a minute. **Large ions** have mean lifetimes of as much as 15 to 20 minutes over the oceans, while in very polluted areas, lifetimes may approach an hour.

ion mobility—(Or ionic mobility.) In gaseous electric **conduction**, the **average velocity** with which a given **ion** drifts through a specified gas under the influence of an **electric field** of unit strength.
  Mobilities are commonly expressed in units of meters per second per volt per meter $[m\ s^{-1}(V\ m^{-1})^{-1}]$. In a vacuum, a single gaseous ion subjected to any nonzero **potential gradient** would accelerate indefinitely; but in the midst of a gas the ion continually experiences collisions with gas molecules. These encounters tend to break up its **trajectory** into a series of short intervals of **acceleration** punctuated by deflections. The net result is that the ion's gross motion resembles **drift** at a uniform **velocity**. The **mobility** depends not only upon the nature of the ion and gas but also upon the **density** of the gas, for the latter controls the **mean free path** of the ion. In **atmospheric electricity**, the mobilities of small and large ions weight their relative importance in atmospheric **conduction**. Small ions have mobilities of about 1.3×10^{-4} m $s^{-1}(V\ m^{-1})^{-1}$ in air at **sea level** with negative small ions exhibiting slightly greater values than do the positive small ions. High humidities suppress **small ion** mobilities slightly. **Large ions** have mobilities of only about 4×10^{-7} m s^{-1} $(V\ m^{-1})^{-1}$ at sea level, their sluggishness being due to their great mass.

Wait, C. R., and W. D. Parkinson, 1951: *Compendium of Meteorology*, 124–125.
Israël, H., 1951: *Compendium of Meteorology*, 146–147.

ion pair—A pair of **ions** of equal and opposite charge formed by **photoionization** or by interaction of matter with any sufficiently energetic **particles** such as **beta particles** or **alpha particles**.

ionic conduction—Any electrical **conduction** where the **current** is sustained by the motion of **ions** (as opposed to **electrons**) within the conductor.
  All electrical conduction in the **atmosphere** is of this type.

ionic mobility—*See* **ion mobility**.

ionization—In **atmospheric electricity**, the process by which neutral atmospheric molecules (**small ions**) or other suspended **particles** (mainly **large ions**) are rendered electrically charged chiefly by collisions with high-energy particles.
  Cosmic rays and radioactive decay are the main sources of **atmospheric ionization**. In the

lower atmosphere, decay electrons of mu-mesons plus **alpha particles** from radioactive gases, as well as **beta particles** and **gamma rays**, serve to ionize air molecules. The rate at which these agents ionize the air is expressed in units of one **ion pair** per cubic centimeter per second, symbolized by I. Cosmic rays at **sea level** yield about $2I$, both over land and at sea. Radioactive gases contribute about $5I$ over land areas at sea level, while radioactive materials in the soil and rocks themselves yield about $4I$. At heights above about 5 km, only cosmic rays provide significant ionization, and this contribution finally reaches a maximum at about 13 km, above which the rate decreases, due to decreasing air density and consequent lack of target molecules for the cosmic rays. *See* **photoionization**.

ionization chamber—An apparatus used to study the production of **ions** in the **atmosphere** by **cosmic ray** and radioactive bombardment of air molecules.
    The chamber is an airtight container usually cylindrical in shape and 25–50 L in volume. An insulated electrode is centrally located in the chamber. In operation a potential is applied between the electrode and the chamber wall. The ions produced in the chamber are collected by the electrode and measured by an **electrometer**. Amplification of ions may occur as the result of gas multiplication.

ionization potential—The amount of **energy** required to remove an **electron** from a given molecule or atom to form an **ion**; usually expressed in units of electron volts.

ionogram—A photographic **display** of **echo** strength as a function of transmitted **frequency** produced by a sweep-frequency **ionosonde**.
    Ionograms are used to infer the structure of the ionospheric **layers** and hence to predict radio transmission concentration.

ionosonde—The traditional technique for measuring the properties of the **ionosphere** by transmission and reception of vertically incident radio pulses at swept frequencies in the **HF** range.

ionosphere—The **atmospheric region** containing significant concentrations of **ions** and **electrons**.
    Its base is at about 70–80 km and it extends to an indefinite height. In terms of the standard upper-atmospheric nomenclature, the ionosphere is collocated with the **thermosphere** and the upper **mesosphere**, while the outer ionosphere also forms part of the **magnetosphere**. Most of the early knowledge of the ionosphere came from **sounding** by ground-based radars known as **ionosondes**, and the sharply defined echoes produced by these instruments led to the definition of ionospheric **layers**, implying well-defined regions with clear maxima of **ionization**. More recent investigations by a number of remote and in situ techniques have shown that such sharply bounded layers do not generally exist (with the exception of the so-called **sporadic-E layers**), and that the layers are better described as regions. The term layer is still frequently used, however. Ionospheric sounding normally shows an **E-layer** in the 100- to 120-km height range, an **F₁-layer** in the 150- to 190-km **range**, and an **F₂-layer** above 200 km. The **D-region**, lying in the upper mesosphere below 100 km, is responsible for daytime **absorption** of high-frequency **radio waves**, but does not usually produce an **echo** on ionosonde recordings. Early suggestions of a C-layer below the D-region, and a **G-layer** above the F₂-layer, have not been generally accepted, and these terms are now obsolete. Over most of the earth, the ionosphere is produced by the action of **solar radiation** of short **wavelength** (extreme **ultraviolet** and **x-ray** radiation) on the atmospheric constituents. At high magnetic latitudes, energetic **particles** of solar or auroral origin become important, and even dominant, sources of ionization. *See* **atmospheric shell**.

ionospheric potential—The **potential** difference between the conductive earth and the conductive **upper atmosphere**, approximately +250 kV.
    This voltage difference is maintained by worldwide **cloud electrification**. Most of the 250 kV is realized well below the formal ionospheric height of maximum **ionization** because of the rapid increase of **electrical conductivity** with height in the **atmosphere**.

ionospheric radar—**Radar** designed for investigating the structure of the **ionosphere** by measuring the properties of the **echoes** returned from the ionosphere.
    Such radars operate in the **HF** part of the **radio frequency band**.

ionospheric recorder—(Also called **ionosonde**.) A radio device for determining the distribution of **virtual height** with **frequency**, and the critical frequencies of the various "layers" of the **ionosphere**.
    A **pulse** at a certain frequency is transmitted vertically, and the time for its return is recorded on an **oscilloscope**; another pulse at a different frequency is then transmitted and timed. The process is repeated until the entire frequency range of interest, usually from about 1 to 25 MHz, has been explored.

ionospheric storm—Term used to denote the major changes that take place in the **F-region** as a result of **solar activity**.

Ionospheric storms are closely associated with **magnetic storms** and can lead to severe disruptions of radio-wave propagation, particularly at high latitudes.

ionospheric tides—Term denoting the system of electrical currents and fields generated in the **ionosphere** by tidal motions in the background **atmosphere**.

These tidal motions are forced mainly by solar heating of atmospheric **water vapor** and **ozone**, and reach large amplitudes at ionospheric heights.

ionospheric trough—A portion of the **F-region**, centered on the **magnetic dip** equator, in which **electron** densities are anomalously low with peaks at 15°–20° latitude on both sides.

The **equatorial trough** appears during daytime and is absent at night. It is attributed to **diffusion** of ionospheric **plasma** down the **magnetic field** lines from the **magnetic equator** as the F-region rises in response to an eastward **electric field** generated by dynamo action in the lower **ionosphere**. One or more troughs in **electron density** are also found at high latitudes and are magnetically linked to the **plasmapause** located at a radial distance of several earth radii in the outer **magnetosphere**.

IPS—Abbreviation for **International Pyrheliometric Scale**.

IR—Abbreviation for **infrared**, or **infrared radiation**.

iridescent clouds—**Ice-crystal clouds** that exhibit brilliant spots or borders of colors, usually red and green, observed up to about 30° from the sun.

This **irisation** results from an optical **diffraction phenomenon**, usually of several orders. The **cloud** particles that occasion this phenomenon are very small (a few micrometers), and locally all are of nearly the same size; they result from local near **adiabatic** lifting and **condensation** in **moist air** often in a **lenticular** (wave) cloud, in **pileus** above a developing **cumulus**, or occasionally in irregular patches of uniform color in a region of shallow **convection** (in this case sometimes called mother-of-pearl cloud).

IRIS—Abbreviation for **Infrared Interferometer Spectrometer**.

irisation—The color exhibited by **iridescent clouds** and at times along the borders of **lenticular clouds**.

Irminger Current—*See* **East Greenland Current**.

iron winds—Northeasterly winds of Central America; they are prevalent during February and March and blow steadily for several days at a time.

irradiance—(Or radiant flux density.) A radiometric term for the rate at which **radiant energy** in a **radiation** field is transferred across a unit area of a surface (real or imaginary) in a hemisphere of directions.

In general, irradiance depends on the orientation of the surface. The radiant energy may be confined to a narrow **range** of frequencies (spectral or **monochromatic** irradiance) or integrated over a broad range of frequencies. Irradiance follows from **radiance** but not, in general, vice versa. The photometric equivalent of irradiance is **illuminance**, obtained by integrating **spectral irradiance** times **luminous efficiency** over the **visible spectrum**. *See* **Poynting vector**.

irradiation—**Radiation** that is incident on a surface.

irregular crystal—(Sometimes called amorphous snow.) A **snow** particle, sometimes covered with **rime**, without visible crystalline facets; sometimes used as a last category for snow identification.

irreversible process—A process in which the total **entropy** of the **universe** (system and its surroundings) increases.

All real processes are irreversible.

irrigation—The artificial application of water to land to promote the growth of crops.

First practiced in arid lands to supplement deficient **rainfall**, it is now used extensively in more humid areas to ensure proper timing of water supply for maximum crop yield. The earliest form of irrigation consisted of diverting **flood** flows of streams onto the cropland. More common practice today is the storage of floodwaters in **reservoirs** from which the water may be withdrawn for use as needed. Extensive irrigation projects have been developed that use **groundwater** pumped from wells.

irrigation requirement—*See* **duty of water**.

irrigation return flow—The amount of **irrigation** water that infiltrates past the root zone and returns fully or partially to the **drainage network** or system.

irrotational—Applied to a **vector field** having zero **vorticity** or **curl** throughout the field.
Two equivalent properties of an irrotational field are that there is no **circulation** about any reducible curve within the fluid, and that a **potential** exists. An **autobarotropic** fluid is irrotational for all time if it is irrotational at any time. Meteorological motions of the smaller scales, for example, **gravity waves**, may be treated as irrotational, but when the **scale** is large enough to take the rotation of the earth into account, only **rotational** motions are of interest. *See* **solenoidal**, **Helmholtz's theorem**.

isabnormal line—(Rare.) On a **chart** or diagram, a line of equal **deviation** from **normal**.
Compare **isanomalous line**.

isallobar—A line of equal change in **atmospheric pressure** during a specified time interval; an **isopleth** of **pressure tendency**.
A common form is drawn for the three-hourly local pressure tendencies on a **synoptic** surface **chart**. Positive and negative isallobars are sometimes referred to as anallobars and katallobars, respectively.

isallobaric—Of equal or constant **pressure** change; this may refer either to the distribution of equal **pressure tendency** in space or to the **constancy** of pressure tendency with time.
The term **allobaric**, or simply, "pressure change," could be used instead. This term, preferably, should not be used to mean "of isallobars."

isallobaric chart—Chart that presents analyses of the changes of **atmospheric pressure** during a specific time interval.

isallobaric high—Same as **pressure-rise center**.

isallobaric low—Same as **pressure-fall center**.

isallobaric maximum—Same as **pressure-rise center**.

isallobaric minimum—Same as **pressure-fall center**.

isallobaric wind—(Also called Brunt–Douglas isallobaric wind.) The **wind** velocity when the **Coriolis force** exactly balances a locally accelerating **geostrophic wind**.
Mathematically, the isallobaric wind V_{is} is defined in terms of the local accelerations but approximated by the **allobaric** gradient as follows:

$$V_{is} = \mathbf{k} \times \frac{1}{f} \frac{\partial \mathbf{V}_g}{\partial t} \doteq -\frac{\alpha}{f^2} \nabla_H \frac{\partial p}{\partial t},$$

where **k** is the vertical unit **vector**, f the **Coriolis parameter**, $\mathbf{V}_g$ the **geostrophic wind**, α the **specific volume**, ∇_H the horizontal **del operator**, and p the **pressure**. The isallobaric wind is thus directed normal to the isallobars toward falling pressure, with magnitude proportional to the allobaric gradient. Because the isallobaric wind is associated with **transient** effects and because of the many assumptions used in deriving its equation, observational evidence of this wind is unsatisfactory.
Haurwitz, B., 1941: *Dynamic Meteorology*, 155–159.

isallohypse—Same as **height-change line**.

isallohypsic wind—Same as **isallobaric wind**, but using height tendency in a **constant-pressure surface** instead of **pressure tendency** in a **constant-height surface**.

isallotherm—A line connecting points of equal change in **temperature** within a given time period.

isametric—(Rare.) *See* **equideparture**.

isanabat—A line drawn through points of equal vertical component of **wind** velocity.
Positive values indicate upward motion, negative values indicate downward motion.

isanakatabar—(Rare.) A line of equal atmospheric-pressure range during a specified time interval.

isanomal—Same as **isanomalous line**.

isanomalous line—(Rare; also called isanomal.) A line drawn through geographic points having equal **anomaly** of some meteorological quantity.
Compare **isabnormal line**.

isanomaly—Line joining points of equal **anomaly** of a **meteorological element**.

isanthesic line—A line drawn through geographical points at which the blossoming of a given plant occurs on the same date.

isarithm—Same as **isopleth**.

ISCCP—Abbreviation for **International Satellite Cloud Climatology Project**.

isentrope—An **isopleth** of **entropy**.
   In meteorology it is usually identified with an isopleth of **potential temperature**.

isentropic—Of equal or constant **entropy** (or, in meteorology, **potential temperature**), with respect to either space or time.

isentropic analysis—Analysis of processes within the **atmosphere** on the basis of the location and configuration of various **isentropic** surfaces, distribution of atmospheric processes, and motion on them.

isentropic chart—A constant-**entropy** chart; a **synoptic chart** presenting the distribution of meteorological **elements** in the **atmosphere** on a surface of constant **potential temperature** (equivalent to an **isentropic surface**).
   It usually contains the plotted data and **analysis** of such elements as **pressure** (or height), **wind**, **temperature**, and moisture at that surface.
   Saucier, W. J., 1955: *Principles of Meteorological Analysis*, chap. 8.

isentropic condensation level—Same as **lifting condensation level**.

isentropic mixing—Any atmospheric **mixing** process that occurs within an **isentropic surface**.
   The fact that many atmospheric motions are reversible **adiabatic processes** renders this type of mixing important, and **exchange coefficients** have been computed therefor.

isentropic surface—A surface in space on which **entropy** (or, in meteorology, **potential temperature**) is everywhere equal; a constant-entropy surface.

isentropic thickness chart—(Also called thick–thin chart.) A **thickness chart** of an **atmospheric layer** bounded by two selected **isentropic** surfaces (surfaces of constant **potential temperature**).
   The **thickness** of such a layer is directly proportional to the **static instability** of that layer; hence, these charts have been called instability charts. *Compare* **isentropic weight chart**; *see* **isentropic chart**.

isentropic vertical coordinate—Vertical coordinate for atmospheric models defined as the **potential temperature** of the **model atmosphere**.
   In practice the isentropic vertical coordinate is replaced by or changed gradually to a terrain-following coordinate near the ground; a change to **pressure coordinates** near the top of the model atmosphere has also been used. *See* **hybrid vertical coordinate**, **vertical coordinate system**, **pressure vertical coordinates**.

isentropic weight chart—A **chart** of atmospheric-pressure difference between two selected **isentropic** surfaces (surfaces of constant **potential temperature**); the greater the **pressure** difference the greater the weight of the air column separating the two surfaces.
   Compare **isentropic thickness chart**; *see* **isentropic chart**.

iso-D—A line of constant **D-value**.

isobar—A line of equal or constant **pressure**; an **isopleth** of pressure.
   In meteorology, it most often refers to a line drawn through all points of equal **atmospheric pressure** along a given reference surface, such as a **constant-height surface** (notably **mean sea level** on surface charts), an **isentropic surface**, the vertical plane of a **synoptic** cross section, etc. The pattern of isobars has always been a main feature of surface-chart **analysis**. Isobars are usually drawn at intervals of one **millibar** or more, depending on the **scale** needed to identify or illustrate a specific meteorological pattern.

isobaric—Characterized by equal or constant **pressure**, with respect to either space or time.

isobaric analysis—An **analysis** depicting contours of constant **pressure**.

isobaric chart—A **weather map** or **chart** displaying meteorological data on a **constant-pressure surface**.
   See **constant-pressure chart**.

isobaric contour chart—A **weather map** or **chart** with **contours** (e.g., contours of **temperature**) analyzed on a surface of constant **pressure** (that is, drawn on an **isobaric surface**).

419

isobaric divergence—The **horizontal divergence** measured on a **constant-pressure surface**, that is, expressed in a system of coordinates with **pressure** as an **independent variable**.

isobaric equivalent temperature—*See* **equivalent temperature**.

isobaric–isosteric solenoid—A **solenoidal** volume bounded by a combination of **isobaric** and **isosteric** surfaces.

isobaric–isosteric tube—*See* **isobaric–isosteric solenoid**.

isobaric mixing—**Mixing** at constant **pressure**.

isobaric surface—A surface of constant **pressure**.
    See **constant-pressure surface**.

isobaric thickness chart—A type of **synoptic chart** showing the **thickness** of a defined layer in the **atmosphere**.
    This layer is the vertical distance between two **constant-pressure surfaces**.

isobaric topography—Same as **height pattern**.

isobaric tube—A **solenoidal** volume bounded by a single **isobaric surface**.

isobaric vorticity—**Relative vorticity** in a **constant-pressure surface**, that is, expressed in a system of coordinates with **pressure** as an **independent variable**.

isobaric wet-bulb temperature—*See* **wet-bulb temperature**.

isobath—(Sometimes called fathom curve.) In **hydrology**, a line on a map connecting all points at which there exists an equal vertical distance between the earth's surface and the **water table**, or equal depths to the upper or lower surface of an **aquifer**.

isobath—A **contour** of equal depth in a body of water, represented on a **bathymetric chart**.

isobathytherm—A line or surface showing the depths in oceans or lakes at which points have the same **temperature**.
    Isobathytherms are usually drawn to show **cross sections** of the **water mass**.

isobront—1. (Sometimes called homobront.) A line drawn through geographical points at which a given phase of **thunderstorm** activity occurred simultaneously.
    2. In **climatology**, a line drawn through geographical points that have the same average number of days with **thunder** in a given period; a type of **isoceraunic line**.

isoceraunic—(Also spelled isokeraunic.) Indicating or having equal **frequency** or **intensity** of **thunderstorm** activity.
    See **isoceraunic line, isobront**.

isoceraunic line—A line drawn through geographical points at which some phenomenon connected with **thunderstorms** has the same **frequency** or **intensity**.
    Its most recent specific use is for lines of equal frequency of **lightning discharges**. *See* **isobront**.

isochasm—(Also called isaurore.) A line connecting points on the earth's surface at which the **aurora** is observed with equal **frequency**.

isochion—Same as **isonival**.

isochoric—Of equal or constant volume, usually applied to a thermodynamic process during which the volume of the system remains unchanged.
    Compare **isosteric**.

isochrone—A line in a **chart** connecting all points having the same time of occurrence of a particular phenomenon or of a particular value of a quantity.
    In meteorology, for example, the past positions of fronts, instability lines, isotherms, etc., on a **continuity chart** constitute an isochrone.

isoclinal line—Same as **isoclinic line**.

isocline—A line connecting points having the same vertical direction (**inclination**) of a particular **vector** quantity.
    This term is not widely used in this sense, probably because of its long-standing geological definition relating to folded rock strata. *See* **isoclinic line**; *compare* **isogon**.

isoclinic line—(Or frequency.) A line drawn through all points on the earth's surface having the same **magnetic inclination**.

The particular isoclinic line drawn through points of zero **inclination** is given the special name of **aclinic line**. *Compare* **isogonic line, isocline.**

isodef—(Rare.) A line of equal percent deficiency from the mean.

The term was introduced by W. R. Baldwin-Wiseman (1934) for a line showing equal differences of the cumulative rainfalls for each month of a **drought** period from the long-period average **rainfall** of the month, expressed as percent of the latter.

> Baldwin-Wiseman, W. R., 1934: The cartographic study of drought. *Quart. J. Roy. Meteor. Soc.*, **60,** 523–532.

isodrosotherm—A line of equal **dewpoint.**

See **isohume.**

isodynamic—In general, a line of equal magnitude of any force.

isoecho—On **radar displays**, a line or **contour** connecting all points of equal **target signal** strength or **radar reflectivity factor**, labeled, for example, in units of dBZ.

Such displays for **radar** or **lidar** (e.g., **PPI, RHI, THI**) can be generated in **real time** using grayshades or colors to show steps in the **reflectivity** pattern. *See* **echo contour.**

isofronts–preiso code—A code in which data on isobars and fronts at **sea level** (or earth's surface) are encoded and transmitted.

It is a modified form of the **international analysis code.**

isogeotherm—(Sometimes called geoisotherm.) A line or surface showing the depths in the ground at which points have the same **temperature.**

Isogeotherms are often drawn against coordinates of time and depth to represent the **diurnal** or annual **variation** of **soil temperature.**

isogon—A line in some given surface joining all points having the same direction of a particular **vector** quantity.

In meteorology, isogons are usually drawn for the **velocity** vector as an aid in constructing **streamlines** for a **wind field**. *Compare* **isotach.**

isogonal map—Same as **conformal map.**

isogonic line—In the study of **terrestrial magnetism**, a line drawn through all points on the earth's surface having the same **magnetic declination**; not to be confused with magnetic **meridian.**

The particular isogonic line drawn through all points having zero **declination** is called the **agonic line**. *Compare* **isoclinic line, isogon.**

isogram—1. (Or isoline.) A line, on a given reference surface, drawn through all points where a given quantity has the same numerical value.

The reference surface can be any **coordinate plane** functionally related to the given quantity (this includes physically defined surfaces in space). This, therefore, is a very general term. Although **isopleth** is used in this same broad sense by meteorologists, it has a more restricted meaning when used in most other sciences. *See also* **isotimic line.**

2. As sometimes restricted, a line drawn through all geographical points that experience the same **frequency** of some meteorological event.

isohaline—1. Of equal or constant **salinity.**

2. A line on a **chart** connecting all points of equal **salinity**; an **isopleth** of salinity.

isoheight—Same as **contour line.**

isohel—Line joining geographical points of equal **insolation** during a specific interval of time.

isohume—A line drawn through points of equal **humidity** on a given surface; an **isopleth** of humidity.

The humidity measures used may be the **relative humidity,** or the actual **moisture content** (**specific humidity** or **mixing ratio**). *See* **isodrosotherm.**

isohyet—A line drawn through geographical points recording equal amounts of **precipitation** during a given time period or for a particular **storm.**

isohyetal map—Map depicting **contours** of equal **precipitation** amounts recorded during a specific time period.

isohypse—Same as **contour line.**

isohypsic—(Rare.) Of equal or constant height; equivalent to "level."

isohypsic chart—Same as **constant-height chart.**

isohypsic surface—Same as **constant-height surface**.

isokeraunic—Same as **isoceraunic**.

isokinetic—British term for **isotach**.

isokinetic sampling—Any technique for collecting airborne particulate matter in which the **collector** is so designed that the airstream entering it has a **velocity** equal to that of the air passing around and outside the collector.

The advantage of isokinetic sampling consists in its freedom from the uncertainties due to selective collection of only the larger, less easily deflected **particulates**. In principle, an isokinetic sampling device has a **collection efficiency** of unity for all sizes of particulates in the sampled air.

isolated system—A thermodynamic system that does not interact in any way with its surroundings.

The **internal energy** and mass of such a system are conserved.

isoline—Same as **isogram**; *see also* **isopleth**.

isomer—A line on a map along which an equal percentage of the total annual **precipitation** falls in a given **season** or month; literally means "equal parts."

isoneph—A line drawn through all points on a map having the same amount of **cloudiness**.

See **nephanalysis**, **nephcurve**.

isonival—(Also called isochion.) Refers to a line of equal **water equivalent of snow** or equal **snow depth**.

isophane—(Rare.) A line drawn through geographical points where a given seasonal biological event occurs on the same date.

isophene—Same as **isophane**.

isopleth—1. In common meteorological usage, a line of equal or constant value of a given quantity, with respect to either space or time; same as **isogram**.

2. (Also called isarithm.) A line drawn through points on a graph at which a given quantity has the same numerical value (or occurs with the same **frequency**) as a function of the two coordinate variables.

Compare **isotimic line**.

3. A straight line along which lie corresponding values of a **dependent** and **independent** **variable**.

isopluvial—A line drawn through geographical points having the same **pluvial index**.

Compare **isomer**.

isoprene—Organic compound, formula C_5H_8; 2–methyl–1,3–butadiene; it is the major hydrocarbon emitted by vegetation, particularly members of the oak genus.

Emission rates depend on the leaf **temperature** and availability of **sunlight**. Isoprene is very reactive with the **hydroxyl radical** and with **ozone**, and its reactivity is thought to contribute to high levels of ozone found over the rural southeastern United States and other areas.

isopycnal—(Or **isopycnic line**, **isopycnic**.) A curve connecting points of equal **density**.

isopycnal mixing—Mixing that occurs along a neutral surface.

Since **particles** can be stirred along **neutral surfaces** with minimal changes in **potential energy**, the majority of horizontal **mixing** in the oceanic interior is **isopycnal** in orientation.

isopycnal surface—Surfaces of constant **density**.

Generally, an isopycnal surface is taken to be a surface of constant **potential density**, so as to compensate for changes of fluid density caused by varying **pressure**.

isopycnic—1. Of equal or constant **density**, with respect to either space or time; equivalent to **isosteric**.

2. An **isopleth** of **density**, equivalent to an **isostere**.

isopycnic line—(Also called, simply, isopycnic.) A line of equal or constant **density**.

It is equivalent to an **isostere**.

isopycnic surface—A surface of equal **density**.

isoshear—A line of equal magnitude of **vertical wind shear**.

isostatic adjustment—The movement of the solid part of the earth until it is in balance; also called isostatic compensation.

The prime example of isostatic adjustment is the continents "floating" on the denser parts of the crust.

isostere—A line of equal or constant **specific volume**.

It is equivalent to an **isopycnic line**.

isosteric—Of equal or constant **specific volume** with respect to either time or space; equivalent to **isopycnic**.

Compare **isochoric**.

isosteric surface—A surface of equal **specific volume**.

isotach—(Also called isovel and, in the United Kingdom, isokinetic.) A line on a given surface connecting points with equal **wind speed**.

Compare **isogon**.

isotach analysis—An **analysis** showing the distribution of **wind speed** by means of **isotachs**, contours of constant wind speed.

isotach chart—A **synoptic chart** showing the distribution of **wind speed** by means of **isotachs**.

isotherm—A line of equal or constant **temperature**.

A distinction is made, infrequently, between a line representing equal temperature in space, **choroisotherm**, and one representing constant temperature in time, **chronoisotherm**. *See* **thermoisopleth**.

isotherm ribbon—A zone of crowded isotherms on a **synoptic** upper-level **chart**.

The **temperature** gradient is many times greater than normally encountered in the **atmosphere**.

isothermal—Of equal or constant **temperature**, with respect to either space or time.

isothermal atmosphere—(Also called exponential atmosphere.) An idealized **atmosphere** in **hydrostatic equilibrium** in which the **temperature** is constant with height and in which, therefore, the **pressure** decreases exponentially upward.

In such an atmosphere the **thickness** between any two heights is given by

$$z_B - z_A = \frac{R_d T_v}{g} \ln \frac{P_A}{P_B},$$

where R_d is the **gas constant** for **dry air**, T_v the **virtual temperature** (°K), g the **acceleration of gravity**, and P_A and P_B the pressures at the heights z_A and z_B, respectively. In the isothermal atmosphere there is no finite height at which the pressure vanishes.

isothermal equilibrium—(Also called conductive equilibrium.) The state of a hypothetical **atmosphere**, at rest and uninfluenced by **radiative heating** or cooling, in which the **conduction** of **heat** from one part to another has, after a sufficient length of time, produced a uniform **temperature** throughout its entire mass.

If such an atmosphere consisted of more than one gas, the **pressure** of each gas would be distributed exponentially according to **Dalton's law**, so that

$$p_n = p_{no}^{\ -(m_n g/kT)^h},$$

where p_{no} is the **surface pressure** and the nth constituent gas, m_n its mean molecular mass, g the **acceleration of gravity**, h the geometric height, k **Boltzmann's constant** (1.3804×10^{-23} J K^{-1}), and T the **absolute** temperature. At a sufficiently great height the lighter gases will predominate. The time necessary for the establishment of isothermal equilibrium in a mixed atmosphere by **diffusion** of one gas through another has been estimated to decrease rapidly with height from about one year near 100 km to a matter of seconds near 200 km. *See* **diffusive equilibrium**, **isothermal atmosphere**.

Lettau, H., 1951: *Compendium of Meteorology,* 320–333.

isothermal process—Any thermodynamic **change of state** of a system that takes place at constant **temperature**.

isotimic—Pertaining to a quantity that has equal value in space at a particular time.

isotimic line—On a given reference surface in space, a line connecting points of equal value of some quantity.

Most of the lines drawn in the **analysis** of **synoptic** charts are **isotimic** lines. *Compare* **isopleth**.

isotimic surface—A surface in space on which the value of a given quantity is everywhere equal.
   Isotimic surfaces are the common reference surfaces for **synoptic charts**, principally **constant-pressure surfaces** and **constant-height surfaces**.

isotope—Chemical term describing alternative forms of the same **element**, differing by the number of **neutrons** in the **nucleus**.
   Isotopes of the same element only differ in their masses and usually have very similar chemical properties. Small differences in chemical reaction rates may be associated with the differing masses—the kinetic isotope effect; such effects are usually of the order of 1% or less but can be as large as a factor of 7 for transfer of **hydrogen**/deuterium.

isotopic tracer—An **isotope** of an **element** that can be used as a **tracer**.

isotropic—Having the same properties in all directions.
   Obtained by combining the Greek iso, meaning alike or same, and tropos, meaning turning. Its antonym is **anisotropic**.

isotropic radiation—**Radiation** with constant **radiance** in all directions.

isotropic radiator—An emitter of constant **radiance** with angle.
   Isotropic radiators obey **Lambert's law**.

isotropic target—In **radar** or **lidar**, a **target** that scatters the same **intensity** of **radiation** in all directions.
   Specifically, for plane-wave incident radiation, an isotropic target scatters in all directions the same **power** per unit solid angle. Real targets are generally not **isotropic**, but **scatter** different intensities in different directions. The concept of an isotropic target is an artifice that allows the **scattering** properties of a real target to be described in terms of those of an equivalent isotropic target. Thus, the **backscattering cross section** of a target is the cross-sectional area of an isotropic target that scatters the same intensity in the direction of the radar receiver as the real target. *See* **phase function**.

isotropic turbulence—**Turbulence** in which the products and squares of the **velocity** components and their derivatives are independent of direction, or, more precisely, invariant with respect to rotation and reflection of the coordinate axes in a **coordinate system** moving with the mean motion of the fluid.
   Then all the normal stresses are equal and the **tangential stresses** are zero. Atmospheric turbulence is generally nonisotropic, although isotropic turbulence is that most easily produced in **wind** tunnel experiments and forms the basis of much of the theoretical analysis of **turbulent flow**. A related but less restricted type of **turbulence** is known as **homologous turbulence**, in which the fluctuations differ only in **scale** at every point in the flow. *See* **stress tensor, Reynolds stresses**.

isovel—Same as **isotach**.

ITcal—Abbreviation for **International Table calorie**.

ITCZ—Abbreviation for **intertropical convergence zone**.

ITOS—Abbreviation for **Improved TIROS Operational System**.

ivory point—A small pointer extending downward from the top of the **cistern** of a **Fortin barometer**.
   The level of the **mercury** in the cistern is adjusted so that it just comes in contact with the end of the pointer, thus setting the zero of the **barometer** scale.

J

J process—(Sometimes called Junction process.) The slowly propagating **electrical breakdown** process in a **cloud** carrying net charge; it can initiate a **K process**.
See **K changes**.

j-value—(Also known as photolysis rate.) A first-order **rate coefficient** (in units of inverse time) for the occurrence of a **photochemical reaction**.
The *j*-value is calculated from the product of the **absorption cross section** of the molecule being photolyzed, the quantum yield for the process, and the **actinic flux**, all integrated over the **wavelength** region of interest. See **actinometer**.

J–W meter—See **Johnson–Williams liquid water probe**.

Jacob's ladder—See **crepuscular rays**.

Jacobian—The determinant formed by the n^2 partial derivatives of n functions of n variables, when the derivatives of each function occupy one row of the determinant.
For the case of two functions $f(x, y)$ and $g(x, y)$, the Jacobian $J(f, g)$ is

$$J(f, g) = \begin{vmatrix} \dfrac{\partial f}{\partial x} & \dfrac{\partial f}{\partial y} \\ \dfrac{\partial g}{\partial x} & \dfrac{\partial g}{\partial y} \end{vmatrix},$$

sometimes also written

$$J\left(\frac{f, g}{x, y}\right) \text{ or } \frac{\partial(f, g)}{\partial(x, y)}.$$

The **geostrophic advection** of any **scalar** ψ may be written

$$(g/f)J(\psi, z),$$

where g is the **acceleration of gravity**, f the **Coriolis parameter**, and z the **isobaric** contour height.

January thaw—In the United States, a period of mild weather popularly supposed to recur each year, in later January; most pronounced in the Northeast and, to a lesser extent, the Midwest.
The daily **temperature** averages at Boston, computed for the years 1873 to 1952, show a well-marked peak on 20–23 January; the same peak occurs in the daily temperatures of Washington, D.C., and New York City. Statistical tests show a high probability that it is a real **singularity**. The January thaw is associated with the frequent occurrence on the above-mentioned dates of southerly winds on the back side of an **anticyclone** off the southeastern United States.

Japan Current—Same as **Kuroshio**.

Japan Sea Deep Water—(Also called Japan Sea Proper Water.) A **water mass** found in the Japan Sea below a depth of 200 m (84% of the volume of the Japan Sea).
Being isolated from all other oceans, it has very uniform **salinity** (34.1) and **temperature** (0°–1°C). See **deep water**.

Japan Sea Middle Water—A **water mass** found at a depth from 25 to 200 m in the Japan Sea in which the **temperature** drops from 17° to 2°C.
It is well ventilated with a high oxygen content of 8 ml l^{-1}.

Japan Sea Proper Water—See **Japan Sea Deep Water**.

jauk—(Also spelled jauch.) A local name for the **foehn** in the Klagenfurt basin of Austria.
It may come from the south but is developed as a **north foehn**.

jet—In meteorology, a common abbreviation for **jet stream**.

jet axis—Same as **jet stream axis**.

jet-effect wind—A local **wind** created by **acceleration** of the airflow through a gap, constriction, or channel in a mountain range or between ranges.
The acceleration can result from a large-scale **pressure gradient**, or by Venturi acceleration through a constricting passage. Pressure gradients from large-scale processes can occur when a large-scale **anticyclone** lies on one side of the barrier, as in the case of canyon or **Wasatch winds**, or

when a **cold front** impinges on a mountain barrier with a gap in it and the cold **air mass** forces its way through the gap, as in the case of the **tehuantepecer**. Other jet-effect winds include the **düsenwind**, the **kossava**, and **gap winds**. *See* also **mountain-gap wind**.

jet streak—Same as **jet stream core**.

jet stream—Relatively strong winds concentrated within a narrow stream in the **atmosphere**.
   While this term may be applied to any such stream regardless of direction (including vertical), it is coming more and more to mean only a quasi-horizontal jet stream of maximum winds embedded in the midlatitude **westerlies**, and concentrated in the high **troposphere**. The question of the maintenance of the jet stream is a cardinal problem of theoretical meteorology. Two such jet streams are sometimes distinguished. The predominant one, the **polar-front jet stream**, is associated with the **polar front** of middle and upper-middle latitudes. Very loosely, it may be said to extend around the hemisphere, but, like the polar front, it is discontinuous and varies greatly from day to day. A **subtropical jet stream** is found, at some longitudes, between 20° and 30° latitude and is strongest off the Asian coast. Currently, in the **analysis** of upper-level charts, a jet stream is indicated wherever it is reliably determined that the **wind speed** equals or exceeds 50 knots. *See* **thermal jet**.

jet stream axis—(Sometimes called jet axis.) The axis of maximum wind speed in a **jet stream**.

jet stream cirrus—**Cirrus** specifically associated with vertical motion with respect to the **jet stream**, which gives a visual indication of the existence and direction of the jet stream across the sky.

jet stream core—(Sometimes called **jet streak**.) The region of a **jet stream axis** with the greatest winds.

jetlet—Sometimes used for "small" **jet stream**.
   The term very likely originated to describe relatively small regions of maximum wind, especially those analyzed as jet streams because the maximum wind falls within an arbitrary range of jet-stream speed.

Jevons effect—The effect of the presence of the **rain gauge** on the **rainfall** measurement.
   In 1861, W. S. Jevons pointed out that the rain gauge causes a **disturbance** in airflow past it, which carries past the gauge part of the **rain** that would normally be captured. The effect is a function of the **wind speed** and the height of the gauge from the ground. **Rain-gauge shields** have been devised to minimize this loss.

jimsphere—A spherical **balloon** made of metallized polyester film.
   It uses a valve to produce a fixed overpressure during ascent. The balloon contains surface protuberances to maintain **turbulent flow** throughout the ascent. It is tracked by a precision radar to obtain accurate **wind profile** data.

jochwinde—The **mountain-gap wind** of the Tauern Pass in the Alps.

Johnson–Williams liquid water probe—A hot-wire type instrument for measuring the liquid **water content** of clouds in situ.
   The **probe** is most often used on research aircraft, but also occasionally at mountain-top installations and in **wind** tunnels. Resistivity changes, which occur as **cloud droplets** in the airflow that impinge on, and evaporate from, an exposed electrically heated wire, are sensed by an electric circuit. The liquid **water content** of the air is estimated from this **signal** using compensations for **air temperature** variations detected by a similar unexposed wire in the probe and knowledge of the **airspeed**.

JONSWAP—Acronym for Joint North Sea Wave Project.

JONSWAP spectrum—A standard spectrum representing a fetch-limited sea.
   A development of the **Pierson–Moskowitz spectrum**, its form depends not only on the **wind speed** but also on the **fetch**.

joran—Same as **juran**.

Jordan sunshine recorder—A **sunshine recorder** of the type in which the timescale is supplied by the motion of the sun.
   It consists of two opaque metal semicylinders mounted with their curved surfaces facing each other. Each of the semicylinders has a short narrow slit in its flat side. Sunlight entering one of the slits falls on light-sensitive paper (blueprint paper) that lines the curved side of the semicylinder. One semicylinder covers morning hours, the other afternoon hours. The sensitivity of the recording paper is variable, and this introduces an uncertainty in the evaluation of the record.

joule—A unit of **energy** in the **mks system** equal to 10^7 ergs, 1 Watt second, or 0.2389 calories.

Joule's constant—(Also called mechanical equivalent of heat.) The amount of mechanical **work** necessary to raise the **temperature** of a given mass of water by 1 **degree** Celsius.

In modern usage it is the **heat capacity** of water, which at room temperature is 4186 Joules kg^{-1}.

jump—1. A **discontinuity** in a function or a derivative of a function such that it assumes different values at a point when the point is approached from different directions.

2. *See* **pressure jump**.

jump model—An idealization for the **atmospheric boundary layer** that assumes uniform (well mixed) values with height within the **boundary layer**, capped by a **discontinuity** (or jump) at the top of the boundary layer.

Another name for this approach is a **slab model** because the uniform part of the boundary layer behaves in a similar way to a uniform slab of material. Such jump or step models are reasonable simplifications for the **convective boundary layer** when vigorous thermals tend to keep the boundary layer well mixed, but are poor idealizations for **statically stable** and neutral boundary layers.

junction streamer—The process by which negative charge centers at successively more distant locations in a **thundercloud** are "tapped" for **discharge** by successive strokes of **cloud-to-ground lightning**.

Junge aerosol layer—A maximum in large-**particle** concentrations observed in the lower **stratosphere** between 15 and 25 km.

junk wind—A south or southeast **monsoon** wind, favorable for the sailing of junks.

The **wind** is known in Thailand, China, and Japan.

junta—A **wind** blowing through Andes Mountain passes, sometimes reaching **hurricane force**.

juran—(Also spelled joran.) A **wind** blowing from the Jura Mountains in Switzerland from the northwest toward Lake Geneva.

It is cold and snowy and may be very turbulent, especially in spring.

jury problem—A differential equation solved numerically by a **method of successive approximations** that fits the solution to given **boundary conditions**.

Elliptic equations, such as the **Poisson equation**, lead to jury problems. A partial differential equation, such as the **barotropic vorticity equation**, may combine a jury problem and a **marching problem**.

juvenile water—Same as **magmatic water**.

K

k–ε closure—A type of one-and-a-half-order **turbulence closure** that retains forecast equations for mean (first-order **statistics**) variables such as **potential temperature** and **wind** components, and also retains equations for **variances** (**turbulence kinetic energy** and **temperature** variance, symbolized by k) and for **molecular dissipation** or destruction of variances (symbolized by ϵ).

Compare **first-order closure, K-theory, second-order closure, nonlocal closure, Reynolds averaging, closure assumptions.**

K band—*See* **radar frequency bands.**

K changes—The **K process** is generally viewed as a recoil streamer or small **return stroke** that occurs when a propagating **discharge** within the **cloud** encounters a pocket of charge opposite to its own.

In this view, the **J process** represents a slowly propagating discharge that initiates the **K process**. This is the case for K changes in cloud discharges. It is reasonable to expect that **cloud discharge** K changes are similar to the in-cloud portion of **ground discharges.**

K-index—*See* **stability index.**

K index—(Also called George's index.) A **stability index** that is a measure of **thunderstorm** potential based on **temperature lapse rate, moisture content** of the lower **troposphere**, and the vertical extent of the moist layer.

The K index is determined by the following equation:

$$K = (850 \text{ mb } T - 500 \text{ mb } T) - 850 \text{ mb } Td - (700 \text{ mb } T - 700 \text{ mb } Td),$$

where T is the **temperature** and Td is the **dewpoint** in degrees Celsius at the **pressure** levels indicated. The higher (positive) the K index, the greater the likelihood of thunderstorm development.

K process—The return propagation of charge along a channel in a **cloud** in response to the initial **breakdown** streamer contacting a region of opposite charge.

In this context it can be viewed as a small **return stroke** in the cloud. *See* **K changes.**

K-theory—*See* **gradient transport theory, eddy viscosity.**

K theory—(Also called mixing-length theory.) A method of describing the movement of **trace** species on the turbulent or subgrid **scale**.

The theory relates the fluxes of the trace species to the **gradient** of the mean quantities via the **eddy diffusivity** (denoted K). *See also* **turbulent flux.**

kaavie—In Scotland, a heavy fall of **snow.**

kachchan—A hot, dry, west or southwest **wind** of **foehn** type in the lee of the Sri Lanka hills during the **southwest monsoon** in June and July.

It is well developed at Batticaloa on the east coast, where it is strong enough to overcome the **sea breeze** and bring maximum temperatures of nearly 38°C.

Kaikias—The Greek word for the cold northeast **wind**, bringing **thunder** and **hail** in winter.

On the **Tower of the Winds** at Athens it is represented by an old man, well clad but with bare arms, tipping hailstones from a **shield** onto the country below.

kal Baisakhi—In India, a short-lived dusty **squall** at the onset of the **southwest monsoon** (April–June) in Bengal.

It is attributed by Bn. Banerji (1938) to a cool dry upper current from the north or northwest meeting a shallow **surface flow** of warm air from the Bay of Bengal along a **quasi-stationary front**. It is similar to the **shamal**. The **front** advances northwestward as the **season** progresses, and similar phenomena termed **loo** or **uala-andhi** extend into the United Provinces and Punjab.

Banerji, Bn., 1938: Nature of 'Nor'westers' of Bengal and their similarity with others. *Beitr. Physik fr. Atmos,* **24**, 231–233.

kaléma—A very heavy **surf** breaking on the Guinea coast of Africa during the winter.

Kalman–Bucy filter—(Or, simply, Kalman filter.) A **four-dimensional data assimilation** method that provides an estimate of the **model** state by evolving explicitly the **error** covariance of the state estimate.

Variants of the Kalman **filter** algorithm are now being applied to atmospheric **data assimilation** problems. The filter estimate is based on all data observed up to and including the current time. Generalizations of the Kalman filter exist for continuum dynamics, for **nonlinear** stochastic systems (e.g., extended or ensemble Kalman filters), for systems that have different types of **noise**, for unknown noise statistics, and for observations beyond the current time (Kalman smoothers).

Kalman, R. E., 1960: A new approach to linear filtering and prediction problems. *Trans. ASME, Ser. D, J. Basic Eng*, **82**, 35–45.

Cohn, S. E., 1997: An introduction to estimation theory. *J. Meteor. Soc. Japan*, **75**, 257–288.

Kamchatka Current—(Also known as the East Kamchatka Current.) The western part of the subpolar **gyre** in the deep (western) part of the Bering Sea.

It flows southward along the Kamchatka Peninsula with speeds of 0.2–0.3 m s⁻¹. Most of its water leaves the Bering Sea and forms one of the two sources of the **Oyashio**.

kamsin—Same as **khamsin**.

kanat—(Also ghanat, ganat.) A mildly sloping tunnel, with the upper end below the **water table** of an **alluvial aquifer**, that is used to collect and transmit water to a lower-elevation surface outlet.

karaburan—(Also called black storm, black buran.) A violent northeast **wind** of Central Asia occurring during spring and summer.

It resembles the white **buran** of winter but, instead of **snow**, it carries clouds of **dust** that darken the sky.

karajol—(Also spelled qarajel, quara.) On the Bulgarian coast, a west **wind** that usually follows **rain** and persists for one to three days.

karema wind—A violent east **wind** on Lake Tanganyika in Africa.

karif—(Also spelled kharif.) A strong southwest **wind** on the southern shore of the Gulf of Aden, especially at Berbera, Somaliland, during the **southwest monsoon**.

It sets in at about 10 P.M. **local time** and continues until about noon on the following day, reaching its greatest strength, averaging 11 m s⁻¹ (25 mph) in July and August, about 8 A.M. It resembles a **land breeze**, but on a much larger **scale** than the **normal** land breeze, owing to the rugged **topography** and consequent great depth of the layer of cooled air.

Kármán constant—*See* **logarithmic velocity profile**.

Kármán vortex street—*See* **vortex street**.

karst—Terrain composed of and underlain by carbonate rocks (usually limestone) that have been significantly altered by dissolution, creating caves and sinkholes.

karst hydrology—The branch of **hydrology** devoted to the study of the movement and occurrence of water in **karst**.

karstbora—The **bora** of the Yugoslavian coast.

katabaric—Same as **katallobaric**.

katabatic front—Frontal surface above which air is descending.

katabatic wind—1. Most widely used in mountain meteorology to denote a downslope flow driven by cooling at the slope surface during periods of light larger-scale winds; the nocturnal component of the **along-slope wind systems**.

The surface cools a vertical column of the **atmosphere** starting at the slope surface and reaching perhaps 10–100 m deep. This column is colder than the column at equivalent levels over the valley or plain, resulting in a **hydrostatic pressure** excess over the slope relative to over the valley or plain. The horizontal **pressure gradient**, maximized at the slope surface, drives an **acceleration** directed away from the slope, or downslope. Although the pressure-gradient forcing is at its maximum at the slope, **surface friction** causes the peak in the katabatic wind speeds to occur above the surface, usually by a few meters to a few tens of meters. The depth of the downslope flow layer on simple slopes has been found to be 0.05 times the vertical drop from the top of the slope. Surface-wind speeds in mountain–valley katabatic flows are often 3–4 m s⁻¹, but on long slopes, they have been found to exceed 8 m s⁻¹. Slopes occur on many scales, and consequently katabatic flows also occur on many scales. At local scales katabatic winds are a component of **mountain–valley wind systems**. At scales ranging from the slopes of individual hills and mountains to the slopes of mountain ranges and massifs, katabatic flows represent the nocturnal component of **mountain–plains wind systems**. Besides diurnal-cycle effects, surface cooling can also result from cold surfaces such as **ice** and **snow cover**. Katabatic flows over such surfaces have been studied as **glacier winds** in valleys and as large-scale **slope flows** in Antarctica and Greenland. The large-scale katabatic wind blowing down the **ice** dome of the Antarctic continent has sometimes reached 50 m s⁻¹ on the periphery of the continent. The **persistence** of the surface forcing and the great extent of the slopes on these great landmasses means that the flows are subject to Coriolis deflection, and thus they are not pure katabatic flows. *See* **downslope wind, gravity wind, drainage wind**.

2. Occasionally used in a more general sense to describe cold air flowing down a slope or incline

on any of a variety of scales, including phenomena such as the **bora**, in addition to thermally forced flows as described above.

From its etymology, the term means simply "going down" or "descending," and thus could refer to any descending flow; some authors have further generalized it to include downslope flows such as the **foehn** or **chinook** even though they do not represent a flow of cold air. This concept has given rise to the expression **katafront**, which indicates flow down a sloped cold-frontal surface.

katafront—A **front** (usually a **cold front**) at which the warm air descends the **frontal surface** (except, presumably, in the lowest layers).
Compare **anafront**.

katallobar—*See* **isallobar**.

katallobaric—(Also called katabaric.) Of, or pertaining to, a decrease in **atmospheric pressure**.
See **allobaric**, **isallobaric**.

katallobaric center—Same as **pressure-fall center**.

katathermometer—A type of **cooling-power anemometer** based upon the principle that the **time constant** of a **thermometer** is a function of its **ventilation**.

The form developed in the early nineteenth century consisted of a **liquid-in-glass thermometer** having two **calibration** markers on the stem corresponding to 38.5° and 35°C. The thermometer was heated to 40°C, and the time required for the column to fall from 38° to 35°C was measured by a stopwatch and used to compute the **wind speed**. It was especially useful for very low wind speeds. The katathermometer was used also, in **human bioclimatology**, to determine **cooling power**.

kaus—(Also spelled quas; also called cowshee, sharki.) A moderate to gale-force southeasterly **wind** in the Persian Gulf; it is accompanied by gloomy weather, **rain**, and **squalls**.

The kaus is most frequent between December and April. It is associated with the passage of a winter **depression**, and is often followed by a strong southwesterly wind, the **suahili**.

kavaburd—Same as **cavaburd**.

kaver—Same as **caver**.

keel—*See* **ice keel**.

kelsher—In England, a heavy fall of **rain**.

Kelvin–Helmholtz billows—1. Cloud forms that arise from **Kelvin–Helmholtz waves**.

2. Vortical structures that result from the growth and **nonlinear** development of unstable waves in a **shear flow**.

The billows get their name from the **instability** responsible for the growth of the unstable waves, **Kelvin–Helmholtz instability**.

Kelvin–Helmholtz instability—An **instability** of the **basic flow** of an incompressible **inviscid fluid** in two parallel infinite streams of different velocities and densities.

If the overlying fluid has **velocity** U_2 and **density** ρ_2, and the underlying fluid has velocity U_1 and density ρ_1, disturbances of the form e^{ikx} (where k is the **wavenumber**) are unstable if

$$g(\rho_1^2 - \rho_2^2) < k\rho_1\rho_2(U_1 - U_2)^2,$$

where g is the **acceleration of gravity**. Thus, the flow is always unstable to short waves (high wavenumber) if $U_1 \neq U_2$.

Drazin, P. G., and W. H. Reid, 1981: *Hydrodynamic Stability*, Cambridge University Press, 14–22.

Kelvin–Helmholtz wave—A **waveform** disturbance that arises from **Kelvin–Helmholtz instability**.

Kelvin's circulation theorem—*See* **circulation theorem**.

Kelvin temperature scale—(Abbreviated K; also called absolute temperature scale, thermodynamic temperature scale). An absolute temperature scale independent of the thermometric properties of the working substance.

On this **scale**, the difference between two temperatures T_1 and T_2 is proportional to the **heat** converted into mechanical **work** by a **Carnot engine** operating between the **isotherms** and **adiabats** through T_1 and T_2. A **gas thermometer** utilizing a **perfect gas** has the same **temperature scale**. For convenience the Kelvin **degree** is identified with the centigrade degree. The **ice point** in the Kelvin scale is 273.16 K. *See* **absolute zero**; *see also* **centigrade temperature scale**.

Kelvin wave—A type of low-frequency **gravity wave** trapped to a vertical boundary, or the **equator**, which propagates anticlockwise (in the Northern Hemisphere) around a basin.

The flow is parallel to the boundary and in **geostrophic balance** with the **pressure gradient** perpendicular to the boundary. The **velocity** normal to the boundary is identically zero. For a homogeneous ocean, the wave is called a **barotropic** or external Kelvin wave, and for a stratified ocean, the wave is called a **baroclinic** or internal Kelvin wave. Near a boundary in a rotating system, a Kelvin wave propagates with **wave crests** perpendicular to the side wall and **wave height** greatest at the side wall to the right of an observer looking in the direction of **wave propagation**. The wave height decreases exponentially from the side wall with an e-folding length **scale** equal to the **Rossby radius of deformation** c/f, in which f is the **Coriolis parameter** and c is the **phase speed** of the wave in the along-boundary direction. In the **shallow water approximation** the waves are nondispersive with **frequency**

$$w = \pm ck,$$

in which k is the along-boundary **wavenumber** and the phase speed

$$c = (gH)^{1/2},$$

with g the **acceleration of gravity** and H the mean fluid depth. Related to Kelvin waves in a channel are **Poincare waves**.

keraunograph—Same as **ceraunograph**.

Kern counter—Same as **dust counter**.

kernel ice—In **aircraft icing**, an extreme form of **rime ice**, that is, very irregular, opaque, and of low density.
> Kernel ice forms at temperatures of $-15°C$ and lower.

ketones—Family of organic **carbonyl compounds** of general formula RC(O)R, where the R represents alkyl groups (not necessarily the same).
> Ketones are formed in the **oxidation** of most of the larger **hydrocarbons**. Their main atmospheric fate is reaction with hydroxyl radicals or **photolysis** in the near **UV**.

Kew barometer—A type of **cistern barometer**.
> No adjustment is made for the **variation** of the level of the **mercury** in the **cistern** as **pressure** changes occur; rather, a uniformly contracting **scale** is used to determine the effective height of the **mercury column**.

key day—Same as **control day**.

khamasseen—Same as **khamsin**.

khamsin—(Also spelled camsin, chamsin, kamsin, khamasseen, khemsin.) A dry, dusty, and generally hot **desert wind** in Egypt and over the Red Sea.
> It is generally southerly or southeasterly, occurring in front of **depressions** moving eastward across North Africa or the southeastern Mediterranean. The deep khamsins occur in spring with depressions traveling east-northeast across the northern Sahara. They are preceded by a **heat wave** lasting about three days and are followed by a **duststorm**. The passage of the depression is marked by a **cold front** bringing Mediterranean air and a sudden drop in **temperature**. See **ghibli, chili, sirocco**.

kharif—Same as **karif**.

khemsin—Same as **khamsin**.

kibli—Same as **ghibli**.

killing freeze—The occurrence of **air temperature** below $0°C$ ($32°F$) that kills annual vegetation without formation of **frost** crystals on surfaces.
> See **freeze, dry freeze, hard freeze, light freeze**.

killing frost—See **frost**.

kilocalorie—Same as **kilogram calorie**.

kilocycle—(Obsolete.) See **kilohertz**.

kilogram calorie—(Abbreviated Kcal, kg-cal, Cal; also called large calorie.) One thousand **calories**.

kilohertz—(Abbreviated kHz; formerly called kilocycle.) A unit of **frequency** equal to one thousand (10^3) cycles per second.
> It is most commonly used in connection with **radio wave** frequencies.

kilojoule—(Abbreviated kj.) A unit of **energy** in the **International System of Units** (SI).

One kilojoule is equal to 10^3 joules, or to 10^{10} ergs, or to 238.9 calories.

kilomegacycle—(Obsolete.) See **gigahertz**.

kilomole—One thousand **gram-moles** of a substance.

kinematic boundary condition—The condition that the fluid **velocity** directed perpendicular to a solid boundary must vanish on the boundary itself.

This may be stated mathematically by the expression

$$\mathbf{n} \cdot \mathbf{u} = 0,$$

where **n** is a unit **vector** normal to a solid surface and **u** is the fluid velocity vector. In meteorology, this boundary condition is often employed in considering flow near the earth's surface. When the boundary is a fluid surface or **interface**, this condition applies to the vector difference of velocities across the interface and requires that the interface, although in motion, will at all times consist of the same fluid parcels. In meteorology, such a condition must be applied at fronts and other surfaces of **discontinuity**. *See also* **dynamic boundary condition**.

kinematic flux—Transport of a **variable** per unit area per unit time (i.e., a **dynamic flux**), but divided by the average air **density** (or in the case of **heat flux**, divided by the product of air density times **specific heat** of air at constant **pressure**).

The resulting kinematic flux has the same units as **velocity** times the variable being transported. For example, a vertical turbulent heat flux is $\overline{w'\theta'}$, while a vertical heat flux associated with mean wind is $\overline{w}\overline{\theta}$, where w is **vertical velocity**, θ is **potential temperature**, an overbar denotes an average, and a prime denotes a **deviation** from the average. Kinematic fluxes are more closely related to meteorological variables that can be easily measured (such as **temperature** and **wind**) than are the associated dynamic fluxes (such as J m^{-2}s^{-1}). Statistically, kinematic fluxes are **covariances**.

kinematic viscosity—A coefficient defined as the ratio of the **dynamic viscosity** of a fluid to its **density**.

The kinematic viscosity of most gases increases with increasing **temperature** and decreasing **pressure**. For **dry air** at 0°C, the kinematic viscosity is about 1.46×10^{-5} m^2 s^{-1}. Common symbols for these variables are μ for dynamic viscosity and ν for kinematic viscosity.

List, R. J., 1951: *Smithsonian Meteorological Tables*, 6th rev. ed., 394–395.

kinematical analysis—Analysis of the **field** of atmospheric flow.

kinematics—The branch of dynamics that describes the properties of pure motion without regard to force, **momentum**, or **energy**.

Translation, **advection**, **vorticity**, and **deformation** are examples of kinematic variables.

kinetic energy—The **energy** that a body possesses as a consequence of its motion, defined as one-half the product of its mass and the square of its speed, $(1/2)mv^2$.

The kinetic energy per unit volume of a **fluid parcel** is thus $(1/2)\rho v^2$, where ρ is the **density** and v the speed of the parcel. See **potential energy**.

kinetic energy equation—*See* **energy equation**.

kinetic theory—The theory that bulk matter, to outward appearances motionless, is composed of huge numbers of atoms and molecules in rapid and incessant motion.

By applying the kinetic theory of gases, relations between some of the bulk properties of gases (e.g., the **ideal gas** law) may be derived and given a molecular interpretation.

Kirchhoff's equation—A relation that equates the **variation** with **temperature** of the **latent heat** of a **phase change** to the difference between the **specific heats** of the two phases.

For the **heat of vaporization** this may be written as

$$\left(\frac{\partial L_v}{\partial T}\right)_p = c_{pv} - c_w,$$

where L_v is the **latent heat** of **vaporization**, c_{pv} is the specific heat at constant **pressure** of **water vapor**, and c_w is the specific heat of liquid water.

Kirchhoff's law—A fundamental **radiation** law that equates the **absorptivity** of matter to its **emissivity** at the same **wavelength**.

Loosely put, this important law asserts that good absorbers of radiation at a given wavelength are also good emitters at that wavelength. For Kirchhoff's law to hold, the matter must be in **local thermodynamic equilibrium**, but the law is otherwise quite general, and applies to both natural and idealized surfaces or volumes. For natural surfaces it is often necessary to make the absorptivities and emissivities functions of direction and **polarization** state before applying the law.

Kirchhoff vortex—An idealized **vortex** in an unbounded fluid with uniform **vorticity** inside an elliptical patch and zero vorticity outside.

For an ellipse with semiaxes a and b and vorticity ω in its interior, it rotates steadily with **angular velocity** $\omega[ab/(a + b)^2]$.

kite observation—An **upper-air observation** by means of instruments carried aloft by a kite.
See **kytoon**.

kite sounding—Soundings by means of kites, sometimes used for measuring certain weather elements in the **lower atmosphere**.

kloof wind—A cold southwest **wind** of Simons Bay, South Africa.

klydonograph—A device attached to electric power lines for estimating certain electrical characteristics of **lightning** by means of the figures produced on photographic film by the lightning-produced **surge** carried over the lines.

The size of the figure is a function of the **potential** and **polarity** of the **lightning discharge**.

klystron—A **power** amplifier tube used to amplify weak **microwave** energy (provided by a radio-frequency exciter) to a high power level for a **radar** transmitter.

A klystron is characterized by high power, large size, high stability, high gain, and high operating voltages. Electrons are formed into a **beam** that is **velocity** modulated by the **input** waveform to produce microwave energy. A klystron is sometimes referred to as a **linear** beam tube because the direction of the **electric field** that accelerates the **electron beam** coincides with the axis of the **magnetic field**, in contrast to a crossed-field tube such as a **magnetron**. Klystrons provide a coherent transmitted **signal** appropriate for **Doppler radar** and pulse-compression applications. They are used in many operational radars, for example, **NEXRAD** (Next Generation Weather Radar) and TDWR (**Terminal Doppler Weather Radar**). *Compare* **magnetron**.

knik wind—Local name for a strong southeast **wind** in the vicinity of Palmer in the Matanuska Valley of Alaska.

The knik wind blows most frequently in the winter, although it may occur at any time of year. In winter the knik winds are accompanied by very pronounced **temperature** rises; cases of more than 10°C in 24 hours have been observed. These winds may last from one to ten days. They result from a **pressure gradient** normal to the Chugach Mountains, causing a pronounced **foehn** effect in the Matanuska Valley.

knoll—Same as **earth hummock**.

Knollenberg probe—Popular term for a set of optical **particle** probes for counting, sizing, and/or imaging **aerosols** or **hydrometeors**, named for the inventor, Robert G. Knollenberg.

knot—The unit of speed in the **nautical system**; one **nautical mile** per hour.
It is equal to 1.1508 statute miles (1.852 km) per hour or 1.687 ft (0.5144 m) per second.

knowledge acquisition—The extraction and formulation of knowledge derived from extant sources, especially from experts.

It includes the development of **knowledge bases**, often by interviewing and observing **domain** experts and extracting rules from their behavior or statements.

knowledge base—The facts, relationships, and procedures that constitute the knowledge about a given **domain** or task; the database of an expert (or **knowledge based**) system.

knowledge-based system—An **expert system**.
Those who prefer the term "expert system" emphasize the necessity of encapsulating the human expert's knowledge and way of utilizing that knowledge. Advocates of "knowledge-based systems" emphasize the importance of the knowledge itself, rather than its source or the type of reasoning employed.

knowledge engineering—The discipline that addresses the planning and programming tasks of building, testing, and deploying **knowledge-based systems**.

It includes the development of **expert systems**, including **knowledge acquisition** and knowledge representation.

Knudsen number—Ratio of **mean free path** of air molecules to **particle** radius; a measure of its ability to behave as a molecule or a particle.

Knudsen's tables—Tables published by Martin Knudsen in 1901 ("Hydrographical Tables"), to facilitate the computation of results of seawater **chlorinity** titrations and **hydrometer** readings, and their conversion to **salinity**, **density**, and **sigma-t**.

koembang—A dry **foehn**like wind from southeast or south in Cheribon and Tegal in Indonesia.

It is caused by the east **monsoon** that develops a jet effect in passing through the gaps in the mountain ranges and descends on the **leeward** side.

Köhler equation—A relationship between the radius of a **cloud droplet** grown on a **hygroscopic nucleus** of a given dry mass, which decreases the **vapor pressure** with respect to a flat surface, and the radius of curvature of the **drop**, which increases the vapor pressure with respect to a flat surface and the equilibrium **relative humidity**.

Kolmogorov constant—The proportionality constant α in Kolmogorov theory, which states that the spectral **energy** S in the **inertial subrange** is $S = \alpha \epsilon^{2/3} k^{-5/3}$, for ϵ representing the **viscous dissipation** rate of **turbulence kinetic energy**, and k the **wavenumber** (inversely proportional to the **wavelength** or **eddy** size).

Measurements of the 1D longitudinal **spectrum** of the **wind** in the **planetary boundary layer** show that this constant is equal to about 0.5.

Kolmogorov microscale—Of the three standard **turbulence length scales**, the one that characterizes the smallest dissipation-scale eddies.

As the **turbulence kinetic energy** cascades from the largest scales down to the smallest scales, the dynamics of the small eddies become independent of the **large-scale** eddies. At the smallest scales, the rate at which **energy** is supplied must equal the rate at which it is dissipated by **viscosity**. Thus, parameters available to form length and **velocity** scales are the **dissipation rate**, ϵ, and the **kinematic viscosity**, ν. The Kolmogorov length, η, and velocity, υ, scales are:

$$\eta = \left(\frac{\nu^3}{\epsilon}\right)^{1/4}$$

$$\upsilon = (\nu\epsilon)^{1/4}.$$

Note that the **Reynolds number** formed from the Kolmogorov microscale is equal to one. Based on the observation that the large eddies lose their energy in about one large **eddy** turnover time, the dissipation rate may be scaled as $\epsilon = u^3/L$ where L is the appropriate length **scale**. Thus, the ratio of the largest to the smallest length scales is

$$\left(\frac{uL}{\nu}\right)^{3/4} = R^{3/4},$$

and the number of **grid** points necessary to resolve a **turbulent flow** in a numerical **model** is therefore proportional to $R^{9/4}$. *Compare* **integral length scales, Taylor microscale**; *see also* **isotropic turbulence, local isotropy, Reynolds number, turbulence kinetic energy, turbulence spectrum, viscous fluid**.

Hinze, J. O., 1975: *Turbulence*, 2d ed., McGraw–Hill, 790 pp.
Tennekes, H., and J. L. Lumley, 1972: *A First Course in Turbulence*, MIT Press, 300 pp.

Kolmogorov's similarity hypotheses—(Also called local similarity hypotheses, universal equilibrium hypotheses.) Statements of the factors determining the **transfer** and **dissipation** of **kinetic energy** at the high **wavenumber** end of the **spectrum** of **turbulence**.

Kolmogorov considers the large **anisotropic** eddies as the sources of **energy**, which is transferred down the size **scale**. At some point the **eddies** lose all structure; they become homogeneous and **isotropic**, that is, "similar." In this region, their energy is determined only by the rate of transfer from the larger eddies and the rate of dissipation by the smaller ones. Kolmogorov stated two similarity hypotheses: 1) At large **Reynolds numbers** the local average properties of the small-scale components of any turbulent motion are determined entirely by **kinematic viscosity** and average rate of dissipation per unit mass. 2) There is an upper subrange (the **inertial subrange**) in this **bandwidth** of small eddies in which the local average properties are determined only by the rate of dissipation per unit mass. It is a consequence of these hypotheses that in the inertial subrange the energy is partitioned among the eddies in proportion to $k^{-5/3}$, where k is the **wavenumber**.

Sutton, O. G., 1953: *Micrometeorology*, 100–101.

Kolmogorov scale—Length **scale** of turbulent motion below which the effects of **molecular viscosity** are nonnegligible.

In three-dimensional **turbulence**, the Kolmogorov scale is $(\nu^2/\epsilon)^{1/4}$, in which ν is the **kinematic viscosity** and ϵ is the **energy** dissipation rate per unit mass.

kona—A stormy, rain-bringing **wind** from the southwest or south-southwest in Hawaii.

It blows about five times a year on the southwest slopes that are in the lee of the prevailing northeast **trade winds**. Kona is the Polynesian word for "leeward." It is associated with a southward or a southeastward swing of the **Aleutian low** and the passage of a **secondary depression** (**kona cyclone**) from northwest to southeast, north of the islands.

kona cyclone—(Also called kona storm.) A slow-moving, extensive **cyclone** that forms in subtropical latitudes during the winter season.

See **kona**.

kona storm—Same as **kona cyclone**.

See **kona**.

konimeter—(Also spelled conimeter.) An instrument for determining the **dust** content of a sample of air.

One form of the instrument consists of a tapered metal tube through which a sample of air is drawn and allowed to impinge upon a glass slide covered with a viscous substance. The **particles** caught are counted and measured with the aid of a microscope. *See* **dust counter**.

koniology—(Also spelled coniology.) The scientific study of **atmospheric dust** and other suspended impurities, such as germs and pollen.

koniscope—(Also spelled coniscope.) An instrument that indicates the presence of **dust** particles in the **atmosphere**.

konisphere—Same as **staubosphere**.

Köppen classification—A **climatic classification** scheme developed by Wladimir Köppen (1846–1940), a German climatologist.

This scheme is based upon annual and monthly means of **temperature** and **precipitation** and also takes into account the vegetation limits. It is a tool for presenting the world pattern of **climate** and for identifying important deviations from this pattern.

Köppen, W. P., and R. Geiger, 1930–1939: *Handbuch der Klimatologie*, Berlin: Gebruder Borntraeger, 6 vols.

Köppen–Supan line—The **isotherm** connecting places that have a **mean temperature** of 10°C (50°F) for the warmest month of the year.

It was adopted by Supan (1879) as the climatic **isopleth** that most nearly corresponded to the **arctic tree line**, and later by Köppen (1900) as indicating the division between **tundra climate** and the **tree climates**.

Supan, A., 1879: Die Temperaturzonen der Erde. *Petermanns Geog. Mitt*, **25**, 349–358.

Köppen, W., 1900: Versuch einer Klassification der Klimate vorsugsweise nach ihren Bezichungen zur Pflanzenwelt. *Geograph. Zeirsehr*, **6**, 593–611, 657–679.

kosava—Same as **kossava**.

koschawa—Same as **kossava**.

Koschmieder's law—*See* **airlight formula**.

kossava—(Also spelled kosava, koschawa.) A cold, very squally **wind**, descending from the east or southeast in the region of the Danube "Iron Gate" through the Carpathians, continuing westward over Belgrade, thence spreading northward to the Rumanian and Hungarian borderlands and southward as far as Nish.

In winter it brings temperatures down to below −29°C and it is cool even in summer, when it is also dusty. It usually occurs with a **depression** over the Adriatic and high pressure over southern Russia, a frequent situation in winter. It is usually explained as a **jet-effect wind** through the Iron Gate, giving speeds well above the **gradient wind**, but J. Küttner (1940) regards it rather as a **katabatic wind** intermediate between **foehn** and **bora**. The kossava has a marked **diurnal** variation, with its maximum occurring between 5 A.M. and 10 A.M.

Küttner, J., 1940: Der Kosava in Serbien. *Meteor. Z*, **57**, 120–123.

Kp—The most widely used of all the indices of geomagnetic activity.

It is intended as a measure of the worldwide ("p" is for planetary) average level of activity; however, it is very sensitive to certain auroral-zone activity and insensitive to some other types of disturbances. It is based on an index called the K index from each of 12 stations between latitudes 48° and 63°, selected for good longitudinal coverage. The Handbook of Geophysics and the Space Environment (1985) has a discussion of this and other geomagnetic indices.

Jursa, A. S., Ed., 1985: The Geomagnetic Field. *The Handbook of Geophysics and the Space Environment*, Air Force Geophysics Laboratory, Hanscom AFB, MA, chap. 4.

Krakatoa winds—(Also spelled Krakatau; formerly called overtrades.) A layer of easterly winds over the **Tropics** at an **altitude** of about 18–24 km.

This layer tops the midtropospheric **westerlies** (the **antitrades**), is at least 6 km deep, and is based at about 2 km above the **tropopause**. This easterly current is more prominent and better defined in the summer hemisphere. It derives its name from the observed behavior of the volcanic **dust** carried around the world after the great eruption of Krakatoa (6°S, 105°E) in 1883.

kriging—An **interpolation** procedure used to estimate a **variable** at unsampled locations using weighted sums of the variable at neighboring **sample** points.

The procedure is designed to minimize the **variance** of the estimation errors. As a meteorological example, kriging can be used for two-dimensional spatial interpolation of irregularly spaced observational data onto a uniform set of **grid** points to provide **input** for a **numerical forecasting** model.

krivu—Same as **crivetz**.

kryptoclimate—*See* **cryptoclimate**.

kryptoclimatology—*See* **cryptoclimatology**.

krypton—(Symbol Kr.) An inert **element**, fourth member of the noble gas family, atomic number 36, atomic weight 83.7; an element found in the **atmosphere** to the extent of only 0.000114% by volume.

Krypton hygrometer—A **hygrometer** technically very similar to the **Lyman-alpha hygrometer**, but with the **hydrogen** radiation source replaced by a Krypton glow tube.

The main **emission** is at 123.58 nm, and the main advantages are a longer lifetime of the Krypton tube and a better **calibration** stability compared to the **Lyman-alpha hygrometer**, which makes this instrument more suitable for continuous **humidity** recordings. On the other hand, the **sensitivity** of the Krypton hygrometer is distinctly lower; therefore it is used mainly for ground-based measurements.

Kuo–Eliassen equation—A **diagnostic equation**, based on **thermal wind** balance, with a solution consisting of **meridional cell** circulations that are forced by **eddy fluxes** of **heat** and **momentum** and by diabatic processes.

> Grotjahn, R., 1993: *Global Atmospheric Circulations: Observations and Theories*, 249–264.

Kurihara grid—A specific finite difference **grid** discretization designed to cover the entire globe with nearly uniform spacing.

> Kurihara, Y., and L. Holloway, 1967: Numerical integration of a nine-level global primitive equations model formulated by the box method. *Mon. Wea. Rev.*, **95**, 509–530 (see especially fig. 5).

Kuroshio—(Also called Japan current.) One of the western **boundary currents** of the North Pacific subtropical **gyre**.

A deep, narrow, and swift current, it continues from the **Philippines Current** in a northeastward direction from Taiwan along the continental rise of the East China Sea, through Tokara Strait, and close to the eastern coast of Japan. At 35°N it separates from the coast and flows eastward into the Pacific as a free jet known as the **Kuroshio Extension**. It forms a marked **temperature** and **salinity** front with the **Oyashio**, which meets the Kuroshio Extension from the north and then flows parallel to it. Like all other western boundary currents, the Kuroshio develops instabilities and sheds **eddies**. Its unique characteristic is that south of Honshu it switches between three quasi-stable paths across the Izu Ridge at irregular intervals of 18 months to several years. Volume **transport** in the Kuroshio increases **downstream** and reaches 57 Sv (57×10^6 m^{3}s^{-1}) in the Kuroshio Extension, increasing seasonally by 15% during summer. The current's path in the extension is characterized by large **meridional** excursions in the so-called First and Second Crest at 145° and 152°E. On approaching the Shatsky Rise at 157°E, the Kuroshio Extension divides into several paths that tend to recombine before the Emperor Seamounts near 170°E cause the current to split again and disintegrate. The flow then continues as the **North Pacific Current**.

Kuroshio countercurrent—Part of the **Kuroshio system**.

Between longitudes 155° and 160°E, considerable water turns south and southwest, forming part of the Kuroshio countercurrent. It runs at a distance of approximately 700 km from the coast as the eastern branch of a large whirl on the right-hand side of the **Kuroshio**.

Kuroshio Extension—*See* **Kuroshio**.

Kuroshio system—The western **boundary current** system of the North Pacific subtropical **gyre** and its eastward continuation, consisting of the **Philippines Current**, the **Kuroshio**, and the **Tsushima Current**, the **Kuroshio Extension** and its continuation in the **North Pacific Current**.

kurtosis—(Symbol β_2 or α_4.) A descriptive measure of a **random variable** in terms of the flatness of its **probability distribution**.

It is defined as follows:

$$\beta_2 = \mu_4/\sigma^4,$$

where μ_4 is the fourth (**statistical**) **moment** about the mean and σ^2 the **variance**. For the **normal distribution**, $\beta_2 = 3$; and it is commonly (though not invariably) found that curves for which $\beta_2 > 3$ are more sharply peaked than the normal, while those for which $\beta_2 < 3$ are flatter than the normal. In particular, the rectangular distribution $f(x) = 1$ $(0 < x < 1)$ has $\beta_2 = 1.8$. The terms **leptokurtic, mesokurtic,** and **platykurtic** refer to curves for which the values of β_2 are, respectively, greater than 3, equal to 3, and less than 3. Excess is a relative expression for kurtosis, and the **coefficient of excess** γ_2 is defined as $\beta_2 - 3$.

kytoon—A captive **balloon** used to maintain meteorological equipment aloft at approximately a constant height.

The kytoon is streamlined, and combines the **aerodynamic** properties of a balloon and a kite.
See **wiresonde**.

L

L band—*See* **radar frequency bands**.

La Niña—The most common of several names given to a significant decrease in **sea surface temperature** ("cold events") in the central and eastern equatorial Pacific.

La Niña is the counterpart to the **El Niño** "warm event," and its spatial and temporal evolution in the equatorial Pacific is, to a considerable extent, the mirror image of El Niño, although La Niña events tend to be somewhat less regular in their behavior and duration. *See also* **El Niño–Southern Oscillation**.

> Philander, S. George, 1990: *El Niño, La Niña, and the Southern Oscillation*, Academic Press, International Geophysics Series, Vol. 46.

la serpe—A long strip of **cloud** that sometimes lies against the southern base of Mount Etna in Sicily. It is said to herald **rain**. *See* **contessa di vento**.

labbé—(Also spelled labé.) A moderate to strong southwest **wind** in Provence (southeastern France), mild, humid, and very **cloudy** or rainy.

On the coast it raises a rough sea. It is not frequent, occurring only in March. In the Swiss–French Alps it is locally termed labech, and is squally with **thunder**, **hail**, and brief torrential downpours; it comes mainly in autumn and winter.

labech—*See* **labbé**.

labile—Unstable; literally, characterized by a tendency to slip.

laboratory tank—A device for physical simulation of the **atmosphere**, using water or other liquids as the working fluid.

Classical examples are 1) rotating tanks colloquially called dishpans, to simulate **long waves** in the **general circulation**; 2) **turbulence** tanks to simulate **boundary layer** evolution and convective **thermals**; and 3) towing tanks to simulate **pollutant** dispersion associated with mean flow about an obstacle. Because of tank size limitations and the characteristics of water compared to air, not all atmospheric flows can be simulated, and care must be taken to **scale** or nondimensionalize the results to make them independent of the working fluid. *Compare* **large eddy simulation models**, **direct numerical simulation**.

Labrador Current—The western **boundary current** of the North Atlantic subpolar **gyre**.

The current receives considerable **input** of **Arctic Surface Water** originating from the **East Greenland Current** and supplied through the West Greenland and Baffin Currents. Its mean transport is close to 35 Sv (35×10^6 m³ s⁻¹). The current is strongest in February when it carries 6 Sv more water than in August. It is also more variable in winter, with a **standard deviation** of 9 Sv in February but only 1 Sv in August. Near the Grand Banks it forms the **polar front**, also known as the **cold wall**, with the northward flowing **Gulf Stream**, with which it shares the shedding of **eddies**. The cold water of the Labrador Current allows **icebergs** from western Greenland to travel as far south as 40°N, the latitude of southern Italy.

Labrador Sea Water—The last **water mass** to contribute to the formation of **North Atlantic Deep Water** before it enters the southward path of the **ocean conveyor belt**.

It occupies the central Labrador Sea with temperatures of 3.0°–3.6°C, **salinities** of 34.86–34.96 **psu**, and consistently high oxygen content.

LAC—Abbreviation for **local area coverage**.

lacunaris—Same as **lacunosus**.

lacunosus—(Formerly called lacunaris.) A **cloud variety** characterized more by the appearance of the spaces between the **cloud** elements than by the elements themselves.

The gaps are generally rounded and often have fringed edges. The overall appearance is that of a honeycomb or net, the negative of that of clouds composed of separate rounded elements. This variety is a modification mainly of the genera **cirrocumulus** and **altocumulus** and may apply to the species **stratiformis**, **castellanus**, or **floccus**. *See* **cloud classification**.

Lafond's tables—A set of tables and associated information for correcting **reversing thermometers** and computing **dynamic height** anomalies, compiled by E. C. Lafond (1951).

> Lafond, E. C., and U.S. Navy Electronics Laboratory, 1951: *Processing Oceanographic Data*, H. O. Pub. No. 614, 1–114.

lag—1. That part of the difference between the **output** of an instrument and its **input** that is due to the failure of the instrument to respond instantaneously to variations of the input signal.

It is a function of the instrument's **time constant**. *See* **time lag**.

2. A time displacement of a **time series**.
See **autocorrelation**.

3. *See* **delay**.

lag coefficient—Same as **time constant**.

lag correlation—Same as **autocorrelation**.

lag time—The time between the center of mass of **rainfall** and the peak of the resulting **hydrograph**.

Lagrangian airshed models—A **model** designed to follow one parcel of air along its **trajectory** through the **atmosphere**.
In the model, the **air parcel** undergoes chemical change, as well as being impacted by dilution and the injection of fresh pollutants.

Lagrangian change—Variations in space and time of a **parcel** identified by its **Lagrangian coordinates**.

Lagrangian coordinates—(Also called material coordinates.) 1. A system of coordinates by which fluid parcels are identified for all time by assigning them coordinates that do not vary with time.
Examples of such coordinates are 1) the values of any properties of the fluid conserved in the motion; or 2) more generally, the positions in space of the parcels at some arbitrarily selected moment. Subsequent positions in space of the parcels are then the dependent variables, functions of time and of the Lagrangian coordinates. Few observations in meteorology are Lagrangian; this would require successive observations in time of the same **air parcel**. Exceptions are the **constant-pressure balloon** observation, which attempts to follow a parcel under the assumption that its **pressure** is conserved, and certain small-scale observations of diffusing **particles**. *Compare* **Eulerian coordinates**; *see also* **Lagrangian equations**.
2. Same as **generalized coordinates**.

Lagrangian correlation—The **correlation** between the properties of a flow following a single **parcel** of fluid through its space and time variations.
Compare **Eulerian correlation**; *see* **correlation coefficient**.

Lagrangian current measurement—In **oceanography**, a current measurement obtained from the determination of the position of a **drifter** at certain time intervals, so that the **mean velocity** over that interval can be calculated.

Lagrangian equations—Any of the **fundamental equations of hydrodynamics** expressed in **Lagrangian coordinates**.
In the Lagrangian description, the independent variables are time and a set of **particle** labels (such as the initial position of each particle). Although the Lagrangian description is employed less frequently than the Eulerian description, it can be particularly useful in combined chemistry/dynamics problems such as **stratosphere/troposphere** exchange along **isentropic** surfaces that intersect the **tropopause**.
Salmon, R., 1998: *Lectures on Geophysical Fluid Dynamics*, Oxford University Press, 5–6.

Lagrangian equations of motion—Differential equations that determine the evolution of a general **dynamical system**.
These are derived using the calculus of variations to minimize the Lagrangian functional equivalent to the time integral of the difference between the kinetic and potential energies with respect to variations in the evolution of the system.

Lagrangian float—A float that is used to obtain Lagrangian (following a fluid **particle**, thus drifting) **current measurements**.
See **Swallow float**, **drifter**.

Lagrangian hydrodynamic equations—Hydrodynamic **equations of motion**, using coordinates of fluid **particles** as the dependent variables.
The independent variables are time and coordinates representing the fluid particles themselves, for example, their positions at time $t = 0$.

Lagrangian mean current—The current determined by dividing the displacement of a fluid **particle** by the time traveled.

Lagrangian timescale—The mean time for an **air parcel** to be displaced from one location to another.

Lagrangian wave—Same as **shallow water wave**.

laheimar—Severe **squalls** during the change of seasons in October and November in Arabia.

lake breeze—A **wind**, similar in origin to the **sea breeze** but generally weaker, blowing from the surface of a large lake onto the shores during the afternoon; it is caused by the difference in **surface temperature** of land and water as in the land and sea breeze system.

In addition to area, the depth of the lake is an important factor; a shallow lake warms up rapidly and is less effective as the source of a lake breeze in summer than is a deep lake. Lake breezes are well developed around the Great Lakes of North America, where they temper the summer heat.

lake effect—Generally, the effect of any lake in modifying the weather about its shore and for some distance **downwind**.

In the United States, this term is applied specifically to the region about the Great Lakes or the Great Salt Lake. More specifically, lake effect often refers to the generation of sometimes spectacular **snowfall** amounts to the lee of the Great Lakes as cold air passes over the lake surface, extracting **heat** and moisture, resulting in **cloud formation** and snowfall downwind of the lake shore.

lake-effect snow—Localized, convective **snow** bands that occur in the lee of lakes when relatively cold airflows over warm water.

In the United States this phenomenon is most noted along the south and east shores of the Great Lakes during arctic cold-air outbreaks.

lake-effect snowstorm—Snowstorm occurring near or **downwind** from the shore of a lake resulting from the warming (destabilization) and moistening of relatively cold air during passage over a warm body of water.

lake evaporation—The **evaporation** from the surface of a lake.

lake ice—Ice formed on the surface of a lake.

lake surface temperature—The **temperature** of the surface layer of a lake.

lambert—A cgs unit of **luminance** (or photometric **brightness**) equal to one **lumen**, or $1/\pi$ **candela** per square centimeter.

This luminance is produce by a **blackbody** source of **luminous** intensity 1 candela at a distance of 1 centimeter. The corresponding **SI** (or mks) unit is the apostilb, a unit 10^4 smaller produced by 1 candela at a distance of 1 m. The sun's disk at **zenith** at **sea level** under **clear** skies has a luminance of about 470 000 lambert, while that of a 60-watt, inside-frosted, tungsten-filament light bulb is about 38 lambert.

Lambert conic projection—A type of **conformal map** in which features on a sphere are projected onto a cone.

The cone can either be tangent to the sphere, for which contact is along one circle, or pass through the sphere, for which contact is along two circles.

Lambert's cosine law—*See* **Lambert's law.**

Lambert's formula—A formula for computing the mean wind-direction from a series of observations.
It may be written

$$\tan \alpha = \frac{E - [\, W\,(NE + SE - NW - SW)\cos 45°]}{N - [\, S\,(NE + NW - SE - SW)\cos 45°]},$$

where α is the mean wind direction, and each point of the **compass** is replaced by the number of observations of **wind** from that direction. *Compare* **resultant wind, wind rose, prevailing wind direction.**

Lambert's law—A law governing the angular dependence of emitted or reflected **radiation** from an idealized surface.

The radiation emitted from or reflected by a surface obeying Lambert's law is unpolarized and has a **radiance** that is constant with angle, or **isotropic**. Such surfaces are variously termed Lambertian or diffuse surfaces, reflectors, or emitters. They may also be termed perfectly diffusing radiators or reflectors. This law is sometimes called **Lambert's cosine law** to distinguish it from the **Bouguer–Lambert law.**

Lambert's law of absorption—Same as **Bouguer–Lambert Law.**

Lambertian surface—A surface that emits or reflects **radiation** isotropically, according to **Lambert's law.**

lambing storm—(Also called lamb-blasts, lamb-showers, lamb storm.) A slight fall of **snow** in the spring in England.

Lambrecht's polymeter—An obsolete device that combines a **mercury barometer** and a **hair hygrometer** in a manner such that the **dewpoint** temperature is derived.

lamellar vector—A **vector** that may be represented as the **gradient** or **ascendent** of a scalar, in symbols, $\nabla\alpha$.
Thus, a lamellar vector field is **irrotational**.

laminar boundary layer—An interfacial region in which flow is smooth and nonturbulent.
Above a surface, a laminar layer will develop and fluid **velocity** will increase with distance from the surface, but not indefinitely. At some point, flow will become turbulent, with the **laminar sublayer** separating the turbulent layer from the surface. In the real world, most laminar boundary layers are extremely thin (order of 1 mm), but can be of biological importance, for example, next to plant leaves or as invertebrate refuges in streams.

laminar flow—(Also called sheet flow, streamline flow.) A flow regime in which fluid motion is smooth and orderly, and in which adjacent layers or laminas slip past each other with little **mixing** between them.
Exchange of material across laminar layers occurs by **molecular diffusion**, a process about 10^6 times less effective than **turbulence**. Laminar flow can be easily predicted as **velocity** increases at a steady rate from a boundary. This contrasts with the chaotic and **random** nature of **turbulent flow**. Laminar flow is not a common occurrence in the statically neutral and unstable **atmosphere** and is confined to a very thin layer (1 mm) adjacent to very smooth surfaces such as **snow** and **ice**. However, in strongly **statically stable** regions such as the the **nocturnal boundary layer**, the **Richardson number** can be large enough that **turbulence** is suppressed, and the flow is laminar over a layer many tens of meters thick.

laminar sublayer—A layer in which the fluid undergoes smooth, nonturbulent flow.
It is found between any surface and a turbulent layer above. *See* **laminar boundary layer**, **laminar flow**.

land breeze—A coastal **breeze** blowing from land to sea, caused by the **temperature** difference when the sea surface is warmer than the adjacent land.
Therefore, it usually blows by night and alternates with **sea breeze**, which blows in the opposite direction by day. *See* **puelche**, **karif**.
Defant, F., 1951: *Compendium of Meteorology*, 655–662.

land evaporation—The **actual evaporation** from a region of land.

land ice—Any part of the earth's seasonal or perennial **ice cover** that has formed over land as the result, principally, of the **freezing** of **precipitation**; opposed to **sea ice** formed by the freezing of seawater.
Thus, an **iceberg** or **tabular iceberg** is land ice as well as its parent **glacier**, **ice sheet**, or **ice shelf**. The two major concentrations of land ice are the ice sheets of Greenland and Antarctica. **Glaciers** and **ice caps** are the other important forms.

land-lash—In England, a heavy fall of **rain**, accompanied by a high **wind**.

land sky—The relatively dark appearance of the underside of a **cloud layer** when it is over land that is not **snow** covered.
This term is used largely in polar regions with reference to the **sky map**; land sky is brighter than **water sky**, but is much darker than **ice blink** or **snow blink**.

landfast ice—Same as **fast ice**.

landing forecast—Aerodrome forecast indicating the weather at a specific aerodrome, intended to meet the requirements of local users and aircraft within one hour's flying time of the aerodrome.

Landsat—A series of U.S. satellites designed for **remote sensing** and mapping of land areas.
These satellites are launched into **sun-synchronous orbits** that return to **image** the same swath, two degrees wide, every 16 days. The primary imaging instrument on the first generation of Landsat satellites was a **multispectral scanner**, a four-**channel** radiometer with a ground **resolution** of 80 meters. *Landsat-4* and *-5* added a second imaging instrument, a **thematic mapper**, a seven-channel instrument with a ground resolution of 30 meters, used to study vegetation, geology, and other surface features. *Landsat-1* (originally called *ERTS*) was launched in 1972, with *Landsat-2* and *-3* following in 1975 and 1978. *Landsat-4* and *-5* were launched in 1982 and 1984.

landslide—Same as **avalanche**.

landspout—1. (Rare.) A **tornado**.

2. Colloquial expression describing tornadoes occurring with a parent **cloud** in its growth stage and with its **vorticity** originating in the **boundary layer**.

The parent cloud does not contain a preexisting midlevel **mesocyclone**. The landspout was so named because it looks like a weak, Florida Keys **waterspout** over land. *See* **nonsupercell tornado**.

Bluestein, H. B., 1985: The formation of a "landspout" in a "broken-line" squall line in Oklahoma. *Preprints, 14th Conf. on Severe Local Storms*, Indianapolis, 267–270.

lane—A narrow, not necessarily navigable, **crack** in **pack ice** that may widen into a **lead**.

Langevin ion—Same as **large ion**.

langkisau—Strong **foehn**like winds during the daytime in Sumatra and the East Indies.

langley—A unit of **energy** per unit area once commonly employed in **radiation** theory; equal to one gram-calorie per square centimeter.

The langley is almost always used in conjunction with some time unit to express a **flux density**; but the time unit has been purposely separated in order that it may be chosen in a manner convenient to each particular problem. The unit was named in honor of the American scientist Samuel P. Langley, 1834–1906, who made many contributions to the knowledge of **solar radiation**. Modern meteorologists tend to use the mks unit $W\ m^{-2}$.

Langmuir cells—The vortices that characterize **Langmuir circulation**.

Langmuir circulation—Roll circulations approximately aligned with the surface **stress** vector that frequently occur in the upper **boundary layer** of oceans or lakes.

Although similar in form to atmospheric **longitudinal roll vortices**, Langmuir circulations are thought to be driven by **nonlinear** interactions between the **surface gravity wave** field and the larger-scale turbulent motions within the **mixed layer**. They are sometimes called windrows because they form lines of surface debris or bubbles in their surface **convergence** zones. Their spatial **scale** is related to the depth of the mixed layer and their **characteristic velocity** is on the order of $8u_*$, where u_* is the **friction velocity** in water. As a result of this scaling, Langmuir circulations generally require surface winds of at least 8 m s^{-1} in order to form. *See* **coherent structures**, **longitudinal rolls**.

Leibovich, S., 1983: The form and dynamics of Langmuir circulations. *Ann. Rev. Fluid Mech.*, **15**, 391–427.

Langmuir layer—A **trace** organic impurity in water with a polar or similar chemical group that disperses the material as a surface near monomolecular layer and influences surface properties as **water vapor** deposition coefficient, **vapor pressure**, and **surface tension**.

Langmuir number—A **nondimensional number** representing the ratio of **viscous** to **inertial forces** in the nondimensionalized governing equations for **Langmuir circulation**:

$$\mathrm{La} = [(v_T\beta/u_*)^{3/2}][(S_0/u_*)^{-1/2}],$$

where v_T is the **eddy viscosity**, $2S_0$ and $1/2\beta$ are the surface value and *e*-folding depth of the **Stokes's drift** current, respectively, and u_* is the **friction velocity**.

This **parameter** can also be interpreted as expressing the balance between the rate of **diffusion** of **streamwise vorticity** by **eddy viscosity** and the rate of production of **streamwise vorticity** by **vortex stretching** accomplished by the Stokes's drift. The Langmuir number is inversely related to the **Reynolds number**.

Langmuir probe—A **probe** used to measure the **electron** temperature of ionized plasmas.

Langmuir probes mounted on spacecraft are often used to measure the electron temperature of the earth's **ionosphere**.

lansan—The strong southeast **trade winds** of the New Hebrides and East Indies.

Laplace equation—The elliptic partial differential equation

$$\nabla^2\phi = 0,$$

where ϕ is a **scalar** function of position, and ∇^2 is the **Laplacian operator**.

In **rectangular Cartesian coordinates** x, y, z, this equation may be written

$$\frac{\partial^2\phi}{\partial x^2} + \frac{\partial^2\phi}{\partial y^2} + \frac{\partial^2\phi}{\partial z^2} = 0.$$

The Laplace equation is satisfied, for example, by the **velocity potential** in an **irrotational** flow, by **gravitational potential** in **free space**, by electrostatic **potential** in the **steady flow** of electric currents in solid conductors, and by the steady-state **temperature** distribution in solids. A solution of the Laplace equation is called an **harmonic function**. *Compare* **Poisson equation**.

Laplace operator—Same as **Laplacian operator**.

Laplace transform—(Also called Laplace transformation.) An integral transform of a function obtained by multiplying the given function $f(t)$ by e^{-pt}, where p is a new **variable**, and integrating with respect to t from $t = 0$ to $t = \infty$.

Thus, the Laplace transform of $f(t)$ is

$$L\{f(t)\} = \int_0^\infty e^{-pt}f(t)dt$$

and may be denoted by the symbol $F(p)$. The Laplace transform is especially useful in solving initial-value problems associated with inhomogeneous **linear** differential equations with constant coefficients. *See* **Fourier transform**.

Laplacian operator—(Or Laplace operator.) The mathematical **operator** $\nabla^2 = \nabla \cdot \nabla$ (or sometimes written Δ) where ∇ is the **del operator**.

In **rectangular Cartesian coordinates**, the Laplacian operator may be expanded in the form

$$\nabla^2 = \frac{\partial^2}{\partial x^2} + \frac{\partial^2}{\partial y^2} + \frac{\partial^2}{\partial z^2}.$$

See **Laplace equation**.

lapse line—A curve showing the **variation** of **temperature** with height.
See **lapse rate**.

lapse rate—The decrease of an atmospheric **variable** with height, the variable being **temperature**, unless otherwise specified.

The term applies ambiguously to the environmental lapse rate and the process lapse rate, and the meaning must often by ascertained from the context.

lard ice—Same as **grease ice**.

large calorie—(Abbreviated Cal; also called kilogram calorie.) A unit of **heat** equal to 1000 **small calories**.

large eddy model—A numerical **model** with an averaging volume sufficiently small that the **eddies** contained entirely within it (i.e., the unresolved eddies) are of **inertial-range** scales and smaller.

The larger **energy-containing eddies** are resolved explicitly away from the boundaries. In the **boundary layer** the energy-containing eddies are not resolved near the surface and near the **inversion** at the top. In these regions a **subgrid-scale parameterization** is needed.

Wyngaard, J. C., 1982: *Atmospheric Turbulence and Air Pollution Modelling*, Nieuwstadt and van Dop, Eds., Reidel, 97–100.

large eddy simulation—1. (Abbreviated LES.) A three-dimensional **numerical simulation** of turbulent flow in which large **eddies** (with scales smaller than the overall dimension of the problem in question) are resolved and the effects of subgrid-**scale** eddies, which are more universal in nature, are parameterized.

Large eddies are important in characterizing one turbulent flow from another. The difference between a large eddy simulation and the traditional phenomenological modeling of **turbulence** is that in the latter case all scales of turbulent motion are parameterized.

2. A modeling technique in which spatial **resolution** extends into the **inertial subrange**, but does not resolve the smallest scales of motion.

The effects of the latter are approximated using subgrid-scale models, which usually draw heavily on the Kolmogorov theory of the inertial subrange. *See* **direct numerical simulation, very large eddy simulation**.

large eddy simulation models—Computer codes that numerically integrate in three dimensions the time-dependent **Navier–Stokes equations** filtered over a **grid** volume much smaller than the size of the **energy-containing eddies** or **wavelengths** of **turbulence**.

The solutions of these models consist of energy-containing large eddies that carry most of the turbulent fluxes. The net effect of small subgrid-scale eddies is treated as locally diffusive and dissipative, typically modeled based on **inertial-subrange** theory.

large halo—Same as **halo of 46°**.

large ion—(Also called slow ion, heavy ion.) An **ion** of relatively large mass and low mobility that is produced by the **attachment** of a **small ion** to an **Aitken nucleus**.

443

Large ions were discovered by P. Langevin and are sometimes referred to as "Langevin ions." Large ions have ion mobilities of the order of 10^{-8} m s^{-1} per volt m^{-1}, or some 10 000 times lower than those of small ions. As a result these atmospheric ions contribute practically nothing to the **conductivity** of the air, except in rare cases where small ions are nearly absent. Typically, they bear only a single electronic charge, as is true of small ions. Large ions move so slowly that they are not destroyed by being neutralized by still other large ions of paired signs, for such collisions are too infrequent. Instead, they are neutralized by union with a small ion of opposite sign. Their mean lifetimes are of the order of 15–20 minutes over the oceans, but may approach 1 h in very polluted air. The **ion density** of large ions varies widely depending upon the degree of **atmospheric pollution**. Representative low-altitude values might be 10^9 m^{-3} in clean country air, 10^{10} m^{-3} in an industrial area, and 10^8 m^{-3} over the oceans.

Wait, G. R., and Parkinson, W. D., 1951: *Compendium of Meteorology*, 120–121.

large Reynolds number flow—The behavior of a fluid with a **Reynolds number** typically greater than 10^4 to 10^6, which usually occurs within the **atmosphere**.

The main property of such flows is a constant **friction** stress within the **surface layer** that depends only on relative roughness but not on the Reynolds number itself. Thus, **molecular viscosity** and qualities occurring in flow descriptions that are dependent on the Reynolds number may be totally ignored.

large scale—In meteorology, a **scale** in which the curvature of the earth is not negligible.

This is the scale of the high tropospheric **long-wave** patterns, with four or five waves around the hemisphere in middle latitudes. These waves are within the province of both the **general circulation** and **synoptic meteorology**, but the terminology should distinguish this scale from that of the **migratory** high and low pressure systems of the lower **troposphere**. **Rossby waves** and other long **barotropic** waves are large-scale disturbances. *See* **cyclonic scale**.

large-scale convection—Organized vertical motion on a larger **scale** than atmospheric **free convection** associated with **cumulus** clouds.

The patterns of vertical motion in hurricanes or in **migratory** cyclones are examples of such **convection**.

large-scale weather situation—State of the **atmosphere** at a particular time over an extensive area.

Larmor frequency—*See* **gyro-frequency**.

laser—(Acronym for light amplification by stimulated emission of radiation.) A device that produces a narrow **beam** of **electromagnetic energy** by recirculating an internal beam many times through an amplifying medium, each time adding a small amount of **energy** to the recirculating beam in a phase-coherent manner.

Typically the **output** beam results when a small amount of recirculating energy is allowed to leak out from the internal "cavity." These devices produce energy at **light** frequencies (which are higher than radio frequencies) in the **infrared**, visible, or **ultraviolet** portions of the **electromagnetic spectrum**, and hence they often use optical technologies.

laser ceilometer—Instrument for measuring **cloud base** height above the surface with **light** detection and ranging (**lidar**), using a pulsed **laser** and **signal** processing to determine the location of strong backscatter from the lower **cloud** boundary.

laser-induced fluorescence—(Abbreviated LIF.) Experimental technique in which the **absorption** of **laser** radiation excites a molecule or atom to an excited **energy level**, followed by **emission** of **radiation** as the system relaxes.

The technique has been particularly useful in facilitating the **detection** of highly reactive **free radical** species such as the **hydroxyl radical** (OH) in either laboratory systems or in field measurements.

laser radar—A device that operates on **radar** principles but that uses a **laser** as its source of transmitted light-frequency (**ultraviolet**, visible, or **infrared**) **energy**, instead of radio-frequency sources. *See* **lidar**.

last glacial—The most recent time (15 000 to 80 000 years ago) during which continental glaciers covered subpolar regions and existed at elevations as much as 1000 m lower than today; corresponding to periods in which **oxygen** isotopes from marine **sediment** cores indicate that global **sea level** was 50–150 m lower and global **temperature** 5°–10°C lower than today.

last interglacial—The most recent time (115 000 to 125 000 years ago) during which global temperatures were as high as or higher than in the postglacial, when continental **glaciers** were limited to the Arctic and Antarctic, and sea levels were near current positions.

latent heat—The **specific enthalpy** difference between two phases of a substance at the same **temperature**.

The latent heat of **vaporization** is the **water vapor** specific enthalpy minus the liquid water specific enthalpy. When the temperature of a system of **dry air** and water vapor is lowered to the **dewpoint** and water vapor condenses, the **enthalpy** released by the **vapor** heats the air–vapor–liquid system, reducing or eliminating the rate of **temperature reduction**. Similarly, when liquid water evaporates, the system must provide enthalpy to the vapor by cooling. The latent heat of **fusion** is the specific enthalpy of water minus that of **ice** and the latent heat of **sublimation** is the specific enthalpy of water vapor minus that of ice. The latent heats of vaporization, fusion, and sublimation of water at 0°C are, respectively,

$$L_v = 2.501 \times 10^6 \text{ J kg}^{-1}$$
$$L_f = 3.337 \times 10^5 \text{ J kg}^{-1}$$
$$L_s = 2.834 \times 10^6 \text{ J kg}^{-1}.$$

It is common to see an expression like "release of latent heat." In other thermodynamic terms in this glossary, such expressions are avoided in favor of others using enthalpy and temperature, which are measurable quantities.

latent instability—Atmospheric conditions above the **level of free convection** when the **lapse rate** is steeper than **moist adiabatic**; has been used more as a quantitative measure than a qualitative condition.

It is becoming an obsolete term, replaced qualitatively by **conditional instability** and quantitatively by **convective available potential energy**.

lateral gustiness—*See* **gustiness components**.

lateral inflow—The addition of water to a **stream**, river, or lake from the sides of the channel or **reservoir**.

Lateral **inflow** can include **groundwater flow, overland flow**, or **interflow**.

lateral mirage—A **mirage** in which the **image** (or images) is displaced laterally from the position of the object.

This is not a difficult mirage to find, especially along the sun-warmed walls of buildings. In many cases, it appears as nothing but an **inferior mirage** turned on its side. However, there are often interesting subtleties. Easiest to find, perhaps, are the high-order multiple images that result from inhomogeneities along the wall. These can arise both from the wall having a slightly wavy surface and from the periodic variations in the internal structure of the wall that alter the **thermal conductivity** and so produce periodic **temperature** variations. Curiously, unlike the inferior mirage, the lateral mirage seems to be capable of producing three images even in the absence of inhomogeneities. The temperature profiles normal to horizontal and vertical surfaces are slightly different. In the case of the inferior mirage, **gravity** acts normal to the surface, while in the case of the lateral mirage, gravity is parallel to the surface. This produces a flow up the wall that results in a temperature profile capable of giving the three-image mirage. Lateral temperature **gradients** in the **free atmosphere**, away from vertical surfaces, are not sufficient to produce lateral mirages; the rare reports of such sightings undoubtedly arose from misinterpretations of observations.

lateral mixing—Horizontal **mixing** perpendicular to the mean flow.

latitude—The angular distance along the **meridian** from the point in question to the **equator**.

Latitude is normally described as so many degrees north or south of the equator. *Compare* **longitude**.

laveche—Same as **leveche**.

law of storms—Historically, the general statement of 1) the manner in which the winds of a **cyclone** rotate about the cyclone's center, and 2) the way that the entire **disturbance** moves over the earth's surface.

The formation of this "law" was largely due to the investigations of Brandes in 1926, Dove in 1828, and Redfield in 1831. This knowledge of the general behavior of storms led to the issuance of rules for seamen instructing them in means of navigating to avoid the dangers of storms at sea.

Lax equivalence theorem—Theorem that states that a consistent **finite-difference approximation** to a well-posed time-dependent **linear** partial differential equation is convergent if, and only if, it is stable.

See **stability**.

Lax–Wendroff differencing scheme—An explicit numerical approximation to the time evolution of fluid flows that is constructed through Taylor expansions to yield second-order **truncation error** using only two consecutive time levels.

The resulting approximation introduces a diffusive **damping** term that helps control **nonlinear** computational **instability** and may be useful for equations with shock discontinuities.

Richtmyer, R., and K. Morton, 1967: *Difference Methods for Initial Value Problems*, 302 pp.

layer cloud—**Stratus** cloud; a continuous cloud sheet capped by an **inversion**.

layer depth—In **oceanography**, the **thickness** of the **mixed layer**; or the depth to the top of the **thermocline**.

layer of no motion—A layer, assumed to be at rest, at some depth in the ocean.

This implies that the **isobaric** surfaces within the layer are level, and hence they may be used as reference surfaces for the computation of absolute **gradient currents**. This same concept can define a **level of no motion** or a surface of no motion.

layers—Any of a number of **altitude** regions found in the **atmosphere**, each of which can be characterized by unique physical or chemical properties.

These layers can exist as the result of chemistry or dynamics. Examples of layers encountered in the **atmosphere** are the **planetary boundary layer**, the D, E, and F layers of the **ionosphere**, and the stratospheric **ozone layer**. *See* **Chapman layer**.

LCL—Abbreviation for **lifting condensation level**.

LDR—Abbreviation for **linear depolarization ratio**.

leaching—1. The removal of materials in solution from soil, rock, or waste.

2. Separation or dissolving out of soluble constituents from a **porous medium** by **percolation** of water.

Dept. of the Interior, U.S. Geological Survey, Office of Water Data Coordination, 1989: *The Federal Glossary of Selected Terms: Subsurface Waterflow and Solute Transport.*

lead—A long fracture or separation between ice **floes** wide enough to be navigated by a ship.

A lead may be covered by thin ice.

leader—(Or leader streamer.) The **electric discharge** that initiates each **return stroke** in a cloud-to-ground **lightning discharge**.

It is a channel of high ionization that propagates through the air by virtue of the electric **breakdown** at its front produced by the charge it lowers. The **stepped leader** initiates the first **stroke** in a **cloud-to-ground flash** and establishes the channel for most subsequent strokes of a lightning discharge. The **dart leader** initiates most subsequent strokes. Dart-stepped leaders begin as dart leaders and end as stepped leaders. The initiating processes in **cloud** discharges are sometimes also called leaders but their properties are not well measured.

leader streamer—Same as **leader**.

leaf area index—The leaf area subtended per unit area of land.

leaf wetness—Liquid moisture on exposed leaf surfaces or other exposed plant parts.

leaf wetness duration—Length of time of continual **exposure** of plant surfaces to liquid moisture.

Leaf wetness duration is often related to plant disease infection periods. *See* **Mills period**.

leakage—The flow of fluid from a **reservoir** or water-bearing unit to an adjacent medium.

leakage coefficient—In a **leaky aquifer**, leakage coefficient for a given **observation well** refers to the **dimensionless parameter**

$$\frac{r}{B} = \frac{r}{\left(\dfrac{KHH'}{K'}\right)^{1/2}}$$

where r is the distance of the observation well from the pumping well; K and K' are the **hydraulic conductivity** of the **aquifer** and the leaky layer, respectively; and H and H' are the thickness of the aquifer and the leaky layer, respectively.

leakance—*See* **leakage coefficient**.

leaky aquifer—An **aquifer** that is overlaid or underlaid by a semipermeable geologic layer.

leapfrog differencing—A **finite-difference approximation** to a time evolution equation in which the time derivative is approximated with values one time step before and one time step ahead of the values that specify other terms of the equation.

The scheme $(f^{n+1} - f^{n-1})/2\Delta t = g(f^n)$ (where superscript n denotes a point in time, separated by step Δt from the prior $[n - 1]$ and subsequent $[n + 1]$ discrete time levels) is a leapfrog approximation to the differential equation $df/dt = g(f)$. *Compare* **implicit time difference**.

least squares—Any procedure that involves minimizing the sum of squared differences.

For example, the **deviation** of the mean from the **population** is less, in the square sense, than any other **linear** combination of the population values. This procedure is most widely used to obtain the constants of a representation of a known **variable** Y in terms of others X_i. Let $Y(s)$ be represented by

$$\sum_{n=0}^{N} a_n f_n[X_i(s)].$$

The a_n's are the constants to be determined, the f_n's are arbitrary functions, and s is a **parameter** common to Y and X_i. N is usually far less than the number of known values of Y and X_i. The system of equations being overdetermined, the constants a_n must be "fitted." The least squares determination of this "fit" proceeds by summing, or integrating when Y and X_i are known continuously,

$$\left\{ Y(s) - \sum_{n=0}^{N} a_n f_n[X_i(s)] \right\}^2 \text{ over } s$$

and minimizing the sum with respect to the a_n's. In particular, for example, if $f_n[X_i(s)] \equiv X_i(s)$, then the **regression function** is being determined; and when $f_n[X_i(s)] \equiv \cos nX_i(s)$, or $\sin nX_i(s)$, then Y is being represented by a multidimensional **Fourier series**. Least squares is feasible only when the unknown constants a_n enter linearly. The **method of least squares** was described independently by Legendre in 1806, Gauss in 1809, and Laplace in 1812.

least-time track—Same as **minimal flight path**.
See **minimal flight**.

lee cyclogenesis—(Also called orographic cyclogenesis.) The synoptic-scale **development** of an atmospheric **cyclonic circulation** on the **downwind** side of a mountain range.

The "lee" side is relative to the mean background airflow. Weak development can occur due to a redistribution of uniform **vorticity** as large-scale flow passes over a mountain barrier (*see* **lee trough**). Stronger cases of lee cyclogenesis occur when the mountain range interacts with a developing **baroclinic wave**. In this instance the mountain acts to position the **cyclone** or generate a **secondary cyclone** in the lee. Often this leaves a weaker parent cyclone that is typically poleward of the development and far from the mountains. Lee cyclogenesis is a multistage process involving a phase of rapid **deepening** followed by a transition to slower **baroclinic** deepening. The mountain barrier disrupts the orderly **advection** of low-level cold air that would occur behind the cyclone over flatter terrain and induces a quasigeostrphic imbalance. As it tends to restore this imbalance (*see* **geostrophic adjustment**), the **atmosphere** produces the beginnings of a **cyclonic** system near the surface. The lack of cool air at low levels and the descending upper-level air make the mountain lee **environment** statically less stable than the surroundings. This enhances the possibility for vertical coupling of **potential vorticity** maxima associated with the approaching **upper-level trough** and the incipient lee **disturbance** at lower levels. Lee cyclogenesis is common on the **leeward** side of the major mountain ranges of the world including the Alps, the Himalayas, the Rockies (both east and west), and the Andes. Many minor ranges support lee cyclogenic activity. *See also* **Genoa cyclone, Colorado low, Alberta clipper, pampero**.

lee depression—Same as **lee trough**.

lee eddies—The small irregular motions or **eddies** produced immediately in the rear of an obstacle in a turbulent fluid.

For sufficiently high **Reynolds number** and for very irregular obstacles, the region of lee eddies may extend a considerable distance **downstream**. As an example in the **atmosphere, mountain waves** may be thought of as lee eddies

lee trough—(Same as dynamic trough.) A pressure **trough** formed on the lee side of a mountain range in situations where the **wind** is blowing with a substantial component across the mountain ridge; often seen on United States weather maps east of the Rocky Mountains, and sometimes east of the Appalachians, where it is less pronounced.

Its formation may be explained thermodynamically by the warming due to **adiabatic** compression of the sinking air on the lee side of the mountain range, or dynamically by generation of **cyclonic circulation** (cyclogenesis) by the horizontal **convergence** associated with **vertical stretching** of air columns passing over the ridge and descending the lee slope. Alternatively, the latter viewpoint is often expressed as the conservation of **potential vorticity**, where the vertical stretching of the columns is compensated by an increase in their **relative vorticity**. *See* **lee cyclogenesis**.

lee wave—1. Any **wave disturbance** that is caused by, and is therefore stationary with respect to, some barrier in the fluid flow.

Whether the **wave** is a **gravity wave**, **inertia wave**, **barotropic wave**, etc., will depend on the structure of the fluid and the dimensions of the barrier. Most research has been devoted to the gravity lee wave (**mountain wave**) in the **atmosphere**, of **wavelength** of order

$$2\pi V[T/g(\gamma_d - \gamma)]^{1/2},$$

where V is the current speed, T the Kelvin **temperature**, g the **acceleration of gravity**, and γ_d and γ the **dry-adiabatic** and **environmental lapse rates**, respectively. This is the wave that is evident in **lenticular** or **Moazagotl** cloud systems and is strikingly exemplified in the **Bishop wave**. Dynamically, the lee wave is the sum of the **free waves** of the system and those wave components forced by the particular shape of the barrier. The **disturbance** is, in general, negligible at any distance **upstream** of the barrier, a result that follows from the dynamics when the system is started from rest, but a point that requires special attention when the steady-state assumption is made. The term lee wave is also applied loosely to nonwave disturbances in the lee of obstacles, such as the **rotor cloud**.

2. A **mountain wave** occurring to the lee of a mountain or mountain barrier.

These waves can become visible in the form of **lenticular** or trapped lee-wave clouds.

Eliassen, A., and E. Kleinschmidt, 1957: Dynamic Meteorology. *Handbuch der Geophysik*, Vol. XLVIII, 59–64.

lee-wave separation—The production of small wavelength **mountain waves** near a mountaintop under conditions of very strong **static stability**.

When the air is very stable and **wind speeds** are slow, the natural **wavelength** of air is often much shorter than the width of the mountain, as indicated by a very small **Froude number**. For this situation, the buoyant restoring force in the air is so strong that the air resists vertical displacement to get over the mountaintop, and instead most of the air flows around the sides of the mountain. The shallow layer of air near the mountaintop that is able to be displaced upward over the mountain will continue in vertical **oscillation** as it blows **downstream**, or separates, from the mountain. *Compare* **lenticular cloud**, **downslope windstorm**.

leeside convergence—Region of **convergence**, often a line, **downwind** of a mountain or mountain ridge during fair-weather daytime conditions that are favorable for the formation of thermally forced upslope flow and deep convective **mixing**.

Convergence forms between upslope (or sometimes light and variable) flow at lower elevations of the lee slopes and downslope flow at higher elevations, which results from the downward convective mixing of ambient **momentum** from the flow above ridgetops. With moist upslope flow and favorable conditions, updrafts produced by the convergence can lead to mountain **cumulus** formation, or trigger **thunderstorm** or **severe weather** activity.

Leeuwin Current—The eastern **boundary current** of the south Indian Ocean.

It occupies the latitude band of the subtropical **gyre** but is not part of it; rather, it is found as a narrow and swift southward flowing current along the west Australian shelf opposing the broad northward flow of the subtropical gyre (the **West Australian Current**) farther offshore. The Leeuwin Current runs against the prevailing **wind**; it is driven by the alongshore **pressure gradient** caused by the connection between the Pacific and Indian Oceans north of Australia. Its water, which is of tropical origin, cools as it proceeds southward, producing **convection** and a continuous deepening of the surface **mixed layer** along the current's path. The associated **heat** loss is a significant heat gain for the **atmosphere**.

leeward—The side of a mountain, ridge, or other flow obstacle away from the large-scale or ridgetop flow direction; the **downwind** side; opposite of **windward**.

leeway—The **leeward** motion of a ship due to **wind**.

It may be expressed as distance, speed, or angular difference between **course** steered and course through the water. This is analogous to **drift** in air navigation.

Leibniz's theorem of calculus—A relationship between the derivative of an integral and the integral of a derivative, that is,

$$\frac{d}{dt}\left[\int_{S_1(t)}^{S_2(t)} A(t,s)ds\right] = \int_{S_1(t)}^{S_2(t)}\left[\frac{\partial A(t,s)}{\partial t}\right]ds + A(t,S_2)\frac{dS_2}{dt} - A(t,S_1)\frac{dS_1}{dt},$$

where S_1 and S_2 are limits of integration, s is a dummy distance or space **variable** such as height

z, t is time, and A is some meteorological quantity, such as **potential temperature** or **humidity**, that is a function of both space and time. If the limits of integration are constant with time, then the last two terms are zero, and the derivative of the integral equals the integral of the derivative. However, there are many atmospheric situations, such as a growing **atmospheric boundary layer**, where the limits of integration can change with time. Namely, if one wishes to integrate over the depth of the **boundary layer** (between limits $z = 0$ to $z = z_i$) to find a boundary layer average, for example, but the top of the boundary layer at height $z_i(t)$ is rising with time, then one must use the full form of Leibniz's theorem to account for this effect.

Lenard effect—(Also called spray electrification, waterfall effect.) The separation of electric charges accompanying the **aerodynamic** breakup of water drops, first studied systematically by the German physicist P. Lenard (1892).

Experiments have shown that the degree of **charge separation** in spray processes depends upon the **drop** temperature, presence of dissolved impurities, speed of the impinging air blast, and contact with foreign surfaces. The largest fragments of the broken drops are observed to carry positive charges and the fine spray of drops carried off in the impinging **air current** carries a net negative charge. Distilled water drops of 4-mm diameter, broken after a 5-cm free fall into an **updraft** of 1 m s^{-1}, were found by Chapman (1953) to yield about 10^{-10} C of separated charge per drop. The Lenard effect was incorporated by Simpson (1927) into his **breaking-drop theory** of **thunderstorm charge** generation, but many critical details are but poorly understood.

Chapman, S., 1953: *Thunderstorm Electricity*, Byers, H. R., ed., 207–213.
Simpson, G. C., 1927: The mechanism of a thunderstorm. *Proc. Roy. Soc. A*, **114**, 376–401.
Lenard, P., 1892: Über die Elektrizität der Wasserfälle. *Ann. Phys., Lpz*, **46**, 584–636.

length of record—The period during which observations have been maintained at a **meteorological station**, and which serves as the frame of reference for climatic data at that station.

The standard length of record for the purpose of a **normal** has been fixed by the World Meteorological Organization as 30 years (i.e., three consecutive 10-year periods), which is a reasonable average for the length of a homogeneous record desirable for most of the meteorological elements. Homogeneous records as long as 50 years are rare due to breaks or gradual changes being introduced by changes in the hours of **observation**, in the observational practices, in the **site** or instruments used, or by a gradual change in the character of the surrounding country, such as the growth of a city. It is often possible, however, to account for these changes and to construct a composite record that may cover a century or more.

lenticular cloud—A commonly used term for clouds of the species **lenticularis.**

lenticularis—(Or lenticular cloud.) A **cloud species** the elements of which have the form of more or less isolated, generally smooth lenses or almonds; the outlines are sharp and sometimes show **irisation**.

These clouds appear most often in formations of **orographic** origin, the result of **lee waves**, in which cases they remain nearly stationary with respect to the terrain (**standing cloud**), but they also occur in regions without marked **orography**. This species is found mainly in the genera **cirrocumulus, altocumulus**, and (rarely) **stratocumulus**. Altocumulus lenticularis differs from cirrocumulus lenticularis in that, when smooth and without elements, it has shadowed parts while the latter is very white throughout. When undulated or subdivided, the altocumulus species differs from stratocumulus lenticularis in that its elements subtend an angle of less than 5° when viewed at an angle of more than 30° above the **horizon**. *See* **cloud classification.**

leptokurtic—*See* **kurtosis.**

leste—Spanish nautical term for east **wind**.

The name is given to a hot, dry, dusty easterly, or southeasterly wind that blows from the Atlantic coast of Morocco out to Madeira and the Canary Islands. It is a form of **sirocco** and occurs in front of **depressions** advancing eastward. *Compare* **levanto**.

levant—Same as **levante.**

levant blanc—*See* **levante**

levante—The Spanish and most widely used term for an east or northeast **wind** occurring along the coast and inland from southern France to the Straits of Gibraltar.

It is moderate or fresh (not as strong as the **gregale**), mild, very humid, **overcast**, and rainy; it occurs with a **depression** over the western **Mediterranean Sea**. In summer it is rare and weak; in January it is inhibited by the Iberian **anticyclone**. It is most frequent from February to May and October to December. A levant (French spelling) with fine weather is a levant blanc; in the Roussillon region of southern France (where, as along the Catalonian coast of Spain, it is called llevant) it often brings floods in the mountain streams. The **levanter** of the Gibraltar Straits is a related phenomenon. *Compare* **leste**, **lombarde**, **levantera**.

levanter—An English name for the **levante**, more specifically applied to winds in the Straits of Gibraltar and on the east coast of Spain.
It blows from east or northeast with high pressure over central Europe and a **depression** over the southwest Mediterranean. It is most frequent and strongest from October to December and February to May, and persists for two or three days.

levantera—A persistent east **wind** in the Adriatic, usually bringing **cloudy** weather.
Compare **levante**, **greco**, **gregale**.

levanto—A hot southeasterly **wind** in the Canary Islands.
Compare **levante**, **leste**.

leveche—(Also spelled laveche.) A name for the **sirocco** in Spain.
It is a hot, sand- and dust-laden **wind** from between southeast and southwest that blows in front of a **depression** on the southeast coast of Spain but extends only a few miles inland.

level ice—Ice that has not been deformed.

level of concern—A **pollutant** concentration level (specific to the averaging time being considered) associated with a condition of exposure, often for sensitive population groups.
Concentrations above the level of concern are related to undesirable health effects, while concentrations below this level do not cause perceptible health effects of significant duration.

level of escape—*See* **critical level of escape**.

level of free convection—(Abbreviated LFC.) The level at which a **parcel** of air lifted **dry-adiabatically** until saturated and **saturation-adiabatically** thereafter would first become warmer than its surroundings in a conditionally unstable **atmosphere**.
On a **thermodynamic diagram** the level of free convection is given by the point of intersection of the process curve, representing the process followed by the ascending parcel, and the **sounding** curve, representing the **lapse rate** of **temperature** in the **environment**. From the level of free convection to the point where the ascending parcel again becomes colder than its surroundings the atmosphere is characterized by **latent instability**. Throughout this region the parcel will gain **kinetic energy** as it rises. *See* **conditional instability**, **convective condensation level**; *compare* **level of free sink**.

level of free sink—(Abbreviated LFS.) The level at which a **parcel** of descending air becomes cooler than its surroundings.
On a **thermodynamic diagram** the level of free sink is given by the point of intersection of the process curve representing the parcel's descent and the **sounding** curve representing the **environment**. The descent can be **moist-** or **dry-adiabatic**, depending on the availability of water substance to evaporate. From the level of free sink to the point where the parcel either reaches the surface or again becomes warmer than its surroundings represents a region of negative **buoyancy** and a descending parcel will be accelerated according to parcel theory. *Compare* **level of free convection**.

level of neutral buoyancy—(Also called equilibrium level.) The level at which an **air parcel**, rising or descending **adiabatically**, attains the same **density** as its **environment**.

level of no motion—An ocean surface at a depth where the current speed is assumed to be zero.
The determination of currents through the dynamic method gives only relative speeds, therefore an ocean surface at depth has to be chosen as a reference of no motion for the water column above it.

level of nondivergence—A level in the **atmosphere** throughout which the horizontal **velocity divergence** is zero.
Although in some meteorological situations there may be several such surfaces (not necessarily level), the level of nondivergence usually considered is that midtropospheric surface that separates the major regions of horizontal **convergence** and **divergence** associated with the typical vertical structure of the **migratory** cyclonic-scale weather systems. Interpreted in this manner, the level of nondivergence is usually assumed to be in the vicinity of 500 mb. The assumption of such a level in theoretical work facilitated the construction of early models in **numerical forecasting**.

level surface—Same as **geopotential surface**.

LEWP—Abbreviation for **line echo wave pattern**.

LF—Abbreviation for low frequency.

See **radio frequency band**.

LFC—Abbreviation for **level of free convection**.

LFM model—Abbreviation for **limited fine-mesh model**.

LFS—Abbreviation for **level of free sink**.

LI—Abbreviation for **lifted index**.

libeccio—Italian name for a southwest **wind**; used especially in northern Corsica for the west or southwest wind that blows throughout the year, and especially in winter when it is often stormy.

On **windward** slopes it brings **rain**, with thunderstorms in summer and autumn. After crossing the mountains it is warm and dry, but may be very turbulent.

liberator—A name sometimes given the west **wind** through the Straits of Gibraltar.

lid—(Also known as cap, capping inversion, capping layer.) A thin layer with enhanced **static stability** separating a layer below possessing large **convective available potential energy** from a layer above with lower static stability.

The presence of a lid is generally accompanied by substantial **convective inhibition. Air parcels** with insufficient **kinetic energy** rising into the bottom of a lid will be unable to penetrate it.

lidar—(Coined word for light detection and ranging.) An instrument combining a pulsed **laser** transmitter and optical **receiver** (usually a telescope) with an electronic **signal** processing unit used for the **detection** and ranging of various distant targets in the **atmosphere**, analogous to the principles of operation of **microwave radar**.

Normally, the **transmitter** and receiver are coaligned and placed closely together to measure the laser energy backscattered by the **target** into the direction of the receiver. The use of lasers allows the **light** pulses to be exceptionally short, highly focused, and **monochromatic**, but laser light suffers from strong, range-limiting **attenuation** in many types of clouds. Depending on the spectral characteristics of the laser and detector and the number of receiver channels, a large variety of lidar applications for atmospheric research are in use. Simple one-channel **laser ceilometers** measure cloud-base heights and internal **cloud** structures; **polarization** lidars measure cloud phase and **hydrometeor** type; differential absorption (DIAL) and Raman lidars measure the concentrations of selected molecular species; high spectral resolution lidars (HSRL) measure the separation of molecular and **aerosol** or cloud constituents; and **Doppler lidars** measure the **radial velocity** of aerosol or cloud targets. Lidar wavelengths **range** from the near-ultraviolet to the midinfrared ($\approx$0.3–12 μm).

lidar constant—The proportionality factor in the **lidar equation**, usually represented by C, relating the **power** returned from a given range to the properties of the **scattering** medium.

The lidar constant depends on characteristics of the particular **lidar** system including the **transmitted power, pulse length**, and **receiver** aperture. *See* **lidar equation**.

lidar equation—An equation, which may appear in different forms depending on the particular system or application, that describes the relation between the **received power** p measured in a **lidar** receiver channel from range r, and the characteristics of the lidar system and the transmission medium (usually the **atmosphere**) through which the **laser** pulse propagates.

The most common form of the equation is for plane-polarized **radiation** and **single scattering**, for which

$$p(r) = \frac{C\beta(r)\,t^2(r)}{r^2},$$

where β is the volume backscattering coefficient at range r, t^2 is the two-way **transmittance** to range r, and C is the **lidar constant**, which depends on such system parameters as the **transmitted power, pulse duration**, and **receiver** characteristics. The transmittance is related to the **volume extinction coefficient** γ by

$$t^2(r) = \exp\left[-2\int_0^r \gamma(s)\,ds\right].$$

Normally **scattering** and **extinction** of the lidar beam are caused by the combined effects of molecules, **aerosols**, and **hydrometeors**, so that β and γ represent the sum of their separate contributions. *Compare* **radar equation**.

lidar ratio—The ratio of the volume backscattering coefficient to the **volume extinction coefficient** of a **scattering** medium.

If this ratio is known or can be assumed, it facilitates inverting the **lidar equation** to solve for the **profile** of the **extinction coefficient** in terms of the measured **power**.

LIF—*See* **laser-induced fluorescence**.

life zone—A **thermal belt**, either of latitude or **altitude**, in which the plant and animal life is of a distinctive type.

The limits of each life zone are mainly fixed by minimum temperatures. The major life zones are boreal, temperate, subtropical, and tropical in both Northern and Southern Hemispheres, with tropical being equatorial. Within each **thermal** life zone, moisture gradients provide further subdivisions into a number of biogeographical provinces. Latitudinal and altitudinal life zones are not strictly equivalent.

lifetimes—The time required for the concentration of a chemical species to decrease to $1/e$ of its original value.

The definition arises from assuming a pseudo-first-order decay of the species concentration, $[A] = [A]_0 \exp(-kt)$. When the time t is equal to the inverse of the first-order **rate coefficient** k, the concentration $[A]$ will have decreased to $1/e$ of the initial concentration $[A]_0$. The atmospheric lifetime of a species in **steady state** can also be defined as the total atmospheric burden divided by either the total loss rate or the total production rate.

lift—The component, perpendicular to the relative **wind** and in the plane of symmetry, of the total force of air on an aircraft or airfoil.

It must be specified whether this applies to a complete aircraft or the parts thereof. In the case of a lighter-than-air craft, this is often called dynamic lift.

lifted index—(Abbreviated LI.) See **stability index**.

lifting condensation level—(Abbreviated LCL; also called isentropic condensation level.) The level at which a **parcel** of **moist air** lifted **dry-adiabatically** would become saturated.

On a **thermodynamic diagram** it is located at the point of intersection of the **dry adiabat** through the point representing the parcel's original **pressure** and **temperature** with the **saturation mixing ratio** line having the same value of the **mixing ratio** as the parcel. The pressure and temperature at the lifting condensation level are usually called the **condensation pressure** and **condensation temperature**, respectively, and the corresponding point on a **thermodynamic diagram** is called either the **characteristic point, adiabatic saturation point,** or **adiabatic condensation point**. *See* **convective condensation level, conditional instability, saturation level**.

lifting condensation level zone—(Abbreviated LCL zone.) The **range** of altitudes within which the **lifting condensation level** (LCL) occurs for different air parcels rising from near the surface.

Due to natural **variability** and land-use heterogeneity, air parcels at different horizontal locations near the ground usually have slightly different temperatures and humidities. As a result, each **air parcel** has its own LCL. Over a town, for example, LCLs might vary from 1000 m over an irrigated park to 1400 m over a paved parking area, thus giving a 400 m thick LCL zone centered at 1200 m. This implies that **cumulus** clouds formed from rising surface-layer air will have slightly different cloud-base altitudes within a region. While it is difficult to see this effect when observed from the ground, it is very obvious to aircraft flying at the average cloud-base **altitude**.

light—Often synonymous with **visible radiation** but sometimes applied to **electromagnetic radiation** well outside the **visible spectrum**.

light air—In the **Beaufort wind scale**, a **wind** with a speed from 1 to 3 knots (1 to 3 mph) or Beaufort Number 1 (Force 1).

light breeze—In the **Beaufort wind scale**, a **wind** with a speed from 4 to 6 knots (4 to 7 mph) or Beaufort Number 2 (Force 2).

light freeze—The occurrence of **air temperature** below 0°C (32°F) that kills some, but not all, annual vegetation.

This often occurs in the 0° to −1°C (32°–30°F) **range**. *See* **freeze**.

light frost—A thin and more or less patchy deposit of **hoarfrost** on surface objects and vegetation. This term is used, inappropriately, for a **light freeze**.

light ion—Same as **small ion**.

light-of-the-night-sky—*See* **airglow**.

light pillar—A general term for what is often called a **sun pillar**, but one that allows for its appearance with such things as street lamps.

light scattering—*See* **scattering**.

lightning—Lightning is a **transient**, high-current **electric discharge** with pathlengths measured in kilometers.

The most common source of lightning is the **electric charge** separated in ordinary **thunderstorm** clouds (**cumulonimbus**). Well over half of all lightning discharges occur within the thunderstorm cloud and are called intracloud discharges. The usual cloud-to-ground lightning (sometimes called **streak lightning** or **forked lightning**) has been studied more extensively than other lightning forms because of its practical interest (i.e., as a cause of injury and death, disturbances in power and communication systems, and ignition of forest fires) and because lightning channels below **cloud level** are more easily photographed and studied with optical instruments. Cloud-to-cloud and cloud-to-air discharges are less common than intracloud or cloud-to-ground lightning. All discharges other than cloud-to-ground are often lumped together and called **cloud discharges**. Lightning is a self-propagating and electrodeless atmospheric **discharge** that, through the **induction** process, transfers the electrical **energy** of an electrified cloud into electrical charges and **current** in its ionized and thus conducting channel. Positive and negative leaders are essential components of the lightning. Only when a **leader** reaches the ground does the ground potential wave (**return stroke**) affect the lightning process. Natural lightning starts as a bidirectional leader, although at different stages of the process unidirectional leader development can occur. Artificially triggered lightning starts on a tall structure or from a **rocket** with a trailing wire. Most of the lightning energy goes into **heat**, with smaller amounts transformed into sonic energy (**thunder**), **radiation**, and **light**. Lightning, in its various forms, is known by many common names, such as streak lightning, forked lightning, **sheet lightning**, and **heat lightning**, and by the less common **air discharge**; also, the rare and mysterious **ball lightning** and **rocket lightning**. An important effect of worldwide lightning activity is the net **transfer** of negative charge from the **atmosphere** to the earth. This fact is of great important in one problem of **atmospheric electricity**, the question of the source of the **supply current**. Existing evidence suggests that lightning discharges occurring sporadically at all times in various parts of the earth, perhaps 100 per second, may be the principal source of negative charge that maintains the earth–ionosphere **potential** difference of several hundred thousand volts in spite of the steady transfer of charge produced by the **air–earth current**. However, there also is evidence that **point discharge currents** may contribute to this more significantly than lightning. *See also* **cloud-to-ground flash**, **intracloud flash lightning discharge**.

Chalmers, J. A., 1957: *Atmospheric Electricity*, 235–255.
Schonland, B. F. J., 1950: *The Flight of Thunderbolts*, 152 pp.
Hagenguth, J. H., 1951: *Compendium of Meteorology*, 136–143.

lightning arrester—Any device designed to carry to the ground (to "ground") the short-duration **surge** currents that appear on power lines and telephone lines during severe thunderstorms.

Early forms of arresters consisted merely of spark gaps to ground, but in cases where the circuit power maintained the spark after termination of the surge, service could only be restored by momentarily shutting off circuit power. Hence, more elaborate arresters have been developed, especially for high-voltage lines, with features to ensure that the circuit power will not maintain the ground circuit after the surge dies out. *Compare* **lightning rod**.

lightning channel—The irregular path through the air along which a **lightning discharge** occurs.

The lightning channel is established at the start of a **discharge** by the growth of a **leader**, which seeks out a path of least resistance between a charge source and the ground or between two charge centers of opposite sign in the **thundercloud** or between a **cloud** charge center and the surrounding air or between charge centers in adjacent clouds.

lightning counter—A device for measuring the number of **lightning** events within a specified time interval.

lightning current—The **current** flowing in a component of the **lightning flash**.

It is usually considered to be the current in the **return stroke**.

lightning damage— Direct damage to property and any indirect damage (due to fire or accidents) caused by **lightning**.

lightning detection network—An integrated array of **lightning** direction finders that provide information for trigonometric location of **cloud-to-ground lightning discharges**.

Timing and direction information from individual receivers are combined to provide evolving maps of lightning occurrences across vast regions that sometimes reach beyond the **range** of **storm** surveillance radars. *See* **sferics receiver**.

lightning direction finder—*See* **sferics receiver**.

lightning discharge—The series of electrical processes taking place within 1 s by which charge is transferred along a **discharge** channel between **electric charge** centers of opposite sign within a **thundercloud (intracloud flash)**, between a **cloud** charge center and the earth's surface (**cloud-to-ground flash** or **ground-to-cloud discharge**), between two different clouds (intercloud or **cloud-to-cloud discharge**), or between a cloud charge and the air (**air discharge**).

It is a very large-scale form of the common **spark discharge**. A single lightning discharge is called a **lightning flash**.

lightning echo—A **radar echo** from **lightning**.

Lightning echoes are explained by the **scattering** of **radar** waves by the high concentration of free electrons created in the narrow channel of a **lightning discharge**. Because the electrons recombine quickly, lightning echoes have short durations, typically 1 s or less.

lightning flash—The total observed **lightning discharge**, generally having a duration of less than 1 s.

A single **flash** is usually composed of many distinct **luminous** events that often occur in such rapid succession that the human eye cannot resolve them.

Lightning Imaging Sensor—(Abbreviated LIS.) An instrument designed to detect **lightning** from space as part of **TRMM**.

lightning mapping system—A **network** of **lightning** detection equipment for locating the electromagnetic sources of a **lightning flash**.

The **flash**, both intracloud and cloud-to-ground, is mapped in three-dimensional space using equipment with a time resolution of less than 1 μs. Since cloud-to-cloud and cloud-to-air are rare lightning phenomena, mapping them has little or no importance.

lightning recorder—Same as **sferics receiver**.

lightning rod—A grounded metallic conductor with its upper extremity extending above the structure that is to be protected from damage due to **lightning**.

The upper extremity, called the air terminal, should be raised well above the top of the structure, to yield an adequate **radius of protection**. The path to ground must consist of a conductor of low total **resistance** and must contain no points of high local resistance. Lower ends should be buried deeply enough in the earth that they will always be in good contact with **soil moisture**. Connection to water pipes usually affords a good ground. *Compare* **lightning arrester**.

lightning stroke—In a **cloud-to-ground discharge**, a **leader** plus its subsequent **return stroke**.

In a typical case, a cloud-to-ground discharge is made up of three or four successive lightning strokes, most following the same **lightning channel**.

lightning suppression—Procedures to prevent the occurrence of **lightning**.

Seeding below **cloud base** with 10-cm fiber **chaff** in a Colorado study resulted in **corona discharges** that caused a discharging **current** to flow within developing or active thunderstorms. Electric fields below thunderstorms seeded with chaff decayed much faster than electric fields below nonseeded storms, and **chaff seeding** of existing thunderstorms greatly reduced **cloud-to-ground flashes** compared to nonseeded storms. Recent evidence suggests that chaff releases may result in a significant decrease in **downwind** cloud-to-ground lightning. Another experimental approach is to use lasers to **discharge** lightning in an overhead **cloud** in order to divert the **flash** from striking people or highly sensitive equipment on the ground; more research is needed to make this a realistic method of lightning suppression. In the 1960s, **seeding** with **silver iodide** was considered in order to produce an excess of **ice crystals** to cause numerous coronal discharges within the **thunderstorm** and reduce the need for the flash to reach the ground, but the test results were complex and difficult to identify. Finally, electric **space charge** was released into the **atmosphere** from a network of high-voltage wires on the ground to produce corona discharge, but a field test showed minimal effects on suppressing lightning.

lightship code—An international code assigned to a lightship to identify it as the source of meteorological observations.

Lightships are considered fixed **observation** locations to distinguish them from ships that are under way. *See* **ship synoptic code**.

lightship station—Surface **synoptic station** situated aboard a lightship.

Liman Current—A cold southward flowing current between Sakhalin and the Asian mainland carrying water from the Sea of Okhotsk into the Sea of Japan.

limb scanning—Probing the **atmosphere** above the earth's **horizon** from a satellite or other space vehicle.

Measuring **radiation** emitted in the horizontal direction at various altitudes above the limb is a useful method for obtaining vertical profiles of atmospheric constituents.

limestone scrubbing—A system for removal of SO_2 from **pollution** sources by injecting calcined limestone into the exhaust gas system.

In a wet **scrubber** (the most common type), the absorbent, a mixture of water and limestone, is sprayed inside a treatment area onto the **flue gas**, which rises from the bottom of the absorber area. Most of the SO_2 in the gas is absorbed by the droplets of absorbent, and is oxidized into calcium sulfate and calcium sulfite compounds. The used absorbent (waste sludge) is disposed of as solid waste or is sold as a gypsum product. The cleaned flue gas (possibly reheated to avoid corrosive **condensation**) is released to the **atmosphere**.

limit cycle—A steady, closed **oscillation** that either attracts or repels neighboring states of the system.

If a limit cycle were to occur in a system represented by the pair of differential equations

$$dx/dt = f(x, y)$$

$$dy/dt = g(x, y),$$

the limit cycle would correspond to a solution $(x(t), y(t))$ the values of which would trace out a closed loop in the x, y-plane as the system evolves with t; the solution represented by the closed loop would either attract or repel neighboring solutions.

limit of convection—(Abbreviated LOC). The **level of neutral buoyancy** for a saturated rising **air parcel**.

If a **cloudy** air parcel is rising under its own **buoyancy**, then the parcel can rise to the **altitude** where the virtual **potential temperatures** of the parcel and **environment** are equal. Neglecting inertial overshoot, the LOC would mark the top of convective clouds. *Compare* **lifting condensation level, level of free convection**.

limit of the atmosphere—The level at which the atmospheric **density** becomes the same as the density of interplanetary space, which is usually taken to be about one **particle** per cubic centimeter.

limited-area forecast model—A numerical **model** with a horizontal domain that does not cover the entire globe.

Such models must be given lateral **boundary conditions** from either a global forecast model or from another limited-area forecast model with a domain that completely encompasses the original one.

limited fine-mesh model—(Abbreviated LFM.) An adaptation of Shuman and Hovermale's (1968) hemispheric six-layer **primitive equations** numerical weather prediction **model**.

It used a finer horizontal **grid** mesh (190.5 km) to reduce **truncation error** and was integrated over a subhemispheric domain that permitted a shorter wait (1 h) for the acquisition of observational data. When combined with the use of **model output statistics** (MOS), the LFM provided a reliable and timely set of guidance products for operational weather forecasts in the United States from 1975 until 1993.

Shuman, F., and J. Hovermale, 1968: An operational six-layer primitive equation model. *J. Appl. Meteor.*, **7**, 525–547.

limiting angle—In **winds aloft** observations using **radio direction finding** (RDF), the **elevation angle** below which errors in **rawinsonde** tracking caused by **reflection** and **refraction** are so great the resulting **wind** computations are unreliable.

Limiting angles may apply to **azimuth** angle measurements if the azimuth angles are measured near structures such as towers or buildings.

limiting wave—A **wave** of maximum height, for a given **wavelength**, beyond which the wave cannot grow without breaking.

line absorption—**Absorption** due to one or more discrete absorption lines.
Compare **band absorption, continuum absorption**.

line average—The **mean value** of observations taken along a straight or curved line, such as along a highway or along an aircraft track.
See **average, ensemble average, area average, time average, ergodic condition**.

line echo wave pattern—(Abbreviated LEWP.) A special configuration in a line of convective storms that indicates the presence of a **low pressure area** and the possibility of damaging winds and tornadoes.

In response to very strong **outflow** winds behind it, a portion of the line may bulge outward forming a **bow echo**.

Nolen, R. H., 1959: A radar pattern associated with tornadoes. *Bull. Amer. Meteor. Soc.*, **40**, 277–279.

line-end vortices—*See* **book-end vortices**.

line gale—Same as **equinoctial storm**.

line integral—The integral of a function along a given curve.

Mathematically, if s is a **linear** coordinate along the curve, the line integral of a function $f(s)$ between points A and B on the curve is

$$\int_A^B f(s)\,ds,$$

where $f(s)$ is evaluated on the curve. The line integral of a **vector field F** along the same curve is

$$\int_A^B \mathbf{F} \cdot d\mathbf{r},$$

where $d\mathbf{r}$ is a **vector** element along the curve of magnitude ds. In general the integral will depend on the curve; in the special case where the **field F** is **irrotational**, it depends only on the points A and B and therefore vanishes for a closed curve. In meteorology, line integrals are of frequent occurrence, for example, in the **circulation**, and in the averaging of a quantity along a latitude circle.

line intensity—A measure of the total effect of an **absorption** or **emission line**.

The line intensity is equal to the integration of the **absorption coefficient** over the entire shape of the absorption line.

line of apsides—In astronomy, the line joining the points where the distance of an orbiting body is greatest and where it is least from the attractive body.

The line extends infinitely in both directions.

line of sight—(Or line of sight path.) An unimpeded path between two points in the **atmosphere** determined by the **trajectory** of a **ray** between them.

Sometimes synonymous with **optical path**, a term that may, however, be a shortened form of **optical pathlength**, which is not the same as line of sight.

line source—*See* **point source**.

line squall—A **squall** that occurs along a **squall line**.

This term is now confined mostly to nautical usage.

line storm—Same as **equinoctial storm**.

line vortex—An idealized, mathematical **vortex** consisting of the limit of the contraction of a **vortex tube** to a curve in space.

The flow surrounding the curve is assumed **irrotational**.

Batchelor, G. K., 1967: *An Introduction to Fluid Dynamics*, Cambridge, p. 93.

line width—A measure of the finite width of a **spectral line** taken as the distance between the two points either side of the line center where the **absorption** has dropped to half of its value at the center.

Transmission of **atmospheric radiation** is affected by line width, especially in the lower **troposphere** where line widths become large due to **pressure broadening**.

linear—Confined to first-degree algebraic terms in the relevant **variables**.

For example, $a + bx + cy$ is linear in x and y; $a \sin x + b \cos y$ is linear in the coefficients a and b, but **nonlinear** in x and y.

linear correlation—*See* **correlation**.

linear depolarization ratio—(Abbreviated LDR.) The ratio of the **power** received in the **orthogonal**, or cross-polarized, channel to that received in the transmission, or copolarized, channel of a **dual-channel radar**, when a linearly polarized **signal** is transmitted.

Because the main component of the linearly polarized, backscattered signal from **hydrometeors** has the same **polarization** as the transmitted polarization, the linear depolarization ratio yields a value less than unity or, equivalently, a negative **decibel** quantity. Note that relative to the **receiver** channels of the radar the linear depolarization ratio is defined inversely to the **circular depolarization ratio**.

linear differential equation—A differential equation that is **linear** in the **dependent variable** and derivatives thereof.

The existence of a wealth of mathematical techniques and tables for the treatment of linear equations guarantees that a physical problem representable by such an equation is very much easier to solve and understand than a **nonlinear** one. The **advection** terms in the **fundamental equations of hydrodynamics** are not linear (strictly, they are "quasi-linear"), and much of **dynamic meteorology** has been an attempt to circumvent this difficulty.

linear instability—An **instability** that can be described (to first-order **accuracy**) by **linear** (or tangent linear) equations.

linear momentum—Same as **momentum**; *see also* **angular momentum**.

linear operator—1. A mathematical **operator** that involves only a **linear** combination of terms or differentials in the **dependent variable**; examples are the **del operator**, the **Laplacian operator**, and the **differential operator**, L where

$$L = \frac{d^n}{dx^n} + a_1(x) \frac{d^{n-1}}{dx^{n-1}} + \cdots + a_{n-1}(x) \frac{d}{dx} + a_n(x);$$

the coefficients $a_1(x) \cdots a_n(x)$ may be functions of the **independent variable** x.

2. A **weighting** function that is applied to past values of a **time series** in order to obtain estimates of future values, determined mathematically in such a way that the **mean-square error** of **prediction** is minimized.

This technique can be extended to more than one **variable**, so that many time series may be considered simultaneously.

linear tangent equations—*See* **tangent linear equation**.

linear water wave theory—The development of **wave equations** from the full **equations of motion**, based on the assumption that the **wave height** to **wavelength** ratio (**wave steepness**) is small.

From this theory it is possible to calculate quantities such as the **phase velocities** and **group velocities** of the waves.

linearity—The property of being **linear**.

linearization—A process of reduction to **linear** form by appropriate change of variables or by approximation.

For example, 1) the equation $Y = Ae^{bx}$ becomes $y = a + bx$ by the transformation $y = \log Y$; $a = \log A$; or 2) the function $\exp(x)$ can be approximated by the linear Taylor **polynomial** $1 + x$ for small $|x|$.

linearized differential equation—A differential equation that has been derived from an original **nonlinear** equation by the treatment of each **dependent variable** as consisting of the sum of an undisturbed or steady component and a **small perturbation** or **deviation** from this mean.

It is assumed that the product of two **perturbation** quantities is negligible compared to the first-order terms in the perturbations or to the undisturbed variables. This process of **linearization**, often called the **method of small perturbations**, leads to a **linear differential equation** with the perturbations of the original dependent variables as the new dependent variables. It has been used successfully to solve problems involving **sound waves**, **gravity waves**, **frontal waves**, **tides**, waves in the upper **westerlies**, and flow over hills and mountains.

Haurwitz, B., 1951: *Compendium of Meteorology*, 401–420.
Panofsky, H., 1954: *Introduction to Dynamic Meteorology*, ch. 4.

lines of force—*See* **electric lines of force**, **magnetic lines of force**.

Linke turbidity factor—A measure of atmospheric **turbidity**, equal to the ratio of total **optical depth** to the **Rayleigh optical depth**.

It is a function of **wavelength**.

Lips—The ancient Greek name for the southwest **wind**, which is the **sea breeze** in Athens.

On the **Tower of the Winds** it is represented by a bare-legged young man carrying a piece of a trireme. This may indicate either that the wind favored home-coming ships or that, when stormy, it caused wrecks. Today the name is applied to any **hot wind**, usually the **sirocco**.

liquid-in-glass thermometer—A **thermometer** in which the thermally sensitive element is a liquid contained in a graduated glass envelope.

The indication of such a thermometer depends upon the difference between the coefficients of **thermal** expansion of the liquid and the glass. Mercury and alcohol are liquids commonly used in meteorological thermometers. *See* **spirit thermometer**.

liquid-in-metal thermometer—A **thermometer** in which the thermally sensitive element is a liquid contained in a metal envelope, frequently in the form of a **Bourdon tube**.

Otherwise, the indicator portion may be liquid-in-glass.

liquid thermometer—A **liquid-in-glass thermometer** or **liquid-in-metal thermometer**.

liquid water content—*See* **water content**.

liquid water loading—The amount of liquid water present within an **air parcel** as **cloud** droplets, **rain**, or **ice**, usually expressed in percent or fraction by weight (e.g., as a **liquid water mixing ratio** r_L) or volume.

The higher the liquid water loading, the greater the average density and colder the **virtual temperature** of the parcel.

liquid water mixing ratio—The ratio of the mass of liquid water to the mass of **dry air** in a unit volume of air.

Units are mass of liquid water per mass of dry air, such as g kg^{-1}. *Compare* **mixing ratio**, **total water mixing ratio**.

liquid water path—A measure of the weight of the liquid water droplets in the **atmosphere** above a unit surface area on the earth, given in units of kg m^{-2}, for example.

The liquid water path may be defined as

$$W_p = \int \rho_{air} r_L dz,$$

where ρ_{air} is the **density** of the (wet) air, r_L is the **liquid water mixing ratio**, and the integral is from the bottom to the top of the column. *See* **liquid water loading**.

liquid water potential temperature—A quantity that is conserved in reversible **adiabatic** motion.

In the simplest approximation it is given by θ_L, where

$$\theta_L \approx \theta - \frac{L_v}{c_{pd}} r_L$$

with θ the **potential temperature**, L_v the **latent heat** of **vaporization**, c_{pd} the **specific heat** of **dry air** at constant **pressure**, and r_l the **liquid water mixing ratio**. A more accurate expression is

$$\theta_L = \theta \left(\frac{\epsilon + r_v}{\epsilon + r_t} \right)^\chi \left(\frac{r_v}{r_t} \right)^{-\gamma} \exp\left[\frac{-L_v r_l}{(c_{pd} + r_t c_{pv}) T} \right],$$

in which χ is the **Poisson constant**, ϵ the ratio of the gas constants of dry air and **water vapor** (0.622), r_v the mixing ratio of water vapor, c_{pv} the specific heat of water vapor, T the **temperature** and $\gamma = r_t R_v / (c_{pd} + r_t c_{pv})$, where r_t is the **total water mixing ratio** and R_v the **gas constant** for water vapor. Three quantities are conserved in reversible adiabatic motion: **equivalent potential temperature**, total water mixing ratio, and liquid water potential temperature. Any two of these may be considered independent, with the third deducible from those two.

Deardorff, J., and K. A. Emanuel, 1994: *Atmospheric Convection*, Oxford Univ. Press, 580 pp.

liquid water static energy—A **thermodynamic variable** (analogous to **liquid water potential temperature**) calculated by first evaporating any liquid water present in the air, and then lowering it **adiabatically** to a reference **altitude** z often taken either as the local surface ($z = 0$), or as the altitude where the **ambient pressure** is 100 kPa:

$$s_L = C_p T + gz - L_v r_L,$$

where g is gravitational **acceleration**, L_v is the **latent heat** of **vaporization**, C_p is the **specific heat** at constant **pressure** for air, T is **absolute** temperature, and r_L is the **liquid water mixing ratio** in the air.

This **variable** is conserved for **moist-adiabatic processes**. *Compare* **dry static energy**, **moist static energy**, **static energy**.

LIS—Abbreviation for **Lightning Imaging Sensor**.

lithometeor—The general term for dry substances suspended in the **atmosphere**, including **dust**, **haze**, **smoke**, and **sand**.

Compare **hydrometeor**, **igneous meteor**, **luminous meteor**; *see* **atmospheric phenomenon**.

lithosphere—1. The solid portion of the earth, as compared to the **atmosphere** and the **hydrosphere**.

2. In plate tectonics, a layer of strength relative to the underlying asthenosphere for deformation at geologic rates.

It includes the crust and part of the upper mantle and is of the order of 100 km in thickness. (Glossary of Geology 1997)

American Geological Institute, 1997: *Glossary of Geology*, 4th ed., J. A. Jackson, Ed., p. 372.

Little Ice Age—A period between approximately A.D. 1550 (or perhaps as early as 1300) and 1850 in which mountain glaciers advanced in many parts of the world.

The precise timing of the advances and retreats varied from region to region. Temperatures were not uniformly colder throughout this period, but rather showed marked variations on decadal timescales.

livestock safety index—(Abbreviated LSI.) An index of animal heat stress.

Categories defined from the **temperature–humidity index** (THI) are based upon an increasing death rate of livestock as the THI value becomes larger. LSI has been related to other heat-related responses of mammals such as weight gain, milk production, and blood chemistry.

THI	LSI	Effect
<75°F (23°C)	No stress	No heat-related problems
75°–78°F (23°–25°C)	Livestock alert	Reduced weight gain, lower milk production, increased respiration rate
79°–83°F (26°–28°C)	Livestock danger	Reduced weight gain, lower milk production, potential mortality if animals further stressed
> 84°F (28°C)	Livestock emergency	Mortality possible if animals further stressed by activity, lack of water, lack of external cooling

living glacier—A **glacier** experiencing **glacier flow**.

Expansion or retreat is determined by the balance between mass **accretion** and mass depletion processes.

Livingstone sphere—A **clay atmometer** in the form of a sphere.

Evaporation indicated by this instrument is supposed to be somewhat representative of that from plant growth.

llebetjado—In Catalonia (northeastern Spain), a hot squally **wind** descending from the Pyrenees and lasting for a few hours.

llevant—*See* **levante**.

LLJ—Abbreviation for **low-level jet**.

lobe—In an **antenna pattern**, a region of local maximum in the emitted **intensity**. The strongest lobe is in the pointing direction of a directional **antenna** and is called the **main lobe**.

The configuration of lobes is determined by three factors: 1) **wavelength**; 2) geometrical properties of the antenna and **feed** system; and 3) mutual **interference** between the direct and reflected rays for an antenna situated above a reflecting surface. The sidelobes or minor lobes are an unavoidable consequence of the finite size of the antenna. Though undesirable, they ordinarily contain much less **power** than the main lobe. *See* **antenna pattern**.

lobe pattern—Same as **antenna pattern**.

LOC—Abbreviation for **limit of convection**.

local acceleration—*See* **acceleration**.

local angular momentum—**Angular momentum** about an arbitrarily located vertical axis that is fixed with respect to the earth.

local area coverage—(Abbreviated LAC.) Ten minutes of **AVHRR** data at full **resolution** (1.1 km at **nadir**) stored on **digital** tape recorders for subsequent playback to **NOAA** Command and Data Acquisition ground stations.

local axis—*See* **curvilinear coordinates**.

local change—The change in a **variable** during a specified interval of time at a fixed point in the **coordinate system** in use.

See **local derivative**.

459

local closure—A method of approximating the effects of **turbulence** at some height z_k that considers only the meteorological state in the immediate vicinity of that height.

For example, if estimating a **turbulent flux** at z_k, one need consider only the meteorological state (**wind speed, wind shear, wind** curvature, **temperature** gradient, etc.) at z_k. The same approach works for higher-order turbulence statistics at z_k, which can be estimated from values and gradients of lower-order **statistics**. Local closure implicitly assumes that turbulence consists of only small **eddies**, and thus it has difficulty approximating those flows where large eddies are important, such as for **thermals** in the **convective boundary layer**. Nonetheless, local closures are easier to use than **nonlocal closures**.

local derivative—The rate of change of a quantity with respect to time at a fixed point of a fluid, $\partial f / \partial t$.

It is related to the **individual derivative** df / dt through the expression

$$\frac{\partial f}{\partial t} = \frac{df}{dt} - \mathbf{V} \cdot \nabla f,$$

where f is a thermodynamic property $f(x, y, z, t)$ of the fluid, $\mathbf{V}$ the **vector** velocity of the fluid, and ∇ the **del operator**. *See* **partial derivative**.

local effects—The production of weather and **climate** idiosyncrasies at a particular location by nearby natural or artificial topographic features.

local establishment—*See* **lunitidal interval**.

local forecast—Generally, any **weather forecast** of conditions over a relatively limited area, such as a city or airport.

local free-convection similarity—Similar to **mixed-layer similarity** as a way to find empirically universal relationships between boundary layer variables, except using a general height z in place of the depth of the **mixed layer** z_i and using local values of fluxes at height z (including in the definition for **Deardorff velocity**).

This approach is designed for use in the statically unstable **surface layer**. *Compare* **local similarity, similarity theory, dimensional analysis, Buckingham Pi theory**.

local horizon—*See* **horizon**.

local inflow—Water entering a **stream** between two gauging stations or, in the case of a lake or **reservoir**, between its principal **inflow** and **outflow** water courses.

local isotropy—The near-isotropic property exhibited by the high-frequency or finescale portion of otherwise **nonisotropic turbulence** that occurs at sufficiently large **Reynolds numbers**.

See **isotropic turbulence**.

local lightning-flash counter—An instrument for counting nearby **lightning** flashes.

local Mach number—The **Mach number** of an isolated section of an airplane or its airframe, as contrasted with the **free-stream Mach number**.

local similarity—Similar to **local free-convection similarity**, except designed for the statically **stable boundary layer**.

Also known as z-less scaling, because the appropriate length **scale** does not depend on height z. *Compare* **similarity theory, dimensional analysis, Buckingham Pi theory**.

local similarity hypothesis—*See* **Kolmogorov's similarity hypotheses**.

local storm—A **storm** of mesometeorological **scale**; thus, **thunderstorms, squalls**, and **tornadoes** are often put in this category.

local thermodynamic equilibrium—(Abbreviated LTE.) A condition under which matter emits **radiation** based on its intrinsic properties and its **temperature**, uninfluenced by the magnitude of any incident radiation.

LTE occurs when the **radiant energy** absorbed by a molecule is distributed across other molecules by collisions before it is reradiated by **emission**. LTE is needed for **Planck's law** and **Kirchhoff's law** to apply, and is typically satisfied at atmospheric pressures higher than about 0.05 mb. Laser radiation is an example of non-LTE emission.

local time—Time, as measured for a given point on the earth.

Local sidereal time, for instance, is by Greenwich sidereal time minus the local longitude of the point (15° longitude is equivalent to a one-hour change in time).

local vorticity—1. The **vorticity** of the earth about the local **zenith**.

2. Same as **relative vorticity**.

local winds—1. **Winds** that, over a small area, differ from those that would be appropriate to the general large-scale **pressure distribution**, or that possess some other peculiarity.

Often these winds have names unique to the area where they occur. Local winds may be classified into three main groups. The first includes diurnally varying airflows that are driven by local gradients of surface **heat flux** (e.g., near the shore of a sea or lake) or by **diurnal heating** or cooling of the ground surface in areas of sloping or mountainous terrain. These include land and **sea breezes**, **mountain–valley circulations**, and **drainage** and **slope winds**. The second group consists of winds produced by the interaction of a **synoptic-scale** flow with **orography**. These may be further subdivided into **barrier jets**, **gap winds**, **downslope windstorms**, and include such local phenomena as the **tehuantepecer**, **Santa Ana**, **foehn**, **mistral**, and **bora**. The third group includes those winds accompanying **convective activity**, more specifically individual **thunderstorms** or **mesoscale convective systems**. These are generally the surface manifestations of precipitation-cooled diverging **outflow** and in some locations are given special names due to the distinctive character of the weather associated with them (e.g., the **haboob**).

2. Local or colloquial names given to frequently occurring or particularly noteworthy winds (sometimes because of the bad weather associated with them), usually from a certain direction.

Often these names reflect the direction from which the **wind** comes (e.g., **sou'wester**, **nor'easter**).

locally generated sea—The wavefield at a given location resulting from the **local effects** of **wind**, without being affected by **swell** generated elsewhere and propagating into the area.

lodos—A southerly **wind** on the Black Sea coast of Bulgaria.

loehis—*See* aloegoe.

lofting—The phenomenon where the upper part of a smoke **plume** diffuses more rapidly upward than the bottom part diffuses downward.

This generally occurs when the **boundary layer** near the ground is more stable than it is aloft. *Compare* **coning, fanning, looping**.

logarithm—The logarithm of any positive number n to the base b is the power l to which that base must be raised in order to satisfy the identity $n = b^l$: $l = \log_b n$.

Logarithms to the base 10 are called common logarithms and written log or $\log_{10}$. Logarithms to the base $e = 2.7182818284\ldots$ are called natural (Napierian, hyperbolic) logarithms, and are often written $\log_e$ or ln. The natural logarithms are the more convenient in any computations involving differentiation

logarithmic differentiation—Finding derivatives by taking the **logarithm** of both sides of an equation and then differentiating.

In meteorology this process is applied, for example, to the **equation of state** for a **perfect gas**, giving

$$\frac{dP}{P} = \frac{d\rho}{\rho} + \frac{dT}{T},$$

or to the **Poisson equation** for **potential temperature**, giving

$$\frac{d\theta}{\theta} = \frac{dT}{T} - \mathrm{K}\,\frac{dP}{P},$$

where P is **pressure**, ρ **density**, T **temperature**, θ potential temperature, and K Poisson's constant.

logarithmic scale—A nonuniform **scale** representing the function $y = \log x$.

See **logarithm, alignment chart**.

logarithmic velocity profile—The **variation** of the mean wind speed with height in the **surface boundary layer** derived with the following assumptions: 1) the mean motion is one-dimensional; 2) the **Coriolis force** can be neglected; 3) the **shearing stress** and **pressure gradient** are independent of height; 4) the **pressure force** can be neglected with respect to the **viscous force**; and 5) the **mixing length** l depends only on the fluid and the distance from the boundary, $l = kz$.

Near aerodynamically smooth surfaces, the result is

$$\frac{\bar{u}}{u_*} = \frac{1}{k} \ln\left(\frac{u_* z}{v}\right) + 5.5,$$

that is, the logarithmic velocity profile, where u_* is the **friction velocity** and v the **kinematic**

461

viscosity. $k \cong 0.4$ and has been called the Kármán constant or **von Kármán's constant**. The equation fails for a height z sufficiently close to the surface. For aerodynamically rough flow, **molecular viscosity** becomes negligible. The profile is then

$$\frac{\overline{u}}{u_*} = \frac{1}{k} \ln \left(\frac{z}{z_0} \right), z \geq z_0.$$

z_0 is a constant related to the average height ϵ of the surface irregularities by $z_0 = \epsilon/30$ and is called the **aerodynamic roughness length**. Another derivation of the logarithmic profile was obtained by Rossby under the assumption that for fully rough flow the roughness affects the **mixing length** only in the region where z and z_0 are comparable. Then $l = k(z + z_0)$ and

$$\frac{\overline{u}}{u_*} = \frac{1}{k} \ln \left(\frac{z + z_0}{z_0} \right).$$

For statically nonneutral conditions, a **stability** correction factor can be included (*see* equation in definition of **aerodynamic roughness length**).

Haugen, D. A., 1973: *Workshop on Micrometeorology*, Amer. Meteor. Soc., 392 pp.
Sutton, O. G., 1953: *Micrometeorology*, sect. 3.9.

lognormal cloud-size distribution—The **logarithm** of **cumulus** cloud diameters and **cloud** thicknesses is normally (Gaussian) distributed, according to many observations.

$$f(x) = \frac{\Delta x}{(2\pi)^{1/2} x s_x} \exp \left[-0.5 \left(\frac{\ln (x/L_x)}{s_x} \right)^2 \right],$$

where x is cloud size, Δx is a small **range** of cloud sizes (i.e., bin width), $f(x)$ is the fraction of clouds with sizes between $x - 0.5\Delta x$ and $x + 0.5\Delta x$, L_x is a location **parameter**, and s_x is a dimensionless spread parameter.

Stull, R. B., 1995: *Meteorology Today for Scientists and Engineers*, 385 pp.

lolly ice—Saltwater **frazil**, a heavy concentration of which is called **sludge**.

lombarde—An easterly **wind** (from Lombardy) that predominates along the French–Italian frontier.
It comes from the High Alps. In winter it is violent and forms **snowdrifts** in the mountain valleys. In the plains it is gentle and very dry. It is associated with an **anticyclone** over France and central Europe, or with high pressure to the southeast of Europe and low pressure to the northwest along with falling **pressure** over western France.

London (sulfurous) smog—Deadly mixture of **smoke** and **fog** peaking in the midtwentieth century in large cities.
A **smog** episode in London in 1952 led to 4000 deaths. The **sulfuric acid** produced from the fossil fuel sources in use at that time led to a choking mixture when incorporated into fog droplets. It is associated with low temperatures, low actinic **flux**, and high humidity. This form of **air pollution** was largely eliminated by legislation in the 1950s that led to reduced emissions of SO_2 and smoke. See **Los Angeles (photochemical) smog**.

long-crested wave—Ocean **surface waves** that are nearly two-dimensional, in that the crests appear very long in comparison with the **wavelength**, and the **energy** propagation is concentrated in a narrow band around the mean wave direction.

long-range forecast—A forecast for a period greater than seven days in advance, although there are no absolute limits to the period embraced by the definition.
Compare **medium-range forecast, extended forecast**.

long shore wind—1. A damp unpleasant **wind** that blows from the south in Madras (India).
2. A **wind** from the northeast at night in Sri Lanka.

long-term hydrological forecast—Prediction of future hydrologic variables such as **streamflow** or snowmelt for periods of at least one week and up to several months in advance.

long-train atmospherics—An extended train of waveform **atmospherics**.

long wave—1. (Or major wave; also called planetary wave.) With regard to atmospheric **circulation**, a **wave** in the major belt of **westerlies** that is characterized by large length and significant **amplitude**.
The wave length is typically longer than that of the rapidly moving individual **cyclonic** and **anticyclonic** disturbances of the lower **troposphere**. The angular **wavenumber** of long waves is generally taken to be from 1 to 5. *Compare* **short wave**; *see* **Rossby wave**.

2. (Also called shallow water wave.) A **wave** with a relatively long **wave length** and **period**. For **ocean waves**, this is typically a wave of period greater than about 10 s and wave length greater than about 150 m.

long-wave formula—*See* **Rossby wave**.

longitude—The angular distance along the **equator** measured from the prime **meridian** to the meridian of the point in question.
Compare **latitude**.

longitudinal gustiness—*See* **gustiness components**.

longitudinal roll vortices—*See* **horizontal convective rolls**.

longitudinal rolls—(Also called **rolls**, roll vortices, organized large eddies.) Atmospheric **coherent structures** in the form of persistent organized counterrotating roll vortices that are approximately aligned with the mean wind and span the depth of the **planetary boundary layer**.
Longitudinal rolls are frequently present in the **atmospheric boundary layer** in near-neutral to moderately unstable **stratification**. They are believed to be the result of **nonlinear** equilibration of mixed convective–dynamic **normal** mode instabilities of the mean boundary flow. Longitudinal rolls produce a nonlocal **transport** not only of **momentum**, but also of **scalar** quantities that mix the **boundary layer** more efficiently than local **turbulent diffusion**. The quasi-two-dimensional longitudinal rolls generate a mean secondary **circulation** that organizes the smaller-scale three-dimensional turbulent **eddies** into **linear** patterns. The existence of longitudinal rolls significantly changes the fluxes within the boundary layer and at the surface. **Flux** profiles also differ between the **updraft** and **downdraft** regions of longitudinal rolls. In favorable thermodynamic conditions **cloud streets** (linear boundary layer cloud patterns) form in the updraft regions between the rolls. *See* **coherent structures, Langmuir circulation, two-dimensional eddies, horizontal convective rolls**.

> Etling, D., and R. A. Brown, 1993: Roll vortices in the planetary boundary layer: A review. *Bound.-Layer Meteor.*, **65**, 215–248.

longitudinal section—A section diagram plotted in the longitudinal dimension of a **stream**.
It is usually used to plot **energy** grade lines.

longitudinal wave—(Also called compressional wave.) An **irrotational** plane **wave A** parallel to the propagation direction in the sense that if $\mathbf{A} = \mathbf{A}_0 \exp(i\mathbf{k} \cdot \mathbf{x} - i\omega t)$, where $\mathbf{k}$ is the **wave vector**, then $\mathbf{k} \times \mathbf{A} = 0$.
An example is an **acoustic wave** in an inviscid medium. *Compare* **transverse wave**.

longitudinal wind—**Head wind** or **tail wind**.

Longmont anticyclone—A **mesoscale** zone of anticyclonically turning winds that develops **downstream** of the Cheyenne Ridge in northeast Colorado and southeast Wyoming, and is often centered just east of the foothills of the Rocky Mountains near the town of Longmont, Colorado.
The cause of the feature is the interaction of the ambient low level northwest flow with the east–west terrain feature known as the Cheyenne Ridge. *See also* **Denver convergence–vorticity zone**.

longshore current—(Also called littoral current.) The resultant current produced by waves being deflected at an angle by the shore.
In this case the current runs roughly parallel to the shoreline. The longshore current is capable of carrying a certain amount of material as long as its **velocity** remains fairly constant; however, any obstruction, such as a submarine rock ridge or a land point cutting across the path of the current, will cause loss of velocity and consequent loss of carrying power.

longwave—Shorthand for **longwave radiation**.

longwave radiation—In meteorology, a term used loosely to distinguish **radiation** at wavelengths longer than about 4 μm, usually of terrestrial origin, from those at shorter wavelengths (**shortwave radiation**), usually of solar origin.
See also **terrestrial radiation**.

loo—(Also called lu, loo marna.) A **hot wind** from the west in India.

Loofah—World War II name for a polar **nephelometer** developed by the British Admiralty Research Laboratory.

lookup table—(Abbreviated LUT.) A mapping of an integer value to another integer value.
This is used in displaying images and represents a mapping of the **pixel value** in the **image** or picture to the value shown on the **display** device. *See also* **color lookup table**.

looming—A **mirage** in which the **image** of distant objects is displaced upward.

Because the displacement increases with distance, a horizontal surface, such as that of a body of water, appears to bend upward and one's perception is that of being inside a broad shallow bowl. Indeed, the upward bending surface results in an (optical) **horizon** that can be much farther from the observer than in the absence of a mirage. Looming is an example of a **superior mirage**. The opposite of looming is **sinking**. Looming occurs when the concave side of **light** rays from a distant object is down, and this in turn occurs when the **refractive index** of the **atmosphere** decreases with height. This is very common, but the displacement is usually sufficiently small as to be unremarkable. However, when there is a **temperature inversion** over the surface, the looming can be striking.

loop antenna—An **antenna** consisting of a conducting coil of any convenient **cross section**, generally circular, that emits and receives radio **energy**.

The principal **lobe** of the **radiation pattern** is wide and is in the direction perpendicular to the plane of the loop. Its primary application in **radio meteorology** is to the **detection** of such **low frequency** radio waves as are employed in **sferics** and in certain radio direction-finding equipment.

Loop Current—The passage of water through the Gulf of Mexico from Yucatan Strait to the Straits of Florida and the connection between the Caribbean and Florida Currents.

The Loop Current is part of the western **boundary current** system of the North Atlantic subtropical **gyre** and as such is swift flowing, extending to great depth, and prone to instabilities. Its path includes a large northward excursion into the gulf beyond 27°N but retreats to 25°N when shedding an **eddy**. Eddies drift slowly westward into the central and western Gulf of Mexico.

loop rating—A **rating curve** that has higher values of **discharge** for a certain **stage** when the river is rising than it does when the river is falling.

Occasionally, the stage discharge curve describes a loop with each rise and fall of the river.

looping—The phenomenon during unstable atmospheric conditions caused by the presence of large convective turbulent motions where a smoke **plume** waves upward and downward like a garden hose.

Compare **coning, fanning, lofting**.

Lorenz–Mie theory—*See* **Mie theory**.

Los Angeles (photochemical) smog—Type of **air pollution** characterized by high levels of **ozone** and low visibility, typically found in cities located in a valley (e.g., Los Angeles, Denver, Mexico City).

Sunlight, oxides of **nitrogen**, and **hydrocarbons** (the latter two of which arise from automobile exhaust) are all required in order for **smog** formation to occur. The most severe episodes occur when a strong **temperature inversion** caps the location and traps the pollutants. The **degradation** in **visibility** is associated with the **light scattering** due to particulate matter. See **London (sulfurous) smog**.

Loschmidt's number—**Avogadro's number** per unit liter, that is, 2.687×10^{22} per liter.

Love wave—A type of seismic **surface wave** having a horizontal motion that is **shear** or transverse to the direction of propagation.

Its **velocity** depends only on **density** and rigidity **modulus**, and not on **bulk modulus**. It is named after A. E. H. Love, the English mathematician who discovered it.

low—(Sometimes called depression.) In meteorology, an "area of low pressure," referring to a minimum of **atmospheric pressure** in two dimensions (closed isobars) on a **constant-height chart** or a minimum of height (closed contours) on a **constant-pressure chart**.

Since a low is, on a **synoptic chart**, always associated with **cyclonic circulation**, the term is used interchangeably with **cyclone**. *Compare* **high**.

low aloft—Same as **upper-level cyclone**.

low cloud—*See* **cloud classification**.

low-flow channel—Stream channel occupied during periods of low flow.

low frequency—(Abbreviated LF.) *See* **radio frequency band**.

low index—A relatively low value of the **zonal index** that, in mid latitudes, indicates a relatively weak westerly component of **wind** flow (usually implying stronger north–south motion), and the characteristic weather attending such motion.

A **circulation pattern** of this type is commonly called a low-index situation.

low-level jet—(Abbreviated LLJ; also called low-level jet stream.) A **jet stream** that is typically found in the lower 2–3 km of the **troposphere**.

At night, sometimes called a **nocturnal jet**. Examples are the **African jet** and the **Somali jet**.

low pressure area—*See* **low**.

low pressure center—*See* **pressure center**.

low pressure system—*See* **low**.

low-temperature hygrometry—The study that deals with the measurement of **water vapor** at low temperatures.

The techniques used differ from those of conventional **hygrometry** because of the extremely small amounts of moisture present at low temperatures and the difficulties imposed by the increase of the time constants of the standard instruments when operated at these temperatures.

low tide—Same as **low water**.

low water—(Popularly called low tide.) The minimum water level reached in a tidal **cycle**.

lower atmosphere—Generally and quite loosely, that part of the **atmosphere** in which most weather phenomena occur (i.e., the **troposphere** and lower **stratosphere**); hence used in contrast to the common meaning for the **upper atmosphere**.

In other contexts, the term implies the lower troposphere.

lowest astronomical tide—The lowest level of **tide** that can be predicted to occur under average meteorological conditions and under any combination of astronomical conditions; often used to define **chart datum** where the tides are semidiurnal.

Lowitz arcs—*See* **arcs of Lowitz**.

LTE—Abbreviation for **local thermodynamic equilibrium**.

lu—Same as **loo**.

luganot—A strong south or south-southeast **wind** of Lake Garda, Italy.

lull—A momentary decrease in the speed of the **wind**.

lumen—A unit of photometric **power**.

The lumen is equal to the amount of photometric power radiated into a unit solid angle (**steradian**) from a small source having a **luminous** intensity of one **candela**. Tungsten-filament light bulbs produces approximately 15 lumens per **watt**.

luminance—The photometric equivalent of **radiance**.

Luminance is obtained by integrating **spectral radiance** weighted by **luminous efficiency** over the **visible spectrum**. *Compare* **illuminance**.

luminance contrast—*See* **contrast**.

luminescence—Any **emission** of **light** at temperatures below that required for incandescence.

luminous—1. In general, pertaining to the **emission** of **visible radiation**.

2. In **photometry**, a modifier used to denote that a given physical quantity, such as luminous emittance, is weighted according to the manner in which the response of the human eye varies with the **wavelength** of the **light**.

luminous cloud—Same as **sheet lightning**.

luminous contrast—*See* **contrast**.

luminous efficiency—The ratio of the **radiant energy** sensed by the average human eye at a particular **wavelength** to that received.

This ratio reaches a maximum inside the visible portion of the **spectrum** and falls to zero outside it. Luminous efficiency is dimensionless but is often given the units of lumens per **watt**. Photometric quantities are obtained by multiplying the corresponding radiometric quantities by the luminous efficiency and so often bear the adjective **luminous**, for example, luminous flux. However, when the radiometric quantities, **radiance** and **irradiance**, are transformed into photometric ones, they are given the special names **luminance** and **illuminance**. The luminous efficiency of cones differs from that of rods.

luminous meteor—Same as **optical meteor**.

lumped chemical mechanism—A simplified atmospheric chemical **model**, most commonly used for urban airshed models.

In a lumped mechanism, organic compounds of similar chemical formula are grouped or "lumped" together and treated as a single representative species. For example, all **aldehydes** (acetaldehyde, propionaldehyde, etc.) might be treated as a single molecule of formula RCHO, which undergoes chemical reactions representative of the aldehydes.

lunar atmospheric tide—An **atmospheric tide** due to the gravitational attraction of the moon.

The only detectable components are the 12-lunar-hour or semidiurnal, as in the oceanic tides, and two others of very nearly the same **period**. The **amplitude** of this atmospheric tide is so small, about 0.06 mb in the **Tropics** and 0.02 mb in middle latitudes, that it is detected only by careful **statistical** analysis of a long record. *See* **tide**.

lunar day—1. (Also called tidal day.) The time required for the earth to rotate once with respect to the moon, that is, the time between two successive upper transits of the moon.

The mean lunar day is approximately 1.035 times as great as the **mean solar day**, or 24 hours 50 minutes.

2. In astronomy, the time required for the moon to revolve once, relative to a fixed star, about its own axis.

lunar rainbow—(Also known as moonbow.) A **rainbow** formed when the **light** source is the moon rather than the sun.

Even though the optics of both is the same, the **luminance** of the lunar rainbow is much lower. As a result of the eye's reduced sensitivity to color at low light levels, the bow may exhibit little color. There are several reasons why the lunar rainbow is seen much less frequently than the solar bow. While the moon and sun spend equal time above the **horizon**, when they are both present, only a solar bow can be seen. This, by itself, means that the lunar rainbow could form only half as often. Further, the moon goes through phases, and so even at night may not contribute enough light to produce a discernible bow. Finally, the **convective showers** in which the rainbow is frequently seen are much less common at night.

lunar tide—That portion of a **tide** that is due to the **tide-producing force** of the moon. *See* **lunar atmospheric tide**.

lunitidal interval—(Also called local establishment.) An old term for the interval of time at a particular location, between the transit (upper or lower) of the moon and the next semidiurnal **high water**.

This varies slightly during a spring-neap tidal **cycle**. The interval at the times of full and new moon is called "high water full and change." *See* **nonharmonic tidal analysis**, **cotidal hour**.

lux—A photometric unit of **illuminance** or **illumination** equal to one **lumen** per square meter.

A level of illumination between 200 and 1000 lux is generally considered to be adequate for homes and offices. *Compare* **foot-candle**.

Lyman alpha emission line—Feature in the **emission spectrum** of **solar radiation**, identified with neutral **hydrogen**, that occurs at a **wavelength** of 121.567 nm. Because of the extremely low **absorption cross section** of molecular **oxygen** at this particular wavelength, this radiation penetrates deep into the earth's **atmosphere**. It is important to the **dissociation** of **trace gases** in the upper **mesosphere** and to the formation of the lower ionospheric **layers** by ionizing certain minor atmospheric constituent gases, especially **nitric oxide**.

Lyman-alpha hygrometer—A **hygrometer** based on the **absorption** of **radiation** by **water vapor** at the Lyman-alpha line, which is an **emission line** of atomic **hydrogen** at 121.567 nm.

Lyman-alpha radiation can be generated by a **glow discharge** in hydrogen, and **detection** is normally accomplished by a **nitric oxide** ion chamber. Two magnesium fluoride windows both at the radiation source and the detector bound the absorption path. Lyman-alpha hygrometers are used on aircraft and on meteorological towers for high-frequency humidity measurements. Inconveniences of the method, like drift of the source intensity or contamination of the windows, are overcome by special **calibration** techniques or by baselining the high-frequency output to the **humidity** values provided by a slower, but stable, hygrometer.

lysimeter—A type of **evaporation gauge** that consists of a tank or pan of soil placed in a field and manipulated so that the soil, water, **thermal**, and vegetative properties in the tank duplicate as closely as possible the properties of the surrounding area.

Relatively sophisticated lysimeters use various methods to determine the reductions in weight of the instruments so that **water loss** due to **evapotranspiration** can be computed. In some cases, the **water table** level in the lysimeter is adjusted to coincide with a high water table outside the tank. Lysimeters are used for several other purposes. In a potential **evapotranspirometer**, the water table in the tank is held sufficiently high so that vegetative evapotranspiration is not limited by **soil moisture** stress. Some types of lysimeters are devoted to studies of chemical **leaching** by water or to the fate of nutrients or potentially toxic substances, for which water is removed from near the bottom of the tank for chemical analysis. Shear-stress lysimeters have been developed to measure the surface **momentum flux** from the **atmosphere** by use of strain gauges or displacement meters attached to the tank.

M

M component—In a **lightning discharge**, an increase in channel luminosity accompanied by a rapid **electric field** variation, itself called an M electric field change.

The M components occur when the channel is already faintly **luminous**. Downward-moving **leaders** have not been observed to precede M components. The M components may be confused with branch components, the increases in channel luminosity that occur between each branch and the ground when the upward-propagating **return stroke** reaches that branch, since the higher branches are obscured by the **cloud**. The M component is a minor surge of **current** that reilluminates the channel in a negative **cloud-to-ground flash**. It may occur within microseconds or up to a few milliseconds of a return stroke. Evidence exists that M components are a result of the sequence of fast in-cloud negative leaders (**K changes**) contacting a conducting ground channel and renewed ground **potential** wave that reilluminate the channel.

M meter—Name applied to a class of instruments that measure the liquid **water content** of the **atmosphere**.
See **capillary collector**.

M-region—Name given to a region of activity on the sun when the details and nature of that activity cannot be established.

This term was used in the 1940s to account for recurrent geomagnetic storms with a **period** the same as the period of solar rotation relative to the earth, that is, 27.3 days. Today these storms are usually associated with high-speed **solar wind** streams emanating from **coronal holes**. *See* **magnetic storm**.

M-unit—*See* **modified refractivity**.

Mach number—A **dimensionless number**, the ratio of a **characteristic velocity** in a fluid to the **speed of sound** in that fluid.

machine learning—The process by which computer systems can be directed to improve their performance over time.
Examples are **neural networks** and genetic algorithms.

mackerel breeze—(Also called mackerel gale.) A **wind** that ruffles the water, favoring the catching of mackerel.

mackerel sky—A sky with considerable **cirrocumulus** or small-element **altocumulus** cloud, resembling the scales on a mackerel; clouds of the variety **vertebratus**.

Maclaurin series—*See* **Taylor's theorem**.

macroburst—A **downburst** on the **misoscale**.
Bow echoes are often associated with macrobursts. *See also* **microburst**.

macroclimate—The general large-scale **climate** of a large area or country, as distinguished from the **mesoclimate** and **microclimate**.

macroclimatology—The study of **macroclimate**.

macrometeorology—The study of the largest-scale aspects of the **atmosphere**, such as the **general circulation** and weather types.

There is a wide gap between this **scale** and the relatively small scale of **mesometeorology**. The gap is bridged by those atmospheric characteristics, referred to as **cyclonic scale**, that are commonly the subject of **synoptic chart** analysis.

macroscale—Meteorological expression referring to **synoptic** events occurring on a **scale** of thousands of kilometers, such as warm and cold fronts.
Compare **mesoscale**, **microscale**.

macroviscosity—A quantity with the **dimensions** of **kinematic viscosity**, defined as $u_* z_0$, where u_* is the **friction velocity** and z_0 the **aerodynamic roughness length**.

In smooth flow the macroviscosity is of the order of one-tenth the kinematic viscosity. In **normal** atmospheric flow, which is fully rough, it is of the order of 10^{-2} m²s⁻¹, or about 1000 times the kinematic viscosity. *See* **logarithmic velocity profile**.

Madden–Julian oscillation—A quasiperiodic **oscillation** of the near-equatorial **troposphere**, most noticeable in the **zonal wind** component in the **boundary layer** and in the upper troposphere, particularly over the Indian Ocean and the western equatorial Pacific.

This phenomenon is named after the codiscoverers. The oscillation can be detected globally in winds near the **tropopause**. The **period** of the oscillation varies between about 30 and 50 days, and appears to represent an eastward-propagating **disturbance** with the structure of a **Kelvin wave** with a vertical half-wavelength of the depth of the troposphere, but with a **phase speed** of only about 8 m s^{-1}, much less than that of an **adiabatic** Kelvin wave. The disturbance is accompanied by strong fluctuations of deep **convection**, easily detectable using satellite observations, and is a major contributor to intraseasonal weather variability in equatorial regions from eastern Africa eastward to the central Pacific.

maelstrom—A tidal whirlpool found between the islands of Moskenesy and Mosken in the Lofoten Islands of northern Norway.

The term is generally applied to other tidal whirlpools.

maestro—A northwesterly **wind** with fine weather that blows, especially in summer, in the Adriatic; it is most frequent on the western shore and is equivalent to the **etesians** of the eastern Mediterranean.

It is also found on the coasts of Corsica and Sardinia. *Compare* **mistral**.

magmatic water—(Also called juvenile water.) Water brought to the earth's surface from great depths by the upward movement of intrusive igneous rocks.

The quantities of neither magmatic water nor **connate water** are appreciable in comparison to **meteoric water**.

magnetic character figure—Same as **C index**.

magnetic crotchet—*See* **sudden ionospheric disturbance**.

magnetic declination—*See* **declination**.

magnetic dip—*See* **inclination**.

magnetic dipole—1. In **geomagnetism**, either of the two points on the earth's surface where a free-swinging magnetic needle points in a vertical direction.

The line connecting these two points does not pass through the center of the earth. These two points constantly move at a slow rate. They are presently in northern Canada and in the Antarctic south of Australia.

2. *See* **dipole**.

magnetic double refraction—(Or magnetic birefringence; also Voigt effect, Cotton–Mouton effect.) Double **refraction** (or linear **birefringence**) induced in a medium as a consequence of a **magnetic field** applied to it; discovered in 1902 by Voigt, who showed that **light** propagating in a **vapor** to which a strong, perpendicular magnetic field is applied exhibits **double refraction**.

The Cotton–Mouton effect (discovered in 1907) is the Voigt effect in a liquid. All double refraction results from **anisotropy**. Magnetic fields applied to otherwise **isotropic** media provide the anisotropy necessary for magnetic double refraction. Radio waves propagating in the **ionosphere** may exhibit magnetic double refraction as a consequence of the geomagnetic field.

magnetic equator—Same as **aclinic line**.

magnetic field—*See* **magnetic induction**.

magnetic field intensity—The magnitude of the **magnetic field** at any point.

Strictly speaking, the magnetic field is the set of all values of the magnetic field intensity, but magnetic field and magnetic field intensity (as well as **magnetic field strength** and magnetic **vector**) are used more or less interchangeably. The trend is to use magnetic field both for the field taken as a whole and for its value at any point, context being sufficient to determine precise meaning.

magnetic field strength—Same as **magnetic field intensity**.

magnetic inclination—*See* **inclination**.

magnetic induction—A **vector field**, usually denoted by **B**, defined as follows.

The **torque N** experienced by a **magnetic dipole** with magnetic **dipole moment m** is

$$\mathbf{N} = \mathbf{m} \times \mathbf{B}.$$

Thus by measuring **N** for **m** oriented in two **orthogonal** directions, the magnetic induction components are obtained as torque components divided by the magnitude of **m**. The fundamental relation linking **electric field E** and magnetic induction **B** to the force on a charge q with **velocity v** is the Lorentz force equation

$$\mathbf{F} = q(\mathbf{E} + \mathbf{v} \times \mathbf{B}).$$

469

Magnetic induction is sometimes called magnetic field, a term usually applied to a different **field** **H**, related to **B** but different from it. In **free space**, **B** and **H** are proportional:

$$\mathbf{H} = \frac{\mathbf{B}}{\mu_0},$$

where μ_0, the **permeability** of free space, is a universal constant. **B** is the primitive field, whereas **H** is secondary, not strictly needed but convenient. Care must be exercised in deciding if, by magnetic field, **B** or **H** is meant. What is usually meant by the electric and magnetic fields (or the **electromagnetic field**) are **E** and **H**, although according to the Lorentz force equation **E** and **B** are the fundamental fields. Moreover, the Lorentz transformation preserves the (**E**, **B**) structure but not the (**E**, **H**) structure.

magnetic lines of force—Same as **electric lines of force**, but for the **magnetic field**.

magnetic north—At any point on the earth's surface, the horizontal direction of the earth's **magnetic lines of force** (direction of a magnetic **meridian**) toward the north magnetic **pole**, that is, a direction indicated by the needle of a magnetic **compass**.
  Because of the wide use of the magnetic compass, magnetic north, rather than **true north**, is the common 0° (or 360°) reference in much of navigational practice, including the designation of airport runway alignment.

magnetic storm—(Or geomagnetic storm.) A worldwide **disturbance** of the earth's **magnetic field**.
  Magnetic storms are frequently characterized by a sudden onset, in which the magnetic field undergoes marked changes in the course of an hour or less, followed by a very gradual return to normalcy, which may take several days. If extreme enough, they may interfere with the operation of electrical power lines and the operation of artificial satellites. Magnetic storms are caused by solar disturbances, though the exact nature of the link between the solar and terrestrial disturbances is not totally understood. They are more frequent during years of high **sunspot number**. Sometimes a magnetic storm can be linked to a particular **coronal mass ejection**. In these cases, the time between the ejection and onset of the magnetic storm is about one or two days. When these disturbances are observable only in the **auroral zones**, they may be termed polar magnetic storms. *See also* **M-region**.

magnetic variation—*See* **declination**.

magnetic wind direction—The direction, with respect to **magnetic north**, from which the **wind** is blowing; distinguish from **true north** direction.
  Magnetic winds are frequently used in aircraft operation, necessitated by the magnetic frame of reference applied to air navigation facilities (such as designation of runway alignment).

magneto anemometer—A **cup anemometer** with its shaft mechanically coupled to a magneto.
  Both the **frequency** and **amplitude** of the voltage generated are proportional to the **wind speed** and may be indicated or recorded by suitable electrical instruments.

magneto-ionic theory—The theory of the propagation of **electromagnetic waves** by an ionized medium in an external **magnetic field**.
  It applies to the propagation of **radio waves** in the **ionosphere**. *See also* **magnetic double refraction**.
    Davies, K., 1965: *Ionospheric Radio Propagation*, Sec. 2.3.

magnetograph—A recording **magnetometer**.

magnetohydrodynamic wave—More or less synonymous with **Alfvén wave**.

magnetohydrodynamics—(Also called hydromagnetics.) The study of the behavior of an electrically conducting fluid in the presence of a **magnetic field**.

magnetometer—General name for an instrument that measures the earth's **magnetic field intensity**.
  See **sine galvanometer**.

magnetopause—The sharp boundary between the earth's **magnetosphere** and the **solar wind** of interplanetary space.
  The magnetopause represents the outer termination of the earth's **atmosphere** and extends from a distance of several earth-radii in the sunward direction to a much larger and rather indefinite distance in the anti-sunward direction, resulting in a cometlike shape with a long tail directed away from the sun.

magnetosphere—The region in which dynamical motions are strongly influenced, or even dominated, by the earth's **magnetic field** as a result of **plasma** effects.

The magnetosphere extends from the **F-region** of the **ionosphere** to the **magnetopause**. *See* **atmospheric shell.**

magnetospheric convection—Motion induced in the **plasma** of the outer **magnetosphere** by interaction with the **solar wind.**

Magnetospheric convection in turn induces motions in the high-latitude **ionosphere** through linkage along the direction of the geomagnetic **field.**

magnetron—A self-excited **oscillator** used as a **radar** transmitter tube.

Magnetrons are characterized by high peak power, small size, efficient operation, and low operating voltage. Emitted electrons interact with an **electric field** and a strong **magnetic field** to generate **microwave** energy. Because the direction of the electric field that accelerates the **electron beam** is perpendicular to the axis of the magnetic field, magnetrons are sometimes referred to as crossed-field tubes. Unlike a **klystron**, a magnetron is not a coherent transmission **source**, but has a randomly changing **phase** from **pulse** to pulse. A coaxial magnetron uses a different architecture and has better stability, higher reliability, and longer life. Magnetrons are used in inexpensive radars and microwave ovens. *Compare* **klystron.**

main lobe—(Or main beam.) The **lobe** in the **radiation pattern** of a directional **antenna** that includes the region of the maximum radiated **power.**

The center of the main lobe defines the **beam** axis. The **beamwidth** in a given plane is usually defined as the angle within which the radiated **intensity** is at least one-half the intensity in the direction of the beam axis. *See* **antenna pattern.**

main meteorological office—(Abbreviated MMO.) An office that provides **meteorological service** for international air navigation in accordance with International Civil Aviation Organization specifications.

MMOs 1) prepare forecasts; 2) supply **meteorological information** and **briefings** to aeronautical personnel; 3) supply meteorological information required by an associated **dependent meteorological office** or **supplementary meteorological office**. *See also* **meteorological watch office.**

main standard time—Synoptic hour when meteorological stations make surface **synoptic** observations that are broadcast on a regional or worldwide scale.

The main standard times are 0000, 0600, 1200, and 1800 **UTC.**

main telecommunications network—(Abbreviated MTN.) The primary means of electronic data transmission between important processing locations; usually applied in a telecommunications system where there are major processing and assimilation locations that have smaller, less capable, or more restricted communications networks providing data to them. These smaller networks are secondary or local telecommunications networks.

major lobe—Same as **main lobe.**

major ridge—A long-wave **ridge** in the large-scale **pressure pattern** of the upper **troposphere.**

major trough—A long-wave **trough** in the large-scale **pressure pattern** of the upper **troposphere.**

major wave—Same as **long wave.**

Maloja wind—A **wind**, named after the Maloja Pass between the Engadine and Bergall, Switzerland, that blows down the valley of the Upper Engadine by day and either up or down by night.

This deviation from the usual nature of mountain and valley winds is attributed to the fact that the stronger daytime **valley wind** from the south overtops the ridge and continues down the Engadine.

Defant, F., 1951: *Compendium of Meteorology,* p. 664.

Malvinas Current—A jetlike northward looping excursion of the **Antarctic Circumpolar Current** east of southern Argentina; also known as the **Falkland Current.**

Somewhere between 33° and 38°S it meets the **Brazil Current** and turns eastward, forming an intense **temperature** and **salinity** front.

mamatele—(Also called mamaliti, mamatili.) A light northwest **wind** of Sicily; a form of **mistral.**

mamma—(Also called mammatus.) Hanging protuberances, like pouches, on the undersurface of a **cloud.**

This supplementary cloud feature occurs mostly with **cirrus, cirrocumulus, altocumulus, altostratus, stratocumulus,** and **cumulonimbus**; in the case of cumulonimbus, mamma generally appear on the underside of the **anvil** (**incus**). *See* **cloud classification.**

mammatus—Same as **mamma.**

man–machine mix—Combination of subjective (human) and computer techniques.

mandatory layer—A layer of the **atmosphere** between two consecutive (or any two) specified **mandatory levels**.

mandatory level—(Or mandatory surface.) One of several constant-pressure levels in the **atmosphere** for which a complete evaluation of data derived from **upper-air observations** is required.

Currently the mandatory **pressure** levels are 1000 mb, 850 mb, 700 mb, 500 mb, 400 mb, 300 mb, 200 mb, 150 mb, 100 mb, 50 mb, 30 mb, 20 mb, 10 mb, 7 mb, 5 mb, 3 mb, 2 mb, and 1 mb. The **radiosonde** code has specific blocks reserved for these data. To have a more complete vertical picture, **significant levels** of **radiosonde observations** are also evaluated. Until just after World War II, mandatory levels were defined by values of height above **sea level**.

mandatory surface—Same as **mandatory level**.

Manning equation—An equation relating the **mean velocity** of flow to channel characteristics.

It is expressed as

$$ V = \left(\frac{1}{n}\right) R^{2/3} S_f^{1/2}, $$

where, at a given section, V is mean velocity, n is the Manning's **roughness coefficient** (indicative of the **resistance** to the flow), S_f is the **gradient** of the **total head line**, and R is the **hydraulic radius**.

manometer—An instrument for measuring differences of **pressure**.

The weight of a column of liquid enclosed in a tube is balanced by the pressures applied at its opposite ends, and the pressure difference is computed from the **hydrostatic equation**. A **mercury barometer** is a type of manometer.

mao mao yuh—(Chinese for "nothing serious.") **Drizzle** in northern China.

map correlation—*See* **pattern correlation**.

map factor—Same as **scale factor**.

map plotting—(Also called map spotting.) In meteorology, the process of transcribing weather information onto maps, diagrams, etc.

It usually refers specifically to decoding **synoptic reports** and entering those data in conventional station-model form on **synoptic charts**. It is done either manually or by computer.

map projection—Any systematic arrangement of meridians and parallels portraying the curved surface of a sphere or spheroid upon a plane.

The methods for generating the **projection** are often classified as geometric or analytic. Geometric projections can be classified according to the type of surface on which the projection is assumed to be developed, such as planes, cylinders or cones. Analytical projections are developed by mathematical computation.

map scale—Same as **scale factor**.

map spotting—Same as **map plotting**.

mapped data—1. A geographic map on which meteorological conditions or **elements** are represented by figures, **symbols**, or **isopleths**.

2. Data values that are projected relative to a precise latitude–longitude **grid** in any specified **projection**, such as Mercator or polar stereographic.

marble crust—An extremely hard **snow crust** occurring in small round patches.

marching problem—A procedure for solving continuous differential equations via stepwise changes in the values of the **independent variables**.

The **dependent variables** are computed at each step. The differential equations are solved for an ordered set of discrete values of the independent variables. For example, in **numerical weather prediction** the state of the **atmosphere** is computed at discrete time steps starting from an observed initial state.

marenco—East-southeast **wind** on Lake Maggiore, Italy.

mares' tails—Long, well-defined wisps of **cirrus** clouds, thicker at one end than the other.

Marfa front—*See* **dryline**.

Margules's equation—An equation for the **equilibrium** inclination of an **interface** separating two homogeneous **air masses** in a steady **geostrophic** motion parallel to the interface,

$$\tan \alpha = \frac{f}{g} \frac{(T_2 v_1 - T_1 v_2)}{(T_2 - T_1)},$$

where α is the angle of inclination of the surface to the horizontal, f the **Coriolis parameter**, g the gravitational **acceleration**, and T_1 and T_2 the **absolute** temperatures of the colder and warmer air masses, respectively, with speeds v_1 and v_2.

This equilibrium condition has been used to calculate the slope of atmospheric frontal surfaces.

marigram—1. The record made by a **marigraph**.

2. Any graphic representation of the rise and fall of **tide**, with time as **abscissa** and height as **ordinate**.

marigraph—A recording **tide gauge**.

marin—A warm moist southeast **wind** from the sea on the French Mediterranean coast and in the Maritime Alps, especially frequent in spring and autumn.

In the Rhône delta it blows also from the south. The marin is associated with depressions that cross southern France or northern Spain and the Gulf of Lions. Generally, it is strong and regular, sometimes violent and turbulent in hilly country as the **ayalas** in the Massif Central; it is very humid, **cloudy** with **hill fog**, and often rainy (unless unaccompanied by fronts, when it is the marin blanc). The heavy rains, which may continue for one or two days on the mountain slopes, cause dangerous river floods. On the western slope of the Cévennes it becomes the **autan**. In the southern Cévennes the marin is called the aygalas. On the coast of Catalonia (northeast Spain) and Roussillon (southern France) it is the marinada and generally occurs with a **depression** centered over or south of the Gulf of Gascony. *Compare* **sirocco**.

marin blanc—*See* **marin**.

marinada—*See* **marin**.

marine barometer—A **mercury barometer** designed for use aboard ship.

The instrument is of the fixed-cistern type (*see* **Kew barometer**). The **mercury** tube is constructed with a wide bore for its upper portion and with a capillary bore for its lower portion. This is done to increase the **time constant** of the instrument and thus prevent the motion of the ship from affecting the reading. The instrument is suspended in gimbals to reduce the effects of **pitch** and **roll** of the ship.

marine climate—(Also called maritime climate, oceanic climate.) A regional **climate** under the predominant influence of the sea, characterized by relatively small seasonal variations and high atmospheric **moisture content**; the antithesis of a **continental climate**.

marine forecast—A forecast, for a specified oceanic and/or coastal area, of weather elements of particular interest to maritime transportation.

These elements include **wind**, **visibility**, the general state of the weather, and **storm warnings**. *Compare* **aviation weather forecast**.

marine meteorology—The part of meteorology that deals mainly with the study of oceanic areas, including island and coastal regions.

In particular it serves the practical needs of surface and air navigation over the oceans. Since there is a close interaction between ocean and **atmosphere**, and oceanic influences upon weather and **climate** can be traced far inland over the continents, modern meteorology uses this name mainly for making regional or administrative distinctions.

marine observation—*See* **marine weather observation**.

marine radar—**Radar** of the type used on ships, can be employed to **image** ocean waves, to determine their directional **spectrum**, and also to determine **ocean currents** by their propagation speed. *See also* **Doppler radar**, **microwave radar**, **synthetic aperture radar**, **high-frequency radar**.

marine rainbow—(Also called sea rainbow.) A **rainbow** seen in sea spray.

It is optically the same as the ordinary rainbow, although the slightly different **index of refraction** of saltwater results in a shift in the angular radius of the bow, which is apparent if accompanied by a bow formed in raindrops.

marine thermometer—*See* **seawater thermometer**, **reversing thermometer**.

marine weather observation—The weather as observed from a ship at sea, usually taken in accordance with procedures specified by the World Meteorological Organization.

The following elements are usually included: total **cloud amount**; **wind direction** and **speed**; **visibility**; current weather; **pressure**; **temperature**; selected cloud-layer data, that is, amount, type, and height; **pressure tendency**; seawater temperature; **dewpoint temperature**; **state of the sea** (waves); **sea ice**; and **icing** onboard ship. Also included are the date and time, and the name, position, **course**, and speed of the ship. The encoded and transmitted marine observations are known as ship reports. *See* **ocean station vessel**.

Mariotte's law—(Also called Boyle–Mariotte law.) Same as **Boyle's law**.

maritime aerosol—Aerosol produced by processes taking place near the ocean surface.

Air entrained by breaking **ocean waves** (visible as whitecaps) forms **bubbles** that break at the water surface. The whitecaps produce a fine **aerosol**; the Rayleigh jet produces coarser **particles** that are carried aloft by the turbulent **eddies** of the winds. Aerosol also forms from chemical reaction of ocean-produced gases. *See* **continental aerosol**.

maritime air—A type of **air mass** the characteristics of which are developed over an extensive water surface and which, therefore, has the basic maritime quality of high humidity content in at least its lower levels.

See **airmass classification**; *compare* **continental air**.

maritime climate—Same as **marine climate**.

maritime cloud—A **cloud** forming in **maritime air** containing relatively low concentrations of **cloud condensation nuclei**.

These clouds are characterized by a broader **droplet** size distribution and low droplet concentrations (some 100 cm^{-3}). *See* **continental cloud**.

Maritime Province Cold Current—*See* **Mid-Japan Sea Cold Current**.

maritime tundra—**Tundra** found along many subarctic coastal belts, usually with a high proportion of arctic plants and animals far south of their **normal** limit.

Markov chain—A **stochastic process** with a finite number of states in which the **probability** of occurrence of a future state is conditional only upon the current state; past states are inconsequential.

In meteorology, Markov chains have been used to describe a **raindrop size distribution** in which the state at time step $n + 1$ is determined only by collisions between pairs of drops comprising the size distribution at time step n.

Markovian—A process for generating a **time series** where the value at any new time depends only on the previous value plus some **random** component.

Marsden chart—A system introduced by Marsden early in the nineteenth century for showing the distribution of meteorological data on a **chart**, especially over the oceans.

A Mercator **map projection** is used, the world between 80°N and 70°S latitudes being divided into **Marsden squares** each of 10° longitude. These squares are systematically numbered to indicate position. Each square may be divided into quarter squares, or into 100 one-degree subsquares numbered from 00 to 99 to give the position to the nearest degree.

Marsden square—A system that divides a Mercator **chart** of the world into squares of 10° latitude by 10° longitude.

Each square is numbered and then subdivided into 100 one-degree squares numbered from 00 to 99. These 100 one-degree squares are applied to the eight octants of the globe. Marsden squares are mainly used for identifying the geographic position of meteorological data over the oceans.

marsh—Periodically or continually inundated areas covered with vegetation of cattails, sedges, rushes, and some woody **hydrophytic plants**.

Marshall–Palmer relation—The Z–R relationship developed by J. S. Marshall and W. M. Palmer (1948) consistent with an exponential **drop-size distribution**.

The relationship is $Z = 200R^{1.6}$, where Z (mm^6 m^{-3}) is the **reflectivity factor** and R (mm h^{-1}) is the **rainfall rate**. The relationship is sometimes generalized to the form $Z = aR^b$, where a and b are adjustable parameters.

Marshall, J. S., and W. McK. Palmer, 1948: The distribution of raindrops with size. *J. Meteor.*, **5**, 165–166.

Marvin sunshine recorder—A **sunshine recorder** of the type in which the timescale is supplied by a **chronograph**.

It consists of two bulbs, one of which is blackened, that communicate through a glass tube of small diameter. The tube is partially filled with **mercury** and contains two electrical contacts. When the instrument is exposed to **sunshine**, the air in the blackened bulb is warmed more than that in the clear bulb. The warmed air expands and forces the mercury through the connecting tube to a point where the electrical contacts are shorted by the mercury. This completes the electrical circuit to the pen on the chronograph. The Marvin sunshine recorder is equally sensitive to the direct rays of the sun and to **diffuse radiation** from the sky (the **heat** from the latter at midday in **overcast** may be more than that from direct sunshine in the early morning); thus, the instrument is not without ambiguities. It is standard equipment at National Weather Service stations.

masked front—A **front** the presence of which is not readily apparent on the **surface synoptic chart** because of local modifying influences such as **radiation**, **topography**, or **mesoscale** processes.

mass absorption coefficient—A measure of the rate of **absorption** of **radiation**, expressed as the **absorption cross section** per unit mass.
Units are $m^2 kg^{-1}$. *See* **absorption coefficient**.

mass accommodation coefficient—The **probability** that reversible uptake of a gas-phase species will occur upon collision of that species with a given surface (liquid or solid).
This is often equated with a "sticking" probability.

mass balance—Usually, a **model** that employs the limitation that the system observed maintains a constant mass, that is, total **mass divergence** for the entire system is zero, but may be nonzero within the system.

mass concentration—Same as **density**.

mass convergence—Negative **mass divergence**.

mass curve—A plotting of the cumulative values of a **variable** as a function of time.
This is applied especially to mass curves of **rainfall** in **storm** studies, to departures of various weather elements from **normal**, and to **streamflow** data for **reservoir** studies.

mass divergence—The **divergence** of the **momentum** field, a measure of the rate of net **flux** of mass out of a unit volume of a system; in symbols, $\nabla \cdot \rho\mathbf{u}$, where ρ is the fluid **density**, $\mathbf{u}$ the **velocity** vector, and ∇ the **del operator**.

mass scattering coefficient—A measure of the rate of **scattering** of **radiation**, expressed as the **scattering cross section** per unit mass.
Units are $m^2 kg^{-1}$. *See* **scattering coefficient**.

mass spectrometry—Technique for the **detection** of chemicals by **ionization**, followed by the use of a **magnetic** or **electric field** to separate the ions according to their mass-to-charge ratio.
The ionization can be achieved by bombardment with a stream of electrons from a heated filament (**electron** impact) or by the **transfer** of charge from a prepared **ion** to the species of interest (chemical ionization). Highly sensitive mass spectrometers have been used to measure small differences in the ratio of **isotopes** in certain molecules to ascertain, for example, the age of the sample or to assess the nature of the source of the sample.

mass-transfer method—Method for estimating the **actual evaporation** from a body of water, assuming it is proportional to the product of **wind** velocity (perhaps raised to a power less than one), the difference between the **saturation vapor pressure** at water surface **temperature** and the **vapor pressure** of the **ambient air**, and an empirical mass-transfer coefficient.

mass transport—The **momentum**, $\rho\mathbf{u}$, where ρ is the fluid **density** and $\mathbf{u}$ the **velocity** vector, considered as the **transport** of fluid mass from one region of space to another.

mast—A tall, usually movable structure, used to mount meteorological sensors on for monitoring near-surface atmospheric parameters.
A mast differs from a tower in that it is designed to have minimal effect on the measurement(s) being made. Circular poles similar to a ship's mast are the most common shape.

master recession curve—*See* **recession curve**.

Matanuska wind—The local name, taken from the Matanuska River, for a strong, gusty, northeast **wind** that occasionally occurs during the winter in the vicinity of Palmer, Alaska.

matched filter—A type of **filter** matched to the known or assumed characteristics of a **target signal**, designed to optimize the **detection** of that **signal** in the presence of **noise**. Typically such filters are implemented in the temporal or spatial **frequency** domain, using **digital** techniques (though **analog** matched filters are found in **radar** receivers).

material coordinates—Same as **Lagrangian coordinates**.

material derivative—Same as **individual derivative**.

mathematical climate—An elementary generalization of the earth's climatic pattern, based entirely upon the **annual cycle** of the sun's inclination.

This early **climatic classification** recognized three basic latitudinal zones (summerless, intermediate, and winterless), which are now known as the **Frigid**, **Temperate**, and **Torrid Zones**, and which are bounded by the **Arctic** and **Antarctic Circles** and the **Tropic of Cancer** and **Tropic of Capricorn**. What we now call "mathematical climate" probably corresponds to what was classically considered to be **climate** (from the Greek word *klima*, meaning "inclination"). It is sometimes used synonymously with **solar climate**, but the latter has a more specific theoretical connotation.

mathematical expectation—Same as **expected value**.

mathematical forecasting—Same as **numerical forecasting**.

matinal—The "morning **wind**," that is, an east wind.
 In the Morvan Mountains and the center of the Massif Central in France, the matinal often blows for several days, especially in summer, and brings fine weather. On winter mornings a northeast or east wind descends the western slopes of the Alps (where it is known as the matinière) bringing cold and generally fine weather. *Compare* **solaire**.

mauka breeze—Night winds of Hawaii of a cool and refreshing nature.

Maunder minimum—*See* **sunspot minimum**, **solar minimum**.

max-wind and shear chart—Same as **maximum-wind and shear chart**.

max-wind level—Same as **maximum-wind level**.

max-wind topography—Same as **maximum-wind topography**.

maximum—The greatest value attained (or attainable) by a function; the opposite of **minimum**.
 An "absolute" maximum is the greatest value within a prescribed interval, while "relative" maxima are the greatest values within arbitrary subintervals, each one of which is "absolute" within its own subinterval, and so on. In records of meteorological observations, "absolute" is with reference to the entire **period of record** for that **station**, and the "relative" values are labeled "annual," "monthly," and "daily."

maximum contaminant level—(Usually abbreviated MCL.) A federal drinking water standard that specifies the maximum concentration level for a regulated chemical in drinking water.

maximum entropy method—A method used to estimate power spectra that contain very sharp peaks.
 Under the assumption that such peaks can be approximated analytically by singularities, the **power spectrum** is written as a rational fraction with a denominator containing a **polynomial**, the zeros of which correspond to frequencies where the spectral density is infinite.

maximum possible sunshine duration—Time interval between **sunrise** and **sunset**.

maximum precipitation—*See* **zone of maximum precipitation**.

maximum probable flood—*See* **probable maximum flood**.

maximum temperature—The highest **temperature** reported for a given location during a given period.

maximum thermometer—A **thermometer** so designed that it registers the **maximum temperature** attained during an interval of time.
 The liquid-in-glass type of maximum thermometer has a bore that is constricted between the bulb and graduated portion of the stem. As the **temperature** rises, a portion of the **mercury** is forced past the constriction and into the graduated section. This mercury is retained when the temperature falls and serves to indicate the highest temperature reached. **Bimetallic thermometers** with a circular **scale** are also used as maximum thermometers. A free **index** mounted concentrically with and driven by the thermometer index is held by **friction** at the maximum temperature. *Compare* **minimum thermometer**; *see* **Townsend support**.

maximum unambiguous range—The maximum range from which a transmitted **radar**, **lidar**, or **sodar** pulse can be reflected and received before the next **pulse** is transmitted. This **range**, r_{max}, is given by $r_{max} = cT/2$, where T is the **interpulse period** and c is the **speed of light** (or **speed of sound** in the case of sodar).
 Range is measured by the time delay between pulse transmission and reception, ordinarily assuming that the received pulse is associated with the most recent transmitted pulse. **Targets** at ranges beyond r_{max} therefore appear at ranges closer than r_{max} because of **range folding**. Special coding of the pulses permits discrimination between echoes from the most recent transmitted pulse and earlier ones, enabling the measurement of ranges beyond r_{max}. *See also* **second-trip echo**, **range aliasing**.

maximum unambiguous velocity—The maximum range of **radial velocity** that can be observed without ambiguity by a **Doppler radar**, **sodar**, or **lidar**.

Velocities outside this interval are folded into the interval. For a **pulsed radar** or sodar operating with an **interpulse period** T, or a lidar with an **A/D converter** sampling interval T, the **unambiguous velocity interval** is $-\lambda/4T$ to $+\lambda/4T$, where λ is the operating **wavelength**. This interval can be extended by the use of dual **PRF** and other techniques. *See* **velocity aliasing**, **Nyquist frequency**.

maximum-wind and shear chart—(Or max-wind and shear chart.) A **synoptic chart** on which are plotted the altitudes of the maximum wind speed, the maximum wind **velocity** (**wind direction** optional), plus the velocity of the **wind** at **mandatory levels** both above and below the level of maximum wind.

These maps can be analyzed for one or more of the quantities plotted.

maximum-wind level—(Or max-wind level.) The height at which the maximum wind speed occurs, typically within the **jet stream**, determined in a **winds-aloft observation**.

See **maximum-wind and shear chart**.

maximum-wind topography—(Or max-wind topography.) The **topography** of the surface of maximum wind speed.

The contours of maximum-wind height form one of the sets of lines on an analyzed **maximum-wind and shear chart**. *See* **maximum-wind level**.

maximum zonal westerlies—The average west-to-east component of **wind** over the continuous 20-degree belt of latitude in which this average is a maximum.

It is usually found, in the winter season, in the vicinity of 40–60°N latitude. *See* **zonal index**.

Maxwell–Boltzmann distribution—(Or Maxwell–Boltzmann law, Maxwell–Boltzmann distribution of kinetic energies or speeds.) The distribution law for kinetic energies (or, equivalently, speeds) of molecules of an **ideal gas** in **equilibrium** at **absolute** temperature T:

$$\frac{2\sqrt{E}}{\sqrt{\pi}\,(kT)^{3/2}}\,e^{-E/kT},$$

where k is **Boltzmann's constant**.

The integral of this function between any two energies is the fraction of the total number of molecules with energies in this **range**. Note the similarity between this function and the **Planck's radiation law**. The former is the distribution law for the energies of an ideal gas of molecules, whereas the latter is the distribution law for the energies of a gas of **photons**.

MCC—Abbreviation for **mesoscale convective complex** or **mesoscale cellular convection**.

MCS—Abbreviation for **mesoscale convective system**.

mean—An arithmetic average.

Types can include mean over an ensemble of experimental realizations, mean over a time period, mean along a line such as a road or flight path, mean in an area such as a farm field, or mean within a volume of air such as can be sampled by a remote **sensor**. For example, the mean wind speed from **anemometer** measurements is the speed at a fixed point averaged over a time period such as 10 minutes.

mean annual range of temperature—The difference between the mean temperatures of the warmest and coldest months of the year.

mean chart—(Or mean map.) Any **chart** on which **isopleths** of the **mean value** of a given **meteorological element** are drawn.

Compare **normal chart**.

mean daily maximum temperature for a month—Mean of the daily maximum temperatures observed during a specific calendar month, either for a specific year or over a specific period of years.

mean daily minimum temperature for a month—Mean of the daily minimum temperatures observed during a specific calendar month, either for a specific year or over a specific period of years.

mean daily temperature—Mean of the temperatures observed at 24 equidistant times in the course of a continuous 24-hour period (normally the **mean solar day** from midnight to midnight according to the **zonal** time of the **station**).

mean day-to-day variation—Mean of the **absolute** values of the differences between the daily means of a **climate** element on two successive days.

mean deviation—The mean of the **absolute** deviation from the mean.

mean Doppler velocity—The **mean velocity** in the **Doppler spectrum**; that is, the first **moment** of the Doppler spectrum expressed in terms of **velocity**.
See **Doppler spectral moments**.

mean free path—1. Average distance a molecule travels in a gas between collisions.
This concept has a meaning only to the extent that the paths of molecules are mostly straight lines interrupted by changes in direction (collisions) over comparatively shorter distances of order the molecular size. Molecules in a liquid are never free in this sense. The mean free path L in a gas of a single molecular species with a **Maxwell–Boltzmann distribution** of speeds is

$$L = \frac{1}{\sqrt{2}\, nS},$$

where n is the number **density** of molecules and S is the mutual collision **cross section**. To the extent that a molecule can be considered a hard sphere with diameter d, $S = \pi d^2$. The concept of molecular diameter is fuzzy and each method for determining it yields different results (Kennard 1938). At **sea level** the mean free path in air is of order 0.1 μm.

2. Average distance a **photon** travels in a turbid medium between **scattering** (scattering mean free path) events, or the average distance a photon travels before being absorbed (**absorption** mean free path).
The scattering mean free path in a medium is the inverse of its **scattering coefficient**; the absorption mean free path is the inverse **absorption coefficient**; and the total mean free path is the inverse of the sum of scattering and absorption coefficients. At visible wavelengths, the scattering mean free path in clouds is of order 10 m.

Kennard, E. H., 1938: *Kinetic Theory of Gases*, 97–114.

mean high water—The average level of all **high waters** at a place over a 19-year period.
See **mean sea level**.

mean high water springs—Average **spring tide** high water level, averaged over a sufficiently long period.

mean higher high water—A tidal **datum**; the average of the higher of the two **high water** heights of each **tidal day**, averaged over the U.S. **National Tidal Datum Epoch**.

mean interdiurnal variability—*See* **mean day-to-day variation**.

mean kinetic energy—The **kinetic energy** associated with the mean wind.
Per unit mass m, the mean kinetic energy (MKE) is

$$\mathrm{MKE}/m = 0.5(\overline{U}^2 + \overline{V}^2 + \overline{W}^2),$$

where (U, V, W) are the Cartesian components of **wind**, and the overbar denotes an average or mean. *Compare* **turbulence kinetic energy**.

mean life—(Or mean lifetime, lifetime.) The average time during which anything can be said to reside within a specified region or to exist unchanged; for example, the mean life of a chemical species in the **atmosphere**, the mean life of a radioactive **nucleus**, the mean life of **photons** in a **cloud**.
For processes following the same exponential law as radioactive decay, mean life is related simply to **half-life**.

mean low water—The average level of **low water** at a place over a 19-year period.

mean low water springs—A tidal **datum**; the average spring **low water** level, averaged over a sufficiently long period.

mean lower low water—A tidal **datum**; the average of the lower of the two **low water** heights of each **tidal day** observed over the U.S. **National Tidal Datum Epoch**.

mean map—Same as **mean chart**.

mean monthly maximum temperature—Mean of the monthly **maximum temperatures** observed during a specific calendar month over a specific period of years.

mean monthly minimum temperature—Mean of the monthly **minimum** observed during a specific calendar month over a specific period of years.

mean sea level—(Abbreviated MSL; popularly called sea level.) The **arithmetic mean** of hourly heights observed over some specified period.

In the United States, mean sea level is defined as the mean height of the surface of the sea for all stages of the **tide** over a 19-year period. Selected values of mean sea level serve as the sea level **datum** for all **elevation** surveys in the United States. In meteorology, mean sea level is used as the reference surface for all altitudes in upper-atmospheric work; in aviation it is the level above which **altitude** is measured by a **pressure altimeter**. Along with **mean high water**, **mean low water**, and **mean lower low water**, mean sea level is a type of tidal datum. *Compare* **half-tide level**, **still-water level**.

mean sea level trends—Changes of **mean sea level** at a **site** over long periods of time, typically decades, also called secular changes.

Global changes due to the increased volume of ocean water are called **eustatic** changes; vertical land movements of regional extent are called eperiogenic changes.

mean solar day—The interval of time between two successive **meridional** transits of the "mean sun," an imaginary point moving with such constant **angular velocity** along the **celestial equator** as to complete one annual circuit in a time period exactly equal to that of the apparent (true) sun in its annual circuit.

The mean solar day is 86 400 seconds, or 1.0027379 sidereal day. This modification of the **apparent solar day** was devised as a means of **smoothing** the irregularities observed in apparent relative motion of sun and earth; the **equation of time** defines the difference between the two. Outside astronomical circles, the mean solar day is the day in common use, but since it is reckoned from noon to noon, it has been modified to the **civil day** for practical use.

mean solar year—*See* **year**.

mean sphere depth—The uniform depth to which the water would cover the earth if the solid surface were smoothed off and were parallel to the surface of the **geoid**.

This depth would be about 2440 m.

mean-square error—(Abbreviated MSE.) The mean square of any **residual**.

In case the mean residual is zero, the mean-square error is the same as the **residual variance**. *See* **regression**.

mean temperature—The average temperature of the air as indicated by a properly exposed **thermometer** during a given time period, usually a day, a month, or a year.

For climatological tables, the mean temperature is generally calculated for each month and for the year. For charts, the observed mean values at **station** level are reduced to **sea level** by adding a **correction** for **elevation**, usually taken as 0.5°C for each 100 m (1°F for 360 ft), but in some mountainous countries different rates are used, based on local observations. *See* **true mean temperature**.

mean tide level—The **arithmetic mean** of **mean high water** and **mean low water**.

This level is not identical with **mean sea level** because of higher harmonics in the tidal constituents.

mean value—Same as **average**.

mean velocity—The **time average** of the **velocity** of a fluid at a fixed point, over a somewhat arbitrary time interval T counted from some fixed time t_0.

For example, the mean velocity of the u component is

$$\bar{u} = \frac{1}{T} \int_{t_0}^{t_0+T} u\,dt.$$

The **time average** of any other quantity can be defined in this manner.

mean virtual temperature—*See* **virtual temperature**.

mean water level—*See* **mean sea level**, **mean tide level**.

mean wind velocity—1. The **vector** average **wind** velocity at a point relative to the earth.

2. For the purpose of **upper air** reports from aircraft, mean wind is derived from the **drift** of the aircraft when flying from one fixed point to another or obtained by flying on a circuit around a fixed observed point and an immediate **wind** deduced from the drift of the aircraft.

meander—Dispersion of **smoke plumes** in the horizontal by means of the **crosswind** component (fluctuations) of the horizontal **wind speed**.

The result is a **plume** that wanders from side to side. When averaged over a finite time period, the result is plume spreading in the horizontal.

measured ceiling—After U.S. weather observing practice, the **ceiling classification** that is applied when the **ceiling** value has been determined by means of 1) a **ceiling light** or **ceilometer**, provided that penetration of the **beam** is not in excess of that normally experienced for the height and type of layer and that the **elevation angle** indicated by the **clinometer** or ceilometer detector does not exceed 84°; 2) the timed disappearance of a **radiosonde balloon** with its height computed; 3) the known heights of unobscured portions of objects, other than natural landmarks, within 1½ nautical miles of any runway of the airport.

A measured ceiling pertains only to clouds or to obscuring phenomena aloft. It is designated M in **aviation weather observations**.

measurement cell—In **radar**, the four-dimensional space (volume × time) from which received **target** signals are sampled and processed to provide a measurement of **reflectivity**, the **Doppler spectrum**, or other properties of the target.

Normally the measurement cell is large compared with the **coherence element** to provide an adequate average of the quantity to be measured or to reduce the amount of data to be stored.

mechanical dispersion—(Also called hydraulic dispersion.) The process whereby solutes are mechanically mixed by **velocity** variations at the microscopic level during advective **transport**.

mechanical energy equation—*See* **energy equation**.

mechanical equivalent of heat—Same as **Joule's constant**.

mechanical instability—Same as **absolute instability**.

mechanical internal boundary layer—(Abbreviated MIBL.) An **internal boundary layer** caused by **advection** of air across a **discontinuity** in **surface roughness**.

When the new surface is rougher than the old one, the MIBL depth grows roughly as the 0.8 power of the ratio of the two **roughness lengths**. In this example, the MIBL grows to include the whole **surface layer**.

mechanical mixing—Any **mixing** process that utilizes the **kinetic energy** of relative fluid motion.

The archetype of mechanical mixing is the **instability** of a vertically sheared flow. Consider a flow with an upper and a lower layer flowing in opposite directions. If the flow is unstable, **perturbations** grow by transferring **momentum** (through **Reynolds stresses**) between the layers and reducing the **velocity** difference between them. The kinetic energy of the **shear flow** (proportional to the mean-square velocity) is therefore reduced, and the energy has gone into the perturbations. These perturbations grow to the point where they become turbulent and cause vertical mixing.

mechanical turbulence—**Turbulence** produced by **shear flow**.

Meddy—A short form for **Mediterranean Eddy**.

Meddies are **salt lenses** containing high amounts of original Gibraltar Outflow Water in their interior. With spatial scales smaller than the internal Rossby radius, they belong to the energetic class of **submesoscale coherent vortices**. They rotate clockwise (anticyclonically) like solid bodies and are encapsulated by strong contrasts (gradients) of **water masses** (properties) and a sharp **vorticity** front at their periphery. Meddies interact with partner vortices, depending on their geographical position and **eddy** population. Spontaneous Meddy release represents **random** salt sources within the Mediterranean salt **tongue** and questions a **large-scale** advection–diffusion salt balance in the North Atlantic. Typical Meddy scales and properties include diameters of ~50–80 km, a vertical extent of ~600–1400 m, a **drift** velocity of ~2–3 cm s^{-1} (with occasional stalls), a rotation **velocity** of ~20–30 cm s^{-1}, a rotation **period** of ~4–10 days, a lifetime of months to two years, and a **salinity** core contrast of 0.2–1 **psu**.

median—One of several measures of **central tendency**.

1) Pertaining to a series of numbers, the median is the middle term when the numbers are arranged in algebraic order. If the number of terms is even, the median is taken as halfway between the two middle terms. 2) Pertaining to a continuous **random variable** x, the median is that value that divides the **probability distribution** into two equal areas. Hence, in terms of the **distribution function** $F(x)$, the median is that value of x for which $F(x) = 1/2$. In case $F(x)$ is discontinuous, the median is defined in such a way as to yield consistent results when the area is cumulated from either end of the distribution.

median volume diameter—A measure of the diameter that contributes most to **cloud** liquid water or mass.

See **drop-size distribution**.

Medieval Warm Period—A period of warmer **climate** in much of northern Europe, the North Atlantic, southern Greenland, and Iceland from about the tenth to thirteenth centuries A.D., coinciding roughly with Europe's Middle Ages.

As the number of well-understood proxy records of climate covering this time period in many parts of the world has increased in recent years, it has become clear that it would be an oversimplification to view this period as uniformly warmer in all seasons, even within the relatively limited North Atlantic/northern Europe region, let alone on a global **scale** (Hughes and Diaz 1994).

> Hughes, M. K., and H. F. Diaz, 1994: Was there a Medieval Warm Period, and if so, where and when?. *Climatic Change*, **26**, 109–142.

medina—A land **wind** during winter at Cadiz, Spain.

mediocris—A **cloud species** peculiar to the genus **cumulus**.
> *See* **cumulus mediocris, cloud classification**.

Mediterranean climate—(Also called etesian climate, dry-summer **subtropical climate**.) Characterized by mild, wet winters and warm to hot, dry summers; typically occurs on the west side of continents between about 30° and 45° latitude.

Mediterranean Deep Water—(Also called Western Mediterranean Deep Water.) A **water mass** that occupies the Eurafrican **Mediterranean Sea** below a depth of 500 m.
> It is formed in winter by **convection** in the Ligurian Sea and the Balearic Basin when very cold Siberian air is channeled through Alpine valleys and descends in a burst known as **mistral** to cool the sea surface. *See* **deep water**.

Mediterranean Eddy—*See* **Meddy**.

Mediterranean front—A **front** that forms in the low pressure zone that covers the Mediterranean between the cold air over Europe and the warm air over the Sahara.

Mediterranean lenses—(Also called Meddies.) Coherent subsurface masses of anticyclonically rotating, warm salty water in the Atlantic Ocean originating from the **Mediterranean Sea**.
> These **mesoscale** lenses have been observed to persist for up to many months. *See* **Meddy**.

Mediterranean outflow—The relatively saline water that flows out of the **Mediterranean Sea** through the Strait of Gibraltar in a subsurface layer just above the **sill depth**.
> The **outflow** consists mainly of Levantine Intermediate Water, which is formed in the extreme eastern basin of the Mediterranean. Once in the Atlantic, the relatively dense **Mediterranean Water** quickly sinks and spreads westward around a depth of 1000 m. The **salinity** anomaly of the Mediterranean Water can be detected several thousand kilometers west of the Strait of Gibraltar. *See* **Mediterranean lenses, Gibraltar outflow water**.

Mediterranean Sea—A part of the World Ocean with little communication with the major ocean basins, in which the **circulation** is controlled by **density** differences.
> Two types of Mediterranean Seas are distinguished. In the arid type, **evaporation** exceeds **precipitation**, increasing the density of the Mediterranean Sea; this produces deep **convection** and an **outflow** of dense **Mediterranean Water** through the connection with the main ocean below an **inflow** of less dense oceanic water. Examples are the Eurafrican Mediterranean and the Red Sea. In the humid type, precipitation exceeds evaporation, lowering the density; this produces a two-layer structure and inflow of denser oceanic water below water of less dense Mediterranean Water. Examples are the Australasian Mediterranean (Indonesian Seas) and the Baltic Sea.

Mediterranean Water—A **water mass** formed in a **Mediterranean Sea**.
> In the Atlantic Ocean the term refers to the water mass that enters from the Eurafrican Mediterranean through the Strait of Gibraltar with a **temperature** of 13.5°C and a **salinity** of 37.8 **psu** and can be traced throughout the Atlantic as a salinity maximum of 36.0–36.2 psu near a depth of 1000 m and progressively deeper and fresher as the distance from the straits increases. *See also* **Australasian Mediterranean Water, Persian Gulf Water, Red Sea Water**.

medium frequency—(Abbreviated MF.) *See* **radio frequency band**.

medium-range forecast—A forecast for a period extending from about three days to seven days in advance; there are no absolute limits to the period embraced by the definition.
> *Compare* **long-range forecast, extended forecast**.

medium-term hydrological forecast—Prediction of hydrologic variables such as **streamflow** or snowmelt for periods three to seven days in advance.

megacycle—(Obsolete.) See **megahertz**.

megahertz—(Abbreviated MHz; formerly called megacycle.) A unit of **frequency** equal to one million (10^6) cycles per second.

megatherm—A type of plant life that requires continuous high temperature and abundant **rainfall** (in excess of **evapotranspiration**)—features that are typical of the **tropical rain forest**.
Compare **mesotherm, microtherm**.

megathermal—Of, or pertaining to, high temperatures.

megathermal climate—A type of **climate** characterized by continuous high temperatures in combination with abundant **rainfall** (in excess of **evapotranspiration**) throughout the year.
This climate is the type required to support the growth of the plant group known as **megatherms**. This is an (A) **climate** under the **Köppen classification** (1931) and, as defined by Thornthwaite's **climatic classification** scheme (1948), this is a climate with annual **potential evapotranspiration** in excess of 114 cm. *Compare* **mesothermal climate, microthermal climate**.

> Trewartha, G. T., 1954: *Introduction to Climate*, 3d ed., McGraw–Hill, 239–266.
> Köppen, W., 1931: *Grundriss der Klimakunde*, Walter die Gruyter Co., Berlin.
> Thornthwaite, C. W., 1948: An approach toward a rational classification of climate. *Geographical Review*, **38**, 55–94.

megathermal period—Same as **Climatic Optimum**.

megathermic—*See* **megathermal**.

mei-yu front—(Also called baiu front). A quasi-persistent, nearly stationary, east–west-oriented weak **baroclinic** zone in the lower **troposphere** that typically stretches from the east China coast, across Taiwan, and eastward into the Pacific, south of Japan. The term "mei-yu" is the Chinese expression for "plum rains."
The mei-yu front generally occurs from mid- to late spring through early to midsummer. This low-level baroclinic zone typically lies beneath a confluent **jet** entrance region aloft situated **downstream** of the Tibetan Plateau. The mei-yu/baiu front is very significant in the weather and **climate** of southeast Asia as it serves as the focus for persistent heavy convective **rainfall** associated with **mesoscale convective complexes** (MCCs) or **mesoscale convective systems** (MCSs) that propagate eastward along the baroclinic zone. The moisture source is typically the South China Sea and sometimes the Bay of Bengal. The usual lifting mechanism is low-level warm-air **advection** in association with a **low-level jet** on the equatorward flank of the baroclinic zone. Deep ascent and resulting organized MCCs/MCSs are especially favored when the low-level warm-air advection is situated beneath the favorable equatorward jet entrance region aloft.

Meinel bands—*See* **hydroxyl emission**.

Mellor–Yamada parameterization—A **parameterization** of a complex **model** for turbulent flows in the **planetary boundary layer**.
A series of simplifications of the full **turbulence** model to remove complex terms and form a closed set of equations leads to a hierarchy of so-called closure models of decreasing complexity labeled level 4 through level 1. The level 2.5 **model** is widely used; it incorporates only one additional equation and produces conventional turbulent fluxes.

melt pond—A pond of liquid water (mostly from melted **snow**) on the surface of **sea ice**, usually occurring in the spring.
Melt ponds are common in the Arctic but less so in the Antarctic.

meltém—1. (Also spelled meltémi.) A strong **wind** from the northeast or east that often sets in suddenly and blows during the day in summer on the Bulgarian coast and in the Bosporus.
2. Same as **monsoon**.

melting band—Same as **melting layer**.

melting layer—The **altitude** interval throughout which ice-phase **precipitation** melts as it descends.
The top of the melting layer is the **melting level**. The melting layer may be several hundred meters deep, reflecting the time it takes for all the **hydrometeors** to undergo the transition from solid to liquid **phase**. The **temperature** of the melting layer is typically 0°C or slightly warmer. *See* **bright band**.

melting level—The **altitude** at which **ice crystals** and snowflakes begin to melt as they descend through the **atmosphere**. In **cloud physics** and in **radar meteorology**, this is the accepted term for the 0°C constant-temperature surface (*see* **bright band**).
It is physically more apt than the corresponding operational term, **freezing level**, for melting of pure **ice** must begin very near 0°C, but **freezing** of liquid water can occur over a broad **range** of temperatures (between 0° and −40°C; *see* **supercooling**). *See also* **freezing point, ice point, melting point**.

melting point—The **temperature** at which a solid substance undergoes **fusion**, that is, melts, changes from solid to liquid form.

The melting point of a substance should be considered a property of its crystalline form only. At the melting point the liquid and solid forms of a substance exist in **equilibrium**. All substances of crystalline nature have their characteristic melting points. For very pure substances the **temperature range** over which the process of fusion occurs is very small. The melting point of a pure crystalline solid is a function of **pressure**; it increases with increasing pressure for most substances. However, in the case of **ice** (and a few other substances) the melting point decreases with increasing pressure (*see* **regelation**). Under a pressure of one **standard atmosphere**, the melting point of pure ice is the same as the **ice point**, that is, 0°C. *Compare* **freezing point**.

meniscus—The crescent-shaped upper surface of a column of liquid.

The curvature of this surface is dependent upon the cross-sectional area of the liquid and the relative ability of the liquid to wet the walls of the enclosure. For liquids that wet the walls of the enclosure, the curvature of the surface is concave. In the case of a **mercury column** enclosed in a glass container, the surface is convex, since **mercury** does not wet glass.

MERCAST—The radio broadcast system for delivery of messages originated by government agencies and addressed to merchant ships when prior arrangements for such delivery have been made.

Navy-issued **storm warnings** are carried on MERCAST. *See* **General Broadcast**, **Fleet Broadcast**.

Mercator projection—*See* **conformal map**.

mercurial barometer—Same as **mercury barometer**.

mercury—(Symbol Hg.) A metallic **element**, atomic number 80, atomic weight 200.61; unique (for metals) in that it remains liquid under all but very extreme temperatures. Its **density** of 13.596 g cm^{-3} and **melting point** of −38.87°C (−37.8°F) make it very useful as the medium for liquid **barometers** and **thermometers**.

Mercury is very poisonous and can be absorbed through the skin. It can form organic derivatives that can enter the food chain, particularly via marine organisms. Atmospheric mercury is predominantly in the elemental form.

mercury barometer—(Or mercurial barometer; formerly called Torricelli's tube.) A glass **manometer**, employing **mercury** in its vertical column, that is used to measure **atmospheric pressure**.

The basic construction, unchanged since Torricelli's experiment in 1643, is a glass tube about three feet long, closed at one end, filled with mercury, and inverted with the open end immersed in a **cistern** of mercury. With the cistern surface exposed to atmospheric pressure, the height of the **mercury column** varies with that **pressure**. Mercury barometers may be classified into three groups according to their construction: **cistern barometers**, **siphon barometers**, and **weight barometers**. *See also* **aneroid barometer**, **inch of mercury**.

mercury column—(Also called barometer column, barometric column.) The column of **mercury** employed in a **mercury barometer**, the height of which (**inches of mercury**) is used as a measure of **atmospheric pressure**.

mercury-in-glass thermometer—A common type of **liquid-in-glass thermometer** used, in meteorology, in **psychrometers** and as a **maximum thermometer**.

mercury-in-steel thermometer—A **liquid-in-metal thermometer** in which **mercury** is enclosed in a steel envelope.

The change in internal **pressure** caused by the **temperature** variation is measured by a **Bourdon tube** that is connected to the mercury by a capillary tube. This instrument is very accurate and has extremely good pen control when arranged as a **thermograph**.

mercury thermometer—A **liquid-in-glass thermometer** or **liquid-in-metal thermometer** using **mercury** as the liquid.

mergozzo—Northwest **wind** on Lake Maggiore, Italy.

meridian—The **great circle** on the earth that passes through the North Pole, the point in question, and the South Pole.

Meridians intersect the **equator** at right angles.

meridional—In meteorology, a flow, average, or functional **variation** taken in a direction that is parallel to a line of longitude; along a **meridian**; northerly or southerly; as opposed to **zonal**.

meridional cell—A very **large-scale convection** circulation in the **atmosphere** or ocean that takes place in a **meridional** plane, with northward and southward currents in opposite branches of the cell, and upward and downward motion in the equatorward and poleward ends of the cell.

There are three annual mean meridional cells in each hemisphere, the strongest of which is the **Hadley cell**. A much weaker **indirect cell** is the **Ferrel cell** located between 30° and 60° latitude. There is a very weak **direct cell** in the polar latitudes. These are integral parts of the **general circulation**. *Compare* **Hadley cell, polar cell**.

meridional circulation—1. An atmospheric **circulation** in a vertical plane oriented along a **meridian**.

It consists, therefore, of the vertical and the **meridional** (north or south) components of motion only. *See* **meridional cell**.

2. *See* **meridional flow**.

meridional exchange—Exchange by the **wind** of properties of air, such as **heat, momentum**, and moisture, between a northern zone and a southern zone in the **atmosphere**.

meridional flow—A type of atmospheric **circulation pattern** in which the **meridional** (north and south) component of motion is unusually pronounced.

The accompanying **zonal** component is usually weaker than **normal**. *Compare* **zonal flow, meridional circulation**; *see* **meridional index**.

meridional front—A **front** in the South Pacific separating successive **migratory** subtropical anticyclones.

Such fronts are essentially in the form of great arcs with meridians of longitude as chords; they have the character of **cold fronts**. *See* **polar trough**.

meridional index—A measure of the component of air motion along **meridians**, averaged, without regard to sign, around a given latitude circle.

When averaged further over five-day periods, the meridional index in the Northern Hemisphere generally varies between 5 and 20 knots in the lower and middle **troposphere** in middle latitudes. *Compare* **zonal index**.

meridional wind—The **wind** or wind component along the local **meridian**, as distinguished from the **zonal wind**.

In a horizontal **coordinate system** fixed locally with the x axis directed eastward and the y axis northward, the meridional wind is positive if from the south, and negative if from the north.

meromictic lake—A lake that does not mix throughout the total water column; the lower part of the water column is perennially isolated.

mesoanalysis—The representation of **temperature**, moisture, **pressure**, and **wind** variations on horizontal scales of 10–100 km.

The **analysis** seeks to define **mesoscale** features of the observed temperature, pressure, moisture, and wind fields that can be related to important local and regional circulations that in turn may have a significant impact on local and regional weather systems. The mesoanalysis differs from the more conventional **synoptic-scale** representation of the wind and pressure features in that smaller-scale features inherent in the wind, pressure, and moisture fields are retained in the analysis.

mesoanticyclone—An anticyclonically rotating **vortex**, around 2–10 km in diameter, in a **convective storm**.

mesochart—A means to illustrate a **mesoanalysis**.

Important **mesoscale** features in the **wind, temperature**, moisture, and **pressure** fields are illustrated on the mesochart by means of **contours** and/or colored (or black-and-white) shading.

mesoclimate—The **climate** of a natural region of small extent, for example, valley, forest, plantation, and park.

Because of subtle differences in **elevation** and **exposure**, the climate may not be representative of the general climate of the region.

mesoclimatology—The study of mesoclimates; the **climatology** of relatively small areas that may not be climatically representative of the general region.

The data used in mesoclimatology are mostly standard observations. The size of the area involved is rather indefinite and may include topographic or landscape features from a few acres to a few square miles, such as a small valley, a forest clearing, a beach, or a village site.

mesocyclone—A cyclonically rotating **vortex**, around 2–10 km in diameter, in a **convective storm**.

The **vorticity** associated with a mesocyclone is often on the order of 10^{-2} s^{-1} or greater. (It should be noted that a mesocyclone is not just any **cyclone** on the **mesoscale**; it refers specifically to cyclones within convective storms.) Mesocyclones are frequently found in conjunction with updrafts in **supercells**. **Tornadoes** sometimes form in mesocyclones. Persistent mesocyclones that have significant vertical extent are detected by **Doppler radar** as mesocyclone signatures. Tornado warnings may be issued when a **mesocyclone signature** is detected.

mesocyclone signature—The **Doppler velocity** pattern of a **mesocyclone** within a severe **thunderstorm**.

In a storm-relative **reference frame**, the idealized **signature** is symmetric about the **radar** viewing direction with marked azimuthal **shear** across the core region between peak Doppler velocity values of opposite sign. Typical signatures consist of Doppler velocity differences of 25–75 m s^{-1} across core diameters of 2–8 km, with resulting azimuthal shear values of 5×10^{-3} s^{-1} to 2×10^{-2} s^{-1}.

mesojet—A **mesoscale** wind maximum.

It typically may have an along-flow length **scale** of tens to hundreds of kilometers and a cross-flow length scale of < 100 km. Mesojets differ from planetary-scale jets, which can have length scales of several thousand kilometers, and synoptic-scale jets, which may have length scales of 1000–2000 km and are commonly found in association with progressive synoptic-scale **troughs** and **ridges**. Larger mesojets may also sometimes be known as **jet streaks**. Mesojets can form adjacent to prominent **orographic** features in association with terrain-channeled flow. Mesojets are also seen in association with organized **mesocale convective systems** as typified by the evaporatively driven **rear-inflow jet** commonly found behind active **squall lines** lines. Mesojets may also be found in conjunction with prominent lower-tropospheric stable layers where the airflow can become decoupled from the **planetary boundary layer**, especially at night. An exceptionally well organized lower-tropospheric mesojet extending over hundreds of kilometers might be known as a **low-level jet**.

mesokurtic—*See* **kurtosis**.

mesolow—A low pressure area on the **mesoscale**.

It has been used to refer both to features observed within **convective storms** and features even larger in **scale**.

mesometeorology—The study of **mesoscale** atmospheric phenomena.

mesonet station—*See* **portable mesonet stations**.

mesopause—The top of the **mesosphere** and the base of the **thermosphere**.

The mesopause is usually located at heights of 85–95 km, and is the site of the coldest temperatures in the **atmosphere**. Temperatures as low as 100 K (− 173°C) have been measured at the mesopause by **rockets**. *See* **atmospheric shell**.

mesopeak—The **temperature** maximum at about 50 km in the **mesosphere**.

mesophyte—A plant that grows under conditions of a medium amount of water; **temperature** is not a limiting factor.

mesopic vision—Same as **parafoveal vision**.

mesoscale—Pertaining to atmospheric phenomena having horizontal scales ranging from a few to several hundred kilometers, including **thunderstorms**, **squall lines**, **fronts**, **precipitation bands** in **tropical** and **extratropical cyclones**, and topographically generated weather systems such as **mountain waves** and **sea** and **land breezes**.

From a dynamical perspective, this term pertains to processes with timescales ranging from the inverse of the **Brunt–Väisälä frequency** to a **pendulum day**, encompassing deep **moist convection** and the full **spectrum** of **inertio-gravity waves** but stopping short of **synoptic-scale** phenomena, which have **Rossby numbers** less than 1.

mesoscale cellular convection—(Sometimes abbreviated MCC.) A regular pattern of convective **cells** that can develop in an **atmospheric boundary layer** heated from below or radiatively cooled from **cloud top**.

This phenomenon is readily observed in satellite imagery during cold air outbreaks when **continental air** passes over the relatively warm coastal ocean. **Cloud lines**, marking **horizontal roll vortices**, form initially in the developing marine atmospheric boundary layer. These lines evolve into **open cells**, which are defined by clouds in the upward motion along the edges of honeycomb-shaped cells, with less **cloudy** subsiding air in their centers. The convective structure further evolves into **closed cells**, which have cloudy centers and cloud-free edges.

mesoscale convective complex—(Abbreviated MCC.) A subset of **mesoscale convective systems** (MCS) that exhibit a large, circular (as observed by satellite), long-lived, cold **cloud shield**.

The cold cloud shield must exhibit the following physical characteristics.

Size: A - Cloud shield with continuously low **infrared** (IR) **temperature** $\leq -32°C$ must have an area $\geq 10^5$ km²; and B - Interior cold **cloud** region with temperature $\leq -52°C$ must have an area $\geq 0.5 \times 10^5$ km².

Initiate: Size definitions A and B are first satisfied

Duration: Size definitions A and B must be met for a period ≥ 6 h.

Maximum extent: Contiguous cold cloud shield (IR temperature $\leq -33°C$) reaches maximum size.

Shape: Eccentricity (minor axis/major axis) ≥ 0.7 at time of maximum extent.

Terminate: Size definitions A and B no longer satisfied.

Alternatively, a dynamical definition of an MCC requires that the system have a **Rossby number** of order 1 and exhibit a horizontal **scale** comparable to the **Rossby radius of deformation**. In midlatitude MCS environments, the Rossby radius of deformation is about 300 km.

mesoscale convective system—(Abbreviated MCS.) A **cloud system** that occurs in connection with an ensemble of **thunderstorms** and produces a contiguous **precipitation area** on the order of 100 km or more in horizontal **scale** in at least one direction.

An MCS exhibits deep, moist convective overturning contiguous with or embedded within a **mesoscale** vertical **circulation** that is at least partially driven by the convective overturning.

mesoscale disturbance—Same as **disturbance** except that the atmospheric feature has a **Rossby number** of order 1 and exhibits a horizontal **scale** comparable to the **Rossby radius of deformation**.

For such systems, both **ageostrophic advection** and **rotational** influences are important.

mesoscale eddies—*See* **mode eddies**.

mesoscale model—A **model** designed to simulate **mesoscale** atmospheric phenomena.

Such models can include analytic solutions of a set of simplified equations governing atmospheric motion, **scale** models of particular geographic regions, and numerical integrations, including **numerical weather prediction** models that can resolve mesoscale circulations. *See* **nonhydrostatic model**.

mesosphere—The region of the **atmosphere** lying above the **stratosphere** and extending from the **stratopause** at about 50 km height to the **mesopause** at 85–95 km.

The mesosphere is characterized by decreasing **temperature** with increasing height, reflecting the decreasing **absorption** of solar **ultraviolet radiation** by **ozone**. Many features of the mesosphere remain poorly understood since in situ measurements are difficult. The region is too high for **balloon** operations and too low for satellites to **orbit**. **Rockets**, while useful, usually travel too rapidly through the region to produce reliable measurements. *See* **atmospheric shell**.

mesosphere–stratosphere–troposphere radar—See **MST radar**.

mesospheric jet—There are two **zonal** jet streams in the layer known as the **mesosphere**.

In January, there is a westerly (blowing from the west) **jet** located at 70 km between 25° and 45°N with a maximum wind of 60 m s⁻¹. There is also an easterly jet of comparable **intensity** located between 30° and 50°S. In July, the direction of the mesospheric jet in each hemisphere is reversed. The westerly jet in the southern hemisphere in July descends to a somewhat lower level (about 50 km) and has a stronger intensity (about 100 m s⁻¹)

mesotherm—A type of plant life that requires moderate temperatures for full growth; moisture is not a limiting factor.

Compare **megatherm**, **microtherm**.

mesothermal—Of, or pertaining to, moderate **temperature**.

mesothermal climate—A type of **climate** characterized by moderate temperatures, that is, a region lacking the constant **heat** of the **Tropics** or the constant cold of the polar caps; middle latitude climate with definite seasonal rhythm in **temperature**, with **amplitude** ranges reaching maximum for the earth in the north intermediate zone.

This is a (C) climate under the **Köppen classification** (1931) and, as defined by Thornthwaite's **climatic classification** scheme (1948), this is a climate with annual **potential evapotranspiration** between 57 and 114 cm. *Compare* **megathermal climate**, **microthermal climate**.

Trewartha, G. T., 1954: *Introduction to Climate*, 3d ed., McGraw–Hill, 289–323.
Köppen, W., 1931: *Grundriss der Klimakunde*, Walter die Gruyter Co., Berlin.
Thornthwaite, C. W., 1948: An approach toward a rational classification of climate. *Geographical Review*, **38**, 55–94.

mesothermic—*See* **mesothermal**.

messenger—A brass weight, usually hinged and with a latch so that it can be fastened around a wire, used to actuate **Nansen bottles** and other oceanographic instruments after they have been lowered to the desired depth.

met—A colloquial contraction for meteorology, occasionally used in a technical sense, especially in aviation.
It can also refer to a **meteorologist**.

metadata—Information that provides enhanced knowledge of the content of data.
For example, the format description of a **METAR** is the metadata about the meteorological data being observed and reported. Literally, metadata are data about data.

metallic barometer—All-metal **barometer**; synonym for **aneroid barometer**.

metamorphosis of snow—The modification of **snow grains** to a less angular, more rounded form accompanied by a gradual increase in **density**.
Metamorphic processes include **sublimation, evaporation**, and vapor **diffusion**.

METAR—1. Abbreviation for Meteorological Terminal Air Report; also known as Aviation Routine Weather Reports.
2. The international standard code for hourly and special observations that took effect on 1 July 1996.
U.S. Department of Commerce, NOAA, 1995: *Federal Meteorological Handbook No. 1 (FMH-1)*, Chapter 9.

meteor—Anything in the air; hence meteorology and its concern with **hydrometeors, lithometeors, igneous meteors, electrometeors**, and **optical meteors**.

Meteor—A series of operational polar-orbiting meteorological satellites launched by the former Soviet Union since 1969.
There have been three series of Meteor satellites, *Meteor-1, Meteor-2*, and *Meteor-3*. Meteor instrumentation includes visible and **infrared** imaging **radiometers**, including an **APT** transmission, an infrared **sounder**, and an instrument to monitor the space environment.

meteor trail—(Or ion column.) The ionized trail left by a meteor or meteoritic **particle** entering the **atmosphere**; a part of the composite phenomenon known as a meteor.
Radar measurements of the **drift** of meteor trails are used to infer **wind** motions in the upper **mesosphere**.

meteoric dust—Atmospheric **dust** that has been produced by meteorites.

meteoric water—**Groundwater** that originates in the **atmosphere** and reaches the **zone of saturation** by **infiltration** and **percolation**; as opposed to **connate** and **magmatic water**.
Most groundwater is meteoric in origin.

meteorogram—1. A record obtained from a **meteorograph**.
2. A **chart** in which meteorological **variables** are plotted against time.

meteorograph—An instrument that automatically records the measurement of two or more meteorological **elements**.
See **aerograph**.

meteorological acoustics—*See* **environmental acoustics**.

meteorological bomb—*See* **bomb**.

meteorological bulletin—A text message composed of **meteorological information** preceded by an appropriate heading that typically identifies its type, point of origin, and issue time.

meteorological cloud chamber—A closed vessel in which the **temperature, pressure**, and **humidity** are controlled in a manner to cause **cloud formation** in the air within the enclosed space.
Cloud chambers were used for early studies of high-energy **particles** such as **cosmic rays** (see **Wilson cloud chamber**). Meteorological versions have been used for studying **cloud** microphysical processes. They usually **range** in size from a few cubic centimeters to a few cubic meters, although cloud chambers as large as hundreds of cubic meters have been constructed.

meteorological code—A code composed of a set of **code forms** and binary codes made up of groups of characters representing meteorological or other geophysical **elements**.

meteorological element—*See* **element**.

meteorological element series—A series of values of any one of the properties or **elements** of the **atmosphere** for a number of places or for a number of instances in time.

meteorological equator—1. The parallel of latitude of 5°N; so named because this is the annual mean latitude of the **equatorial trough**.

2. (Also called **equatorial trough**.) The axis of the **barotropic** current that characterizes the low **troposphere** in equatorial regions.

This axis is marked by the presence of a **convergence line** (the **intertropical convergence zone**).

3. *See* **heat equator, hyetal equator, thermal equator**.

meteorological forecast—A **prediction** of the future state of the **atmosphere**, typically in terms of **state variables** such as **temperature**, moisture, and **momentum**.

See **forecast**; *compare* **weather forecast**.

meteorological information—**Meteorological report, analysis**, forecast, and any other statement relating to existing or expected meteorological conditions.

meteorological institute—National or regional technical, scientific, and administrative organization with activities concerned with the different theoretical and practical branches of meteorology.

meteorological instrument—A measuring device for determining the present value of a meteorological quantity under observation.

meteorological message—A **meteorological bulletin** preceded by a starting line and followed by an end-of-message signal.

World Meteorological Organization, 1992: *Manual on the Global Telecommunication System, Vol. I, Global Aspects*, WMO-No. 386, p. A.II-1.

meteorological network—A group of **meteorological observing stations** spread over a given area for a specific purpose.

meteorological noise—Originally, the small-scale, high-frequency solutions to the **fundamental equations of hydrodynamics**, which may obscure the solution required for **numerical forecasting**.

However, the meaning must be extended to include unwanted frequencies in general, and the term "noise" may be applied legitimately by the investigator of **atmospheric tides** to the moving **cyclonic-scale** weather patterns. *See* **noise filtering, Kalman–Bucy filter**.

meteorological observation—Evaluation or measurement of one or more meteorological **elements**.

meteorological observatory—1. A scientific establishment devoted to making particularly precise and detailed meteorological, geophysical, and related astronomical phenomena.

2. (Obsolete.) A location where only **synoptic** and **radar** meteorological observations were made.

meteorological observer—Member of a **meteorological service** or a volunteer approved by a meteorological service who makes and transmits **meteorological observations**.

meteorological observing station—Place where hydrometeorological observations are made with approval of the WMO member or members concerned.

meteorological office—Office designated to provide hydrometeorological services.

See **main meteorological office, dependent meteorological office, supplementary meteorological office**.

meteorological optics—That part of **atmospheric optics** concerned with the study of patterns observable with the naked eye.

This latter restriction is often relaxed slightly to allow the use of simple aids such as binoculars or a polarizing **filter**. Topics included in meteorological optics are sky color, **mirages, rainbows, halos, glories, coronas**, and **shines**.

meteorological range—(Also called standard visibility, standard visual range.) An empirically consistent measure of the **visual range** of a **target**; a concept developed to eliminate from consideration the **threshold contrast** and **adaptation luminance**, both of which vary from **observer** to observer.

The meteorological range is the distance V' in the black target form of the **visual-range formula**,

$$V' = \frac{1}{\sigma} \ln \frac{1}{\epsilon},$$

when ϵ, the threshold contrast, is set equal to 0.02. Thus, V' is a function only of the **extinction coefficient** σ of the **atmosphere** at the time and place in question.

meteorological reconnaissance flight—An aircraft flight made for the specific purpose of obtaining weather information in a region inadequately served by surface observations (generally over the sea). Such flights are normally made to gather information about **tropical storms**.

meteorological report—Statement of observed meteorological conditions related to a specific time and location.

meteorological rocket—A **rocket** designed primarily for routine **upper-air observation** (as opposed to research) in the lower 80 km of the **atmosphere**, but especially that region above 30 km inaccessible to **balloons**.
See **rocketsonde**.

meteorological satellite—(Acronym: metsat.) Environmental and weather satellites (such as **GOES**, **Meteosat, GMS, NOAA, DMSP**) that carry instruments to remotely sense portions of the **electromagnetic spectrum** radiated from the earth and the surrounding **atmosphere** for use in the preparation of various meteorological observations and forecasts.

meteorological service—National or regional technical, scientific, and administrative organization with activities concerned with the different theoretical and practical branches of meteorology.

meteorological station—*See* **meteorological observing station**.

meteorological symbol—A letter, number, diagrammatic sign, or character used in weather records or on weather maps to indicate meteorological phenomena, both past and present, in a concise and accurate form.
Symbols were first suggested and used by J. A. Lambert in 1771; his symbols included only **clouds, rain, snow, fog**, and **thunder**. The symbology pertaining to surface observations in the United States is contained in the "Manual of Surface Observations," 7th ed., 1955; that pertaining to the preparation of **weather maps** and **weather analysis** is contained in "Preparation of Weather Maps," U.S. Weather Bureau, 1942, and "Weather Analysis Symbols," U.S. Weather Bureau, 1950.

meteorological tide—*See* **radiational tide**.

meteorological transmission—Use of electronic means to transmit meteorological data.

meteorological visibility—*See* **visibility**.

meteorological visibility at night—*See* **visibility**.

meteorological watch office—A **meteorological office** specified under International Civil Aviation Organization procedures to maintain watch over meteorological conditions within a defined area or along designated routes or portions thereof for the purpose of supplying **meteorological information**, in particular, meteorological warnings.
A meteorological watch office may be an independent office or may be part of a **main meteorological office** (MMO) or **dependent meteorological office**. In the United States, this office is either a National Center for Environmental Prediction (an independent office) or a Warning and Forecast Office (an MMO).

meteorologist—A person who is professionally employed in the study or practice of meteorology.
It often refers to individuals who have completed the requirements for a college degree in meteorology or **atmospheric science**.
Amer. Meteor. Soc., 1987: *Bull. Amer. Meteor. Soc.,* **68**, p. 1570.
Amer. Meteor. Soc., 1991: *Bull. Amer. Meteor. Soc.,* **72**, p. 61.

meteorology—1. The study of the physics, chemistry, and dynamics of the earth's **atmosphere**, including the related effects at the air–earth boundary over both land and the oceans. Fundamental topics include the composition, structure, and motion of the atmosphere. The goals ascribed to meteorology are the complete understanding and accurate **prediction** of atmospheric phenomena.
See also **atmospheric science**; *compare* **climatology**.

2. In popular usage, the underlying science of weather and weather forecasting.

Meteosat—A series of European geostationary satellites, initiated with the launch of *Meteosat-1* in November 1977.
Meteosat-1 was the first **geostationary satellite** to provide 6.7-μm **water vapor imagery**. Meteosat is positioned above the prime **meridian** (0° longitude).

meter—In the **International System of Units**, the fundamental unit of length; equal to 100 cm, 3.2808399 ft, or 39.370079 in.

meter–tonne–second system—(Abbreviated mts system.) A system of physical units based upon the use of the meter, the metric ton (or tonne: 10^6 grams), and the second as elementary quantities of length, mass, and time, respectively.

In this system, **density** is expressed in tonne m^{-3}, **velocity** or speed in m s^{-1}, force in tonne m^{-2} (or sthene), **pressure** in centibars (or pieze), and **energy** in kilojoules. The mts system is used primarily by European engineers.

methacrolein—Unsaturated aldehyde, 2-formyl **propene**, formed in the atmospheric **oxidation** of **isoprene** in about 30 percent yield.

methane—(Also called **marsh** gas.) Colorless, inflammable gas of formula CH_4; the simplest hydrocarbon.

Methane enters the **atmosphere** as a result of the anaerobic decay of organic matter in, for example, swamps and rice paddies, and is also produced in large quantities by cattle and termites. It is formed along with coal and oil in fossil fuel deposits, and released to the atmosphere on mining. Methane is itself burned as a fuel, being the major constituent of natural gas. The atmospheric **mixing ratio** of methane is currently about 1.7 parts per million and has been rising gradually since the industrial era began. The atmospheric lifetime of methane is about eight years. As well as influencing the chemistry of the atmosphere, methane is a strong greenhouse gas and an important source of stratospheric **water vapor**, and it contributes to global warming.

methanesulfonic acid—Acidic gas, formula CH_3SO_3H, formed in the **oxidation** of **dimethyl sulfide** in the marine **atmosphere**.

As a result of its low volatility it is mainly found in the liquid (**aerosol**) **phase**.

methanol—Alcohol, also known as methyl alcohol (common name wood alcohol), formula CH_3OH, formed in small quantities in the **oxidation** of **methane**; possibly emitted in large amounts from various species of vegetation.

method of characteristics—A method of solving systems of **nonlinear** differential equations by constructing the **characteristics** for the equations over the region of known initial data and proceeding along these lines to determine the solutions for later times or for new regions of space.

This method has been used successfully in the study of gas flows and has been applied to the meteorological problem of one-dimensional **unsteady flow** of air under an **inversion** surface.
>   Freeman, J. C., 1951: *Compendium of Meteorology*, 421–433.

method of images—Analytical technique for solving potential problems by superposition of solutions of individual sources and sinks.

Commonly, one **image** is placed opposite a second one, with each having equal magnitude and being equally spaced from a boundary plane.

method of least squares—*See* **least squares**.

method of perturbations—*See* **method of small perturbations**.

method of small perturbations—(Also called method of perturbations, perturbation method.) The **linearization** of the appropriate equations governing a system by the assumption of a **steady state**, with departures from that steady state limited to **small perturbations**.

See **linearized differential equation**.

method of successive approximations—The solution of an equation or a set of simultaneous equations by proceeding from an initial approximation to a series of repeated trial solutions, each depending upon the immediately preceding approximation, in such a manner that the discrepancy between the newest estimated solution and the true solution is systematically reduced.

Newton's method for determining the roots of an algebraic equation is an example of the method of successive approximations. For partial differential equations, the **relaxation method** is a widely applied example of the method of successive approximations.

methyl bromide—An organic compound, formula CH_3Br, present in the **atmosphere** as the result of both natural (oceanic production) and **anthropogenic** (use as a soil fumigant) sources; this compound is the largest single source of bromine to the **stratosphere**, with a tropospheric **mixing ratio** of about 10 parts per trillion (by volume).

methyl chloride—The most abundant single halocarbon, formula CH_3Cl, found in the **atmosphere**, with a **mixing ratio** of about 600 parts per trillion (by volume) in the **troposphere**.

This compound is mostly of natural origin, as a result of production in the oceans. It provides the natural background amount of chlorine that was believed to be present in the preindustrial **stratosphere** and that will likely be present in the future following the phaseout of other chlorine source compounds (**chlorofluorocarbons**, etc.).

methyl chloroform—A chlorine-containing organic compound, formula CH_3CCl_3.

It is used in industrial applications as a solvent and a degreasing agent. The production of this compound is now forbidden as a result of the **Montreal Protocol**. Because its sources to the **atmosphere** (**anthropogenic** activity) and its losses (mostly through reaction with the **hydroxyl radical**) are well understood, the atmospheric abundance of this compound has been used to infer the average hydroxyl concentration in the **troposphere**.

methyl iodide—An organic compound, formula CH_3I, present in the **atmosphere** as the result of its production in the oceans by marine algae.
> It is subject to rapid **photolysis** in the **lower atmosphere** (with a lifetime of a few days).

methylvinyl ketone—Unsaturated carbonyl compound, formula $CH_3C(O)CH=CH_2$, or 3-oxybutene.
> One of the major products of **isoprene** oxidation in the **atmosphere**.

metro—A contraction for "meteorology" often heard among personnel in the military weather services.

metsat—Acronym for **meteorological satellite**.

MF—Abbreviation for medium frequency.
> *See* **radio frequency band**.

MIBL—Abbreviation for **mechanical internal boundary layer**.

Michael-riggs—(Also called rig.) Autumn **gales** occurring about Michaelmas (29 September).

Michaelson actinograph—A **pyrheliometer** of the bimetallic type used to measure the **intensity** of **direct solar radiation**.
> The **radiation** is measured in terms of the angular deflection of a blackened **bimetallic strip** that is exposed to the direct solar beams. In its original form the instrument was also responsive to **ambient temperature** variations. However, modifications have been introduced that eliminate this source of **error**.

microbarm—That portion of the record of a **microbarograph** between any two (or a specified small number of) successive crossings of the average **pressure** level (in the same direction); analogous to **microseism**.

microbarogram—The record or **trace** made by a **microbarograph**.

microbarograph—An **aneroid barograph** designed to record **atmospheric pressure** variations of very small magnitude.

microbarovariograph—Sensitive **barometer** used to record continuously on an enlarged **scale** the short period variations of **pressure**.

microburst—A **downburst** that covers an area less than 4 km along a side with peak winds that last 2–5 minutes.
> Differential **velocity** across the **divergence** center is greater than 10 m s^{-1}. The strong **wind shears** associated with a microburst can result in aircraft accidents.

microclimate—The fine climatic structure of the air space that extends from the very surface of the earth to a height where the effects of the immediate character of the underlying surface no longer can be distinguished from the general local **climate** (**mesoclimate** or **macroclimate**).
> The microclimate varies with and in turn is superimposed upon the larger-scale conditions. While some rigid limits have been placed on the thickness of the layer concerned, it is more realistic to consider variable thicknesses. (Observe the microclimate of a putting green versus that of a redwood forest.) Generally, four times the height of surface growth or structures defines the level where microclimatic overtones disappear. Microclimate can be subdivided into as many different classes as there are types of underlying surface. With sufficient detail, this could be almost limitless. Currently, the most studied broad types are the "urban microclimate," affected by pavement, buildings, **air pollution**, dense inhabitation, etc., the "vegetation microclimate," concerned with the complex nature of the air space occupied by vegetation, and its effects upon the vegetation (*see* **phytoclimatology**); and the microclimate of confined spaces (the **cryptoclimate**) of houses, greenhouses, caves, etc.
> > Geiger, R., 1951: *Compendium of Meteorology*, 993–1003.

microclimatology—The study of **microclimate**.
> It includes the study of profiles of **temperature**, moisture, and **wind** in the lowest stratum of air, the effect of the vegetation and of **shelterbelts**, and the effect of towns and buildings in modifying the **macroclimate**. The study of the "microclimate" of confined spaces is termed **cryptoclimatology**, and that of plant communities, **phytoclimatology**. *Compare* **macroclimate**, **mesoclimate**.

Geiger, R., 1951: *Compendium of Meteorology*, 993–1003.

microfront—A sharp horizontal **temperature** contrast of a few degrees Celsius within a width of tens of centimeters to a few meters along the trailing edge of a **thermal** plume in the **surface layer** of the **atmospheric boundary layer**.

microlayer—1. A thin **interface** between two other layers.

2. The very thin (order of 1 mm) layer of air adjacent to the surface for which molecular **transport** (**conduction, diffusion, viscosity**) dominates over **turbulent transport**.
See **laminar boundary layer**.

micrometeorology—A part of meteorology that deals with observations and processes in the smallest scales of time and space, approximately smaller than 1 km and less than a day (i.e., local processes).
Micrometeorological processes are limited to shallow layers of frictional influence (slightly larger-scale phenomena like convective **thermals** are not part of micrometeorology). Therefore, the subject of micrometeorology is the bottom of the **atmospheric boundary layer**; namely, the **surface layer**. Exchange processes of **energy**, gases, etc., between the **atmosphere** and the surface (water, land, plants) are important topics. Therefore, micrometeorology is closely connected with most of the human activities in the atmosphere. Microclimatology describes time averaged (long-term) micrometeorological processes, and micrometeorologists are interested in their fluctuations. *Compare* **mesometeorology, macrometeorology**.

micrometer—(Also called micron; abbreviated μm.) A unit of length equal to one-millionth of a meter or one-thousandth of a millimeter.
The micrometer is a convenient length for measuring wavelengths of **infrared radiation**, diameters of atmospheric **particles**, etc.

micron—Same as **micrometer**.

micropluviometer—(Same as **ombrometer**.) A **rain gauge** capable of measuring very small amounts of **precipitation**, amounts that are less than could be measured by an ordinary rain gauge.

micropulsations—Geomagnetic micropulsations are cyclic fluctuations of the earth's **magnetic field** in the **amplitude** range of a fraction of an **nT** (or gamma) to, on rare occasions, as much as a few tens of nT.
Periods **range** from about 0.1 sec to 10 minutes.

microscale—Atmospheric motions with Lagrangian **Rossby numbers** greater than 200 or spatial scales 2 km or less.

Emanuel, K. A., 1986: Overview and definition of mesoscale meteorology. *Mesoscale Meteorology and Forecasting*, P. Ray, Ed., Amer. Meteor. Soc., P. 13.
Orlanski, I., 1975: A rational subdivision of scales for atmospheric processes. *Bull. Amer. Meteor. Soc.*, **56**, 527–530.

microseism—A collective term for small motions in the earth that are unrelated to an **earthquake** and that have a **period** of 1.0–9.0 s.
They are caused by a variety of natural and artificial agents. Certain types of microseisms seem to be closely correlated with **pressure** disturbances and can be used to locate such disturbances, especially in the case of **tropical cyclones**. In addition, traffic, industrial activities, and **wind** flexure of trees and tall structures can create microseisms.

microstrip antenna—A thin, flat, printed circuit board **antenna**. The radiating elements of the antenna are conducting strips or patches printed on the upper surface of a thin dielectric substrate that is backed by a conducting ground plate.
This type of antenna has been used in **wind profilers** (particularly **boundary layer radars**) operating in the **UHF** radar **band**.

microstructure—Small-scale structure in the fields of **temperature, salinity, density**, and/or **velocity**.
The term microstructure refers in the ocean to structures of vertical **scale** smaller than 10 m, usually associated with overturning motions and **diapycnal mixing** events.

microtherm—In botany, a type of plant life that requires an annual **mean temperature** of 0°–14°C for **normal** growth.
Such flora are typical of **tundra** and **Boreal woodland**.

microthermal—Of, or pertaining to, low **temperature**.

microthermal climate—A type of **climate** characterized by low annual mean temperatures (between

0° and 14°C), that is, a region of genuine winter emphasized by the usual **snow** mantle, and a true, although many times short, summer to produce a characteristic annual **climate** cycle.

This is a (D) climate under the **Köppen classification** (1931) and, as defined by Thornthwaite's **climatic classification** scheme (1948), this is a climate with annual **potential evapotranspiration** between 14 and 43 cm. *Compare* **megathermal climate, mesothermal climate.**

Trewartha, G. T., 1954: *Introduction to Climate*, 3d ed., McGraw–Hill, 324–357.

Köppen, W., 1931: *Grundriss der Klimakunde*, Walter die Gruyter Co., Berlin.

Thornthwaite, C. W., 1948: An approach toward a rational classification of climate. *Geographical Review*, **38**, 55–94.

microthermic—*See* **microthermal.**

microvariation of pressure—Minute **pressure** variation identifiable only with ultrasensitive equipment.

microwave—**Electromagnetic radiation** having **wavelengths** between approximately 1 mm and 1 m (corresponding to 0.3- and 300-GHz **frequency**) bounded on the short-wavelength side by **far infrared** (< 1 mm) and on the long-wavelength side by **very high frequency** radio waves (> 1 m).

Passive systems operating at these wavelengths are sometimes called passive microwave systems. Active systems operating at these wavelengths are called **radar**, although the definition of radar requires a capability to measure distance that is not always included in active microwave systems. The limits of the microwave region are not precisely fixed.

microwave probing—Sensing **electromagnetic radiation** having wavelengths between approximately 1 mm and 1 m.

The **Microwave Sounding Unit** (MSU) and the Special Sensing Microwave/Imager (SSM/I) are examples of systems that utilize microwave probing.

microwave radar—A type of **radar** that employs **microwave** scatterometers deployed aboard aircraft and satellites that can compute **wind speed** using algorithms relating the wind speed to the **backscattering cross section.**

See also **Doppler radar, marine radar, synthetic aperture radar, high-frequency radar.**

microwave radiation—**Electromagnetic radiation** generally in the **frequency** range between 300 MHz and 300 GHz (free-space **wavelengths** between 1 and 1000 mm).

Within these frequencies lie the UHF, SHF, and EHF **radio frequency bands**. Radars operate at microwave frequencies.

microwave refractometer—An instrument that measures the **refractive index** of air by determining changes in **humidity.**

The refractive index of the **moist air** is measured either by measuring its **dielectric constant** using special capacitors or by measuring **frequency** shifts. For the latter method a measuring cell containing the moist air is compared with a reference cell containing **dry air** or an inert reference gas. The frequencies used are about 10 GHz, which corresponds to a **wavelength** of about 30 cm.

Microwave Sounding Unit—(Abbreviated MSU.) One of three **sounders** that comprise the **TIROS-N Operational Vertical Sounder** package on **NOAA** polar orbiter satellites.

The primary function is to make **temperature** soundings in the presence of clouds.

Mid-Japan Sea Cold Current—The slow southward movement of cold water in the central Japan Sea that, when meeting the northward flowing **Tsushima Current**, establishes the **Arctic Polar Front** that stretches from southern Korea to Hokkaido.

middle atmosphere—The region lying between the **troposphere** and the **thermosphere**, comprising the **stratosphere** and the **mesosphere.**

middle clouds—*See* **cloud classification.**

middle infrared—The portion of the **electromagnetic spectrum** lying between the **near-infrared** and the **far infrared.**

This covers the **wavelength** range approximately from 4 to 15 μm, but usage varies. *See also* **thermal infrared.**

middle-latitude westerlies—Same as **westerlies.**

middle-level cloud—Same as **middle clouds.**

midget tropical cyclone—A **tropical cyclone** with a radius to the outermost closed **isobar** of 100–200 km.

These cyclones can support hurricane-force winds with central pressures significantly higher than larger storms.

midnight wind—In Germany, a local **wind** from the south that sets in regularly, under **anticyclonic** conditions, over the Upper Bavarian lakes Würm and Ammer, soon after midnight.

It is a **mountain wind** due to **nocturnal radiation** and reaches a **velocity** of only 1–2 m s^{-1} (3–4 mph).

midsection method—Method for computing the **discharge** of a **stream** by dividing the **cross section** by verticals into sections.

The total discharge is the sum of the products of the **average velocity** in each vertical by the depth of that vertical and by the mean width of the two adjacent sections.

Mie scattering—Scattering of **electromagnetic waves** by homogeneous spheres of arbitrary size, named after Gustav Mie (1868–1957), whose theory of 1908 explains the process.

See **Mie theory**; *compare* **Rayleigh scattering**.

Mie theory—The theory, within the framework of continuum **electromagnetic theory**, of **scattering** and **absorption** of a plane, **harmonic** wave with arbitrary **frequency** and state of **polarization** by a homogeneous sphere of arbitrary size and composition.

Mie theory describes in detail atmospheric optical phenomena such as **rainbows, glories**, and **coronas**. Because Gustav Mie in 1908 was not the first to treat scattering by an arbitrary sphere, the term Lorenz-Mie theory has come into use in recent years.

> Kerker, M., 1969: *The Scattering of Light and Other Electromagnetic Radiation*, 54–59.
> Logan, N., 1965: *Proc. IEEE*, **53**, 773–785.
> van de Hulst, H. C., 1957: *Light Scattering by Small Particles*, ch. 9.
> Bohren, C. F., and D. R. Huffman, 1983: *Absorption and Scattering of Light by Small Particles*, ch. 4.
> Bohren, C. F., 1992: *Opt. Phot. News*, **3**, 18–19.

miejour—A warm, moist **sea breeze** from the south that sets in at midday in Provence, France, south of Mount Ventoux.

In the Roussillon region the midday south **wind** (mitgjorn) is irregular and generally **light**, and is dry after crossing the Pyrenees.

migratory—In meteorology, commonly applied to **pressure systems** embedded in the **westerlies** and, therefore, moving in a general west-to-east direction.

mil—1. A unit of angular measurement, sometimes used in **radar** antennas, equal to 1/6400 of the circumference of a circle.

2. A unit of length, equal to 0.001 in., used in measuring the diameter of wire.

Milankovitch hypothesis—*See* **Milankovitch theory**.

Milankovitch oscillations—Variations in **climate** that are due to variations in the receipt of **solar radiation** associated with 1) the precession of the **equinoxes** and **solstices**; 2) the varying tilt of the earth's rotational axis; and 3) the varying eccentricity of the earth's orbit.

Milankovitch Pleistocene climate variation—The major **oscillation** in the **Pleistocene epoch**, which, according to Milankovitch, was astronomical in origin.

Milankovitch solar radiation curve—A **radiation** curve that combines the systematic effects of the precession of the **equinoxes**, the tilt of the earth's rotational axis, and the eccentricity of the earth's **orbit**.

Early in the twentieth century, a Serbian mathematician and physicist, Milutin Milankovitch (1879–1958), calculated the composite **solar radiation** curve and used it to account for the variations of **climate**. He postulated that the effects of seasonal and latitudinal distribution of **incoming solar radiation** influenced climatic fluctuations of the order tens to hundreds of thousands of years, with each period of radiation minimum causing an **ice age**.

Milankovitch theory—The theory, introduced by the Serbian mathematician M. Milankovitch during the first half of the twentieth century, that variations in the precession of the **equinoxes** and **solstices**, the varying tilt of the earth's rotational axis, and the varying eccentricity of the earth's **orbit** are responsible for the sequence of **ice ages** during the **Pleistocene era**.

> Milankovitch, M., 1941: Canon of insolation and the ice age problem (in Yugoslavian). *K. Serb. Acad. Beorg.*, Spec. Publ. 132 (English translation by Israel Program for Scientific Translations, Jerusalem, 1969).

mile—A unit of distance equal to 5280 ft or 1.609 km; commonly known as **statute mile**.

Compare **nautical mile**.

milky ice—A form of **aircraft icing** intermediate in all respects between **clear ice** and **rime ice**.

It forms in the **temperature range** between $-4°$ and $-15°C$.

milky weather—*See* **whiteout**.

millibar—(Abbreviated mb.) A **pressure** unit of 1000 dynes cm^{-2}, convenient for reporting atmospheric pressures.

The millibar does not fit into any commonly employed system of physical units. One millibar equals one **hectopascal**.

millimeter of mercury—*See* **inch of mercury**.

Mills period—(Also called Mills infection period.) Environmental conditions favorable to promoting development of a vector (spores in the air or soil and insect populations) or disease.

Mindanao Current—One of the western **boundary currents** of the Pacific Ocean.

It flows southward with great speed along the island of Mindanao and carries a **transport** of 25–35 Sv (25–35 × 10^6 m^{3}s^{-1}) from the **North Equatorial Current** into the **North Equatorial Countercurrent**. It is part of the **Mindanao Eddy**.

Mindanao Eddy—A **recirculation** system between the **North Equatorial Current** and the **North Equatorial Countercurrent** in the extreme west of the Pacific Ocean, east of Mindanao.

The **circulation** is counterclockwise, about 800 km in diameter, and reaches to a depth of 250 m. Its western part forms the **Mindanao Current**.

mini-supercell—**Convective storm** that contains similar **radar** characteristics to those of a **supercell** (e.g., **hook echo**, **WER**, **BWER**), but is significantly smaller in height and width.

The diameter of the radar-detected rotation is 1–8 km. This is a relatively new storm type, the existence of which has been confirmed by data from the recently installed **WSR–88D** radars in the United States. Mini-supercells occur in areas where the height of the **equilibrium** level is low, most often in the northern United States, but possibly under certain weather conditions in any area of the world. They are sometimes found in landfalling **tropical cyclones**.

minimal flight—Ideally, an aircraft flight so planned and navigated that is completed in the least possible time.

Such planning should take into consideration the complete three-dimensional **wind** pattern en route as related to aircraft operating characteristics. To date, however, minimal flight is still largely a two-dimensional (**pressure-pattern flight**) concept. One practical method of determining a **minimal flight path** is the **wave-front method**. *Compare* **optimum flight**.

minimal flight path—(Also called least-time track, minimum time track.) In air navigation, the route that yields the least **travel time** between two points, for the expected weather conditions.

minimal headings—The set of headings for which the flight time from **departure** to destination is minimum.

See **minimal flight**, **wave-front method**.

minimum—The least value attained (or attainable) by a function; the opposite of maximum.

(*See* further discussion under **maximum**.)

minimum annual flow—The smallest observed **streamflow** value during a **water year**.

May be either an instantaneous value, the smallest 24-hour average, or an average over a longer period of time.

minimum detectable signal—Same as **threshold signal**.

minimum deviation—The minimum total **deviation**, relative to neighboring deviations, of an incident **wave** (or **ray**) transmitted by a bounded, **optically homogeneous** body (e.g., prism, **ice crystal**, **raindrop**).

The corresponding angle between the direction of the incident wave and that transmitted by the body is the **angle of minimum deviation**. **Rainbows** and **arcs of 22°** are, according to geometrical optics, formed at angles of minimum deviation.

minimum temperature—The lowest **temperature** attained at a specific location during a specified period.

minimum thermometer—A **thermometer** that automatically registers the lowest **temperature** attained during an interval of time.

The alcohol-in-glass minimum thermometer contains a dumbbell-shaped **index** that is kept on the bulb side of the **meniscus** by **surface tension**. The thermometer is installed in a horizontal mounting (*see* **Townsend support**) so that as the temperature falls, the index is pulled toward the bulb and remains at the **minimum** point as the temperature rises. A **bimetallic thermometer** with a circular dial is also used as a minimum thermometer. A free index, mounted concentrically with and driven by the thermometer index, is held by **friction** at the **minimum temperature**. *Compare* **maximum thermometer**.

minimum time track—Same as **minimal flight path**.

minimums—Same as **operational weather limits**.

minor lobe—Same as **sidelobe**.

minor ridge—A **ridge** of smaller **scale** than a **long-wave** ridge.
  It ordinarily moves rapidly and is associated with a **migratory** anticyclonic **disturbance** in the lower **troposphere**. *See* **short wave**.

minor trough—A pressure trough of smaller **scale** than a **long-wave** trough.
  It ordinarily moves rapidly and is associated with a **migratory** cyclonic **disturbance** in the lower **troposphere**. *See* **short wave**.

minor wave—Same as **short wave**.

mintra—Highest **temperature**, for a particular **pressure**, at which a **condensation trail** can form.

minuano—A cold southwesterly **wind** blowing in the winter (June through September) in the coastal region of southern Brazil, the country of the Minuano Indians.
  It is a weak **pampero**. *Compare* **reboyo**.

mirage—An **image** formed when the **atmosphere** behaves as a lens.
  Mirages have a very small angular extent compared with that of the sky: They are normally seen near the **horizon** and involve image displacements and distortions of less than half a **degree**. Consequently, even though visible with the naked eye, they are easiest to see with the aid of binoculars or when photographed with a telephoto lens. The name applied to a particular type of mirage is dependent upon the way in which the appearance of the image differs from that of the object. The simplest distinction for the observer is that between a mirage that exhibits but a single image and one showing multiple images. If there is only a single image, and if that image is displaced down from the position of the object, it is said that there is **sinking**; if up, **looming**. If the image appears vertically enlarged, there is **towering**; if vertically shrunken, there is **stooping**. Recognition of these states depends critically on one's knowledge or memory of the appearance of the scene in the absence of a mirage, because all that is seen is the image. However, the change is often so striking as to make classification fairly easy. Mirages are explained by the **refraction** of **light** through an atmosphere with a **gradient** of **refractive index**. The refractive index of air depends mainly on the molecular number **density** of air, but as the layer through which the majority of the refractive bending occurs is often fairly thin, this density variation is primarily dependent upon **temperature**. Indeed, it a simple matter to associate a particular type of mirage with the shape of a temperature profile. Because the observer is located inside this atmospheric lens, the mirage can change its appearance markedly as a result of slight changes in position, say changing the height of the observer above a surface. *See* **fata bromosa, fata morgana**.

mirror nephoscope—(Also called reflecting nephoscope, cloud mirror.) A **nephoscope** in which the motion of the **cloud** is observed by its **reflection** in a mirror.
  A representative instrument consists of a black mirror disk, engraved with special concentric circles calibrated in degrees, and mounted on a tripod stand fitted with leveling screws. An eyepiece is arranged so that it can be rotated about the center of the mirror and adjusted to various distances above the mirror surface. The **observer** orients the mirror so that its zero corresponds to **true north**, and then adjusts the eyepiece until the cloud is observed at the center of the mirror. The cloud's direction of motion is indicated by the **azimuth** at which the **image** leaves the mirror.

miscible displacement—In a **porous medium**, the advance of one fluid into the pore space occupied by another fluid when the two fluids fully dissolve into each other.
  Compare **immiscible displacement**.

misocyclone—A horizontal **vortex** with a width of between 40 m and 4 km.
  It is often used to refer to 1) a vortex within a **convective storm** with a horizontal **scale** of less than 4 km, and 2) a near-surface vortex with a horizontal scale of less than 4 km along a **convergence line**.
    Fujita, T., 1981: Tornadoes and downbursts in the context of generalized planetary scales. *J. Atmos. Sci.*, **38**, 1511–1534.

misoscale—On a **scale** of 40 m to 4 km.
    Fujita, T., 1981: Tornadoes and downbursts in the context of generalized planetary scales. *J. Atmos. Sci.*, **38**, 1511–1534.

mist—1. A **suspension** in the air consisting of an aggregate of microscopic water **droplets** or wet **hygroscopic** particles (of diameter not less than 0.5 mm or 0.02 in.), reducing the **visibility** at the earth's surface to not less than 1 km or 5/8 mi.

The term mist is used in weather reports when there is such obscurity and the associated visibility is 1000 m or more, and the corresponding **relative humidity** is 95% or more, but is generally lower than 100%. These hydrometeors form a thin greyish veil that covers the landscape. It also reduces visibility, but to a lesser extent than **fog**.

2. In popular usage in the United States, same as **drizzle**.

mistbow—Same as **cloudbow**.

mistral—A north **wind** that blows down the Rhone valley south of Valence, France, and into the Gulf of Lions.

It is strong, squally, cold, and dry, the combined result of the basic **circulation**, a **fall wind**, and **jet-effect wind**. It blows from the north or northwest in the Rhône Delta, where it is strongest, from northwest in Provence and from northeast in the valley of the Durance below Sisteron. A general mistral usually begins with the **development** of a **depression** over the Tyrrhenian Sea or Gulf of Genoa with an **anticyclone** advancing from the Azores to central France. It often exceeds 27 m s^{-1} (60 mph) and reaches 38 m s^{-1} (85 mph) in the lower Rhône valley and 22 m s^{-1} (50 mph) at Marseilles, decreasing both east and west and out to sea. It remains strong to a height of 2–3 km. In the absence of a strong **pressure gradient**, a weaker katabatic local mistral develops in the Rhône valley. A general mistral usually lasts for several days, sometimes with short lulls. It is most violent in winter and spring, and may do considerable damage. Market gardens and orchards are protected from it by windbreaks, and rural houses are built with only a few openings on the side exposed to it. The mistral has a variety of local names: mangofango in Provence; sécaire, maistrau, maistre, or magistral in Cévennes; dramundan in Perpignan; **cierzo** in Spain; **cers** in the Pyrenees, etc. South of Mont Ventoux a similar wind is named **bise**. A local west wind of mistral type that descends from Mt. Canigou to the plains of Roussillon is called canigonenc. *Compare* **bora, tramontana, maestro**; *see also* **cavaliers**.

Defant, F., 1951: *Compendium of Meteorology*, p. 670.

mitigation—An action to make something less severe.

A mitigation plan for a **source** of pollutants is a plan to reduce the amount of pollutants emitted. A mitigation plan could be voluntary or in response to a regulation or law.

mixed cloud—A **cloud** containing both water drops (supercooled at temperatures below 0°C) and **ice crystals**, hence a cloud with a composition between that of a **water cloud** and that of an **ice-crystal cloud**.

A mixed cloud is unstable in the sense that the **equilibrium vapor pressure** difference between water drops and ice crystals at subfreezing temperatures promotes growth of the ice crystals at the expense of the water drops, which is the basis for the **Bergeron–Findeisen theory** of **precipitation**. Most convective clouds extending into air colder than about −10°C are mixed clouds, though the proportion of ice crystals to water drops may be small until the cloud builds to levels of still lower **temperature**. Some clouds (**lenticularis**, **mountain wave** clouds) form at temperatures near −35°C and contain only very small amounts of ice crystals.

mixed distribution—(Also called compound distribution.) A **frequency distribution** that is composed of a mixture of several unlike **populations** of data or populations with different parameters.

mixed layer—1. (Abbreviated ML; sometimes called convective mixed layer, convective boundary layer, or mixing layer in air-pollution meteorology.) A type of **atmospheric boundary layer** characterized by vigorous **turbulence** tending to stir and uniformly mix, primarily in the vertical, quantities such as conservative **tracer** concentrations, **potential temperature**, and **momentum** or **wind speed**.

Moisture is often not so well mixed, showing a slight decrease with height. The vigorous turbulence can be caused by either strong winds or **wind shears** that generate **mechanical turbulence** (called **forced convection**), or by buoyant turbulence (called **free convection**) associated with large **thermals**. The buoyantly generated mixed layers are usually **statically unstable**, caused by heating at the bottom boundary such as the earth's surface or **radiative cooling** at the tops of **cloud** or **fog** layers within the mixed layer. The terms mixed layer, convective mixed layer, and convective boundary layer commonly imply only the buoyantly stirred layer. During fair weather over land, mixed layers are usually daytime phenomena generated buoyantly, with growth caused by **entrainment** of free-atmosphere air into the **mixed-layer top**. *See* **mixed-layer depth, entrainment zone, radix layer, uniform layer**.

2. In **oceanography**, a fully turbulent region of quasi-isopycnal water (i.e., virtually uniform

potential density) that, in the case of the surface mixed layer, is bounded above by the air–sea **interface** and below by the **transition layer**.

Mixed **layer depth** is often defined as the depth at which potential density differs from that of the surface by 0.01 kg m^{-1}.

mixed-layer capping inversion—The statically stable layer of air at the top of the **atmospheric boundary layer**.

Because the **troposphere** is statically stable on the average (i.e., **potential temperature** increases with height), and because **turbulence** in the **boundary layer** causes potential temperatures to become somewhat well mixed there, conservation of **heat** requires that there be a potential temperature increase (i.e., a temperature step or **inversion**) at the top of the boundary layer. It is this inversion that separates the boundary layer from the rest of the troposphere by limiting the domain of turbulence. It is also responsible for trapping **pollutants** near the ground during **fair** weather.

mixed-layer depth—(Also called mixed-layer height, mixed-layer top, mixing height.) The **thickness**, z_i, of the **mixed layer**, defined as the location of a capping **temperature inversion** or **statically stable** layer of air.

Often associated with, or measured by, a sharp increase of **potential temperature** with height, a sharp decrease of water-vapor **mixing ratio**, a sharp decrease in **turbulence intensity**, a sharp decrease in **pollutant** concentration, a change of **wind speed** to **geostrophic**, a minimum of turbulent **heat flux**, and a maximum of **signal intensity** from remote sensors such as **sodars** and **wind profilers**. Quite variable in space and time, the mixed-layer depth typically increases during fair-weather daytime over land from tens of meters shortly after **sunrise** to 1–4 km before **sunset**, depending on the location and **season**.

mixed-layer evolution—The three-part change of the **atmospheric boundary layer** that typically occurs during **fair** weather over land on sunny days.

In the early morning, the **mixed layer** is shallow, slowly deepening, cool (in a **potential temperature** sense), and is capped by the remains of the **stable boundary layer** from the previous night. In mid- to late morning, the top of the mixed layer exhibits rapid rise as heating eliminates the nocturnal **inversion**, and the mixed layer grows through the **residual layer**. The third stage in late morning and afternoon is that of a deep (order of 1–2 km) **convective boundary layer** of relatively constant depth.

mixed-layer height—Same as **mixed-layer depth**.

mixed-layer models—(Also called slab models.) An approximation that treats the **atmospheric boundary layer** as though variables such as **potential temperature**, **momentum**, **pollutants**, and **humidity** were uniform with height.

This type of **model** is popular because of its simplicity, requiring forecasts of only the average variables in the **mixed layer** and of the change of **mixed-layer depth**. During sunny days over land, the actual **boundary layer** is often sufficiently well mixed that a uniform **slab approximation** is a fairly good approximation.

mixed layer models—Models of the upper ocean (usually one-dimensional) that take advantage of the concept of a **mixed layer** by assuming 1) the **mean temperature**, **salinity**, and horizontal **velocity** are quasi-uniform within the mixed layer; and 2) a quasi-discontinuous distribution (**jump**) exists for the same variables just below the mixed layer.

This permits integrating the **momentum** and **tracer** equations from the bottom of the mixed layer to the surface to get equations for the bulk mixed layer velocity, temperature, and salinity, and adding an equation for the evolution of the **mixed-layer depth** derived from the vertically integrated (across the mixed layer) **turbulence kinetic energy** (TKE) equation.

mixed-layer similarity—An empirical method of finding universal relationships between **boundary layer** variables that are made dimensionless using the **Deardorff velocity** w_*, the **mixed-layer depth** z_i, and the mixed-layer **temperature scale** $\theta_* = \overline{w'\theta'}_s / w_*$, where $\overline{w'\theta'}_s$ is the surface kinematic **heat flux**.

The resulting universal relationships are valid only for convective **mixed layers**. An example is

$$\frac{\overline{w'\theta'}(z)}{w_*\theta_*^{ML}} = 1 - 1.2\,\frac{z}{z_i}.$$

Compare **local free-convection similarity, local similarity, similarity theory, dimensional analysis, Buckingham Pi theory**.

Stull, R. B., 1988: *An Introduction to Boundary Layer Meteorology*, 666 pp.

mixed-layer spectra—**Fourier analysis** spectra of **velocity** variance for the atmospheric **mixed layer** that exhibit universal **similarity** characteristics when properly normalized by **mixed-layer** scaling variables.

mixed-layer top—Same as **mixed-layer depth**.

mixed-layer venting—Removal of **pollutants** out of the top of the **atmospheric boundary layer** through the **mixed-layer capping inversion**.

Normally pollutants cannot escape through the **capping inversion**. However, penetrating cumulus clouds, thunderstorms, mountain circulations, and frontal circulations can force polluted air through the **inversion** to vent pollutants into the **free atmosphere**.

mixed nucleus—A **cloud condensation nucleus** (CCN) composed of both soluble **hygroscopic** matter and insoluble, possibly wettable, matter.

Its mixed composition probably results from a **coagulation** process or from cycles of **condensation** and **evaporation**, possibly involving chemical reactions. When applied to natural **ice nuclei** or **cloud seeding** nuclei (as **silver iodide**), the term implies the presence of a soluble component that leads to condensation prior to initiation of **freezing**.

mixed rain and snow—*See* **rain and snow mixed**.

mixed Rossby–gravity wave—(Also called Yanai wave.) An eastward propagating **equatorial wave** that has a **meridional** velocity component symmetric about the **equator** and a **zonal** velocity component anti-symmetric about the MCC.

For large positive (eastward) zonal **wavenumbers** its **dispersion relation** is **Kelvin wave**–like and for large but negative (westward) zonal wavenumbers its dispersion relation is **Rossby wave**–like. *See also* **Rossby–gravity wave**.

mixed tide—A tidal regime where both the **diurnal** and semidiurnal components are significant.

mixing—1. The result of irregular fluctuations in fluid motions on all scales from the molecular to large **eddies**.

Gradients of conservative properties such as **potential temperature**, **momentum**, **humidity**, and concentrations of **particles** and gaseous constituents are reduced by mixing, tending toward a state of uniform distribution. *See* **turbulence, eddy flux, diffusion**.

2. In electronics, the nonlinear (nonadditive) combining of signals.

The common mixing element is a diode or set of diodes. The common desired result of mixing two sinusoidal signals is the multiplicative product, with terms at the sum and difference frequencies. Mixing is used to shift signals to different **carrier** frequencies.

mixing cloud—A **cloud** formed when two subsaturated volumes of **moist air** with different **temperatures** and **vapor pressures** mix **isobarically** and **adiabatically** to form a volume of moist air with an intermediate temperature and vapor pressure above the **saturation** value at that temperature.

Popular but misleading terms for mixing clouds are **steam**, **steam fog**, and steam clouds. These terms obscure the essential mechanism by which mixing clouds are formed and confuse **water vapor** with liquid water. Mixing clouds are the consequence of neither cooling nor heating because both volumes enter into the mixture symmetrically. They are a consequence of the nonlinearity of the **Clausius–Clapeyron equation** together with the **linearity** of the **first law of thermodynamics** for moist air. An example of a mixing cloud is a **condensation trail**.

mixing condensation level—Same as **condensation level**.

mixing efficiency—*See* **Richardson number**.

mixing fog—Fog, light and of short duration, produced by the **mixing** of two moist but nonsaturated **air masses** with different **temperatures**.

mixing height—Same as **mixed-layer depth**.

mixing layer—Same as **mixed layer**.

mixing length—1. An average distance of **air parcel** turbulent movement toward a reference height, where the average is a root-mean-square distance.

It is also known as Prandtl's mixing length, l, after Ludwig Prandtl who devised it in 1925 to explain **turbulent fluxes** such as the **Reynolds stress**, τ. Prandtl started with Boussinesq's **first-order turbulence closure** hypothesis that $\tau = \rho K(d\overline{U}/dz)$, where ρ is **density**, $\overline{U}$ is average horizontal **velocity**, and K is kinematic **eddy viscosity**. He further recognized that **exchange coefficient** K has units of length times velocity, and proposed that $K = lw$, where w is a representative average turbulent vertical velocity. Prandtl also suggested that turbulent vertical motions are caused by the collision of air parcels moving horizontally at different speeds. This results in turbulent vertical velocity being proportional to turbulent horizontal velocity. From this, it can be shown that eddy viscosity can be approximated by $K = l^2|d\overline{U}/dz|$, which can be used in Boussinesq's first-order closure.

2. A mean length of travel over which an **air parcel** maintains its identity before being mixed with the surrounding fluid; analogous to the **mean free path** of a molecule.

Stull, R. B., 1988: *An Introduction to Boundary Layer Meteorology*, 666 pp.

mixing-length theory—Same as *K* **theory**.

mixing line—A method of thermodynamic analysis where the conserved variables for two different states of air (i.e., **air parcels**) are plotted on a **thermodynamic diagram**, and the ending state of a mixture of the two parcels is found on the straight line connecting the two initial states, with relative distance along the line proportional to the relative amounts of each parcel in the mixture.

Mixing line analysis can help determine the origin of air within clouds. *See* **conserved variable diagram**.

mixing potential—The amount of turbulent **mixing** necessary to eliminate any static and dynamic **instabilities** in the **atmosphere**.

Because atmospheric circulations that include turbulent **eddies** have finite velocities, the actual amount of mixing that can occur during a finite time interval might be less than the mixing potential. Some nonlocal **turbulence** models parameterize the amount of turbulence as proportional to the mixing potential, where local as well as nonlocal instabilities are considered. *See* **responsive parameterization**.

mixing ratio—The ratio of the mass of a variable atmospheric constituent to the mass of **dry air**.

If not otherwise indicated, the term normally refers to **water vapor**. For many purposes, the mixing ratio may be approximated by the **specific humidity**. Either r or w is commonly used to symbolize **water vapor** mixing ratio, with r used for thermodynamic terms in this glossary. In terms of the **pressure** p and **vapor pressure** e, the mixing ratio r is

$$r = \frac{0.622e}{p - e}.$$

Compare **absolute humidity, relative humidity, dewpoint**.

mixing zone—Zone separating regions of two different fluids through which a **gradient** exists to mix the fluid properties.

mizzle—*See* **Scotch mist**.

mks system—A system of units consisting of meters, kilograms, and seconds.

This natural, easily understood system is used around the world in commerce and by the public, but is not yet used extensively in the United States except in science and in some engineering disciplines.

Compare **meter–tonne–second system**.

ML—Abbreviation for **mixed layer**.

MMO—Abbreviation for **main meteorological office**.

moat—A ring of reduced **precipitation** and **radar reflectivity** just outside the **eyewall** of a **tropical cyclone** or between inner and outer **concentric eyewalls**.

In many tropical cyclones, this is a distinct minimum of precipitation, with higher precipitation rates occurring farther from the center.

Moazagotl—A stationary bank of **cirriform** cloud marking the upper portion of the system of lenticular clouds formed in the **lee wave** produced by flow across the Sudeten Mountains in southeastern Germany; a type of **foehn cloud**.

The Moazagotl reaches its maximum development in the colder months, but especially in the autumn. It usually occurs when the air is conditionally unstable and **wind speeds** exceed certain critical values. *Compare* **Bishop wave, chinook arch**.

Hewson, E. W., and R. W. Longley, 1951: *Meteorology, Theoretical and Applied*, 449–450.

Moazagotl wind—The strong **wind** blowing across a mountain crest, responsible for the formation of the **Moazagotl** cloud.

mobile ship—A ship equipped to take **upper-air soundings**.

mobile ship station—**Weather station** onboard a moving ship.

mobile source—A category of **pollution** sources that release emissions as they travel, for example, automobiles, ships, etc.
Compare **stationary source**.

mobile weather station—A vehicle used to provide meteorological observations at a place where no fixed **station** exists, or to study the **mesoscale** processes or the **microclimate** of a region.

mobility—Quantitative measure of the **velocity** of ions in an **electric field** of unit strength.
The mobility can be expressed as

$$\mu = \frac{v}{E},$$

where v is the velocity of the **ion** and E is the **electric field strength**. The units are square meters per volt per second. Heavier ions will have lower mobilities than lighter ones and values of mobility vary inversely with **density**, producing slower mobilities near the earth's surface.

mock moon—Same as **paraselene**.
See **parhelion**.

mock sun—A locally bright spot of **light** in the sky other than the sun itself.
As such, it is usually applied to a **parhelion**, but has also been used for other spots such as **paranthelia**.

mode—One of several accepted measures of **central tendency**.
It is the most probable value of a discrete **variate**, or the point of maximum **probability** density in the case of a continuous variate.

mode eddies—(Also known as mesoscale eddies.) In oceanography, densely packed, irregularly oval-shaped high and low pressure centers roughly 400 km (240 miles) in diameter in which current intensities are typically tenfold greater than the local means.

mode water—A term for water of exceptionally uniform properties over an extensive depth **range**, caused in most instances by **convection**.
Mode waters represent regions of **water mass formation**; they are not necessarily **water masses** in their own right but contribute significant volumes of water to other water masses. Because they represent regions of deep sinking of **surface water**, mode water formation regions are atmospheric **heat** sources. **Subantarctic Mode Water** is formed during winter in the **subantarctic zone** just north of the **subantarctic front** and contributes to the lower **temperature range** of **central water**; only in the extreme eastern Pacific Ocean does it obtain a **temperature** low enough to contribute to **Antarctic Intermediate Water**. **Subtropical Mode Water** is mostly formed through enhanced **subduction** at selected locations of the **subtropics** and contributes to the upper temperature range of central water. Examples of Subtropical Mode Water are the 18°C water formed in the **Sargasso Sea**, Madeira Mode Water formed at the same temperature but in the vicinity of Madeira, and 13°C water formed not by surface processes but through mixing in **Agulhas Current** eddies as they enter the **Benguela Current**.

model—A tool for simulating or predicting the behavior of a **dynamical system** like the **atmosphere**.
Models can be based on subjective **heuristic** methods, **statistics** (*see* **statistical dynamical model**, **model output statistics**), numerical methods (*see* **numerical forecasting**), simplified physical systems (*see* **dishpan experiments**), analogy (*see* **analogs**), etc. The term is now most commonly applied to numerical models.

model atmosphere—Any theoretical representation of the **atmosphere**, particularly of vertical **temperature** distribution.
See **adiabatic atmosphere**, **homogeneous atmosphere**, **isothermal atmosphere**, **thermotropic model**, **equivalent barotropic model**, **barotropic model**, **standard atmosphere**.

model calibration—Adjustment of the parameters of a mathematical or numerical **model** in order to optimize the agreement between observed data and the model's predictions.

model output statistics—(Abbreviated MOS.) For a **numerical weather prediction** model, **statistical** relations between model-forecast **variables** and observed weather variables, used for either **correction** of model-forecast variables or **prediction** of variables not explicitly forecast by the **model**.
They have often taken the form of multilinear **regression** equations derived by screening potential model-forecast variables as predictors. The method produces forecasts of weather variables that to some extent account for the **random** and **systematic errors** in the **numerical weather prediction** model. *Compare* **perfect prognosis method**.

moderate breeze—In the **Beaufort wind scale**, a **wind** with a speed from 11 to 16 knots (13 to 18 mph) or Beaufort Number 4 (Force 4).

moderate gale—In the **Beaufort wind scale**, a **wind** with a speed from 28 to 33 knots (32 to 38 mph) or Beaufort Number 7 (Force 7).

modified refractive index—In radio engineering, the **refractive index** of air on a hypothetical flat earth, given by

$$n + \frac{z}{R},$$

where z is the height above the surface, R is the earth's radius, and n is the actual refractive index of air.

For a spherical earth with an **atmosphere** having a uniform refractive index, rays are curved relative to the earth. This is approximately equivalent to a flat earth with an atmosphere in which the refractive index varies with height as z/R. *See* **modified refractivity**.

modified refractivity—Related to **modified refractive index** by

$$M = (n + \frac{z}{R} - 1) \times 10^6$$

and said to be measured in M units.

The term modified refractive index is sometimes carelessly used for M.

modon—A **dipole** eddy structure of counterrotating flows, such as neighboring **cyclonic** and **anticyclonic** vortices that propagate through the mean flow together.

modular flow—Flow across a measuring device in which the **water level** upstream is not influenced by the **downstream** water level.

modulation—1. In general, the modification of some property of a phenomenon by another distinct phenomenon.

2. In radio, **radar, sodar,** and **lidar,** the modification of the **amplitude, phase,** or **frequency** of a **carrier** to convey information or to permit resolved measurements at particular **range,** frequency, or **velocity** intervals.

modulator—A device for effecting the process of **modulation**.

The **amplitude, frequency,** or **phase** of a **carrier** signal is varied with time by an applied information **signal**.

modulus—A real, positive quantity that measures the magnitude of some number.

For instance, the modulus of a complex number is the square root of the sum of the squares of its components. Often it means, simply, the numerical ("absolute") value $|x|$ of an algebraic quantity x.

modus ponens—(Latin for affirmative mode.) An **inference** rule that states that if A is true, and A implies B, then B is also true.

modus tollens—(Latin for denial mode.) An **inference** rule that states that if B is false and A implies B, then A is also false.

This is considered an "unsafe" approach compared to **modus ponens** because proving a negative is always weaker than proving a positive.

moist adiabat—(Also called saturation adiabat.) On a **thermodynamic diagram**, an **isopleth** of **equivalent potential temperature** or **pseudoequivalent potential temperature**.

Moist adiabats on most diagrams are drawn for the **pseudoadiabatic process**, in which liquid water is removed as soon as it is condensed.

moist-adiabatic lapse rate—(Or saturation-adiabatic lapse rate.) The rate of decrease of **temperature** with height along a **moist adiabat**.

It is given approximately by Γ_m in the following:

$$\Gamma_m = g \frac{1 + \frac{L_v r_v}{RT}}{c_{pd} + \frac{L_v^2 r_v \epsilon}{RT^2}},$$

where g is gravitational **acceleration**, c_{pd} is the **specific heat** at constant **pressure** of **dry air,** r_v

is the **mixing ratio** of **water vapor**, L_v is the **latent heat** of **vaporization**, R is the **gas constant** for dry air, ϵ is the ratio of the gas constants for dry air and water vapor, and T is temperature. This expression is an approximation to both the reversible moist **adiabatic lapse rate** and the **pseudoadiabatic lapse rate**, with more accurate expressions given under those definitions. When most of the condensed water is frozen, this may be replaced by a similar expression but with L_v replaced by the latent heat of **sublimation**.

moist-adiabatic process—(Also known as saturation-adiabatic process.) An **adiabatic process** for which the air is saturated and may contain liquid water.

A distinction is made between the **reversible process**, in which total water is conserved, and the **pseudoadiabatic** or irreversible moist adiabatic process, in which liquid water is assumed to be removed as soon as it is condensed.

moist air—1. In **atmospheric thermodynamics**, air that is a mixture of dry air and any amount of **water vapor**.

Compare **dry air, saturation.**

2. Generally, air with a high **relative humidity**.

moist climate—In C. W. Thornthwaite's 1948 **climatic classification**, any type of **climate** in which the seasonal **water surplus** counteracts seasonal water deficiency; thus it has a **moisture index** greater than zero. Included are the **moist subhumid, humid,** and **perhumid climates.** In contrast are the **dry climates.** The moist climates are further subdivided according to values of **aridity index** into the following: little or no moisture deficiency; moderate summer water deficiency; moderate winter water deficiency; large summer water deficiency; and large winter water deficiency.

Compare **rainy climate.**

Thornthwaite, C. W., 1948: An approach toward a rational classification of climate. *Geogr. Rev.*, **38**, 55–94.

moist convection—Atmospheric **convection** in which the **phase** changes of water play an appreciable role.

All **cumuliform** clouds are manifestations of moist convection. The **enthalpy** exchange between condensing **water vapor** or **freezing** liquid water and air (*see* **latent heat**) is a major contributor to the positive **buoyancy** of updrafts, while the reverse exchange between air and evaporating water or melting **ice** contributes strongly to the negative buoyancy of downdrafts.

moist enthalpy—*See* **enthalpy.**

moist static energy—A **thermodynamic variable** (analogous to **equivalent potential temperature**) calculated by hypothetically lifting air adiabatically to the top of the **atmosphere** and allowing all **water vapor** present in the air to condense and release **latent heat**:

$$s_e = C_p T + gz + L_v r,$$

where g is gravitational **acceleration**, L_v is the latent heat of **vaporization**, C_p is the **specific heat** at constant **pressure** for air, T is **absolute** temperature, z is height above some reference level (either the local surface at $z = 0$ or the height where the **ambient pressure** is 100 kPa), and r is the water vapor **mixing ratio** in the air. *Compare* **dry static energy, liquid water static energy, saturation static energy.**

moist subhumid climate—*See* **subhumid climate.**

moist tongue—An extension or protrusion of **moist air** into a region of lower **moisture content.** Cloudiness and **precipitation** are closely related to moist tongues.

moisture—(Or moisture content.) In meteorology, a general term usually referring to the **water vapor** content of the **atmosphere**, or to the total water substance (gaseous, liquid, and solid) present in a given volume of air.

In **climatology**, moisture refers more specifically to quantities of **precipitation** or to **precipitation effectiveness.** *See* **humidity;** *see also* **soil moisture.**

moisture adjustment—The adjustment of observed **precipitation** in a **storm** by the ratio of the estimated probable maximum **precipitable water** over the basin under study to the actual precipitable water calculated for the particular storm.

The procedure is one step in determining the **probable maximum precipitation** for design purposes.

moisture content—Same as moisture.

moisture-continuity equation—The **water vapor** storage equation as applied to the **atmosphere**.

The general form of the equation is written

$$\frac{dS}{dt} = I + E - O - P,$$

where I is the atmospheric moisture inflow, E the **evapotranspiration** from the ground, O the atmospheric moisture outflow, P **precipitation**, and dS/dt the time rate of change of moisture storage in the portion of the atmosphere under consideration. In practice the equation is more commonly applied to a finite interval of time and the various terms become mean values in this interval.

moisture equivalent—The weight percentage of water (with respect to dry weight) retained by an initially saturated 1-cm-thick soil sample after it has been subjected to a **centrifugal force** 1000 times the **force of gravity** for 30 minutes.

moisture factor—One of the simplest measures of **precipitation effectiveness**, given by Lang as

$$\text{moisture factor} = \frac{P}{T},$$

where P (cm) is **precipitation** and T (°C) **mean temperature** for the period in question.

This index recognizes only that as **temperature** increases, the effective moisture decreases due to greater **evaporation**. A number of greater refinements of this concept exist: De Martonne's **index of aridity**; Angström's **humidity** coefficient; Gorczyński's aridity coefficient; Thornthwaite's **precipitation-effectiveness index** and **moisture index**; and Köppen's formulas for outlining **steppe climate** and **desert climate**.

moisture flux—*See* **heat flux**.

moisture index—1. That portion of total **precipitation** used to satisfy plant (vegetation) needs.

2. As used by C. W. Thornthwaite in his 1948 **climatic classification**: an overall measure of **precipitation effectiveness** for plant growth that takes into consideration the weighted influence of **water surplus** and water deficiency as related to **water need** and as they vary according to **season**. For a given **station**, it is calculated by the formula

$$I_m = \text{humidity index} - 0.6(\text{aridity index}),$$

which becomes

$$I_m = \frac{100s - 60d}{n},$$

where I_m is the moisture index, s the water surplus, d the water deficiency, and n the water need. The calculation of s and d is made on a normal month-to-month basis, with s being the total surplus from all months having a water surplus, and d the total of all monthly deficiencies; each is represented by the difference between monthly **precipitation** and monthly **potential evapotranspiration** (in centimeters or inches). Here n is the annual potential evapotranspiration.

The moisture index replaced Thornthwaite's previously used (1931) **precipitation-effectiveness index**. *Compare* **Palmer Drought Severity Index**.

Thornthwaite, C. W., 1948: An approach toward a rational classification of climate. *Geographical Review*, **38**, 55–94.

Thornthwaite, C. W., 1931: Climates of North America according to a new classification. *Geographical Review*, **21**, 633–655.

moisture inversion—An increase with height of the **moisture content** of air; specifically, the layer through which this increase occurs, or the **altitude** at which the increase begins. *See* **inversion**.

moisture pooling—The development of an area in the **boundary layer** (e.g., often observed on surface or 850-mb charts) where moisture values become higher than in the surrounding region.

Moisture pooling typically occurs in an area of low-level **convergence** during the warm (growing) **season**, and can have a significant effect on **convection** initiation and evolution.

Johns, R. M., 1993: Meteorological conditions associated with bow echo development in convective storms. *Weather and Forecasting*, **8**, 294–299.

moisture profile—Curve representing air or **soil moisture** as a function of height or depth.

moisture tension—The **energy** with which water is held in soil; equivalent to the negative **pressure** to which water must be exposed to so that it is in hydraulic **equilibrium**, through a permeable membrane, with soil water.

moisture volume percentage—Expression of **soil moisture content** obtained by taking the ratio of the volume of water in the soil to the total bulk soil volume and multiplying by 100.

moisture weight percentage—Expression of **soil moisture content** obtained by taking the ratio of the weight of water in the soil to the oven-dry weight of the soil and multiplying by 100.

molality—A measure of the concentration of a dissolved species in a solution.
> The molality is given as the number of **moles** of the compound dissolved per kilogram of solution. *Compare* **molarity**.

molan—**Breeze** blowing from Arve toward Geneva in Switzerland.

molar specific heat capacity—(Or molar specific heat.) The **heat capacity** of a system divided by the number of **moles** in that system.

molarity—A measure of the concentration of a dissolved species in a solution.
> The molarity is defined in terms of the number of **moles** of the compound dissolved per liter of solution. *Compare* **molality**.

mole—A unit of mass numerically equal to the **molecular weight** of the substance.
> The **gram-mole** or gram-molecule is the mass in grams numerically equal to the molecular weight; for example, a gram-mole of **oxygen** is 32 grams.

mole fraction—In a gas mixture, the ratio of the **partial pressure** of a gas to the **total pressure** of the mixture.
> *Compare* **mixing ratio**, **specific humidity**.

mole fraction of water vapor—The ratio of **vapor pressure**, e, to **total pressure**, p, related to **vapor** mixing ratio by

$$\frac{e}{p} = \frac{\epsilon r_v}{1 + \epsilon r_v},$$

where r_v is **mixing ratio** and ϵ is the ratio of the gas constants of **dry air** and **water vapor**.

mole number—*See* **Raoult's law**.

molecular conduction—*See* **conduction**.

molecular diffusion—*See* **diffusion**.

molecular dissipation—*See* **dissipation**.

molecular-scale temperature—A **temperature** parameter T_M defined by the following relation:

$$T_M = T\,\frac{M_0}{M},$$

where T is the actual "kinetic" temperature, M_0 the **molecular weight** of air at **sea level** (28.966), and M the molecular weight of air at the point where the temperature is being specified.
> Molecular-scale temperature has application in specifying temperatures at extremely high altitudes (*see* **standard atmosphere**). Below about 90 km, $T_M = T$, but above that level, T_M becomes increasingly greater than T.

molecular viscosity—*See* **viscosity**.

molecular viscosity coefficient—Same as **dynamic viscosity**.

molecular weight—The weight of 1 **mole** (equal to **Avogadro's number** of **particles**) of a particular molecule.
> The mass of 1 mole of **carbon** atoms is defined as 12 grams.

Moll thermopile—A **thermopile** used in some types of **radiation** instruments.
> Alternate junctions of series-connected manganan–constantan thermocouples are embedded in a shielded nonconducting plate having a large **heat capacity**. The remaining junctions, which are blackened, are exposed directly to the radiation. The voltage developed by the **thermocouple** is proportional to the **intensity** of radiation. *See* **solarimeter**.

mollisol—Same as **active layer**, but in U.S. Department of Agriculture soil taxonomy, mollisol has many more specific properties.

mollition—The act or process of thawing the **active layer** (**mollisol**).

moment—1. The product of a distance and another **parameter**.

The moment may be about a point, line, or plane; if the parameter is a **vector**, the moment is the **vector product** of the vector distance from the point, line, or plane, into the parameter. Thus, the moment of the **momentum** of a **fluid parcel** per unit volume about an axis is $\mathbf{r} \times \rho\mathbf{u}$, where $\mathbf{r}$ is the vector from axis to the parcel, ρ the **density**, and $\mathbf{u}$ the **velocity** vector of the parcel; this is also called the **angular momentum**. The moment of a force $\mathbf{F}$ about an axis is $\mathbf{r} \times \mathbf{F}$, called the **torque**. The second moment of a parameter is the moment of the first moment, and so on, for higher moments.

2. By analogy, in **statistical** terminology, the **mean value** of a power of a **random variable**. The symbol μ_n' (or ν_n) is used for a **raw moment** as distinguished from the corresponding **central moment** μ_n taken about the mean μ. Thus the raw moments are

$$\mu_n' \equiv \nu_n \equiv E(x^n), \ (n = 0, 1, 2, \cdots),$$

where $E(x^n)$ is the **expected value** of the **variate** x to the nth power. In particular, $\mu_0' = 1$ and $\mu_1' \equiv \nu_1 \equiv \mu$. The central moments are

$$\mu_n \equiv E[(x - \mu)^n], \ (n = 0, 1, 2, \cdots),$$

where $E[(x - \mu)^n]$ is the expected value of the nth power of the **deviation** of the variate from its mean. In particular, $\mu_0 \equiv 1$, $\mu_1 \equiv 0$, $\mu_2 \equiv \sigma^2$, where σ^2 is the **variance**.

moment of inertia—For a system of point masses, the sum of the product of each mass and the square of its perpendicular (minimum) distance from a given axis.

For a continuous mass distribution, the sum becomes an integral. In general, moment of inertia is a **tensor** because moments of inertia are different for three **orthogonal** axes. Moment of inertia plays a similar role in **angular momentum** to that which mass (inertia) plays in **linear momentum**.
Goldstein, H., 1959: *Classical Mechanics*, Ch. 5.

momentum—(Usually means **linear momentum** as opposed to **angular momentum**.) In **Newtonian mechanics** the (**linear**) momentum $\mathbf{p}$ of a body with mass m and **velocity v** is the product of these two quantities:

$$\mathbf{p} = m\mathbf{v}.$$

In the absence of forces, momentum is conserved. But momentum is a more fundamental quantity than simply the product of mass and velocity. For example, photons have momenta, which can be transferred to objects (as evidenced by **radiation pressure**), and yet the **photon** has zero **rest mass**. Thus, momentum is best looked upon as a single entity, complete in itself, governed by the dynamical law

$$\mathbf{F} = \frac{d\mathbf{p}}{dt}$$

where $\mathbf{F}$ is the force acting on the body with momentum $\mathbf{p}$.

momentum flux—The vertical **flux** of horizontal **momentum**, equal to the force per unit area, or **stress**.

The **Reynolds stress** (τ_R) can be determined from the **covariance** of the fluctuations of the horizontal (u', v') and vertical (w') **wind** components, by

$$\tau_R = \rho(\overline{u'w'}^2 + \overline{v'w'}^2)^{1/2},$$

where ρ is air density. A direct measurement is possible with eddy-correlation techniques, or an indirect determination can be made using **Monin–Obukhov similarity** flux-profile relationships (also called **universal functions**). The relation between the **velocity** scale (**friction velocity** u_*) and the momentum flux is

$$u_* = (\tau_R/\rho)^{1/2} = (\overline{u'w'}^2 + \overline{v'w'}^2)^{1/4}.$$

Momentum flux can be associated with either **mean velocity** components, internal **gravity waves**, or with turbulent velocity fluctuations. For **turbulence**, the momentum flux is also called the Reynolds stress. For waves, it is related to **mountain wave** drag.

momentum transfer—Transport of a property of air known as **momentum**, which is the product of mass and **velocity** of the air, from one location to another.

Monin–Obukhov equation—Same as **Monin–Obukhov similarity theory**.

Monin–Obukhov scaling length—*See* **Monin–Obukhov similarity theory**.

Monin–Obukhov similarity theory—A relationship describing the vertical behavior of nondimensionalized mean flow and **turbulence** properties within the atmospheric **surface layer** (the lowest 10% or so of the **atmospheric boundary layer**) as a function of the Monin–Obukhov key parameters.

These key parameters are the height z above the surface, the **buoyancy** parameter ratio g/T_v of inertia and buoyancy forces, the kinematic surface **stress** τ_0/ρ, and the surface **virtual temperature** flux

$$Q_{v0} = H_{v0}/(\rho C_p) = \overline{w' T_{v0}},$$

where g is gravitational **acceleration**, T_v is virtual temperature, τ_0 is turbulent stress at the surface, ρ is air density, Q_{v0} is a kinematic virtual **heat flux** at the surface, H_{v0} is a dynamic virtual heat flux at the surface, C_p is the **specific heat** of air at constant **pressure**, and $\overline{w' T_{v0}}$ is the **covariance** of **vertical velocity** w with virtual temperature near the surface. The key parameters can be used to define a set of four dimensional scales for the **surface layer**: 1) the **friction velocity** or shearing **velocity**, a velocity scale,

$$u_* = (\tau_0/\rho)^{1/2};$$

2) a surface-layer **temperature scale**,

$$T_{*SL} = Q_{v0}/u_*;$$

3) a length **scale** called the **Obukhov length**,

$$L = \frac{-u_*^3 T_v}{k g Q_{v0}},$$

where k is the **von Kármán constant**; and 4) the height above ground scale, z. These key scales can then be used in **dimensional analysis** to express all surface-layer flow properties as dimensionless **universal functions** of z/L. For example, the mean **wind shear** in any quasi-stationary, locally homogeneous surface layer can be written as

$$\frac{\partial U}{\partial z} = \frac{u_*}{z} f\left(\frac{z}{L}\right),$$

where f is a universal function of the dimensionless height z/L. The forms of the universal functions are not given by the Monin–Obukhov theory, but must be determined theoretically or empirically. Monin–Obukhov similarity theory is the basic **similarity** hypothesis for the horizontally homogeneous surface layer. With these equations and the hypothesis that the fluxes in the surface layer are uniform with height, the **momentum flux**, **sensible heat** flux, and fluxes of **water vapor** and other gases can be determined. *Compare* **aerodynamic roughness length, Richardson number.**

monitor well—*See* **observation well.**

monochromatic—1. Characterized by a single **frequency.**

For example, a monochromatic (or time harmonic) **electromagnetic wave** is one with a single frequency. Although monochromatic originally meant characterized by a single hue, the term has been extended to **electromagnetic radiation** beyond the **visible spectrum** and even to waves that are not electromagnetic.

2. In **radar**, radiometry, and **lidar**, of or pertaining to a single **wavelength.**
See **coherence.**

monochromatic radiation—**Radiation** taken over a sufficiently small **spectral interval** that the **radiance** is invariant with **wavelength.**
Compare **narrowband radiation, broadband radiation.**

monostatic radar—The most common **radar** system configuration, with the radar **receiver** at the same location as the radar **transmitter.**

In such a system, surfaces of constant **range** are spheres centered at the radar site, and only the radial component of **target** velocity causes a **Doppler frequency shift**. *Compare* **bistatic radar.**

monoterpenes—*See* **terpenes.**

monsoon—(Derived from Arabic mausim, a **season**.) A name for seasonal **winds.**

It was first applied to the winds over the Arabian Sea, which blow for six months from northeast and for six months from southwest, but it has been extended to similar winds in other parts of the world. Even in Europe the prevailing west to northwest winds of summer have been called the "European monsoon." The primary cause is the much greater annual **variation** of **temperature** over large land areas compared with neighboring ocean surfaces, causing an excess of **pressure** over the continents in winter and a deficit in summer, but other factors such as the relief features of the land have a considerable effect. The monsoons are strongest on the southern and eastern sides of Asia, the largest landmass, but monsoons also occur on the coasts of tropical regions wherever the **planetary circulation** is not strong enough to inhibit them. They have been described in Spain, northern Australia, Africa except the Mediterranean, Texas, and the western coasts of the United States and Chile. In India the term is popularly applied chiefly to the **southwest monsoon** and, by extension, to the rains which it brings. *See* **brisa**, **elephanta**; *compare* **etesians**, **meltém**.

monsoon circulation—*See* **monsoon**.

monsoon climate—Type of **climate** found in regions subject to **monsoons** and characterized by a dry winter and a wet summer.
    The monsoon climate is best developed on the fringes of the **Tropics** (e.g., India). The Indian subcontinent has a long winter–spring **dry season** that includes a "cold season" followed by a short "hot season" just preceding the rains; a summer and an early autumn rainy season that is usually very wet but varies greatly from year to year; and a secondary maximum of **temperature** right after the **rainy season**.

monsoon current—A seasonal, eastward flowing **ocean current** of the Indian Ocean.
    See **North Equatorial Current**, **North Equatorial Countercurrent**, **South Equatorial Countercurrent**.

monsoon depression—A **depression** that forms within the **monsoon trough**.
    The term is most frequently used to describe weak **cyclonic** disturbances that form over the Bay of Bengal and generally track northwestward over the Indian subcontinent. These occasionally intensify into **tropical cyclones** if they remain over warm ocean water long enough. The term is also used to describe depressions that form within the monsoon trough near Australia and in the western North Pacific region. The term has gained ascendancy in use to refer to a broad tropical cyclonic **vortex** characterized by 1) its large size, where the outermost closed **isobar** may have a diameter on the order of 600 n mi (1000 km); 2) a loosely organized cluster of deep convective elements, which may form an elongated band of deep **convection** in the east semicircle; 3) a low-level **wind** distribution that features a 100 n mi (200 km) diameter light-wind core, which may be surrounded by a band of gales or contain a highly asymmetric **wind field**; and 4) a lack of a distinct **cloud system** center. Most **monsoon** depressions that develop in the western North Pacific eventually acquire persistent central convection and accelerated core winds, marking their transitions into conventional **tropical cyclones**.

monsoon fog—(Rare.) An **advection fog** produced as a **monsoon circulation** that transports warm **moist air** over a colder surface.

monsoon gyre—A convection of the **summer monsoon** circulation of the western North Pacific characterized by 1) a very large nearly circular low-level **cyclonic** vortex (not the result of the expanding **wind field** of a preexisting **monsoon depression** or **tropical cyclone**) that has an outermost closed **isobar** with a diameter on the order of 1200 n mi (2500 km); 2) a **cloud band** bordering the southern through eastern periphery of the **vortex**/surface **low**; and 3) a relatively long (two week) life span.
    Initially, a subsequent regime exists in its core and western and northwestern quadrants with light winds and **scattered** low **cumulus** clouds; later, the area within the outer closed isobar may fill with deep **convective cloud** and become a isobar or tropical cyclone. Note: a series of midget tropical cyclones may emerge from the "head" or leading edge of the peripheral tropical cyclone of a monsoon gyre.

monsoon low—A seasonal **low** found over a continent in the summer and over the adjacent sea in the winter.
    Examples are the lows over the southwestern United States and India in summer and those located off lower California and in the Bay of Bengal in winter. Palmer (1951) points out that, while the winter and summer monsoon lows appear similar on mean charts, they are dynamically quite different. *Compare* **thermal low**.
        Palmer, C. E., 1951: *Compendium of Meteorology*, p. 873.

monsoon surge—The temporary extension of deep **monsoon** flow into a region not normally dominated by persistent monsoon flow.
    This temporary extension or surge may last from a few days to three weeks. These surges most commonly occur eastward across the Philippine Sea into the western North Pacific and east of Australia into the western South Pacific. The establishment of a **reverse-oriented monsoon trough**

is accompanied by an eastward surge in the monsoon flow. Monsoon surges are often precursors to the **development** of **tropical cyclones**.

monsoon trough—The line in a **weather map** showing the locations of relatively minimum sea level **pressure** in a **monsoon** region.

Most of the active transient disturbances producing the monsoon **rain** develop and move along the monsoon trough region.

MONT code—(Short for mountain code.) A **synoptic code** in which observations of clouds at and below **station elevation** of mountain stations are encoded and transmitted.

Monte Carlo method—A **numerical modeling** procedure that makes use of **random numbers** to simulate processes that involve an element of chance.

In Monte Carlo simulation, a particular experiment is repeated many times with different randomly determined data to allow **statistical** conclusions to be drawn.

Monte Carlo model—In **radiation**, the solution of **radiative transfer** by **random** walks of simulated **photon** trajectories through an absorbing, **scattering**, or emitting medium.

Repetition of a large number (a million repetitions may be needed) of independent simulations can produce accurate results even for very complicated problems.

Montgomery streamfunction—The quantity $gz + c_p T$ measured on an **isentropic surface** where g is the **acceleration of gravity**, z the height of the isentropic surface, c_p the **specific heat** of air at constant **pressure**, and T the Kelvin **temperature**; or it is sometimes defined as the same quantity divided by the **Coriolis parameter**.

This is the **streamfunction** for the **geostrophic wind** on the isentropic surface.

monthly climatological summary—A **climatological summary** issued on a monthly basis containing data for the month for a single **station** or an area.

monthly maximum temperature—Highest **temperature** recorded during a particular calendar month in a specified year.

monthly minimum temperature—Lowest **temperature** recorded during a particular calendar month in a specified year.

monthly record—Form on which **meteorological observations** are recorded daily to produce a monthly summary.

Montreal Protocol—(on Substances that Deplete the Ozone Layer.) An international agreement, signed in 1987, that serves to eliminate the use of compounds that have been linked to destruction of stratospheric **ozone** (including the **chlorofluorocarbons** and **halons**).

Subsequent amendments and adjustments have been made to the Protocol (London, 1990; Copenhagen, 1992; Vienna, 1995) that have increased the number of compounds under regulation and accelerated the phaseout of the production of these compounds.

moon illusion—The moon seen near the **horizon** appears larger than the moon seen high in the sky.

This difference is illusory, for there is no difference in the angular widths of moon from one situation to the other. (There is normally a small difference in the angular heights of the moon due to **refraction** in the **atmosphere**, but this serves to lessen the height on the horizon rather than increase it). Yet the illusion is sufficiently compelling to cause most observers to be convinced that the moon actually has a significantly larger angular size when near the horizon, and that this has a physical origin in the optics of the atmosphere. But the phenomenon is perceptual and its explanation lies in the realm of psychology. No single explanation has been found that accounts for all aspects of what people claim to see. One explanation that accounts for some aspects of the phenomenon does relate to **meteorological optics**. The **clear sky** is not perceived to be a hemisphere, but a variety of shapes, for example, a flattened dome; the horizon being seen as significantly farther from the observer than the **zenith**. Further, the moon appears to be pasted on the horizon and so shares its distance. The perceptual phenomenon of size constancy will then cause something of fixed angular size but apparently varying in distance to appear larger at the greater distance of the horizon. Suffice it to say that there are aspects of the illusion that are consistent with this explanation and others that are at variance with it.

Hershenson, M., Ed., 1989: *The Moon Illusion*, Laurence Erlbaum Associates, Publisher, Hillsdale, New Jersey, 421 pp.

Neuberger, H., 1957: *Introduction to Physical Meteorology*, Pennsylvania State University, 149–157.

moonbow—Same as **lunar rainbow**.

moor-gallop—In England, a sudden **squall** across the moors.

moraine—Ridges or deposits of rock debris transported by a **glacier**.

Moraines are left after a glacier has receded, providing evidence of its former extent. Common forms are ground moraine, formed under a glacier; lateral moraine, along the sides; medial moraine, down the center; and end moraine, deposited at the terminus.

> Armstrong, T., B. Roberts, and C. Swithinbank, 1973: *Illustrated Glossary of Snow and Ice*, Scott Polar Research Institute, Special Pub. No. 4, 2d ed.; The Scholar Press Ltd., Menston, Yorkshire, UK.

morgeasson—Same as **morget**.

morget—(Also called morgeasson.) The night **land breeze** on Lake Geneva, Switzerland.

It blows from the north from 5–7 P.M. until 7–9 A.M. as a powerful **breeze**. In the late fall and winter it blows almost throughout the day. Its complementary **lake breeze** is the **rebat**.

morning glory—A wind **squall** or succession of wind squalls, frequently accompanied by a spectacular low **roll cloud** or a series of such clouds, that occurs early in the morning, mainly in the late **dry season** (September–October) at places around the southern coast of the Gulf of Carpentaria region of northern Australia.

The cloud lines may be up to a few kilometers across in the direction of travel, usually from the east or northeast, and are of considerable lateral extent, often stretching from **horizon** to horizon with a remarkably uniform **cross section**.

morphology—A study of the structure, form, or shape of meteorological phenomena, such as **clouds** or **ice crystals**; includes the classification of these phenomena.

MOS—Abbreviation for **model output statistics**.

mother-cloud—A **cloud** from which another cloud has grown or been formed.

See **cloud classification**.

mother-of-pearl clouds—*See* **iridescent cloud, nacreous clouds**.

motorboating—(Obsolete.) The **sound** heard through the monitoring speaker of an audio-modulated **radiosonde** when the audio **signal** became so low in **frequency** that it resembled the sound of a motorboat.

It corresponded to unmeasurable low humidities. Modern sondes do not have such stringent limitations or produce such a signal.

Mount Rose snow sampler—A **snow** water content sampler consisting of a hollow tube of steel or duralumin and having an internal diameter of 1.485 in. so that each inch of water in the sample weighs one ounce.

This sampler is used almost exclusively in sampling deep snow in the mountains of the western United States. Its design includes a cutting lip for penetrating **ice** layers in the snow, longitudinal slots for cleaning after a measurement, and accessory wrenches for assembling and driving. The tube is made in short sections with threaded couplings so that it can be disassembled for transportation.

mountain barometer—Any conventional **barometer** fitted with an extended **scale** so that **atmospheric pressure** measurements may be made at both high and low altitudes.

mountain breeze—A nocturnal component of the **mountain–plains** or **mountain–valley wind systems** encountered during periods of light **synoptic** flow.

mountain climate—(Also called highland climate.) Generally, the **climate** of high elevations.

Mountain climates are distinguished by the **departure** of their characteristics from those of surrounding lowlands, and the one common basis for this distinction is that of atmospheric rarefaction. Aside from this, great variety is introduced by differences in latitude, **elevation**, and **exposure** to the sun. Thus, there exists no single, clearly defined, mountain climate. The most common climatic results of high elevation are those of decreased **pressure**, reduced **oxygen** availability, decreased **temperature**, and increased **insolation**; the last two combine to produce a typical "hot sun and cold shade" condition. Precipitation is heavier on the **windward side** of a mountain barrier than on the **leeward (orographic precipitation)**, and on the windward side it increases upward to the **zone of maximum precipitation**, then decreases again. On many tropical mountains the forest zone extends into the level of average cloud height, which causes an excessively damp climate and produces the so-called **fog forest**. The **orography** gives rise to many **local winds**, chief among which are the **foehn**, **mountain** and **valley winds**, **mountain-gap winds**, and **downslope winds** of many sorts. Great interest in mountain climate has centered in the relatively well-populated, equatorial Andes. There, four zones of **elevation** are delimited: tierra caliente (hot land); tierra tamplada (temperate land); tierra fria (cool land); and tierra helada (land of **frost**).

> Trewartha, G. T., 1954: *An Introduction to Climate*, 367–377.

Landsberg, H. E., 1950: *Physical Meteorology*, 212–218.
Miller, A. A., 1943: *Climatology*, 271–288.

mountain fog—Fog formed by **orographic lifting** to **condensation** of **moist air** up a mountain slope.
See **orographic fog**.

mountain-gap wind—A local **wind** blowing through a gap between mountains, a **gap wind**.
This term was introduced by R. S. Scorer (1952) for the surface winds blowing through the Strait of Gibraltar. When air stratification is stable, as it usually is in summer, the air tends to flow through the gap from high to low pressure, emerging as a "jet" with large **standing eddies** in the lee of the gap. The excess of **pressure** on the **upwind** side is attributed to a **pool of cold air** held up by the mountains. Similar winds occur at other gaps in mountain ranges, such as the **tehuantepecer** and the **jochwinde**, and in long channels, such as the Strait of Juan de Fuca between the Olympic Mountains of Washington and Vancouver Island, British Columbia. *Compare* **jet-effect wind**, **canyon wind**, **mountain wind**

Scorer, R. S., 1952: Mountain-gap winds: A study of surface wind at Gilbraltar. *Quart. J. R. Meteor. Soc.*, **78**, 53–61.

mountain lee wave—*See* **lee wave**, **mountain wave**.

mountain observation—A collection of simultaneous meteorological measurements taken and recorded in a mountainous location.
The harshness of the high-mountain **environment**, the inaccessibility of sites, and the remoteness of these regions are special problems that have limited the availability of long-term records of mountain weather. These difficulties are compounded by the issue of representativeness of a measurement. Over flatter, simpler terrain, care is taken to place instrumentation in exposed locations where the measurement can be considered as representing a larger area. Barry (1992) defines at least three types of situations in the mountains: "summit, slope, and valley bottom— apart from considerations of slope orientation, slope angle, topographic screening, and irregularities of small-scale relief." Thus it is very difficult, perhaps inappropriate, to claim representativeness for a single **observation**, and one must interpret mountain observations with great caution. Barry further states, "These factors necessitate either a very dense network of stations or some other approach to determining **mountain climate**. In the future, the use of ground-based and satellite remote sensors combined with intensive case studies of particular phenomena, may provide the best solution."

Barry, R. G., 1992: *Mountain Weather and Climate*, 2nd ed., Routledge, London, 9–10.

mountain–plains wind systems—The **diurnal** cycle of **local winds** between a mountain or a mountain range and the adjacent or surrounding plains during periods of weak **synoptic** flow.
Winds at lower elevations blow from the plains toward the mountains during daytime and from the mountains toward the plains at night. An upper return branch of the **circulation** at higher levels is sometimes present, blowing in the direction opposite the surface winds and completing the circulation. The mountain-plains wind system is most apparent on individual days when skies are **clear** and the general prevailing winds are weak, but it is also seen in climatological averages. The term is usually used to represent the larger massif-scale circulations, of which the embedded **along-valley wind systems** and **along-slope wind systems** are key components. On the smaller mountain and mountain-range scale, the components of the diurnal cycle are a nocturnal mountain (**katabatic**) wind and a daytime wind from the plains toward the mountains (**anabatic wind**).

mountain station—*See* **high-altitude station**.

mountain torque—The **torque** about a given axis exerted on or by the earth's surface due to the force associated with a difference of **pressure** on two sides of a mountain.
For example, if the axis is the earth's **polar axis** and the air pressure is higher (level for level) on the west side of a mountain than on the east, there exists a mountain torque in the vicinity tending to speed up the earth's rotation. *See also* **frictional torque**.

mountain tundra—Same as **alpine tundra**.

mountain–valley wind systems—The **diurnal** cycle of **local winds** in a mountain valley during **clear** or mostly clear periods of weak **synoptic** flow.
The traditional components of the **cycle** are upslope (**anabatic**) winds, the daytime **upvalley wind**, downslope (**katabatic**) winds, and the nighttime **downvalley wind** (Defant,1951). In this traditional view, each component has corresponding compensatory currents aloft, presumably to form a **closed circulation**. For example, the downvalley wind would lie beneath a wind aloft directed up the valley. Observationally, these compensatory currents have been verified in some cases but not found in others. The classic **model** has the upvalley wind continuing until after

sunset, but in many semiarid regions (or during dry periods), when the **Bowen ratio** is large and the surface **heat flux** is strong, boundary layer **convection** interrupts this model by **mixing** ridgetop winds down to the surface for much of the mid- to late afternoon. Where the valley opens onto a plain, a **valley outflow jet**, which represents a continuation of the downvalley wind, often extends many kilometers over the plain at night. Variations from this basic scheme arise from valley orientation, especially affecting the transition periods. In a north–south valley, for example, sunset occurs first on the east-facing slopes, and katabatic flows begin earlier there, which can result in **cross-valley winds** connecting with anabatic flow components on the west-facing sidewalls still in the sun. Similarly, at **sunrise** the east-facing slope is exposed to **sunshine** earlier than the rest of the valley, and therefore anabatic flows begin earlier there.

Defant, F., 1951: *Compendium of Meteorology*, 663–665.

Banta, R. M., and W. R. Cotton, 1981: An analysis of local wind systems in a broad mountain basin. *J. Climate Appl. Meteor.*, **20**, 1255–1266.

Whiteman. C. D., 1990: Observations of thermally developed wind systems in mountainous terrain. *Meteorol. Monographs*, **45**, 5–42.

mountain wave—An atmospheric **gravity wave**, formed when **stable air** flow passes over a mountain or mountain barrier.

Mountain waves are often standing or nearly so, at least to the extent that **upstream** environmental conditions (and **diurnal** forcing) are stationary. Two divisions of mountain wave are recognized, vertically propagating and trapped **lee waves**. Vertically propagating mountain waves over a barrier may have horizontal **wavelengths** of many tens of kilometers or more, usually extend upward into the lower **stratosphere**, and in pure form, **tilt** upwind with height. They can accompany **foehn**, **chinook**, or **bora** wind conditions. They have the capability to concentrate **momentum** on the lee slopes, sometimes in structures resembling a **hydraulic jump**, leading to occasionally violent downslope windstorms. When sufficient moisture is present in the upstream flow, vertically propagating mountain waves produce interesting **cloud** forms, including **altocumulus** standing **lenticular** (ACSL) and other foehn clouds. Intense waves can present a significant hazard to aviation by producing severe or even extreme **clear air turbulence**. Trapped lee waves generally have horizontal wavelengths of 5–35 km. They occur within or beneath a layer of high **static stability** and moderate **wind** speeds at low levels of the **troposphere** (the lowest 1–5 km) lying beneath a layer of low **stability** and strong winds in the middle and upper troposphere. These conditions are often diagnosed using a **vertical profile** of the **Scorer parameter**, a sharp decrease in midtroposphere indicating conditions favorable to trapped lee wave formation. Trapped lee waves assume the form of a series of waves running parallel to the ridges, and the crests of these waves often contain altocumulus, **stratocumulus**, **wave clouds**, or **rotor clouds** in parallel bands that can be very striking in satellite pictures. Because **wave** energy is trapped within the stable layer, these waves (and accompanying **cloud bands**) may dissipate only very slowly **downwind**, and they can continue **downstream** for many wavelengths spanning many tens of kilometers. Flow beneath the **wave crests**, occasionally made visible by rotor clouds, is often turbulent, thus presenting a significant hazard to low-level aviation. Vertically propagating mountain waves and trapped lee waves can coexist, and sometimes lee waves are incompletely trapped or "leaky," leading to a variety of complex rotor interactions. This complexity of rotor patterns often produces interesting variations in **cloud** forms. As mountain waves propagate upward, the rotor's **amplitude** can grow to the point that the rotor "breaks," that is, the rotor becomes convectively unstable and overturns. Wave breaking can have an important role in vertically redistributing horizontal atmospheric momentum, as it slows the **atmosphere** by **turbulent transport** of the earth's momentum upward.

mountain-wave cloud—A **cloud** that forms in the rising branches of mountain waves and occupies the crests of the waves.

The most distinctive are the sharp-edged, lens-, or almond-shaped **lenticular** clouds, but a variety of **stratocumulus**, **altocumulus**, and **cirrocumulus** forms appear in both the main, vertically propagating waves and in the lee waves. *See* **mountain wave**, **foehn cloud**.

mountain wind—A nocturnal, thermally forced **wind** from the direction of the mountains, generated by cooling along the mountain slopes; a **downvalley wind**, or the nighttime downslope (**katabatic**) component of a **mountain–plains wind system**.

moutonnée—1. Ice-molded **hummocks** in relatively resistant bedrock.

The characteristic streamline form of this **glaciation** is related to the direction of movement of the former **glacier**.

2. An **ice field** of **polar ice** in which there are streamlined **hummocks**.

movable-scale barometer—A **mercury barometer** of the fixed **cistern** type in which a movable **scale** terminating in an **ivory point** is used to compensate for the variations in height of the **mercury** in the cistern.

The position of the scale is adjusted so that the ivory point just touches the mercury in the cistern.

moving average—Same as **consecutive mean**.

moving fetch—The resultant fetch (ocean area where waves are generated by a **wind** having a constant direction and speed) if a **wind field** is moving together with a **wave** field.

moving-target indication—(Abbreviated MTI.) An incoherent Doppler technique for eliminating stationary targets from **radar displays**.

For each transmitted **pulse**, the **frequency** or **phase** of the returned **signal** is compared with that of the transmitted signal, and signals are rejected for which there is no change in frequency or phase. MTI was an early technique for detecting airplanes or other moving targets in the presence of **clutter**. It is not used in modern **weather radars**.

Mozambique Current—One of the western **boundary currents** of the subtropical **gyre** in the southern Indian Ocean.

It flows southward along the east coast of Mozambique through the Mozambique Channel and contributes about 30 Sv (30 × 10⁶ m³s⁻¹) to the **Agulhas Current**. Only about 6 Sv of this **transport** enter the Mozambique Channel from the north, fed from the northern branch of the **South Equatorial Current**; the remainder comes from the **East Madagascar Current**, which makes a northward loop south of Madagascar before it joins the Mozambique Current north of 15°S.

MSL—Abbreviation for **mean sea level**.

MST radar—(Abbreviation for mesosphere–stratosphere–troposphere radar.) A type of **wind profiler** designed to measure **winds** and other atmospheric parameters up to altitudes of 100 km or more.

In the **troposphere** and lower **stratosphere** (up to about 30 km) the **radar** signal is returned from **refractive index** fluctuations produced by **turbulence** in the **neutral atmosphere** (*see* **clear-air echo**). In the upper stratosphere and lower **mesosphere** (between about 60 and 100 km) refractive index variations are strengthened by the strong vertical **gradient** in **electron density**. Because the **scale** size of the refractive index fluctuations must be of the order of one-half the radar **wavelength** (**Bragg scattering**), and the minimum scale size of turbulence increases with height from a few centimeters in the lower troposphere to several meters in the upper stratosphere and lower mesosphere, most MST radars have operated in the **VHF** band (typically 30–60 MHz, 5–10 m wavelengths). Very sensitive **UHF** radars can detect echoes from **incoherent scattering** (**thermal** or Thomson **scatter**) by electrons above 60 km. MST radars are characterized by high-powered transmitters and large antennas. (VHF antennas range from 100 to 300 m across.) Similar wind profilers that lack the **transmitter** power and **antenna** area to detect returns from the upper stratosphere and the mesosphere have often been called ST (stratosphere–troposphere) radars.

MSU—Abbreviation for **Microwave Sounding Unit**.

MTI—Abbreviation for **moving-target indication**.

MTN—Abbreviation for **main telecommunications network**.

mts system—Abbreviation for **meter–tonne–second system**

mud rain—**Rain** containing a noticeable concentration of **particles** of sand or **dust** that may originate from very distant regions.

Compare **blood rain, sulfur rain**.

mudflow—A dense, highly viscous flowing mass of predominantly fine-grained earthy material, with **water content** approximately 20% by weight (or approximately 40% by volume).

Their high density and **viscosity** allow mudflows to travel at speeds of more than 10 m s⁻¹ in mountain canyons, and to carry very large boulders on low gradients.

muerto—A summer **norther** of Mexico.

muggy—Colloquially descriptive of warm and especially humid weather.

multiannual storage—Storage of water in a **reservoir** that can be carried over multiple years with variations in **inflow** and demand.

multiband system—(Term used in **remote sensing**.) A system for simultaneously observing the same **target** with several filtered bands.

Usually applied to imaging **radiometers** that use dispersant optics to separate **wavelength** bands for viewing by several filtered detectors. Also applied to cameras.

multicell convective storm—A **convective storm** system usually composed of a cluster of ordinary convective cells at various stages of their life **cycle**.

New cells within the convective system are generated primarily by either low-level **convergence** along a preexisting boundary, or by lifting at the leading edge of the system-scale **cold pool** that was produced by the previous cells. A multicell **storm** may have a lifetime of several hours, and may also have supercells incorporated as a part of the system as well. *See also* **cell**, **ordinary cell**, **supercell**, **thunderstorm**.

multichannel sea surface temperature—An **algorithm** designed to produce improved estimates of **sea surface temperature** using multispectral measurements from **AVHRR**.

McClain, E. P., W. G. Pichel, and C. C. Walton, 1985: Comparative performance of AVHRR-based multichannel sea surface temperatures. *J. Geophys. Res.*, **90**, 11 587–11 601.

multichannel system—A **scanning** system capable of observing and recording several **channels** of data simultaneously through the same **aperture**.

multilevel model—A **model** that represents the vertical structure of the **atmosphere** with more than one discrete level or basis function.

Compare **barotropic model**, **equivalent barotropic model**; *see* **baroclinic model**.

multiparameter radar—A **radar** capable of deriving more than one quantity from observations of a **target**. In meteorological applications, the term is usually applied to radars capable of measuring either 1) **reflectivity** and at least one polarization-dependent quantity, or 2) reflectivity and at least one wavelength-dependent quantity.

The term is not ordinarily applied to radars that operate with a single **wavelength** and measure only reflectivity and **Doppler velocity**. *Compare* **polarimetric radar**.

multipath—(Coined word for multiple-path radio propagation.) The process or condition in which **radiation** travels between **source** and **receiver** via more than one propagation path.

Because there can be only one "direct" path, some process of **reflection, refraction**, or **scattering** must account for multipath propagation.

multipath transmission—Propagation of acoustic or **electromagnetic radiation** between two points along two or more paths because of the nonuniformity of the propagating medium.

multiple correlation—The **correlation** between a **random variable** and its **regression function**.

If Y denotes the regression function of a random variable (**variate**) y with respect to certain other variates $x_1, x_2 \ldots, x_n$ then the **coefficient of multiple correlation** between y and the x's is defined as the coefficient of simple, **linear correlation** between y and Y. However, the constants of the regression function automatically adjust for algebraic sign, with the result that the **coefficient of correlation** between y and Y cannot be negative; in fact, its value is precisely equal to the ratio of their two **standard deviations**, that is, $\sigma(Y)/\sigma(y)$. Therefore, the coefficient of multiple correlation ranges from 0 to 1, and the square of the coefficient of multiple correlation is equal to the **relative reduction** (or **percent reduction**), that is, the ratio of **explained variance** to **total variance**. Since, in practice, the true regression function Y is seldom known, it is ordinarily necessary to hypothesize its mathematical form and determine the constants by **least squares**, thus obtaining the approximation Y'. In that case, the conventional estimate of the multiple correlation is the **sample** value of the simple linear correlation (symbol R) between y and Y', although a better estimate is obtained by incorporating a correction for **degrees of freedom**. Such a corrected value R' is given as follows:

$$R' = \frac{[(N - 1) R^2 - n]^{1/2}}{[N - (n + 1)]^{1/2}},$$

where N denotes the sample size and $n + 1$ equals the total number of constants (including the **absolute** term) determined from the data. In case $(N - 1) R^2 < n$, the value of R' is taken as zero. *See* **regression**.

multiple correlation coefficient—*See* **multiple correlation**.

multiple drift correction—In air navigation, a method of **pressure-pattern flight** utilizing a set of drift-correction angles applied over successive segments of the flight route.

Multiple **drift** corrections are necessary to maintain a prescribed **course** over the earth when a **single drift correction** would result in a relatively meandering course. *See* **minimal flight**.

multiple incursion theory—(Obsolete.) Hailstone growth by several ascents and descents within a **cloud**.

At present, this is considered a common occurrence, but not a necessary one for the formation of **hail**.

multiple-phase flow—(Also called multiphase flow.) In a **porous medium**, the simultaneous flow of at least two immiscible fluids.
> *Compare* **miscible displacement**.

multiple-purpose reservoir—A **reservoir** designed and operated to serve two or more purposes, such as **flood** control, hydroelectric **power, navigation, irrigation, pollution** control, water supply, and recreation.

multiple register—A **chronograph** used to make a time record of certain measured meteorological **elements**.
> The most common type, the **triple register**, records **wind direction** and **speed**, duration of **sunshine**, and amount of **rainfall** (sensed, respectively, by a contact **anemometer, Marvin sunshine recorder**, and **tipping-bucket rain gauge**). The register consists of a rotating, clock-driven drum on a helical axis, a separate pen for each element, and the actuating mechanism for the pens. Double registers are also used. Multiple registers of this type are becoming obsolete.

multiple-scattering—Radiative **transfer** in which more than one **scattering** event may be of importance before transmission, **reflection**, or **absorption**.
> Multiple-scattering is the dominant effect on the transfer of **solar radiation** within clouds, giving rise to **diffuse radiation**. *Compare* **single-scattering**.

multiple scattering—**Scattering** of **radiation**, usually electromagnetic but possibly acoustic, by an array of objects (e.g., atoms, molecules, **particles**) each of which is excited to **scatter** (radiate) not only by an external **source** but also by the **scattered radiation** from the other objects in the array.
> Multiple scattering is distinguished from **single scattering**, an idealization strictly realized only with a single object excited by an infinitely distant source. Scattering as a consequence solely of excitation by the external source is sometimes referred to as **primary scattering**, the remaining scattering as **secondary scattering**, which is misleading in that it can be decomposed into an infinite series of primary, secondary, tertiary, and higher-order scattering. Multiple scattering can be classified according to the **coherence** properties of the array and the external source. For incoherent multiple scattering, **phase** differences of scattered waves are **random**, and scattered powers are additive. For coherent multiple scattering, phase differences of scattered waves are not random, and scattered fields are additive. Incoherent and coherent multiple scattering are idealizations. An example of (primarily) incoherent multiple scattering is scattering of **sunlight** by thick clouds. An example of (mostly) coherent multiple scattering is **specular reflection** by a glass of water. Scattering by a single **cloud droplet** is an example of scattering by a coherent array—the water molecules in the droplet stick together (cohere) in the sense that the phase differences between their individual scattered waves are fixed—whereas scattering by the entire **cloud** is incoherent in the sense that for droplets separated by random distances large compared with the **wavelength**, the phase differences between waves scattered by individual droplets are essentially random.

multiple-trip echo—In **pulsed radar**, an **echo** from a given transmitted **pulse** that is not received until transmission of one or more additional pulses.
> Such echoes appear on a **display** at the wrong **range** because of **range folding**. *See* **maximum unambiguous range**.

multiple tropopause—A frequent condition in which the **tropopause** appears not as a continuous single "surface" of **discontinuity** between the **troposphere** and **stratosphere** but as a series of quasi-horizontal "leaves" that are partly overlapping in steplike arrangement.
> The multiple tropopause is most common above regions of large horizontal **temperature** contrast in the troposphere. In the more extreme conditions the components of the multiple tropopause are not distinct, and the tropopause is then just a deep zone of transition between troposphere and stratosphere.

multiplexor—A device to collect a variety of data and then interleave the data into a single record or onto a telecommunications link.
> The reverse device is a demultiplexor.

multispectral remote sensing—**Remote sensing** in two or more spectral bands simultaneously, such as visible and **infrared**.

multispectral scanner—A **remote sensing** device that is capable of recording data in several bands of the **electromagnetic spectrum**.

multivariate objective analysis—An analysis procedure based on derivation of an analytic formula containing two or more variables to approximate values of a physical quantity at specified data points.
> The resulting formula serves as an interpolating function to estimate the physical quantity

between data points. When two **independent variables** are used, the result is the equation of a surface that allows objective determination of the **isopleths**. In a typical meteorological application, **objective analysis** could be used to derive from scattered observations an expression for **contour** height as a continuous function of latitude and longitude for the purpose of generating a **weather map**.

multivariate signal—A **signal** that consists of several distinguishable components.

A multivariate signal may contain information that describes both the temporal and spatial **variability** of a single physical quantity.

multiyear ice—**Sea ice** that has survived more than two summer melt seasons.

Such **ice** is typically 3 m or more thick, is less saline, and has smoother **hummocks** and **ridges** than does younger **ice**. Undeformed multiyear ice is distinguished by its undular surface (remnants of drained or refrozen melt ponds). Multiyear ridges are distinct from first-year ridges in that they are typically smaller, more rounded, nearly solid ice and are therefore a serious impediment to surface ships.

Munk boundary layer—A horizontal **boundary layer** in which lateral **transport** of **momentum** exerts a **torque** on fluid parcels, thereby allowing them to cross **isolines** of background **potential vorticity**.

In many ocean general **circulation** models, the western **boundary currents** are Munk boundary layers.

Munsell color chart—A radiometrically traceable color system based on the Commission International de l'Eclairage (CIE).

muskeg—A swamp or **bog** occurring in depressions in poorly drained **alluvial** or **glacial** terrain in northern Canada or the United States.

The depression usually accumulates a saturated, highly compressible mixture of mineral particles and decaying vegetal matter, topped by a hummocky surface of sphagnum moss, and incapable of supporting heavy loads or traffic. In the colder and wetter parts of Alaska, these accumulations spread widely over low-amplitude terrain and are not confined to depressions.

Muskingum method—An approximate hydrological method of **flood routing** through a **reach** of river, based on the **equation of continuity** and a **storage equation** expressing the **linear** dependence of the water volume in the reach on the weighted **inflow** and **outflow**.

mutatus—A term used when the whole or a large part of a **cloud** undergoes a change from one genus to another.

The new cloud is then given the name of the appropriate genus, followed by the name of the genus of the **mother-cloud** with the addition of the suffix "mutatus" (e.g., stratus stratocumulomutatus). *Compare* **genitus**.

Myers rating—A relative measure of **flood** productivity.

It is a numerical factor used to compare extreme floods on different streams, given by the extreme discharge (in cubic feet per second) on a **stream** divided by 100 times the square root of the **drainage area** (in square miles). Typical values in the eastern United States range from 30 to 40, but values as **low** as 1 or 2 occur in the Great Basin, and some as high as 300 occur on some basins in west Texas.

N

n'aschi—(Also spelled nashi.) The Arabic name for a northeasterly **wind** that occurs in winter on the Iranian coast of the Persian Gulf, especially near the entrance to the gulf and also on the Makran coast.

It is probably of the **bora** type, though less strong, representing the **outflow** of cold air from central Asia. The n'aschi is part of the Asiatic **monsoon** system.

N.T.P.—Abbreviation for **normal temperature and pressure**.
See **standard temperature and pressure**.

N-unit—*See* **refractivity**.

N weather—(Obsolete.) Abbreviation for **instrument weather**.

n-year event—Same as **T-year event**.

NAAQS—Abbreviation for **National Ambient Air Quality Standards**.

NACA Standard Atmosphere—See **standard atmosphere**.

nacreous clouds—(Also called mother of pearl clouds; rarely, luminous clouds.) *See* **polar stratospheric clouds**.

nadir—1. Satellite **subpoint** on the earth's surface that is centered directly below the satellite.
This is also referred to as the point of zero nadir angle for a satellite in earth orbit.

2. The point on a given observer's **celestial sphere** diametrically opposite his **zenith**, that is, directly below the observer.

NADW—Abbreviation for **North Atlantic Deep Water**.

Nansen bottle—A device used by oceanographers to obtain subsurface samples of **seawater**.
The "bottle" is lowered by wire with its valves open at both ends. It is then closed in situ by allowing a weight (called a **messenger**) to slide down the wire and strike the reversing mechanism. This causes the bottle to turn upside down, closing the valves and reversing the reversing thermometers, which are mounted on the bottle in a special **thermometer** case. If, as is usually the case, a series of bottles are lowered, then the reversal of each bottle releases another messenger to actuate the bottle beneath it.

Nansen cast—A series of **Nansen bottle** water samples and associated **temperature** observations resulting from one release of a **messenger**.

nappe—The profile assumed by water flowing over a **weir** in a vertical drop.

narboné—Same as **narbonnais**.

narbonnais—(Also spelled narboné.) In France, a **wind** coming from Narbonne; a north wind in the Roussillon region of southern France resembling the **tramontana**.
If associated with an influx of **arctic air**, it may be very stormy with heavy falls of **rain** or **snow**. It is especially violent in the region of Perpignan where it blows in a succession of squalls for several days. In Provence it is rarer and blows from the west. In lower Languedoc and the southern Cévennes, the narbonnais is an infrequent, mild, moist, moderate southwest wind in winter and early spring, sometimes bringing **thunderstorms**.

narrow-beam radiogoniometer—Same as **narrow-sector recorder**.

narrow-sector recorder—A **radio direction-finder** with which **atmospherics** are received from a limited sector related to the position of the **antenna**.
The antenna is usually rotated continuously and the bearings of the atmospherics recorded automatically. *See* **sferics**.

narrowband radiation—Radiation over a **range** of wavelengths for which the Planck function does not change significantly, but for which the spectral **absorption coefficient** may be highly variable
Transmission functions for narrowband radiation, unlike the exponential function for **monochromatic radiation** (see **Bouguer's law**), require the application of **band** models or numerical techniques such as the **correlated-k** method. Spectral intervals for narrowband radiation have widths that are typically 100 cm^{-1}.

NASA Scatterometer—(Acronym NSCAT.) A specialized **microwave radar** designed to measure surface winds over the oceans from an orbiting satellite.

nashi—Same as **n'aschi**.

NAT—*See* **nitric acid trihydrate**.

National Ambient Air Quality Standards—(Abbreviated NAAQS.) Air quality standards regarding **air pollution** for the United States that are required to be met by industry and are enforceable by law.
> *See* **emission standard**.

national standard barometer—**Barometer** designated by a WMO member as the reference **standard** barometer for its own country.

National Tidal Datum Epoch—The specific 19-year period adopted by the National Ocean Service as the official time segment over which **sea level** observations are taken and reduced to obtain mean values for **datum** definition.
> The present epoch is 1960–78. It is reviewed annually for revision and must be actively considered for revision every 25 years.

natural control—Natural conditions in a channel that make the **water level** at, or **upstream** of, that location a stable indicator of the **discharge**.
> Commonly, this control will be a condition where **critical depth** occurs due to a change from a mild to a steep slope.

natural convection—Same as **free convection**.

natural coordinates—An **orthogonal**, or mutually perpendicular, system of **curvilinear coordinates** for the description of fluid motion, consisting of an axis t tangent to the instantaneous **velocity** vector and an axis n normal to this velocity vector to the left in the horizontal plane, to which a vertically directed axis z may be added for the description of three-dimensional flow.
> Such a **coordinate system** often permits a concise formulation of atmospheric dynamical problems, especially in the Lagrangian system of **hydrodynamics**.
> > Holton, J. R., 1992: *An Introduction to Dynamic Meteorology*, 3d edition, Academic Press, 61–63.

natural flow—Streamflow that occurs from a **basin** under natural conditions, that is, without regulation by **hydraulic structures**.

natural frequency—(Also called characteristic frequency.) Any **frequency** of small-amplitude **oscillation** for a system with a position of stable **equilibrium** and in the absence of external forces.
> One must be careful about what is meant by "external forces." A simple pendulum of length h, when disturbed slightly from its equilibrium position, oscillates with natural (circular) frequency

$$\omega = \left(\frac{g}{h}\right)^{1/2},$$

where g is the **acceleration** due to **gravity**. Yet this oscillation requires an external force, namely, gravity. In general, a mechanical system with a position of stable equilibrium has a set of distinct natural frequencies, one for each degree of freedom.

natural remanent magnetization—(Abbreviated NRM.) The permanent magnetism of a rock.
> A rock may also have other types of magnetism, such as induced magnetism that may change when the rock is moved to another location in the geomagnetic **field**. The NRM does not change with another location but is "frozen" into the rock.

natural sulfur cycle—Conceptualization of the processes associated with the production and loss of sulfur gases in the **atmosphere**.
> The **cycle** consists of 1) the **emission** of reduced sulfur gases (such as **dimethyl sulfide**, **carbonyl sulfide**) by biological material; 2) the **oxidation** of these gases to **sulfur dioxide** or **sulfuric acid** in the atmosphere; and 3) the subsequent **deposition** of these acidic species to the earth's surface.

natural synoptic period—Period of time during which the essential characteristics of a particular **synoptic situation** persist over a large area of the globe.
> The term is used in long-range forecasting in the former USSR.

natural synoptic region—A large area of the globe in which **synoptic** processes possess well-defined characteristics that are independent of processes occurring in other areas of the globe.
> The term is used in long-range forecasting in the former USSR.

natural synoptic season—Period of the year characterized by weather conditions of a particular type over a large area of the globe.
> The term is used in long-range forecasting in the former USSR.

natural variability of climate—*See* **climate variability**.

naulu—In Hawaii, an intense **shower**.

nautical mile—The length of one minute of arc along any **great circle** on the earth's surface.

Since this actual distance varies slightly with latitude, a nautical mile by international agreement is defined as 1852 meters (6076.103 feet or 1.1508 statute miles). *Compare* **mile**.

nautical system—A system for expressing distance, speed, and **acceleration** in which 1) the distance of one minute of arc along a **meridian** or **great circle** is one **nautical mile**; 2) a nautical mile per hour is a **knot**; 3) a nautical mile per hour per hour is the acceleration in knots per hour.

Although the nautical system originated with marine operations, it has been adopted to report winds and aircraft speeds.

nautical twilight—The **twilight** stage during which the sun's unrefracted center is at **elevation** angles $-6° > h_0 > -12°$.

During a **clear** evening's nautical twilight, horizontal **illuminance** decreases from ~3.5–2 **lux** to ~0.008 lux. At nautical twilight's bright upper limit, the brightest stars are visible and the (ocean) **horizon** is distinct. At its dark lower limit, the horizon is generally invisible.

Navier–Stokes equations—The **equations of motion** for a **viscous fluid** that may be written

$$\frac{d\mathbf{u}}{dt} = -\frac{1}{\rho}\nabla p + \mathbf{F} + \nu\nabla^2\mathbf{u} + \frac{1}{3}\nu\nabla(\nabla \cdot \mathbf{u}),$$

where p is the **pressure**, ρ the **density**, $\mathbf{F}$ the total external force, $\mathbf{u}$ the fluid **velocity**, and ν the **kinematic viscosity**.

For an **incompressible fluid**, the term in $\nabla \cdot \mathbf{u}$ (**divergence**) vanishes and the effects of **viscosity** then play a role analogous to that of **temperature** in **thermal** conduction and to that of density in simple **diffusion**. Solutions of the Navier–Stokes equations have been obtained only in a limited number of special cases; in atmospheric motion, the effects of **molecular viscosity** are usually overshadowed by the action of turbulent processes and the Navier–Stokes equations have been of little direct application. The use of the concept of **eddy viscosity** has overcome this limitation in certain problems. The equations are derived on the basis of certain simplifying assumptions concerning the **stress tensor** of the fluid; in one dimension they represent the assumption referred to as the **Newtonian friction law**. *See also* **Ekman spiral**, **logarithmic velocity profile**.

navigable semicircle—The side of a **tropical cyclone** to the left of the direction of movement of the **storm** in the Northern Hemisphere (to the right in the Southern Hemisphere), where the winds are weaker because the **cyclone**'s translation and rotation speeds subtract.

Compare **dangerous semicircle**.

navigation—The process of calculating the earth coordinates (latitude and longitude) of remotely sensed data.

Navigation requires an accurate knowledge of the position of the satellite in its **orbit**, the orientation of the satellite, and the **scanning** geometry of the instrument sensors.

NDVI—Abbreviation for Normalized Difference Vegetation Index.

See **vegetation index**.

neap range—The average semidiurnal **tidal range** occurring at the time of **neap tide**.

neap tide—A **tide** of decreased **amplitude**, occurring semimonthly one or two days after **quadrature**.

Compare **spring tide**, **tropic tide**, **equatorial tide**.

near gale—**Wind** with a speed between 28 and 33 knots or force 7 on the Beaufort wind scale.

near-infrared—Pertaining to or the same as **near-infrared radiation**.

near-infrared radiation—The preferred term for the shorter wavelengths in the **infrared** region of the **electromagnetic spectrum** extending from about 0.75 μm (visible red) to around 3 μm.

The term usually emphasizes the **radiation** reflected from plant materials, which peaks around 0.85 μm. Near-infrared is sometimes called **solar infrared**, since the **flux of radiation** in these wavelengths is a maximum during the daylight hours.

near-polar-orbiting satellite—*See* **polar-orbiting satellite**.

nebelwind—Same as **fog wind**.

NEBUL code—A code in which data on **cloud** systems are encoded and transmitted; it is a modified form of the **international analysis code**.

nebule—A unit of atmospheric **opacity**.

One nebule is the opacity of a screen having such **transmissivity** T_n that

$$T_n^{100} = 0.001,$$

that is, such that 100 of the screens placed in optical series would transmit only one-thousandth of the incident **light**. One speaks of a given **atmosphere** or atmospheric stratum as possessing an opacity of a certain number of nebules per kilometer. The transmissivity T of an **optical path** r km long through an atmosphere of opacity n nebules per km is given by

$$T = T_n^{nr},$$

where T_n is defined above. An opacity of one nebule per km is equivalent to an **extinction coefficient** of 0.069 per km.

Middleton, W. E. K., 1952: *Vision through the Atmosphere*, p. 14.

nebulosus—A **cloud species** with the appearance of a nebulous veil, showing no distinct details.

This species is found principally in the genera **cirrostratus** and **stratus**. Stratus nebulosus is the most common species of stratus. Cirrostratus nebulosus produces **halo** phenomena. *See* **cloud classification**.

NEC—Abbreviation for **North Equatorial Current**.

NECC—Abbreviation for **North Equatorial Countercurrent**.

needle—*See* **ice needle**.

needle ice—1. Same as **candle ice**.

2. Same as **frazil**.

Neel temperature—Same as **Curie temperature**.

negative cloud-to-ground lightning—A **lightning flash** or **stroke** between a **cloud** and the ground that lowers negative charge to the ground.

negative feedback—A sequence of interactions that damps or reduces the response to an initial **perturbation**.

For example, consider a surface that is subjected to an increase in incoming **radiation**. This change in the **energy balance** produces an increase in **temperature** which, by virtue of the **Stefan–Boltzmann law**, results in an increase in the radiation emitted by the surface. Thus, the interaction by temperature and radiation acts to partially counteract the original perturbation. *Compare* **positive feedback**.

negative ground flash—Same as **negative cloud-to-ground lightning**.

negative isothermal vorticity advection—The **advection** of negative **vorticity** along an **isothermal** surface.

negative rain—**Rain** that exhibits a net negative electrical charge.

negative viscosity—A characteristic of a system in which **momentum** is transferred from a region of lower **velocity** toward a flow of higher velocity, increasing the mean **wind shear**.

Compare **viscosity**.

nemere—In Hungary, a stormy, cold **fall wind**.

Neoglacial—A period of general expansion of glaciers variously defined as spanning from approximately 3000 to 2000 years ago or covering the last 4000–5000 years.

neon—(Symbol Ne.) An **inert gas** that is the second member of the noble gas family, atomic number 10, atomic weight 20.183.

It is a colorless, odorless, monatomic **element** found in the **atmosphere** to the extent of about 0.0018% by volume of **dry air**.

neper—The analogue of the **decibel**, utilizing natural logarithms rather than logarithms to the base 10.

It is given by the relation $\ln[(S_1/S_2)^{1/2}]$, where S_1 and S_2 are **flux** densities. One neper is equal to 8.686 decibels.

neph chart—*See* **nephanalysis**.

nephanalysis—The **analysis** of a **synoptic chart** focusing on the types and amount of clouds and **precipitation**.

Cloud systems (or nephsystems) are identified both as entities and in relation to analysis features determined from other information. A **chart** of this type is sometimes termed a **neph chart**. *See* **isoneph**, **nephcurve**.

nephcurve—In **nephanalysis**, a line bounding a significant portion of a **cloud system** that permits the analyst to extract information.

Examples are clear-sky lines, **precipitation** lines, cloud-type lines, ceiling-height lines. *Compare* **isoneph**.

nephelometer—1. General name for instruments that measure, at more than one angle, the **scattering function** of **particles** suspended in a medium.

Information obtained from such instruments may be used to determine the size of the suspended particles and the **visual range** through the medium.

2. Same as **nephometer**.

Middleton, W. E. K., 1952: *Vision through the Atmosphere*, 200–212.

nephelometry—From the Greek "nephele," **cloud**, usually applied to measurement of the angular dependence of **scattering** of **electromagnetic radiation** by a **suspension** of **particles**.

nepheloscope—1. A laboratory instrument for the production of **clouds** by the **condensation** process.

2. An instrument for demonstrating the **temperature** changes that occur in air that is rapidly expanded or compressed.

nephology—The study of **clouds**.

nephometer—(Also called nephelometer.) 1. A general term for instruments designed to measure the amount of **cloudiness**.

An early type consists of a convex hemispherical mirror mapped into six parts. The amount of **cloud** coverage on the mirror is noted by the **observer**.

2. Same as **nephelometer**.

nephoscope—An instrument for determining the direction and relative speed of **cloud** motion.

There are two basic designs of nephoscope: the **direct-vision nephoscope** and the **mirror nephoscope**.

nephsystem—Same as **cloud system**.

nested grids—A high-resolution region of **discretization** embedded within a low-resolution region of a numerical **model** or **analysis** system.

Fine **resolution** may be desired to focus on a **mesoscale** feature such as a **tropical cyclone** or on a geographical area of interest. In two-way nested grids, information is passed back and forth between the high- and low-resolution regions, in contrast with one-way grids in which information is passed merely from low- to high-resolution regions. *See* **grid telescoping**, **variable resolution model**.

net balance—The **mass balance** at the end of the **balance year**.

It represents the annual addition or loss of mass at a point on a **glacier**. *See also* **average net balance**.

net outgoing IR—Same as **net terrestrial radiation**.

net primary production—The part of the **gross primary production** that remains after deducting the amount released by plant (autotrophic) respiration.

It is the amount of **energy** that is stored in plant tissues and is often expressed as the amount of **carbon** or dry organic matter (biomass) produced over the time period.

net pyranometer—**Net radiometer** designed to measure only **solar radiation** wavelengths.

net pyrgeometer—**Net radiometer** designed to measure only **thermal infrared** wavelengths of **electromagnetic radiation**.

net pyrradiometer—**Net radiometer** designed to measure both solar and **thermal** wavelengths of **radiation** as a single sum of **energy**.

net radiation—In general, the difference between absorbed and emitted **radiation**.

Commonly used to refer to the **surface radiation budget**, but may be used more generally.

net radiometer—A **radiometer** designed to measure the difference in **irradiance** coming from two opposing hemispheric fields of view.

Typically, this instrument would be used to measure the difference of upwelling and downwelling irradiance at some level in the **atmosphere**, including the surface.

net solar radiation—The difference between the **solar radiation** fluxes directed downward and upward; net **flux** of solar radiation.

net storm rain—*See* **effective rainfall**.

net terrestrial radiation—(Also called net outgoing IR.) The net **flux density** of **terrestrial radiation**. This is the difference between upwelling and downwelling longwave flux density, which may be referenced to any specified **altitude**, but is most commonly referenced to the surface. The long-term global average for the net terrestrial radiation leaving the earth's surface is approximately 50 W m^{-2}.

network—System of small and mottled areas of enhanced **brightness** in the solar **chromosphere**. Their size varies from 5 to 20 arc-seconds. These features evolve from the **plage** areas and, like the plage, their number and **intensity** varies with the **solar cycle**. Also referred to as enhanced network, active network, and bright network.

network density—Measure of the proximity of observing stations in a network established for some specific meteorological studies or research.

Neuhoff diagram—*See* **thermodynamic diagram**.

neural network—Usually used to mean "artificial neural network," a computer program inspired by simple models of the brain.
It consists of a network of nodes connected by weighted links that establish the relationships between nodes. Each node sums the weighted inputs entering it and compares the result to a (usually) **nonlinear** function to produce its own **output**. Most neural networks have a training **rule** establishing how the weights are adjusted to bring the average output closest to that desired. The term can also mean the computer chip containing the fully trained (unmodifiable) system used in automatic sensors and controls. *See also* **machine learning**.

neutercane— Large **storm** system possessing both tropical and **extratropical** characteristics.

neutral atmosphere—Term used in an ionospheric context to describe the nonionized component of the **atmosphere**.

neutral equilibrium—A condition of a system for which a **small perturbation** of a **parcel** of the system causes it to neither depart from its new position nor return to its previous one.

neutral mode—(Also called neutral wave.) In **hydrodynamic instability** theory, a **wave** solution the **amplitude** of which does not change with time; it neither grows nor decays.
In contrast, the amplitude of a **growing mode** (or wave) increases with time; that of a **decaying mode** (or wave) decreases with time. The latter two are **unstable waves**.

neutral occlusion—*See* **occluded front**.

neutral oscillation—*See* **oscillation**.

neutral point—1. In the **clear sky**, one of several points in the **principal plane** (or very near it) where the degree of **linear** polarization $P = 0$ (i.e., **polarization** is neutral).
Stated another way, a neutral point occurs where Stokes parameter $Q = U = 0$. In fact, P often differs indistinguishably from zero within a small area ($\sim 3°$ radius) around the nominal **principal point**, and thus neutral polarization routinely occurs slightly outside the principal plane. For a **single-scattering** molecular **atmosphere**, the neutral points would coincide with the solar and antisolar points. In the real atmosphere, **scattering** by large **particles** (e.g., **haze**) and **multiple scattering** displace the observed neutral points. The Arago neutral point occurs $\sim 10°–30°$ above the **antisolar point**. The Brewster neutral point occurs below the sun and the **Brewster point** occurs above it. When the sun is low in the sky, the Brewster and Babinet points can be as much as $25°–35°$ from it, although their angular distances need not be equal. In principle, the Brewster and Babinet points coincide with the **zenith** sun.
2. Same as **hyperbolic point**.

neutral point—Same as **col**.

neutral stability—Same as **neutral equilibrium**.

neutral surfaces—A neutral surface is that surface along which fluid **particles** can be exchanged without doing any **work** against **gravity**.
For many oceanographic purposes it is sufficiently accurate to assume that this surface is defined

by a constant value of **potential density**. In **ocean mixing** studies, however, this definition will often not be sufficiently accurate due to lateral variations of **temperature**, **salinity**, and **pressure** along a potential density surface. These variations will cause fluid particles from different points within the potential density surface to migrate slightly off that surface when they are displaced laterally.

neutral wave—Same as **neutral mode**.

neutron—An uncharged **subatomic particle**.
The **rest mass** of the neutron is slightly greater than that of the **proton**. Atomic nuclei are composed of protons and neutrons bound together by nuclear forces. The common term for neutron or proton is nucleon. Although the neutron is electrically neutral it does possess a charge structure and a **magnetic dipole** moment. The existence of the neutron was first demonstrated by James Chadwick in 1932.
> Boorse, H. A., and L. Motz, 1966: *The World of the Atom*, Vol. II, 1288–1308.

neutron logging—Assessment of the fluid properties in a borehole by bombardment with fast-moving **neutrons** and measurement of resultant **radiation**.

neutron-scattering method—Technique used for the indirect measurement of **soil moisture content** based on the thermalization (slowing) and **scattering** of emitted **neutron** radiation upon encountering **hydrogen** atoms, which occurs primarily in the water in moist soil.

neutrosphere—The **atmospheric shell** from the earth's surface upward in which the atmospheric constituents are for the most part not ionized, that is, electrically neutral.
The region of transition between the neutrosphere and the **ionosphere** is somewhere between 70 and 90 km, depending on latitude and **season**.

nevada—A cold **wind** descending from a mountain **glacier** (**glacier wind**) or **snowfield**, for example, in the higher valleys of Ecuador.

névé—1. Same as **firn**.
2. Same as **accumulation area**.

new candle—*See* **candela**.

New Guinea Coastal Current—A seasonal western boundary **current** through Vitiaz Strait and along the northern coast of New Guinea.
The current flows northwestward from April to November and is then a link between the northern branch of the **South Equatorial Current** and the **Halmahera Eddy**. It reverses during December to March, opposing the **New Guinea Coastal Undercurrent** underneath, when the Philippines experience **monsoon** winds blowing from the northeast in the Northern Hemisphere and from the northwest south of the **equator**.

New Guinea Coastal Undercurrent—A western boundary **current** and part of the **equatorial current system** in the Pacific Ocean.
It flows northwestward through Vitiaz Strait and along the northern coast of New Guinea. It receives its water from the northern branch of the **South Equatorial Current** and is the major **source** of the **Equatorial Undercurrent**.

new snow—1. Recently fallen **snow** in which the original form of the **snow crystals** is recognizable.
2. The amount of **snow** fallen within the previous 24 hours.

newton—A unit of force that, when applied to a body of mass one kilogram, gives it an **acceleration** of one meter per second squared ($1 \text{ N} = 1 \text{ kg m s}^{-2}$).

Newton's law of cooling—As stated by Newton, the exponential decrease of **temperature** with time of a body cooling in a fluid.
Newton's law of cooling is an approximation valid under restricted conditions, not a universal law of nature.
> Bohren, C. F., 1991: *Amer. J. Phys.*, **58**, 956–960.

Newton's laws of motion—A set of three postulates first set forth by Sir Isaac Newton in the middle of the seventeenth century.
According to the first law (the law of inertia), a body (point mass) remains at rest or in a state of uniform motion unless acted on by a force, where the position of the body is specified relative to an **inertial reference frame**. The second law states that the time rate of change of (**linear**) **momentum** of a body is equal to the force on the body; Newton took **momentum** to be the product of mass and **velocity**, which is only an approximation valid at speeds much less than that

of **light**. Finally, the third law states that if two bodies exert forces on each other, they are equal in magnitude and opposite in direction. Although apparently simple and unambiguous, Newton's laws have engendered considerable discussion over the extent to which some of them are mere definitions (e.g., of inertial reference frame, mass, and force) or are true laws in the sense of being subject to experimental **verification**. Despite the ambiguity of Newton's laws, they have proven to be an efficient way of describing the physical world (within limits) and form the basis for the **equations of motion** of fluids.

Newtonian fluid—A fluid in which the **stress tensor** is proportional to the rate of **deformation**, that is, a fluid satisfying the **Navier–Stokes equations**.
    See **viscous force**.

Newtonian friction law—(Also called Newton's formula for the **stress**.) The statement that the tangential force (i.e., the force in the direction of the flow) per unit area acting at an arbitrary level within a fluid contained between two rigid horizontal plates, one of which is motionless and the other which is in **steady motion**, is proportional to the **shear** of the fluid motion at that level.
    Mathematically, the law is given by

$$\tau = \mu \, \frac{\partial u}{\partial z},$$

where τ is the tangential force per unit area, usually called the **shearing stress**; μ a constant of proportionality called the **dynamic viscosity**; and $\partial u / \partial z$ the shear of the fluid flow normal to the resting plate. In deriving this expression Newton assumed that either the speed u of the moving plate or the distance between the plates was so small that, once a **steady state** was reached, the speed of the fluid increased linearly from zero at the resting plate to the speed u at the moving plate. In this case both the shear of the motion and the shearing stress are constant throughout the fluid.

Newtonian mechanics—(Also called classical mechanics.) A system of mechanics based on **Newton's laws of motion**.
    The salient characteristics of Newtonian mechanics are that mass and **energy** are separately conserved, all physical variables can take on a continuous set of values, the state of a system at any instant uniquely determines its state at any later instant (determinism), and interactions at a distance are instantaneous. *Compare* **quantum theory**.

NEXRAD—1. Acronym for Next Generation Weather Radar program, a joint program of the U.S. Departments of Commerce, Defense, and Transportation during the 1980s and 1990s to develop and deploy a network of operational **Doppler weather radars**.
    2. The popular name for the **radar** itself, the **WSR-88D** weather radar, which became the operational network radar for the U.S. National Weather Service, U.S. Air Force, and Federal Aviation Administration during the early and middle 1990s. The nominal transmitted **wavelength** of the WSR-88D is 10.5 cm (**S band**) and the nominal circular **beamwidth** is 0.95°.

nieve penitente—Same as **penitent ice**.

night-sky light—Same as **airglow**.

night visual range—(Also called nighttime visual range, transmission range.) The maximum distance at which a normal observer can see a particular point light **source** under given atmospheric conditions.
    For a **light** source of **luminous** power P_v (in lumens), uniform atmospheric **extinction coefficient** σ along the viewing direction, and **threshold illuminance** E_{thresh} for the **observer**, the night visual range x is

$$x = \frac{1}{\sigma} \ln \left(\frac{P_v}{E_{thresh} x^2} \right).$$

Unlike the **daytime visual range**, the night visual range is determined by both the **inverse-square law** and atmospheric **extinction**. Note that for a light to be seen at a distance $x > 0$, its **illuminance** at the observer (P_v / x^2) must exceed E_{thresh}. Even if σ is known, there is no unique night visual range because x also depends on P_v and the observer's **dark adaptation**. *See* **Allard's law**

night wind—Dry **squalls** that occur at night in southwest Africa and the Congo.
    It is likely that this term is loosely applied to other **diurnal** local winds such as **mountain wind**, **land breeze**, **midnight wind**, etc.

nightglow—*See* **airglow**.

nilas—A thin elastic crust of **ice** up to 10 cm thick that, under pressure, may deform by **finger rafting**.

nimbostratus—(Abbreviated Ns.) A principal **cloud** type (**cloud genus**), gray colored and often dark, rendered diffuse by more or less continuously falling **rain, snow, sleet**, etc., of the ordinary varieties and not accompanied by **lightning, thunder,** or **hail**.

In most cases the **precipitation** reaches the ground, but not necessarily. Nimbostratus is composed of suspended water droplets, sometimes supercooled, and of falling **raindrops** and/or **snow crystals** or **snowflakes**. It occupies a layer of large horizontal and vertical extent. The great **density** and **thickness** (usually many thousands of feet) of this cloud prevent observation of the sun; this, plus the absence of small droplets in its lower portion, gives nimbostratus the appearance of dim and uniform lighting from within. It also follows that nimbostratus has no well-defined base, but rather a deep zone of **visibility** attenuation. Frequently a false base may appear at the level where snow melts into rain. Nimbostratus usually results from the thickening of **altostratus** to the point where the sun becomes totally indiscernible (Ns altostratomutatus); this point in time usually coincides with the beginning of relatively continuous precipitation. Rarely, it may evolve in like manner from **stratocumulus** or **altocumulus** (Ns stratocumulomutatus or Ns altocumulomutatus). Nimbostratus sometimes forms by the spreading of **cumulonimbus** or **cumulus congestus** when these clouds produce **rainfall** (Ns cumulonimbogenitus or Ns cumulogenitus). By definition, nimbostratus is always accompanied by the complementary features **praecipitatio** or **virga**. The accessory **cloud, pannus,** also is a common feature. At first the pannus consists of separate units, but later they may merge into a continuous layer and extend upward into the nimbostratus. Nimbostratus is most easily confused, in identification, with thick masses of altostratus, **stratus,** or stratocumulus. Altostratus, however, is lighter in color, appears less uniform from below, and does not completely hide the sun. In case of further doubt, a cloud is called nimbostratus if precipitation from it reaches the ground. Stratus also may have precipitation, but only of very small-sized **particles**. Stratocumulus shows clear relief and a well-defined limit of its base. *See* **cloud classification**.

nimbus—Former name for any rain-producing **cloud**.

It is not now recognized in the international **cloud classification**.

Nimbus—An experimental series of meteorological satellites, launched between August 1964 and October 1978.

Nimbus 1 was the first sun-synchronous, **three-axis stabilized** meteorological satellite. Nimbus satellites became the nation's principal platform for **remote sensing** research. Many of the instruments tested on Nimbus evolved into operational instruments used on other satellites.

nine-light indicator—A remote **indicator** for **wind speed** and direction used in conjunction with a contact **anemometer** and a **wind vane**.

The indicator consists of a center light, connected to the contact anemometer, surrounded by eight equally spaced lights that are individually connected to a set of similarly spaced electrical contacts on the wind vane. Wind speed is determined by counting the number of flashes of the center wind vane during an interval of time. Direction, indicated by the position of the illuminated outer bulbs, is given to 16 points of the **compass**.

Nipher shield—A conically shaped, copper, **rain-gauge shield**.

It is used to prevent the formation of vertical wind eddies in the vicinity of the mouth of the gauge, thereby making the rainfall **catch** a representative one. *Compare* **Alter shield**.

nirtra—*See* **aloegoe**.

nitrate—Any chemical compound containing ONO_2.

These can be either **organic nitrates** (e.g., CH_3ONO_2) or inorganic nitrates (eg., $ClONO_2$). *See also* **nitrate ion, nitrate radical**.

nitrate ion—The ionized form of **nitric acid** in solution, formula NO_3^-.

Nitrate is a common, if not universal, component of rainwater and is often used as a marker for **acid precipitation**, since its production is usually associated with combustion sources. It is an important plant nutrient.

nitrate radical—Free radical, formula NO_3, formed from the reaction of **nitrogen dioxide** with **ozone**.

NO_3 has a very strong optical **absorption** in the visible portion of the **spectrum**. Absorption of **solar radiation** in this region leads to **photodissociation** and thus a very short atmospheric lifetime during the day. However, at night NO_3 can build up to substantial levels and react with some families of organic molecules, allowing **oxidation** chemistry that would not otherwise occur at night. Nitrate reacts to form **dinitrogen pentoxide**, which can be taken up into aqueous droplets and increase acidity. Optical absorption has been used to detect NO_3 in the **atmosphere**.

nitric acid—A very soluble, acidic gas, formula HNO_3, the end product of the **oxidation** of emitted gases.

It is a major component of acidic **precipitation** in continental regions. In the clean background **troposphere**, its removal in precipitation acts as a **sink** for **odd hydrogen** and **nitrogen** compounds and limits the formation of **ozone**.

nitric acid trihydrate—(Also known as NAT.) A stable crystalline form of **nitric acid** and water, consisting of three molecules of water for every molecule of nitric acid. **Polar stratospheric clouds** (type I) may be in the form of NAT at least part of the time.

nitric oxide—A colorless gas, formula NO, the most common form of **nitrogen** emitted into the **atmosphere**, either by fuel combustion or due to natural emissions.

Nitric oxide is interconverted with **nitrogen dioxide** fairly readily in the atmosphere, resulting in **catalytic cycles** leading to **ozone** formation in the **troposphere** and ozone loss in the **stratosphere**.

nitrification—The process of converting reduced **nitrogen** (as **ammonia** or ammonium) to its more oxidized forms (nitrite or **nitrate** ions).

It is a two-step process, with each step carried out by distinct groups of bacteria. The first step is the **oxidation** of ammonia to nitrite, and the second is oxidation of nitrite to nitrate.

nitrogen—(Symbol N.) A colorless, tasteless, odorless gaseous **element**, atomic number 7, atomic weight 14.007.

It is the most abundant constituent of the **atmosphere**, amounting to 78.09% by volume of **dry air**. The molecular formula for nitrogen gas is N_2; its **molecular weight** is 28.016. Nitrogen enters the atmosphere from volcanoes and from the decay of organic matter. It is removed from the atmosphere by certain natural nitrogen-fixing bacteria for use in plant life processes. Free nitrogen is very inactive, but can be broken down by high-energy reactions such as occur in **lightning**, high **temperature** combustion, or in the **upper atmosphere**. Nitrogen-containing compounds are very reactive and play integral roles in the production and destruction of **ozone** in the atmosphere. Atomic nitrogen, N, occurs in significant quantities only at altitudes above about 100 km.

nitrogen cycle—A continuous series of natural processes by which **nitrogen** successively passes through air, soil, and organisms involving principally organism decay, **nitrogen fixation**, **nitrification**, and **denitrification**.

See **nitrogen-fixing plants**.

nitrogen dioxide—A brown gas, formula NO_2, found at all levels in the **atmosphere**.

In the **troposphere** it photodissociates to give free **oxygen** atoms, which then form **ozone**, and is thus a key player in local and **regional air pollution** events. In the **stratosphere** it participates in catalytic ozone destruction cycles, but also forms stable **nitrate** reservoir species that ameliorate ozone loss.

nitrogen fixation—The incorporation of gaseous molecular **nitrogen**, N_2, into nitrogenous compounds.

Abiotic fixation of N_2 occurs via **lightning** and photochemical conversion in the **atmosphere**. Biotic fixation of N_2 is done by specialized bacteria that construct the hemoglobin-like enzymes necessary to cleave the strong triple bond of molecular nitrogen.

nitrogen-fixing plants—Certain plants (mostly leguminous) that are capable of biological conversion of molecular dinitrogen (atmospheric **nitrogen**, or N_2) to organic combinations (**nitrogen fixation**), or to forms usable in biological processes, through the action of symbiotic rhizobia (bacteria) in the root nodules.

nitrogen oxides—Family of compounds in which **nitrogen** is bound to **oxygen**.

The most abundant is **nitrous oxide**, formula N_2O, which is relatively unreactive. Nitric oxide (NO) and **nitrogen dioxide** (NO_2) are highly reactive, and are present in much lower amounts in the **atmosphere**. Together, NO and NO_2 are classed as **odd nitrogen**, or **active nitrogen**.

nitrogen pentoxide—*See* **dinitrogen pentoxide**.

nitrous acid—1. Weak acid formed in solution by neutralization of the nitrite **ion**.
Also formed when **nitrogen dioxide**, NO_2, dissolves in water.

2. Gaseous compound, formula HONO, that builds up in polluted, typically urban air, containing **nitrogen oxides**.

The only known gas-phase reaction leading to its formation, that between hydroxyl radicals (OH) and **nitric oxide** (NO), is not rapid enough to account for observed levels, and the formation is thus thought to involve **particles**. Nitrous acid photodissociates to give OH + NO, and can thus serve as an OH source, particularly in the early morning.

nitrous oxide—Colorless gas, formula N_2O, released by bacterial activity at the earth's surface.

It has an atmospheric lifetime of about 160 years and is currently present at a level of about 330 ppb. Its atmospheric significance is that it is transported into the **stratosphere**, where its reaction with excited **oxygen** atoms (O^1D) is the major source of **active nitrogen**; it is also a major greenhouse gas. In large amounts, it has anesthetic properties (laughing gas).

nival—Characterized by **snow** or a snowy **environment**.

NMHCs—*See* **nonmethane hydrocarbons**.

90° halo—Same as **Hevel's halo**.

NOAA—An operational series of meteorological satellites, named for the National Oceanic and Atmospheric Administration.

See **POES**.

noble gases—(Also called rare gases, inert gases.) The elements **helium** (He), **neon** (Ne), **argon** (Ar), **krypton** (Kr), **xenon** (Xe), and **radon** (Rn) that make up the eighth period of the periodic table.

Because of their completely filled atomic orbitals, these gases are extremely nonreactive, and build up to measurable concentrations in the **atmosphere**.

noctilucent clouds—(Rarely called luminous clouds.) Thin silvery-blue cirrus-like clouds frequently seen during summer twilight conditions at high latitudes (above 50°) in both hemispheres.

They are the highest visible clouds in the **atmosphere**, occurring in the upper **mesosphere** at heights of about 85 km, and are closely related to the **polar mesospheric clouds** seen in satellite observations at similar altitudes over the summer polar cap. Noctilucent clouds are now known to consist of tiny **ice** particles with **dimensions** of the order of tens of nanometers, growing in the extreme cold of the summer polar **mesopause** region. The **condensation nuclei** on which the **particles** grow are thought to be either **smoke** and **dust** particles of meteoric origin or large hydrated positive ions. Strong upwelling of air from below, associated with a pole-to-pole **meridional circulation** in the upper mesosphere, is responsible for both the extreme cold and the upward **flux** of **water vapor**. Although water-vapor **mixing ratios** are very low (less than 10 parts per million by volume) in the region, the temperatures are also low enough to produce a high degree of **supersaturation** at times. Anomalously strong radar **echoes** from the region, known as polar summer mesospheric echoes, are also associated with the clouds. *Compare* **nacreous clouds**, **polar stratospheric clouds**.

nocturnal boundary layer—The cool layer of air adjacent to the ground that forms at night.

At night under **clear** skies, **radiation** to space cools the land surface, which in turn cools the adjacent air through processes of **molecular conduction**, **turbulence**, and **radiative transfer**. This causes a **stable boundary layer** to form and grow to depths of a few hundreds of meters, depending on the **season**. Many interacting processes can occur within the statically stable nocturnal boundary layer: patchy sporadic turbulence, internal **gravity waves**, **drainage flows**, **inertial oscillations**, and **nocturnal jets**.

nocturnal cooling—The lowering of **temperature** during nighttime, due to a net loss of **radiant energy**.

See **radiational cooling**.

nocturnal jet—Usually, a **low-level jet** that occurs at night.

nocturnal minimum temperature—Lowest **temperature** recorded between **sunset** and **sunrise**.

nocturnal radiation—(Obsolete.) Same as **terrestrial radiation**.

nodal factors—Small adjustments to the amplitudes and phases of **harmonic** tidal constituents to allow for modulations over an approximate 18.61-year **cycle**, the **period** of a nodal tidal cycle.

nodal increment—The difference in degrees longitude between successive **ascending nodes** of a satellite in polar **orbit**.

It is the amount of turning of the earth under the satellite, measured in degrees longitude, that takes place during one **nodal period**.

nodal period—The time that elapses between successive passages of a satellite through successive **ascending nodes**.

node—One of the two points of intersection of the **orbit** of a satellite with the plane of the **equator** of the earth.

The **ascending node** of equatorial crossing refers to that point on the plane of the **equator** at which the satellite crosses from the Southern to Northern Hemisphere. The **descending node** of equatorial crossing denotes that point at which the satellite crosses the plane of the **equator** from the Northern to the Southern Hemisphere.

noise—1. Any function, often of time but possibly of any **variable**, with a substantial lack of **correlation** between values at successive times.

For example, the **autocorrelation function** of **white noise** is a sharp spike: Its value at any instant is correlated only with its value at that instant.

2. Any change in a **signal** that degrades its capability to transmit information (e.g., the audible static on AM radio resulting from **lightning discharges**).

3. *See* **meteorological noise**.

noise factor—Same as **noise figure**.

noise figure—(Or noise factor.) A number by which the performance of a radio **receiver** can be specified.

Essentially, the noise figure is the ratio of the **noise** generated by the actual receiver to the noise output of an "ideal" receiver with the same overall **gain** and **bandwidth** when the receivers are connected to a room **temperature** load. The **noise power** from a simple load is equal to kTB, where k is **Boltzmann's constant**; T the **absolute** temperature of the load, for example, resistor; and B the measurement bandwidth. *See also* **noise level**, **signal-to-noise ratio**.

noise filtering—The removal of components of a numerical solution of the hydrodynamical equations that correspond to time or space scales not of immediate interest.

See **Kalman–Bucy filter**, **meteorological noise**.

noise level—Roughly, the amount of **noise power** in a **signal**.

The noise level is a limit where simple **power** measurements start to have trouble measuring the **signal**.

noise power—The **power** of the **noise** component of a **signal**.

noise temperature—A measure of the **noise power** of a device or circuit.

The noise temperature is the **temperature** of a resistor that has noise power equal to that of the device or circuit. Specifically, the noise temperature is defined by $T = N/kB$, where N is the noise power within **bandwidth** B, and $k = 1.38 \times 10^{-23}$J K^{-1} is **Boltzmann's constant**. A **radar** system is characterized by several noise temperatures: the **antenna** temperature T_a, the **receiver** temperature T_r, and the transmission line temperature T_l. The transmission line temperature is a measure of the noise power within the receiver bandwidth generated by the resistive losses in the transmission line or **waveguide** between the antenna and the receiver. The transmission line temperature is frequently combined with either the antenna or the receiver temperature, depending on the reference point for the measurement. The total system temperature is $T = T_a + T_l + T_r$.

noise threshold—The power level of a **signal** below which **noise** is likely to obscure the signal and above which the signal is discernible.

nomogram—(Or alignment chart; also called nomograph, nomographic chart.) The graphical representation of an equation of three variables $f(u, v, w) = 0$, by means of three graphical scales (not necessarily straight), arranged in such a manner that any straight line, called an index line, cuts the scales in values of u, v, and w satisfying the equation.

By introducing auxiliary variables and constructing auxiliary scales, equations containing more than three variables may also be represented by nomograms.

nomograph—Same as **nomogram**.

nomographic chart—Same as **nomogram**.

nonadiabatic process—Same as **diabatic process**.

noncoherent echo—(Obsolete.) **Incoherent echo**.

noncoherent target—(Obsolete.) A radar **target**, such as **rain**, that produces an **incoherent echo**.

noncontributing area—Same as **closed drainage**.

nondeterministic—Unpredictable in terms of observable antecedents and known laws.

This is a relative term pertaining to a given state of knowledge but not necessarily implying ultimate unpredictability. *Compare* **deterministic**, **random**.

nondimensional equation—An equation that is independent of the units of measurement as it only involves **nondimensional numbers**, parameters, and variables.
This is usually the result of **dimensional analysis**.

nondimensional number—A number with a value that is independent of the units of measurement.
Dimensionless numbers often arise from **dimensional analysis** and include common quantities such as the **Reynolds number, Rossby number, Rayleigh number, Richardson number**, etc. *See* **dimensionless group**.

nondimensional parameter—Parameter of a problem with a value that is independent of the units of measurement.
See **nondimensional equation, dimensional analysis**.

nondispersive infrared spectrometry—Any form of **analysis** by **infrared** absorption that does not rely on a monochromator to separate the wavelengths of the **radiation**.
If the **target** molecule is the only one known to absorb strongly in the **wavelength** region of interest, a broadband optical **filter** may be used, such as in instruments to measure ambient **carbon dioxide**. If the target molecule absorbs less strongly, or if the spectral region is congested, more specific techniques such as **gas filter correlation spectroscopy** must be used.

nondivergence level—*See* **level of nondivergence**.

nonfrontal squall line—Same as **prefrontal squall line**.

nonharmonic tidal analysis—Analysis and **prediction** using traditional methods such as the **local establishment**.
See **lunitidal interval**.

nonhydrostatic model—An **atmospheric model** in which the **hydrostatic approximation** is not made, so that the vertical **momentum** equation is solved.
This allows nonhydrostatic models to be used successfully for horizontal scales of the order of 100 m, resolving small-scale **mesoscale** circulations such as **cumulus** convection and sea-breeze circulations. In recent years, computer power has made mesoscale weather **prediction** with nonhydrostatic models feasible, and several such models are in routine use by major meteorological modeling groups and operational centers. *See* **hydrostatic model**.

noninductive charging mechanism—A charging mechanism whereby the **electric field** increase is independent of the existing electric field.

nonisotropic turbulence—*See* **isotropic turbulence**.

nonlinear—Not a **linear** function of the relevant **variables**.

nonlinear instability—The **instability** of a physical or mathematical system that arises from the **nonlinear** nature of relevant **variables** and their interactions within the system.

nonlinearity of the equation of state—The **nonlinear** dependence of seawater **density** on temperature, **salinity**, and **pressure** in the **equation of state**.

nonlocal closure—A method of approximating unknown **turbulence** quantities, such as covariances, by sums or integrals (over the whole domain of turbulence) of known quantities.
This mimics the effects of a **spectrum** of **eddy** sizes causing **mixing** from various distances. Examples are **transilient turbulence theory** and **spectral diffusivity** theory. *See* **nonlocal flux, nonlocal mixing**.

nonlocal flux—The vertical **turbulent transport** of a quantity such as **heat** per unit area per time across any height index k, between **source** heights j to destination heights i, where i and j are on opposite sides of k vertically.
It is nonlocal because i and j need not be neighbors. Using kinematic **heat flux** as an example,

$$\overline{w'\theta'}(t, \Delta t) = \frac{\Delta z}{\Delta t} \sum_{i=1}^{k} \sum_{j=k+1}^{n} [c_{ji}(t, \Delta t)\theta_i(t) - c_{ij}(t, \Delta t)\theta_j(t)],$$

where n is the total number of **grid** cells in a vertical column, Δz is the vertical thickness of each grid cell (i.e., spacing between grid points), Δt is the time increment over which **eddies** transport heat starting from time t, θ_a is the **potential temperature** at grid cell a, and c_{ab} is the **transilient matrix** specifying the fraction of air that ended in destination cell a originated at source cell b, for a and b dummy indices. *See* **transilient turbulence theory, nonlocal mixing**.

nonlocal mixing—The vertical movement and intermingling of fluid from all possible **source** locations (neighboring and nonneighboring), to produce a mixture at some other location.

The fraction of air mixing into each destination height index i from any source height j is given by a **transilient matrix** c_{ij}. The resulting state of the mixture is given by **transilient turbulence theory**. *See* **nonlocal flux**.

nonlocal static stability—A term emphasizing that the whole **vertical profile** of **virtual potential temperature** θ_v, not just the the local vertical **gradient** $\partial\theta_v/\partial z$, must be used when determining whether flow will become laminar or turbulent.

See also **static stability**.

nonmethane hydrocarbons—(Abbreviated NMHCs.) Collectively, all the **hydrocarbons** other than **methane**.

Methane is fairly long-lived in the **atmosphere** and has a large and relatively constant **mixing ratio** in the **troposphere**. The other hydrocarbons have **lifetimes** factors of 30–30 000 times shorter than methane, thus showing much greater **variability**, and tend to have more localized sources. Since such sources are often **anthropogenic** in nature, the total concentration of NMHCs is often used as a measure of the degree of **pollution** of an **air mass**.

nonrecording rain gauge—Any member of the class of **rain gauges** in which **rain** amount as a function of time is not automatically recorded.

See **eight-inch rain gauge**.

nonsupercell tornado—A **tornado** that occurs with a parent **cloud** in its growth stage and with its **vorticity** originating in the **boundary layer**.

The parent cloud does not contain a preexisting midlevel **mesocyclone**. **Landspouts** and **gustnadoes** are examples of the nonsupercell tornado.

Roberts, R. O., and J. W. Wilson, 1995: The genesis of three nonsupercell tornadoes observed with dual-Doppler radar. *Mon. Wea. Rev.*, **123**, 3408–3436.

nonviscous fluid—*See* **inviscid fluid**.

nonwetting liquid—The fluid in a pair of immiscible fluids that has a **contact angle** with the solid surface greater than 90°.

It is the fluid with the lower affinity for the solid surface.

nor'easter—Common contraction for **northeaster**.

See also **northeast storm**.

nor'wester—Contraction for **northwester**.

Nordenskjöld line—The line connecting all places at which the **mean temperature** (°C) of the warmest month is equal to $9 - 0.1k$, where k is the mean temperature of the coldest month. (In degrees Fahrenheit the relationship becomes $51.4 - 0.1k$.)

This line fits the **arctic tree line** better than any other purely climatic **isopleth** hitherto tried.

Hare, F. K., 1951: *Compendium of Meteorology*, p. 955.

normal—1. Referring to a **normal distribution**.

2. Regular or typical in the sense of lying within the limits of common occurrence, but sometimes denoting a unique value, as a measure of **central tendency**.

Either sense presupposes a stable **probability distribution**.

3. As usually used in meteorology, the average value of a **meteorological element** over any fixed period of years that is recognized as a **standard** for the country and element concerned.

Often erroneously interpreted by the general public as meaning the weather patterns that one should expect. In the broadest sense, "normals" should consist of a suite of descriptive **statistics**, including measures of **central tendency** (e.g., mean, **median**), **range** (e.g., **standard deviation**, **interquartile range**, **extremes**), **variation**, and **frequency** of occurrence. At the International Meteorological Conference at Warsaw in 1935, the years 1901–30 were selected as the international standard period for normals. Recommended international usage is to recalculate the normals at the end of every decade using the preceding 30 years. This practice is used to take account of the slow changes in **climate** and to add more recently established stations to the network with observed normals. Normals should be based on actual observations if available; otherwise a recognized method should be used to "reduce" shorter series to the normal period by comparison with neighboring stations. Recognized methods of adjusting for inhomogeneities should be used to account for breaks or gradual changes introduced into the data record by changes in the hours of **observation**, in the observational practices, in the **site** or instruments used, or by a gradual change in the character of the surrounding country, such as the growth of a city. The years covered by a normal should always be clearly stated, since averages for different periods of the same length are rarely the same. *See* **climatological standard normals**.

normal aeration—The complete renewal of **soil air** to a depth of 20 cm about once each hour.

normal barometer—A **barometer** of such **accuracy** that it can be used for the determination of **pressure** standards.
    An instrument such as a large-bore **mercury barometer** is usually used as a normal barometer.

normal chart—(Or normal map.) In meteorology, any **chart** that shows the distribution of the official **normal** values of a **meteorological element**.
    Compare **mean chart**.

normal circular distribution—*See* **normal distribution**.

normal curve of error—*See* **normal distribution, error distribution**.

normal dispersion—*See* **dispersion**.

normal distribution—(Or Gaussian distribution.) The **fundamental frequency** distribution of **statistical** analysis.
    A continuous **variate** x is said to have a normal distribution or to be normally distributed if it possesses a **density function** $f(x)$ that satisfies the equation

$$f(x) = \frac{1}{\sigma\sqrt{2\pi}}\, e^{-(x-\mu)^2/2\sigma^2} \ (-\infty < x < \infty),$$

where μ is the **arithmetic mean** (or first **moment**) and σ is the **standard deviation**. About two-thirds of the total area under the curve is included between $x = \mu - \sigma$ and $x = \mu + \sigma$. The corresponding **frequency distribution** of vectors is the **normal circular distribution** in which the frequencies of **vector** deviations are represented by a series of circles centered on a vector mean. When applied to **error distribution**, this function is the **normal law of errors**, and the distribution is called the **normal curve of error**. Although discovered by DeMoivre, the normal distribution is usually called the Gaussian distribution. In early anthropometric studies and also investigations of **random errors** in physical measurements, the variates exhibited the normal distribution so faithfully that this distribution was mistakenly assumed to be the governing principle of nearly all random phenomena and was therefore given the name "normal." While less universal than formerly believed, the normal distribution does have remarkable breadth of application, inasmuch as the distribution of averages computed from repeated random samples of almost any **population** tends more and more nearly toward the "normal" form as the **sample** size increases. Formulated in precise terms, this proposition is known as the **central limit theorem**.

normal functions—*See* **orthogonal functions**.

normal law of errors—*See* **normal distribution**.

normal map—Same as **normal chart**.

normal mode initialization—A method that uses **model** normal modes to adjust the initial conditions of an **NWP** model so that high-frequency oscillations are removed from the subsequent forecast.
    The most successful procedure is known as nonlinear normal mode initialization, in which the **normal modes** are used to **filter** out high frequencies from the time derivative of the initial conditions.
        Daley, R., 1991: *Atmospheric Data Analysis*, 263–321.

normal modes—A solution of the **tangent linear model** of a set of differential equations.
    Usually a **wave** type structure is assumed of the form $e^{ik(x-t)}$ and solutions with imaginary k are considered.

normal population—In **statistical** terminology, a collection of quantities having a **normal distribution**.

normal stresses—The components of the **stress tensor** that are normal to the faces of the fluid element.

normal temperature and pressure—(Abbreviated N.T.P.) Same as **standard temperature and pressure**.

normal water—(Also called Copenhagen water, standard seawater.) A **seawater** preparation, the **chlorinity** of which lies between 19.30 and 19.50 grams per kilogram (or **per mille**) and has been determined to within ±0.001 per mille.

Normal water is used as a convenient comparison **standard** for chlorinity measurements of seawater samples by titration. It is prepared by the Hydrographical Laboratories, Copenhagen, Denmark.

normalize—1. To change in **scale** so that the sum of squares, or the integral of the square, of the transformed quantity is unity.
See **orthogonal functions**.

2. To transform a **random variable** so that the resulting random variable has a **normal distribution**.

Normalized Difference Vegetation Index—(Abbreviated NDVI.) *See* **vegetation index**.

Normand's theorem—A theorem in **psychrometry** that states that the **dewpoint temperature** is always less than or equal to the **wet-bulb temperature**, which is always less than or equal to the **dry-bulb temperature**.
It is not, however, true in supersaturated air or for air temperatures below **freezing** if the wet bulb is frozen and the air is supersaturated with respect to **ice**.

nortada—A strong, persistent northerly **wind** in the Philippines.

norte—1. The winter north **wind** in Spain.

2. A strong cold northeasterly **wind** in Mexico and on the shores of the Gulf of Mexico.
It results from an **outbreak** of cold air from the north; actually, the Mexican extension of a **norther**. *See* **chocolatero**.

3. A **norther** in Central America.
See **papagayo**.

North American anticyclone—Same as **North American high**.

North American high—(Or North American anticyclone.) The relatively weak general **area of high pressure** that, as shown on mean charts of **sea level pressure**, covers most of North America during winter.
This **pressure system** is not nearly as well-defined as the analogous **Siberian high**.

North Atlantic Current—(Also known as North Atlantic Drift, West Wind Drift.) The eastward flowing **current** that originates from the **Gulf Stream Extension** east of the Grand Banks (about 40°N, 50°W).
It initially forms part of the Atlantic subtropical **gyre** but separates from it after less than 500 km near 45°W, turning northeastward and following the **Arctic Polar Front**, also known as the **North Wall**, with a **transport** of some 30 Sv (30×10^6 m³s⁻¹). Some of this water enters the subpolar gyre through **mixing** across the polar front and feeds the **Irminger Current**, but most of it is delivered to the **Norwegian Current**. The North Atlantic Current carries warm subtropical water much farther north than any other current of the Northern Hemisphere. As a result the **climate** of northern Europe is much milder than the climate of Alaska or northern Siberia, both of which are located at comparable latitude.

North Atlantic Deep Water— (Abbreviated NADW.) A **water mass** found in the Atlantic at depths between 1000 and 4000 m that can be traced from there into most other ocean basins.
It is formed in the North Atlantic from some 5 Sv (5×10^6 m³s⁻¹) of Atlantic Bottom Water entering through Denmark Strait and across the Scotland–Faeroe–Iceland Ridge. This water flows toward the Labrador Sea, entraining another 5 Sv from the eastern North Atlantic on its way. Another 5 Sv is added in the Labrador Sea by winter convection, giving a total of 15 Sv of NADW formation. The NADW formation process is the engine of the **ocean conveyor belt**, which makes NADW one of the most important water masses for today's **climate**.

North Atlantic Drift—*See* **North Atlantic Current**.

North Atlantic oscillation—*See* **oscillation**.

North Brazil Current—One of the western **boundary currents** of the Atlantic Ocean and part of the pathway for water from the Southern into the Northern Hemisphere in the global **ocean conveyor belt**.
Flowing northward along the coast of northern Brazil with speeds up to 0.8 m s⁻¹, it originates from the **South Equatorial Current** and continues as the **Guyana Current**.

North Equatorial Countercurrent—(Abbreviated NECC.) A band of eastward flow between the westward flowing North and South Equatorial Currents.
The location and strength of the NECC is determined by the **intertropical convergence zone**

(ITCZ) of the **atmosphere**. In the Pacific Ocean it is strongest in May–January when it flows between 5° and 10°N with 0.4–0.6 m s⁻¹; in February–April it is restricted to 4°–6°N with speeds below 0.2 m s⁻¹ and disappears east of 110°W. In the Atlantic Ocean the NECC is observed between 5° and 10°N with speeds of 0.1–0.3 m s⁻¹; it is strongest during August when it flows from South America into the Gulf of Guinea and weakest in February when it is restricted to the region east of 20°W. In the Indian Ocean the NECC exists during the northeast **monsoon** season only and is then the only **countercurrent**; it is therefore mostly called the **Equatorial Countercurrent**. It is centered on 5°S, again the location of the ITCZ.

North Equatorial Current—(Abbreviated NEC.) The broad region of uniform westward flow that forms the southern part of the Northern Hemisphere subtropical **gyres** driven by the **trade winds**.

Being directly **wind** driven, the NEC responds quickly to variations in the **wind field** and is therefore strongest in winter (February). In the Atlantic Ocean it is found between 8° and 30°N with speeds of 0.1–0.3 m s⁻¹. In the Pacific Ocean it has similar speed but is limited to 8°–20°N. In the Indian Ocean it exists only during December–April when the northeast **monsoon** produces the same wind forcing as the Northern Hemisphere trade winds. It then runs as a narrow **current** of 0.3 m s⁻¹ from Malacca Strait to Sri Lanka where it bends southward and accelerates in the region 60°–75°E to 0.5–0.8 m s⁻¹ between 2°S and 5°N and continues along the **equator**.

north foehn—A northerly **foehn** wind blowing down the Italian side of the Alps.

The northern slopes are normally cooler than the southern slopes, and the dynamic warming is often insufficient to overcome the difference of **temperature**. Hence a warm dry northerly **wind** of foehnlike character occurs less frequently than the **south foehn**.

North Frigid Zone—*See* **Frigid Zone**.

North Korea Cold Current—A southward flowing **current** along the northern Korean coast fed from the **Liman Current**.

At 37°–38°N it meets the East Korean Warm Current to establish the **Arctic Polar Front** of the Japan Sea. It is the **source** of the **Mid-Japan Sea Cold Current**, which withdraws water from it along its way to reinforce the polar front across the entire length of the Japan Sea.

North Pacific Current—(Also known as the West Wind Drift.) The eastward **current** that forms the northern part of the North Pacific subtropical **gyre**.

It originates from the **Kuroshio Extension** east of the Emperor Seamounts (170°W) and maintains the **Arctic Polar Front** with the Aleutian or **Pacific Subarctic Current**, with which it experiences much **mixing** as it proceeds eastward. A broad band of flow some 2000 km wide, it feeds its water into the **California Current** on approaching the North American coast.

North Pacific oscillation—*See* **oscillation**.

North Wall—(Also called cold wall.) The steep water-temperature **gradient** between the **Gulf Stream** and 1) the **Slope Water** inshore of the Gulf Stream or 2) the **Labrador Current** north of the Gulf Stream.

northeast storm—(Also called **northeaster**, nor'easter.) A **cyclonic** storm of the east coast of North America, so called because the winds over the coastal area are from the northeast.

They may occur at any time of year but are most frequent and most violent between September and April. Northeast storms usually develop in lower–middle latitudes (30°–40°N) within 100 miles east or west of the coastline. They progress generally northward to northeastward and typically attain maximum intensity near New England and the Maritime Provinces. They nearly always bring **precipitation**, winds of **gale** force, rough seas, and, occasionally, coastal **flooding** to the affected regions.

northeast trades—The **trade winds** of the Northern Hemisphere.

northeaster—(Commonly contracted nor'easter.) A northeast **wind**, particularly a strong wind or **gale**.

Two well-known examples are the **black northeaster** of Australia and New Zealand and the **northeast storm** of the east coast of North America.

norther—A northerly **wind**; in general, a cold windstorm from the north.

The term has several specific applications: 1) In the southern United States, especially in Texas (Texas norther), in the Gulf of Mexico, in the Gulf of Panama away from the coast, and in Central America (**norte**), the norther is a strong cold wind from between northeast and northwest. It occurs between November and April, freshening during the afternoon and decreasing at night. It is a cold air outbreak associated with the southward movement of a **cold anticyclone**. It is usually preceded by a warm and **cloudy** or rainy spell with southerly winds. The norther comes as a rushing blast and brings a sudden drop of **temperature** of as much as 25°F in one hour or 50°F in three hours in winter. 2) The **California norther** is a strong, very dry, dusty, northerly wind that blows in late spring, summer, and early fall in the valley of California or on the West Coast when **pressure** is high over the mountains to the north. It lasts from one to four days. The dryness is due to **adiabatic** warming during descent. In summer it is very hot. 3) The Portuguese norther is the beginning of the **trade** wind west of Portugal. 4) Norther is used for a strong north wind on the coast of Chile that blows occasionally in summer. 5) In southeast Australia, a hot dry wind from the **desert** is called a norther. *See also* **chocolatero, tehuantepecer**; *compare* **burster, pampero**.

northern lights—Same as **aurora borealis**; *see* **aurora**.

northern nanny—A cold **storm** of **hail** and **wind** from the north in England.

northwester—(Often contracted nor'wester.) A northwesterly **wind** (as **Canterbury northwester**). *See also* **kal Baisakhi**.

Norwegian Current—A warm **current** flowing northeastward along the Norwegian coast.
It is the continuation of the **North Atlantic Current** and discharges about 10 Sv (10×10^6 m³ s⁻¹) into the Arctic Ocean. Because its waters are of subtropical origin it has a significant impact on the **climate** of northern Europe; ports in northern Norway located at 70°N are ice-free throughout the year.

Norwegian Sea Deep Water—One of the **water masses**, with a **temperature** of −0.95°C, that contributes to the formation of **Arctic Bottom Water**.

NOTAM—The aviation communications code word for "notice to airmen."
Compare **GENOT**.

Notos—The Greek name for the south **wind**, sultry and rainy.
On the **Tower of the Winds** in Athens it is represented by a lightly clad young man carrying an inverted jar from which water is falling.

Nova Zemlya—This phenomenon owes its name to an event in 1596 when explorers (in search of the northeast passage) wintering on the island of Nova Zemlya saw a distorted **image** of the sun two weeks before astronomical calculation would have had it rise.
Since that time, the term has been used generically for any such observation of an image of the sun when the actual sun was substantially below the **horizon**. In the original case, the angular difference between the image and the object was 5°. This is explained by the ducting of the **sunlight** between the surface and a lifted **inversion**.

nowcast—A short-term **weather forecast**, generally for the next few hours.
The U.S. National Weather Service specifies zero to three hours, though up to six hours may be used by some. *Compare* **short-range forecast, very short-range forecast, long-range forecast**.

NO$_x$—Abbreviation for various oxides of **nitrogen**, including NO (**nitric oxide**), NO₂ (**nitrogen dioxide**), or N₂O (**nitrous oxide**).
These gaseous pollutants can be transformed in the presence of water to **nitric acid** HNO_3, one of the acids in **acid deposition**. They are formed anthropogenically inside combustion engines, where the ambient nitrogen N_2 and **oxygen** O_2 gases in the **atmosphere** can react at high temperatures. They are also produced naturally by **lightning** and by biological activity in soils. *See* **active nitrogen**.

Noxon cliff—Phenomenon found in the **upper atmosphere** during winter and early spring in which the total column amount of **active nitrogen** is reduced due to the removal of **nitric acid**.
The mechanism is closely related to **denitrification** in polar regions.

NO$_y$—*See* **total reactive nitrogen**.

NRM—Abbreviation for **natural remanent magnetization**.

NRM wind scale—A **wind scale** adapted by the U.S. Forest Service for use in the forested areas of the northern Rocky Mountains (sometimes abbreviated NRM).
It is an adaptation of the **Beaufort wind scale**. The difference between these two scales lies in the specification of the visual effects of **wind**; the force numbers and the corresponding wind speeds are the same in both.

NSCAT—Acronym for **NASA Scatterometer**.

nT—The nanotesler (nT); a measure of geomagnetic **field strength**.
There are 10^9 nanoteslers in a tesler, which is the unit of measure of **magnetic field strength**

in the **SI** system of units. An nT is equal to a gamma. There are 10^5 gammas in an **oersted**, which is the unit of measure of magnetic field strength in the **cgs system**.

nuclear winter—A term used to describe the surface cooling that might result from the emission of extensive clouds of **smoke** (from burning cities, fuel sites, and forests) following the detonation of hundreds of warheads in a nuclear exchange.

nucleation—The process of initiation of a new **phase** in a supercooled (for liquid) or supersaturated (for solution or **vapor**) **environment**; the initiation of a **phase change** of a substance to a lower thermodynamic **energy** state (vapor to liquid **condensation**, vapor to solid **deposition**, liquid to solid **freezing**).

In nature, heterogeneous nucleation is the more common where such a change takes place on small **particles** of different composition and structure. Homogeneous nucleation occurs when the **change of state** centers upon embryos that exist in the same initial state as the changing substance. In this case, the nucleation system contains only one component, and it is termed homogeneous nucleation. In meteorology, particularly in **cloud physics**, a number of types of nucleation are of interest. The process by which **cloud condensation nuclei** initiate the phase change from vapor to liquid is important in all **cloud formation** problems. The physical nature of **freezing nuclei** that may be responsible for the conversion of drops of **supercooled water** into **ice crystals** is critically important in **precipitation** theory, as is the clarification of the role of homogeneous nucleation near $-40°C$. Thermodynamically, all nucleation processes involve free energy decrease associated with the bulk phase change and the free energy increase associated with the creation of new interfaces between phases.

nuclei counter—*See* **nucleus counter**.

nucleus—1. The positively charged core of an atom about which its **electrons** orbit.

Almost all the mass of an atom resides in its nucleus, the diameter of which is about 10^4 times smaller than that of the atom. The nucleus of an (electrically neutral) atom is made up of **protons**, equal in number to its electrons, and **neutrons** bound together by nuclear forces.

2. In **physical meteorology**, a **particle** of any nature that initiates a **phase** transition in an **environment** supersaturated or supercooled with respect to a phase with lower chemical (**Gibbs**) potential, for example, a solid or liquid particle or gas/**vapor** bubble in a supercooled/supersaturated environment.

See **Aitken nucleus**.

nucleus counter—1. Any of several devices for determining the number of **condensation nuclei** or **ice nuclei** in a sample of air.

2. Same as **dust counter**.

nudging—A **four-dimensional data assimilation** method that uses dynamical relaxation to adjust toward **observations** (observation nudging) or toward an **analysis** (analysis nudging).

Nudging is accomplished through the inclusion of a forcing term in the **model** dynamics, with a tunable coefficient that represents the **relaxation time** scale. Computationally inexpensive, nudging is based on **heuristic** and physical considerations.

numerical dispersion—The separation of different Fourier components of a **finite-difference approximation** into a train of oscillations that travel with different speeds.

Numerical dispersion occurs whenever the **dispersion relation** for the difference approximation is **nonlinear**.

numerical forecasting—(Also called mathematical forecasting, dynamical forecasting, **physical forecasting**, numerical weather prediction.) The integration of the governing equations of **hydrodynamics** by numerical methods subject to specified initial conditions.

Numerical approximations are fundamental to almost all dynamical weather prediction schemes since the complexity and nonlinearity of the hydrodynamic equations do not allow exact solutions of the continuous equations. *See* **numerical integration**, **numerical simulation**.

Haltiner, G. J., and R. T. Williams, 1980: *Numerical Prediction and Dynamic Meteorology*, Wiley, New York, 477 pp.

numerical instability—*See* **stability**.

numerical integration—1. The integration of an analytical expression or of discrete or continuous data by approximate numerical methods.

These methods usually involve fitting sample curves to successive groupings or sets of the data and performing the integration step-wise.

2. A solution of the governing equations of **hydrodynamics** by numerical methods.

The numerical solutions are carried out with the aid of computers ranging from desktop workstations to the most powerful computers available. The latter are required, in particular, for the timely production of operational global **weather forecasts**. *See* **finite differencing, finite-element model, spectral model.**

numerical modeling—In **oceanography**, the **prediction** of flow evolution via numerical construction of approximate solutions to the governing equations.

Solutions are obtained by assigning discrete values to temporal and spatial derivatives in order to convert the governing differential equations into algebraic equations that can be solved by using computational methods. Because computational resources are finite, no one technique is ideal for all applications. Some models define the equations on very fine spatial intervals (*see* **direct numerical simulation**). This approach furnishes solutions that are very accurate, but that span only small spatial regions (spatial scales of a few meters, at present). At the other extreme, some models span entire ocean basins by using large spatial intervals (hundreds of kilometers). Here, approximation of unresolved motions is a crucial and difficult issue (*see* **very large eddy simulation**). Similar trade-offs must be made with respect to temporal solutions. Numerical models also differ in the equations and **boundary conditions** that are employed. The most general **model** commonly used in oceanography includes **momentum** conservation via the incompressible **Navier–Stokes equations** with the **Boussinesq approximation**, mass conservation via the **incompressibility** condition, and equations expressing conservation of **heat** energy and salt (e.g., Gill 1982). For large-scale applications, the **hydrostatic approximation** is usually made. The vertical coordinate may be the geometric height, or a convenient substitute such as **density, pressure, logarithm** of pressure, or **potential temperature**. Surface boundary conditions generally express fluxes of momentum, **heat**, and **freshwater** from the **atmosphere**. Basin-scale models use boundary conditions that approximate the effects of bottom **topography**. Smaller-scale models typically specify periodic conditions at the side boundaries and an **energy** radiation condition at the bottom. *See also* **column model, mixed layer models, coupled model.**

Gill, A. E., 1982: *Atmosphere–Ocean Dynamics*, Academic Press.

numerical simulation—A **numerical integration** in which the goal is typically to study the behavior of the solution without regard to the initial conditions (to distinguish it from a numerical forecast).

Thus the integrations are usually for extended periods to allow the solution to become effectively independent of the initial conditions.

numerical weather prediction—Same as **numerical forecasting**.

Nusselt number—(Sometimes called **heat transfer coefficient**.) A **nondimensional number** arising in the problem of **heat transfer** in fluids.

It may be written

$$\text{Nu} = QL/kS\Delta T,$$

where Q is the quantity of **heat** transferred in unit time from an immersed body across an area S, ΔT a characteristic **temperature** difference, L a characteristic **length**, and k the **thermal conductivity**. With suitable choices of characteristic quantities, it may be interpreted as a ratio of heat actually transferred to heat that would be transferred under the circumstances by pure **conduction**.

NWP—A popular abbreviation for numerical weather prediction.
See **numerical forecasting**.

Nyquist frequency—(Also called **turnover frequency**.) The highest **frequency** that can be determined in a **Fourier analysis** of a discrete sampling of data.

If a **time series** is sampled at interval Δt, this frequency is $1/2\Delta t$ cps.

O

O'Brien cubic polynomial—An approximation for the **eddy diffusivity** K as a function of height z in a **boundary layer** of depth h with **surface layer** (SL) of depth z_{SL}:

$$K(z) = K(h) + \left[\frac{h-z}{h-z_{SL}}\right]^2 \left\{K(z_{SL}) - K(h) + (z - z_{SL})\left[\frac{\partial K}{\partial z}\bigg|_{SL} + 2\frac{K(z_{SL}) - K(h)}{h - z_{SL}}\right]\right\},$$

where the tunable parameters in the equation are eddy diffusivities at the top of the surface layer and at the top of the boundary layer, $K(z_{SL})$ and $K(h)$, respectively, the heights of those two layers, and the **gradient** of eddy diffusivity at the top of the surface layer $\partial K/\partial z$.

Above the top of the boundary layer the eddy diffusivity is assumed to be constant at $K(h)$, while at the surface it is assumed to be zero. *See* **K-theory, gradient transport theory, first-order closure, closure assumptions**.

O'Brien, J. J., 1970: A note on the vertical structure of the eddy exchange coefficient in the planetary boundary layer. *J. Atmos. Sci.*, **27**, 1213–1215.

oasis effect—Evaporative cooling effect due to **heat** advection when a **source** of water exists in an otherwise arid area.

In addition to true **desert** oases, the oasis effect is also characteristic of natural bodies of water in arid surroundings, melting **snow** patches, irrigated fields in arid areas, or irrigated urban lawns and parks. Latent **heat flux** from such an oasis can exceed the locally available radiative **flux** twofold; **advection** of **sensible heat** from surrounding warmer surfaces and airmass **subsidence** over the cooler area provides the remainder. **Evaporation** also exceeds the local **precipitation**, the extra water coming from wells, river flow, and **irrigation**.

ob—In meteorology, a commonly used abbreviation for **weather observation**.

oberwind—A **night wind** from mountains or the upper ends of lakes; a **wind** of Salzkammergut in Austria.

objective analysis—An **analysis** that is free from any direct subjective influences resulting from human experience, interpretation, or bias.

See **objective forecast**.

objective forecast—Typically used to denote that a forecast is free from any direct subjective influences resulting from human experience, interpretation, or bias.

Examples are **numerical** and **statistical forecasts**. *See* **subjective forecast**.

oblique Cartesian coordinates—*See* **Cartesian coordinates**.

oblique visibility—Same as **oblique visual range**.

oblique visual range—(Also called oblique visibility, slant visibility.) The greatest distance at which a specified **target** can be perceived when viewed along a **line of sight** inclined to the horizontal.

One must distinguish upward from downward oblique visual range because of the quite different background **luminance** prevailing in the two cases. Furthermore, a **range** can only be considered with respect to some given type of target, as is also true of the ordinary (horizontal) **visual range**. In view of the great importance of the downward oblique visual range in air to ground visual contact and the upward oblique visual range in visual detection of aircraft, it is unfortunate that no satisfactory theory for this has yet been developed. The principal obstacle in treating this problem lies in the typically nonuniform height variation of the **extinction coefficient**.

obliquity of the ecliptic—The angle between the **plane of the ecliptic** (or the plane of the earth's **orbit**) and the plane of the earth's **equator**; the "tilt" of the earth.

The obliquity of the ecliptic is computed from the following formula:

$$23°27'08.26'' - 0.4684\,(t - 1900)'',$$

where t is the year for which the obliquity is desired. For 1999, the value was 23°26′21.89″. It is the oblique orientation of the earth's axis relative to its orbit that accounts for the seasons, for, in the period of a year, the **angle of incidence** of **incoming solar radiation** varies by nearly 47° at any one place. Particularly at high latitudes, this results in a great seasonal **temperature** contrast. M. Milankovitch has calculated that the obliquity of the ecliptic varies between 24.5° and 22° in the course of 40 000 years. This **variation** may be considered as a long-period **climatic control** and is included in the astronomical theory of **ice ages**.

obscuration—1. (Also called obscured sky cover.) In U.S. weather observing practice, the designation for the **sky cover** when the sky is completely hidden by surface-based obscuring phenomena.

It is encoded "X" in aviation weather observations; it always constitutes a **ceiling**, the height of which is the value of **vertical visibility** into the **obscuring phenomenon**. *Compare* **partial obscuration**.

2. A surface-based **obscuring phenomenon**.

obscured sky cover—Same as **obscuration**.

obscuring phenomenon—(Also called **obscuration**.) Any collection of **particles**, aloft or in contact with the earth's surface, dense enough to be discernible to the **observer**.

Examples are **haze**, **dust**, **smoke**, **fog** or **ice fog**, spray or **mist**, **drifting** or **blowing snow**, **duststorms** or **sandstorms**, **dust whirls** or **sand whirls**, and volcanic **ash**. Potentially, all **hydrometeors** and **lithometeors** may be obscuring phenomena.

observation—*See* **weather observation**.

observation well—A well drilled into an **aquifer** for the purpose of obtaining **water level**, water temperature, or water quality data.

observational day—Any 24-hour period selected as the basis for climatologic or hydrologic observations.

Where observations are recorded automatically, the observational day is commonly taken as the calendar day. Where only one **observation** is made in 24 hours, the observational day is assumed to be the 24 hours ending at the time of observation.

observational error—The difference between the true value of some quantity and its observed value.

Every **observation** is subject to certain errors, as follows. 1) **Systematic errors** affect the whole of a series of observations in nearly the same way. For example, the **scale** of an instrument may be out of adjustment. These instrument errors can be detected and corrected by comparison with a **standard**. The **personal equation** of an **observer** may lead him to make small systematic errors in his readings; for example, if the scale is not at eye level. 2) **Random errors**, which appear in any series of observations, are generally small and as likely to be positive as negative; their magnitudes are usually distributed according to the **error distribution**. 3) Mistakes are widely discrepant readings.

observational network—A group of **meteorological observing stations** spread over a given area for a specific purpose.

observer—In meteorology, anyone who takes a **weather observation**.

observing systems simulation experiment—(Abbreviated OSSE.) An experiment in which a **model** is assumed to represent the exact behavior of the physical system (the **atmosphere**) of interest.

The model results are then sampled in a fashion to mimic some real or hypothetical observing system. OSSEs are an inexpensive way to guide the design of new or enhanced observing systems or to evaluate the usefulness of existing observations.

obstacle flow signature—The **Doppler velocity** signature of middle-altitude flow around the strong **updraft** region of a severe **thunderstorm**. The **signature** indicates stronger flow around the edges of the updraft region and weaker (or reversed) flow in the **wake** region immediately **downwind**.

Though present for all **radar** viewing directions, it is most pronounced when a significant component of **storm** motion is toward or away from the radar.

obstruction to vision—In U.S. weather observing practice, one of a class of atmospheric phenomena, other than the weather class of phenomena, that may reduce **horizontal visibility** at the earth's surface.

Obstructions to vision are listed in the Manual of Surface Observations (**WBAN**), Circular N, as follows: **fog**, **ground fog**, **blowing snow**, **blowing sand**, **blowing dust**, **ice fog**, **haze**, **smoke**, **dust** and **blowing spray**. These are encoded as a part of an **aviation weather observation**.

Obukhov length—A **parameter** with dimension of length that gives a relation between parameters characterizing dynamic, **thermal**, and buoyant processes.

At altitudes below this length scale, **shear production** of **turbulence kinetic energy** dominates over buoyant production of **turbulence**. It is defined by

$$L = \frac{-u_*^3 T_v}{kg Q_{v0}},$$

where k is **von Kármán's constant**, u_* is the **friction velocity** (a measure of turbulent surface **stress**), g is gravitational **acceleration**, T_v is **virtual temperature**, and Q_{v0} is a kinematic virtual temperature flux at the surface. The parameter was first described by Obukhov in 1946, and therefore should not be called the Monin–Obukhov length, even though there is a **Monin–Obukhov similarity theory** that uses it. The Obukhov length, of order one to tens of meters, is the characteristic height scale of the dynamic sublayer. The Obukhov length is zero for neutral **stratification**, and positive (negative) for stable (unstable) stratifications. The dimensionless Obukhov length z/L (where z is height above the surface) is used as a **stability parameter**, with $z/L = 0$ for statically **neutral stability**, and is positive (negative) in a typical **range** of 1 to 5 (-5 to -1) for stable (unstable) **stratification**. *Compare* **similarity theory, dimensional analysis, Buckingham Pi theory, surface layer.**

occluded cyclone—Any **cyclone** (or **low**) within which there is little or no **temperature** advection; typically associated with the formation of an **occluded front**.

occluded depression—A **depression** in which there has developed an **occluded front**.

occluded front—(Commonly called **occlusion**; also called frontal occlusion.) A **front** that forms as a **cyclone** moves deeper into colder air.
  This front will separate air behind the **cold front** from air ahead of the **warm front**. This is a common process in the late stages of wave-cyclone **development**, but is not limited to occurrence within a **wave cyclone**. There are three basic types of occluded front, determined by the relative coldness of the air behind the original cold front to the air ahead of the warm (or stationary) front. 1) A **cold occlusion** results when the coldest air is behind the cold front. The cold front undercuts the warm front and, at the earth's surface, coldest air replaces less cold air. 2) When the coldest air lies ahead of the warm front, a **warm occlusion** is formed, in which case the original cold front is forced aloft at the warm front surface. At the earth's surface, coldest air is replaced by less cold air. 3) A third and frequent type, a **neutral occlusion**, results when there is no appreciable **temperature** difference between the cold air masses of the cold and warm fronts. In this case frontal characteristics at the earth's surface consist mainly of a pressure **trough**, a **wind-shift line**, and a band of **cloudiness** and **precipitation**. *See* **bent-back occlusion.**

occlusion—1. (Or frontal occlusion.) In meteorology, the process of formation of an **occluded front**.
  Some persons restrict the use of this term to the usual case where the process begins at the apex of a **wave cyclone**; when the process begins at some distance from the apex, they call it **seclusion**.
  2. Same as **occluded front**.

ocean—1. The intercommunicating body of saltwater occupying the depressions of the earth's surface.
  2. One of the major primary subdivisions of the above, bounded by continents, the **equator**, and other imaginary lines.
  See **sea**.

ocean conveyor belt—The global **recirculation** of **water masses** that determines today's **climate**.
  The conveyor belt is driven by the sinking of **North Atlantic Deep Water** (NADW) through cooling of the **surface water** in the Greenland and Labrador Seas. NADW flows southward through the Atlantic below a depth of 3000 m. When it reaches the **Antarctic Circumpolar Current** (ACC), some of it continues into the Indian and Pacific Oceans at depth, enters the Atlantic through the Drake Passage and returns to the North Atlantic. Most NADW, however, rises very close to the surface in the ACC, where it freshens considerably through contact with surface water and enters all three oceans as **Antarctic Intermediate Water** at depths of 700–1000 m. Antarctic Intermediate Water penetrates into the Northern Hemisphere, being slowly entrained by **central water**, the water mass above it. Pacific central water enters the Indian Ocean through the Indonesian Seas. It then joins Indian central water to flow eastward and then southward in the subtropical **gyre**. **Agulhas Current** eddies carry it into the Atlantic, where it moves northward with the **Benguela** and **Brazil Currents** and in the **Gulf Stream** system toward the Greenland and Labrador Seas to cool and sink again, thus completing the conveyor belt circulation.

ocean current—A movement of ocean water characterized by regularity, either of a cyclic nature or, more commonly, as a continuous stream flowing along a definable path.
  Three general classes, by cause, may be distinguished: 1) currents related to **seawater** density gradients, comprising the various types of **gradient current**; 2) wind-driven currents, which are those directly produced by the **stress** exerted by the **wind** upon the ocean surface; and 3) currents produced by long-wave motions. The latter are principally tidal currents, but may include currents associated with **internal waves**, **tsunamis**, and **seiches**. The major ocean currents are of continuous, stream-flow character, and are of first-order importance in the maintenance of the earth's thermodynamic balance.

ocean mixing—Any process or series of processes by which parcels of ocean water with different properties are brought into intimate small-scale contact, so that **molecular diffusion** erases the differences between them.

There is a distinction between stirring, which moves the water parcels into intimate contact, and **mixing**, the final process of molecular diffusion that blends the water parcels together. The term "mixing" is currently used to describe all of the processes, including molecular diffusion.

ocean station—As defined by the International Civil Aviation Organization, a specifically located area of ocean surface, roughly square, and 200 nautical miles on a side.

An **ocean station vessel** on patrol is said to be "on station" when it is within the perimeter of the area. *Compare* **ocean weather station**; *see* **station**.

ocean station vessel—(Abbreviated OSV.) An oceangoing vessel assigned to patrol an **ocean station**.

These ships are specially equipped to take comprehensive **meteorological observations** of conditions both at the surface and aloft. The U.S. vessels are provided by the U.S. Coast Guard, and the meteorological personnel and equipment are provided by the National Weather Service.

Ocean Topography Experiment—(Acronym TOPEX.) A cooperative project between the United States and France to develop an advanced satellite system dedicated to observing the earth's oceans.

The TOPEX/Poseidon satellite, launched in September 1992, uses sophisticated altimeters to make **sea level measurements** in combination with a **microwave** radiometer to correct the **altimeter** measurements for changes in total **water vapor** content of the **atmosphere**, resulting in an **absolute** accuracy of about 4 cm.

ocean waves—Waves, on the surface of the ocean, generated by **wind** action.

Their dynamics are governed by the influence of **gravity** and, for short waves of **wavelength** less than about 20 cm, **surface tension**. The waves are dispersive, with **angular frequency**, ω, being given by

$$\omega = (gk + Tk^3)^{1/2},$$

where g is the **acceleration** due to gravity, $k = 2\pi/\lambda$ is the **wavenumber**, λ is the wavelength, and T is the surface tension divided by the water density.

ocean weather ship—Same as **ocean station vessel**.

ocean weather station—As defined by the World Meteorological Organization, a specific maritime location occupied by a ship equipped and staffed to observe weather and sea conditions and report the observations by international exchange.

Compare **ocean station**; *see* **station**.

oceanic anticyclone—Same as **subtropical high**.

oceanic climate—Same as **marine climate**.

oceanic high—Same as **subtropical high**.

oceanic meteorology—Study of the interaction between the sea and the **atmosphere**.

oceanic mixed layer—*See* **mixed layer**.

oceanic surface mixed layer—*See* **mixed layer**.

oceanicity—(Or oceanity.) The degree to which a point on the earth's surface is in all respects subject to the influence of the sea; the opposite of **continentality**.

Oceanicity usually refers to **climate** and its effects. One measure for this characteristic is the ratio of the frequencies of **maritime** to **continental** types of **air mass**.

oceanity—Same as **oceanicity**.

oceanography—The study of the sea, embracing and integrating all knowledge pertaining to the sea's physical boundaries, the chemistry and physics of **seawater**, and marine biology.

octa— A fraction equal to one-eighth of the celestial dome, used in the coding of **cloud** amounts in **synoptic** meteorological observations.

octane—Family of alkane molecules, formula C_8H_{18}.

Several different isomers exist with different structures but the same chemical formula. One of the isomers, 2,3,4-trimethylpentane, is the **standard** for knock properties in automobile engines. The octane rating measures the percentage of 2,3,4-trimethylpentane in a mixture with heptane that would give the same knock characteristics as the fuel under test.

odd chlorine—A collective term for the suite of inorganic **chlorine compounds** found in the at-

mosphere, including chlorine atoms (Cl), chlorine monoxide (ClO), **hydrogen chloride** (HCl), **chlorine nitrate** ($ClNO_3$), **hypochlorous acid** (HOCl), and **chlorine monoxide dimer** (Cl_2O_2). The interconversion of these compounds in the **stratosphere** leads to **ozone** depletion.

odd hydrogen—Chemical family, comprising **hydrogen** atoms (H), hydroxyl radicals (OH), and hydroperoxyl radicals (HO_2), that participates in **ozone** destruction in the **stratosphere**.

Due to the rapid interconversion of these three species, they are normally treated as one lumped species and partitioned according to the local chemical **environment**.

odd nitrogen—*See* **active nitrogen**.

odd oxygen—The sum of the concentrations of **oxygen** atoms and **ozone** (O_3).

In the **lower atmosphere**, the concentration of O_3 greatly exceeds that of O, and the odd oxygen level can be approximated by the O_3 concentration. *See* **Chapman mechanism**.

odd-oxygen system—The chemical species that contain an odd number of **oxygen** atoms, that is, chiefly atomic oxygen (O) and **ozone** (O_3).

oe—A violent **whirlwind** of the Faeroe Island region (north of the British Isles).

oersted—A unit of **magnetic field intensity** equal to one **dyne** per cgs unit magnetic **pole**.
See **gauss**.

off-airways—Any aircraft **course** or track that does not lie within the bounds of prescribed airways.

off-center PPI scope—An obsolete type of **radar** display in which the center of a **PPI** display is offset from the geometrical center of the **oscilloscope**.

offshore wind—Wind blowing from land to sea.

During **synoptic** conditions of light winds, offshore winds near the surface often occur at night as a component of the **land breeze**.

Ogasawara high—A lower tropospheric **anticyclone** in the **subtropics** of the western Pacific that often occurs during summer and influences conditions over Japan.

Named after the Ogasawara Islands (27°N, 141°E), over which it is often centered.

ogive—1. A band or **wave** on the surface of a **valley glacier**, stretching from side to side and arched in the direction of flow.

2. The graph of a cumulative **frequency distribution**.

ohmic current—A **current** that is proportional to the voltage and inversely proportional to the **resistance**.

oil slick—A layer of oil on the water surface, usually as a result of an accidental or deliberate spill of crude oil or other petroleum product.

The thickness of an oil slick can range from a few molecules thick up to many millimeters. Thin oil slicks appear as a "blue sheen", due to optical **interference** effects. Even when very thin, slicks cause significant **damping** of centimeter-scale surface waves and thus appear as low-backscatter regions in airborne or satellite radar images. After spillage, oil slicks spread by the effects of the winds, tides, **gravity**, **surface tension**, and **ocean current** shear and **turbulence**. They eventually disperse by **evaporation**, dissolution, and downmixing by breaking waves. Wave action can also lead to the formation of oil-in-water emulsions.

old snow—Deposited **snow** in which the original crystalline forms are no longer recognizable, such as **firn**, **spring snow**.

Old Wives' summer—A period of **calm**, **clear** weather, with cold nights and misty mornings but fine warm days, which sets in over central Europe toward the end of September; comparable to **Indian summer**.

It has been explained as a transition between the summer and winter pressure types. In summer, central Europe is dominated by the **Azores high**, from which a **wedge** of high pressure extends to southwestern Germany. In winter, the dominant feature is the **Siberian high**, from which a **ridge** extends across Switzerland. Between these two stages there is often a period, on the average occurring between 18 and 22 September, during which an independent **anticyclone** forms over Germany. As this gradually drifts away eastward, the Old Wives' summer tends to be delayed until October in the western part of the former Soviet Union. The term itself probably stems from the widespread existence of "old wives' tales" concerning this striking feature of autumn weather.

Compare **St. Luke's summer**, **St. Martin's summer**.

olefins—*See* **alkenes**.

Olland cycle—A modulating system used in some forms of **chronometric radiosonde**.

The meteorological transducing elements are so designed that each of them controls an electrical contact arm that sweeps a sector of a rotating wire helix or spiral. During a rotation of the helix or spiral each electrical contact momentarily touches the wire at a point in the **cycle** determined by the value of the meteorological **parameter** being measured. The chronometric pulses are translated into their meteorological equivalents at the receiving **station** by a **chronograph** operating in synchronism with the rotation of the helix or spiral.

OLS—Abbreviation for **Operational Linescan System**.

ombrometer—1. (Also called **micropluviometer**, trace recorder.) Specifically, a **rain gauge** capable of measuring very small amounts of **precipitation**.

2. Any **rain gauge**.

ombroscope—An instrument that indicates the presence of **precipitation**.

The ombroscope consists of a heated, water-sensitive surface that indicates by mechanical or electrical techniques the occurrence of precipitation. The **output** of the instrument may be arranged to trip an alarm, to record on a time chart, to raise the top on a convertible automobile, etc.

omega equation—A **diagnostic equation** by which the **vertical velocity** in **pressure coordinates** ($\omega = Dp/Dt$) may be calculated according to **quasigeostrophic theory**:

$$\sigma \nabla_H^2 \omega + f^2 \frac{\partial^2 \omega}{\partial p^2} = f \frac{\partial}{\partial p}\left[\mathbf{v}_g \cdot \nabla_H \left(f + \zeta_g\right)\right] - \nabla_H^2\left(\mathbf{u}_g \cdot \nabla_H \frac{\partial \phi}{\partial p}\right),$$

for which f is the **Coriolis parameter**, σ is the **static stability**, $\mathbf{v}_g$ is the **geostrophic** velocity vector, ζ_g is the geostrophic **relative vorticity**, ϕ is the **geopotential**, ∇_H^2 is the horizontal **Laplacian operator**, and ∇_H is the horizontal **del operator**.

The right-hand side of the omega equation can also be expressed in terms of the **divergence** of the **Q vector**.

Holton, J. R., 1992: *An Introduction to Dynamic Meteorology*, 3d edition, Academic Press, 166–175.

omega sun—A two-image **inferior mirage** of the setting or rising sun in which the erect **image** of the solar disc, which forms the top of the omega, touches the inverted segment of the lower image, which forms the base of the omega.

When the sun is even lower, all that is seen are two images of the upper segment of the sun floating back to back above the optical **horizon** and having an outline similar to that of an American football. The stage at which only two images of the upper rim are seen is one way in which the **green flash** is formed.

on and off instruments—A description used by aircraft crews for flight conditions wherein the encountered flight weather alternately allows **visual flight** and requires **instrument flight**.

on instruments—Same as **instrument flight**.

on top—In aviation terminology, descriptive of in-flight weather conditions that allow the use of modified **visual flight** techniques at **flight level**, but below which level the clouds and/or obscuring phenomena necessitate **instrument flight**.

Under certain limited conditions **visual flight rules** may be used while flying on top (**VFR on top**).

one-and-a-half order closure—A **higher-order closure** for **turbulence** where forecast equations are retained for mean variables (e.g., first-order or first **moment** mean values of **potential temperature** and **wind** components), and for **variance** of selected variables (e.g., selected second **statistical** moments such as **turbulence kinetic energy** or potential temperature variance).

It is not **second-order closure** because forecast equations are not retained for covariances, which are also second moments statistically. Any other higher-order **statistics** remaining in the equations are approximated by the mean and variance values. One type of one-and-a-half-order closure is k-ϵ closure. *Compare* **first-order closure**, **K-theory**, **second-order closure**, **nonlocal closure**, **Reynolds averaging**, **closure assumptions**.

one-dimensional cloud probe—*See* **optical imaging probe**.

one-dimensional model—A mathematical **model** that simulates variations in chemical composition along one spatial coordinate, usually **altitude**, as a function of time.

One-dimensional models are usually used to examine the effects of a change in chemistry on the vertical distribution of **trace atmospheric constituents**, for example, **ozone**, at a given latitude. The models are fairly cheap and easy to run, but do not represent atmospheric **meridional** or **zonal** transport of atmospheric constituents.

one-way entrainment process—At the **interface** between a **turbulent boundary layer** and a **laminar boundary layer**, laminar air is incorporated into the turbulent layer, but none of the turbulent air is incorporated into the laminar layer.

This causes the turbulent layer to get thicker as it erodes the laminar layer. Such a process explains why turbulent boundary layers increase **thickness** (because laminar air from the **free atmosphere** and **capping inversion** are incorporated into the **mixed layer**) and why **air pollutants** are trapped in the mixed layer (because none of the turbulent mixed layer air can carry pollutants into the laminar air).

onshore wind—A **wind** blowing from water onto land; the wind may be a result of heating differences between land and water or related to **synoptic** weather patterns.

opacity—The same as the **optical thickness**.

opacus—A variety of **cloud** (sheet, layer, or patch), the greater part of which is sufficiently dense to obscure the sun (betweeen 10 and 20 optical depths).

This variety is found in the genera **altocumulus**, **altostratus**, **stratocumulus**, and **stratus**. (Note: **cumulus** and **cumulonimbus** clouds are inherently opaque.) In the case of **altocumulus opacus** or **stratocumulus opacus** the elements stand out in true relief at the **cloud base**, rather than as a consequence of varying degrees of **opacity**. Also, in these cases, this variety usually modifies the species **stratiformis**. *See* **cloud classification**.

opaque sky cover—In U.S. weather observing practice, the amount (in tenths) of **sky cover** that completely hides all that might be above it; opposed to **transparent sky cover**.

In the case of an **obscuration** or **partial obscuration**, the **total sky cover** attributed thereto must be opaque.

open-cell cumulus—**Cumulus** clouds that align themselves in polygonal or elliptical shapes with clear center areas surrounded by clouds.

Open-cell cumulus are found mainly over oceans in areas of strong cold-air **advection**.

open cells—The regular array of clear patches with a connecting lattice of **cloud** that comprises a **layer cloud**.

open channel flow—Flow of a fluid with its surface exposed to the **atmosphere**.

open system—A thermodynamic system so chosen that there may be **transfer** of mass across the boundaries; for example, an **air parcel** undergoing a **pseudoadiabatic expansion**.
Compare **closed system**.

Operational Linescan System—(Abbreviated OLS.) The visible and **infrared** imager carried aboard the **DMSP** satellites, which provides a variety of moderate (2.7 km at **nadir**) and high-resolution (0.6 km at nadir) imagery.

During daytime passes, the OLS is normally configured to collect high-resolution **visible imagery** in conjunction with a single **channel** of moderate-**resolution** IR imagery. During nighttime passes the **sensor** configuration is usually reversed. Nighttime visible data are obtained with a high-performance photomultiplier tube that allows viewing in visible wavelengths of clouds and land areas illuminated by a quarter moon or brighter. City lights and the **aurora** can also be viewed quite clearly, with or without a moon.

operational numerical model—A numerical formulation of the governing equations of **hydrodynamics** applied to operational (as opposed to research) problems; typically used in reference to operational weather prediction models.

operational weather limits—(Also called minimums.) The limiting values of **ceiling**, **visibility** and **wind**, or **runway visual range**, established as safety minimums for aircraft landings and takeoffs.

Civil aircraft operate under limits stated in Civil Air Regulations and military aircraft operate under limits established by the respective military organizations. Limits for day and night operations usually differ. Also, the limits vary according to airport **environment**, navigational aids, and **type** of aircraft.

operative temperature—In the study of **human bioclimatology**, one of several parameters devised to measure the cooling effect of the air upon a human body. It is equal to the **temperature** at which a specified hypothetical **environment** would support the same **heat** loss from an unclothed, reclining human body as the actual environment.

In the hypothetical environment, the wall and air temperatures are equal and the air movement is 7.6 cm s^{-1}. From experiments it has been found that

$$\text{operative temperature} = 0.48 T_r + 0.19 [v^{1/2} T_a - (v^{1/2} - 2.76) T_s],$$

where T_r is the mean radiant temperature, T_a the mean air temperature, T_s the mean skin temperature (all in degrees Celsius), and v the air speed in centimeters per second. *Compare* **cooling power, cooling temperature, effective temperature**.

Newburgh, L. H., 1949: *Physiology of Heat Regulation and the Science of Clothing*, W. B. Sanders, Philadelphia, p. 281.
Buettner, K. J. K., 1951: *Compendium of Meteorology*, p. 1115.

operator—A mathematical symbol that stands for a specific operation upon a **variable** or function. *See* **linear operator, del operator, Laplacian operator**.

opposing wind—Generally, same as **headwind**; the opposite of **following wind**.

opposition—In astronomy, the arrangement of the earth, sun, and one of the other planets or the moon, in which the angle subtended at the earth between the sun and the third body, in the **plane of the ecliptic**, is 180°.

When this is the case, a superior planet is on the opposite side of the earth from the sun and is most easily observed. *Compare* **conjunction, quadrature**.

optical air mass—(Originally called air mass.) A measure of the length of the path through the **atmosphere** to **sea level** traversed by **light** rays from a celestial body, expressed as a multiple of the pathlength for a light source at the **zenith**.

It is approximately equal to the secant of the **zenith distance** of the given celestial body for zenith distances up to about 70°. **Bemporad's formula** must be used for more accurate determination. To get a representative value at high elevation, the above values must be multiplied by the ratio of the actual **atmospheric pressure** to the **sea level pressure**.

List, R. J., Ed., 1951: *Smithsonian Meteorological Tables*, 6th rev. ed., p. 422.

optical counter—Device for counting the concentration of **particles** in air that takes advantage of the **light scattering** caused by particles.

Modern instruments use **laser** light sources and are able to give information on the shape of the particle or its physical state.

optical depth—The **optical thickness** measured vertically above some given **altitude**.

Optical depth is dimensionless and may be used to specify many different radiative characteristics of the **atmosphere**. *See* **aerosol optical depth, cloud optical depth, Rayleigh optical depth**.

optical haze—*See* **scintillation**.

optical hygrometer—Same as **spectral hygrometer**; a group of hygrometers based on the spectral **attenuation** of **radiation** by specific **water vapor** absorption lines or bands.

optical imaging probe—(Also called optical array probe.) An **optical particle probe** that records the size and shape of the **shadow** of each **particle** that intercepts and attenuates the **illumination** by a **laser** beam.

Of these, shadowing probes differentiate particle shadows from the **light** of the unobstructed **beam** with **linear** arrays of optically activated diodes; for example, the **one-dimensional cloud probe** normally for sizing 10-μm to either 300- or 600-μm cloud hydrometeors, the **two-dimensional cloud probe** normally for sizing and imaging 25–800-μm **cloud** hydrometeors, and the **two-dimensional precipitation probe** normally for sizing and imaging 200-μm to 6.4-mm **precipitation** hydrometeors. The electronics record either the maximum one-dimensional width of the shadow or two dimensions and shape of the shadow, to estimate particle size and type (**raindrop, crystal** growth habit, etc.). Another type, the **cloud particle imager**, illuminates each particle with a pulsed laser and records size, shape, and detailed structure of each **hydrometeor** with a high-resolution solid-state digital imaging camera, normally to size 5-μm–2.3-mm hydrometeors.

optical mass—The vertical integral of the **density** of absorbers between two altitudes; used mainly in determining the transmission through an absorbing gas.

Dimensions are mass per unit area (e.g., kg m^{-2}).

optical meteor—A rarely used term describing any of the phenomena of **meteorological optics**, such as a **halo, rainbow**, or **mirage**.

optical particle probe—(Also called electro-optical particle probe.) Any of a class of instruments for sizing and counting large numbers of individual **aerosols** or **hydrometeors** to characterize populations, or also for imaging and classifying the shapes of individual hydrometeors by measuring the optical **illumination, scattering**, or **attenuation** of a **laser** beam by each **particle** and recording the result electronically.

Included are **optical scattering probes** and **optical imaging probes**. Primarily airborne instruments, they can be adapted for stationary use at the surface for **precipitation** or **fog** monitoring, in some cases with forced **ventilation** to create a particle flux.

optical path—1. Same as **line of sight**.
See also **optical pathlength**.

2. The integral of the **volume extinction coefficient** over the path traveled by **monochromatic radiation**.
See **optical thickness**.

optical pathlength—(Sometimes called **optical path** or optical length.) The **line integral** of the **refractive index** (real part) along a **ray** connecting two points in an **optically homogeneous** medium.

Because refractive index depends on **frequency**, so does optical pathlength. Although optical pathlength may be shortened to optical path, there is a difference between a path and its length. *Compare* **optical thickness**.

optical rain gauge—A gauge that measures the **scintillation** in an optical **beam** produced by **rain-drops** falling between a **light** source and an optical **receiver**.

By measuring the **energy** in selected **frequency** bands of the scintillation spectrum and comparing their ratios, the occurrence of precipitation, type of precipitation, and **intensity** of precipitation can be estimated.

Nystuen, J. A., et al., 1996: A comparison of automatic rain gauges. *J. Atmos. Oceanic Technol.*, **13**, 62–73.

optical scattering probe—A type of **optical particle probe** that measures the forward **scattering** caused by each **particle** or ensemble of particles intercepting a **laser** beam.

The scattering is converted to an equivalent size via **calibration** and scattering theory for spheres. Examples are the passive cavity aerosol **spectrometer** probe, normally for sizing 0.1–3.0-μm diameter **aerosols**, the **forward scattering spectrometer probe**, normally for sizing either 0.3–20-μm or 0.5–47-μm aerosols and **cloud droplets**, and the optical **cloud drop** spectrometer, normally for sizing 2–200-μm cloud droplets.

optical thickness—1. (Also **absorption optical thickness**, scattering optical thickness, total optical thickness, **optical depth**.) The (dimensionless) **line integral** of the **absorption coefficient**, or of the **scattering coefficient**, or of the sum of the two, along any path in a **scattering** and absorbing medium.

More often than not, the qualifiers "absorption," "scattering," and "total" are omitted (context sometimes being sufficient to determine which is meant). Optical thickness also depends on the **wavelength** of the **radiation** of interest. In a uniform medium, optical thickness has a simple physical interpretation as the length of a path in units of **mean free path**. Optical thickness and optical depth are used more or less synonymously; if there is a distinction between them it is that optical thickness is applied to an entire path through a medium. Normal optical thickness is optical thickness along a vertical path. *Compare* **optical pathlength**.

2. The degree to which a **cloud** modifies the **light** passing through it.

Optical thickness depends on the physical constitution (**crystals**, **drops**, **droplets**), and the form; the overall effect depends on the **scatter** parameter and the **phase function** for the **particles** as well as their concentration and the vertical extent of the cloud.

Optical Transient Detector—(Abbreviated OTD.) An instrument, designed to detect **lightning** from space, that was launched on 3 April 1995 aboard the *MicroLab-1* satellite into a near polar **orbit** at an **inclination** of 70°.

The OTD was an engineering prototype of the **Lightning Imaging Sensor** that was subsequently launched on the **TRMM** satellite.

optically homogeneous—Homogeneous on the **scale** of the **wavelength** of the **electromagnetic radiation** of interest.

Pure liquid water is optically homogeneous over the **visible spectrum** because one cubic wavelength of water contains many molecules, whereas a **cloud** of water droplets is not optically homogeneous. Optical homogeneity is more general than transparency, usually restricted to visible wavelengths. A body is said to be transparent if it transmits images. Optical homogeneity is necessary for transparency but not sufficient. A sufficiently thick sample of an absorbing optically homogeneous material would be described as opaque rather than transparent.

optically smooth—Smooth on the **scale** of the **wavelength** of the **electromagnetic radiation** of interest.

No surface is absolutely smooth, if for no other reason than matter is composed of molecules in motion. An approximate criterion for smoothness is the Rayleigh criterion. A surface is reckoned to be optically smooth if $d < \lambda/(8 \cos \theta)$, where d is the **surface roughness** (e.g., root-mean-square roughness height measured from a reference plane), λ is the wavelength of the incident **illumination**, and θ is the **angle of incidence** of this illumination. Thus, a surface that is smooth at some wavelengths is rough at others, or that is rough at some angles of incidence and smooth at others (e.g., near-grazing angles).

optimal perturbation—A **perturbation** that is intended to produce the largest (optimal) change to a forecast measure J.

For example, if J is a measure of **forecast error**, an optimal perturbation δx could represent a change to the forecast's initial conditions designed to minimize forecast error, given a constraint that δx has magnitude comparable to the typical **analysis** errors of operational forecast models. *See* **adjoint sensitivity**.

optimal yield—Amount of annual withdrawal from an **aquifer** or **reservoir** that best meets a set of economic, environmental, and/or social objectives, subject to a set of prescribed constraints.

optimum flight—An aircraft flight so planned and navigated that it is completed under the optimum conditions of minimum time and minimum exposure to dangerous flying weather.

An optimum flight often is not a minimal flight.

optimum interpolation—Commonly known as OI, this procedure provides an estimate of the state of the **atmosphere** by a weighted least squares fit to observations and a **background field**, usually provided by a **NWP** model forecast.

The weights are the inverse of the **error** covariance matrices for the observations and the background field. The word "optimum" is misleading, because in practice it is difficult to define the error covariances accurately. A more appropriate term is "statistical interpolation."

Daley, R., 1991: *Atmospheric Data Analysis*, 98–184.

ora—(Also spelled aura.) A regular **valley wind** at Lake Garda in Italy.

See also **suer**.

orbit—The path that a satellite follows in its motion through space relative to the attracting body.

orbit number—A particular circular path beginning at the satellite's **ascending node**.

The orbit number from launch to the first ascending node is designated as zero.

orbital elements—A collection of quantities that, together, describe the size, shape, and orientation of an **orbit**.

The classical orbital elements include the semimajor axis, eccentricity, **inclination**, argument of **perigee**, **right ascension** of **ascending node**, mean anomaly, and epoch time.

orbital motion—In **hydrodynamics**, the motion of a fluid **particle** induced by the passage of a progressive **gravity wave**.

When the **wave height** is small and the fluid depth is great, the orbit is a circle the radius of which decreases exponentially with depth. In shallow fluids, the orbit is an ellipse, which degenerates into a horizontal line at the bottom boundary of the fluid.

orbital velocity—As an ocean wave propagates, the water itself moves in an approximately circular motion (in **deep water**), with the **amplitude** of the motion decaying with depth below the sea surface.

The **velocity** associated with the fluid motion is known as the orbital (or **particle**) velocity.

ordinary cell—The most basic component of a **convective storm**, consisting of a single main **updraft** that is usually quickly replaced by a **downdraft** once **precipitation** begins.

Ordinary cells are especially observed in environments with weak **vertical wind shear**, and typically have lifetimes of 30–50 minutes. Ordinary cells are the primary component of multicell storms. *See also* **convective cell**, **multicell convective storm**, **thunderstorm**.

ordinary ray—(Or ordinary wave.) When a **plane wave** is incident on a uniaxial medium (an **anisotropic** medium with two equivalent **orthogonal** directions), two transmitted waves result: the ordinary ray (or wave), the **phase velocity** of which does not depend on its direction, and the **extraordinary ray** (or wave), the phase velocity of which depends on the angle it makes with the optic axis (the unique direction along which waves propagate as if the medium were optically **isotropic**).

ordinate—The vertical coordinate in a two-dimensional system of **rectangular Cartesian coordinates**; usually denoted by y. Also, the vertical axis of any graph.

See **abscissa**.

organic acids—Group of organic compounds containing the carboxylic group -C(O)OH, formed by the **oxidation** of **aldehydes**.

Acids such as formic, acetic, and pyruvic acids are often found in **air masses** containing the oxidation products of organic matter. Other multifunctional acids such as oxalic and malonic acid are often found in liquid droplets. The mechanism of their formation is not fully understood at the moment. The organic acids are relatively weak and do not contribute much to the acidity of the atmospheric **aerosol**.

organic nitrates—Nitrogen-containing compounds, general formula $RONO_2$, where R is an alkyl (or organic) group; formed in a minor (2%–30%) channel of the reaction of **nitric oxide** with the corresponding alkylperoxy radical.

The extent of formation of organic nitrates in an **air mass** has been used as a measure of the age of the air mass. They are a reservoir species for **NO_x**.

organic peroxides—Organic molecules containing an oxygen–oxygen linkage, general formula ROOR or ROOH, where R is an organic group.

In the **atmosphere**, ROOH is most commonly encountered, and the compound is an organic hydroperoxide.

organized large eddies—*See* **longitudinal rolls, coherent structures**.

oriented overgrowth—Same as **epitaxis**.

orographic—Relating to mountains and mountain effects.

Often refers to influences of mountains or mountain ranges on airflow, but also used to describe effects on other meteorological quantities such as **temperature**, **humidity**, or **precipitation** distribution. A major effect is **orographic lifting**.

orographic cloud—Mountain clouds produced by **orographic lifting** of **moist air** to **saturation**.

Clouds formed by upslope winds are generally **stratiform**, those formed by **mountain wave** updrafts are often **lenticularis**-type or **wave clouds**, and those formed by heating, such as elevated **heat** source or **leeside convergence** effects, are generally **cumuliform**. Upslope and wave clouds are clouds with form and extent determined by the disturbing effects of **orography** upon the passing flow of air. Because these clouds are linked to the **topography**, they are generally **standing clouds**, even though the winds at the same level may be very strong. Orographic upslope clouds include stratiform **cap** or **crest clouds** and the **foehn wall**. Convective orographic clouds are also strongly tied to the topography. Banta (1990) finds that mountain flows, which are driven by the topography interacting with the large-scale winds and the **diurnal heating** cycle, "play a significant role in determining where convective cells will initiate, and often how **precipitation** from the showers will be distributed spatially . . . they regulate not only the location of **storm** initiations, but also the timing."

Banta, R. M., 1990: The role of mountain flows in making clouds. *Meteor. Monogr.*, **45**, p. 283.

orographic depression—Same as **lee trough**.

orographic fog—**Fog** formed as **moist air** blows up a mountain slope and becomes saturated.

Flow can result from large-scale upslope winds or thermally forced upslope (**anabatic**) winds. *See* **anabatic wind, orographic cloud**.

orographic isobar—A **distortion** of an **isobar** or isobars (surface or upper level) due to a mountain range or other obstruction.

orographic lifting—Ascending air flow caused by mountains.

Mechanisms that produce the lifting fall into two broad categories: 1) the upward deflection of horizontal larger-scale flow by the **orography** acting as an obstacle or barrier; or 2) the daytime heating of mountain surfaces to produce **anabatic** flow along the slopes and updrafts in the vicinity of the peaks. The first category includes both direct effects, such as forced lifting and vertically propagating waves, and indirect effects, such as upstream **blocking** and **lee waves**. Even though this term strictly refers only to lifting by mountains, it is sometimes extended to include effects of hills or long sloping **topography**. When sufficient moisture is present in the rising air, **orographic fog** or clouds may form.

orographic occlusion—An **occluded front** in which the **occlusion** process has been hastened by the retardation of the **warm front** along the **windward** slopes of a mountain range.

orographic precipitation—**Precipitation** caused or enhanced by one of the mechanisms of **orographic lifting** of **moist air**.

Examples of precipitation caused by mountains include **rainfall** from **orographic** stratus produced by forced lifting and precipitation from orographic cumuli caused by daytime heating of mountain slopes. Many of the classic examples of locations having excessive annual precipitation are located on the **windward** slopes of mountains facing a steady **wind** from a warm ocean. As another example, wintertime orographic stratus (**cap clouds**) often produce the major water supply for populated semiarid regions such as the mountainous western United States, and as a result these **cloud** systems have been a target of precipitation enhancement, cloud-seeding projects intended to produce **snowpack** augmentation. Orographic precipitation is not always limited to the ascending ground, but may extend for some distance windward of the base of the barrier (**upwind effect**), and for a short distance to the lee of the barrier (**spillover**). The lee side with respect to prevailing moist flow is often characterized as the dry **rain shadow**. *See* **seeder–feeder**.

orographic rainfall—As commonly used, same as **orographic precipitation**.
  See **rainfall**.

orographic snow line—The **freezing level** for **precipitation** formed by **orographic lifting**; the **elevation** above which **rain** or **drizzle** turns to **snow**.

orographic storm—A **storm** that is produced or significantly enhanced by mountain effects such as **orographic lifting** or **lee cyclogenesis**.

orographic vortex—Atmospheric **vortex** or **whirlwind** produced by flow over or past mountains and other obstacles.
  Orographic vortices exist over a wide **range** of scales and orientations, from **eddies** of a few tens to a few hundreds of meters across, oriented in any direction, and shed by individual peaks or other topographic obstacles, to **synoptic-scale** cyclones, vertical vortices that form or intensify in the **lee trough** downwind of mountain-range scale barriers (*see* **lee cyclogenesis**). Eddies of several hundred meters to a few tens of kilometers across (the larger scale representing approximately the scale of the mountain producing them) contribute to **aircraft turbulence** and enhance damage during **downslope windstorm** conditions. They often are the result of periodic shedding from the obstacle that produced them. Especially strong vertical vortices have been called "mountain-adoes," indicating a resemblance to mountain tornadoes. Under strong convective heating conditions vortices spawned in the mountains sometimes continue **downwind** over the heated plains and participate in the initiation of **dust devils**.

orographic wind flow—Wind flow caused, affected, or influenced by mountains.
  Orographic effects include both dynamic, in which mountains disturb or distort an existing approach flow, and thermodynamic, in which heating or cooling of mountain-slope surfaces generates flow. Dynamic effects include **aerodynamic** obstacle effects, **mountain waves**, **channeling**, orographic **blocking**, and processes leading to **foehn**, **bora**, and **gap winds**. Thermodynamic effects include **anabatic winds**, **katabatic winds**, **valley breezes**, and **mountain breezes**. *See also* **mountain–valley wind systems**, **mountain–plains wind systems**.

orography—1. The nature of a region with respect to its elevated terrain.
  2. That branch of **geomorphology** that deals with the disposition and character of hills and mountains.

oroshi—In Japan, a dry, northwesterly **bora** descending the mountains of central Honshu onto the Kanto Plain in winter.
  This **wind** is strong, usually exceeding 10 m s^{-1}.

orsure—A stormy north to northeast **wind** in the Gulf of Lions off the southern coast of France.

orthogonal—1. Originally, at right angles; later generalized to mean the vanishing of a sum (or integral) of products.
  The cosine of the angle between two vectors, $\mathbf{V}_1$ and $\mathbf{V}_2$ with respective components, (x_1, y_1, z_1) and (x_2, y_2, z_2), is proportional to the sum of products, $x_1 x_2 + y_1 y_2 + z_1 z_2$. Hence, if the vectors are perpendicular, the latter sum equals zero. For this reason any two series of numbers, $(x_1, x_2, \cdots, x_n)$ and $(y_1, y_2, \cdots, y_n)$ is said to be orthogonal if

$$\sum_i x_i y_i = 0.$$

  See **orthogonal functions**.

  2. On an ocean-wave **refraction diagram**, a ray drawn everywhere at right angles to **wave crests**.

orthogonal antennas—In **radar**, a pair of transmitting and receiving antennas, or a single **antenna** used for transmission and reception, designed for measuring the **copolarized signal** and the **cross-polarized signal**.

See **dual-channel radar**.

orthogonal curvilinear coordinates—*See* **curvilinear coordinates**.

orthogonal functions—A set of functions, any two of which, by analogy to **orthogonal** vectors, vanish if their product is summed by integration over a specified interval.
For example, $f(x)$ and $g(x)$ are orthogonal in the interval $x = a$ to $x = b$ if

$$\int_a^b f(x)g(x)dx = 0.$$

The functions are also said to be normal if

$$\int_a^b [f(x)]^2 dx = \int_a^b [g(x)]^2 dx = 1.$$

The most familiar examples of such functions, many of which have great importance in mathematical physics, are the sine and cosine functions between zero and 2π.

orthogonal lines—Perpendicular lines.

orthomorphic map—Same as **conformal map**.

oscillation—1. A swinging, as of a pendulum.
Often applied to periodic motion or **variation** in time of any quantity, although may mean any more or less regular variation between fixed bounds. Fluctuation is more suggestive of irregular variation. Vibration is a near synonym except that oscillation is applied to variations in space as well as time. The **amplitude** of a **damped oscillation** steadily decreases. Oscillations are said to be forced or free according to whether the oscillating system is or is not acted upon by an external force, although what constitutes such a force is a matter of convention. An ordinary pendulum is acted on by the external **force of gravity**, and yet the pendulum probably would be described as undergoing free oscillation.
2. As used by Sir Gilbert Walker, a single number, empirically derived, that represents the distribution of **pressure** and **temperature** over a wide ocean area.
Basically, the process is one of **weighting** pressure and temperature values for selected island and coastal stations, and algebraically combining them. These numbers were originally employed in correlations with single **station** values. Three such "oscillations" were derived: the **North Atlantic oscillation**; the **North Pacific oscillation**; and the **Southern Oscillation**.

oscillator—The general term for an electrical device that generates alternating currents or voltages.
The oscillator is classified according to the **frequency** of the generated **signal**.

oscillatory wave—Same as **wave of oscillation**.

oscilloscope—Same as **cathode-ray oscilloscope**.

osmometer—A device for measuring soil water amounts, especially under roads; seldom used.

osmosis—The process by which a solvent is allowed to diffuse through a semipermeable membrane from a solution of higher concentration to one of lower concentration.

Osos wind—In California, a strong northwest **wind** blowing from the Los Osos valley to the San Luis valley.

OSSE—Abbreviation for **observing systems simulation experiment**.

ostria—(Also called auster.) A warm southerly **wind** on the Bulgarian coast; it is considered a precursor of bad weather.
See **austru**; *compare* **lodos**.

OSV—Abbreviation for **ocean station vessel**.

OTD—Abbreviation for **Optical Transient Detector**.

ouari—A south **wind** of Djibouti in Africa; it is similar to the **khamsin**.

outbreak—*See* **polar outbreak**.

outer atmosphere—Very generally, the **atmosphere** at a great distance from the earth's surface; possibly best used as an approximate synonym for **exosphere**.

outer eyewall—*See* **concentric eyewalls**.

outer layer—For flow over a hill, the top layer in the **boundary layer** that accelerates relative to its **upstream** value due to the Bernoulli effect.

outer product—Same as **vector product**.

outer vortex—The outermost region of a **tropical cyclone**, encompassing all of the region affected by the **storm** outside its **eyewall**.

outflow—A **current** exiting through a strait or passage.

outflow boundary—A surface boundary formed by the horizontal spreading of thunderstorm-cooled air.

Outflow boundaries may intersect with each other or with other features (**fronts, low-level jets**) and act to focus new **convection**. Outflow boundaries may be short-lived, or last for longer than a day. *See* **gust front**.

outflow jet—Nocturnal cold-air jet flowing out of the mouth of a valley or canyon as it opens onto a plain.

When the flow is fully developed, its depth near the exit is about the same as the height of the valley or canyon sidewalls there. It represents a continuation or extension of the mountain (down-valley or downcanyon) **breeze** generated in the valley by surface cooling. The jet often achieves peak **wind speeds** outside the valley. This **acceleration** is probably due to the conversion of **potential** to **kinetic energy**, as the column of cold air flowing out of the valley, freed of the confining sidewalls, fans out in the horizontal and compresses vertically. Another factor is the release of the flow from **surface friction** along the sidewalls. Maximum speeds in the jet begin near the surface early in the evening, but for well-developed jets at exits from deep valleys the highest speeds may be at 300–500 m or more above the ground. In deep, extensive valley drainage systems, peak speeds may exceed 10 m s^{-1} on **clear** nights.

outgoing longwave radiation—The radiative character of **energy** radiated from the warmer earth to cooler space, often derived from **window** channel measurements from satellites in polar **orbit**.

outlet glacier—A **valley glacier** that drains an **ice sheet** or **ice cap** and flows through a gap in peripheral mountains.

outlook—Issued to indicate that a hazardous weather or hydrologic event may develop.

It is intended to provide information to those who need considerable lead time to prepare for the event.

outo—*See* **autan**.

output—(Or output signal.) The quantity that is delivered by an instrument or a component of an instrument; used in contradistinction to the **input**.

overcast—1. Descriptive of a **sky cover** of 1.0 (95% or more) when at least a portion of this amount is attributable to **clouds** or **obscuring phenomena** aloft; that is, when the **total sky cover** is not due entirely to surface-based obscuring phenomena.

In **aviation weather observations**, an overcast sky cover is denoted by the **symbol** "⊕"; it may be explicitly identified as **thin** (predominantly transparent); otherwise a predominantly opaque status is implicit. An opaque overcast sky cover always constitutes a **ceiling**. *See* **obscuration**.

2. Popularly, the **cloud layer** that covers most or all of the sky.

It generally suggests a widespread layer of clouds such as that considered typical of a **warm front**.

overflow—1. Excess water that spills over the ordinary confines of a body of water.

2. A **current** exiting across the sill of a **basin**.

overland flow—(Also called surface flow.) Water flowing over the ground surface toward a **stream** channel or water body.

Upon reaching the channel or water body, it is called **surface runoff**.

overlap—The area common to two successive satellite images or scan swaths along the same or adjacent flight or orbital strips.

The amount of overlap is expressed as a percentage of **image** area or scanned area.

overlapping average—Same as **consecutive mean**.

overlapping mean—Same as **consecutive mean**.

overrunning—A condition existing when an **air mass** aloft is in motion relative to another air mass of greater **density** at the surface.

This term is usually applied in the case of warm air ascending the surface of a **warm front** or **quasi-stationary front**.

overseeding—A **cloud seeding** operation in which an excess of nucleating material is released over that required to produce the desired effect.

As the term is normally used, the excess is relative to that amount of nucleating material expected to produce the maximum amount of **precipitation** received at the ground. For example, in **seeding** a **supercooled water** cloud with **dry ice** or **silver iodide** in order to increase precipitation, the addition of too much seeding material will create too many **ice crystals** insofar as they compete for the available **water vapor**, and few of them will grow large enough to fall as precipitation. There are unresolved questions about what composes the optimal amount of seeding material to produce the desired effect. These questions include how to estimate the right amount of material and when and where to deliver it to the **cloud**; it is difficult to assign quantitative values that will result in overseeding. *See* **cloud seeding**.

overshooting top—(Or anvil dome, penetrating top.) A domelike protrusion above a **cumulonimbus** anvil, representing the intrusion of an **updraft** through its **equilibrium** level (**level of neutral buoyancy**).

It is usually a **transient** feature because the rising **parcel's momentum** acquired during its buoyant ascent carries it past the point where it is in equilibrium; the air within it rapidly becomes negatively buoyant and descends. Tall and persistent overshooting tops are frequently observed with strong or severe **thunderstorms** in which there is a nearly continuous stream of buoyant updrafts.

overtrades—Obsolete term for **Krakatoa winds**.

overturn—(Also called convective overturn.) The **mixing** of a stratified body of water due to **density** changes, usually caused by **temperature** changes.

Owens dust recorder—An instrument for rapidly obtaining samples of airborne **dust**; a type of **dust counter**.

Particles pass through a cylindrical chamber and are drawn at high velocity through a narrow slit in front of a microscope cover glass. The fall in **pressure** due to expansion of the air passing through the slit causes the formation of a moisture film on the glass to which the dust adheres. An analysis for quantity and size of the **particles** on the cover glass is made with the aid of a microscope. The vacuum required to operate the instrument is developed by an attached hand pump.

ox's eye—A nautical term of Guinea for **hurricane**.

oxidant—Substance capable of causing **oxidation** of, for example, an atmospheric species.

The most common oxidant in the **troposphere** is **ozone**, which can be detected by its reaction with potassium iodide (KI). Thus, by extension, any species that oxidizes KI was historically classed as an atmospheric oxidant.

oxidation—Reaction of a substance with **oxygen**, or incorporation of oxygen into a molecule.

Examples are combustion and rust formation. In solution-phase ionic chemistry, oxidation is formally defined as a process that reduces the **electron density** at a given site. Since oxygen is a very electronegative species, incorporation of oxygen into a molecule results in a reduction in electron density. Due to the high proportion of oxygen, the **atmosphere** is intrinsically oxidative in nature, and photochemical degradation of **pollutants** usually involves an increased degree of oxygenation, and a consequent reduction in the number of **hydrogen** atoms in a molecule, for example,

$$CH_4 \rightarrow HCHO \rightarrow HCOOH \rightarrow CO_2.$$

Hence, the driving force in the atmosphere is from reduced emissions to oxidized products.

oxidizing capacity of atmosphere—The ability of the **atmosphere** to cleanse itself of terrestrial, often organic, emissions.

The degradation of most organic compounds is an **oxidation** process initiated by attack by a **hydroxyl radical**. Thus, the ability of the atmosphere to sustain a concentration of OH sufficient to remove current levels of **hydrocarbons**, etc., is usually referred to as its oxidizing capacity. The expression is qualitative, and no quantitative measure of oxidizing capacity has been adopted to date.

oxygen—(Symbol O.) An **element**, atomic number 8, atomic weight 16.0; molecular oxygen, formula O_2, **molecular weight** 32, is the second most abundant species in the **atmosphere**, with an abundance of approximately 21% at **sea level**.

The atmospheric abundance of O_2 remains fairly constant up to about 80 km, above which substantial **photodissociation** to atomic oxygen occurs. Oxygen is a prerequisite to almost all forms of terrestrial life. Oxygen was probably released from minerals such as carbonates resulting in the evolution from a reducing to an oxidizing atmosphere. The general tendency is for reduced emissions from the earth's surface to be oxidized to simpler, oxygen-containing species. Atomic oxygen is formed in the **photolysis** of molecular oxygen, O_2; **ozone**, O_3; or **nitrogen dioxide**, NO_2, in the atmosphere. Below about 40 km, its predominant fate is **recombination** with molecular oxygen to form ozone. Above that **altitude** it can participate in other chemical reactions, which may lead to ozone destruction. Both molecular and atomic oxygen have low-lying electronically excited states that are important in the atmosphere. The $^1\Delta$ and $^1\Sigma$ states of O_2 are relatively long-lived, and **fluorescence** from these states contributes to the **airglow**. The O^1D state of atomic oxygen, formed in ozone photolysis, reacts to form the **hydroxyl radical**, which is the primary **oxidant** in the atmosphere.

oxygen band—Regions of the **electromagnetic spectrum** in which molecular oxygen, O_2, absorbs **solar radiation** but plays an insignificant role in the direct heating of the **atmosphere**.
This **band** is strong between 0.13 and 0.17 μm, and is of special importance in the **absorption** of **ultraviolet radiation**.

Oyashio—The western **boundary current** of the North Pacific subpolar **gyre**.
Fed by the **Alaskan Stream** and **Kamchatka Current**, the Oyashio flows southward along the Kuril Islands and the east coast of northern Japan to 39°–40°N and occasionally as far south as 36°N. Near southern Hokkaido (43°N) the current splits into two branches known as the First and Second Oyashio Intrusion. In the region east of Tsugaru Strait (41°N) it is dominated by eddies; one or two **cyclonic** (warm core) eddies are formed each year. Every six years or so one of them grows into a giant **eddy** that then dominates the region for a year. Farther **downstream** the Oyashio meets the northward flowing **Kuroshio** to form the **Arctic Polar Front**. Both currents maintain their own frontal systems, the Oyashio Front being characterized by temperatures of 2°–8°C and feeding into the **Aleutian Current**, while the Kuroshio Front has temperatures of 10°–15°C and feeds into the **North Pacific Current**.

Ozmidov scale—A length **scale** for the description of turbulent flows under stable **stratification**, defined as the square root of the ratio between the **dissipation rate** of turbulent **kinetic energy** and the third power of the **Brunt–Väisälä frequency**.
In flows where **turbulence** and **wave motion** are simultaneously present, the inverse of the Ozmidov scale defines the **buoyancy wavenumber**, which separates the **buoyancy subrange** from the **inertial subrange**.

ozone—A nearly colorless gas, formula O_3, **molecular weight** 48, that appears blue in the condensed **phase** or at high concentration, with a characteristic odor like that of weak chlorine.
It is formed in the reaction between atomic **oxygen** and molecular oxygen: $O + O_2 \overset{M}{\rightarrow} O_3$. It is a very strong absorber of **ultraviolet radiation**, and the presence of the **ozone layer** in the **upper atmosphere** provides an **ozone shield** that prevents dangerous **radiation** from reaching the earth's surface and allows the existence of life in its present forms. Ozone, produced by photochemical reactions, is found at all altitudes in the **atmosphere**. The total amount of ozone in the atmosphere would correspond to less than 1 part per million if uniformly distributed, or a column amount of about 3 mm if compressed to **sea level pressure**. In the **troposphere**, it is regarded as a **pollutant**, and its presence in high concentrations can lead to respiratory stress and crop damage. Ozone is an important component of **photochemical smog** and can also be formed locally by the action of electrical discharges on the air. Ozone in the free troposphere often results from downward **transport** from the **stratosphere**. In the stratosphere, ozone is formed following the **absorption** of radiation by molecular oxygen. Its **mixing ratio** there can reach several parts per million, and the **temperature inversion** characteristic of the stratosphere is due to the strong absorption of **energy** by ozone molecules in this region. In the stratosphere, ozone is destroyed predominantly by **catalytic cycles** involving free radicals, many of which are formed as products of human activity. Ozone has several radiation absorption bands that are atmospherically important: the very intense Hartley **band**, between 200 and 300 nm, which is responsible for much of the heating of the upper atmosphere; the Huggins bands, between 320 and 360 nm; the **Chappuis bands**, between 450 and 650 nm; and **infrared** bands, centered at 4.7, 9.6, and 14.1 μm. All the above bands have been used for the **detection** of ozone using various **remote sensing** techniques. Absorption by ozone in the infrared is responsible for its effectiveness as a **greenhouse gas**. *See* **Dobson unit**.

Finlayson–Pitts, B. J., and J. N. Pitts, 1986: *Atmospheric Chemistry*, Wiley–Interscience, New York, 1098 pp.
Seinfeld, J. H., and S. N. Pandis, 1998: *Atmospheric Chemistry and Physics*, Wiley–Interscience, New York, 1326 pp.

Volz, A., and D. Kley, 1988: Evaluation of the Montsouris series of ozone measurements made in the nineteenth century. *Nature,* **332,** 240–242.

ozone-depleting potential—A measure of the ability of a halocarbon source gas to destroy stratospheric **ozone**, determined relative to an equal mass of chlorofluorocarbon-11.

The ozone-depleting potential is usually determined by examining the effects on the **ozone layer** of the addition of the compound under investigation using atmospheric models.

ozone hole—A characteristic severe depletion of stratospheric **ozone** that occurs each spring over the Antarctic continent.

The depletion is caused by the catalytic destruction of ozone by chlorine, released from fluorocarbons and activated by the presence of polar stratospheric **cloud** particles in the extreme cold of the Antarctic **stratosphere**.

ozone isopleth plot—A plot in which contours of **tropospheric ozone** are plotted as a function of the **active nitrogen** mixing ratio versus total hydrocarbon **mixing ratio**.

Regions where **ozone** production is limited by the availability of **nitrogen oxides** or by **hydrocarbons** are readily apparent in such a plot.

ozone layer—1. Same as **ozonosphere**.

2. Generally, any layer in the **atmosphere** in which there is a maximum of **ozone** concentration.

ozone shield—Popular expression, referring to the fact that the strong **absorption** of **ultraviolet** light by stratospheric **ozone** limits the amount of dangerous **radiation** reaching the earth's surface.

ozone spectrophotometer—*See* **Dobson spectrophotometer**.

ozonesonde—A **radiosonde** equipped with an instrument to measure the atmospheric concentration of **ozone** (O_3).

ozonosphere—(Also called ozone layer.) A region of the **atmosphere** from about 15 to 60 km (roughly the extent of the **stratosphere**) that contains large concentrations of **ozone**.

The ozone concentration peaks at about 10^{13} molecules per cubic centimeter near 20 km, while the **mixing ratio** peaks at slightly higher **altitude** (about 9–10 ppm at 30 km). Ozone in this region is produced from **photolysis** of O_2 molecules and is destroyed by reactions involving the oxides of **nitrogen**, chlorine, and **hydrogen**. Because of the strong **UV** absorption **spectrum** of ozone, the ozone layer effectively limits penetration of UV radiation to the earth's surface to wavelengths longer than 290 nm.

Crutzen, P. J., 1971: The influence of nitrogen oxides on the atmospheric ozone content. *Quart. J. Roy. Meteor. Society,* **96 (408),** 320–325.

Crutzen, P. J., 1971: Ozone production rates in the oxygen–hydrogen–nitrogen atmosphere. *J. Geophys. Res,* **76,** 7311–7327.

Molina, M. J., and F. S. Rowland, 1974: Stratospheric sink for chlorofluorocarbons: Chlorine atom catalyzed destruction of ozone. *Nature,* **249,** p. 890.

Stolarski, R. S., and R. J. Cicerone, 1974: Stratospheric chlorine: A possible sink for ozone. *Can. J. Chem.,* **52,** 1610–1615.

Johnston, H. S., 1971: Reduction of stratospheric ozone by nitrogen oxide catalysts from supersonic transport exhaust. *Science,* **173,** p. 517.

P

P-E index—Abbreviation for **precipitation-effectiveness index**.

P–E ratio—Abbreviation for **precipitation–evaporation ratio**.

p system—*See* **pressure coordinates**.

Pacific Deep Water—A **water mass** found between 1000 and 3000 m in the Pacific Ocean.
   It is characterized by very sluggish flow and correspondingly low oxygen content. *See* **deep water**.

Pacific high—(Or Pacific anticyclone.) The nearly permanent **subtropical high** of the North Pacific Ocean centered, in the mean, at 30°–40°N and 140°–150°W.
   On mean charts of **sea level pressure**, this **high** is a principal **center of action**.

Pacific–North American pattern—(Abbreviated PNA pattern.) A wave-train **signal** that spans from the equatorial Pacific through the northwest of North America to the southeastern part of North America.
   This is the strongest teleconnection pattern of **low-frequency** variability of the atmospheric **circulation** in winter in response to changes in the **sea surface temperature**.

Pacific Subarctic Current—*See* **Aleutian Current**.

pack ice—(Also called ice pack.) All **sea ice** other than **fast ice**; thus, sea ice that is capable of substantial motion and **deformation**.

packed tower—A **pollution** control device for reducing (scrubbing) sulfur dioxide (SO_2) emissions.
   This device consists of a column or tower that is packed with absorption material such as **sodium** or calcium carbonate, which removes the SO_2 from the **flue gas** as it passes through the tower.

paesa—A violent north-northeast **wind** of Lake Garda in Italy.

paesano—A northerly night **breeze**, blowing down from the mountains, of Lake Garda in Italy. *See* **sover**.

PAGES—(Acronym for Past Global Changes.) A core project of the International Geosphere–Biosphere Program of the International Council of Scientific Unions that is focused on two temporal streams—the last 2000 years studied at interannual to interdecadal resolution, and the longer timescale of the cycles of **glaciation**/deglaciation.

PAHs—*See* **polycyclic aromatic hydrocarbons**.

painter—(Also called Callao painter, Peruvian paint.) Blackening of ship's paint by **hydrogen sulfide** from decaying marine organisms, especially during **El Niño**.

pair production—*See* **antiparticle**.

paleoclimate—(Or geological climate.) **Climate** for periods prior to the development of measuring instruments, including historic and **geologic time**, for which only **proxy climate records** are available.

paleoclimatic sequence—The sequence of climatic changes in **geologic time**.
   It shows a succession of **oscillations** between warm periods and **ice ages**, but superimposed on this are numerous shorter oscillations. The tendency to regard the whole of a **geologic period**, lasting for 20 million years and more, as having a single type of **climate** is a great oversimplification, as is shown by the succession of **glacial** and **interglacial** periods in the **Quaternary**. Even the warm periods are known to have been made up of successions of climates of different degrees of warmth; but until much more information is available, it will not be possible to set out in detail the sequence of changes of the earlier paleoclimates.

paleoclimatology—The study of past **climates** throughout geologic and historic time (paleoclimates), and the causes of their variations.

paleocrystic ice—Old **sea ice**, generally considered to be at least ten years old; it is nearly always a form of **pressure ice** and is often found in **floebergs** and in the **pack ice** of the central Arctic Ocean.

paleomagnetism—The study of natural remanent magnetization in order to determine the **intensity** and direction of the earth's **magnetic field** in the geologic past.

Palmer Drought Severity Index—(Abbreviated PDSI.) An index formulated by Palmer (1965) that compares the actual **amount of precipitation** received in an area during a specified period with the **normal** or average amount expected during that same period.

The PDSI is based on a procedure of hydrologic or water balance accounting by which excesses or deficiencies in moisture are determined in relation to average climatic values. Values taken into account in the calculation of the index include **precipitation, potential** and **actual evapotranspiration, infiltration** of water into a given **soil zone**, and **runoff**. This index builds on Thornthwaite's work (1931, 1948), adding 1) soil depth zones to better represent regional change in soil water-holding capacity; and 2) movement between soil zones and, hence, plant moisture stress, that is, too wet or too dry.

Palmer, W. C., 1965: *Meteorological Drought Research Paper 45*, U.S. Weather Bureau, Washington, D.C.

Thornthwaite, C. W., 1948: An approach toward a rational classification of climate. *Geogr. Rev.*, **38**, 55–94.

Thornthwaite, C. W., 1931: Climates of North America according to a new classification. *Geogr. Rev.*, **21**, 633–655.

palouser—A **duststorm** of northwestern Labrador.

Paluch diagram—*See* **conserved variable diagrams**.

pampas—The **steppe** of South America, especially Argentina.

pampero—A cold squally south or southwest **wind** over the **pampas** of the Argentine and Uruguay, which sets in with the passage of a **cold front**; the South American counterpart of a **norther**.

It is often accompanied by **squalls, thunderstorms**, and a sudden drop of **temperature**. The pampero seco is rainless; the pampero sucio brings a **duststorm**. *Compare* **minuano**; *see* **zonda**.

PAN—Common abbreviation for **peroxyacetyl nitrate**.

pan coefficient—The ratio of the amount of **evaporation** from a large body of water to that measured in an **evaporation pan**.

pan evaporation—Measured **water loss** from **free water** surface of **class-A evaporation pan**.

panas oetara—A strong, warm and dry north **wind** in February in Indonesia.

pancake ice—Roughly circular accumulations of **frazil ice**, usually less than about 3 m in diameter, with raised rims caused by collisions.

This form of **ice** is common in the Antarctic. Pancake ice may develop from **grease ice** or **shuga**.

pannus—Numerous **cloud** shreds below the main cloud.

These shreds may constitute a layer, which may be separated from the main part of the cloud, or attached to it. This **accessory cloud** occurs mostly with **nimbostratus, cumulus**, and **cumulonimbus**. *See* **scud, cloud classification**.

papagayo—(Also spelled popogaio.) A violent northeasterly **fall wind** on the Pacific coast of Nicaragua and Guatemala.

It consists of the cold **air mass** of a **norte** that has overridden the mountains of Central America and, being a descending **wind**, brings fine **clear** weather. Papagayos are most frequent and strongest in January and February, often lasting three or four days. They weaken between 7 and 10 A.M. and freshen again, sometimes to **gale** force, in the evening and early night. *Compare* **tehuantepecer**.

PAR—Abbreviation for **photosynthetically active radiation**.

parabolic antenna—An **antenna** employing a **reflector** in the form of a **paraboloid**.

Common in **weather radar**, such an antenna produces collimated **radiation** from a **feed** at the focus, providing high **gain** and narrow **beamwidth**.

paraboloid—(Or equilibrium paraboloid.) The geometrical figure obtained by rotating a parabola about its axis.

In experimental meteorology and **oceanography** such a figure is sometimes used as the lower boundary of a **model atmosphere** or ocean, and is especially important because, at a proper rotation rate, it is an **equipotential surface** for **apparent gravity** in the **model**. If ω is the rotation rate of the apparatus, g the **acceleration** of local earth's **gravity**, z the coordinate parallel to the vertical rotation axis, and r the radial coordinate, an equilibrium paraboloid is given by

$$z = \omega^2 r^2 / 2g.$$

This surface is the laboratory equivalent of the spheroidal equipotential surfaces of the earth's apparent gravity field.

paraffins—General name for **alkanes**; usually used with reference to alkanes with a large number of **carbon** atoms that are liquids or waxy solids.

parafoveal vision—(Also called mesopic vision.) That part of an **image** focused on the region of the retina surrounding the fovea.

This region contains a mixture of cones and rods and does not provide as high a **resolution** as does the fovea. Although rods do not permit color vision, they respond to much lower **illuminance**. Nighttime vision is performed primarily with the rods. *See* **dark adaptation**.

parameter—1. In general, any quantity of a problem that is not an **independent variable**.

More specifically, the term is often used to distinguish, from **dependent variables**, quantities that may be more or less arbitrarily assigned values for purposes of the problem at hand.

2. In **statistical** terminology, any numerical constant derived from a **population** or a **probability distribution**.

Specifically, it is an arbitrary constant in the mathematical expression of a probability distribution. For example, in the distribution given by

$$f(\mathbf{x}) = \alpha e^{-\alpha x},$$

the constant α is a parameter.

parameterization—The representation, in a dynamic **model**, of physical effects in terms of admittedly oversimplified parameters, rather than realistically requiring such effects to be consequences of the dynamics of the system.

See **Mellor–Yamada parameterization, subgrid-scale process, convective adjustment**.

parametric equations—A set of equations in which the **independent variables** or coordinates are each expressed in terms of a **parameter**.

For example, instead of investigating $y = f(x)$, or $F(x, y) = 0$, it is often advantageous to express both x and y in terms of a parameter u: $x = g(u)$; $y = G(u)$. The parameter may or may not have a useful geometric or physical interpretation.

paranthelion—A **halo** in the form of a faint spot at the same angular **elevation** as the sun, but located in the vicinity of the **anthelion** (a point at the same elevation as the sun but on the opposite side of the sky).

The one at 120° from the sun (60° from the anthelion) is explained by internal reflections in an oriented hexagonal **plate crystal**.

paraselene—*See* **parhelion**.

paraselenic circle—*See* **parhelic circle**.

parcel—An imaginary volume of fluid to which may be assigned various thermodynamic and kinematic quantities.

The size of a parcel is arbitrary but is generally much smaller than the characteristic **scale** of **variability** of its **environment**.

parcel method—(Also called path method.) A method of testing for **instability** in which a displacement is made from a **steady state** under the assumption that only the **parcel** or parcels displaced are affected, the **environment** remaining unchanged.

Although this method has been applied to various problems (e.g., **inertial instability**), its most familiar context is with vertical displacements from **hydrostatic equilibrium**, in which the parcel displaced is assumed to undergo **adiabatic temperature changes**, and the **buoyant force** resulting from its contrast with the unchanged environment leads to the criterion for **stability**, $\gamma < \Gamma$, where γ is the **lapse rate** of **virtual temperature** with height and Γ is the **dry-** or **saturation-adiabatic lapse rate**, according to the condition of the parcel. *Compare* **slice method**.

parhelic circle—A **halo** in the form of a faint, white, horizontal arc at the **elevation** of the sun.

Small segments are more frequently seen than the complete circle. When caused by the moon, it is called the **paraselenic circle**. The parhelic circle is explained by **reflection** by the vertical faces of oriented **ice crystals**, such as the sides of large hexagonal plates, which also produce **parhelia** and **paranthelia**.

parhelion—(Also called sun dog.) A **halo** in the form of a colored spot at the same angular **elevation** as the sun.

The lunar counterpart is a paraselene. The most common parhelia are seen about 22° on either side of the sun. That angular distance increases from 22° when the sun is on the **horizon** to over twice that when the sun climbs to 60°. The sun side of a parhelion is reddish. The parhelia of 22° can be explained by **refraction** in hexagonal crystals falling with principal axes vertical. The effective prism angle is 60° when the sun is on the horizon, but this increases as the sun climbs, resulting in greater displacement of the parhelion.

parrey—Same as **perry**.

parry—Same as **perry**.

Parry arcs—A **halo** in the form of arcs that appear outside the upper and lower **tangent arcs** to the halo of 22°.
 The columnar crystals have their long axes horizontal, which by itself would produce the tangent arcs, and a prism face horizontal. *See* **minimum deviation, anthelic arcs**.
  Humphreys, W. J., 1940: *Physics of the Air*, 3d ed., 524–526.

Parseval's theorem—A theorem relating the product of two functions to the products of their **Fourier series** components.
 If the functions are $f(x)$ and $F(x)$, and their Fourier series components have respective amplitudes a_n, b_n and A_n, B_n, Parseval's theorem states that under certain general conditions

$$\frac{1}{\pi}\int_{-\pi}^{\pi} f(x)F(x)dx = \tfrac{1}{2}A_0 a_0 + \sum_{n=1}^{\infty}(A_n a_n + B_n b_n).$$

There is an analogous theorem for **Fourier transforms**.

partial correlation—The **correlation** between the residuals of two **random variables** (variates) with respect to common regressors.
 Denoting the **regression function** of two variates y and z with respect to a common set of regressors $x_1, x_2, \cdots x_n$ by Y and Z, the coefficient of partial correlation between y and z is defined as the coefficient of simple **linear correlation** between $(y - Y)$ and $(z - Z)$. To estimate the partial correlation, it is usually necessary to resort to **sample** approximations Y' and Z' of Y and Z. In that case, the estimate of the partial correlation is the sample value of the coefficient of simple, linear correlation between $(y - Y')$ and $(z - Z')$. In the simplest case in which Y' and Z' are taken as **linear** functions of a single **variable** x, the sample estimate $r_{yz.x}$ of the partial correlation coefficient is given by the formula

$$r_{yz.x} = \frac{r_{yz} - r_{yx}r_{zx}}{[(1 - r_{yx}^2)(1 - r_{zx}^2)]^{1/2}},$$

where the symbol r_{uv} denotes the sample coefficient of linear correlation between any pair of variates u, v. *See* **regression**.

partial derivative—The ordinary derivative of a function of two or more **variables** with respect to one of the variables, the others being considered constants.
 If the variables are x and y, the partial derivatives of $F(x, y)$ are written $\partial f/\partial x$ and $\partial f/\partial y$, or $D_x f$ and $D_y f$, or f_x and f_y. The partial derivative of a variable with respect to time is known as the **local derivative**.

partial drought—In British **climatology**, a relative **drought** period of at least 29 consecutive days during which the average daily **rainfall** does not exceed 0.01 in.
 Compare **absolute drought**.

partial-duration series—A series composed of all events during the **period of record** that exceed some set criterion, for example, all **floods** above a selected base, or all daily **rainfalls** greater than a specified amount.
 Such series are used in **frequency analysis** to determine return periods, etc.

partial obscuration—A situation when a part of the celestial dome is hidden by a weather phenomenon.
 In U.S. weather observing procedures, this is the designation for **sky cover** when part (0.1–0.9) of the sky is completely hidden by surface-based obscuring phenomena. It is encoded "−X" in **aviation weather observations**; it never, by itself, constitutes a **ceiling**, for the overhead **vertical visibility** is not restricted. *Compare* **obscuration, thin**.

partial potential temperature—Same as **potential temperature**.

partial pressure—The **pressure** that a component of a gaseous mixture would have if it alone occupied the same volume at the same **temperature** as the mixture.
 See **Dalton's law**.

partial tide—*See* **tidal constituent**.

partially penetrating well—A well in which the screened portion through which the water enters the well does not penetrate the full thickness of the formation.

particle—An aggregation of sufficiently many atoms or molecules that it can be assigned macroscopic properties such as volume, **density**, **pressure**, and **temperature**.

But sometimes by particle, without qualification, is meant a **subatomic particle** such as the **proton** or **neutron** (which themselves are composed of other "elementary particles") or the **electron**. *See also* **particles**.

particle charge—The **electric charge** associated with a **particle**, for example, a **raindrop, ice crystal, hailstone,** or **aerosol**.

particle derivative—Same as **individual derivative**.

particle velocity—In ocean wave studies, the instantaneous **velocity** of a water particle undergoing **orbital motion**.

It has the **scalar** value

$$\frac{\pi}{T} \, He^{-2\pi z/L},$$

where T is the **wave period**, H the **wave height**, z the depth below **still-water level**, and L the **wave length**. At the crest, its direction is the same as the direction of progress of the **wave**; at the trough it is in the opposite direction. *See* **orbital velocity**.

particles—Components of the **atmosphere** composed of solid or liquid matter.

Particles may be both released from the earth's surface, such as **dust** or **smoke**, or formed in the atmosphere, as in **rain** or **ice** particles or sulfate **aerosol**. The particles in the atmosphere are usually defined in terms of their size, or diameter. Particles less than 100 μm in diameter are referred to as aerosol particles. These are aerodynamically stable and settle out only slowly (strictly speaking, the term aerosol refers to the gas–particle **colloidal system**, not just the particulate **phase**). The aerosol is usually divided into three modes: the Aitken mode (diameter less than 0.5 μm), the **accumulation mode** (0.5–2.0 μm), and the coarse mode (greater than 2 μm). The Aitken and accumulation modes are collectively referred to as **fine particles**. The larger particles of the coarse mode compose **clouds** and **hydrometeors** such as rain and **sleet**, which precipitate out. *See also* **particle**.

particulate loading—1. The mass of suspended particulate **aerosols** per mass or volume of air.

2. A measure of the amount of mass of **particulates** per unit time that activities in a region generate in the **atmosphere**.

If a simple "box model" is used to estimate the associated particulate concentration, the result is proportional to the total **emission rate** and is inversely proportional to the **wind speed** and the vertical cross-sectional area of the airflow through a plane perpendicular to the **wind**.

particulates—The term for solid or liquid **particles** found in the air.

Some particles are large or dark enough to be seen as **soot** or **smoke**. Others are so small they can be detected only with an electron microscope. Because particles originate from a variety of mobile and stationary sources, their chemical and physical compositions vary widely. Particulate matter can be directly emitted or can be formed in the **atmosphere** when gaseous pollutants such as sulfur dioxide (SO_2) and NO_x react to form **fine particles**. See **aerosol**.

partly cloudy—1. In U.S. climatological practice, the character of a day's weather when the average cloudiness, as determined from frequent observations, has been from 0.4 to 0.7 for the 24-hour period.

Compare **clear, cloudy**.

2. In popular usage, the state of the weather when **clouds** are conspicuously present but do not completely dull the day or the sky at any moment.

In **weather forecast** terminology, this term may be used when the expected **cloudiness** is from about 0.3 to 0.6.

pascal—The **SI** derived unit of **pressure**.

One pascal (Pa) is equal to 1 **newton** m^{-2}. The kilopascal (kPa) is the preferred unit for **atmospheric pressure**, but the more familiar **millibar** (mb) is the unit of pressure generally used by meteorologists, by international agreement; 1 mb = 1 hPa (**hectopascal**). For a typical **sea level pressure**, 102.345 kPa = 1023.45 hPa = 1023.45 mb.

Pascal's law—A hydrostatic principle that **pressure** applied to an enclosed fluid is transmitted undiminished to every portion of the fluid and to the walls of the containing vessel.

Pasquill–Gifford categories—*See* **stability categories**.

Pasquill stability classes—A method of categorizing the **stability** of a region of the **atmosphere** in terms of the horizontal **surface wind**, the amount of **solar radiation**, and the fractional **cloud cover**.

passive cavity aerosol spectrometer probe.—*See* **optical scattering probe.**

passive cloud—A **cumulus** cloud that is no longer dynamically connected with the **atmospheric boundary layer** via updrafts or downdrafts.
These clouds decay and dissipate just above the **boundary layer** top and can leave behind a concentration of **pollutants**. The clouds are radiatively coupled with the boundary layer and shade the ground.

passive front—Same as **inactive front.**

passive permafrost—(Also called fossil permafrost.) Permanently **frozen ground (permafrost)** that, under present climatic conditions, will not refreeze if thawed; opposed to **active permafrost.**

passive system—A **remote sensing** system that relies on the **emission** of natural levels of **radiation** from the target.
Most satellite-borne meteorological instruments are passive systems.

past weather—*See* **weather.**

pastagram—*See* **thermodynamic diagram.**

patchy turbulence—**Turbulence** that is not continuous in space, but is separated by regions of **stability** and **laminar flow.**
Examples include the turbulence in a horizontal plane across the **entrainment zone** of the **convective boundary layer**, or a stable, **nocturnal boundary layer** with patches of **turbulent flow** caused by windbreaks scattered across the landscape.

path—Same as **trajectory.**

path line—A curve along which a fluid **particle** travels.

path method—Same as **parcel method.**

pattern correlation—The Pearson **product-moment** coefficient of **linear correlation** between two **variables** that are respectively the values of the same variables at corresponding locations on two different maps.
The two different maps can be for different times, for different levels in the vertical direction, for forecast and observed values, etc. Occasionally referred to as **map correlation**. *See* **anomaly correlation**, a special case of pattern correlation.

patterned ground—A general term for any ground surface exhibiting a discernibly ordered, more-or-less symmetrical, morphological pattern of ground and, when present, vegetation.

Pavlovsky's approximation—Similar to the **Dupuit–Forchheimer assumptions**, except that the **streamlines** are assumed to be parallel to the sloping **aquifer** base rather than horizontal.

payload—1. In referring to vehicles for scientific research and/or **observation**, the dimensions (often only the weight) of the scientific equipment carried by a **rocket**, aircraft, ship, etc.
This usually includes sensors, data storage and telemeter gear, and instrument power supply, and sometimes includes special auxiliary equipment and recovery gear.
2. The environmental sensors and instruments carried on an aircraft, **rocket**, or satellite.

PBL—Abbreviation for **planetary boundary layer.**
See **atmospheric boundary layer.**

PCA events—Abbreviation for **polar cap absorption events.**

PDSI—Abbreviation for **Palmer Drought Severity Index.**

PE—In **radar**, abbreviation for **permanent echo.**

pea-soup fog—A thick, nearly opaque **fog.**
The term is often encountered as a simile, for example, "The fog was as thick as pea soup."

peak current—Usually refers to the maximum current in a lightning **return stroke.**

peak discharge—For a given flow record, the maximum instantaneous **discharge.**

peak frequency—The **frequency** (period/wavelength) of waves represented by a peak (maximum energy) in the **wave spectrum**; sometimes known as the dominant frequency.

peak gust—After U.S. weather observing practice, the highest "instantaneous" **wind speed** recorded at a **station** during a specified period, usually the 24-hour **observational day.**

Therefore, a peak gust need not be a true **gust** of **wind**. *Compare* **fastest mile**.

peak power—1. The **power** averaged over that carrier-frequency **cycle** that occurs when the power is maximum (usually one-half the maximum instantaneous power).

In **radar**, the **transmitted power** averaged over one **cycle** of the **carrier** at the position in the **pulse** where the **power** is at a maximum.

For typical **weather radars**, the peak power is in the **range** from several kilowatts to a megawatt.

pearl lightning—Same as **beaded lightning**.

pearl necklace lightning—Same as **beaded lightning**.

Pearson distribution—A skewed **probability distribution function** of the form

$$f(x) = \frac{1}{\beta\Gamma(p)} \left(\frac{x - \alpha}{\beta}\right)^{p-1} e^{-(x-\alpha)/\beta},$$

where β and p are positive constants, α is any real-value constant, and $\Gamma(p)$ represents the gamma function with argument p.

Abramowitz, M., and I. A. Stegun, 1965: *Handbook of Mathematical Functions*, p. 930.

Péclet number—A **nondimensional number** arising in problems of **heat transfer** in fluids.

It is the ratio of **heat** advection to heat diffusion and may be written

$$\text{Pé} = UL/K,$$

where U is a **characteristic velocity**, L a **characteristic length**, and K the **thermometric conductivity**. Also,

$$\text{Pé} = \text{Re} \times \text{Pr},$$

the product of the **Reynolds number** and **Prandtl number**.

pedestal—In **radar**, a device for supporting and positioning the **antenna**.

Typically, the pedestal allows the **azimuth** and **elevation** angles of the antenna to be controlled separately or in a coordinated way to permit different methods of **scanning**.

pedestal cloud—*See* **wall cloud**.

peesash—(Also called peshash, pisachee, pisachi.) A hot, dry, dust-laden **wind** of India.

peesweep storm—(Also called peaseweep, peesweip, peewit, teuchit, swallow storm.) An early-spring **storm** in Scotland and England.

pellicular water—*See* **field capacity**.

pendulum day—The time required for the plane of a freely suspended (Foucault) pendulum to complete an apparent rotation about the local vertical.

This **period** τ is given by the formula

$$\tau = \frac{2\pi}{\Omega\sin\phi} = \frac{\text{sidereal day}}{\sin\phi},$$

where Ω is the angular speed of the earth and ϕ the latitude. *See* **inertial flow**.

penetrating top—*See* **overshooting top**.

penetration range—Same as **night visual range**.

penetrative convection—The result of buoyant **thermals** that encounter a capping stable layer and have sufficient **energy** to travel some distance into the stable layer.

This penetration may lead to **mixing** of fluid between the stable and convective layers. A common example is the penetration of thermals from the atmospheric **convective boundary layer** into the **potential temperature** inversion capping the **mixed layer**. *See* **entrainment zone**.

penetrometer—A pointed device that indicates the amount of **resistance** encountered when it is forced into a material such as **snow** or soil.

See **ram penetrometer**.

penitent ice—A spike or pillar of compacted **snow**, **firn**, or **glacier ice** caused by differential melting and **evaporation**.

Necessary for this formation are 1) **air temperature** near **freezing**; 2) **dewpoint** much below freezing; and 3) strong **insolation**. Consequently, penitent ice is most developed on low-latitude mountains, especially the Chilean Andes, but has been found in polar regions. Penitents are oriented individually toward the noonday sun, and usually occur in east–west lines. The term is derived from the Spanish nieve penitente (penitent snow), which is still widely used throughout the literature.

penitent snow—Same as **penitent ice**.

penknife ice—Same as **candle ice**.

Penman–Monteith method—A micrometeorological method to determine the rate of moisture transport (i.e., **evaporation rate** E or **latent heat** flux) away from a surface employing the concept of **aerodynamic resistance** and requiring measurements of **net radiation** R_n, ground **heat flux** G, **temperature**, **humidity** and **wind speed**.

The latent heat **flux** is given by

$$\lambda E = \frac{S(R_n - G) + \rho C_p [e_s(T_z) - e(z)]/r_a}{S + \gamma[(r_a + r_c)/r_a]},$$

where λ is the latent heat of **vaporization**; r_c and r_a are **aerodynamic** resistances; C_p and ρ are the **specific heat** and **density** of air, respectively; S is the rate of change of **saturation vapor pressure** with temperature; γ is the **psychrometric constant**; e_s is the saturation vapor pressure; T_z is the temperature at a height z above the surface; and e is the **vapor pressure**. The product λE is the latent heat flux.

Stull, R. B., 1988: *An Introduction to Boundary Layer Meteorology*, 666 pp.

pennant—A means of representing **wind speed** in the **plotting** of a **synoptic chart**; it is a triangular flag, or pennant, drawn pointing toward lower **pressure** from a **wind-direction shaft**.

In the **five-and-ten system**, a pennant represents 50 knots and replaces five **barbs**.

pentad—A group of five.

In **climatology**, it is applied to a period of five consecutive days. It is often preferred to the week for climatological purposes since it is an exact factor of the 365-day year.

per mille—(Symbol ‰.) Per thousand or 10^{-3}; used in the same way as percent (%, per hundred, 10^{-2}).

Per mille (by weight), is commonly used in **oceanography** for **salinity** and **chlorinity**; for example, a salinity of 0.03452 (or 3.452 percent), is commonly stated as 34.52 per mille.

percent reduction—*See* **multiple correlation, regression**.

percentage of possible sunshine—Ratio of the actual duration of **bright sunshine** to the geographically or topographically possible duration.

percentile—One of a set of numbers on the random-variable axis that divides a **probability distribution** into 100 equal areas; it is a **quantile** equal to one one-hundredth of a total **population**.

perched groundwater—A body of (often temporarily) stored water, generally of moderate dimensions, supported by a relatively impermeable stratum, and located between a **water table** and the ground surface.

perched stream—**Stream** that is separated from the underlying **groundwater** by a zone of unsaturated material.

Compare **insulated stream**.

percolation—The **gravity flow** of water within soil.

pereletok—(Also called intergelisol.) A frozen layer of ground, at the base of the **active layer**, that may persist for one or several years.

The term is Russian meaning "survives over the summer." Pereletok may easily be mistaken for **permafrost**.

perennial stream—A **stream** that contains water at all times except during extreme **drought**.

Compare **ephemeral stream, intermittent stream**.

perennially frozen ground—Same as **permafrost**.

perfect fluid—1. Sometimes used for an inviscid, incompressible, **homogeneous fluid**.

2. Same as **inviscid fluid**.

3. *See* **ideal gas**.

perfect gas—Same as **ideal gas**.

perfect-gas laws—Same as **gas laws**.

perfect prognosis method—See **perfect prognostic**.

perfect prognostic—(Often called perfect prog, perfect prognosis method.) A method or technique of developing objective forecasting aids.

Suitable **statistical** relationships are found between a **predictand** and one or more observed **variables** that can be forecast by one or more numerical (dynamic) **prediction** models. The relationships can be determined by **linear** or nonlinear **regression**, multiple **discriminant analysis**, or other statistical methods. In practice, the relationships are applied to the appropriate **output** of numerical prediction model(s) to yield forecasts of the predictand. In essence, the output of the **model**(s) is considered perfect, hence the name. The difference between **model output statistics** (**MOS**) and perfect prognostic is that in MOS the predictand is related to the actual model output, while in perfect prog, the predictand is related to observations or representations of them at (nearly) concurrent times.

perfectly diffuse radiator—A body that emits **radiance** isotropically.
See **Lambert's cosine law**.

perfectly diffuse reflector—A body that reflects **radiance** isotropically.
See **Lambert's cosine law**.

perfluorocarbons—Collective term for a series of totally fluorinated **hydrocarbons** (e.g., CF_4, C_2F_6), which are present in the **atmosphere** as a result of **anthropogenic** activities.

These species are very resistive to destruction in the atmosphere and thus have extremely long **lifetimes** (thousands of years). As a consequence of their extremely long lifetimes and strong **infrared** absorption features, the large greenhouse warming potentials of these compounds is of some concern.

pergelation—The act or process of forming **permafrost** (pergelisol).

pergelisol—Same as **permafrost**.

pergelisol table—Same as **permafrost table**.

perhumid climate—As defined by C. W. Thornthwaite in his 1948 **climatic classification**, a type of **climate** that has **humidity index** values of +100 and above.

This is his wettest type of climate (letter code *A*) and compares closely to the **wet climate** that heads his 1931 grouping of **humidity provinces**.

> Thornthwaite, C. W., 1948: An approach toward a rational classification of climate. *Geogr. Rev.*, **38**, 55–94.
> Thornthwaite, C. W., 1931: Climates of North America according to a new classification. *Geogr. Rev.*, **21**, 633–655.

pericenter—That point of any **orbit** nearest to the center of attraction; the opposite of **apocenter**.
See **perigee, perihelion**.

perigean range—*See* **perigean tide**.

perigean tide—**Tide** of increased **range** when the moon is near **perigee**, its nearest approach to the earth in its elliptical orbit.

perigee—The point in an **orbit** at which any orbiting object, for example, planet, moon, artificial satellite, is closest to the attracting body; the opposite of **apogee**.
Compare **perihelion**.

periglacial—Of, or pertaining to, the outer perimeter of a **glacier**, particularly to the fringe areas surrounding the great **continental glaciers** of the geologic **ice ages**.

Thus, "periglacial weathering" is said to have produced certain characteristic land forms.

periglacial climate—The **climate** characteristic of the regions immediately bordering the outer perimeter of an **ice cap** or **continental glacier**.

The principal climatic feature is the high frequency of very cold and dry winds off the ice area. These regions have been thought to create ideal conditions for the maintenance of a belt of intense **cyclonic** activity.

perihelion—The point on the earth's **orbit** that is nearest the sun.

At present, the earth reaches this point on about 1 January, but the date varies irregularly from year to year (due to leap years) and also has a slow secular change. For a more complete discussion, *see* **aphelion**; *see also* **pericenter, perigee**.

period—Any function of time $f(t)$ is periodic with period t if $f(t) = f(t + \tau)$ for all times t, where τ is the smallest number for which this equality holds.

Without qualification, period often means temporal period, but could mean spatial period (**wavelength**), the repeat distance of a spatially periodic function. *See* **frequency**.

period of record—The length of time that continuous, reliable observations of any of the weather elements are available at a particular location.

period of validity—Specified period of time in which the conditions stated in a **weather forecast** apply.

periodicity—A state or condition characterized by regular repetition in time or space.
See **period**.

periodogram—A plot of **amplitude** or squared amplitude against **frequency** for the **wave** components of a periodic function represented by a **Fourier series**.

perlucidus—A **cloud variety**, usually of the species **stratiformis**, in which distinct spaces between its elements permit the sun, moon, blue sky, or higher clouds to be seen.
These openings may be very small. This variety is found only in the genera **altocumulus** and **stratocumulus**. *See* **cloud classification**.

permafrost—1. (Also called perennially frozen ground, pergelisol, permanently frozen ground.) A layer of soil or bedrock at a variable depth beneath the surface of the earth in which the **temperature** has been below **freezing** continuously from a few to several thousands of years.
Permafrost exists where the summer heating fails to descend to the base of the layer of **frozen ground**. A continuous stratum of permafrost is found where the annual **mean temperature** is below about $-5°C$ ($23°F$). *Compare* **pereletok**; *see* **active layer**.
2. As limited in application by P. F. Svetsov, soil that is known to have been frozen for at least a century.
Muller, S. W., 1947: *Permafrost, or Permanently Frozen Ground, and Related Engineering Problems.*
Hare, F. K., 1951: *Compendium of Meteorology*, p. 958, and map, p. 956.

permafrost island—An isolated area of **permafrost**.

permafrost table—(Also called pergelisol table.) The more or less irregular surface in the ground that marks the upper limit of the **permafrost**; not to be confused with **frost table**.

permanent anticyclone—A **high** that exists in approximately the same location most of the year, every year, and thus is evident on long-term average annual mean charts.

permanent control—A stable **cross section** of a channel, often man-made or on bedrock, that is used for monitoring measurements of depth and **velocity**.

permanent depression—A **low** that exists in approximately the same location most of the year, every year, and thus is evident on long-term average annual mean charts.

permanent echo—(Abbreviated PE.) A **radar echo** from a stationary object, such as a building, a hill, or power lines, that may interfere with the **detection** or measurement of other kinds of echoes.
See **ground clutter**.

permanent gas—A gaseous component of the **atmosphere** with its **mixing ratio** nearly constant with time and height.

Permanent Service for Mean Sea Level—(Abbreviated PSMSL.) The organization responsible for collection, analysis, interpretation, and publication of **mean sea level** data from a global network of gauges.
PSMSL is one of the Federation of Astronomical and Geophysical Services, under the auspices of the International Council of Scientific Unions.

permanent wave—A **wave** (in a fluid) moving with no change in **streamline** pattern, and which, therefore, is a **stationary wave** relative to a **coordinate system** moving with the wave.

permanently frozen ground—Same as **permafrost**.

permeability—The portion of the **hydraulic conductivity** of a **porous medium** that is dependent on the pore structure only:

$$K = \frac{k\rho g}{\mu},$$

where K is hydraulic conductivity, k is permeability, ρ is **density**, μ is **dynamic viscosity**, and g is **acceleration of gravity**.

permeability coefficient—*See* **permeability**.

permeameter—One of any number of devices used to measure the **permeability** of porous media.

permittivity—*See* **relative permittivity**.

peroxides—*See* **organic peroxides, hydroperoxides, hydrogen peroxide**.

peroxyacetyl nitrate—(Abbreviated PAN; also called peroxyacetic nitric anhydride.) Organic compound formed in the **atmosphere** from the addition of **nitrogen dioxide**, NO_2, to the peroxyacyl radical formed in the **oxidation** of **acetaldehyde**.

PAN and its larger homologs are irritants to the eyes and breathing system. They are formed in dangerous levels in **photochemical smogs**. The PANs are thermally quite stable "reservoir" species and can **transport** active **nitrogen** from polluted to pristine regions.

peroxyalkyl radical—More correctly called **alkylperoxy radicals**.

perpetual frost climate—(Also called frost climate, ice-cap climate.) In general, the **climate** of the **ice cap** regions of the world; thus, it requires temperatures sufficiently cold so that the annual **accumulation** of **snow** and **ice** is never exceeded by **ablation**.

The perpetual frost climate is one of the **polar climates** in W. Köppen's **climatic classification** and is characterized by a warmest-month **mean temperature** of less than 0°C (32°F). It is designated by letter code *EF* and is equivalent to the **frost climate** of C. W. Thornthwaite's classifications and to Nordenskjöld's (1928) **high arctic climate**.

> Thornthwaite, C. W., 1931: Climates of North America according to a new classification. *Geogr. Rev.*, **21**, 633–655.
> Thornthwaite, C. W., 1948: An approach toward a rational classification of climate. *Geogr. Rev.*, **38**, 55–94.
> Nordenskjöld, O., and L. Meeking, 1928: *The Geography of the Polar Regions*, American Geographical Society, Special Pub. No. 8, New York.
> Köppen, W. P., and R. Geiger, 1930–1939: *Handbuch der Klimatologie*, Berlin: Gebruder Borntraeger, 6 vols.

perry—(Also spelled parrey, parry, pirrie, pirry.) A sudden, heavy fall of **rain**; a **squall** in England, sometimes referred to as "half a **gale**."

Pers sunshine recorder—A **sunshine recorder** of the type in which the timescale is supplied by the motion of the sun.

The instrument, which is pointed at the **celestial pole**, consists of a hemispherical mirror mounted externally on the optical axis of a camera. The lens of the camera forms an **image** of the sun that is reflected by the hemispherical mirror so that as the sun moves across the sky, the image traces an arc of a circle on the photographic paper.

Persian Gulf Water—A **Mediterranean Water** mass originating in the Persian Gulf, where it has a **temperature** above 20°C and a **salinity** above 38 **psu**.

It spreads, much diluted, at 250 m in the Arabian Sea and is found at a depth of 500–600 m south of Madagascar.

persistence—1. The previous value in a **time series**.

Thus, if $x(t)$ denotes the present value, the value of persistence would be $x(t - 1)$, whence the latter value is regarded as "persisting." It is used as an objective **standard** in the **verification** of **weather forecasts**. Sometimes it is extended to mean $x(t - h)$ where h is arbitrary.

2. (Also called constancy, steadiness.) With respect to the long-term nature of the **wind** at a given location, the ratio of the magnitude of the main **wind vector** to the average speed of the wind without regard to direction.

3. (Also called inertial forecast.) In general, the tendency for the occurrence of a specific event to be more probable, at a given time, if that same event has occurred in the immediately preceding time period.

persistence forecast—In meteorology, a forecast that the future weather condition will be the same as the present condition.

The persistence forecast is often used as a **standard** of comparison in measuring the degree of **skill** of forecasts prepared by other methods, especially for very short projections. *See* **persistence**; *compare* **random forecast, probability forecast**.

persistence tendency—The limited tendency for existing weather conditions to persist.

It is reflected by the positive **correlation** between successive values of most meteorological **elements** when arranged in their order of occurrence.

persistent oscillation—*See* **oscillation**.

personal equation—A systematic **observational error** due to the characteristics of the **observer**.

The **uncertainty** in a reading made by an observer may be ascertained by a **statistical** analysis of his readings.

perturbation—Any departure introduced into an assumed **steady state** of a system.

The magnitude is often assumed to be small so that product terms in the **dependent variables** may be neglected; the term "perturbation" is therefore sometimes used as synonymous with "small perturbation." The perturbation may be concentrated at a point or in a finite volume of space; it may be a **wave** (sine or cosine function); in the case of a rotating system, it may be symmetric about the axis of rotation; or it may be a displacement by the **parcel method**. The mathematical work in an **instability** problem may be facilitated by the **perturbation technique**, whether or not the equations are linearized. In **synoptic meteorology**, this term is used for any departure from **zonal flow** within the major **zonal** currents of the **atmosphere**. It is especially applied to the wavelike disturbances within the **tropical easterlies**. *See* **easterly wave**; *compare* **disturbance**.

perturbation equation—Any equation governing the behavior of a **perturbation**.
This may or may not be a **linearized differential equation**.

perturbation method—Same as **method of small perturbations**.

perturbation motion—The motion of a **disturbance** (usually, but not necessarily, assumed infinitesimal), as opposed to the motion of the **steady state** of the system on which the **perturbation** is superimposed.

perturbation quantity—Any **parameter** of a system, for example, **velocity** components or **temperature**, that may or may not have been assumed to be **small perturbations** from a mean or steady-state value.

perturbation technique—A mathematical technique to eliminate **linear** terms in an equation in order to retain the **nonlinear** (**turbulence**) terms.
Variables such as **potential temperature** (θ) or **velocity** (U) can be partitioned into mean (slowly varying) and **perturbation** (rapidly varying) components. Mean components or averages are often represented with an overbar, while perturbation quantities are indicated with a prime:

$$\theta = \overline{\theta} + \theta' \text{ or } U = \overline{U} + u'.$$

When substituted in the **equations of motion** or other budget equations, the resulting equations have terms that explicitly describe the mean and turbulence components, and the interaction between these components. Next, the whole **perturbation equation** can be averaged, which eliminates the linear terms (terms having only one perturbation variable, such as $\overline{u'\partial\theta/\partial x}$). The remaining **nonlinear** terms (terms that have products of two or more perturbation quantities, such as $\overline{u'\partial\theta'/\partial x}$) represent **turbulent fluxes**, **variances**, or **correlations**. *Compare* **ensemble average**, **area average**, **Reynolds averaging**.
Starr, V. P., 1966: *Physics of Negative Viscosity Phenomena*, McGraw-Hill Pub. Co., 256 pp.
Stull, R. B., 1988: *An Introduction to Boundary Layer Meteorology*, 666 pp.

perturbation theory—The systematic derivation of linearized equations of systems by the **method of small perturbations**, exhibiting the assumptions involved; or any **model** derived by use of this method.

Peru/Chile Current—The eastern **boundary current** of the south Pacific subtropical **gyre**, also known as the **Humboldt Current**.
The Peru/Chile current originates where part of the water that flows toward the east across the subantarctic Pacific Ocean is deflected toward the north as it approaches South America. It flows northward along the coast of Chile, Peru, and Ecuador, and is associated with the economically most important **coastal upwelling** of the World Ocean. The upwelling region extends from 40°S into the equatorial region, where the current separates from the coast and turns toward the west, joining the **South Equatorial Current**, and the coastal upwelling blends into the **equatorial upwelling** belt.

Peruvian paint—Same as **painter**.

peshash—Same as **peesash**.

petit St. Bernard—A **mountain wind** of Haute Tarentaire in France.

pH—In aqueous solution, the **logarithm** (base 10) of the inverse of the concentration of **hydrogen** ions expressed in moles per liter; a measure of solution acidity. Thus, $pH = -\log[H^+]$.
The pH of neutral water is 7, since it contains 10^{-7} moles hydrogen ions and 10^{-7} moles hydroxyl **ions** per liter. Lower pH values correspond to more acidic solution. The pH of natural waters in the **atmosphere** is 5.5 or less, due to the dissolution of acidic gases such as **carbon dioxide** and **sulfur dioxide**. In polluted continental areas the pH can drop to as low as 3.0, a condition known as **acid rain**, due to the production of sulfuric and nitric acids from **anthropogenic** activity.

phase—1. For any type of periodic motion (e.g., rotation, **oscillation**) a point or stage in the **period** to which the motion has advanced with respect to a given initial point.

Specifically, the phase or **phase angle** is the angular measure along a **simple harmonic wave**, the **linear** distance of one **wavelength** being 360° of phase measure. This is often generalized by equating one **cycle** of any oscillation to 360°. *See* **delay, interference, surface of constant phase**.

2. The state of **aggregation** of a substance, for example, solid, liquid, or gas.

phase angle—*See* **phase**.

phase averaging—Averaging a **wave** characteristic over one full **cycle**.

Buoyant **oscillations** in the form of **gravity waves** can exist in the **atmosphere** and ocean, and these oscillations can **transport** momentum and **energy**. However, the amount of transport often oscillates too, so an instantaneous measure of these **momentum** or energy fluxes would not be representative. A better value is obtained by measuring the transport at many closely spaced intervals during the passage of any one **wave**, and then finding the **mean value** of the measurements.

phase change—(Or phase transformation.) A thermodynamic process in which a substance changes from one **phase** to another.

A phase change entails **discontinuity**. At a given **temperature**, two phases of a pure, homogeneous substance are characterized by different enthalpies (and entropies), and the **enthalpy** difference is called **latent heat**. The phase transitions of greatest importance to meteorology are those between **water vapor**, liquid water, and **ice**.

phase delay—1. In the **transfer** of a single-frequency **signal** from one point to another in a system, the time delay of a part of the **wave** identifying its **phase**.

2. The difference in **phase** between one **signal** and another.

phase distortion—The **distortion** that occurs in an instrument when the relative phases of the **input** signal differ from those of the **output** signal.

phase front—Same as **surface of constant phase**.

phase function—A function describing the dependence of scattered **radiance** on **scattering** angle.

The phase function is a dimensionless and normalized version of the **scattering function**, such that the integral of the phase function over 4π steradians equals 4π. The phase function for a given **wavelength** is a property of the medium, not of the incident **radiation**, provided the incident radiation is unpolarized. As the **size parameter** of a scatterer increases, its phase function becomes more **anisotropic**, with progressively more of the scattered radiance being concentrated into a **diffraction peak**. *See also* **Rayleigh phase function, Henyey–Greenstein phase function**.

phase instability—The **instability** associated with a supercooled or supersaturated **phase** of matter.

A small **fluctuation** (on a molecular **scale**) or foreign **particle** may cause change to a more stable phase. For example, at a **temperature** below the equilibrium **melting point** of ice, the ice phase is stable, whereas liquid water is supercooled and is unstable because its chemical potential is higher than that of ice. A small fluctuation such as collision with ice may cause the **supercooled water** to change into ice.

phase matrix—A more complicated form of the **phase function** that is needed to treat the **scattering** of polarized **radiation**.

phase modulation—**Modulation** in which the **phase** of a **signal** is caused to depart from its reference value by an amount proportional to the instantaneous value of the modulating function.

A signal phase modulated by a given function can be regarded as the same as a signal **frequency** modulated by the time derivative of that function.

phase shift—1. Change in **phase** of a **signal** as it passes through a **filter**, some other system component, or a transmission medium.

2. The method employed for electronic **beam** steering in phased-array **radar**.

3. For **polarimetric radar**, the **differential phase shift** between the copolarized and cross-polarized components, a measure of the **propagation effect** interpreted in terms of the degree of preferred orientation of the scatterers in the propagation medium.

phase space—1. A Fourier spectral representation of physical space.

By computing the **Fourier transform** of a physical **signal**, such as **temperature** measurements frequently sampled during some period, the resulting amplitudes of the sine and cosine waves are an alternative way of describing the data. These amplitudes represent the signal in phase space.

2. A technique used in **nonlinear** dynamics and **chaos** theory to examine processes by the

evolving relationship between **dependent variables**, rather than by the relationship between dependent and **independent variables**.
See **fast Fourier transform** (FFT).

phase spectrum—A measure of the relative **phase** between two meteorological **variables**, segregated by **wavelength**.
It is the phase difference for any **frequency** between two **time series** that yields the greatest **correlation**. The phase spectrum is computed as the arctangent of the ratio of the **quadrature spectrum** to the **cospectrum**. For example, **turbulence** usually consists of **vertical velocity** and **potential temperature** either in phase (for daytime **convection**) or 180° out of phase (for turbulence in a **statically stable** environment). However, **gravity waves** usually consist of vertical velocity and potential temperature that are 90° out of phase. Thus, if one were analyzing the cross spectra at night in a **stable boundary layer**, and found 90° **phase shift** for the longer wavelengths, but 180° for the shorter wavelengths, then one could infer that long-wavelength gravity waves are propagating through a region of small-eddy turbulence.

phase speed—(Or wave speed; also called phase velocity, wave velocity.) The speed of propagation of a mathematical **surface of constant phase** (or phase angle) of a time-harmonic **wave**.
Compare **group velocity**.

phase transformation—*See* **phase change**.

phase velocity—1. Same as **phase speed**.
2. The **velocity** at which a **wave** of given **frequency** and **wave length** advances across the ocean.
See **dispersion relationship**.

phased-array antenna—An **antenna** that has a **radiation pattern** determined by the relative phases and amplitudes of the currents on the individual antenna elements.
The direction of the **antenna pattern** can be steered by properly varying the relative phases of those elements.

phenogram—Chart showing elapsed time between plant or crop development stages through its life **cycle**.
The unit of time may be in calendar units such as days or in other measures such as **thermal** time (**growing degree-days**), fraction of life cycle, etc. *See* **phenology**.

phenological observation—A record of a plant or crop describing the physiological development stage or physical stage.
For example, date of tillering (rapid vegetative plant growth stage in wheat), date of flowering, date of physiological maturity, date of planting, etc. *See* **phenology**.

phenology—The science that treats the periodic biological phenomena with relation to **climate**, especially seasonal changes.
Phenological events are stages of plant growth. From a climatological viewpoint, these phenomena serve as bases for the interpretation of progress in local seasons and the climatic zones, and are considered to integrate the effects of a number of bioclimatic factors on rate of plant development. Phenology may be considered a branch of the science of bioclimatics, the sequence of plant or crop development stages through its life **cycle**. Growth stages may be defined by stage of physiological development such as germination, first true leaf, flowering, maturity, etc., and/or by physical stage such as planting, emergence, harvest, etc.

Philippines Current—One of the western **boundary currents** of the North Pacific subtropical **gyre**.
It flows northward from just north of Mindanao (10°N) to Taiwan where it continues as the **Kuroshio**. It is fed from the **North Equatorial Current**, from which it continues to entrain water along its way.

phosphorescence—*See* **luminescence**.

photo cell—Same as **photoelectric cell**.

photochemical air pollution—Type of **air pollution**, such as Los Angeles **smog**, associated with the buildup of **oxidation** products formed from the **degradation** of **hydrocarbons**, etc.
The term arises because **sunlight** is required to initiate **photolysis** reactions. **Ozone** and **nitrogen dioxide** are always present in photochemically produced mixtures; often other species such as **peroxyacetyl nitrate** and **formaldehyde** are also produced.

photochemical oxidants—Products of the **degradation** of atmospheric emissions resulting from **photooxidation** processes.

Products include **ozone, nitrogen dioxide**, and **peroxyacetyl nitrate**, all of which contribute to the total **oxidant** concentration.

photochemical reaction—A chemical reaction that involves either the **absorption** or **emission** of **radiation**.

The absorption of an **ultraviolet** photon often provides the **energy** required to break chemical bonds and initiate a reaction sequence. Examples of photochemical reactions are the **photolysis** of **nitrogen dioxide**, $NO_2 \rightarrow NO + O$, or **ozone**, $O_3 \rightarrow O_2 + O$. The latter reaction leads to the initiation of chain reactions that cause the breakdown of **hydrocarbons** and other **pollutants** in the **troposphere**. *See also* **photolysis, photodecomposition**.

photochemical smog—Air contaminated with **ozone, nitrogen oxides**, and **hydrocarbons**, with or without natural **fog** being present.

In the presence of **sunlight, hydrocarbons** and NO_x arc involved in a complex series of chemical reactions that eventually creates ozone and other **oxidants** as **secondary pollutants**. However, ozone is also destroyed by NO_x. Photochemical **air pollution** levels are generally proportional to concentrations of nitrogen oxides and hydrocarbons; they also increase with strong solar **intensity** and high ambient temperatures, which increase biogenic volatile organic emissions to the **atmosphere** from vegetation. The pollutant levels are inversely proportional to **wind speed** and **inversion** height. *See also* **smog**.

photochemistry—The study of the chemical and physical changes occurring when a molecule or atom absorbs **light**.

These can include **photodecomposition, fluorescence, chemiluminescence**, and quenching of the **excited state**. The **atmosphere** can be considered as a complex photochemical reactor, the chemistry of which is driven by the visible and **ultraviolet radiation** emanating from the sun.

photoconductive cell—A **photoelectric cell** with electrical **resistance** that varies with the amount of **illumination** falling upon the sensitive area of the cell.

$$\text{Conductivity} = 1/(\textbf{resistivity}).$$

photodecomposition—The chemical destruction of a molecule following the **absorption** of **light** energy.

See also **photolysis, photochemical reaction**.

photodissociation—Fragmentation of a molecule into two or more components, which may or may not be charged, as a consequence of **absorption** of a **photon** (interaction with **electromagnetic radiation**).

photoelectric cell—(Also called photo cell.) A **transducer** that converts **electromagnetic radiation** in the **infrared**, visible, and **ultraviolet** regions of the **electromagnetic spectrum** into electrical quantities such as voltage, **current**, or **resistance**.

See **photoelectric effect**.

photoelectric effect—Any process in which **illumination** of a material by **electromagnetic radiation** causes **electrons** to be separated from atoms or molecules.

Photoelectric effect is often synonymous with surface photoelectric effect, "the release of electrons by **light** at the boundary between a solid or liquid . . . and usually a gas" (Hughes and Dubridge 1932). The photoelectric effect was discovered in 1887 by Heinrich Hertz. The fundamental law of photoelectricity is the Einstein law

$$e = h\nu - p,$$

where e is the (maximum) **kinetic energy** of the emitted **photoelectron**, ν is the **frequency** of the source of illumination, h is **Planck's constant**, and p is the (minimum) difference between the **potential energy** of an electron inside and outside the material to which it is bound. An implicit assumption underlying this equation is that the initial kinetic energy of the electron is negligible compared with its maximum kinetic energy. The Einstein law played a fundamental role in establishing the **quantum theory** of light (Leighton 1959).

Hughes, A. L., and L. A. DuBridge, 1932: *Photoelectric Phenomena*, 1–2.
Leighton, R. B., 1959: *Principles of Modern Physics*, 67–69.

photoelectric photometer—A device that uses a **photoelectric cell** to measure **electromagnetic radiation** in the visible portion of the **electromagnetic spectrum**.

photoelectric photometry—In contrast to the methods of visual **photometry**, an objective approach to the problems of photometry, wherein any of several types of photoelectric devices are used to replace the human eye as the **sensing element**.

photoelectric transmittance meter—(Or photoelectric transmissometer.) An instrument for measuring the **transmissivity** of the **atmosphere**; a type of transmissometer.

It consists of a constant-intensity collimated **light** source located at a suitable distance from a **photoelectric cell**. Variation in the **turbidity** of the atmosphere causes changes in the **intensity** of the light received by the photo cell, thereby varying its electrical **output**.

photoelectron—A **free electron** produced by **photoionization**.

photogrammetry—The analysis of data using photogrammetric procedures and techniques.

photographic barograph—A **mercury barometer** arranged so that the position of the upper or lower **meniscus** may be measured photographically.

In one design the **image** of the meniscus is formed on a rotating drum covered with sensitized paper so that a continuous record is obtained of **pressure** as a function of time. *See* **barograph**.

photoionization—Fragmentation of an atom or molecule into two or more components, some of which are charged (**electrons, ions**), as a consequence of **absorption** of a **photon** (interaction with **electromagnetic radiation**).

Photoionization is an example of the **photoelectric effect**.

photolabile—A chemical that is easily dissociated by the **absorption** of **light**, usually in the visible or **ultraviolet** region.

An example of a photolabile species is **nitrous acid**, HONO, which has a very short lifetime in **sunlight**.

photolysis—The process by which a chemical species undergoes a chemical change as the result of the **absorption** of a **photon** of **light** energy.

See also **photodecomposition, photochemical reaction, photooxidation**.

photometer—An instrument for measuring the **luminance, luminous** intensity, or **illuminance** of a **light** source.

Analogous to **radiometer**, but with an **output** weighted by the **wavelength** response of the human eye (i.e., scaled by **luminous efficiency**). It is common in historical literature to see this term used for any radiometer that responds primarily to wavelengths that are visible to the human eye, with or without scaling for luminous efficiency. A photometer used to measure the **intensity** of a distant light is sometimes referred to as a **telephotometer** or **transmissometer**. *See* **hydrophotometer, radiometer**.

photometry—The measurement of the visual aspect of **radiant energy (light)**.

As such, it is distinguished from **radiometry** in that photometry takes into account the varying **sensitivity** of the eye to different **wavelengths** of light.

photon—The massless **particle** that, according to the **quantum theory** of **radiation**, carries the smallest discrete amount of **electromagnetic energy** $h\nu$, where h is **Planck's constant** and ν is the **frequency** of the associated **electromagnetic wave**.

Although the photon has no mass it does have **(linear) momentum** $h\nu/c$, where c is the free-space **speed of light**, and intrinsic **angular momentum** (spin) $h/2\pi$, as evidenced by **radiation pressure** and **radiation torque**.

photon flow rate—The rate of flow of **photons**.

photon flux—The **photon flow rate** per unit plane surface area.

photooxidation—The general term used to describe the conversion of a reduced molecule to an oxidized form in the presence of molecular **oxygen** via a set of chemical reactions that are initiated by **photolysis**.

In the **atmosphere, hydrocarbons** are converted to **carbon monoxide**, CO, and **carbon dioxide**, CO_2, via photooxidation.

photoperiod—Day length or duration of daily **exposure** to light and dark periods, either natural or artificially manipulated.

photoperiodism—Response of an organism to the relative duration of dark and light periods.

In plants, photoperiodism may affect flower or seed development, vegetative growth, formation of bulbs and tubers, leaf shape, character and extent of branching, abscission (dropping of vegetative growth, i.e., protective seed sheath) and leaf fall, root development, dormancy, and death.

photopic vision—Vision mediated by cones at **normal** levels of **luminance**.

Cones allow color vision. *See* **foveal vision**.

photopolarimeter—A **polarimeter** operating at optical **wavelengths**, especially in the **band** of wavelengths visible to the human eye.

photosphere—The intensely bright portion of the sun (or any other star) visible to the unaided eye.
  Although the sun appears to have a surface, the photosphere is actually a gaseous layer many hundreds of kilometers thick from which the **light** is emitted.

photostationary state relation—A relationship that determines the ratio of the concentrations of **nitric oxide**, NO, and **nitrogen dioxide**, NO_2, in the **troposphere**.
  In its simplest form, the ratio is controlled by the following chemical reactions, which interconvert NO and NO_2 without any change in the **ozone** concentration:

$$NO + O_3 \rightarrow NO_2 + O_2,$$

$$NO_2 + h\nu (+ O_2) \rightarrow NO + O_3.$$

In the real **atmosphere**, the ratio is perturbed by the presence of other **oxidants** (mostly hydroperoxyl and organic peroxyl radicals), which also convert NO to NO_2 and lead to net ozone production. The photostationary state relation is also occasionally referred to as the Leighton relationship, after Philip Leighton.
  Leighton, P. A., 1961: *Photochemistry of Air Pollution*, Academic Press, New York.

photosynthesis—Process by which cells containing chlorophyll in green plants convert incident **light** to **chemical energy** and synthesize organic compounds from inorganic compounds, especially carbohydrates from **carbon dioxide** and water, accompanied by the simultaneous release of **oxygen**.

photosynthetically active radiation—(Abbreviated PAR.) The **electromagnetic energy** in the 400–700 nm **wavelength** range.
  Measured as the photosynthetic **photon flux** (PPF) in quanta per second per square meter, or mole of quanta per second per square meter, or photosynthetic **irradiance** (PI) in watts per square meter for the specified wavelength band.

phreatic cycle—The time of the successive rise and decline of a **water table**, over a day, year, or longer time period.

phreatic line—*See* **water table**.

phreatic surface—Same as **water table**.

phreatic water—Same as **groundwater**, in an **unconfined aquifer**.

phreatic zone—Same as **zone of saturation**.

phreatophyte—A plant that habitually obtains its water supply either directly from the **zone of saturation** or through the **capillary fringe**.
  Salt cedar, mesquite, greasewood, sycamore, willow, and cottonwood are examples.

physical climate—The actual **climate** of a place, as distinguished from a hypothetical climate, such as that simulated by a mathematical **model**.

physical climatology—The major branch of **climatology** that deals with the explanation of **climate**, rather than with presentation of it (**climatography**).

physical equation—Same as **equation of piezotropy**.

physical forecasting—Forecasts based on the governing equations of **hydrodynamics**.
  The terminology is typically used to distinguish the forecasts from those based on **statistical** models.

physical hydrodynamics—*See* **hydrodynamics**.

physical meteorology—A subfield of meteorology generally restricted to that part of meteorology not explicitly devoted to atmospheric motions.
  There is no real distinction between **atmospheric physics** and physical meteorology. Physical meteorology usually deals with optical, electrical, acoustic, and thermodynamic phenomena of the **troposphere**, its chemical composition, the laws of **radiation**, and the physics of **clouds** and **precipitation**.

Physical Oceanographic Real-Time System—(Abbreviated PORTS.) A system of **current**, **water level**, and **meteorological stations** telemetering their data to a central location for real-time dissemination, processing, and archival.

Available to pilots, mariners, the U.S. Coast Guard, and all marine interests in voice, graphical, and **digital** form and via the World Wide Web. The PORTS concept was first introduced in the United States by NOAA's National Ocean Service in Tampa Bay, Florida, and subsequent systems have been installed in the Port of New York and New Jersey, in Galveston Bay, and in San Francisco Bay.

physical oceanography—The study of the physical properties of the ocean, including but not limited to **temperature**, **salinity**, and **ocean currents**, and of the physical processes that determine those properties.

physics of the atmosphere—*See* **physical meteorology**.

physiography—Same as **geomorphology**.

phytoclimate—1. The climatic characteristics of the air spaces occupied by plant communities within the **canopy**.

2. Description of **climate** characteristics of a region defined largely by distribution of plant species.

phytoclimatology—1. The study of the **microclimate** in the air space occupied by plant communities within the **canopy**, on the surfaces of the plants themselves, and, in some cases, in air spaces within the plants.

2. The study of climatic regions defined largely by distribution of plant species.

phytometer—A device (similar to a **potometer**) for measuring **transpiration**.

It consists of a vessel containing soil in which one or more plants are rooted and sealed so that water can escape only by transpiration from the plant.

phytotoxic—Poisonous to plants.

Sensitive plants react to **pollutant** chemicals by suppression of plant growth or reduced productivity.

phytotron—A controllable artificial **environment** used for the study of plants under well-defined conditions.

Pi theorem—(Or **Buckingham Pi theory**.) The basis for **dimensional analysis**.

The theorem states that an equation for a physical system that can be written $f(Q_1, Q_2, \ldots, Q_m) = 0$ can also be written as $g(\pi_1, \pi_2, \ldots, \pi_{m-n}) = 0$ where Q_i are m dimensional parameters, numbers, and **variables**; π_i are $m - n$ nondimensional quantities; and n is the number of fundamental dimensional units.

pibal—Contraction for **pilot-balloon observation**.

Piché evaporimeter—A porous paper wick **atmometer**.

The instrument consists of a graduated tube, closed at one end, that is filled with distilled water and then covered with a larger circular piece of filter paper held in place by a disc and collar arrangement. In operation the instrument is inverted so that the distilled water is in contact with the filter paper. The amount of **evaporation** that occurs during an interval of time is determined by noting the change in level of the **meniscus** of the water. *See* **Dines anemometer**, **Pitot-tube anemometer**; *compare* **Venturi tube**.

picture element—(Acronym **pixel**.) The smallest **element** (spectral or spatial) sensed by a satellite sensor.

A pixel generally appears as a rectangle with a uniform shade or color. A row of pixels forms a **scan line**, and a series of scan lines forms an **image**.

piedmont glacier—The lobe-shaped, expanded, terminal part of a **valley glacier** spread out over broad lowlands at the base of mountains.

Armstrong, T., B. Roberts, and C. Swithinbank, 1973: *Illustrated Glossary of Snow and Ice*, Scott Polar Research Institute, Special Pub. No. 4, 2d ed.; The Scholar Press Ltd., Menston Yorkshire, UK.

Pierson–Moskowitz spectrum—A standard **wave spectrum** representing a **fully developed sea**.

The form of the spectrum is dependent only on the **wind speed**. The Pierson–Moskowitz **frequency** would be the minimum peak frequency a **wind sea** can attain for a given wind speed.

piezometer—A device (tube or pipe) that allows one to determine the **elevation** of **hydraulic head** in an **aquifer** at a given point.

piezometric head—The **pressure** that exists in a **confined aquifer**.

Specifically, it is the **elevation** above a **datum** plus the **pressure head**.

piezometric surface—The imaginary surface that everywhere coincides with the **piezometric head** of the water in the **aquifer**.

In areas of **artesian ground water**, it is above the land surface.

piezotropy—*See* **equation of piezotropy.**

pileus—(Also called **cap cloud**, scarf cloud.) An **accessory cloud** of small horizontal extent, often **cirriform**, in the form of a cap, hood, or scarf, which occurs above or attached to the top of a **cumulus** or **cumulonimbus** (less often **stratocumulus**) **cloud** that often pierces it.

Sometimes several pileus clouds are observed above each other. Pileus is formed as a moist layer locally lifted due to rising cloud below. *See* **cloud classification.**

pilmer—In England, a heavy **shower** of **rain.**

pilot balloon—A small ascending balloon observed with a **theodolite** in order to obtain data for the computation of the speed and direction of **upper-air** winds.

pilot-balloon observation—(Commonly contracted pibal.) A method of **winds-aloft observation**, that is, the determination of **wind** speeds and directions in the **atmosphere** above a **station**.

This is done by reading the **elevation** and **azimuth** angles of a **theodolite** while visually tracking a **pilot balloon**. The ascension rate of the balloon is approximately determined by careful inflation to a given total **lift**. After release from the ground, periodic readings (usually at one-minute intervals) of elevation and azimuth angles of the balloon are recorded. These data are transferred to a **winds-aloft plotting board**, and the wind speed and direction at selected levels are calculated by trigonometric methods. *See* **graphing board, winds-aloft plotting board;** *compare* **rabal, rawin, rawinsonde, double-theodolite observation.**

pilot-balloon station—Station at which **upper winds** are measured by optical tracking of a free **balloon.**

pilot briefing—(Or flight-weather briefing, flight briefing.) Oral commentary on the observed and forecast weather conditions along a route, given by a **forecaster** to the pilot, navigator, or other air crew member prior to takeoff.

See **debriefing.**

pilot chart—A **chart** of a major ocean area that presents in graphic form averages obtained from weather, **wave**, **ice**, and other marine data gathered over many years in meteorology and **oceanography** to aid the navigator in selecting the quickest and safest routes; published in the United States by the Defense Mapping Agency Hydrographic/Topographic Center from data provided by the U.S. Naval Oceanographic Office and the National Environmental, Satellite Data and Information Service of the National Oceanic and Atmospheric Administration.

pilot report—(Commonly contracted PIREP; also called aircraft report.) A report of in-flight weather by an aircraft pilot or crew member.

A complete report includes the following information in this order: location and/or extent of reported weather phenomena; time of **observation**; description of phenomena; **altitude** of phenomena; **type** of aircraft (only with reports of **turbulence** or **icing**). This term is also applied to more informal and less complete reports of weather aloft as well as to post-flight reports filed after landing. *Compare* **aircraft observation.**

pilot streamer—A relatively slow-moving, nonluminous **lightning** streamer, the existence of which has been postulated but not verified, to provide a physical explanation for the observed intermittent mode of advance of a **stepped leader** as it initiates a cloud-to-ground **lightning discharge**.

Whereas the stepped leader descends at an average speed of the order of 10^5 m s^{-1} during its downward motion, it advances only about 50 m at a time with a speed greater than average and then pauses for 50–100 µs before resuming its downward movement. The average downward speed has been associated with an invisible **streamer**, the pilot streamer, which is postulated to descend at a uniform speed only slightly in excess of the ionizing speed of **electrons** in air and lay down a trail of weak residual ionization along which the stepped leader moves very rapidly in a pulsating manner. The idea of a pilot **leader** has been supplanted by more modern theory based on studies of sparks over long distances in the laboratory.

pinene—Structural isomers, formula $C_{10}H_{16}$; α- and β-pinene are members of the monoterpene family.

Stored in the needles of many conifer trees and emitted under stress, they are responsible for the characteristic odor of pine forests. The pinenes are very reactive **hydrocarbons.**

piner—In England, a rather **strong breeze** from north and northeast.

pingo—A large **frost mound** of more than one year's duration.

While this Eskimo term is used in several related senses, the above meaning is becoming increasingly accepted.

pink snow—*See* **red snow**.

pip—1. In **radar** terminology, an **echo** from a **point target**.

It originally stemmed from the appearance of these types of echoes on **A-scopes** and **R-scopes**.

2. On **weather maps**, the triangles or half-circles along **fronts** that point in the direction of frontal movement.

piping—The progressive development of internal **erosion** by seepage, appearing **downstream** as a hole or seam discharging water that contains soil particles.

PIREP—(Or pirep.) The aviation communications code word and commonly used contraction for **pilot report**.

pirry—Same as **perry**.

pisachee—Same as **peesash**.

pitch—The **oscillation** of a ship about the lateral axis, that is, alternate rising and falling of bow and stern.

See **roll, yaw, ship motion**.

Pitot-static tube—Same as **Pitot tube**.

Pitot tube—(Or Pitot-static tube.) A pressure **anemometer** consisting of two concentric tubes that are oriented parallel to the flow.

The inner tube is open at the **upstream** end to sense the **total pressure**, while the outer tube is closed with a rounded contour and has a ring of small **static pressure** ports a short distance **downstream**. Each tube is connected to a **manometer** and the difference between the two pressures, the **dynamic pressure**, is proportional to the square of the fluid speed.

Pitot-tube anemometer—A pressure-tube **anemometer**, consisting of a **Pitot tube** mounted on the **windward** end of a **wind vane**.

See also **Dines anemometer**.

pivot arm anemometer—An instrument for measuring the **wind speed** by determining the force (usually by measurement of mechanical deflection) on a suspended plate exposed to the airstream.

In sensitive devices for rapid response **turbulence** measurements, a **feedback** system may be used to generate an appropriate restoring force to maintain the plate at its **equilibrium** position.

pixel—A **picture element**, that is, a single **element** of an **image** or picture.

A pixel is the smallest element of a picture. Other elements are rows, columns, and subframes, which are rectangular regions of the image or picture.

pixel value—The **intensity** of a **pixel**, usually an integer.

For grayscale images, the pixel value is typically an 8-bit data value (with a **range** of 0 to 255) or a 16-bit data value (with a range of 0 to 65535). For color images, there are 8-bit, 16-bit, 24-bit, and 30-bit colors. The 24-bit colors are known as true colors and consist of three 8-bit pixels, one each for red, green, and blue intensity.

plage—Areas of intensified **brightness** in the solar **chromosphere** that are usually observed in an **emission line** of calcium.

These regions are usually broad and irregular and approximately coincide with **faculae** and **sunspots** in the underlying **photosphere**. The number and size of these regions varies during the **solar cycle**, causing enhanced chromospheric **emission** during **solar maximum** relative to a lower emission during **solar minimum**.

plan position indicator—(Abbreviated PPI.) A type of **radar** display on which the **reflectivity** or other properties of **echoes** are plotted in plan position, forming a maplike **display**, with radial distance from the center representing **range** and the angle of the **radius vector** representing **azimuth** angle.

The PPI is usually produced by **scanning** the **antenna** in azimuth with the **elevation** fixed. It may be an **intensity**-modulated display or may use a **scale** of colors to represent the values of the function displayed.

Planck function—See **Planck's radiation law**.

Planck's constant—A universal constant, denoted by h, with the value 6.626075 x 10^{-34} J s, in the **quantum theory** of matter and **radiation**.

Planck's constant is the bridge between the **wave** and **particle** descriptions of **light**, an electromagnetic **wave** of **frequency** v alternatively described as a stream of photons each with **energy** hv. According to the **quantum theory**, when the energy of an atom, molecule, or **nucleus** changes from one discrete **energy level** E_1 to another E_2, **conservation of energy** requires the **emission** (creation) or **absorption** (annihilation) of a **photon** with energy given by the Einstein frequency condition

$$hv = E_2 - E_1.$$

Planck's constant is a fundamental scaling **parameter** of the **universe**, determining, among other things, the sizes of atoms and molecules.

Planck's radiation law—The distribution law of **photon** energies for **radiation** in **equilibrium** with matter at **absolute** temperature T:

$$\frac{2\pi}{h^2 c^2} \frac{E^3}{e^{E/kT} - 1},$$

where E is photon energy, c is the free-space **speed of light**, h is **Planck's constant**, and k is **Boltzmann's constant**.

The integral of this function between any two energies is the contribution to the total **irradiance** from all photons with energies in this **range**. This law is mathematically similar and physically analogous to the Maxwell–Boltzmann distribution for the **kinetic energies** of (ideal) gas molecules in equilibrium. One of the salient differences between these two distributions is that the number of gas molecules within a container is conserved whereas the number of photons is not, increasing with the absolute temperature of the container. Within an enclosure the walls of which are opaque (but not necessarily black) and at **temperature** T, the distribution of photon energies is given by Planck's radiation law.

plane albedo—The **albedo** of a flat surface.

In contrast to the **spherical albedo**, this usually refers to the albedo of a flat surface illuminated only by **direct solar radiation** from a specified direction.

plane atmospheric wave—An **atmospheric wave** represented in two-dimensional **rectangular Cartesian coordinates**, in contrast to a **wave** considered on the spherical earth.

plane-dendritic crystal—(Or plane dendrite; sometimes called stellar crystal.) An **ice crystal** exhibiting an elaborately branched (dendritic) structure of hexagonal symmetry, with its much larger dimension lying perpendicular to the principal (c axis) of the **crystal**.

Such crystals usually form by vapor **deposition** at atmospheric temperatures between about $-14°$ and $-18°C$. In sufficient concentration they readily aggregate to **snowflakes** by interlocking of their branches. *See* **dendrite**.

plane flow—(Also called two-dimensional flow, planar flow.) A flow in which the **streamlines** all lie in parallel planes.

plane of the ecliptic—The plane of the **great circle** on the celestial plane that is the apparent annual path of the sun around the earth.

plane parallel atmosphere—An approximation used in many **radiation** models that depict the **atmosphere** as being only one-dimensional and bounded at the top and bottom by horizontal plane surfaces.

plane polarization—(Archaic.) Linear **polarization**.

plane source—*See* **point source**.

plane wave—Any **wave** for which the **surface of constant phase** is a plane.

planetary albedo—The ratio of reflected-to-incident shortwave **flux density** at the top of the atmosphere.

Planetary albedos are a function of time and space, depending in particular on the nature of **cloud** and ground cover. For the earth as a whole, the long term average planetary albedo is about 0.31.

planetary atmosphere—*See* **atmosphere**.

planetary boundary layer—(Abbreviated PBL.) Same as **atmospheric boundary layer**.

planetary circulation—1. The instantaneous **large-scale** flow in the **troposphere** when viewed as a hemispheric or worldwide system.

2. The mean or time-averaged hemispheric **circulation** of the **atmosphere**; in this sense, almost synonymous with **general circulation**.

3. Same as **general circulation**.

planetary-geostrophic—An approximation to the **equations of motion** for very large length scales, of the order of the earth's radius, obtained by suppressing **nonlinear** advection terms from the **momentum** equations.

The remaining equations may be viscous, are hydrostatic, and include full **thermodynamics**.

planetary vorticity effect—The effect of the **variation** of the earth's **vorticity** with latitude in altering the **relative vorticity** of a flow with a **meridional** component.

A fluid with a **free surface** in a rotating cylinder will exhibit a corresponding effect, owing to the shrinking or stretching of radially displaced columns. *See* **beta plane**, **vorticity equation**.

planetary wave—1. Same as **long wave**.

2. Same as **Rossby wave**.

planetary-wave formula—*See* **Rossby wave**.

plant climate—Same as **phytoclimate**.

plant climate zone—1. Area in which a common set of **temperature** ranges, **humidity** patterns, and other geographic and seasonal characteristics combine to create a particular plant distribution by allowing certain plants to succeed and causing others to fail.

2. A region in a **climatic classification** system that places greatest importance on plant distribution to identify climatic characteristics.

plasma—An ionized gas composed of positive and negative charges (and possibly neutral atoms and molecules) of almost equal charge **density**.

At least one kind of charge is mobile. The term was coined by Langmuir and Tonks (1929) "to designate that portion of an arc-type **discharge** in which the densities of **ions** and **electrons** are high but substantially equal." A more quantitative definition can be given in terms of the Debye shielding distance, the distance over which the density of negative charges can be appreciably different from that of positive charges: A plasma is an ionized gas for which the Debye shielding distance is small compared with a **characteristic length** (Spitzer 1962). According to this definition the **ionosphere** is a plasma, and so is a slab of aluminum, but in atmospheric usage it is limited to an ionized gas.

Langmuir, I., and L. Tonks, 1929: *Phys. Rev.*, **33**, p. 196.
Spitzer, L., 1962: *Physics of Fully Ionized Gases*, 2d ed., p. 22.

plasmapause—A shell surrounding the earth at a radial distance of a few earth radii in the outer **magnetosphere**, marking a rapid decrease in **electron density** with increasing distance.

The plasmapause is the outer boundary of the **plasmasphere** and is usually connected along **magnetic field** lines to an **ionospheric trough**.

plasmasphere—The region of the **magnetosphere** in which the **plasma** (composed of **electrons** and positive **ions**) rotates with the earth.

Beyond the **plasmapause**, **magnetospheric convection** driven by the interaction of the **solar wind** with the outer magnetosphere dominates the motion.

plate crystal—(Or platelet.) An **ice crystal** usually having an **aspect ratio** greater than 2 and as much as 100 and with hexagonal symmetry, although trigonal (threefold) and occasionally other symmetries occur.

As such crystals fall through the clouds in which they form, they may encounter conditions causing them to develop dendritic extensions, that is, become **plane-dendritic crystals**.

plateau glacier—*See* **glacier**.

plateau station—*See* **high-altitude station**.

platelet—Same as **plate crystal**.

platykurtic—*See* **kurtosis**.

playa—In the western United States, the sandy, salty or mudcaked flat floor of a **desert** basin having interior drainage, usually occupied by a shallow lake during or after prolonged **heavy rains**.

Pleistocene climate—The **climate** from about 2 500 000 to 10 000 years ago, differing from earlier (**Pliocene**) climate in being generally colder and with greater extremes of **glacial** (cold) to **interglacial** (warm) climate.

Characterized in the last 875 000 years by repeated glacials, each lasting approximately 110 000 years, punctuated by interglacials lasting 10 000 to 15 000 years. Before that, the predominant **periodicity** was 41 000 years rather than 110 000 years. The distribution of continents and oceans has been relatively stable during the Pleistocene, but the **oscillation** between glacials and inter-glacials has been characterized by major changes in atmospheric concentrations of optically active gases such as **carbon dioxide** and **methane** (higher in interglacials), and global changes in **sea level** (lower in glacials) associated with changes in the volume of **ice** on land (lower in interglacials). Changes in the amount, and the seasonal and latitudinal distribution, of **insolation** resulting from the evolving characteristics of the earth's orbit around the sun play a major role as the pacemaker of these changes. *See* **Milankovitch theory**.

Pleistocene epoch—The period from 2 500 000 to 10 000 years ago, during which **continental glaciers** periodically expanded to cover subpolar regions in both hemispheres.

Pleistocene glaciation—The periodic expansion of **continental ice** sheets to cover much of Canada and northeastern Asia during the interval 2 500 000–10 000 years ago.

Pleistocene Ice Age—*See* **Pleistocene epoch**.

Pliocene climatic optimum—The period from about 3.2 to 2.5 million years ago during which polar **temperature** was 5°–10°C warmer than today, the Antarctic and Arctic **ice caps** nearly melted, and **sea level** was 35 m higher.

plotting—Same as **map plotting**.

plotting board—*See* **winds-aloft plotting board**.

plotting model—Conventional pattern for the **plotting** of symbols around a **station location** on a **synoptic chart**.

plotting position—An empirical distribution, based on a **random sample** from a (possibly unknown) **probability distribution**, obtained by plotting the exceedance (or cumulative) **probability** of the **sample** distribution against the sample value.
The **exceedance probability** for a particular sample value is a function of sample size and the rank of the particular sample. For exceedance probabilities, the sample values are ranked from largest to smallest. The general expression in common use for plotting position is

$$P = \frac{r - b}{n + 1 - 2b},$$

where r is the ordered rank of a sample value, n is the sample size, and b is a constant between 0 and 1, depending on the plotting method. *See* **California plotting position, Cunnane plotting position, Gringorten plotting position, Hazen plotting position, Weibull plotting position**.

plotting symbols—Conventional ideograms on a meteorological **chart** or in a message that represent various meteorological observations.

plow wind—(Or plough wind.) Strong straight-line winds associated with nontornadic **outflow** from strong **thunderstorms**.
Used by Canadian meteorologists, particularly in Manitoba; no longer used in the United States. *See also* **derecho, downburst**.

plowshares—*See* **foam crust**.

plum rains—See **bai-u**.

plume—1. Buoyant jet in which the **buoyancy** is supplied from a **point source**; the buoyant region is continuous.
See **thermal**.

2. A mostly horizontal (sometimes initially vertical) stream of air **pollutant** that is being blown **downwind** from a smokestack.
Typical smoke-plume diameters are of order 1–10 m initially, gradually expanding to 100 m or more, while lengths can be order of 1–100 km. The path and shape of the **smoke** plume can indicate the nature of **turbulence** in the **atmospheric boundary layer**, such as **looping** plumes, **fanning** plumes, and **coning** plumes.

Emanuel, K. A., 1994: *Atmospheric Convection*, Oxford University Press, p. 16.

plume rise model—An **algorithm** for calculating the **altitude** that a **plume** will rise due to **momentum** and **buoyancy** forces before reaching an **equilibrium** height.
Plume rise increases with higher buoyancy or momentum of the plume and decreases with

increasing **wind speed** or vertical **temperature** gradient in the **atmosphere**. The rate of rise is fastest at the point of emission and decreases due to the **entrainment** of **ambient air**, which has minimal momentum and generally lower temperatures than the original plume. The plume is considered to be at its final height when the rate of rise decreases to at point where it is equivalent to vertical velocities generated by **turbulence** in the atmosphere.

pluvial—1. Pertaining to **rain**, or more broadly, to **precipitation**, particularly to an abundant amount thereof.
> *Compare* **hyetal, glacial.**

> 2. Pertaining to an interval of **geologic time** that was marked by a relatively large **amount of precipitation**; the opposite of **interpluvial**.
> This is usually applied to those periods of heavy **rainfall** in the lower latitudes associated with the equatorward advance of the **glacierization** of an **ice age**. Thus, a **pluvial period** of low latitudes generally coincides with a **glacial** stage of higher latitudes. During pluvial periods the surfaces of existing lakes rose to high levels. In some cases several lakes combined into great sheets of water, such as Lakes Lahontan and Bonneville in the Great Basin of the western United States.

pluvial index—(Rare.) The **amount of precipitation** falling in one day, or other specified period, that is likely to be equaled or exceeded at a given place only once in a century, that is, a precipitation amount that has a **return period** of 100 years.

pluvial period—Intervals of **wet climate** within the **Pleistocene** during which lake levels were high; generally corresponding to **glacials** in regions with **Mediterranean climate** and to **interglacials** in regions with **monsoonal climate**.

pluviograph—Same as **recording rain gauge**.

pluviometer—Same as **rain gauge**.

pluviometric coefficient—(Also called hyetal coefficient.) For any month at a given **station**, the ratio of the monthly **normal** precipitation to one-twelfth of the annual normal precipitation.
> Seen collectively, the 12 pluviometric coefficients describe the normal month-to-month distribution of the normal annual **precipitation** in terms of each month's "share" of the annual amount. *See* **isomer**.

pluviometric quotient—The ratio of the actual **amount of precipitation** collected at a specific place, during a specific month, to the amount that would have been obtained if the mean annual amount had been equally distributed over every day of the year.

pluviometry—Study of **precipitation**, including its nature and distribution, and techniques of measurement.

PM-10—Particulate matter smaller than 10 μm that is suspended in the air.
> *See* **aerosol, airborne particulates, criteria pollutants**.

PM-2.5—Particulate matter smaller than 2.5 μm that is suspended in the air.
> *See* **aerosol, airborne particulates, criteria pollutants**.

PMC—Abbreviation for **polar mesospheric clouds**.

PMSE—Abbreviation for **polar mesospheric summer echoes**.

PNA—Abbreviation for **Pacific–North American pattern**.

poëchore—**Steppe** region.
> *See* **biochore**.

POES—Abbreviation for **Polar-Orbiting Operational Environmental Satellite**.

pogonip—Same as **ice fog**; a Native American word applied particularly to ice fogs occurring in the mountain valleys of the western United States.

Poincare wave—A **gravity wave** that is slow enough (**low frequency**) to feel the effects of the earth's rotation, so that the **Coriolis parameter** appears in the **dispersion relation**.
> Within a channel in a rotating system, a Poincare wave has sinusoidally varying cross-channel **velocity** with an integral or half integral number of cross-channel waves spanning the channel. In the **shallow water approximation** the waves have dispersion relationship with squared **frequency**

$$\omega^2 = f^2 + c^2(k^2 + \pi^2 n^2/L^2),$$

in which f is the Coriolis parameter, k is the **wavenumber** along the channel, L is the width of the channel, n is any positive integer, and c is the **phase speed** for shallow water gravity waves;

$$c = (gH)^{1/2},$$

in which g is the **acceleration** due to **gravity** and H is the mean depth of the fluid. Related to Poincare waves are **Kelvin waves**, which take the role of the mode with $n = 0$.

point—1. Position or time of occurrence, as in **boiling point, freezing point, compass** point, **point rainfall**, etc.

2. In Australia, a unit of **precipitation** amount; equal to one one-hundredth of an inch.

point data—Station **observation** data as opposed to area information as might be derived from analyses of data from multiple points or remotely sensed.

point discharge—A silent, nonluminous, gaseous **electrical discharge** from a pointed conductor maintained at a **potential** that differs from that of the surrounding gas.

In the **atmosphere**, trees and other grounded objects with points and protuberances may, in disturbed weather, be sources of **point discharge current**. Close to a pointed and grounded conductor that extends above surrounding objects, the local **electric field strength** may be many times greater than that existing at the same level far from the elevated conductor. When this local **field** reaches such a value that a **free electron**, finding itself acted upon by this field, can be accelerated (in one **mean free path**) to a sufficiently high velocity to ionize neutral air molecules, point discharge will begin. Different structures will yield point discharge under quite different gross field conditions, for geometry is critically important. Point discharge is recognized as a major process of charge **transfer** between electrified clouds and the earth, and is a leading item in the charge balance of the global electrical circuit. *Compare* **corona discharge, spark discharge**.

point discharge current—The electrical **current** accompanying any specified source of **point discharge**.

In the electrical budget of the earth–atmosphere system, point discharge currents are of considerable significance as a major component of the **supply current**. Estimates made by Schonland (1928) of the point discharge current from trees in arid southwest Africa suggest that this process accounts for about 20 times as much delivery of negative charge to the earth during typical **thunderstorms** as do **lightning discharges**. Although the great height of **thundercloud** bases in arid regions, such as that referred to in Schonland's study, tends to favor point discharge over lightning charge transfer, point discharge still seems more significant than lightning even in England, where Wormell (1953) found for Cambridge a ratio of about 5:1 in favor of point discharge over lightning charge transfer.

Chalmers, J. A., 1957: *Atmospheric Electricity*, 156–175.
Wormell, T. W., 1953: Atmospheric electricity: some recent trends and problems. *Quart. J. Roy. Meteor. Soc.*, **79**, 3–50.
Schonland, B. F. J., 1928: The polarity of thunderclouds. *Proc. Roy. Soc. A*, **118**, 233–251.

point gauge—A device used to measure water levels.

It consists of an adjustable, pointed metallic rod that is lowered until it just pierces the water surface. A **hook gauge** is a similar device, except that the rod is U-shaped, and the submerged hook is raised until the point just breaks the water surface.

point of occlusion—On a **synoptic chart**, the point at which a **warm front, cold front**, and **occlusion** meet.

point precipitation—**Precipitation** at a particular **site**.

point rainfall—**Rainfall** measured at a particular **site**.

point source—1. With respect to **radiation**, a single point in space emitting radiation.

The radiation from such a source may be expressed as **flux** per unit solid angle (W sr^{-1}).

2. In experimental studies of **atmospheric turbulence** and **diffusion**, a source of particulate matter from a single fixed point.

There are continuous and instantaneous point sources. The analogous concepts of **line source** and **plane source** are also frequently encountered.

3. In **hydrodynamics**, a source of mass, that is, a **singular point** in the **field** where the **equation of continuity** fails.

point target—A radar target that is small compared with the **pulse volume**, which is the cross-sectional area of the **radar beam** multiplied by half the length of the radar pulse.

Typically, a point target is an object such as an aircraft, ship, or building that reflects the radar signal from relatively simple, discrete surfaces. *See* **coherent target**; *compare* **distributed target**.

point-to-multipoint communication—Also known as a multicast, this is a communications system involving one source that transmits a **signal** that is received at multiple destinations.

For example, television broadcasts are an example of a point-to-multipoint communication.

point-to-point communication—A communication between two and only two locations.

A **normal** phone call is an example of a point-to-point communication (excluding conference calls).

point vortex—A straight **line vortex**, the flow of which can be modeled in two dimensions, with a point concentration of **vorticity** surrounded by **irrotational** flow.

See **line vortex**.

poise—The unit of **viscosity** in the centimeter-gram-second system, one gm cm^{-1}s^{-1}, named in honor of Poiseuille.

Poiseuille flow—According to Poiseuille's law, the **laminar flow** of a liquid through a circular tube as determined by

$$q = \frac{\pi r^4}{8\mu} \left(\frac{p_1 - p_2}{L} \right),$$

where q = **discharge**; L = length of pipe; p_1, p_2 = **pressure** at pipe ends; μ = **dynamic viscosity**; and r = radius of pipe.

Compare **Couette flow**.

Poiseuille–Hagen law—The relation between **velocity** and **pressure gradient** in **Poiseuille flow**.

Poisson constant—The ratio κ of the **gas constant** R to the **specific heat** at constant **pressure**, c_p.

For **dry air**, $\kappa = 0.2854$. For **moist air**,

$$\kappa = \frac{R_d}{c_{pd}} \frac{1 + r_v/\epsilon}{1 + r_v c_{pv}/c_{pd}} \approx 0.2854(1 - 0.24 r_v),$$

where R_d and c_{pd} are the gas constant and specific heat of dry air, ϵ is the ratio of the gas constants of **water vapor** and dry air, c_{pv} is the specific heat of water vapor, and r_v is the water vapor **mixing ratio**. The Poisson constant appears in the **Poisson equation** and in the definition of **potential temperature**.

Poisson distribution—A one-parameter, discrete **frequency distribution** giving the **probability** that n points (or events) will be (or occur) in an interval (or time) x, provided that these points are individually independent and that the number occurring in a subinterval does not influence the number occurring in any other nonoverlapping subinterval.

It has the form $P(n, x) = e^{-\kappa x}(\kappa x)^n/n!$. The mean and **variance** are both κx, and κ is the average density (or rate) with which the events occur. When κx is large, the Poisson distribution approaches the **normal distribution**. The **binomial distribution** approaches the Poisson when the number of events n becomes large and the **probability** of success p becomes small in such a way that $np \to \kappa x$. The Poisson distribution arises in such problems as radioactive and photoelectric emissions, **thermal** noise, service demands, and telephone traffic.

Poisson equation—1. The partial differential equation

$$\nabla^2 \phi = F,$$

where ∇^2 is the **Laplacian operator**, ϕ a **scalar** function of position, and F a given function of the independent space **variables**.

For the special case $F = 0$, the Poisson equation reduces to the **Laplace equation**. *See* **relaxation method**.

2. The relationship between **temperature** T and **pressure** p of an **ideal gas** undergoing an **adiabatic process**; given by

$$\frac{T}{T_0} = \left(\frac{p}{p_0} \right)^\kappa,$$

where T_0 and p_0 are initial state values and κ is the **Poisson constant**. With p_0 given as a reference pressure of 100 kPa, T_0 is equal to the **potential temperature**.

polacke—(Also spelled polake.) A cold, dry, northeasterly **wind** in Bohemia descending from the Sudeten Mountains (from the direction of Poland).

polar air—A type of **air mass** with characteristics developed over high latitudes, especially within the subpolar highs.

Continental polar air (cP) has low surface **temperature**, low moisture content, and, especially in its **source** regions, great **stability** in the lower layers. It is shallow in comparison with **arctic air**. Maritime polar air (mP) initially possesses similar properties to those of continental polar air, but in passing over warmer water it becomes unstable with a higher **moisture content**.

polar air depression—A weak **low pressure system** that forms in a **polar air** mass.

polar amplification—In **climate** modeling studies, the tendency for simulated **temperature** changes to be larger at high latitudes, as in the case of the warming induced by increased **greenhouse gases**.

polar angle—*See* **spherical coordinates**.

polar anticyclone—1. Same as **arctic high**.

2. Same as **subpolar high**.

Polar Atlantic Water—*See* **intermediate water**.

polar axis—*See* **polar coordinates**.

polar blackout—*See* **radio blackout**.

polar cap absorption events—(Abbreviated PCA events.) Episodes of intense **absorption** of **HF** and **VHF** radio waves over the polar caps caused by excess ionization resulting from an influx of solar energetic **particles** into the **upper atmosphere**.

PCA events have a pronounced **diurnal** variation, being much more intense during daytime than at night, and typically last for two to three days. *See also* **solar proton event**.

polar-cap ice—Same as **polar ice**.

polar cell—A weak **meridional circulation** in the high-latitude **troposphere** characterized by ascending motion in the subpolar latitudes (50°–70°), descending motion over the pole, poleward motion aloft, and equatorward motion near the surface.

As a residual of many **transient** weather systems, the polar cell is barely detectable in means with respect to time of latitude–height **cross sections**.

polar climate—(Also called frost climate, snow climate.) A **climatic zone** located in the polar latitudes marked by conditions too harsh to support vegetation.

See **Frigid zone**.

polar continental air—A **continental air** mass that develops over or near the polar region.

polar coordinates—1. In the plane, a system of **curvilinear coordinates** in which a point is located by its distance r from the origin (or pole) and by the angle θ which a line joining the given point and the origin makes with a fixed reference line, called the polar axis.

The relations between **rectangular Cartesian coordinates** and polar coordinates are

$$x = r \cos \theta, \, y = r \sin \theta,$$

where the origins of the two systems coincide and the polar axis coincides with the x axis.

2. In space, same as **spherical coordinates**.

See also **cylindrical coordinates**.

polar cyclone—Same as **polar vortex**.

polar easterlies—The rather shallow, irregular, and diffuse easterly winds located poleward of the **subpolar low pressure belt**.

In the mean in the Northern Hemisphere, these **easterlies** exist to an appreciable extent only north of the **Aleutian low** and the **Icelandic low**.

polar-easterlies index—A measure of the strength of the easterly **wind** between the latitudes of 55° and 70° N.

The index is computed from the average **sea level pressure** difference between these latitudes and is expressed as the east to west component of **geostrophic wind** to a tenth of a meter per second. *Compare* **zonal index, temperate-westerlies index, subtropical-easterlies index**.

polar front—1. According to the **polar-front theory**, the semipermanent, semicontinuous **front** separating air masses of tropical and polar origin.

This is the major front in terms of **air mass** contrast and susceptibility to **cyclonic** disturbance. *Compare* **arctic front**.

2. In oceanography, see **Antarctic Polar Front, Arctic Polar Front**.

polar-front jet stream—*See* **jet stream**.

polar-front theory—A theory originated by the Scandinavian school of meteorologists whereby a **polar front**, separating **air masses** of polar and tropical origin, gives rise to **cyclonic** disturbances that intensify and travel along the **front**, passing through various phases of a characteristic life history.

This theory ushered in a new era of atmospheric **analysis** and remains an important basis of practical **synoptic meteorology** and weather forecasting even today.

> Bjerknes, J., and Solberg, H., 1922: Life cycles of cyclones and polar front theory of atmospheric circulation. *Geofys. Publ.*, **3**, 3–18.

polar frontal zone—In the Southern Hemisphere, the region of low **salinity** water between the **Antarctic Polar Front** and the **subantarctic front**.

polar glacier—In Ahlmann's **glacier** classification, a glacier with an **accumulation area** covered by **firn** and with subsurface temperatures below **freezing** throughout the year.

The two subtypes are the **high polar glacier** and the **subpolar glacier**.

polar high—1. Same as **arctic high**.

2. Same as **subpolar high**.

polar ice—1. The thickest form of **sea ice**, one to several years old (perhaps paleocrystic), and sometimes more than 3 m (10 ft) thick.

2. (Also called polar-cap ice.) The **pack ice** of the central Arctic Ocean.

polar invasion—A vigorous thrust of **polar air** behind a **polar front**.

polar low—A small but intense **cyclone** that forms in cold **polar air** advected over warmer water.

These vortices often form in the subpolar North Pacific and subpolar North Atlantic equatorward of the **sea ice** margin. Horizontal scales **range** from several tens to several hundreds of kilometers. Because of strong winds and intense **precipitation**, these cyclones are sometimes referred to as "arctic hurricanes."

polar magnetic storm—*See* **magnetic storm**.

polar maritime air—A **maritime air** mass that develops over or near the polar region.

polar mesospheric clouds—(Abbreviated PMC.) **Cirrus**-like clouds seen from spacecraft over the polar regions during summer in both hemispheres.

They occur near the **mesopause**, at heights of roughly 85 km, and are closely related to **noctilucent clouds**.

polar mesospheric summer echoes—(Abbreviated PMSE.) Anomalously strong **radar echoes** received from the **mesopause** region during summer at high latitudes.

They are loosely associated with the occurrence of **noctilucent clouds** and **polar mesospheric clouds**.

polar night jet stream—A core of strong westerly winds that develop during autumn and winter in the upper **stratosphere** and **mesosphere** near the boundary of the polar night.

Radiative cooling in the polar night appears to maintain the required **baroclinicity**.

Polar-Orbiting Operational Environmental Satellite—(Abbreviated POES.) A general name given the U.S. series of polar-orbiting satellites beginning in 1966 with the launch of *ESSA-1*.

The series continued with eight additional satellites in the **TOS** series (*ESSA-1* through -9), six satellites in the **ITOS** series between 1970 and 1976, and evolving into the current series of TIROS-N satellites, beginning 1978. An improved version of TIROS-N was introduced in 1983, termed **Advanced TIROS-N**. With the launch of NOAA-K (renamed *NOAA-15* after launch) in May 1998, POES moved to another new generation of spacecraft with a full complement of new and improved instruments. POES satellites are sometimes referred to as the **NOAA** series of satellites, based on the name (e.g., *NOAA-15*) given to POES satellites since 1970.

polar-orbiting satellite—A satellite with an **orbit** that lies in a plane passing through the center of the earth and is inclined to the equatorial plane such that the **subpoint** track traverses polar latitudes on every orbit.

polar outbreak—(Or cold-air outbreak.) The movement of a cold **air mass** from its **source region**; almost invariably applied to a vigorous equatorward thrust of cold **polar air**, a rapid equatorward movement of the **polar front**.

polar stratospheric clouds—(Abbreviated PSC; also called nacreous clouds, mother-of-pearl clouds; rarely, luminous clouds.) Clouds are **cirrus** or **altocumulus lenticularis**, and show very strong

irisation similar to that of mother-of-pearl, especially when the sun is several degrees below the **horizon**.

They occur at heights about 20–30 km above the earth. These clouds are rarely seen, and it would appear that they can be observed only in certain regions. They have been observed mainly over Antarctica due to the colder temperatures present there in the **circumpolar vortex**. They also form over Scotland and Scandinavia in winter during periods with an intense, broad, deep, and homogeneous westerly to northwesterly flow of air over northern Europe; they are also observed in Alaska. Less frequent sightings have been reported at lower latitudes. The simultaneous occurrence of various colors of the **spectrum** in more or less irregular patterns strongly suggests **diffraction** by spherical **particles**. The exact physical constitution of **cloud** particles has been determined by aircraft (e.g., the NASA ER2 aircraft) penetration showing the presence of **nitric acid** hydrates (in particular nitric acid tri-hydrate, type I) with the addition of water ice at temperatures a few degrees colder (type II). Nuclei for clouds are thought to be **sulfuric acid** aerosol, possibly of volcanic origin. The clouds form in regions where dynamic lifting or **radiational cooling** lowers the air to temperatures below **saturation** for these different constituents (about −95°C). PSC are thought to play a major role in the formation of the "**ozone hole**" because they absorb **odd nitrogen** from the **atmosphere**, which allows the catalytic destruction of **ozone** to occur. Nacreous clouds appear stationary and, by day, often resemble pale cirrus. At **sunset**, all the colors of the spectrum appear; as the sky darkens after sunset, they increase in brilliance. As the sun drops lower and lower below the horizon and the clouds are lighted by last rays, the various colors are replaced by a general coloration that is first orange and then becomes pink, contrasting vividly with the darkening sky. The clouds next become gray and the colors of the spectrum reappear but very weakly, then fade out rapidly. Later, up until about two hours after sunset, the nacreous clouds can still be distinguished standing out against the starry sky as tenuous and gray clouds. They can even be observed all night if there is moonlight. Before **dawn**, the same series of aspects appear, but in reverse order.

polar trough—In **tropical meteorology**, a **wave trough** in the **circumpolar westerlies** having sufficient **amplitude** to reach the **Tropics** in the **upper air**.

At the surface it is manifest as a **trough** in the **tropical easterlies**, but at moderate elevations it is characterized by westerly winds. It moves generally from west to east, accompanied by considerable **cloudiness**. **Cumulus congestus** and **cumulonimbus** clouds are usually found in and around the trough lines. Early- and late-season (June and October) **hurricanes** of the western Caribbean frequently form in polar troughs. *See* **meridional front**.

polar vortex—(Also called polar cyclone, **polar low**, circumpolar whirl.) The planetary-scale **cyclonic circulation**, centered generally in the polar regions, extending from the middle **troposphere** to the **stratosphere**.

The westerly airflow is largely a manifestation of the **thermal wind** above the **polar frontal zone** of middle and subpolar latitudes. The **vortex** is strongest in winter when the pole-to-equator **temperature** gradient is strongest. In the Northern Hemisphere, the vortex has two centers in the mean, one near Baffin Island and the other over northeast Siberia.

polar wandering—The steady motion of the rotational axis relative to coordinates fixed in the earth.

The **pole** moves at a rate of about 10 cm per year in the general direction of Philadelphia, Pennsylvania. Polar wander is believed to be due to the redistribution of matter within the earth. It is sometimes called "apparent polar wander" because it is difficult to separate from **continental drift**.

polar westerlies—Same as **westerlies**.

polar wind—The permanent **outflow** of **ionization** from the polar regions of the **magnetosphere**.

polar wind divide—A somewhat antiquated term for the very diffuse boundary of low pressure between the midlatitude **westerlies** and the **polar easterlies**.

polarimeter—An instrument for determining the **degree of polarization** of **light**.
See **photopolarimeter**.

polarimetric radar—A **radar** capable of measuring any or all of the polarization-dependent attributes of a **target** or **backscattering** medium.

The term may denote a radar capable of measuring the full **polarization matrix** by means of variable transmitted **polarization** and dual-channel reception. It also may denote a simpler radar that transmits a single polarization and receives separately the copolarized and cross-polarized components of the returned **signal**. *Compare* **dual-channel radar, dual-polarization radar, polarization-diversity radar**.

polariscope—An instrument for studying, or examining substances in, polarized **light**.

See **Savart polariscope.**

polarity—May be applied to any property of a physical system that can take on only two values, usually opposite in some sense (e.g., sign or direction).

Thus, the **electron** and **positron** are said to have opposite polarity because their charges are equal in magnitude but opposite in sign. Similarly, anodes and cathodes are electrodes of opposite polarity.

polarizability—The proportionality factor between the **dipole moment** (usually electric but also magnetic) induced in an atom, molecule, or even a **particle** and the inducing **electric** or **magnetic field**.

If the induced electric dipole moment is **p**,

$$\mathbf{p} = \alpha\mathbf{E},$$

where **E** is the electric field and α the electric polarizability. In general, α is a **tensor** (because **p** and **E** are not necessarily parallel) and depends on the **frequency** of the (periodic) field **E**.

polarization—1. With respect to a transverse **electromagnetic wave**, the **correlation** between two **orthogonal** components of its **electric** (or, equivalently, **magnetic**) field.

If the ratio of the **amplitudes** of these two components and the difference in their **phases** is constant in time (completely correlated), the **wave** is said to be polarized (or completely polarized or 100% polarized). If these two amplitudes and phases are uncorrelated, the wave is said to be unpolarized (or 0% polarized). These are two extreme degrees of correlation, never strictly realized in nature, all real waves being partially polarized (or partially correlated). Associated with a polarized wave is its vibration ellipse traced out in time by the oscillating electric field at a given point in space. A line (circle) is a special ellipse, and a wave with such a vibration ellipse is said to be linearly (circularly) polarized, but the general state of (complete) polarization is elliptical. A vibration is characterized by its handedness (the sense in which it rotates; clockwise or counterclockwise), the ratio of its minor to major axis (ellipticity), and its orientation (**azimuth**). Any **beam** may be decomposed uniquely as an incoherent superposition of two beams, one unpolarized and one polarized. Thus, the ratio of the **transmitted power** of the polarized component to the total transmitted power may be taken as a measure of the **degree of polarization** of the beam. The vibration ellipse of the polarized component and the degree of polarization define the state of polarization of the beam. Polarization would be an uninteresting (indeed, unmeasurable) property of **electromagnetic radiation** were it not for the fact that two beams, identical in all respects except their state of polarization, may interact with matter differently. **Skylight** (for a molecular **atmosphere**) is, in general, partially linearly polarized, the degree of polarization being greatest approximately 90° from the sun.

2. With respect to **particles** in an **electric field**, the displacement of the charge centers within a particle in response to the electric force acting thereon.

See **polarizability.**

polarization-diversity radar—A **radar** capable of measuring polarization-dependent attributes of a **target**.

McCormick and Hendry (1975) used this term to describe a radar that transmits a fixed (or slowly variable) **polarization** while receiving signals of identical and **orthogonal** polarization. The term is equivalent to **dual-channel radar**. *Compare* **dual-polarization radar, polarimetric radar**.

McCormick, G. C., and A. Hendry, 1975: Principles for the radar determination of the polarization properties of precipitation. *Radio Sci.*, **10**, 421–434.

polarization matrix—*See* **covariance matrix, scattering matrix.**

polder—A tract of low-lying land, especially in the Netherlands, that has been reclaimed from the sea or other body of water, and is protected by dikes.

pole—1. A point in an **electromagnetic field** at which **electric** or **magnetic field** lines converge. *See also* **dipole, magnetic dipole.**

2. For any circle on the surface of a sphere, the point of intersection of the surface of the sphere and the normal line through the center of the circle.

The North and South geographic poles are the poles of the **equator** or of any other latitude circle.

3. The origin of a system of **polar coordinates.**

pole of inaccessibility—The approximate center of the most consolidated portion of the arctic **pack ice** near 83°S or 84°N, and 160°W.

pole-star recorder—An instrument used to determine approximately the amount of **cloudiness** during the dark hours.

It consists of a fixed long-focus camera positioned so that Polaris is permanently within its **field of view**. The apparent motion of the star appears as a circular arc on the photograph and is interrupted as clouds come between the star and the camera.

pole tide—Small variations in **sea level** due to the Chandler wobble of the axis of rotation of the earth.

This has a **period** close to 436 days. Maximum **amplitudes** of more than 30 mm are found in the Gulf of Bothnia, but elsewhere amplitudes are only a few millimeters.

pollen analysis—Analysis of the distribution of pollen grains of various species contained in **surface layer** deposits, especially peat bogs and lake sediments, from which a record of past **climate** may be inferred.

pollutant—*See* **air pollution**.

pollutant standards index—(Abbreviated PSI.) A relative **scale** developed by the U.S. Environmental Protection Agency that applies to specified **air pollutants** (ozone, **carbon monoxide, nitrogen dioxide, sulfur dioxide**, and **particulates**), designed to clearly and simply inform the public of the current air quality.

The scale is normalized such that an index value of 100 corresponds to the U.S. federal air-quality **standard** for each specified pollutant. The minimum and maximum values are 0 and 500, corresponding to perfectly clean and extremely dirty air. Instead of presenting to the public different numbers for different pollutants, usually only the worst value is presented. The relationship between the pollutant standards index and health advisory level are: 0–50 Good; 50–100 Moderate; 100–200 Unhealthful; 200–300 Very unhealthful; >300 Hazardous. *See* **air pollution episode**.

pollution—*See* **air pollution**.

polycyclic aromatic hydrocarbons—(Abbreviated PAHs.) Class of large aromatic molecules composed of several **benzene** rings fused together.

Some PAHs show very high carcinogenic and mutagenic activity. They are found in organic residues, such as **soot**, coal tar, and combustion exhaust. Due to their low volatility, PAHs are usually taken up onto organic aerosols, which facilitates their inhalation.

polyn'ya—(Also called ice clearing.) A Russian term meaning an area of open water, possibly containing some thin ice, within the **ice pack**.

A polyn'ya is distinguished from a **lead** by being a broad opening rather than a long, narrow fracture.

polynomial—A mathematical expression of the form

$$a_0 + a_1x + a_2x^2 + \ldots + a_nx^n,$$

where the a's are real or complex numbers and n is a positive integer, called the degree of the polynomial; or, in the multivariable case, a weighted sum of products of integral powers of two or more **variables**.

polytropic atmosphere—A **model atmosphere** in **hydrostatic equilibrium** with a constant nonzero **lapse rate**.

The vertical distribution of **pressure** and **temperature** is given by

$$\frac{p}{p_0} = \left(\frac{T}{T_0}\right)^{g/R\gamma},$$

where p is the pressure, T the Kelvin temperature, g the **acceleration of gravity**, R the **gas constant** for air, and γ the **environmental lapse rate**, the subscript zeros denoting values at the earth's surface.

polytropic process—A thermodynamic process in which changes of **pressure** p and **density** ρ are related according to the formula

$$p\rho^{-\lambda} = p_0\rho_0^{-\lambda},$$

where λ is a constant and subscript zeros denote initial values of the **variables**.

Therefore pressure and **temperature** are similarly related:

$$\frac{p}{p_0} = \left(\frac{T}{T_0}\right)^k,$$

where k is the **coefficient of polytropy**. For **isobaric** processes, $k = 0$; for **isosteric** processes, $k = 1$; for **adiabatic processes** $k = c_p/R$, where c_p is the **specific heat** at constant pressure and R is the **gas constant**; sometimes applied to circumstances when adiabatic heating or cooling combine with slow ascent or descent to produce a particular **lapse rate**. In meteorology this formula is applied to individual **air parcels** and should be distinguished from that for a **polytropic atmosphere**, which describes a distribution of pressure and temperature in space. *See also* **equation of piezotropy**.

POMAR code—A code (part **synoptic code** and part aviation operations) in which certain observable meteorological **elements** and certain aircraft data are encoded and transmitted from transport aircraft.

pondage—Water held in a **reservoir** for short periods to regulate **natural flow**, usually for hydro-electric power.

ponente—A west **wind** on the Côte d'Azur (French Mediterranean coast), the northern Roussillon region, and Corsica.
   On the Côte d'Azur it is a weakened **mistral** and brings **clear** skies. In northern Roussillon it is the **land breeze** of early morning, changing to southeast during the day, and generally preceding the **tramontana**. *Compare* **poniente**.

Ponentino—Local name for the westerly **sea breeze** on the west coast of Italy, and especially in Rome.

poniente—The west **wind** in the Strait of Gibraltar.
   See **mountain-gap wind**, **liberator**; *compare* **ponente**.

pontias—A **mountain wind** at Lyons, Department of Drôme, in southeastern France.

pool of cold air—A relatively shallow region of cold air of limited horizontal extent.
   For example, during winter at high latitudes in mountainous regions, cold air can become trapped in valleys, analogous to pools of water. *See also* **cold-air pool**.

pooling—*See* **moisture pooling**.

poorga—Same as **purga**.

pop-up thunderstorm—An **airmass thunderstorm** that forms rapidly in an otherwise rain-free **environment**.
   This most often occurs on warm, humid days, in unstable meteorological conditions.

popogaio—Same as **papagayo**.

population—(Also called universe.) In **statistical** usage, any definite class of individuals or objects.
   Compare **sample**.

poriaz—Violent northeast **winds** along the western shore of the Black Sea near the Bosporus in Russia.

Porlezzina—An east **wind** on Lake Lugano (Italy and Switzerland), blowing from the Gulf of Porlezza.

porosity—The volume of void space per total volume.
   See also **void ratio**.

porous medium—A medium that has numerous interstices, whether connected or isolated.

portable mesonet stations—Portable **observation** systems that can be set up and operated unattended at field sites to measure, for example, **temperature**, **winds**, **humidity**, **surface pressure**, **solar** and **infrared radiation**, **rainfall**, and sometimes more specialized **variables** such as **sensible** and **latent heat** fluxes, **momentum flux**, surface radiation temperature, **net radiation**, **soil moisture**, and soil **heat flux**.
   They are often self-contained, powered by solar cells and batteries, and able to transmit data to a base **station** from remote locations. Typically they are deployed in a network from a few kilometers to tens of kilometers apart to provide baseline data for **mesoscale** multifaceted observational studies.

PORTS—Abbreviation for **Physical Oceanographic Real-Time System**.

Portugal Current—One of the eastern **boundary currents** of the North Atlantic subtropical **gyre**.
   Flowing southward along the Portuguese coast, it is the continuation of the **Azores Current** and continues as the **Canary Current**. Because of its location outside the **trade-wind belt**, it is not associated with significant **upwelling**.

position vector—A **vector** that leads from the origin of a given **coordinate system** to a given point in space, thus specifying the position of the point relative to the chosen coordinate system.

positive cloud-to-ground lightning—A **lightning flash** or **stroke** between a **cloud** and the ground that lowers positive charge from the cloud to the ground.

positive discharge—A positive discharge lowers positive charge to the ground via a **lightning flash**. The flash may be initiated in the **cloud** or from the ground.

positive feedback—A sequence of interactions that amplifies the response to an initial **perturbation**. The **snow** and **ice** albedo–temperature **feedback** is an example of a positive feedback. *Compare* **negative feedback**.

positive ground flash—Same as **positive cloud-to-ground lightning**.

positive isothermal vorticity advection—Positive **vorticity advection** along an **isothermal** surface.

positive rain—**Rain** having a net positive electrical charge.

positron—(Or antielectron.) A positively charged **subatomic particle** with the same mass and charge magnitude as that of the **electron**.
The electron's **antiparticle**, the positron was first observed in 1932 in **cosmic rays** by Carl Anderson using a Wilson **cloud chamber**.
Boorse, H. A., and L. Motz, 1966: *The World of the Atom*, Vol. II, 1261–1267.
Anderson, C. D., 1961: *Amer. J. Phys.*, **29**, 825–830.

postfrontal fog—**Fog** that forms behind or after a **frontal passage**, usually a **cold front**.
It forms from the **evaporation** of **precipitation** into a cold **air mass** that causes the air mass to reach its **wet-bulb temperature**.

potamology—1. (From the Greek potamos, "river.") The science of rivers.
2. More specifically, the interdisciplinary branch of **hydraulics**, **hydrology**, and fluid dynamics dealing with surface **streams** and their regimes. Potamology generally focuses on issues of fluvial **erosion**, **transport**, and **sedimentation**; **fluvial dynamics**; and river metamorphosis or change through time.

potential—1. A function of space, the **gradient** of which is equal to a force.
In symbols,

$$\mathbf{F} = -\nabla \phi,$$

where $\mathbf{F}$ is the force, ∇ the **del operator**, and ϕ the potential. A force that may be so expressed is said to be "conservative," and the **work** done against it in motion from one given **equipotential surface** to another is independent of the path of the motion. In meteorology, the **force of gravity** has a potential, the **geopotential**, which, if the **acceleration of gravity** g is taken as constant, may be written $\phi = gZ$, where Z is the height coordinate. The **pressure force** has in general no potential, nor do the **Coriolis** or **viscous forces**. By extension and analogy, the **velocity potential**, **acceleration** potential, and **Gibbs function** (**thermodynamic potential**) are defined.
2. Applied to the value that an atmospheric **thermodynamic variable** would attain if processed **adiabatically** from its initial **pressure** to a **standard pressure**, typically 100 kPa.
See **potential density**, **potential temperature**.

potential density—The **density** an **air parcel** would attain if compressed **adiabatically** by descent to the **standard pressure** of 100 kPa.
The potential density ρ' is most easily defined in relation to the **potential temperature** θ as

$$\rho' = p/R\theta,$$

where p is a **pressure** of 100 kPa and R the **gas constant**, in appropriate units.

potential drop—The difference in hydraulic **potential** between two equipotential lines.
Analogous to the **contour interval** on topographic maps.

potential energy—The **energy** a system has by virtue of its position; the negative of the **work** done in taking a system from a reference configuration, where the potential energy is assigned the value zero, to a given configuration, with no change in **kinetic energy** of the system.
An example of potential energy is the **gravitational potential energy** of a point mass m at a distance r from the center of a spherically symmetric body with mass M (e.g., a planet):

$$-\frac{GMm}{r},$$

where G is the **universal gravitational constant** and the reference potential energy is taken as zero at infinity. At distances z above the surface of the body that are small compared with its radius, the potential energy is approximately

$$mgz,$$

where g is the **acceleration** due to **gravity** at the surface and the zero of potential energy is taken at the surface ($z = 0$). Molecular potential energies, arising from short-range forces much stronger than **gravitation**, are involved in all chemical reactions, are responsible for the cohesiveness of liquids and solids, and influence a host of processes such as **evaporation** and **condensation**.

potential equivalent temperature—Same as **equivalent potential temperature**.

potential evaporation—(Also called evaporative power, evaporation power, evaporative capacity, evaporation capacity, evaporativity.) A measure of the degree to which the weather or **climate** of a region is favorable to the process of **evaporation**.
    It is usually considered to be the rate of evaporation, under existing atmospheric conditions, from a surface of water that is chemically pure and has the **temperature** of the lowest layer of the **atmosphere**. *See* **Bowen ratio, relative evaporation**.

potential evaporation rate—Same as **potential evaporation**.

potential evapotranspiration—1. The amount of water evaporated (both as **transpiration** and **evaporation** from the soil) from an area of continuous, uniform vegetation that covers the whole ground and that is well supplied with water.
    Generally, the amount of moisture that, if available, would be removed from a given land area by **evapotranspiration**; expressed in units of water depth. It can be measured in a dry basin by determining the amount of **irrigation** water used, and in wetter regions, by the difference between **rainfall** and **runoff**, or by the supply of water required to maintain a constant amount of **soil moisture** in an isolated block of the soil. *See* **evapotranspirometer**.
    2. The quantity of water evaporated (both as **transpiration** and **evaporation** from the soil) per unit area, per unit time from an extensive stretch of continuous, uniform vegetation that covers the whole ground and that is well supplied with water.; an empirical index of the above.
    As given in Thornthwaite's 1948 **climatic classification**, it is equal to the summation of the 12 successive monthly values of the expression ct^a, where t is the monthly **mean temperature** in degrees Celsius, and a and c are coefficients that depend upon the annual heat index.

potential flow theory— **Irrotational, inviscid**, free-slip flow for which **Bernoulli's equation** applies along a **streamline**.

potential gradient—In general, the local space rate of change of any **potential**, as the **gravitational potential** gradient or the **velocity potential** gradient.
    In **atmospheric electricity**, the **electric potential gradient** (**electric field strength**) of the **atmosphere** is commonly referred to as the **atmospheric electric field**. The **electric field** is the negative of the potential gradient.

potential index of refraction—*See* **refractive index**.

potential instability—(Also called convective instability, thermal instability.) The state of an unsaturated layer or column of air in the **atmosphere** with a **wet-bulb potential temperature** (or **equivalent potential temperature**) that decreases with **elevation**.
    If such a column is lifted bodily until completely saturated, it will become unstable (i.e., its **temperature lapse rate** will exceed the **saturation-adiabatic lapse rate**) regardless of its initial **stratification**.
    Saucier, W. J., 1955: *Principles of Meteorological Analysis*, 76–78.

potential predictability—A hypothetical level of **predictability** that would be obtained under certain assumptions regarding errors in the knowledge of current and past states of the system, and the **model** used to predict the future states of the system.
    For instance, one common usage defines the potential predictability as the predictability assuming that the forecast model being used is perfect and that the only source of **error** arises from uncertainties in the **initial conditions**.

potential refractive index—In radio engineering, the **refractive index** for an **atmosphere** in which **potential temperature** and **specific humidity** are constant with height.
    The **gradient** of potential refractive index depends on the values of potential temperature and specific humidity, but over the **range** of values commonly encountered in the earth's atmosphere, this gradient does not vary much and is given approximately by

$$\frac{dn}{dz} = -\frac{1}{4R},$$

where R is the earth's radius. *See* **potential refractivity**.

potential refractivity—An index, expressed in B-units, obtained by subtracting the **gradient** of **potential refractive index** from the gradient of **refractive index** and multiplying by 10^6:

$$B = \left[(n-1) + \frac{z}{4R}\right] \times 10^6.$$

The gradient of B is zero in an (approximately) **neutral atmosphere**.

potential temperature—1. The **temperature** that an unsaturated **parcel** of **dry air** would have if brought **adiabatically** and reversibly from its initial state to a **standard pressure**, p_0, typically 100 kPa.

Its mathematical expression is

$$\theta = T(p_0/p)^\kappa,$$

where θ is the potential temperature, T is temperature, and κ is the **Poisson constant**. This exponent is often assumed to be 2/7, the ratio of the **gas constant** to the **specific heat capacity** at constant pressure for an ideal diatomic gas. *See* **virtual potential temperature**, **liquid water potential temperature**, **equivalent potential temperature**, **wet-bulb potential temperature**.

2. In **oceanography**, the **temperature** that a water sample would attain if raised **adiabatically** to the sea surface.

For the deepest points of the ocean, which are just over 10 000 m, the **adiabatic cooling** would be less than 1.5°C.

potential transpiration—The **water loss**, often expressed as a rate of **flux**, by a plant (or vegetated surface) when soil water is not limiting.

Usually determined empirically for a plant or vegetation type and referenced to **evaporation** from a **class-A evaporation pan**. Potential transpiration may exceed **pan evaporation** (soybean at peak vegetative stage is about 105% of pan). *See* **evapotranspiration**.

potential vorticity—(Sometimes called absolute potential vorticity.) The **specific volume** times the **scalar product** of the **absolute vorticity** vector and the **gradient** of **potential temperature**:

$$P = \alpha\,(2\Omega + \nabla \times \mathbf{u}) \cdot \nabla\theta,$$

where α is the specific volume, Ω the **angular velocity** vector of the earth's rotation, $\mathbf{u}$ the three-dimensional **vector** velocity relative to the rotating earth, and θ the potential temperature.

In the absence of **friction** and **heat** sources, the **Ertel potential vorticity** P is a materially **conservative property** (it remains constant for each **particle**). In **spherical coordinates** (λ, ϕ, r), where λ is longitude, ϕ is latitude, and r is the distance from the center of the earth, the above expression for P becomes

$$P = \alpha\left[\left(\frac{\partial w}{r\partial\phi} - \frac{\partial(rv)}{r\partial r}\right)\frac{\partial\theta}{r\cos\phi\partial\lambda} + \left(2\Omega\cos\phi + \frac{\partial(ru)}{r\partial r} - \frac{\partial w}{r\cos\phi\partial\lambda}\right)\frac{\partial\theta}{r\partial\phi}\right.$$
$$\left. + \left(2\Omega\sin\phi + \frac{\partial v}{r\cos\phi\partial\lambda} - \frac{\partial(u\cos\phi)}{r\cos\phi\partial\phi}\right)\frac{\partial\theta}{\partial r}\right].$$

This nonhydrostatic version is not necessary for the **analysis** of large-scale weather systems, and an approximate hydrostatic version is usually used. This approximate version neglects terms involving the **vertical velocity** w, neglects the Coriolis terms proportional to the cosine of the latitude, and makes selective use of $r \approx a$, where a is the constant radius of the earth. In this way we obtain the approximate form

$$P = \alpha\left[-\frac{\partial v}{\partial z}\frac{\partial\theta}{a\cos\phi\partial\lambda} + \frac{\partial u}{\partial z}\frac{\partial\theta}{a\partial\phi} + \left(2\Omega\sin\phi + \frac{\partial v}{a\cos\phi\partial\lambda} - \frac{\partial(u\cos\phi)}{a\cos\phi\partial\phi}\right)\frac{\partial\theta}{\partial z}\right].$$

The potential vorticity has the **SI** units $m^2\ s^{-1}K\ kg^{-1}$. It has become accepted to define 1.0×10^{-6} $m^2\ s^{-1}K\ kg^{-1}$ as one potential vorticity unit (1 PVU). *See* **vorticity equation**.

Gill, A. E., 1982: *Atmosphere–Ocean Dynamics*, Academic Press, 237–241.

potentiometer—An instrument for measuring differences in **electric potential**.

Essentially, this instrument balances the unknown voltage against a known, adjustable voltage. If the balancing is accomplished automatically, the instrument is called a self-balancing potentiometer. Potentiometers are frequently used in conjunction with **thermocouples** for measuring **temperature**.

potentiometric surface—Same as **piezometric surface**.

potometer—A device (similar to a **phytometer**) for measuring **transpiration**.

It consists of a small vessel containing water and sealed so that the only escape of moisture is by transpiration from a leaf, twig, or small plant with its cut end inserted in the water.

poudrin—(Rare.) Same as **ice crystal**.

poultry stress index—(Abbreviated PSI.) An index of poultry **heat** stress.

Categories defined from the **temperature–humidity index** (THI) are based upon an increasing death rate of poultry as the THI becomes larger.

THI	PSI	Effect
<82°F (27°C)	No stress	No heat-related increase in daily mortality
82°–84°F (27°–29°C)	Moderate stress	0.5%–1% increase in daily mortality
84°–86°F (29°–30°C)	Severe stress	1%–2% increase in daily mortality
>86°F (30°C)	Very severe stress	>2% increase in daily mortality

powder snow—A skiing term for a cover of **dry snow** that has not been compacted in any way.

power—The rate, often expressed in watts, at which **energy** is exchanged or transmitted.

In **radar** it usually refers to the rate at which **electromagnetic energy** is radiated from or received at the **antenna**.

power density—In an **electromagnetic wave**, the rate of **power** flow in a specific direction at a particular point in a transmission medium, expressed as **energy** per unit time (power, or **radiant flux**) per unit cross-sectional area normal to the direction of propagation.

The power density generally diminishes with increasing distance from the source as a result of **absorption**, **reflection**, **scattering**, and possibly other effects, as well as geometric spreading of the **beam**. For surfaces or objects that intercept the **radiation** at a sufficiently long distance from the source, the propagating energy may be regarded as plane-wave or parallel-beam radiation. Then the power density is the same as the **irradiance** at a surface normal to the beam.

power density spectrum—(Sometimes called power spectrum.) A measure of the contribution to the **total variance** from a given **frequency band** in the generalized Fourier representation of a **random** function.

If $f(t)$ is a random function, the total **energy** $\int_{-\infty}^{\infty} f^2 dt$ is infinite, so the **Fourier integral** representation is inadequate. If a transform is defined over a finite interval

$$F_T(\omega) = \frac{1}{2\pi} \int_{-T}^{T} f(t)e^{-i\omega t} dt,$$

under suitably restrictive conditions the power density spectrum may be defined as

$$\lim_{T \to \infty} \frac{\pi}{T} |F_T(\omega)|^2.$$

The theorem, proved by N. Wiener, establishing the analogy between the analysis of random functions and ordinary **Fourier analysis**, is that the power density spectrum is the **Fourier transform** of the **autocorrelation function**, which is defined for random functions as

$$\lim_{T \to \infty} \frac{1}{2T} \int_{-T}^{T} f(t)f(t + \tau) dt.$$

See **power spectrum**.

power-law profile—A formula for the **variation** of **wind** with height in the **surface boundary layer**.

It is an alternative to the **logarithmic velocity profile**, and the assumptions are the same, with the exception of the form of the dependence of **mixing length** l on height z. Here

$$l = l_1 z^p, \, p \neq 1.$$

Then

$$\frac{\bar{u}}{u_*} = q\left(\frac{u_* z}{v}\right),$$

where $\bar{u}$ is the **mean velocity**, u_* the **friction velocity**, v the **kinematic viscosity**, and

$$q = \left(\frac{v}{u_*}\right)^{1-p} \frac{1}{l_1(1-p)}.$$

For moderate **Reynolds numbers**, $p = 6/7$ (the seventh-root profile) is empirically verified, but for large Reynolds numbers p is between this value and unity. It is to be noted that if $\bar{u}$ is proportional to z^m, and if the **stress** is assumed independent of height, then the **eddy viscosity** v_e is proportional to z^{1-m}. These relations are known as Schmidt's conjugate-power laws.

Sutton, O. G., 1953: *Micrometeorology*, 78–85.

power series—An infinite series of increasing powers of the **variable** x, of the form

$$\sum_{n=0}^{\infty} a_n x^n \equiv a_0 + a_1 x + a_2 x^2 + \cdots + a_n x^n + \cdots.$$

Both the x and the coefficients may taken on complex values. The totality of values of x for which a power series is convergent is called the interval of **convergence** of the series.

power spectrum—1. The square of the **amplitude** of the (complex) Fourier coefficient of a given periodic function.

Thus if $f(t)$ is periodic with **period** T, its **Fourier coefficients** are

$$F(n) = \frac{1}{T} \int_0^T f(t)\, e^{-in\omega t} dt,$$

where $\omega = 2\pi/T$, and the power spectrum of $f(t)$ is $|F(n)|^2$. Here n takes integral values and the **spectrum** is discrete. The total **energy** of the periodic function is infinite, but the **power**, or energy per unit period, is finite. In the case of the **aperiodic** function containing finite total energy, the **energy density spectrum** is the corresponding **spectral function**. This is a continuous function of **frequency** and therefore has **dimensions** of energy/frequency (energy density). In the case of a **random** function containing infinite total energy but not periodic, the **power density spectrum** is the corresponding spectral function. The mathematical conditions governing analogous theorems in these three classes of functions are different. However, when actual computations of observational data are involved, a finite number of discrete values are used, and the effect is the same as if the function were assumed to be periodic outside the interval of computation. Thus, it is the power spectrum that is exhibited. But all types of spectra referred to may be considered as measures of the contribution of given frequencies in the Fourier representation of the original function. The terms "power" and "energy" are usually retained to indicate relative dimensions regardless of the actual dimensions of the functions analyzed, which may be functions of space as well as time. Computation of the power spectrum in practice may be facilitated by use of the theorem that it is the Fourier coefficient of the **autocorrelation function**.

2. Same as **power density spectrum**.

Poynting vector—The **vector** cross product of the **electric field E** and the **magnetic field H**:

$$\mathbf{S} = \mathbf{E} \times \mathbf{H}.$$

This vector specifies the instantaneous rate of propagation of **electromagnetic energy**. The amount of **energy** transmitted per unit time across a unit area, with normal **n**, is **n · S**. In optics, **irradiance** is the Poynting vector averaged over a time sufficiently long compared with the **period** but short when compared to the characteristic response times of radiometric instruments.

PPI—Abbreviation for **plan position indicator**.

PPI scope—Same as **PPI**.

practical salinity units—(Abbreviated psu.) See **salinity**.

praecipitatio—A **cloud** supplementary feature for **precipitation** falling from a cloud and apparently reaching the earth's surface.

See **cloud classification**; *compare* **virga**.

prairie—A flat or gently undulating plain that is grassy and generally treeless; specifically, such an area in southern Canada and the northern and central United States where it extends from the foothills of the Rocky Mountains to about 88°W longitude.

Its **climate**, with light summer rains and high summer temperatures, is highly favorable for the growth of cereals, but there is a considerable risk of **drought** especially in certain portions, where a semiarid climate (**steppe climate**) prevails. The prairie is similar, but not completely analogous, to the **steppe** regions of Europe and Asia.

prairie climate—*See* **subhumid climate**.

Prandtl number—The nondimensional ratio between the product of **heat** advection and **viscous forces** and the product of heat **diffusion** and **inertial forces** in a given fluid.
It may be given as

$$\mathrm{Pr} = \frac{C_p\mu}{k},$$

where C_p is the **specific heat** at constant **pressure**, μ the **dynamic viscosity**, and k the **thermal conductivity**. The Prandtl number may also be defined as the ratio of the **Reynolds** to the **Péclet numbers**, or as the ratio of **kinematic viscosity** to **thermometric conductivity**.

pre-cold frontal squall line—Same as **prefrontal squall line**.

preactivation—The ability of an **ice** nucleating **particle** to act at a higher **temperature** or a lower **supersaturation** once it has activated an **ice crystal** that has subsequently evaporated, provided that critical conditions of temperature and undersaturation are not exceeded.

prebaratic chart—*See* **surface forecast chart**.

precession rate—1. The angular motion of the orbital line of nodes in fixed space; positive to the east, negative to the west.
The precession rate for a **sun-synchronous orbit** is $-0.986°$ per day, or about $360°$ per year. The net effect is that the satellite's orbital plane rotates slowly around the earth at the same rate and direction that the earth rotates around the sun, hence, sun-synchronous.
2. The period of time needed for a positional shift in the location of astronomical bodies.

precipitable water—(Or precipitable water vapor.) The total atmospheric **water vapor** contained in a vertical column of unit cross-sectional area extending between any two specified levels, commonly expressed in terms of the height to which that water substance would stand if completely condensed and collected in a vessel of the same unit **cross section**.
The total precipitable water is that contained in a column of unit cross section extending all of the way from the earth's surface to the "top" of the **atmosphere**. Mathematically, if $x(p)$ is the **mixing ratio** at the **pressure** level, p, then the precipitable water vapor, W, contained in a layer bounded by pressures p_1 and p_2 is given by

$$W = \frac{1}{g} \int_{p_1}^{p_2} x\,dp,$$

where g is the **acceleration of gravity**. In actual rainstorms, particularly thunderstorms, amounts of **rain** very often exceed the total precipitable water vapor of the overlying atmosphere. This results from the action of **convergence** that brings into the rainstorm the water vapor from a surrounding area that is often quite large. Nevertheless, there is general **correlation** between **precipitation** amounts in given storms and the precipitable water vapor of the **air masses** involved in those storms.

precipitable water vapor—Same as **precipitable water**.

precipitation—1. All liquid or solid **phase** aqueous **particles** that originate in the **atmosphere** and fall to the earth's surface.
2. The amount, usually expressed in millimeters or inches of liquid water depth, of the water substance that has fallen at a given point over a specified period of time.
As this is usually measured in a fixed **rain gauge**, small amounts of **dew**, **frost**, **rime**, etc., may be included in the total. The more common term **rainfall** is also used in this total sense to include not only amounts of **rain**, but also the water equivalents of **frozen precipitation**. For obvious reasons, precipitation is the preferred general term.

precipitation aloft—Vertical or inclined trails or wisps of water or **ice** particles falling out of a **cloud** but evaporating before reaching the earth's surface in the dry layer beneath the cloud.
See **virga**.

precipitation area—1. Specifically, on a synoptic **surface chart**, an area over which **precipitation** is falling.

2. In **radar**, the region in space containing **precipitation** echoes.

precipitation attenuation—**Attenuation** of **electromagnetic waves** propagating through **precipitation**. Depending on the **wavelength** of the **radiation**, the attenuation is accounted for by some combination of **absorption** and **scattering** by the precipitation particles.

The relative importance of scattering tends to increase as the wavelength becomes shorter. For **radar**, the **specific attenuation** Y (dB km^{-1}) due to **rain** is described by empirical power relations of the form $Y = aR^b$, where R (mm h^{-1}) is the **rainfall rate**, and a and b are empirical constants that depend on wavelength and **temperature**. The specific attenuation of **snow** is less than that of rain, and for wavelengths of 10 cm and longer is usually negligible. For **lidar**, the precipitation attenuation is approximately proportional to the cross-sectional area of the precipitation particles per unit volume. *See* **rain attenuation**; *compare* **cloud attenuation**.

precipitation band—*See* **banded structure**.

precipitation ceiling—After U.S. weather observing practice, a **ceiling classification** applied when the **ceiling** value is the **vertical visibility** upward into **precipitation**.

This is necessary when precipitation obscures the **cloud base** and prevents a determination of its height. All precipitation ceilings are estimations, but the following are used as guides: the height corresponding to the upper limit of a **ceilometer** reaction; the top of a **ceiling-light** projector **beam**; or the height at which a **ceiling balloon** or **pilot balloon** completely disappears. These guides usually indicate values that are lower than the actual **vertical visibility**. Precipitation ceilings are designated P in **aviation weather observations**.

precipitation cell—*See* **cell**.

precipitation current—The downward **transport** of charge, from **cloud** region to earth, that occurs in a fall of electrically charged **rain** or other **hydrometeors**; a particular case of a **convection current**.

Observations of the charge on individual **raindrops** during thunderstorms have revealed a complex picture. On average, more positive than negative charge is brought to earth by **precipitation** currents, but wide deviations occur both within individual storms and from one **storm** to another. The reasons for these wide fluctuations are not understood. Precipitation currents in continuous rain generally vary from about 10^{-12} to 10^{-10} A m^{-2}, while **thunderstorm** currents become as large as 10^{-8} A m^{-2}.

precipitation day—A day on which **precipitation** is observed.

The minimum amount of precipitation considered necessary to constitute a precipitation day varies from country to country but is generally 0.1 mm or 0.01 in.

precipitation duration—The period of time in which continuous **precipitation** is observed, or occurs at a specific point or within a specific area.

precipitation echo—A radar echo from **rain**, **snow**, or **hail**.

precipitation effectiveness—1. That portion of total **precipitation** used to satisfy vegetation needs.

2. The actual availability of **precipitation** used in plant development.

Availability is affected by such factors as **precipitation intensity**, **season**, **temperature**, ground cover, sod type, etc. The dependence of precipitation effectiveness on temperature and/or **evaporation** has been expressed in many ways: Köppen's formulas for defining **desert climate**, Lang's **moisture factor**, De Martonne's **index of aridity**, Gorczyński's **aridity coefficient**, Angström's humidity coefficient, Transeau's precipitation–evaporation quotient, and Thornthwaite's **precipitation-effectiveness index**.

Transeau, E. N., 1905: Forest Center of Eastern America. *American Naturalist,* **39**, 875–899.
Thornthwaite, C. W., 1948: An approach toward a rational classification of climate. *Geogr. Rev.,* **38**, 55–94.
Köppen, W., 1931: *Grundriss der Klimakunde,* Walter die Gruyter Co., Berlin.

precipitation-effectiveness index—(Commonly abbreviated P-E index; also called precipitation–evaporation index) For a given location, a measure of the long-range effectiveness of **precipitation** in promoting plant growth:

$$\text{P-E index} = 10 \sum_{n=1}^{12} (\text{P-E index})_n,$$

that is, it is equal to 10 times the sum of the monthly **precipitation–evaporation ratios** (monthly **precipitation** amounts divided by monthly **evaporation** amounts).

See **moisture index**, **temperature-efficiency index**.

precipitation-effectiveness ratio—Same as **precipitation–evaporation ratio**.

precipitation electricity—1. That branch of the study of **atmospheric electricity** concerned with the electrical charge carried by **precipitation** particles and with the manner in which these charges are acquired.

2. The electrical charge borne by **precipitation** particles.

A very complex and highly variable picture is obtained when charges are measured on individual **raindrops** or **snow crystals** and no present theory approaches a complete explanation of all details. In general, more raindrops are positively than negatively charged. Sometimes the prevailing sign of the charges even shifts in the course of a given **storm**'s lifetime. *See* **precipitation current**; *compare* **ion-capture theory**.

> Chalmers, J. A., 1957: *Atmospheric Electricity*, 176–199.

precipitation–evaporation index—Same as **precipitation-effectiveness index**.

precipitation–evaporation ratio—(Abbreviated P–E ratio; also called precipitation-effectiveness ratio.) For a given locality and month, an empirical expression devised for the purpose of classifying climates numerically on the basis of **precipitation** and **evaporation**:

$$\text{P–E ratio} = 11.5 \left(\frac{P}{T - 10} \right)^{10/9},$$

where P is the **normal** monthly precipitation in inches and T the normal monthly **temperature** in degrees Fahrenheit. All temperatures below 28.4°F are given the value of 28.4°, and P–E ratios greater than 40 are counted as 40.

See **precipitation-effectiveness index**.

precipitation gauge—General term for any instrument that measures **precipitation**, usually meaning a **rain gauge** or **snow gauge**.

precipitation-generating element—Same as **generating cell**.

precipitation intensity—The rate of **precipitation**, usually expressed in millimeters or inches per hour.

precipitation inversion—An elevation band along mountain slopes where **precipitation** decreases with increasing **elevation**.

In the lowest 1–2 km above the base of a mountain, precipitation typically increases with height until the **zone of maximum precipitation** is reached, above which precipitation decreases with height. This zone of decrease is the precipitation inversion.

precipitation physics— The study of the formation and **precipitation** of liquid and solid **hydrometeors** from **clouds**; a branch of **cloud physics** and of **physical meteorology**.

precipitation regime—Characteristics of the seasonal distribution of **precipitation** at a particular place.

precipitation scavenging—Same as **scavenging by precipitation**.

precipitation shadow—Same as **rain shadow**.

precipitation shield—On satellite images, an area of clouds, often within a larger **cloud shield**, that, based upon surface or **radar observations**, is resulting in **precipitation**.

precipitation station—A **station** where only observations of **precipitation** are made.

See **third-order climatological station**.

precipitation trails—Same as **virga**.

See **precipitation trajectory**.

precipitation trajectory—In **radar meteorology**, a characteristic pattern observed on **RHI** or **THI** displays caused by **snow** falling from an isolated region in space or a **generating cell**.

The shape of the **trajectory** depends on the **wind speed** at the **altitude** of origin of the snow, the fall speed of the snow, and the **vertical profile** of the ambient **wind** through which the snow falls.

precipitators—Devices for removing **particulates** from pollutant-laden gas streams.

The most common type, an **electrostatic precipitator**, functions by charging the **dust** to be removed with **ions** and then collecting the ionized particulates onto a special surface. The **collector** surface is cleaned by mechanical means.

precision—The quality of being exactly defined.

Sometimes indicated by the minimum number of significant digits required for an adequate representation of a quantity. Not the same as **accuracy** but often confused as such. A measurement having small **random error** is said to have high precision; a measurement having small **systematic error** or bias is said to have high accuracy.

precision aneroid barometer—*See* **aneroid barometer**.

predictability—The extent to which future states of a system may be predicted based on knowledge of current and past states of the system.

Since knowledge of the system's past and current states is generally imperfect, as are the **models** that utilize this knowledge to produce a **prediction**, predictability is inherently limited. Even with arbitrarily accurate models and observations, there may still be limits to the predictability of a physical system. *See* **chaos**.

predictability limit—The time beyond which it is no longer possible to predict the state of a system, given knowledge of current and past states, with a desired level of **accuracy**.

predictand—*See* **regression**.

prediction—*See* **numerical weather prediction, climate prediction**; *compare* **weather forecast**.

predictor—*See* **regression**.

preexponential factor—The factor A in the **Arrhenius expression** for a **rate coefficient**, $k = A \exp(-E_a/RT)$.

prefrontal fog—*See* **frontal fog**.

prefrontal squall line—A **squall line** less than about 100 km ahead of a **cold front**, in the **warm sector**, having an orientation more or less parallel to the cold front.

prefrontal thunderstorm—Same as **thunderstorm**, except that the **storm** occurs in the **warm sector** ahead of a **cold front**.

preliminary phase—*See* **foehn phase**.

present weather—*See* **weather**.

pressure—1. A type of **stress** characterized by uniformity in all directions.

As a measurable on a surface, the net force per unit area normal to that surface exerted by molecules rebounding from it. In dynamics, it is that part of the **stress tensor** that is independent of **viscosity** and depends only upon the molecular motion appropriate to the local **temperature** and **density**. It is the negative of the mean of the three normal stresses. The concept of pressure as employed in **thermodynamics** is based upon an **equilibrium** system, where tangential forces vanish and normal forces are equal.

2. In meteorology, commonly used for **atmospheric pressure**.

3. In mechanics, same as **stress**.

4. *See* **radiation pressure**.

pressure altimeter—(Also called barometric altimeter.) An **aneroid barometer** calibrated to convert **atmospheric pressure** into **altitude**.

Altimeters use **standard atmosphere** pressure–height relations in converting **pressure** into altitude. Therefore, the **altimeter** shows **indicated altitude**, which may, and frequently does, differ from the actual altitude. An altimeter may be set to measure altitude from an arbitrarily chosen level. It is common practice to use **mean sea level**; the level of the **constant-pressure surface** of 29.92 in. of **mercury** is also used; and, less frequently, the constant-pressure surface of the pressure at **airport height**. *See* **altimeter setting, pressure altitude**.

pressure altitude—The **altitude** that corresponds to a given value of **atmospheric pressure** according to the ICAO **standard atmosphere**.

For aircraft flying above 18 000 feet **MSL**, it is the **indicated altitude** of a **pressure altimeter** at an **altimeter setting** of 1013.2 mb (29.92 in. of **mercury**); therefore, it is the indicated altitude above the 1013.2-mb **constant-pressure surface**. Aircraft flying below 18 000 feet MSL use the current altimeter setting as measured at the nearest airport and reported by air traffic control.

pressure anomaly—Difference between the measured **pressure** value at a given location and the mean pressure value, sometimes given at the parallel of the latitude for a given location.

pressure broadening—Same as **collision broadening**.

pressure capsule—Same as **aneroid capsule**.

pressure center—On a **synoptic chart** (or on a **mean chart** of **atmospheric pressure**), a point of local minimum or maximum pressure; the center of a **low** or **high**.

Pressure center is also a center of **cyclonic** or **anticyclonic circulation**. *See* **center of action**, **central pressure**.

pressure-change chart—(Also called pressure-tendency chart.) A **chart** indicating the change in **atmospheric pressure** of a **constant-height surface** over some specified interval of time; comparable to a **height-change chart**.

See **pressure tendency**, **isallobar**, **differential chart**.

pressure coordinates—A **coordinate system** in which **atmospheric pressure** is taken as the vertical coordinate.

Compare θ **coordinate system**.

pressure correlation—The **covariance** between **pressure** and another **variable**, usually **velocity**.

An example is $\overline{w'p'}$, where p is pressure, w is **vertical velocity**, the prime denotes a **deviation** or **perturbation** from the mean, and the overbar denotes an average. This particular example is related to the vertical redistribution of **kinetic energy** and can be associated with either **turbulence** or **waves**.

pressure distribution—*See* **distribution**.

pressure drag—*See* **drag**.

pressure-fall center—(Or center of falls; also called katallobaric center, isallobaric minimum, isallobaric low.) A point of maximum decrease in **atmospheric pressure** over a specified interval of time; on **synoptic charts**, a point of greatest negative **pressure tendency**; opposed to a **pressure-rise center**.

pressure field—*See* **field**.

pressure fluctuation spectrum—*See* **pressure spectrum**.

pressure force—(Or pressure-gradient force.) The force due to differences of **pressure** within a fluid mass.

The (**vector**) force per unit volume is equal to the **pressure gradient**, $-\nabla p$, and the force per unit mass (**specific** force) is equal to the product of the volume force and the **specific volume**, $-\alpha\nabla p$. In the **atmosphere**, the vertical component of this force is of the order of 10^4 times the horizontal component; in meteorological literature the term "pressure force" usually refers only to the latter **horizontal pressure force**.

pressure gradient—(In meteorology, also called barometric gradient.) The rate of decrease (**gradient**) of **pressure** in space at a fixed time.

The term is sometimes loosely used to denote simply the magnitude of the gradient of the **pressure field**.

pressure-gradient force—Same as **pressure force**.

pressure head—Pressure in **head** units, for example, meters of fluid, equal to the force per unit area divided by the product of the **density** of the fluid and the **acceleration** due to **gravity**.

It is the depth of fluid that would exert an equivalent **pressure**.

pressure ice—**Sea ice** (or river ice or **lake ice**) that has been deformed or altered by the lateral stresses of any combination of **wind**, water currents, **tides**, **waves**, and **surf**.

This may include **ice** pressed against the shore, or one piece of ice upon another. Its two major forms are **rafted ice** and **tented ice**, which, individually or in combinations, may form **pressure ridges** or **hummocked ice**.

pressure jump—Dynamically the same as the **hydraulic jump**, but in meteorological contexts the equations are applied to a **temperature inversion** or to a system of two inversions.

The phenomenon of a pressure jump is thus a steady-state propagation of a sudden finite change of **inversion** height, in analogy to the **shock wave** in a compressible fluid. The **prefrontal squall line** has been interpreted as a pressure jump, with the **cold front** providing the initial pistonlike impetus. *See also* **compression wave**.

pressure jump detector—A **sensor** designed to detect the rapid rise in **pressure** associated with the leading edges of **outflow** boundaries (**pressure jump**) or beneath strong **downdrafts** (pressure nose).

pressure melting—The melting of **ice** due to applied **pressure**.

The **melting point** of pure ice is lowered 0.0072 K per **atmosphere** of applied pressure. Pressure melting is responsible for **regelation**.

pressure pattern—(Sometimes called baric topography.) In meteorology, the general geometric characteristics of **atmospheric pressure** distribution as revealed by **isobars** on a **constant-height chart**; usually applied to **cyclonic-scale** features of a **surface chart**.
Compare **height pattern, circulation pattern**.

pressure-pattern flight—In general, an aircraft flight so planned and navigated as to take advantage of the flight-altitude winds (**pressure pattern**) to reduce flying time.
As techniques have advanced, this concept has become increasingly referred to as **minimal flight**. Probably the most widely used method today is based upon the determination of **D-values** with **pressure** and radio altimeters while in flight. This provides for the continual adjustment of flight **course** to take fullest advantage of winds in long flights over water. Flight plans made out by the navigator are based on meteorological forecasts of **wind** distribution. These forecasts are usually prepared in the form of maps analyzed in terms of D-values at the flight **pressure altitude**. During flight, observations of actual D-values can conveniently be used to make relatively small corrections. *See* **constant-pressure-pattern flight, aerologation, single-heading navigation, 4-D chart, wave-front method, single drift correction, multiple drift correction**.

pressure reduction—*See* **reduction**.

pressure ridge—1. *See* **ridge**.
2. A **ridge** of ice, up to 35 m (100 ft) high and sometimes several kilometers long, in **pressure ice**.

pressure-rise center—(Or center of rises; also called anallobaric center, isallobaric maximum, isallobaric high.) A point of maximum increase in **atmospheric pressure** over a specified interval of time; on **synoptic charts**, a point of greatest positive **pressure tendency**; opposed to a **pressure-fall center**.

pressure spectrum—Within the **inertial subrange**, a **spectrum** of the form

$$S_p(k_1) = c\epsilon^{4/3}k_1{}^{-7/3},$$

where ϵ is **dissipation** of **turbulence energy** and k_1 is **wavenumber** in longitudinal direction. At smaller wavenumbers the pressure spectrum increases considerably due to **mesoscale** motions and **synoptic** weather systems.

pressure-sphere anemometer—An **anemometer** that measures the three components of the **wind vector** by means of an array of **pressure** ports bored in a sphere.
A **Pitot tube** is located in a **venturi** bored through the center of the sphere and up to 12 auxiliary ports are arranged on the surface of the sphere.

pressure stress—*See* **pressure, stress tensor**.

pressure surge—A sudden rise in **pressure** due to a **mesoscale** high pressure area, often formed by the **outflow** of a **thunderstorm**; typically associated with a **prefrontal squall line**.
The decrease in pressure after the passage of the mesoscale **high** is known as a **wake depression** or **wake low**.

pressure system—An individual **cyclonic-scale** feature of atmospheric **circulation**, commonly used to denote either a **high** or a **low**, less frequently a **ridge** or a **trough**.

pressure tendency—(Also called barometric tendency.) The character and amount of **atmospheric pressure** change during a specified period of time, often a three-hour period preceding an **observation**.
Pressure tendency is composed of two parts, the pressure change and the pressure characteristic. The pressure change is the net difference between pressure readings at the beginning and ending of a specified interval of time. The pressure characteristic is an indication of how the pressure has been changing during that specified period of time, for example, decreasing then increasing, or increasing and then increasing more rapidly. *See* **tendency**.

pressure-tendency chart—Same as **pressure-change chart**.

pressure tendency equation—*See* **tendency equation**.

pressure tide gauges—Instruments that measure the **pressure** below the sea surface; this pressure may be converted to **sea level** if the air pressure, the gravitational **acceleration**, and the water density are known.

pressure topography—Same as **height pattern**.

pressure vertical coordinates—A **vertical coordinate system** based on the **pressure** of the atmosphere.

pressure wave—A short-period **oscillation** of **pressure** such as that associated with the propagation of **sound** through the **atmosphere**; a type of **longitudinal wave**.

Pressure waves are usually recorded on sensitive **microbarographs** capable of measuring pressure changes of amounts down to 10^{-4} mb. Typical values for the **period** and **wavelength** of pressure waves are ½ to 5 s and 100 to 1500 m, respectively. Pressure waves produced by explosions in the **upper atmosphere** are of value in determining the high-altitude temperatures and winds. *See* **sound wave**, **compression wave**, **microbarm**.

Gutenberg, B., 1951: *Compendium of Meteorology*, 366–375.

prester—A **whirlwind** or **waterspout** accompanied by **lightning** in the Mediterranean and Greece.

prevailing visibility—In U.S. weather observing practice, the greatest **horizontal visibility** that is equaled or surpassed throughout half of the **horizon** circle; it need not be a continuous half.

In the case of rapidly varying conditions, it is the average of the prevailing visibility while the **observation** is being taken. This value is entered in a **surface weather observation**. If the value is less than seven miles, it must be explained by reporting a type of weather or **obstruction to vision**.

prevailing wind direction—(Or prevailing wind.) The **wind direction** most frequently observed during a given period.

The periods most frequently used are the **observational day**, month, **season**, and year. Methods for determination vary from a simple count of periodic observations to the computation of a **wind rose**. *Compare* **resultant wind**.

PRF—Abbreviation for **pulse repetition frequency**.

Price meter—The **current meter** in common use in the United States.

Six conical cups, mounted around a vertical axis, rotate and cause a **signal** in a set of headphones with each rotation. Tail vanes and a heavy weight stabilize the instrument.

primary circulation—The prevailing fundamental atmospheric **circulation** on a planetary **scale** that must exist in response to 1) **radiation** differences with latitude, 2) the rotation of the earth, and 3) the particular distribution of land and oceans; and that is required from the viewpoint of **conservation of energy**.

Primary circulation and **general circulation** are sometimes taken synonymously. They may be distinguished, however, on the basis of approach; that is, primary circulation is the basic system of winds, of which the secondary and **tertiary circulation** are perturbations, while general circulation encompasses at least the **secondary circulations**. *See* **macrometeorology**.

primary cyclone—(Or primary low.) Any **cyclone** (or **low**), especially a **frontal cyclone**, within which a **circulation** of one or more secondary cyclones have developed.

The term is sometimes applied to the stronger of a system of two cyclones that together form a multilobed cyclone.

primary depression—Same as **primary cyclone**; usually used in the context of tropical **lows**.

primary front—The principal, and usually original, **front** in any **frontal system** in which **secondary fronts** are found.

primary low—Same as **primary cyclone**.

primary pollutants—**Pollutants** released directly into the **atmosphere**, such as **hydrocarbons**, **sulfur dioxide**, or **nitric oxide**.

primary rainbow—A **rainbow** that is distinguished from other rainbows by its angular radius, color order, and **brightness**.

This bow is seen between about 40° and 42° from the antisolar point (shadow of the observer's head) or equivalently, between 140° and 138° from a **light** source (such as the sun). Reds are found to the outside of the bow (closest to the sun) with the blues to the inside. The primary bow is usually brighter than any of the other bows. The primary rainbow is certainly the most frequently noticed bow, but the purity and **range** of its colors fall a long way short of that assumed by the popular dictum: all the colors of the fall. Frequently accompanying the primary bow are the **secondary bow** (lying about 8° outside the primary bow) and the **supernumerary bows** (immediately inside the primary bow, and often confined to the upper portions of the arc). Infrequently seen are the **reflection bows**. A theory of the bow that approximates the behavior of light as a

ray is able to account for the difference in position and color order of the primary and secondary bows. In this theory, the position of each bow is determined by the minimum angle of **deviation** of the light passing through a **drop**. The difference is that the light that forms the primary bow has undergone one internal reflection, while the light that forms the secondary bow has undergone two internal reflections. This is a useful approximation to reality, but it fails to capture many important features of observable bows. *Compare* **secondary rainbow**.

primary scattering—*See* **multiple scattering**.

primitive equations—The **Eulerian equations** of motion of a fluid in which the primary **dependent variables** are the fluid's **velocity** components.
   These equations govern a wide variety of fluid motions and form the basis of most hydrodynamical **analysis**. In meteorology, these equations are frequently specialized to apply directly to the **cyclonic-scale** motions by the introduction of the so-called **filtering approximations**. *See* **equations of motion**.

principal band—A prominent **spiral band** in a **tropical cyclone**, generally on the downshear side, where **convergence** occurs as air moving with the **vortex** overtakes the environmental flow.

principal components analysis—**Regression analysis** to determine from a set of **independent variables** (predictors) those contributing most to the **explained variance**.
   See **regression**.

principal front—*See* **primary front**.

principal land station—Surface **synoptic station** on land, suitably equipped and staffed, at which are observed specific **elements** and from which **meteorological observations** for international exchange are normally transmitted.

principal plane—(Also called sun's meridian or sun's vertical.) In clear-sky optics, the vertical plane defined by sun, **zenith**, and observer.
   The **clear** daytime sky's **radiance**, **polarization**, and chromaticity patterns are largely symmetric about the principal plane.

principal point—The point on the earth where a satellite sensor is focused at any time during its **orbit**.
   If the **sensor** vertical axis is perpendicular to the earth's surface, the principal point coincides with the **subpoint**.
   2. A term used in **remote sensing**; the point where the optical axis intersects the **principal plane**.

principle of geometric association—A principle used in forecasting the **thickness** of a layer between two given **constant-pressure surfaces**; more commonly used prior to the advent of **numerical forecasting** techniques.
   In practical use of the principle, specific thicknesses are considered to be associated with corresponding points on the lower of the two constant-pressure surfaces (the upper one can also be used). A **prognostic chart** is prepared for the lower surface and the assumption is made that the thickness of the layer at the corresponding points is unchanged at the same corresponding points on the prognostic map. The prognostic **thickness chart** is then constructed from thickness values at the corresponding points.

probabilistic process—(Also called **stochastic process**.) A **variable**, $X(t)$, with a value over time (or space) described by probabilistic laws.

probability—The chance that a prescribed event will occur, represented as a pure number p in the **range** $0 \leq p \leq 1$.
   The probability of an impossible event is zero and that of an inevitable event is unity. Probability is estimated empirically by **relative frequency**, that is, the number of times the particular event occurs divided by the total count of all events in the class considered. *See* **probability theory**.

probability density function—(Or density function; also called frequency function.) The **statistical** function that shows how the density of possible observations in a **population** is distributed.
   It is the derivative $f(x)$ of the **distribution function** $F(x)$ of a **random variable**, if $F(x)$ is differentiable. Geometrically, $f(x)$ is the **ordinate** of a curve such that $f(x)dx$ yields the **probability** that the random variable will assume some value within the **range** dx of x. The density function is nonnegative, and its total integral is unity. Sometimes the probability density function is called the distribution function, but this practice causes confusion and is not recommended.

probability distribution—The mathematical description of a **random variable** in terms of its admissible values and the **probability** associated, in an appropriate sense, with each value.

The probability distribution of a continuous **variate** is defined by stating the mathematical equation of the **distribution function** $F(x)$ or the **probability density function** $f(x)$ (if it exists) together with the **range** over which the equation holds. The probability distribution of a discrete variate is commonly defined by stating the equation for the probability $p(x)$ that the variate will assume any particular value x, and indicating what values are possible.

probability distribution function—Same as **distribution function**.

probability forecast—A forecast of the **probability** of occurrence of one or more categorical events. *Compare* **persistence forecast, random forecast**.

probability integral—The classical form (still widely used in engineering work) of the definite integral of the special **normal distribution** for which the mean $\mu = 0$ and **standard deviation** $\sigma = \frac{1}{\sqrt{2}}$.

Geometrically, the probability integral equals the area under this density curve between $-z$ and z, where z is an arbitrary positive number. Often denoted by the symbol erf z (read "error function of z") the probability integral is defined thus:

$$\text{erf } z \equiv \frac{2}{\sqrt{\pi}} \int_0^z e^{-x^2} dx.$$

Modern **statistical** usage favors the unit normal **variate** u, which is such that $\mu = 0$ and $\sigma = 1$. The relation between the probability integral erf z and the **distribution function** $F(u)$ of the unit normal variate u is as follows:

$$u \text{ positive: } F(u) = \tfrac{1}{2} + \tfrac{1}{2} \text{ erf}\left(\frac{u}{2}\right),$$

$$u \text{ negative: } F(u) = \tfrac{1}{2} - \tfrac{1}{2} \text{ erf}\left(\frac{-u}{2}\right).$$

See **unit normal distribution**.

probability of detection—*See* **skill**.

probability paper—A type of graph paper on which the cumulative or **exceedance probability** of **random variables** for a specified distribution plot is a straight line when plotted.

Traditionally, values of the random variable are on the vertical axis, and **probability** values on the horizontal axis. The empirical distribution obtained by plotting data points is used for **return period** determination. *See* **plotting position**.

probability theory—The mathematical theory of **random (nondeterministic)** phenomena.

probable error—The magnitude of a **deviation** from a **statistic** that will be exceeded with **probability** of 0.50, or on half the occasions.

For a **normal distribution** it is 0.6745 times the **standard deviation**. The probable error is not "probable" in any peculiar sense and should have no more significance attached to it than the above.

probable maximum flood—(Also called maximum possible flood.) **Flood** that can be expected from the most severe combination of critical meteorologic and hydrologic conditions that are reasonably possible in a region.

probable maximum precipitation—[Also called maximum probable precipitation, maximum possible precipitation (rare).] Theoretically, the greatest depth of **precipitation** for a given duration that is physically possible over a given size **storm** area at a particular geographical location at a certain time of year.

probe—In **geophysics**, the device used to make a **sounding**.

process lapse rate—The rate of decrease of the **temperature** of an **air parcel** as it is lifted, $-dT/dz$, or occasionally dT/dp, where p is **pressure**.

The concept may be applied to other atmospheric **variables**, for example, the process lapse rate of **density**. The process lapse rate is determined by the character of the fluid processes and should be carefully distinguished from the **environmental lapse rate**, which is determined by the distribution of temperature in space. In the **atmosphere** the process lapse rate is usually assumed to be either the **dry-adiabatic lapse rate** or the **moist-adiabatic lapse rate**.

product-moment—The **expected value** of a product.

production rule—An **IF–THEN rule** for representing knowledge in a **rule-based system**.

production system—A problem-solving system using a rule-based architecture consisting of a **knowledge base** of **rules** and general facts, a **working memory** of facts concerning the current case, and an **inference engine** for manipulating both.
 It is a type of **expert system** or **knowledge-based system**.

profile—In meteorology, a graph of the value of a **scalar** quantity versus a horizontal, vertical, or timescale.
 It usually refers to a vertical representation. *Compare* **contour, cross section, time section**.

profile matching—A method of merging the shape or mathematical representation of a meteorological **variable** versus height in one layer with a different shape or representation in an adjacent layer.
 An example in the **atmospheric boundary layer** would be the matching of an exponential **wind profile** at the top of a forest **canopy** with a logarithmic wind profile just above it in the **surface layer**.

profile similarity—The assumption that the height dependence of **scalar** quantities and **wind speed** in the **atmospheric boundary layer**, after appropriate scaling, exhibits universal behavior.
 The appropriate **similarity** scales depend on the height above the surface and the **static stability**.

profiler—In general, a **remote sensing** device that receives **electromagnetic** (or **acoustic**) waves transmitted through, emitted by, or reflected from the **atmosphere** in order to produce a **vertical profile** of one or more atmospheric quantities.
 Often used more specifically to refer to a **wind profiler**.

prog—Common contraction for **prognostic chart**.

prog chart—Common contraction for **prognostic chart**.

prognosis—*See* **forecast**.

prognostic chart—(Commonly contracted prog, prog chart.) A **chart** showing, principally, the expected **pressure pattern** (or **height pattern**) of a given **synoptic chart** at a specified future time.
 Usually, positions of **fronts** are also included, and the forecast values of other meteorological **elements** may be superimposed.

prognostic clouds—A set of **prognostic equations** containing parameterizations of various **cloud microphysics** processes, governing the evolution of **cloud** water.
 The term "prognostic clouds" is usually reserved for numerical models that predict weather and **climate**. Prognostic cloud schemes may differ as to the number of prognostic equations and parameterized microphysics processes (the sources and sinks of cloud water), and as to whether or not they advect cloud water. In current **climate models**, spectral size distributions of cloud water **particles** (and falling condensate) are almost always assumed, rather than predicted, and their (radiative) effective radii are specified. Most often, the cloud volume fraction is parameterized by diagnostic relationships, as in purely empirically based, diagnostic cloud parameterizations, but it can be prognostic. Key cloud radiative properties, such as **cloud optical depth** and **emissivity**, may be parameterized in terms of the prognostic cloud **variables** and the (radiative) effective radii of the cloud particles.

prognostic contour chart—*See* **prognostic chart, contour chart**.

prognostic equation—Any equation governing a system that contains a time derivative of a quantity and therefore can be used to determine the value of that quantity at a later time when the other terms in the equation are known (e.g., **vorticity equation**).
 Compare **diagnostic equation**; *see also* **regression**.

prograde orbit—*See* **inclination**.

progressive wave—1. A **wave** that moves relative to a fixed **coordinate system** in a fluid; or, in meteorology, a **wave** or wavelike **disturbance** that moves relative to the earth's surface.
 Progressive waves are to be distinguished from **stationary waves**, which show no relative translation. **Standing waves** can be treated mathematically as two equal and oppositely directed progressive waves superimposed upon each other.
 2. In oceanography, a **wave** the travel of which can be followed by monitoring the movement of the crest.
 Energy is transmitted; that is, the wave form travels significant distances, but the water particles perform oscillatory motions. *See also* **Kelvin wave**.

projection—The correspondence between a domain of the earth's surface and a plane surface (map) such that each point on one corresponds to one and only one point on the other.

Typical projections used on **weather charts** include **stereographic**, **Lambert conic**, and **Mercator**.

prominence—A filament-like protuberance from the **chromosphere** of the sun.

Prominences can be observed visually (optically) whenever the sun's disk is masked, as during an eclipse or by using a **coronagraph**; and can be observed instrumentally by **filtering** in certain **wavelengths**, as with a **spectroheliograph**. A typical prominence is 6000 to 12 000 km thick, 60 000 km high, and 200 000 km long. These features appear as **filaments** when they are seen against the solar disk.

propagation constant—For a given **frequency** of **radiant energy**, a complex quantity describing the medium through which the **radiation** is propagating. The real part is the **specific attenuation**, usually measured in decibels per unit path length, and the imaginary part is the **phase** constant or change in phase, in radians per unit path length.

propagation effect—In **radar**, a change in the **polarization** state of a **signal** because of propagation through and interaction with an **anisotropic** medium, for example, a **cloud** of **precipitation** particles of nonspherical shape that have principal axes with a common alignment or preferred orientation.

See **depolarization, differential attenuation, differential phase shift.**

propane—Third member of the **alkane** family, formula C_3H_8.

Emissions of propane are usually associated with **anthropogenic** activity, and its presence in an **air mass** can be a good indicator of **pollution**.

propeller anemometer—A rotation **anemometer** that has a fixed axis upon which a propeller consisting of helicoidal vanes is mounted.

propeller-type current meter—A device for **streamflow** measurement, using the rotation rate of a propeller as an **indicator** of **flow velocity**.

propeller-vane anemometer—An instrument for the measurement of both horizontal **wind speed** and direction, consisting of a **propeller anemometer** mounted on the **windward** end of a **wind vane**.

The wind vane serves the dual purposes of measuring **wind direction** and aligning the propeller axis in the direction of the **wind**.

propene—A member of the **alkene** family, formula C_3H_6.

Major sources include fuel combustion, **biomass burning**, and the oceans.

protected thermometer—A **reversing thermometer** that is encased in a strong glass outer shell that protects it against **hydrostatic pressure**.

Compare **unprotected thermometer.**

proton—A positively charged **subatomic particle** with a **rest mass** of 1.67262×10^{-27} kg, slightly less than that of the **neutron** and about 1836 times that of the **electron**.

Atomic nuclei are composed of protons and neutrons bound together by nuclear forces. The common term for neutron or proton is nucleon.

proton precipitation—The **flux** of energetic **protons** into the **upper atmosphere** from the **radiation belts**, the outer **magnetosphere**, or the sun.

proton–proton chain—(Or proton–proton cycle.) A chain of **fusion** reactions the net effect of which is that four **protons** yield a **helium** nucleus.

This chain accounts for about 99% of the **energy conversion** that keeps the sun at its high **temperature**. We owe our existence to the proton–proton chain, the source of the sun's **energy**.

protonosphere—The region of the **outer atmosphere** in which the dominant gas is **hydrogen**.

proxy climate record—A biologic or geologic structure of known age from which information on past **climate** may be extracted; may also be used with reference to climate information extracted from historical documents.

Proxy climate records are of use for times and places lacking instrumental records. Examples include fossil species assemblages from rocks, geomorphic structures, the isotopic or species composition of marine or lacustrine sediments, the isotopic and other chemical composition of polar and very high elevation **ice**, and the structure and chemical composition of annual growth rings in trees and growth bands in corals.

PSC—Abbreviation for **polar stratospheric clouds.**

pseudo cold front—Same as **pseudo front**.

pseudo front—(Also called pseudo cold front.) A small-scale **front**, formed in association with organized severe **convective activity**, between a mass of rain-cooled air from the **thunderstorm** clouds and the warm surrounding air.

It may be the leading edge of a **bubble high**. *See* **outflow boundary**.

pseudo wet-bulb potential temperature—Same as **wet-bulb potential temperature**.

pseudo wet-bulb temperature—*See* **wet-bulb temperature**.

pseudoadiabat—On a **thermodynamic diagram**, a line representing a **pseudoadiabatic expansion** of an **air parcel**.

pseudoadiabatic chart—(Or pseudoadiabatic diagram.) *See* **thermodynamic diagram**.

pseudoadiabatic diagram—*See* **thermodynamic diagram**.

pseudoadiabatic expansion—A **saturation-adiabatic process** in which the condensed water substance is removed from the system, and therefore best treated by the **thermodynamics** of open systems.

Meteorologically, this process corresponds to rising air from which the moisture is precipitating. Descent of air so lifted becomes by definition a **dry-adiabatic process**. *See* **pseudoadiabatic process**.

pseudoadiabatic lapse rate—The rate of decrease of **temperature** with height of a **parcel** undergoing a **pseudoadiabatic process**.

It is given by

$$\Gamma_{ps} = g \, \frac{(1 + r_v)\left(1 + \dfrac{L_v r_v}{RT}\right)}{c_{pd} + r_v c_{pv} + \dfrac{L_v^2 r_v(\epsilon + r_v)}{RT^2}},$$

where Γ_{ps} is the pseudoadiabatic lapse rate, g is gravitational **acceleration**, r_v is the **mixing ratio** of **water vapor**, c_{pd} and c_{pv} are the **specific heats** at constant **pressure** of **dry air** and water vapor, L_v is the **latent heat** of **vaporization**, R is the dry air gas constant, $\epsilon \approx 0.62$ is the ratio of the gas constants of dry air and water vapor, and T is temperature. The above **lapse rate** is usually within 1 percent of those shown under **moist-adiabatic lapse rate** and **reversible moist-adiabatic lapse rate**.

pseudoadiabatic process—(Also called irreversible moist-adiabatic process.) A **moist-adiabatic process** in which the liquid water that condenses is assumed to be removed as soon as it is formed, by idealized instantaneous **precipitation**.

The pseudoadiabatic process is only defined for expansion, since a **parcel** that is compressed after such expansion will follow the **dry-adiabatic lapse rate**. A process similar to pseudoadiabatic descent can occur, however, if **drizzle** is evaporated into a relatively slow **downdraft**.

pseudocoloring—The creation of a **color lookup table** for a grayscale **image**, or the creation of a color lookup table for a color image that is not the standard color mapping.

For example, in a grayscale image with levels 0, 1, 2, 3, and 4, pseudocoloring is a color lookup table that maps 0 to black, 1 to red, 2 to green, 3 to blue, and 4 to white.

pseudoequivalent potential temperature—The **temperature** a sample of air would have if it were expanded by a **pseudoadiabatic process** to zero **pressure** and then compressed to a reference pressure of 100 kPa by a **dry-adiabatic process**.

This quantity is conserved in a pseudoadiabatic process and is given approximately by

$$\theta_{ep} = T \left(\frac{p_0}{p}\right)^{0.2854\,(1 - 0.28^{r_v})} \exp\left[r_v(1 + 0.81 r_v)\left(\frac{3376}{T_c} - 2.54\right)\right],$$

where T is the temperature, p is the pressure, T_c is the **condensation temperature** (obtainable from the **dewpoint formula**) and r_v is the **water vapor** mixing ratio. When $r_v = 0$, $\theta_{ep} = \theta$, the **potential temperature**.

Bolton, D., 1980: The computation of equivalent potential temperature. *Mon. Wea. Rev.*, **108**, 1046–1053.

pseudoequivalent temperature—*See* **equivalent temperature**.

pseudorandom numbers—Sequences of algorithmically generated numbers that pass certain tests for **randomness** but that are not truly **random** due to the inherently **deterministic** nature of the procedure.

PSI—1. [Abbreviation for pounds (force) per square inch.] A nonstandard unit of **pressure**, where 1 PSI = 6.895238 kPa.

2. Abbreviation for **pollutant standards index**.

3. Abbreviation for **poultry stress index**.

PSMSL—Abbreviation for **Permanent Service for Mean Sea Level**.

psu—Abbreviation for practical salinity units. See **salinity**.

psychrometer—An instrument used to measure **humidity**. It consists of two thermometers exposed side by side, one of which (the dry bulb) is an ordinary glass **thermometer**, while the other (the wet bulb) has its bulb covered with a jacket of clean muslin that is saturated with distilled water prior to an **observation**.

The temperature measured by the wet-bulb thermometer is generally lower (due to **evaporation** of water from the wet bulb) than that measured by the dry bulb. The difference in the temperatures is a measure of the humidity of the air; the lower the ambient humidity, the greater the rate of evaporation and, consequently, the greater the depression of the wet-bulb temperature. The size of the **wet-bulb depression** is related to the ambient humidity by the **psychrometric formula**. *See* **aspiration psychrometer, Assmann psychrometer, sling psychrometer**.

psychrometric chart—A **nomograph** for graphically obtaining **relative humidity, absolute humidity**, and **dewpoint** from **wet-** and **dry-bulb thermometer** readings.

psychrometric constant—The ratio of **specific heat** (C_p) of **moist air** at constant **pressure** to **latent heat** (L_v) of **vaporization** of water.

This constant has a value of $\gamma = C_p/L_v \cong 0.4$ (g_{water}/kg_{air}) K^{-1}. Latent **heat flux**, when multiplied by this constant, yields a **moisture flux**.

psychrometric formula—(Also known as hygrometric formula.) The semi-empirical relation giving the **vapor pressure** in terms of the **barometer** and **psychrometer** readings.

For temperatures in degrees Celsius, this formula is

$$e = e_s(T_w) - 6.60 \times 10^{-4}(1 + 0.00115 T_w)p(T - T_w),$$

where T and T_w are the **dry-bulb** and **wet-bulb temperatures**, respectively; e is the vapor pressure; $e_s(T_w)$ is the **saturation vapor pressure** at the wet-bulb temperature; and p is **pressure**. For temperatures below **freezing**, with the wet bulb covered with **ice**, 6.60×10^{-4} is replaced by, approximately, 5.82×10^{-4}.

psychrometric tables—Tables prepared from the **psychrometric formula** and used to obtain **vapor pressure, relative humidity**, and **dewpoint** from values of **wet-bulb** and **dry-bulb temperature**.

psychrometry—The science and techniques associated with psychrometric measurements. *See* **psychrometer**.

puelche—An east **wind** that has crossed the Andes; the Andean **foehn** of the South American west coast.

This term is sometimes used for a **land breeze** in areas where the Andes descend sharply into the Pacific Ocean. The corresponding **sea breeze** is the **virazon**. *See* **fog wind**.

pulse—In **radar, sodar**, or **lidar** a single short-duration transmission (or burst) of **energy**.

A pulse is characterized by its **radio frequency, pulse repetition frequency, pulse duration**, and **peak power**.

pulse amplitude modulation—*See* **pulse modulation**.

pulse coding—The use of **modulation** within a **radar** or **sodar** pulse to increase the **range resolution** beyond that normally achievable with a **pulse** of the same length.

In a common technique, the **phase** of the **transmitter** carrier is inverted several times in an optimized sequence during the transmitted pulse. This modulation is undone by a complementary process in the **receiver** to produce an effective **pulse length** narrower by a factor equal to the number of inversions in the transmitted pulse sequence. *See* **pulse compression**.

pulse compression—The use of special forms of **frequency, phase**, or **amplitude modulation** to permit a **radar** system to achieve higher **range resolution** than that normally permitted by a given **pulse duration**. A suitably modulated **transmitter** pulse of duration τ (and hence range resolution $c\tau/2$, where c is the **velocity** of **light**) may be processed after reception to obtain a higher range resolution $c\tau/2n$, where n is the pulse-compression ratio.

Compression ratios of 10–100 are commonly achieved through the use of **linear** FM, **nonlinear** FM, or phase-coded **modulation**, implemented by **analog** or **digital** means. The advantage of pulse compression over simply transmitting shorter pulses is that high range resolution is achieved while maintaining the benefits of high pulse **energy**.

pulse duration—In **radar**, **sodar**, or **lidar**, the period of time during which a **pulse** is being transmitted.

pulse duration modulation—*See* **pulse modulation**.

pulse frequency—Same as **pulse repetition frequency**.

pulse frequency modulation—*See* **pulse modulation**.

pulse integration—The estimation of **signal** parameters from a sequence of **pulses** in a **radar** or **lidar** system.

Incoherent integration, in which the **signal intensity** from successive pulses is added, is used in simple lidar systems, while **coherent integration**, with the **phase** of the signal taken into account, is used in coherent radars.

pulse integrator—An electronic device for the measurement of the average received **power** from a **target** illuminated by a pulsed transmission.

pulse length—In **radar**, **sodar**, or **lidar**, the extent of a transmitted **pulse**, measured in units of length.

So defined, the pulse length is the **pulse duration** times the **velocity** of propagation of the **energy**. However, the term pulse length is sometimes used in place of pulse duration.

pulse length modulation—*See* **pulse modulation**.

pulse modulation—1. Modulation of a **carrier** signal by a train of **pulses**, as in **pulsed radar**.

2. Use of a series of **pulses** modulated to carry information. The **modulation** may involve changes of pulse amplitude, position, **phase**, or duration.

3. Modulation of the **waveform** characteristics within an individual **pulse** during the duration of the pulse, as in **pulse compression**.

pulse-pair processing—An efficient computational **algorithm** for **digital** radar data processing that provides estimates of the **mean Doppler velocity** and the **Doppler spread**.

Based on the properties of the **autocorrelation function** of the **radar** signal in successive pulses, this procedure is usually much faster than the **fast Fourier transform** algorithm, but it does not yield the complete **Doppler spectrum**.

pulse period—Same as **interpulse period**.

pulse phase modulation—*See* **pulse modulation**.

pulse position modulation—*See* **pulse modulation**.

pulse radar—Same as **pulsed radar**.

pulse repetition frequency—(Abbreviated PRF.) The rate at which pulses of radio **energy** are transmitted by a **pulsed radar**.

The PRF determines the **maximum unambiguous range** and the maximum unambiguous **Doppler velocity** measurable by a particular **radar**. For typical **weather radars**, the PRF is in the **range** from several hundred **hertz** to several kilohertz.

pulse-time-modulated radiosonde—(Also called time-interval radiosonde.) A **radiosonde** that transmits the indications of the meteorological sensing elements in the form of pulses spaced in time.

The meteorological data are evaluated from the intervals between the pulses. The first meteorological telemeter was developed by Olland in 1875 and made use of this principle.

pulse time modulation—*See* **pulse modulation**.

pulse volume—The volume in space within which are located the scatterers that contribute to the **radar echo** arriving at the **receiver** at a particular instant.

Individual scatterers within the pulse volume contribute to the instantaneous **signal** and cannot be resolved. The pulse volume is therefore sometimes called the **resolution volume**. The extent in **range** of the pulse volume is determined by the **convolution** of the transmitted **pulse** with the receiver filter response and the extent in **azimuth** by the **antenna pattern**. Ordinarily, the radial extent of the volume is one-half the **pulse length** and the transverse extent is the 3-dB **beamwidth**.

pulse width—Same as **pulse length**.

pulse width modulation—*See* **pulse modulation**.

pulsed-light cloud-height indicator—A remote-sensing instrument used for the determination of **cloud** heights.

It operates on the principle of **pulse radar**, employing visible **light** rather than **radio waves**. *See* **ceilometer, laser ceilometer, cloud-height indicator**.

pulsed radar—A **radar** that transmits and receives individual pulses of radio **energy**, as opposed, for example, to a **continuous-wave radar**.

The **range** to a **target** is measured by the time for a **pulse** to travel from the **transmitter** out to the **receiver** and back.

pumping—Rapid vertical **oscillations** of the column of a **mercury barometer**.

The oscillations are caused by the variations of **ambient pressure** due to the occurrence of wind **gusts**. Aboard ship, they may also be caused by the vessel's motion.

pumping head—The sum of the **static** head plus **friction**, as well as all other **head** losses on a pump, for a given **discharge**.

pumping test—A method for determining the **transmissivity** and **storativity** of a **confined** or **unconfined aquifer**.

pumping water level—The **water level** in a well that is being pumped at a given rate.

purga—(Also spelled poorga.) In Russia, a **severe storm**, similar to the **blizzard** and **buran**, that rages in the **tundra** regions of northern Siberia in winter.

See **burga**.

purl—In **radar meteorology**, the looping flight path described by an aircraft's track during **storm** investigation.

> Mapes, B. E., 1995: Diabatic divergence profiles in Western Pacific mesoscale convective systems. *J. Atmos. Sci.*, **52**, p. 1810.

purple light—A faint band, purple in color, seen over much of the solar sky during a **clear** twilight.

The purple light exists when sun **elevation** $-2° < h_0 < -6°$. Its azimuthal width is $\sim 40°–80°$, and its vertical or elevation-angle width is $\sim 10°–15°$. Maximum **luminance** occurs at $h_0 \sim -4°$, and the area of purple light steadily descends toward the solar point during evening **twilight**. At $h_0 \sim -7°$, the **bright segment** replaces the purple light.

pycnocline—A vertical **density** gradient (as determined by the vertical **temperature** and **salinity** gradients and **equation of state**) in some layer of a body of water, which is appreciably greater than the gradients above and below it; also a layer in which such a **gradient** occurs.

The principal pycnoclines in the ocean are either seasonal, due to heating of the **surface water** in summer or **freshwater** inputs, or permanent.

pyranogram—(Or solarigram.) Record made by a **pyranograph** (solarigraph).

pyranograph—(Or solarigraph.) Recording **pyranometer** (solarimeter).

pyranometer—(Sometimes called **solarimeter**.) General name for the class of **actinometers** that measure the combined **intensity** of incoming **direct solar radiation** and **diffuse sky radiation**.

The pyranometer consists of a recorder and a **radiation** sensing element that is mounted so that it views the entire sky (radiation from the solid angle 2π on a plane surface). *See* **pyrheliometer, Robitzsch actinograph, albedometer**.

pyrgeometer—Instrument for measuring **radiation** in the **longwave** range between 2 and 60 mm. As a horizontal downward-facing black surface, it measures the **terrestrial radiation**, and as a horizontal upward-facing black surface, it measures the **atmospheric radiation**.

pyrheliogram—(Or actinogram.) Record made by a **pyrheliograph** (actinograph).

pyrheliograph—(Or actinograph.) Recording **pyrheliometer** (actinometer).

pyrheliometer—General term for the class of **actinometers** that measure the **intensity** of **direct solar radiation**.

The instrument consists of a **radiation** sensing element enclosed in a casing that is closed except for a small **aperture** through which the direct solar rays enter. Pyrheliometers can be classified on the basis of the sensing elements employed. In one form the **sensing element** is a blackened water **calorimeter**. The rise in the **temperature** of the water gives a measure of the amount of **radiant energy** absorbed during the **exposure** of the instrument. Another type of sensing element consists of a blackened plate of high **heat capacity**. When radiation is allowed to fall on the plate for a period short compared to the **thermal** time constant, the temperature rise of the plate is proportional to the intensity of the incoming radiation. A third type of sensing element consists of a pair of plates, one blackened and one reflecting, that are continuously exposed to the incoming radiation. The temperature differential between the plates is proportional to the intensity of the incoming radiation. See Hand (1946) for descriptions of various types of pyrheliometers, for example, silver-disc pyrheliometer, **water-flow pyrheliometer**, Eppley pyrheliometer, **spectropyrheliometer**, **Michaelson actinograph**.

Hand, I. F., 1946: *Pyrheliometers and Pyrheliometric Measurements*, U.S. Weather Bureau.

pyrheliometric scale—Scale of measurement of **irradiance** as determined by an absolute standard type **pyrheliometer**; now superseded by the World Radiometric Reference.

pyrheliometry—The science and study of pyrheliometric measurements.
See **pyrheliometer**.

pyrhenerwind—A **foehn** of the Austrian Alps.

pyrocumulus—A **cumulus** cloud formed by a rising **thermal** from a fire, or enhanced by buoyant **plume** emissions from an industrial combustion process.
See also **venting mixed layer air**.

pyrolysis—The destruction of a chemical compound by heating or burning.

pyrotechnic flare—A **cloud seeding** device in which the cloud seeding agent is first vaporized at a high **temperature** and then condensed as an **aerosol** as the **vapor** cools.
Flares have been made for producing **ice** forming or **hygroscopic** seeding materials. Typical flares burn for 20 seconds to several minutes, and release 10–100 g of **seeding** material. Airborne flares are usually mounted in external racks on cloud seeding aircraft. Large flares are burned **upstream** or within clouds; smaller flares may be ignited and dropped through clouds.

pyrradiometer—Instrument for measuring the total **irradiance** (solar and atmospheric) on a horizontal surface from a solid angle of 2π.

Q

Q band—*See* **radar frequency bands**.

Q burst—An electromagnetic **transient** launched by a **lightning discharge** of large **amplitude** within the earth–ionosphere cavity.

Originally named by Toshio Ogawa (1967), the "Q" connotes "quiet" and characterizes a transient in which the fundamental 8-Hz mode of the **Schumann resonances** is the dominant contributor. In general, a mix of Schumann modes makes up a Q burst.

> Ogawa, T., et al., 1967: Worldwide simultaneity of occurrence of a Q-type ELF burst in the Schumann resonance frequency range. *J. Geomag. Geoelectr.*, **19**, 377, 384.

Q channel—*See* **I and Q channels**.

Q code—A letter code used by aircraft in requests for information; it is also used in the supply of information to aircraft.

Certain items in the code relate to **meteorological information**, for example, QFE refers to **station pressure**, QNH to **altimeter setting**.

Q noise—Quasi-continuous sequences of VHF (very high frequency) **radiation** emanating from **lightning** that is associated with high-speed **K-change** activity along pre-ionized lightning channels. The process was named by David Proctor (1974).

> Proctor, D., 1974: VHF radio pictures of lightning. *CSIR Special Report, No. TEL 120*, Pretoria, South Africa.

Q vector—A horizontal **vector**, arising in **quasigeostrophic** and **semigeostrophic theory**, the **divergence** of which appears on the right-hand side of the **omega equation**.

In the context of *f*-**plane** (i.e., the **Coriolis parameter** *f* is assumed constant) quasigeostrophic theory, the **Q** vector is defined as

$$\mathbf{Q} = -\frac{g}{\theta_0}\left(\frac{\partial \mathbf{v}_g}{\partial x} \cdot \nabla_p\theta, \frac{\partial \mathbf{v}_g}{\partial y} \cdot \nabla_p\theta\right),$$

where g is the **acceleration of gravity**, θ_0 a constant reference value of the **potential temperature**, $\mathbf{v}_g$ the horizontal **geostrophic wind**, ∇_p the horizontal **gradient** operator on a **constant-pressure surface**, and θ the potential temperature. In the context of *f*-plane semigeostrophic theory, the definition of the **Q vector** is identical except that the physical coordinates (x, y) are replaced by the **geostrophic coordinates** (X, Y). The **Q** vector tends to point in the direction of rising air. If **Q** points toward warm air, the **geostrophic flow** is **frontogenetic**. If **Q** points toward cold air, the geostrophic flow is **frontolytic**.

qarajel—Same as **karajol**.

qaus—Same as **kaus**.

qibla—*See* **ghibli**.

QPF—Abbreviation for **quantitative precipitation forecast**.

quadrant analysis—A method of **analysis** that determines whether two different **perturbation** (turbulent) variables are correlated by plotting their joint frequency of occurrence.

quadrant electrometer—A very sensitive **electrostatic electrometer** for measuring small **potential** differences.

It consists of a metal needle in the form of a lemniscate suspended horizontally inside a short metal container that has been divided into four electrically insulated quadrants. In the most frequent mode of operation, alternate quadrants are electrically connected to the terminals of a battery that has a grounded midpoint. The unknown potential is applied to the needle, causing a rotation proportional to this potential.

quadrature—1. In astronomy, the arrangement of the earth, sun, and another planet or the moon in which the angle subtended at the earth between the sun and the third body, in the **plane of the ecliptic**, is 90°.

The first and third quarters of the moon are positions of quadrature. *See also* **conjunction**, **opposition**.

2. In **radar** systems, an **orthogonal** relationship between two coherent signals in which the **phase** of one **signal** is offset by 90° from the phase of the other. Two signals in quadrature may be regarded as a single **complex signal**.

quadrature spectrum—The imaginary part of the **cross spectrum** of two functions.

quality of snow—The amount of **ice** in a **snow** sample expressed as a percent of the weight of the sample.

quality of water—The physical, chemical, and biological characteristics of water with respect to its suitability for a particular use.

quantile—A generic term for any fraction that divides a collection of observations arranged in order of magnitude into two specific parts.

Thus, the upper **quartile** is the quantile that separates the upper one-quarter from the lower three-quarters of the observations. *See also* **percentile, decile**.

quantitative precipitation estimation—The estimation of the amount of **rainfall** or **rainfall rates** based on **radar** measurements or satellite data.

quantitative precipitation forecast—(Abbreviated QPF.) A **prediction** of the **amount of precipitation** that will fall at a given location in a given time interval.

quantum theory—(Also quantum mechanics.) A theory of matter and **radiation**, developed in its essentials mostly between 1900 and 1930, distinguished from classical theory (or classical mechanics, **Newtonian mechanics**) in two important respects: discreteness and indeterminism.

According to classical theory, measurable physical variables such as **energy** and **momentum** can have a continuous set of values; according to quantum theory, however, these variables can have only a discrete set of values (and hence are said to be quantized). The dynamical laws of classical theory are **deterministic**: Given the exact initial state of a system, its future state is in principle exactly determined by dynamical laws. Quantum theory is inherently probabilistic: Its dynamical laws yield only the probabilities that certain discrete values for physical variables will be measured for an ensemble of similarly prepared systems. *See* **energy level, photon**.

quara—Same as **karajol**.

quartile—One of a set of numbers (a **quantile**) on the **random-variable** axis that divides a **probability distribution** into four equal areas.

The three quartile points that lie between the extremes of the distribution are designated as Q_1, Q_2, Q_3 and are defined in terms of the **distribution function** $F(x)$ as follows:

$$F(Q_1) = 0.25; \ F(Q_2) = 0.50; \ F(Q_3) = 0.75.$$

Thus, Q_2 coincides with the **median**. In empirical **relative frequency** tables, the quartiles are estimated by **interpolation**. The **interquartile range** $2Q$ is the distance from Q_1 to Q_3; half of this distance Q is called the **semi-interquartile range** (or quartile deviation) and is sometimes used as a crude measure of **variability** or **spread**.

quartile deviation—Same as **semi-interquartile range**.

quasi-biennial oscillation—An **oscillation** in the **zonal** winds of the equatorial **stratosphere** having a **period** that fluctuates between about 24 and 30 months.

This oscillation is a manifestation of a downward propagation of winds with alternating sign. This phenomenon is sometimes referred to as the stratospheric quasi-biennial oscillation to distinguish it from other atmospheric features that also have spectral peaks near two years.

quasi-biennial periodicity—Same as **quasi-biennial oscillation**.

quasi-equilibrium—In meteorology, used to describe a state of **statistical** equilibrium between ensembles of **cumulus** convective clouds and larger-scale processes.

In such an **equilibrium** state, the rate of generation of **convective available potential energy** by the larger-scale processes is balanced by **frictional dissipation** within the convective clouds.

quasi-hydrostatic approximation—(Or quasi-hydrostatic assumption; also called hydrostatic approximation.) The use of the **hydrostatic equation** as the vertical **equation of motion**, thus implying that the vertical accelerations are small without constraining them to be zero.

This compromise takes advantage of the smallness of organized vertical accelerations in **cyclonic-scale** motions while allowing theoretically for a realistic distribution of vertical velocities, which may be computed from the other equations of a **closed system**. Dynamically, the effect of the quasi-hydrostatic approximation is to eliminate or **filter** out the higher frequencies, corresponding to **sound waves** and certain (but not all) **gravity waves**, from the fundamental equations, while retaining the frequencies corresponding to cyclonic-scale motions. Combined often with the **quasigeostrophic approximation**, this assumption is much used in theoretical work associated with **numerical forecasting**. An example of phenomena to which it is inapplicable is the **lee wave**. For the discussion of this and other types of gravity waves, it is common to assume **hydrostatic equilibrium** in the **basic flow** but not in the **perturbation**. *See* **filtering approximations**.

quasi-Lagrangian coordinates—A system of mixed **Eulerian** and **Lagrangian coordinates**.

At least one coordinate of each **fluid parcel** must therefore be unvarying with time. Such a system takes advantage of the fact that in many atmospheric models there is one property (but not three properties) conserved in the motion. The most frequently used system of this type is the (x, y, θ) system under **adiabatic** motions, where x and y are **Cartesian coordinates**, and θ the **potential temperature**. If **water vapor** in all phases is admitted to the system, **wet-bulb potential temperature** or a similar conservative **temperature** may be used.

quasi-nondivergent—In analogy to **quasigeostrophic**, characterizing a **model** in which a nondivergent **velocity** field is used in every context except in the **divergence** term in the **vorticity equation**.

This velocity field is computed from some steady-state equation such as the **balance equation** (thus including nonlocal accelerations) and is more general than the **geostrophic approximation**. The **horizontal divergence** must, of course, be computed from the **equation of continuity** or from another equation.

quasi-stationary front—(Commonly called stationary front.) A **front** that is stationary or nearly so.

Conventionally, a front that is moving at a speed less than about five knots is generally considered to be quasi-stationary. In **synoptic chart** analysis, a quasi-stationary front is one that has not moved appreciably from its position on the last previous synoptic chart (three or six hours before).

quasi-stationary perturbation—A **departure** from the long-term average state that is apparent in averages of a week or longer.

quasi-stationary time series—*See* **stationary time series**.

quasi steady state—A situation that is changing slowly enough that it can be considered to be constant.

For example, **atmospheric turbulence** has a fast **response time**, while the **atmospheric boundary layer** depth that controls the **turbulence** grows with a slower timescale. Thus, turbulence can be analyzed as if it were unchanging because it quickly reaches **equilibrium** for any instantaneous **boundary layer** depth.

quasigeostrophic—A system or flow that evolves slowly in time compared to the rotation **period** of the earth, has a length scale of the **deformation radius** or larger, and undergoes only limited vertical excursions.

Formally, the quasigeostrophic system of equations is derived from an expansion of terms in powers of the **Rossby number**, which is presumed small.

quasigeostrophic approximation—(Also called geostrophic approximation, pseudogeostrophic approximation.) A form of the **primitive equations** in which an approximation to the actual winds is selectively used in the momentum and thermodynamic equations.

Specifically, horizontal winds are replaced by their **geostrophic** values in the horizontal **acceleration** terms of the momentum equations, and **horizontal advection** in the thermodynamic equation is approximated by **geostrophic advection**. In addition, the quasigeostrophic approximation neglects **vertical advection** of **momentum** and replaces the four-dimensional **static stability** parameter with a basic-state static stability, which is a function of the vertical coordinate only. The quasigeostrophic approximation is used in the **analysis** of extratropical **synoptic-scale** systems, in which winds can be closely approximated by their geostrophic values. This approximation is not accurate in situations in which the **ageostrophic wind** plays an important advective role, for example, around fronts.

quasigeostrophic current—A **current** in which an approximate **geostrophic balance** between **Coriolis** and **pressure gradient** forces holds, but for which other terms, generally the inertial terms involving temporal change or **advective acceleration**, play a key dynamic role (through **vortex stretching** effects) despite their relatively small magnitude.

For the **geostrophic approximation** to hold, the flow must be nearly steady (timescale $\gg$ **pendulum day**), weak, and large-scale (small **Rossby number**), with negligible **friction** (small **Ekman number**).

quasigeostrophic equations—The system of momentum and thermodynamic equations, derived through the **quasigeostrophic approximation**, that fulfill the requirements of **quasigeostrophic theory**.

At the **momentum** level and using the pseudoheight coordinate $z = [1 - (p/p_0)^\kappa]c_p\theta_0/g$ the equations are

$$\frac{\partial u_g}{\partial t} + u_g \frac{\partial u_g}{\partial x} + v_g \frac{\partial u_g}{\partial y} - fv_{ag} = 0,$$

$$\frac{\partial v_g}{\partial t} + u_g \frac{\partial v_g}{\partial x} + v_g \frac{\partial v_g}{\partial y} + fu_{ag} = 0,$$

$$\frac{\partial \phi}{\partial z} = \left(\frac{g}{\theta_0}\right)\theta,$$

$$\frac{\partial u_{ag}}{\partial x} + \frac{\partial v_{ag}}{\partial y} + \frac{\partial(\rho w)}{\rho \partial z} = 0,$$

$$\frac{\partial \theta}{\partial t} + u_g \frac{\partial \theta}{\partial x} + v_g \frac{\partial \theta}{\partial y} + \frac{\theta_0}{g} N^2 w = 0,$$

in which x, y, and z form a Cartesian **coordinate system** with u_g and v_g the eastward and northward **geostrophic wind** components, u_{ag} and v_{ag} the eastward and northward **ageostrophic wind** components, w the **vertical velocity** in the pseudoheight coordinate, ϕ the **geopotential**, θ the **potential temperature**, g the **acceleration of gravity**, f the **Coriolis parameter**, ρ the **density** in z space, θ_0 the reference potential temperature, and N the **Brunt–Väisälä frequency**.

quasigeostrophic equilibrium—Said of a flow that fulfills the requirements of the **quasigeostrophic equations**.

See **quasigeostrophic approximation**.

quasigeostrophic motion—Atmospheric motions that fulfill the principles and requirements of **quasigeostrophic theory**.

quasigeostrophic theory—A theory of **atmospheric dynamics** that involves the **quasigeostrophic approximation** in the derivation of the **quasigeostrophic equations**.

Quasigeostrophic theory is relatively accurate for **synoptic-scale** atmospheric motions in which the **Rossby number** is less than unity. However, it cannot accurately describe some atmospheric structures such as fronts or small strong low pressure cells as well as **semigeostrophic theory**.

quasiperiodic—A term with a variety of meanings usually used to describe motion, a series of data, or a mathematical function the behavior of which suggests a recurring pattern but fails to conform to the strict meaning of **periodicity** in that an ordered set of values does not repeat at regular intervals of time or space.

Quaternary climate—The **climate** of the last 2 500 000 years, including the alternating glacial–interglacial climate of the **Pleistocene** and the comparatively warm climate of the Holocene or postglacial (the last 10 000 years).

Quaternary Ice Age—See **Quaternary period**.

Quaternary period—The last two million years of **geologic time**, comprising the Pleistocene and Holocene **glacial epochs**. Estimates of the date of the beginning of the Quaternary vary between 2.5 and 1.6 million years ago.

quintile—For a set of data arranged in order, values that partition the data into five groups, each containing one-fifth of the total number of observations.

Also, values of a **random variable** that partition the **variable** range into five subintervals in such a way that the **probability** taken over each subinterval is equal to one-fifth.

R

R-meter—Analog **radar** device used in the 1950s to measure the **fluctuation** rate of the detected **target signal** from which may be estimated the root-mean-square **relative velocity** among the scatterers in the **pulse volume**.

This approach was made obsolete when pulsed **Doppler radar** was introduced to meteorology in the late 1950s.

R-scope—A **radar** display with coordinates of received **signal** amplitude versus **range** but differing from the **A-display** by starting from a range offset from zero.

rabal—A method of **winds-aloft observation**; that is, the determination of **wind** speeds and directions in the **atmosphere** above a **station**.

It is accomplished by recording the **elevation** and **azimuth** angles of the balloon at specified time intervals while visually tracking a **radiosonde balloon** with a **theodolite**. A rabal is basically the same as a **pilot-balloon observation**, except that the height data are derived from the **radiosonde observation** rather than from an assumed balloon ascension rate.

racoon—A **zero-pressure balloon** flying high above a very cold **tropopause** in tropical or summer midlatitudes.

When the balloon cools and descends at night, its **radiation** temperature does not change, but its lift increases as it descends to the colder levels. The lost lift is overcome and the balloon floats at a lower **altitude** without the need for ballast. The flight altitude of the balloon is radiation-controlled.

radar—(Coined word for radio detection and ranging.) An electronic instrument used for the **detection** and ranging of distant objects of such composition that they **scatter** or reflect radio **energy**.

A radar consists of a **transmitter, receiver, antenna, display**, and associated equipment for control and **signal** processing. The most common radars are monostatic radars, which use the same antenna for both transmission and reception. These radars depend on backscattering to produce a detectable **echo** from a **target**. **Bistatic radars** have the transmitter and its antenna at one location and the receiver and its antenna at a remote location. These radars depend upon forward **scattering** to produce a detectable signal. Radio energy emitted by the transmitter and focused by the antenna of a **monostatic radar** propagates outward through the **atmosphere** in a narrow **beam**. Objects lying in the path of the beam reflect, scatter, and absorb the energy. A small portion of the reflected and **scattered** energy, called the **target signal**, travels back along the same path through the atmosphere and is intercepted by the receiving antenna. The time delay between the transmitted signal and the target signal is used to determine the distance or **slant range** of the target from the radar. The direction in which the focused beam is pointing at the instant the target signal is received (i.e., the **azimuth** and **elevation** angles of the antenna) determine the direction and height of the target. This information is presented visually as echoes on different types of **radar displays**. Because **hydrometeors** scatter radio energy, **weather radars**, operating in certain **radar frequency bands**, can detect the presence of **precipitation** and other weather phenomena at distances up to several hundred kilometers from the radar, depending upon meteorological conditions and the type of radar. **MST radars** and **wind profilers**, which operate at longer **wavelengths** than weather radars, are able to detect echoes from optically **clear air** that are caused by spatial fluctuations of **refractivity**. Additional information provided by a radar about a target may include the **radial velocity** or rate of change of range, as measured by a **Doppler radar**, or the depolarizing characteristics of the target, as measured by a **polarimetric radar**.

radar algorithm—A computer program that automatically detects the presence of desired features or signatures in **radar** data.

radar altimeter—(Also called radio altimeter.) An onboard **radar** for determining the **altitude** of an aircraft above an underlying surface.

Pulse-radar techniques measure altitude in terms of the transit time of the radar **pulse**; **continuous-wave radar** measures altitude in terms of the **phase** difference between the transmitted and received **signals**.

radar altitude—The **altitude** of an aircraft as determined by radar-type radio **altimeter**; thus, the actual distance from the nearest terrain feature.

For all practical purposes, the radar altitude over oceans is equal to the height above **mean sea level**.

radar band—1. A **radar frequency band**.

2. A **precipitation band**.

radar beam—The focused electromagnetic emissions from a **radar** antenna.

The **beam** is defined by the **main lobe** of the **antenna pattern**.

radar calibration—1. The act of determining the proportionality factor or **radar constant** that relates the **radar reflectivity** of a **target** to the **power** measured at the **output** of a radar **receiver**.

2. The numerical value of the **radar constant** that associates **target** reflectivity and measured **power**.

See **radar equation**.

radar climatology—The temporal and spatial **statistics** of **radar** weather **echoes**.

See **radar meteorology**, **radio climatology**.

radar coded message—The encoded and transmitted report of **radar** features observed by a **WSR–88D** radar. The message consists of three main parts.

Part A contains a **display** and tabular listing of **composite reflectivity** data; the display may also contain annotations or graphical overlays generated by various radar algorithms. Part B contains a **vertical profile** of horizontal **wind** obtained from the **velocity–azimuth display** (VAD) **algorithm**. Part C contains the locations and characteristics of features automatically generated by radar algorithms (e.g., **tornadic vortex signature**, **mesocyclone signature**, **storm** centroid, storm top, and **hail** indices) as well as optional manual entries (e.g., **precipitation** type, height of **melting level**, location and movement of individual storms and **tropical storms**, and location of **line echo wave patterns**).

radar constant—The combination of **radar** system parameters and physical constants appearing in the **radar equation** that determines the proportionality factor between the **reflectivity** of a radar target at a given **range** and the **power** measured at the receiving **antenna**.

These parameters include **peak power**, **antenna gain** or **aperture**, **beamwidth**, **pulse duration**, and **wavelength**. *See* **radar equation**.

radar cross section—For a **scattering** object or **target** at a certain **range**, the cross-sectional area of an **isotropic** scatterer at the same range that would return the same **power** to a **radar** as the actual target.

The radar cross section σ may be defined by

$$\sigma = \frac{4\pi r^2 S_r}{S_i},$$

where S_i is the **power density** incident on the target at range r and S_r is the backscattered power density at the receiving **antenna**. *See* **radar equation**; *compare* **backscattering cross section**.

radar displays—Maps, plots, or graphical presentations of **weather radar** observations designed to facilitate interpretation of the observations.

Some displays are maps of such basic quantities as **reflectivity factor** or **mean Doppler velocity** on polar or other coordinate systems (*see* **PPI** and **RHI**). Others are maps of derived quantities, such as **rainfall rate** or **vertically integrated liquid** water; or of quantities or phenomena based on computer algorithms, such as **vortex signatures**, **downbursts**, **fronts**, or regions where **hail** is likely. Another class of radar display is employed for special analyses, for example, the **VAD** used for kinematic analysis of the **wind field**.

radar duct—*See* **duct**, **radio duct**.

radar echo—*See* **echo**.

radar equation—An equation expressing the **power** of a **radar echo** at the **input** of the receiving **antenna** of a **radar** as a function of the **range** and **radar cross section** of a **target**.

For a **point target** and plane-polarized **radiation**, it is written

$$P_r = P_t \frac{G^2 \lambda^2 \sigma}{64\pi^3 r^4},$$

where P_r is the **received power**, P_t the **peak power** transmitted by the radar, G the **gain** of the antenna, λ the **wavelength**, r the distance to the target, and σ the **radar cross section** of the target. For a **distributed target** such as **precipitation**, which fills the **radar beam**, the radar equation may be written

$$\overline{P_r} = P_t \frac{G^2 \lambda^2 \theta \phi h}{1024\pi^2 \ln 2} \frac{\eta}{r^2},$$

where θ and ϕ denote the antenna **beamwidths** in the horizontal and vertical planes, h is the

pulse length of the transmitted **signal**, and η is the **radar reflectivity** of the target. This equation assumes that the **antenna pattern** has a Gaussian shape and that the **scattering** volume is uniformly filled. The radar signals received from distributed targets fluctuate; the overbar on P_r indicates that it is a **time average** over a period equal to several multiples of the **coherence time** of the received signal. From the radar equation, the fundamental measurable property of precipitation is the radar reflectivity η, which depends on the sizes and concentration of the **hydrometeors** and their thermodynamic **phase**. For hydrometeors small enough for the **Rayleigh scattering** approximation, it is given by

$$\eta = \frac{|K|^2 \pi^5}{\lambda^4} Z,$$

where $|K|^2$ is a **dielectric factor**, approximately equal to 0.93 for water and 0.21 for **ice**. The factor Z is called the **radar reflectivity factor** of the precipitation. It equals the sum of the sixth-powers of the diameters of the water drops in a unit volume of space or of the melted diameters of the **snow** and ice particles in a unit volume. It may be expressed in terms of the **drop-size distribution** as

$$Z = \int_0^\infty N(D)D^6 dD,$$

where $N(D) \, dD$ is the number of drops per unit volume with diameters in the interval dD. (For ice-phase precipitation, $N(D)$ is the distribution of melted diameters.)

radar frequency band—A frequency band of **microwave radiation** within which radars operate. The radar frequency bands were first designated by code letters for secrecy during World War II; these letters are still in common use, although the exact **frequency** intervals to which they apply have undergone some redefinition.

They all fall within the **UHF**, **SHF**, and **EHF** radio frequency bands. The bands normally used for **radar** detection of **precipitation** and **clouds** are the following.

Frequency band	Frequency range (GHz)	Wavelength range (cm)
L band	1–2	15–30
S band	2–4	7.5–15
C band	4–8	3.75–7.5
X band	8–12	2.5–3.75
Ku band	12–18	1.67–2.5
K band	18–27	1.11–1.67
Ka band	27–40	0.75–1.11
V band	40–75	0.4–0.75
W band	75–110	0.27–0.4

Radars operating in the S, C, and X bands are the ones mainly used for precipitation measurements. Attenuation of the transmitted **radio frequency** energy by atmospheric gases, precipitation, and cloud particles is severe for all frequency bands higher than X band, and even X band can suffer severe **attenuation** in **heavy rain**. Nevertheless, because radars operating in the K, Ka, and W bands are able to detect clouds, they are used for cloud observations even though they are not able to penetrate far through precipitation. Radar **wind profilers** and **MST radars** operate at lower frequencies than those included in this table, namely, in the **UHF** and **VHF** bands. *Compare* **radio frequency band**.

radar horizon—*See* **radio horizon**.

radar hydrology—The use of meteorological **radar** measurements for hydrological purposes, particularly for estimating the current **precipitation intensity** as a function of location over a region and the total **precipitation** during a prescribed time interval, and for deriving estimates of **runoff** and **streamflow**.

radar hygrometer—*See* **microwave refractometer**.

radar interferometry—Techniques for measuring properties of a **scattering** medium by comparing the signals received at separate antennas or at separate points on the same **antenna**, or by comparing signals received using two slightly different **carrier frequencies**.

For example, in spatial **interferometry** the **angle of arrival** of **signals** is found by comparing the phases of signals received at two or more separated or spaced antennas. This information, combined with Doppler **frequency** measurements, can be used to estimate the **velocity** of the

target. In frequency domain interferometry, the **phases** of signals at two slightly different carrier frequencies are compared to improve the **range resolution.**

radar meteorological observation—(Or **radar weather observation.**) An evaluation of the **echoes** that appear on the **display** of a **weather radar** in terms of the orientation, **intensity**, tendency of intensity, height, movement, and unique characteristics of echoes, which may be indicative of certain types of severe storms (such as **hurricanes**, **tornadoes**, or **thunderstorms**).

Such an **observation** is commonly abbreviated ROB (for **radar observation**).

radar meteorology—The study of the **atmosphere** and weather using **radar** as the means of observation and measurement.

One of the branches of **physical meteorology**, it shares much in common with **cloud physics** and, more generally, atmospheric **remote sensing**. For essays on the history of radar meteorology and research reviews, see Atlas (1990).

Atlas, D., Ed., 1990: *Radar in Meteorology*, American Meteorological Society, Boston, 806 pp.

radar mile—(Obsolete.) The round-trip time for a radar **pulse** to travel out and back to a **target** one **statute mile** away, or 10.75 µs.

radar observation—(Abbreviated ROB.) The encoded and transmitted report of a **radar meteorological observation.**

These reports usually give the **azimuth**, distance, **altitude**, shape, **intensity**, movement, and other characteristics of **echoes** observed by **radar.**

radar polarimetry—Radar measurements based on comparing the **polarization** properties of the transmitted and received **signals.**

See **polarimetric radar.**

radar–rain gauge comparison—The comparison of hydrological measurements (principally average rain rate or total **precipitation** accumulation over an interval) made with one or more meteorological radars with measurements made over the same spatial and temporal interval by in situ instruments.

Such comparisons are usually made to calibrate the **radar** measurements, though sampling problems (spatial inhomogeneity of precipitation, sparseness of the **rain gauge** network, and the **vertical profile** of **radar reflectivity**) are formidable.

radar range equation—Same as **radar equation.**

radar reflectivity—In general, a measure of the efficiency of a **radar** target in intercepting and returning radio **energy**. It depends upon the size, shape, **aspect**, and **dielectric** properties of the **target.**

It includes not only the effects of **reflection** but also **scattering** and **diffraction**. In particular, the radar reflectivity of a meteorological target depends upon such factors as 1) the number of **hydrometeors** per unit volume; 2) the sizes of the hydrometeors; 3) the physical state of the hydrometeors (**ice** or water); 4) the shape or shapes of the individual elements of the group; and 5) if asymmetrical, their aspect with respect to the radar. The radar reflectivity η has **dimensions** of area per unit volume (e.g., cm^2m^{-3}, or, more commonly, cm^{-1} or m^{-1}) and is defined by

$$\eta = \sum_i N_i \sigma_i,$$

where N_i is the number of hydrometeors per unit volume with backscattering cross section σ_i and the summation is over all the hydrometeors in a unit volume. For spherical hydrometeors small enough compared with the **wavelength** for the **Rayleigh scattering** approximation to be valid, the **radar cross section** is related to **particle** size by

$$\sigma = \frac{\pi^5 |K|^2 D^6}{\lambda^4},$$

where λ is the wavelength, D the diameter of the hydrometeor, and K a **dielectric factor** defined by

$$K = \frac{m^2 - 1}{m^2 + 2},$$

where m is the **complex index of refraction** of the hydrometeor. *See* **radar equation**, **radar reflectivity**

radar reflectivity factor—A quantity determined by the **drop-size distribution** of **precipitation**, which is proportional to the **radar reflectivity** if the precipitation particles are spheres small compared with the **radar** wavelength.

Given the drop-size distribution of a sample of **rain**, the radar reflectivity factor may be computed by summing the sixth-powers of the diameters of all the drops contained in a unit volume of space. Or, regarding the drop-size distribution $N(D)$ as a continuous function of **drop** size, the **reflectivity factor** Z may be written as

$$Z = \int_0^\infty N(D)D^6 dD.$$

For ice-phase precipitation, $N(D)$ is the distribution of melted diameters. Conventional units of Z are mm^6 m^{-3} and it is sometimes measured on a **logarithmic scale** in units of dBZ. The **equivalent reflectivity factor** Z_e may be estimated from measurements of the radar reflectivity η of precipitation and is defined by

$$Z_e = \frac{\eta\lambda^4}{0.93\pi^5},$$

where λ is the radar **wavelength** and 0.93 is the **dielectric factor** for water. Either the reflectivity factor or the equivalent reflectivity factor is frequently used to estimate **rainfall rate** using relationships of the form $Z = aR^b$, where a and b are empirical constants and R is the rainfall rate. For R in millimeters per hour and Z or Z_e in mm^6 m^{-3}, values of a range from 200 to 600 and those of b range from 1.5 to 2. The particular combination of $a = 200$ and $b = 1.6$ defines the **Marshall–Palmer relation.** *See* **radar equation.**

radar report—(Abbreviated RAREP; obsolete.) *See* **radar observation.**

radar resolution—The minimum separation between two **targets** that permits them to be distinguished by a **radar.**
　　The minimum separation in **range** along a given direction from the radar is one-half the transmitted **pulse length.** The minimum angular separation at a given range is approximately the 3-dB **beamwidth** of the radar. These minimum dimensions determine the **radar resolution volume.** Targets within the volume are not resolved and appear to be merged into a single target on a **radar display.** Radar **resolution** is usually of more concern for **point targets** than for weather **echoes,** because the latter are beam-filling targets that are considerably larger than the **resolution volume.** *Compare* **pulse volume.**

radar resolution volume—*See* **radar resolution.**

radar return—An **echo** received by a **radar.**

radar scatterometer—An active **microwave** instrument designed to infer **wind speed** and direction by precisely measuring the **backscattering cross section** (or normalized **radar cross section**). The **Seasat-A** satellite scatterometer was an example.

radar signal spectrograph—(Abbreviated RASAPH; obsolete.) **Analog** radar device used in the 1950s to provide the **power spectrum** of the fluctuations in the received **signal intensity,** which may be interpreted in terms of the relative velocities among the scatterers in the **pulse volume.**
　　This approach was made obsolete when pulsed **Doppler radar** was introduced to meteorology in the late 1950s.

radar sounding—An **upper-air observation** that uses **radar** to obtain meteorological measurements of **hydrometeors** and **winds aloft.**

radar storm detection—The process of detecting storms or **precipitation** areas using a **weather radar.**
　　See **radar, radar meteorology.**

radar theodolite—(Obsolete.) A **radar** that is used to obtain the **azimuth, elevation,** and **slant range** of an airborne **target.**
　　See **theodolite.**

radar volume—Same as **pulse volume.**

radar weather observation—Detection and interpretation of **echoes** from **precipitation.**
　　Such observations are used for **analysis** and forecasting of precipitation, for research in **cloud** and **precipitation physics,** and for **prediction** of and research on **severe storms.** Observations may be reported in a standard format as a **radar coded message.** *See* **radar meteorology.**

radar wind sounding—Upper-level **wind** estimates derived from the movement of a free balloon equipped with a passive or active detector .

radar wind system—Use of **radar** for determining the **azimuth, elevation,** and **range** of a balloon-borne **target** and, from this position information as a function of time, the **wind speed** and **direction**.

radarscope—A **cathode-ray oscilloscope** on which radar **echoes** are displayed.
　　See **radar displays**.

radarsonde—Same as **radar wind system**.

radial inflow—Component of the **wind** velocity directed inward toward the center of an atmospheric system.

radial symmetry—The symmetry of a configuration the properties of which are functions only of radial distance from an origin, and thus independent of the azimuthal coordinate in two dimensions and of azimuthal and latitudinal coordinates in three dimensions.
　　Radial symmetry in two dimensions is often called **circular symmetry**; in three dimensions, **spherical symmetry**.

radial velocity—The component of a three-dimensional **velocity** vector oriented along the radial direction from the origin point or axis in **polar**, **cylindrical**, or **spherical coordinates**.
　　In connection with **Doppler radar**, the radial velocity component is called **Doppler velocity**. *See also* **curvilinear coordinates**.

radial wind—Component of the **wind** at a point in a direction along a **radius vector** from the center of a circulating wind system.

radian—A unit of angular measure; one radian is that angle with an intercepted arc on a circle equal in length to the radius of the circle.
　　Thus, π radians equals 180°. *See also* **degree**.

radiance—A radiometric term for the rate at which **radiant energy** in a set of directions confined to a unit solid angle around a particular direction is transferred across unit area of a surface (real or imaginary) projected onto this direction.
　　Unlike **irradiance**, radiance is a property solely of a **radiation** field, not of the orientation of the surface. The **SI** units of radiance are $W\ m^{-2}\ sr^{-1}$. In general, radiance depends on time, position, and direction as well as **frequency** (**monochromatic** or **spectral radiance**) or **range** of frequencies. Irradiance for any surface is the integral of radiance over a hemisphere of directions above or below that surface. The photometric equivalent of radiance is **luminance**, obtained by integrating spectral radiance weighted by **luminous efficiency** over the **visible spectrum**.

radiance temperature—*See* **brightness temperature**.

radiant—1. Pertaining to the **emission** or the measurement of **electromagnetic radiation**.
　　Compare **luminous**.
　　2. In describing auroras, a projected point of intersection of lines drawn coincident with auroral streamers, that is, the point from which the **aurora** seems to originate.

radiant density—The **radiant energy** passing through a unit volume from all directions; units are **joules** per cubic meter $(J\ m^{-3})$.

radiant energy—Infrequently, any **energy** propagated by a physical quantity governed by a **wave equation**; units are **joules**.

radiant-energy thermometer—An instrument that determines the **blackbody** temperature of a substance by measuring its **thermal radiation**.
　　The substance need not be thermally black over the whole **spectrum**, since it is possible to limit the measurement to those frequencies where it is black.

radiant exitance—**Radiant flux** per unit area leaving a plane surface; no longer used as a substitute for **emittance**.
　　In current usage the equivalent term **flux density** is preferred, or **irradiance** is adapted to include **emission** from a surface; units are **watts** per square meter $(W\ m^{-2})$.

radiant exposure—**Radiant energy** incident on a unit surface over some specified time period; units are **joules** per square meter $(J\ m^{-2})$.

radiant flux—(Sometimes called hemispheric flux.) **Radiant energy** per unit time passing some specified area from one side; units are **watts** (W).

radiant flux density—(Also called **irradiance**.) **Radiant energy** crossing unit area from one side in unit time; units are **watts** per square meter $(W\ m^{-2})$.

radiant intensity—The **radiant flux** per unit solid angle, as from a **point source**; units are **watts** per **steradian** (W sr^{-1}). *See* **intensity**.

radiant power—*See* **radiant flux**.

radiation—1. The process by which **electromagnetic radiation** is propagated through **free space**. The propagation takes place at the **speed of light** (3.00 x 10^8 m s^{-1} in vacuum) by way of joint (**orthogonal**) **oscillations** in the **electric** and **magnetic fields**. This process is to be distinguished from other forms of **energy transfer** such as **conduction** and **convection**.

2. Propagation of **energy** by any physical quantity governed by a **wave equation**.

3. *See* **alpha ray**, **beta ray**.

radiation balance—*See* **net radiation**.

radiation belts—(Also called Van Allen radiation belts.) Belts of energetic **electrons**, **protons**, and heavier **ions** encircling the earth and trapped in the geomagnetic **field**.

The radiation belts are characterized by intense fluxes of high-energy **radiation**, creating a dangerous **environment** for spacecraft. The **particle** density and **energy spectrum**, as well as the physical characteristics of the belts, depend on the level of **solar activity**.

radiation chart—Any **chart** or diagram providing a graphical solution to the **flux** integrals arising in problems of atmospheric infrared **radiative transfer**.

Radiation charts have been superseded by computer models.

radiation equilibrium—*See* **radiative equilibrium**.

radiation flux—*See* **radiant flux**.

radiation fog—A common type of **fog**, produced over a land area when **radiational cooling** reduces the **air temperature** to or below its **dewpoint**.

Thus, a strict radiation fog is a nighttime occurrence, although it may begin to form by evening **twilight** and often does not dissipate until after **sunrise**. Factors favoring the formation of radiation fog are 1) a shallow **surface layer** of relatively **moist air** beneath a dry layer and **clear skies**, and 2) light surface winds. It can be most confusing near sea coasts with cold coastal water. It can be difficult at times to differentiate between this and other types of fog, especially since nighttime cooling intensifies all fogs. Radiation fog is frequently and logically called **ground fog**, but in U.S. weather observing practice, the latter term is defined only with respect to the amount of sky that is obscured by the fog.

radiation frost—**Freezing** conditions that typically occur on **clear** nights with little or no **wind**, when the outgoing is greater than the incoming **radiation** and cooling **air temperature** near the surface creates a stable **temperature inversion** near the ground.

radiation inversion—A relatively cool layer of air, usually adjacent to a ground surface cooled by net loss of **radiation**, in which the **air temperature** increases with height.

Sometimes this term is also loosely extended to include layers in which the **potential temperature** increases with height, that is, to all **statically stable** layers of air formed by radiatively cooled ground. **Radiational cooling** is responsible for forming nocturnal stable **boundary layers**. It is also associated with formation of **dew**, **frost**, and **fog** if the **humidity** is sufficiently high. It is rare for elevated radiation inversions to form because cooling that creates a relative minimum in potential temperature at some height above the surface would cause cold air to descend from the **inversion**, creating **turbulence** that would destroy the inversion.

radiation laws—1. The four physical laws or equations that are most commonly used to explain the **emission** of **radiation**: 1) **Kirchhoff's law**; 2) **Planck's law**; 3) **Stefan–Boltzmann law**; and 4) **Wien's displacement law**.

Of these, 1) and 2) are fundamental. The remaining two can be derived from 2).

2. All of the more inclusive assemblage of empirical and theoretical laws describing all manifestations of radiative phenomena, for example, **Bouguer–Lambert law**, **Lambert's law**, **Stefan–Boltzmann law**, etc.

radiation model—A mathematical or computational description of **radiative transfer** that produces **radiation** products for specified atmospheric conditions.

They may be relatively simple approximations, as in the **two stream** and **delta Eddington** models, which are basically **monochromatic** models. More complicated broadband models include the **correlated-k**, and more complicated **scattering** models include **Monte Carlo models**. Use of radiation models has replaced earlier use of radiation charts.

radiation pattern—(Also called **antenna pattern**, lobe pattern, coverage diagram.) A diagram showing the **intensity** of the **radiation** field in all directions from a transmitting radio or **radar** antenna at a given distance from the **antenna**.

It usually refers to the geometrical solid in space that encompasses all of the **lobes** of the antenna. For a receiving antenna, it is the response of the antenna to a **signal** having unit **field strength** and arriving from all directions. The transmitting and receiving radiation patterns for a single antenna are identical. Two types of radiation patterns should be distinguished: 1) the **free space** radiation pattern that is the complete lobe pattern of the antenna and is a function of the **wavelength**, **feed** system, and **reflector** characteristics; and 2) the **field** radiation pattern that differs primarily from the free space pattern by the formation of **interference** lobes whenever direct and reflected wavetrains interfere with each other, as is found in most surface-based radars. The envelope of these interference lobes has the same shape, but, for a perfectly reflecting surface, it has up to twice the **amplitude** of the free-space radiation pattern.

radiation point—*See* **radiatus**.

radiation pressure—The **pressure** (or force) on a body illuminated by **electromagnetic radiation**.

By ordinary standards radiation pressure is extremely small. For example, the radiation pressure on an object exposed to intense **sunlight** is about 10^{11} times smaller than **sea level atmospheric pressure**. But for small (comparable to or smaller than the **wavelengths** of visible and near-visible **radiation**) **particles** subjected to only the gravitational attraction of the sun and the repulsive radiation pressure of its radiation (e.g., cometary particles), radiation pressure is not negligible, which accounts for the curvature of the tails of comets.

radiation shield—A device used on certain types of instruments to prevent unwanted **radiation** from biasing the measurement of a quantity.

The radiation shield for a **thermometer**, for example, usually consists of a short length of brightly polished metal tubing that encloses the thermosensitive element. The tubing is usually perforated or opened at the bottom to allow proper **ventilation** and is sometimes artificially ventilated, as in the **Assmann psychrometer**.

radiation thermometer—*See* **radiant-energy thermometer**.

radiation torque—The **torque** on a body illuminated by **electromagnetic radiation**.

By ordinary standards **radiation** torques are extremely small, but the **photon** possesses an intrinsic **angular momentum** (or spin) that can be transferred to matter when it interacts with **light**.

radiational cooling—In meteorology, the result of **radiative cooling** of the earth's surface and adjacent air.

Radiational cooling occurs, as is typical on **calm**, **clear** nights, whenever the **longwave** emission from the surface is not balanced by significant amounts of absorbed **shortwave radiation** or downwelling longwave from the **atmosphere** above the surface, and there are no nonradiative sources of sufficient **energy** to make up the difference.

radiational tide—**Tides** generated by regular periodic meteorological forcing.

These are principally at annual (seasonal changes in solar **insolation**), daily (solar-driven **land** and **sea breeze**), and twice-daily (solar-driven **cycle** in **barometric pressure**) periods.

radiative–convective equilibrium—The **equilibrium** state of an atmospheric column for which any net loss or gain of **radiant energy** is balanced by the vertical **transport** of latent or **sensible heat**.

Radiative–convective gain provides an approximate description of the long-term global average of the vertical **temperature** structure of the **atmosphere**, and is useful in some types of **climate models**.

radiative–convective model—A type of **climate model** that simulates the **vertical profile** of atmospheric **temperature** under the assumption of **radiative–convective equilibrium**.

Useful for theoretical studies of **climate sensitivity**, these models are capable of sophisticated treatments of **radiative transfer**. They typically ignore the effects of horizontal **transport**.

radiative cooling—The process by which **temperature** decreases due to an excess of emitted **radiation** over absorbed radiation.

radiative equilibrium—The balance between the gain and loss of **radiant energy** by a specified system.

In radiative equilibrium, there is no net **radiative heating** or cooling of the system. Examples of systems in, or approximately in, radiative equilibrium include the interior of thick water clouds and the long-term average of the planet Earth.

radiative flux density—(Or radiative flux.) The radiative **energy** per unit time and unit area.

The radiative flux density is obtained through integration over all **radiances** in a hemisphere.

radiative flux divergence—The differential increase of radiative **flux** leaving a region compared to that entering it.

A positive flux divergence is usually directly related to the **radiative cooling** rate, and a negative flux divergence to the **radiative heating** rate.

radiative forcing—1. In **radiation**, the net **flux of radiation** into or out of a system.

As a consequence of radiative forcing there must be some change to the nonradiative **energy** states of the system (e.g., its **temperature** may change).

2. In **climatology**, a systematic **perturbation** to the climatological value of the net **radiant flux density** at some point in the earth's **climate system**.

For example, this perturbation may be due to a change in concentration of the radiatively active gases, a change in **solar radiation** reaching the earth, or changes in **surface albedo**.

radiative heating—The process by which **temperature** increases due to an excess of absorbed **radiation** over emitted radiation.

radiative transfer—The physics and mathematics of how **radiation** passes through a medium that may contain any combination of scatterers, absorbers, and emitters.

In meteorology, the study of radiative transfer dates back at least to Lord Rayleigh's investigations in 1871 on the nature of **scattering** by air molecules. Since that time, sophisticated mathematical solutions to radiative transfer have been developed, especially for one-dimensional **monochromatic** problems. The more general problem of solving broadband radiative transfer in three-dimensional **cloudy** atmospheres requires a computational solution.

radiatively active gas—A term used to describe a gas occurring naturally or produced **anthropogenically** that affects **atmospheric radiation** by **absorption** or **emission**.

See also **greenhouse gases**.

radiator—Any source of **radiant energy**.

radiatus—A **cloud variety**, the elements of which are arranged in straight parallel bands.

Owing to the effect of perspective, these bands seem to converge toward a point on the **horizon**, or, when the bands cross the entire sky, toward two opposite points (radiation points, V points, vanishing points, radiant points). This variety occurs in the genera **cirrus**, **altocumulus**, **altostratus**, and **stratocumulus**, and may modify many of the species, but principally **stratiformis**. *See* **cloud classification**, **cloud streets**.

radio acoustic sounding system—(Abbreviated RASS.) A ground-based **remote sensing** method that combines **radar** and acoustic techniques to determine the **vertical profile** of **virtual temperature**, often used in conjunction with **wind profilers**.

With the RASS technique, **sound waves** generated by an acoustic source located near the radar produce perturbations in the atmospheric **refractive index**. The perturbations propagate upward in the air at the local **speed of sound** and serve as a **target** for the radar. When the acoustic **wavelength** is matched to half the radar wavelength (the **Bragg scattering** condition), a **resonance** condition develops that leads to the production of a detectable **radar echo**. The **Doppler shift** of this echo is a measure of the sum of the **speed of sound** and any vertical air motion in the **radar volume**. Early work with RASS either ignored the vertical air motion or used time averaging to reduce its effect. Some modern RASS processors have the ability to correct for vertical air motion. The local speed of sound is related to the virtual temperature by

$$C_s = (\gamma R T_v)^{1/2} \approx 20 T_v^{1/2},$$

where C_s (m s^{-1}) is the speed of sound, R is the **gas constant** of air, γ ($= c_p/c_v$) is the ratio of the **specific heats** of air, and T_v is the virtual temperature in degrees Kelvin. Using this relation, the **vertical profile** of sound speed can be converted to a profile of virtual temperature. The RASS technique depends on close matching of the acoustic wavelength to half the radar wavelength (called Bragg matching). To ensure that Bragg matching occurs throughout the **range** of temperature expected in the vertical profile, the acoustic source is swept in **frequency** over an interval broad enough to contain all the necessary spatial wavelengths. Bragg matching also requires that different acoustic frequency bands be used for different radar frequencies. Acoustic frequencies range from around 100 Hz for 50-MHz radars up to 2 kHz for radars operating around 1 GHz. Because acoustic **attenuation** in the **atmosphere** is strongly dependent on frequency, the height coverage or reach of RASS varies greatly with radar frequency. Typical small UHF **boundary layer radars** can measure RASS temperature profiles to about 1.5 km, whereas large VHF **MST radars** can measure the **temperature** up to 10 km or higher. *See also* **electromagnetic acoustic probe**.

radio atmometer—An instrument designed to measure the effect of **sunlight** upon **evaporation** from plant foliage.

It consists of a porous **clay atmometer** with a blackened surface so that it absorbs **radiant energy**.

radio beacon—A type of radio **transmitter** system having wide-angle coverage.

It may emit signals continuously or, like the **transponder**, may respond to a received **signal** before operating.

radio blackout—(Also called arctic blackout, blackout, polar blackout.) A prolonged period of **fading** of radio communications that occurs naturally in the polar regions.

An arctic blackout may last for days during periods of intense **auroral** activity.

radio climatology—The study of regional and seasonal variations in the manner of propagation of radio **energy** through the **atmosphere**.

radio direction finder—(Also called radio theodolite.) A ground-based, steerable radio **antenna** that tracks a moving radiotransmitter such as that contained in a **radiosonde**.

See **radio direction finding**.

radio direction finding—(Abbreviated RDF.) The use of a **radio direction finder** to track precisely the angular position (**azimuth** and **elevation**) of a **transmitter**; often used as a method of determining **upper-level winds** represented by a **radiosonde**'s movement during flight.

See also **wind finding**.

radio duct—A rather shallow, almost horizontal layer in the **atmosphere** through which vertical **temperature** and moisture gradients produce a **refractive index** lapse rate greater than 157 **N-units** per kilometer. Conditions necessary for the formation of ducts are strongly increasing temperature and decreasing **relative humidity** with height.

The resulting **superstandard propagation** causes the curvature of radio rays to be greater than that of the earth. Radio **energy** that originates within a **duct** and leaves the **antenna** at angles near the horizontal may thus be trapped within the layer. The effect is similar to that of a **mirage** (sometimes called "radio mirage"), and radar targets may be detected at phenomenally long ranges if both **target** and **radar** are in the duct. The greater the **elevation angle** between radar and target, the less the possibility of serious **distortion** due to transmission through ducts. Ducts may be surface-based or elevated, with typical thickness ranging from about 10 to 300 m. When elevated ducts occur, they are generally associated with **subsidence** or **frontal inversions**. Elevated ducts are rarely found above 5 km. *See* **anomalous propagation**, **skip effect**.

radio fadeout—*See* **fading**.

radio frequency—1. (Abbreviated RF.) The number of oscillations per second of the electric and magnetic fields in the radio portion of the **electromagnetic spectrum**, generally that portion between 10^4 and 10^{12} Hz; specifically, the **frequency** of a given radio **carrier wave**.

2. In radio or **radar**, pertaining to a **signal** at the transmitted or received **frequency**, as opposed to a signal translated to a different frequency (**IF signal**) or detected (**video signal**).

radio frequency band—A specified **range** of **frequencies** of **electromagnetic waves**.

Radio frequency bands are classified as follows.

Frequency band	Frequency range
Very low frequency (VLF)	< 30 kHz
Low frequency (LF)	30–300 kHz
Medium frequency (MF)	300–3000 kHz
High frequency (HF)	3000–30 000 kHz
Very high frequency (VHF)	30–300 MHz
Ultra high frequency (UHF)	300–3000 MHz
Super high frequency (SHF)	3000–30 000 MHz
Extremely high frequency (EHF)	30–300 GHz

These bands, based on multiples of 3, are a consequence of the **speed of light** being very nearly equal to 3×10^8 m s^{-1}. Thus, VHF corresponds to free-space **wavelengths** 1–10 m and so on. *Compare* **radar frequency band**.

radio goniometer—Same as **radio direction finder**.

radio hole—Fundamentally no different from a **shadow zone** (at radio frequencies), although the term is often restricted to one associated with a **duct**.

radio horizon—The locus of points at which direct rays from a radio **transmitter** become tangential to the earth's surface.

The radio horizon extends beyond the geometric and visible horizons in conditions of **normal** atmospheric **refraction**. It may be decreased or increased in particular cases as **standard propagation** is replaced by **substandard propagation** or **superstandard propagation**, respectively. Beyond the radio horizon, surface targets cannot be detected under VHF atmospheric conditions although significant radio **power** is sometimes detected in the **diffraction zone** below the **horizon**. This power is a result of **scattering** by turbulence-produced atmospheric inhomogeneities. Assuming a **smooth surface**, the distance of the radio horizon is given approximately by the equation $R = (17h)^{1/2}$, where R is the distance in kilometers and h is the height in meters of the antenna above the surface. *See* **effective earth radius**, **scatter propagation**.

radio meteorology—That branch of meteorology embracing the propagation of **radio waves** in the **atmosphere** and the use of such waves for the **remote sensing** of **clouds**, **storms**, **precipitation**, **turbulence**, **winds**, and various physical properties of the atmosphere.

Radio meteorology encompasses **radar meteorology**.

> Bean, B. R., and E. J. Dutton, 1996: *Radio Meteorology*, National Bureau of Standards Monograph No. 92, U.S. Government Printing Office, 435 pp.

radio navigation aid—A general term to include all means of radio signals used for **navigation**, typically for aircraft, ships, and other transportation systems; for example, the nondirectional Beacon (NDB), long-range navigation (LORAN), tactical air navigation (TACAN), distance measuring equipment (DME), very high frequency omnirange tactical air navigation (VORTAC), the Omega system, and the **Global Positioning System** (GPS).

radio sounding—Observations of meteorological variables such as **temperature**, **pressure**, and **relative humidity** by a **radiosonde** in the **upper air**.

The data are transmitted instantaneously to the observing **station**.

radio source—A natural source of **electromagnetic radiation** in the **radio frequency band**.

Lightning would be a common example.

radio theodolite—*See* **radio direction finder**.

radio wave—**Electromagnetic radiation** with a **frequency** lying within a **radio frequency band**.

radio waves—That portion of the **electromagnetic spectrum** having **wavelengths** longer than about 10 cm.

radio wind observation—Determination of the **upper-level wind** by the tracking of a free balloon by electronic means other than **radar**.

radioactive carbon—Name given to the **isotope** ^{14}C, which has a **half-life** of about 5500 years and is formed in the **upper atmosphere** by the reaction of neutrons on **nitrogen**.

It has also been formed locally by nuclear detonations in the upper atmosphere. Thus, it is useful as a radioactive **tracer** for distinguishing stratospheric **transport** and for dating organic matter. As a result of the gradual decay of this isotope, carbon-containing compounds that have been deposited at the earth's surface have a different isotopic **signature** than the **carbon** in the **atmosphere**.

radioactive dating—Use of radioactive **isotopes** to estimate the age of a sample of water or rock; more precisely, radiometric dating.

See also **groundwater dating**.

radioactive decay—In reference to **conductivity** near the ground, the primary source of **ions** in the lowest 1 or 2 km of the **atmosphere** above land. (**Cosmic radiation** is dominant over the oceans and at higher altitudes.)

The ionizing **radiation** can take the form of 1) **emission** of α, β, or γ radiation by radioactive materials in the soil; or 2) emission from gaseous radioactive daughter products (**radon**) that emanate from the soil. The α radiation emitted by materials in the ground is absorbed in the lowest few centimeters of the **atmosphere**, β radiation penetrates a few meters, and γ radiation a few hundred meters. The **ionization** produced by radioactive gases in the atmosphere is highly variable and depends on the rate of emission from the soil and also on atmospheric **dispersion**. ^{222}Rn, which has a **half-life** of 3.8 days, is produced by the uranium decay series. ^{220}Rn (**thoron**) is produced by the thorium decay series and has a half-life of 54 seconds.

radioactive fallout—The eventual descent to the earth's surface of radioactive matter placed in the **atmosphere** by an atomic or thermonuclear explosion.

Fallout is commonly separated into three classes. 1) Local or close-in fallout occurs in the vicinity (mostly **downwind**) of the detonation in a matter of hours. It consists mainly of **particles** greater than 25 microns in diameter, mostly earthen matter rendered radioactive by attachment. Local fallout occurs for an estimated 20%–80% of the total fission products (depending on location and type of detonation). 2) Intermediate or tropospheric fallout remains in the atmosphere for weeks to months, falling out principally by **precipitation scavenging**, but also by gravitational settling and impingement upon vegetation. Only a few percent of total fission products fall out in this manner. 3) Delayed or stratospheric fallout results from high-yield detonations that thrust their clouds into the **stratosphere**. Storage times vary from less than one year to as high as ten years, depending upon latitude and altitude of detonation. The particles are of submicron size.

radioactive gas—1. In **atmospheric electricity**, any one of the three radioactive inert gases, **radon**, **thoron**, and **actinon**, that contributes to **atmospheric ionization** by virtue of the ionizing effect of the **alpha particles** that each emits on disintegration.

These three gases are isotopic to each other, all having **atomic number** 86. They are members of distinct families of radioactive elements, but each is formed as a result of alpha **emission** and each decays by that process. They form in the interstices of soil or porous rocks containing their respective parent atoms in the forms of salts or minerals. By the process of **exhalation**, they enter the surface layers of the **atmosphere** and are then carried upward by **turbulence** and **convection**.

2. Any gaseous material containing radioactive atoms.

Israël, H., 1951: *Compendium of Meteorology*, 155–161.

radioactive-particle ionization—The production of **ions** by radioactive **particles**, typically near the surface where radioactive substances exist in the soil.

radioactive precipitation—Washout of radioactive **aerosol** particles following release into the **atmosphere** from 1950s weapon tests above ground or from nuclear reactor leaks, as at Chernobyl.

radioactive snow gauge—A device that automatically and continuously records the **water equivalent of snow** on a given surface as a function of time.

A small sample of a radioactive salt is placed in the ground in a lead-shielded collimator that directs a **beam** of radioactive **particles** vertically upward. A Geiger–Müller counting system (located above the **snow** level) measures the amount of depletion of **radiation** caused by the presence of the snow.

radioactivity—1. The spontaneous transition of an atomic **nucleus** to a lower **energy** state (radioactive decay) accompanied by the **emission** of an **alpha particle**, a **beta particle**, or **gamma radiation**.

Alpha emission results in a reduction of **atomic number** by two and mass number by four; (negative) beta emission results in an increase of atomic number by one but no change in mass number; in gamma emission, atomic number and mass number are unchanged. Several naturally occurring **isotopes** are radioactive, including carbon-14 and potassium-40, which reside in the human body. **Radon** (strictly, radon-222) is a natural **radioactive gas** originating ultimately from the radioactive decay of uranium-238. Artificial radioactivity, as opposed to natural radioactivity, is a consequence of bombardment of isotopes that are not radioactive with **neutrons**, **protons**, and other subatomic particles. *See* **half-life**.

2. (Often simply activity.) Rate of decay of a radioactive **isotope**.

The unit of radioactivity is the **curie**.

radiomaximograph—A device for measuring the **field strength** of sferics.

radiometer—An instrument that measures radiated electromagnetic **power**.

Radiometers operating at **infrared** and **microwave** frequencies are used in passive atmospheric **remote sensing**.

radiometric dating—*See* **radioactive dating**.

radiometry—The technique of measuring **radiant energy**, especially radiant energy in that portion of the total **electromagnetic spectrum** lying adjacent to the visible region.

Radiometry is to be distinguished from the closely related subject of **photometry**, the latter being specifically concerned with the quantitative response to **visible radiation** of the human eye. *See* **actinometer**.

radiosonde—An expendable **meteorological instrument** package, often borne aloft by a free-flight **balloon**, that measures, from the surface to the **stratosphere**, the vertical profiles of atmospheric variables and transmits the data via radio to a ground receiving system.

Radiosondes typically measure **temperature**, **humidity**, and, in many cases, **pressure**. Radiosonde temperature sensors generally measure temperature-induced changes in the electrical re-

sistance, capacitance, or voltage of a material. Radiosonde humidity sensors can be substances that respond in a known way to changes in ambient humidity or instruments that directly measure a characteristic of the air that is dependent on its **water vapor** content. Radiosonde pressure sensors are typically **aneroid** cells, a part of which flexes in proportion to pressure changes. Some radiosondes do not measure pressure, but pressure data are calculated from the **hypsometric equation** using temperature, humidity, and height data. Some radiosondes also measure **wind speed** and **direction**. *See* **rawinsonde**.

radiosonde balloon—A balloon used to carry a **radiosonde** aloft.
    Radiosonde balloons are larger than **pilot balloons** and **ceiling balloons**, are generally filled with **hydrogen** or **helium** to achieve lift, and burst at altitudes of about 30 km in daytime and 25 km at night.

radiosonde ground equipment—The nonflight equipment necessary to track a **radiosonde**, receive its **telemetry** signal, convert the **signal** into meteorological data, process the data into calculated parameters and graphical displays, and transmit coded messages representing the upper-air **radiosonde observation**, or raob.

radiosonde observation—(Acronym raob.) An evaluation of **pressure**, **temperature**, and **relative humidity** data received from a balloon-borne **radiosonde**.
    The processed data are usually presented in terms of **geopotential height**, temperature, and **dewpoint** at **mandatory** and **significant pressure levels**. If the position of the radiosonde is measured to determine **winds aloft**, then the **observation** is called a **rawinsonde observation**.

radiosonde station—Station where **radiosonde** observations are routinely made by use of packages with instrument and transmitting equipment carried aloft by **balloons**.

radiosonde system—The collective ground and flight equipment, along with computer software, necessary to make a **radiosonde** flight, collect and process its data, and transmit the associated **upper-air** radiosonde message.

radius of protection—The radius of the circle within which a **lightning discharge** will not strike due to presence of an elevated **lightning rod** at the center.
    A rule of thumb is that this radius is equal to the height of the rod, though cases of **lightning damage** inside this distance are well established.

radius vector—*See* **spherical coordinates**.

radix layer—The bottom fifth of the convective **atmospheric boundary layer** where **virtual potential temperature**, **wind speed**, and **humidity** vertical gradients are nonzero.
    A **convective boundary layer** comprises three layers: the radix layer touching the ground, a **uniform layer** in the interior of the convective boundary layer, and an **entrainment zone** at the top. The classical **surface layer**, where **Monin–Obukhov similarity theory** applies to small shear-driven turbulent **eddies**, is a bottom subdomain of the radix layer.

RADOB—Report of ground **radar weather observation**.
    These encoded reports and transmitted reports usually give **azimuth**, distance, **altitude**, **intensity**, shape, movement, and other characteristics of precipitation **echoes** observed by the **radar**. *See also* **RAREP**.

radome—A dome used to cover the **antenna** assembly of a **radar** to protect it from the effects of weather.
    It may refer to either a surface-based or an airborne radar installation. Radomes must be made of material transparent to radio **energy**.

radon—(Symbol Rn.) A **radioactive gas**, **atomic number** 86, atomic weight 222; an inert gaseous **element** with the property of emitting highly ionizing **alpha particles** during its radioactive disintegration, which makes it important in the atmospheric **ion** economy.
    Radon is a member of the uranium–radium family of radioactive elements. It forms by alpha **emission** from radium, which has a **half-life** of 1590 years. Radon itself decays by alpha emission in a half-life of only 3.82 days. Short as this radon half-life is, it is much longer than that of its two radioactive isotopes, **thoron** and **actinon**, so radon has a much greater opportunity to be carried by **turbulence** and **convection** to considerable heights and contribute to the **ionization** of the **atmosphere** before decaying. The oceans contain such slight concentrations of dissolved radium salts that radon production is very small over the oceans. Thus, its presence in air samples is often used to indicate the time since the air was over land. Its effects on human health are currently cause for concern since it can accumulate in inadequately ventilated underground areas where it decomposes into radioactive polonium that forms particles.

Israël, H., 1951: *Compendium of Meteorology*, 155–157.

raffiche—(Also refoli.) In the Mediterranean region, gusts from the mountains; violent gusts of the **bora**.

RAFOS technology—A Lagrangian method for the acoustic observation of **ocean currents** in the interior of the ocean.

The main components of a RAFOS system are a minimum of three fixed **sound** sources moored at intermediate depths and free-drifting floats labeling a certain **water mass**. The term RAFOS (SOFAR spelled backwards) was selected for this technology to indicate that the sound transmission direction is the opposite of the **SOFAR technology**. In both cases, the minimum in vertical profiles of sound velocity in the ocean is utilized for communication between the sources and receivers. RAFOS floats are expendable receivers that listen for coded signals from the moored RAFOS sound sources. After their underwater mission is terminated (duration up to two years), RAFOS floats drop a ballast weight, ascend to the surface, and transmit their internally stored data via satellite link to a shore-based receiving **station**. The RAFOS technology enables acoustic observations of ocean currents to within the millimeter per second **range**.

rafted ice—(Also called telescoped ice.) Deformed **sea ice** in which one piece has overridden another.

rageas—*See* **bora**.

ragged ceiling—(Obsolete.) **Indefinite ceiling**; the term still enjoys some popular use because of its descriptive quality.

raggiatura—Land **squalls** descending with great force from ravines and valleys in high land in Italy; they extend only a short distance off the west coast.

ragut—*See* **bora**.

rain—**Precipitation** in the form of liquid water drops that have diameters greater than 0.5 mm, or, if widely scattered, the **drops** may be smaller.

The only other form of liquid precipitation, **drizzle**, is to be distinguished from rain in that **drizzle drops** are generally less than 0.5 mm in diameter, are very much more numerous, and reduce **visibility** much more than does light rain. For observing purposes, the **intensity** of **rainfall** at any given time and place may be classified as 1) "light," the rate of fall varying between a **trace** and 0.25 cm (0.10 in.) per hour, the maximum rate of fall being no more than 0.025 cm (0.01 in.) in six minutes; 2) "moderate," from 0.26 to 0.76 cm (0.11 to 0.30 in.) per hour, the maximum rate of fall being no more than 0.076 cm (0.03 in.) in six minutes; 3) "heavy," over 0.76 cm (0.30 in.) per hour or more than 0.076 cm (0.03 in.) in six minutes. When **rain gauge** measurements are not readily available to determine the **rainfall intensity**, estimates may be made according to a descriptive system set forth in observing manuals.

rain and snow mixed—Precipitation consisting of a mixture of **rain** and **wet snow**.

It usually occurs when the **temperature** of the air layer near the ground is slightly above **freezing**. The British term for this mixture is **sleet** (which has a different meaning in the United States).

rain area—*See* **precipitation area**.

rain attenuation—The depletion of **electromagnetic energy** during propagation through **rain**, caused by **raindrop** scattering and **absorption**.

The effect is described by **Beer's law** as the fractional depletion of **radiance** per unit pathlength,

$$\frac{dL}{L} = -\gamma \, ds,$$

where L is the **monochromatic** radiance at a given **wavelength**, γ is the volume **attenuation coefficient** (or **extinction coefficient**) due to **absorption** and **scattering** by **rain**, and ds is an increment of pathlength. In **radar**, the depletion of **power** from incident plane-wave **radiation** is sometimes described by the **specific attenuation** Y, usually expressed in units of decibels per kilometer, related to the volume attenuation coefficient by

$$Y = 4.343\gamma,$$

where Y is in decibels per kilometer when γ is in inverse kilometers. The proportionality factor is $10 \log_{10} e$. The specific attenuation for a given radar wavelength depends on the **temperature** and the **drop-size distribution**. If this distribution is known or can be approximated as a function of **rainfall rate**, the specific attenuation can be estimated as a function of rainfall rate.

rain cloud—Any **cloud** from which **rain** falls; a popular term having no technical denotation.

In older **cloud classification** systems, any cloud from which rain or **snow** fell was called **nimbus**.

rain crust—A type of **snow crust** formed by refreezing after surface **snow** crystals have been melted and wetted by liquid **precipitation**.

This type of crust is composed of individual **ice** particles such as **firn**. **Rain** may also help to form **film crust** or **ice crust**.

rain day—(Also called day of rain.) In British **climatology**, a period of 24 hours, normally commencing at 0900 **UTC**, in which at least 0.01 in. or 0.2 mm of **precipitation** is recorded.

It is the practice of the U.S. National Weather Service to use phrases that are more self-defining, such as "day with measurable precipitation" or "day with 0.01 inch or more" (of precipitation).

rain factor—A coefficient designed to measure the combined effect of **temperature** and moisture on the formation of soil humus.

It is obtained by dividing the annual **rainfall** (mm) by the mean annual temperature (°C). When the rain factor is below 50, salty and sandy soils form; above this value the soil type passes through laterite, brown earth, and black earth to earth humus and leached soil with the rain factor above 160.

rain forest—Generally, a forest that grows in a region of heavy annual **precipitation**, such as in a **humid climate**.

Two types are distinguished: 1) the **tropical rain forest** (often simply called the "rain forest"); and 2) the **temperate rain forest**.

rain forest climate—*See* **wet climate, tropical rain forest climate**.

rain gauge—Any instrument designed to measure **rain** amount; includes **recording, nonrecording,** and **rain-intensity gauges**.

rain-gauge shield—(Also called wind shield.) A device that surrounds a **rain gauge** and acts to maintain horizontal flow in the vicinity of the funnel so that the **catch** will not be influenced by **eddies** generated near the gauge.

Types in use include the **Alter shield, Nipher shield,** and **Wild fence**.

rain-gauge wind shield—Same as **rain-gauge shield**.

rain gush—Same as **cloudburst**.

rain gust—Same as **cloudburst**.

rain-intensity gauge—(Also called rain-rate gauge, rate-of-rainfall gauge.) An instrument that measures the rate at which **rain** is falling.

In the Jardi design, water from the rain collector enters a chamber containing a float with a tapered needle that controls the rate of **outflow**. The higher the rain rate, the higher the float, and the greater the outflow.

rain making—Popular term applied to all activities designed to increase, through any artificial means, the **amount of precipitation** released from a **cloud**.

The techniques of **cloud seeding** have been studied carefully and extensively, especially since V. J. Schaefer's discovery of the effect of **dry ice** on **supercooled clouds**.

Schaefer, V. J., 1946: The production of ice crystals in a cloud of supercooled water droplets. *Science*, **104**, N.Y., p. 457.

rain-out—1. Same as **rain washout**.

2. A process in a dissipating **cumulonimbus** that results in all the large **hydrometeors** falling out as **precipitation**, leaving a nonprecipitating **cloud** composed of small **ice crystals**.

rain-rate gauge—Same as **rain-intensity gauge**.

rain shadow—A region of sharply reduced **precipitation** on the lee side of an **orographic** barrier, as compared with regions **upwind** of the barrier.

Slopes facing **windward** with respect to prevailing or seasonal moisture-bearing flows typically experience heavy **orographic precipitation**. To the lee of the barrier, however, the sinking air warms, dries, and becomes more stable, suppressing precipitation. Two dramatic and often-cited examples are the Ghat Mountains of western India, which receive annually more than 600 cm of **rainfall** at locations on their western slopes but 60 cm or less on their eastern slopes, and the island of Hawaii, where up to 450 cm of **rain** falls on the slopes facing the northeast **trade winds**, but less than 100 cm falls at locations on the lee side of the island. A good example of rain shadow in the United States is the region east of the Sierra Nevadas; there the prevailing westerly winds deposit most of their moisture on the western slopes of the range, whereas to the east lies the Great Basin **desert**.

rain shower—A brief period of **rainfall** in which **intensity** can be variable and may change rapidly. It is often convective in nature.

rain simulator—(Also called sprinkler **infiltrometer**.) A device used for rainfall-runoff studies, **erosion** plot studies, and **infiltration** that applies water in the form and at a rate comparable with natural **rainfall**.

rain stage—Part of an obsolete conceptual **model** of an **air parcel** ascent referring to that portion of the ascent during which the parcel undergoes **pseudoadiabatic expansion** at temperatures above **freezing**.
Other portions of the ascent were described as the **dry stage**, **snow stage**, and **hail stage**.

rain stimulation by artificial means—Any form of **weather modification**, usually involving **cloud seeding**, where the objective is to induce the formation of **precipitation** or to enhance the amount.

rain washout—Capture of **particulates** and gaseous **pollutants** by falling **raindrops**.
In both cases, a "washout coefficient" (W_p) can be estimated that expresses the fractional depletion (X/X_0) of the pollutant during a **washout** period t:

$$X/X_0 = \exp(-W_p t),$$

where X_0 is the initial pollutant concentration and X is the concentration at time t later.

rainband—The complete **cloud** and **precipitation** structure associated with an area of **rainfall** sufficiently elongated that an orientation can be assigned.
The band is either nonconvective or only weakly convective.

rainbow—Any one of a family of large, colored, circular (or nearly circular) arcs formed by **light** (usually **sunlight**) falling on a **population** of water **drops** such as provided by **rain, cloud, fog**, or spray.
The apparent center of the arcs is normally the shadow of the observer's head, so the rainbow is a personal phenomenon with each person seeing a slightly different bow. Although the rainbow can form a circle, when caused by rain the bottom part of the circle is usually cut off by the ground, leaving an arc, the extent of which depends upon the **elevation** of the light. The term rainbow is not applied to the small, nearly circular arcs seen around the sun and antisolar point in clouds. These are the **corona** and **glory**. Nor is the term applied to the large circular arcs formed by light falling on **ice crystals**. These are the **halos**. Rainbows are seen as related groups of arcs, such as the **primary rainbow** (with red on the outside, blue on the inside), the (larger) **secondary rainbow** (with red on the inside), the **supernumerary rainbows** (seen to the inside of the primary bow), and the reflection bows (the centers of which are above the **horizon**). However, the appearance of these bows can vary markedly depending upon whether they formed in rain, **drizzle**, or cloud, as both the radius and color purity of the arcs depend on drop size. Certainly, the brightest and most frequently seen of the bows is the primary rainbow, but whether the whole arc above the horizon is seen or not depends upon the location of the rain. It is not uncommon for someone to report seeing two rainbows, when these were merely two unconnected portions of the same bow. There is a hierarchy of theories of the rainbow with simplicity being purchased at the expense of verisimilitude. Theories that treat light as a series of rays do a good job of explaining the approximate positions and colors of the primary and secondary bows but fail to account for the supernumerary bows. To account for such easily observable features of the natural bow as the variations of color purity, **brightness**, and distribution of the supernumeraries around the arc, such things as the wave nature of light, the **drop-size distribution** in a cloud, and the size-dependent shape of **raindrops** must be taken into account.

raindrop—A **drop** of water of diameter greater than 0.5 mm falling through the **atmosphere**.
In careful usage, falling drops with diameters between 0.2 and 0.5 mm are called **drizzle drops** rather than raindrops, but this distinction is frequently overlooked and all drops with diameters in excess of 0.2 mm are called raindrops. The limiting diameter of 0.2 mm is rather arbitrary, but has been employed because drops of this size fall rapidly enough (about 0.7 m s^{-1}) to survive evaporative **dissipation** for a distance of the order of several hundred meters, the exact survival distance being a function of the **relative humidity**. Drops much smaller than this limiting size fall so slowly from most clouds that they evaporate before reaching the ground. **Virga** is almost always composed of drops with diameters just below the limiting size assigned to drizzle drops. Raindrops are very much larger than **cloud drops**. A typical raindrop might have a diameter of 1–2 mm, while a typical cloud drop diameter is of the order of 0.01–0.02 mm. Raindrops fall between 2 and 12 meters per second (depending on **altitude**); those larger than about 1 mm are increasingly deformed by airflow (with flatter bases), the largest raindrops having a height to width ratio of 1:2. Raindrops may form by **coalescence** of cloud drops or from melting **ice** precipitation.

Any given **rainfall** is characterized by a certain **drop-size distribution** of its raindrops, and even within a given **storm** this distribution may change its characteristics. The largest drops observed in heavy **thunderstorms** may have equivalent spherical diameters of 5–8 mm. Raindrops of such large size are rare, but occasionally form in the **warm rain process** by **accretion** of cloud water or can result from melting **hail**.

raindrop size distribution—The **spread** of sizes for raindrops falling at a given location; distributions change significantly with the process of formation (from melting **snow** or drop **coalescence**) and **shear** of horizontal and vertical **wind**, which locally sorts drops through their differential **fall velocity**.
> *See* **raindrop**, **drop-size distribution**.

raindrop spectrograph—An instrument that automatically determines the **raindrop size distribution**.
> *See* **disdrometer**.

rainfall—The **amount of precipitation** of any type (including the liquid equivalent of frozen **hydrometeors**); usually taken as that amount measured by means of a **rain gauge** (thus a small, varying amount of direct **condensation** is included).
> A more accurate term would be **precipitation** or precipitation amount. However, the broad use of "rainfall" is firmly established in meteorology, especially in hydrologic and climatological literature. Its best utilization would confine it to liquid precipitation, and so would provide a distinction between precipitation immediately accessible to soil and streams and that delayed in storage as **snow** or **ice** on the earth's surface.

rainfall distribution—Manner in which the depth of **rainfall** varies in space and time.

rainfall duration—The interval of time elapsed between the beginning and ending of a **rainfall** event.

rainfall effectiveness—Same as **precipitation effectiveness**.

rainfall excess—The volume of **rainfall** available for direct **surface runoff**.
> It is equal to the total amount of rainfall minus all abstractions including **interception**, **depression storage**, and **infiltration**.

rainfall frequency—1. (Also called frequency curve.) The **probability distribution** specifying the **exceedance probability** of different **rainfall** depths for a given duration (such as 1 hour, or a 24-hour day).
> The exceedance **frequency** is often reported as a **return period** (in years), which is the reciprocal of the annual exceedance frequency.
> 2. The expected number of times, during a specified time period, that a given **precipitation** depth will be exceeded.

rainfall intensity—Same as **precipitation intensity**.

rainfall intensity pattern—Distribution of **rainfall rate**, in time, during a **storm**.
> *See also* **hyetograph**.

rainfall intensity return period—For a given duration (such as 24 hours), the average time in years between the occurrence of **rainfall rates** of a given **intensity**.
> *See also* **return period**.

rainfall inversion—Same as **precipitation inversion**.

rainfall loss—(Also called rainfall abstraction, storm loss.) **Rainfall** that is retained in a **basin** during a **storm** event and does not contribute to **direct runoff**.
> Retention occurs due to the processes of **infiltration**, **interception**, and **depression storage**.

rainfall rate—A measure of the **intensity** of **rainfall** by calculating the amount of **rain** that would fall over a given interval of time if the **rainfall intensity** were constant over that time period.
> The rate is typically expressed in terms of length (depth) per unit time, for example, millimeters per hour, or inches per hour.

rainfall regime—The character of the seasonal distribution of **rainfall** at any place.
> The chief rainfall regimes, as defined by W. G. Kendrew (1961), are equatorial, tropical, monsoonal, oceanic and continental **westerlies**, and Mediterranean.
>> Kendrew, W. G., 1961: *Climates of the Continents*, 5th ed., Oxford, Clarendon Press, 608 pp.

rainfall station—A **station** at which the only regular measurements made are those of **rainfall**.

rainstorm—A **storm** characterized by the fall of liquid **precipitation**.

rainy climate—In W. Köppen's 1936 **climatic classification**, any **climate** type other than the **dry climates**. However, it is generally understood that this refers principally to the **tree climates** and not the **polar climates**.
Compare **moist climate, wet climate**.

Köppen, W. P., and R. Geiger, 1930–1939: *Handbuch der Klimatologie*, Berlin: Gebruder Borntraeger, 6 vols..

rainy season—(Also called wet season.) In certain types of **climate**, an annually recurring period of one or more months during which **precipitation** is a maximum for that region; the opposite of **dry season**.

ram penetrometer—(Also called **ramsonde**.) A cone-tipped metal rod designed to be driven downward into deposited **snow** or **firn**.
The measured amount of force required to drive the rod a given distance is an indication of the physical properties of the snow or firn.

ramp structure—A sawtooth **temperature** pattern in a **time series** measured at the bottom of the **atmospheric boundary layer**.
Each tooth or ramp (gradual **linear** increase in temperature followed by a rapid decrease) is associated with the passage of a surface-layer **convective plume**.

ramsonde—British term for **ram penetrometer**.

random—Eluding precise **prediction**, completely irregular.
In connection with **probability** and **statistics**, the term random implies collective or long-run regularity; thus a long record of the behavior of a random phenomenon presumably gives a fair indication of its general behavior in another long record, although the individual observations have no discernible system of progression. *Compare* **nondeterministic, stochastic**.

random error—The inherent imprecision of a given process of measurement; the unpredictable component of repeated independent measurements on the same object under sensibly uniform conditions.
It is found experimentally that, given sufficient refinement of reading, a series of independent measurements $x_1, x_2, \ldots, x_n$ will vary one from another even when conditions are most stringently controlled. Hence, any such measurement x_i may be regarded as composed of two terms:

$$x_i = \mu + v_i,$$

where μ (ordinarily the true value) is a numerical constant common to all members of the series and v_i, the random error, is an unpredictable **deviation** from μ. The principal conclusion of classical investigations of errors of measurement (by Gauss and Laplace) was that, as a consequence of the **central limit theorem**, repeated measurements under controlled conditions usually follow the **normal distribution**, and the corresponding distribution of the random error is known as the **error distribution**.

random forecast—A "forecast" in which one of a set of meteorological contingencies is selected on the basis of chance.
The random forecast is often used as a standard of comparison in determining the degree of **skill** of another forecast method. *Compare* **persistence forecast, probability forecast**.

random numbers—A set of numbers arranged in **random** order.

random process—A series of procedures or operations that produces **random** output.

random sample—A **sample** selected at **random** from a **population**.

random sea—A sea in which a large number of **wave** components are present, a result of the **random** processes involved in **wave generation**.
The behavior of the waves may be represented statistically in terms of a **wave spectrum**. Most sea states may be described as random seas.

random variable—(Or variate.) A **variable** characterized by **random** behavior in assuming its different possible values.
Mathematically, it is described by its **probability distribution**, which specifies the possible values of a random variable together with the **probability** associated (in an appropriate sense) with each value. A random variable is said to be continuous if its possible values extend over a continuum, discrete if its possible values are separated by finite intervals. *See* **probability theory, statistical independence**.

random walk model—An example of a **stochastic process** that is defined by $X(0) = X_0$, and

$$X(t) = X_0 + \sum_{i=1}^{t} R_i = X(t-1) + R_t,$$

where R_i are independent and identically distributed **random variables**.

randomization—Arrangement of data in such a way as to simulate chance occurrence.

randomized seeding trial—A series of field tests expected to yield unbiased **statistical** data of the efficiency of a given agent used in **cloud seeding**.

Attempts are made to simulate chance distributions in order to reduce the effects of uncontrolled natural **variability** in meteorological and other external conditions. Also required are the delineations of a defined target area and an undisturbed, but otherwise similar, control area.

randomness—The property of being **random**.

Computer-generated, pseudorandom numbers are sometimes used to **model** the behavior of certain meteorological processes. The utilized algorithms can be tested for randomness through application of **statistical** procedures designed to detect unwanted correlations in the sequences of numbers produced.

range—1. The difference between the maximum and minimum of a given set of numbers; in a periodic process it is twice the **amplitude**, that is, the **wave height**.

2. The distance between two objects, usually an **observation** point and an object under observation.

See **slant range**.

3. A maximum distance attributable to some process, as in **visual range** or the range of an aircraft.

4. The difference between **high** and **low water** in a tidal **cycle**.

Ranges greater than 4 m are sometimes termed macrotidal and those less than 2 m are termed microtidal. Intermediate ranges are termed mesotidal.

5. In **radar**, **lidar**, and **sodar**, the radial distance measured outward from the location of the **transmitter**; ordinarily the distance to a **target**.

range aliasing—(Also called range folding.) In **radar meteorology**, a sampling problem that arises when **echoes** located beyond the **maximum unambiguous range** (r_{max}) are received as if they were within this **range** of the **radar**.

A radar ordinarily computes range to targets by measuring the time interval between the transmission of a **pulse** and the receipt of the returned **signal**, assuming that the signal was associated with the pulse just transmitted. However, depending on the **pulse repetition frequency**, the returned signal may be associated with one of several pulses transmitted prior to the latest one. Therefore, a returned signal, indicated as originating at range r, could have originated at $r + r_{max}$ (**second-trip echo**), or $r + 2r_{max}$ (third-trip echo), etc. A range-aliased echo from a weather target is sometimes recognizable by a distorted shape. It may appear elongated radially or shrunk in **azimuth** extent because the radial length is unaffected by **aliasing** and is a correct measure of the **target** size while the azimuthal width decreases with increasing range from the radar.

range attenuation—In **radar** terminology, the decrease of **power density** with **range** in accordance with the **inverse-square law** as a result of the **divergence** of the **beam**.

For one-way transmission, this **attenuation** is proportional to $1/r^2$ where r is the range from the **antenna**. For a **monostatic radar** and a **point target**, the range attenuation is proportional to $1/r^4$ because the beam diverges on the outward path and the return path. For monostatic radar and a **distributed target**, the range attenuation varies between these two limits, depending on the extent to which the **pulse volume** is filled with **precipitation** or other scatterers. Range attenuation only arises from the spreading of the attenuation with distance and should not be confused with attenuation caused by **scattering** and **absorption**. *See* **radar equation, beam filling, range normalization**.

range distortion—The **distortion** or lack of **resolution** in measuring the distance to a **target** by **radar**, **lidar**, or **sodar** caused by the finite duration of the transmitted **pulse**.

A **point target** returns an **echo** that is spread over a **range** of approximately half the **pulse length**. *See* **radar resolution**.

range error—The **error** in **range**, generally negligible, caused by propagation of **radar** waves through an inhomogeneous **atmosphere** in which the **refractivity** is sufficiently variable to make the propagation path depart significantly from a straight line.

range folding—Same as **range aliasing**.

range gate—A selectable interval of **range** (or of time delay from transmission) within which returning **radar** signals are measured.

Gating is used to isolate the **echoes** from different regions of distributed targets. Contiguous range gates of narrow width, separated by a distance equal to half the transmitted **pulse length**, are often used in **lidar** and **weather radar** systems. Some systems employ as many as a thousand range gates to measure the signals returned along each pointing direction. *See* **gating**.

range gating—*See* **gating**.

range–height indicator—(Abbreviated RHI.) A type of **radar** display on which the **reflectivity** or other properties of **echoes** are displayed as a function of **range** and **elevation angle** in **polar coordinates**.

The RHI may be produced by **scanning** the **antenna** in **elevation** with the **azimuth** fixed, or it may be composed of data from successive **PPI** displays at different elevation angles. It may be an **intensity** modulated **display** or may use a **scale** of colors to represent the values of the function displayed.

range marker—On a radar **display**, graphic marks that facilitate the determination of the **range** of a **target** from the **radar**.

Range markers may be in the form of a **scale** on a range axis or, as on a **PPI** display, in the form of concentric circles with the position of the radar at the center.

range normalization—In **radar** and **lidar**, adjustment of the **received power** to compensate for the **divergence** of **radiant flux density** with **range**. The **received power** is multiplied by $(r/r_0)^2$, where r is the range and r_0 is an arbitrary reference range (e.g., 1 km).

This normalization assumes that a **distributed target** fills the **pulse volume** at range r (*see* **beam filling**) so that the size of the **scattering** volume increases with r^2 even as the incident **flux density** decreases with r^2. The normalization in effect corrects for the **flux** divergence on the return trip from the scattering volume to the **receiver**. *See* **lidar equation**, **radar equation**.

range resolution—The least radial separation between two targets in the same direction from a **radar** that allows them to be distinguished.

This separation equals one-half the transmitted **pulse length**. Targets closer together than this distance are not resolved and appear as a single **target** on the **display**. *See* **radar resolution**.

Rankine temperature scale—A **temperature scale** with the **degree** of the **Fahrenheit temperature scale** and the zero **point** of the **Kelvin temperature scale**.

The **ice point** is thus 491.69 degrees Rankine and the **boiling point** of water is 671.69 degrees Rankine.

Rankine vortex—A two-dimensional circular flow in which an inner circular region about the origin is in **solid rotation**:

$$\frac{V}{r} = \text{constant},$$

where V is the tangential speed and r the distance from the origin.

The outer region is free of **vorticity**, the speed being inversely proportional to the distance from the origin (as in the **V–R vortex**),

$$Vr = \text{constant}.$$

This vortex has occasionally been used as a **model** for the **wind** distribution in a **hurricane**.

Milne-Thomson, L. M., 1955: *Theoretical Hydrodynamics*, 3d ed., 318–323.

RAOB—Acronym for **radiosonde observation**.

Raoult's law—Physical law relating the change in **vapor pressure** of a liquid to the amount of solute dissolved in it.

The law states that $P_0 - P = P_0 x_i$, where x_i is the **mole fraction** of the dissolved solute. The quantity $P_0 - P$ is sometimes referred to as the **vapor tension** of the solution. Consequences of Raoult's law are the so-called colligative properties of solutions, that is, the depression of **freezing** and **melting points** of solutions relative to those of the pure solvent and osmotic **pressure**. Raoult's law is observed in everyday situations every winter when we put salt on sidewalks to melt the **ice**. This is simply a depression of the melting point associated with the dissolution of salt in water.

rapid distortion theory—A theory that assumes **eddy** vorticity is conserved in turbulent air undergoing rapid stretching, while **shear production** of **turbulence** remains zero or unchanged.

It is used to help predict the flow of turbulent air over hills and predicts that the **vertical**

velocity variance should increase in the **outer layer** of accelerated winds flowing over a hill, while longitudinal velocity variance should decrease.

rapid interval imaging—The collection of satellite imagery from geostationary satellites over limited geographical areas with unusually short time intervals between images, sometimes as little as one minute.
See **rapid interval scan**.

rapid interval scan—The mode of data gathering by **GOES** whereby an **image** is produced at intervals of 1–10 minutes instead of the **normal** intervals of 15 (formerly 30) minutes.
Rapid interval **scanning** is used primarily for research and to observe rapidly developing **severe weather**, **tropical storm**, or **flash flood** situations.

rapid scan mode—Same as **rapid interval scan**.

rare gases—Same as **noble gases**.

rare optical phenomenon—Literally, an optical phenomenon that occurs only rarely.
As such, this would hardly seem to merit discussion. Yet the literature is replete with discussions of frequently occurring optical phenomena that are nevertheless described by authors as being rare. The explanation for this discrepancy is to be found in the difference between a rare event and a rarely observed event. The authors who characterize, say, a **green flash** as rare are undoubtedly doing so on the basis of their having rarely seen one. Yet green flashes can be seen frequently if one knows when and how to look for them. Clearly, a rarely observed event owes as much to the observer as it does to the event. For example, if one were to characterize the frequency of sunrises on the basis of most observer's frequency of observation, one would have to conclude that sunrises are very much rarer than sunsets. There certainly are rare optical phenomena, but there are many more common optical phenomena that are merely rarely observed due to personal habits.

rarefaction wave—Same as **expansion wave**.

RAREP—(Or rarep.) Aviation weather communications code word for **radar report**.
See **radar meteorological observation**.

RASAPH—Abbreviation for **radar signal spectrograph**.

RASS—Abbreviation for **radio acoustic sounding system**.

rate coefficient—A constant of proportionality relating the rate of a chemical reaction to the concentrations of the chemical species involved in the reaction.
For a second-order reaction, A + B → products, the units of the rate coefficient are (concentration)$^{-1}$(time)$^{-1}$.

rate constants—More correctly known as **rate coefficients**.

rate of accretion—Volume of water added to an **aquifer** per unit area and per unit time.

rate-of-climb indicator—An instrument that shows the rate at which an aircraft is gaining or losing **altitude**.
It is in actuality a rate-of-pressure-change **indicator**, a sensitive aircraft **aneroid barometer** with a slow leak to the **aneroid capsule** calibrated in terms of altitude change.

rate-of-rainfall gauge—Same as **rain-intensity gauge**.

rating curve—For a given point on a **stream**, a graph of **discharge** versus **stage**.
See **loop rating**; *compare* **gauge relation**.

rating flume—A **flume** used for the purpose of calibrating flow measuring devices.

rational horizon—*See* **horizon**.

rational method—(Also known as the rational formula.) Empirical equation relating the **peak discharge** from a **basin** to the product of a dimensionless **runoff coefficient** that is estimated from basin hydrologic characteristics, the **catchment area**, and a **rainfall intensity**.

raw—Colloquially descriptive of uncomfortably cold weather, usually meaning cold and damp, but sometimes cold and windy.

raw moment—In **statistics**, a **moment** taken about the origin.

raw water—Untreated water, or water received for further treatment.

rawin—A method of **winds-aloft observation**; that is, the determination of **wind** speeds and directions in the **atmosphere** above a **station**.

It is accomplished by tracking a balloon-borne **radar** target, **responder**, or **radiosonde** transmitter with either radar or a **radio direction finder**. With a radio direction finder, the height data must be supplied by other means, normally by concurrent **radiosonde observation**. With radar, if height data are not otherwise supplied, the **slant range** must be recorded in addition to the angles of **elevation** and **azimuth**.

rawin target—A special type of **radar** target tied beneath a free balloon and designed to be an efficient **reflector** of radio **energy**.
Such targets usually consist of a **corner reflector** and are made of some reflecting material stretched over styrofoam, or light wooden or metal struts.

rawinsonde—An **upper-air sounding** that includes determination of **wind** speeds and directions.
Historically, wind data were obtained by tracking a balloon-borne **radiosonde** with a **radio direction finder**. Contemporary methods include measuring position or radiosonde velocity from **GPS** or Loran radio navigation signals.

rawinsonde observation—A combined **radiosonde** and **radio wind observation**.

ray—An imaginary bundle of propagating electromagnetic or acoustic **energy**, the lateral **dimensions** of which are negligible.
It is impossible to isolate a ray. Nevertheless, rays are useful conceptual devices if used with a knowledge of their limitations. For example, the **rainbow** can be described by imagining **sunlight** incident on a **raindrop** to be subdivided into many rays, each of which obeys the laws of (specular) **reflection** and **refraction**. Because rays do not exist, ray optics (or geometrical optics) is an approximation.

ray tracing—A graphical or mathematical approximation scheme for determining the propagation of **electromagnetic** or **sound waves** by following the path of rays obeying the laws of **reflection** and **refraction**.
Applications of ray tracing to atmospheric problems include estimating **rainbow** angles, determining the characteristics of **mirages**, and mapping the propagation of **sound** in the **atmosphere** and oceans.

Rayleigh atmosphere—A **model** of the **clear** atmosphere used for **light scattering** calculations.
A pure Rayleigh atmosphere contains only the permanent atmospheric gases that **scatter** light by **Rayleigh scattering**. It excludes the effects of **water vapor**, **clouds**, and **aerosols**.

Rayleigh–Bénard convection—A special case of **Bénard cells** with an **aspect ratio** of 3:1.
It was first observed in the laboratory as **convection** in a shallow liquid, heated from below (as in a skillet) or cooled from above by **evaporation** (as in metallic paint). Such observations have similarities to convection in shallow cells in atmospheric layers as in **cirrocumulus**, with upward motion at the center, where **cloud** forms, and descending motion at the edges, which are **clear**.

Rayleigh number—The nondimensional ratio between the product of **buoyancy** forces and heat **advection** and the product of **viscous forces** and heat **conduction** in a fluid.
It is written as

$$\mathrm{Ra} = \frac{g \, |\Delta_z T| \, \alpha d^3}{\nu k},$$

where g is the **acceleration of gravity**, $\Delta_z T$ a characteristic vertical **temperature** difference in the characteristic depth d, α the **coefficient of expansion**, ν the **kinematic viscosity**, and k the molecular **conductivity**. The Rayleigh number is equal to the product of the **Grashoff** and **Prandtl numbers**, and is the critical **parameter** in the theory of **thermal instability** for laboratory flows.

Rayleigh optical depth—The vertical **optical thickness** of a **Rayleigh atmosphere**.
The Rayleigh optical thickness is directly proportional to **surface pressure** and inversely proportional to the fourth power of the **wavelength**. At a wavelength of 0.5 μm, the Rayleigh optical depth is about 0.145 at **sea level**.

Rayleigh phase function—The **phase function** applicable to **Rayleigh scattering** of unpolarized **radiation**.
The Rayleigh phase function is expressed as $p(\theta) = (3/4)(1 + \cos^2\theta)$, where θ is the **scattering** angle.

Rayleigh's formula—*See* aerodynamic force, drag, drag coefficient.

Rayleigh's scattering law—(Also called **Rayleigh scattering**.) An approximate law of **scattering** of

electromagnetic waves by molecules and **particles** small compared with the **wavelength** of the **illumination** at wavelengths for which **absorption** is sufficiently small.

According to this law, first derived in 1871 by Lord Rayleigh using simple dimensional arguments, scattering in all directions by an object is inversely proportional to the fourth power of the wavelength of the illumination. Scattering of **sunlight** by air molecules does not obey this law exactly, although it is a good approximation. Rayleigh's scattering law also predicts that scattering by a particle is proportional to the square of its volume. *Compare* **Mie theory**.

> Young, A. T., 1982: *Phys. Today*, Jan., 2–8.

Rayleigh scattering— Approximate theory for electromagnetic **scattering** by small **particles** named for Lord Rayleigh (John William Strutt, 1842–1919), who in 1871 showed that the blue color of the **clear sky** is explained by the scattering of **light** by molecules in the **atmosphere**.

The Rayleigh approximation to the more complex **Mie theory** requires that the particles be very small in comparison to the **wavelength** of the **radiation**. The range of applicability of the Rayleigh theory depends on the **refractive index** of the particles. For water drops the criterion is usually stated as $D < \lambda/10$, where D is the **drop** diameter and λ is the wavelength of the incident **radiation**. Characteristics of Rayleigh scattering are that the **scattering cross section** of a sphere of diameter D is proportional to D^6/λ^4 and that the **phase function** is proportional to $(1 + \cos^2\theta)$, where θ is the scattering angle. In **radar**, Rayleigh scattering theory is usually employed to interpret the observations of **echoes** from **precipitation**, even though the Rayleigh criterion is not satisfied by **raindrops** for wavelengths much shorter than 10 cm or by **hailstones** for even longer wavelengths. Corrections based on Mie theory are sometimes applied to observations at wavelengths shorter than 3 cm. Large raindrops and hailstones can deviate from spherical shape; radar scattering by nonspherical particles has been approximated by the Gans theory for small ellipsoids.

> Bohren, C. F., and D. R. Huffman, 1983: *Absorption and Scattering of Light by Small Particles*, Wiley, New York, 3–223.
> McCartney, E. J., 1976: *Optics of the Atmosphere*, Wiley, New York, 176–262.

Rayleigh smoothness criterion—*See* **optically smooth**.

Rayleigh wave—1. A two-dimensional **barotropic disturbance** in a fluid having one or more discontinuities in the **vorticity** profile.

2. A **wave** propagated along the surface of a semi-infinite elastic solid and bearing certain analogies to a **surface gravity wave** in a fluid.

RDF—Abbreviation for **radio direction finding**.

reach—1. A specified portion of a **stream** channel, commonly taken between two stream **gauging stations**, but may be taken between any two specified endpoints.

2. In hydraulic and **sediment** transportation calculations, a reach of **stream** is a specified length of a stream channel used for computational purposes such as **flood routing**.

reaction kinetics—The branch of physical chemistry that deals with the determination of the rates and mechanisms of chemical reactions.

The measurement of the **rate coefficients** for a large number of gas-phase chemical reactions has improved our ability to **model** the impact of human activities on the **atmosphere**. For description of some of the techniques used, *see* **flash photolysis systems, laser-induced fluorescence, resonance fluorescence**.

reaction order—In an elementary chemical reaction, the reaction order is the number of species that actually participate in the reaction.

In a complex reaction scheme, the overall apparent reaction order may be fractional or may change with time.

reaction rate—(Sometimes referred to simply as rate.) The change with time in the concentration of a reactant or product involved in a chemical reaction.

The rate of the reaction is given by the product of the **rate coefficient** for the particular reaction and the concentrations of the reactants.

real horizon—*See* **horizon**.

real time—Data or observations for which the reporting or recording of events is nearly simultaneous with their occurrence.

It is distinguished from archival retrieval of the data.

real-time processing—A universally used term for the processing of data immediately upon their receipt.

real-time transmission—A universally used term for the immediate transmission of observed data.

reanalysis—Same as **analysis**, except for two important practical differences.

First, it is not done in **real time**, and second, the **background field** is made by an **NWP** model that does not change over the entire **period** of the reanalysis. A serious problem with **climate change** studies made from standard analyses results from frequent changes in the **model** used to generate the **background field**. These changes (including changes in **resolution** and **orography**) lead to discontinuities in a **time series** of real-time analyses. A reanalysis yields complete, global gridded data that are as temporally homogeneous as possible. Reanalysis data include many derived fields (heating, **soil moisture** over land, etc.) for which direct observations are nearly absent.

rear-inflow jet—A **mesoscale** circulation feature in which a system-relative current of air enters and flows through the **stratiform** precipitation region of **mesoscale convective systems** from the rear.

The rear-inflow jet forms in response to the upshear-tilting of the convective **circulation**, as the horizontal **buoyancy** gradients along the back edge of the system create a circulation that draws midlevel air in from the rear. The rear-inflow jet supplies potentially cold and dry midlevel air that aids in the production of convective and system-scale **downdrafts**.

rear of a depression—A British term for the sector of advancing cold air of a midlatitude **depression** in its mature or decaying state.

Reaumur temperature scale—(Abbreviated R.) A **scale** with the **ice point** at 0° and the **boiling point** of water at 80°, with **pressure** of one **atmosphere**.

rebat—The **lake breeze** of Lake Geneva, Switzerland; it blows from about 10 A.M. to 4 P.M.

Its complementary **land breeze** is the **morget**.

reboyo—A persistent (daylong) **storm** from the southwest during the **rainy season** on the Brazilian coast.

Compare **minuano**.

recco—Common contraction for **reconnaissance**.

See **aircraft weather reconnaissance**.

RECCO code—An **international synoptic code** for communicating aircraft observations.

received power—In **radar**, the amount of **power** received at the **antenna** after being scattered by a **target**.

This power is normally on the order of microwatts as compared with the kilowatts or megawatts of **transmitted power**. By convention, received power is measured in **decibels** relative to a reference value of 1 mW, that is, in dBm. The average received power is proportional to the **radar reflectivity** of the target, while fluctuations in the received power are caused by relative motions among the individual **scattering** elements that make up the target.

receiver—An instrument used to detect the presence of and to determine the information carried by **electromagnetic radiation**.

A receiver includes circuits designed to detect, amplify, rectify, and shape the incoming radio-frequency **signals** received at the **antenna** in such a manner that the information-containing component of this received **energy** can be delivered to the desired indicating or recording equipment.

receiver (noise) temperature—A measure of the **noise power** density at the **output** of a radio or **radar** receiver due to internally generated **noise** and passive losses in the **receiver**. In particular, the receiver temperature T_r is the **temperature** of a resistor having noise power per unit **bandwidth** equal to that of the receiver output at a given **frequency**.

It is given in terms of the power output P and bandwidth B of the receiver by $T_r = P/kB$, where k denotes **Boltzmann's constant**, 1.38×10^{-23} J K^{-1}. The receiver temperature is related to the receiver noise factor F by $T_r = T_o(F - 1)$ with $T_o = 290$ K. *See* **noise temperature, noise figure**.

receiver gain—The ratio of the **output** signal **power** to the **input** signal power of a **receiver**, usually expressed in **decibels**.

recession—1. The decrease in **streamflow**, at a point along a **stream** channel, following the passage of a crest.

2. (Also called falling limb.) That portion of a **hydrograph** that shows the rate of decrease of **stage** or **discharge** following the passage of a crest; the opposite of **rising limb**.

recession curve—(Or master recession curve.) A smoothed composite of the **recessions** of several observed **hydrographs**, drawn to represent the characteristic time graph of decreasing total **runoff** for a **drainage area** after passage of a peak flow.

Curves of this type, designed to characterize the nature of a drainage area, may be constructed as a plot of flow (**stage** of **discharge**) versus time, or as a plot of flow versus flow at some fixed later interval; also, separate recession curves may be derived for **surface runoff**, **groundwater runoff**, and even for **interflow**.

recharge—The process of increasing the water stored in the **saturated zone** of an **aquifer**.

recharge area—The area that contributes to the increase of the water stored in the **saturated zone** of an **aquifer**.

recharge capacity—The capability of the **unsaturated zone** to transmit water to the **saturated zone**.

recirculation—A **closed circulation** system within a larger oceanic **circulation**, in which water is returned to its place of departure through a much shorter route than by following the main circulation path.

recombination—The process by which a positive and a negative **ion** join to form a neutral molecule or other neutral **particle**.

In the literature of **atmospheric electricity** this term is applied both to the simple case of capture of **free electrons** by positive atomic or molecular ions, and to the more complex case of neutralization of a positive **small ion** by a negative small ion or a similar (but much rarer) neutralization of **large ions**. Recombination is, in general, a process accompanied by **emission** of **radiation**. The **light** emitted from the channel of a **lightning stroke** is recombination radiation. The much less concentrated recombinations steadily occurring in all parts of the **atmosphere** where ions are forming and disappearing do not yield observable radiation. The intermediate case of **glow discharge** may be thought of as the most diffuse case of visibly detectable recombination. The rate at which electrons, small ions, and large ions recombine is a function of their respective mobilities and of their concentrations. The former dependence is expressed in terms of the **recombination coefficient** of the particular ion type. *See also* **airglow**, **aurora**.

recombination coefficient—A measure of the specific rate at which oppositely charged ions join to form neutral **particles** (a measure of **ion** recombination).

Mathematically, if q is the **ionization** rate expressed in terms of ion pairs created per unit volume per unit time, n_1 the number of positive **small ions** per unit volume at time t, and n_2 the number of negative small ions per unit volume at the same instant, then the time rate of change in, for example, positive small-ion **density** is

$$\frac{dn_1}{dt} = q - a n_1 n_2,$$

if we ignore effects due to **Aitken nuclei** and **large ions**. In this expression a is the recombination coefficient for small ions and is of the order of 10^{-6} cm^3 s^{-1} at **sea level**. The recombination coefficient between large and small ions is of the same magnitude; the recombination coefficient for large ions neutralizing large ions is much smaller, of the order of 10^{-9} cm^3 s^{-1}. *See* **combination coefficient**.

recombination energy—The **energy** released as **heat** or **light** when two oppositely charged portions of an atom or molecule rejoin to form a neutral atom or molecule.

recon—Common contraction for **reconnaissance**.
See **aircraft weather reconnaissance**.

reconnaissance—(Often contracted recco, recon.) *See* **aircraft weather reconnaissance**.

record observation—(Commonly called hourly observation.) A type of **aviation weather observation**; the most complete of all such observations, usually taken at regularly specified and equal intervals (hourly, usually on the hour). This type of **observation** has been replaced by the **METAR**.

recording albedometer—An instrument that chronicles the ratio of the **radiation** reflected by a surface to the radiation incident upon it.

recording instrument—A device that makes a record (written or by other means) of **observations** of a meteorological quantity.

recording potentiometer—An instrument that automatically records as a function of time the voltage applied to it.

It is frequently used in meteorology in conjunction with thermocouples to record **temperature**.

recording rain gauge—An instrument that automatically records the **amount of precipitation** collected as a function of time.

The most common types, according to their principle of operation, are 1) **tipping-bucket rain gauge**, 2) **weighing rain gauge**, 3) **capacitance rain gauge**, and 4) **optical rain gauge**.

recovery test—A type of **pumping test** in which, after the pumping stops, the rise of the **water table** or piezometric level in the pumped well, or in surrounding **observation wells**, is measured at predetermined time intervals.

recovery time—In **radar**, the time after the transmission of a **pulse** for the **receiver** to regain its full **sensitivity**.

Ordinarily the receiver is turned off during pulse transmission to protect it from the strong transmitted **signal** and is turned back on as soon as the **transmitter** is turned off. The recovery time is the transient time during which full receiver sensitivity is restored. Because the receiver is not operating at its full sensitivity at times shorter than the recovery time, quantitative **reflectivity** measurements at close ranges are difficult or impossible. Through the use of fast switches, recovery times may be in the order of a fraction of a microsecond, so that the minimum range for quantitative measurements can be as small as a few tens of meters.

rectangular Cartesian coordinates—*See* **Cartesian coordinates**.

rectangular curvilinear coordinates—*See* **curvilinear coordinates**.

rectilinear current—*See* **reversing current**.

rectilinear wind shear—Same as **unidirectional vertical wind shear**.

recurrence formula—A formula that relates successive members of a sequence of terms or formulas to earlier members of the sequence, so that all members may be determined in order.

recurrence frequency—*See* **pulse repetition frequency**.

recurrence interval—Same as **return period**.

recurvature—The change in direction of **tropical cyclone** movement from westward and poleward to eastward and poleward, under the influence of midlatitude **westerlies**.

Such recurvature of the path frequently occurs as storms move into midlatitudes and is a major concern in tropical cyclone forecasting.

recycled water—Used water that has had its quality sufficiently upgraded so that it can be used for another purpose or restored for the same purpose.

red flash—A flash of red **light** seen on or (seemingly) adjacent to the lower rim of the low sun (at either **sunrise** or **sunset**).

Much of what is said about the **green flash** applies to the red flash, only transferred to the base of the sun and applied to multiple images of the lower red rim. During a sunset, an island of red light will sometimes form just below the rest of the sun, proceed to join the main disc, and then progress up the sides of the sun as serrations. The red flash is not perceived as being as striking an event as the green flash if for no other reason than that its color is more characteristic of the low sun.

red noise—Noise with **energy** monotonically decreasing as the **frequency** increases; unlike **white noise** where the same energy is found at all frequencies.

Red Sea Water—A **Mediterranean Water** mass originating in the Red Sea, where it has a **temperature** of 22°C and a **salinity** above 39 **psu**.

It spreads, much diluted, at a depth between 600 and 800 m in the Arabian Sea and is found at a depth between 1000 and 1100 m south of Madagascar.

red snow—A **snow** surface of reddish color due to the presence within it of certain microscopic algae (**cryoplankton**) or **particles** of red **dust**.

red tide—The term applied to toxic algal blooms caused by several genera of dinoflagellates (Gymnodinium and Gonyaulax) that turn the sea red and are frequently associated with a deterioration in water quality.

The color occurs as a result of the reaction of a red pigment, peridinin, to **light** during **photosynthesis**. These toxic algal blooms pose a serious threat to marine life and are potentially harmful to humans. The term has no connection with astronomical tides. However, its association with the word "tide" is from popular observations of its movements with tidal currents in **estuaries**.

reduced gravity—The effective change in the **acceleration of gravity** acting on one fluid in contact with a fluid of different **density** due to **buoyancy** forces.

Explicitly, the reduced gravity is

$$g' = g\Delta\rho/\rho_0,$$

in which g is the **acceleration of gravity**, ρ_0 is the reference density, and $\Delta\rho$ is the difference in density between the two fluids.

reduced grid—A term generally used for grids with points that lie along lines of latitude and longitude when some of the points at higher latitudes are removed.

Due to the convergence of the meridians, points along a parallel of latitude placed at regular angular intervals of longitude will become closer and closer together in geometric distance as the poles are approached. If points are removed at higher latitudes, the geometric distance between the remaining points does not become extremely small. *See* **variable resolution model**.

reduced pressure—The calculated value of **atmospheric pressure** at **mean sea level** or some other specified level, as derived (reduced) from **station pressure** or **actual pressure**.

Thus, **sea level pressure** is nearly always a reduced pressure. *See* **reduction**.

reduced temperature—*See* **reduction**.

reduction—In general, the transformation of data from a "raw" form to some usable form.

In meteorology, this often refers to the conversion of the observed value of an element to the value that it theoretically would have at some selected or standard level, usually **mean sea level**. The most common reduction in weather observing is that of **station pressure** to **sea level pressure**. Temperature is sometimes reduced to a **sea level** value, mostly in climatological work. Most calculations of this sort are based upon approximate actual atmospheric conditions or on the **standard atmosphere**.

reduction of pressure to a standard level—Calculation of the **pressure** at a standard level from the pressure measured at another level by taking into account, according to theory, the weight of a column of air between the two levels.

reduction of temperature to mean sea level—Calculation by which **temperature** observed at a particular **altitude** is reduced to value for **mean sea level**, assuming a mean **lapse rate** of the **atmosphere**.

Reech number—The reciprocal Lg/M^2 of the **Froude number**, where g is the **acceleration of gravity**, L a **characteristic length**, and M a characteristic speed.

reference atmosphere—*See* **standard atmosphere**.

reference crop evaporation—The quantity of water evaporated per unit area, per unit time from a continuous, uniform area of grass with a fixed height of 0.12 m, an **albedo** of 0.23, and a surface **resistance** of 69 s m^{-1}; approximated by the rate of **evaporation** for an extensive area of 0.08 to 0.15 m high green grass cover of uniform height, actively growing, completely shading the ground, and not short of water.

reference frame—Same as **coordinate system**.

reflectance—A general term referring to the **radiation** reflected from, or **scattered** back through, a given surface in response to radiation incident on the surface with the same **wavelength** or wavelength range.

Different definitions arise depending on the directionality of the radiation under consideration. Reflectance in terms of the ratio of reflected to incident **irradiance** is generally termed **albedo**, which may also be described in terms of the **directional-hemispherical reflectance** (when the incident irradiance is from a specific direction), or the **bihemispherical reflectance** (when the incident irradiance is diffuse). Reflectance of **radiance** from a specified incident direction into a specified reflected direction is termed the bidirectional reflectance, which may also be described by the **bidirectional reflection function** (BDRF), the **bidirectional reflectance distribution function** (BRDF), or the **bidirectional reflectance factor**. *See* **reflectivity**.

reflection—A change of direction and possibly **amplitude** of an electromagnetic, acoustic, or any other **wave** propagating in a material medium, as a consequence of spatial **variation** in the properties of the medium.

In specular (mirrorlike) reflection, the spatial variation is abrupt (on the **scale** of the **wavelength**), as at an **interface** between water and air. **Specular reflection** is described by the law of reflection, according to which incident and reflected waves lie in the plane of incidence, defined by the normal to the interface and the direction of the incident wave, and make the same **angle of incidence** with this normal. Specular reflection is distinguished from **refraction** in that the direction of propagation of the reflected wave has a component opposite the direction of the incident wave. Although the law of (specular) reflection is often a good approximation, it is not exact: **Diffuse reflection** in directions not accounted for by the law of reflection always accompanies specular reflection because matter is not homogeneous on all scales. Light reflected by a **cloud** illuminated by **sunlight** is an example of diffuse reflection. Reflection may also refer to the change of direction of a **beam** of **particles**, in the broadest sense of this term.

reflection coefficient—(Also called Fresnel coefficient.) The ratio of the **amplitude** of the reflected **electric field** to the amplitude of the field incident at the **optically smooth** planar **interface** between two **optically homogeneous** media.

The incident and reflected fields are plane **harmonic** and the interface is large in lateral extent compared with the **wavelength** of the **illumination**. May also be used for the ratio of reflected to incident irradiances, that is, the ratio of the normal (to the **interface**) component of the reflected **Poynting vector** to that of the incident normal. A better term for this quantity is **reflectivity**. The reflection coefficient (and hence reflectivity) depends on the **angle of incidence** of the illumination, its wavelength (by way of the wavelength-dependence of the relative **refractive index** of the two media), and its state of **polarization**. These coefficients taken together are sometimes called the "Fresnel formulae" or "Fresnel relations." Reflection coefficient may mean the ratio of any reflected to incident **irradiance** (reflectivity). *See also* **transmission coefficient**.

reflection nephoscope—Same as **mirror nephoscope**.

reflection rainbow—A **rainbow** formed by a **light** source reflected by an extended water surface; not to be confused with a rainbow seen reflected in a still body of water.

When the water is calm, the center of a reflection rainbow is at the same **elevation** as the sun but in the opposite part of the sky. Such a bow intersects the ordinary rainbow (primary or secondary) at the **horizon**. When the water is rough (which it often is under the windy conditions in the vicinity of a **shower**) a family of reflection bows is formed, the envelope of which produces an essentially vertical bow, which intersects the ordinary bow at the horizon.

Humphreys, W. J., 1940: *Physics of the Air*, 3d ed., 498–499.

reflectivity—Often the ratio of any reflected to incident **irradiance**.

This term is sometimes qualified, as in bidirectional (or biangular) reflectivity, meaning the ratio of reflected **radiance** (in a particular direction) to the incident radiance (in a particular direction). *See also* **reflection coefficient**, **radar reflectivity**, **radar reflectivity factor**.

reflectivity factor—See **radar reflectivity factor**.

reflector—1. In general, any object that reflects incident **energy**; usually a device designed for specific **reflection** characteristics.

2. The part of a radio or **radar** antenna system that focuses and directs the transmitted **wave**.

refoli—Same as **raffiche**.

refraction—A change of direction and possibly **amplitude** of an electromagnetic, acoustic, or any other **wave** propagating in a material medium, homogeneous on the **scale** of the **wavelength**, as a consequence of spatial **variation** in the properties of the medium.

This variation can be abrupt (on the scale of the wavelength), as in refraction of **electromagnetic waves** because of an air–water **interface**, or gradual, as in refraction of electromagnetic waves by the **atmosphere**. Refraction is distinguished from (specular) **reflection** in that the direction of propagation of the refracted wave does not have a component opposite the direction of the incident wave. *See* **atmospheric refraction**, **refractive index**; *compare* **diffraction**, **scattering**.

refraction coefficient—In **oceanography**, the square root of the ratio of the spacing between orthogonals in **deep water** and in shallow water.

It is a measure of the effect of **refraction** in diminishing **wave height** by increasing the length of the **wave crest**.

refraction diagram—A diagram showing the trajectories of **wave crests** in shallow water under the influence of **refraction**.

refraction index—Same as **index of refraction**.

refraction of water waves—The process by which the direction of a **wave crest** changes due to different propagation speeds along the wave crest.

In shallow water, **refraction** causes the crests of surface waves to bend towards the direction of the bottom contours. Also, currents can cause refraction.

refractive index—(Or **index of refraction**; also **absolute refractive index**.) The ratio of the free-space **speed of light** c, a universal constant, to the **phase velocity** of a plane harmonic **electro-magnetic wave** in an **optically homogeneous**, unbounded medium.

The refractive index, often denoted as n, of a given material in a given state depends on the **frequency** of the **plane wave**. A plane **harmonic** wave incident on an **optically smooth** interface between two dissimilar, negligibly absorbing media undergoes a change in direction specified by Snel's law (attributed to Willebrord Snel, almost always misspelled as Snell):

$$\frac{\sin \theta_i}{\sin \theta_t} = \frac{n_t}{n_i},$$

where θ_i is the angle between the normal to the **interface** and the **wave vector** of the (incident) wave in the medium with refractive index n_i, and θ_t is the angle between the normal to the interface and the wave vector of the (transmitted or refracted) wave in the medium with refractive index n_t. The ratio n_t/n_i is the relative refractive index. Strictly speaking, plane harmonic wave propagation in an unbounded, homogeneous medium is specified by a **complex refractive index**, sometimes written as

$$n \pm ik,$$

where n and k are nonnegative but otherwise unrestricted. The choice of sign depends on the convention for harmonic time dependence,

$$\exp(\mp i\omega t),$$

where ω is **circular frequency**. Here the real part n is as defined previously, and the imaginary part k, sometimes called the **absorption** index, is related to the **absorption coefficient** by

$$\frac{4\pi k}{\lambda},$$

where λ is the free-space **wavelength**. The inverse of the absorption coefficient is the distance over which the **irradiance** of a plane wave is attenuated (by absorption) by the factor e. Although the real part of the complex refractive index is often denoted as n, no symbol is widely used for the imaginary part. In particular, the complex refractive index is sometimes written

$$n(1 \pm ik),$$

in which instance the absorption coefficient is

$$\frac{4\pi k}{\lambda},$$

where λ is the wavelength in the medium. Two widespread myths about n are that it can never be less than 1 and that it stands in a one-to-one relation with (mass) **density**. For a refutation of the first see Brillouin (1960); for a refutation of the second see Barr (1955).

Brillouin, L., 1960: *Wave Propagation and Group Velocity.*
Barr, E. S., 1955: *Amer. J. Phys.*, 623–624.

refractive modulus—*See* **refractivity**.

refractivity—(Also refractive modulus.) In radio engineering, related to the **refractive index** n of air by

$$N = (n - 1) \times 10^6,$$

and given in N-units.

N also sometimes denotes **modified refractive index**.

refractometer—An instrument for measuring the **index of refraction** of a liquid, gas, or solid.

Refractometers in general use in meteorology operate in the **microwave** region and are based on the principle that the resonant **frequency** of a cavity depends on the **dielectric constant** of its contents. In the Crain refractometer, a microwave oscillator is stabilized by the cavity at its resonant frequency. Two such stabilized systems are used, one with a sealed cavity and one with a perforated cavity open to the **atmosphere**. These are spaced in frequency by 30–50 MHz, and the difference frequencies are measured. The change in difference frequency is almost linearly related to the change in **refractive index** of the perforated cavity's contents. The Birnbaum refractometer is similar in its operating principles.

Refsdal diagram—*See* **thermodynamic diagram**.

regelation—A twofold process in which a localized region on the surface of a piece of **ice** melts when **pressure** is applied to that region (**pressure melting**) and then refreezes when pressure is reduced.

Regelation was discovered by Faraday, who found that two pieces of ice at 0°C would **freeze** together if pressed against each other and then released. Regelation occurs only for substances, such as ice, that have the property of expanding upon freezing, for the melting points of those substances decrease with increasing external pressure. The **melting point** of pure ice decreases with pressure at the rate of 0.0072°C per **atmosphere**. Since this rate is very small, regelation only occurs at ice temperatures of 0°C or very slightly less. The fact that snowballs can be packed well at near 0°C but not at much colder temperatures is a consequence of regelation.

regeneration of a depression—**Deepening** of a **depression** that had been **filling**.

regime—*See* **rainfall regime, Rossby regime**.

regime channel—A term originating in design of stable **irrigation** canals.

A **stream** channel is said to be in regime if it is transporting water and **sediment** in **equilibrium** such that there is neither scour of the channel bed and banks nor sediment deposition in the channel.

region of escape—*See* **exosphere**.

region of influence—The domain of air that is modified by an embedded object.

For example, consider a steady **wind** with straight streamlines well **upstream** of an isolated building. As the air nears the building it will diverge to flow around the building, eventually converging behind the building and becoming straight again. The domain where the streamlines are affected by the building is the region of influence. Similar regions can be found for flow around mountains, airplanes, and weather instruments.

regional air pollution—**Pollutants** that have been emitted from all sources in a region and have had time to mix, diffuse from their peak concentration, and undergo physical, chemical, and photo-chemical reactions.

The size of a region is indeterminate, but usually incorporates one or more cities, and is on the order of 100 to 10 000 km².

regional analysis—1. Involves the use of **rainfall, flood** flow, or other records at a number of sites to obtain a general description of the behavior of such phenomena in an area. Regional analyses are used to improve upon the description of the distribution of rainfall or **runoff** that can be obtained using only the limited records available at most sites.

2. The development of general predictive models of mean rainfall or flood flows, flood flow quantiles, or other **statistics**, perhaps making use of physiographic information such as **altitude, drainage basin** area, or channel slope.

regional basic synoptic network—A WMO network consisting of **synoptic stations** with a specified observational program satisfying minimum regional requirements that enable members to fulfill their responsibilities within World Weather Watch and in the application of meteorology.

regional broadcast—A radio or radio-facsimile broadcast covering a region or part of an ocean basin such as an offshore **forecast area** or the Gulf of Mexico.

regional forecast—In general, a **weather forecast** for a specified geographic area.

Regional Meteorological Center—A center for the **Global Data Processing System** that has the primary purpose of issuing meteorological analyses and prognoses on a regional scale for a specified geographic area.

Regional Meteorological Telecommunications Network—(Abbreviated RMTN.) A broadcast system for distributing various National Weather Service **weather charts** and information to an area that is smaller than an entire hemisphere.

register—The writing component of a **recording instrument**.

It is frequently located at some distance from the sensing portion of the instrument. *See* **multiple register**.

regressand—*See* **regression**.

regression—The **statistical** counterpart or analogue of the functional expression, in ordinary mathematics, of one **variable** in terms of others.

A **random variable** is seldom uniquely determined by any other variables, but it may assume a unique **mean value** for a prescribed set of values of any other variables. The **variate** y is statistically dependent upon other variates $x_1, x_2, \cdots, x_n$ when it has different **probability distributions** for different sets of values of the x's. In that case its mean value, called its **conditional mean**, corresponding to given values of the x's will ordinarily be a function of the x's. The **regression**

function Y of y with respect to $x_1, x_2, \cdots, x_n$ is the functional expression, in terms of the x's, of the conditional mean of y. This is the basis of statistical estimation or **prediction** of y for known values of the x's. From the definition of the regression function, we may deduce the following fundamental properties:

$$E(Y) = E(y), E(y - Y) = 0;$$

$$E[Y(y - Y)] = 0, E(Y^2) = E(yY);$$

$$\sigma^2(y) = \sigma^2(Y) + \sigma^2(y - Y),$$

where $\sigma^2(w)$ denotes the **variance** of any variate w, and $E(w)$ denotes the **expected value** of w. The variate y is called the regressand, and the associated variates $x_1, x_2, \cdots, x_n$ are called regressors; or, alternatively, y is called the predictand, and the x's are called predictors. When it is necessary to resort to an approximation Y' of the true regression function Y, the approximating function is usually expanded as a series of terms $Y_1, Y_2, \cdots, Y_m$, each of which may involve one or more of the basic variates $x_1, x_2, \cdots, x_n$. By extension of the original definitions, the component functions $Y_1, Y_2, \cdots, Y_m$ are then called regressors or predictors. Various quantities associated with regression are referred to by the following technical terms: The variance $\sigma^2(y)$ of the regressand is called the **total variance**. The quantity $y - Y$ is variously termed the **residual**, the **error**, the **error of estimate**. Its variance $\sigma^2(y - Y)$ is called the **unexplained variance**, the **residual variance**, the **mean-square error**; and its positive square root $\sigma(y - Y)$ is called the residual **standard deviation**, the **standard error of estimate**, the **standard error**, the **root-mean-square error**. The variance $\sigma^2(Y)$ of the regression function is called the **explained variance** or the **variance reduction**; the ratio $\sigma^2(Y)/\sigma^2(y)$ of explained to total variance is called the **relative reduction**, or, expressed in percent, the percent reduction.

regression analysis—The representation of a **variable** by a **regression function**.

regression coefficient—A coefficient of a **predictor** or an additive constant in a **regression equation**. The term is usually used in the context of **linear** or multiple linear regression. *See* **regression**.

regression equation—The equation of the **regression function**. It may be of any functional form and the terms may be **orthogonal** or not.

regression function—*See* **regression**.

regression line—The graph of a **linear** regression function. *See* **regression**.

regressor—*See* **regression**.

regular broadcast—A radio or radio-facsimile **HF** broadcast covering a high-seas oceanic area.

regular reflector—Same as **specular reflector**.

rejection region—*See* **significance test**.

relative angular momentum—The **moment** of the **relative momentum** about a point. *See* **angular momentum**.

relative contour—Same as **thickness line**.

relative coordinate system—Any **coordinate system** that is moving with respect to an **inertial coordinate system**. In practice, atmospheric motion is always referred to a relative system fixed to the surface of the earth. Referred to a relative system, various apparent forces arise in Newton's laws owing to motion of the system. *See*, for example, **centrifugal force**, **Coriolis force**.

relative current—The **vector** difference in water velocity between two locations.

relative evaporation—The ratio of the **actual evaporation** to **potential evaporation**.

relative frequency—Proportionate **frequency** per **observation**. If an event occurs N' times in N trials, its relative frequency is N'/N. Relative frequency is the empirical counterpart of **probability**.

relative gradient current—*See* **gradient current**.

relative humidity—The ratio of the **vapor pressure** to the **saturation vapor pressure** with respect to water.

This quantity is alternatively defined by the World Meteorological Organization as the ratio of the **mixing ratio** to the **saturation mixing ratio**. These two definitions yield almost identical numerical values. Relative humidity is usually expressed in percent and can be computed from psychrometric data. Unless specified otherwise, relative humidity is reported with respect to water rather than **ice** because most **hygrometers** are sensitive to relative humidity with respect to water even at subfreezing temperatures, and because the air can easily become supersaturated with respect to ice, which would require three digits in coded messages for **relative humidity with respect to ice**.

Retallach, B. J., 1974: *Physical Meteorology*, p. 83.

relative humidity with respect to ice—The ratio of the **vapor pressure** to the **saturation vapor pressure** with respect to **ice**.

relative hypsography—Same as **thickness pattern**.

relative ionospheric opacity meter—(Or riometer.) A device for measuring the electromagnetic capacity of the **ionosphere** using the strength of a cosmic **radio source** relative to that during minimums of ionospheric disturbances.

relative isohypse—Same as **thickness line**.

relative momentum—The product of the mass of a **particle** and its **relative velocity**; or, in the case of a fluid, the product of **density** and relative velocity.
  See **momentum**.

relative permeability—The dimensionless ratio of the **effective permeability** of a **porous medium** to the **intrinsic permeability**.

relative permittivity—(Also called dielectric function, dielectric constant, specific inductive capacity.) The frequency-dependent response of **optically homogeneous** matter to excitation by a time-harmonic **electric field**.
  If the electric **polarization P** (average **dipole moment** per unit volume) of a material satisfies the (assumed) constitutive relation

$$\mathbf{P} = \epsilon_0 \chi \mathbf{E},$$

where ϵ_0 is the **permittivity** of **free space** (a universal constant), **E** is the electric field, and χ is the electric susceptibility, then the (dimensionless) relative permittivity is

$$\epsilon = 1 + \chi.$$

ϵ is a complex-valued function of **frequency**, its imaginary part related to **absorption** of **electromagnetic waves**. Although dielectric constant is often used as a synonym for relative permittivity (or dielectric function), the former term is misleading given that it may vary by nearly a factor of 100 over the **electromagnetic spectrum**, and hence can hardly be said to be "constant." By the dielectric constant (or, better, static dielectric constant) of a material is often meant the ratio of the capacitance of a parallel-plate capacitor with that material between the plates to the capacitance with a vacuum between them. The **refractive index** of a nonmagnetic material is the square root of its relative permittivity.

relative reduction—*See* **multiple correlation, regression**.

relative sea level—**Sea level** measured relative to a local **tide gauge benchmark**.
  Changes include both local vertical land movements and absolute sea level changes.

relative sunspot number—(Also called sunspot number, Wolf number, Zurich number.) A measure of **sunspot** activity, computed from the formula $R = k(10g + f)$, where R is the relative sunspot number, f the number of individual spots, g the number of groups of spots, and k a factor that varies with the **observer's** **personal equation**, the **seeing**, and the observatory (location and instrumentation).

relative topography—Same as **thickness pattern**.

relative velocity—**Velocity** as measured in a **relative coordinate system**.

relative vorticity—(Also called local vorticity.) The **vorticity** as measured in a system of coordinates fixed on the earth's surface.
  Usually, only the vertical component of the vorticity is meant. *See also* **geostrophic vorticity, vorticity equation**.

relaxation method—An iterative numerical method for solving elliptic partial differential equations.

For example, a **Poisson equation**, $\nabla^2\phi = F(x, y)$, where ∇^2 is the **Laplacian operator** and the function $F(x, y)$ is given.

Panofsky, H., 1956: *Introduction to Dynamic Meteorology*, 168–174.

relaxation of a trough—A decrease in the **amplitude** of a **trough**.

In middle latitudes, it implies a poleward movement of an **isopleth** that defines a trough (i.e., height or **thickness** contour). It is the opposite of **meridional** or equatorward extension of a trough.

relaxation time—In general, the time interval required for a system exposed to some discontinuous change of **environment** to undergo the fraction $(1 - e^{-1})$, or about 63%, of the total **change of state** that it would exhibit after an infinitely long time.

For example, a **thermometer** initially at **equilibrium** in a bath at **temperature** T_1 will exhibit an exponential change of temperature with time after being suddenly plunged into a bath at temperature T_2, theoretically assuming the new temperature T_2 only after an infinitely long time. The finite time interval required for the thermometer to undergo a change of amount $(T_1 - T_2)(1 - e^{-1})$ is called the thermal relaxation time of the thermometer. Occasionally, the fraction $9/10$ is used in place of $(1 - e^{-1})$, so contexts must always be checked to be certain of the definition employed in a given case. The definition may also change for an underdamped device. The change of state of such a device may oscillate several times while approaching its final value.

relaxed eddy accumulation—Method of measuring fluxes of **trace gases**; similar to **eddy accumulation** except that the volume rate at which the **sample** is collected is constant.

The **flux** is estimated as the product of the **standard deviation** of the **vertical velocity**, the difference in the concentration in each **reservoir**, and a scaling **parameter**. The technique has been applied to the measurement of the fluxes of **heat**, **water vapor** and some chemical species.

Businger, J. A., and S. P. Oncley, 1990: Flux measurement with conditional sampling. *J. of Atmos. Oceanic Technol.*, **7**, 349–352.

remote sensing—A method of obtaining information about properties of an object without coming into physical contact with that object.

remote sensing of the atmosphere—*See* **atmospheric sounding**.

removal correction—*See* **barometric corrections**.

repetition frequency—*See* **pulse repetition frequency**.

representative basin—A characteristic **basin** in a natural region, where hydrologic observations are taken, so that the measurements would represent a broad area, rather than taking measurements on all basins in a given region.

representative meteorological observation—**Observation** considered valid for a more or less extended area around a point (**station**) where an observation is made.

representativeness—Property of an **air mass** that is typical of the air mass as a whole and thus may be used in **airmass analysis**.

réseau—A network, **grid**, or system.

In meteorology, this term has been adapted by the World Meteorological Organization for the worldwide network of meteorological stations that have been chosen to represent the meteorology of the globe: the réseau mondial.

reservoir—1. A space capable of storing a fluid.

2. A supply of a substance, especially a reserve or extra supply.

3. A natural or man-made lake that serves to store water; often the release is controlled so that withdrawals can be managed.

reservoir compounds—Any of a series of relatively long-lived compounds that sequester more reactive oxides of **nitrogen**, **hydrogen**, and halogen in the **stratosphere**.

Examples include 1) **chlorine nitrate**, $ClNO_3$, which ties up both ClO and NO_2, and 2) HO_2NO_2, which is a HO_2 and NO_2 reservoir. The presence of these compounds serves to mediate the ability of the more reactive oxides to destroy stratospheric **ozone**, or to produce **tropospheric ozone**.

reshabar—(Also spelled rushabar.) Literally "black wind"; a strong, very turbulent, dry, northeast **wind** of **bora** type that blows down mountain ranges in southern Kurdistan in Iran.

The reshabar is dry and hot in summer and cold in winter.

residence time—A measure of the length of time that molecules will stay in a chamber or region.

It can be estimated by dividing the volume of a chamber or region by the volumetric flow rate through it.

For example, the residence time of air in clouds is important for aqueous-phase chemical reactions and for the size to which **hydrometeors** can grow.

residual—(Also called error of estimate.) In general, the difference between any quantity and an approximation to it; in particular, the difference $(y - Y)$ between any **random variable** y and its **regression function** Y.

See **regression**.

residual depression storage—The accumulation of water in depressions on the **basin** surface, ranging in size from lakes and swamps to cavities the size of soil grains, with no possibility for escape as **runoff**.

Compare **depression storage**.

residual layer—The middle portion of the nocturnal **atmospheric boundary layer** characterized by weak sporadic **turbulence** and initially uniformly mixed **potential temperature** and **pollutants** remaining from the **mixed layer** of the previous day.

Below the statically neutral residual layer is the **stable boundary layer** in contact with the radiatively cooled ground, and above it is the **capping inversion** that separates boundary layer air from free-atmosphere air.

residual mass curve—Graph of the cumulative departures from a given reference such as the arithmetic average as a function of time or date.

residual variance—(Also called unexplained variance.) In general, the **variance** of any **residual**; in particular, the variance $\sigma^2 (y - Y)$ of the difference between any **variate** y and its **regression function** Y.

See **mean-square error**.

resistance—1. In general, any force that tends to oppose motion.

2. Same as **drag**.

3. In electricity, the opposition offered by a substance to the passage of an electric current; the reciprocal of **conductance**.

By virtue of the resistance, a portion of the electrical **energy** is converted into **heat**. *See* **conductivity**.

resistance thermometer—(Also called electrical resistance thermometer) A type of **electrical thermometer** in which the **thermal** element is a substance with an electrical **resistance** that varies with the **temperature**.

Such thermometers can be made with very short time constants and are capable of very accurate measurements. They are commonly used in **radiosondes**.

resistivity—(Also called specific resistance.) The electrical **resistance** per unit length and per unit reciprocal cross-sectional area of a given material at a specified **temperature**.

It is also possible to define the resistivity of a substance as the resistance of a cube of that substance having edges of unit length, with the understanding that the **current** flows normal to opposite faces and is distributed uniformly over them. Resistivity is commonly expressed in units of ohm centimeters. The reciprocal of resistivity is **conductivity**.

resolution—1. The degree to which nearly equal values of a quantity can be discriminated.

2. The smallest measurable change in a quantity.

3. The least value of a measured quantity that can be distinguished.

4. A formal **inference** rule permitting computer programs to reason logically.

5. The ability of an optical system to render visible separate parts of an object or to distinguish between different sources of **light**.

resolution volume—*See* **pulse volume**.

resonance—The condition that results when a system is acted upon by a periodic driving force the **frequency** of which coincides with one of the natural frequencies of the system.

The steady-state **amplitude** of the system, for fixed amplitude of the driving force, is a local maximum at a resonance frequency.

resonance fluorescence—Analytical technique in which an atom or molecule absorbs and then emits **radiation** of a specific frequency, followed by **detection** of the emitted radiation.

The radiation is usually provided by a **plasma** containing the atom or molecule to be detected, ensuring that radiation of the correct **frequency** is available (hence the term **resonance**). The technique has been adapted with great success to atmospheric measurements, for example, the measurement of ClO radicals in the **antarctic ozone hole**.

Anderson, J. G., D. W. Toohey, and W. H. Brune, 1991: Free radicals within the Antarctic vortex: The role of CFCs in Antarctic ozone loss. *Science*, **251**, p. 39.

resonance theory—Theory developed by Lord Kelvin in 1855 explaining the relatively large **amplitude** of the solar semidiurnal component of the **atmospheric tides** on the basis of the **resonance** of a hypothetical natural **atmospheric oscillation** of the same **period**.
(From WMO International Meteorological Vocabulary.)

resonance trough—(Obsolete.) A **large-scale** pressure **trough** that forms in a location that is consistent with the position of other dominant troughs and the distance between troughs expected from dynamical considerations.
For example, the mean trough over the Mediterranean in winter is referred to as a resonance trough because of its relationship to the two orographically forced troughs of the east coasts of North America and Asia.

resonance waves over hills—**Mountain waves** with very large amplitudes produced when the natural **wavelength** of the air matches the width of the hill. The **Froude number** is near 1 in this type of flow.

resonance waves over thermals— A situation similar to **resonance waves over hills**, except that the wavy air is flowing over the tops of convective **thermals** or **cumulus** clouds.
Some of the **kinetic energy** of the rising thermals, when they reach and overshoot into a **statically stable capping inversion**, is converted into **wave** energy.

resonator—In radio and **radar** applications, a circuit that will resonate at a certain **frequency**, or over a **range** of frequencies, when properly excited.
An important type of resonator is the cavity resonator, a closed hollow volume having conducting walls. The frequency at which such cavities resonate is a function of their volume and shape. They are used for making accurate frequency comparisons and for generating radio frequencies, usually in the **microwave** region.

responder—In general, an instrument that indicates reception of an electric or electromagnetic **signal**.
Compare **transponder**.

response analysis—The representation of observed tidal variations in terms of the frequency-dependent **amplitude** and **phase** responses to **input** or forcing functions, usually the **gravitational potential** due to the moon and sun, and the radiational meteorological forcing.

response function—Mathematical function describing the response of a system to a given **input**, for example, the **streamflow** response to a given input of **rainfall**; usually applied to **linear** systems.

response time—Same as **relaxation time**.

responsive parameterization—A method to approximate the amount of **turbulence** by assuming that it is proportional to the amount of **static** or **dynamic instability** in the air, since turbulence responds to instabilities and reduces or removes them by causing turbulent **mixing**. An example of such a **parameterization** would be to relate the **eddy viscosity** to the **Richardson number**, or the **transilient matrix** to the **mixing potential**.

rest mass—(Also called proper mass, invariant mass, velocity-independent mass.) The mass of an object measured in a **coordinate system** in which the object is at rest.
To retain within special relativity the classical form of Newton's dynamical law of motion for a body with **momentum p** and **velocity v** acted upon by a force **F**

$$\mathbf{F} = \frac{d\mathbf{p}}{dt},$$

momentum may be written as

$$\mathbf{p} = m_R\mathbf{v}.$$

The relativistic mass is defined as

$$m_R = \frac{m}{\left(1 - \frac{v^2}{c^2}\right)^{1/2}},$$

where m is the rest mass and c is the free-space **speed of light**. For meteorological applications, rest mass and relativistic mass are so nearly identical that it is largely pointless to make a distinction between the two.

Helliwell, T. M., 1966: *Introduction to Special Relativity*, 148–150.

restoring—A technique used in **numerical modeling**, particularly ocean modeling, in which predicted variables are slowly changed toward prescribed values.

The prescribed values are typically based on observations. An arbitrary **time constant** determines the degree to which the predicted variables can deviate from the prescribed values.

resultant—The sum of **vectors**.

resultant wind—In **climatology**, the vectorial average of all **wind** directions and speeds for a given level at a given place for a certain period, such as a month.

It is obtained by resolving each wind observation into components from south and west, summing over the given period, obtaining the averages, and reconverting the average components into a single **vector**. *Compare* **Lambert's formula**, **wind rose**, **prevailing wind direction**.

retarded flow—Unsteady flow in which the **velocity** decreases at a point over time.

retarding basin—A **reservoir** that reduces peak **floods** through temporary storage.

retention curve—(Also called soil moisture characteristic curve.) In a **porous medium**, the relationship between the volumetric **water content** and the soil water **pressure head**.

retention storage—Water retained in the capillary pores (micropores) of the soil where there is sufficient tension to retain the moisture against the **force of gravity**, so that it is not free to **drain**, and is in large part available to plants.

See also **detention storage**, **field capacity**, **gravitational water**.

retreater—A defective **maximum thermometer** of the **liquid-in-glass** type in which the **mercury** flows too freely through the constriction.

Such a **thermometer** will indicate a **maximum temperature** that is too low.

retrograde depression—A **depression** that moves in a direction opposite to the mean flow in which it is embedded.

retrograde orbit—*See* **inclination**.

retrograde system—*See* **retrogression**

retrograde wave—In **meteorology**, an **atmospheric wave** that moves in a direction opposite to that of the flow in which the wave is embedded.

Retrogression of a particular wave on daily charts is rarely seen, but it is frequently observed on five-day or monthly mean charts.

retrogression—In meteorology, the motion of an **atmospheric wave** or **pressure system** in a direction opposite to that of the **basic flow** in which it is embedded.

retroreflection—**Reflection** in which there is a pronounced maximum in the backward direction.

retroreflector—Any instrument used to cause reflected **radiation** to return along paths parallel to those of their corresponding incident rays.

One type, the **corner reflector**, is an efficient **radar** target. Trihedral prisms are used to retroreflect **light** in measuring atmospheric **extinction coefficients**.

return flow—Water that, having been diverted from a water body, usually for **irrigation**, is not consumed in the process and flows back to the original or some other water body.

return period—(Also called recurrence interval.) The average time until the next occurrence of a defined event.

When the time to the next occurrence has a geometric distribution, the return period is equal to the inverse of **probability** of the event occurring in the next time period, that is, $T = 1/P$, where T is the return period, in number of time intervals, and P is the **probability** of the next event's occurrence in a given time interval.

return stroke—The intense luminosity that propagates upward from earth to **cloud base** in the last phase of each **lightning stroke** of a **cloud-to-ground discharge**.

In a typical **flash**, the first return stroke ascends as soon as the descending **stepped leader** makes electrical contact with the earth, often aided by short ascending **ground streamers**. The second and all subsequent return strokes differ only in that they are initiated by a **dart leader** and not a stepped leader. It is the return stroke that produces almost all of the luminosity and charge **transfer** in most cloud-to-ground strokes. Its great speed of ascent (about 1×10^8 m s^{-1}) is made

possible by residual ionization of the **lightning channel** remaining from passage of the immediately preceding **leader**, and this speed is enhanced by the convergent nature of the **electric field** in which channel electrons are drawn down toward the ascending tip in the region of the **streamer**'s **electron avalanche**. Current peaks as high as 3×10^5 A have been reported, and values of 3×10^4 A are fairly typical. The entire process of the return stroke is completed in a few tens of microseconds, and even most of this is spent in a long decay period following an early rapid rise to full current value in only a few microseconds. Both the current and propagation speed decrease with height. In negative cloud-to-ground flashes the return stroke deposits the positive charge of several coulombs on the preceding negative leader channel, thus charging earth negatively. In positive cloud-to-ground flashes, the return stroke deposits the negative charge of several tens of coulombs on the preceding positive leader channel, thus increasing positive charge on the ground. In negative cloud-to-ground flashes, multiple return strokes are common. Positive cloud-to-ground flashes, in contrast, typically have only one return stroke. The return streamer of cloud-to-ground discharges is so intense because of the high electrical **conductivity** of the ground, and hence this type of streamer is not to be found in **air discharges**, **cloud discharges**, or **cloud-to-cloud discharges**.
> Hagenguth, J. H., 1951: *Compendium of Meteorology*, 137–141.

returning polar maritime air—Air that has originated over subpolar oceans and is returning poleward after an equatorward excursion around an **anticyclone**.

reversal of the monsoon—The twice-yearly change in sign of some **monsoonal** circulations.
> For example, the Indian monsoon reverses in October–November and again in April–May.

reversal temperature—The observed **temperature** at which the charge transferred to riming **graupel** particles during collisions with **ice crystals** in **thunderstorms** reverses sign.
> The reversal temperature is affected by the **cloud** liquid **water content** such that graupel charges positively (negatively) at high (low) liquid water contents and **temperatures** above (below) the reversal temperature. The process can account for the positive charging of ice crystals carried to the top of the clouds, for the negative charge region typically at temperatures between $-10°$ and $-20°C$, and for the lower positive charge center.

reverse cell—A circulating fluid system in which the **circulation** in a vertical plane is **thermally indirect**, that is, cooler air rises relative to warmer air.

reverse flow thermometer housing—A **thermometer** housing that prevents the direct airstream from affecting the thermometer.
> The air intake is reversed so that inertial forces prevent the heavier water and **ice** particles from following the reversed airstream. This allows the thermometer to operate free from moisture effects.

reverse-oriented monsoon trough—A **monsoon trough** with a southwest–northeast orientation (which is the reverse of the **normal** northwest–southeast orientation of the **trough** axis).
> This type of monsoon trough often penetrates into subtropical areas (normally the province of easterly flow).

reversed tide—*See* **reversing current**.

reversible moist-adiabatic process—A **moist-adiabatic process** in which the air is maintained at **saturation** by the **evaporation** or **condensation** of water substance, the **enthalpy** of **water vapor** formed or removed being supplied by or to the air, respectively.
> In contrast to a **pseudoadiabatic expansion**, the liquid water that condenses in an **air parcel** expanding through a reversible moist-adiabatic process is carried with the parcel, so that subsequent compression occurs with moist-**adiabatic** warming, leading to the original state. This can only happen if the condensed water drops are small enough to have negligible fallout velocities. The **moist-adiabatic lapse rate** for the **reversible process** is given by

$$\Gamma_{rm} = g \; \frac{(1 + r_t)\left(1 + \dfrac{L_v r_v}{RT}\right)}{c_{pd} + r_v c_{pv} + r_l c + \dfrac{L_v^2 r_v(\epsilon + r_v)}{RT^2}},$$

where Γ_{rm} is the reversible moist-adiabatic lapse rate; g is **gravity**; r_v, r_l, and r_t are the **mixing** of water vapor, liquid water, and total water; c_{pd}, c_{pv}, and c are the **specific heats** at constant **pressure** of **dry air**, water vapor, and liquid water; L_v is the **heat of vaporization**; R is the dry air gas constant; ϵ is the ratio of the **gas constants** of dry air and water vapor (≈ 0.622); and T is **temperature**.

reversible process—A conceptual process in which the **entropy** of the **universe** (system plus surroundings) is constant.

No real process is reversible.

reversing current—(Also called rectilinear current.) A **tidal current** that flows alternately in approximately opposite directions with a **slack water** at each reversal of direction.

reversing thermometer—A **mercury-in-glass thermometer** that records **temperature** upon being inverted and thereafter retains its reading until returned to the first position.

It consists of a conventional bulb connected to a capillary in which a constriction is placed so that upon reversal the **mercury column** breaks off in a reproducible manner. The **mercury** runs down into a smaller bulb at the other end of the capillary, which is graduated to read temperature. A 360° turn in a locally widened portion of the capillary serves as a trap to prevent further addition of mercury if the **thermometer** is warmed and the mercury expands past the break-off point. The remote-reading potentialities of reversing thermometers make them particularly suitable for use in measuring temperatures at depths in the sea. In this application, both protected thermometers and unprotected thermometers are used, each of which is provided with an **auxiliary thermometer**. They are generally used in pairs in **Nansen bottles**. They are usually read to 0.01°C, and after the proper corrections have been applied, their readings are considered reliable to 0.02°C. Details of the **correction** procedure are given in **Lafond's Tables**.

revised local reference—A **datum level** defined by the **Permanent Service for Mean Sea Level** for each **station**, relative to which the **mean sea level** is approximately 7 m above.

Measurements at the location, over different periods, and to different **tide gauge** benchmarks are all related to this defined **datum**. The value of 7 m was chosen to avoid confusion with other local datum definitions.

revolving storm—(Obsolete.) **Cyclone**.

Reynolds averaging—An averaging procedure applied to **variable** quantities such as **wind speed** and **temperature** in a **turbulent flow**.

If the variable is $\tilde{S}$, the averaging procedure may be given by

$$\overline{S} \equiv \frac{1}{T} \int_0^T \tilde{S}(t)dt \equiv S$$

and $\tilde{S} - S = s$ is the fluctuating part so that $\overline{s} = 0$ where S is the average value, s the fluctuating part, and the bar indicates the averaging process. The average is usually taken over a period of time but it may be taken over space or over an ensemble of realizations. This decomposition is called Reynolds averaging. Reynolds averaging has been applied to the **Navier–Stokes equations** to formulate, for example, turbulent fluxes and the **turbulence kinetic energy** equation. The drawback of this procedure is that it leads to the problem of **turbulence closure**.

Reynolds effect—A process of **drop** growth (primarily of historical interest) suggested originally by O. Reynolds, which involves net **evaporation** from warmer **cloud** drops and **condensation** on cooler drops.

Reynolds number—The dimensionless ratio of the **inertial force** $(\sim U^2/L)$ to the **viscous force** $(\sim \nu U/L^2)$ in the **Navier–Stokes equations**, where U is a **characteristic velocity**, L is a **characteristic length**, and ν is the **kinematic viscosity** of the fluid; thus,

$$\text{Re} = UL/\nu.$$

The Reynolds number is of great importance in the theory of **hydrodynamic stability** and the origin of **turbulence**. The inertia force generates **vortex stretching** and **nonlinear** interactions and hence creates **randomness**. Turbulence occurs when the inertia term dominates the viscous term, that is, when the Reynolds number is large. For many engineering flows, turbulence occurs when $\text{Re} > \text{Re}_c$, where the critical Reynolds number is roughly $\text{Re}_c = 2100$. See **large Reynolds number flow**.

Reynolds stresses—The mean forces (per unit area) imposed on the mean flow by turbulent fluctuations.

They arise from the **nonlinear** advection term when the **Navier–Stokes equations** are Reynolds decomposed and Reynolds averaged. The general form of the Reynolds **stress tensor** is

$$\tau_{ij} = -\rho \overline{u_i u_j}.$$

Kinematically, the **velocity** correlation $-\overline{u_i u_j}$ represents the **momentum flux** in direction $\hat{x}_i$ across a plane perpendicular to direction $\hat{x}_j$ or the momentum flux in direction $\hat{x}_j$ across a plane perpendicular to direction $\hat{x}_i$. In a **turbulent flow**, the **divergence** of the Reynolds stresses are of leading order in the mean momentum budgets. Typically they are several orders of magnitude larger than the **viscous stresses**. In the **boundary layer**, the most important Reynolds stresses are the vertical fluxes of horizontal **momentum**, $-\rho\,\overline{uw}$ and $-\rho\,\overline{vw}$.

RF—Abbreviation for **radio frequency**.

RF signal—(Abbreviation for radio frequency signal.) In radio or **radar**, the **signal** at the transmitted or received **frequency**, as opposed to a signal translated to a different frequency (**IF signal**) or a detected signal (**video signal**).

RHI—Abbreviation for **range–height indicator**.

RHI scope—Same as **range–height indicator**.

rhumb-line course—A **course**, route, or track over the earth along which the angle between the course and **true north** is everywhere the same.

The rhumb-line course is longer than a **great-circle course**, and its main justification lies in the simplification of navigation problems. *See* **grid navigation**.

ribbon lightning—Ordinary **cloud-to-ground lightning** that appears to be spread horizontally into a ribbon of parallel **luminous** streaks when a very strong **wind** is blowing at right angles to the **observer**'s **line of sight**.

Successive strokes of the **lightning flash** are then displaced by small angular amounts and may appear to the eye or camera as distinct paths. The same effect is readily created artificially by rapid transverse movement of a camera during film exposure.

ribut—Sharp, short **squalls** during comparatively **calm** winds from May to November in Malaya.

Richardson number—The dimensionless ratio of buoyant suppression of **turbulence** to **shear** generation of turbulence.

It is defined as

$$Ri = \frac{g\beta}{(\partial u/\partial z)^2},$$

where g is the **acceleration of gravity**, β a representative **vertical stability** (commonly $\partial\theta/\partial z$, where θ is **potential temperature**), and $\partial u/\partial z$ is a characteristic vertical shear of the **wind**. It is used as a **dynamic stability** measure to determine if turbulence will exist. *See* **gradient Richardson number, bulk Richardson number, critical Richardson number, flux Richardson number**.

ridge—1. (Sometimes called wedge.) In meteorology, an elongated area of relatively high atmospheric **pressure**, almost always associated with and most clearly identified as an area of maximum anticyclonic curvature of **wind** flow.

The locus of this maximum curvature is called the **ridge line**. Sometimes, particularly in discussions of atmospheric waves embedded in the **westerlies**, a ridge line is considered to be a line drawn through all points at which the anticyclonically curved **isobars** or contour lines are tangent to a latitude circle. The most common use of this term is to distinguish it from the **closed circulation** of a **high** (or **anticyclone**); but a ridge may include a high (and an **upper-air ridge** may be associated with a surface high) and a high may have one or more distinct ridges radiating from its center. The opposite of a ridge is a **trough**.

2. Also used as reference to other meteorological quantities such as **equivalent potential temperature, temperature**, and **mixing ratio**.

That is, an elongated area of relatively high values of any particular **field** emanating from a maximum.

3. In **oceanography**, a **linear** accumulation of broken ice blocks projecting upward, formed by **ice** deformation, often at the edge of a **floe**.

A ridge is distinguished from a **hummock** by being much longer than it is wide. The term ridge is often used to describe an entire ridged ice feature, in which case the portion above the water line is termed the sail and the portion below the water line is termed the keel.

ridge aloft—Same as **upper-level ridge**.

ridge line—(Also called ridge axis.) A line connecting the points of maximum anticyclonic curvature in a **ridge**.

ridge of high pressure—*See* **ridge**.

riefne—An intense **storm** of Malta in the Mediterranean.

rig—Same as **Michael-riggs**.

right ascension—In astronomy, the angular distance measured eastward along the **celestial equator** from the **vernal equinox** to the hour circle (celestial **meridian**) of the celestial body in question.

Right ascension and **declination** are the two principal coordinates used in positional astronomy.

right-hand rotation—In general, the sense of rotation indicated by curling the fingers of your right hand about the axis denoted by your thumb.

When used to describe horizontal **circulation** in the **atmosphere**, right-hand rotation is **cyclonic** in the Northern Hemisphere, **anticyclonic** in the Southern Hemisphere. *See* **angular velocity**.

right-handed rectangular coordinates—Rectangular **Cartesian coordinates** in which the coordinates x, y, and z are measured along three mutually perpendicular axes such that a rotation of the positive x axis into the positive y axis will drive a right-handed screw in the direction of the positive z axis.

The Cartesian coordinates most often used in meteorology form a right-handed system: x increasing to east, y to north, z to **zenith**.

rime—A white or milky and opaque granular deposit of **ice** formed by the rapid **freezing** of **supercooled water** drops as they impinge upon an exposed object.

It is denser and harder than **hoarfrost**, but lighter, softer, and less transparent than **glaze**. Rime is composed essentially of discrete ice granules and has densities as low as 0.2–0.3 g cm^{-3}. Glaze is generally continuous but with some air pockets and has much higher densities. Factors that favor rime formation are small **drop** size, slow **accretion**, a high degree of **supercooling**, and rapid **dissipation** of **latent heat** of **fusion**. The opposite effects favor glaze formation. Both rime and glaze occur when supercooled water drops strike an object at a **temperature** below freezing. Such formation on terrestrial objects constitutes an **ice storm**; on aircraft, it is called **aircraft icing** (where rime is known as **rime ice**). Either rime or glaze may form on **snow crystals**, **droxtals**, or other ice particles in the **atmosphere**. When such a deposit is wholly or chiefly of rime, **snow pellets** result; when most or all of the deposit is glaze, ordinary **hail** or **ice pellets** result. The alternating clear and opaque layers of some hailstones represent glaze and rime, deposited under varying conditions around the growing **hailstone**. *See also* **hard rime**, **soft rime**.

rime fog—1. A **supercooled fog** that deposits **rime** on exposed objects.

See **fog deposit**.

2. Same as **ice fog**.

rime ice—Same as **rime**, but especially applied to rime formation on aircraft.

Flight through an extremely **supercooled cloud** ($-10°C$ or colder) is very conducive to rime icing. This type of **ice** weighs less than **clear ice**, but it may seriously distort airfoil shape and thereby diminish the **lift**. In aviation parlance, ice that has the ideal rime character may be called **kernel ice**, and that intermediate between rime and clear ice may be called **milky ice**.

rime rod—A simple, cooled cylindrical rod, usually of metal or glass, that is exposed to the airflow in a **cloud** to collect **supercooled cloud** droplets for chemical **analysis**.

The **collection efficiency** of the rod, which is dependent on the size of the drops, can be calculated.

rind ice—*See* **ice rind**.

Rinehart projection—A **map projection** consisting of a set of overlays, used with **radar displays**, to compensate for the possibility of **second-trip echoes**.

The overlays, shifted in **range** by multiples of the **maximum unambiguous range**, make it possible to determine where range-aliased echoes are in relation to cities, political boundaries, or other features commonly plotted on radar displays. If **radar** echoes are automatically dealiased, the Rinehart projection need not be used.

Ringelmann chart—A **chart** used in making subjective estimates of the amount of solid matter emitted by smokestacks.

The **observer** compares the grayness of the **smoke** with a series of shade diagrams formed by horizontal and vertical black lines on a white background.

Ringelmann shades—*See* **Ringelmann chart**.

riometer—*See* **relative ionospheric opacity meter**.

rip—The agitation of water caused by the interaction of water currents or by a rapid **current** setting in over an irregular bottom; for example, a tide rip.

rip current—A narrow **current** in the **surf zone** flowing seaward from the shore.

It usually appears as a visible band of agitated water and is the return movement of water piled up on the shore by incoming waves and winds.

ripe—Descriptive of **snow** that is in a condition to discharge meltwater.

Ripe snow usually has a coarse crystalline structure, a **snow density** near 0.5 kg m⁻³, and a **temperature** near 0°C.

ripening of snow—The process by which a **snow pack** reaches a state where it can yield meltwater, including warming of the snowpack to 0°C, wetting of the **snow**, and coarsening of the snow texture.

ripple—A **wave** on a fluid surface, of sufficiently short **wavelength**, in which **gravity** is the dominant influence.

Same as **capillary wave**.

ripple wave—A **wave** that is controlled to a significant degree by both **surface tension** and **gravity**.

rise time—Usually, the time required for a **pulse** to increase from 10% of its final value to 90% of its final value.

Rise time is less frequently measured between the 5% and 95% points or the 1% and 99% points. Rise time is used to specify the **transient** response of an instrument, and is similar to its **time constant**, **relaxation time**, or **response time**, although these latter terms use $(1 - e^{-1})$, or about 63%, as the fractional change in state over which time is measured, beginning at an initial value. For example, if a step increase of 10° is applied to a **thermometer** registering 0° with a rise time of 50 s, the thermometer would increase from 1° to 9° in 50 s. Note that nothing is specified concerning the time required for the instrument to respond to the first or last 10%. *See also* **time lag**.

rising limb—The rising portion of the **hydrograph** resulting from **runoff** of **rainfall** or **snowmelt**.
See **recession**.

rising tide—(Sometimes called **flood tide**.) The portion of the **tide** cycle between **low water** and the following **high water**.

river—A natural **stream** of water of considerable volume.

river basin—The total area drained by a river and its tributaries.
Compare **watershed**, **drainage area**.

river forecast—A forecast of the expected **stage** or **discharge** at a specified time, or of the total volume of flow within a specified time interval, at one or more points along a **stream**.

river gauge—(Also called stream gauge.) A device for measuring the **river stage**.

river stage—The level of the water surface in a river measured with reference to some **datum**.

river surface temperature—Temperature of the water at the surface of a river.

river system—The aggregate of **stream** channels draining a **river basin**.

rmse—Abbreviation for **root-mean-square error**.

RMTN—Abbreviation for **Regional Meteorological Telecommunications Network**.

roaring forties—A popular nautical term for the stormy ocean regions between 40° and 50° latitude.

It nearly always refers to the Southern Hemisphere, where there is an almost completely uninterrupted belt of ocean with strong prevailing westerly winds.

ROB—Abbreviation for **radar observation**; aviation weather communications code word.
See **radar meteorological observation**.

Robin Hood's wind—Saturated air with temperatures near **freezing**; it is **raw** and penetrating.

robin sphere—An inflatable sphere released at **apogee** from a sounding **rocket**.

The sphere is inflated by **evaporation** of a contained liquid on release. A **corner reflector** is ejected as the sphere inflates, permitting track during descent by a precision radar. Winds and **density** data are derived from the position and **fall velocity** of the sphere. ("Robin" is an acronym for "rocket-balloon-instrument.")

Robitzsch actinograph—A **pyranometer** developed by M. Robitzsch.

Its design utilizes three bimetallic strips that are exposed horizontally at the center of a hemispherical glass bowl. The outer strips are white reflectors, and the center strip is a blackened absorber. The bimetals are joined in such a manner that the pen of the instrument deflects in proportion to the difference in **temperature** between the black and white strips and is thus proportional to the **intensity** of the received **radiation**. This instrument must be calibrated periodically.

rockair—A high-altitude **sounding** system consisting of a small solid-propellant research **rocket** carried aloft by an aircraft.

The rocket is fired while the aircraft is in vertical ascent.

rocket—A vehicle designed or adapted for high-altitude research.

The ideal meteorological research rocket has a low thrust and long burning time to avoid high acceleration while attaining high altitude, has a large and recoverable **payload**, and is a stable platform for experimental equipment.

rocket lightning—A form of **cloud discharge**, generally horizontal and at **cloud base**, with a **luminous** channel appearing to advance through the air with visually resolvable speed, often intermittently.

rocket-triggered lightning—A form of artificial **lightning discharge** initiated with a **rocket** trailing wire that may or may not be connected to the ground.

The first phase of the **discharge** is a unidirectional **leader** starting from the tip of the wire. When the low end of the wire is not connected to ground, bidirectional leader development occurs from both ends of the wire, similar to **lightning** initiation from aircraft. In the case of negative **space charge** overhead (usual summer **thunderstorm** condition), a triggered lightning may only be a positive leader or may become a sequence of **dart leader–return stroke** processes following the initial positive leader. The latter is analogous to the subsequent return stroke process in a negative **cloud-to-ground flash** with the initial positive leader being analogous to the first return stroke. In the case of positive space charge overhead (usual winter storm condition), the triggered lightning is a single negative leader.

rocketsonde—A **meteorological instrument** package that measures vertical profiles of atmospheric winds and either **temperature** or **density** upon descent after ejection from a **rocket** at or near **apogee**.

Rocketsondes reach altitudes above those typical for balloon-borne **radiosondes**.

ROFOR—An international code word used to indicate a **route forecast** (along an air route).
See also **ROFOT, ROMET**.

ROFOT—An international code word used to indicate a **route forecast**, units in the English system.
See also **ROFOR, ROMET**.

rogue wave—*See* **freak wave**.

roll—Transverse **oscillation** of a ship about its longitudinal axis.
See **pitch, yaw, ship motion**.

roll cloud—1. The popular term for **arcus**.

2. A low-level, horizontal, tube-shaped **arcus** cloud associated with a **gust front** of a **convective storm** (or occasionally a **cold front**).

Roll clouds are relatively rare; they are completely detached from the convective storm's **cloud base**, thus differentiating them from the more familiar **shelf clouds**. Roll clouds appear to be rolling about a horizontal axis because of the shearing effects and horizontal **vorticity** provided by the differing **air masses**. *See also* **rotor cloud, morning glory**.

roll vortex—*See* **horizontal convective rolls**.

rollers—Large breakers formed on a beach, usually as a result of an incoming **swell** that has traveled a long distance across the ocean.
See also **long-crested wave**.

rolls—1. Overturning, quasi-two-dimensional circulations parallel to mean wind in the layer they occupy in which individual **particles** move **downwind** in a helical motion.

In the **atmosphere, boundary layer rolls** usually consist of alternating counterrotating helices, aligned nearly parallel to the mean boundary layer **wind**. When clouds are present, they form over the upwelling parts of the roll circulation. Called **cloud streets**, they are a good measure of roll wavelength. Typically, cloud streets are spaced at about two to three times the depth of the rolls, although larger spacings are not unusual. Several mechanisms have been proposed for forming rolls; formation and maintenance of atmospheric rolls is thought to involve both **buoyancy** and **shear** effects. In some cases, rolls are thought to result from the action of **gravity waves** on the **boundary layer**. Rolls occur in the **convective boundary layer** and have been observed with both stronger winds and midboundary layer wind maxima. They have also been observed for lighter winds with weaker buoyant forcing.

2. The overturning motion that results from breaking **Kelvin–Helmholtz waves**.

ROMET—An international code word denoting a **route forecast**, units in the metric system.
See also **ROFOR, ROFOT**.

rondada—A nautical term of Spain denoting **wind** that shifts diurnally from northwest through north, east, south, and west.

Röntgen ray—(Or Roentgen ray.) Same as **x-ray**; rarely used these days.
X-ray was the coinage of Wilhelm Röntgen, who discovered them.
Boorse, H. A., and L. Motz, 1966: *The World of the Atom*, Vol. I, 385–401.

root-mean-square error—(Abbreviated rmse; also called standard error of estimate, residual standard deviation.) The positive square root of the **mean-square error**.
It is equal to the **standard error** only when the mean error is zero.

rope cloud—A very narrow, long, sometimes meandering, **cumulus** cloud formation that is frequently visible in satellite imagery.
It is generally associated with a **cold front** or a land–**sea breeze** front.

ropes of Maui—Same as **crepuscular rays**.

rosau—West or southwest **breeze** by day in the Rhone Valley, France.

Rosemount temperature housing—A **thermometer** housing that directs the ambient airflow to the thermometer around a sharp right-angle bend in the intake channel.
See **reverse flow thermometer housing**.

Rossby diagram—*See* **conserved variable diagrams**.

Rossby formula—*See* **Rossby wave**.

Rossby–gravity wave—(Also called mixed Rossby–gravity wave, Yanai wave.) An **equatorial wave** with a **dispersion relation** asymptotic to that for **equatorial Kelvin waves** for large positive (eastward) **zonal** wavenumbers and asymptotic to that for equatorial **Rossby waves** for large negative (westward) zonal **wavenumbers**.
In the **shallow water approximation**, the dispersion relationship is given by **frequency**

$$\omega = kc[1 - \{1 + 4\beta/(k^2 c)\}^{1/2}]/2,$$

in which k is the zonal wavenumber, β is the **meridional** gradient of the **Coriolis parameter** at the **equator**, and

$$c = (gH)^{1/2},$$

for which g is the **acceleration of gravity** and H is the mean fluid depth.

Rossby number—A **dimensionless number** relating the ratio of **inertial** to **Coriolis forces** for a given flow of a rotating fluid.
Explicitly, the Rossby number is

$$\text{Ro} = \frac{U}{fL},$$

where U is the **velocity** scale, f is the **Coriolis parameter**, and L is the horizontal length **scale**. This number plays a fundamental role in defining the **regime** of large-scale geophysical fluid dynamics. Large-scale flows are defined as those that are significantly influenced by the earth's rotation and with sufficiently large L for Ro to be order one or less (e.g., flows with sufficiently small Rossby number are in **geostrophic balance**).

Rossby number similarity—A modeling technique that describes the height dependence of meteorological parameters in the upper part of the **atmospheric boundary layer** by $(z/z_0)(\text{Ro})^{-1}$, where z_0 is the **surface roughness** length and $\text{Ro} = u_*/fz_0$ is called the friction **Rossby number**, where f is the **Coriolis parameter** and u_* is the **friction velocity**. Other Rossby numbers are also used, always in the form of a **velocity** scale divided by the product of a length times the Coriolis parameter.

Rossby parameter—(Also called Rossby term.) The northward **variation** of the **Coriolis parameter**, arising from the sphericity of the earth.
In **spherical coordinates**,

$$\beta = \frac{1}{a} \frac{d}{d\phi} (2\Omega\sin\phi) = \frac{2\Omega\cos\phi}{a},$$

where Ω is the angular speed of the earth, ϕ the latitude, a the mean radius of the earth, and β the Rossby parameter. The Rossby parameter, usually treated as a constant, is of importance dynamically in producing Rossby waves. *See* **Rossby wave, barotropic instability, beta plane**.

Rossby radius of deformation—1. The distance that cold pools of air can spread under the influence of the **Coriolis force**.

A **cold pool** will initially spread out toward and under warmer air because of higher **pressure** under the cold, denser air. However, as the spreading **velocity** increases, the Coriolis force will increasingly turn the velocity vector until it is parallel, rather than perpendicular, to the **pressure gradient**. At this point, no further spreading will occur and the winds will be in **geostrophic equilibrium**. The final **equilibrium** distance traveled by the edge of the cold air equals the **external Rossby radius of deformation**, λ_R:

$$\lambda_R = \frac{(gH\Delta\theta/\theta_0)^{1/2}}{f_c},$$

where g is gravitational **acceleration**, H is the initial depth of the cold pool, $\Delta\theta$ is the **potential temperature** contrast between the cold and surrounding warm air, θ_0 is the **absolute** potential temperature of the warm air, and f_c is the **Coriolis parameter**.

2. An internal Rossby radius of deformation can be defined for fluids with a **gradient** of **potential temperature** rather than a **temperature** interface:

$$\lambda_R = \frac{N_{BV}Z_T}{f_c},$$

where N_{BV} is the average **Brunt–Väisälä frequency** within the **troposphere** and Z_T is the depth of the troposphere.

This radius is important for determining the **phase speed** and **wavelength** of **baroclinic waves** (**Rossby waves**) in the **general circulation**. An alternative definition for internal Rossby radius of deformation is

$$\lambda_R = \frac{GZ_T}{f_c z_i},$$

where G is the **geostrophic wind** speed and z_i is the depth of the **atmospheric boundary layer**, approximated here as $z_i = G/N_{BV}$. This form is useful in determining boundary layer (Ekman) pumping through the top of the **boundary layer**.

Rossby regime—A type of flow pattern in a rotating fluid with differential radial heating in which the major radial **transport** of **heat** and **momentum** is effected by horizontal **eddies** of low **wavenumber**.

This regime occurs for low values of the **Rossby number** (on the order of 0.1). The term has been used in connection with **dishpan experiments**, but applies as well to the **atmosphere** of the earth and other planets.

Rossby term—Same as **Rossby parameter**.

Rossby wave—(Also called planetary wave.) A **wave** on a uniform **current** in a two-dimensional nondivergent fluid system, rotating with varying angular speed about the local vertical (**beta plane**).

This is a special case of a **barotropic disturbance**, conserving **absolute vorticity**. Applied to atmospheric flow, it takes into account the **variability** of the **Coriolis parameter** while assuming the motion to be two-dimensional. The **wave speed** c is given by

$$c = \overline{u} - \frac{\beta}{K^2},$$

where $\overline{u}$ is the mean westerly flow, β is the **Rossby parameter**, and $K^2 = k^2 + l^2$, the total **wavenumber** squared. (This formula is known as the Rossby formula, long-wave formula, or planetary-wave formula.) A stationary Rossby wave is thus of the order of the distance between the large-scale semipermanent **troughs** and **ridges** in the middle **troposphere**. The Rossby wave moves westward relative to the current, in effect slowing the eastward movement of long-wave components relative to the short-wave components in a **barotropic** flow. This effect is important in a numerical forecast with a **barotropic model**, but attempts to apply the formula to actual **contour** patterns considered as waves have less dynamic justification and correspondingly less success. *See* **long wave**.

Holton, J. R., 1992: *An Introduction to Dynamic Meteorology*, 3d edition, Academic Press, 216–222.

rotary current—A **tidal current** that flows continuously, with the direction of flow changing through all points of the **compass** during a tidal **cycle**; found away from coastal or shallow water flow restrictions, where reversing currents are more probable. *Compare* **reversing current**; *see* **current ellipse**.

rotating-beam ceilometer—*See* **ceilometer**.

rotating multicylinder—An instrument consisting of a series of graduated cylinders possessing selective **collection efficiencies**.

It is used for the measurement of quantities relating to the size distribution of **cloud droplets**. *See* **impactor**.

rotating Reynolds number—(Or rotation Reynolds number.) A **nondimensional number** arising in problems of a rotating **viscous fluid**.

It may appear either as $\Omega h^2/\nu$, in which case it equals one-half the square root of the **Taylor number**, or as $\Omega r^2/\nu$, where r is a suitable radius, h a representative depth, Ω the **absolute** angular speed, and ν the **kinematic viscosity**.

rotation Reynolds number—Same as **rotating Reynolds number**.

rotational—Possessing **vorticity**.

See also **irrotational**.

rotational instability—In general, any **instability** of a rotating fluid system; usually synonymous with **inertial instability**.

rotenturm wind—A warm south **wind** blowing through Rotenturm Pass in the Transylvanian Alps.

rotor—Circulation of flow about a horizontal or nearly horizontal axis that is usually associated with flow over the lee side of a barrier, such as a mountain range.

See **rotor cloud**.

rotor cloud—(Sometimes called **roll cloud**.) A turbulent, altocumulus-type **cloud formation** found in the lee of some large mountain barriers, particularly in the Sierra Nevada near Bishop, California.

The air in the **cloud** rotates around an axis parallel to the range. The term was first applied to clouds of this type in Europe, especially in the Riesengebirge and on Crossfell. The rotation may extend to the ground, cause hazards to aircraft, and carry large amounts of **dust** aloft. Rotor clouds are often associated with **lee wave** (lenticular) clouds that may be present above. *See* **Bishop wave**.

rotoscope—1. A device used in experimental meteorology for viewing the relative motions in a rotating fluid system.

It consists of a Dove prism (Hecht 1974) mounted in a rotating barrel and aligned along the axis of rotation of the vessel containing the fluid. Since the **image** in the prism rotates twice during each revolution of the prism itself, the rotating vessel will appear stationary if the rate of rotation of the barrel is one-half the rotation of the vessel, and in the correct sense.

2. An instrument that measures the **flow velocity** of a gas through a tube.

It consists of a lightweight piston, externally threaded, that fits loosely inside a vertically mounted glass tube. The gas flows in at the bottom of the tube, lifting the piston and imparting a rotary motion to it as air flows through the threaded section. The height to which the piston rises is a measure of the flow velocity of the particular gas.

Hecht, E., 1974: *Geometric Optics*, Addison–Wesley Publishing Co., 167–168.

rotten ice—Any piece, body, or area of **ice** that is in the process of melting or disintegrating.

It is characterized by honeycomb structure, weak bonding between crystals, or the presence of meltwater or **seawater** between grains. Rotten ice may appear transparent (and thus dark) when saturated with seawater and so may be easily confused with newly forming **black ice**.

rough air—An aviation term for **turbulence** encountered in flight.

rough surface—*See* **aerodynamically rough surface**.

roughness coefficient—Same as **roughness length**.

roughness length—(Also called roughness coefficient.) A measure of the roughness of a surface over which a fluid is flowing, defined as follows:

$$z_0 = \frac{\epsilon}{30},$$

where z_0 is the roughness length and ϵ is the average height of surface irregularities.

See **logarithmic velocity profile.**

roughness length—*See* **aerodynamic roughness length.**

roughness parameter—Same as **roughness length.**

roughness sublayer—(Same as **transition layer.**) The lowest **atmospheric layer** immediately adjacent to a surface covered with relatively large roughness elements such as stones, vegetation, trees, or buildings.

The roughness sublayer extends from the surface up to about two to five times the height of the roughness elements and includes the **canopy** layer. Within the roughness sublayer the flow is three-dimensional, since it is dynamically influenced by length scales of individual roughness elements and **surface layer** scaling cannot be expected to apply. *Compare* **aerodynamic roughness length.**

roundoff error—(Also called truncation error.) The **error** in a numerical computation due to the finite **precision** in the representation of numbers in computers.

route component—The average forecast **wind** component parallel to the flight path at **flight level** for an entire route.

Route component is positive if helping (**tailwind**) and negative if retarding (**headwind**).

route cross section—Graphic representation of the meteorological conditions observed simultaneously in a vertical selection of the **atmosphere**, taken along a specific route of flight.

Compare **flight cross section.**

route forecast—**Aviation forecast** for a specific air route or for a specific portion of an air route.

See also **ROFOR, ROFOT, ROMET, flight cross section, route cross section.**

rubber ice—Newly formed **sea ice** that is weak and elastic.

rule—A law or regulation that governs behavior, actions, or operations.

In **rule-based systems**, only those rules with true antecedents are used. For example, a rule that begins "IF the **temperature** is less than 0°C . . . " is ignored whenever the measured temperature is 0°C or higher. *See* **IF–THEN rule, production rule, antecedent, consequent.**

rule-based system—A **knowledge-based system** that uses **rules** as its knowledge representation.

See also **expert system.**

rule-of-thumb—*See* **heuristic.**

rules of inference—**Rules** for deriving truths from previously stated or proven truths.

It is the basis of the **inference engine** in most **expert systems.** *See also* **chain rule, modus ponens, modus tollens, resolution.**

running mean—Same as **consecutive mean.**

runoff—The water, derived from **precipitation**, that ultimately reaches **stream** channels.

runoff coefficient—A coefficient relating the amount of **runoff** to the **amount of precipitation** received.

See **rational method.**

runoff cycle—The part of the **hydrologic cycle** undergone by water between the time it reaches the land as **precipitation** and its subsequent **evapotranspiration** or **discharge** through **stream** channels.

runoff plot—A small, experimental plot of land used to study **surface runoff.**

runway elevation—1. Same as **airport elevation.**

2. The **actual elevation** of the touchdown zone of a runway.

It is published on the **IFR** Approach Plates.

runway observation—An evaluation of certain meteorological elements observed at a specified point on or near an airport runway.

Temperature, wind speed and **direction, ceiling,** and **visibility** are among the elements frequently observed at such locations because of the importance of these data to aircraft landing and takeoff operations. *See* **runway temperature, runway visibility, runway visual range.**

runway temperature—The **temperature** of the air just above the runway at an airport (usually at about 4 ft but ideally at engine and/or wing height), used in the determination of **density altitude.**

Therefore, runway temperature observations are made and reported at airports when critical values of density altitude prevail.

runway visibility—The **visibility** along an identified runway, determined from a specified point on the runway with the **observer** facing in the same direction as a pilot using the runway.

Compare **runway visual range**.

runway visual range—(Abbreviated RVR.) The **range** over which the pilot of an aircraft on the center line of a runway can see the runway surface markings or the lights delineating the runway or identifying its center line.

RVR may be determined by an **observer** located at the end of the runway, facing in the direction of landing, or by means of a **transmissometer** installed near the end of the runway.

rush of snowmelt flood—The rate of change in time of **discharge** and water surface elevations during spring snowmelt.

rushabar—Same as **reshabar**.

RVR—Abbreviation for **runway visual range**.

S

S-band—Describes radars operating at a **wavelength** of 10 cm.

S band—*See* **radar frequency band**.

S-curve—The **hydrograph**, for a given **river basin**, that would theoretically result from a continuous, constant excess **rainfall rate** per specified period T resulting in a **runoff** volume of 1.00 unit depth (e.g., 1.00 cm or 1.00 in.) over the basin for each period T.

The curve has a characteristic S-shaped **rising limb**, reaching a constant final **discharge** rate equivalent to the constant excess rainfall rate multiplied by the area of the basin. An S-curve is constructed from a series of unit hydrographs of period T by the simple process of successive displacement by T and summing up the ordinates (discharge values). In theory, the unit hydrographs for storms of any duration may be derived from the resulting S-curve.

S-curve method—A method of deriving a **unit hydrograph** for a desired duration from one of another duration through the use of an **S-curve**.

S-values—Specific **temperature** anomalies given by the relation

$$S = \frac{T - T_p}{T_p},$$

where T is the actual temperature and T_p is the standard-atmosphere temperature at a point where the **pressure altitude** is z_p.

S-values are usually expressed in terms of tenths of degrees per hundred degrees **absolute** temperature. S-value lines are drawn on 4D charts because they are related to D-values by

$$\Delta_v D = S\Delta z_p,$$

where $\Delta_v D$ is the **variation** in **D-value** for a given Δz_p variation in pressure altitude.

saddle-back—In aviation terminology, the cloudless air between the "towers" of two **cumulus congestus** or **cumulonimbus** clouds and above a lower **cloud** mass.

saddle point—Same as **col**.

Saffir–Simpson hurricane scale—A classification scheme for **hurricane** intensity based on the maximum surface **wind speed** and the type and extent of damage done by the **storm**.

The wind speed categories are as follows: 1) 33–42 m s^{-1} (65–82 knots); 2) 43–49 m s^{-1} (83–95 knots); 3) 50–58 m s^{-1} (96–113 knots); 4) 59–69 m s^{-1} (114–134 knots); and 5) 70 m s^{-1} (135 knots) and higher. These categories are used routinely by weather forecasters in North America to characterize the **intensity** of hurricanes for the public.

SAGE—Abbreviation for **Stratospheric Aerosol Gas Experiment**.

sahel—1. The region of poor and intermittent **rains** south of the Sahara Desert in Africa.

2. A strong dust-bearing **desert wind** in Morocco.

sail—*See* **ridge**.

sailing directions—A descriptive book for the use of mariners, containing detailed information on a wide variety of material important to navigators of coastal and intracoastal waters.

Most of this information cannot be shown graphically on the standard nautical charts and is not readily available elsewhere. This information includes navigation regulations, outstanding landmarks, channel and anchorage peculiarities, dangers, weather, **ice**, currents, and port facilities. They are compiled and issued by national **hydrographic** authorities. For waters of the United States and its possessions, sailing directions are published by the National Ocean Service and are called **United States Coast Pilots**.

Saint Elmo's fire—*See* **St. Elmo's fire**.

salinity—A measure of the quantity of dissolved salts in **seawater**.

It is formally defined as the total amount of dissolved solids in seawater in parts per thousand (0/00) by weight when all the carbonate has been converted to oxide, the bromide and iodide to chloride, and all organic matter is completely oxidized. These qualifications result from the chemical difficulty in drying the salts in seawater. In practice, salinity is not determined directly but is computed from **chlorinity**, **electrical conductivity**, **refractive index**, or some other property with a relationship to salinity that is well established. The relationship between chlorinity Cl and salinity S as set forth in **Knudsen's tables** is

$$S = 0.03 + 1.805 \ Cl.$$

In 1940, however, a better expression for the relationship between total dissolved salts Σ and chlorinity was found to be

$$\sum = 0.07 + 1.811 \ Cl.$$

In more recent times, with the advent of devices that measure continuous records of conductivity electronically (e.g., **CTD** or **conductivity–temperature–depth profiler**), a new "practical salinity scale" has been determined. It is defined in terms of its electrical conductivity relative to a prescribed **standard** and it is given the units psu, for "practical salinity units." For most purposes one can assume that the new unit, psu, and the older unit, 0/00, are synonymous.

salinity bridge—An instrument for determining **salinity** of water (a **salinometer**) by measuring **electrical conductivity** of the water sample with a Wheatstone bridge.

salinometer—Any device or instrument for determining **salinity**, especially one based on **electrical conductivity** methods.

salt content—Concentration of dissolved salts in a body of water.

salt fingering—A form of **double-diffusive convection** that occurs when warm, salty water overlies cold **freshwater**.
   A **parcel** of freshwater moved upward will gain heat more quickly than it gains salt and so will become lighter than the surrounding water. It will rise unstably as a result. Salt fingering has been observed in the Caribbean Sea where it gives rise to stable layers hundreds of kilometers in extent. The process transports salt much more efficiently than **heat**. *See also* **Turner angle**.

salt haze—A **haze** created by the presence of finely divided **particles** of sea salt in the air, usually derived from the **evaporation** of sea spray.

salt lens—*See* **Mediterranean lenses, Meddy.**

salt seeding—A form of **cloud seeding** in which salt **particles** are released into the **atmosphere** with the objective of increasing **precipitation**.
   The salt particles are **hygroscopic** and readily condense water and grow large enough to become centers for **coalescence** growth of precipitation, or participate in the **ice accretion** process. *See* **cloud seeding, weather modification.**

saltation—*See* **sandstorm.**

saltwater intrusion—Movement of saline water into an **aquifer**.
   See also **seawater intrusion.**

SAM—Abbreviation for **Stratospheric Aerosol Measurement.**

sample—In **statistics**, a group of observations selected from a statistical **population** by a set procedure.
   Samples may be selected at **random** or systematically. The sample is taken in an attempt to estimate the population. *See* **random sample.**

SAMS—Abbreviation for **Stratospheric and Mesospheric Sounder.**

San Marco satellites—A series of satellites from the 1960s developed jointly by the United States and Italy.

sand auger—*See* **dust whirl, dust devil.**

sand devil—*See* **dust devil.**

sand haze—Reduced **visibility** in the **atmospheric boundary layer** caused by suspended **particles** of soil, mixed into the air during strong winds.
   Sand haze is particularly prevalent in **desert** regions where there is little moisture and few plants to hold the sand grains to the surface. After a **sandstorm** the larger sand grains will fall out of the air quickly, leaving a sand haze of medium size particles (1–100 μm diameters, including **silt** and fine sand) and small particles (< 1 μm diameters, including clay particles). *See* **sandstorm, haboob, aerosol, airborne particulates.**

sand mist—Same as **bai.**

sand pillar—*See* **dust devil.**

sand snow—**Snow** that has fallen at very cold temperatures (of the order of −25°C).
   A surface cover of this snow has the consistency of **dust** or light, dry sand. *See* **wild snow.**

sand wall—The leading edge of a **gust front** that is filled with suspended sand **particles** as is often associated with **thunderstorm** winds over **desert** regions.

The sand wall is the leading edge of the **haboob**, and looks like a knobby, dun-colored wall of turbulent air. *See* **saltation, aerosol, airborne particulates, downburst**.

sand whirl—(Obsolete.) A **dust devil**.

sandstorm—A strong **wind** carrying sand through the air.

The diameter of most of the **particles** ranges from 0.08 to 1 mm. In contrast to a **duststorm**, the sand particles are mostly confined to the lowest 3.5 m (10 ft), rarely rise more than 15 m (50 ft) above the ground, and proceed mainly in a series of leaps (**saltation**). Sandstorms are best developed in **desert** regions where there is loose sand, often in dunes, without much admixture of **dust**. Sandstorms are due to strong winds caused or enhanced by surface heating and tend to form during the day and die out at night.

sansar—(Also called sarsar, shamsir.) "Icy **wind** of death"; a northwest **wind** of Iran.

Santa Ana—In the United States, a dry, foehnlike **desert wind** in southern California, generally blowing from the northeast or east, especially in the pass and river valley of Santa Ana, California, and other nearby passes, where it is further modified as a **mountain-gap wind**.

It is driven by strong **pressure gradients** from an **anticyclone** over the Great Basin of the western United States. It blows, often hot and sometimes with great force, from the deserts to the east of the Sierra Nevada Mountains and may carry a large amount of **dust**. The combination of **heat**, dryness, and strong winds make it an especially hazardous **fire weather** condition. It most frequently occurs in late fall and winter (October–March); when it comes in spring, however, it can do great damage to fruit trees.

Santa Rosa storm—In Argentina, an "annual" **storm** near the end of August.

saoet—*See* **aloegoe**.

SAR—Abbreviation for **synthetic aperture radar**.

sàrca—A violent north **wind** of Lake Garda in Italy.

Sargasso Sea—The region (actually in the **horse latitudes**) of the North Atlantic Ocean to the east and south of the **Gulf Stream** system.

This is a region of **convergence** of the surface waters, and is characterized by clear, warm water, a deep blue color, and large quantities of floating sargassum or "gulf weed."

sarsar—Same as **sansar**.

sastruga—A single **sastrugi** ridge (in some English-language articles).

sastrugi—Sharp irregular ridges formed on a **snow** surface by **wind erosion** and deposition.

The ridges are parallel to the direction of the prevailing **wind**. The Russian word (zastrugi) is a collective noun and lacks a singular form. An individual ridge in a field of sastrugi has been called a sastruga in some English-language articles.

satan—*See* **dust whirl**.

satellite—1. An artificial platform placed into **orbit** around the earth, often carrying instruments to gather environmental data.

2. Any natural or man-made object that **orbits** about an astronomical body.

satellite image navigation—Identifying the geographic frame of reference and latitude and longitude coordinates that correspond to the scan coordinates of a satellite image.

Satellite Infrared Spectrometer—(Abbreviated SIRS.) The first space-based, grating **spectrometer** used for vertical **temperature** soundings of the **atmosphere**.

SIRS was flown on *Nimbus-3* and *-4* launched in April 1969 and April 1970, and was a forerunner of the *HIRS-2* operational instrument on the **TIROS**-N satellites since 1978.

satellite meteorology—The use of artificial earth-orbiting satellites for the purposes of imaging the atmospheric, land, and oceanic systems; providing atmospheric profiling; and collecting and relaying environmental data.

Satellite meteorology involves the sampling of weather and **climate** features in time and space, as well as the development of new algorithms and interpretation methods, satellite sensors, and products for weather and weather applications.

Satellite pour l'Observation de la Terre—(Abbreviated SPOT.) A series of French satellites using visible and **near-infrared** wavelength instruments to gather high spatial **resolution** images of the earth.

The primary SPOT imaging instrument consists of two identical visible sensors that can record **digital** images in either a multispectral mode (3 channels, each with a ground resolution of 20 m) or a panchromatic mode (with ground resolution of 10 m).

satellite sounding—Measurements of various atmospheric parameters (such as **temperature** or **water vapor**) at various heights or **pressure** levels, as derived from satellite-based sensors.

satellite subpoint—*See* **subpoint**.

satellite wind estimate—*See* **cloud motion vector**.

satellite winds—*See* **cloud motion vector**.

satellite zenith angle—The angle between a straight line from a point on the earth's surface to the satellite and a line from the same point on the earth's surface that is perpendicular to the earth's surface at that point (the **zenith** point).
Compare **solar zenith angle**.

satin ice—Same as **acicular ice**.

saturated soil—Soil that has all its micropore and macropore spaces filled with water.

The soil cannot retain additional water until water is deleted from some of these pores. Saturated soils promote anaerobic conditions.

saturated zone—A geological formation where the pore spaces are fully saturated with water, that is, **groundwater**.

saturation—The condition in which **vapor pressure** is equal to the **equilibrium vapor pressure** over a plane surface of pure liquid water, or sometimes **ice**.

saturation adiabat—Same as **moist adiabat**.

saturation-adiabatic lapse rate—Same as **moist-adiabatic lapse rate**.

saturation-adiabatic process—Same as **moist-adiabatic process**.

saturation deficit—1. The amount by which the **water vapor** in the air must be increased to achieve **saturation** without changing the environmental **temperature** and **pressure**.

The saturation deficit may be expressed in terms of a **vapor pressure** deficit, an **absolute humidity** deficit, or a **relative humidity** deficit.

2. The physiological saturation deficit: "The difference between the amount of **vapor** actually present in the air (i.e., the **absolute humidity**) and amount that saturated air at body **temperature** contains (viz., about 45 gm per cubic m)."
Kendrew, W. G., 1930: *Climate*, p. 189.

saturation equivalent potential temperature—A thermodynamic variable θ_{es} defined as

$$\theta_{es} = \theta + \left(\frac{L_v\theta}{C_pT}\right)r_{sat}$$

where θ is **potential temperature**, L_v is the **latent heat** of **vaporization**, C_p is **specific heat** at constant **pressure**, T is **absolute** temperature, and r_{sat} is the theoretical value of the **saturation mixing ratio** of water in air of **temperature** T at pressure p.
Compare **saturation static energy**, **equivalent potential temperature**.

saturation level—The **altitude** (and its corresponding **pressure** P_{sat}) to which an **air parcel** must be lifted **dry-adiabatically** or lowered **moist-adiabatically** to be just saturated (100% **relative humidity** with no liquid water present).

For unsaturated air, this is commonly known as the **lifting condensation level**. Saturation level is a conserved **variable** that does not change during **adiabatic** lifting or lowering of saturated or unsaturated air and can thus be used as a **tracer** for that air parcel. When paired with the corresponding **saturation** air **temperature** at that altitude, the result is a **saturation point** that can be represented on a **thermodynamic diagram**.

saturation mixing ratio—A thermodynamic **function of state**; the value of the **mixing ratio** of saturated air at the given **temperature** and **pressure**.

This value may be read directly from a **thermodynamic diagram**. Without specific qualification, the saturation mixing ratio refers to **saturation** with respect to a plane surface of pure water. Saturation mixing ratios may also be specified with respect to a plane **ice** surface.

saturation point—A point on a **thermodynamic diagram** representing the state of the air defined by a pair of variables: P_{sat} and T_{sat}, where P_{sat} is the **saturation pressure** and T_{sat} is the **saturation temperature**.

These latter variables are the **pressure** and **temperature** of an **air parcel** that was raised **dry-adiabatically** or lowered **moist-adiabatically** until it was just at **saturation**. The **saturation point** is a conserved **variable** for **adiabatic** motion and thus serves as a **tracer** for the air parcel on a thermodynamic diagram.

saturation pressure—The **pressure** at the **saturation point**.

saturation signal—A received **signal** with a power level that exceeds the **dynamic range** of the **receiver**. For such a signal, any increase in the power level causes no appreciable change in the **output** of the receiver.

saturation specific humidity—The **specific humidity** of **water vapor** corresponding to the **saturation mixing ratio**.

saturation static energy—A thermodynamic variable s_{es} defined as

$$s_{es} = C_p T + gz + L_v r_{sat}$$

where g is gravitational **acceleration**, z is height above some reference level (often taken as the height where the **pressure** is 100 kPa), L_v is the **latent heat** of **vaporization**, C_p is **specific heat** at constant pressure, T is **absolute** temperature, and r_{sat} is the theoretical value of **saturation mixing ratio** of water in air of **temperature** T at pressure p.

Compare **saturation equivalent potential temperature, dry static energy, Montgomery streamfunction, moist static energy, liquid water static energy**.

saturation temperature—The **temperature** at the **saturation point**.

saturation vapor pressure—1. The **vapor pressure** of a system, at a given **temperature**, for which the **vapor** of a substance is in **equilibrium** with a plane surface of that substance's pure liquid or solid **phase**; that is, the vapor pressure of a system that has attained **saturation** but not **supersaturation**.

The saturation vapor pressure of any pure substance, with respect to a specified parent phase, is an intrinsic property of that substance, and is a function of temperature alone. For **water vapor**, the **saturation pressure** over supercooled liquid differs appreciably from that over **ice**. *Compare* **equilibrium vapor pressure, vapor tension;** *see* **Clausius–Clapeyron equation**.

2. Same as **equilibrium vapor pressure**.

savanna—(Also spelled savannah.) A tropical or subtropical region of grassland and other drought-resistant (xerophilous) vegetation.

This type of growth occurs in regions that have a long **dry season** (usually "winter-dry") but a heavy **rainy season**, and continuously high temperatures.

savanna climate—Same as **tropical savanna climate**.

Savart polariscope—A **polariscope** consisting of a specially constructed double plate polarizer and a tourmaline plate analyzer.

Polarized **light** passing through the instrument is indicated by the presence of parallel colored fringes, while unpolarized light results in a uniform **field**. *See* **Voss polariscope**.

sawtooth wave—A **waveform** that increases linearly with time for a fixed interval, returns abruptly to the original level, and repeats the process periodically, producing a shape resembling the teeth of a saw.

SBL—Abbreviation for **stable boundary layer**.

SBL strength—The **air temperature** decrease near the surface when the surface is colder than the air.

For a nocturnal **SBL** over land, it can be measured as the **potential temperature** difference between the **residual layer** and the air temperature near the ground. **Inversion** strength increases as greater cooling continues over longer times and decreases with increasing **wind speed** and turbulent **mixing**.

SBLI—Abbreviation for **surface-based lifted index**.

SBUV—Abbreviation for **Solar Backscatter Ultraviolet Radiometer**.

scalar—Any physical quantity with a **field** that can be described by a single numerical value at each point in space.

A scalar quantity is distinguished from a **vector** quantity by the fact that a scalar quantity possesses only magnitude, whereas a vector quantity possesses both magnitude and direction. Thus, **pressure** is a scalar quantity and **velocity** is a vector quantity.

scalar product—(Also called dot product, direct product, inner product.) A **scalar** equal to the product of the magnitudes of any two **vectors** and the cosine of the angle θ between their positive directions.

For two vectors **A** and **B**, the scalar product is most commonly written **A** · **B**, read "**A** dot **B**," and occasionally as (**AB**). If the vectors **A** and **B** have the components A_x, B_x, A_y, B_y, and A_z, B_z along rectangular Cartesian x, y, and z axes, respectively, then

$$\mathbf{A} \cdot \mathbf{B} = A_x B_x + A_y B_y + A_z B_z = |\mathbf{A}||\mathbf{B}| \cos \theta = AB \cos \theta.$$

If a scalar product is zero, one of the vectors is zero or else the two are perpendicular. *See* **vector product**.

scale—1. Regular markings on an instrument used to allow the reading of the measured quantity or setting.

2. A factor that relates the indication of the measuring instrument to the value of the quantity.

3. An order of magnitude aid in estimating meteorological parameters (e.g., **mesoscale**).

scale analysis—An analysis method usually using the nondimensional equations to determine which terms are dominant for a particular phenomenon or situation so that the smaller terms can be neglected, resulting in a simplified set of equations.

For example, the **quasigeostrophic equations** were derived by a scale analysis.

scale factor—(Also called map scale, map factor.) The ratio between the distance separating two points on a map of the earth's surface at a standard latitude (or latitudes) and the distance between the corresponding points on the earth's surface itself.

For most map projections, this factor is a slowly varying function of latitude, and for synoptic-chart base maps it is usually of the order of 10^{-6} to 10^{-7}. The scale factor is sometimes defined as the reciprocal of the above ratio. *Compare* **conformal map**.

scale height—The height within which some **parameter**, such as **pressure** or **density**, decreases by a factor $1/e$ in an **isothermal atmosphere**.

The term is most often used in an ionospheric context, but is equally applicable to the **neutral atmosphere**. It is a measure of the effective "thickness" of an **atmospheric layer** and is expressed mathematically as

$$H = \frac{kT}{mg} = \frac{R^*T}{Mg},$$

where T, m, and g are, respectively, the **absolute** temperature, the molecular mass, and the **acceleration** due to **gravity**; R^* is the **universal gas constant**; and k is **Boltzmann's constant**.

scale of turbulence—*See* **turbulence length scales**.

SCAMS—Abbreviation for **Scanning Microwave Spectrometer**.

scan line—1. One sweep of a **radiometer** across the area or volume being viewed.

Many radiometers on satellites scan across the earth from **horizon** to horizon.

2. The narrow strip that is swept by the **instantaneous field of view** of a detector in a **scanner** system mounted on the ground or on a satellite or aircraft.

scanner—A **sensor** with a narrow **field of view** that must scan to enlarge the portion of the **atmosphere** or earth being measured.

scanning—In **radar meteorology**, the motion of the **radar** antenna during data collection.

Scanning usually follows a systematic pattern involving one of the following: 1) In horizontal scanning, used to generate **PPI** displays, the **antenna** is continuously rotated in **azimuth** around the **horizon** or is rotated back and forth in a sector (sector scanning); at the completion of each 360° or **sector scan**, the **elevation angle** of the scan typically is increased; 2) Vertical scanning, used to generate **RHI** displays, is accomplished by holding the azimuth constant while continuously varying the elevation angle of the antenna; at the completion of each vertical scan, the azimuth typically is incremented and the vertical scan proceeds in the opposite direction. *See also* **volume scan**.

Scanning Microwave Spectrometer—(Abbreviated SCAMS.) A five **channel** scanning **radiometer** on *Nimbus-6* (launched June 1975) used to measure temperatures over ocean surfaces, **water vapor**, and liquid water.

SCAMS was a precursor to the **MSU** used operationally on the **TIROS**-N (**NOAA** series) satellites since 1978.

Scanning Multifrequency Microwave Radiometer—(Abbreviated SMMR.) An instrument designed to measure **precipitation** rate, column-integrated liquid water and **water vapor**, **sea surface temperature** and **wind speed**, **soil moisture**, and **sea ice** concentration.
    This **radiometer** flew on **Seasat** (launched in June 1978) and *Nimbus-7* (launched in October 1978).

scanning radiometer—A **radiometer** that used a rotating or an oscillating plane mirror to scan a path normal to the movement of the instrument.

SCAPE—Abbreviation for **slantwise convective available potential energy**.

scarf cloud—Same as **pileus**.

scatter—1. Same as **scattering**; or, sometimes used in referring to the **scattered radiation**.
    2. The relative **dispersion** of points on a graph, especially with respect to a **mean value**, or any curve used to represent the points.
    See **scatter diagram**, **spread**.

scatter angle—(Or scattering angle.) The angle between any given **ray** of **scattered radiation** and the incident ray.
    Convention varies as to whether this angle is measured with respect to the direction in which the incident **radiation** was advancing or with respect to the direction from scatterer to radiation source. *See* **scattering**.

scatter diagram—(Also called scattergram.) A plot representing corresponding observed values of two variables x and y as points in **Cartesian coordinates**.
    If the two variables are functionally related, the points will be bunched, but if they are not functionally related, the points will be scattered uniformly over the plane. Scatter diagrams are used to explore the influence of one **variable** upon another, strong relationships being revealed as a concentration around a definite curve.

scatter propagation—Propagation of **electromagnetic waves** by irregularities in the **refractive index** of the propagation medium.
    See **tropospheric scatter**.

scattered—1. In **radiation**, *see* **scattering**.
    2. A sky coverage of 1/8 through 4/8.
    In U.S. weather observing procedures, this is reported with the contraction "SCT."

scattered power—In **radar**, the **power** scattered by a **target**, or the amount of this power arriving at the **antenna** of the **receiver**.

scattered radiation—Radiation that has undergone one or more **scattering** processes.
    See also **diffuse radiation**, **single-scattering**, **multiple-scattering**.

scattergram—Same as **scatter diagram**.

scattering—In a broad sense, the process by which matter is excited to radiate by an external source of **electromagnetic radiation**, as distinguished from **emission** of **radiation** by matter, which occurs even in the absence of such a source.
    By this definition, **reflection**, **refraction**, and even **diffraction** of **electromagnetic waves** are subsumed under scattering. Sometimes scattering is applied in a restricted sense to that radiation not accounted for by the laws of **specular reflection** and refraction, which are approximate because matter is not continuous on all scales. Often the term **scattered radiation** is applied to that radiation observed in directions other than that of the source and may also be applied to **acoustic** and other waves. If there is no change in **frequency** between the incident and scattered radiation, the scattering is sometimes said to be elastic; the converse is inelastic. Scattering is also applied to any interaction between **particles** that results in a change in direction. *See* **multiple scattering**, **Mie theory**, **Rayleigh's scattering law**.
        Born, M., and E. Wolf, 1965: *Principles of Optics*, 3d rev. ed., 98–108.
        Doyle, W. T., 1958: *Am. J. Phys.*, **53**, 463–468.

scattering area coefficient—(Obsolete.) The **scattering efficiency factor**.

scattering coefficient—A measure of the **extinction** due to **scattering** of **monochromatic radiation** as it traverses a medium containing scattering **particles**.

Usually expressed as a **volume scattering coefficient** with units of reciprocal length (i.e., area per unit volume), but also as a **mass scattering coefficient** with units of area per unit mass.

scattering cross section—The hypothetical area normal to the incident **radiation** that would geometrically intercept the total amount of radiation actually scattered by a **scattering** particle.

scattering efficiency factor—The ratio of the **scattering cross section** to the geometrical cross section.

Scattering efficiency factors increase with the sixth power of the radius for **Rayleigh scatterers** and tend to a value of 2 for nonabsorbing scatterers in the **geometric optics** limit.

scattering function—A function characterizing the angular distribution of **scattered radiation** in terms of the **scattering** angle; usually used in its normalized form as the **phase function**.

scattering matrix—A 2 × 2 matrix comprising four **signal** amplitudes that characterizes the **scattering** from a **target** in terms of a **polarization** basis.

The basis is defined by two **orthogonal** vectors, for example, horizontal and vertical or right and left circular. The diagonal terms are the amplitudes of the copolarized signals corresponding to the two transmitted polarizations, and the off-diagonal terms are the amplitudes of the cross-polarized signals. Each term carries a double subscript that denotes the transmitted and received polarization. *Compare* **covariance matrix**.

scatterometer—A **radar** system that infers **wind speed** by measuring the **backscattering cross section** (or normalized **radar cross section**).

scavenging by precipitation—Removal of **pollutants** from the air by either **rain** or **snow**.

Rainout (or snowout), which is the in-cloud capture of **particulates** as **condensation nuclei**, is one form of scavenging. The other form is washout, the below-cloud capture of **particulates** and gaseous pollutants by falling **raindrops**. Large **particles** are most efficiently removed by **washout**. Small particles (especially those less than 1 μm in diameter) more easily follow the airstream flowing around raindrops and generally avoid capture by raindrops except in **heavy rain** events.

scavenging coefficient—A **parameterization** of the rate of loss of gases or **aerosol** particles from the **atmosphere** by their incorporation into larger drops, for example, **rain** or other forms of **precipitation**.

scharnitzer—A cold, northerly **wind** of long duration in Tirol.

Schlernwind—East **wind** blowing down from the Schlern near Bozen in Tirol.

schlieren—Parcels or strata of air having densities sufficiently different from that of their surroundings that they may be discerned by means of **refraction** anomalies in transmitted **light**.

All of the natural **scintillation** phenomena in the **atmosphere** result from the presence of density schlieren developed by turbulent processes. The schlieren method is an experimental technique for optically detecting the presence of slight **density**, and hence **temperature** and/or **pressure**, variations in gases and liquids by virtue of refraction effects.

Schmidt number—A **dimensionless number** relating the ratio of inertial to molecular diffusive forces.

Explicitly,

$$Sc = UL/\kappa_D,$$

in which U and L are **characteristic velocity** and length scales, respectively, and κ_D is the **diffusion** constant of a solute in solution, such as salt in water.

Schuman–Ludlam limit—Critical liquid **water content** for growth by **accretion** of a **hailstone** or other object in a **supercooled cloud** such that for a given **temperature** and air speed all the accreted water is frozen and the **surface temperature** is above 0°C.

Above this limit more water is accreted than can be frozen and either is shed as liquid or is retained as **spongy ice**.

Schumann resonance—**Electromagnetic waves** in the **ELF** range trapped in the spherical cavity formed by the conductive earth and the conductive **ionosphere**.

The fundamental **resonance** mode represents one **wavelength** around the earth with a corresponding **frequency** of approximately 8 Hz. The resonances are maintained continuously by global **lightning** activity. Five to seven higher-order modes are frequently discernible.

scintillation—Rapid fluctuations in the **amplitude** and **phase** of electromagnetic or acoustic waves that have propagated through a medium containing fluctuations in **refractive index**, such as the **atmosphere**.

The most common example of optical scintillation is the "twinkling" of stars observed through the atmosphere. Scintillation arises as a result of **random** angular **scattering** produced by refractive index fluctuations. For **electromagnetic wave** propagation, these result from fluctuations in **temperature** and, especially at **far infrared** and radio frequencies, **humidity**. Scintillation in **acoustic wave** propagation arises from **velocity** and temperature fluctuations. Fluctuations in the amplitude of different **frequency** components in the **spectrum** of an object can give rise to apparent changes in its color (**chromatic scintillation**); an example is the **random** red and blue twinkling of bright stars near the **horizon**. Scintillation **statistics** have been used to study **turbulence** in regions ranging from the **planetary boundary layer** to the **ionosphere**, as well as interplanetary and interstellar space. Scintillation is important for astronomical imaging, optical and radio communications, **laser** and acoustical propagation, active and passive **remote sensing**, and the performance of the **Global Positioning System**.

scintillometer—(Also called scintillation meter.) A type of **photoelectric photometer** used in a method of determining high-altitude winds on the assumption that **stellar scintillation** is caused by atmospheric inhomogeneities (**schlieren**) being carried along by the **wind** near **tropopause** level.

In one design a star is observed with a single telescope that looks through a four-inch long rotatable slit. Two **scintillation** frequencies are monitored, one high and one low. The slit is rotated until the ratio of high-frequency to low-frequency scintillation is a minimum. At this point, the slit is oriented parallel to the **wind**. The magnitude of the **frequency** ratio is a function of the **wind speed**. In another design, two telescopes mounted four inches apart are focused on a star. Only one frequency need be monitored. By direct observation of **phase** relationships as the telescopes are rotated about a mutual axis, both direction and speed can be determined with less ambiguity and uncertainty than with the single telescope type.

scirocco—Same as **sirocco**.

Sciron—Same as **Skiron**.

SCMR—Abbreviation for **Surface Composition Mapping Radiometer**.

Scofield–Oliver technique—A manual technique designed to estimate **rainfall** from convective precipitating systems using consecutive enhanced **infrared** imagery from **GOES**.

Scofield, R. A., and V. J. Oliver, 1977: A scheme for estimating convective rainfall from satellite imagery. *NOAA Technical Memorandum NESS 86*, U.S. Dept. of Commerce, Wash., D.C., 47 pp.
Scofield, R. A., 1987: The NESDIS operational convective precipitation technique. *Mon. Wea. Rev.*, **115**, 1773–1792.

scope—The general abbreviation for an instrument of viewing, such as telescope, microscope, or **oscilloscope**.

In **radar** installations, the **cathode-ray oscilloscope** indicators are commonly referred to as scopes or radarscopes.

Scorer parameter—The quantity $l(z)$ arising from the **wave equation** for atmospheric **gravity waves** describing flow over a mountain barrier:

$$l^2(z) = N^2/U^2 - (\partial^2 U/\partial z^2)/U,$$

where $N = N(z)$ is the **Brunt–Väisälä frequency** and $U = U(z)$ is the **vertical profile** of the horizontal **wind**, both quantities determined from an **atmospheric sounding** upstream of the barrier.

The first term on the right-hand side usually dominates, but occasionally the second term, the velocity-profile curvature term, can be of similar magnitude. When l^2 is nearly constant with height, conditions are favorable for vertically propagating mountain waves. This **parameter** is most often used, however, as an indicator of when trapped lee waves (*see* **mountain wave**) can be expected; they occur when $l^2(z)$ decreases strongly with height. This is especially true if this decrease occurs suddenly in mid **troposphere**, dividing the troposphere into two regions, a lower layer of large $l^2(z)$ (high stability) and an upper layer of small $l^2(z)$ (low stability). l, the square root of the parameter, has units of **wavenumber** (inverse length), and the wavenumber of the resonant lee wave lies between l of the upper layer and l of the lower layer—the equivalent **wavelength** generally lying between 5 and 25 km in the **atmosphere**. Mountain ranges wide enough to force wavelengths long relative to l_{upper} (the l in the upper layer) produce vertically propagating waves with wavenumbers less than l_{upper}. Small objects (that force wavenumbers greater than l_{lower}) produce waves that are evanescent, or vanishing with height.

Scotch mist—A combination of thick **mist** (or **fog**) and heavy **drizzle** occurring frequently in Scotland and in parts of England.

In the Devon-Cornwall peninsula, the same phenomenon is referred to as "mizzle."

scotopic vision—Vision mediated by rods alone at very low levels of **luminance**.

Rods do not allow color vision. *See* **dark adaptation**.

SCR—Abbreviation for **Selective Chopper Radiometer**.

screen—1. *See* **cathode-ray tube**.

2. *See* **instrument shelter**.

screened pan—An **evaporation pan** the top of which is covered by wire mesh or screening to prevent animals from drinking the water and debris from falling into it.

The screening reduces air circulation and **insolation**, and increases the **pan coefficient**.

screening layer—Space charge layer on the surface of a **cloud** opposite in **polarity** to the main charge inside the cloud.

This can lead to the underestimation of the magnitude of the main charge when measured remotely.

scrubber—A **flue gas** system that removes most of the sulfur dioxide, SO_2, from exhaust gases; generally used for large coal-fired boilers.

See also **limestone scrubbing**.

scud—Ragged low **clouds**, usually **stratus fractus** or **cumulus fractus**, that occur below the main **cloud base**.

They are often found in the vicinity of rainshafts. Several mechanisms may explain their occurrence. They may represent rising air that has greater-than-average **humidity** due to the **evaporation** of **rain** or water on the ground. They may evaporate before reaching the main cloud due to **mixing** with the intervening **dry air**. Scud may also be attributed to the breakup of **raindrops**. The breakup fragments that are able to accumulate in regions of high humidity in rainshafts may be the source of some scud. *See* **pannus**.

SCV—Abbreviation for **submesoscale coherent vortices**.

sea—1. Same as ocean.

2. A subdivision of an ocean.

All seas except "inland seas" are physically interconnected parts of the earth's total saltwater system. Two types are distinguished, mediterranean and adjacent. Mediterraneans are groups of seas, collectively separated from the major water body as an individual sea. Adjacent seas are those connected individually to the larger body.

3. Same as **state of the sea**.

4. Sea **surface waves** within their **fetch**; opposed to **swell**.

See **fully developed sea**.

sea breeze—A coastal local **wind** that blows from sea to land, caused by the **temperature** difference when the sea surface is colder than the adjacent land.

Therefore, it usually blows on relatively **calm**, sunny, summer days, and alternates with the oppositely directed, usually weaker, nighttime **land breeze**. As the sea breeze regime progresses, the wind develops a component parallel to the coast, owing to the **Coriolis deflection**. *See* **lake breeze, brisa, doctor, virazon, sea-breeze front**.

Defant, F., 1951: *Compendium of Meteorology*, 655–672.

sea-breeze front—The horizontal **discontinuity** in **temperature** and **humidity** that marks the leading edge of the intrusion of cooler, more moist marine air associated with a **sea breeze**.

It might also have a **wind** shift, **pollution** confluence, and enhanced **cumulus** clouds along its leading edge.

sea clutter—(Also called sea echo, sea return.) Radar **echoes** from the surface of the sea caused by **scattering** from waves, ripples, and spray.

sea echo—Same as **sea clutter**.

sea fog—A type of **advection fog** formed when air that has been lying over a warm water surface is transported over a colder water surface, resulting in cooling of the lower layer of air below its **dewpoint**.

sea horizon—*See* **horizon**.

sea ice—1. Specifically, **ice** formed by the **freezing** of **seawater**; as opposed, principally, to **land ice**.

In brief, it forms first as **lolly ice** (**frazil** crystals), thickens into **sludge**, and coagulates into **sheet ice**, **pancake ice**, or into **floes** of various shapes and sizes. Thereafter, sea ice may develop into **pack ice** and/or become a form of **pressure ice**.

2. Generally, any **ice** floating in the sea.

sea level—The level of the sea after averaging out the short-term variations due to **wind** waves.

It is used loosely as a synonym for **mean sea level**.

sea level chart—Same as **surface chart**.

sea level horizon—*See* **horizon**.

sea level measurements—Averaging out the effects of waves to determine sea level.

The measurements can be made in several ways, for example, for the reading of **tide** poles, the averaging is by eye, in **stilling wells** the waves are damped out by a narrow constriction, acoustic and **pressure** gauges may apply electronic averaging to rapid samples, and satellite altimetry is corrected for general **wave** conditions within the **footprint** of the transmission.

sea level pressure—The **atmospheric pressure** at **mean sea level**, either directly measured or, most commonly, empirically determined from the observed **station pressure**.

In regions where the earth's surface is above **sea level**, it is standard observational practice to reduce the observed **surface pressure** to the value that would exist at a point at sea level directly below if air of a **temperature** corresponding to that actually present at the surface were present all the way down to sea level. In actual practice, the **mean temperature** for the preceding 12 hours is employed, rather than the current temperature. This "reduction of **pressure** to sea level" is responsible for many anomalies in the **pressure field** in mountainous areas on the **surface synoptic chart**.

sea level pressure chart—Same as **surface chart**.

sea level rise—Long-term increases in **mean sea level**.

The expression is popularly applied to anticipated **eustatic sea level changes** due to the **greenhouse effect** and associated global warming.

sea rainbow—Same as **marine rainbow**.

sea return—Same as **sea clutter**.

sea-salt nucleus—A **condensation nucleus** of a very **hygroscopic** nature produced by partial or complete **desiccation** of **particles** of sea spray or of **seawater** droplets derived from breaking **bubbles**.

That such nuclei are important in **condensation** processes over the oceans and near coasts is fairly well established, but it appears that they are not in general the source of nuclei in condensation over continental interiors.

sea smoke—(Also antarctic sea smoke, arctic sea smoke, arctic smoke.) Same as **steam fog**, but often specifically applied to steam fog rising from small areas of open water within **sea ice** in the Arctic or Antarctic region.

sea state—A description of the properties of sea **surface waves** at a given time and place.

This might be given in terms of the **wave spectrum**, or more simply in terms of the **significant wave height** and some measure of the **wave period**.

sea surface temperature—The **temperature** of the ocean surface.

The term sea surface temperature is generally meant to be representative of the upper few meters of the ocean as opposed to the skin temperature, which is the temperature of the upper few centimeters.

sea turn—A **wind** coming from the sea, often bringing **mist**.

The use of this term is limited mainly to New England. *See* **back-door cold front**.

Sea-Viewing Wide Field-of-View Sensor—(Acronym SeaWiFS.) A multispectral ocean color **sensor** developed by NASA and carried on the *SeaStar* satellite.

The **scanner** produces imagery from six visible and two **near-infrared** spectral bands optimized for the study of ocean color.

search radar—A **radar** designed to determine the approximate location of objects, usually aircraft or ships.

The beams of such radars, called fan beams, are usually wider in the vertical than in the horizontal, making it possible to scan large volumes of space quickly. *Compare* **tracking radar**.

search space—In **expert systems**, the set of all possible rules, facts, and relationships as set out in the rules.

searchlight—In meteorology, same as **ceiling light**.

searchlighting—The procedure of continuously pointing a **radar beam** at a particular object or in a particular direction.

Seasat—A satellite designed to demonstrate the feasibility of global ocean monitoring from space.

After a successful launch on 28 June 1978, *Seasat* produced promising results, but the mission was terminated after only three months following a failure in the satellite's electrical system. Instrumentation flown on *Seasat* included a **radar altimeter**, a microwave **scatterometer** for measuring surface **wind speed** and **direction**, a **scanning** multichannel microwave **radiometer** for **sea surface temperature**, a visible and **infrared** radiometer, and the first spaceborne **synthetic aperture radar**.

Seasat **surface wind**—Surface **winds** above the ocean derived using the **Seasat**-A **scatterometer**.

season—A division of the year according to some regularly recurrent phenomena, usually astronomical or climatic.

> *See* **winter, spring, summer, autumn**.

SeaStar—A satellite system jointly developed by NASA and private industry to measure ocean color for research and operations.

Ocean color reveals the presence and concentration of phytoplankton, sediments, and dissolved organic chemicals in the uppermost layers of the ocean. Measurements are made with **SeaWiFS**, a multispectral imaging **spectrometer**. *SeaStar* was successfully launched on 1 August 1997.

seawater—The water of the seas, distinguished from **freshwater** by its appreciable **salinity**.

The distinction in usage of saltwater and seawater is not very sharply drawn. Commonly, seawater is used as the antithesis of specific types of freshwater, as river water, lake water, rainwater, etc., whereas saltwater is merely the antithesis of freshwater in general.

seawater intrusion—Movement of ocean water into a coastal **aquifer**, often caused by pumping of the aquifer.

> *See also* **saltwater intrusion**.

seawater thermometer—A **thermometer** designed for use in measuring the **temperature** of **seawater**.

One form of this instrument consists of a **mercury-in-glass thermometer** protected by a perforated metal case. This is used to measure the temperature of a sample of seawater. Another form consists of a mercury-in-glass thermometer surrounded by a metal case that forms a well around the bulb of the thermometer. When the thermometer is raised from the water, a sample is retained in the well for temperature measurement. *See* **reversing thermometer, bucket thermometer**.

SeaWiFS—Acronym for **Sea-Viewing Wide Field-of-View Sensor**.

SEC—Abbreviation for **South Equatorial Current**.

seca—A **drought**, or dry wind, in Brazil.

sécaire—*See* **mistral**.

SECC—Abbreviation for **South Equatorial Countercurrent**.

Secchi disk—A white disk, 30 cm (12 in.) or more in diameter, that is lowered into the sea to estimate transparency of the water.

The depths are noted at which it first disappears when lowered and reappears when raised. *See* **Forel scale**.

sechard—A dry, warm **foehn** wind over Lake Geneva in Switzerland.

seclusion—A special (and rare) case of the process of **occlusion**, where the point at which the **cold front** first overtakes the **warm front** (or **quasi-stationary front**) is at some distance from the apex of the **wave cyclone**.

An isolated mass of warm air is completely surrounded by colder air.

second-foot—A unit of water measure equal to 1 cubic foot per second, or about 449 gallons per minute.

It is the unit of **stream** discharge commonly used in the United States.

second-foot day—The volume of water represented by a flow of 1 cubic foot per second for 24 hours; equal to 86 400 cubic feet.

This is used extensively as a unit of **runoff** volume or **reservoir** capacity and is closely equivalent to 2 acre-feet.

second law of thermodynamics—An inequality that is fundamentally different from the **first law** because it specifies the direction in which a natural process will evolve rather than merely requiring that certain quantities are conserved.

As formulated by Planck, the second law asserts that a thermodynamic state function, S, known as **entropy**, exists for all physical systems. For the universe and for a system isolated from its surroundings,

$$\frac{dS}{dt} \geq 0.$$

Equality prevails only for reversible processes or when the system is in a **steady state**. When the universe of a system is a maximum, no further evolution of the system is possible. The second law is often asserted in other forms, including the following.

When two bodies at different temperatures interact, the **temperature** of the hotter body can only decrease and that of the colder body can only increase unless **work** is done.

No device can continuously deliver mechanical work and produce no effect other than cooling a **reservoir**.

In the neighborhood of every state that can be reached reversibly, there exist states that cannot be reached by a reversible, **adiabatic process**, or, in other words, that can be reached only irreversibly or cannot be reached at all.

Dutton, J. A., 1995: *Dynamics of Atmospheric Motion*, Dover Press, 45–51, 406–410.

Sommerfeld, A., 1964: *Thermodynamics and Statistical Mechanics*, Academic Press, 26–36, 39.

second-order climatological station—A **station** at which observations of **atmospheric pressure**, **temperature, humidity, winds, clouds**, and weather are made at least twice daily at fixed hours, and at which the daily maximum and minimum of **temperature**, the daily **amount of precipitation**, and the duration of **bright sunshine** are observed.

Compare **second-order station**; *see also* **first-order climatological station, third-order climatological station, climatological substation, precipitation station**.

second-order closure—A type of higher-order **turbulence closure** in which **prognostic equations** are retained for mean variables (such as **potential temperature** and **wind**), as well as for some of the higher-order **statistics**, including **variance** (such as **turbulence kinetic energy** or **temperature** variance) and **covariance** (kinematic fluxes such as for **heat** and **momentum**).

Any third- and higher-moment statistics remaining in the equations are approximated as functions of lower-order statistics and **independent variables**. *Compare* **first-order closure, K-theory, one-and-a-half order closure, nonlocal closure, closure assumptions**.

second-order reaction—A chemical reaction for which the rate of disappearance of reactants or the rate of appearance of the reaction product(s) is proportional to the product of the concentration of two reacting species, which may be the same.

The constant of proportionality is the second-order **rate coefficient** for the chemical reaction and has the units of inverse concentration multiplied by inverse time.

second-order station—After U.S. National Weather Service practice, a **station** staffed by personnel certified to make **aviation weather observations** and/or **synoptic weather observations**.

Compare **second-order climatological station**; *see also* **first-order station**.

second-trip echo—In pulsed radar, an **echo** from a given **pulse** that is not received until after the transmission of the next pulse.

See **maximum unambiguous range**.

second-year ice—Sea ice that has survived only one summer melt **season**.

secondary bands—**Squall lines** in the form of spirals accompanying **tropical cyclones**.

Secondary bands generally lack local swirling **wind** maxima and are fixed to the cyclone motion or the environmental **shear**, in contrast with the **principal band**.

secondary circulation—1. Atmospheric **circulation** features of **cyclonic scale**.

Use of the term is usually reserved for distinguishing between the various dimensions of atmospheric circulation, that is, **primary circulation, tertiary circulation**. *See also* **general circulation**.

2. A **circulation** induced by the presence of a stronger **circulation** as a result of dynamical constraints. A **frictional secondary flow** is an example.

3. Organized flow superimposed on a larger-scale mean circulation.

For example, **roll vortices** are a **secondary circulation** in the **atmospheric boundary layer**. They "fill" the **boundary layer** vertically, but have a width of only two to three times the boundary layer depth, while the mean wind **profile** extends over a much broader region.

secondary cold front—A **secondary front** that forms behind a **frontal cyclone** and within a cold **air mass** characterized by an appreciable horizontal **temperature** gradient.

These are not uncommon phenomena; however, they often appear more as weak **troughs** or **instability lines** embedded in the cold-air flow. It may be difficult to determine when or if these become true fronts.

secondary cyclone—(Or secondary low.) A **cyclone** that forms near, or in association with, a **primary cyclone**.

For example, secondary cyclones often form along the east coast of the United States when a primary cyclone is present in the Great Lakes region. Similarly, secondary cyclones often occur over the Baltic when a primary cyclone is present near the coast of Norway. *See also* **center jump**.

secondary depression—Same as **secondary cyclone**.

Usually used in the context of tropical **lows**.

secondary emission—Emission of subatomic **particles** (and/or **photons**) stimulated by primary **radiation**, for example, **cosmic rays** impinging on other particles and causing them, by disruption of their **electron** configurations or even of their nuclei, to emit particles and/or photons in turn.

secondary flow—*See* **secondary circulation**.

secondary front—A **front** that may form within a **baroclinic** cold **air mass**, which itself is separated from a warm air mass by a primary **frontal system**.

The most common type is the **secondary cold front**.

secondary hydrometric station—A temporary **hydrometric station** supplementing the data from permanent stations; usually used for specific projects.

secondary ice crystal—An **ice crystal** formed by a process other than homogeneous or **heterogeneous nucleation**, as by shatter of a **drop** freezing on **accretion**, or breakup of an **ice** particle on **evaporation**.

See **ice multiplication**.

secondary instrument—An instrument with **calibration** determined by comparison with an **absolute instrument**.

secondary low—Same as **secondary cyclone**.

secondary pollutants—**Pollutants** that are formed in the **atmosphere** as a result of chemical reactions.

Secondary pollutants are often **photochemical oxidants** such as **ozone** or **nitrogen dioxide**, or components of **acid rain** such as **sulfuric acid** or **nitric acid**.

secondary rainbow—Distinguished from other rainbows by its angular radius, color order, and **brightness**.

The bow is seen between about 50° and 54° from the antisolar point (shadow of the **observer's** head) or, equivalently, between 130° and 134° from a **light** source (such as the sun). Reds are found to the inside of the bow (closest to the primary bow) with the blues to the outside. The secondary bow is usually dimmer than the primary bow. Any theory of the bow that approximates the behavior of light as a **ray** attributes the secondary bow to light that has undergone two internal reflections. The losses accompanying the additional **reflection** account for the bow being fainter than the primary bow. This may mean that the bow has insufficient **contrast** to be distinguished from the background. It is sometimes incorrectly asserted that the extra reflection, by itself, is responsible for the reversal of colors from those of the primary bow. *See also* **rainbow, supernumerary rainbows**; *compare* **primary rainbow**.

secondary scattering—*See* **multiple scattering**.

sector scan—A **radar** scanning procedure in which the **antenna** sweeps through an interval of **azimuth** less than the full 360° at a fixed **elevation angle**.

Often the azimuth extent of a sector scan may be only a few degrees in order to concentrate the observations on a particular **storm** or region of a storm.

sector wind—The average observed or computed **wind** (direction and speed) at **flight level** for a given sector of an air route.

Sectors for over-ocean flights usually consist of 10 degrees of longitude. *Compare* **track wind**, **spot wind**.

sectorized image—Satellite imagery that is reformatted into subsections covering specified geographic areas.

secular trend—Same as **secular variation**.

secular variation—In general, a slow trend, the actual timescale implied being different for different contexts.

sediment—1. An accumulation of rock and mineral particles transported by water (fluvial sediment) or by **wind** (**aeolian** sediment).

2. A collective term for rock and mineral particles that 1) are being transported by a fluid (sediment in **transport**, **suspension**, or motion) caused by the fluid motion or 2) have been deposited by the fluid (i.e., sediment deposits).

sediment-carrying capacity—A conceptual term related to the ability of flowing water, as a **stream**, to **transport** sediment.

Literally, the carrying capacity of a stream is the amount of **sediment** it can transport under the given flow conditions. This is generally a descriptive term not used for quantitative statements of rates and amounts of sediment discharge, and its use is generally discouraged.

sediment concentration—A term used to describe the mass of **sediment** in a fluid per unit volume of the fluid.

For fluvial sediment, less commonly as the mass of sediment per mass of water as expressed by percent by weight. The sediment concentration is often expressed as kilograms of sediment per cubic meter of water, or more commonly, in terms of milligrams (mg) of sediment per liter of water.

sediment discharge rating—(Also called silt discharge rating.) The graphic or functional relationship between **sediment** discharge rate (mass per unit time) in flowing water and the corresponding water discharge rate in volume per unit time or mass per unit time.

The graphical relationship is called a rating curve and the functional relationship is called a rating equation or rating function. *Compare* **capacity of the wind**.

sediment grading—Literally, the grouping of **sediment** particles into size classes or grades.

sediment sampler—The apparatus or equipment used to take samples of the mixture of fluid and **sediment**.

sediment yield—1. Total quantity of **sediment** moving out of a **watershed** in a given time interval, expressed in units of mass per unit time.

2. Total **sediment** discharge from a **watershed** relative to the watershed area, expressed in units of mass per unit time per unit area.

sedimentation—The process of depositing material by water, **wind**, or **glaciers**.

sedimentation diameter—1. Because of irregular shapes of natural **sediment** particles, defined in terms of similar characteristics and behavior to "equivalent diameter" spherical particles.

2. Operationally, the diameter of a sphere of the same specific weight (weight per unit volume) and the same terminal **velocity** falling in the same fluid as the given **sediment** particle.

sedimentology—The science of the production, composition, **transport**, and deposition of **sediment**.

seeder–feeder—Orographic precipitation-enhancement mechanism, in which **precipitation** from an upper-level precipitating **cloud** (seeder) falls through a lower-level **orographic** stratus cloud (feeder) capping a hill or small mountain.

Precipitation droplets or **ice** particles fall from the higher seeder cloud and collect cloud water as they pass through the lower feeder cloud by collision and **coalescence** or **accretion**, thus producing greater precipitation on the hill under the **cap cloud** than on the nearby flat land. The effectiveness of the process depends on sufficiently strong low-level moist flow to maintain the cloud water content in the orographic feeder cloud and the continuing availability of precipitation particles from the seeder cloud.

seeding—*See* **cloud seeding**.

seeding agent—Material used for **cloud seeding**.

seeding devices—A generator used for producing **cloud seeding** particles and injecting them into **clouds**.

The type is selected for the particular **seeding** material, application, and delivery system. Different types of seeding devices have been used, including combustion burners, pyrotechnic flares, atomizers, dry powder dispersal systems, dry-ice crushers, and dispensers of liquids that cool substantially on **evaporation** (such as **propane** or liquid CO_2).

seeding rate—In a **cloud seeding** operation, seeding rate refers to the amount of seeding material released per unit time, per unit distance traveled, or per amount of air.

The amount of seeding material can be characterized by mass, volume, or number of **particles**.

seeing—(Or astronomical seeing.) A term long used by astronomers for the **degradation** of images by the **atmosphere**.

seepage face—A boundary between the **saturated zone** and the **atmosphere** along which **ground-water** discharges at practically **atmospheric pressure**.

seepage spring—Spring that discharges from a **saturated zone** over a relatively large area.

seepage velocity—(Also known as average actual flow velocity, average linear velocity.) The **discharge** (volume per unit time) per unit area of void space in a **porous medium**.

seiche—1. The **oscillation** of a body of water at its natural **period**.

Coastal measurements of **sea level** often show seiches with amplitudes of a few centimeters and periods of a few minutes due to oscillations of the local harbor, **estuary**, or bay, superimposed on the **normal** tidal changes.

2. In the Great Lakes area of the United States, any sudden rise in the water of a harbor or a lake, whether or not it is oscillatory.

Although inaccurate in a strict sense, this usage is well established in the Great Lakes area.

seismic sea wave—Same as **tsunami**.

seismograph—An instrument used to measure and record **earthquake** vibrations and other earth tremors.

It consists essentially of a heavy pendulum hung so that it can swing freely in one direction. A pen is attached to the end of the pendulum so that the pendulum's motion, relative to a recorder fixed solidly to the earth, may be determined. A seismological observatory requires three seismographs in order to measure the three component directions of the tremors and to fix approximately their origin. A study of the records of sensitive seismographs has shown that some of the tremors may be attributed to oceanographic and atmospheric phenomena. *See* **microseism**.

seistan—(Also called bad-i-sad-o-bistroz.) A strong **wind** of **monsoon** origin that blows from between the northwest and north-northwest and sets in about the end of May or early June in the historic Seistan district of eastern Iran and Afghanistan.

It continues almost without cessation until about the end of September. From its duration it is known as the wind of 120 days (bad-i-sad-o-bistroz). It sometimes reaches a **velocity** of more than 31 m s^{-1} (70 mph) and carries much sand and **dust**. This sand blast is very erosive; buildings are eaten away and undercut near the ground. In some places the sand is deposited as wandering dunes, which overwhelm buildings and choke water supplies. All buildings in the region have blank walls on the **windward side**. The seistan is associated (along with the European **etesians**, and the **shamal**) with the deep summertime **low** over northwest India.

selatan—(Also spelled slatan.) Strong, dry, southerly winds of the southeast **monsoon** in Indonesia.

selection rules—A set of rules that determine whether the **absorption** or **fluorescence** of **radiation** will be favorable.

The "rules" derive from a quantum mechanical description of **light** absorption. A given transition may be "allowed" or "forbidden." **Forbidden transitions** can occur, but with very much lower **intensity** than allowed ones. See also **allowed transitons**.

selective absorption—Absorption that varies with the **wavelength** of **radiation** incident upon the absorbing substance.

A substance that absorbs in such fashion is called a selective absorber and is to be contrasted with a **blackbody** or **graybody** absorber. **Water vapor, carbon dioxide**, and **ozone** are significant contributors to selective absorption in the earth's **atmosphere**.

Selective Chopper Radiometer—(Abbreviated SCR.) A 16-channel nonscanning **radiometer** flown on *Nimbus-5* (launched December 1972) for **sounding** the **atmosphere** in the **infrared** spectrum.

selective scattering—Scattering that varies with the **wavelength** of **radiation** incident upon the **scattering** particles.

In general, the largest and most complex degree of selectivity is found for wavelengths nearly equal to the diameter of the scattering **particles**.

selene—Suffix meaning moon or lunar in terms related to **meteorological optics**. The plural is selena. *Compare* **helion**.

self-collection—(Also called large **hydrometeor** self-collection.) A **cloud** modeling term for the growth of large **drizzle drops** by collision and **coalescence** with smaller drizzle drops.

This process is identified as being responsible for the rapid increase in the mode size of the **precipitation** liquid water spectrum. Self-collection, **autoconversion**, and **accretion** have been identified as the primary governors of the **collision–coalescence process**.

self-healing of atmosphere—Effect in which depletion of **ozone** at high altitudes is partially compensated for by an increased penetration of **solar radiation** to lower altitudes, thus increasing the rate of ozone production at the lower altitudes.

selsyn—1. Designation for a self-synchronous motor system, consisting of a driver motor and one or more remote followers (or repeaters) with armatures remaining in synchronization with that of the driver.

2. Informal Navy designation for a wind-measuring system consisting of a **wind vane** and a **bridled-cup anemometer**, both of which are coupled to selsyn drivers and remotely record or indicate by means of the repeaters of the selsyn system.

selvas—The **tropical rain forest** of South America.

SEM—Abbreviation for **Space Environmental Monitor**.

semantic network—A labeled, directed graph with nodes representing physical or conceptual objects and labeled arcs representing relations between objects.

This permits the use of generic rules, inheritance, and object-oriented programming, which in turn allow the development of meteorological **expert systems** that can be adapted to more than one location.

semi-implicit method—A finite difference approximation in which some terms producing time change are specified at an earlier time level.

The approximation $(f^{n+1} - f^{n-1})/2\Delta t = g(f^{n+1}) + h(f^n)$ (where superscript n denotes a point in time, separated by step Δt from the prior $[n - 1]$ and subsequent $[n + 1]$ discrete time level) is an example of a semi-implicit approximation to $df/dt = g(f) + h(f)$. Semi-implicit approximations may increase computational efficiency when g produces relatively higher frequencies or more rapid time changes in f than does h. *See* **implicit time difference**.

semi-interquartile range—*See* **quartile**.

semi-Lagrangian method—A physically motivated numerical technique for solving the **advection** (**transport**) equation.

In the advective form, $DQ/Dt = 0$, where $D(\ )/Dt$ is the **total derivative**, and "mixing ratio" Q is an invariant along a flow **trajectory**. By tracing (along the flow trajectory) backward in time to the "departure point", the value at the "arrival point" can be obtained by an **interpolation** or a remapping procedure (between the fixed **Eulerian grid** and a time-dependent distorted **Lagrangian grid**). Because of the discrete particle–like approach, total mass is generally not conserved. To ensure mass conservation, the semi-Lagrangian method can be formulated with the conservative flux form. The singular **particle** discretization is replaced by a finite control-volume **discretization**. Analogous to an Eulerian flux-form formulation, total **flux** from the **upstream** direction, computed in the Lagrangian fashion, is used for the **prediction** of the volume-averaged quantity, which can be the **density** or a density-weighted **mixing ratio**–like quantity. Because the size of the time step is not limited by the **CFL** condition, both the advective-form and the flux-form semi-Lagrangian methods are computationally efficient, particularly in spherical geometry.

Staniforth, A., and J. Cote, 1991: Semi-Lagrangian integration schemes for atmospheric models – A review. *Mon. Wea. Rev.*, **119**, 2206–2223.

Lin, S.-J., and R. B. Rood, 1996: Multidimensional flux-form semi-Lagrangian transport schemes. *Mon. Wea. Rev.*, **124**, 2046–2070.

semiannual oscillation—That component of the **annual cycle** of a **variable** that consists of a sinusoidal **oscillation** with a **period** of six months.

Strong semiannual signals in **thermal** and **momentum** fields of the **troposphere** are found in both the **Tropics** (especially in the eastern hemisphere) and in the oceanic regions of Southern Hemisphere middle and high latitudes.

semiarid climate—In Thornthwaite's 1931 **climatic classification**, a **humidity province** in which the principal plant life is drought-resistant short grasses.

Köppen called these conditions the **steppe climate**. Semiarid regions are very susceptible to severe **drought**.

Thornthwaite, C. W., 1931: The climates of North America according to a new classification. *Geogr. Rev.*, **21**, 633–655.

Köppen, W. P., and R. Geiger, 1930–1939: *Handbuch der Klimatologie*, Berlin: Gebruder Borntraeger, 6 vols.

semiarid zone—Aridity is a climatological condition in which the **amount of precipitation** received (supply) is exceeded, on average, by **potential evapotranspiration** (demand).

A number of physically based indices have been proposed to describe this deficiency that relate **precipitation** to **temperature** and **humidity**. For practical purposes, in the temperate and tropical zones, **semiarid climates** generally receive between 200 and 500 mm of precipitation per year on average, and **arid climates** receive less than 200 mm. However, definitions based on mean precipitation are not always satisfactory, because they do not express **variability** or the likelihood of **drought**. **Rainfall** in arid climates is extremely variable, with coefficients of **variation** for arid climates exceeding 50%, and coefficients for semiarid climates ranging between 30% and 50%. In broad economic terms, variability restricts the potential use of these lands. Arid climates are unsuitable for growing crops using rainfall alone, because crops may fail three or more years out of ten. Semiarid climates often support grasses that are suitable for grazing animals. Thus, livestock raising is often more appropriate, less risky, and more common than rainfed agriculture in semiarid areas.

semiconfined aquifer—*See* **leaky aquifer**.

semidiurnal tide—A **tide** having two high waters and two low waters each **lunar day**, with little or no **diurnal inequality**.

This applies equally to **solar tides** and to **atmospheric tides**. Worldwide, semidiurnal tides are the most important because the global ocean is near to **resonance** at the **period** of semidiurnal gravitational forcing.

semigeostrophic equations—The system of **momentum** and thermodynamic equations in **geostrophic coordinates** X, Y, Z, T, derived through the **geostrophic momentum approximation**, that fulfills the requirements of **semigeostrophic theory**.

At the momentum level, and using the pseudoheight coordinate $Z = [1 - (p/p_0)^\kappa] c_p \theta_0 / g$, the equations are

$$\frac{\partial u_g}{\partial T} + u_g \frac{\partial u_g}{\partial X} + v_g \frac{\partial u_g}{\partial Y} - f v^*_{ag} = 0,$$

$$\frac{\partial v_g}{\partial T} + u_g \frac{\partial v_g}{\partial X} + v_g \frac{\partial v_g}{\partial Y} + f u^*_{ag} = 0,$$

$$\frac{\partial \Phi}{\partial Z} = \left(\frac{g}{\theta_0}\right)\theta,$$

$$\frac{\partial u^*_{ag}}{\partial X} + \frac{\partial v^*_{ag}}{\partial Y} + \frac{\partial(\rho w^*)}{\rho \partial Z} = 0,$$

$$\frac{\partial \theta}{\partial T} + u_g \frac{\partial \theta}{\partial X} + v_g \frac{\partial \theta}{\partial Y} + \frac{\theta_0}{g} q_g \rho w^* = 0,$$

where u_g, v_g are the **geostrophic wind** components; u^*_{ag}, v^*_{ag} the transformed **ageostrophic wind** components; w^* the transformed **vertical velocity**; $\Phi = \phi + (1/2)(u^2_g + v^2_g)$ a **potential** function derived from the **geopotential** ϕ; q_g the geostrophic **potential vorticity**; θ the **potential temperature**; g the **acceleration of gravity**; f the **Coriolis parameter**; ρ the **density** in Z-space; and θ_0 a reference potential temperature.

semigeostrophic motion—Atmospheric motion that fulfills the principles and requirements of **semigeostrophic theory**.

semigeostrophic theory—A more accurate alternative to **quasigeostrophic theory** that involves the **geostrophic momentum approximation** to the quasi-static **primitive equations**.

In semigeostrophic theory the full effects of **ageostrophic advection** are included. Atmospheric structures such as fronts, small strong low pressure cells, and broad weak high pressure cells are more accurately represented in semigeostrophic theory than in quasigeostrophic theory.

semipermanent anticyclone—A high pressure **circulation** feature on monthly mean charts that is often evident in the same location during different years.
> Examples include the **North American high** and the **Siberian high**.

semipermanent depression—A low pressure **circulation** feature on monthly mean charts that is often evident in the same location during different years.
> Examples include the **Aleutian low** and the **Icelandic low**.

sensible heat—A somewhat archaic term that is typically the outcome of heating a surface without evaporating water from it.
> Sensible **heat** per unit mass can be roughly identified with **specific enthalpy** of unsaturated air, that is, approximately $c_{pd}T$, where c_{pd} is the **specific heat** of **dry air** at constant **pressure** and T is **temperature**. Sensible heat is often compared with **latent heat**, which is the difference between the **enthalpy** of **water vapor** and that of liquid water.

sensible heat flow—The **transfer** of **sensible heat** (**enthalpy**) from one region of a fluid to another by fluid motion.
> In the **atmosphere**, the poleward **transport** of sensible heat across a given latitude belt is given by the integral $\int c_p \rho Tv ds$, where c_p is the **specific heat** of air at constant **pressure**, ρ the air density, v the northward component of the **wind speed**, T the **air temperature**, and ds an element of the vertical boundary at the given latitude. It must be assumed that the net **mass transport** $\int \rho v ds$ must vanish or the question of the zero-point **internal energy** will arise. *See also* **energy equation**.

sensible horizon—*See* **horizon**.

sensible temperature—(Obsolete.) Same as **effective temperature**; now usually replaced by **windchill index** for winter conditions and **heat index** for summer conditions.

sensing element—(Or sensor.) The component of an instrument that converts an **input** signal into a quantity that is measured by another part of the instrument.

sensitivity—In radio terminology, the degree to which a **receiver** will respond to an **input** of given strength.
> The greater the sensitivity, the weaker are the signals detected. *See* **threshold signal**.

sensitivity time control—(Abbreviated STC.) The term, now somewhat obsolete, used to denote the procedure of reducing the **sensitivity** of a **radar** receiver at close ranges to compensate for the limited **dynamic range** of the **receiver** and avoid saturation.
> The primary disadvantage of STC is that weak **echoes** at close **range** may be undetected because of the lowered sensitivity. The term is sometimes used erroneously to denote range correction of the **received power** to compensate for the inverse-square falloff of the **power** from distributed targets with range. *See* **radar equation**.

sensor—Same as **sensing element**.

separation of flow—*See* **boundary layer separation**.

separation of variables—The assumption that the solution of a partial differential equation is equal to a product of functions, each being a function of only one of the **independent variables**.
> Each function then satisfies an ordinary differential equation and the original equation is said to be separable. This method has been widely applied in **linear** boundary value problems admitting permanent **waves** as solutions.

sequence—With respect to **aviation weather observations**, same as **collective**.

serac—A large block of **ice**, generally taller than broad, formed by the fracturing of ice.
> Most commonly found within an **icefall**, at the edge of an ice cliff, or at the margins of fast-moving ice.

serein—The doubtful phenomenon of fine **rain** falling from an apparently **clear sky**, the **clouds**, if any, being too thin to be visible.
> Frequently fine rain is observed with a clear sky overhead, but clouds to **windward** clearly indicate the source of the drops. Also, it has been defined as fine rain falling from a clear sky after **sunset**, probably actually referring to **dew**, a survival of the old belief that dew falls.

serial communications—Communications over a single wire, where only one **bit** is transmitted at a time.
> In the past this has typically meant a communication line to a printing device, such as a terminal or printer, with the expectation that the characters are all printable or printer/display characters. Today, however, a serial communication does not guarantee use of only printable characters.

serial correlation—Usually same as **autocorrelation**, but occasionally used to designate lagged correlations between two separate series of observations, for example, the **correlation** between x_i, and y_{i+1} in the series $x_1, x_2, \ldots$ and $y_1, y_2, \ldots$. The serial correlation coefficient is the **product-moment** correlation coefficient for the lagged correlations or for the correlations between series.

serial correlation coefficient—*See* **serial correlation**.

serial station—(Also called hydrographic station.) An oceanographic **station** consisting of one or more **Nansen casts**.

servo loop—In general, an automatic **feedback** control system in which the controlled **variable** is mechanical position or any of its derivatives. In **radar**, it is a method for controlling the positioning of an **antenna**.
The servo loop control system, or servomechanism, employs an alignment error voltage to command the antenna to move to a position that will reduce the alignment error to zero. The system consists of a computer, amplifiers, filters, and motors that move the antenna about its vertical and horizontal axes.

sesquiterpenes—*See* **terpenes**.

set—The direction toward which an **ocean current** flows.

severe local storm—*See* **severe storm**.

severe storm—In general, any destructive **storm**, but usually applied to severe local storms in particular, that is, intense **thunderstorms**, **hailstorms**, and **tornadoes**.

severe-storm observation—An **observation** (and report) of the occurrence, location, time, and direction of movement of **severe storms**.

severe weather—Generally, same as **severe storm**.

severe weather threat index—(Abbreviated SWEAT index, SWI.) See **stability index**.

SFAZI code—(Short for spherics azimuth code.) An international code used to report the direction of **sferics** azimuth in terms of bearings from the observing **station**.
See **direction finder**.

SFAZU code—An international code used to report the direction of **sferics** azimuth for the previous 24 hours in terms of bearings from the observing **station**.

sferics—1. (Also spelled spherics.) The study of **atmospherics**, especially from a meteorological point of view.
This involves techniques of locating and tracking atmospherics sources and evaluating received signals (**waveform**, **frequency**, etc.) in terms of source.
2. Same as **atmospherics**.

sferics fix—The determination of the **bearing** to the **lightning** source usually based on the measurement of the horizontal **magnetic field** with **orthogonal** coils or **loop antennas**.

sferics observation—The **detection** of **electromagnetic radiation** from **lightning** generally in the **frequency** range 10–30 kHz.
The physical measurement can include the **electric field**, the **magnetic field**, or both. **Sferics** are generally attributed to the high current phases of source, that is, **return strokes** and **K changes**.

sferics receiver—(Also called lightning direction finder.) An instrument that measures, electronically, the direction of arrival, **intensity**, and rate of occurrence of **atmospherics**; a type of **radio direction finder**, it is most commonly used to detect and locate **cloud-to-ground lightning discharges** from distant thunderstorms.
In its simplest form the instrument consists of two orthogonally crossed **antennas** that measure the **electromagnetic field** changes produced by a **lightning discharge** and determine the direction from which the changes arrived. Negative and positive **polarity** cloud-to-ground discharges can be distinguished. **Cloud-to-cloud discharges** can be distinguished based on characteristics of the received **signal**, and the geometry of nearby discharge channels may be determined. *See also* **narrow-sector recorder**, **lightning detection network**.

sferics source—That portion of a **lightning discharge** that radiates strongly in the **frequency** interval 10–30 kHz.
The physical source is generally identified with the **return stroke** in flashes to ground and the **K change** in the case of **intracloud flashes**.

677

shadow—In **radar**, an **azimuth** sector with no **echoes** because the transmitted **signal** is masked by local prominences such as hills or buildings.

shadow band pyranometer—An instrument for measuring the **diffuse sky radiation**.

A shadow band is added to a conventional **pyranometer** at such an angle that it blocks out the **direct solar radiation** throughout the course of a day.

shadow bands—1. Same as **crepuscular rays**.

2. A **scintillation** phenomenon wherein dark bands appear to move rapidly across the **luminous** surface of any distant **light** source.

shadow zone—(Often called diffraction zone.) May be used for any region that would not be illuminated by a given source of **electromagnetic** (or acoustic) radiation if it propagated strictly according to **ray** optics in a homogeneous (on the **scale** of the **wavelength**) medium.

Some radiation, however, does penetrate shadow zones because of **scattering** by the propagating medium or by obstacles within it. Because of the widespread misconception that **diffraction** is fundamentally different from scattering, the term **diffraction zone** is also used, especially by radio engineers, who might say that the earth diffracts **radio waves** into this zone (and probably would ignore scattering by the **atmosphere**). Yet all **radiation** that penetrates a shadow zone does so because of scattering by the atmosphere and by solid and liquid bodies (including the earth) within the atmosphere or at its boundary.

shaitan—Same as **dust whirl**.

shale ice—Mass of thin and brittle plates of **freshwater** ice formed by breaking up of thin, skin **ice** into small pieces lumped together.

shallow convection parameterization—The representation in a numerical **model** of the turbulent transports of **heat** and moisture by nonprecipitating **cumulus** clouds with **cloud** tops below 3000 m above the surface.

These shallow cumulus clouds are found all over the globe but they, and their associated turbulent transports, are of particular significance in the **trade wind** region where they provide the heat and moisture that maintains the thermodynamic structure of the lower **troposphere**. *See* **parameterization**.

shallow fog—In weather-observing terminology, low-lying **fog** that does not obstruct **horizontal visibility** at a level 2 m (6 ft) or more above the surface of the earth.

This is, almost invariably, a form of **radiation fog**. *Compare* **ground fog**.

shallow low—1. A region of low **pressure** with a weak horizontal **gradient**.

2. Low **pressure** at the earth's surface that may be reflected aloft as a **trough** in only the lower levels of the **atmosphere**.

shallow water approximation—An approximation to the **equations of motion** whereby it is assumed that the fluid is homogeneous and horizontal scales of interest are much larger than the depth of the fluid.

See **shallow water wave**.

shallow water wave—(Also called **long wave**, Lagrangian wave.) An ocean wave with its length sufficiently large compared to the water depth (i.e., 25 or more times the depth) that the following approximation is valid:

$$c = (gH)^{1/2},$$

where c is the **wave velocity**, g the **acceleration of gravity**, and H the water depth.

Thus the **velocity** of shallow water waves is independent of **wavelength** L. In water depths between $0.5L$ and $0.04L$ it is necessary to use the more precise expression

$$c = [(gL/2\pi) \tanh(2\pi H/L)]^{1/2}.$$

See **deep-water wave**, **gravity wave**.

shaluk—Any hot **desert wind** other than the **simoom**.

shamal—(Also called barih; also spelled shemaal, shimal, shumal.) The northwest **wind** in the lower valley of the Tigris and Euphrates and the Persian Gulf.

It may set in suddenly at any time, and generally lasts from one to five days, dying down at night and freshening again by day; however, in June and early July it continues almost without cessation (the "great" or "forty-day" shamal). Although the wind rarely exceeds 13 m s⁻¹ (30 mph), it is very hot, dry, and dusty. The sky is cloudless but the **haze** is often so thick as to obscure the land, making navigation dangerous.

shamsir—Same as **sansar**.

sharki—(Also spelled sherki, shuquee, shurgee, shurkiya.) Same as **kaus**.

sharp-edged gust—A wind **gust** that results in an instantaneous change in direction or speed.
See **effective gust velocity, derived gust velocity**.

shear—The **variation** (usually the directional derivative) of a **vector field** along a given direction in space.
The most frequent context for this concept is **wind shear**.

shear flow—Fluid motion having a **velocity** field characterized by the presence of **velocity shear**.

shear–gravity wave—A combination of **gravity waves** and a **Helmholtz wave** on a **surface of discontinuity** of **density** and **velocity**.
If the densities of the lower and upper layers respectively are ρ and ρ' and the velocities U and U', the **phase speed** c of the shear–gravity wave is

$$c = \frac{\rho U + \rho' U'}{\rho + \rho'} \pm \left[\frac{gL}{2\pi} \frac{\rho - \rho'}{\rho + \rho'} - \frac{\rho\rho'(U - U')^2}{(\rho + \rho')^2} \right]^{1/2},$$

where g is the **acceleration of gravity** and L the **wavelength**. The motion is unstable if and only if the bracketed quantity is negative; the **density** difference thus contributes to **stability** and the velocity difference to **instability**. Applications have been made to atmospheric **frontal surfaces** and **inversions**; perhaps the most successful of these is to the phenomenon of **billow clouds**. Reasonable atmospheric values for the parameters yield stationary wavelengths of the order of 1 km.

Drazin, P. G., and W. H. Reid, 1981: *Hydrodynamic Stability*, Cambridge University Press, 14–22.

shear hodograph—*See* **hodograph**.

shear instability—*See* **Kelvin–Helmholtz instability**.

shear layer—A layer in the **atmosphere** across which there is a change of **wind speed** or **wind direction**.

shear line—In meteorology, a line or narrow zone across which there is an abrupt change in the horizontal **wind** component parallel to this line; a line of maximum horizontal **wind shear**.

shear production—The generation of **turbulence kinetic energy** caused by **wind shear**.
This mechanical production term S of the turbulence kinetic energy budget equation is (written in **Einstein's summation notation**)

$$S = -\overline{u_i' u_j'}\, \frac{\partial \overline{U_i}}{\partial x_j}$$

and is almost always positive, that is, a source. In the fair-weather, turbulent **atmospheric boundary layer**, the terms associated with horizontal wind usually dominate, allowing the shear production term to be approximated as

$$S = -\overline{u'w'}\, \frac{\partial \overline{U}}{\partial z} - \overline{v'w'}\, \frac{\partial \overline{V}}{\partial z},$$

where ($\overline{u'w'}$, $\overline{v'w'}$) are the **Reynolds stresses** (turbulent kinematic **momentum fluxes**) in the (x, y) directions, (U, V) are the respective wind components, the prime denotes **deviation** from the mean, and the overbar denotes an average. The ratio of the buoyant to the negative of the shear production terms is also known as the **flux Richardson number** and is a measure of **dynamic stability**.

shear ridge—A ridge formed primarily by shear **deformation**, usually found along the boundary between **fast ice** and **pack ice**.

shear vector—The derivative of any **vector** in any spatial direction.
The particular direction should be clear from the context.

shear wave—1. Same as **Helmholtz wave**.
2. A **wave** propagating within an elastic solid.

shearing deformation—*See* **deformation**.

shearing instability—1. Same as **Helmholtz instability**.
2. A complex **hydrodynamic instability** phenomenon exhibited by a stratified **shear flow**.

A necessary condition for this type of **instability** is that the local **Richardson number** is somewhere less than one-quarter.

Drazin, P. G., and W. H. Reid, 1981: *Hydrodynamic Stability*, Cambridge University Press, 325–333.

shearing stress—Any of the tangential components of the **stress tensor**.

In meteorology, the shearing stress is often that on a horizontal surface in the direction of the **wind**:

$$\tau = \mu \, \frac{du}{dz},$$

where μ is the dynamic (molecular) **viscosity**, u the **wind speed**, and z the vertical coordinate. A corresponding **eddy** stress, or **Reynolds stress**, may be defined, with an **eddy viscosity** in place of the **molecular viscosity**. *See also* **surface friction**, **Newtonian friction law**.

shearing wave—*See* **transverse wave**.

sheet erosion—(Also called sheet wash.) Erosion of thin layers of earth-surface material, more or less evenly, from extended areas of gently sloping land by broad continuous sheets of running water, without the formation of rills, gullies, or other channelized flow.

sheet flow—Same as **laminar flow**.

sheet frost—A thick coating of **rime** formed on windows and other surfaces.

sheet ice—**Ice** formed in a smooth thin layer on a water surface by the **coagulation** of **frazil** through rapid **freezing**.

sheet lightning—(Also called luminous cloud.) A diffuse, but sometimes fairly bright, **illumination** of those parts of a **thundercloud** that surround the path of a **lightning flash**, particularly a **cloud discharge** or **cloud-to-cloud discharge**.

Thus, sheet lightning is no unique form of **lightning** but only one manifestation of ordinary lightning types in the presence of obscuring **clouds**. *Compare* **heat lightning**.

sheet wash—Same as **sheet erosion**.

shelf cloud—A low-level, horizontal, wedge-shaped **arcus** cloud associated with a **convective storm**'s **gust front** (or occasionally a **cold front**).

The shelf cloud is attached to the convective storm's **cloud base**. Rising motion can be seen in the leading (outer) part of the shelf cloud, while the underside appears turbulent and tattered. *Compare* **roll cloud**.

shelf ice—Same as **ice shelf**.

shelf waves—**Topographic waves** generated over a shelf, typically by the **wind** acting in the presence of a coastal boundary.

For an observer traveling with the **wave**, the coast is on the right in the Northern Hemisphere and on the left in the Southern Hemisphere. *See also* **coastally trapped waves**, **Kelvin wave**.

shell ice—(Also called cat ice.) Ice, on a body of water, that remains as an unbroken surface when the **water level** drops so that a cavity is formed between the water surface and the **ice**.

shelter—*See* **instrument shelter**.

shelterbelt—A belt of trees and/or shrubs arranged as a protection against strong winds; a type of **windbreak**.

The trees may be specially planted or left standing when the original forest is cut. A shelterbelt decreases the force of the **wind** near the ground, both **upwind** for a distance of up to six times the height of the barrier, and **downwind** for a distance of fifteen to twenty times the height. These ratios are roughly constant, irrespective of the height of the belt. The lowest **wind speed** is found downwind at a distance of three or four times the height of the belt. If the trees are too dense, the air beyond this quiet zone is often turbulent, with downdrafts that may flatten crops. The best form is a moderately dense belt of mixed conifers and deciduous trees, five to ten yards wide, containing at least three rows of trees at right angles to the prevailing winds. A system of shelterbelts to give protection to crops over a large area should be planted at intervals of about twenty-five times their expected height.

U.S. Dept. of Agriculture, 1941: *Climate and Man, 1941 Yearbook of Agriculture*, 484–485.
Brooks, C. E. P., 1951: *Climate in Everyday Life*, 270–272.

sheltering coefficient—The constant of proportionality in Jeffreys's expression for the **work** done by the **wind** on the sea surface.

It was assumed by Jeffreys that this work is proportional to the product of the **density** of air, the slope of the water surface, and the square of the wind velocity relative to the **wave**. Although the value of the sheltering coefficient (0.27) found by Jeffreys gives a satisfactory explanation of the initial formation of **wind waves**, observations of the increase of wind waves shows that this value is about ten times too high for the subsequent **transfer** of **energy** from wind to waves.

shemaal—Same as **shamal**.

sherki—Same as **sharki**; *see* **kaus**.

SHF—Abbreviation for **super high frequency**; *see* **radio frequency band**.

shield—*See* **rain-gauge shield**, **radiation shield**.

shimal—Same as **shamal**.

shimmer—*See* **scintillation**.

shine—A general term for a dry **heiligenschein**, moist heiligenschein, or **sylvanshine**.

ship motion—The complex motion imparted to a ship upon encountering waves.
   All the motions can be regarded as combinations of three oscillations about horizontal or vertical axes (**roll**, **pitch**, and **yaw**), and three **linear** displacements of the **center of gravity** (**heave**, **surge**, and **sway**).

ship report—The encoded and transmitted report of a **marine weather observation** by a ship at sea.

ship routing—Advice to shipping companies and ship captains of the best routes to be taken between any two ports by a particular ship.
   Criteria for selection include avoidance of navigational hazards such as **fog** or **ice**, fuel economy, and avoidance of damage to ship or cargo (e.g., from a major **storm**).

ship synoptic code—A **synoptic code** for communicating **marine weather observations**.
   It is a modification of the **international synoptic code**. Abbreviated forms of the ship synoptic code are 1) **abbreviated ship code**, 2) short ship code, and 3) **lightship code**.

shoal—Shallow water usually associated with the presence of sand bars below the surface.
   Sometimes these sand bars are exposed during low tides.

shoaling—The increase in the **wave height** of **wind waves** in shallow water because of the **divergence** in **wave group** velocity.
   This happens when the depth decreases near the coast.

shock wave—Propagation of a shock front, a narrow region in a fluid over which its thermodynamic state changes markedly.
   The thickness of the front is of order a few **mean free paths** in the medium ahead of the **wave**.
      Bradley, J. N., 1962: *Shock Waves in Chemistry and Physics*, 1–2.

shore ice—Compact **sea ice** that is attached to the shore or to anchoring points on the sea bed.
   It is a type of **fast ice**, and may sometimes be **rafted ice** or **ice** that has been beached by **wind**, **tides**, **currents**, or by ice pressure.

shore lead—A **lead** between **pack ice** and **fast ice**, or between **floating ice** and the shore.
   It may be closed by refreezing or by **wind** or **currents** so that only a **tide crack** remains.

short-crested wave—A **wave** that has a small extent in the direction perpendicular to the direction of propagation.
   Most waves in the ocean are short-crested.

short-range forecast—1. A **weather forecast** made for a time period up to 48 hours.
   The U.S. National Weather Service issues short-range forecasts by part of the day, for example, today, tonight, etc. *Compare* **extended forecast**, **long-range forecast**, **nowcast**.
   2. Same as **daily forecast**.

short-term hydrologic forecast—Forecast of hydrologic events on a short-term basis, such as a few hours, up to a few days.

short wave—1. (Or minor wave.) With regard to atmospheric **circulation**, a **progressive wave** in the horizontal pattern of air motion with dimensions of **cyclonic scale**, as distinguished from a **long wave**.

A short wave moves in the same direction as that of the prevailing basic **current** through the **troposphere**. The **angular wavenumber** of short waves ranges between eight and twenty. *See* **cyclone wave**.

2. A **wave** with a relatively short **wavelength** and **period**.

For ocean wind waves, this usually means waves with periods shorter than about 60 s.

shortwave radiation—1. Energy in the visible and near-visible portion of the **electromagnetic spectrum** (0.4–1.0 μm in **wavelength**).

2. In meteorology, a term used loosely to distinguish **radiation** in the visible and near-visible portions of the **electromagnetic spectrum** (roughly 0.4–4.0 μm in **wavelength**), usually of solar origin, from that at longer wavelengths (**longwave radiation**), usually of terrestrial origin.

Showalter stability index—*See* **stability index**.

shower—Precipitation from a **convective cloud**.

Showers are characterized by the suddenness with which they start and stop, by the rapid changes of **intensity**, and usually by rapid changes in the appearance of the sky. In weather observing practice, showers are always reported in terms of the basic type of **precipitation** that is falling, that is, **rain** showers, **snow** showers, **sleet** showers. In **aviation weather observations**, these are encoded RW, SW, and EW, respectively. *See* **airmass shower**.

shuga—An accumulation of spongy, whitish chunks of **ice** a few centimeters across formed from **grease ice** or **slush**.

It forms instead of **pancake ice** if the **freezing** takes place in **seawater** that is considerably agitated.

shumal—Same as **shamal**.

shuquee—Same as **sharki**; *see* **kaus**.

shurgee—Same as **sharki**; *see* **kaus**.

shurkiya—Same as **sharki**; *see* **kaus**.

SI—Acronym for the **Systeme Internationale**, or the **International System of Units**.

si giring giring—*See* **aloegoe**.

Siberian anticylcone—Same as **Siberian high**.

Siberian high—(Or Siberian anticyclone.) An **area of high pressure** that forms over Siberia in winter and that is particularly apparent on mean charts of **sea level pressure**.

Its center is near Lake Baikal, where the average sea level pressure exceeds 1030 mb from late November to early March. This **anticyclone** is enhanced by the surrounding mountains that prevent the cold air from flowing away readily. In the center of the anticyclone the **normal** clockwise **circulation** is replaced by **katabatic winds** down the river valleys, but to the east along the Pacific Coast there is a belt of very strong northerly winds. The offshore flow is known as the **winter monsoon**. In summer the Siberian high is replaced by a low pressure area.

SID—Abbreviation for **sudden ionospheric disturbance**.

side-scanning sonar—A method of surveying the bottom of the ocean or other body of water to obtain detailed acoustic images, frequently used to locate debris, such as from aircraft accidents or sunken ships.

The acoustic **beam** is directed perpendicularly to the direction of travel. The ship moves at a constant **velocity**, and each sonar ping insonifies a slightly different wedge of bottom. The apparent depth is actually the **slant range** from the ship to the **target**. The **display** is a map of slant range and distance along the track.

sideband—The **frequency band**, above or below the **carrier frequency**, that is produced by the process of **modulation**.

sidelobe—*See* **lobe**.

sidelooking radar—An airborne **radar** used for high-resolution ground mapping that employs a fixed **antenna** beam pointing out of the side of the aircraft.

Narrow **azimuth resolution** is obtained with a long **aperture** mounted along the side of the aircraft or by use of **synthetic aperture radar** processing.

sidereal year—The time it takes the earth to complete one full orbit around the sun relative to the fixed stars.

siffanto—A southwest **wind** of the Adriatic Sea; it is often violent.

sigma-n—(Symbol σ_n, where $n = 1, 2$, etc.) The **density** (in kg m^{-3}) that a **parcel** of water would have if moved **adiabatically** to n kilometers depth, less 1000 kg m^{-3}; analogous to **sigma-theta** but with a deeper reference **pressure**.

sigma-STp—(Symbol σ_{STp}.) The **density** ρ (in kg m^{-3}) that seawater of **salinity** S and **temperature** T would have at **pressure** p, less 1000 kg m^{-3}.
In symbols,

$$\sigma_{STp} = \rho(S,T,p) - 1000 \text{ kg m}^{-3},$$

where p is the **atmospheric pressure** relative to that at the sea surface. This definition includes the effects of **compressibility** in calculating the density but does not include the effects of **adiabatic** warming due to compression.

sigma-t—(Symbol σ_t.) The **density** ρ (in kg m^{-3}) that seawater, of **salinity** S and **temperature** T, would have at **sea level** (i.e., at **atmospheric pressure**), less 1000 kg m^{-3}.
In symbols,

$$\sigma_t = \rho(S,T,0) - 1000 \text{ kg m}^{-3},$$

where $p = 0$ is the atmospheric pressure relative to that at the sea surface.

sigma-t,p—(Symbol $\sigma_{t,p}$.) **Sigma-t**, corrected to the **hydrostatic pressure** in situ.

sigma-theta—(Symbol σ_θ.) The **density** ρ (less 1000 kg m^{-3}) that a **parcel** of water of **salinity** S and **temperature** T would have if it were raised **adiabatically** to the sea surface, where its **temperature** would change from T to **potential temperature** θ due to **adiabatic** decompression.
In symbols,

$$\sigma_\theta = \rho(S,\theta,0) - 1000 \text{ kg m}^{-3}.$$

sigma vertical coordinate—A vertical coordinate for atmospheric models defined as **pressure** normalized by its surface value, or as the difference in pressure and its value at the top of the **model atmosphere** normalized by the surface value of this difference.
Thus, $\sigma = (p - p_T)/(p_S - p_T)$ where p is pressure, and the subscripts T and S stand for the top and the ground surface values of the model atmosphere, respectively. With the sigma coordinate, the lowest **coordinate surface** follows the **model** terrain, resulting in significant simplification of the equations compared to pressure or to an unmodified geometric height coordinate. *See* **eta vertical coordinate, hybrid vertical coordinate, isentropic vertical coordinate, pressure vertical coordinates, height vertical coordinate**.

sigma-zero—**Sigma-t** at 0°C.
Knudsen's tables give values of sigma-zero as a function of **salinity** or **chlorinity**, as well as corrections to be applied to obtain sigma-t.

SIGMET information—(Contraction for significant meteorological information.) Information issued by a **meteorological watch office** concerning the occurrence or expected occurrence of en route phenomena that may affect the safety of aircraft operations, such as **convection, turbulence,** and **icing**.

signal—Any **carrier** of information; opposed to **noise**.
See **carrier wave**.

signal generator—An electronic instrument used for the production of electromagnetic or acoustic signals with certain desired characteristics.
It is useful in testing and **calibration**.

signal intensity—*See* **signal strength**.

signal strength—1. In radio, a measure of the received **radio frequency** power; generally expressed in decibels relative to some **standard** value, normally either one milliwatt or that **power** that would have resulted at the same distance under **free space** transmission.
Could also be applied to the strength of an optical **signal** transmitted or received by a **lidar** or an acoustic signal transmitted or received by a **sodar**. The term **field strength** is commonly used as a synonym.
2. In **radar**, the strength of the received **signal**, usually expressed in logarithmic **power** units, in particular **decibels** relative to 1 mW, that is, dBm.

signal-to-noise ratio—(Abbreviated SNR.) A ratio that measures the information content of a **signal**, usually defined as the ratio of the **power** of the signal unaffected by **noise** to the power of the noise.

Commonly measured in **decibels**, the signal-to-noise ratio is sometimes defined for a specified **bandwidth**.

signal velocity—The speed of propagation of a hydrodynamic influence.

For the cyclonic-scale **quasigeostrophic** disturbances, the effective signal velocity is given approximately by the maximum **group velocity** and is usually of the order of 30° longitude per day.

signature—1. Any characteristic or series of characteristics by which a material, phenomenon, or change may be recognized in an **image** or **dataset**.

2. In **radar**, a term used to designate recognizable identifying characteristics of any one of many **cloud** properties or **mesoscale** phenomena, such as **gust fronts**, **vortices**, the **melting layer**, **microbursts**, and **convergence lines**.

signature analysis techniques—1. Techniques that use the **variation** in the spectral **reflectance** or **emittance** of objects as a method of identification.

2. Methods, such as the **Dvorak** and **Scofield–Oliver techniques**, that use satellite imagery to estimate intensities or trends of meteorological features.

significance level—*See* **significance test**.

significance test—A test of the reliability of estimate of **statistical** parameters.

Such tests proceed by assuming that the estimates are not significant and are those to be expected from sampling a particular **population**, and then, from the properties of the population, determining the probabilities of such occurrences. The hypothesis (that the estimates are not significant) is rejected only when an observational result is found to be significant, that is, when the obtained result belongs to an objectively specified unfavorable class (**critical region** or **rejection region**) having a fixed, small **probability** of occurrence in **random** samples from the hypothesized population. When the result falls in the **acceptance region**, it is not significant and the hypothesis cannot be rejected. The boundaries of the classes are set in such a way that the total probability (unity) is appropriately divided between them, say 0.95, 0.05 or 0.99, 0.01. The probability assigned to the critical region, commonly either 0.05 or 0.01, is called the **significance level**. *See* **chi-square test**, **Student's t-test**, **analysis of variance**.

significant level—In a **radiosonde observation**, a level (other than a **mandatory level**) for which values of **pressure**, **temperature**, and **humidity** are reported because temperature and/or moisture-content data at that level are sufficiently important or unusual to warrant the attention of the **forecaster**, or they are required for the reasonably accurate reproduction of the **radiosonde observation**.

There are definite rules governing the selection of significant levels, set forth in the Manual of Radiosonde Observations, WBAN Circular P, 7th ed. rev., 1957.

significant wave—In ocean wave forecasting, a fictitious **wave** with a height and **period** equal to the average height and period of the highest one-third of the actual waves that pass a fixed point.

significant wave height—Defined traditionally as the mean height of the highest third of the **waves**, but now usually defined as four times the root-mean-square of the surface **elevation** (or equivalently as four times the square root of the first **moment** of the **wave spectrum**).

significant weather—In aviation weather, the occurrence or expected occurrence of specified en route weather phenomena that may affect the operation of aircraft.

significant weather chart—Chart displaying the observed or forecast significant weather phenomena at different flight levels that may affect the operation of the aircraft.

sigua—A straight-blowing **monsoon** gale of the Philippines.

sikussak—Very old **sea ice** trapped in fjords.

It resembles **glacier ice** because **snowfall** and **snowdrifts** contribute to its formation.

sill depth—(Also called threshold depth.) The maximum depth at which there is horizontal communication between an ocean basin and the open ocean.

silt—Sediment **particles** in the size **range** of 1/256 to 1/16 mm (or 0.0039 to 0.0625 mm).

Traditionally, **sediment** particle sizes in the silt-sized range are defined by their **sedimentation diameter**.

silt-discharge rating—(Archaic.) Term formerly used synonymously with **sediment discharge rating**.

silver-disk pyrheliometer—An instrument used for the measurement of **direct solar radiation**.

It consists of a silver disk located at the lower end of a tube containing a diaphragm that serves as the **radiation** receiver for a **calorimeter.** Radiation falling on the silver disk is periodically intercepted by means of a shutter located in the tube, causing **temperature** fluctuations of the calorimeter that are proportional to the **intensity** of the radiation. The instrument is normally used as a **secondary instrument** and is calibrated against the **water-flow pyrheliometer.** It was used by the U.S. Weather Bureau as a standard instrument. Currently, this instrument's function has been replaced by an **absolute cavity radiometer.** *See* **pyrheliometer, cavity radiometer.**

silver frost—(Also called silver thaw.) Colloquial expression for a deposit of **glaze** built up on trees, shrubs, and other exposed objects during a fall of **freezing precipitation**; the product of an **ice storm.**

silver iodide—An inorganic chemical compound, AgI, that has a crystalline structure (symmetry; lattice spacing) similar to **ice** and a very low solubility in water, and can be easily generated as an **aerosol.**

It was discovered by Bernard Vonnegut in 1947 after a search of the crystallographic tables as an effective nucleating agent for **supercooled water.** In **cloud seeding** applications it is usually combined with small amounts of other materials (e.g., bromine, chlorine, copper) to enhance nucleating properties through change of lattice dimension to approach more closely that of ice.

Vonnegut, B., 1947: The nucleation of ice formation by silver iodide. *J. Appl. Phys.*, **18**, p. 593.

silver iodide seeding—*See* **cloud seeding.**

silver storm—Same as **ice storm.**

silver thaw—Same as **silver frost.**

similarity—(Also called similitude.) *See* **dynamic similarity.**

similarity relationship—The formula relating **dimensionless groups** in **similarity theory.**

While **Buckingham Pi theory** allows one to identify which dimensionless groups of variables might be relevant to a particular flow situation, it does not give a formula relating these groups. Such a formula is found empirically.

similarity theory—An empirical method of finding universal relationships between **variables** that are made dimensionless using appropriate scaling factors.

The **dimensionless groups** of variables are called Pi groups and are found using a **dimensional analysis** method known as **Buckingham Pi theory.** Similarity methods have proved very useful in the **atmospheric boundary layer,** where the complexity of turbulent processes precludes direct solution of the exact governing equations. *See* **mixed-layer similarity, local free-convection similarity, local similarity.**

Stull, R. B., 1988: *An Introduction to Boundary Layer Meteorology,* 666 pp.

similarity theory of turbulence—*See* **Kolmogorov's similarity hypotheses.**

simm—Same as **simoom.**

simoom—(Many variant spellings.) A strong, dry, dust-laden **desert wind** that blows in the Sahara, Palestine, Jordan, Syria, and the **desert** of Arabia.

Its **temperature** may exceed 54°C and the **humidity** may fall below 10%. The name means "poison wind" and is given because the sudden onset of a simoom may cause **heat** stroke. This is attributed to the fact that the **hot wind** brings more heat to the body than can be disposed of by the **evaporation** of perspiration.

simple average—Same as **arithmetic mean.**

simple harmonic wave—An **oscillation** translating with constant speed and **amplitude,** and represented mathematically by a trigonometric or complex exponential function.

Thus,

$$A \sin\left(\frac{2\pi}{\lambda}x - vt + \phi\right)$$

or

$$A \exp\left[i\left(\frac{2\pi}{\lambda}x - vt + \phi\right)\right]$$

represents a simple harmonic wave of amplitude A, **wavelength** λ, **frequency** v, and **phase angle** ϕ. In ocean wave studies, a simple **harmonic** progressive **wave** is an idealized wave characterized by constant speed of propagation and a straight crest of indefinite length. *See also* **wavenumber, phase speed**.

simple linear correlation—*See* **correlation**.

simple reflector—Same as **specular reflector**.

simple vortex—*See* **point vortex**.

simple wave—The solution of a set of quasi-linear differential equations describing a fluid flow that possesses a family of straight-line **characteristics**.

In such a fluid motion the **dependent variables**, or simple combinations thereof, are constant along the characteristics, and may be used as the basis of the integration of the equations by the **method of characteristics**. The simple waves of most interest to meteorology are the **expansion wave** and the **compression wave**. *See* **simple harmonic wave**.

sine galvanometer—A **magnetometer** of the electromagnetic type that is used to measure the horizontal **intensity** of the earth's **magnetic field**.

The instrument serves as a national standard for measurement of this quantity.

single drift correction—In air navigation, a method of **pressure-pattern flight** utilizing a single value of **drift-correction angle** applied to an entire flight from point of origin to destination.

The value is based upon a forecast of **mean wind velocity** en route. *See* **multiple drift correction**; *compare* **aerologation, single-heading navigation**.

single-grid-heading navigation—Same as **aerologation**.

single-heading navigation—1. A system of air navigation whereby a flight is accomplished using a single value of **compass** heading.

The **heading** must be so determined that it compensates for the net effect of the **wind** en route. It should also be noted that a single-heading **course** rarely coincides with a **great-circle course**, except in **grid navigation**. Single-heading navigation is not necessarily economical in terms of least time, but its advantage lies in its simplicity. *Compare* **single drift correction**.

2. Same as **aerologation**.

single observer forecasting—Local weather forecasting that is based purely on **observation** of the weather elements for the same locality.

The application of experience in **synoptic meteorology**, combined with physical reasoning, is capable of producing reasonably reliable local forecasts for a few hours ahead; in some situations reliability is possible for an appreciable time ahead. The observations chiefly used in such forecasting are those of **pressure** and **pressure tendency**, **wind** velocity at surface and higher levels, **cloud**, and **temperature**.

single-scatter albedo—The ratio of the **scattering coefficient** to the **extinction coefficient**.

The single-scatter albedo affects **radiative transfer** in a **scattering** medium and is typically a function of **wavelength**. Throughout much of the **visible spectrum**, the single-scatter albedo of **clouds** and most atmospheric gases is very close to unity, and it typically varies considerably throughout the **infrared** spectrum.

single-scattering—A useful approximation to **radiative transfer** in which the **radiation** undergoes at most one **scattering** event.

Single-scattering is most appropriate for scattering **optical thicknesses** less than about 0.1 when there is no **absorption**, but can be applied to thicker media as the **single-scatter albedo** is lowered.

single scattering—1. In a propagation medium consisting of a **dispersion** of **scattering** particles, the situation whereby the **electromagnetic field** in the vicinity of any **particle** is unaffected by scattering from the other particles.

Then the total scattered field is just the sum of the fields scattered by the individual particles, each of which is acted on by the external **field** in isolation from the other particles.

2. The assumption that the total scattered field from a **dispersion** of **scattering** particles is the sum of the fields scattered by the individual **particles**.

Whether single scattering is a good approximation depends on the characteristics of the particles, the **wavelength** of the **radiation**, and the way the scattered field is measured. It is more likely to be a good approximation for particles in dilute concentration that are small compared with the wavelength, and for experiments in which the **beamwidth** of the detector is narrow. The assumption of single scattering underlies the interpretation of most **weather radar** and **wind profiler** observations. For **lidar**, single scattering is said to occur when a transmitted **photon** experiences just one scattering event before returning to the **receiver**. *Compare* **multiple scattering**.

single-station analysis—The **analysis** (or reconstruction) of the weather pattern from more or less continuous meteorological observations made at a single geographic location; or the body of techniques employed in such an analysis.

The extension of these techniques to produce a **weather forecast** is known as **single-station forecasting**.

Berry, F. A., E. Bollay, and N. R. Beers, 1945: *Handbook of Meteorology*, 858–879.

single-station forecasting—Weather forecasting based on meteorological observations made at a single station.

See **single-station analysis**.

single-theodolite observation—The usual type of **pilot-balloon observation**, that is, using one theodolite.

Compare **double-theodolite observation**.

singular corresponding point—A center of elevation or depression on a **constant-pressure chart** (or a center of high or low pressure on a **constant-height chart**) considered as a reappearing characteristic of successive charts.

See **corresponding point, principle of geometric association**.

singular point—1. Of a differential equation, a point at which the coefficients are not expandable in a **Taylor series**.

2. Of a function of a complex **variable**, a point at which the function does not have a derivative.

3. (Also called singularity.) Of a flow **field**, a point at which the direction of flow is not uniquely determined, hence, a point of zero speed, for example, a **col**.

singular value decomposition—A numerical procedure to bring a matrix on diagonal form with its singular values on the diagonal.

For a given matrix $\mathbf{M}$ the method finds **orthogonal** matrices $\mathbf{U} = [\mathbf{u}_1, \ldots, \mathbf{u}_n]$ and $\mathbf{V} = [\mathbf{v}_1, \ldots, \mathbf{v}_n]$ so that $\mathbf{U}^T \mathbf{M} \mathbf{V} = \Sigma$ with $\Sigma = diag(\sigma_1, \ldots, \sigma_n)$. The columns of $\mathbf{U}$ and $\mathbf{V}$ are the left and right **singular vectors** of $\mathbf{M}$, respectively, the σ_i are the singular values, and $\mathbf{M}\mathbf{u}_i = \sigma_i \mathbf{v}_i$.

singular vectors—**Vectors** that maximize the ratio between the initial norm of a **perturbation** and the final norm of the perturbation after integration with the **tangent linear model**.

See **singular value decomposition**.

singularity—1. Same as **singular point**.

2. A characteristic meteorological condition that tends to occur on or near a specific date more frequently than chance would indicate.

See **January thaw**.

sink—A route or **reservoir** by which a measurable quantity may exit a system, such as by accumulation (in a reservoir) or chemical conversion.

Compare **source**.

sink processes for trace gases—Any process (chemical or physical) that removes a particular chemical species from the **atmosphere**.

The abundance of a species in the atmosphere can be determined from a consideration of the various sources of the compound to the atmosphere and the sink processes that remove it.

sinkhole—1. A natural **depression** or cavity in a land surface, generally occurring in limestone regions and formed by solution or by collapse of a cavern roof.

2. Popularly, a sudden slumping of the surface due to overpumping of the **aquifer**.

sinking—A **mirage** in which the **image** of distant objects is displaced downward.

Because the displacement increases with distance, a horizontal surface, such as that of a body of water, **desert**, or road, appears to bend downward and one's perception is that of being on top of an inverted bowl or possibly on a planet with a very much smaller radius. Indeed, the downward bending surface results in an (optical) **horizon** that can be very much closer to the observer than in the absence of a mirage. Sinking is an example of an **inferior mirage**. The opposite of sinking is **looming**. Sinking occurs when the concave side of **light** rays from a distant object is up, and this in turn occurs when the **refractive index** of the **atmosphere** increases with height. This only happens near a surface when the **heat flux** is upward and so the **temperature** gradient decreases with height. This is common over a warm surface (such as might occur over sun-warmed ground or a lake at night). Sinking is often accompanied by a two-image inferior mirage. *Compare* **stooping**.

sinusoidal pattern—A **synoptic** pattern, approximating to a sine curve, that is formed by alternate **troughs** and **ridges** of about equal **amplitude**.

siphon barograph—A recording **siphon barometer**.

siphon barometer—A **mercury barometer** so constructed that the upper and lower **mercury** surfaces have the same diameter.

It is not a standard meteorological instrument, but is sometimes utilized for special purposes. *See* **float barograph**.

sirocco—(Also spelled scirocco.) A warm south or southeast **wind** in advance of a **depression** moving eastward across the southern **Mediterranean Sea** or North Africa.

The air comes from the Sahara (as a **desert wind**) and is dry and dusty, but the term is not used in North Africa, where it is called chom (hot) or arifi (thirsty). In crossing the Mediterranean the sirocco picks up much moisture because of its high temperature, and reaches Malta, Sicily, and southern Italy as a very enervating, hot, humid wind. As it travels northward, it causes **fog** and **rain**. In some parts of the Mediterranean region the word may be used for any warm southerly wind, often of **foehn** type. In the extreme southwest of Greece a warm foehn crossing the coastal mountains is named sirocco di levante. There are a number of local variants of the spelling such as xaroco (Portuguese), jaloque or xaloque (Spanish), xaloc or xaloch (Catalonian). In the Rhône delta the warm rainy southeast sirocco is called eissero. On Zakynthos Island it is called lampaditsa. *See* **solano**, **ghibli**, **chili**, **simoom**, **leveche**, **marin**.

sirocco di levante—*See* **sirocco**.

siroeang—*See* **aloegoe**.

SIRS—Abbreviation for **Satellite Infrared Spectrometer**.

site—*See* **site of station**.

site characterization—1. A general term applied to the investigation activities at a specific location that examine natural phenomena and human-induced conditions important to the **resolution** of environmental, safety, and water-resource issues.

2. The program of exploration and research, both in the laboratory and in the field, undertaken to establish the geologic conditions and the ranges of the geologic parameters of a particular **site** relevant to the program. It includes borings, surface excavations, excavation of exploratory shafts, limited subsurface lateral excavations and borings, and in situ testing at depths needed to determine the suitability of the site for a geologic repository, but does not include preliminary borings and geophysical testing needed to decide whether site characterization should be undertaken.

site of station—The location of a **meteorological station** from the point of view of geography, orientation, and position of **shelter** and various instruments to include latitude, longitude, and **elevation**.

Six's thermometer—A combination **maximum thermometer** and **minimum thermometer**.

The tube is shaped in the form of a "U" having a bulb at either end. One bulb is filled with a clear liquid that expands or contracts with **temperature** variation, forcing before it a short column of **mercury** having iron indices at either end. The indices remain at the extreme positions reached by the **mercury column**, thus indicating the maximum and minimum temperatures. The indices can be reset with the aid of a magnet.

size parameter—The ratio of the size of a spherical **scattering** particle to the **wavelength** of the **radiation** being scattered, that is, $\alpha = \pi d / \lambda$, where d is the diameter of the **particle**, and λ is the wavelength of the incident radiation.

Mie theory describes the general process of **extinction**, scattering, and **absorption** of spherical particles as a function of the size parameter and **refractive index**. For $\alpha < \approx 0.1$, the **Rayleigh scattering** approximation becomes valid, whereas for $\alpha > \approx 100$, geometrical optics becomes valid.

skavler—A Norwegian word generally equivalent to **sastrugi**.

skew T–log$p$ diagram—An **emagram** (**temperature** and **logarithm** of **pressure** as coordinates) with the isotherms rotated 45° clockwise to produce greater separation of **isotherms** and **dry adiabats**. *See* **thermodynamic diagram**

skewness—Departure from symmetry.

In **statistics**, the **coefficient of skewness** γ_1 of a **random variable** or of a **probability distribution** is defined as $\gamma_1 = \mu_3 / \sigma^3$, where μ_3 is the third **moment** about the mean and σ is the **standard deviation**. Where $\gamma_1 > 0$ the typical curve trails off toward the right and hence is said to be skewed to the right; when $\gamma_1 < 0$ the longer tail is on the left, and the curve is said to be skewed to the left.

skill—A **statistical** evaluation of the **accuracy** of forecasts or the effectiveness of **detection** techniques.

Several simple formulations are commonly used in meteorology. The skill score (SS) is useful for evaluating predictions of temperatures, pressures, or the numerical values of other parameters. It compares a **forecaster**'s root-mean-squared or mean-absolute **prediction** errors, E_f, over a period of time, with those of a reference technique, E_{refr}, such as forecasts based entirely on **climatology** or **persistence**, which involve no **analysis** of **synoptic** weather conditions:

$$SS = 1 - (E_f / E_{refr}).$$

If SS > 0, the forecaster or technique is deemed to possess some skill compared to the reference technique. For binary, yes/no kinds of forecasts or detection techniques, the probability of detection (POD), false alarm rate (FAR), and critical success index (CSI) may be useful evaluators. For example, if A is the number of forecasts that **rain** would occur when it subsequently did occur (forecast = yes, observation = yes), B is the number of forecasts of no rain when rain occurred (no, yes), and C is the number of forecasts of rain when rain did not occur (yes, no), then

$$POD = A /(A + B)$$

$$FAR = C/(A + C)$$

$$CSI = A /(A + B +C).$$

For perfect forecasting or detection, POD = CSI =1.0 and FAR = 0.0. POD and FAR scores should be presented as a pair.

skill score—*See* **skill**.

skin drag—A retarding force on the **wind** caused by viscous **friction** as air flows parallel to a **smooth surface** that it touches.
> *Compare* **form drag**, **wave drag**; *see* **drag**.

skin friction—(Or viscous drag.) *See* **drag**.

skin-friction coefficient—(Also called friction coefficient, **drag coefficient**.) A dimensionless drag coefficient expressing the proportionality between the **frictional force** per unit area, or the **shearing stress** τ_0 exerted by the **wind** at the earth's surface, and the square of the **surface wind** speed M:

$$\tau_0 = \tfrac{1}{2} C_D \rho M^2,$$

where C_D is the skin-friction coefficient and ρ the air density.
> The skin-friction coefficient is independent of **wind speed** except where the wind modifies the underlying surface, for example, a sheet of water or field of tall grass. It appears to be independent of **thermal** stability and, over a fully **rough surface**, of the wind speed at the reference level, but it does depend on the height of the reference level. Values derived for the **atmosphere** have almost the same magnitudes as those used in **aerodynamics**, ranging from about 0.005 over smooth water to 0.015 over grassland. Some writers define the skin-friction coefficient as twice the value given here. *See also* **drag**; *compare* **form drag**, **wave drag**.
>> Sutton, O. G., 1953: *Micrometeorology*, 255–264.

skip effect—A phenomenon in which **sound** or radio **energy** may be detected only at various distance intervals from the energy source, as the result of the presence of an energy reflecting or refracting layer in the **atmosphere**.
> For long **radio waves**, the **ionosphere** acts as the reflecting layer. For shorter wavelengths, the effect may be produced by strong **superstandard propagation** in elevated **layers** of the **troposphere**. Skip effects make it possible on occasion to detect targets at distances far greater than the **normal** radio **horizon** while closer targets remain undetected.

Skiron—(Also spelled Sciron.) The Greek name for the northwest **wind**, which is cold in winter but hot and dry in summer.
> On the **Tower of the Winds** in Athens it is represented as an old man warmly clad and carrying what appears to be a fire-basket.

sky—1. The apparent surface against which all **aerial** objects are seen from the earth.
> 2. Same as sky condition, **sky cover**, **state of the sky**.

sky cover—In **surface weather observations**, a term used to denote one or more of the following: 1) the amount of sky covered but not necessarily concealed by **clouds** or by **obscuring phenomena** aloft; 2) the amount of sky concealed by obscuring phenomena that reach the ground; or 3) the amount of sky covered or concealed by a combination of 1) and 2).
> Opaque sky cover is the amount of sky completely hidden by clouds or obscuring phenomena,

while **total sky cover** includes this plus the amount of sky covered but not concealed (transparent). Sky cover, for any level aloft, is described as **thin** if the ratio of transparent to total sky cover at and below that level is one-half or more. Sky cover is reported in tenths, so that 0.0 indicates a **clear sky** and 1.0 (or 10/10) indicates a completely covered sky. Amount of sky cover for any given level is determined according to the **summation principle**. The following classifications of sky cover are used in **aviation weather observations: clear, scattered, broken, overcast, partial obscuration, obscuration.**

sky map—A pattern of variable brightness on the underside of a **cloud layer** as a consequence of the variable reflectivity of the surface immediately beneath, whether **ice**, **snow**, land, or water.
　　This term is used mainly in polar regions. *See* **snow blink, ice blink**.

sky radiation—Same as **diffuse sky radiation**.

sky slightly clouded—Sky with a **total cloud cover** equal to one or two **octas**.

sky sweeper—A dry **foehn** wind from the northwest descending the coastal hills of Majorca.

sky wave—In radio terminology, or for **lightning** transients, **electromagnetic waves** received after they have been reflected by the **ionosphere**.
　　Compare **ground wave**.

skyhook balloon—A large plastic **constant-level balloon** for duration flying at very high altitudes, used for determining **wind** fields and measuring upper-atmospheric parameters.

skylight—Same as **diffuse sky radiation**.

slab approximation—A simplifying approximation regarding the horizontal **momentum** in the **mixed layer**, based on the assumption that vertical **shears** across the mixed layer are much smaller than the average shear between the mixed layer and the **transition layer** below it; hence the mixed layer can be viewed as moving like a slab.

slab layer—In **oceanography**, a **mixed layer** for which the **slab approximation** can be made.

slablike motion—A horizontal **velocity** field (usually that of the **mixed layer**), for which the **slab approximation** can be made.

slab model—A **model** of a surface or bottom **mixed layer** in which all quantities (**scalar** and **vector**) are assumed to be completely and instantaneously homogenized.
　　The assumption is that the processes affecting the mixed layer as a whole act slowly in comparison to the **mixing** time of the large **eddies** that stir the layer. *See also* **mixed-layer models, jump model**.

slack water—The state of a **tidal current** when its speed is near zero, especially the moment when a reversing (rectilinear) current changes direction and its speed is zero.
　　For a theoretical standing **tidal wave**, slack water occurs at the times of **high water** and **low water**, while for a theoretical progressive tidal wave, slack water occurs midway between high and low water.

slake—The crumbling and disintegration of soils due to wetting or drying.

slant range—The **line-of-sight** distance between two objects.

slant visibility—1. Same as **approach visibility**.
　　2. Same as **oblique visual range**.

slantwise convection—A form of **convection** driven by a combination of gravitational and centrifugal forces.
　　Slantwise convection can occur in **baroclinic** flows in which the slantwise-upward displacement of **air parcels**, elongated in the direction of the **thermal wind**, results in a **vector** combination of **buoyancy** and **Coriolis** (or centrifugal) and **pressure-gradient** accelerations that drive the parcel in the same direction as the displacement. *See also* **symmetric instability**.

slantwise convective available potential energy—(Abbreviated SCAPE.) The total amount of **potential energy** available for conversion into **kinetic energy** in the slantwise-upward displacement of a particular sample of air in an **atmosphere** that is unstable to **slantwise convection**.
　　It is given by

$$\text{SCAPE} = \int_{PNB}^{P} |_M |R_d \left(T_\rho - T_{\rho e} \right) d \ln p$$

where R_d is the **gas constant** for **dry air**; T_ρ is the **density temperature** of the sample in question, lifted to a **pressure** p by a reversible **adiabatic process**; $T_{\rho e}$ is the density temperature of the **environment** through which the sample is lifted, PNB is the pressure at which the lifted sample has the same density temperature as its environment; and $|_M|$ denotes that the sample should be lifted slantwise along a surface of constant **absolute momentum**, or, in circularly symmetric flows such as **tropical cyclones**, constant **angular momentum**.

slatan—Same as **selatan**.

sleeper—*See* **etesians**.

sleet—1. *See* **ice pellets**.

2. In British terminology, and colloquially in some parts of the United States, **precipitation** in the form of a mixture of **rain** and **snow**.

slice method—A method of evaluating the **static stability** over a limited area at any reference level in the **atmosphere**.

Unlike the **parcel method**, the slice method takes into account **continuity** of mass by considering both upward and downward motion but not **mixing**. The **temperature** excess of the rising air will be 1) negative (stable conditions), 2) zero (neutral or indifferent conditions), or 3) positive (unstable conditions) according to whether the expression

$$\frac{A_d}{A}\,(\gamma_d - \gamma') - (\gamma_d - \gamma)$$

is 1) less than zero, 2) equal to zero, or 3) greater than zero, respectively, where A is the total area of the region under consideration, A_d the area covered by the descending current, γ the **environmental lapse rate** of temperature, and γ' and γ_d the **adiabatic process** lapse rates followed, respectively, by the ascending and descending currents. Since the ratio A_d/A can be evaluated only after **convection** has set in, the slice method is used mainly as a diagnostic, rather than a prognostic, tool.

Emanuel, K., 1994: *Atmospheric Convection*, Oxford University Press, 175–178.

sliding scale—A set of combinations of **ceilings** and **visibilities** that constitute the **operational weather limits** at an airport; as the observed value of one **element** increases, the limiting value of the other element decreases, and vice versa.

sling psychrometer—A **psychrometer** in which the **wet-** and **dry-bulb thermometers** are mounted upon a frame connected to a handle at one end by means of a bearing or length of chain.

Thus, the psychrometer may be whirled by hand in order to provide the necessary **ventilation**.

sling thermometer—A **thermometer** mounted upon a frame connected to a handle at one end by means of a bearing or length of chain, so that the thermometer may be whirled by hand in order to provide the necessary **ventilation**.

slope area method—An indirect method of **discharge** measurement, especially after **floods**, by determining the water surface slope during the peak flow, by using the **high water** marks, a survey of two or more river cross sections along a stretch of the river, and by estimating the river **roughness coefficient**.

The discharge is then calculated by the **Manning equation** or **Chezy equation**.

slope current—*See* **gradient current**.

slope flow—Along-slope flows generated by heating or cooling of the slope.

A warm slope surface, as produced by daytime heating, generates **anabatic** or upslope winds, whereas a cool surface, as from **nocturnal cooling**, generates **katabatic** or downslope winds. Slope flows are a component of **mountain–valley** and **mountain–plains wind systems**.

slope of a front—Tangent of the angle formed by the **frontal surface** with a horizontal plane.

slope of an isobaric surface—Tangent of the angle formed by an **isobar** with a horizontal plane.

Slope Water—A **water mass** found in the permanent **thermocline** between the **Gulf Stream** and the **continental shelf** north of Cape Hatteras (35°N); similar properties as North Atlantic Central Water but about 0.8 **psu** lower in **salinity**.

slope winds—Cross-valley winds that are driven up or down sloping terrain by buoyant forces.

Examples are **anabatic** (upslope) winds caused by rising warm air during daytime and **katabatic** (downslope) winds caused by descending cold air at night. *Compare* **slope flows**.

slope windstorm—(Also called **downslope windstorm**.) Occurs when strong **synoptic-scale** winds blow over a mountain ridge top, where the winds are trapped below a strong **temperature inversion** located at an **altitude** just above the ridge top.

The resulting **fall winds** on the lee side of the ridge are trapped close to the surface and can destroy buildings and blow down trees. *See also* **chinook, foehn**.

slow ion—Same as **large ion**.

slow manifold—The set of all atmospheric states or solutions to a forecast model that evolve slowly with time.

slow tail—The **extremely low frequency** (ELF) component of some **sferics** that lags the initial **very low frequency** (VLF) arrival because of the lower **phase velocity** at low frequencies.

The cause of the slow tail is still controversial. One explanation attributes it to the **continuing current**. The **lightning** radiation arrives after the initiating ground **stroke** because 1) it has components below about 3 kHz in frequencies (necessary for the propagation in the earth–iono-sphere **waveguide** cavity), and 2) it suffers **dispersion** as a result of the waveguide propagation that broadens the original waveguide impulse, making it look like a tail.

> Wait, J. R., 1960: On the theory of the slow-tail portion of atmospheric waveforms. *J. Geophys. Res.*, **65**, 1939.

sludge—(Also called slush.) A dense accumulation of **frazil** (or lolly ice); an early stage in the **freezing** of a body of water.

The sea surface becomes thick and soupy and sometimes greasy in appearance. Sludge depth seldom exceeds one foot. *See* **grease ice**.

sludging—Same as **solifluction**.

sluff—A small downhill movement of **snow**.

slug test—The determination of **hydraulic conductivity** in the vicinity of a well by rapidly displacing the water in the well and calculating the amount of time necessary for the well water to return to its previous level.

slush—1. **Snow** or **ice** on the ground that has been reduced to a soft watery mixture by **rain**, warm **temperature**, and/or chemical treatment.

2. Same as **sludge**.

slush icing—The accumulation of **ice** and water on exposed surfaces of aircraft when the craft is flown through **wet snow** and liquid drops at temperatures near 0°C.
  See **aircraft icing**.

small calorie—(Abbreviated cal.) Same as **calorie**.

small circle—A line formed by the intersection of the surface of a sphere and a plane that does not pass through the center of the sphere.
  See **great circle**.

small-craft warning—A **warning**, for marine interests, of impending winds up to 28 knots (32 mph).

After 1 January 1958, the storm-warning signals for this condition became 1) one triangular red pennant by day and 2) a red lantern over a white lantern by night. *See* **storm warning**.

small eddy closure—*See* **gradient transport theory, K-theory, eddy viscosity**.

small eddy theory—*See* **gradient transport theory, K-theory, eddy viscosity**.

small hail—**Hail** with a diameter less than 0.64 cm (0.25 in.).
  See **ice pellets**.

small halo—Same as **halo of 22°**.

small ion—(Also called light ion, fast ion.) An atmospheric **ion** of the type that has the greatest **mobility** and hence, collectively, is the principal agent of atmospheric **conduction**.

The exact physical nature of the small ion has never been fully clarified, but much evidence indicates that each is a singly charged atmospheric molecule (or, rarely, an atom) about which a few other neutral molecules are held by the electrical attraction of the central ionized molecule. Estimates of the number of satellite molecules are as high as twelve. When freshly formed by any of several **atmospheric ionization** processes, small ions are probably singly charged molecules, but after a number of collisions with neutral molecules they acquire (actually, in a fraction of a second) their cluster of satellites. Even with these satellites clustering about the central charged molecules, the **ion mobility** of the resulting complex is of the order of 10^4 times greater than that of large ions. Negative small ions exhibit slightly greater mobilities than positive small ions,

1.9×10^{-4} m s^{-1} per volt cm^{-1} being typical of negative, and 1.4×10^{-4} m s^{-1} per volt m^{-1} being typical of positive small ions in **dry air** at **sea level**. Small ions may disappear either by direct **recombination** with oppositely charged small ions or by combination with neutral **Aitken nuclei** to form new **large ions**, or by combination with large ions of opposite sign. The concentration of small ions near sea level is typically about 5×10^{-4} of each sign per m^3, both over the oceans and over land. This concentration increases with **altitude**, and at 18 km it is about 10^{-3} per m^3.

small-ion combination—Either of two processes by which small ions disappear.
The first of these processes is the union of a **small ion** and a neutral **Aitken nucleus** to form a new **large ion**; the second is the neutralization of a large ion by the small ion. The rate at which these processes occur is expressed by the magnitude of the **combination coefficient** for each process.

small perturbation—In the context of **tangent linear** and **adjoint models**, **perturbations** are considered small if their development can be described with acceptable **accuracy** by **tangent linear equations**.
In many cases, this class of perturbations includes those that are comparable in magnitude to the typical analysis errors of operational forecast models. *See* **tangent linear approximation**.

smaze—(Rare.) A combination of **smoke** and **haze**; or, a very light smoke condition that resembles haze.

SMMR—Abbreviation for **Scanning Multifrequency Microwave Radiometer**.

SMO—Abbreviation for **supplementary meteorological office**.

smog—As originally coined in 1905 by Des Voeux: a natural **fog** contaminated by industrial pollutants, a mixture of **smoke** and fog.
Today, it is the common term applied to problematical, largely urban, **air pollution**, with or without the natural fog; however, some visible manifestation is almost always implied. Smogs are constituted in great variety, but a major dichotomy exists between the photochemical smogs of **nitrogen oxides** and **hydrocarbons** emitted mainly by automobile engines and, on the other hand, the sulfur-laden, sometimes deadly, smogs produced by the large-scale combustion of fuel oil and coal. Both types contain **carbon monoxide, carbon dioxide**, and a variety of **particulates**. *See* **Los Angeles (photochemical) smog, London (sulfurous) smog**.

smog chambers—Large reaction chambers in which chemical reactions associated with **smog** chemistry are studied.
Typically, mixtures of a hydrocarbon in air containing traces of **nitric oxide** are irradiated, and the rate of buildup of **ozone, nitrogen dioxide**, and organic products is measured and recorded.

smoke—1. Foreign particulate matter in the **atmosphere** resulting from combustion processes; a type of **lithometeor**.
When smoke is present, the disk of the sun at **sunrise** and **sunset** appears very red, and during the daytime has a reddish tinge. Smoke that has come a great distance from its source, such as from forest fires, usually has a light grayish or bluish color and is evenly distributed in the **upper air**. *See* **smog, haze**.
2. Applied to some types of **fog**.
See **sea smoke**.

smoke horizon—The top of a **smoke** layer that is confined by a low-level **temperature inversion** in such a way as to give the appearance of the **horizon** when viewed from above against the sky.
In such instances the **true horizon** is usually obscured by the smoke layer. *See also* **dust horizon, fog horizon, haze horizon**.

smoke pall—A condition of **visibility** restriction caused by an overspreading mass of **smoke** that produces a gloomy effect.

smoke plumes—Trails of emissions, advected by the **wind**, that are laden with **particulates** and gaseous **pollutants** from a continuous source.
Plumes are generally visible for at least a short distance **downwind** due to condensed **water vapor, particulate loading**, or visible gaseous compounds.

smokes—Dense, white **fogs** common in the **dry season** on the Guinea coast of Africa.
They often precede a **harmattan**.

smooth surface—*See* **aerodynamically smooth surface**.

smoothing—An averaging of data in space or time, designed to compensate for **random errors** or fluctuations of a **scale** smaller than that presumed significant to the problem at hand.

Thus, for example, a **thermometer** smooths the **temperature** reading on the scale of its **time constant**; the **analysis** of a sea level **weather map** smooths the **pressure field** on a space scale more or less systematically determined by the analyst by taking each **pressure** as representative not of a point but of an area about the point. *See* **consecutive mean, curve fitting, filtering, bloxam**.

SMS—Abbreviation for **Synchronous Meteorological Satellite**.

smudging—A frost-preventive measure used in orchards.

Properly, it means the production of heavy **smoke** from the burning of **fossil fuels** intended to prevent **radiational cooling**, but the term is applied to both heating and smoke production.

snap—A colloquial term for a brief period of extreme (generally cold) weather setting in suddenly, as in a "cold snap."

snappy—Colloquially descriptive of cold (and often dry and windy) weather.

Snel's law—(Often misspelled Snell's law.) *See* **refractive index**.

Snell's law—*See* **Snel's law**.

sno—Same as **elvegust**.

SNOTEL—(Acronym for snow telemetry.) In remote areas of the western United States, SNOTEL sites, comprising a **snow pillow**, a shielded **standpipe storage precipitation gauge**, and a radio **transmitter**, are used to telemeter **precipitation** data to a satellite.

snow—**Precipitation** composed of white or translucent **ice crystals**, chiefly in complex branch hexagonal form and often agglomerated into **snowflakes**.

For weather-observing purposes, the **intensity** of snow is characterized as 1) light when the **visibility** is 1 km (5/8 **statute mile**) or more; 2) moderate when the visibility is less than 1 km (5/8 statute mile) but not more than 1/2 km (5/16 statute mile); and 3) heavy when the visibility is less than 1/2 km (5/16 statute mile).

snow accumulation—(also called snow depth.) A measurement of the **depth of snow** on the ground made either since the **snow** began falling or since a previous **observation**.

The total accumulation is equivalent to the total snow depth during a **storm**, or after any single snowstorm or series of storms. Snow accumulation can vary due to settling and melting and will therefore vary depending on how often it is measured. For example, if **new snow** is measured every hour during a relatively long duration storm, it is likely that the summed accumulations may exceed a total snow accumulation measured only once at the end of the storm.

snow banner—(Also called snow smoke, snow plume.) **Snow** being blown from a mountain crest.

It is sometimes mistaken for volcanic **smoke** or a **banner cloud**.

snow blindness—Impaired vision or temporary blindness caused by **sunlight** reflected from **snow** surfaces.

The medical name is niphablepsia. "Symptoms of snow blindness are a gritty sensation under the eyelids, excessive watering, double vision. First aid is to place the casualty in the dark or bandage the eyes; application of cool compresses alleviates pain. Most cases will recover in 18 hours without medical treatment"(from *Glossary of Arctic and Subarctic Terms* 1955).

Arctic, Desert, Tropic Information Center (ADTIC) Research Studies Institute, 1955: *Glossary of Arctic and Subarctic Terms*, ADTIC Pub. A-105, Maxwell AFB, AL, 90 pp.

snow blink—(Also called snow sky.) A relatively bright region on the underside of **clouds** produced by the **reflection** of **light** by **snow**.

This term is used in polar regions where it contributes to the **sky map**; snow blink is brighter than **ice blink**, or reflection of light by land or water.

snow board—Specially constructed board used to identify the surface of **snow** or **ice** that has been covered by more recent **snowfall**.

Snow boards are used as an aid in obtaining representative samples of solid **precipitation** at times when the **catch** in the gauge is considered erroneous, for example, in windy weather and **wet snow**. They provide the reference level for the measurement of new snowfall and ice. They are constructed of thin metal, wood, or other light material so as to not sink in the snow. They should be at least 40 cm by 40 cm and covered with white cloth or plastic.

NASA/National Weather Service, 1972: *National Weather Service Observing Handbook No. 2, Substation Observations*, NOAA, US GPO.

snow cap—Snow covering the ridges and peaks of mountains when no **snow** exists at lower elevations.

snow climate—Same as **polar climate**.
   Compare **forest climate, snow forest climate**.

snow cloud—Any **cloud** from which **snow** falls; a popular term having no technical connotation.

snow concrete—**Snow** that has been compressed at low temperatures and that sets into a tough substance of considerably greater strength than uncompressed snow (from *Glossary of Arctic and Subarctic Terms* 1955).

   Arctic, Desert, Tropic Information Center (ADTIC) Research Studies Institute, 1955: *Glossary of Arctic and Subarctic Terms*, ADTIC Pub. A-105, Maxwell AFB, AL, 90 pp.

snow course—An established line or transect of measurements of snow **water equivalent** across a **snow** field in representative mountainous terrain, where appreciable snow accumulates, to monitor seasonal snowpack.
   Although intended to be measured over a distance of several hundred meters, in practice snow courses may consist of four or so individual measurements. The measurements are used for **runoff** prediction and for assessing the potential for **flooding** or **drought**, and they are also used in longer-term climatological studies. Along the snow course, core samples are taken periodically (often monthly) throughout the snow season. Approximately 300 snow courses are operated over the western United States.

snow cover—1. The areal extent of snow-covered ground, usually expressed as percent of total area in a given region.
   2. In general, a layer of **snow** on the ground surface.
   Compare **snowfield, snowpack**.
   3. The **depth of snow** on the ground, usually expressed in inches or centimeters.

snow-cover chart—A **synoptic chart** showing areas covered by **snow** and showing **contour lines** of **snow depth**.

snow crust—A crisp, firm, outer surface upon **snow**.
   Basically, three types of snow crusts exist, formed by 1) the refreezing of surface snow, after melting and/or wetting, to form a hard layer of snow (**sun crust, rain crust, spring crust**); 2) the packing of snow into a hard layer by **wind** action (**wind crust, wind slab**); and 3) the **freezing** of **surface water**, however derived, to form a continuous layer of **ice** on top of snow (**film crust, ice crust**). A snow crust is designated as "breakable" or "unbreakable" according to its ability to support a person on skis.

snow crystal—Any of several types of **ice crystal** found in **snow**.
   A snow crystal is a single **crystal**, in contrast to a **snowflake**, which is usually an aggregate of many single snow crystals.

snow day—A day on which a fall of **snow** is observed.

snow density—The ratio of the volume of meltwater that can be derived from a sample of **snow** to the original volume of the sample; strictly speaking, this is the **specific gravity** of the snow sample.
   Freshly fallen snow usually has a snow density of 0.07 to 0.15; **glacial** ice formed from compacted snow (or **firn**) has a maximum density of about 0.91. Values as low as 0.004 have been measured.
   Compare **water content, water equivalent**.

snow depth—Same as **snow accumulation**.

snow eater—1. Any warm **wind** blowing over a **snow** surface; usually applied to a **foehn** wind, that is, schneefresser.
   See also **chinook**.
   2. A **fog** over a **snow** surface; so called because of the frequently observed rapidity with which a **snow cover** disappears after a fog sets in.
   As **water vapor** from the air condenses on the snow, the **latent heat** of **condensation** causes the snow to become warmer and melt faster.

snow fence—1. An open, slatted board fence usually 1–3 m high, placed **upwind** of a railroad track or highway.
   The fence serves to create **eddies** in the **downstream** airflow, resulting in a reduced **wind speed** such that **snow** is deposited close to the fence on its **leeward** side. The intent is to provide a comparatively clear zone along the railroad track or highway. A snow fence is also used to accumulate **drifting snow** in a flat windswept area to reduce the depth of **ground frost** and increase **soil moisture** as the snow melts.
   2. Same as **Wild fence**.

snow flurry—Common term for a light **snow shower**, lasting for only a short period of time.

snow forest climate—A major category (the *D* climates) in W. Köppen's **climatic classification**, defined by a coldest-month **mean temperature** of less than −3°C (26.6°F) and a warmest-month mean temperature of greater than 10°C (50°F).

The first limit separates it from **temperate rainy climates**, and the second from **tundra climates**. It is distinguished from the **dry climates** by a function of annual **temperature** and **precipitation** (see formulas under **steppe climate**). The outstanding feature of these climates is the cold winters with at least a month of snow-covered ground. These are the coldest of the **tree climates**. In C. W. Thornthwaite's classifications, this general type of **climate** would be 1) in the 1931 system, **humid** or **subhumid** and **microthermal climate**, or **taiga climate**; and 2) in the 1948 system, humid or moist subhumid and microthermal climate.

Köppen, W. P., and R. Geiger, 1930–1939: *Handbuch der Klimatologie*, Berlin: Gebruder Borntraeger, 6 vols.
Thornthwaite, C. W., 1931: The climates of North America according to a new classification. *Geogr. Rev.*, **21**, 633–655.
Thornthwaite, C. W., 1948: An approach toward a rational classification of climate. *Geogr. Rev.*, **38**, 55–94.

snow garland—A rare and beautiful phenomenon in which **snow** is festooned from trees, fences, etc., in the form of a rope of snow, several feet long and several inches in diameter, formed and sustained by **surface tension** acting in thin films of water bonding individual crystals.

Such garlands form only when the **surface temperature** is close to the **melting point**, for only then will the requisite films of slightly **supercooled water** exist.

snow gauge—An instrument for measuring the vertical **depth of snow**.

The **eight-inch rain gauge** is adapted for **snow** measurement by removing the funnel and measuring tube so that snow is collected in the **overflow** can. The **weighing rain gauge** is also used for measuring **snowfall** by removing the funnel portion of the **collector**. Other instruments used for measuring **snow depth** include the **snow sampler** and **snow stake**.

snow-generating level—A layer in the middle or upper **troposphere** in widespread **precipitation** in which **ice crystals** form in small convective cells and fall to lower altitudes.

Such layers are thought to be characterized by **convective instability**. Small convective cells called generating cells that develop within the layer may produce ice crystals that then settle into lower altitudes. The base of the convectively unstable layer is called the snow-generating level. On time–height displays from vertically pointing radars, generating cells are typically observed to extend about 1–2 km above the snow-generating level. **Snow trails**, or streamers, are terms used to describe the trails of **precipitation** that subsequently emerge from the base of the generating cells.

snow geyser—Fine powdery **snow** blown upward by a **snow tremor**.

snow grains—(Also called granular snow.) **Precipitation** in the form of very small, white opaque **particles** of ice; the solid equivalent of **drizzle**.

They resemble **snow pellets** in external appearance, but are more flattened and elongated, and generally have diameters of less than 1 mm; they neither shatter nor bounce when they hit a hard surface. Descriptions of the physical structure of snow grains vary widely and include very fine, simple **ice crystals**; tiny, complex **snow crystals**; small, compact bundles of **rime**; and particles with a rime core and a fine **glaze** coating. It is agreed that snow grains usually fall in very small quantities, mostly from **stratus** clouds or from **fog**, and never in the form of a **shower**.

snow ice—Ice formed by **freezing** of a mixture of **snow** and water.

snow line—1. In general, the outer boundary of a snow-covered area.

It has at least two specific applications: 1) the actual lower limit of the **snow cap** on high terrain at any given time; 2) the ever-changing equatorward limit of **snow cover**, particularly in the Northern Hemisphere winter.

2. Same as **firn line**. *See* **climatic snow line**.

snow mat—A special device used to mark the surface between **old** and **new snow**.

It consists of a piece of white duck 28 inches square, having in each corner triangular pockets in which are inserted slats placed diagonally to keep the mat taut and flat. *See* **snow board**, **snow gauge**.

snow mist—Same as **ice crystals**.

snow patch—An isolated area of **snow cover**.

snow pellets—(Also called soft hail, **graupel**, tapioca snow.) **Precipitation** consisting of white, opaque, approximately round (sometimes conical) **ice** particles having a snowlike structure, and about 2–5 mm in diameter.

Snow pellets are crisp and easily crushed, differing in this respect from **snow grains**. They rebound when they fall on a hard surface and often break up. In most cases, snow pellets fall in **shower** form, often before or together with **snow**, and chiefly on occasions when the **surface temperature** is at or slightly below 0°C (32°F). It is formed as a result of **accretion** of supercooled droplets collected on what is initially a falling **ice crystal** (probably of the spatial aggregate type).

snow pillow—An instrument designed to provide a direct estimate of water equivalence of a **snowpack** by measuring the **pressure** due to the mass of overlying **snow**.

snow plume—Same as **snow banner**.

snow ripple—Same as **wind ripple**.

snow roller—A rolled-up, cylindrical mass of **snow**, rather common in mountainous or hilly regions.
 It occurs when snow, moist enough to be cohesive, is picked up by **wind** blowing down a slope and rolled onward and downward until it either becomes too large or the ground levels off too much for the wind to propel it farther. Snow rollers vary in size from very small cylinders to some as large as 1.5 m long and more than 2 m in circumference.

snow sampler—(Also called snow tube.) A hollow tube for collecting a sample of **snow** in situ.
 See **Mount Rose snow sampler**

snow scale—Same as **snow stake**.

snow shed—A protective structure erected over railroad tracks to prevent deep **snow** from accumulating on the tracks.
 They are used in deep cuts in mountains where **avalanches** are common.

snow shower—A brief period of **snowfall** in which **intensity** can be variable and may change rapidly.
 A snow shower in which only light **snow** falls for a few minutes is typically called a **snow flurry**.

snow sky—Same as **snow blink**.

snow smoke—Same as **snow banner**.

snow stage—Part of an obsolete conceptual **model** of **air parcel** ascent referring to conditions under which the **condensation level** is at a **temperature** below **freezing** and it is assumed that all condensed **vapor** immediately freezes.
 Other portions of the ascent were described as the **dry stage**, the **rain stage**, and the **hail stage**.

snow stake—(Also called snow scale.) A wooden **scale**, calibrated in inches or centimeters, used in regions of deep **snow** to measure its depth.
 The scale is bolted to a wood post or angle iron set in the ground.

snow survey—The process of determining depth and **water content** of **snow** at representative points, for example, along a **snow course**.

snow trails—(Also called streamers.) The trails of **precipitation** that emerge from the base of generating cells typically observed on time–height displays from vertically pointing radars.
 Snow trails are also commonly observed on **range–height indicator** displays. Snow trails emerge from a layer of **convective instability** that often exists in the middle or upper **troposphere** in widespread storms. Small convective cells developing within this layer produce the **ice crystals** that then fall to lower altitudes. The base of the convectively unstable layer is called the **snow-generating level**. The shape and vertical extent of the streamers depend on the vertical profiles of **wind** and **relative humidity** in the layer through which the precipitation falls.

snow tremor—A disturbance in a **snowfield** caused by the simultaneous settling of a large area of thick **snow crust** or surface layer.
 It occurs when **wind** action has maintained the top foot or more as closely packed, fine-grained **snow** rather impervious to air movement. Meanwhile, at lower depths, **firnification** has caused the larger crystals (**depth hoar**) to grow at the expense of smaller ones, creating air pockets and a weak structure. The collapse of this structure (a perceptible drop, but rarely as much as 1 in.) may be accompanied by a loud report; over a large, level field, adjacent patches may settle as a series of tremors. Occasionally, a **snow geyser** may be blown upward through the crack of a settling patch.

snow tube—Same as **snow sampler**.

snowbreak—Any barrier designed to shelter an object or area from **snow**; analogous to **windbreak**.
 See **snow fence, snow shed**.

snowcreep—An extremely slow, continuous, downhill movement of a mass of **snow**.

snowcrete—Same as **snow concrete**.

snowdrift—A mound or bank of **snow** deposited as sloping surfaces and peaks, often behind obstacles and irregularities, due to **eddies** in the **wind field**.

snowdrift glacier—A semipermanent mass of **firn** formed by drifted **snow** in depressions in the ground or behind obstructions.

snowdrift ice—Permanent or semipermanent masses of **ice**, formed by the accumulation of drifted **snow** in the lee of projections or in depressions of the ground.

snowfall—1. In **surface weather observations**, usually expressed as centimeters or inches of **snow depth** per six-hourly period.

2. The accumulation of **snow** during a specified period of time.

snowfield—1. Generally, an extensive area of snow-covered ground or **ice**, relatively smooth and uniform in appearance and composition.

This term is often used to describe such an area in otherwise coarse, mountainous, or **glacial** terrain.

2. In **glaciology**, a region of permanent **snow cover**, more specifically applied to the **accumulation area** of **glaciers**.

snowflake—Colloquially an **ice crystal**, or more commonly an **aggregation** of many crystals that falls from a **cloud**.

Simple snowflakes (single crystals) exhibit beautiful variety of form, but the symmetrical shapes reproduced so often in photomicrographs are not found frequently in snowfalls. Broken single crystals, fragments, or clusters of such elements are much more typical of actual **snow**. Snowflakes made up of clusters of crystals (many thousand or more) or **crystal** fragments may grow as large as three to four inches in diameter, often building themselves into hollow cones falling point downward. In extremely still air, flakes with diameters as large as 10 inches have been reported.

snowmelt—The water resulting from the melting of **snow**.

snowmelt flood—A substantial rise in **stream** or river discharge caused by snowmelt **runoff**.

snowpack—A laterally extensive accumulation of **snow** on the ground that persists through winter and melts in the spring and summer.

snowpack yield—The amount of water that drains from the **snowpack** as a result of **precipitation** and/or **snowmelt**.

snowquake—Same as **snow tremor**.

snowslide—Same as **avalanche**.

snowstorm—A **storm** characterized by a fall of **frozen precipitation** in the form of **snow**.

SNR—Abbreviation for **signal-to-noise ratio**.

socked in—In the early days of aviation, commonly used to describe weather at an airport when **ceiling** and/or **visibility** were of such low values that the airport was effectively closed to aircraft operations.

The term probably originated from reference to the **wind sock**. When the sock was in the **clouds** or when the visibility was so badly impaired that the sock was not visible, the airport was "socked in." The expression is still widely used.

sodar—(Coined word from sound detection and ranging.) **Sound-wave** transmitting and receiving equipment operated on principles analogous to those of **radar**.

Irregularities in atmospheric **temperature** and **wind** velocity constitute sources of **scattering** for **acoustic waves**. Sodars measure vertical profiles of the mean and turbulent properties of the sound to heights of several hundred meters by transmitting acoustic waves upward and measuring the **Doppler shift** in the backscattered acoustic signals. Sodar is sometimes inappropriately called **acoustic radar**.

sodium—(Symbol Na.) A very light, soft, metallic **element**, **atomic number** 11, atomic mass 23; one of the most common metals on earth.

It is alkaline, very reactive, and, due to its high activity, sodium is almost always found in chemical combination with other elements. It is a major component of minerals, ocean water, and marine **aerosol** particles. Sodium is present in atomic form in the **upper atmosphere** due to the

ablation of meteors and forms a sodium layer at the approximately 80-km **altitude**. Sodium has a very strong, yellow, optical transition at 593 nm (sodium D-line). **Emission** from this line is seen in the **airglow**, following the reaction of sodium with **ozone**. The transition has also been used in ground-based **lidar** sounding applications to detect the passage of **gravity waves**.

SOFAR technology—(Acronym for Sound Fixing and Ranging technology.) Technology developed during World War II when it was discovered that there was a "channel" in the deep ocean within which the acoustic energy from a small explosive charge (deployed in the water by a downed aviator, for example) could transmit over long distances.

An array of hydrophones could be used to range on and roughly locate the source of the charge thereby allowing rescue of downed pilots far out to sea. This "channeling" of **sound** occurs because there is a minimum in the vertical sound speed **profile** in the ocean caused by changes in the **density** of the water column. The density is affected by water **temperature**, **pressure** (depth), and **salinity**. Changes in the **speed of sound** in the water are largely due to changes in temperature and pressure, with salinity offering only a minor effect. *See also* **RAFOS technology**.

soft hail—Same as **snow pellets**.

soft radiation—Quantum **radiation** (**particles** or **photons**) of limited penetrating power, typically of lower **frequency** or longer **wavelength**.

A 10-cm thickness of lead is usually used as the criterion upon which the relative penetrating power of various types of radiation is based. Hard radiation will penetrate such a **shield**; soft radiation will not. *See* **hard radiation**.

soft rime—A white, opaque coating of fine **rime** deposited chiefly on vertical surfaces, especially on points and edges of objects, generally in **supercooled fog**.

On the **windward side**, soft rime may grow to very thick layers, long feathery cones, or needles pointing into the **wind** and having a structure similar to that of **frost**.

soft water—Water without significant hardness, that is, low (< 60 mg/l) in concentration of magnesium or calcium salts.

soil air—The air and other gases in spaces in the soil; specifically, that which is found within the **zone of aeration**.

The **aeration** of soil by exchange with the **free air** above is a continuing process, but is accelerated by **pressure** and **temperature** changes (*see* **normal aeration**). The composition of soil air is modified by bacterial activity and by chemical processes due to the action of **oxygen** and **carbon dioxide** in the presence of **soil moisture**. Generally speaking, soil air contains more carbon dioxide than does free air. In addition to the gases originating in the soil, the following may be present in the **soil atmosphere** in measurable quantities: **ammonia**, **hydrogen sulfide**, **methane**, and other **hydrocarbons** resulting from decay of organic materials. Traces of **helium**, **radon**, **thoron**, and **actinon** from radioactive processes also diffuse through the soil layers into the **atmosphere** by the process of **exhalation**.

soil atmosphere—Same as **soil air**.

soil creep—Slow downhill movement, usually over relatively short distances, of near-surface masses of soil and loose rock material on hillslopes under the influence of **gravity**, soil dynamics (shrink–swell or freeze–thaw action), and soil-water movements.

Same as **surficial creep**.

soil flow—Same as **solifluction**.

soil moisture—The total amount of water, including the **water vapor**, in an unsaturated soil.

soil moisture content—The amount of water in an unsaturated soil, expressed as a volume of water per unit volume of porous media, or as a mass of water per unit oven-dry mass of soil.

soil moisture deficit—The difference between the amount of water actually in the soil and the amount of water that the soil can hold.

The amount of water the soil can hold is generally called **field capacity**, which is the amount of water that remains after gravitational forces have drained water from the soil macropores. *See* **field water-holding capacity**, **gravitational water**.

soil moisture profile—A graph of soil **water content** as a function of depth in the soil.

soil moisture tension—A measure of the amount of **work** that must be done (suction that would need to be applied) to move water in the soil.

Also the **energy** of retention at the outer edge of the moisture film of the soil **particles**. Tensions are generally expressed in **pascals**, **bars**, or **atmospheres**. *See* **water potential**.

soil stripe—(Same as earth stripe, soil strip.) A zone of sorted soil **particles** oriented as one of several alternating bands of coarse and fine material appearing on the soil surface and commonly aligned parallel to downslope **transport** on hillslopes.

soil temperature—The **temperature** measured at a given soil depth, typically at 2, 4, 8, and sometimes 20 and 40 in.

Many biological processes, including seed germination, plant emergence, microbial activity, and soil respiration are a function of soil temperature.

soil thermograph—A remote-recording **thermograph** with a **sensing element** that may be buried at various depths in the earth.

See **soil thermometer**.

soil thermometer—(Also called earth thermometer) A **thermometer** used to measure the **temperature** of the soil.

Two forms of the **mercury-in-glass thermometer** are used for this purpose. For measurement at small depths, a thermometer with a right-angle bend in the stem is used. The bulb is inserted into a hole in the ground with the stem lying along the surface. A thermometer that has been fused into an outer protecting glass shield is used for measurement at greater depths. Wax is inserted between the bulb and the shield to increase the **time constant**. To obtain a measurement, the instrument is lowered into a steel tube that has been driven into the soil to the desired depth.

soil-water pressure—The **pressure** of the water in a soil expressed relative to **atmospheric pressure** (gauge pressure).

soil zone—A vertical or horizontal region of soil with common characteristics.

soilclime—Expression of **soil temperature** and moisture status.

sol–air temperature—The **temperature** that, under conditions of no **direct solar radiation** and no air motion, would cause the same **heat transfer** into a house as that caused by the interplay of all existing atmospheric conditions.

It is given by the formula

$$\theta = T + \frac{aI - E}{c},$$

where θ is the sol–air temperature, T the outdoor **air temperature**, a the radiational **absorptivity** of the wall surface, I the incident **radiation**, c the coefficient of convective heat transfer between air and building material, and E the difference between the **longwave radiation** emitted and received by the surface. In practice, this equation has been reduced to

$$\theta = T + \frac{a'I}{c'},$$

where θ, T, and I are the same as above; a' is a constant for construction material; and c' is a constant for **climate**.

> Landsberg, H. E., 1954: Bioclimatology of housing. *Recent Studies in Bioclimatology, Meteor. Monogr.*, **2(8)**, p. 86.

solaire—A name generally applied to winds from an easterly direction (i.e., from the rising sun) in central and southern France. Local variants are soulédras, soulèdre.

In the Morvan Mountains the solaire blows from the south. *Compare* **matinal, solano**.

solano—A southeasterly or easterly **wind** on the southeast coast of Spain in summer; usually an extension of the **sirocco**.

It is hot and humid and sometimes brings **rain**; when dry it is dusty. *Compare* **solaire**.

solar activity—Any type of **variation** in the appearance or **energy** output of the sun; usually associated with the variation of sunspots and other features over the 11-year **solar cycle**.

The number and magnitude of these features vary with time, from minimum values to maximum values, that is, from **solar minimum** to **solar maximum**. *See* **faculae, flare, granules, plage, prominence, spicules, sunspot**.

solar atmospheric tide—An **atmospheric tide** due to the **thermal** or gravitational action of the sun.

Six- and eight-hour components of small **amplitude** have been observed. They are primarily thermal in origin. The 12-hour component has by many times the greatest amplitude of any atmospheric **tidal component**, about 1.5 mb at the **equator** and 0.5 mb in mid-latitudes. This relatively large amplitude is often explained as a **resonance** effect. The 24-hour component is a **thermal tide** with great local **variability**.

solar aureole—(Also called circumsolar radiation.) **Radiation** scattered out of the direct solar **beam** is often forward peaked, especially when the **scattering** is caused by **dust**.

Therefore, the **scattered radiation** is concentrated at small angles close to the sun. This aureole phenomenon provides a sensitive measure of dust loading in the **atmosphere**.

Solar Backscatter Ultraviolet Radiometer—(Abbreviated SBUV.) A nonscanning, **nadir** viewing instrument that measures backscattered **solar radiation** to estimate global **ozone** distribution.

SBUV-1 was tested on *Nimbus-7*, launched in October 1978, with a modified version, SBUV-2, becoming an operational instrument on most of the **Advanced TIROS-N** satellites.

solar climate—The hypothetical **climate** that would prevail on a uniform solid earth with no **atmosphere**.

Thus, it is a climate of **temperature** alone and is determined only by the amount of **solar radiation** received, that is, determined by the **elevation** of the sun as it varies with **season**. Solar climatic zones, therefore, are parallel to lines of latitude. The solar climate and the nearly synonymous **mathematical climate** have both grown out of the earliest approach to **climatology**, as is indicated by the fact that "climate" takes its name from the **inclination** (Greek *klima*) of the sun's rays.

solar constant—*See* **total solar irradiance**.

solar corona—*See* **corona**.

solar cycle—Most indicators of **solar activity** vary in a quasiperiodic manner, with successive maxima separated by an average interval of about 11 years—the so-called solar cycle.

If reversal of the magnetic-field **polarity** in a given hemisphere in successive 11-year periods is taken into account, the complete solar cycle may be considered to average some 22 years.

solar day—The time interval from midnight to midnight, with midnight being the sun's transit across the lower **meridian**.

solar flare—*See* **flare**.

solar infrared—That portion of the **infrared** spectrum containing significant **radiant energy** from the sun.

Wavelengths range from 720 nm to about 4 μm. *See also* **near-infrared**.

solar maximum—*See* **sunspot maximum**.

solar minimum—*See* **sunspot minimum**.

solar prominence—*See* **prominence**.

solar proton event—The episodic **flux** of high-energy protons and other energetic **particles** from the sun.

The **ionization** created by these particles as they enter the **upper atmosphere** is responsible for **polar cap absorption events** and for chemical effects in the neutral upper atmosphere.

solar radiation—The total **electromagnetic radiation** emitted by the sun.

To a first approximation, the sun radiates as a **blackbody** at a **temperature** of about 5700 K; hence, about 99.9% of its **energy** output falls within the **wavelength** interval from 0.15 to 4.0 μm, with peak **intensity** near 0.5 μm. About one-half of the total energy in the solar **beam** is contained within the **visible spectrum** from 0.4 to 0.7 μm, and most of the other half lies in the **near-infrared**, a small additional portion lying in the **ultraviolet**. *See* **insolation, direct solar radiation, diffuse sky radiation, global radiation, extraterrestrial radiation, solar constant, total solar irradiance**.

Fritz, S., 1951: *Compendium of Meteorology*, 17–19.

solar radiation observation—An evaluation of the **radiation** from the sun that reaches the **observation** point.

The observing instrument is usually a **pyrheliometer** or **pyranometer**. Two types of such observations are taken. The more common consists of measurements of the radiation reaching a horizontal surface, consisting of both radiation from the sun (**direct solar radiation**) and that reaching the instrument indirectly by **scattering** in the **atmosphere** (**diffuse sky radiation**). The second type of observation involves the use of an equatorial mount that keeps the instrument pointed directly at the sun at all times. The sensitive surface of the instrument is normal to the path of **solar radiation** and is shielded from indirect radiation from the sky.

solar radio emission—**Energy** emitted by the sun at wavelengths detected by radio receivers; in particular, radio **noise** produced by solar disturbances, especially solar **flares**.

solar signal—**Electromagnetic energy** from the sun measured by a **radar** and used as a reference **signal** for **calibration**.

The solar signal may also be regarded as **noise** when a **radar beam** scans across the sun or when it interferes with communications. *See* **sun pointing**.

solar spectrum—That part of the **electromagnetic spectrum** occupied by the wavelengths of **solar radiation**.

The shape is well represented by the continuum **emission** from a **blackbody** with **temperature** near 5700 K superimposed with **Fraunhofer absorption lines**.

solar–terrestrial relationships—The **statistical** or physical links between events that occur on the sun and a response observed on the earth, primarily in the **atmosphere**.

solar tide—A **partial tide**, with a **period** of 12 hours, caused by the **tide-producing force** of the sun.

See **solar atmospheric tide**.

solar–topographic theory—The theory that the changes of **climate** through **geologic time** (the paleoclimates) have been due to changes of land and sea distribution and **orography** combined with fluctuations of **solar radiation** of the order of 10%–20% on either side of the mean.

Topographic changes may account for the occurrence of **ice ages** as a whole, but not for the more rapid alternations of **glacial** and **interglacial** periods. A combination of the topographic with the solar fluctuations seems to offer a nearly complete explanation of climatic changes, but it is not yet clear which plays the main part.

solar wind—Stream of ionized gas, mainly **hydrogen**, continuously flowing outward from the sun, at very high velocity (≈ 400 km s^{-1} on the average) and with variable intensity.

While passing near the earth the solar wind interacts with the **magnetic field** to produce various effects in the **upper atmosphere** (e.g., **aurora**).

solar zenith angle—Angle measured at the earth's surface between the sun and the **zenith**.

solarimeter—1. A **pyranometer** developed by W. Gorczyński, consisting of a **Moll thermopile** shielded from the **wind** by a bell glass.

2. Name sometimes used as a generic term in place of **pyranometer**.

solaure—Same as **solore**.

solenoid—A tube formed in space by the intersection of unit-interval **isotimic** surfaces of two **scalar** quantities.

The number of solenoids enclosed by a space curve is therefore equal to the **flux** of the **vector product** of the two gradients through a surface bounded by the curve, or

$$\int_A \int (\nabla \phi_1 \times \nabla \phi_2) \cdot d\mathbf{s} = \oint \phi_1 d\phi_2,$$

where $d\mathbf{s}$ is the **vector** element of area of a surface bounded by the given curve. Solenoids formed by the intersection of surfaces of equal **pressure** and **density** are frequently referred to in meteorology. A **barotropic atmosphere** implies the absence of solenoids of this type, since surfaces of equal pressure and density coincide. *See also* **baroclinity, vorticity equation, circulation theorem**.

solenoidal—Applied to a **vector field** having zero **divergence**, hence, one that may be expressed as the **curl** of another **vector**:

$$\mathbf{a} = \nabla \times \mathbf{b},$$

where $\mathbf{a}$ is the solenoidal vector and $\mathbf{b}$ is a vector field (sometimes called the **vector potential** of $\mathbf{a}$) that can be determined from the differential equation.

See **irrotational, Helmholtz's theorem**.

solenoidal index—The difference between the **mean virtual temperature** from the surface to some specified **upper level** averaged around the earth at 55° latitude, and the mean virtual temperature for the corresponding layer averaged at 35° latitude.

solid rotation—The rotation of a system as though it were a solid or rigid body rotating about a fixed axis, all points within the body having the same **angular velocity**.

solifluction—1. (Sometimes called soil fluction.) The viscous downslope flow of water-saturated soil and other surficial materials, particularly in regions underlaid by **frozen ground** (not necessarily **permafrost**) acting as a barrier to downward water percolation.

2. Any slow soil flowage on slopes in which the near-surface unconsolidated material is saturated by water.

solitary wave—A **gravity wave** consisting of a single elevation of finite **amplitude** that propagates without change of form.

First described in 1844 by Scott Russell in a British Association Report, its existence is a result of a balance between nonlinearity, which tends to steepen the **wave front** in consequence of the increase of **wave speed** with amplitude, and **dispersion**, which tends to spread the wave front as the wave speed of any spectral component decreases with increasing **wavenumber**. Most extensively studied are solitary waves on the **free surface** of a homogeneous, nonrotating fluid of **finite depth**. Surface solitary waves are also the easiest to observe. However, there also exist internal solitary waves, as the balance between nonlinearity and dispersion may be possible in the absence of a free surface by virtue of any or all of **stratification**, **shear**, **compressibility**, and rotation. *See also* **envelope soliton**.

Miles, J. W., 1980: Solitary waves. *Annu. Rev. Fluid Mech.*, **12**, 11–43.

soliton—A portion (one-half) of a **modon**, such as a **blocking** event in the midlatitude **planetary wave** pattern.

solore—(Also spelled solaure.) A cold, **night wind** of the mountains following the course of the Drôme River in southern France.

solstice—1. Either of the two points on the sun's apparent annual path where it is displaced farthest, north or south, from the earth's **equator**, that is, a point of greatest **deviation** of the **ecliptic** from the **celestial equator**.

The **Tropic of Cancer** (north) and **Tropic of Capricorn** (south) are defined as those parallels of latitude that lie directly beneath a solstice.

2. Popularly, the time at which the sun is farthest north or south; the "time of the solstice."

In the Northern Hemisphere the **summer solstice** falls on or about 21 June, and the **winter solstice** on or about 22 December. The reverse is true in southern latitudes. *Compare* **equinox**.

solubility—The total amount of solute species in **equilibrium** with the solid crystals from which the solutes were derived at constant **temperature** and **pressure**.

Somali Current—A prominent western **boundary current** in the northern Indian Ocean.

During the northeast **monsoon** season the Somali Current flows southward from 5° to 1°N in December, expanding to 10°N–4°S in January–February and contracting again to 4°N–1°S in March. It is then fed from the **North Equatorial Current** and discharges into the **Equatorial Countercurrent**. During all these months its speed is 0.7–1.0 m s^{-1}. During the **southwest monsoon** the current develops into an intense northward jet with extreme surface speeds; 2 m s^{-1} have been reported for May and 3.5 m s^{-1} for June. The jet is fed from the **South Equatorial Current** and flows along the eastern coast of the Horn of Africa; part of it continues along the Arabian Peninsula as the **East Arabian Current**. South of 5°N the jet is shallow; southward flow continues below a depth of 150 m. North of 5°N the jet deepens and embraces the permanent **thermocline**. During its northward phase the Somali Current is associated with strong **upwelling** between 2° and 10°N. The upwelled cold water turns offshore near Ras Hafun (11°N), forming a large **anticyclonic** eddy with a diameter of about 500 km known as the **Great Whirl**. Eventually the water from the Somali Current enters the **Southwest Monsoon Current**.

Somali jet—A low-level southwesterly **jet** over the Arabian Sea in the summer months, off the coast of Somalia.

It is the northern branch of a cross-equatorial flow, giving rise to a major supply of moisture in support of the Asian **summer monsoon**.

sondo—*See* **zonda**.

sonic anemometer—An **anemometer** that measures **linear** components of the **wind vector** by determining the effect of the **wind** on transit times of acoustic pulses transmitted in opposite directions across known paths.

It operates on the principle that the propagation **velocity** of a **sound wave** in a moving medium is equal to the **velocity of sound** with respect to the medium plus the velocity of the medium. The wind velocity **vector** is determined by measuring the **wind speed** along three independent linear paths, for example, three **orthogonal** paths. The sonic anemometer is an **absolute instrument** and has the advantages of a very short **time constant** and an absence of moving mechanical parts.

sonic thermometer—A **thermometer** based on the principle that the **velocity** of a **sound wave** is a function of the **virtual temperature** of the medium through which it passes.

By transmitting acoustic pulses in opposite directions between two transducers, the **travel time** difference can be used to infer the **virtual temperature**, provided that the **wind speed** across the acoustic path is known. Three-axis sonic anemometers, in addition to measuring three-dimensional winds, are also used as sonic thermometers. Sonic thermometers possess very short **time constants** and eliminate **radiation** error.

sonic wave—Same as **sound wave**.

sonora—A summer **thunderstorm** in the mountains and deserts of southern California and Baja California, Mexico.

sonora weather—A term used by old-timers living in the coastal plain of Southern California to describe summer episodes of hot, humid weather associated with widespread middle and high cloudiness and **thunderstorm** activity, sometimes intense, over the mountains and deserts to the east.

Summer thunderstorms, unusual over the coastal plain of southern California, occur during such episodes. Sonora weather is a particular manifestation of the Southwest **monsoon**, occurring when the more typical summer monsoonal flow of midlevel moisture into Arizona and New Mexico (from Sonora State, Mexico, and points south) is shunted westward.

soot—Fine particulate mass, mostly **carbon**, that is emitted as a result of incomplete fuel combustion.

sopero—Same as **sover**.

sorbtion nucleus—An older term for **deposition nucleus**.

soroche—Spanish term for the mountain sickness experienced on the plateaus of the Andes.

sorption—The taking up of one substance by another, either **absorption** or **adsorption**.

sou'easter—Contraction for **southeaster**.

sou'wester—Contraction for **southwester**.

soulaire—*See* **solaire**.

soulèdre—*See* **solaire**.

sound—A **pressure** fluctuation, usually in the **range** of audible frequencies, resulting from a displacement of a gas, liquid, or solid, that can be detected by a mechanical or electromechanical **transducer** (e.g., a **barometer**, microphone, or the human ear).

sound channel—The depth range within the ocean where the **speed of sound** in **seawater** takes on a local minimum value.

Sound waves generated within this sound channel or **waveguide** will, through continual **refraction** in the vertical due to **temperature** effects above and **pressure** effects below, remain within it. These sounds can propagate large horizontal distances within this waveguide with relatively little loss in **amplitude**.

sound pressure level—The (usually logarithmic) units for **sound** pressure measurements.

The sound pressure level is equal to 20 ($\log_{10}[p/p_0]$), in decibels, where p is the measured acoustical **pressure**, and p_0 is a reference pressure that must be given explicitly.

sound wave—(Also called acoustic wave, sonic wave.) A mechanical disturbance advancing with finite **velocity** through an elastic medium and consisting of longitudinal displacements of the ultimate **particles** of the medium, that is, consisting of compressional and rarefactional displacements parallel to the direction of advance of the disturbance; a **longitudinal wave**.

Sound waves are small-amplitude **adiabatic** oscillations. The **wave equation** governing the motion of sound waves has the form

$$\nabla^2 \phi = \frac{1}{c^2}\, \frac{\partial^2 \phi}{\partial t^2},$$

where ∇^2 is the **Laplace operator**, ϕ the **velocity potential**, t the time, and c the **speed of sound**, the **density** variations and velocities being taken small. As so defined, this includes waves outside the **frequency** limits of human hearing, which limits customarily define **sound**. Gases, liquids, and solids transmit sound waves, and the propagation velocity is characteristic of the nature and physical state of each of these media. In those cases where a steadily vibrating sound generator acts as a source of waves, one may speak of a uniform **wave train**; in other cases (explosions,

lightning discharges) a violent initial disturbance sends out a principal **wave**, followed by waves of more or less rapidly diminishing **amplitude**. *See* **ultrasonic, infrasonic, pressure wave**.

sounder—An instrument that acquires multispectral measurements from which vertical profiles of atmospheric **temperature** and **humidity** can be derived.

sounding—1. In **geophysics**, any penetration of the natural **environment** for scientific **observation**.

2. In meteorology, same as **upper-air observation**.

However, a common connotation is that of a single complete set of **radiosonde** observations.

3. The measurement of the depth of water beneath a vessel.

sounding balloon—A free, unmanned balloon instrumented and/or observed for the purpose of obtaining a **sounding** of the **atmosphere**.

See **balloon** for list of types.

sounding pole—(Also called sounding rod.) Graduated rigid pole or rod for measuring the depth of water.

sounding weight—Weight of streamlined shape attached to a **sounding** line or to the suspension of a **current meter** when observing depth and/or velocities in streams.

soundings—*See* **atmospheric sounding**.

source—In **hydrodynamics**, a point, line, or area at which mass or **energy** is added to a system, either instantaneously or continuously.

Conversely, a **sink** is a point where mass or energy is removed from the system. An **incompressible fluid** will possess sources or sinks of mass only at points where the **divergence** of its **velocity** vector is nonzero; a source is associated with positive divergence and a sink with negative divergence (**convergence**). The fluid is usually assumed to pass outward from a source or inward to a sink equally in all directions along radial lines. The strength of a source, for example, the rate of mass flow of fluid of unit **density** across a curve enclosing the source, is given by

$$Q = 2\pi \, r v_n,$$

where r is the distance from the source and v_n the radial speed.

source region—Extensive region of the earth's surface characterized by essentially uniform surface conditions, and so located with respect to the general atmospheric **circulation pattern** that a volume of air remains in contact with the surface long enough to acquire properties that distinguish it as an **air mass** (e.g., central Canada for continental **polar air**).

See **airmass source region**.

source strength—*See* **source**.

source terms—In spectral **wave** modeling, the contributions to net **energy** increase and decrease within each spectral **band**, due to such effects as energy input from the **wind**, **nonlinear** transfer of energy between different spectral components, energy dissipation by **wave breaking**, bottom **stress**, etc.

These contributions do not include the spatial **flux** of energy due to **wave propagation** or the flux in **wavenumber** space due to changes in water depth or **current**.

South Atlantic convergence zone—A local maximum in **cloudiness, precipitation**, and low-level **convergence** that occurs in a northwest–southeast oriented swath extending across southeastern Brazil and the western South Atlantic.

This counterpart to the **South Pacific convergence zone** is strongest in summer.

South Atlantic Current—The eastward current that forms the southern part of the South Atlantic subtropical **gyre**.

It is fed by the **Brazil Current** and follows the **subtropical front**, gradually losing water to the subtropical gyre and Brazil Current recirculation. About 20–25 Sv (20–25×10^6 m^{3}s^{-1}) reach the African coast and continue as the **Benguela Current**. Cooling of the warm Brazil Current water along its path makes the South Atlantic Current a **heat** source for the **atmosphere**.

South Atlantic high—One of the semipermanent **highs** of the **general circulation**, centered at about 25°S, 15°W.

South Equatorial Countercurrent—(Abbreviated SECC.) A band of weak eastward flow in the Atlantic and Pacific Oceans embedded in the **South Equatorial Current** near 8°S, caused by a **wind stress** minimum in the Southern Hemisphere **trade winds**.

In the Pacific Ocean the SECC is controlled by the Asian–Australian **monsoon** and is strongest during the northwest monsoon (December–April), with speeds approaching 0.3 m s^{-1}; it is barely seen during the remainder of the year. East of the date line it decreases rapidly in strength and is absent from the eastern Pacific during most of the year. In the Atlantic Ocean it is weak, narrow, and variable and has its largest speed of little more than 0.1 m s^{-1} often below the surface at a depth of 100 m.

South Equatorial Current—(Abbreviated SEC.) The broad region of uniform westward flow driven by the **trade winds** that forms the northern part of the Southern Hemisphere subtropical **gyres**.

Being directly **wind** driven, the SEC responds quickly to variations in the **wind field** and is therefore strongest in winter (August). In the Atlantic Ocean it is found between 3°N and 25°S with speeds of 0.1–0.3 m s^{-1}. In the Pacific Ocean it covers the same latitude band but attains 0.6 m s^{-1} and a **transport** of about 27 Sv (27 × 10^6 m^{3}s^{-1}) in August; this decreases to 7 Sv in February. In the Indian Ocean it occupies the latitude band 8°–30°S during the northeast **monsoon** (December–April) and expands northward to 6°S in September during the **southwest monsoon**, with speeds close to 0.3 m s^{-1} throughout the year.

south foehn—A **foehn** wind blowing from south (northern Italy) to north (Switzerland, southern Germany, and western Austria) over the Alps.

Because these winds blow from a warmer region to a cooler region, they are often accompanied by dramatic **temperature** and **wind speed** increases. The south foehn often results from flow in the **warm sector** of a **cyclone** system and thus it often occurs ahead of a **cold front**.

South Indian Ocean Current—The eastward current that forms the southern part of the subtropical **gyre** in the Indian Ocean.

It is fed by the **Agulhas Current** and follows the **subtropical front**, gradually losing water to the subtropical gyre and Agulhas Current recirculation. East of Africa it begins with 60 Sv (60 × 10^6 m^{3}s^{-1}) and arrives off the coast of western Australia with 10 Sv, which continues as the **West Australian Current**. Cooling of the warm Agulhas Current water along its path makes the South Indian Ocean Current a **heat** source for the **atmosphere**.

South Java Current—A seasonal current along the south coast of Java.

The current flows southeastward during December–April and northwestward during June–October, when it is associated with **coastal upwelling**.

South Pacific convergence zone—(Abbreviated SPCZ.) A persistent and greatly elongated zone of low-level **convergence** extending from approximately 140°E near the **equator** to approximately 120°W at 30°S.

The zone is not quite **linear**, but is oriented more west to east near the equator and has a more diagonal orientation (northwest to southeast) at higher latitudes. The low-level convergence of moisture leads to a persistent **cloud band** along the SPCZ.

> Trenberth, K. E., 1991: *Teleconnections Linking Worldwide Climate Anomalies*, H. Glantz, R. Katz, and N. Nichols, Eds., 22–23.

South Pacific Current—The eastward current that forms the southern part of the South Pacific subtropical **gyre**.

It is fed by the **East Australian Current** and its continuation, the **East Auckland Current** and **East Cape Current**, and follows the **subtropical front**. It is much weaker than the **South Atlantic** and **South Indian Ocean Currents**, carrying little more than 5 Sv (5 × 10^6 m^{3}s^{-1}). Nevertheless, cooling of the warm water from the western **boundary currents** along its path makes the South Pacific Current a **heat** source for the **atmosphere**. It feeds its water into the **Peru/Chile Current**.

southeast-storm warning—*See* **storm warning**.

southeast trades—(Or southeast trade winds.) *See* **trade winds**.

southeaster—(Often contracted sou'easter.) A southeasterly **wind**, particularly a strong wind or **gale**, for example, the winter southeast storms of the Bay of San Francisco.

A specific application to a local wind is the southeaster of Table Bay, South Africa (**Table Mountain southeaster**). When there are dark **clouds** over Table Mountain, it is a black southeaster; when there are no clouds, it is a blind southeaster. *See* **Cape doctor**.

souther—A south **wind**, especially a strong wind or **gale**.

southerly burster—Same as **burster**.

southern lights—*See* **aurora australis**.

Southern Oscillation—Originally defined in 1924 by Gilbert Walker as a low-latitude, planetary-scale "seesaw" in **sea level pressure**, with one pole in the eastern Pacific and the other in the western Pacific–Indian Ocean region.

The **pressure** seesaw is associated with a global pattern of atmospheric anomalies in **circulation**, **temperature**, and **precipitation**. The primary timescale of the **oscillation** is interannual–multiyear, and it is now recognized to be primarily a response to basin-scale **sea surface temperature** variations in the equatorial Pacific arising from coupled ocean–atmosphere interactions, the opposite extremes of which are the **El Niño** and **La Niña** warm and cold events. *See also* **ENSO**.

> Philander, S. George, 1990: *El Niño, La Niña, and the Southern Oscillation*, Academic Press, International Geophysics Series, Vol. 46.
> Walker, G. T., 1924: Correlation of seasonal variations in weather IX: A further study of world weather. *Mem. Indian Meteor. Dep.*, **24**, 275–332.

Southern Oscillation index—A measure of the state of the **Southern Oscillation**.

A common index used for this purpose is the **sea level pressure** at Tahiti minus the sea level pressure at Darwin, Australia, divided by the **standard deviation** of that quantity. It is closely associated with **El Niño** and is thus often referred to as **ENSO** (El Niño–Southern Oscillation).

southwest monsoon—The **rainy season** in India.

See also **monsoon**.

Southwest Monsoon Current—The broad eastward flow that dominates the Indian Ocean north of 5°S during the **southwest monsoon** season (June–October) and extends into the Arabian Sea and Bay of Bengal.

Current speeds are generally 0.2–0.3 m s^{-1} but increase to 0.5–1.0 m s^{-1} south of Sri Lanka. The Southwest Monsoon Current replaces the westward flowing **North Equatorial Current** and absorbs the eastward flowing **Equatorial Countercurrent** of the northeast **monsoon** season.

southwest-storm warning—*See* **storm warning**.

southwester—(Often contracted sou'wester.) A southwest **wind**, particularly a strong wind or **gale**.

sover—(Also called sopero.) A night **breeze** blowing down-valley near Lake Garda in Italy; it sometimes reaches **gale** force.

See **paesano**.

SO$_x$—Abbreviation for various oxides of sulfur, including SO$_2$ (**sulfur dioxide**) and SO$_3$ (sulfur trioxide).

These gaseous and solid **pollutants** can be transformed in the presence of water to **sulfuric acid** H_2SO_4, one of the acids in **acid rain**. They are formed anthropogenically by combustion of coal or coal that contains sulfur impurities. They are also produced naturally by volcanoes and by biological activity in swamps, tidal areas, oceans, and by some soil bacteria.

Soya Warm Current—*See* **Tsushima Current**.

space charge—Any net electrical charge that exists in a given region of space.

In electronics, this usually refers to the electrons in the space between the filament and plate of an **electron** tube. In **atmospheric electricity**, space charge refers to a preponderance of either negative or positive **ions** within any given portion of the **atmosphere**. A net positive space charge is found in **fair** weather at all altitudes in the atmosphere and is largest near the earth's surface. The general downward **flux** of this positive space charge is known as the **air–earth conduction current**.

Space Environmental Monitor—(Abbreviated SEM.) A multichannel, charged particle **spectrometer** that measures the population of charged **particles** that compose the earth's **radiation belts** and the related phenomena resulting from **solar activity**.

SEM packages are flown on the **GOES**, **GMS**, and **POES** series of satellites.

space weather—(Modern designation for cosmical meteorology.) Designation for the study of those phenomena that lie wholly or in part outside the earth's **atmosphere**.

The term is analogous to terrestrial meteorology in that it includes the study of planetary atmospheres and solar–weather relationships.

spark discharge—That type of gaseous **electrical discharge** in which the charge **transfer** occurs transiently along a relatively constricted path of high **ion** density, resulting in high luminosity.

It is of short duration and to be contrasted with the nonluminous **point discharge**, the **corona discharge**, and the continuous **arc discharge**. The exact meaning to be attached to the term "spark discharge" varies somewhat in the literature. It is frequently applied to just the **transient** phase of the establishment of any arc discharge. A **lightning discharge** can be considered a large-scale spark discharge.

spatial average—The mathematical **mean value** over multiple points in space.

These can be points along a line or path (**line average**), along a surface (**area average**), or within a volume (**volume average**). *Compare* **time average, ensemble average**; *see also* **ergodic condition**.

spatial dendrite—A complex **ice crystal** with dendritic arms that extend in many directions (spatially) from a central **nucleus**, probably a frozen **droplet** that originally froze as a polycrystal.
Its form is roughly spherical. *See* **dendritic crystal**.

SPCZ—Abbreviation for **South Pacific convergence zone**.

SPECI—In the **METAR** observation program, a **surface observation** issued on a nonroutine basis as dictated by changing meteorological conditions.

> National Oceanographic and Atmospheric Administration (NOAA), 1995: *Federal Meteorological Handbook No. 1 (FMH-1)*, chap. 9.

special observation—A category of **aviation weather observation** taken to report significant changes in one or more of the observed elements since the last **record observation**.
The specific criteria demanding a special observation change from time to time. Basically they cover 1) significant changes in **ceiling** and **visibility**, particularly in the low ranges; 2) the appearance of low **clouds**; 3) the appearance and disappearance of a **tornado** or **thunderstorm**; 4) the beginning and ending of frozen or **freezing precipitation**; 5) **pressure jumps**; and 6) significant changes in the **wind**.

Special Sensor Microwave Imager—(Abbreviated SSM/I.) A passive conical-scanning multichannel **microwave** radiometer on **DMSP** satellites that is used to derive parameters such as land and **sea surface temperature**, ocean surface **wind speed, precipitable water, rainfall rate**, and **snow/ice cover**.

Special Sensor Microwave Temperature—(Abbreviated SSM/T.) The **microwave** temperature and **water vapor** sounders on the **DMSP** polar-orbiting satellites.
SSM/T is a **temperature** sounder, while SSM/T2 is a water vapor **profiler**. Many of the capabilities of the SSM/T and SSM/T2 instruments will be duplicated on the **AMSU**.

special weather report—The encoded and transmitted report of a **special observation**. This has been replaced by the **SPECI** observation in the **METAR** code.

special weather statements—Statements issued by a **weather service** that discuss weather situations including long-term events that describe phenomena of interest or concern to a number of users.
These are used when a **warning** is in progress to provide additional detail, describe events, and provide appropriate response recommendations. These statements may refer to inclement or hazardous weather, but are not so limited.

specific—Applied to a physical quantity, especially a **thermodynamic variable**, usually means per unit mass.
For example, **specific volume** is volume divided by mass, **specific enthalpy** is **enthalpy** divided by mass, and so on. Specific quantities are common in meteorology, and hence specific is often omitted as a qualifier, it being understood.

specific absorption—Same as **specific capacity**.

specific attenuation—*See* **attenuation coefficient**.

specific capacity—The ratio of **discharge** into a **recharge** well to the rise of **head** at **equilibrium**.

specific conductance—Measurement of the ability of soil water to conduct electricity, usually expressed in units of siemens per meter or millimhos per centimeter.
Specific conductance is related to the concentration of **ions** in the water and is used to estimate the concentration of dissolved solids in the water.

specific discharge—1. In **surface water hydrology, discharge** per unit area of an **upstream** watershed.
2. In **open channel flow, discharge** per unit width of channel.
3. In **groundwater**, the volumetric flow rate per unit area of a porous media, also known as **flow velocity** or **specific flux**; if computed using **Darcy's law**, known as Darcy's **flux**.

specific energy—In **hydrology**, the total **energy** per unit weight of fluid, at any **cross section** of a channel, measured above the channel bottom.
The equation for specific energy, E, is

$$E = y + \frac{v^2}{2g},$$

where y is the depth of flow, v is the **average velocity**, and g is the gravitational attraction.

specific enthalpy—*See* **enthalpy**.

specific entropy—*See* **entropy**.

specific flux—Same as **specific discharge**.

specific gravity—The ratio of the **density** of a substance to the density of water, usually determined at 4°C.

specific heat—*See* **specific heat capacity**.

specific heat capacity—(Or specific heat.) The **heat capacity** of a system divided by its mass.

It is a property solely of the substance of which the system is composed. As with heat capacities, specific heats are commonly defined for processes occurring at either constant volume (c_v) or constant **pressure** (c_p). For an **ideal gas**, both are constant with **temperature** and related by $c_p = c_v + R$ with R the **gas constant**. For **dry air** at 273 K,

$$c_p = 1005.7 \pm 2.5 \text{ J kg}^{-1}\text{K}^{-1}$$

$$c_v = 719 \pm 2.5 \text{ J kg}^{-1}\text{K}^{-1}.$$

For **moist air**, the specific heat capacities of the dry air and **water vapor** must be combined in proportion to their respective mass fractions.

Dutton, J. A., 1995: *Dynamics of Atmospheric Motion*, Dover Press, 41–45, 406–410.

Sommerfeld, A., 1964: *Thermodynamics and Statistical Mechanics*, Academic Press, p. 45.

specific humidity—In a system of **moist air**, the (dimensionless) ratio of the mass of **water vapor** to the total mass of the system.

The specific humidity q is related to the **mixing ratio** r_v by

$$q = \frac{r_v}{1 + r_v}.$$

For many purposes they can be equated. *See also* **absolute humidity**, **relative humidity**, **dewpoint**.

specific intensity—Same as **radiance**.

specific retention—The ratio of the volume of water that an initially saturated porous-medium sample retains after being drained by **gravity** to the volume of the **porous medium**.

specific storage—(Applies to aquifers.) The volume of water released from storage per unit volume of a saturated **porous medium** per unit decrease in the **hydraulic head**.

specific volume—Volume per unit mass of a substance, and hence the reciprocal of **density**; commonly symbolized by v or α.

specific-volume anomaly—(Or anomaly of specific volume; also called steric anomaly; symbol δ.) In **oceanography**, the excess of the actual specific volume of the **seawater** at any point in the ocean over the **specific volume** of seawater of **salinity 35 psu** and **temperature** 0°C at the same **pressure**.

The integral of specific-volume anomaly with depth is the **dynamic-height anomaly**. *See* **thermosteric anomaly**.

specific yield—(Applies to unconfined aquifers.) The volume of water released from storage from a unit area of the **aquifer** per unit decline of the **hydraulic head**.

specification—*See* **perfect prognosis method**.

spectral diffusivity—A form of **eddy viscosity** or **eddy diffusivity** that varies with **eddy** size.

For a **spectrum** of turbulent eddy sizes there is a spectrum of diffusivities. By integrating over all eddy sizes and diffusivities, the total effect of **turbulence** can be approximated. This is a form of nonlocal **turbulence closure** and can be shown to be related to **transilient turbulence theory**.

spectral function—(Or **spectrum**.) The Fourier representation of a given function, that is, the **Fourier transform** if the given function is **aperiodic**, or the set of coefficients of the **Fourier series** if the given function is periodic.

The existence of the spectral function depends on the mathematical behavior of the given function. If it exists, the spectral function will in general be a complex function, having both **amplitude** and **phase**. *See also* **continuous spectrum**, **discrete spectrum**.

spectral gap—A **wavenumber**, **wavelength**, or **frequency band** within a Fourier **energy spectrum** that has a relative minimum of spectral **energy**.

Much of the theoretical **development** of **turbulence** in the **atmosphere** is based on the assumption of a spectral gap between larger-wavelength motions (called mean motions) and small-scale motions (called turbulence). However, a growing body of experimental evidence indicates that there is often not a spectral gap in the **atmospheric boundary layer**, thereby raising questions about the **Reynolds averaging** approach that has formed the basis for turbulence theory for the past century.

spectral hygrometer—(Also called **optical hygrometer**.) A **hygrometer** that determines the amount of precipitable moisture in a given region of the **atmosphere** by measuring the **attenuation** of **radiant energy** caused by the **absorption bands** of **water vapor**.

The instrument consists of a collimated **energy** source, separated by the region under investigation from a detector that is sensitive to those frequencies that correspond to the absorption bands of water vapor. The basis for determining the water vapor concentration is **Beer's Law**: $I / I_0 = \exp(-kx)$, where I is the **light** intensity after passing through the **sample**, I_0 is the incident **intensity**, x is the pathlength reduced to some **absolute** standard like **STP**, and k is the **absorption coefficient**. The most useful regions of the **electromagnetic spectrum** for this purpose lie in the **ultraviolet** and **infrared** regions. The most widespread application is the monitoring of very-high-frequency variations in **humidity**, as the **time constant** of a spectral hygrometer is typically just a few milliseconds. The use of spectral hygrometers remains mostly restricted to research applications. *See* **Krypton hygrometer, Lyman-alpha hygrometer, differential absorption hygrometer**.

spectral interval—(Or spectral band, spectral channel.) The width, generally expressed in **wavelength**, **wavenumber**, or **frequency**, of a particular portion of the **electromagnetic spectrum**.

A given **sensor** is designed to measure or be sensitive to **energy** received at the satellite from that part of the **spectrum**.

spectral irradiance—The **irradiance** per unit **wavelength** or **wavenumber** interval.

Units are typically $W\ m^{-2}\ \mu m^{-1}$ or $W\ m^{-2}(cm^{-1})^{-1}$.

spectral line—A bright, or dark, line found in the **spectrum** of some **radiant** source.

The **wavelength** location of the line is controlled by the physics of the **emission** (bright line) or **absorption** (dark line) process involved. *See* **absorption line, emission line**.

spectral method—A numerical method for solving differential equations in which the **dependent variables** are expanded as series of **orthogonal** basis functions and the original equations are reduced to a set of algebraic equations for the modal amplitudes.

spectral model—A **model** in which the prognostic **field** variables are represented as sums of a finite set of spectral modes rather than at gridpoints.

The spectral modes may be Fourier modes in the one-dimensional case or double Fourier modes or **spherical harmonics** in the two-dimensional case. The advantage of a spectral model is that horizontal derivatives can be calculated exactly for the spectral modes represented in the model and thus the model error is confined only to the unrepresented higher spectral modes beyond the model's spectral truncation.

spectral numerical analysis—The expression of a **field** as a sum of spectral modes.

The modes may be Fourier modes or **spherical harmonics**. While a continuous field can be expressed exactly as a sum of an infinite number of modes, in practice it is approximated as a sum of a finite number of modes with the **error** confined to the higher unrepresented spectral modes.

spectral numerical prediction—The forecasting of atmospheric phenomena using a **spectral model**.

The **model** may be a simple shallow water model or a complex three-dimensional system with full parameterized physics. The model predicts the atmospheric variables in spectral space, but they are generally transformed to physical space for display, dissemination, and **verification**.

spectral radiance—The **radiance** per unit **wavelength** or **wavenumber** interval.

Units are typically $W\ m^{-2}\ sr^{-1}\ \mu m^{-1}$ or $W\ m^{-2}\ sr^{-1}(cm^{-1})^{-1}$.

spectral similarity—Universal behavior of Fourier spectra of **turbulence** in the **atmospheric boundary layer** when the spectral **energy** and the **frequency** (or **wavenumber**) are made dimensionless using the appropriate boundary layer scales.

Separate spectral similarity relationships have been found for the **surface layer** and for the **convective mixed layer**. *See* **similarity theory, dimensional analysis, fast Fourier transform, Buckingham Pi theory, inertial subrange, Kolmogorov's similarity hypotheses**.

spectral transform model—A **spectral model** in which the **nonlinear** terms in the governing equations are computed in physical space.

Thus, for every **model** time step, some fields must be transformed to physical space where nonlinear terms are computed and then transformed back to spectral space where they can be applied. These nonlinear terms may consist of **advection** terms, **pressure gradient** terms, and **energy conversion** terms, as well as parameterized physics terms. This is the most common kind of spectral model, as it avoids the expensive computation of interactions among modes in spectral space.

spectre of the Brocken—Same as **Brocken spectre**.

spectrobolometer—A **bolometer** that has some capability to resolve the **electromagnetic spectrum**.

spectrograph—A **spectroscope** with a photographic or other recording device to capture an **image** of the entire **spectrum**, or portions thereof, at one instant in time.

spectroheliograph—An instrument for taking photographs (spectroheliograms) of an **image** of the sun in **monochromatic** light.
   The **wavelength** of **light** chosen for this purpose corresponds to one of the **Fraunhofer lines**, usually the light of **hydrogen** and/or ionized calcium. A similar instrument used for visual, instead of photographic, observations is the spectrohelioscope.

spectrohelioscope—*See* **spectroheliograph**.

spectrometer—A generic term for a device that measures the **intensity** of **radiation** as a function of **frequency** (or **wavelength**) of the radiation or that breaks down any **signal** into its frequency components.

spectrophotometer—A **photometer** that measures the **intensity** of **radiation** as a function of **frequency** (or **wavelength**) of the radiation.
   In one design, radiation enters the spectrophotometer through a slit and is dispersed by means of a prism. A **bolometer** having a fixed **aperture** scans the dispersed radiation so that the intensity over a narrow **wave** band is obtained as a function of frequency. *See* **Dobson spectrophotometer**.

spectropyrheliometer—An instrument that measures the spectral distribution of the **intensity** of **direct solar radiation**.
   See **pyrheliometer, spectrophotometer**.

spectroscope—An apparatus to effect **dispersion** of **radiation** and visual display of the **spectrum** obtained.

spectroscopy—The study of the interaction between matter and **radiation**.
   Since the **energy** levels in all atoms and molecules are discrete (quantized), the **absorption** or **emission** of radiation occurs at distinct energies (**frequencies**) characteristic of a given chemical. The energy levels probed may be rotational (**microwave** region), vibrational (**infrared** region), or electronic (visible or **ultraviolet**). In mass spectroscopy, molecules are ionized and the different masses selected by use of an **electric** or **magnetic field**. Many spectroscopic techniques have been used to identify molecules or atoms in the **atmosphere**.

spectrum—1. **Radiant energy** (e.g., electromagnetic, acoustic) per unit **frequency** (or **wavelength**) interval over some **range** of frequencies; may also be applied to any **distribution function**.
   For example, the distribution of **droplet** sizes in **clouds** is sometimes called a size spectrum.
   2. Mathematically, same as **spectral function**.

specular reflection—**Reflection** as from a mirror.

specular reflector—A mirrorlike surface for which the **angle of reflection** equals the **angle of incidence**; to be contrasted with a **diffuse reflector**.
   Most natural surfaces are too irregular to act as good specular reflectors of **solar radiation**, the main exception being calm water surfaces.

speed of light—Without qualification usually means the speed of propagation of **electromagnetic radiation** of any **frequency** in **free space**, a universal constant with the value 2.99792×10^8 m s^{-1}.
   Could also mean the **phase velocity** or **group velocity** of an **electromagnetic wave** in a material medium.

speed of sound—1. (Also velocity of sound.) Usually taken to mean the **phase speed** of an acoustic (or **sound**) wave.
   In an ideal, stationary gas the speed of sound c is a thermodynamic property depending only on the **equilibrium** state of the gas and is given by

$$c = (\gamma RT)^{1/2},$$

where γ is the ratio of **specific heat** capacities, at constant **pressure** and volume, respectively; R is the **gas constant**; and T is **absolute** temperature.

2. In **seawater**, a function of **temperature**, **salinity**, and **pressure**.

speed-up height—The height above hilltop experiencing the greatest increase in **wind speed** due to the Bernoulli effect, as compared to the **upstream** value of the wind speed at the same height.

speed-up ratio—The increase in speed of air that accelerates over the tops of hills due to the Bernoulli effect divided by the ambient **wind speed** well **upwind** of the hill.

speed-up wind—The increase in **wind speed** above the top of a hill due to the Bernoulli effect.

sphere calibration—The procedure of calibrating a **radar** by measuring the **power** reflected by and returned to the radar from a conducting sphere of a known **radar cross section**. In principle, this method establishes the **radar constant**, including the effects of the **antenna gain** and any **attenuation** of the **signal** in the radar system between the **antenna** and the **receiver**, effects that are sometimes difficult to estimate by other means.

In practice, the method is sometimes difficult to apply. The approach is usually to suspend the sphere beneath a tethered (sometimes free-floating) **balloon**. The problem is to keep the moving sphere at the center of the **radar beam** during the time required for a measurement of the returned power.

spherical albedo—The **albedo** of a surface when the incident **radiation** is **isotropic**.

Spherical albedos are also used as the average of the **plane albedo** over all sun angles, or as the effective albedo of an entire planet.

spherical coordinates—(Also called polar coordinates in space, geographical coordinates.) A system of **curvilinear coordinates** in which the position of a point in space is designated by its distance r from the origin or pole along the **radius vector**, the angle ϕ between the radius vector and a vertically directed **polar axis** called the **cone angle** or colatitude, and the angle θ between the plane of ϕ and a fixed **meridian** plane through the polar axis, called the **polar angle** or longitude.

A constant-amplitude radius vector $\mathbf{r}$ confines a point to a sphere of radius r about the pole. The angles ϕ and θ serve to determine the position of the point on the sphere. The relations between the spherical coordinates and the **rectangular Cartesian coordinates** (x, y, z) are $x = r \cos \theta \sin \phi$; $y = r \sin \theta \sin \phi$; $z = r \cos \phi$.

spherical harmonic—An analytic basis function on the sphere that is commonly used in a **spectral model**.

A spherical harmonic is defined for each total **wavenumber** n and **zonal wavenumber** m as the following function of sine of latitude μ and longitude λ:

$$Y_{m,n}(\mu, \lambda) = P_{m,n}(\mu)e^{im\lambda}$$

where $P_{m,n}$ is the associated Legendre function defined as

$$P_{m,n}(\mu) = \left[\frac{(2n + 1)(n - m)!}{(n + m)!} \right]^{1/2} \frac{(1 - \mu^2)^{m/2}}{2^n n!} \frac{d^{n+m}}{d\mu^{n+m}} (\mu^2 - 1)^n.$$

The spherical harmonic basis functions satisfy the **orthogonal** relationship

$$\frac{1}{4\pi} \int_0^{2\pi} \int_{-1}^{1} Y_{m,n} Y^*_{m',n'} d\mu \, d\lambda = \left(\begin{smallmatrix} 1 & \text{for} & (m',n') & = & (m,n) \\ 0 & \text{for} & (m',n') & \neq & (m,n) \end{smallmatrix} \right)$$

and they satisfy the elliptic equation on the sphere:

$$\nabla^2 Y_{m,n} + \frac{n(n + 1)}{a^2} Y_{m,n} = 0.$$

spherical harmonics—In analogy to **harmonic** functions in the plane, the solutions of the **Laplace equation** in **spherical coordinates**.

Spherical surface harmonics are special sets taken over the surface of a sphere; therefore, the harmonic components are restricted to an integral number of waves over the sphere. Spherical harmonics have been applied in the study of the large-scale oscillations of the **atmosphere**.

spherical pyranometer—**Radiometer** designed to measure only **solar radiation** incident on the surface of an infinitesimally small sphere.

Practically, the sphere of the radiometer is at least 2 cm in diameter.

spherical pyrgeometer—**Radiometer** designed to measure only **thermal infrared** radiation incident on the surface of an infinitesimally small sphere.

spherical pyrradiometer—**Radiometer** designed to measure both **solar** and **thermal infrared** radiation incident on the surface of an infinitesimally small sphere.

spherical shell apparatus—An experimental apparatus intended to represent, for modeling purposes, the geometric and dynamic effects of the earth's sphericity.

It is composed of two concentric spheres (usually glass) of different radii. The space between them is filled with some working fluid and the entire shell is rotated. The inner sphere represents the surface of the earth, and the outer sphere is an artificial upper limit of the **model atmosphere**.

spherical symmetry—Having properties independent of latitude and longitude; properties that depend only on radial distance.

spherical wave—Any **wave** for which the **surface of constant phase** is a sphere.

spherics—*See* **atmospherics**.

spicule—1. A spike of **ice** formed during the **freezing** of a water drop or a contained volume of water (as in a puddle or freezer container); expansion of the inward freezing ice expels any remaining water through a weak point in the shell which then freezes as a spike in the colder **environment**.

2. Bright spike of **luminous** gas extending from the **chromosphere** into the **corona** of the sun.

Spicules are several hundred kilometers in diameter and extend outward 5000–10 000 km. Observed in photographs of the limb, these features have a lifetime of several minutes.

spider lightning—**Lightning** with extraordinary lateral extent near a **cloud base** where its dendritic structure is clearly visible.

This lightning type is prevalent beneath the stratiform **anvil** of **mesoscale convective systems** and is often associated with positive **ground flashes**. This **discharge** form is also referred to as **sheet lightning**.

spillover—That part of **orographic precipitation** that is carried over the peaks by the **wind** so that it reaches the ground on the lee side of the barrier.

spillway—A structure that allows release of excess water from a dam or other hydraulic structure.

spillway capacity—The maximum discharge rate that a **spillway** has been designed to convey.

spillway design flood—**Flood** used to design a dam **spillway**; also, maximum flow that could pass through a spillway without causing serious structural damage.

spin-scan cloud camera—A camera system developed by scientists at the University of Wisconsin and flown on the first **ATS** in 1966.

This camera made use of the spin-stabilization of the satellite to produce high-resolution **cloud** imagery. The spin-scan cloud camera was the forerunner of the **VISSR** imaging instrument used in later **geostationary satellites**.

spindown time—The length of time required for **friction** to reduce the **relative vorticity** of a column of fluid to $1/e = 37.8\%$ of its initial value.

This timescale is used to characterize the importance of friction and presumes an exponential decay of **relative vorticity** as would occur for **linear** friction processes.

spinup time—The length of time required for forcing effects such as **wind** to induce **velocities** and/or **vorticities** of $(e - 1)/e = 63.2\%$ of the value obtained at long times.

This characterizes the time required for forcing to be felt and is usually the same as the **spindown time**.

spiral band—*See* **hurricane band**.

spiral scanning—In **radar**, coordinated **scanning** in **azimuth** and **elevation** so that much of the space in the hemisphere above a ground-based radar is surveyed.

In a common form of spiral scanning, the **antenna** rotates continuously in azimuth around a full circle while the elevation is slowly increased from the horizontal to a high angle.

spirit thermometer—A **liquid-in-glass thermometer** that uses an organic substance such as alcohol as the thermometric liquid.

This type of **thermometer** has a very low **freezing point** and a high coefficient of expansion. The **accuracy** of thermometers using an organic liquid, however, is much less than that of thermometers using **mercury**. Spirit thermometers are used when the low freezing point is important, and in such special application as the **minimum thermometer**.

Middleton, W. E. K., and A. F. Spilhaus, 1953: *Meteorological Instruments*, 3d ed., rev., 73–75.

spissatus—A **cloud species** unique to the genus **cirrus**.
See **cirrus spissatus, cloud classification**.

spit—Colloquially, a fall of **rain** or **snow** in very light amount; a **sprinkle**, or **snow flurry**.

Spitzbergen Current—See **West Spitzbergen Current**.

splash—A dispersion of **raindrop** water rebounding after reaching the earth's surface.

spline function—A function composed of **polynomial** pieces defined on adjoining subintervals and with **continuity** conditions imposed on the function and its derivatives at all points including those connecting the subintervals.

The most commonly used spline function is the cubic spline that is made up of polynomial pieces of degree three, where the polynomial coefficients are determined in such a way that the composite function and its first and second derivatives are continuous at the connecting points. The graph of the cubic spline is a smooth, nonoscillatory curve. An interpolating spline is a spline function that passes through a finite set of data points.

splintering—A process from which **ice** splinters are produced due to the **pressure** buildup in the ice shell when a **drop** is **freezing**, resulting from the ice-water expansion on **phase change**.

This causes the shell to deform and crack and eject ice splinters. It has been suggested as one of the **ice multiplication** mechanisms.

split cold front—A **cold front** in which the **upper level** portion of the **front** moves at a faster speed than the low-level portion.

Eventually, the front evolves into two distinct parts rather than one continuous front. As the **upper front** moves ahead of the lower front, deep **clouds** and embedded **rain** bands form well ahead of the surface frontal position. See **katafront**.

split-explicit method—An **explicit time difference** approximation in which terms contributing to high frequencies of the total solution are integrated using relatively small time steps, while terms contributing to lower frequencies are integrated using relatively long time steps.
See **implicit time difference**.

split window—1. Segmentation of a **remote sensing** window **channel** into smaller increments of **wavelength**.

2. A technique used to calculate land and **sea surface temperatures**, where corrections are made for the atmospheric modification of **upwelling** radiation from the surface.

splitting convective storm—The process by which a single **convective cell** splits into two **supercells**, one dominated by **cyclonic rotation** and the other by **anticyclonic rotation**, their paths then deviating substantially from each other and other nearby convective cells.

Splitting storms require strong environmental **vertical wind shear**, with **unidirectional vertical wind shear** promoting the **development** of mirror-image supercells, and clockwise or counterclockwise **hodograph** curvature in the lowest few kilometers above ground level promoting the preferred development of a cyclonic or anticyclonic supercell, respectively. The cyclonic supercell is often observed to propagate to the right of the mean wind (in the Northern Hemisphere), while the anticyclonic supercell propagates to the left of the mean wind.

spongy ice—Ice that contains liquid water; a composite of **ice** and liquid water.

Spongy ice is a form of **accretion** of supercooled drops in which the **heat transfer** is inadequate to **freeze** all the water. The excess water is included within the growing ice. Spongy ice may also form during the melting of low-density accretions by the meltwater soaking into the pores.

spontaneous nucleation—Old term for **homogeneous nucleation**.
See **nucleation**.

sporadic E-layer—A narrow layer of enhanced **electron** and **ion density** in the lower **ionosphere**.

Most sporadic E-layers are thought to be formed by the action of dynamo forces on metallic positive **ions** deposited by incoming meteors.

SPOT—Abbreviation for **Satellite pour l'Observation de la Terre**.

spot wind—In air navigation, **wind direction** and **speed**, either observed or forecast if so specified, at a designated **altitude** over a fixed location.
Compare **track wind, sector wind**.

spotting—See **map plotting**.

spout—See **waterspout**.

spray electrification—A process of **charge separation** associated with the mechanical disruption of liquid drops and believed to be based on the electrical **double layer** at the air–water **interface**.

Spray electrification is the common explanation for the **electric field** generation in the vicinity of waterfalls.

spread—1. (Also called variability.) The general **departure** of individual values from **central tendency**.

Spread is reflected geometrically in the **probability** curve as the width of the region over which the probability density is appreciable. *See* **scatter, variance**.

2. Popular contraction for **dewpoint spread**.

spread-F—Term derived from ionospheric **sounding** when the **echo** from the **F-region** of the **ionosphere** becomes blurred, with multiple echoes occurring over a **range** of distances and frequencies.

Spread-F is basically caused by the existence of many small-scale irregularities in **electron density** within the echoing region, instead of the **normal** smooth **variation** of electron density with height.

spread skill correlation—The **correlation** between the **spread** of an **ensemble forecast** and the **skill** of the **ensemble mean**.

Even in a perfect **model**, the spread skill correlation may be rather low since the spread is an estimate of the **expected value** of the difference between the ensemble mean and the **verification**, not an estimate of this difference in individual cases. *See* **ensemble spread**.

spring—The **season** of the year composing the transition period from winter to summer; the **vernal** season, during which the sun is approaching the **summer solstice**.

In popular usage and for most meteorological purposes, spring is customarily taken to include the months of March, April, and May in the Northern Hemisphere, and September, October, and November in the Southern Hemisphere. Except in the **Tropics**, spring is a season of rising temperatures and decreasing **cyclonic** activity over continents. In much of the Tropics, neither spring nor fall is recognizable, and in polar regions, both are very short-lived.

spring crust—A type of **snow crust**, formed when loose **firn** is recemented by a decrease in **temperature**.

It is most common in late winter and spring.

spring equinox—Same as **vernal equinox**.

spring sludge—Same as **rotten ice**.

spring snow—(Also called corn snow, granular snow.) A coarse, granular, **wet snow**, resembling finely chopped melted **ice**, generally found in the spring.

spring tapping—Process of gathering springwater into pipes and/or canals.

spring tide—**Tide** near the time of **syzygy**, when ranges between **high water** and **low water** are greatest.

Compare **neap tide, tropic tide, equatorial tide**.

sprinkle—Popular term for a very light shower of **rain**.

sprite—Weak **luminous** emissions that appear directly above an active **thunderstorm** and are coincident with **cloud-to-ground** or **intracloud lightning flashes**.

Their spatial structures range from small single or multiple vertically elongated spots, to spots with faint extrusions above and below, to bright groupings that extend from the **cloud** tops to altitudes up to about 95 km. Sprites are predominantly red. The brightest region lies in the **altitude** range 65–75 km, above which there is often a faint red glow or wispy structure that extends to about 90 km. Below the bright red region, blue tendril-like filamentary structures often extend downward to as low as 40 km. High-speed **photometer** measurements show that the duration of sprites is only a few milliseconds. Current evidence strongly suggests that sprites preferentially occur in decaying portions of thunderstorms and are correlated with large positive cloud-to-ground flashes. The optical **intensity** of sprite clusters, estimated by comparison with tabulated stellar intensities, is comparable to a moderately bright **auroral arc**. The optical **energy** is roughly 10–50 kJ per event, with a corresponding optical **power** of 5–25 MW. Assuming that optical energy constitutes 10^{-3} of the total for the event, the energy and power are on the order of 10–100 MJ and 5–50 GW, respectively. Early research reports for these events referred to them by a variety of names, including upward lightning, upward discharges, cloud-to-stratosphere discharges, and cloud-to-ionosphere discharges. Now they are simply referred to as sprites, a whimsical term that evokes a sense of their fleeting nature, while at the same time remaining nonjudgmental about physical processes that have yet to be determined. *Compare* **blue jets**.

Sq—The solar daily magnetic variations on quiet days.

It is determined from the records at geomagnetic stations for days that are believed free of noncyclic disturbances.

squall—1. A strong **wind** characterized by a sudden onset, a duration of the order of minutes, and then a rather sudden decrease in speed.

In U.S. observational practice, a squall is reported only if a **wind speed** of 16 knots or higher is sustained for at least two minutes (thereby distinguishing it from a **gust**). *See* **line squall**, **white squall**, **williwaw**.

2. In nautical use, a **severe local storm** considered as a whole, that is, **winds** and **cloud** mass and (if any) **precipitation**, **thunder** and **lightning**.

See **squall line**, **arched squall**, **black squall**, **sumatra**, **Abrolhos squalls**.

squall line—A line of active **thunderstorms**, either continuous or with breaks, including contiguous **precipitation** areas resulting from the existence of the thunderstorms.

The squall line is a type of **mesoscale convective system** distinguished from other types by a larger length-to-width ratio.

squall-line thunderstorm—An individual **thunderstorm** included within a **squall line**.

square wave—A **waveform** that increases suddenly from a base level to another level, remains at that level for a certain time interval, then abruptly returns to the base level for the same or a different time interval, and repeats this behavior periodically.

The modulating waveform of a **pulsed radar** approximates a square wave.

SREH—Abbreviation for **storm-relative environmental helicity**.

SSM/I—Abbreviation for **Special Sensor Microwave Imager**.

SSM/T—Abbreviation for **Special Sensor Microwave Temperature**.

SSU—Abbreviation for **Stratospheric Sounding Unit**.

St. Elmo's fire—(Also called Elmo's fire, corposant.) **Corona** or **point discharges** that occur when the environmental **electric field** is high, typically at the tips of sharp conductors that enhance the electric field.

This name was given the phenomenon by Mediterranean sailors who regarded it as a visitation of their patron saint, Elmo (Erasmus). Their superstition also led to the equivalent term corposant. An appearance of St. Elmo's fire was regarded as a good omen, for it tends to occur in those latter phases of a violent **thunderstorm** when most of the surface wind and wave disturbance is over.

Schonland, B. F. J., 1950: *The Flight of Thunderbolts*, 44–47.

St. Luke's summer—In English folklore, a period of fine, calm weather, similar to **Indian summer**, occurring in October.

St. Luke's Day is 18 October. *Compare* **St. Martin's summer**, **All-hallow summer**, **Old Wives' summer**.

St. Martin's summer—In English folklore, a period of fine, **calm** weather, similar to an **Indian summer**, occurring in November.

St. Martin's Day (Martinmas) is 11 November. *Compare* **St. Luke's summer**, **All-hallow summer**, **Old Wives' summer**.

St. Swithin's Day—In English folklore, a day that is popularly supposed to govern the weather of the succeeding 40 days; specifically, if it rains on St. Swithin's Day, 15 July, it will continue to **rain** for 40 days.

Not even a tendency toward this is borne out by weather records. Similar legends relate to other saints' days in many countries. *See* **control day**.

stability—1. The characteristic of a system if sufficiently small disturbances have only small effects, either decreasing in **amplitude** or oscillating periodically; it is asymptotically stable if the effect of small disturbances vanishes for long time periods.

A system that is not stable is referred to as unstable, for which small disturbances may lead to large effects. Some authors also distinguish a neutral or marginally stable case, in which disturbances do not vanish, but also do not grow without bound. Classically, stability was defined only with respect to systems in **equilibrium**. More recently it has been extended to apply to evolving systems, for which an unstable **disturbance** leads to an evolution that becomes uncorrelated with the undisturbed evolution. From this standpoint stability and **predictability** can be equated.

2. Same as **static stability**.

3. The property that each computed solution (in exact arithmetic) of a finite difference approximation remains bounded for all possible choices of the time step.
See **Lax equivalence theorem**.

4. The ability of **laminar flow** to become turbulent in a fluid.

stability categories—The **range** of stabilities organized in categories designed to provide semiquantitative measures of the **mixing** capabilities of the **lower atmosphere**; usually termed the Pasquill–Gifford categories.

> Pasquill, F., 1962: *Atmospheric Diffusion*, Van Nostrande, p. 209.

stability chart—A **synoptic chart** that shows the distribution of a **stability index**.

stability index—Any of several quantities that attempt to evaluate the potential for **convective storm** activity and that may be readily evaluated from operational **sounding** data.

In the definitions, T_p and D_p correspond to the **temperature** and **dewpoint** at **pressure** levels p. Those most commonly used are the following.

Showalter stability index: An index given by $S = (T_{500} - T_L)$, where T_L is the temperature (°C) of a **parcel** lifted from 850 to 500 mb, **dry-adiabatically** to **saturation** and **moist-adiabatically** above that. As the index decreases to zero and below, the likelihood of **showers** and **thunderstorms** is considered to increase (Showalter 1947).

K-index: This index is due to George (1960) and is defined by $K = (T_{850} - T_{500}) + D_{850} - (T_{700} - D_{700})$. The first term is a **lapse rate** term, while the second and third are related to the moisture between 850 and 700 mb, and are strongly influenced by the 700-mb temperature–dewpoint **spread**. As this index increases from a value of 20 or so, the likelihood of showers and thunderstorms is expected to increase.

Totals Indices: The Total Totals index is attributable to Miller (1972). It is defined as the sum of two indices: $TT = VT + CT$ where VT is the Vertical Totals index, defined by $VT = T_{850} - T_{500}$, A value of about 40 corresponds to a **dry-adiabatic lapse rate**. For a **moist-adiabatic lapse rate** it is about 20 for $T_{850} = 15°C$, about 30 for $T_{850} = 0°C$. The Cross Totals index, CT, is defined by $CT = D_{850} - T_{500}$, so is strongly influenced by the 850-mb moisture. Showers and thunderstorms become increasingly likely from TT values of about 30, and severe thunderstorms are considered likely for values of 50 or more.

Lifted index: This index, developed by Galway (1956), is $L = (T_L - T_{500})$, so is nominally identical to the Showalter index, except that the parcel being lifted (dry-adiabatically to saturation and then moist-adiabatically to 500 mb) is defined by the **dry adiabat** running through the predicted surface afternoon temperature maximum and the mean **mixing ratio** in the lowest 900 m of the sounding. If no further heating is expected, as with a sounding taken in the late afternoon, then the mean **potential temperature** in the lowest 900 m of the sounding defines the dry adiabat used for the parcel. Numerous variations, focused on how the lifted parcel is defined, have been used since the original definition. The values of this index tend to be somewhat lower than those of Showalter, and the interpretation depends to some extent on how the lifted parcel is defined.

SWEAT index: (Or severe weather threat index; also abbreviated SWI.) Another index attributable to Miller (1972), used mainly for analyzing the potential for severe thunderstorms, is defined as $SW = 20(TT - 49) + 12D_{850} + 2V_{850} + V_{500} + 125[\sin(\Delta V_{500 - 850}) + 0.2]$ where TT is the Total Totals index (set to zero if less than 49), V_{850} and V_{500} are the 850- and 500-mb **wind speeds**, and $\Delta V_{500 - 850}$ is the 500-mb **wind direction** minus the 850-mb wind direction, in degrees. The last term is set to zero if any of the following conditions are not met: 1) 850-mb wind direction is in the range from 130 to 250 degrees; 2) 500-mb wind direction is not in the range 210 to 310 degrees; 3) the difference in wind directions is positive, or 4) both 850- and 500-mb wind speeds are at least 15 knots. No term in the formula is allowed to be negative. The severe thunderstorm threat is considered to increase from values of about 300 and higher; **tornadoes** are considered to increase in likelihood from values of about 400 and up.

> Showalter, A. K., 1947: A stability index for forecasting thunderstorms. *Bull. Amer. Meteor. Soc.*, **34**, 250–252.
>
> George, J. J., 1960: *Weather Forecasting for Aeronautics*, Academic Press, 673 pp.
>
> Miller, R. C., 1972: Notes on analysis and severe storm forecasting procedures of the Air Force Global Weather Central. *Tech. Rept. 200(R)*, Headquarters, Air Weather Service, USAF, 190 pp.
>
> Galway, J. G., 1956: The lifted index as a predictor of latent instability. *Bull. Amer. Meteor. Soc.*, 528–529.

stability parameter—*See* **stability index**.

stable air—Air in which **static stability** prevails, a condition that depends on the vertical gradients of **air temperature** and **humidity**.

stable air mass—**Air mass** having a stable **stratification** in its lower layer, and consequently free from **convection**, having a low degree of **turbulence**, and containing either **stratiform** clouds, **fog**, or no **clouds** at all.

stable boundary layer—(Abbreviated SBL.) A cool layer of air adjacent to a cold surface of the earth, where **temperature** within that layer is statically stably stratified.

SBLs can form at night over land when the earth is cooled by net loss of **radiation**, and they can form at any time when air moves over a relatively cooler land or water surface. Many interacting processes can occur within the SBL: patchy sporadic **turbulence**, **internal gravity waves**, drainage flows, **inertial oscillations**, and nocturnal jets. *See* **nocturnal boundary layer**.

stable motion—A motion in which **small perturbations** or disturbances do not grow.

stable oscillation—*See* **oscillation, stability**.

stable profile—An **atmospheric sounding** that indicates a consistent positive change of **temperature** with height (a **temperature inversion**) or zero **convective available potential energy** (**CAPE**) at any height in the **atmosphere**.

See also **Richardson number, barotropic instability**.

stable wave—Generally, any **wave** that exhibits **stable motion** and is not an **unstable wave**.

stack effect—An effect due to **heat** sources within a local exhaust enclosure (stack) producing convective air currents with vertical velocities proportional to the rate of heat transferred to the surrounding air and to the height of rise of the heated air.

When hot gases rise through a stack, the vertical stack exit **velocity** is proportional to the square root of the difference in the densities of the heated air column and that of an equal column of the surrounding **ambient air**.

stacking—The holding pattern of aircraft awaiting their turn to approach and land at an airport.

Stacking aircraft one above the other has to be employed when weather conditions interfere with normal **VFR** direct landing approach to the extent that aircraft arrive at a more rapid rate than they are able to approach and land. Stacked aircraft are usually separated by a fixed **altitude**, and when the bottom aircraft lands all those in the stack drop down one altitude separation. Newly arrived aircraft take a place at the top of the stack.

staff gauge—A graduated **scale** placed in a position so that the **stage** of a **stream** may be read directly therefrom; a type of **river gauge**.

Staff gauges may be painted on bridge piers and pilings, and painted wooden scales or enameled metal scales may be mounted on bridge piers, docks, trees, or specially constructed supports to be partially submerged by the stream at all times.

stage—The level of the water surface in a **stream**, river, or **reservoir**, measured with reference to some **datum**.

stage-discharge relation—Relationship between **water level** and **discharge** at a certain location in a **stream** or river, also referred to as a **rating curve** or **uniform flow** rating curve, expressed as a single valued function.

Dynamic stage-discharge relationships, which form **loop rating** curves, occur when the **stage** is not just a function of the discharge, but is also a function of the variable energy slope.

stage relation—Same as **gauge relation**.

staggered grid—A **grid** on which all of the **variables** are not predicted at all of the points but rather are interspersed at alternate points.

Grids using certain types of staggering, such as between mass and the components of **momentum** in the horizontal, have been shown to possess superior numerical properties to those of unstaggered grids.

stagnant glacier—Same as **dead glacier**.

stagnation area—1. For air blowing toward an object such as the wall of a building, the point that marks the center of **divergence** of air along that wall as the streamlines split to flow around the building.

The stagnation area corresponds to a relative maximum of **static pressure**.

2. During strongly **statically stable** conditions, the region just **upstream** of a mountain where the air is blocked by the mountain.

This blocked flow could also contain a **cavity** of reverse flow.

3. An **air pollution** term for an **anticyclonic** region of **subsidence** and light winds that tends to trap pollutants near the ground where concentrations can become large.

stagnation pressure—1. Usually, same as **total pressure**.

2. Sometimes, same as **dynamic pressure**.

stalling Mach number—The **Mach number** of an aircraft when the coefficient of **lift** of the **aerodynamic** surfaces is the maximum obtainable for the **pressure altitude, true airspeed**, and angle of attack under which the craft is operated.

stalo—(Coined word from stable local oscillator.) A very stable **radio frequency** local **oscillator** used in radio and **radar** for mixing received signals to produce an **intermediate frequency** signal. *See* **heterodyne**.

stand—The condition at **high water** or **low water** when there is no change in the height of the **tide**. *Compare* **slack water**.

standard—An established unit of measure, reference instrument, or method appropriate for the **calibration** of other instruments.

standard artillery atmosphere—A set of values describing atmospheric conditions on which ballistic computations are based, namely, no **wind**; a **surface temperature** of 15°C; a **surface pressure** of 1000 millibars; a surface **relative humidity** of 78%; and a **lapse rate** that yields a prescribed **density altitude** relation.

standard artillery zone—A vertical subdivision of the **standard artillery atmosphere**; it may be considered a layer of air of prescribed **thickness** and **altitude**.

standard atmosphere—1. A hypothetical vertical distribution of atmospheric **temperature, pressure**, and **density** that, by international agreement, is taken to be representative of the **atmosphere** for purposes of **pressure altimeter** calibrations, aircraft performance calculations, aircraft and missile design, ballistic tables, etc.

The air is assumed to obey the **perfect gas** law and the **hydrostatic equation**, which, taken together, relate temperature, pressure, and density variations in the vertical. It is further assumed that the air contains no **water vapor** and that the acceleration of **gravity** does not change with height. This last assumption is tantamount to adopting a particular unit of **geopotential height** in place of a unit of geometric height for representing the measure of vertical displacement, for the two units are numerically equivalent in both the metric and English systems, as defined in connection with the standard atmosphere. The current standard atmosphere is that adopted in 1976 and is a slight modification of one adopted in 1952 by the International Civil Aeronautical Organization (ICAO), which, in turn, supplanted the NACA Standard Atmosphere (or U.S. Standard Atmosphere) prepared in 1925. It assumes **sea level** values as follows:

—Temperature 288.15 K (15°C)

—Pressure 101 325 Pa (1013.25 mb, 760 mm of Hg, or 29.92 in. of Hg)

—Density 1225 g m^{-2} (1.225 g L^{-1})

—Mean molar mass — 28.964 g **mole**$^{-1}$.

The parametric assumptions and physical constants used in preparing the current standard atmosphere are as follows.

1) Zero **pressure altitude** corresponds to that pressure that will support a column of **mercury** 760 mm high. This pressure is taken to be 1.013250 × 10^6 dynes cm^{-2}, or 1013.250 mb, and is known as one standard atmosphere or one atmosphere.

2) The **gas constant** for **dry air** is 2.8704 × 10^6 ergs gm^{-1}K^{-1}.

3) The **ice point** at one standard atmosphere pressure is 273.16 K.

4) The **acceleration of gravity** is 980.665 cm s^{-2}.

5) The temperature at zero pressure altitude is 15°C or 288.15 K.

6) The density at zero pressure altitude is 0.0012250 gm cm^{-3}

7) The **lapse rate** of temperature in the **tropopause** is 6.5°C km^{-1}.

8) The pressure altitude of the tropopause is 11 km.

9) The temperature at the tropopause is −56.5°C.

The **ARDC Model Atmosphere**, 1959, extended the above standard approximately as follows:

1) The lapse rate from 11 to 25 km is 0°C km^{-1}.

2) The lapse rate from 25 to 47 km is +3.0°C km^{-1}; temperature at 47 km is +9.5°C.

3) The lapse rate from 47 to 53 km is 0°C km^{-1}.

4) The lapse rate from 53 to 75 km is −3.9°C km^{-1}; temperature at 75 km is −76.3°C.

5) The lapse rate from 75 to 90 km is 0°C km^{-1}.

6) The lapse rate from 90 to 126 km is +3.5°C km^{-1}; temperature at 126 km is +49.7°C (molecular-scale temperatures).

7) The lapse rate from 126 to 175 km is +10.0°C km^{-1}; temperature at 175 km is 539.7°C (molecular-scale temperatures).

8) The lapse rate from 175 to 500 km is +5.8°C km⁻¹; temperature at 500 km is 2424.7°C (molecular-scale temperatures).

The U.S. Extension to the ICAO Standard Atmosphere is essentially a recomputation of the above data from the surface to 300 km. *See* **air**.

2. A standard unit of **atmospheric pressure**, the 45° **atmosphere**, defined as that **pressure** exerted by a 760-mm column of **mercury** at 45° latitude at **sea level** at **temperature** 0°C (**acceleration of gravity** = 980.616 cm s⁻²).

One 45° atmosphere = 760 mm Hg (45°) = 29.9213 in Hg (45°) = 1013.200 mb = 101.320 kPa.

3. A standard unit of **atmospheric pressure**, defined as that **pressure** exerted by a 760-mm column of **mercury** at **standard gravity** (980.665 cm s⁻² at **temperature** 0°C).

This is a unit recommended for meteorological use. One standard atmosphere = 760 mm Hg = 29.9213 in Hg = 1013.250 mb = 101.325 kPa.

4. With respect to radio propagation, that hypothetical **atmosphere** in which **standard propagation** exists, that is, one in which the **index of refraction** decreases with height at a rate of 12 **N-units** per 1000 ft.

COESA, 1976: *U.S. Standard Atmosphere*, U.S. Gov. Print. Off., Wash., D.C.

Jursa, A. S., ed., 1985: *Handbook of Geophysics and the Space Environment*, Air Force Geophysics Laboratory.

Minzner, R. A., K. S. W. Champion, and H. L. Pond, 1959: *The ARDC Model Atmosphere, 1959*, Air Force Surveys in Geophysics No. 115 (AFCRC-TR-59-267), Air Force Cambridge Research Center.

standard atmospheric pressure—That **pressure** (force per unit area) recorded when the height of a column of **mercury** is reduced to **standard gravity** and **temperature**.

It is the pressure exerted by a column of mercury 760 mm high, of **density** 13 595.1 kg m⁻³, subject to a gravitational **acceleration** of 9.80665 m s⁻², and equals 1013.25 hPa. This is a **sea level** consideration at 45°N latitude.

standard density—In physics, usually the **density** of water at the **ice point** (0°C), 0.99984 g cm⁻³; or the maximum density of water (at 4°C), 0.99997 g cm⁻³.

In the U.S. Standard Atmosphere–1976, the density of air at a **standard temperature** of 15°C and a **standard pressure** of 1013.25 mb is 1225.00 g cm⁻³.

Jursa, A. S., Ed., 1985: *Handbook of Geophysics and the Space Environment*, Air Force Geophysics Laboratory, p. 14-1.

standard density altitude—**Altitude** in the **standard atmosphere** corresponding to a particular **density**.

standard depth—A depth below the sea surface at which water properties should be measured and reported, either directly or by **interpolation**, according to the proposal by the International Association of Physical Oceanography in 1936; analogous to the mandatory levels of meteorological **upper-air observations**.

standard deviation—The positive square root σ of the **variance** σ².

This is a measure of the **scatter** or spread in a series of observations.

standard error—The **standard deviation** (positive square-root of the **variance**) of the errors associated with physical measurements of an unknown quantity, or **statistical** estimates of an unknown **parameter** or of a **random variable**.

See **root-mean-square error**, **regression**.

standard error of estimate—Same as **root-mean-square error**.

standard gravity— (Symbol $g_{45,0}$.) The **acceleration of gravity** at 45° latitude and **mean sea level**, equal to 9.80616 m s⁻².

standard isobaric surface—A surface of constant **pressure**, defined by international agreement, used for representing and analyzing conditions of the **atmosphere**.

standard precipitation index—An index developed by McKee et al. (1993) to quantify **precipitation** deficit at a given location for multiple timescales.

Standardized precipitation is the difference of precipitation from the mean for a specified time divided by the **standard deviation**, where the mean and standard deviation are determined from the climatological record. The fact that precipitation is not normally distributed is overcome by applying a transformation (i.e., **gamma function**) to the distribution.

McKee, T. B., N. J. Doesken, and J. Kleist, 1993: The relationship of drought frequency and duration to time scales. Preprints, *Eighth Conference on Applied Climatology*, Anaheim, CA, Amer. Meteor. Soc., 179–184.

standard pressure—1. A **pressure** of one **atmosphere**, in **SI** units, 101 325 N m⁻².

In older works standard pressure sometimes means the arbitrary reference pressure in the expression for **potential temperature**; but this use of standard pressure should be avoided, given the widespread use of standard pressure in many fields of science to mean a pressure of one atmosphere. *Compare* **standard temperature.**

In meteorology, the arbitrarily selected **atmospheric pressure** of 1000 mb to which **adiabatic** processes are referred for definitions of **potential temperature, equivalent potential temperature,** etc.

Other pressures may be used as **standard** for specific purposes.

standard pressure altitude—The **altitude** that corresponds to a given value of **atmospheric pressure** according to the ICAO **standard atmosphere.**

It is the **indicated altitude** of a **pressure altimeter** at an **altimeter setting** of 29.92 in. of **mercury** (1013.2 hPa); therefore, it is the indicated altitude above the 1013.2-hPa **constant-pressure surface.**

standard project flood—The **discharge** expected to result from the most severe combination of meteorological and hydrological conditions that are reasonably characteristic of the geographic region involved.

standard propagation—The propagation of radio **energy** over a smooth spherical earth of uniform **dielectric constant** and **conductivity** under conditions of **standard refraction** in the **atmosphere,** that is, an atmosphere in which the **refractive index** decreases uniformly with height at a rate of approximately 40 **N-units** per kilometer.

Standard propagation leads to **ray** curvature due to **refraction** with a value approximately one-fourth that of the earth's curvature, giving a **radio horizon** that is about 15% farther than the distance to the geometric **horizon.** This is equivalent to straight-line propagation over a fictitious earth with radius of four-thirds the radius of the actual earth. *See* **effective earth radius, super-standard propagation, substandard propagation, standard atmosphere.**

standard refraction—*See* **standard propagation.**

standard seawater—Same as **normal water.**

standard target—A **radar** target of known **radar cross section** used to calibrate a radar or monitor its performance.

Metal spheres or corner reflectors of known dimensions are examples.

standard temperature—In physics, usually the **ice point** (0°C); less frequently, the **temperature** of maximum water **density** (4°C).

In meteorology, this has no generally accepted meaning, except that it may refer to the temperature at zero **pressure altitude** in the **standard atmosphere** (15°C).

standard temperature and pressure—(Abbreviated STP; also called normal temperature and pressure.) A phrase used in physics to indicate a **temperature** of 0°C and a **pressure** of one **standard atmosphere.**

standard time of observation—A time specified in the manual on the **Global Observing System** for making meteorological observations.

standard visibility—Same as **meteorological range.**

standard visual range—Same as **meteorological range.**

standardized variate—A dimensionless **random variable,** usually that which measures the distance from the mean in units of the **standard deviation.**

Thus, if x is a random variable with mean μ and standard deviation σ, the **variable** z defined by the formula

$$z = \frac{x - \mu}{\sigma}$$

is a standardized variate.

standing cloud—Any stationary **cloud** maintaining its position with respect to a mountain peak or ridge, such as a **banner cloud, cap cloud, crest cloud,** or **cloud** of the species **lenticularis.**

standing eddies—Same as **stationary eddies.**

standing wave—1. (Also called **stationary wave.**) A **wave** that is stationary with respect to the medium in which it is embedded, for example, two equal **gravity waves** moving in opposite directions.

The antinodes (points of maximum amplitude) oscillate while the nodes remain stationary.

2. In oceanography, **wave** motion in an enclosed or semienclosed sea, where the incident and reflected progressive waves combine to give a node of zero tidal **amplitude**.

Maximum tidal amplitudes are found at the head of the **basin** where **reflection** occurs. No **energy** is transmitted in a standing wave, nor is there any progression of the wave pattern.

standpipe storage precipitation gauge—A **precipitation gauge** comprising a tall pipe 30.5 cm (12 in.) in diameter designed especially to measure the **water content** of **snow** in remote mountainous areas for unattended operation; part of a **SNOTEL** site.

The bottom contains a known amount of an oil–antifreeze mixture. The oil prevents **evaporation** and the antifreeze melts the snow. A pressure **transducer** converts the oil–antifreeze–water mixture into a calibrated water depth. Daily changes in depth provide seasonal and annual accumulations.

Stanton number—The reciprocal of the **Prandtl number**.

Stark effect—A shift in the **energy** levels of an isolated atom or molecule as a consequence of an external **electric field**.

Energy levels depend on the internal forces that electrons and nuclei exert on each other as well as on any external forces (e.g., an **electromagnetic field**). An observable consequence of the Stark effect is the shifting and broadening of **spectral lines**. The Stark effect is the electrical analogue of the **Zeeman effect**.

starting plume—**Plume** with a well-defined, advancing upper edge.

state of ground—State of soil with regard to soil firmness and **moisture content**; ability to support equipment to work soil; **soil temperature** suitability for seed germination.

state of the sea—(Or **sea state**; also called sea.) A description of the properties of the wind-generated **waves** on the surface of the sea.

state of the sky—The aspect of the sky in reference to the **cloud cover**.

The state of the sky is fully described when the amounts, kinds, directions of movement, and heights of all **clouds** are given.

state parameter—Same as **state variable**.

state variable—(Or state parameter, thermodynamic variable.) A term for **pressure, temperature**, and **specific volume**; other thermodynamic quantities that depend on these or other variables are often referred to as thermodynamic functions of state.

In this sense, the **equation of state** specifies the relation among the state variables.

static—Audio **frequency** signals, usually regarded as **noise**, that are detected by radio receivers.

Radio noise emitted by **lightning** is the most common natural source of static. There are also many common man-made sources, such as **corona discharges** from high-voltage transmission lines, defective vehicular ignition systems, and high-power switching relays for large motors.

static energies—Measures of the thermodynamic state of the air, similar to **potential temperature**.

For example, the **dry static energy** (also known as the **Montgomery streamfunction**) is

$$s = C_p T + gz,$$

where g is **acceleration of gravity**, z is height above some reference level (often taken as the height where the **pressure** is 100 kPa), C_p is **specific heat** at constant pressure, and T is **absolute** temperature. Typical atmospheric values are on the order of 300 kJ kg^{-1}. Compare this **static energy** to one definition for **potential temperature**,

$$\theta = T + \frac{gz}{C_p} = \frac{s}{C_p}.$$

See **moist static energy, liquid water static energy, saturation static energy**.

static energy—The sum of the **enthalpy** and **potential energy** per unit mass of **dry air**, given by

$$h = (c_{pd} + r_t c_w)T + L_v r_v + (1 + r_t)gz + \text{constant},$$

where h is the static energy (unfortunately using the same symbol as that conventionally used for **specific enthalpy**), c_{pd} is the **specific heat** at constant **pressure** of dry air, c_w is the specific heat of liquid water, L_v is the **latent heat** of **vaporization**, r_v and r_t are the **vapor** and total water **mixing ratios**, T is the **temperature**, z is the height above the surface, and g is the **acceleration of gravity**. Static energy plus **kinetic energy** is conserved along steady-state trajectories under **adiabatic** reversible conditions.

static instability—*See* **instability, static stability**.

static level—In **hydrology**, the height to which water will rise in an **artesian well**.
The static level of a flowing well is above the ground surface.

static pressure—In engineering fluid mechanics, the **pressure** in a homogeneous **incompressible fluid** in **steady flow** along a level **streamline** at points other than the stagnation point.
Thus if p is the static pressure, **Bernoulli's equation** gives

$$p + \frac{1}{2}\rho V^2 = p_1,$$

where ρ is the **density** of the fluid, V the speed, and p_1 the pressure at the stagnation point, called the **total pressure**. The **kinetic energy** per unit volume $(1/2)\rho V^2$ is also called the **dynamic pressure**. The static pressure is that measured by a **barometer** moving with the fluid. Since the static pressure is the pressure in the moving fluid and is distributed along the streamline exactly as the **hydrodynamic pressure**, the terminology is most unfortunately chosen. Since it is rigorously defined only when Bernoulli's equation applies, meteorologists do well in avoiding the term. The unqualified term "pressure" is quite satisfactory in this connection. However, the instrumental precautions taken in measuring the static pressure in fluid mechanics must also be applied to meteorological barometers so that it is the pressure and not the **wind speed** that is being measured. The measured meteorological pressure is in approximate **hydrostatic equilibrium** because of the relatively small vertical accelerations in the **atmosphere**, but this condition does not ordinarily obtain in those studies in which the concept of static pressure is used. Thus static pressure and **hydrostatic pressure** must be distinguished.

static pressure port—A termination device permitting **pressure** sensors to measure the true **atmospheric pressure** in the presence of **winds**.

static stability—(Also called **hydrostatic stability, vertical stability**.) The ability of a fluid at rest to become turbulent or laminar due to the effects of **buoyancy**.
A fluid, such as air, tending to become or remain turbulent is said to be statically unstable; one tending to become or remain laminar is statically stable; and one on the borderline between the two (which might remain laminar or turbulent depending on its history) is statically neutral. The concept of static stability can also be applied to air not at rest by considering only the buoyant effects and neglecting all other **shear** and inertial effects of motion. However, if any of these other **dynamic stability** effects would indicate that the flow is dynamically unstable, then the flow will become turbulent regardless of the static stability. That is, **turbulence** has physical priority, when considering all possible measures of flow **stability** (e.g., the air is turbulent if any one or more of static, dynamic, inertial, **barotropic**, etc., effects indicates **instability**). Turbulence that forms in statically **unstable air** will act to reduce or eliminate the instability that caused it by moving less dense fluid up and more dense fluid down, and by creating a neutrally buoyant mixture. Thus, turbulence will tend to decay with time as static instabilities are eliminated, unless some outside forcing (such as heating of the bottom of a layer of air by contact with the warm ground during a sunny day) continually acts to destabilize the air. This latter mechanism is one of the reasons why the **atmospheric boundary layer** can be turbulent all day. *Compare* **dynamic stability, lapse rate, Brunt–Väisälä frequency, nonlocal static stability, adiabatic equilibrium**; *see also* **slice method, buoyant instability**

Stull, R. B., 1991: Static stability—An update. *Bull. Amer. Meteor. Soc.*, **72**, 1521–1529.

station—1. In science generally, a permanent or temporary location where scientific observations and measurements are made.
In meteorology, several types of stations are officially defined, including **first-order station; first-order climatological station; second-order station; second-order climatological station; third-order climatological station; climatological substation; ocean weather station**.

2. In **oceanography**, the geographic location at which any set of oceanographic observations was taken; also, the observations recorded at the location.
The appropriate verbal phrase is "occupy a station." *See also* **serial station**; *compare* **ocean station, ocean weather station**.

station circle—A small circle on a **weather chart** marking the location of a **synoptic station**.

station continuity chart—A **chart** or graph on which time is one coordinate and one (or more) of the observed meteorological elements at that **station** is the other coordinate.
Compare **continuity chart**.

station designator—(Also called station identifier.) Group of five digits (IIiii) used in meteorological messages to identify the **station** of origin of an **observation** report.

It consists of the block number (II), defining the area in which the **station** is located, and the **station index number** (iii). *See* **international index numbers**.

station elevation—The vertical distance above **mean sea level** that is adopted as the reference **datum level** for all current measurements of **atmospheric pressure** at the **station**.

It may or may not be the same as either **climatological station** elevation or **elevation of ivory point**. This term is denoted by the symbol H_p in international usage. The atmospheric pressure with respect to station elevation is called the **station pressure**.

station identifier—Same as **station designator**.

station index number—For U.S. purposes, the **WBAN** number used by the National Weather Service to identify a **station**.

For international purposes, the WMO five-digit **station designator** is used. *Compare* **international index numbers**.

station location—Geographical coordinates of a **meteorological station** to include **elevation**.

station model—A specified pattern for entering, on a **weather map**, the meteorological symbols that represent the state of the weather at a particular observing **station**.

station pressure—The **atmospheric pressure** computed for the level of the **station elevation**.

This may or may not be the same as either the **climatological station pressure** or the **actual pressure**, the difference being attributable to the difference in reference elevations. Station **pressure** is usually the base value from which **sea level pressure** and **altimeter setting** are determined.

station-year method—The combination of records from several sites so as to obtain a single long record (with a length the sum of the lengths of the individual records).

The composite record is used to obtain an average **frequency distribution** describing the phenomena at sites in the region. *See also* **regional analysis**.

stationary eddies—(Also called standing eddies.) In studies of the **general circulation**, the eddies are the departures of a **field** (e.g., **temperature** or **relative vorticity**) from the **zonal** mean of that field; the stationary eddies are the time-averaged, or time-invariant, component of the **eddy** field.

stationary front—Same as **quasi-stationary front**.

stationary Gaussian process—Same as **Gaussian process**.

stationary Gaussian time series—Same as **Gaussian process**.

stationary motion—Same as **steady state**.

stationary orbit—The **orbit** in which an equatorial satellite revolves around the earth at the same rate as the earth rotates on its axis.

As a result, viewed from a point on the earth's surface, the satellite appears to be stationary.

stationary phase—*See* **foehn phase**.

stationary process—A **stochastic process**, $X(t)$, with properties that do not change over time or space.

This means that the marginal distributions of the **variate** (its mean, **variances**, and similar characteristics) are time independent. Furthermore, the joint distribution of the values of the process at two (or more) times, $X(t)$ and $X(s)$, can only depend on the time difference(s), $t - s$.

stationary source—A type of **air pollution** source that releases emissions from a specific location and is permanent or semipermanent structures at that location.

Examples are smokestacks, vents, power plants, mines, farms, buildings, and trees. *Compare* **mobile source**.

stationary-state hypothesis—(Also called steady-state hypothesis.) A chemical species is said to be in **steady state** when the rate of its formation and the rate of its destruction are approximately equal, such that the concentration of the species remains nearly constant. The lifetime is then the atmospheric burden divided by the **source** or **sink** rate.

stationary time series—A **time series** having stable **statistical** properties in the following sense.

Let $x(t)$ denote the value of the **variable** at time t. Hold t fixed and imagine an indefinite series of repetitions of essentially the same generating process, giving rise to a **population** (ensemble) of values of $x(t)$. For a stationary time series, the ensemble **probability distribution** of $x(t)$ is independent of t. When the probability distribution changes very gradually with t, the time series is called quasi-stationary.

stationary wave—A **wave** with nodes that are stationary relative to the given **coordinate system**.

Any **permanent wave** may be rendered stationary by appropriately chosen coordinates. In meteorology the coordinate system is usually fixed with respect to the earth, so that a stationary wave usually refers to one that is stationary relative to the earth's surface. *See* **standing wave**.

stationary-wave length—The length given by the **long-wave formula** for a large-scale atmospheric **stationary wave**.

It is given by

$$L_s = 2\pi \left(\frac{U}{\beta}\right)^{1/2},$$

where U is the **zonal wind** speed and β the **Rossby parameter**. *See* **Rossby wave**.

statistic—Generally, a number describing some characteristic of a **population** or samples therefrom.

Specifically, an estimate of a **statistical parameter** computed from a **sample**.

statistical—Pertaining to or characterizing **random** phenomena, or referring to **statistics**.

statistical dependence—*See* **statistical independence**.

statistical–dynamical model—A **climate model** in which the **statistical** behavior of **synoptic-scale** dynamical processes is represented parametrically.

In contrast to **general circulation models**, statistical–dynamical models offer computational advantages, but the parameterizations adopted in these models can only roughly approximate the effects of synoptic-scale **eddies**.

statistical forecast—A forecast based on a systematic **statistical** examination of data representing past observed behavior of the system to be forecast, including observations of useful predictors outside the system.

In short-term **climate** forecasting, CCA (canonical **correlation** analysis), as described by Barnston (1994), is a good example of a statistical forecast. Depending on method and scope, the limitations of statistical forecasts are related to shortness of record, danger of overfitting, assumptions of **linearity** (often), absence (often) of physical considerations, etc. Purely statistical forecasts in weather forecasting have become rare; however, a combination of dynamical **model** output and **statistics** is very common in weather forecasting. Some statistical methods are guided by physical principles to such an extent that they resemble dynamical models. An example of the latter is empirical **wave propagation** described by Qin and van den Dool (1996). *See* **perfect prognosis method**, **MOS**.

Barnston, A., 1994: Linear statistical short-term climate predictive skill in the Northern Hemisphere. *J. Climate*, **7**, 1513–1564.

Qin, J., and H. M. van den Dool, 1996: Simple extensions of an NWP model. *Mon. Wea. Rev.*, **124**, 277–287.

statistical independence—The relationship between two or more **random variables** (variates) when their joint **probability density function** can be expressed as the product of the individual functions,

$$F(x,y) = H(x)\, G(y),$$

and the ranges of **variation** of x and y are independent.

Otherwise, variates are statistically dependent.

statistical interpolation—*See* **optimum interpolation**.

statistical parameter—*See* **parameter**.

statistical significance test—*See* **significance test**.

statistics—The systematic **analysis** of **random** phenomena.

Primarily, it is the application of **probability theory** to specific data, but includes special techniques and principles not subsumed under **probability**. Statistics is concerned with collecting and processing data, summarizing information, estimating descriptive constants (parameters), discovering empirical laws, testing hypotheses, and designing experiments in such a way that valid inferences can be drawn from empirical evidence.

statute mile—*See* **mile**.

staubosphere—(Or konisphere; both rare.) The "dust atmosphere" of the earth.

steadiness—With respect to the **wind**, same as **persistence**.

steady flow—A system in which the flow characteristics, such as depth of flow or **mean velocity**, at any given point, do not change with time.

steady motion—Same as **steady state**.

steady state—(Also called steady motion, stationary motion.) A fluid motion in which the velocities at every point of the **field** are independent of time; **streamlines** and **trajectories** are identical.

Sometimes it is further assumed that all other properties of the fluid (**pressure**, **density**, etc.) are also independent of time. All local derivatives in the fundamental equations then vanish. A steady-state solution to a theoretical problem suggests two further questions: how the steady state came to exist (the **initial-value problem**), and whether it will persist (the **instability** problem).

steady-state hypothesis—Same as **stationary-state hypothesis**.

steam—1. A popular term for **mixing cloud**.

2. Water **vapor** at a **temperature** greater than the **boiling point**.

steam devils—Gentle whirls of foggy air that form within **steam fog** when cold air advects over a warmer body of water or saturated surface; analogous to **dust devils**, except that water droplets are the **tracers** that make the whirl visible.

These are usually shallow phenomena in the **surface layer** of a **thermal internal boundary layer**.

steam fog—(Or sea smoke; also called arctic sea smoke, antarctic sea smoke, **frost smoke**, water smoke, sea mist, steam mist.) **Fog** formed when **water vapor** is added to air that is much colder than the **vapor**'s source; most commonly, when very cold air drifts across relatively warm water.

No matter what the nature of the vapor source (warm water, industrial combustion exhaust, exhaled breath), its **equilibrium vapor pressure** is greater than that corresponding to the colder air; thus, the water vapor, upon becoming mixed with and cooled by the cold air, rapidly condenses. It should be noted that this mechanism never allows the fog to actually reach the vapor source. Also, upon further **mixing** in sufficiently turbulent or convective flow (only a slight degree is needed), the fog particles evaporate at a more or less well defined upper limit of the fog. Note, also, that although **advection** of air is necessary to produce steam fog, it differs greatly from an **advection fog** in the usual sense, which is caused by warm, **moist air** moving over a cold surface. Steam fog is commonly observed over lakes and streams on cold autumn mornings as well as in polar regions. It is sometimes confused with **ice fog**, but its **particles** are entirely liquid. At temperatures below $-29°C$ ($-20°F$), these may **freeze** into **droxtals** and create a type of **ice fog** that may be known as frost smoke.

steam mist—Same as **steam fog**.

steepness—Usually defined as the ratio of **wave height** to **wavelength**, but occasionally defined as the **amplitude** multiplied by the **wavenumber** of the **wave**.

See **wave steepness**.

steering—In meteorology, loosely used for any influence upon the direction of movement of an **atmospheric disturbance** exerted by another aspect of the state of the **atmosphere**.

Thus, it might be said that a surface **pressure system** tends to be steered by **isotherms**, **contour lines**, **streamlines** aloft, warm-sector **isobars**, the orientation of a **warm front**, etc. Nearly always this principle is applied to the relationship between the **velocity** of a **cyclone** and the velocity of the **basic flow** in which it is embedded.

steering current—Same as **steering flow**.

steering flow—(Or steering current.) A **basic flow** that exerts a strong influence upon the direction of movement of disturbances embedded in it.

This is the concept of **steering** in meteorology.

steering level—A level in the **atmosphere** where the **velocity** of the **basic flow** bears a direct relationship to the **velocity** of movement of an **atmospheric disturbance** embedded in the flow.

This concept is most often applied as a subjective tool in forecasting the motion of surface **pressure systems**.

steering line—According to Bjerknes's **cyclone model**, the line of **convergence** (corresponding to the **warm front** of a **wave cyclone**) that tends to be parallel to the direction of motion of the **cyclone** at the line's point of juncture with the cyclone center.

Stefan–Boltzmann constant—A constant of proportionality that relates the **radiant** emittance of a **blackbody** to the fourth power of the body's **absolute** temperature.

Its value is 5.670×10^{-8} W m^{-2} K^{-4}.

Stefan–Boltzmann law—One of the **radiation laws**, which states that the amount of **energy** radiated per unit time from a unit surface area of an ideal **blackbody** is proportional to the fourth power of the **absolute** temperature of the blackbody.

The law is written

$$E = \sigma T^4,$$

where E is the **emittance** of the blackbody, σ the **Stefan–Boltzmann constant**, and T the absolute temperature of the blackbody. This law was established experimentally by Stefan and was given theoretical support by thermodynamic reasoning due to Boltzmann. This law may be deduced by integrating **Planck's law** over the entire **frequency** spectrum.

stellar crystal—A kind of **plane-dendritic crystal**, often referring to those with relatively thin branches without elaborate patterns.

stellar scintillation—*See* **scintillation**.

stem flow—**Rainfall** intercepted by tree canopies and transported directly to the ground via the stems and trunks.

step function—(Or stepped function.) A function that has different constant values over adjacent subintervals; thus it has discontinuities at the ends of each interval.
The **distribution function** of a discrete **variate** and a **histogram** have this shape.

steppe—An area of grass-covered and generally treeless plains, with a **semiarid climate**, which forms a broad belt over southeastern Europe and the southwestern part of the former Soviet Union.

steppe climate—(Also called **semiarid climate**.) The type of **climate** in which **precipitation** is very slight but sufficient for the growth of short, sparse grass.
This is typical of the **steppe** regions of south-central Eurasia. In his 1936 **climatic classification**, W. Köppen assigns maximum values of annual precipitation to separate this **dry climate** from the rainy climates as follows. For precipitation chiefly in the winter,

$$p = 0.44(t - 32).$$

For precipitation evenly distributed during the year,

$$p = 0.44(t - 19.4).$$

For precipitation chiefly in the summer,

$$p = 0.44(t - 6.8).$$

In the above, p is the mean annual precipitation in inches, and t the mean annual **temperature** in degrees Fahrenheit. (For minimum values of precipitation, *see* formulas under **desert climate**.) This scheme was modified by Bailey to

$$p = 0.41 \left(T - \frac{R}{4} \right),$$

where R is the precipitation falling during the six colder months. In Köppen's system the steppe climate is designated letter code BS. The semiarid climate of Thornthwaite (1931) corresponds closely to the steppe climate.

Köppen, W. P., and R. Geiger, 1930–1939: *Handbuch der Klimatologie*, Berlin: Gebruder Borntraeger, 6 vols.
Thornthwaite, C. W., 1931: The climates of North America according to a new classification. *Geogr. Rev.*, **21**, 633–655.

stepped function—Same as **step function**.

stepped leader—The initial **leader** of a **lightning discharge**; an intermittently advancing column of high ionization and charge that establishes the channel for a first **return stroke**.
The peculiar characteristic of this type of leader is its stepwise growth at intervals of about 50–100 μs. The **velocity** of growth during the brief intervals of advance, each only about 1 μs in duration, is quite high (about 5×10^7 m s^{-1}), but the long stationary phases reduce its effective speed to only about 5×10^5 m s^{-1}. To help explain its mode of advance, the concept of a **pilot streamer** was originally suggested, but has been supplanted by analogy to recent work on long laboratory sparks.

steradian—The **SI** unit of solid angle that, having its vertex in the center of a sphere, cuts off an area of the surface of the sphere equal to that of a square with sides of length equal to the radius of the sphere.
There are 4π steradians in a sphere.

stereo image—Same as **stereographic image**.

stereographic image—(Or stereo image.) A depiction of a three-dimensional object that results from stereoscopic viewing.

A stereo **image** is obtained by simultaneously viewing a **cloud**, for example, from satellites at different locations from which the height of the cloud may be determined.

stereographic projection—A type of **conformal map** in which features on a sphere are projected onto a plane tangent to the sphere.

The source of projecting rays is a point diametrically opposite the tangent point. *See* **conformal map**.

stereoscopic image—Two pictures taken with a spatial or time separation that are then arranged to be viewed simultaneously.

When so viewed they provide the sense of a three-dimensional scene using the innate capability of the human visual system to detect three dimensions.

steric anomaly—Same as **specific-volume anomaly**.

Stevenson screen—A type of **instrument shelter**.

The shelter is a wooden box painted white with double louvered sides and mounted on a stand 122 cm (4 ft) above the ground. In addition to the **dry-** and **wet-bulb thermometers**, it usually contains **maximum** and **minimum thermometers**.

sticking coefficient—*See* **mass accommodation coefficient**.

stiff system—A system of differential equations with solutions that contain a rapidly **damping** component (as would describe the displacement of a stiff spring when stretched and then released).

Such systems are used to describe rapid photochemical reactions that take place in the **upper atmosphere**. Special numerical techniques are often required to solve stiff systems in order to maintain **stability** of the computation.

Stikine wind—The severe, gusty east-northeast **wind** in the vicinity of the Stikine River near Wrangell, Alaska.

It is produced under the same conditions as a **Taku wind**.

stilb—A unit of **luminance** (or **brightness**) equal to one **international candle** per cm^2.

still-water level—The level that the sea surface would assume in the absence of **wind waves**; not to be confused with **mean sea level** or **half-tide level**.

still well—A device, used in **evaporation pan** measurements, that provides an undisturbed water surface and support for the **hook gauge**.

The National Weather Service model consists of a brass cylinder, 8 in. high and 3.5 in. in diameter, mounted over a hole in a triangular galvanized iron base that is provided with leveling screws.

stilling-well gauges—An instrument system for measuring **sea levels**; tidal changes of levels are detected by the movement of a float in a well, which is connected to the open sea by a restricted hole or narrow pipe.

Wind waves are eliminated by the constriction of the connection.

stirring—*See* **mixing**.

stochastic—Conjectural; in **statistical** analysis, a synonym for **random**.

stochastic hydrology—The science that pertains to the probabilistic description and modeling of the value of hydrologic phenomena, particularly the dynamic behavior and the **statistical** analysis of records of such phenomena.

stochastic model—A **model** of a system that includes some sort of **random** forcing.

In many cases, stochastic models are used to simulate **deterministic** systems that include smaller-**scale** phenomena that cannot be accurately observed or modeled. As such, these small-scale phenomena are effectively unpredictable. A good stochastic model manages to represent the average effect of unresolved phenomena on larger-scale phenomena in terms of a random forcing.

stochastic particle modeling—A Lagrangian approach for the simulation of atmospheric **dispersion** of passive, nonbuoyant, and nonreactive **tracers** in a **turbulent flow**.

Particle trajectories are calculated by assuming the evolution of turbulent velocities to be **Markovian** (thus experiencing an **acceleration** of partly **stochastic** nature). One-particle models assume the **particles** to be independent of each other and thus are able to predict ensemble-average **statistics** only. Two-particle models take into account the relative movement of particle pairs and can therefore predict concentration fluctuations and hence provide information on the actual realization of a **plume** of **pollutants**.

stochastic process—A system that evolves in time according to probabilistic equations, that is, the behavior of the system is determined by one or more time-dependent **random variables**.

stoke—The unit of **kinematic viscosity** in the centimeter-gram-second system, one cm^2s^{-1}, named in honor of Sir George Gabriel Stokes (1819–1903).

Stokes's drift—The Lagrangian **drift** associated with **irrotational** surface **gravity waves**.
   See **Stokes's drift velocity**.

Stokes's drift velocity—The **average velocity** of a fluid **particle** in **wave motion** of infinitesimal **amplitude**.
   To the order of the square of the **wave amplitude**, it is

$$\mathbf{u}_s(x) = <\int_{t_0}^t \mathbf{u}(\mathbf{x},t')dt' \cdot \nabla(\mathbf{x},t)>,$$

where $<\cdot>$ represents the average over a **wave period**, t is time, $\mathbf{u}$ is the local (Eulerian) **velocity** of the wave motion, and $\mathbf{x}$ is the position of the fluid particle at $t = t_0$. As an example, for surface **gravity waves** of a homogeneous ocean with a flat bottom,

$$\mathbf{u}_s = ([a^2\omega\mathbf{k}]/[2 \sinh^2 kH])/(\cosh [2k\{z + H\}]),$$

where a is the wave amplitude, $\mathbf{k}$ is the **wavenumber** vector, $k = |\mathbf{k}|$, ω is the **wave frequency**, z is the vertical coordinate (negative below the ocean surface), and H is the depth of the ocean.

Stokes's law—An approximate relation, valid at low Reynolds numbers, for the **drag** force on a sphere moving relative to a fluid.
   According to this law the drag on a sphere of radius a with speed v in a fluid with **dynamic viscosity** μ is

$$F_D = 6\pi a \mu v.$$

Stokes's law is a good approximation for the drag on **cloud droplets** falling in air, but not on **raindrops**.

Stokes's streamfunction—*See* **streamfunction**.

Stokes's theorem—The statement that if s is a surface in three dimensions having a closed curve c as its boundary, then the **circulation** of a **vector V** around c is equal to the **flux** of the **vorticity** (**curl** of **V**) through s, under certain mathematical conditions on these fields and boundaries:

$$\int\int_s \mathbf{n} \cdot (\nabla \times \mathbf{V})ds = \oint_c \mathbf{V} \cdot d\mathbf{r},$$

where $\mathbf{n}$ is the unit vector normal to s on that side of s arbitrarily taken as the positive side. The positive direction along c is defined as the direction along which an observer, traveling on the positive side of s, would proceed while keeping the enclosed area to his left. $d\mathbf{r}$ is a vector line element of c.
   For two-dimensional flow in the x, y plane, Stokes's theorem becomes

$$\int\int_s \zeta ds = \oint \mathbf{V} \cdot d\mathbf{r},$$

where ζ is the vertical component of the vorticity. This states that the circulation around a given curve per unit area enclosed by the curve is equal to the average vorticity within the area.

Stokes wave—A **nonlinear** irrotational long-crested ocean **wave**, neglecting **surface tension** effects.
   It was first described by Sir George Gabriel Stokes (1819–1903), who produced a series expansion in powers of the **wave amplitude**. Its steepest form has a crest with a 120° angle.

Stokesian wave—Same as **deep-water wave**.

stomatal resistance—The opposition to **transport** of quantities such as **water vapor** and **carbon dioxide** to or from the stomata (pores) on the leaves of plants.
   For water vapor, stomatal resistance r is defined as

$$r = \frac{\rho(q_i - q_s)}{E},$$

where q_i and q_s are the **specific humidities** in the interior of the stomate cavity and at the exterior surface of the leaf, ρ is air density, and E is the **moisture flux**. The dimension of r is time per distance, that is, inverse **velocity**. This equation is analogous to that for an electrical **resistance**, proportional to the ratio of voltage **potential** divided by **current** flow. *See* **transpiration**.

> Oke, T. R., 1987: *Boundary Layer Climates*, chap. 4.

stone ice—Same as **ground ice**.

stooping—A **mirage** in which the angular height of the **image** is less than that of the object.

As the width is unaffected (the angular width of image width remains that of the object because the **refractive index** gradient is vertical), the **aspect ratio** is altered and distant images appear vertically squashed. Stooping often accompanies **looming**: Distant features appear lifted and squashed. *Compare* **towering, sinking**.

storage coefficient—(Also known as storativity.) The volume of water released from storage in a **confined aquifer** per unit surface area per unit decrease in the **hydraulic head**.

The storage coefficient is the product of the **specific storage** and the aquifer thickness.

storage curve—In a **surface water** body, a curve depicting the volume of stored water as a function of **stage**.

storage equation—In **hydrology**, the **equation of continuity** applied to **unsteady flow**.

It states that the fluid **inflow** to a given space during an interval of time minus the **outflow** during the same interval is equal to the change in storage. It is applied in hydrology to the routing of **floods** through a **reservoir** or a **reach** of a **stream**. The **moisture-continuity equation** applied to the **atmosphere** is a modification of this.

storage ratio—Ratio of the net available storage of a **reservoir** to the annual mean inflow.

storage routing—Same as **streamflow routing**.

storage term—The local (Eulerian) rate of change of a quantity with time, $\partial/\partial t$, also known as a **tendency** term.

It indicates the increase or decrease in a quantity.

storativity—Same as **storage coefficient**.

storm—1. Any disturbed state of the **atmosphere**, especially as affecting the earth's surface, implying inclement and possibly destructive weather.

There are at least three somewhat different viewpoints of storms. 1) In **synoptic meteorology**, a storm is a complete individual **disturbance** identified on **synoptic charts** as a complex of **pressure, wind, clouds, precipitation**, etc., or identified by such mesometeorological means as **radar** or sferics. Thus, storms range in scale from **tornadoes** and **thunderstorms**, through **tropical cyclones**, to widespread **extratropical cyclones**. 2) From a local and special interest viewpoint, a storm is a **transient** occurrence identified by its most destructive or spectacular aspect(s). In this manner we speak of **rainstorms, windstorms, hailstorms, snowstorms**, etc. Notable special cases are **blizzards, ice storms, sandstorms**, and **duststorms**. 3) To a hydrologist, "storm" alludes primarily to the space- and time-distribution of **rainfall** over a given region. *See* **local storm, severe storm**.

2. *See* **magnetic storm**

3. (Also called storm wind, violent storm.) In the **Beaufort wind scale**, a **wind** with a speed from 56 to 63 knots (64 to 72 mph) or Beaufort Number 11 (Force 11).

storm chaser—One who intercepts, by car, van, or truck, **severe convective storms** for sport or for scientific research.

Storm chasing is particularly popular in the Plains of the United States.

storm duration—Period of time between the onset and end of **precipitation**.

storm loss—*See* **rainfall loss**.

storm model—A physical, three-dimensional representation of the **inflow, outflow**, and vertical motion of air and **water vapor** in a **storm**.

It is used in **hydrometeorology** as the basis for computing the effective **precipitable water** from the surface **dewpoint** by application of the **equation of continuity**.

storm-relative environmental helicity—(Abbreviated SREH.) A measure of the **streamwise vorticity** within the **inflow** environment of a **convective storm**.

It is calculated by multiplying the storm-relative inflow velocity **vector** by the **streamwise vorticity** and integrating this quantity over the inflow depth. Geometrically, the storm-relative environmental helicity is represented by the area on a **hodograph** swept out by the storm-relative

wind vectors between specified levels (typically the surface and 3 km to represent the primary **storm** inflow). It is thought to be a measure of the tendency of a **supercell** to rotate.

storm surge—(Also called **storm tide, storm wave, hurricane tide.**) A rise and onshore **surge** of **seawater** as the result primarily of the winds of a **storm**, and secondarily of the **surface pressure** drop near the storm center.

The magnitude of the surge depends on the size, **intensity**, and movement of the storm; the shape of the coastline; nearshore underwater **topography**; and the state of the astronomical tides. The storm surge is responsible for most loss of life in **tropical cyclones** worldwide.

storm tide—1. Same as **storm surge.**

2. The height of a **storm surge** (or **hurricane wave**) above the astronomically predicted level of the sea.

storm track—1. A region in which the **synoptic** eddy activity is statistically and locally most prevalent and intense.

It also roughly corresponds to the mean trajectory of the **cyclones** in winter. In winter, there are two **storm** tracks in the Northern Hemisphere centered at about 45°N, one over the Pacific Ocean and the other over the Atlantic Ocean. There is one storm track in the Southern Hemisphere over the South Indian Ocean region at about 50°S prevalent throughout most of the year.

2. The path followed by a meteorological phenomenon, for example, a center of low atmospheric **pressure**, a severe **thunderstorm**, a **tornado**.

storm transposition—Technique used to establish a definite lower bound for **probable maximum precipitation** by moving an observed **depth–area–duration pattern** from where it occurred to another place where it may reasonably be expected to occur.

storm warning—1. A **warning** of sustained winds of 48 knots (55 mph) or more either predicted or occurring, not associated with **tropical cyclones.**

The storm-warning signals for this condition are 1) one square red flag with black center by day and 2) two red lanterns at night.

2. Prior to 1 January 1958, a **warning**, for marine interests, of impending winds of from 28 to 63 knots (32 to 72 mph).

3. A message in plain language broadcast to all Navy ships and merchant ships over appropriate **Fleet Broadcasts**.

The **warning** gives information on the position, movement, **intensity**, etc., of a **low pressure center**. *See* **small-craft warning, gale warning, whole-gale warning, hurricane warning.**

storm-warning signal—An arrangement of flags or pennants (by day) and lanterns (by night) displayed on a coastal **storm-warning tower.**

storm-warning tower—A tower, generally constructed of steel, for displaying coastal **storm-warning signals.**

storm wave—1. A **wave** generated during a **storm**, usually taken to refer to a wave of great height.

2. Same as **storm surge.**

storm wind—See **storm.**

STP—Abbreviation for **standard temperature and pressure**, that is, 0°C and 1 **atmosphere** (1013.3 mb).

straight-line wind—Current of air in which the ground-relative motion does not have any significant curvature.

Used in the context of **surface winds** that inflict damage; to be distinguished from winds in **tornadoes**, which have significant curvature.

strain tensor—(Or rate-of-strain tensor.) The 3×3 array of coefficients representing the **deformation** of a nonrigid body subjected to **stress**.

Batchelor, G. K., 1967: *An Introduction to Fluid Dynamics*, Cambridge University Press, 79–84.

stratification—The existence or formation of distinct layers or laminae in a body of water identified by differences in **thermal** or **salinity** characteristics (e.g., densities) or by **oxygen** or nutrient content.

stratified fluid—A fluid having **density** variation along the axis of **gravity**, usually implying upward decrease of density, that is, a **stratification** characterized by **static stability**.

stratiform—Descriptive of **clouds** of extensive horizontal development, as contrasted to the vertically developed **cumuliform** types.
>*See* **stratus, altostratus, cirrostratus, nimbostratus, stratocumulus**.

stratiform cloud—*See* **cloud classification, stratiform, stratiformis**.

stratiform precipitation area—A region of **precipitation** from a **nimbostratus** cloud, which may or may not be an outgrowth of a **cumulonimbus** cloud, in which the air motions are strong enough for **vapor** to be condensed or deposited on **particles** but weak enough that the particles cannot grow effectively by collection of **cloud** water **droplets**.
>Generally the vertical motions are weak enough that the growing particles fall relative to the ground. In especially well-developed **stratiform** precipitation, precipitating **ice** particles falling and growing by vapor deposition aggregate to form large **snowflakes**; then the snowflakes melt and produce a **bright band** on **radar**.

stratiformis—A **cloud species** consisting of a very extensive horizontal layer or layers that need not be continuous.
>This species is the most common form of the genera **altocumulus** and **stratocumulus** and is occasionally found in **cirrocumulus**. *See* **cloud classification**.

stratocumulus—(Abbreviated Sc.) A principal **cloud** type (**cloud genus**), predominantly **stratiform**, in the form of a gray and/or whitish layer or patch, which nearly always has dark parts and is nonfibrous (except for **virga**).
>Its elements are tesselated, rounded, roll-shaped, etc.; they may or may not be merged, and usually are arranged in orderly groups, lines, or undulations, giving the appearance of a simple (or occasionally a cross-pattern) **wave system**. These elements are generally flat-topped, smooth, and large; observed at an angle of more than 30° above the **horizon**, the individual stratocumulus element subtends an angle of greater than 5°. When a layer is continuous, the elemental structure is revealed in true relief on its under surface. Stratocumulus is composed of small water droplets, sometimes accompanied by larger **droplets**, **soft hail**, and (rarely) by **snowflakes**. When the cloud is not very thick, the **diffraction** phenomena **corona** and **irisation** appear. Under ordinary conditions, **ice crystals** are too sparse even to give the cloud a fibrous aspect; however, in extremely cold weather, ice crystals may be numerous enough to produce abundant virga, and sometimes even **halo** phenomena. **Mamma** may be a supplementary feature of stratocumulus, in which case the mammiform protuberances may develop to the point where they seem about to detach themselves from the main cloud. Virga may form under the cloud, particularly at very low temperatures. **Precipitation** rarely occurs with stratocumulus. Stratocumulus frequently forms in **clear air**. It may also form from the rising of **stratus**, and by the convective or undulatory transformation of stratus, or **nimbostratus**, with or without change of height (Sc stratomutatus or Sc nimbostratomutatus). Stratocumulus is analogous to **altocumulus** and forms directly from the latter when the elements grow to a sufficient size (Sc altocumulomutatus). The further humidification, accompanied by **turbulence** and/or **convection**, of an already humid layer of air near the base of nimbostratus or even **altostratus** can form stratocumulus (Sc nimbostratogenitus or Sc altostratogenitus). If the ascending currents that produce **cumulus** or **cumulonimbus** approach an upper layer of **stable air**, they slow down, and all or a portion of the **mother-cloud** tends to diverge gradually and spread horizontally, often producing stratocumulus (Sc cumulogenitus or Sc cumulonimbogenitus). A particular form of Sc cumulogenitus, previously called stratocumulus vesperalis, often occurs in the evening when convection decreases, resulting in the gradual **dissipation** of both bases and tops of **cumuliform** clouds. Since stratocumulus may be transformed directly from or into altocumulus, stratus, and nimbostratus, all transitional stages may be observed. By convention, altocumulus is composed of apparently smaller elements (often simply because of its higher **altitude**); stratus and nimbostratus do not show regular subdivisions or **wave** form, and they have a more fibrous aspect. When the base of stratocumulus is rendered diffuse by precipitation, the cloud becomes nimbostratus. *See* **cloud classification**.

stratocumulus castellanus—*See* **castellanus**.

stratocumulus floccus—*See* **floccus**.

stratocumulus lenticularis—*See* **lenticularis**.

stratocumulus mammatus—*See* **mamma**.

stratocumulus opacus—*See* **opacus**.

stratocumulus perlucidus—*See* **perlucidus**.

stratocumulus radiatus—*See* **radiatus**.

stratocumulus stratiformis—See **stratiformis**.

stratocumulus translucidus—*See* **translucidus**.

stratocumulus undulatus—*See* **undulatus**.

stratocumulus vesperalis—*See* **stratocumulus**.

stratopause—The top of the **stratosphere**.

stratosphere—The region of the **atmosphere** extending from the top of the **troposphere** (the **tropopause**), at heights of roughly 10–17 km, to the base of the **mesosphere** (the **stratopause**), at a height of roughly 50 km.

The stratosphere is characterized by constant or increasing temperatures with increasing height and marked **vertical stability**. It owes its existence to heating of **ozone** by solar **ultraviolet radiation**, and its **temperature** varies from −85°C or less near the tropical tropopause to roughly 0°C at the stratopause. While the major constituents of the stratosphere are molecular **nitrogen** and **oxygen**, just as in the troposphere, the stratosphere contains a number of minor chemical species that result from photochemical reactions in the intense ultraviolet radiation environment. Chief among these is ozone, the presence of which shelters the underlying atmosphere and the earth's surface from exposure to potentially dangerous ultraviolet radiation. *See* **atmospheric shell**.

Stratospheric Aerosol Gas Experiment—(Abbreviated SAGE.) A limb **sounding** instrument that was first flown as a Small Applications Explorer Mission in 1979.

Since this initial mission, a series of SAGE instruments have been flown on a variety of satellites. In all cases, the SAGE instruments measure vertical profiles of **aerosol** and gas concentrations by measuring the **extinction** of **sunlight** during satellite sunrises and sunsets. SAGE II was launched in October 1984 on the **ERBS**, and SAGE III will be flown on a number of satellites as part of NASA's **EOS** program. Current plans call for placing SAGE III on a Russian **Meteor** platform, as well as on the International Space Station.

Stratospheric Aerosol Measurement—(Abbreviated SAM.) A single **channel** sun **photometer** centered at 1 **micron** used to measure **aerosol** extinction above the earth tangent point by looking at the sun during spacecraft **sunrise** and **sunset**.

SAM was initially tested as a hand-held sun photometer on *Apollo–Soyuz* in 1975. A follow-on SAM-II instrument was carried onboard *Nimbus-7* (launched October 1978).

Stratospheric and Mesospheric Sounder—(Abbreviated SAMS.) A multichannel **infrared** limb **sounder** on *Nimbus-7* (launched October 1978) designed to extract vertical **temperature** soundings from upper atmospheric **emissions**.

stratospheric coupling—The interaction between disturbances in the **stratosphere** and those in the **troposphere**.

The interaction can take the form of dynamic forcing and/or an exchange of mass.

stratospheric oscillation—Traveling **wave** in the **stratosphere** with a **period** of between four and five days and forced by diabatic heating in the **troposphere**.

The diabatic forcing is manifested as cloud-brightness oscillations that are particularly prominent at 5°–10°N in the northern winter, indicating that they are oscillations in the **intertropical convergence zone**.

Stratospheric Sounding Unit—(Abbreviated SSU.) A three-channel **infrared** sounder on operational **NOAA** polar-orbiting satellites.

The three channels are used to determine profiles of **temperature** in the **stratosphere**.

stratospheric steering—The **steering** of lower-level atmospheric disturbances along the **contour lines** of the **tropopause**, which lines are presumably roughly parallel to the direction of the **wind** at tropopause level.

stratospheric sulfate layer—(Also called Junge aerosol layer.) A region in the lower **stratosphere** (approximately 15–25 km in **altitude** and worldwide) where submicrometer-sized **particles** composed of aqueous **sulfuric acid** are present.

Oxidation of sulfur compounds (primarily **carbonyl sulfide**, OCS, and **sulfur dioxide**, SO_2) is believed to be the source of sulfuric acid (H_2SO_4). Volcanic eruptions directly inject large quantities of H_2S or SO_2 into the stratosphere, causing large increases in the amount of particulate sulfuric acid. The particles undergo a slow falling-out process (timescale of years) into the **lower atmosphere**. Because these sulfuric acid particles are present in the **ozone layer**, heterogeneous chemical reactions that occur on them can have a significant effect on **ozone** by influencing the abundance of **trace** species that affect ozone concentrations.

Junge, C. E., C. W. Chagnon, and J. E. Manson, 1961: Stratospheric aerosols. *J. Meteor.*, **18**, p.81.

stratospheric warming—(Also called sudden warming.) A rise in **temperature** of the **stratosphere** in the polar region in late winter resulting from enhanced propagation of **energy** from the **troposphere** by planetary-scale waves.

Temperatures near 50 mb can increase by as much as 40°C in just a few days.

Andrews, D. F., J. R. Holton, and C. B. Leovy, 1987: *Atmosphere Dynamics*, 259–294.

stratus—(Abbreviated St.) A principal **cloud** type (**cloud genus**) in the form of a gray layer with a rather uniform base.

Stratus does not usually produce **precipitation**, but when it does occur it is in the form of minute **particles**, such as **drizzle, ice crystals**, or **snow grains**. Stratus often occurs in the form of ragged patches, or cloud fragments (**stratus fractus**), in which case rapid transformation is a common characteristic. Stratus clouds have characteristically low vertical velocities, usually less than 1 m s^{-1}. When the sun is seen through the cloud, its outline is clearly discernible, and it may be accompanied by **corona** phenomena. In the immediate area of the solar disk, stratus may appear very white. Away from the sun, and at times when the cloud is sufficiently thick to obscure it, stratus gives off a weak, uniform **luminance**. The particulate composition of stratus is quite uniform, usually of fairly widely dispersed water **droplets** and, at lower temperatures, of ice crystals (although this is much less common). **Halo** phenomena may occur with this latter composition. Dense stratus often contains particles of precipitation. The prior existence of any other cloud in the low or middle levels is seldom required for the formation of stratus. A common mode of stratus development is the transformation of **fog**, the lower part of which evaporates while the upper part may rise (St nebulomutatus). As can be expected by its close relationship to fog, stratus follows a **diurnal** cycle with a maximum (over land) in the night and early morning. **Insolation** tends to dissipate this cloud rapidly, and often brings about the transformation of stratus fragments to **cumulus** clouds. Fog arriving from the sea frequently becomes stratus over the adjacent land. Coastal regions also provide the low-level moisture and frequently the lapse-rate **stability** conducive to its formation, and therefore these areas have the greatest stratus status. Stratus also develops from **stratocumulus** when the undersurface of the latter descends or for any reason loses its relief or apparent subdivisions (St stratocumulomutatus). **Nimbostratus** and **cumulonimbus** often produce stratus fractus, as precipitation from these clouds causes low-level **condensation** (St fractusnimbostratogenitus or St fractus cumulonimbogenitus). Stratus fractus in this form constitutes the accessory feature **pannus** of these **mother-clouds**. Stratocumulus and nimbostratus are the clouds most difficult to distinguish from stratus. Stratus is lower and lacks the uniform undulations or relief of stratocumulus. More difficulty is encountered when differentiating it from nimbostratus. Their modes of formation are different, nimbostratus usually having been formed from a preexisting mid- or low-level cloud; nimbostratus is more dense and has a wetter aspect, and its precipitation is of the ordinary varieties. As a final distinction, the **wind** accompanying nimbostratus is usually stronger than that with stratus. *See* **cloud classification**.

stratus fractus—(Previously called fractostratus.) *See* **fractus**.

stratus nebulosus—*See* **nebulosus**.

stratus opacus—*See* **opacus**.

stratus translucidus—*See* **translucidus**.

stratus undulatus—*See* **undulatus**.

strays—In radio terminology, same as **atmospherics**.

streak lightning—Ordinary **lightning**, of a **cloud-to-ground discharge**, that appears to be entirely concentrated in a single, relatively straight **lightning channel**.

Compare **forked lightning, zigzag lightning**

stream—Body of **water** moving under the influence of **gravity**, to lower levels, in a well-defined natural channel.

stream frequency—A measure of topographic texture based on the ratio of the number of **stream** segments per unit area of the **basin**.

stream gauge—(Also spelled stream gage.) Same as **river gauge**.

stream gauging—The process of measuring river **discharge**.

stream hydrograph—A plot of stream **discharge** as a function of time.

stream network—(Also called drainage network.) The geomorphologic patterns of hierarchical connections formed by streams as they **drain** the **watershed**.

stream order— (Also called channel order.) The designation by a dimensionless integer series (1, 2, 3, . . .) of the relative position of **stream** segments in the network of a **drainage basin**, with 1 designating a stream with no tributaries.

stream piracy—The natural movement of **groundwater** from one **river basin** to another (sometimes distant) basin.

stream profile—**Stream** bed **profile** along the direction of flow.

stream segment—A portion of a **stream** defined either between its beginning and a **confluence**, or between two confluences.

stream tube—A region of space bounded by arbitrarily selected **streamlines** of a fluid.

streamer—1. A sinuous channel of very high **ion** density that propagates itself though a gas by continual establishment of an **electron avalanche** just ahead of its advancing tip.

In **lightning** discharges, the **stepped leader**, **dart leader**, and **return stroke** all constitute special types of streamers.

2. *See* **aurora**.

3. A term used to describe a **dropsonde observation** when, because of either partial or complete parachute failure, the descent speed of the **dropsonde** exceeds 1500 feet per minute.

4. *See* **snow trails**.

streamflow—The water flowing in a **stream** channel.

streamflow routing—(Also called flood routing, storage routing.) In **hydrology**, a procedure used to derive a **downstream** hydrograph from an **upstream** hydrograph (and/or **tributary** hydrographs) and considerations of **local inflow** by solving the **storage equation**.

streamfunction—1. A **parameter** of two-dimensional, nondivergent flow, with a value that is constant along each **streamline**.

For flow in the (x,y) plane, the streamfunction ψ is related to the respective coordinate velocities u and v by the equations

$$u = - \frac{\partial \psi}{\partial y} \text{ and } v = \frac{\partial \psi}{\partial x}.$$

In meteorology the most common application of the streamfunction is in the assumption of **geostrophic equilibrium**. If variations in the **Coriolis parameter** f are ignored, the streamfunction in a **constant-pressure surface** is proportional to the **geopotential** gz, that is, $\psi = gz/f$; in an **isentropic surface**, it is $\psi = (gz + c_p T)/f$, called the **Montgomery streamfunction**, where c_p is the **specific heat** at constant **pressure** and T the Kelvin **temperature**.

2. Stokes's streamfunction (also called current function): If the flow is three-dimensional but is **axisymmetric** (i.e., the same in every plane containing the axis of symmetry), a Stokes's streamfunction ψ will exist such that

$$v_s = - \frac{1}{r} \frac{\partial \psi}{\partial n},$$

where v_s is the speed in an arbitrary direction s, r is the distance from the axis of symmetry, and n is normal to the direction s, increasing to the left.

Note that Stokes's streamfunction has **dimensions** of volume per time. Streamfunctions can also be defined for more complex three-dimensional flows.

streamline—A line with its tangent at any point in a fluid parallel to the instantaneous **velocity** of the fluid at that point.

The differential equations of the streamline may be written $d\mathbf{r} \times \mathbf{v} = 0$, where $d\mathbf{r}$ is an element of the streamline and $\mathbf{v}$ the **velocity** vector; or in **Cartesian coordinates**, $dx/u = dy/v = dz/w$, where u, v, and w are the fluid velocities along the **orthogonal** x, y, and z axes, respectively. In steady-state flow the streamlines coincide with the trajectories of the fluid **particles**; otherwise, the streamline pattern changes with time. A two-dimensional wind-vector **field** may be completely specified by streamlines and **isotachs**. *See* **free streamline**; *compare* **trajectory**.

streamline flow—Same as **laminar flow**.

streamwise vorticity—The component of **vorticity** that is parallel to the ambient **velocity** vector.

See also **helicity**.

Streeter–Phelps equation—A mathematical **model** of the **variation** of dissolved **oxygen** in a **stream** due to a point load of waste **discharge**.

The waste discharge can be characterized by **biochemical oxygen demand** (BOD) or chemical oxygen demand (COD) and its decay coefficient. The natural **aeration** of the stream is also included. The equation gives the sag in the dissolved oxygen in the water as a function of the distance from the point load.

Streeter, H. W., and E. Phelps, 1925: *Bulletin 146*, U.S. Public Health Service.

stress—Generally, a surface force, measured per unit area.

In meteorology the term refers to some particular component of the **stress tensor** evident from the context, often to a **Reynolds stress**, $-\overline{\rho u' w'}$, where ρ is the **density**, u' and w' are **eddy** velocities, and the overbar indicates a **time average**. *See also* **surface friction**, **pressure**, **shearing stress**.

stress tensor—The complete set of **stress** components in a medium, which are written as a tensor τ_{ij}.

It has nine components, one for each of the coordinate faces of an imaginary fluid **element** upon which the stress acts ($j = x, y, z$) and for each direction in which the stress is directed ($i = x, y, z$) By definition, an **inviscid fluid** is one in which the six **tangential stresses** ($i \neq j$) are zero, and the three normal stresses ($i = j$) are equal to the negative of the **pressure**.

stretched gridded data—(Becoming obsolete.) A simple remapping scheme for polar-orbiting satellites designed to remove some of the **distortion** caused by the increasing **sensor** field of view towards the extreme limits of a scan swath.

In modern processing systems it is more common to just remap the **digital** imagery to an extreme **map projection**.

stretched VISSR data—GOES VISSR **image** data that have been stretched in time by the temporary storage and retransmission of those data.

The data rate of the **VISSR** data is reduced from 28 Mbits per second to 2 Mbits per second, allowing ground **station** users to receive the data more easily. Since the launch of *GOES-8*, stretched VISSR data have been replaced by **GVAR** data.

stretching deformation—*See* **deformation**.

striations—Narrow, straight, or curved streaks that appear in extensive **overcast** cloud regions.

A low sun angle creates the striation effect by differential shadowing of varying **cloud** levels.

stroke—*See* **lightning stroke**.

stroke density—The areal **density** of **lightning** discharges over a given region during some specified period of time, as number per square **mile** or per square kilometer.

strong breeze—In the **Beaufort wind scale**, a **wind** with a speed from 22 to 27 knots (25 to 31 mph) or Beaufort Number 6 (Force 6).

strong gale—In the **Beaufort wind scale**, a **wind** with a speed from 41 to 47 knots (47 to 54 mph) or Beaufort Number 9 (Force 9).

Strouhal number—A **nondimensional number** occurring in the study of periodic or quasiperiodic variations in the **wake** of objects immersed in a fluid **stream**.

The phenomena are due to the more or less regular formation of vortices or **vortex streets** in the wake (especially at **Reynolds numbers** around 100 in the case of circular cylinders):

$$S = nD/U,$$

where n is a **frequency**, D a representative length, and U a representative **velocity** of the stream. Common examples are **aeolian sounds** such as those produced by electric power or telephone lines in a **wind**.

structural stability—For a **dynamical system**, the property that small changes in the parameters do not qualitatively alter the system's behavior.

structure constant—*See* **structure function**.

structure function—The **variance** of the difference of a **turbulence** quantity in space or time, given by

$$D_q(r) = \overline{[q(x) - q(x + r)]^2},$$

where q is a **scalar** or a **velocity** component.

For small separations ($r \leq z$, where z is the height above the surface), this equation may be written

$$D_q(r) = 4.01b\chi_q\epsilon^{-1/3}r^{2/3},$$

where b is about 0.7 for **temperature** and 0.76 for **humidity**, χ_q is the rate of destruction of $q^2/2$ by **molecular diffusion**, and ϵ is the rate of **dissipation** of **turbulence energy** into **heat**. The coefficient

$$C_q^2 = 4.01b\chi_q\epsilon^{-1/3}$$

is called the **structure constant** of q. This formulation eliminates the mean component of a fluctuating quantity and provides a measure of **fluctuation** variance within a **band** of **wavelengths** (or frequencies) determined by the separation distance (or **lag time**).

Panofsky, H. A., and J. A. Dutton, 1984: *Atmospheric Turbulence*, Wiley, 182–183.

Student's t-test—(Also called t test.) A **significance test** applicable to a **sample** mean of a sample estimate of a **regression coefficient** or a **correlation coefficient** (the latter only when the hypothetical value is zero).

It is used to test an assumed value of the corresponding **parameter**. It is also applicable to the comparison of two (but not more than two) sample means or two regression coefficients, but not of two correlation coefficients. The t test has different formulas for different uses.

sturmpause—Same as **foehn pause**.

Stüve diagram—*See* **thermodynamic diagram**.

suahili—*See* **ghibli**.

subantarctic front—The northern front of the **Antarctic Circumpolar Current** that separates the **polar frontal zone** in the south from the **subantarctic zone** in the north.

It is characterized by **sea surface temperatures** near 7°–9°C and a **salinity** minimum of 33.8–34.0 **psu** produced by high **rainfall**.

Subantarctic Mode Water—*See* **mode water**.

Subantarctic Upper Water—The **water mass** of the **subantarctic zone**.

It has a **temperature** of 4°–10°C in winter (August) and 4°–14°C in summer (January) and a **salinity** of 33.9–34.9 **psu** in winter, which can decrease to 33.0 in summer from the melting of **ice**.

subantarctic zone—The region between the **subantarctic front** and the **subtropical front**.

subarctic climate—Same as **taiga climate**.

subarctic current—Same as **Aleutian Current**.

Subarctic Intermediate Water—A **water mass** identified by a **salinity** minimum found at a depth of about 600 m in the North Pacific Ocean.

It is formed at various locations along the **Arctic Polar Front** and spreads toward the **equator** between the **central water** and the **deep water**. The equivalent water mass in the Atlantic Ocean is known as **Arctic Intermediate Water**.

Subarctic Upper Water—The **water mass** found north of the North Pacific **subtropical convergence** region between 50°N and the Alaskan coast.

It has a **temperature** of 2°–6°C and a **salinity** below 33 **psu**.

subatomic particle—Any of the individual, elemental **particles** that comprise an atom.

The ones most commonly encountered in atmospheric sciences are the **proton**, **neutron**, and **electron**. Other, less common particles such as neutrinos may also penetrate the **atmosphere** as constituents of the **solar wind**.

SubBoreal—A period of more **continental climate** 3000 to more than 2000 years ago identified as part of the Blytt–Sernander sequence of inferred climates in northern Europe.

subcloud layer—The portion of the **boundary layer** extending from the surface to the average altitude of the base of **clouds** with updrafts originating in the boundary layer.

It encompasses the **mixed layer** and the **surface layer**.

subcooling—Same as **supercooling**.

subduction—A process of **water mass formation** through a combination of **wind** action and cooling.

A **convergence** in the wind-driven **surface current** field pushes water down along constant-density surfaces. During winter this water is convectively mixed as a result of surface cooling. When spring warming sets in, the mixed water is isolated under a thin surface layer of warmer water but continues to move downward on constant-density surfaces in response to the **current** convergence, eventually moving out of reach of next winter's **mixing**. It is then injected into the ocean interior as a new **water mass** contribution.

subfrontal cloud—A **cloud** that forms in the cold **air mass** below the **elevation** of the **frontal zone**. Such clouds can be associated with either a **warm front** or a **cold front**.

subgelisol—Unfrozen ground beneath **permafrost**.
Compare **talik**.

subgeostrophic wind—Any **wind** of lesser speed than the **geostrophic wind** required by the existing **pressure gradient**.
A subgeostrophic wind is not necessarily a **subgradient wind**.

subgradient wind—A **wind** of lower speed than the **gradient wind** required by the existing **pressure gradient** and **centrifugal force**.

subgrid-scale parameterization—*See* **large-eddy model**.

subgrid-scale process—Atmospheric processes that cannot be adequately resolved within a **numerical simulation**.
Examples can include **turbulent fluxes, phase changes** of water, chemical reactions, and **radiative flux divergence**. Such processes are often parameterized in numerical integrations and even neglected in some applications. *See* **parameterization, convective adjustment, Mellor–Yamada parameterization, shallow convection parameterization**.

subhumid climate—A **humidity province** of Thornthwaite's 1931 **climate** classification, defined by **precipitation-effectiveness index** values of 32–63, and designated letter code *C*.
Based on its typical vegetation, the subhumid climate is sometimes called grassland, or prairie, climate. In 1948, Thornthwaite divided this category into **moist subhumid** and **dry subhumid climate**, with **moisture index** values of 0 to +20 and -20 to 0, respectively. The zero value of moisture index also divides, overall, the **moist climates** from the **dry climates**. This type of climate and the **semiarid climate** are very susceptible to **drought** conditions.

> Thornthwaite, C. W., 1931: The climates of North America according to a new classification. *Geogr. Rev.*, **21**, 633–655.
> Thornthwaite, C. W., 1948: An approach toward a rational classification of climate. *Geogr. Rev.*, **38**, 55–94.

subjective forecast—A forecast based on decisions made by an individual.
Conversely, an **objective forecast** is based solely on the results from a set of mathematical equations or numerical **model**.

sublimation—The process of **phase** transition from solid directly to **vapor** in the absence of melting.
Thus an **ice crystal** or icicle sublimes under low **relative humidity** at temperatures below 0°C. The process is analogous to **evaporation** of a liquid. Colloquially the terms are used interchangeably for the solid–vapor transition (**evaporation**). For growth, the term sublimation has been replaced by **deposition** since the 1970s. There is evidence that deposition nucleation does occur, although there may be an adsorbed layer prior to **nucleation**. It appears that most **nuclei** in the **atmosphere** require near–water **saturation** before they initiate **ice**.

sublimation nucleus—(Obsolete.) Any **particle** upon which an **ice crystal** may grow by the process of **sublimation**.
A. Wegener (1911) was the first to suggest that the **atmosphere** might contain nuclei upon which sublimation (**deposition**) of **water vapor** directly to **ice** occurs in a manner similar to **condensation** upon **condensation nuclei**. In experiments where various nuclei are added to **supercooled clouds**, it is generally difficult to determine whether these nuclei function as **deposition nuclei** or as **freezing nuclei**. It has never been demonstrated that natural deposition nuclei (as distinguished from freezing nuclei) exist in the atmosphere. In the laboratory, cases of artificial preparation of true sublimation nuclei have been reported that activate ice crystals without reaching anywhere close to water **saturation**; lack of such nuclei in the **atmosphere** results from the predominance of **mixed nuclei**.

> Wegener, A., 1911: *Thermodynamik der Atmosphäre*, Barth, Leipzig.

sublimation of ice—The **change of phase** of water from solid to gas without passing through the liquid **phase**.

submesoscale coherent vortices—(Abbreviated SCV.) Isolated, long-lived **eddies**, having spatial scales smaller than **mesoscale eddies**; usually **anticyclonic**, having a **geopotential** maximum and **stratification** minimum in the center, with a **velocity** field having an interior maximum.
SCV types include **thermocline** SCVs, subthermocline SCVs, **Meddies**, Arctic Ocean SCVs, and others.

McWilliams, J. C., 1985: Submesoscale, coherent vortices in the ocean. *Rev. Geophys.*, **23**, 165–182.

suborbital track—Vertical **projection** of the actual flight path of an aerial or space vehicle onto the surface of the earth or other body.
See **ground track**.

subpoint—The point on the earth centered directly below a satellite.
This is also called the point of zero **nadir** angle for a satellite in earth orbit.

subpolar anticyclone—Same as **subpolar high**.

subpolar glacier—In Ahlmann's **glacier** classification, a **polar glacier** with 10–20 m of **firn** in the **accumulation area** where some melting occurs.

subpolar high—(Also called subpolar anticyclone, polar high, polar anticyclone.) A **high** that forms over the cold continental surfaces of subpolar latitudes, principally in Northern Hemisphere winter.
These highs typically migrate eastward and equatorward.

subpolar low pressure belt—A band of low pressure located, in the mean, between 50° and 70° latitude.
In the Northern Hemisphere, this belt consists of the **Aleutian low** and the **Icelandic low**. In the Southern Hemisphere, it is supposed to exist around the periphery of the Antarctic continent.

subpolar westerlies—*See* **westerlies**.

subrefraction—*See* **substandard propagation**.

subregional broadcast—A radio broadcast covering an area smaller than that of a **regional broadcast**.

subsidence—A descending motion of air in the **atmosphere**, usually with the implication that the condition extends over a rather broad area.

subsidence inversion—An increase in **temperature** with height produced by the **adiabatic** warming of a layer of subsiding air.
This **inversion** is enhanced by vertical **mixing** in the air layer below the inversion.

subsoil ice—Same as **ground ice**.

substandard propagation—The propagation of radio **energy** under conditions of substandard **refraction** in the **atmosphere**; that is, refraction by an atmosphere or section of the atmosphere in which the **refractive index** decreases with height at a rate of less than 40 **N-units** per kilometer.
Substandard propagation produces less than the **normal** downward bending, or even upward bending, of **radio waves** as they travel through the atmosphere, giving closer radio horizons and decreased **radar** and radio coverage. It occurs primarily when propagation takes place through a layer in which the **specific humidity** is constant or increases with height. *See* **standard propagation, superstandard propagation**.

substantial derivative—Same as **individual derivative**.

subsun—A **halo** in the form of a bright white spot as far below the **horizon** as the sun is above it.
The subsun is explained by **reflection** by the horizontal faces of oriented **ice crystals** (e.g., large hexagonal plates). Such crystals are never perfectly horizontal with the result that the shape of the subsun changes slightly with solar **elevation**, being circular for high suns and increasingly teardrop-shaped for low suns.

subsurface current—A **current** below a **surface current**, therefore not in direct contact with the **atmosphere**.

subsurface float—A centrally buoyant device, designed to drift at a specific depth below the water's surface, used to measure water **velocity** at that depth.

subsurface flow—All water flowing below the land surface, which may eventually (after a long time) contribute to **base flow** or become **deep percolation**.

subsurface storm flow—Part of the storm **rainfall** that flows below the land surface for a certain distance before eventually reaching the surface channel after a fairly short time.

subsurface water—Water in the **lithosphere**, that is, below the ground surface.

subterranean ice—Same as **ground ice**.

subterranean stream—A body of **subsurface water** flowing through interconnected caves or caverns or through large interconnected interstices.

subtrack—The locus of subpoints made by a satellite as it moves through its **orbit**.

subtropical anticyclone—Same as **subtropical high**.

subtropical calms—The light and variable **winds** that occur in the centers of the **subtropical high pressure belts** over the oceans.
See **horse latitudes, calms of Cancer, calms of Capricorn**.

subtropical climate—In general, a **climatic zone** with a **climate** typical of the **subtropics**, with warm **temperatures** and meager **precipitation**.

subtropical convergence—The region characterized by a **convergence** of the **current** in the surface **mixed layer** (**Ekman layer** convergence) where water is subducted into the permanent or oceanic **thermocline** as a result.
This region reaches typically from 15°–20° latitude to 45° latitude. Historically, the term subtropical convergence has also been used for the **subtropical front**.

subtropical countercurrent—An eastward flowing narrow **current** in the center of a subtropical **gyre** where the general water movement is weakly westward.
The three subtropical countercurrents found in the North Pacific at 20°–26°N between Hawaii and Asia extend to depths of at least 800 m and have speeds of 0.15 m s^{-1}. Other subtropical countercurrents have been reported north of the Hawaiian Ridge and in the Coral Sea.

subtropical cyclone—A **cyclone** in tropical or subtropical latitudes (from the **equator** to about 50°N) that has characteristics of both **tropical cyclones** and midlatitude (or **extratropical**) cyclones.
They occur in regions of weak to moderate horizontal **temperature** gradient and extract the associated **available potential energy**, as do **baroclinic cyclones**, but they also receive some or most of their **energy** from convective redistribution of **heat** acquired from the sea, as do tropical cyclones. These storms usually have a radius of maximum winds that is larger than what is observed in purely tropical systems, and their maximum sustained winds have not been observed to exceed about 32 m s^{-1} (64 knots). Subtropical cyclones sometimes become true tropical cyclones, and likewise, tropical cyclones occasionally become subtropical storms. Subtropical cyclones in the Atlantic **basin** are classified by their maximum sustained surface winds: Subtropical depressions have surface winds less than 18 m s^{-1} (35 knots), while subtropical storms have surface winds greater than or equal to 18 m s^{-1}. While these storms are not given names, forecasters do issue warnings for them.

subtropical easterlies—Same as **tropical easterlies**.

subtropical easterlies index—A measure of the strength of the easterly **surface wind** between the latitudes of 20° and 35°N.
The index is computed from the average **sea level pressure** difference between these latitudes and is expressed as the east to west component of the corresponding **geostrophic wind** to the tenth of a meter per second. *Compare* **zonal index, polar-easterlies index, temperate-westerlies index**.

subtropical front—A zone of enhanced **meridional** gradients of **sea surface temperature** and **salinity** in the poleward part of the **subtropical convergence**.
In the Southern Hemisphere, the subtropical front can be traced from 40°S at the east coast of South America across the Atlantic into the Indian Ocean and across the Great Australian Bight, where it shifts to 45°S to pass south of Tasmania and reach the southern tip of New Zealand's South Island. It continues in the Pacific Ocean from the Chatham Rise east of New Zealand near 40°S and reaches the west coast of South America near 30°S. In the Northern Hemisphere a well-developed subtropical front exists in the Pacific Ocean between 25°N, 135°E and 30°N, 140°W.

subtropical high—(Or subtropical anticyclone; also called oceanic anticyclone, oceanic high.) One of the semipermanent highs of the **subtropical high pressure belt**.
These highs appear as centers of action on mean charts of **sea level pressure**, generally between 20° and 40° latitude. They lie over the oceans and are best developed in the summer season. *See* **Azores high, Bermuda high, Pacific high**.

subtropical high pressure belt—(Or subtropical ridge.) One of the two bands of high atmospheric **pressure** that are centered, in the mean, near 30°N and 30°S latitudes.
These belts are formed by the **subtropical highs**.

subtropical jet stream—A **band** of relatively strong **winds** concentrated between 20° and 40° latitude in the middle and upper **troposphere**.
It can be present at any longitude but is generally strongest off the Asian coast. *See* **jet stream**.

Subtropical Mode Water—*See* **mode water**.

subtropical ridge—Same as **subtropical high pressure belt**; or, the axis thereof.

subtropical westerlies—*See* **westerlies**.

subtropics—The indefinite belts in each hemisphere between the regions of **tropical** and **temperate climates**.

    The polar boundaries are considered to be roughly 35°N and 35°S latitudes, but vary greatly according to continental influence, being farther poleward on the west coasts of continents and farther equatorward on the east coasts.

suchovei—Same as **sukhovei**.

suction vortices—Smaller-scale secondary vortices within a **tornado** core that orbit around a central axis.

    The transition of a one-celled **vortex** into secondary vortices in laboratory and numerical simulations occurs at high swirl ratios. The vortices produce cycloidal swaths within tornado damage tracks and are often used to explain the gradation of **wind** damage caused by a tornado. Structures in the path of a suction vortex are damaged while others are spared.

sudden change report—Special **meteorological report** transmitted by a **station** to indicate a deterioration or improvement in the weather significant for a particular need, especially aeronautics.

sudden ionospheric disturbance—(Abbreviated SID.) A complex combination of sudden changes in the condition of the **ionosphere**, and the effects of these changes.

    A sudden ionospheric disturbance usually occurs in association with a **solar flare** and is seen only on the sunlit side of the earth. The return of the ionosphere to its "normal" condition following a pronounced sudden ionospheric disturbance usually takes from half an hour to an hour. The following are the most important effects accompanying a sudden ionospheric disturbance: 1) shortwave fadeout, a condition in which there is a marked and abrupt increase in **absorption** in the **D-region** for high-frequency (**HF**) **radio waves**, and a consequent loss of long-distance radio reception in this **range** of frequencies; 2) **magnetic crotchet**, a sudden change in the horizontal component of the earth's **magnetic field** due to an increase in the **conductivity** of the lower ionosphere, the changes being in the nature of an augmentation of the **normal** quiet-day magnetic change; 3) sudden enhancements of long-wave **atmospherics** recorded in the **frequency** range between 10 and 100 kHz, due to the improved **reflectivity** at oblique incidence of the D-region for such low-frequency radio waves; 4) sudden **phase** anomalies of discrete low-frequency radio waves (10–100 kHz), due to a lowering of the D-region; and 5) sudden field-strength anomalies of distant low-frequency radio signals (10–100 kHz), due to **interference** between the **ground wave** and the **sky wave**.

sudden warming—Same as **stratospheric warming**.

sudois—A southwest **wind** on Lake Geneva, Switzerland.

suer—A violent north-northwest **wind** at Lake Garda in Italy.

suestada—Strong southeast **winds** occurring in winter along the coast of Argentina, Uruguay, and southern Brazil.

    They cause heavy seas and are accompanied by **fog** and **rain**; the counterpart of the **northeast storm** in North America.

sugar berg—An **iceberg** of porous **glacier ice**.

sugar snow—Same as **depth hoar**.

sukhovei—(Also spelled suchovei.) Literally "dry wind"; in Russia, a dry, hot, dusty **wind** in the southern steppes.

    It blows principally from the east and frequently brings a prolonged **drought** and crop damage.

sulfates—Chemical derivatives of **sulfuric acid**; can be regarded as being formed by neutralizing sulfuric acid by an appropriate base, for example, reaction with NaOH forms **sodium** sulfate.

    Sulfates are a large component of **seawater** and, hence, sea-salt **aerosol**.

sulfur dioxide—Acidic gas, formula SO_2, formed in the combustion of many fuels and in the **oxidation** of naturally occurring sulfur gases.

    It is the primary sulfur gas emitted from combustion sources and is a precursor to **sulfuric acid**, which is a major constituent of **acid rain**.

sulfur hexafluoride—Extremely unreactive gas, formula SF_6, which is of entirely **anthropogenic** origin.

    It is used as a **dielectric** insulator or to provide a chemically inert **atmosphere**. Due to its inertness, SF_6 has a very long atmospheric lifetime (>1000 years) and has been used to estimate the age of stratospheric air. It is also used as a chemical **tracer** in tropospheric field experiments.

sulfur rain— **Rain** of yellowish color caused by foreign matter (e.g., pollen, yellow **dust**) picked up by **raindrops** during descent.

A dust-filled subcloud layer is required to yield this effect, and the **particles** must contain sufficient **sulfur dioxide** to be yellow in color. *Compare* **blood rain, mud rain.**

sulfuric acid—Strong acid, formula H_2SO_4, formed as the end product in the **oxidation** of most sulfur compounds.

Sulfuric acid is a major constituent of **acid rain** associated with fossil fuel combustion. As a consequence of its low volatility, it can lead to **particle** formation, and is a major constituent of **cloud condensation nuclei** and the background **aerosol**, including the **stratospheric sulfate layer.**

sulfurous smog—*See* **London (sulfurous) smog.**

sultriness—An oppressively uncomfortable state of the weather that results from the simultaneous occurrence of high temperature and high humidity. It is often enhanced by **calm** air and **cloudiness.**

This term is gaining technical recognition in light of the increased study of human comfort and air conditioning. Sultriness has been expressed in terms of the values of various measures of comfort. Some of the lower limits are **water vapor** pressure of 14 mm; **dewpoint** of 19°C (65°F); **equivalent temperature** of 56°C (133°F); **effective temperature** of 24°C (75°F); and the following pairs of **temperature** and **relative humidity**, 35°C (95°F) and 25%, 30°C (86°F) and 40%, 25°C (77°F) and 65%. *See* **comfort chart, comfort zone.**

sumatra—A **squall**, with **wind speeds** occasionally exceeding 13 m s^{-1} (30 mph), in the Malacca Strait between Malay and Sumatra during the **southwest monsoon** (April through November).

It usually blows from the southwest, sometimes from the west or northwest, raising a heavy sea on the Malay coast. The **wind** veers and strengthens and a heavy bank or arch of cumulonimbus **arcus** passes overhead (**arched squall**) with **heavy rain** and often **thunder**. Sumatras usually occur at night; they bring a sudden drop in **temperature** and are generally due to the descent of air cooled by **radiation** on the high ground of northern Sumatra. In a few cases they mark an **air mass** boundary during the advance of the **monsoon.** They are said to occur simultaneously along a line of 320 km (200 miles) or more that advances in a direction between southeast and northeast at about 9 m s^{-1} (20 mph).

summation notation—*See* **Einstein's summation notation.**

summation principle—The principle that the **sky cover** at any level is equal to the summation of the sky cover in the lowest layer, plus the additional sky cover in all successively higher layers up to and including the layer in question.

summer—Astronomically, between the **summer solstice** and **autumnal equinox** in the Northern Hemisphere, and beteen the **winter solstice** and **vernal equinox** in the Southern Hemisphere; the warmest **season** of the year everywhere except in some tropical regions.

Popularly and for most meteorological purposes, summer is taken to include June, July, and August in the Northern Hemisphere, and December, January, and February in the Southern Hemisphere, the opposite of winter. *See* **Indian summer, Old Wives' summer, St. Luke's summer, St. Martin's summer.**

summer monsoon—The summer phase of the **annual cycle** of **winds** driven by the land–sea **thermal** contrast.

The heating of the landmasses during summer leads to a deficit of **atmospheric pressure** over the continents, resulting in an onshore flow. Other factors such as the **topography** of the land have a considerable effect. The summer monsoon is strongest on the southern and eastern sides of Asia, but it can also occur over northern Australia, parts of Africa, the southwestern United States, and the Mediterranean. *See* **monsoon;** *compare* **winter monsoon.**

summer solstice—For either hemisphere, the **solstice** at which the sun is above that hemisphere. In northern latitudes, this occurs approximately on 21 June.

sun cross—A **halo** in which vertical and horizontal streaks of **light** extend from the sun. Most of those observed probably result from the combination of a **parhelic circle** and a **sun pillar.**

sun crust—A type of **snow crust** formed by refreezing after surface **snow crystals** have been melted by the sun.

742

This type of crust is composed of individual **ice** particles such as **firn**, but the sun's action may also produce a **film crust**.

sun dog—Same as **parhelion**.

sun drawing water—Same as **crepuscular rays**.

sun pillar—A **halo** in the form of a pillar of **light** extending above or below the sun and usually seen when the sun is low in the sky.
It is explained by **reflection** by the sides of columnar **ice crystals** falling with their long axes horizontal. The term **light pillar** is sometimes used when the source of light is artificial, such as street lamps.

sun pointing—For **radar**, the procedure of pointing a radar at the sun to receive and measure the **energy** radiated by the sun at the **radio frequency** to which the radar receiver is sensitive.
Using the sun as a **signal** source, it is possible to check the **azimuth** and **elevation** alignments of the **antenna** and to estimate the **antenna gain**. Antenna gain measurements require knowledge of the solar **irradiance** in the **frequency band** of the **receiver**. This information is available from several observatories around the world that observe the **solar spectrum** at local noon each day. *See* **solar signal**.

sun-synchronous orbit—An **orbit** that precesses 360° during the course of the year, permitting the satellite to obtain views of a given geographical area at the same **local time** each day.
A **sun-synchronous satellite** always crosses the equator on the **ascending node** at the same local solar time.

sun-synchronous satellite—A satellite in an **orbit** that precesses 360° during the course of the year, permitting the satellite to obtain views of a given geographical area at the same **local time** each day.
A sun-synchronous satellite always crosses the equator on the **ascending node** at the same local solar time.

sun–weather correlation—A relationship claimed to exist between solar phenomena and tropospheric weather phenomena.

sunglint—The portion of **shortwave radiation** illuminating a water surface that is reflected back to space.
This **reflection** appears in **visible imagery** as a bright area on an otherwise dark water area when winds are light and the water surface is smooth. The rougher the water, the more diffuse the glint pattern.

sunlight—Light from the sun; also called **sunshine**.
The majority of the **energy** from the sun is divided into three parts: **ultraviolet radiation**, **wavelengths** less than about 0.4 μm; **visible radiation**, wavelengths between about 0.4 and 0.7 μm; **infrared radiation**, wavelengths greater than about 0.7 μm. There are conflicting conventions as to whether all three regions are referred to as **light**, or whether that term should only be applied to the visible portion of the **spectrum**. The ultraviolet radiation is sometimes called short waves and the infrared region, long waves.

sunlit aurora—An **aurora** that occurs in the sunlit part of the **upper atmosphere** above the **earth's shadow**.
Sunlit auroras have been observed to extend up to about 1000 km.

sunrise—1. The phenomenon of the sun's daily appearance on the eastern **horizon** as a result of the earth's rotation.
Under **normal** conditions the upper limb of the sun appears to be on the **horizon** of an observer at **sea level**. The parallax of the sun is ignored. The observed time may differ from the tabular time because of a **variation** of the **atmospheric refraction** from the adopted value (34') and because of a difference in height of the observer and the actual horizon.
2. Contraction for "time of sunrise," which is defined by the U.S. Naval Observatory.
See also **twilight**; *compare* **sunset**.
United States Naval Observatory and her Majesty's Nautical Almanac Office, Annual: *The Astronomical Almanac*, U.S. Government Printing Office, Washington, D.C. and HMSO, London, England.

sunscald—An injury to woody plants that results in local death of the plant tissues.
In summer it is due to excessive heating by the sun's rays; in winter, it is caused by the great **variation** of **temperature** on the side of trees exposed to the sun during cold weather.

sunset—1. The phenomenon of the sun's daily disappearance below the western **horizon** as a result of the earth's rotation.

2. Contraction for "time of sunset," which is defined by the U.S. Naval Observatory. *Compare* **sunrise**.

> United States Naval Observatory and her Majesty's Nautical Almanac Office, Annual: *The Astronomical Almanac*, U.S. Government Printing Office, Washington, D.C. and HMSO, London, England.

sunshine—Direct **radiation** from the sun, as opposed to the shading of a location by **clouds** or by other obstructions.

Because of variations in atmospheric **turbidity** (alone and in conjunction with the **optical air mass**), and because of the variable **optical thickness** of clouds, the distinction between sunshine and lack of sunshine is an arbitrary one that is largely dependent upon the type of **sunshine recorder** in use or upon the quality of subjective estimates.

sunshine recorder—An instrument designed to record the duration of **sunshine** without regard to **intensity** at a given location.

Sunshine recorders may be classified into two groups according to the method by which the recorder timescale is obtained. In one class of instruments the timescale is obtained from the motion of the sun in the manner of a sundial (*see* **Campbell–Stokes recorder**, **Jordan sunshine recorder**, **Pers sunshine recorder**). In the second class of instruments the timescale is supplied by a **chronograph** (*see* **Marvin sunshine recorder**). *Compare* **actinometer**; *see also* **heliograph**, **twilight correction**.

sunspot—A relatively dark area on the surface of the sun consisting of a dark central umbra surrounded by a penumbra, which is intermediate in **brightness** between the umbra and the surrounding **photosphere**.

Sunspots are often nearly circular with a typical dimension of 20 000 km. The strongest solar magnetic fields, up to 4000 gauss, are found within the umbra. Sunspots usually occur in pairs with opposite magnetic polarities. They have a lifetime ranging from a few days to several months. Their occurrence exhibits approximately an 11-year **period** (the **sunspot cycle**). *See* **relative sunspot number**.

sunspot cycle—A **cycle** with an approximate length of 11 years, but varying between about 7 and 17 years, in the number and area of sunspots, as given by the **relative sunspot number**.

This number rises from a minimum of 0–10 to a maximum of 50–140 in about four years time, and then declines more slowly. An approximate 11-year cycle has been found or suggested in **geomagnetism**, **frequency** of **aurora**, and other ionospheric characteristics. The **u index** of geomagnetic **intensity** variation shows one of the strongest known correlations to **solar activity**. *See also* **solar cycle**.

sunspot maximum—Periods of time when the **relative sunspot number** is high. These periods of time occur approximately every 11 years and represent the maximum in the **sunspot cycle**. *See also* **solar cycle**; *compare* **sunspot minimum**.

sunspot minimum—Periods of time when the **relative sunspot number** is low.

These periods of time occur approximately every 11 years and represent the minimum in the **sunspot cycle**. There have been prolonged periods of very low sunspot activity from AD 1100 to 1250, from 1460 to 1550, and from 1645 to 1715. The last period is usually called the "Maunder minimum" after E. W. Maunder, the British astronomer who, along with Gustav Sporer of Germany, first called attention to it. The **Maunder minimum** coincides very closely with the coldest part of the **Little Ice Age**. See also **solar cycle**.

sunspot number—Same as **relative sunspot number**.

sunspot relative number—Same as **relative sunspot number**.

super high frequency—(Abbreviated SHF.) *See* **radio frequency band**.

super pressure balloon—A sealed **balloon** made from a material of high modulus of elasticity.

The balloon has an excess of lifting gas, producing an overpressure at the **altitude** of neutral **buoyancy**. As the balloon gas heats and cools, the excess pressure stresses the balloon film. If the balloon material has a high enough modulus, strain is minimal. With system mass constant and balloon volume almost constant, the balloon floats at a near-constant **density** level night and day in the absence of **convection** or wetting.

superadiabatic lapse rate—An **environmental lapse rate** greater than the **dry-adiabatic lapse rate**, such that **potential temperature** decreases with height.

superbolt—A **lightning discharge** of extraordinary peak luminosity when observed from space.

supercell—An often dangerous **convective storm** that consists primarily of a single, quasi-steady rotating **updraft**, which persists for a period of time much longer than it takes an **air parcel** to rise from the base of the updraft to its summit (often much longer than 10–20 min).

Most rotating updrafts are characterized by cyclonic **vorticity** (*see* **mesocyclone**). The supercell typically has a very organized internal structure that enables it to propagate continuously. It may exist for several hours and usually forms in an **environment** with strong **vertical wind shear**. Supercells often propagate in a direction and with a speed other than indicated by the mean wind in the environment. Such storms sometimes evolve through a splitting process, which produces a **cyclonic**, right-moving (with respect to the mean wind), and **anticyclonic**, left-moving, pair of supercells. **Severe weather** often accompanies supercells, which are capable of producing high winds, large **hail**, and strong, long-lived **tornadoes**. *See also* **convective storm, thunderstorm, splitting convective storm, cell, bulk Richardson number**.

supercell tornado—A **tornado** that occurs within **supercells** that contain well-established midlevel **mesocyclones**.

supercooled cloud—A **cloud** composed of supercooled liquid water drops.

The importance of such clouds lies in their unstable composition, since natural or artificial addition of **ice crystals** or other **ice nuclei** will initiate the rapid **phase change** to a **mixed cloud** or to an ice crystal cloud as described by the **Bergeron–Findeisen theory**. This type of cloud constitutes the principal source of **aircraft icing**.

supercooled fog—**Fog** containing water **droplets** in liquid form that exists at temperatures colder than 0°C; a **supercooled cloud** immediately above the surface.

supercooled rain—Liquid **precipitation** at temperatures below **freezing**.

At midlatitudes, supercooled rain often forms first as an **ice crystal** or **snow** in the **clouds**, which then melts as it falls through an elevated layer of air warmer than freezing before reaching the thick bottom layer of cold air that cools the **drop** below freezing. A supercooled drop freezes instantly on contact with surfaces such as electrical power lines, trees, and roads during an **ice storm**. *See* **glaze**.

supercooled water—Liquid water at a **temperature** below the **freezing point**.

supercooling—(Also called subcooling or undercooling; see note below.) The reduction of **temperature** of any liquid below the **melting point** of that substance's solid **phase**; that is, cooling beyond its nominal **freezing point**.

A liquid may be supercooled to varying degrees, depending upon the relative lack of **freezing nuclei** or solid boundary irregularities within its **environment**, and freedom from agitation. This supercoolability of a substance with a crystalline solid form (as opposed to amorphous matter) stems from the unique **energy** transformations necessary for the formation of the first **crystal** nucleus, whereafter all adjacent liquid immediately becomes solid unless or until the **latent heat** released elevates the system's **temperature** sufficiently to arrest the process. It should be noted that the reverse process is not possible; a crystalline solid cannot be "superheated"; therefore, a substance's **melting point** is very conservative, only slightly dependent upon **pressure**. Supercooled **clouds** are quite common. In extreme cases, they have been observed at temperatures as low as −40°C (−40°F). The smaller and purer the water **droplets**, the more likely is supercooling. *Compare* **supersaturation**; *see* **nucleation**.

(Note: The choice of terminology for this concept has been variable and controversial for many years, largely based on conceptual views of the prefixes super, sub, and under. Supercooling remains, however, the most frequently utilized term.)

supercritical flow—Flow with a **celerity** greater than that of a **gravity wave**.

Equivalently, flow for which the **Froude number** is greater than 1.

supergeostrophic wind—Any **wind** of greater speed than the **geostrophic wind** required by the existing **pressure gradient**.

A supergeostrophic wind is not necessarily a **supergradient wind**.

supergradient wind—A **wind** of greater speed than the **gradient wind** required by the existing **pressure gradient** and **centrifugal force**.

superimposed ice—Ice exposed at the surface of a **glacier** that was formed by the **freezing** of melted **snow** after deposition; usually located below the **snow line** and above the **equilibrium line**.

Although it is **ice**, it is part of the **accumulation area**.

superior air—An exceptionally dry mass of air formed by **subsidence** and usually found aloft but occasionally reaching the earth's surface during extreme subsidence processes.

It is often found above **tropical maritime air**, bounded by the **trade-wind inversion**.

superior mirage—A **mirage** in which the **image** or images are displaced upward from the position of the object.

If only a single image of distant objects is seen, then the term **looming** is often applied: A horizontal surface appears to curve upward with increasing distance and terminate in a relatively distant optical **horizon**. The looming might be accompanied by either **stooping** or **towering**. The superior mirage is most striking when it exhibits three or more images. The upper and lower images are always erect, while a single middle image will be inverted. No matter the number, images will alternate between erect and inverted, although sometimes a pair will appear back to back and might be interpreted as a single image. Although textbooks sometimes suggest that what is seen is an object and an even number of images, all are images, and have positions and magnifications that differ from that of the object. What is seen is dependent upon both the distance to objects and the height of the eye. A change of either can produce markedly different image characteristics. Superior mirages occur over a surface when (molecular number) **density** decreases with height, but are always most striking when **temperature** increases with height. Then, what is seen (for a particular distance and observing height) depends critically on the shape of the temperature (and thus, the **refractive index**) **profile**. Everything from stooping to towering to multiple images is a possible result of fairly simple profiles. A common profile, a lifted **temperature inversion**, can not only produce the three-image mirage but, since **internal gravity waves** often occur on such an inversion, the resulting periodic horizontal inhomogeneity can also produce a higher number of images. Such lifted inversions are common over, but hardly confined to, enclosed bodies of water on warm afternoons when the warmer air from the surrounding land flows over the colder water. *Compare* **inferior mirage, sinking**.

supernumerary rainbows—(Or supernumerary bows.) A series of fine weakly colored bows that can frequently be seen just inside the **primary rainbow**.

When formed in rain showers, where there is a broad distribution of **drop** sizes, these bows are mainly seen near the top of the **rainbow** arch, but fade toward the vertical portions of the primary bow. They owe their name (beyond the prescribed number) to the fact that an explanation of rainbows based upon a treatment of **light** as a series of rays is incapable of accounting for them. However, when light is treated as a **wave**, the supernumerary bows become higher-order **interference** maxima, for which the primary bow is but the first maximum. In this sense, the supernumerary bows are as much a part of the primary bow as are, say, its colors.

superposed-epoch method—A technique used to study the relationship between two **time series**, where one series records the occurrence or nonoccurrence of a discrete event of particular interest and the other consists of data for a hypothetically related **variable**.

Relationships are established or rejected by comparison of variable behavior before or after the occurrence of like events. As a meteorological example, the method might be used to examine the influence of **sunspot maxima** (discrete events) on **cloud cover** (the variable).

superrefraction—*See* **superstandard propagation**.

supersaturation—1. In meteorology, the condition existing in a given portion of the **atmosphere** (or other space) when the **relative humidity** is greater than 100%, that is, when it contains more **water vapor** than is needed to produce **saturation** with respect to a plane surface of pure water or pure **ice**.

Such supersaturation does develop because frequently there is no "plane surface of pure water (or ice)" available. In the absence of water surfaces and in the absence of **condensation nuclei** or any wettable surfaces, **phase change** from **vapor** to liquid cannot occur due to the free **energy** barrier imposed by the **surface free energy** of the embryonic **droplets** that would then have to form by **spontaneous nucleation**. Humid air, purified of all foreign nuclei, can be expanded in **cloud** chambers to relative humidities of the order of 400% without any **condensation** taking place. Cloud condensation occurs in our atmosphere at relative humidities near 100% only because nature provides an abundance of condensation nuclei.

2. In physical chemistry, the condition existing in a solution when it contains more solute than is needed to cause **saturation**.

Thermodynamically, this type of supersaturation is closely allied to supersaturation of a **vapor** since the solute cannot crystallize out in solutions free from impurities or seed crystals of the solute.

supersonic—Referring usually to an object moving faster than the **speed of sound** in the gas or liquid surrounding it.

superstandard propagation—The propagation of **radio waves** under conditions of superstandard **refraction** (superrefraction) in the **atmosphere**; that is, refraction by an atmosphere or section of the atmosphere in which the **refractive index** decreases with height at a rate of greater than 40 **N-units** per kilometer.

Superstandard propagation produces greater than **normal** downward bending of **radio waves** as they travel through the atmosphere, giving extended **radio horizons** and increased **radar** coverage. It is caused primarily by propagation through layers near the earth's surface in which the **dewpoint** temperature is rapidly decreasing or the **temperature** is increasing with height. Such conditions are commonly observed near coastlines when a layer of warm **dry air** overlies a cool moist layer adjacent to the ocean surface. A layer in which the downward bending is greater than the curvature of the earth is a **radio duct**. Frequently, the general term, **anomalous propagation**, is used for superstandard propagation. *See* **standard propagation, substandard propagation**.

supertyphoon—A **typhoon** with maximum sustained 1-min mean surface winds of 67 m s⁻¹(130 knots) or greater.

supplemental observations—Additional measurements added to the standard **observational network** taken to minimize some measure of **forecast error**.
See **adaptive observations**.

supplementary features—A term used to refer to **cloud** features (such as **mamma, virga, pileus**, etc.) that are supplementary to other clouds.

supplementary land station—A **surface synoptic station** on land, other than a **principal land station**.

supplementary meteorological office—(Abbreviated SMO.) An office that, according to International Civil Aviation Organization specifications, is competent to supply aeronautical personnel with 1) **meteorological information** received from a **main meteorological office** or **dependent meteorological office**; 2) **meteorological reports** otherwise available.

supplementary observation—**Meteorological observation** made, in addition to those fixed for a scheduled time, to meet special requirements, such as **tropical cyclone** forecasting.

supplementary station—A **surface synoptic station** other than a principal **synoptic station**.

supply current—The electrical **current** in the **atmosphere** that is required to balance the observed **air–earth current** of fair-weather regions by transporting positive charge upward or negative charge downward.

Accounting for the supply current has been for many years a key problem of the field of **atmospheric electricity** and has received much attention. A quasi-steady current of about 1800 A for the earth as a whole is estimated to be required to balance the air–earth current. Wilson (1920) suggested that the **thunderstorms** present in widely scattered regions of the earth at any one time might be responsible for the supply current. Although this suggestion has not been fully confirmed, there is growing conviction that this is correct. When one considers an average over many storms, thunderstorm **lightning** transports negative charge downward to earth, as does **point discharge** in the regions below thunderstorms. Also, positive **ions** flow upward above active thunderstorms. *See* **air–earth conduction current, point discharge current**.

Gish, O. H., 1951: *Compendium of Meteorology*, 113–118.
Wilson, C. T. R., 1920: Investigations on lightning discharges and on the electric field of thunderstorms. *Phil. Trans. A*, **221**, 73–115.

supragelisol—Same as **suprapermafrost layer**.

suprapermafrost layer—The layer of ground above **permafrost**; it includes the **active layer** and possibly occurrences of **talik** and **pereletok**.

sur—A cold **wind** in Brazil.

suraçon—A very chilling, stormy, rainy **wind** in Bolivia.

surf—The **wave** activity occurring on a beach inshore of the point at which incoming waves break.
Generally, surf consists of waves that have broken and therefore have air to some degree mixed in with the water.

surf beat—A nearshore phenomenon of the long **period** disturbance in **surf zone** associated with the occurrence of **wave groups**.
One example is the release and **reflection** of bounded long waves by breaking wave groups in the surf zone.

surf zone—The area in the **coastal zone** with breaking waves; also the collective term for **breakers**.

surface air temperature—*See* **surface temperature**.

surface albedo—The ratio, expressed as a percentage, of the amount of **electromagnetic radiation** reflected by the earth's surface to the amount incident upon it.

Value varies with **wavelength** and with the surface composition. For example, **snow** and **ice** vary from 80% to 85% and bare ground from 10% to 20%. *Compare* **albedo**.

surface-based lifted index—(Abbreviated SBLI.) A **stability index** that is identical to the **lifted index** (LI) except that it modifies the standard 1200 and 0000 **UTC** soundings or numerical-model **soundings** by incorporating hourly surface observational data, and the characteristics of the lifted **parcel** are from surface values of **temperature** and **dewpoint**.

The advantage of this index is that it can be computed at hourly intervals and it typically reflects short-term changes in the **boundary layer**. The surface-based lifted index is typically more unstable than the LI. As with the LI, **thunderstorms** become increasingly likely the further the surface-based lifted index decreases below a threshold of zero.

> Hales, J. E., Jr., and C. A. Doswell III, 1982: High-resolution diagnosis of instability using hourly surface-lifted parcel temperatures. *Preprints, 12th Conf. on Severe Local Storms*, 172–175.

surface boundary layer—(Also called constant flux layer, surface layer.) A layer of air of order tens of meters thick adjacent to the ground where mechanical (**shear**) generation of **turbulence** exceeds buoyant generation or **consumption**.

In this layer **Monin–Obukhov similarity theory** can be used to describe the **logarithmic wind profile**. The **friction velocity** u_* is nearly constant with height in the surface layer. *Compare* **atmospheric boundary layer, radix layer, Obukhov length, aerodynamic roughness length**.

surface chart—(Also called surface map, sea level chart, sea level pressure chart.) An **analyzed chart** of **surface weather observations**.

Essentially, a surface chart shows the distribution of **sea level pressure**, including the positions of **highs, lows, ridges**, and **troughs** and the location and character of **fronts** and various boundaries such as **drylines, outflow boundaries, sea-breeze fronts**, and **convergence lines**. Often added to this are **symbols** of occurring weather phenomena, **analysis** of **pressure tendency** (isallobars), indications of the movement of **pressure systems** and fronts, and perhaps others, depending upon the intended use of the chart. Although the **pressure** is referred to **mean sea level**, all other elements on this chart are presented as they occur at the surface point of **observation**. A chart in this general form is the one commonly referred to as the weather map. When the surface chart is used in conjunction with **constant-pressure charts** of the **upper atmosphere** (e.g., in **differential analysis**), sea level pressure is usually converted to the height of the 1000-mb surface. The chart is then usually called the 1000-mb chart.

Surface Composition Mapping Radiometer—(Abbreviated SCMR.) A three-**channel** scanning **radiometer** on *Nimbus-5* (launched December 1972) measuring **radiation** in the visible and **infrared** spectrum to determine the composition of the earth's surface.

surface current—The **current** at the sea surface, which is partly due to the effects of **wind** and **waves**.

Empirically, it is found that the **drift current** at the surface is 2.5%–3.0% of the **wind speed**, and it decreases rapidly in the uppermost meter.

surface detention—(Also called detention storage.) The portion of the storm **rainfall** that flows on the land surface toward the channel, but has not yet reached it.

It does not include **depression storage**.

surface energy—*See* **surface tension**.

surface energy balance—A statement of the **conservation of energy** applied to a given surface.

The main terms involved include the vertical fluxes of **energy** into or out of the surface due to **net radiation, sensible heat**, and **latent heat**, as well as the net horizontal fluxes of energy that may take place below the surface (e.g., due to **ocean currents**). Any nonzero residual flux is typically applied as a **storage term**, increasing or decreasing the **internal energy** below the surface, usually resulting in an associated change of **surface temperature**.

surface float—A buoyant device subject to **wind** effects used to measure water velocity on **surface water** bodies.

surface flow—Same as **overland flow**.

surface forecast chart—**Prognostic chart** for a given time, that is, 24 hours ahead of the surface **synoptic situation**.

surface free energy—Same as **surface tension**.

surface friction—Resistance to movement of air flowing along the surface of the earth or other surface such as an airplane wing.

The total **drag** at the surface is a combination of **skin drag** due to **viscosity**, **form drag** due to **pressure** forces created as the **wind** hits roughness elements, and **gravity wave drag** in the case of **statically stable** air. In the **atmosphere**, surface friction is related to turbulent drag. *Compare* **friction velocity, drag coefficient, Reynolds stress.**

surface front—*See* **front.**

surface gravity wave—A **wave** that propagates, typically, on the surface of water under the influence of **buoyancy** forces.

In water of uniform depth H, the **dispersion relationship** is given by squared **frequency**

$$\omega^2 = gk \tanh(kH),$$

in which k is the **wavenumber** and g is the **acceleration of gravity**. In shallow water, when the **wavelength** is much larger than the fluid depth, the waves are nondispersive; the dispersion relationship is then

$$\omega = ck,$$

in which **phase speed** $c = (gH)^{1/2}$. In **deep water** the waves have the dispersion relationship

$$\omega = \pm(gH)^{1/2}.$$

surface hoar—1. Fernlike **ice crystals** formed directly on a **snow** surface by **deposition**; a type of **hoarfrost.**

2. **Hoarfrost** that has grown primarily in two dimensions, as on a window or other smooth surface.

surface integral—In **rectangular Cartesian coordinates**, the integral

$$\int_S \int f(x, y, z)dxdy = \int_R \int f[x, y, \phi(x, y)]dxdy,$$

where $f(x, y, z)$ is a single-valued continuous function of x, y and z in a region in which S is a surface given by $z = \phi(x, y)$, and R is the projection of S on the xy-plane.

Similar surface integrals exist over surfaces $x = \phi(y, z)$ and $y = \phi(z, x)$. In **vector** notation the surface integral of a single-valued continuous vector **F** over the surface s can be written

$$\int_S \mathbf{F} \cdot \mathbf{n}ds,$$

where **n** is an outward directed unit vector normal to the surface. The surface integral for a closed surface may be related to a volume integral by the **divergence theorem**. *See* **line integral.**

surface inversion—(Or ground inversion.) A **temperature inversion** based at the earth's surface; that is, an increase of **temperature** with height beginning at the ground level.

This condition is due primarily to greater radiative loss of **heat** at and near the surface than at levels above. Thus, surface inversions are common over land prior to **sunrise** and in winter over high-latitude continental interiors.

surface layer—Same as **surface boundary layer.**

surface map—Same as **surface chart.**

surface observation—*See* **surface weather observation.**

surface of constant amplitude—*See* **surface of constant phase.**

surface of constant phase—If a solution to any **wave equation** has a time-harmonic (sinusoidal) solution (exact or approximate) of the form

$$Ae^{i\phi},$$

where the **amplitude** A and **phase** ϕ are functions of position, the equation

$$\phi(x, y, z) = \text{constant}$$

defines a surface of constant phase.

If A is a **scalar**, the equation

$$A(x, y, z) = \text{constant}$$

defines a **surface of constant amplitude**.

surface of discontinuity—Same as **interface**, but more usually applied to the **atmosphere**.

An atmospheric **front** is represented ideally by a surface of discontinuity of **velocity**, **density**, **temperature**, and **pressure gradient**; the **tropopause** is represented ideally by a surface of discontinuity of, for example, the derivatives: **lapse rate** and **wind shear**.

surface ozone—Refers to the presence of **ozone** at the earth's surface.

The term is also used in reference to the episodic observation of near **total ozone** depletion that occurs in certain regions of the arctic marine **boundary layer** in springtime. This ozone depletion (not to be confused with the stratospheric polar **ozone hole**) likely occurs as the result of bromine- and chlorine-catalyzed chemistry. The sources of the **halogens** are not well understood, but it is likely that they emanate from sea salt. *See* **photochemical air pollution**.

surface pressure—In meteorology, the **atmospheric pressure** at a given location on the earth's surface.

This expression is applied loosely and about equally to the more specific terms: **station pressure** and **sea level pressure**.

surface radiation balance—*See* **surface radiation budget**.

surface radiation budget—(Also called surface radiation balance or **net radiation**, which is the preferred term.) The net **radiative flux density** across a given point on the earth's surface, averaged over a specified time interval.

Major components of the surface radiation budget are downward **shortwave** (**direct** plus **diffuse solar radiation**), upward shortwave (reflected), downward **longwave** (emitted from different levels of the **atmosphere**), and upward longwave (emitted from the surface). On a global annual average, the radiation budget for the earth's surface is positive, indicating an excess of solar heating over longwave loss, and is approximately 100 watts per square meter. Instantaneous values, however, may be positive or negative, and **range** through many hundreds of watts per square meter. In the annual global mean, the positive surface radiation budget is assumed to be matched equally by a negative **atmospheric radiation budget** so that the planet as a whole is in **radiative equilibrium**.

surface renewal model—A **turbulent exchange** model that assumes that fluid near a surface is intermittently replaced by well-mixed fluid from adjacent layers, thereby disrupting the viscous sublayer at the surface itself.

Near-surface **gradients** are assumed to be restored by **conduction** according to

$$\frac{\partial T}{\partial t} = \kappa \frac{\partial^2 T}{\partial z^2},$$

where T is **temperature** and κ is **thermal diffusivity**.

Kraus, E. B., and J. A. Businger, 1994: *Atmosphere–Ocean Interaction*, Oxford University Press, 138–147.

surface retention—(Also called surface storage.) The portion of storm **rainfall** that is intercepted, stored in depressions, or otherwise sufficiently delayed that it fails to reach the **basin outlet** within the time interval of the storm **hydrograph**.

surface roughness—The geometric characteristic of a surface associated with its efficiency as a **momentum** sink for **turbulent flow**, due to the generation of **drag** forces and increased **vertical wind shear**.

In **micrometeorology**, the surface roughness is usually measured by the **roughness length**, a length **scale** that arises as an integration constant in the derivation of the **logarithmic wind profile** relation. In **neutral stability** the logarithmic wind profile extrapolates to zero **wind** velocity at a height equal to the surface roughness length. Several formulas exist to parameterize this length scale as a function of the roughness element geometry (e.g., spacing and silhouette area). Tabulated values for various surface types are published in most micrometeorological texts. *See* **momentum flux, zero-plane displacement**.

Stull, R. B., 1988: *An Introduction to Boundary Layer Meteorology*, 666 pp.

surface runoff—The water that reaches streams (ranging from the large permanent streams to the tiny rills and rivulets that carry water only during rains) by traveling over the surface of the soil.

Thus, surface runoff takes place only over the relatively short distance to the nearest minor channel. *See* **overland flow**.

surface storage—Same as **surface retention**.

surface synoptic chart—*See* **surface chart**.

surface synoptic station—*See* **synoptic station**.

surface temperature—1. In meteorology, the **temperature** of the air near the surface of the earth; almost invariably determined by a **thermometer** in an **instrument shelter**.

2. In **oceanography**, the **temperature** of the layer of **seawater** nearest the **atmosphere**. *See* **sea surface temperature**.

surface tension—(Also called surface energy, surface free energy, capillary forces, interfacial tension.) The tangential force acting at the **interface** between a liquid and air (or, more correctly, its own **vapor**) caused by the difference in attraction between liquid molecules and gaseous molecules.

Expressed either as a force per unit length of interface or as an **energy** per unit area of interface.

surface thermometer—In **oceanography**, a **thermometer** used in a bucket of **seawater** to measure **sea surface temperature**.

surface trough—An elongated area with relatively low **pressure** values when reduced to **sea level**.

surface velocity—Fluid **velocity** at or very near the fluid surface.

surface visibility—The **visibility** determined from a point on the ground, as opposed to **control-tower visibility**.

surface water—All bodies of water on the surface of the earth.

surface water hydrology—Branch of **hydrology** of the land surface dealing with **rainfall–runoff** relationships and the general **water budget** on the surface of the earth.

surface wave—1. A **gravity wave** formed on the **free surface** of a fluid.

In classical **hydrodynamics**, to distinguish surface waves from **tidal waves**, the condition is imposed that vertical accelerations are not negligible. Dynamically, this **wave** is similar to that on an **interface** separating two fluids, becoming identical in the case of zero **density** in the upper fluid.

2. *See* **ground wave**.

3. A **wave** at the **interface** of a light and a much heavier fluid, in particular of the air–sea interface.

See **surface gravity wave**.

surface weather observation—An evaluation of the state of the **atmosphere** as observed from a point at the surface of the earth, as opposed to an **upper-air observation**.

This term is applied mainly to observations that are taken for the primary purpose of preparing **surface synoptic charts**. Major types of surface observation are **synoptic weather observation**, **aviation weather observation**, and **marine weather observation**. **Climatological observations** may also be included.

surface wetness—Same as **leaf wetness**.

surface wetness duration—Same as **leaf wetness duration**.

surface wind—The **wind** measured at a surface observing **station**.

This wind is customarily measured at a standard distance above the ground to minimize the distorting effects of local obstacles and terrain. *See* **anemometer**, **wind vane**, **instrument exposure**, **winds aloft**.

surface wind stress—The force per unit area on a surface due to the airflow above it.

For a mean wind speed U_r at a reference height z_r, **drag coefficient** $C_D(z_r)$ for that same reference height, and air density ρ, the surface wind stress τ is given by $\tau = C_D(z_r)U_r^2$.

surfactant film—A thin film on the water surface, usually due to long-chain molecules with one hydrophilic end and one hydrophobic end.

Such a film has a **surface tension** that varies with the surfactant concentration, and the resulting effect on the near-surface viscous **boundary layer** causes considerably enhanced **damping** of centimeter-scale waves.

surficial creep—Same as **soil creep**.

surge—1. See **storm surge**.

2. *See* **surge current**.

3. *See* **surge line**.

4. In **hydrology**, a sudden change in **discharge** resulting from the opening or closing of a

gate that controls the flow in a channel, or by the sudden introduction of additional water into the channel.

    5. The fore and aft movement of the **center of gravity** of a ship. *See* **heave**, **sway**, **ship motion**.

    6. Water transported up a beach by breaking waves.

surge current—A short-duration, high-amperage, electric **current** impulse that may sweep through an electrical network, as a power transmission network, when some portion of it is strongly influenced by the electrical activity of a **thunderstorm**.

    This activity may take the form of a direct **lightning** strike, or it may be simply the release of previously induced charge on the line when a **thundercloud** overhead suddenly discharges itself. Such current surges flow rapidly through the lines until they find a path to the ground through arc-over at a weak insulator or by entering terminal equipment at the end of the line. Use of grounded guard wires above the power lines of a transmission system reduces the frequency of surge current difficulties, and installation of lightning arresters at sensitive terminal equipment protects the system against damage there.

surge line—In meteorology, a line along which a **discontinuity** in the **wind speed** occurs.

    Usually, but not always, the wind speed is strongest **upstream** from the surge line. Sometimes it is also accompanied by a change in **wind direction**.

suroet—A persistent, rain-bearing, southwest (sud-ouest) **wind** on the west coast of France.

survey datum—The **datum** to which levels on land surveys are related; often defined in terms of a **mean sea level**.

    A survey datum is a horizontal surface to within the **accuracy** of the survey methods.

suspended phase—*See* **suspension**.

suspending phase—*See* **suspension**.

suspension—In physical chemistry, a system composed of one substance (suspended phase, suspensoid) dispersed throughout another substance (suspending phase) in a moderately finely divided state, but not so finely divided as to acquire the **stability** of a **colloidal system**.

    Given sufficient time, a suspension will, by definition, separate itself by gravitational action into two visibly distinct portions, whereas a colloidal system, by definition, is stable. Dust in the **atmosphere** is an example of a suspension of a solid in a gas.

suspensoid—*See* **suspension**.

Swallow float—In **oceanography**, a special **drifting buoy** (named after Mary Swallow) that can be used in adjustable **density** (thus subsurface) horizons.

swallow storm—Same as **peesweep storm**.

swash—The intermittent landward flow of water up a beach face driven by the action of breaking **waves**; the opposite of **backwash**.

sway—Lateral movement of the **center of gravity** of a ship.
    See **heave**, **surge**, **ship motion**.

SWEAT index—*See* **stability index**.

sweep—1. A single traversal of the **electron beam** along any **coordinate axis** on the face of a **cathode-ray oscilloscope**.

    2. A single rotation of an **antenna** at fixed **azimuth** or **elevation**.

sweep length—For a given **radarscope**, the maximum distance from the **radar** that is represented on the face of the scope.

    The sweep length, therefore, specifies the **scale** of the scope **display**.

swell—Surface **gravity waves** on the ocean that are not growing or being sustained any longer by the **wind**.

    Generated by the wind some distance away and now propagating freely across the ocean away from their area of generation, these waves can propagate in directions that differ from the direction of the wind, in contrast to a **wind sea**. *See* **generating area**.

SWI—Abbreviation for severe weather threat index.
    See **stability index**.

sylphon—Same as **aneroid capsule**.

sylvanshine—An optical phenomenon in which some species of dew-covered plants become strongly retroreflective.

On a warm summer evening when a **beam** of **light**, such as car headlights, illuminates dew-covered trees, the scene is reminiscent of snow-covered trees in the moonlight. Four things must be present for the sylvanshine to be seen: the proper species of tree, the proper **season** of the year, dew-covered leaves, and a view looking directly down the beam of light illuminating the tree in an otherwise dark scene. The sylvanshine is closely related to the **heiligenschein**.

Fraser, A. B., 1994: The Sylvanshine: retroreflection in dew-covered trees. *Appl. Optics*, **33**, 4539–4547.

symbol—*See* **plotting symbols**.

symmetric instability—Similar to **inertial instability** but caused by the imbalance between **pressure gradient** and **inertial forces** for infinitesimal disturbances that meridionally displace fluid along **isentropes** (in the **atmosphere**) or **isopycnals** (in the ocean).

For **geostrophic** motion in the Northern Hemisphere, symmetric instability may occur only if the **potential vorticity** is negative.

symmetry point—In meteorology, a point in a **time series** an equal distance on either side of which the rate of change of the element has the same magnitude but opposite sign.

synchronous communication—Communication where the sender and receiver share a clock signal, and data are constantly being transmitted, even if the data are only an idle pattern (indicating no data are present).

synchronous detection—The procedure of using a reference **waveform** in a **radar** receiver that is exactly synchronized with the **carrier frequency**.

Specifically, a coherent local **oscillator** provides a reference **signal** that is mixed with the incoming signals to recover the **amplitudes** and **phases** of the **echoes**.

Synchronous Meteorological Satellite—(Abbreviated SMS.) The first generation of semioperational geostationary **meteorological satellites**.

SMS-1, launched in May 1974, and *SMS-2*, launched in February 1975, carried the first **VISSR** instruments.

SYNOP code—Contraction for **synoptic code**.

synoptic—1. In general, pertaining to or affording an overall view.

In meteorology, this term has become somewhat specialized in referring to the use of meteorological data obtained simultaneously over a wide area for the purpose of presenting a comprehensive and nearly instantaneous picture of the state of the **atmosphere**. Thus, to a **meteorologist**, "synoptic" takes on the additional connotation of simultaneity.

2. A specific scale of atmospheric motion with a typical **range** of many hundreds of kilometers, including such phenomena as **cyclones** and **tropical cyclones**.

Compare **mesoscale**.

synoptic analysis—The **analysis** of **synoptic charts**, or the body of techniques so employed.

synoptic chart—In meteorology, any **chart** or map on which data and analyses are presented that describe the state of the **atmosphere** over a large area at a given moment in time.

The possible variety of such charts is almost limitless, but in meteorological history there has been a more or less standard set of synoptic charts, including **surface charts** and the **constant-pressure charts** of the **upper air**. Other synoptic charts include **isentropic charts** and **constant-height charts**, both used for **upper-air analysis**. There are a number of auxiliary and special-purpose synoptic charts, including **thickness charts**, **tropopause charts**, **stability charts**, **change charts**, **continuity charts**, etc., that have useful applications for preparing forecasts of weather events at various locations. *See also* **mean chart**, **prognostic chart**, **cross section**, **profile**.

synoptic climatology—The study of **climate** from the perspective of atmospheric **circulation**, with emphasis on the connections between circulation patterns and climatic differences.

The common approach is to determine distinct categories of **synoptic** weather patterns and then to assess statistically the weather conditions associated with these patterns. The term, which refers to synchronous weather systems such as **cyclones** and **anticyclones**, originated during the 1940s when military services became concerned with the **climatology** of **weather types** and its implications for transportation. While **dynamic climatology** is often global in scope, synoptic climatology generally deals with hemispheric or local climates.

synoptic code—1. In general, any code by which **synoptic weather observations** are communicated.

Among the synoptic codes in use are the **international synoptic code**, **ship synoptic code**, **U.S. airways code**, and **RECCO code**.

2. Same as **international synoptic code**.

synoptic forecasting—Forecast methods based upon **analysis** of a set and/or series of **synoptic charts**; the most common means of arriving at a **weather forecast**.

These techniques usually contain elements of a physical, kinematic, and climatological nature and are, to an appreciable degree, subjective. This term is used primarily to distinguish these methods from others such as **mesoscale** forecasting, **numerical forecasting**, **statistical** forecasting, and **climatological forecasting**.

Bundgaard, R. C., 1951: *Compendium of Meteorology*, 766–795.

synoptic hour—Hour (**UTC**), determined by international agreement, at which **meteorological observations** are made simultaneously throughout the world.

synoptic meteorology—The study and **analysis** of **synoptic** weather information (**synoptic charts, synoptic weather observations**).

synoptic model—Any **model** specifying a space distribution of some meteorological **elements**.

The distribution of **clouds, precipitation, wind, temperature**, and **pressure** in the vicinity of a **front** is an example of a synoptic model.

synoptic observation—*See* **synoptic weather observation**.

synoptic report—An encoded and transmitted **synoptic weather observation**.

synoptic scale—Used with respect to weather systems ranging in size from several hundred kilometers to several thousand kilometers, the **scale** of **migratory** high and low pressure systems (frontal **cyclones**) of the lower **troposphere**.

See **cyclonic scale**.

synoptic situation—The general state of the **atmosphere** as described by the major features of **synoptic charts**.

synoptic station—A **station** at which **meteorological observations** are made for the purposes of **synoptic analysis**.

The observations are made at the main **synoptic** times of 0000, 0600, 1200, 1800 **UTC** and normally at the intermediate synoptic hours of 0300, 0900, 1500, 2100 UTC and are entered into a coded format for dissemination.

synoptic type—A characteristic atmospheric **circulation** that occurs in a region from time to time.

synoptic wave chart—A **chart** of an ocean area on which are plotted **synoptic** wave reports from vessels, along with computer-generated **wave heights** for areas where reports are lacking.

Atmospheric **fronts, highs**, and **lows** are also shown. Isolines of wave height and the boundaries of areas having the same dominant **wave** direction are drawn.

synoptic weather observation—A **surface weather observation**, made at periodic times (usually at three-hourly and six-hourly intervals specified by the World Meteorological Organization), of **sky cover, state of the sky, cloud height, atmospheric pressure** reduced to **sea level, temperature, dewpoint, wind speed** and **direction, amount of precipitation, hydrometeors** and **lithometeors**, and special phenomena that prevail at the time of the observation or have been observed since the previous specified observation.

Compare **aviation weather observation**.

synthetic aperture radar—(Abbreviated SAR). **Radar** deployed aboard aircraft and satellites that produces a two-dimensional **image** of the **target** surface.

The position of an object along the direction parallel to the movement of the observing platform is determined by the **Doppler shift** of the received **signal**. Phenomena observed by SAR include **swell** waves, **current** patterns imaged because of the varying sea surface roughness due to **wave–current interaction**, and oil spills and natural films that appear as areas of low image **intensity** as a result of their **damping** effect on centimeter-scale **surface waves**. *See also* **Doppler radar, marine radar, microwave radar, high-frequency radar**.

synthetic hydrograph—An empirical **unit hydrograph**, developed for ungauged watersheds, that is based upon measurable physical **basin** characteristics.

systematic error—That part of the **inaccuracy** of a measuring instrument, or **statistical** estimate of a **parameter**, that is due to a single cause or small number of causes having the same sign, and hence, in principle, is correctable; a bias or constant offset from the true value.

In the absence of **random errors**, the true value is equal to the instrumental reading or statistical mean estimate minus the systematic error.

systematic observations—Observations taken at standard, preset times and places.

Système Internationale—(Or **International System of Units**; abbreviated SI.) The preferred system for measuring most quantities in scientific disciplines.

syzygy—1. The astronomical condition of alignment of the earth, moon, and sun at new and full moon, the time of maximum spring tidal forcing.

At these times the **range** of **tide** is greater than average.

2. A west **wind** on the seas between New Guinea and Australia preceding the **summer (northwest) monsoon**.

T

θ coordinate system—A coordinate system in **atmospheric dynamics** in which θ (**potential temperature**) is used as the vertical coordinate.

θ system—A system of the **equations of motion** expressed in the **θ coordinate system**.

T-E index—Abbreviation for **temperature-efficiency index**.

T-E ratio—Abbreviation for **temperature-efficiency ratio**.

T–S curve—Abbreviation for **temperature–salinity curve**.

T–S diagram—Abbreviation for **temperature–salinity diagram**.

T–S relation—Same as **T–S curve**; *see* **temperature–salinity diagram**.

t **test**—Same as **Student's t-test**.

T-year event—The T-year event is the magnitude of a phenomenon that recurs on average every T years.

For rare events, the T-year event has the **probability** 1/T of being exceeded in any year. For commonly occurring events, the probability of the T-year event being exceeded at least once in T years is

$$P = 1 - e^{1/T}.$$

tabetisol—Same as **talik**.

table iceberg—Same as **tabular iceberg**.

Table Mountain southeaster—*See* **southeaster**.

tablecloth—*See* **Cape doctor**.

tabular iceberg—(Also called table iceberg; formerly called barrier iceberg.) An **iceberg** that has broken off from an **ice shelf**.

Newly formed tabular icebergs have nearly vertical sides and flat tops. In the Antarctic, they may be tens of kilometers wide, up to 160 km (100 miles) long, and as much as 300 m (1000 ft) thick, with about 30 m (100 ft) exposed above the sea surface. In the Arctic, large icebergs of this type are called **ice islands**, but they are considerably smaller than the largest of the antarctic variety.

TAF—The international standard code format for terminal forecasts issued for airports, which took effect on I July 1996.

> U.S. Department of Commerce, NOAA/NWS, 1996: *Weather Service Operations Manual*, Chapter D-31.

taiga—The open northern part of the **boreal forest**.

It consists of open woodland of coniferous trees growing in a rich floor of lichen (mainly reindeer moss or caribou moss), and is generally cold and swampy. The taiga lies immediately south of the **tundra**. In spring, it is often flooded by water from northward flowing rivers, the lower reaches of which are still frozen.

taiga climate—(Also called subarctic climate.) In general, a **climate** that produces **taiga** vegetation; that is, one that is too cold for prolific tree growth but milder than the **tundra climate** and moist enough to promote appreciable vegetation.

This climate type appears as a subdivision of Köppen's 1936 **snow forest climate** and Thornthwaite's 1931 **microthermal climate**.

> Köppen, W. P., and R. Geiger, 1930–1939: *Handbuch der Klimatologie*, Berlin: Gebruder Borntraeger, 6 vols.
> Thornthwaite, C. W., 1931: The climates of North America according to a new classification. *Geogr. Rev.*, **21**, 633–655.

tail cloud—Colloquial expression for a horizontal **cloud band**, often laminar and tube-shaped, but sometimes ragged and turbulent, that is attached to a **wall cloud**.

> Fujita, T., 1959: A detailed analysis of the Fargo tornadoes of June 20, 1957. *U.S. Wea. Bur. Res. Paper 42*, p. 15.

tail Doppler radar—A scanning **Doppler radar** mounted in the tail of an airplane, usually employed for **helical scanning**.

See **airborne weather radar**.

tailwater—Water located just **downstream** of a hydraulic structure.

tailwind—(Also called **following wind**.) A **wind** that assists the intended progress of an exposed, moving object, for example, rendering an airborne object's groundspeed greater than its **airspeed**; the opposite of a **headwind**.

The tailwind component is directed along the **heading**, not the **course**. *See* **wind factor, crosswind**.

taino—Name given a **tropical cyclone** (**hurricane**) in parts of the Greater Antilles.

Taiwan Warm Current—A northward flowing current that carries warm subtropical water into the East China Sea.

It originates partly as an offshoot of the **Kuroshio** just northeast of Taiwan, partly as a current through Taiwan Strait. When it meets the southward flowing **China Coastal Current** near 30°S, it submerges and continues northward below a depth of 5 m.

Taku wind—A strong, gusty, east-northeast **wind**, occurring in the vicinity of Juneau, Alaska, between October and March.

It sometimes attains **hurricane** force at the mouth of the Taku River, after which it is named. *See* **Stikine wind**.

talik—(Also called tabetisol.) A Russian term applied to permanently unfrozen ground in regions of **permafrost**.

This usually applies to a layer that lies above the permafrost but below the **active layer**, that is, when the **permafrost table** is deeper than the depth reached by winter freezing from the surface. Talik is also found within and beneath permafrost; when it occurs beneath the permafrost it is equivalent to **subgelisol**.

tamboen—*See* **aloegoe**.

tangent arcs—A **halo** in the form of an arc tangent to a circular halo.

The most common of these are the upper and lower tangent arcs to the **halo of 22°**. They form separate arcs when the sun is low, but as the sun climbs they join to produce the **circumscribed halo**. These arcs are explained by **refraction** through the 60° prism sides of columnar crystals oriented with their long axes horizontal. *See* **46° lateral arcs**.

tangent linear approximation—An assumption used in applications of **tangent linear models** and **adjoint models** that the evolution of **small perturbations** in **nonlinear** models may be approximated by tangent linear (and adjoint) equations for finite time intervals.

In forecasts of **extratropical** synoptic-scale weather systems, this approximation is generally valid for two to three days with dry tangent linear and adjoint models, although qualitatively useful information may be obtained for longer forecast intervals. The **accuracy** implied by the tangent linear approximation applies equally to the **perturbation** forecast of a tangent linear model and to the sensitivity provided by the corresponding adjoint model.

tangent linear equation—An equation of the form $\mathbf{x}_1 = L\mathbf{x}_0$, in which L is a **linearization** about the time-varying forecast trajectory of a **nonlinear** model ($M(\overline{\mathbf{x}}, t)$), and $\mathbf{x}_0$ and $\mathbf{x}_1$ are first-order perturbations of nonlinear vector $\mathbf{x}$.

See **tangent linear model, adjoint equation**.

tangent linear model—(Abbreviated TLM.) A **model**, comprising **tangent linear equations**, that maps a **perturbation** vector, $\delta\mathbf{x}(t_1) = L\delta\mathbf{x}(t_0)$, from initial time t_0 to forecast time t_1.

Here, L is the tangent **linear operator** and $\mathbf{x}$ is the model state **vector**. A TLM provides a first-order approximation to the evolution of perturbations in a **nonlinear** forecast **trajectory**. *See* **tangent linear approximation, small perturbation, adjoint model**.

tangential acceleration—The component of the **acceleration** directed along the **velocity** vector (**streamline**), with magnitude equal to the rate of change of speed of the **parcel** dV/dt, where V is the speed.

In horizontal, frictionless atmospheric flow, the tangential acceleration is balanced by the tangential **pressure force**,

$$\frac{dV}{dt} = -\alpha\frac{\partial p}{\partial s},$$

where α is the **specific volume**, p the **pressure**, and s a coordinate along the streamline. Thus, flow without tangential acceleration is along the **isobars**, and the **wind** is the **gradient wind**.

tangential stresses—The components of the **stress tensor** that are tangential to the faces of the fluid element.

tangential wind—The component of the **velocity** directed along the velocity vector (**streamline**), with magnitude equal to the speed of the **parcel** at that point.

tapioca snow—Same as **snow pellets**.

tarantata—A **strong breeze** from the northwest in the Mediterranean region.

target—Any of the many types of objects detected by **radar**.

A radar target must have an **index of refraction** sufficiently different from that of the **atmosphere** to return a **target signal** to the radar by **reflection**, **refraction**, or **scattering**. Also, it must be near enough and have a large enough **radar cross section** that the target signal will exceed the threshold of detectability of the radar receiver. The target is then said to produce a detectable **echo**.

target signal—The **electromagnetic radiation** returned to a **radar** by a **target**. The target signal is characterized by its **amplitude, phase, frequency,** and **polarization**.

These are the properties that, with suitable equipment, are available for measurement and may be used to infer such target properties as **reflectivity** or **Doppler velocity**. A Cartesian component of the **electric field** vector of the target signal at a given distance from the radar varies with time t and may be written

$$E(t) = A(t) \cos[\phi(t) - \omega_0 t],$$

where A is the **signal** amplitude, ϕ the phase in radians, and $\omega_0 = 2\pi f_0$ with f_0 the **carrier frequency** in **hertz**. The amplitude varies with time because of changes in the **radar cross section** of the target or, for distributed targets, because of movements of the separate **scattering** elements relative to each other. The phase will vary with time because of motion of the target toward or away from the radar. The Doppler **frequency** in hertz is $1/2\pi(d\phi/dt)$. The **intensity** I of the target signal is defined as the average value of the square of E, evaluated over one **cycle** of the **carrier**. Thus, $I(t) = A^2(t)/2$.

target volume—Same as **pulse volume**.

targeted observations—*See* **adaptive observations**.

taryn—Russian term for **frozen ground** that lasts for more than one **season**.

TAS—Abbreviation for **true airspeed**.

tau-value—The time rate of changes of D-values at a fixed point defined by relation,

$$\tau = \frac{\Delta_t D}{\Delta t},$$

where Δt is the change in time and $\Delta_t D$ is the change in **D-value** during this time interval.

Tau-values are expressed in terms of feet per hour. Tau-value lines are drawn on 4D charts and compose the time dimension of these charts.

Taylor effect—A phenomenon investigated experimentally and theoretically by G. I. Taylor in which the relative motion of a homogeneous rotating liquid tends to be the same in all planes perpendicular to the axis of rotation.

For example, if a sphere is introduced as an obstruction, the fluid flows around it as if it were a cylinder extending the entire depth of the fluid parallel to the axis of the system. It is due essentially to the presence of very **geostrophic**, homogeneous conditions so that the **thermal wind** is zero. It occurs at very low values of the **Rossby number**.

Taylor microscale—Of the three standard **turbulence length scales**, the one for which **viscous dissipation** begins to affect the **eddies**.

Thus it marks the transition from the **inertial subrange** to the **dissipation** range. *Compare* **integral length scales, Kolmogorov microscale**.

Tennekes, H., and J. L. Lumley, 1972: *A First Course in Turbulence*, MIT Press, 65–66.
Townsend, A. A., 1976: *The Structure of Turbulent Shear Flow*, 2d ed., Cambridge University Press, 46–47.

Taylor number—A **nondimensional number** arising in problems of a rotating **viscous fluid**.

It may be written

$$\mathrm{T} = f^2 h^4/\nu^2,$$

where f is the **Coriolis parameter** (or, for a cylindrical system, twice the rate of rotation of the system), h is representative of the depth of the fluid, and ν is the **kinematic viscosity**. The square root of the Taylor number is a **rotating Reynolds number**, and the fourth root is proportional to the ratio of the depth h to the depth of the **Ekman layer**.

Taylor polynomial—The **polynomial** obtained by truncation of a **Taylor series** after a finite number of terms.
See **Taylor's theorem**.

Taylor's diffusion theorem—The theorem of G. I. Taylor in the **statistical** theory of **atmospheric turbulence**:

$$\overline{x^2} = 2\overline{u^2} \int_0^T \int_0^t R(\xi)\, d\xi\, dt,$$

where x is the distance traveled by a **particle** in the time interval T, u is the **fluctuation** or **eddy velocity** of the particle, and $R(\xi)$ is the **Lagrangian correlation** coefficient between the particle's **velocity** at time t and $t + s$.

Taylor's hypothesis—An assumption that **advection** contributed by turbulent circulations themselves is small and that therefore the advection of a **field** of **turbulence** past a fixed point can be taken to be entirely due to the mean flow; also known as the Taylor "frozen turbulence" hypothesis.
It only holds if the relative **turbulence intensity** is small; that is, $u/U << 1$, where U is the **mean velocity** and u the **eddy velocity**. Then the substitution $t = x/U$ is a good approximation.

Taylor's theorem—If all the derivatives of a function $f(x)$ are continuous in the vicinity of $x = a$, then $f(x)$ can be expressed in an infinite series (the **Taylor series**):

$$f(x) = f(a) + f'(x)(x - a) + \frac{1}{2!}f''(x)(x - a)^2 + \cdots + \frac{1}{n!}f^{(n)}(a)(x - a)^n + \cdots.$$

The case $a = 0$ is called a **Maclaurin series**.

Taylor series—*See* **Taylor's theorem**.

TDWR—Abbreviation for **Terminal Doppler Weather Radar**.

teardrop balloon—A **sounding balloon** that, when operationally inflated, resembles an inverted teardrop.
This shape was determined primarily by **aerodynamic** considerations of the problem of obtaining maximum stable rates of **balloon** ascension. Such balloons are not commonly used for operational **soundings**.

teeth of the gale—An old nautical term for the direction from which the **wind** is blowing (**upwind**, **windward**).
To sail into the teeth of the gale or into the eye of the wind is to sail to windward.

Tehuano—A Spanish term frequently used to denote a burst of strong offshore (southward) **wind**, lasting a day or more, that blows from the Gulf of Mexico across the **Gulf of Tehuantepec**.
Tehuano events are associated with winter surges of cold air that spread southward from the United States across the Gulf of Mexico behind strong weather fronts. Tehuanos produce much local cooling of the gulf waters and frequently cause **anticyclonic** warm-core ocean eddies to form and propagate west-southwestward near 12°–14°N. *See* **papagayo**.

tehuantepecer—A violent squally **wind** from north or north-northeast in the **Gulf of Tehuantepec** (south of southern Mexico) in winter.
It originates in the Gulf of Mexico as a **norther** that crosses the isthmus and blows through the gap between the Mexican and Guatemalan mountains. It may be felt up to 160 km (100 miles) out to sea. *Compare* **papagayo**; *see* **mountain-gap wind**, **jet-effect wind**.

telecommunication—A communication using an electronic (radio, telephone, telegraph) transmission system.
Used to imply rapid and distant communication.

teleconnection—1. A linkage between weather changes occurring in widely separated regions of the globe.
2. A significant positive or negative **correlation** in the fluctuations of a **field** at widely separated points.
Most commonly applied to **variability** on monthly and longer timescales, the name refers to the fact that such correlations suggest that information is propagating between the distant points through the **atmosphere**.

telegraphic equation—(Also called telegrapher's equation.) A partial differential equation usually written in the form

$$\frac{\partial^2 u}{\partial t^2} = a \frac{\partial^2 u}{\partial x^2} + bu,$$

where t is a time coordinate, x is a space coordinate, and a and b are positive constants.

This is the equation governing the flow of electricity in a cable. When $b < 0$, the equation applies to the motion of long gravitational waves in a rotating ocean.

telemetry—The process of measuring a quantity or quantities, transmitting the measured value to a distant **station**, and there interpreting, indicating, or recording the quantities measured.

Typically, spacecraft telemetry includes parameters, such as spacecraft **temperature**, **power** supply, and orbital information, that are transmitted to earth to monitor the status of spacecraft.

telephotometer—A **photometer** that measures the amount of **light** received from a distant light source.

When specifically used to measure the **transmissivity** of the intervening **atmosphere** (or other medium), it is usually termed a **transmissometer**. *See* **visibility meter**.

telephotometry—The body of principles and techniques concerned with measuring atmospheric **extinction** using various types of **telephotometer**.

See **photometry**, **radiometer**

telescoped ice—Same as **rafted ice**.

telethermometer—A **temperature**-measuring system in which the thermally sensitive element is located at a distance from the indicating element.

telethermoscope—A **temperature** telemeter, frequently used in a **weather station** to indicate the temperature at the **instrument shelter** located outside.

See **telethermometer**.

Television and Infrared Observation Satellite—(Abbreviated TIROS.) The first satellite series dedicated to meteorological observations.

TIROS-1, launched in April 1960, carried a vidicon television camera to provide visible images of the earth. *TIROS-8*, launched in December 1963, introduced **APT**, while *TIROS-9*, launched in January 1965 in a sun-synchronous **orbit**, produced the first global composite imagery. The first ten satellites in the TIROS series, launched from 1960 to 1965, were experimental systems. From 1966 to 1969 TIROS satellites were used for the first series of operational polar-orbiting satellites, the **TIROS Operational System**. These satellites were named *ESSA-1* through *ESSA-9*. A second series of operational satellites, the **Improved TIROS Operational System**, was launched between 1970 and 1976 and includes the *ITOS-1* satellite, as well as *NOAA-1* through 5. The current polar-orbiting series began with *TIROS-N*, launched on 13 October 1978, and continuing with *NOAA-6, -7,* and *-12. NOAA-8* through *-11* and *NOAA-14* were modified versions of *TIROS-N*, designated **Advanced TIROS-N** satellites. Like **GOES**, the polar-orbiting satellites are typically given a letter designation as they are built and then renamed with a number after they are successfully launched. Thus NOAA-J became *NOAA-14* after launch. *TIROS-N* is an exception to this rule.

telluric lines—Dark **absorption** lines observed in a **spectrum** of **solar radiation** obtained with an instrument located near the earth's surface.

These lines are produced by constituents of the **atmosphere** of the earth itself. They appear relatively narrow due to the cool **temperature** of the atmospheric gases, and are quite distinct from the broad **Fraunhofer lines** caused by absorption by high-temperature solar gases. The terrestrial nature of these narrow telluric lines is revealed by their **intensity** variation with **solar zenith angle** and by their freedom from any **Doppler broadening** due to solar rotation. **Water vapor** produces the strongest of the telluric lines in the **visible spectrum**.

TEMPAL code—A code in which data on the 0° and −10°C isotherms are encoded and transmitted.

It is a modification of the **international analysis code**.

temperate belt—As given by A. Supan in 1879, a belt around the earth within which the annual **mean temperature** is less than 20°C (68°F) and the mean temperature of the warmest month is higher than 10°C (50°F).

These limits separate this belt from the **hot belt** and **cold caps**, respectively. *See* **climatic classification**.

Supan, A., 1879: Die Temperaturzonen der Erde. *Petermanns Geog. Mitt.*, **25**, 349–358.

temperate climate—Very generally, the **climatic zone** of the "middle" latitudes; the **variable** climates between the extremes of **tropical climate** and **polar climate**.

temperate glacier—In Ahlmann's **glacier** classification, a glacier that, at the end of the melting **season**, is composed of **firn**, and **ice** at the **melting point**.

temperate rain forest—A type of forest that exists in cool but generally frost-free regions of heavy annual **precipitation**.

It consists mainly of mixed deciduous trees, usually with one dominant species. With the onset of winter the forest becomes dormant and remains so until spring, when it resumes active growth. Temperate rain forests are found principally near the west coasts of southern South America and northern North America, in New Zealand, and in northern Japan. *Compare* **tropical rain forest**.

temperate rainy climate—One of the major categories (the C climate) in W. Köppen's 1936 **climatic classification**. These climates have a coldest-month **mean temperature** of less than 18°C (64.4°F) and greater than −3°C (26.6°F), with a warmest-month mean temperature of more than 10°C (50°F).

These limits distinguish it, respectively, from **tropical rainy climates**, **snow forest climates**, and **tundra climates**. It is separated from the **dry climates** by a function of annual **temperature** and **precipitation**. In C. W. Thornthwaite's 1931 classifications, the humid or moist subhumid and **mesothermal climates** would very closely correspond to the above. *See* formulas under **steppe climate**.

> Köppen, W. P., and R. Geiger, 1930–1939: *Handbuch der Klimatologie*, Berlin: Gebruder Borntraeger, 6 vols.
> Thornthwaite, C. W., 1931: The climates of North America according to a new classification. *Geogr. Rev.*, **21**, 633–655.

temperate westerlies—*See* **westerlies**.

temperate-westerlies index—A measure of the strength of the westerly **surface wind** between 35° and 55°N.

The index is computed from the average **sea level pressure** difference between these latitudes and is expressed as the west to east component of **geostrophic wind** to a tenth of a meter per second. *Compare* **zonal index**, **polar-easterlies index**, **subtropical-easterlies index**.

Temperate Zone—Either of the two latitudinal zones on the earth's surface that lie between 23°26′ and 66°34′N and S (the North Temperate Zone and South Temperate Zone, respectively).

It is one of three subdivisions of the **mathematical climate**, which, in turn, is the earliest and simplest form of **climatic classification**. The other two divisions are the **Frigid Zone** and the **Torrid Zone**.

temperature—The quantity measured by a **thermometer**.

Bodies in **thermal** equilibrium with each other have the same temperature. In gaseous fluid dynamics, temperature represents molecular **kinetic energy**, which is then consistent with the **equation of state** and with definitions of **pressure** as the average force of molecular impacts and **density** as the total mass of molecules in a volume. For an **ideal gas**, temperature is the ratio of **internal energy** to the **specific heat capacity** at constant volume.

temperature anomaly—*See* **anomaly**.

temperature belt—1. *See* **temperature zone**.

2. (Rare.) The "belt" that may be drawn on a **thermograph** trace or other **temperature** graph by connecting the daily maxima with one line and the daily minima with another.

temperature coefficient—The ratio of the speeds of a chemical reaction at two temperatures differing by 10°C.

It varies with different materials but is generally near 2 (i.e., the **reaction rate** doubles for each rise of **temperature** by 10°C), especially for reactions involving **water vapor**. This value may be connected with the fact that **equilibrium vapor pressure** nearly (but not quite) doubles for a rise of temperature by 10°C. The temperature coefficient for bacterial action is also near 2.

temperature correction—The **correction** applied to an instrument to account for the effect of **temperature** upon its response characteristics.

temperature–dewpoint spread—*See* **dewpoint depression**.

temperature efficiency—Same as **thermal efficiency**.

temperature-efficiency index—(Abbreviated T-E index.) For a given location, a measure of the long-range effectiveness of **temperature (thermal efficiency)** in promoting plant growth.

Numerically, the T-E index is equal to the sum of the 12 monthly **temperature-efficiency ratios** (T-E ratio).

temperature-efficiency ratio—(Abbreviated T-E ratio.) For a given location and month, a measure of **thermal efficiency** equal to the **departure** of the **normal** monthly **temperature** above 32°F divided by 4:

$$\text{T-E ratio} = \frac{T - 32}{4},$$

where T is the normal monthly temperature in degrees Fahrenheit, except that all values below 32° are counted as 32°. No **dimensions** are assigned to this ratio.

temperature extremes— The highest and lowest values of temperatures attained at a specified location during a given time interval, for example, daily, monthly, or seasonal.

In **climatology**, if this is the whole period for which observations are available, it is called the "absolute extreme."

temperature–humidity index—(Abbreviated THI; also known as discomfort index, **effective temperature**.) An index to determine the effect of summer conditions on human comfort, combining **temperature** and **humidity**.

Several equations have been used to calculate the index, dependent on the data availability:

$$\text{THI} = 0.4 \, (T_d + T_w) + 15$$

$$\text{THI} = 0.55 T_d + 0.2 T_{dp} + 17.5$$

$$\text{THI} = T_d - (0.55 - 0.55 \text{RH}) \, (T_d - 58),$$

where T_d is the **dry-bulb temperature** in °F, T_w is the **wet-bulb temperature** in °F, T_{dp} is the **dewpoint** temperature in °F, and RH is the **relative humidity** in percent. In the equation, RH is used as a decimal; in other words, 50% relative humidity is indicated as 0.50. Studies have shown that relatively few people in the summer will be uncomfortable from heat and humidity while THI is 70 or below; about half will be uncomfortable when THI reaches 75; and almost everyone will be uncomfortable when THI reaches 79. There are portions of the United States in which the THI has reached values around 90.

Temperature–Humidity Infrared Radiometer—(Abbreviated THIR.) A two-channel (6.5 and 11.5 μm) **scanning radiometer** flown on *Nimbus*- *4* through -*7* used to provide information on moisture and high-level clouds in the upper **troposphere** and **stratosphere**.

temperature inversion—A layer in which **temperature** increases with **altitude**.

The principal characteristic of an **inversion layer** is its marked **static stability**, so that very little **turbulent exchange** can occur within it. Strong **wind shears** often occur across inversion layers, and abrupt changes in concentrations of atmospheric **particulates** and atmospheric **water vapor** may be encountered on ascending through the **inversion**. When an inversion is mentioned in meteorological literature and discussion, a temperature inversion is usually meant. *See* **frontal inversion, subsidence inversion, trade-wind inversion**.

temperature lapse rate—*See* **lapse rate**.

temperature microscale—Similar to the **Taylor microscale**, which is based on the **dissipation** of **turbulence kinetic energy**, a **microscale** λ_θ formulated on the dissipation of the **variance** of the **temperature** fluctuations θ by

$$\epsilon_\theta = \kappa \, \overline{\frac{\partial \theta}{\partial x_j} \frac{\partial \theta}{\partial x_j}} = 6\kappa \, \frac{\overline{\theta^2}}{\lambda_\theta^2}, \text{ or}$$

$$\lambda_\theta = (6\kappa \overline{\theta^2})^{1/2},$$

where κ is the **thermal diffusivity**.

temperature–moisture index—An index indicating the initiation of **convective cloud**.

This is defined as

temperature–moisture index = $[150.0 - \text{sfc}/100 - 2(\text{T850} + \text{T700} + \text{T500})] - 100\text{PW700}$,

where sfc is the surface **elevation** of the **station** in m; T850, T700, and T500 are the 850-, 700-, and 500-hPa temperatures in °C; and PW700 is the **precipitable water** in cm between the surface and the 700-hPa level. The more negative the index, the greater the **probability** of **convective precipitation**.

temperature of the soil surface—Temperature at the soil–atmosphere interface.

temperature province—A major division of C. W. Thornthwaite's 1931 schemes of **climatic classification**, determined as a function of the **temperature-efficiency index** or the **potential evapotranspiration**.

In the 1931 system, six main temperature provinces (**climates**) are distinguished: 1) **tropical**; 2) **mesothermal**; 3) **microthermal**; 4) **taiga**; 5) **tundra**; and 6) **frost**. In the 1948 system they are 1) **megathermal**; 2) **mesothermal**; 3) **microthermal**; 4) tundra; and 5) **frost**. *Compare* **climatic province**; *see* **humidity province**.

> Thornthwaite, C. W., 1931: The climates of North America according to a new classification. *Geogr. Rev.*, **21**, 633–655.
> Thornthwaite, C. W., 1948: An approach toward a rational classification of climate. *Geogr. Rev.*, **38**, 55–94.

temperature range—The difference between the maximum and minimum temperatures or between the highest and lowest mean temperatures during a specified time interval, for example, daily, monthly, or seasonal.

temperature reduction—*See* **reduction**.

temperature–salinity curve—(Abbreviated T–S curve.) Given measurements at a single location of $T(z)$ and $S(z)$, oceanographers often plot $T(z)$ versus $S(z)$, showing the depth dependence only parametrically.

The main reason for doing this is that **water masses** generally are characterized by their location on the *T–S* plane, and so can be identified by plotting a *T–S* curve.

temperature–salinity diagram—(Abbreviated T–S diagram.) A graph with **temperature** as **ordinate** and **salinity** as **abscissa**, on which the points observed at a single oceanographic **serial station** are joined by a curve (the **T–S curve**).

temperature scale—(Or thermometric scale.) *See* **Celsius temperature scale, centigrade temperature scale, Fahrenheit temperature scale, Kelvin temperature scale, Rankine temperature scale, Reaumur temperature scale, molecular-scale temperature**.

temperature-sensing element—That part of a thermometric instrument that is directly affected by its **thermal** state.

Thus, in a **resistance thermometer**, the resistor is the temperature-sensing element.

temperature zone—Very generally, a portion of the earth's surface defined by relatively uniform **temperature** characteristics, and usually bounded by selected values of some measure of temperature or temperature effect.

All of the following may be considered "temperature zones": A. Supan's 1879 **hot belt, temperate belt**, and **cold cap**; W. Köppen's 1936 **tropical rainy climates, temperate rainy climates, snow forest climates**, and **polar climates**; C. W. Thornthwaite's 1931 **temperature provinces**. It is occasionally used for a vertical subdivision of **thermal belts** in mountainous terrain. *See also* **mathematical climate, solar climate**.

> Supan, A., 1879: Die Temperaturzonen der Erde. *Petermanns Geog. Mitt.*, **25**, 349–358.
> Köppen, W. P., and R. Geiger, 1930–1939: *Handbuch der Klimatologie*, Berlin: Gebruder Borntraeger, 6 vols.
> Thornthwaite, C. W., 1931: The climates of North America according to a new classification. *Geogr. Rev.*, **21**, 633–655.

tempestada—*See* **colla**.

temporale—A rainy **wind** from the southwest to west resulting from a deflection of the **southeast trades** of the eastern South Pacific onto the Pacific coasts of Central America.

Temporales are most frequent in July and August, when they may reach **gale** force and raise a heavy sea.

tendency—The local rate of change of a **vector** or **scalar** quantity with time at a given point in space.

Thus, in symbols, $\partial p / \partial t$ is the **pressure tendency**, $\partial \zeta / \partial t$ is the **vorticity** tendency, etc. Because of the difficulty of measuring instantaneous variations in the **atmosphere**, variations are usually obtained from the differences in magnitudes over a finite period of time and the definition of tendency is frequently broadened to include the **local time** variations so obtained. An example is the familiar three-hourly pressure tendency given in **surface weather observations**; in fact, the term "tendency" alone often means the pressure tendency.

tendency chart—Same as **change chart**.

tendency equation—An equation for the **local change** of **pressure** at any point in the **atmosphere**, derived by combining the **equation of continuity** with an integrated form of the **hydrostatic equation**.

In its basic form it is

$$\left(\frac{\partial p}{\partial t}\right)_h = -\int_h^\infty g\left(\frac{\partial(\rho u)}{\partial x} + \frac{\partial(\rho v)}{\partial y}\right) dz + (\rho g w)_h,$$

where $(\partial p/\partial t)_h$ represents the **pressure tendency** at a height h, g is the **acceleration of gravity**, ρ is the air density, and u, v, and w are, respectively, the x, y, and z components of the **wind** velocity. The subscript h indicates that the quantity is to be measured at the level $z = h$. The first term on the right represents the local change of pressure due to the net horizontal **mass convergence** above the level h. The second term represents the local change of pressure due to vertical motion through the level h. In general, these two terms balance one another so that the pressure tendency is obtained as a small difference between two quantities of large magnitude, a very undesirable feature for computational purposes. This difficulty is partially eliminated by making use of a form of the tendency equation that assumes that individual **density** changes are **adiabatic**, that $\partial p/\partial z = dp/dz$, and that the **advection** of pressure by the wind is negligible. It may be written

$$\left(\frac{\partial p}{\partial t}\right)_h = \int_0^{p_h} \frac{1}{T}\left(u\frac{\partial T}{\partial x} + v\frac{\partial T}{\partial y}\right) dp + \int_0^{p_h} \frac{1}{T}(\gamma_d - \gamma)w\,dp,$$

where T is the Kelvin **temperature**; $\gamma_d = -dT/dz$ the **dry-** or **saturation-adiabatic lapse rate**, depending on whether the air is unsaturated or saturated, respectively; and $\gamma = -\partial T/\partial z$, the **environmental lapse rate**. This equation is used most often to estimate the integrated vertical motion, given the pressure tendency and the advection.

Panofsky, H., 1956: *Introduction to Dynamic Meteorology*, 124–130.

tendency interval—The finite increment of time over which a change of the value of a **meteorological element** is measured in order to estimate its **tendency**.

The most familiar example is the three-hour time interval over which local **pressure** differences are measured in determining **pressure tendency**.

tenggara—A strong, dry, hazy, east or southeast **wind** during the east **monsoon** in the Spermunde Archipelago.

See **tongara, broeboe**.

tension saturated zone—*See* **capillary fringe**.

tensor—An array of functions that obeys certain laws of transformation.

The motivation for the use of tensors in some branches of physics is that they are invariants, not depending on the particular **coordinate system** employed. The tensors (e.g., **stress tensor**) that enter into meteorology are of a particularly simple kind, called Cartesian tensors, so the important theorems of tensor analysis play no role, but notational convenience justifies their use. A one-row or one-column tensor array is a **vector**.

tented ice—**Pressure ice** in which two floes have been pushed into the air, leaving an air space underneath.

tephigram—*See* **thermodynamic diagram**.

tercile—1. For a set of data arranged in order, values that partition the data into three groups, each containing one-third of the total number of observations.

2. Values of a **random variable** that partition the **variable** range into three subintervals in such a way that the **probability** taken over each subinterval is equal to one-third.

tereno—Same as **terrenho**.

Terminal Doppler Weather Radar—(Abbreviated TDWR.) **Doppler radar** installed during the 1990s by the Federal Aviation Administration at U.S. airports with high traffic density or susceptibility to **wind shear**.

Radars are located 10–20 km from the airports in order to detect **microbursts, gust fronts**, and **convective storms** along arrival and departure paths. The nominal transmitted **wavelength** of the TDWR is 5.3 cm (**C band**) and the nominal circular **beamwidth** is 0.5°.

terminal fall velocity—(Or terminal velocity.) The particular falling speed, for any given object moving through a fluid medium of specified physical properties, at which the **drag** forces and buoyant forces exerted by the fluid on the object just equal the gravitational force acting on the object. It falls at constant speed, unless it moves into air layers of different physical properties.

In the **atmosphere**, the latter effect is so gradual that objects such as raindrops, which attain terminal **velocity** at great heights above the surface, may be regarded as continuously adjusting their speeds to remain at all times essentially in the terminal fall condition. The terminal fall velocity of water droplets in still air can be computed from **Stokes's law** for drops smaller than 80 μm in diameter. Above that size, empirical values must be used.

terminal forecast—An **aviation weather forecast** for one or more specified air terminals.
See also **TAF**.

terminator—The **great circle** of transition from daylight to darkness on the earth's surface.
The terminator appears in visible environmental satellite imagery as a gradual change from easily detectable **cloud** features to nearly black imagery.

terpenes—Naturally occurring **hydrocarbons**, emitted by many trees and plants.
They mostly have very strong smells and are responsible for the aromas of the vegetation in which they are found. Terpenes can be thought of as being built from units of **isoprene**, C_5H_8, joined together into chains and rings. Monoterpenes, formula $C_{10}H_{16}$, constitute the major emissions from conifers and fruit trees. Sesquiterpenes, formula $C_{15}H_{24}$, are commonly found in citrus trees. Terpenes are very reactive and may contribute to diminution of air quality in forested areas. The name comes from turpentine, a liquid consisting of several terpene compounds distilled from the resin of pine trees.

terpenoids—Any organic compound the structure of which is derived from the **isoprene** entity.

terral—*See* **virazon**.

terral levante—A **land breeze** of Spain and Brazil, sometimes a northwest **squall** of **foehn** character.

terrenho—(Also spelled tereno.) A cold, dry, land **wind** of India.

terrestrial magnetism—Same as **geomagnetism**.

terrestrial radiation—**Longwave radiation** originating by **thermal** emission from the earth's surface and/or its **atmosphere**; to be distinguished from **solar radiation**.

terrestrial refraction—A variety of phenomena that result from the **refraction** of terrestrial **light** by the earth's **atmosphere**.
It is distinguished from **astronomical refraction**, which is used when the source is outside our atmosphere.

terrestrial scintillation—*See* **scintillation**.

terrestrial surface radiation—**Longwave radiation** originating by **thermal** emission from the earth's surface only.

tertiary circulation—(Obsolete.) The generally small, localized atmospheric circulations.
They are represented by such phenomena as **local winds**, **thunderstorms**, and **tornadoes**.
Compare **primary circulation**, **secondary circulation**; *see* **mesometeorology**.

Tetens's formula—An analytic expression for **saturation vapor pressure**, e_s, given as a function of **temperature** by

$$e_s(T) = 0.611 \times 10^{7.5T/(T + 237.3)},$$

where the temperature T is given in °C and the **vapor pressure** is in kPa.
A slightly improved version is shown under **Clausius–Clapeyron equation**.
Tetens, O., 1930: Uber einige meteorologische Begriffe. Z. Geophys., **6**, 297–309.

tethered balloon—A positively buoyant unmanned balloon attached to a cable that is used to raise and lower the balloon.
Instrument packages are suspended from the balloon to measure characteristics such as **air temperature**, **winds**, **humidity**, **refractive index**, **trace gases**, and **aerosols** as functions of time (by keeping the height constant) or height (by raising and lowering the balloon). They have been used for measuring both mean values and turbulent fluctuations. Balloon heights of over a kilometer have been reached, and instrument packages are sometimes deployed at more than one height below the balloon. Tethered balloons have been deployed from fixed locations over land and from ships.

tethersonde—A **radiosonde** attached to a fixed or **tethered balloon**.
The balloon is usually larger than a balloon used for **upper-air soundings**, and the tether usually limits the sounding to the **boundary layer**. The radiosonde is typically moved up and down the tether to get multiple, high-resolution profiles of the boundary layer.

tetroon—A balloon of tetrahedron shape made from a cylinder of plastic film.

The cylinder is sealed orthogonally at the two ends, producing a low-cost balloon that simulates a sphere.

teuchit—Same as **peesweep storm**.

Teweles–Wobus index—A **verification** system designed for the 30-hour sea level **prognostic chart** produced by the U.S. **WBAN** Analysis Center in the 1950s.

texture—In a photo **image**, the frequency of change and arrangement of tones, often due to highlights and shadows created by the irregular surface being viewed.

thaw—1. To melt a substance, **ice** for example, by warming it to a **temperature** greater than the **melting point** of the substance, or to have frozen contents melted.

2. To free something from the binding action of **ice** by warming it to a **temperature** above the **melting point** of ice.

3. A warm spell when **ice** and **snow** melt, for example, "January thaw."

thawing index—As used by the U.S. Army Corps of Engineers, the number of Fahrenheit **degree-days** (above and below 32°F) between the lowest and highest points on the cumulative degree-days **time curve** for one **thawing season**.

The thawing index determined from air temperatures at 4.5 ft above the ground is commonly designated the **air thawing index**, while that determined from temperatures immediately below a surface is called the surface thawing index. *Compare* **freezing index**; *see also* **growing season**.

thawing season—As used by the U.S. Army Corps of Engineers, the period of time between the lowest point and the succeeding highest point on the **time curve** of cumulative Fahrenheit **degree-days** above and below 32°F; the opposite of **freezing season**.

See also **growing season**.

thematic mapper—A seven-**channel** instrument on **Landsat** series satellites used to make maps of **infrared** emission and **reflection** from the earth.

Most thematic mapper images are used to study vegetation, geology, and other surface features.

theodolite—An optical instrument, similar to a surveyors's transit telescope, used to visually track a **radiosonde balloon** and determine its **azimuth** and **elevation** angles while in flight.

thermal—1. Pertaining to **temperature** or **heat**.

2. A discrete buoyant element in which the **buoyancy** is confined to a limited volume of fluid. *See* **plume**.

3. A relatively small-scale, rising current of air produced when the **atmosphere** is heated enough locally by the earth's surface to produce **absolute instability** in its lowest **layers**.

The use of this term is usually reserved to denote those currents either too small and/or too dry to produce convective clouds; thus, thermals are a common **source** of low-level **clear-air turbulence**. It is generally believed that the term originated in glider flying, and it is still very commonly used in this reference.

thermal band—*See* **thermal infrared**.

thermal belt—1. (Also called thermal zone.) Any one of several possible horizontal belts of vegetation type found in mountainous terrain.

These belts are primarily the result of vertical **temperature** variation. See, for example, **frostless zone, timber line**.

2. An elevation band along mountain and other terrain slopes where nighttime surface temperatures remain relatively mild compared with temperatures above and below.

Drainage winds carry the coldest air down the slopes to the bottom of the valley. The belt of warmer air (thermal belt) lies above this **pool of cold air**. Above the warm belt, **temperature** exhibits its **normal** decline with **elevation**, augmented by increased **radiation** loss from lower air **density** and lower **moisture content** at higher altitudes. The impact of this milder slope **climate** is a longer **growing season**, an earlier leafing out and blossoming of trees and other vegetation, and the ability to grow crops that could not survive at lower or higher elevations (e.g., vineyards). Geiger (1965) suggests that this effect influenced early settlement locations: "In Germany this area was preferred for the earliest villages, monasteries, and country houses."

Geiger, R., 1965: *The Climate Near the Ground*, Harvard University Press, Cambridge, Mass., p. 437.

thermal capacity—Same as **heat capacity**.

thermal climate—**Climate** as defined by **temperature**.

thermal conductivity—The proportionality factor between **energy** flux and **temperature** gradient (*see* **conduction**).

Thermal conductivity is to an extent an intrinsic property of a medium but may depend on temperature. The thermal conductivity of air is about 50% greater than that of **water vapor** and that of both increases (approximately) as the square root of **absolute** temperature. The thermal conductivity of liquid water is about 25 times that of air. Thermal conductivities of solids, especially metals, are thousands of times greater than that of air.

thermal constant—(Also called thermometric constant.) The quantity of **heat** required to complete some stage, or the whole, of a plant's growth.

It was first investigated by Réaumur about 1735, using the sum of the mean daily temperatures. It may also be measured by the sum of air temperatures above some standard such as 42°F (*see* **degree-day**). In modern **agricultural climatology** this concept has been largely abandoned in favor of complex influence factors such as **evapotranspiration**. It has also been established that soil temperatures at or somewhat below 10 cm are more important for plant growth than **air temperature**.

thermal core—The interior of a convective **thermal** that is relatively undiluted by lateral **entrainment** or **intromission**.

thermal current—Same as **thermal**.

thermal diffusivity—(Also called heat conductivity, thermometric conductivity.) The ratio of the **thermal conductivity** k of a substance to the product of its **specific heat** c and its **density** ρ:

$$K = k/c\rho.$$

For a gas, c is the specific heat at constant **pressure**, c_p. The thermal diffusivity determines the rate of heating given the **temperature** distribution according to an equation, which for one spatial dimension is

$$\frac{dT}{dt} = K\frac{\partial^2 T}{\partial z^2}.$$

At a temperature of 0°C, the thermal diffusivity for air is 1.9×10^{-5} m^2 s^{-1}. At other temperatures, the ratio μ/K is nearly constant, where μ is the **dynamic viscosity**. *See also* **diffusivity**, **kinematic viscosity**.

List, R. J., 1951: *Smithsonian Meteorological Tables*, 6th rev. ed., 394–395.

thermal efficiency—1. Same as **thermodynamic efficiency**.

2. In **climatology**, an expression of the effectiveness of **temperature** in determining the rate of growth, assuming sufficient moisture.

The idea was introduced by B. E. and G. J. Livingston (1913). It was applied by C. W. Thornthwaite (1948) in his system of **climatic classification**. The recognition of this general concept led to one of the first uses of the degree-day, that is, application to plant growth and relationship to the phenological **effective temperature** of about 42°F. *Compare* **precipitation effectiveness**.

Livingston, B. E., and G. J. Livingston, 1913: Temperature coefficients in plant geography and climatology. *Botanical Gazette*, **56**, 349–375.
Thornthwaite, C. W., 1948: An approach toward a rational classification of climate. *Geogr. Review*, **38**, 55–94.

thermal-efficiency index—Same as **temperature-efficiency index**.

thermal-efficiency ratio—Same as **temperature-efficiency ratio**.

thermal energy—Same as **heat**.

thermal equator—1. Same as **heat equator**.

2. (Rare.) The belt around the earth that is bounded by the mean annual isotherms of 27°C (80°F); also, the middle line of this belt.

3. (Rare.) The belt of hottest temperatures around the earth as it annually migrates with the inclination of the sun.

thermal gradient—1. (Or geothermal gradient.) According to Smithsonian Physical Tables, the rate of **variation** of **temperature** in soil and rock from the surface of the earth down to depths of the order of kilometers.

It varies greatly from place to place, depending on the geological history of the region, the **radioactivity** of the underlying rocks, and the **conductivity** of the upper rocks. An average is about +10°C per km.

2. Same as **temperature** gradient.
See also **lapse rate**.

thermal high—An area of high atmospheric **pressure** near the surface resulting from the cooling of air by a cold underlying surface, and remaining relatively stationary over the cold surface.
Compare **thermal low, cold high**; *see* **glacial anticyclone**.

thermal infrared—The middle and **far infrared** portions of the **spectrum**; mainly used to distinguish **infrared radiation** of terrestrial origin from the **solar infrared**.

thermal instability—1. Same as **potential instability**.

2. (Obsolete.) The **instability** of a fluid layer heated from below.

thermal internal boundary layer—An **internal boundary layer** caused by **advection** of air across a **discontinuity** in **surface temperature**.
Important in coastal plains, this layer increases in depth to merge eventually with the **convective boundary layer** some distance from the coastline. In this example of cool air advection, the thermal internal boundary layer grows in depth as the square root of distance from the discontinuity.

thermal inversion—*See* **temperature inversion**.

thermal jet—(Rare.) A region in the **atmosphere** where **isotherms** or **thickness lines** are closely packed; therefore, a region of very strong **thermal wind**.

thermal load—1. Amount of **heat** transported by a river or canal per unit time (Joules per second).

2. Amount of **heat** discharged by an outlet into a water body per unit time (Joules per second).

thermal low—(Or heat low.) An area of low atmospheric **pressure** near the surface resulting from heating of the lower **troposphere** and the subsequent lifting of **isobaric surfaces** and **divergence** of air aloft.
Thermal lows are common to the continental **subtropics** in summer; they remain stationary over the warm surface areas that produce them; their **cyclonic circulation** is generally weak and diffuse; they are nonfrontal. *Compare* **monsoon low**.

thermal neutrality—(Also called thermoneutrality.) The condition in which the **thermal** environment of a homeothermic animal is such that its **heat** production (metabolism) is not increased either by cold stress or heat stress.
The **temperature range** in which this minimum occurs is called the zone of thermal neutrality. For humans, this zone is 29°–31°C (84°–88°F).

thermal pollution—Generally, the **discharge** of water heated by industrial processes into natural water bodies.
Also, the "waste heat" generated by industrial processes, such as those associated with petrochemical facilities, coke ovens, and flares, that result in large "plumes" of heated **ambient air** released into the **atmosphere**.

thermal precipitator—A device often used to sample airborne **particles** by passing the particle-laden air between hot and cold plates.
The **thermal** or phoretic forces deposit the particles on the cold surface.

thermal quality of snow—Same as **quality of snow**.

thermal radiation—Electromagnetic **radiation** emitted by all matter in **thermal** equilibrium as the result of the thermal excitation of its molecules.

thermal Rossby number—The nondimensional ratio of the **inertial force** due to the **thermal wind** and the **Coriolis force** in the flow of a fluid heated from below:

$$\mathrm{Ro}_T = \frac{U_T}{fL},$$

where f is the **Coriolis parameter**, L a **characteristic length**, and U_T a characteristic thermal wind. The characteristic thermal wind is

$$U_T = \frac{g\epsilon(\Delta,\theta)\delta}{f\Delta r},$$

where g is the **acceleration of gravity**, ϵ the **coefficient of thermal expansion**, $\Delta,\theta/\Delta r$ a characteristic radial **temperature** gradient, and δ depth of the fluid. *Compare* **Rossby number**; *see* **thermal instability**.

thermal roughness—*See* **roughness length.**

thermal steering—The **steering** of an **atmospheric disturbance** in the direction of the **thermal wind** in its vicinity; equivalent to steering along **thickness lines.**
   For this purpose the thermal wind is usually taken from the earth's surface to a level in the middle **troposphere.**

thermal tide—A **variation** in atmospheric pressure due to the **diurnal** differential heating of the **atmosphere** by the sun; so called in analogy to the conventional **gravitational tide.**
   See **solar atmospheric tide.**

thermal vorticity—The **vorticity** of the **thermal wind,** defined by analogy with the **geostrophic vorticity,**

$$\zeta_T = \frac{g}{f}\,\nabla_p^2 h,$$

where ζ_T is the thermal vorticity, g the **acceleration of gravity,** f the **Coriolis parameter,** ∇_p the **del operator** in the **isobaric surface,** and h the **thickness** of a layer bounded by two isobaric surfaces.

thermal vorticity advection—The **advection** or **transport** of the **thermal vorticity** by the **thermal wind,** in analogy to the advection of the **vorticity** by the **wind.**
   If $\mathbf{v}_T$ denotes the thermal wind and ζ_T the thermal vorticity, the thermal vorticity advection may be written

$$\mathbf{v}_T \cdot \nabla_H \zeta_T,$$

where ∇_H is the horizontal **del operator.** In a two-level **baroclinic model** of the **atmosphere,** the thermal vorticity advection contributes to the development of **geostrophic vorticity** and is a measure of the conversion of **potential energy** into **kinetic energy** of cyclonic-scale flow.

thermal wave—A **gravity wave** in the free **troposphere** excited by boundary layer **thermals** overshooting into the **capping inversion.**

thermal wind—The mean **wind-shear** vector in **geostrophic balance** with the **gradient** of **mean temperature** of a layer bounded by two **isobaric surfaces.**
   Its analogy to the **geostrophic wind** is best demonstrated by expressing the thermal wind $\mathbf{v}_T$ in terms of the **thickness** h of the layer:

$$f\mathbf{k} \times \mathbf{v}_T = -g\nabla_p h,$$

where f is the **Coriolis parameter,** $\mathbf{k}$ the vertical unit **vector,** g the **acceleration of gravity,** and ∇_p the **del operator** in the isobaric surface. The thermal wind is directed along the isotherms with cold air to the left in the Northern Hemisphere and to the right in the Southern Hemisphere. Its magnitude V_T is given by the formula

$$V_T = \left| \frac{g}{f}\,\frac{\partial h}{\partial n} \right|,$$

where n is a coordinate normal to the thickness lines. See **thermal wind equation.**

thermal wind equation—An equation for the vertical **variation** of the **geostrophic wind** in **hydrostatic equilibrium,** which may be written in the form

$$-\frac{\partial \mathbf{v}_g}{\partial p} = \frac{R}{pf}\,\mathbf{k} \times \nabla_p T,$$

where $\mathbf{v}_g$ is the **vector** geostrophic wind, p the **pressure** (used here as the vertical coordinate), R the **gas constant** for air, f the **Coriolis parameter,** $\mathbf{k}$ a vertically directed unit vector, and ∇_p the isobaric **del operator.**
   This equation shows that the geostrophic wind shear $-\partial \mathbf{v}_g/\partial p$ is a vector parallel to the isotherms in such a sense that the cold air is on the left of the **shear** (in the Northern Hemisphere). The geostrophic wind at the top of an **atmospheric layer** may be considered the sum of the **wind** at the bottom of the layer and the **thermal wind** determined by the mean **isotherms** within the layer, that is, by the **thickness pattern** of the layer.

thermal zone—1. Same as **temperature zone.**
   2. Same as **thermal belt.**

thermally direct—Refers to a process by which internal and/or **potential energy** is transferred to **kinetic energy**, as for example in **buoyant convection**.

thermally indirect—The reverse of **thermally direct**, so that **kinetic energy** is transferred to internal and/or **potential energy**.

An example occurs in the **instability** of **shear flow** in a stably stratified **environment**. Although the term may be considered confusing, it is widely used in contemporary literature.

thermistor—A device with electrical **resistance** that varies markedly and monotonically and that possesses a negative **temperature coefficient** of **resistivity**.

The thermistors commonly used in meteorology are composed of solid semiconducting materials with resistance that decreases 4% per °C. They are constructed in a variety of sizes and may be obtained with **thermal** time constants of a millisecond or less. Meteorological applications include **thermometers**, **anemometers**, and **bolometers**.

thermistor thermometer—A device for estimating **temperature** based on a known relationship of its electrical **resistance** to temperature.

The term "thermistor" is more specific than the term "resistance thermometer" because it implies that the material has been selected for a strong temperature dependence to simplify the determination of temperature. *See* **thermistor**.

thermo-integrator—An apparatus, used in studying soil temperatures, for measuring the total supply of **heat** during a given period.

It consists of a long nickel coil forming a 100-ohm **resistance thermometer** and a 6-volt battery, the **current** used being recorded on a galvanometer. The coil is attached to a rod for insertion. A **mercury thermometer** can be used. The instrument can also measure the **heat balance** in a plant cover.

thermo-osmosis—The movement of liquid in a **porous medium** due to differences in **temperature**.

thermocline—A vertical **temperature** gradient, in some layer of a body of water, that is appreciably greater than the gradients above and below it; also a layer in which such a **gradient** occurs.

The permanent thermocline refers to the thermocline not affected by the seasonal and **diurnal** changes in the surface forcing; it is therefore located below the yearly maximum depth of the **mixed layer** and the influence of the **atmosphere**. The seasonal thermocline refers to the thermocline not affected by the diurnal changes in the surface forcing. In general, it is established each year by heating of the **surface water** in the summer, and is destroyed the following winter by cooling at the surface and wind-driven **mixing**. The diurnal thermocline refers to the thermocline that, in general, is established each day by heating of the surface water and is destroyed the following night by cooling and/or mixing. *See also* **transition layer**.

thermoclinicity—The spatial **gradient** of **potential temperature** on an **isopycnal surface**.

If thermoclinicity is zero, all fluid parcels in an isopycnal surface have the same potential temperature and are interchangeable. If thermoclinicity is nonzero, fluid parcels in the same isopycnal surface can have different potential temperature T and **salinity** S. Exchange of such parcels will lead to large T/S differences within the surface.

thermocouple—A **temperature-sensing element** that converts **thermal energy** directly into electrical **energy**.

In its basic form it consists of two dissimilar metallic electrical conductors connected in a closed loop. Each junction forms a thermocouple. One thermocouple is maintained at a known **temperature** (usually 0°C or a measured temperature) and the other thermocouple is used to measure the unknown temperature. The **signal** voltage is a function of the temperature, and the smooth curve can be handled with a simple **linear** fit over a moderate **temperature range**. Different materials have different curves. Popular thermocouples (and change in voltage per °C) include iron-constantan (50 mv per °C), copper-constantan (38 mv per °C), and various platinum alloys. Thermocouples are also important in home furnaces to detect the pilot light or that the fuel has ignited. A chain of thermocouples, called a **thermopile**, can be used as a power supply if a source of heat and cold is available.

thermocouple thermometer—Same as **thermoelectric thermometer**.

thermodynamic diagram—(Sometimes called adiabatic chart, adiabatic diagram.) Any **chart** or graph representing values of **pressure**, **density**, **temperature**, **water vapor**, or functions thereof, such that the **equation of state**, the Clapeyron–Clausius equation, and the **first law of thermodynamics** for **adiabatic** and **saturation** or **pseudoadiabatic processes** are satisfied.

A very large number of individual diagrams fall under this general description. These common diagrams are mathematical transformations of each other. They display (reading counterclockwise about a point) **isobars** (approximately horizontal), **isotherms**, **vapor lines**, **saturation** or **pseudoadiabats**, and **dry adiabats**. When an **air parcel** undergoes a **reversible process**, the succession of states is represented on the thermodynamic diagram by a curve. A cyclic process is represented by a closed curve. On some charts the area so enclosed is directly proportional to the **work** done in the process, and some authors prefer to restrict the use of "thermodynamic diagram" to such charts. See table below.

Common Thermodynamic Diagrams and Some of Their Properties

Chart	X (x-axis)	Y (y-axis)	Equation for $T = $ const	Equation for $\theta = $ const	Thermodynamic work by system	Remarks
Aerogram or **Refsdal diagram**	$\ln T$	$-T \ln p$	$X = \ln T$	$Y = \dfrac{c_p}{R}(\ln \theta - X) + \text{const}$	$-R\int Y dX$	Only isotherms are straight lines.
Clapeyron $(\alpha, -p)$	α	$-p$	$Y = -\dfrac{RT}{X}$	$Y = \text{const}\left(\dfrac{\theta}{X}\right)^{c_p/c_v}$	$-\int Y dX$	Classic physics diagram; isentropes curved and nearly parallel to isotherms.
Emagram or **Neuhoff diagram**	T	$-\ln p$	$X = T$	$Y = \text{const} - \dfrac{c_p}{R}\ln\left(\dfrac{X}{\theta}\right)$	$-R\int Y dX$	Isentropes slightly curved.
Pastagram	T_0	$T_0\left[1 - \left(\dfrac{p}{p_0}\right)^{R/c_p}\right]$	$Y = X - T$	$Y = X\left[1 - \left(\dfrac{\theta}{X}\right)^{1/(1 - g/c_p\gamma)}\right]$	Not relevant	T_0, p_0 are surface values, $\gamma = 6.5$ K km^{-1}; used mainly for hydrostatic calculations.
Stüve or Pseudo-adiabatic	T	$-p^{R/c_p}$	$X = T$	$Y = \text{const}(X/\theta)$	$c_p\int\ln(-Y)dX$	Isentropes are straight, inclined lines; proportionality of area and work sacrificed.
Skew T–log p or Skewed emagram	$T - c \ln p$	$-\ln p$	$Y = \dfrac{X - T}{c}$	$\dfrac{R}{c_p}Y + \ln(X - cY) = \ln \theta + \text{const}$	$-R\int Y dX$	Isotherms straight, but tilted according to $T = X - cY$; isentropes curved.
Tephigram	T	$\ln \theta$	$X = T$	$Y = \ln \theta + \text{const}$	$-c_p\int Y dX$	Isobars are curved, inclined lines.
Thetagram	T	$-p$	$X = T$	$Y = X \ln(-Y) + \text{const}$	$-R\int(X/Y)dY$	Isentropes curved; area not proportional to work.

thermodynamic efficiency—(Also called thermal efficiency, Carnot efficiency.) In **thermodynamics**, the ratio of the **work** done by a **heat engine** to the total **heat** supplied by the heat source.

thermodynamic energy equation—*See* **energy equation**.

thermodynamic function of state—Often, in **thermodynamics**, the same as **state variable** or **thermodynamic variable**.

Some authors, however, use state variable for **pressure**, **temperature**, and **specific volume**, and then refer to other thermodynamics quantities depending on these or other variables as functions of state.

thermodynamic method for mixed layer growth—*See* **encroachment method**.

thermodynamic potential—Same as **Gibbs function**.

thermodynamic probability—Proportional to the number of equally likely states that a system may attain.

Boltzmann proposed that the **entropy** S and the thermodynamic probability W were proportional, and Planck subsequently showed that $S = k \ln W$ with k **Boltzmann's constant**. This result is of such fundamental significance in physics that it was carved as an adornment on Boltzmann's gravestone in Vienna. *See* **third law of thermodynamics**.

thermodynamic speed limit—A theoretical limit to the windspeed of a **tornado**, based on the assumption of a tornado core with the thermodynamic characteristics of the **pseudoadiabatic lapse rate** and **hydrostatic balance**.

The thermodynamic speed limit may be expressed in terms of CAPE as $(2\ \text{CAPE})^{1/2}$, where CAPE is the **convective available potential energy**. The thermodynamic speed limit is often exceeded, illustrating that nonhydrostatic effects are important.

thermodynamic temperature scale—Same as **Kelvin temperature scale**.

thermodynamic variable—Same as **state variable**.

thermodynamics—A collection of ideas and axioms, leading to differential equations specifying rates of change, that describes our experience with processes that involve fluxes of **heat** and changes in **energy** content.

Thermodynamics introduces a new concept—**temperature**—absent from classical mechanics and other branches of physics. Classical thermodynamics deals with **equilibrium** states, concentrating on initial and final configurations, not on the processes involved in evolution.

Dutton, J. A., 1995: *Dynamics of Atmospheric Motion*, Dover Press, 35–36, 406–410.
Richards, P. I., 1959: *Manual of Mathematical Physics*, Pergamon Press, p. 30.
Sommerfeld, A., 1964: *Thermodynamics and Statistical Mechanics*, Academic Press, v, 1.

thermoelectric thermometer—(Also called thermocouple thermometer.) A type of **electrical thermometer** consisting of two thermocouples that are series connected with a **potentiometer** and a constant-temperature bath.

One couple, called the reference junction, is placed in a constant-temperature bath, while the other is used as the measuring junction. The measuring junction can be made physically very small in order to have practically a negligible **thermal** time constant. *See* **bottle thermometer**.

Middleton, W. E. K., and A. F. Spilhaus, 1953: *Meteorological Instruments*, 3d ed., rev., 88–91.

thermogram—The record of a **thermograph**.

thermograph—A self-recording **thermometer**.

The thermometric element is most commonly either a bimetal strip or a **Bourdon tube** filled with a liquid. In the first case the bimetal element has the form of a helical coil with one end rigidly fastened to the instrument and the other to the recording pen. In the second case the tube is made with an elliptical **cross section** so that an expression of the liquid caused by a **temperature** increase will cause the radius or curvature of the bend to increase, thus moving the instrument pen, which is fastened to the tip of the tube. A **resistance thermometer** and a **thermoelectric thermometer** may be converted into thermographs if provision is made to record their **output**. *See* **aspiration thermograph**, **hygrothermograph**, **mercury-in-steel thermometer**.

thermohaline circulation—That part of the large-scale ocean circulation driven by the fluxes of **heat** and **freshwater** at the ocean surface.

The freshwater **flux** affects **salinity**, and both **temperature** and salinity changes cause **density** changes that drive the thermohaline circulation. The present-day forcing consists of cooling and net **precipitation** in high latitudes, warming and **evaporation** in subtropical latitudes; note the opposing effects on density. The present-day thermohaline circulation consists of 1) sinking of strongly cooled, moderately saline water in relatively small regions located in areas of relatively strong winter cooling; 2) deep flow throughout the global ocean basins; and 3) slow **upwelling** toward the surface. Its **transport** is small compared to wind-driven transport, but it is believed that the thermohaline circulation is responsible for much of the heat transported by the ocean. *See* **gradient current.**

thermohygrogram—*See* **hygrothermogram.**

thermohygrograph—*See* **hygrothermograph.**

thermohygrometer—*See* **hygrothermometer.**

thermoisopleth—An **isopleth** of **temperature**; specifically, a line on a graph connecting values of constant temperature as a function of two coordinates.
　　See **isotherm.**

thermokarst topography—A pattern of waterfilled sinks, cave-ins, and other small depressions caused by the local melting of **permafrost.**

thermometer—An instrument for measuring **temperature** by utilizing the **variation** of the physical properties of substances according to their **thermal** states.
　　Thermometers may be classified into six types according to their construction: **gas thermometer, liquid-in-glass thermometer, deformation thermometer, electrical thermometer, liquid-in-metal thermometer,** and **sonic thermometer.** *See also* **thermograph.**
　　　　Middleton, W. E. K., and A. F. Spilhaus, 1953: *Meteorological Instruments*, 3d ed., rev., 67–92.

thermometer screen—Same as **instrument shelter.**

thermometer shelter—Same as **instrument shelter.**

thermometer support—A device used to hold liquid-in-glass **maximum** and **minimum thermometers** in the proper recording position inside an **instrument shelter,** and to permit them to be read and reset.
　　See **Townsend support.**

thermometric conductivity—Same as **thermal diffusivity.**

thermometric constant—Same as **thermal constant.**

thermometric scale—Same as **temperature scale.**

thermometry—The science of **temperature** measurement.

thermoneutral zone—The range of **ambient temperature** in which normal metabolism provides enough **heat** to maintain an essentially constant body **temperature** in homeothermic animals. The limits of the zone depend on the species and breed of an animal and its age, sex, degree of **acclimatization,** how it is fed, and even the time of day.
　　The zone is narrow for the young of a species (e.g., 29°–30°C for a chick) and wide for a well-fed large adult (e.g., −30° to +25°C for a cow).

thermoneutrality—Same as **thermal neutrality.**

thermoperiodicity—Response of an organism to periodic (annual or **diurnal**) changes of **temperature.**

thermopile—A **transducer** for converting **thermal energy** directly into electrical **energy.**
　　It is composed of pairs of **thermocouples** that are connected either in series or in parallel. The **output** voltage of N pairs of series-connected thermocouples is N times the voltage developed by a single pair, while the **current** developed by N pairs of parallel connected thermocouples is N times the current developed by a single pair. Thermopiles are used in thermoelectric **radiation** instruments when the output of a single pair of thermocouples is not large enough. *See* **moll thermopile, pyrheliometer.**

thermoscope—An instrument that measures **temperature** changes, in contrast with a **thermometer,** which measures the **absolute** temperature.
　　　　Middleton, W. E. K., 1969: *Invention of the Meteorological Instruments*, Johns Hopkins Press, Baltimore, p. 44.

thermoscreen—In U.S. Navy terminology, same as **instrument shelter.**

thermosphere—The **atmospheric shell** extending from the top of the **mesosphere** to outer space.
　　It is a region of more or less steadily increasing **temperature** with height, starting at roughly 100 km. The thermosphere includes, therefore, the **exosphere** and most of the **ionosphere.**

thermosteric anomaly—The **specific-volume anomaly** (steric anomaly) that the **seawater** at any point would attain if the seawater were brought isothermally to a **pressure** of one **standard atmosphere**.

In other words, thermosteric anomaly is the specific-volume anomaly calculated for the given **salinity** and **temperature** but for a **standard pressure**.

thermotropic model—A **model atmosphere** used in **numerical forecasting** in which the parameters to be forecast are the height of one **constant-pressure surface** (usually 500 mb) and one **temperature** (usually the **mean temperature** between 1000 and 500 mb).

Thus, a surface **prognostic chart** can also be constructed. The **quasigeostrophic approximation** is employed and the **thermal wind** is assumed constant with height.

thetagram—*See* **thermodynamic diagram**.

THI—1. Abbreviation for **temperature–humidity index**.

2. Abbreviation for **time–height indicator**.

thick–thin chart—Same as **isentropic thickness chart**.

thickness—In **synoptic meteorology**, the vertical depth, measured in geometric or **geopotential** units, of a layer in the **atmosphere** bounded by surfaces of two different values of the same physical quantity, usually **constant-pressure surfaces**.

See **thickness chart**.

thickness chart—A type of **synoptic chart** showing the **thickness** of a certain physically defined layer in the **atmosphere**.

Currently it almost always refers to an **isobaric thickness chart**, that is, a **chart** of vertical distance between two **constant-pressure surfaces** and is often proportional to the **temperature** of that layer. This chart consists of a pattern of **thickness lines** either drawn directly to data plotted on the chart or, more commonly, drawn by the single graphical process of **differential analysis**. *See* **isentropic thickness chart**, **vertical differential chart**.

thickness line—(Also called relative contour, relative isohypse.) A line drawn through all geographic points at which the **thickness** of a given **atmospheric layer** is the same; an **isopleth** of thickness.

The pattern of thickness lines constitutes a **thickness chart**.

thickness pattern—(Also called relative hypsography, relative topography.) The general geometric distribution of **thickness** contour lines on a **thickness chart**.

Thiessen polygon method—A method of assigning areal significance to **point rainfall** values.

Perpendicular bisectors are constructed to the lines joining each measuring **station** with those immediately surrounding it. These bisectors form a series of polygons, each polygon containing one station. The value of **precipitation** measured at a station is assigned to the whole area covered by the enclosing polygon.

thin—As used in **aviation weather observations**, descriptive of a **sky cover** that is predominantly transparent.

According to the **summation principle**, at any level, if the ratio of the **transparent sky cover** to the **total sky cover** (opaque plus transparent) is one-half or more, then the **cloud layer** at that level must be classified as "thin." It is denoted by the **symbol** "−" preceding the appropriate sky cover symbol.

thin line echo—Same as **fine line**.

THIR—Abbreviation for **Temperature–Humidity Infrared Radiometer**.

third law of thermodynamics—The statement that every substance has a finite positive **entropy**, and the entropy of a crystalline substance is zero at the **temperature** of **absolute zero**.

Modern **quantum theory** has shown that the entropy of crystals at 0 K is not necessarily zero. If the **crystal** has any asymmetry, it may exist in more than one state; and there is, in addition, an entropy residue deriving from nuclear spin. *See* **thermodynamic probability**.

third-order climatological station—As defined by the World Meteorological Organization in 1956, a **station**, other than a **precipitation station**, at which the observations are of the same kind as at a **second-order climatological station**, but are 1) not so comprehensive; or 2) made once a day only; or 3) made at other than the specified hours.

This designation is not used officially in the United States, but types of stations that would fit under this category include **climatological substations** and certain **aeronautical weather reporting stations**.

third-order closure—A technique for approximating **turbulence** by retaining forecast equations for first-, second-, and third-order **statistical** moments, and parameterizing all other higher **moments**.

Compare **higher-order closure, second-order closure, first-order closure, nonlocal closure, local closure.**

thirty-day forecast—A **weather forecast** for a period of 30 days.

As issued by the U.S. National Weather Service, the forecast items are information concerning expected departures of **temperature** and **precipitation** from **normal**.

Thornthwaite moisture index—A method for monitoring the soil **water budget** for relatively large areas based on the pioneering work of C. W. Thornthwaite (1948) using the concept of **potential evapotranspiration.**

The **Palmer Drought Severity Index** and the companion "**crop moisture index**," developed by W. C. Palmer, are modifications of the earlier Thornthwaite work.

> Thornthwaite, C. W., 1948: An approach toward a rational classification of climate. *Geogr. Review*, **38**, 55–94.

thoron—(Symbol Tn.) A **radioactive gas, atomic number** 86, atomic weight 220; an inert gaseous **element** with radioactive decay, accompanied by emission of **alpha particles**, responsible for a portion of the **ionization** observed in the **lower atmosphere**.

It is a member of the thorium family of radioactive elements and is the immediate descendant of the radium **isotope** of atomic number 88 and atomic weight 224 (called thorium X at one time), which decays by alpha emission to thoron in a **half-life** of 3.6 days. Thoron, in turn, decays by alpha emission in a half-life of only 54.5 seconds, yielding an isotope of polonium. Thoron, like each of the other two radioactive gases (**radon** and **actinon**) that are its isotopes, enters the **atmosphere** by the process of **exhalation** after its formation by radioactive disintegration within soil or rocks. The very short half-life of thoron allows it little time to be carried from the earth's surface to higher levels, so its contribution to **atmospheric ionization** is made largely in the lowest few meters of the atmosphere.

three-axis stabilized—Refers to a spacecraft that rotates once per **orbit**, allowing its instruments to point constantly toward the earth.

three-dimensional model—Mathematical **model** used to simulate the variations in chemical composition over the earth's surface as a function of **altitude** with time.

Such models are very expensive to run, since they contain detailed descriptions of **atmospheric transport** and usually only contain a limited number of trace chemical species. More detailed chemistry is usually encountered in two-dimensional models, where the **variation** with longitude is not considered.

three-dimensional variational analysis—A **variational objective analysis** performed at a single time in the three space dimensions to create an estimate of the atmospheric state.

threshold contrast—The minimum **contrast** C_{thresh} at which an observer can just distinguish a **target** object from its surroundings.

Threshold contrast depends on target angular size θ, the surrounding **luminance** L_b, and the desired **probability of detection** (usually 50%–99%). For a given L_b, C_{thresh} has a minimum value at some optimal θ (i.e., the target size that is most easily detected). Threshold contrast also varies among observers and across time for a given observer. Because **detection** outdoors primarily depends on luminance rather than chromaticity differences, C_{thresh} is usually calculated from spectrally integrated luminances. For alerted observers (i.e., those expecting a target to appear) and daytime luminance levels, C_{thresh} ranges from ~0.005 to 5.0 or more. Unalerted observers may require C_{thresh} values that are five or more times greater than those for alerted observers.

threshold depth—Same as **sill depth**.

threshold illuminance—For a point **light** source at night, the smallest **illuminance** E_{thresh} at the observer that can be seen for a given surrounding **luminance** and state of **dark adaptation**.

Although the threshold illuminance is not a constant, a dark-adapted observer can usually see a nonflashing light that produces $\sim1.5\times10^{-7}$ **lux** at the eye. *See also* **Allard's law, night visual range**.

threshold of audibility—The level above which **sound** intensity, at any specified **frequency**, must rise in order to be detected by the average human ear.

This threshold value decreases with increasing frequency from a value of about one microwatt m^{-2} at 50 Hz to slightly less than 10^{-6} microwatts m^{-2} at about 2000 Hz, and then increases with increasing frequency to about 10^4 microwatt m^{-2} near 30 000 Hz. As the sound intensity rises above the threshold of audibility at any frequency, continued increase will finally raise the **intensity** to a second type of limit, the **threshold of pain** (threshold of discomfort, threshold of feeling). This threshold is rather insensitive to frequency, being found at about 10^6 microwatts m^{-2} for nearly all frequencies in the audible **range**. It represents the limit above which appreciable increase in sound intensity will lead to sensible pain in the average human ear. *See* **decibel**.

threshold of hearing—The smallest root-mean-square **sound** pressure that can be perceived by an average person without special aids.
  Usually specified as 0 dB, it corresponds to 2×10^{-5} N m^{-2} or 0.0002 microbar.

threshold of nucleation—The thermodynamic condition (**temperature, supersaturation/super-cooling**) at which the rate of formation of the nascent phase first becomes appreciable.
  See also **nucleation**.

threshold of pain—(Or threshold of discomfort, threshold of feeling.) The root-mean-square **sound** pressure at which an average person will begin to experience physical pain from ambient sound.
  It corresponds to a **sound pressure level** of 134 dB, that is, 100 N m^{-2} or 1 **millibar**.

threshold signal—(Also called minimum detectable signal.) A received radio **signal** (or **radar echo**) with **power** just above the **noise level** of the **receiver**.
  Compare **saturation signal**.

throughfall—Part of **rainfall** that reaches the ground directly through the vegetative **canopy**, through intershrub spaces in the canopy, and as **drip** from the leaves, twigs, and stems.

thrust anemometer—Same as **drag anemometer**.

thunder—The **sound** emitted by rapidly expanding gases along the channel of a **lightning discharge**.
  Some three-fourths of the electrical **energy** of a lightning discharge is expended, via ion–molecule collisions, in heating the atmospheric gases in and immediately around the **luminous** channel. In a few tens of microseconds, the channel rises to a local **temperature** of the order of 10 000°C, with the result that a violent quasi-cylindrical **pressure wave** is sent out, followed by a succession of rarefactions and compressions induced by the inherent elasticity of the air. These compressions are heard as thunder. Most of the sonic energy results from the return streamers of each individual **lightning stroke**, but an initial tearing sound is produced by the **stepped leader**; and the sharp click or crack heard at very close range, just prior to the main crash of thunder, is caused by the **ground streamer** ascending to meet the stepped leader of the first **stroke**. Thunder is seldom heard at points farther than 15 miles from the lightning discharge, with 25 miles an approximate upper limit, and 10 miles a fairly typical value of the **range** of audibility. At such distances, thunder has the characteristic rumbling sound of very low pitch. The pitch is low when heard at large distances only because of the strong **attenuation** of the high-frequency components of the original sound. The rumbling results chiefly from the varying arrival times of the sound waves emitted by the portions of the sinuous **lightning channel** that are located at varying distances from the **observer**, and secondarily from echoing and from the multiplicity of the strokes of a composite **flash**. *See* **electrometeor**.

thunderbolt—In mythology, a **lightning flash** accompanied by a material "bolt" or dart; this is the legendary cause of the damage done by **lightning**.
  It is still used as a popular term for a **lightning discharge** accompanied by **thunder**.
  Schonland, B. F. J., 1950: *The Flight of Thunderbolts*, 1–8.

thundercloud—A convenient and frequently used term for the **cloud** mass of a **thunderstorm**, that is, a **cumulonimbus**.

thunderhead—A popular term for the **incus** ("anvil") of a **cumulonimbus** cloud; or, less appropriately, the upper portion of a swelling **cumulus**, or the entire cumulonimbus.

thundersquall—Strictly, the combined occurrence of a **thunderstorm** and a **squall**, the squall usually being associated with the **downrush** phenomenon typical of a well-developed thunderstorm.

thunderstorm—(Sometimes called electrical storm.) In general, a **local storm**, invariably produced by a **cumulonimbus** cloud and always accompanied by **lightning** and **thunder**, usually with strong gusts of **wind, heavy rain**, and sometimes with **hail**.
  It is usually of short duration, seldom over two hours for any one **storm**. A thunderstorm is a consequence of atmospheric **instability** and constitutes, loosely, an overturning of air layers in order to achieve a more stable **density** stratification. A strong convective **updraft** is a distinguishing feature of this storm in its early phases. A strong **downdraft** in a column of **precipitation** marks its dissipating stages. Thunderstorms often build to altitudes of 40 000–50 000 ft in midlatitudes and to even greater heights in the **Tropics**; only the great **stability** of the lower **stratosphere** limits their upward growth. A unique quality of thunderstorms is their striking electrical activity.

The study of thunderstorm electricity includes not only lightning phenomena per se but all of the complexities of **thunderstorm charge separation** and all charge distribution within the realm of thunderstorm influence. In U.S. weather observing procedure, a thunderstorm is reported whenever thunder is heard at the **station**; it is reported on regularly scheduled observations if thunder is heard within 15 minutes preceding the **observation**. Thunderstorms are reported as light, medium, or heavy according to 1) the nature of the lightning and thunder; 2) the type and **intensity** of the precipitation, if any; 3) the speed and **gustiness** of the wind; 4) the appearance of the clouds; and 5) the effect upon **surface temperature**. From the viewpoint of the **synoptic** meteorologist, thunderstorms may be classified by the nature of the overall weather situation, such as **airmass thunderstorm**, **frontal thunderstorm**, and **squall-line thunderstorm**.

> Byers, H. R., and R. R. Braham Jr., 1949: *The Thunderstorm*, U.S. Government Printing Office, 287 pp.
> Byers, H. R., 1951: *Compendium of Meteorology*, p. 681.

thunderstorm cell—The **convective cell** of a **cumulonimbus** cloud having **lightning** and **thunder**.

thunderstorm charge—The existence of regions of net charge in a **thunderstorm**.

During **transient** collisions of **ice crystals** with riming **graupel** pellets, charge is transferred. The separating **particles** then carry equal and opposite charges; the larger (often negative) particles fall while the smaller ones (often positively charged ice crystals) are carried up in the **updraft** to produce a vertical **electric field** that eventually produces **lightning**. The charge **transfer** process is not completely understood, but possible processes include charges on the surface layers of the particles, charges on dislocations in the **ice** lattice, **temperature** differences along surface features that may be broken off during collisions, and contact **potential** differences between the surfaces of the interacting particles. *See* **breaking-drop theory, ion-capture theory**.

thunderstorm charge separation—The process by which the large electric fields found within thunderclouds are generated; the process by which **particles** bearing opposite electrical charge are given those charges and transported to different regions of the **active cloud**.

Accounting for the rapid and extensive separation of **electric charge** within thunderstorms is still one of the central problems in the study of **thunderstorm** electricity. Many theories have been proposed to explain **charge separation**, including the **breaking-drop theory**, the **ion-capture theory**, a theory involving the **Workman–Reynolds effect**, and a mechanism involving the bounce of **ice crystals** from growing **graupel**. None is entirely satisfactory in being able to account fully for the observed charge separation required to maintain a very active thunderstorm producing one **discharge** per second or so. Much evidence points toward particle-size difference and hence falling-speed difference as a necessary factor in the transportation of the oppositely charged particles in opposite directions in the updrafts of convective clouds, in regions where ice crystals are produced in the presence of graupel in regions between **updraft** and **downdraft**.

thunderstorm cirrus—Same as **cirrus spissatus**.

thunderstorm day—An **observational day** during which **thunder** is heard at the **station**. **Precipitation** need not occur.

thunderstorm dipole—The simplest representation of the electrostatic structure of an electrified **cloud** with overall charge neutrality.

Ordinary **thunderstorms** are characterized by upper positive charge and lower negative charge.

thunderstorm electrification—The process by which regions of net positive and negative **electric charges** are produced in **clouds**.

thunderstorm initiation mechanism—Rising motion associated with an atmospheric feature capable of releasing **convective instability** leading to **thunderstorm** development.

Examples are rising motion associated with **fronts**, the **dryline**, **gust fronts**, **upper-air disturbances**, heated elevated terrain, terrain–airflow interaction, and **sea breezes** and other mesocale circulations resulting from horizontal gradients in radiative properties of the underlying surface.

thunderstorm outflow—The relatively cool pool of air that results when a **thunderstorm** downdraft reaches the earth's surface and spreads horizontally as a **density current**.

See also **outflow boundary, gust front**.

thunderstorm tripole—A refinement of the simpler **dipole** representation for the electrostatic structure of isolated electrified clouds.

The tripole structure includes the lower positive charge center that appears in many observations.

thunderstorm turbulence—*See* **convective turbulence**.

TID—Abbreviation for **traveling ionospheric disturbances**.

tidal bore—*See* **bore**.

tidal component—Same as **partial tide**.

tidal constituent—*See* **harmonic analysis**.

tidal current—Same as **tidal stream**.

tidal day—Same as **lunar day**.

tidal excursion—The **Lagrangian** movement of a water particle during a tidal **cycle**.

tidal glacier—A **glacier** with its terminus in the ocean or sea.

tidal marsh—A low, flat marshland traversed by interlacing channels and tidal sloughs, and usually inundated by **tides**.

tidal prism—The volume of water exchanged between a lagoon or **estuary** and the open sea in the course of a complete tidal **cycle**.

tidal range—(Or tide range.) The difference in height between **low** and **high water** tidal levels, equal to twice the **tidal amplitude**.

tidal river—River or section of river (usually near the mouth) in which **water levels** and flow may be affected by the **tides**.

tidal stream—(Also called tidal current.) Horizontal water movements due to **tidal forcing**.

tidal wave—1. The **wave motion** of the tides.
2. In popular usage, any unusually high (and therefore destructive) **water level** along a shore. It usually refers to either a **storm surge** or **tsunami**.

tidal wind—A very **light breeze** that occurs in **calm** weather in inlets where the **tide** sets strongly; it blows onshore with **rising tide** and offshore with ebbing tide.

tide—1. The periodic rising and falling of the earth's oceans and **atmosphere**.
It results from the **tide-producing forces** of the moon and sun acting upon the rotating earth. This **disturbance** actually propagates as a **wave** through the **atmosphere** and along the surface of the waters of the earth. Atmospheric tides are always so designated, whereas the term "tide" alone commonly implies the oceanic variety. Sometimes, the consequent horizontal movement of water along the coastlines is also called "tide," but it is preferable to designate the latter as **tidal current**, reserving the name tide for the vertical wavelike movement. *See* **equatorial tide, neap tide, spring tide, tropic tide**.
2. *See* **rip current, red tide, storm tide**.

tide amplitude—One-half of the difference in height between consecutive **high water** and **low water**; hence, half of the **tidal range**.

tide crack—A crack between the moving **sea ice** and the unmoving **ice foot**.
It may widen to form a **shore lead**.

tide gauge—A device for measuring the height of **tide**.
It may be simply a graduated staff in a sheltered location where visual observations can be made at any desired time, or it may consist of an elaborate **recording instrument** (sometimes called a **marigraph**) making a continuous graphic record of tide height against time. Such an instrument is usually activated by a float in a pipe communicating with the sea through a small hole that filters out shorter waves.

tide gauge benchmark—A stable **benchmark** near a gauge to which **tide gauge** datum is referred.
It is connected to local auxiliary benchmarks to check local **stability** and to guard against accidental damage. The tide gauge datum is a horizontal plane defined at a fixed arbitrary level below a tide gauge benchmark.

tide-producing force—The slight local difference between the gravitational attraction of two astronomical bodies and the **centrifugal force** that holds them apart.
These forces are exactly equal and opposite at the **center of gravity** of either of the bodies, but, since gravitational attraction is inversely proportional to the square of the distance, it varies from point to point on the surface of the bodies. Therefore, gravitational attraction predominates at the surface point nearest to the other body, while centrifugal "repulsion" predominates at the surface point farthest from the other body. Hence there are two regions where tide-producing forces are at a maximum, and normally there are two tides each **lunar day** and **solar day**.

tide range—Same as **tidal range**.

tide tables—Annual tabulations of daily predictions of the times and heights of **high water** and **low water** at various places.

Such tables are constructed from astronomical data and from the results of **harmonic analyses** of previous observations at the desired point. They are compiled and issued by national **hydrographic** authorities, for example, the U.S. Coast and Geodetic Survey. The heights in tide tables are usually measured from **chart datum** rather than **mean sea level**.

tilt—In **synoptic meteorology**, the inclination to the vertical of a significant feature of the **circulation** (or **pressure**) pattern or of the **field** of **temperature** or moisture.

For example, **troughs** in the **westerlies** usually display a westward tilt with **altitude** in the lower and middle **troposphere**.

tilting term—Same as **twisting term**.

timber line—1. The poleward limit of tree growth.
See **arctic tree line**, **tree line**.

2. In mountainous regions, the line above which climatic conditions do not allow the upright growth of trees.

time—Duration as measured by some clock.

Atomic clocks give the most accurate measure of time. Less regular timekeepers are those based on the rotation of the earth and other bodies of the solar system.

time average—The mean of values measured over different times.
Compare **ensemble average**, **spatial average**, **volume average**, **ergodic condition**.

time constant—(Also called lag coefficient.) Generally, the time required for an instrument to indicate a given percentage of the final reading resulting from an **input** signal; the **relaxation time** of an instrument.

In the general case for instruments such as **thermometers**, with responses exponential in character to step changes in an applied **signal**, the time constant is equal to the time required for the instrument to indicate 63.2% of the total change, that is, the time to respond to all but $1/e$ of the original signal change.

time cross section—Same as **time section**.

time curve—(Also called time front, hour-out line.) In the **wave-front method** of **minimal flight** planning, a line through positions attainable in an equal flight time from a given origin.

time-dependent flow—A flow with a **velocity** field that changes with time.

time–distance graph—A graph used to determine the ground distance for air-route legs of a specified time interval.

Time–distance relationships are often simplified by considering air, **wind**, and ground distances for flight legs of 1-h duration; thus, the **hourly distance scale**.

time-domain averaging—*See* **coherent integration**.

time front—Same as **time curve**.

time–height indicator—(Abbreviated THI.) A type of **radar** display on which the **reflectivity**, **Doppler velocity**, or other properties of **echoes** are displayed as a function of time and height in rectangular coordinates.

The THI is ordinarily composed of successive observations with a vertically pointing **radar beam**. It may be an **intensity**-modulated **display** or may use a **scale** of colors to represent the values of the function displayed.

time–height section—A facsimile **trace** of a vertically directed **radar**, specifically, a **cloud-detection radar**.
See **time–height indicator**, **time section**.

time lag—The total time between the application of a **signal** to an instrument and the full indication of that signal within the **uncertainty** of the instrument.
Compare **time constant**, **rise time**, **relaxation time**.

time-of-arrival technique—The time-of-arrival technique refers to locating the source of an emitted **signal** from a precise recording of the time that a signal is observed.

For example, the time interval between an observed **lightning flash** and the arrival of the **thunder** can be used to estimate the distance to the lightning flash. On the average, a time arrival difference of five seconds indicates that a lightning flash occurred one **mile** away from the observer, since the **speed of sound** in air is approximately 1000 ft s^{-1}.

time of concentration—An expression of the length of time it takes for water to travel from some designated geographical location of the **watershed** to another, or from some identifiable time on the **histogram** of the runoff-causing event to the time of peak or centroid of the storm **hydrograph**.

time section—(Also called time cross section.) A diagram in which one coordinate is time and the other is distance (usually height, in which case it is a vertical time section).
> *Compare* **cross section**, **profile**, **time–height section**.

time series—The values of a **variable** generated successively in time.
> A continuous **barograph** trace is an example of a continuous time series, while a **sequence** of hourly pressures is an example of a discrete time series. Graphically, a time series is usually plotted with time as the **abscissa** and the values of the function as the **ordinate**. Time series may be either stationary or nonstationary. For **stationary time series** the actual dynamics that motivate the series are constant from one period to the next. For nonstationary time series the dynamics are continually changing and such series are less susceptible to **statistical** analysis.
> > Wadsworth, G. P., 1951: *Compendium of Meteorology*, 849–855.

tipping-bucket rain gauge—A **recording rain gauge** in which the water collected continuously drains through a funnel into one of a pair of chambers or buckets that are balanced bistably on a horizontal axis.
> When a predetermined amount of water has been collected, commonly 0.25 mm (0.01 in.) of **rain**, the bucket tips, spilling the water and placing the other bucket under the funnel. An electronic switch is excited each time the bucket tips so that both **rain rate** and accumulation can be determined from the record of tips.

TIROS—Abbreviation for **Television and Infrared Observation Satellite**.

TIROS-N Operational Vertical Sounder—(Abbreviated TOVS.) An **atmospheric sounding** system composed of three instruments carried on the **TIROS**-*N* and *NOAA*-*6* through -*14* **polar-orbiting satellites**.
> The **HIRS**, **MSU**, and **SSU** instruments compose the TOVS system.

TIROS Operational System—(Abbreviated TOS.) The name given to the initial system of operational **polar-orbiting meteorological satellites** in the United States, specifically *ESSA*-*1* to -*9*.
> *See* **TIROS**.

tivano—A night **breeze** blowing down the valley at Lake Como in Italy.

TKE—Abbreviation for **turbulence kinetic energy**.

tofan—(Also spelled tufon, tufan.) A violent spring storm common in the mountains of Indonesia.

tomography—The reconstruction of the **temperature** structure of the ocean from acoustic signals in multiple vertical planes.
> *See* **acoustic tomography**.

TOMS—Abbreviation for **Total Ozone Mapping Spectrometer**.

tongara—A hazy, southeast **wind** in the Macassar Strait.
> *See* **tenggara, broeboe**.

tongue—In **oceanography**, a three-dimensional, tonguelike intrusion of finite extent in the along-front direction.
> *See* **interleaving**.

top-down/bottom-up diffusion—An **atmospheric boundary layer** theory that splits **turbulent transport** into two components, one that is entrained into the **boundary layer** top and then diffused downward, and one that is injected from the earth's surface and brought upward into the boundary layer.
> These two linearly decomposed components of a passive **scalar** assume that the top-down component has zero **flux** at the surface while the bottom-up component has zero flux at the top. This theory allows some aspects of vertical **transport** across a **mixed layer** to be modeled, even though the actual vertical **gradient** of the passive scalar is zero, which is often the case for vigorous **buoyant convection**.

TOPEX—Acronym for **Ocean Topography Experiment**.

topographic amplification factor—A ratio, calculated solely from topographic (terrain elevation) data, of the **diurnal** temperature **amplitude** or **range** in a valley to that at the same **altitude** over the adjacent plain.
> The diurnal amplitude is larger (or "amplified") over the valley than over the plain because the

volume of air is less in the valley, but the amount of heating or cooling is approximately the same in both locations. The significance of the **temperature** differential caused by this amplification is that it produces a horizontal **pressure gradient** between the valley and the surrounding plains, or along the axis of the valley itself, that reverses twice per day, driving the diurnally varying **along-valley wind system**, upvalley during the day and downvalley at night. *See* **upvalley wind, down-valley wind**.

McKee, T. B., and R. D. O'Neal, 1989: The role of valley geometry and energy budget in the formation of nocturnal valley winds. *J. Appl. Meteor.*, **28**, 445–456.

Steinacker, R., 1984: Area-height distribution of a valley and its relation to the valley wind. *Contrib. Atmos. Phys.*, **57**, 64–71.

Whiteman, C. D., 1990: Observations of thermally developed wind systems in mountainous terrain. *Meteor. Monogr.*, No. 45, 9–13.

topographic waves—**Waves** with a restoring force arising from variations in depth.

The stretching or compression of displaced columns of water generates anomalous **vorticity** tending to drive them back to their original position.

topography—1. Generally, the disposition of the major natural and man-made physical features of the earth's surface, such as would be entered on a map.

This may include forests, rivers, highways, bridges, etc., as well as **contour lines** of **elevation**, although the term is often used to denote elevation characteristics (particularly **orographic** features) alone.

2. The study or process of topographic mapping.

topside sounder—A satellite designed to determine **ion concentration** within the **ionosphere** as measured from above the ionosphere.

tornadic vortex signature—(Abbreviated TVS.) The **Doppler velocity** signature of a **tornado** or of an incipient tornado-like **circulation** aloft.

As the **signature** occurs when the **radar beam** is wider than the **vortex**, the measured Doppler velocities are weaker than the **rotational** velocities within the vortex and the apparent core diameter is larger than that of the vortex. The signature, which may extend throughout a considerable vertical depth, is ideally characterized by extreme Doppler velocity values of opposite sign separated in **azimuth** by the equivalent of one **beamwidth**. However, since most radars display and record Doppler velocity values at discrete azimuthal intervals, the extreme Doppler velocity values are usually at azimuthally adjacent positions that are roughly one beamwidth apart. If the centers of the radar beam and the vortex coincide, the signature includes a zero Doppler velocity value that separates the extreme values.

tornado—1. A violently rotating column of air, in contact with the ground, either pendant from a **cumuliform** cloud or underneath a cumuliform cloud, and often (but not always) visible as a **funnel cloud**.

When tornadoes do occur without any visible funnel cloud, debris at the surface is usually the indication of the existence of an intense **circulation** in contact with the ground. On a local **scale**, the tornado is the most intense of all atmospheric circulations. Its **vortex**, typically a few hundred meters in diameter, usually rotates cyclonically (on rare occasions anticyclonically rotating tornadoes have been observed) with **wind speeds** as low as 18 m s^{-1} (40 mph) to wind speeds as high as 135 m s^{-1} (300 mph). Wind speeds are sometimes estimated on the basis of wind damage using the **Fujita scale**. Some tornadoes may also contain secondary vortices (**suction vortices**). Tornadoes occur on all continents but are most common in the United States, where the average number of reported tornadoes is roughly 1000 per year, with the majority of them on the central plains and in the southeastern states (*see* **Tornado Alley**). They can occur throughout the year at any time of day. In the central plains of the United States they are most frequent in spring during the late afternoon. *See also* **supercell tornado, nonsupercell tornado, gustnado, landspout, waterspout**.

2. A violent **thundersquall** in West Africa and adjacent Atlantic waters.

Tornado Alley—A term often used by the media to denote a zone in the Great Plains region of the central United States, often a north–south oriented region centered on north Texas, Oklahoma, Kansas, and Nebraska, where tornadoes are most frequent.

Since **statistics** are variable on all timescales, the term has little scientific value.

tornado belt—(Obsolete.) Same as **Tornado Alley**.

tornado cyclone—A term coined by Brooks (1949) to describe a surface low pressure area in a **convective storm** that, with its attendant winds, has a radius of about 8–16 km and is associated with, but is larger than, a **tornado**.

Starting with Agee (1976), the tornado cyclone has been redefined as a distinct **circulation** with a **scale** larger than that of the tornado but smaller than that of the **mesocyclone** (although embedded within it and smaller in scale than the **mesolow**). The intermediate-scale tornado cyclone is sometimes inferred from high-resolution **Doppler radar** observations, but at other times it is not apparent.

Brooks, E. M., 1949: The tornado cyclone. *Weatherwise*, **2**, 32–33.
Agee, E. M., 1976: Multiple vortex features in the tornado cyclone and the occurrence of tornado families. *Mon. Wea. Rev.*, **104**, 552–563.

tornado echo—A type of **precipitation echo** observed in connection with large, strong tornadoes located at relatively close ranges from a **radar**.
It appears on **PPI** displays in the form of a doughnut-shaped **echo** at the end of a **hook echo** in a **supercell** thunderstorm. A quasi-vertical tube having minimum radar **reflectivity** in the center may be observed on an **RHI** display or on a vertical sequence of PPI displays.

tornado outbreak—Multiple **tornado** occurrences associated with a particular **synoptic-scale** system.
In recent years, Galway (1977) has defined ten or more tornadoes as constituting an **outbreak**.
Galway, J. G., 1977: Some climatological aspects of tornado outbreaks. *Mon. Wea. Rev.*, **105**, 477–484.

torque—The **moment** of a force about a given point; that is, the **vector product** of the **position vector** (from the given point to the point at which the force is applied) and the force.
See **mountain torque, frictional torque.**

torrent—Flow with high **velocity** and great **turbulence.**

Torricelli's tube—An early and once universal name for the **mercury barometer.**

Torricellian vacuum—The "vacuum" above the column in a **mercury barometer.**

Torrid Zone—The zone of the earth's surface that lies between the **Tropic of Cancer** and the **Tropic of Capricorn.**
This is one of the three subdivisions of the **mathematical climate**; the other two are the **Temperate Zone** and the **Frigid Zone.**

torsion hygrometer—A **hygrometer** in which the rotation of the hygrometric element is a function of the **humidity.**
Such hygrometers are constructed by taking a substance in which length is a function of the humidity and twisting or spiraling it under tension in such a manner that a change in length will cause a further rotation of the element. *See* **hygroscope.**

TOS—Abbreviation for **TIROS Operational System.**

toscà—A southwest **wind** on Lake Garda in Italy.

total acidity—Amount of both weak and strong acids in water, usually expressed in milliequivalents of a strong base necessary to neutralize one liter of a water sample, using methyl-red or phenolphthalein as an indicator.

total cloud cover—Fraction of the sky hidden by all visible **clouds.**

total conductivity—In **atmospheric electricity**, the sum of the electrical conductivities of the positive and negative **ions** found in a given portion of the **atmosphere.**

total derivative—The rate of change of a function of two or more **variables** with reference to a **parameter** on which these variables are dependent.
If $z = f(x, y, t)$, $x = \phi(t)$, and $y = \theta(t)$, the total derivative of z with reference to t, written dz/dt, is given by

$$\frac{dz}{dt} = \frac{\partial f(x, y)}{\partial x} \frac{d\phi(t)}{dt} + \frac{\partial f(x, y)}{\partial y} \frac{d\theta(t)}{dt} + \frac{\partial f}{\partial t},$$

where $\frac{\partial f(x, y)}{\partial x}$ and $\frac{\partial f(x, y)}{\partial y}$ are the partial derivatives of $f(x, y)$ with respect to x and y, respectively; $\frac{d\phi(t)}{dt}$ and $\frac{d\theta(t)}{dt}$ are the total derivatives of $\phi(t)$ and $\theta(t)$, respectively, with respect to t. For a function of one variable, the **partial derivative** equals the total derivative. The **individual derivative** is the total derivative when the parameter t represents time.

total differential—The total differential of a function $f = f(x_1, x_2, \cdots, x_n)$ of n **variables** is defined by the equation

$$df = \frac{\partial f}{\partial x_1}\, dx_1 + \frac{\partial f}{\partial x_2}\, dx_2 + \cdots + \frac{\partial f}{\partial x_n}\, dx_n,$$

whether or not x_1, x_2, $\cdots$, and x_n are independent of each other. $\frac{\partial f}{\partial x_1}$, $\frac{\partial f}{\partial x_2}$, $\cdots$, and $\frac{\partial f}{\partial x_n}$ denote the partial derivatives of f with respect to x_1, x_2, $\cdots$, and x_n respectively. The total differential df represents the change in f associated with simultaneous infinitesimal changes in x_1, x_2, $\cdots$, and x_n given by dx_1, dx_2, $\cdots$, dx_n, respectively, to the order of dx_1, dx_2, $\cdots$, dx_n.

total emissive power—Same as **emittance**.

total energy equation—*See* **energy equation**.

total evaporation—*See* **evapotranspiration**.

total head—The sum of the elevation **head**, **pressure head**, and **velocity head**.

total head line—The line connecting the **total head** at different points along the flow.

total intensity—The integral of **radiance** over area, with units of W sr^{-1}.
Alternatively, the integral of **spectral radiance** over **wavelength**, with units of W m^{-2} sr^{-1}.

total ozone—The total amount of **ozone** present in a column of the earth's **atmosphere**, often expressed in **Dobson units**.

Total Ozone Mapping Spectrometer—(Abbreviated TOMS.) A satellite instrument that measures backscattered **ultraviolet radiation** to infer total column **ozone**.
TOMS was first flown on *Nimbus-7*, launched in November 1978, and has subsequently flown on *Meteor-3*, *ADEOS*, and other international satellites. The **antarctic ozone hole** was first discovered through analysis of TOMS data.

Total Ozone Monitoring Satellite—Satellite-borne **spectrometer** for measuring the total column amount of **ozone** above the earth.
The spectrometer works by measuring the amount of **ultraviolet radiation** scattered back into space at several wavelengths. Data from TOMS have been invaluable in evaluating long-term trends in global ozone and also annual events such as the **antarctic ozone hole**.

total potential energy—*See* **potential energy**.

total pressure—(Also called stagnation pressure.) The sum of the **static pressure** and the **dynamic pressure** when these concepts are applicable.
Since this is the **pressure** at the stagnation point of a **streamline**, it is measured by an ideal **Pitot tube** directed exactly **upstream**. The total pressure satisfies the **hydrostatic equation**.

total radiation—The integral of spectral **radiation** over all relevant wavelengths. This is most commonly applied to the sum of **shortwave** and **longwave** irradiances, but usage varies somewhat with context.

total reactive nitrogen—Collective name for oxidized forms of **nitrogen** in the **atmosphere** such as **nitric oxide** (NO), **nitrogen dioxide** (NO$_2$), **nitric acid** (HNO$_3$), and **organic nitrates**; usually designated by NO$_y$.
The species can all be readily converted to NO by a heated catalyst and detected by an NO **chemiluminescence** detector.

total scattering cross section—The integrated **scattering cross section** for all ϕ and θ.

total sky cover—*See* **sky cover**.

total solar irradiance—(Abbreviated TSI.) The amount of **solar radiation** received outside the earth's **atmosphere** on a surface normal to the incident **radiation**, and at the earth's mean distance from the sun.
Reliable measurements of solar radiation can only be made from space and the precise record extends back only to 1978. The generally accepted value is 1368 W m^{-2} with an **accuracy** of about 0.2%. Variations of a few tenths of a percent are common, usually associated with the passage of sunspots across the solar disk. The **solar cycle** variation of TSI is on the order of 0.1%.

Total Totals index—See **stability index**.

total variance—*See* **regression**.

total vorticity—Usually, the magnitude of the **vorticity** vector, all components included, as opposed to only the vertical component of the vorticity.

total water mixing ratio—The **mixing ratio** of the sum of vaporous, liquid, and solid water.

totalizer rain gauge—A **nonrecording rain gauge** designed to be used at stations that can be visited only infrequently.

The collection can may contain a solution with a low **freezing point** and a thin layer of oil to prevent **evaporation**. The **amount of precipitation** is found after subtracting the amount of solution.

> Middleton, W. E. K., and A. F. Spilhaus, 1953: *Meteorological Instruments*, 3d ed., rev., Univ. of Toronto Press, 286 pp.

Totals Indices—*See* **stability index**.

touriello—A south **wind** of **foehn** type descending from the Pyrenees in the Ariège valley, France.

It is especially violent in February and March when it melts the **snow**, **flooding** the rivers, and sometimes causing avalanches. It causes a premature spring that forces the buds on fruit trees, which are subsequently killed by **frost**. In August and September it comes as a strong drying wind, generally lasting three to four days.

Toussaint's formula—A rule proposed by Toussaint for the **linear** decrease of **temperature** with height in an **atmosphere** for which the temperature at **mean sea level** is 15°C.

It is given by

$$t = 15 - 0.0065 \, z,$$

where t is the temperature in degrees Celsius and z is the geometric height in meters above mean sea level. Toussaint's formula is used to determine temperature below 11 000 m in the **ICAO Standard Atmosphere**.

TOVS—Abbreviation for **TIROS-N Operational Vertical Sounder**.

Tower of the Winds—An octagonal marble building in Athens erected not later than 35 BC and still standing.

The sides face the points of the Athenian **compass** and carry a frieze of male personifications of the **winds** from those directions: **Boreas** (N), **Kaikias** (NE), **Apheliotes** (E), **Euros** (SE), **Notos** (S), **Lips** (SW), **Zephyros** (W), and **Skiron** (NW). These figures are reproduced on the tower of the Radcliffe Observatory, Oxford, United Kingdom, and in the Library of the Blue Hill Observatory near Boston, Massachusetts. The Tower was not a **meteorological observatory**, though it originally carried a **wind vane** on the roof, but was built to measure time, the walls bearing sundials, with a water clock inside for use during **cloudy** weather.

tower visibility—*See* **control-tower visibility**.

towering—A **mirage** in which the angular height of the **image** is greater than that of the object.

As the width is unaffected (the angular width of image width remains that of the object because the **refractive index** gradient is vertical), the **aspect ratio** is altered and distant images appear vertically enlarged. Towering often accompanies **sinking**—distant features appear depressed and enlarged—but it can also accompany **looming**. *Compare* **stooping**.

towering cumulus—A descriptive term, used mostly in weather observing, for **cumulus congestus**.

Townsend support—A fixed support for mounting **maximum** and **minimum thermometers** of the **liquid-in-glass** type.

The support holds the thermometers at the correct operating attitude and also permits their rotation for resetting when desired.

toxic pollutants—Substances that are known or suspected to cause cancer or other serious health problems, such as birth defects.

Regulations concern the concentration of these substances at the source of emission. *Compare* **criteria pollutants**.

TR tube—Abbreviation for **transmit–receive tube**.

trace—1. In general, an unmeasurable (less than 0.01 in.) quantity of **precipitation**.

2. An insignificantly small quantity.

3. The record made by any self-registering instrument.

Thus, one may speak of the **barograph** trace, the **hygrothermograph** trace, etc.

trace atmospheric constituents—Same as **trace gas**.

trace gas—Chemical present in the **atmosphere** at a very low level (typically parts per million or less), usually because of its very reactive nature, or because of a very low production or **emission rate**.

Examples include **ozone, nitrogen oxides**, and **carbon monoxide**.

trace recorder—Same as **ombrometer**.

tracer—1. A chemical or thermodynamic property of the flow that is conserved during **advection**.

It can be used to track air-parcel movement and to identify the origins of **air masses**. Examples are **absolute humidity, equivalent potential temperature, radioactivity**, and **CCN** composition.

2. Any substance in the **atmosphere** that can be used to track the history of an **air mass**.

It can be chemical or radioactive in nature. The main requirement for a tracer is that its lifetime be substantially longer than the **transport** process under study. An example of an inert chemical tracer is SF_6, which is often released during a field experiment and measured at a later time to assess the extent of dilution of the air mass. Chemicals such as **methane** (CH_4) and **nitrous oxide** (N_2O), which are released at the earth's surface and destroyed slowly in the atmosphere, can be used to infer vertical rates of transport. CO released in the **boundary layer** can be used to trace transport in **convection**. Radioactive tracers such as ^{14}C and ^{90}Sr have been used to test models of stratospheric **circulation**. Certain atmospheric gases have also been used as tracers in ocean waters, for example, the **chlorofluorocarbons**.

track of a depression—A plot of the locations of a **depression**, or area of low pressure.

track wind—In air navigation, **wind direction** and **speed**, either observed or forecast if so specified, over a fixed air route or segment of a route for a designated **altitude**.

Compare **spot wind, sector wind**.

tracking radar—A **radar** that is primarily used to automatically track the position of nonmeteorological targets that are usually small relative to the radar pulse volume, for example, an aircraft or **balloon**.

tracking system—General name for apparatus, such as a **tracking radar**, used in following and recording the position of objects in the sky.

A **theodolite** and an **observer** form an optical tracking system used in **pilot balloon** runs.

tractive force—Force, parallel to the streambed, exerted by flowing water on a **sediment** particle at rest.

trade—1. *See* **trade winds**.

2. Of or pertaining to the **trade winds** or the region in which the trade winds are found.

trade air—The type of air, usually warm and moist, of which the **trade winds** consist.

Its chief thermodynamic characteristic is the presence of the **trade-wind inversion**. *See* **tropical air**.

trade cumulus—Same as **trade-wind cumulus**.

trade inversion—A characteristic **temperature inversion** usually present in the trade-wind streams over the eastern portions of the tropical oceans.

It is found in large-scale subsiding flows constituting the descent branches of the **Hadley cell** and **Walker circulation**. The **subsidence** warming in the **inversion layer** is balanced by **radiative cooling** and **evaporation** from the tops of **trade** cumuli. The height of the base of this **inversion** varies from about 500 m at the eastern extremities of the subtropical highs to about 2000 m at the western and equatorial extremities. In the **equatorial trough** zone and over the western portions of the **trade-wind belt**, the inversion does not exist as a mean condition, although it appears in certain weather patterns. The strength of the inversion varies enormously, occasionally being more than 10°C over 1 km, but sometimes being absent altogether, especially in the Northern Hemisphere. The inversion is generally strongest when the height of its base is lowest, and vice versa. The thickness of the inversion layer varies from only a few meters to more than 1000 m. On the average its **thickness** is about 400 m. The airflow below the inversion is very moist and filled with **cumulus** clouds (trade cumuli). Above it, the air is warm and exceedingly dry; this structure is so characteristic of the trade current that tropical analysts think of the tropical **troposphere** as consisting of a lower moist and an upper dry layer.

Riehl, H., 1954: *Tropical Meteorology*, ch. II.

trade-wind belt—The latitudes occupied by the **trade winds**, generally between about 30° in the winter hemisphere and 35° in the summer hemisphere.

trade-wind cumulus—(Or trade cumulus.) The characteristic **cumulus** cloud of the **trade winds** over the oceans in average, undisturbed weather conditions.

These clouds are generally 5000–7000 ft thick at peak development and are based at about 2000–2500 ft **altitude**. The individual **cloud** usually exhibits a blocklike appearance since its vertical growth ends abruptly in the lower stratum of the **trade-wind inversion**. A group of fully grown clouds show considerable uniformity in size and shape.

trade-wind desert—An area of very little **rainfall** and high temperature that occurs where the **trade winds** or their equivalent (such as the **harmattan**) blow over land.

The best examples are the Sahara and Kalahari deserts. The trade winds, blowing from higher latitudes, are very drying, and **cloudiness** is almost absent in these **desert** regions.

trade-wind front—**Front** occurring in the warm **season** between a flow of oceanic **trade winds** and warm air from the continents.

trade-wind inversion—*See* **trade inversion**.

trade winds—1. (Commonly called trades.) The **wind** system, occupying most of the **Tropics**, that blows from the **subtropical highs** toward the **equatorial trough**; a major component of the **general circulation of the atmosphere**.

The winds are northeasterly in the Northern Hemisphere and southeasterly in the Southern Hemisphere; hence they are known as the **northeast trades** and **southeast trades**, respectively. The trade winds are best developed on the eastern and equatorial sides of the great subtropical highs, especially over the Atlantic. In the Northern Hemisphere they begin as north-northeast winds at about latitude 30°N in January and latitude 35°N in July, gradually **veering** to northeast and east-northeast as they approach the **equator**. Their southern limit is a few degrees north of the equator. The southeast trades occupy a comparable region in the Southern Hemisphere and similarly change from south-southeast on their poleward side to southeast near the equator. In the Pacific, the trade winds are properly developed only in the eastern half of that ocean, and in the Indian Ocean, only south of about 10°S. They are primarily surface winds, their usual depth being from 3000 to 5000 ft, although they sometimes extend to much greater altitudes. They are characterized by great **constancy** of direction and, to a lesser degree, speed; the trades are the most consistent wind system on earth. *See* **antitrades**, **tropical easterlies**, **equatorial easterlies**.

2. A name given to the prevailing **westerlies** in California and to the northwest winds that blow in Oregon in summer.

trades—Common contraction for **trade winds**.

traersu—A violent east **wind** of Lake Garda in Italy.

trailing flare—A pyrotechnic device used in **weather modification** or **cloud seeding** to produce a controlled release of **seeding agent** from an aircraft.

trailing front—A **cold front** with a large longitudinal extent.

trajectory—(Or path.) A curve in space tracing the points successively occupied by a **particle** in motion.

At any given instant the **velocity** vector of the particle is tangent to the trajectory. In steady-state flow, the trajectories and streamlines of the fluid parcels are identical. Otherwise, the curvature of the trajectory K_T is related to the curvature of the **streamline** K_s by

$$K_T = K_s - \frac{1}{V}\frac{\partial \psi}{\partial t},$$

where V is the **parcel** speed and $\partial \psi / \partial t$ is the **local change** of the **wind direction**. The curvatures and **wind** change are positive for the **cyclonic** sense of flow.

tramontana—A cold **wind** from the northeast or north, particularly on the west coast of Italy and northern Corsica, but also in the Balearic Islands and the Ebro Valley in Catalonia.

Like the **mistral**, it is associated with the advance of an **anticyclone** from the west following a **depression** over the Mediterranean. Weather is fine with occasional **instability showers**. In Languedoc and Roussillon (southern France) a similar wind (tramontane) blows from the northwest, but the name is also applied to an invasion of **polar air** from the northwest, which is squally or tempestuous, dry, and cold except south of the Cévennes where it becomes **foehn**-like. This type occurs during the **filling of a depression** in the Gulf of Genoa and persists for eight to twelve days, mainly in winter and early spring; it rises to a peak at midday and weakens at night. On the Côte d'Azur and in eastern Provence, the tramontane is sometimes called the montagnère or montagneuse.

transducer—A device for converting **energy** from one form to another.

For example, a **thermocouple** transduces **heat** energy into electrical energy.

transfer—*See* **energy transfer, conduction, mixing, exchange coefficients, transport.**

transfer velocity—A **velocity** scale used in the **transport** of **trace** constituents across the air–sea **interface**, given by

$$F = v(C_w - C_a),$$

where F is the **flux** of constituent C across the interface and v is the transfer velocity.

The indices a and w refer to air and water, respectively. The constituent C_w must be considered the concentration in the air in **equilibrium** with the water.

Kraus, E. B., and J. A. Businger, 1994: *Atmosphere–Ocean Interaction*, Oxford University Press, p. 163.

transient—Varying in time, as opposed to **steady state.**

transient climate response—The time-dependent response of the **climate system** to a change in external forcing as simulated by a **climate model**.

The transient climate response differs from the **equilibrium** response because of the varying time constants of different components of the climate system.

transient eddies—In contrast to stationary or **standing eddies**, the component of the **eddy** field that varies with time.

In this context "eddy" refers to departures from the **zonal** mean, such as the **migratory** cyclones and anticyclones in the extratropics.

transient problem—The study of the evolution of a time-dependent dynamic system from an **equilibrium** state A to another equilibrium state B.

The different equilibrium states may be two different equilibrium states of a **dynamical system** having multiequilibrium states, or different equilibrium states that are due to change(s) in either the internal or the external characteristics of the dynamical system.

transient thermocline—A thermally stratified layer with a location in the water column that changes with time.

Transient thermoclines are formed during daytime periods and are subsequently mixed down to the permanent or main **thermocline**.

transilient matrix—*See* **transilient turbulence theory.**

transilient turbulence theory—A method for parameterizing **turbulence** that allows nonlocal vertical **mixing** between every pair of **grid** points in a vertical column, even between nonneighboring points.

This method can account for the advective-like **turbulent transport** within large coherent turbulence structures such as **thermals**, where large diameter (1 km) **updraft** cores **transport** air from near the surface to the top of the **mixed layer** with little or no dilution. The method also parameterizes the mixing effects of medium and small size **eddies**, so it gives a physical-space representation of a **spectrum** of turbulence wavelengths. The framework for this **parameterization** is a matrix equation:

$$S_i(t + \Delta t) = \sum_{j=1}^{n} c_{ij}(t, \Delta t) \, S_j(t),$$

where S_j is the initial value at time t of any **scalar** such as **potential temperature, specific humidity,** or **wind** velocity component at any source grid point j, and S_i is the final value after timestep Δt at destination grid point i. The matrix c_{ij} is called a transilient matrix and indicates the fraction of the air ending at destination grid cell i that came from source grid cell j. The equation is summed over all grid points n representing a column of air. Transilient turbulence theory is called a nonlocal, **first-order turbulence closure**. *See* **nonlocal mixing, nonlocal flux;** *compare* **K-theory.**

Stull, R. B., 1993: Review of nonlocal mixing in turbulent atmospheres. *Bound.-Layer Meteor.*, **62**, 21–96.

transition layer—1. The thin layer that separates thicker layers of different characteristics.

2. The **capping inversion** or **entrainment zone** at the top of the **convective mixed layer**.

3. The **statically stable** layer near the base of convective clouds in the **Tropics**.

4. A stratified layer of a body of water between the **mixed layer** and the undisturbed fluid beneath it; it refers to the uppermost (closest to the surface) **thermocline** at any given time.

transition state theory—Same as **activated complex theory.**

transition to turbulence—The evolution from a **laminar** to a **turbulent** state, usually via a sequence of distinct bifurcations.

transitional flow—A flow in which the **viscous** and **Reynolds stresses** are of approximately equal magnitude.

It is transitional between **laminar flow** and **turbulent flow**. *See* **laminar boundary layer**, **turbulent boundary layer**.

translatory wave—Same as **wave of translation**.

translucidus—A **cloud variety** occurring in a layer, patch, or extensive sheet, the greater part of which is sufficiently translucent to reveal the position of the sun, or through which higher clouds may be discerned.

This variety is found in the genera **altocumulus, altostratus, stratocumulus,** and **stratus,** and is usually a modification of the species **stratiformis** or **lenticularis.** (Note: With the exception of **cirrus spissatus,** all **cirriform** clouds are inherently translucent.) *See* **cloud classification.**

transmissibility—*See* **transmissivity.**

transmission coefficient—(Also Fresnel coefficient.) Sometimes the ratio of the **amplitude** of the transmitted **electric field** to the amplitude of the field incident at the **optically smooth** planar **interface** between two **optically homogeneous** media.

The incident and transmitted fields are plane **harmonic** waves and the interface is large in lateral extent compared with the **wavelength** of the **illumination.** May also be the ratio of transmitted to incident **irradiances,** that is, the ratio of the normal (to the interface) component of the transmitted **Poynting vector** to that of the incident Poynting vector. A better term for this quantity is **transmissivity.** The transmission coefficient (and hence transmissivity) depends on the **angle of incidence** of the illumination, its wavelength (by way of the wavelength-dependence of the relative **refractive index** of the two media), and its state of **polarization** (*see also* **reflection coefficient**). These coefficients taken together are sometimes called the "Fresnel formulae" or "Fresnel relations." Transmission coefficient may mean the ratio of any transmitted to incident irradiance (transmissivity). Transmission coefficient, transmissivity, **transmittance,** and **transmission function** are used more or less synonymously but not always consistently. Within the same work two or more of these terms may mean the same physical quantity. Context is a guide to the exact meaning of different authors in different fields and on different occasions, but is not always sufficient to decipher them. Moreover, these terms are not restricted to **electromagnetic waves** but may be applied to transmission of **acoustic** and other waves.

transmission function—*See* **transmission coefficient, transmissivity.**

transmission loss—A general term for the reduction of **power** in a transmitted radio **signal** resulting from any or all of such effects as **range attenuation, precipitation attenuation, multipath transmission,** etc.

transmission range—Same as **night visual range.**

transmissivity—1. Often the ratio of any transmitted to incident **irradiance.**
See also **transmission coefficient.**

2. (Also called transmissibility.) Measure of **aquifer** permeability, defined as the volume of water passing through a vertical surface of unit width per unit **head** gradient across the surface.

transmissometer—(Or telephotometer; also called transmittance meter, hazemeter.) An instrument for measuring the **extinction coefficient** of the **atmosphere** and for the determination of **visual range.**
See **runway visual range, visibility meter.**

transmissometry—The technique of determining the **extinction** characteristics of a medium by measuring the transmission of a **light** beam of known initial **intensity** directed into that medium.

transmit–receive tube—(Abbreviated TR tube.) A gas-filled **waveguide** cavity that acts as a short circuit when ionized by high-power **energy** but is transparent to low-power emission when not ionized.

It is used as a switch to protect the **receiver** of a **radar** from the high power of the **transmitter** while passing the low-power signals received at the **antenna.** *See* **recovery time.**

transmittance—Often synonymous with **transmissivity,** but may also be used for transmitted **irradiance.**

transmittance meter—Same as **transmissometer.**

transmitted power—In radar, the **power** that is transmitted from the **antenna** into space.

For a **pulsed radar,** the **peak power** transmitted is usually much higher than the **average**

power transmitted. The ratio of the average power to the peak power equals the duty cycle, which is the product of the **pulse duration** and the **PRF**. *Compare* **received power**.

transmitter—A device used for the generation of signals of any type and form that are to be transmitted.

In radio and **radar**, it is that portion of the equipment that includes electronic circuits designed to generate, amplify, and shape the **radio frequency** energy that is delivered to the **antenna** where it is radiated out into space. *See* **receiver**.

transosonde—The flight of a **constant-level balloon**, the **trajectory** of which is determined by ground tracking equipment.

Thus, it is a form of **upper-air**, quasi-horizontal "sounding." The most usual observations are of successive positions of the balloon located by radio direction-finding or radio-navigation equipment, giving trajectory, **wind speed**, and **wind direction**. Instrumentation can also be added to the balloon to sense and transmit **pressure**, **temperature**, **relative humidity**, and other meteorological elements.

transparent sky cover—In U.S. weather observing practice, that portion of **sky cover** through which higher clouds, blue sky, etc., may be observed; opposed to **opaque sky cover**.
Compare **thin**.

transpiration—The process by which water in plants is transferred as **water vapor** to the **atmosphere** from a single leaf, or the amount of water so transferred.

transpiration ratio—The ratio of weight of water transpired by a plant during its **growing season** to the weight of dry matter produced (usually exclusive of roots).

transponder—A device that relays electrical signals not necessarily in the same form or on the same **frequency** as received.

transport—The movement of a substance or characteristic.

Characteristics that can be transported in the **atmosphere** are **heat** (**temperature**), moisture, **momentum**, chemicals, **turbulence**, etc. The transport is sometimes interpreted as a **flux density** (characteristic per unit area per time), or as a flow rate (characteristic per time). *See* **transport processes**.

transport processes—Physical methods of moving substances or characteristics.

Methods of **transport** in the **atmosphere** include **turbulent transport** by **eddy** winds and advective transport by the mean wind.

transverse cirrus banding—Irregularly spaced bandlike **cirrus** clouds that form nearly perpendicular to a **jet stream axis**.

They are usually visible in the strongest portions of the subtropical **jet** and can also be seen in **tropical cyclone** outflow regions.

transverse velocity—The component of the **wind vector** that is normal to the radial viewing direction of a **Doppler radar** or **Doppler lidar** and therefore is not measurable.

Various assumptions and techniques are employed to estimate this missing component. *See also* **Doppler velocity**.

transverse wave—A solenoidal plane **wave A** perpendicular to the propagation direction in the sense that if $\mathbf{A} = \mathbf{A}_0 \exp(i\mathbf{k} \cdot \mathbf{x} - i\omega t)$, where $\mathbf{k}$ is the **wave vector**, then $\mathbf{k} \times \mathbf{A} = 0$.

Electromagnetic waves are examples of transverse waves. **Acoustic waves** in solids have both **irrotational** and **solenoidal** components that propagate with different **phase velocities**. *Compare* **longitudinal wave**.

travel time—The time of flow of a **parcel** or packet of water, **contaminant**, or **tracer** from one point to another.

traveling ionospheric disturbances—(Abbreviated TID.) Disturbances in the **electron density** of the **ionosphere** that are observed to propagate for large distances (thousands of kilometers).

They are detected by ionospheric **sounding** techniques and are now attributed to the effects of **gravity waves** that have grown to large amplitudes in propagating upward into the ionosphere.

traveling wave tube—(Abbreviated TWT.) A **linear** beam amplifier tube used in medium-power **radar** transmitters.

In a TWT, a stream of electrons interacts continuously or repeatedly with a guided **electromagnetic wave** moving substantially in synchronism with it and in such a way that there is a net **transfer** of **energy** from the stream to the wave. A TWT is characterized by high bandwidth but has somewhat lower **gain**, **power**, and efficiency than a **klystron**.

traverse—A westerly **wind** in central France.

It is moderate to strong, generally squally, humid, and thundery in summer, especially on slopes facing west; it is cold in winter and spring and brings **snow** or **hail** showers. In Auvergne it brings continuous **rain** (the Grand Vent or plouazaou). In the Alps it is cold and squally. Houses in regions subject to the traverse are protected by roughcast or double walls, and in the Mont Cantal part of Auvergne, where it is especially violent, zinc sheets are used on the walls; roofs are built of large stone slabs.

traversia—A South American nautical term (especially in Chile) for a west **wind** from the sea; a "side" wind.

traversier—In the Mediterranean, dangerous **winds** blowing directly into port.

tree climate—In W. Köppen's 1936 **climatic classification**, any type of **climate** that supports the growth of trees. This includes the **tropical rainy climates**, **temperate rainy climates**, and **snow-forest climates**. Excluded are the **dry climates** and **polar climates**.

Tree climates are separated from polar climates by the **isotherm** representing the warmest month **mean temperature** of 10°C (50°F). They are separated from the dry climates by a certain value of annual **precipitation** as a function of **temperature**. In this latter sense they are also known as **rainy climates**. *See* **Köppen–Supan line, arctic tree line**.

> Köppen, W. P., and R. Geiger, 1930–1939: *Handbuch der Klimatologie*, Berlin: Gebruder Borntraeger, 6 vols.

tree line—The poleward limit of tree growth; the botanical boundary between **tundra** and **boreal forest**.

The **arctic tree line** has been studied extensively, but in the Southern Hemisphere, a "tree line" can be inferred only by comparison of vegetation on islands in the southern oceans. *See also* **timber line**.

tree ring—Variable width of rings produced by seasonal growth as observed in the horizontal cross section cut from a tree trunk.

The number of rings observed corresponds to the age of the tree. Tree rings are used to date past **climate** events in **dendrochronology** work.

tree-ring climatology—*See* **dendrochronology**.

trend line—In **climatology**, a plot of monthly mean values.

triboelectrification—A process of **charge separation** that involves the rubbing together of dissimilar material surfaces.

The triboelectric series is a classification scheme for the ordering of the tendency for positive charge acquisition in rubbing. The detailed physical mechanism in triboelectrification is a long unsolved problem.

tributary—A **stream** that flows into a larger stream or lake.

trihedral reflector—*See* **corner reflector**.

triple correlation—A third-moment **statistic** found by the product of three perturbations (deviations from mean values) averaged over all observations.

This is used in **turbulence** theory to describe the **turbulent transport** of turbulent quantities. For example, the vertical **turbulent flux** of temperature **variance** is $\overline{w'T'T'}$, where w' is perturbation vertical **velocity**, and T' is **perturbation** temperature.

triple-Doppler analysis—A **radar** analysis technique that makes use of **radial velocity** measurements by three or more **Doppler radars** to deduce the three-dimensional **velocity** vector of radar **echoes**.

Such observations allow more accurate determination of the three-dimensional velocity than **dual-Doppler analysis**, which requires additional assumptions about **boundary conditions** to estimate the **vertical velocity** component.

triple point—1. Same as **triple state**.

2. A junction point within the **tropics** of three distinct **air masses**, considered to be an ideal point of origin for a **tropical cyclone**.

triple register—*See* **multiple register**.

triple scalar product—The **scalar** $\mathbf{A} \cdot (\mathbf{B} \times \mathbf{C})$ written $(\mathbf{ABC})$ or $[\mathbf{ABC}]$, where $\mathbf{A}$, $\mathbf{B}$, and $\mathbf{C}$ are any three vectors.

The dot denotes a **scalar product** and the cross a **vector product**. When $\mathbf{A}$, $\mathbf{B}$, and $\mathbf{C}$ are written in terms of their components along the x, y, and z axes of the **rectangular Cartesian**

coordinates, that is, $\mathbf{A} = a_1\mathbf{i} + a_2\mathbf{j} + a_3\mathbf{k}$, $\mathbf{B} = b_1\mathbf{i} + b_2\mathbf{j} + b_3\mathbf{k}$, and $\mathbf{C} = c_1\mathbf{i} + c_2\mathbf{j} + c_3\mathbf{k}$, the triple scalar product is the determinant

$$(\mathbf{ABC}) = (\mathbf{CAB}) = (\mathbf{BCA}) = |a_1 a_2 a_3 b_1 b_2 b_3 c_1 c_2 c_3|.$$

Any cyclic change among the vectors in a triple product does not alter its value.

triple state—(Or triple point.) The thermodynamic state at which three phases of a substance exist in **equilibrium**.

For water substance, the triple state occurs at $e_s = 0.611$ kPa, and $T = 273.16$ K, where e_s is the **saturation vapor pressure** of the **water vapor** and T the Kelvin **temperature**. At this state the **specific volumes** of the water vapor, water, and **ice** are, respectively,

$$\alpha_v = 206.3 \text{ m}^3\text{kg}^{-1},$$

$$\alpha_w = 0.00100 \text{ m}^3\text{kg}^{-1},$$

$$\alpha_i = 0.001091 \text{ m}^3\text{kg}^{-1}.$$

tritium—A radioactive **isotope** of **hydrogen**, symbol H^3 or T, with a **half-life** of about 12 years.

Tritium is formed by **cosmic rays** at levels near the **tropopause** and diffuses slowly into the **lower atmosphere**. It is also deposited in the **atmosphere** by nuclear detonations. Its **radioactivity** and relatively short half-life make it useful in certain geochronologic studies.

TRMM—Abbreviation for **Tropical Rainfall Measuring Mission**.

tromba—A **whirlwind** of Malta.

Tropic of Cancer—The northern parallel of maximum solar **declination**, approximately 23°27′N latitude.

See **obliquity of the ecliptic**.

Tropic of Capricorn—The southern parallel of maximum solar **declination**, approximately 23°27′S latitude.

See **obliquity of the ecliptic**.

tropic tide—**Tide** occurring when the moon is near maximum **declination**; the **diurnal inequality** is then at a maximum.

Compare **equatorial tide, spring tide, neap tide**.

tropical air—A type of **air mass** with characteristics developed over low latitudes.

Maritime tropical air (mT), the principal type, is produced over the tropical and subtropical seas. It is very warm and humid and is frequently carried poleward on the western flanks of the **subtropical highs**. Continental tropical air (cT) is produced over subtropical arid regions and is hot and very dry. *See* **airmass classification**, **trade air**; *compare* **polar air**.

tropical climate—In general, a **climatic zone** with a **climate** typical of equatorial and tropical regions; that is, one with continually high **temperatures** and with considerable **precipitation**, at least during part of the year.

See **tropical rainy climate, tropical rain forest climate, tropical monsoon climate, tropical savanna climate, megathermal climate, climatic classification**.

tropical continental air—A **continental air** mass that develops over or near tropical regions, typically equatorward of 30° latitude.

tropical cyclone—The general term for a **cyclone** that originates over the tropical oceans.

This term encompasses tropical depressions, tropical storms, hurricanes, and typhoons. At maturity, the tropical cyclone is one of the most intense and feared storms of the world; winds exceeding 90 m s^{-1}(175 knots) have been measured, and its rains are torrential. Tropical cyclones are initiated by a large variety of disturbances, including **easterly waves** and **monsoon troughs**. Once formed, they are maintained by the extraction of **latent heat** from the ocean at high temperature and **heat** export at the low temperatures of the tropical upper **troposphere**. After formation, tropical cyclones usually move to the west and generally slightly poleward, then may "recurve," that is, move into the midlatitude **westerlies** and back toward the east. Not all tropical cyclones recurve. Many dissipate after entering a continent in the **Tropics**, and a smaller number die over the tropical oceans. Tropical cyclones are more nearly circularly symmetric than are **frontal**

cyclones. Fully mature tropical cyclones range in diameter from 100 to well over 1000 km. The surface winds spiral inward cyclonically, becoming more nearly circular near the center. The **wind field** pattern is that of a circularly symmetric spiral added to a straight current in the direction of propagation of the cyclone. The winds do not converge toward a point but rather become, ultimately, roughly tangent to a circle bounding the **eye** of the storm. Pressure gradients, and resulting winds, are nearly always much stronger than those of **extratropical storms**. The **cloud** and **rain** patterns vary from storm to storm, but in general there are spiral bands in the **outer vortex**, while the most intense rain and winds occur in the **eyewall**. Occasionally, multiple eyewalls occur and evolve through a **concentric eyewall cycle**. Tropical cyclones are experienced in several areas of the world. In general, they form over the tropical oceans (except the South Atlantic and the eastern South Pacific) and affect the eastern and equatorward portions of the continents. They occur in the tropical North Atlantic (including the Caribbean Sea and Gulf of Mexico), the North Pacific off the west coast of Mexico and occasionally as far west as Hawaii, the western North Pacific (including the Philippine Islands and the China Sea), the Bay of Bengal and the Arabian Sea, the southern Indian Ocean off the coasts of Madagascar and the northwest coast of Australia, and the South Pacific Ocean from the east coast of Australia to about 140°W. By international agreement, tropical cyclones have been classified according to their **intensity** as follows: 1) **tropical depression**, with winds up to 17 m s^{-1}(34 knots); 2) tropical storm, with winds of 18–32 m s^{-1}(35–64 knots); and 3) severe tropical cyclone, **hurricane** or **typhoon**, with winds of 33 m s^{-1}(65 knots) or higher. It should be noted that the **wind speeds** referred to above are 10-min average wind speeds at **standard** anemometer level (10 m), except that in the United States, 1-min average wind speeds are used.

tropical cyclone classification system from satellite imagery—*See* **Dvorak technique**.

tropical cyclone twins—Two counterrotating **tropical cyclones** straddling the **equator**.
    These occur in the tropical western Pacific and Indian Oceans, and are usually accompanied by westerly **wind** bursts on and near the equator.

tropical depression—A **tropical cyclone** with a closed **wind** circulation and maximum surface winds up to 17 m s^{-1} (34 knots).

tropical disturbance—A **migratory**, organized region of **convective showers** and **thunderstorms** in the **Tropics** that maintains its identity for at least 24 hours but has no closed **wind** circulation.
    The system may or may not be associated with a detectable **perturbation** of the low-level wind or **pressure field**.

tropical easterlies—(Also called subtropical easterlies.) A term applied to the **trade winds** when they are shallow and exhibit a strong vertical **shear**.
    With this structure, the **easterlies** give way at about 1.5 km to the upper **westerlies (antitrades)**, which are sufficiently strong and deep to determine **cloudiness** and weather. The tropical easterlies occupy the poleward margin of the **Tropics** in summer and can cover most of the tropical belt in winter. *Compare* **equatorial easterlies**.

tropical front—*See* **intertropical front**.

tropical maritime air—A **maritime air** mass that develops over or near tropical regions, typically equatorward of 30° latitude.

tropical meteorology—The study of the tropical **atmosphere**.
    The dividing lines, in each hemisphere, between the **tropical easterlies** and the midlatitude **westerlies** in the middle **troposphere** roughly define the poleward boundaries of this region. Whereas many **circulation** systems in middle and high latitudes are nearly **adiabatic** and **quasi-geostrophic**, tropical systems are often strongly influenced by **cumulus** convection and surface heating, and can be less often dealt with using quasigeostrophic techniques. Many tropical circulations are driven or strongly influenced by coupling with the ocean. Examples of important tropical systems include the **Hadley** and **Walker circulations**, **monsoons**, **tropical cyclones**, the **Madden–Julian oscillation**, **easterly waves**, and **El Niño–Southern Oscillation**. The stratospheric circulation is dominated by the **quasi-biennial oscillation** and also contains the ascent branch of the Dobson circulation. Although tropical meteorology may be said to be a distinct endeavor, there are strong interactions between tropical and **extratropical** circulation systems.

tropical monsoon climate—One of Köppen's 1936 **tropical rainy climates**; it is sufficiently warm and rainy to produce **tropical rain forest** vegetation, but it does exhibit the **monsoon climate** influences in that it has a winter **dry season**.
    Some authors do not recognize this as a separate climatic type, but rather include it within the **tropical rain forest climate**.

Köppen, W. P., and R. Geiger, 1930–1939: *Handbuch der Klimatologie*, Berlin: Gebruder Borntraeger, 6 vols.

tropical rain forest—(Also called equatorial forest, selvas.) A type of **rain forest** that exists in tropical regions where **precipitation** is heavy, generally more than 250 cm (98 in.) per year.

It consists mainly of a wide variety of lofty trees, which carry a profusion of parasitic or climbing plants, and, in some portions, a "jungle" of dense undergrowth near the ground. For lack of marked climatic seasons, growth proceeds throughout the year. *Compare* **temperate rain forest**.

tropical rain forest climate—(Also called tropical wet climate.) In general, the **climate** that maintains **tropical rain forest** vegetation; that is, a climate of unbroken warmth, high humidity, and heavy annual **precipitation**.

Tropical Rainfall Measuring Mission—(Abbreviated TRMM.) A joint scientific satellite project between the United States and Japan designed to monitor and study tropical **rainfall** and the associated release of **energy** to the **atmosphere**.

The TRMM instrument package includes the first spaceborne **precipitation** radar, the TRMM microwave imager, a visible and infrared **scanner**, a **cloud** and earth **radiant energy** system, and a **lightning** imaging **sensor**. The TRMM satellite was successfully launched from the Tanegashima Space Center in Japan on 27 November 1997.

tropical rainy climate—A major category (the *A* climates) in W. Köppen's 1936 **climatic classification**; in order to be so classified, a **climate** must have these two characteristics: 1) the **mean temperature** of the coldest month must be 18°C (64.4°F) or higher, separating it from **temperate rainy climates**; 2) the annual **precipitation** must be in excess of a certain amount to distinguish it from **dry climates** (*see* formulas under **steppe climate**).

The three principal types of climate included in this category are the **tropical rain forest climate**, **tropical savanna climate**, and **tropical monsoon climate**
Köppen, W. P., and R. Geiger, 1930–1939: *Handbuch der Klimatologie*, Berlin: Gebruder Borntraeger, 6 vols.

tropical savanna climate—In general, the type of **climate** that produces the vegetation of the tropical and subtropical **savanna**; thus, a **climate** with a winter **dry season**, a relatively short but heavy summer **rainy season**, and high year-round **temperatures**.

tropical storm—*See* **tropical cyclone**.

tropical upper-tropospheric trough—A semipermanent **trough** extending east-northeast to west-southwest from about 35°N in the eastern Pacific to about 15°–20°N in the central west Pacific.

A similar structure exists over the Atlantic Ocean, where the mean trough extends from Cuba toward Spain.
Sadler, J. C., 1963: *Rocket and Satellite Meteorology*, Wesler, H., and J. E. Caskey Jr., Eds., North–Holland Publ., Amsterdam, 333–356.

tropical wet and dry climate—Same as **tropical savanna climate**.

tropical wet climate—Same as **tropical rain forest climate**.

Tropical Wind Energy Conversion and Reference Level Experiment—(Abbreviated TWERLE.) An observing system utilizing lightweight balloons to record weather data that were transmitted through *Nimbus-6*, launched in June 1975, to a ground **station**.

tropical year—The time that it takes the sun to travel from one **vernal equinox** to the next.

Tropics—1. Any portion of the earth characterized by a **tropical climate**.

2. Same as **Torrid Zone**.
See **Tropic of Cancer**, **Tropic of Capricorn**.

tropopause—The boundary between the **troposphere** and **stratosphere**, usually characterized by an abrupt change of **lapse rate**.

The change is in the direction of increased atmospheric **stability** from regions below to regions above the tropopause. Its height varies from 15 to 20 km (9 to 12 miles) in the **Tropics** to about 10 km (6 miles) in polar regions. In polar regions in winter it is often difficult or impossible to determine just where the tropopause lies, since under some conditions there is no abrupt change in lapse rate at any height. It has become apparent that the tropopause consists of several discrete, overlapping "leaves," a **multiple tropopause**, rather than a single continuous surface. In general, the leaves descend, step-wise, from the **equator** to the poles.

tropopause break line—On a **tropopause chart**, a line drawn to show the edge of a tropopause "leaf."
See **tropopause**, **multiple tropopause**.

tropopause chart—A **synoptic chart** showing the **contour lines** of the **tropopause** and **tropopause break lines**.

tropopause fold—Local folding of the **tropopause** over an intense **cyclone**.

tropopause funnel—A sharp depression in the level of the **tropopause**; found occasionally above strong frontal zones and intense cyclones.
A **linear** fold of the tropopause often has a funnel-like appearance in a vertical **cross section** through the fold. *See* **tropopause fold**.

tropopause inversion—(Sometimes called upper inversion.) The decrease in the **lapse rate** of temperature encountered at the level of the **tropopause**.

tropopause leaf—*See* **tropopause**.

tropopause wave—A **wave** that occurs along the surface of the **tropopause**, as an **internal wave**.

troposphere—That portion of the **atmosphere** from the earth's surface to the **tropopause**; that is, the lowest 10–20 km (6–12 mi) of the atmosphere; the portion of the atmosphere where most weather occurs.
The troposphere is characterized by decreasing **temperature** with height, appreciable vertical **wind** motion, appreciable **water vapor**, and weather. Dynamically, the troposphere can be divided into the following **layers: surface boundary layer, Ekman layer**, and **free atmosphere**. *See* **atmospheric shell**.

tropospheric ozone—A term used to distinguish **ozone** present in the **troposphere** from the more commonly talked about stratospheric **ozone layer**.
Tropospheric ozone results from **transport** from the **stratosphere** and from photochemical production (**oxidation** of **carbon monoxide, methane**, and other **hydrocarbons**). In remote regions of the troposphere, production and loss of ozone are nearly in balance, while in regions impacted by **anthropogenic** NO_x and hydrocarbon emissions, net ozone production occurs.

tropospheric scatter—(Also called troposcatter.) Propagation of **radio waves** through the **atmosphere** caused by **scattering** from inhomogeneities in the **refractive index** of the air.
Troposcatter enables propagation beyond the **radio horizon**.

trough—In meteorology, an elongated area of relatively low atmospheric **pressure**; the opposite of a **ridge**.
The axis of a trough is the **trough line**. This term is commonly used to distinguish the previous condition from the **closed circulation** of a **low** (or **cyclone**), but a large-scale trough may include one or more lows, an **upper-air trough** may be associated with a lower-level low, and a low may have one or more distinct troughs radiating from it. *See* **front, dynamic trough, easterly wave, equatorial wave**.

trough aloft—Same as **upper-level trough**.

trough line—A line along which **pressures** are lower than in the surroundings and where the **cyclonic** curvature of the isobars is a maximum.

trowal—In Canadian weather terminology, the **projection** on the earth's surface of a tongue of warm air aloft.

true airspeed—(Abbreviated TAS; also called corrected airspeed.) The **indicated airspeed** corrected for **temperature** and **altitude**.
These corrections are approximate.

true altitude—1. The true vertical distance above **mean sea level**.
2. Same as **corrected altitude**.

true freezing point—Same as **melting point**.

true horizon—*See* **horizon**.

true mean temperature—As adopted by the International Meteorological Organization, a monthly or annual mean of **air temperature** based upon hourly observations at a given place, or on some combination of less frequent observations designed to represent this mean as nearly as possible. *Compare* **mean temperature**.

true north—The direction from any point on the earth's surface toward the geographic North Pole; the northerly direction along any **projection** of the earth's axis upon the earth's surface, for example, along a longitude line.
Except for much of navigational practice (which uses **magnetic north**), true north is the universal 0° (or 360°) mapping reference. True north differs from magnetic north by the **magnetic declination** at that geographic point.

true solar day—Same as **apparent solar day**.

true wind direction—The direction, with respect to **true north**, from which the **wind** is blowing; distinguished from **magnetic wind direction**.

In all standard upper-air and surface weather observations, it is the true wind direction that is reported, usually in terms of tens of degrees in the 360° **compass**.

truncation error—*See* **roundoff error**.

Tsugaru Warm Current—*See* **Tsushima Current**.

tsunami—(Also called seismic sea wave.) **Waves** generated by seismic activity.

Tsunami are also popularly, but inaccurately, called **tidal waves**. When they reach shallow coastal regions, amplitudes may increase to several meters. The Pacific Ocean is particularly vulnerable to tsunami.

Tsushima Current—A warm current flowing northward through the Japan Sea along the west coast of Kyushu and Honshu.

The Tsushima Current is a branch of the western **boundary current** of the North Pacific subtropical **gyre**, which is split by the Japanese islands. It branches off the **Kuroshio** near 30°N to enter the Japan Sea through Korea Strait, where it carries 1.3 Sv (1.3×10^6 m^3 s^{-1}, about 2% of the total Kuroshio transport) with speeds near 0.4 m s^{-1} in summer (August) but only 0.2 Sv with less than 0.1 m s^{-1} in winter (January). Most of the summer transport is fed into the **East Korea Warm Current** but rejoins the Tsushima Current after the East Korea Warm Current separates from the coast at 36°–38°N. The seasonal **variability** of the Tsushima Current effects the **hydrography** of the southern Japan Sea greatly, reducing surface **salinity** from 35 **psu** in winter to below 32.5 psu in summer when the current carries low salinity water from the Yellow Sea. Most of the Tsushima Current rejoins the Kuroshio through the eastward flowing **Tsugaru Warm Current**, which passes through Tsugaru Strait (the passage between Honshu and Hokkaido). This current runs into the **Oyashio** near 42°N, which forces it to flow southward on the shelf along the east coast of Honshu to meet and join the northward flowing Kuroshio near 35°N. Another part of the Tsushima Current continues farther north, pushing the **polar front** to its most northern position in the Pacific, to enter the Sea of Okhotsk between Hokkaido and Sakhalin. This water traverses the Sea of Okhotsk as the **Soya Warm Current**, a rapid current with speeds reaching 1 m s^{-1} that stays close to the coast. Current **shear** between the fast-flowing coastal water and the offshore region persistently produces **eddies** of between 10 and 50 km in diameter. The water leaves the Sea of Okhotsk near 46°–47°N to flow south between the Oyashio and the east coast of Hokkaido.

tsuyu—Same as **bai-u**.

TT—Abbreviation for **Total Totals index**.
See **stability index**.

tuba—(Obsolete; commonly called **funnel cloud**; also called pendant cloud, tornado cloud.) In meteorology, a **cloud** column or inverted cloud cone, pendant from a **cloud base**. This supplementary feature occurs mostly with **cumulus** and **cumulonimbus**; when it reaches the earth's surface it constitutes the **cloudy** manifestation of an intense **vortex**, namely, a **tornado** or **waterspout**.
See **cloud classification**.

tufan—1. *See* **tofan**.

2. (From the Arabic, meaning "smoke.") Same as **typhoon**.

Tulipan radiometer—A calorimetric **radiation** instrument of historic interest used for the measurement of the amount of outgoing **heat** radiation from the earth during an interval of time.

The time integration is performed by allowing the radiation to fall on an uninsulated vessel containing a volatile liquid. The amount of liquid distilled into a connected insulated vessel is a measure of the incident radiation. *See* **radiometer**.

tunable laser spectroscopy—A technique based upon the measurement of **light** absorption upon irradiating a sample using a tunable **laser** source.

This technique is generally used for quantitative measurements of the sample concentration or studying the spectral characteristics of the sample. Specific applications of this technique include **saturation** spectroscopy, **heterodyne** spectroscopy, opto-acoustic **spectroscopy**, and **trace gas** measurements in combustion, atmospheric, and kinetic studies. *See also* **diode laser**, **dye laser**.

tundra—(Also called **arctic desert**.) Treeless plains that lie poleward of the **tree line**.

The plants thereon are sedges, mosses, lichens, and a few small shrubs. Tundra is mostly underlaid by **permafrost**, with the result that drainage is bad and the soil may be saturated for long periods. It does not have a permanent snow–ice cover.

tundra climate—Generally, the **climate** that produces **tundra** vegetation; it is too cold for the growth of trees but does not have a permanent snow–ice cover.

In W. Köppen's 1936 **climatic classification** it is one of the **polar climates**, defined as having a **mean temperature** for the warmest month of less than 10°C (50°F) (in contrast to **snow forest** and **temperate rainy climates**) but higher than 0°C (32°F) (which is the limit of **perpetual frost climate**). Tundra is designated *ET*. Tundra climate appears as a **temperature province** in C. W. Thornthwaite's 1931 classification.

> Köppen, W. P., and R. Geiger, 1930–1939: *Handbuch der Klimatologie*, Berlin: Gebruder Borntraeger, 6 vols.
> Thornthwaite, C. W., 1931: The climates of North America according to a new classification. *Geogr. Rev.*, **21**, 633–655.

tundra desert—See **arctic desert, tundra**.

turbidimeter—An instrument that measures the reduction in transmission of **light** that is caused by interposing a solution containing solid **particles** between the light source and the eye.

By using a known volume of solution in comparison with a **standard**, this instrument makes it possible to determine the mass effect, attributable to the number and size of the particles in the solution, and thus the quantitative amount of material present.

turbidity—The effect of (primarily) **aerosols**, through their total **optical depth**, in reducing the transmission of **direct solar radiation** to the surface below that through a purely molecular **atmosphere**.

Measures of turbidity refer to the total **aerosol optical depth**, either directly at a specified **wavelength** (e.g., the **Volz turbidity factor** or the **Ångström turbidity coefficient**, which is referenced to a wavelength of 1 μm), or indirectly by the ratio of aerosol to **Rayleigh optical depth** (e.g., the **Linke turbidity factor**).

turbidity factor—A measure of **turbidity**, related to **aerosol optical depth**, as in the **Ångström turbidity coefficient, Linke turbidity factor**, or **Volz turbidity factor**.

turbidity maximum—A region in an **estuary** where the tidally driven interaction between freshwater and saltwater generates a relative maximum concentration of suspended **sediment** and thus **turbidity**.

turbonada—A short **thundersquall** on the north Spanish coast, sometimes accompanied by **waterspouts**.

turbopause—The surface that separates the **homosphere**, in which the constituents of the **atmosphere** are well mixed by **turbulence**, from the **heterosphere**, in which constituents adopt their individual distributions with height as the result of **molecular diffusion**.

The turbopause is not very clearly marked, but usually lies at a height of about 100 km, near the base of the **thermosphere**.

turbosphere—(Rare.) The region of the **atmosphere** in which **turbulence** frequently exists; thus, it is the **troposphere**, in contrast to the supposedly nonturbulent **stratosphere**.

turbulence—1. Irregular fluctuations occurring in fluid motions.

It is characteristic of turbulence that the fluctuations occur in all three **velocity** components and are unpredictable in detail; however, statistically distinct properties of the turbulence can be identified and profitably analyzed. Turbulence exhibits a broad **range** of spatial and temporal scales resulting in efficient **mixing** of fluid properties. Analysis reveals that the **kinetic energy** of turbulence flows from the larger spatial scales to smaller and smaller scales and ultimately is transformed by molecular (viscous) **dissipation** to **thermal energy**. Therefore, to maintain turbulence, kinetic energy must be supplied at the larger scales. *See also* **ocean mixing**.

2. Random and continuously changing air motions that are superposed on the mean motion of the air.

See **aircraft turbulence**.

> Fleagle, R. G., and J. A. Businger, 1980: *An Introduction to Atmospheric Physics*, 2d ed., Academic Press, p. 264.
> Frisch, U., 1995: *Turbulence: The Legacy of A. N. Kolmogorov*, Cambridge University Press, 1–22.
> Hinze, J. O., 1975: *Turbulence*, 2d ed., McGraw–Hill, 790 pp.

turbulence closure—The problem in **turbulence** analysis that occurs when **Reynolds averaging** is applied to **Navier–Stokes equations**; the result is that there are more unknowns than equations.

In order to solve the problem, assumptions have to be made concerning the unknown quantities in the equations. These unknowns appear as correlations between the fluctuating quantities. The simplest closure is the so-called **K theory**. A more advanced form is the **direct interaction approximation**.

turbulence component—One of the constituent fluctuating **velocities** of a two- or three-dimensional **field**.
See **turbulence intensity**.

turbulence energy—Same as **turbulence kinetic energy**.

turbulence intensity—The ratio of the root-mean-square of the **eddy velocity** to the mean **wind speed**.
In general, it is a quantity that characterizes the **intensity** of **gusts** in the airflow.

turbulence kinetic energy—(Abbreviated TKE.) The **mean kinetic energy** per unit mass associated with **eddies** in **turbulent flow**.
A budget equation for TKE can be formed from the **Navier–Stokes equations** for **incompressible fluid** flow through Reynolds decomposition and **Reynolds averaging**. For most meteorological applications it is assumed that **buoyancy** effects are in the z direction only, that the mean vertical **velocity** is zero (hydrostatic scaling), and that **molecular diffusion** is neglected. *See also* **hydrostatic balance, isotropic turbulence, local isotropy, Reynolds number, Reynolds stresses, turbulence length scales, turbulence spectrum, viscous fluid**.
Hinze, J. O., 1975: *Turbulence*, 2d ed., McGraw–Hill, 790 pp.

turbulence length scales—Measures of the **eddy** scale sizes in **turbulent flow**.
The separation between the largest and smallest sizes is determined by the **Reynolds number**. The largest length scales are usually imposed by the flow geometry, for example, the **boundary layer** depth. Because **turbulence kinetic energy** is extracted from the mean flow at the largest scales, they are often referred to as the "energy-containing" **range**. The smallest scales are set by the **viscosity** and the rate at which **energy** is supplied by the largest-**scale** eddies. Intermediate between these scales are the **inertial subrange** scales for which turbulence kinetic energy is neither generated nor destroyed but is transferred from larger to smaller scales. Smaller-scale eddies are generated from the larger eddies through the **nonlinear** process of **vortex stretching**. Typically, energy is transferred from the largest eddies to the smallest ones on a timescale of about one large-eddy turnover. There are standard turbulence length scales for each of the eddy scale sizes; **integral length scales** for the **energy-containing eddies**, **Taylor microscale** for the inertial subrange eddies, and **Kolmogorov microscale** for the **dissipation** range eddies.

turbulence shear stresses—Same as **Reynolds stresses**.

turbulence spectrum—A plot of the **energy** distribution of turbulent **eddies** versus **wavelength** or **frequency**.
Wyngaard, J. C., 1973: On surface-layer turbulence. *Workshop on Micrometeorology*, D. E. Haugen, Ed., AMS, p. 132.

turbulent boundary layer—An **atmospheric boundary layer** containing a **range** of quasi-random **eddies** or swirls that tend to cause **mixing** and **dispersion** of **tracers** within it.
Most atmospheric boundary layers form on earth because of **turbulence** acting in a **statically stable** troposphere.

turbulent diffusion—(Or eddy diffusion.) **Diffusion** or **dispersion** due to the effects of turbulent motions.
When quantifying turbulent diffusion, it is important to distinguish between two types of measurement: 1) single-particle diffusion, where we measure the distances between diffusing **particles** at a fixed point or a point moving with the mean fluid **velocity**, and where the mean-square particle displacement initially increases rapidly with time (e.g., linearly); and 2) two-particle diffusion, where we measure the distances between particle pairs or the dimensions of a cloud of diffusing particles. Here the mean-square distance initially increases more slowly (e.g., as the 5/2 power of the time).

turbulent dissipation—*See* **dissipation**.

turbulent exchange—*See* **eddy flux**.

turbulent flow—A fluid flow characterized by **turbulence**.
See **eddy, laminar flow, transitional flow**.

turbulent flux—Transport of a quantity by quasi-random **eddies** or swirls; the **covariance** between a **velocity** component and any **variable**.
For example, **time series** of **potential temperature** θ and **vertical velocity** w can be used to find the turbulent vertical kinematic **heat flux**, F_H:

$$F_H = \frac{1}{N} \sum_{i=1}^{N} (w_i - \overline{w})(\theta_i - \overline{\theta}) = \frac{1}{N} \sum_{i=1}^{N} w_i'\theta_i' = \overline{w'\theta'},$$

where there are N data points in each **time series**, i is the time-series data index, primes denote deviations from the mean (i.e., the turbulent gusts), and the overbar denotes an average over the whole time series.

turbulent shear stresses—Same as **Reynolds stresses**.

turbulent transfer coefficients—*See* **exchange coefficients**.

turbulent transport—The movement of a substance or characteristic by **eddy** motions of the **wind**. *See* **turbulent flux**.

Turner angle—A **parameter** used to define the local **stability** of an inviscid water column to double-diffusive **convection**.
   Defined as

$$\text{Tu} = \tan^{-1}(\alpha\partial_z T - \beta\partial_z S, \; \alpha\partial_z T + \beta\partial_z S),$$

where $\tan^{-1}$ is the four-quadrant arctangent, α is the **coefficient of thermal expansion**, β is the coefficient of saline contraction, T is **temperature**, and S is **salinity**. If $-45° < \text{Tu} < 45°$, the column is **statically stable**. If $-90° > \text{Tu}$ or $\text{Tu} > 90°$, the column is statically unstable to Rayleigh–Taylor **instability**. If $-90° < \text{Tu} < -45°$, the column is unstable to **diffusive convection**. If $45° < \text{Tu} < 90°$, the column is unstable to **salt fingering**.

turning latitude—The latitude below which the **meridional** structure of an equatorial **gravity** or **Rossby wave** is wavelike and beyond which the meridional structure is decaying.
   With the β-plane approximation, the turning latitude (one in each hemisphere) is defined as

$$y_c = \pm (2n + 1)c/\beta,$$

where n is the meridional mode number, c is the **phase speed** of a given vertical mode **Kelvin wave**, and β is the derivative of **Coriolis parameter** f with respect to latitude.

turnover frequency—Same as **Nyquist frequency**.

TVS—Abbreviation for **tornadic vortex signature**.

TWERLE—Abbreviation for **Tropical Wind Energy Conversion and Reference Level Experiment**.

twilight—The period after **sunset** or before **sunrise** when all or part of the sky is visibly bright because of **sunlight** scattered by clouds or the **clear sky**.
   Twilight also refers to the sky's appearance during this period. By convention, there are three sequential stages of twilight: **civil twilight**, **nautical twilight**, and **astronomical twilight**. (Some definitions of twilight set its upper limit as high as sun elevations $h_0 = 5°–10°$.) Regardless of **cloud cover**, **illuminance** at the earth's surface decreases steadily during evening twilight and increases during morning twilight. However, under **partly cloudy** or **overcast** skies, some minor **brightness** fluctuations can occur as twilight progresses. The color and **luminance** patterns of clear skies change in complex ways throughout twilight (*see* **bright segment**, **dark segment**, **purple light**), while overcast skies usually grow bluer during evening civil twilight. The length of nautical, astronomical, and civil twilight varies greatly with latitude and time of year. The **annual range** of the duration of twilight increases with latitude. For example, at polar latitudes twilight may last as long as 24 hours or may not occur at all. *Compare* **dawn**, **dusk**.
   List, R. J., Ed., 1951: *Smithsonian Meteorological Tables*, 6th rev. ed., 506–520.

twilight arch—Same as **bright segment**.

twilight correction—In the interpretation of the records of **sunshine**, the difference between the time of **sunrise** and the time at which a record of sunshine first began to be made by the **sunshine recorder**; and conversely at **sunset**.
   This **correction** is added only when the **horizon** is **clear** during the period.

twilight glow—A faint, constant glow seen in the **twilight** sky and associated with **airglow**.

twin-gauge station—(Also twin-gage station.) Gauging **station** at which two **water level** gauges are used to define the water surface slope for developing a **stage–discharge** relationship.

twinkling—See **scintillation**.

twister—In the United States, a colloquial term for **tornado**.

twisting term—(Also called tilting term, tipping term.) The term in the **vorticity equation** that represents the generation of **vertical vorticity** by the twisting of horizontal **vorticity** into the vertical through the agency of **shear** in the **vertical velocity**.

In symbols this term is

$$\left(\frac{\partial w}{\partial y} - \frac{\partial v}{\partial z}\right)\frac{\partial w}{\partial x} + \left(\frac{\partial u}{\partial z} - \frac{\partial w}{\partial x}\right)\frac{\partial w}{\partial y},$$

where u, v, and w are the **velocity** components along the coordinate directions x, y, and z, respectively.

two-dimensional cloud probe—*See* **optical imaging probe**.

two-dimensional eddies—*See* **two-dimensional turbulence**.

two-dimensional model—A mathematical **model** of the **atmosphere** that simulates variations of concentration with two spatial coordinates (usually vertical and latitudinal) as a function of time.

Since the prevailing winds tend to move zonally (around circles of latitude), variations in chemical concentrations tend to be less pronounced in that coordinate, and the simplification resulting from using a zonally averaged model leads to a considerable saving in computation time.

two-dimensional precipitation probe—*See* **optical imaging probe**.

two-dimensional turbulence—The special case of **turbulence** in which the scales of the turbulent velocities in two dimensions (often the horizontal plane) are much larger than in the third dimension, and the horizontal **eddies** can be treated separately from the vertical.

In consequence, the equations of geophysical fluid dynamics can be formulated in especially simple and productive forms for applications to atmospheric and oceanic flows. Applications include especially study of large-scale atmospheric and oceanic disturbances, the **general circulation**, and **climate change**. *See also* **two-dimensional eddies**.

Frisch, U., 1995: *Turbulence: The Legacy of A. N. Kolmogorov*, Cambridge University Press, 240–241.
Tennekes, H., and J. L. Lumley, 1972: *A First Course in Turbulence*, MIT Press, p. 91.

two stream—An approximate technique for handling the complexity of monochromatic **radiative transfer** in atmospheric models.

The technique reduces the directionality of the problem to an upwelling and a downwelling **irradiance**, and provides moderate **accuracy** for reflected and transmitted irradiances with little computational expense. Several variations of the two-stream technique exist, suited to different problems.

TWT—Abbreviation for **traveling wave tube**.

Tyndall flowers—Small water-filled cavities, often of basically hexagonal shape, that appear in the interior of **ice** masses upon which **light** is falling.

Their formation results from the melting ice by radiative **absorption** at points of defect in the **ice crystal** lattice.

type—1. A specific classification of aircraft having the same basic design, including all modifications that result in a change in handling or flight characteristics.

2. *See* **weather type**.

type-α leader—A **stepped leader** that exhibits very little branching and with individual steps that are short and so weakly **luminous** as to be difficult to discern on high-speed streak photographs. *Compare* **type-β leader**.

type-β leader—A **stepped leader** with the upper portion of the channel characterized by longer and brighter steps than those found in the lower portion of the channel.

Many researchers report that a preliminary **breakdown** precedes the stepped leader in **electric field** records. The beta leader may be this preliminary stepped leader. *Compare* **type-α leader**.

typhoon—(Also spelled typhon.) A severe **tropical cyclone** in the western North Pacific.

The name is derived either from Cantonese t'ai fung (a "great wind"), from Arabic **tufan** ("smoke"), or from Greek typhon (a "monster"). Aristotle used typhon for a wind-containing **cloud** (*Meteorologica*, III, 1). For a more complete discussion, *see* **tropical cyclone**.

typhoon bar—*See* **cloud bar**.

typhoon eye—*See* **eye**.

typhoon warning—*See* **hurricane warning**.

typical year—The 12 months, January–December, selected from the entire **period of record** span for a given location as the most representative, or typical, for that month.

Thus, for a given **station**, a typical year will be a composite year with, possibly, each month of data from a different **calendar year**. As applied to the building construction and energy management industries, a typical meteorological year (TMY) consists of a full set of 8760 hourly **weather observations** (365 days × 24 h) containing real weather **sequences** that represent the long-term climatic modal (i.e., most frequent) conditions for a particular location. The hourly data include extraterrestrial, global and **direct solar radiation** estimates and observed **ceiling, sky cover, visibility**, weather (type), **sea level** and **station pressure**, **dry-bulb** and **dewpoint temperatures, cloud amounts**, and **total** and **opaque sky cover**. TMY data have been utilized, for example, by engineers in the design and evaluation of energy systems.

U

U figure—Same as **U index**.

U index—(Also called U figure.) The difference between consecutive **daily mean** values of the horizontal component of the geomagnetic **field**.

Each value is derived from a 48-hour interval covering two GMT days and is assigned to the second day of the pair. The monthly U index, the mean of the daily values, is the most frequently used. *See* **u index**.

u index—The value

$$u = U/\sin A \cos B,$$

where U is the U index, A the magnetic **co-latitude**, and B the angle between the magnetic **meridian** and the horizontal component of the magnetic **intensity**. The annual and longer-period mean values of the u index exhibit one of the strongest **solar-terrestrial relationships** known.

U.S. airways code—(Or airways code.) A **synoptic code** for communicating **aviation weather observations**.

U.S. Extension to the ICAO Standard Atmosphere—*See* **standard atmosphere**.

U.S. Standard Atmosphere—*See* **standard atmosphere**.

uala-andhi—See **kal Baisakhi**.

UARS—Abbreviation for **Upper Atmosphere Research Satellite**.

ubac—The shady (usually poleward, or north in the Northern Hemisphere) side of a mountain.

The term originated and is most often used in reference to mountains in the Alps. It is characterized by a lower **timber line** and **snow line** than the sunny side (the **adret**).

UHF—Abbreviation for ultra high frequency.
See **radio frequency band**.

Ulloa's ring—(Also called Bouguer's halo.) An infrequently observed, faint white, circular arc or complete ring of **light** that has a radius of 39° and is centered on the **antisolar point**.

When observed, it is usually in the form of a separate outer ring around an **anticorona**.
Tricker, R. A. R., 1970: *An Introduction to Meteorological Optics*, 192–193.

ultimate infiltration capacity—Steady or near-steady **infiltration rate** when water at the surface is maintained at **atmospheric pressure**.

ultra high frequency—(Abbreviated UHF.) *See* **radio frequency band**.

ultrafine particles—Very small atmospheric **particles**, less than 10 nm in diameter.

Such particles are thought to be formed by binary **nucleation** of compounds, such as **sulfuric acid** and water, and lead to an increase in the total number of particles present. The increased particle number **density** has consequences for **climate** effects such as **light scattering**.

ultrasonic—Referring to **sound waves** with frequencies higher than those at the upper limit of unimpaired human hearing, usually between 16 and 20 KHz.

ultraviolet—(Abbreviated UV.) Pertaining to or same as **ultraviolet radiation**.

ultraviolet radiation—(Abbreviated UV.) **Electromagnetic radiation** of shorter **wavelength** than **visible radiation** but longer than x-rays.

Wavelengths of UV radiation **range** from 5 to 400 nm, which may be further subdivided into the **UV-A**, **UV-B**, and **UV-C** ranges. UV radiation contains about 9% of the total **energy** of the solar **electromagnetic spectrum**. Such radiation has marked **actinic** and bactericidal action, and produces **fluorescence** in a number of substances. Ultraviolet radiation from the sun is responsible for many complex photochemical reactions characteristic of the **upper atmosphere**, for example, the formation of the **ozone layer** through ultraviolet dissociation of **oxygen** molecules followed by **recombination** to form **ozone**. The **absorption** of UV by stratospheric ozone and upper atmospheric oxygen is sufficiently strong that very little ultraviolet radiation with wavelengths shorter than about 300 nm reaches the earth's surface.

Umkehr effect—An **anomaly** of the relative **zenith** radiances of **scattered** sunlight at certain wavelengths in the **ultraviolet** as the sun approaches the **horizon**, due to the presence of the **ozone layer**.

The ratio of radiances measured in the zenith direction at two **wavelengths**, one strongly absorbed by **ozone**, the other not so strongly absorbed, shows a steady decrease with **solar zenith angle** until a zenith angle of about 86°, whereupon the ratio increases. The details of this reversal can be used to determine the vertical distribution of ozone concentration.

unambiguous range interval—Same as **maximum unambiguous range**.

unambiguous velocity interval—Same as **maximum unambiguous velocity**.

uncertainty—The **standard deviation** of a sufficiently large number of measurements of the same quantity by the same instrument or method.

Hence, the noncorrectable part of the **inaccuracy** of an instrument; it represents the limit of measurement **precision**. The uncertainty of an instrument is caused by the unpredictable effects upon its performance of such factors as **friction**, **backlash**, and electronic **noise**.

uncinus—A **cloud species** unique to the genus **cirrus**.
See **cloud classification**, **cirrus uncinus**.

unconfined aquifer—(Also known as water table aquifer.) Aquifer with an upper boundary defined by the **water table** and the surface at **atmospheric pressure**.

undamped oscillation—See **oscillation**, **damping**.

undercast—A **cloud layer** of ten-tenths (1.0) coverage as viewed from an observation point above the layer.
The term is most generally used in pilot reporting of in-flight weather conditions.

undercooling—Same as **supercooling**.

undercurrent—A **current** flowing underneath another current at a different speed or in the opposite direction.

underground ice—Same as **ground ice**.

undermelting—The melting from below of any **floating ice**.

undertow—See **rip current**.

undisturbed motion—The **steady state** of a system before **perturbations** are introduced.

undular bore—A propagating **disturbance** that is characterized by a sudden and relatively permanent change in the height of a horizontal fluid **interface** and in the **velocity** of the fluid beneath the interface with oscillations of its depth, and of **wind** and **temperature** characteristics within it.
A classic example in the **atmosphere** is the **morning glory** phenomenon in Australia triggered by the **sea breeze**, which is characterized by a smooth band of **cloud** along the leading edge reminiscent of a gust-front **arc cloud**, and other interface bands along the **wave crests** of the following lower-amplitude waves. Undular bores can also occur in advance of cold-air outflows (**gust fronts**) from thunderstorms. Compare **gravity current**.

undulatus—(Also called **billow cloud**, windrow cloud, wave cloud.) A **cloud variety** composed of merged or separate elements that are elongated and parallel, either suggestive of **ocean waves** or arranged in ranks and files.
Sometimes two distinct **wave** systems are apparent (biundulatus). The formation is by **gravity waves** that exhibit broad, nearly parallel lines of **cloud** oriented normal to the **wind direction**, with cloud bases near an **inversion** surface. See **cloud classification**; compare **cloud streets**.

unexplained variance—Same as **residual variance**.

unfiltered model—A numerical **model** with equations that accommodate the full range of dynamic modes of the hydrodynamic system being simulated.
See **filtered equations**.

unfreezing—The upward movement of stones to the surface as a result of repeated **freezing** and thawing of the containing soil; a result of **congeliturbation**.

unidirectional vertical wind shear—(Sometimes referred to as **rectilinear wind shear**.) The situation in which the **vertical wind shear** vector does not change direction with height.
Presented on a **hodograph**, unidirectional **shear** appears as a straight line. Unidirectional vertical wind shear does not always indicate unidirectional winds, since a straight-line hodograph can result from many different **wind profiles**.

unifilar electrometer—An electrostatic type of **electrometer** that utilizes a single fiber as the sensitive element.
A quartz fiber, coated with conducting material, is mounted midway between two electrical

plates, across which a **potential** is applied. Then a charge is placed on the fiber and its deflection is noted against a **scale** in the eyepiece of the observing telescope. *Compare* **bifilar electrometer**.

uniform flow—Flow in open channels or closed conduits where flow characteristics such as depth or **mean velocity** do not change with distance.

uniform layer—The middle portion of the daytime **atmospheric boundary layer** characterized by convective **thermals** creating vigorous **turbulence** in a well-mixed region of vertically uniform **potential temperature**, **wind speed**, and **pollutants**.

This layer is above the **radix** and **surface layers**, and below the **entrainment zone** or **capping inversion**. Within it, **smoke plumes** are observed to loop up and down as they blow **downwind**. It is **statically unstable**, in spite of the **adiabatic lapse rate**.

unimodal spectrum—In **radar**, a **Doppler spectrum** that has only one mode or peak.

unit hydrograph—The **discharge** at a point, expressed as a function of time, due to a unit of **effective rainfall** that is applied uniformly over the contributing area during a specified period of time.

The volume of the unit hydrograph is, by definition, one unit. The unit hydrograph is the **response function** of a **river basin** approximated as a **linear** system.

unit normal distribution—A **normal distribution** such that the mean $\mu = 0$ and **standard deviation** $\sigma = 1$; hence, the **probability density function** $f(x)$ is given by

$$f(x) = \frac{1}{\sqrt{2\pi}} \, e^{-x^2/2} \, (-\infty < x < \infty).$$

This **density function** and its definite integral (usually between 0 and arbitrary upper limits) have been extensively tabulated.

unit storm—For a specified duration, one unit depth of net **rainfall** that occurs over an entire **watershed** at a uniform **intensity**.

United States Coast Pilot—A multivolume set of **sailing directions** published by the National Ocean Service that covers a wide variety of information important to navigators of U.S. coastal and intracoastal waters, and waters of the Great Lakes.

Most of this information cannot be shown graphically on the standard nautical charts and is not readily available elsewhere. This information includes navigation regulations, outstanding landmarks, channel and anchorage peculiarities, dangers, weather, **ice**, currents, and port facilities. Each Coast Pilot is corrected through Notices to Mariners issued subsequent to the date of original publication.

universal equilibrium hypothesis—*See* **Kolmogorov's similarity hypotheses**.

universal functions—According to the **Monin–Obukhov similarity theory**, the dimensionless **shear**

$$\phi_M \equiv \frac{\partial M}{\partial z} \frac{kz}{u_*},$$

temperature gradient

$$\phi_H \equiv \frac{\partial \theta}{\partial z} \frac{kz}{\theta_*},$$

and other gradients in the **surface layer** are proportional to dimensionless universal **stability** functions, where M is **wind speed**, θ is **potential temperature**, u_* is **friction velocity**, and θ_* is the surface kinematic **heat flux** divided by **friction velocity**.

These functions have a value of nearly 1 for neutral **stratification**, **range** over $0 < (z/L) < 1$ for unstable stratification, and $z/L > 1$ for stable stratification, where z is height above the surface and L is the **Obukhov length**. For strong stability $(z/L >> 1)$ the universal functions are nearly constant. Presently the most-used universal functions are based on an experiment conducted in Kansas in 1968, with some corrections made in the succeeding 30 years. The **accuracy** is about 10% for unstable and 20% for stable stratification. The following empirical forms of the universal functions, based on a von **Kármán constant** of k = 0.4, are currently used. For unstable stratification:

$$\phi_M(z/L) = [1 - 19.3(z/L)]^{-1/4} \text{ for momentum flux;}$$

$$\phi_H(z/L) = 0.95[1 - 11.6(z/L)]^{-1/2} \text{ for sensible heat flux.}$$

For neutral stratification:

$$\phi_M(z/L) = 1 \text{ for momentum flux;}$$

$$\phi_H(z/L) = 0.95 \text{ for sensible heat flux.}$$

For stable stratification:

$$\phi_M(z/L) = 1 + 6.0(z/L) \text{ for momentum flux;}$$

$$\phi_H(z/L) = 0.95 + 7.8(z/L) \text{ for sensible heat flux.}$$

These universal functions are also called **flux-profile relationships**.

universal gas constant—*See* **gas constant**.

universal gravitational constant—*See* **gravitation**.

universal rain gauge—A **weighing rain gauge** in which the weight of the **catch** is converted to centimeters or inches of **precipitation** and recorded by pen on a clock-driven **chart**.
The clock cylinder usually rotates one revolution in 24 hours.
> NOAA/National Weather Service, 1972: *National Weather Service Observing Handbook No. 2, Substation Observations*, NOAA, US GPO.

universal time—(Abbreviated UT.) The basis for civil timekeeping.
It is formally defined by a mathematical formula which relates UT to Greenwich mean sidereal time. Depending on the context, universal time and UT are commonly used to mean 1) UT0, which is dependent on the **observer**'s location; 2) UT1, which removes the effect of the motion of the geographic pole; or 3) coordinated universal time (UTC). Since 1 January 1972, **weather services** have used UTC as the standard of time. *See also* **zone time**
> United States Naval Observatory and her Majesty's Nautical Almanac Office, Annual: *The Astronomical Almanac*, U.S. Government Printing Office, Washington, D.C. and HMSO, London, England.

universe—In **statistical** terminology, synonymous with **population**.

unlimited ceiling—A **ceiling** condition that exists 1) when the **total sky cover** is less than 0.6; 2) when the total **transparent sky cover** is 0.5 or more; or 3) when surface-based obscuring phenomena are classed as **partial obscuration** (i.e., obscures 0.9 or less of the sky) and no layer aloft is reported as **broken** or **overcast**.

unprotected thermometer—A **reversing thermometer** (for seawater **temperature**) that is not protected against **hydrostatic pressure**.
The **mercury** bulb is therefore squeezed, and the amount of mercury broken off on reversal is a function both of temperature and of hydrostatic pressure. When compared with the simultaneous reading of a **protected thermometer**, which is affected by temperature only, the unprotected thermometer reading can be converted to **pressure**, and then, by applying the mean density of the water, to depth.

unrestricted visibility—The **visibility** when no obstructions to vision exist in sufficient quantity to reduce the visibility to less than seven miles.
Otherwise, in an **aviation weather observation**, an **obstruction to vision** must be entered.

unsaturated flow—The flow of water in soil or rock that is not water-saturated.

unsaturated hydraulic conductivity—The proportionality constant between the volumetric **flux** of water and the **hydraulic gradient** in a **porous medium** for cases of **water content** less than **saturation**.
Often expressed as a function of soil-water pressure **head** and/or water content, it includes the effects of the water-filled pore structure and the **viscosity** and **density** of the water.

unsaturated zone—The zone between the ground surface and the top of the **saturated zone** (**capillary fringe**) associated with the regional **water table**.

unsharp masking—An **image processing** technique where a second **image** is created that is an out-of-focus (blurred, low-pass filtered) version of the original.
The second image's **pixel** values are subtracted from the corresponding pixel values in the original image, enhancing detail in the image. Its use (and terminology) dates back to early photographic techniques.

unstable air—Air in which **static instability** prevails.
This condition is determined by the vertical gradients of **air temperature** and **humidity**.

unstable air mass—Air mass having **static instability** in the lowest layer.
Convective clouds and **precipitation** occur if **moisture content** is sufficiently high and there is a mechanism present, for example, surface heating or **orographic lifting**, to initiate **convection**.

unstable channel—A **stream** channel experiencing significant morphologic changes, such as aggradation or degradation.

unstable equilibrium—The state of a system that may not be restored after the introduction of a **perturbation**.
See **equilibrium**, **instability**.

unstable oscillation—*See* **oscillation**.

unstable wave—A **wave motion** with an **amplitude** that increases with time, or total **energy** that increases at the expense of its **environment**.
See **instability**.

unsteady flow—Flow in which the channel depth and **velocity** change with time.

unterwind—A **breeze** blowing up-valley on the lakes of Salzkammergut in Austria.

upbank thaw—The precedence of a **thaw** in a valley, sometimes by many hours, by a thaw or marked rise in **temperature** at mountain level in the same vicinity.
The phenomenon is usually caused by the arrival at higher levels of the warm air in advance of a surface **warm front**. It may also be caused by the **subsidence** and dynamical heating of air at the higher level. The associated **inversion** of the **normal** temperature **lapse rate** is a contributory cause of **glaze**.

updraft—*See* **draft**.

updraft curtains—Long narrow sheets of warm air rising from a heated surface.
These are also called microfronts because of the rapid **temperature** changes measured when these advect past fixed sensors. Updraft curtains are **surface layer** phenomena that gradually merge and change shape into **boundary layer** thermals as they rise higher into the **mixed layer**.

upglide cloud—Cloud formed by **condensation** within a mass of **moist air** that is subject to ascending motion above a **frontal surface** of **discontinuity**.

upgradient flux—(Also called countergradient flux.) The **transport** of any quantity S from small to large values, contrary to that expected by **diffusion** theory.
Upgradient flux is typically observed in the top fifth of the **atmospheric boundary layer**, where **heat flux** is often observed to flow from colder to warmer values of **potential temperature** caused by nondiffusion **transport processes** such as convective thermals. Some theories that account for this countergradient flux include **top-down/bottom-up diffusion** and **transilient turbulence theory**.

upper-air—(Also aerological.) Having to do with the **free atmosphere**, including the **troposphere** and **stratosphere**.
Upper-air observations are distinguished from surface observations, even though an **upper-air observation** may include data from the surface.

upper air—In **synoptic meteorology** and in weather observing, that portion of the **atmosphere** that is above the lower **troposphere**.
No distinct lower limit is set but the term can be generally applied to the levels above 850 mb. *Compare* **upper atmosphere**.

upper-air analysis—*See* **upper-level chart**.

upper-air anticyclone—*See* **upper-level anticyclone**.

upper-air chart—Same as **upper-level chart**.

upper-air climatology—The scientific study of the **atmosphere** above the **boundary layer**.
In addition to the presentation of **upper-air** data (**temperature**, **wind**, **pressure**, etc.), it includes **analysis** of the causes of spatial and temporal variations of upper-air variables. Among the more common presentations are spatial maps of means over specified time periods (months, seasons, years). *See* **climatology**.

upper-air cyclone—*See* **upper-level cyclone**.

upper-air disturbance—(Or upper-level disturbance.) A **disturbance** of the **flow pattern** in the **upper air**, particularly one that is more strongly developed aloft than near the ground.

upper-air observation—(Also called **sounding, upper-air sounding.**) A measurement of atmospheric conditions aloft, above the effective **range** of a **surface weather observation**.
    This is a general term, but is usually applied to those observations that are used in the **analysis** of **upper-air** charts (as opposed to measurements of upper-atmospheric quantities primarily for research). Among the elements evaluated are **pressure, temperature, relative humidity** (e.g., by **radiosonde** aircraft observations), and **wind speed** and **direction** (e.g., by **rawinsonde**, aircraft, or **wind profiling radars**). Also, some mountain stations are high enough and exposed enough so that their observations may be included in the upper-air network at their **elevation**. *See also* **meteorological rocket, radiosonde balloon**.

upper-air ridge—Same as **upper-level ridge**.

upper-air sounding—(Also called **upper-air observation**.) A measurement of the **vertical profile** of the thermodynamic and kinematic state of the **atmosphere**.
    A **radiosonde** makes an in situ point measurement of the atmosphere that it passes through. In contrast, a **radar** or **lidar** profiler makes a remotely sensed volumetric measurement of the atmosphere above the **profiler** location. *See also* **radiosonde observation, rawinsonde, rocket-sonde, sounding**.

upper-air station—A surface location from which **radiosonde observations** are made.

upper-air synoptic station—*See* **upper-air station**.

upper-air trough—Same as **upper-level trough**.

upper anticyclone—Same as **upper-level anticyclone**.

upper atmosphere—The general term applied to the **atmosphere** above the **troposphere**. *See* **atmospheric shell**.

Upper Atmosphere Research Satellite—(Abbreviated *UARS*.) A satellite designed to study the chemistry, dynamics, and **energetics** of the **stratosphere, mesosphere,** and lower **thermosphere**.
    UARS was launched in September 1991. *UARS* instruments include the cryogenic limb array etalon **spectrometer**, an improved stratospheric and mesospheric **sounder**, a **microwave** limb sounder, a halogen occultation experiment, a high-resolution Doppler imager, a **wind** imaging **interferometer**, a solar–stellar **irradiance** comparison experiment, a solar **ultraviolet** spectral irradiance monitor, a **particle** environment monitor, and the active **cavity radiometer** irradiance monitor.

upper cold front—*See* **upper front**.

upper cyclone—Same as **upper-level cyclone**.

upper front—A **front** that is present in the **upper air** but typically does not extend to the ground.

upper high—Same as **upper-level anticyclone**.

upper inversion—Same as **tropopause inversion**.

upper level—*See* **upper air**.

upper-level anticyclone—(Also called upper-level high, **upper anticyclone**, upper high, high-level anticyclone, high aloft.) An **anticyclonic circulation** existing in the **upper air**.
    This often refers to such anticyclones only when they are much more pronounced at upper levels than at and near the earth's surface.

upper-level chart—(Or upper-air chart.) A **synoptic chart** of meteorological conditions in the **upper air**, almost invariably referring to a standard **constant-pressure chart**.

upper-level cyclone—(Also called upper-level low, upper cyclone, upper low, high-level cyclone, low aloft.) A **cyclonic circulation** existing in the **upper air**; specifically as seen on an upper-level **constant-pressure chart**.
    This term is often restricted to such **cyclones** associated with relatively little cyclonic circulation in the **lower atmosphere**.

upper-level disturbance—Same as **upper-air disturbance**.

upper-level high—Same as **upper-level anticyclone**.

upper-level low—Same as **upper-level cyclone**.

upper-level rainband—A **rainband** associated with **upper-level** features, but not with any surface-level feature.

upper-level ridge—(Also called upper ridge, upper-air ridge, high-level ridge, ridge aloft.) A **pressure ridge** existing in the **upper air**, especially one that is stronger aloft than near the earth's surface.

upper-level trough—(Also called upper trough, upper-air trough, high-level trough, trough aloft.) A pressure **trough** existing in the **upper air**.

This term is sometimes restricted to those troughs that are much more pronounced aloft than near the earth's surface. These troughs are often described as either **short-wave** or **long-wave** features.

upper-level winds—Same as **winds aloft**.

upper low—Same as **upper-level cyclone**.

upper ridge—Same as **upper-level ridge**.

upper trough—Same as **upper-level trough**.

upper wind chart—A **chart** depicting winds in the upper levels of the **atmosphere**, often including **wind speed**, **wind direction**, or both.

upper winds—Same as **winds aloft**.

uprush—A term sometimes applied to the strong upward-flow **air current** in **cumulus** clouds during their stage of rapid **development**, often preceding a **thunderstorm**.

upslope fog—A type of **fog** formed when air flows upward over rising terrain and is, consequently, **adiabatically** cooled to or below its **dewpoint**.

upslope wind—1. A **wind** directed up a slope, often used to describe winds produced by processes larger in **scale** than the slope.

Because this flow produces rising atmospheric motion, upslope winds experience cooling, increasing **relative humidity**, decreasing **stability**, and, if sufficient moisture is present, the formation of **fog**, **clouds**, and **precipitation**.

2. Flow directed up a mountain slope and driven by heating at the earth's surface under light larger-scale flow conditions; a component of the mountain–valley or **mountain–plains wind systems**; same as **anabatic wind**.

upstream—In the direction from which a fluid is flowing.

upvalley wind—A daytime, thermally forced, along-valley component of **along-valley wind systems** from the direction of the plains or valley toward the mountains, produced by **diurnal heating** of the valley air; a daytime along-valley component of the **mountain-valley wind systems** encountered during periods of light synoptic or other larger-scale flow.

The mechanism of the upvalley wind is as follows. Air in the valley heats faster than an equivalent vertical column of air over the adjacent plains as a result of 1) the difference in the ratio of the volume of air heated in each location to the horizontal area intercepting **radiation** and 2) **subsidence** over the middle of the valley compensating for the flow up the slopes. The more effective heating in the valley produces a **pressure** drop in the valley relative to the plains, and this in turn produces an upvalley wind, called the **valley breeze**, beginning in late morning one to four hours after **sunrise**. It often reaches 3–5 m s^{-1} at the surface and >5 m s^{-1} above the surface. Mature upvalley flows tend to fill the valley, that is, the depth of the upvalley flow is approximately equal to the depth of the valley. *See also* **topographic amplification factor**.

upwelling—An ascending motion of **subsurface water** by which water from deeper layers is brought into the surface layer and is removed from the area of upwelling by divergent horizontal flow.

See also **coastal upwelling, equatorial upwelling, downwelling**.

upwind—In the direction from which the **wind** is blowing.

upwind effect—The effect of an **orographic** barrier in producing **orographic precipitation** to the **windward** of the base of the barrier as a result of **orographic lifting** mechanisms that produce rising motion **upwind** of the barrier.

Orographic blocking of moist, stable approach flow is an example of one such mechanism.

urban boundary layer—The **internal boundary layer** formed when air flows over a city.

It is a **mesoscale** phenomenon, the characteristics of which are affected by the nature of the urban surface. When a new rural **boundary layer** downwind of the city forms at the surface, the urban boundary layer is isolated aloft and is then called the urban plume. *Compare* **urban heat island, urban canopy layer**.

urban canopy—The assemblage of buildings, trees, and other objects composing a town or city and the spaces between them.

The concept is roughly analogous to that of a vegetative **canopy** except that the built part is open to the sky and has no stem or trunk zone. Together with the air layer beneath rooftop and treetop level, it forms the **urban canopy layer**.

urban canopy layer—The layer of air in the **urban canopy** beneath the mean height of the buildings and trees.

Its **climate** is dominated by **microscale** processes due to the complex array of surfaces (their orientation, **albedo**, **emissivity**, **thermal** properties, wetness, etc.). It is a zone of multiple **reflection** and **emission**, wakes and vortices, especially in the **urban canyons**. *Compare* **urban boundary layer**.

urban canyons—The characteristic geometry formed by a city street and its flanking buildings.

Together with the intervening roofs they provide a first-order repetitive structure to cities with characteristic patterns of **sunlight** and shade and cross-street vortices. Commonly described by their nondimensional **aspect ratio** H/W, where H is the mean height of the buildings and W the width of the street.

urban climate—The **climate** affected by the presence of a town or city.

Urban development greatly modifies the radiative, **thermal**, moisture, and **aerodynamic** properties of the surface. This change alters the fluxes and balances of **heat**, mass, and **momentum**, producing a distinct **urban boundary layer**.

urban climatology—The study of **urban climate**.

urban effects—*See* **urban heat island**, **urban climate**.

urban heat island—(Or heat island.) Closed isotherms indicating an area of the surface that is relatively warm; most commonly associated areas of human disturbance such as towns and cities.

The physiographic analogy derives from the similarity between the pattern of isotherms and height contours of an island on a topographic map. Heat islands commonly also possess "cliffs" at the urban–rural fringe and a "peak" in the most built-up core of the city. The annual **mean temperature** of a large city (say 10^6 inhabitants) may be 1°–2°C warmer than before development, and on individual **calm**, **clear** nights may be up to 12°C warmer. The warmth extends vertically to form an urban heat dome in near calm, and an urban heat plume in more windy conditions.

urban hydrology—Study of the effects of urban conditions on **rainfall–runoff** relationships.

urban plume—*See* **urban boundary layer**.

urban runoff—**Runoff** generated from urban areas.

urban–rural circulation—(Also called city–country wind breeze system.) A **circulation** system driven by urban–rural differences of **temperature** (i.e., the **urban heat island**); a toroidal circulation system of country breezes at low level converging on the city center, resulting in uplift with **divergence** and return flows aloft.

The country breezes arrive in pulses and the strength of the circulation depends on the **static stability** as well as horizontal temperature differences.

UTC—*See* **universal time**.

UV—1. Abbreviation for **ultraviolet radiation**.

2. Abbreviation for **ultraviolet**.

UV-A—**Ultraviolet radiation** in the **wavelength** band of 0.32–0.40 μm; a component of the **spectrum** of **solar radiation** that can tan human skin or cause redness in sensitive skin.

UV and IR hygrometers—Instruments using **absorption** of **electromagnetic radiation** to sense **humidity**.

Ultraviolet (UV) hygrometers use a single **wavelength** absorbed by molecular **water vapor**, such as the Lyman-alpha line produced by a **hydrogen** discharge tube. **Infrared** (IR) hygrometers can also operate on a single **absorption line** if the source is an infrared-emitting tunable **diode laser** (TDL), though broad-band instruments, which integrate the absorption across many individual lines, are also used. These instruments utilize **Beer's law** to estimate the **density** of absorption across the path between the source and a detector (e.g., an **ionization** tube for ultraviolet, or a photodiode for infrared), which contains the air sample. In many cases, these hygrometers have sufficient speed of response that they can resolve rapid turbulent fluctuations of humidity.

UV-B—**Ultraviolet radiation** in the **wavelength** band of 0.29–0.32 μm; a component of the **spectrum** of **solar radiation** that can cause sunburn and skin cancer.

This **band** is only partially blocked by the **atmospheric ozone** layer.

UV-C—**Ultraviolet radiation** in the **wavelength** band of 0.20–0.29 μm; a very harmful component of the **spectrum** of **solar radiation** that can cause chromosome mutations, death of single-cell organisms, and damage to the cornea of the eye.

This **band** is almost completely blocked by the **atmospheric ozone** layer.

V

V-band—*See* **radar frequency bands**.

V–R vortex—A two-dimensional circular flow in which an infinite **vorticity** is concentrated at the origin, the rest of the fluid being free of vorticity.

The (tangential) speed V is inversely proportional to the distance R from the origin:

$$VR = \text{constant}.$$

If the flow is considered identical in all parallel planes, the vertical axis through the origin is a **vortex filament**. See **Rankine vortex**.

V-shaped depression—On a **surface chart**, a **low** or **trough** about which the isobars display a pronounced "V" shape, with the point of the "V" usually extending equatorward from the parent low.

This term is most frequently found in European literature, where it is applied mainly to the southern extension of frontal troughs associated with deep lows that have migrated across the North Atlantic Ocean. *See* **depression**.

V-shaped isobars—Isobars that display a pronounced "V" or kinked shape due to local increases in **pressure gradients**, which are typically found in the vicinity of strong **fronts**.

vacillation—**Oscillation**, usually of a small **amplitude**, about a reference state.

vacuum correction—The **correction** to the reading of a **mercury barometer** required by the imperfections in the vacuum above the **mercury column**, due to the presence of **water vapor** and air.

This correction is a function of both **temperature** and **pressure**. *See* **barometric corrections**.

vacuum-tube electrometer—An **electrometer** that makes use of the amplifying properties of specially designed vacuum tubes.

These instruments are more rugged than the **electrostatic electrometer** and possess the additional advantage of being easily converted into self-registering instruments.

VAD—Abbreviation for **velocity–azimuth display**.

VAD wind profile—(Abbreviated VWP.) Time–height **profile** of horizontal **wind** vectors derived from a scanning **Doppler radar** using a **velocity–azimuth display** (VAD) **algorithm**.

vadose water—Water present within the zone between the regional **water table** and the ground surface, that is, within the **vadose zone**.

vadose zone—The zone between ground surface and the regional **water table**.

It includes the **unsaturated zone** and the **capillary fringe** (**tension saturated zone**).

vaguio—Same as **baguio**.

valais wind—The notable **valley wind** that blows along the Rhône valley from the upper end of Lake Geneva (Valais Canton).

It is sufficiently strong and regular to distort the growth of trees.

validation—1. In **expert systems**, the determination of how well the task is performed.

For **weather forecast** expert systems, it corresponds to **forecast verification**. Consequently, there is often confusion among meteorologists between validation and **verification** as used in **artificial intelligence**.

2. Comparison of a measurement from a new instrument or technique with older, established measurements of the same property or **parameter**.

valley breeze—Same as **valley wind**.

valley exit jet—*See* **outflow jet**.

valley glacier—A **glacier** lying primarily in a mountain valley.

valley outflow jet—*See* **outflow jet**.

valley storage—Volume of water stored between any two specified points along a **stream**, including both the channel and the **flood plain**.

valley wind—(Or valley breeze.) A **wind** that ascends a mountain valley (**upvalley wind**) during the day; the daytime component of a **mountain–valley wind system**.

Van Allen radiation belts—*See* **radiation belts**.

Van der Waal's equation—The best known of the many laws that have been proposed to describe the thermodynamic behavior of real gases and their departures from the **ideal gas** laws.

It states

$$\left(p + \frac{a}{\alpha^2}\right)(\alpha - b) = RT,$$

where a and b are constants dependent upon the gas, p the **pressure** of the gas, α its **specific volume** (measured in units of the specific volume of the gas at **normal temperature and pressure**), R the **universal gas constant**, and T the Kelvin **temperature**. Gas–liquid change characteristics and the properties of a substance in these two phases are predicted with some success by Van der Waal's equation, but not at all by the **equation of state** for perfect gases. When p is small and α is large, the correction terms are small and the two equations become identical.

Van't Hoff factor—The effective number of **moles** of **ions** in solution per molecule of salt.

For dilute aqueous solutions, it is the number of ions in a molecule; it influences colligative properties such as solution **freezing point** depression in **mixed cloud** nuclei.

vane—Same as **wind vane**.

vanishing point—*See* **radiatus**.

vapor—Any substance existing in the gaseous state at a **temperature** lower than that of its **critical point**; that is, a gas cool enough to be liquefied if sufficient **pressure** were applied to it.

If any vapor is cooled sufficiently, say at constant pressure, it ultimately reaches a state of **saturation** such that further removal of **heat** is accompanied by **condensation** to the liquid **phase**. Except for states quite close to that of saturation, vapors exhibit the general properties of all gases. Quantitatively, however, vapors exhibit measurable departures from **perfect-gas laws** even in states well removed from that of saturation. Since the **critical temperature** for water (374°C) is far above any atmospheric temperatures (except for the extreme **upper air**), all water substance found in the **atmosphere** in the gaseous state is appropriately called **water vapor**.

vapor concentration—Same as **vapor density**.

vapor density—(Also called absolute humidity.) In a system of **moist air**, the ratio of the mass of **water vapor** present to the volume occupied by the mixture; that is, the **density** of the water vapor component.

Because this measure of atmospheric **humidity** is not conservative with respect to **adiabatic expansion** or compression, it is not commonly used by meteorologists. *Compare* **mixing ratio**, **specific humidity**, **relative humidity**, **dewpoint**.

vapor line—On a **thermodynamic diagram**, an **isopleth** representing **saturation mixing ratio**, or **saturation specific humidity**, or other moisture variable.

vapor pressure—(Also called vapor tension.) The **pressure** exerted by the molecules of a given **vapor**.

For a pure, confined vapor, it is that vapor's pressure on the walls of its containing vessel; for a vapor mixed with other vapors or gases, it is that vapor's contribution to the **total pressure** (i.e., its **partial pressure**). In meteorology, vapor pressure is used almost exclusively to denote the partial pressure of **water vapor** in the **atmosphere**. Care must be exercised in interpreting the term's meaning as used in other branches of science. *See* **saturation vapor pressure**, **equilibrium vapor pressure**.

vapor tension—1. (Obsolete.) Same as **vapor pressure**.

2. The maximum possible **vapor pressure** that can be exerted, at a given **temperature**, by a system composed of a plane surface of a liquid or solid substance in contact with that substance's **vapor**.

As in the case of solutions, the vapor need not be of the same chemical composition as the parent substance. As so used, vapor tension is an intrinsic property of a substance, usually attributed to its liquid or solid state. *Compare* **equilibrium vapor pressure**, **saturation vapor pressure**.

vapor trail—Same as **condensation trail**.

vaporization—Same as **evaporation**.

vardar—(Also called vardarac.) A cold **fall wind** blowing from the northwest down the Vardar valley in Greece to the Gulf of Salonica.

It occurs where **atmospheric pressure** over eastern Europe is higher than over the Aegean Sea, as is often the case in winter. It persists for two or three days with a **mean velocity** of 5–7 m s⁻¹(10–15 mph), rising to 16 m s⁻¹ (35 mph) in **squalls**. It is strongest where the Vardar River leaves the mountains, but it extends for some distance out to sea. A similar **wind**, the Struma fall wind, blows in the Struma valley.

vardarac—Same as **vardar**.

variability—Mathematically, same as **spread**.

variable—Something that can assume different values or states.
See **dependent variable**, **independent variable**, **random variable**.

variable ceiling—After U.S. weather observing practice, a condition in which the **ceiling** rapidly increases and decreases while the ceiling observation is being made.
The average of the observed values is used as the reported ceiling, and it is reported only for ceilings of less than 3000 ft. Variable ceiling is reported in the remarks section of the **observation**.

variable resolution model—A numerical **model** with horizontal **resolution** arranged to be non-uniform.
This enables finer details of the solution in a chosen region of interest to be captured in a way that preserves computational economy elsewhere. *See* **reduced grid**, **nested grids**.

variable visibility—In U.S. weather observing procedures, a condition when the **prevailing visibility** is less than three statute miles and is rapidly increasing or decreasing by one-half **mile** or more during the period of **observation**.
Variable visibilities are reported in the remarks section of the observations.

variable wind—**Wind** that changes direction frequently.

variance—A measure of variability (or **spread**).
It is denoted by σ^2 and defined as the mean-square **deviation** from the mean, that is, the mean of the squares of the differences between individual values of x and the **mean value** μ:

$$\sigma^2 \equiv E[(x - \mu)^2] \equiv E(x^2) - \mu^2,$$

where E denotes **expected value**. The positive square root σ of the variance is called the **standard deviation**. An unbiased estimate s^2 of the variance σ^2 is obtained from n independent observations $x_1, x_2, \cdots, x_n$ and their **sample** average $\overline{x}$ as follows:

$$s^2 = \left[\sum_{i=1}^{n} (x_i - \overline{x})^2\right]/(n - 1),$$

and the positive square root s of s^2 is taken as an estimate of the standard deviation σ.

variance analysis—*See* **analysis of variance**.

variance ratio—Same as **F-ratio**.

variance reduction—*See* **regression**.

variate—Same as **random variable**.

variation—1. The **range** within which values of a **variable** lie, as in the **diurnal** or annual variation.
2. Same as **declination**.

variational objective analysis—An **objective analysis** technique used to create an estimate of the atmospheric state that maximizes (or minimizes) a mathematical measure of desirable (or undesirable) characteristics.
The analysis characteristics usually include measures of the fit to data, **background field**, and dynamical constraints.

varied flow—Used to describe nonuniform flow conditions in open channels and conduits, with changing **cross section** or slope with distance.
When **discharge** is constant, the **velocity** changes with changes in slope and cross section. Relatively slow changes are known as gradually varied flow and are strongly influenced by surface **resistance**. Dramatic changes in velocity and depth occurring over a very short distance are known as rapidly varied flow and surface resistance effects are small.

vario—Same as **baguio**.

variograph—A recording **variometer**.

variometer—An instrument designed to study very small fluctuations of some quantity.

The **microbarograph** is an example of a recording **pressure** variometer. A common form is an instrument designed to measure small changes in **altitude** by sensing small pressure changes.

varves—The annual layers of **sediment** deposited in lakes and fiords by meltwater from **glaciers**.
Each layer consists of two parts deposited at different seasons and differing in color and texture so that the layers can be measured and counted. If the series is complete, the number of layers gives the date on which the ground was vacated by the retreating **ice**. Varves were discovered by G. de Geer at Lake Ragunda in Sweden in 1905, and have since been found in many other regions. They differ in thickness from year to year. These differences were attributed by de Geer to variations of **solar radiation** and hence are supposed to be similar in all parts of the world. It was proposed that in this way dates of varve series could be determined even if the upper layers were missing, but this method of dating has been doubted.

VAS—Abbreviation for **VISSR Atmospheric Sounder**.

vaudaire—(Also called vauderon.) A violent south **wind**; a **foehn** of Lake Geneva in Switzerland.

vault—Same as a **bounded weak echo region** (BWER).

vector—Any quantity, such as force, **velocity**, or **acceleration**, that has both magnitude and direction at each point in space, as opposed to a **scalar** that has magnitude only.
Such a quantity may be represented geometrically by an arrow of length proportional to its magnitude, pointing in the assigned direction. A unit vector is a vector of unit length; in particular, the three unit vectors along the positive x, y, and z axes of **rectangular Cartesian coordinates** are denoted, respectively, by $\mathbf{i}$, $\mathbf{j}$, and $\mathbf{k}$. Any vector $\mathbf{A}$ can be represented in terms of its components a_1, a_2, and a_3 along the coordinate axes x, y, and z, respectively; for example, $\mathbf{A} = a_1\mathbf{i} + a_2\mathbf{j} + a_3\mathbf{k}$. A vector drawn from a fixed origin to a given point (x, y, z) is called a **position vector** and is usually symbolized by $\mathbf{r}$; in rectangular Cartesian coordinates,

$$\mathbf{r} = x\mathbf{i} + y\mathbf{j} + z\mathbf{k}.$$

Equations written in vector form are valid in any **coordinate system**. Mathematically, a vector is a single-row or single-column array of functions obeying certain laws of transformation. *See* **scalar product, vector product, Helmholtz's theorem**.

vector field—*See* **field**.

vector potential—*See* **solenoidal**.

vector product—(Also called cross product, outer product.) A **vector** with magnitude equal to the product of the magnitudes of any two given vectors and the sine of the angle between their positive directions.
For two vectors $\mathbf{A}$ and $\mathbf{B}$, the vector product is often written $\mathbf{A} \times \mathbf{B}$ (read "$\mathbf{A}$ cross $\mathbf{B}$"), and defines a vector perpendicular to both $\mathbf{A}$ and $\mathbf{B}$ and so directed that a **right-hand rotation** about $\mathbf{A} \times \mathbf{B}$ through an angle of not more than 180° carries $\mathbf{A}$ into $\mathbf{B}$. The magnitude of $\mathbf{A} \times \mathbf{B}$ is equal to twice the area of the triangle of which $\mathbf{A}$ and $\mathbf{B}$ are coterminous sides. If the vector product is zero, either one of the vectors is zero or the two are parallel. When $\mathbf{A}$ and $\mathbf{B}$ are written in terms of their components along the x, y, and z axes of the **rectangular Cartesian coordinates**, that is,

$$\mathbf{A} = a_1\mathbf{i} + a_2\mathbf{j} + a_3\mathbf{k},$$

$$\mathbf{B} = b_1\mathbf{i} + b_2\mathbf{j} + b_3\mathbf{k},$$

then the vector product is the determinant

$$\mathbf{A} \times \mathbf{B} = -\mathbf{B} \times \mathbf{A} = \begin{vmatrix} \mathbf{i} & a_1 & b_1 \\ \mathbf{j} & a_2 & b_2 \\ \mathbf{k} & a_3 & b_3 \end{vmatrix}.$$

See **scalar product**.

veering—1. According to general international usage, a change in **wind direction** in a clockwise sense (e.g., south to southwest to west) in either hemisphere of the earth; the opposite of **backing**.
2. According to widespread usage among U.S. meteorologists, a change in **wind direction** in a clockwise sense in the Northern Hemisphere, counterclockwise in the Southern Hemisphere; the opposite of **backing**.

veering wind—In the Northern Hemisphere, a **wind** that rotates in a clockwise direction with increasing height; the opposite of **backing wind**.

vegetation index—A numerical value used to predict or assess vegetative characteristics such as plant leaf area, total biomass, and general health and vigor of the surface vegetation.

Vegetation indices are usually derived from **multispectral remote sensing** observations. Since growing plants strongly reflect the wavelengths of **light** in the **near-infrared**, combinations of measurements in the near-infrared and visible-red portions of the spectra are used to generate a variety of different indices. Perhaps the most common vegetation index is the Normalized Difference Vegetation Index (NDVI), defined as

$$(NIR - RED) / (NIR + RED),$$

where NIR and RED are the **radiance** or **reflectance** measured in the near-infrared and red portions of the **spectrum**, respectively. This, and other indices, can be adapted to a variety of **remote sensing** satellites and instruments.

velocity—The time rate of change of a **position vector**; that is, a change of position expressed in terms of speed and direction.

If $\mathbf{x}$ is the position vector of a given point in space and t is time, the velocity, $\mathbf{u}$, is given by

$$\mathbf{u} = \frac{D\mathbf{x}}{Dt}.$$

See also **relative velocity**, **absolute velocity**.

velocity aliasing—(Also called velocity folding.) A basic sampling problem that arises when the unambiguous **velocity** sampling interval is less than the full **range** of naturally occurring velocities, causing the erroneous appearance of higher velocities within the sampling interval.

This phenomenon occurs in **Doppler velocity** measurements when the **maximum unambiguous velocity** interval ($\pm V_{max}$) is less than the full range of velocities being measured. Any true velocity, V, appears within the interval from $-V_{max}$ to $+V_{max}$, with the value V', which is related to the true velocity by $V = V' \pm 2nV_{max}$ where n is an integer. Therefore a given measured velocity V' may be caused by many values of the true velocity V. For example, suppose $V_{max} = 25$ m s^{-1} and the measured velocity $V' = -15$ m s^{-1}. Then the values of true velocity that could account for this measurement are the following: -15 m s^{-1} (for $n = 0$); $+35$ or -65 m s^{-1} (for $n = 1$); $+85$ or -115 m s^{-1} (for $n = 2$); etc. In some instances the erroneous velocities can be recognized and ambiguities resolved by additional considerations, such as the requirement of spatial **continuity** of the velocity field. *See also* **aliasing**, **Nyquist frequency**.

velocity–area method—A method of **stream gauging** in which depth and vertically averaged **velocity** measurements are taken at successive locations along a stream **cross section** and are numerically integrated to give the **streamflow** discharge.

velocity–azimuth display—(Abbreviated VAD.) A **radar** display of the **mean Doppler velocity** at a given **range** as a function of the **azimuth** angle as the radar antenna rotates through (usually) a complete 360° azimuth scan at a constant **elevation angle**.

Harmonic analysis of a VAD is used to determine spatially averaged kinematic properties of the **wind field**, namely, the **wind speed** and **direction**, the **horizontal divergence**, the horizontal **deformation**, and the orientation of the **axis of dilatation**. By repeating this process at many elevation angles or ranges, a **vertical profile** of the horizontal **wind** (**VAD wind profile**) above the radar site can be constructed.

velocity–contour method—A method of measuring **stream** discharge in which point **velocity** measurements are translated into average cross-sectional flow velocities by contouring the point velocities; these averages are then multiplied by the areas of the cross sections to give the **discharge**.

velocity curve—*See* **velocity distribution**.

velocity defect law—The **momentum** integral relation that expresses the mean stress between any two stations in terms of the difference in the **velocity** profiles between the stations.

Also known as von Kármán's momentum integral.

Brown, R. A., 1974: *Analytical Methods in PBL Modelling*, Halsted Press, appendix B, 148 pp.

velocity distribution—(Also known as velocity profile, velocity curve.) Usually the curve of point velocities along a vertical line normal to the flow direction that denotes the **variation** of **velocity** with depth.

In general flow, it may be a set of curves describing the flow variation along a **cross section**.

velocity divergence—*See* divergence.

velocity entrainment—The incorporation of laminar air having high velocity into a turbulent region of lower mean velocity.

velocity folding—Same as velocity aliasing.

velocity head—The kinetic energy of flow per unit weight of flowing liquid. ($h_v = v^2/2g$, where v is the flow velocity and g is the acceleration of gravity.)

velocity of approach—*See* approach velocity.

velocity of escape—*See* escape velocity.

velocity of sound—*See* speed of sound.

velocity potential—A scalar function with its gradient equal to the velocity vector **u** of an irrotational flow.

If $\chi(x, y, z)$ is the velocity potential,

$$\mathbf{u} = -\nabla\chi.$$

If the flow is also nondivergent, the velocity potential satisfies the Laplace equation

$$\nabla^2\chi = 0.$$

The velocity is everywhere normal to the surfaces of constant velocity potential. If a velocity potential exists, it is simpler to describe the motion by means of the potential rather than the vector velocity, since the former is a single scalar function whereas the latter is a set of three scalar functions.

velocity pressure—Same as dynamic pressure, wind pressure.

velocity profile—1. The variation of horizontal wind speed and direction (or of horizontal components of wind) with height such as would be measured with a sounding.
  2. The equations describing the variation of wind with height.
  A classic example is the logarithmic wind speed profile in the surface layer of the atmospheric boundary layer. *Compare* shear.
  3. *See* velocity distribution.

velocity rod—Floating rod weighted at the base so that it travels in an almost vertical position; the immersed portion may be adjustable.

velocity shear—The local variation of a velocity vector in a given direction.
  In oceanography, this frequently refers to the variation of the ocean current with depth. *See also* shear flow.

velocity spectra—Spectra that make up the components of the turbulence kinetic energy.
  See also spectral gap.
      Kaimal, J. C., and J. J. Finnigan, 1994: *Atmospheric Boundary Layer Flows*, Oxford University Press, 37–38.

velopause—The level at which the wind speed reaches a maximum or minimum (including zero) in a vertical profile of wind; analogous to the tropopause in a vertical profile of temperature.

velum—An accessory cloud veil of great horizontal extent draped over or penetrated by cumuliform clouds.
  Velum occurs with cumulus and cumulonimbus. *See* cloud classification.

Vema Channel—A deep trough at 31.3°S, 39.4°W in the Rio Grande Rise, South Atlantic Ocean, with a depth of 4646 m and a width of approximately 18 km. The Vema Channel acts as a conduit for Weddell Sea Bottom Water and Antarctic Bottom Water (potential temperature below 2°C) between the Argentine and the Brazil Basins, with a mean transport at 4 Sv (4×10^6 m³s⁻¹). Named after the research vessel "Vema" of the former Lamont–Doherty Geological Observatory, Palisades, New York (now Lamont–Doherty Earth Observatory).

vena contracta—The ratio of the area measured at the most contracted section of a jet of liquid issuing from an opening to the area of the opening.

vendaval—A stormy southwest wind on the southern Mediterranean coast of Spain and in the Strait of Gibraltar.
  It occurs with a low advancing from the west in late autumn, winter, or early spring, and is often accompanied by thunderstorms and violent squalls.

vent da Mùt—A strong, "wet," wind of Lake Garda in Italy.

vent des dames—A daily **sea breeze** of about 7 m s⁻¹ (15 mph) from the southwest in summer on the Mediterranean coast east of the Rhône delta, extending some 32 km (20 miles) inland.

vent du midi—In France, a south **wind** in the center of the Massif Central and the southern Cévennes. It is warm, moist, and generally followed by a southwest wind with **heavy rain**.

VENTAL-PREVENTAL code—A code, used primarily in French North Africa, in which data on **contours** of constant **pressure** or prognostic contours are encoded and transmitted.

Except in key words, it is identical with the **international analysis code** insofar as symbolic form is concerned.

ventifact—A stone with its form modified by the sand-blasting effect of wind-carried sediments.

Ventifacts are common in **periglacial** zones as well in deserts. *See* **dreikanter**.

ventilated psychrometer—A **hygrometer** that is artificially aspirated; in this way proper readings can be obtained within about three minutes.

In the case of a naturally aspirated wet bulb, the correct **wet-bulb temperature** will be attained after approximately 15 minutes provided the water reservoir has about the same **temperature** as the air. If the water temperature differs substantially from that of the air, it may be necessary to wait up to 30 minutes.

ventilated thermocline—A **model** of the oceanic **thermocline** in which **parcels** are advected horizontally while maintaining their **potential vorticity**.

The potential vorticity is set at the time of **subduction**, when the parcel leaves the **mixed layer** and enters the thermocline.

ventilation—1. In the equation for the **time constant** of a **thermometer**, a quantity equal to the product of **wind speed** and air **density**.

The time constant varies inversely with ventilation. The concentration of an **air pollutant** is inversely proportional to the ventilation, the mass **flux** of "clean air" moving past the observer. Stagnation, a condition caused by the lack of ventilation, is historically associated with major **air pollution** episodes. It occurs due to the lack of horizontal wind speed and the lack of vertical wind speed, for example, caused by an **inversion**.

2. The exchange of properties with the **surface layer** such that property concentrations are brought closer to **equilibrium** values with the **atmosphere**.

Such exchange may occur without **water mass formation**.

3. In weather-observing terminology, the process of causing "representative" air to be in contact with the sensing elements of observing instruments; especially applied to producing a flow of air past the bulb of a **wet-bulb thermometer**.

venting mixed layer air—The action of venting pollutants from the **convective boundary layer** to the **atmosphere** above by **cumulus** clouds that break through the stably stratified **entrainment zone** at the top of the **mixed layer**.

vento di sotto—**Breezes** blowing up-lake on Lake Garda in Italy.

Venturi effect—An increase in **wind speed** and decrease of **pressure** as air flows through a constriction or mountain gap.

The relationship between wind speed and pressure is given by **Bernoulli's equation**, which states that the sum of **kinetic energy** plus pressure is constant along a **streamline**. *Compare* **gap wind**, **funneling**.

Venturi tube—A tube designed to measure the rate of flow of fluids.

It consists of a tube having a constriction or throat at its midsection. The difference between the **pressure** measured at the inlet and at the throat is a function of the fluid **velocity**. The instrument is frequently used in wind-tunnel work and as a speed **indicator** for aircraft. *Compare* **Pitot tube**.

veranillo—The lesser **dry season**, made up of a few weeks of hot dry weather, that breaks up the summer **rainy season** on the Pacific coast of Mexico and Central America.

See **verano**.

verano—In Mexico and Central America, the main **dry season**, generally occurring from November through April.

See **veranillo**.

verdant zone—Same as **frostless zone**.

verglas—Same as **glaze**.

verification—1. *See* **forecast verification**.

2. In expert systems, the process of ensuring that the **knowledge base** is correct and complete, adequately modeling human **expertise**.
See **validation**.

vernal—Pertaining to spring.
The corresponding adjectives for summer, fall, and winter are **aestival**, **autumnal**, and **hibernal**.

vernal equinox—The **equinox** at which the sun crosses the **celestial equator** from the Southern Hemisphere to the Northern Hemisphere.
The time of this occurrence is approximately 21 March.

vernalization—Exposure of seed or plants to low temperature to induce or accelerate the development of the ability to form flowers.

vernier scale—A small, movable, graduated **scale**, adjacent and parallel to the main scale of an instrument, which provides a means for interpolating between the graduations of the main scale.
Generally historical, like the slide rule; commonly replaced by **digital** electronics.

vertebratus—A **cloud variety** (applied mainly to the genus **cirrus**), the elements of which are arranged in a manner suggestive of vertebrae, ribs, or a fish skeleton.
See **cloud classification**.

vertical advection—*See* **advection**.

vertical anemometer—General name for an instrument designed to measure the vertical component of the **wind speed**.
See **anemoclinometer**; *compare* **vertical-axis anemometer**.

vertical-axis anemometer—General name for an instrument designed to measure the vertical component of the **wind vector**.
Compare **vertical anemometer**.

vertical-beam radar—A **radar** employing a vertically pointing **beam**.

vertical coordinate system—The system used to define the **finite-difference approximation** applied in the direction roughly perpendicular to the earth's surface in a numerical **model**.
The most obvious example is the use of geometric height; however, there are numerical or physical advantages to many other vertical coordinate systems. For example, a "terrain following" vertical coordinate system is one with a lowest **coordinate surface** that follows the terrain. *See* **sigma vertical coordinate, eta vertical coordinate, hybrid vertical coordinate, isentropic vertical coordinate, pressure vertical coordinates**.

vertical-current recorder—General term for an instrument that records the vertical electric **current** in the **atmosphere**.

vertical differential chart—A **synoptic chart** showing the difference in value of a **meteorological element** between two levels in the **atmosphere**.
A common example is the **thickness chart**. *See* **differential analysis, differential chart**.

vertical gustiness—*See* **gustiness components**.

vertical jet—Same as **uprush**.

vertical modes—A complete set of eigenfunctions of a vertical **eigenvalue** problem derived from the linearized governing equations of fluid motions of given mean stratification.
These modes are often referred to as natural or dynamic modes.

vertical profile—A graph showing the **variation** of a meteorological event with height.

vertical section—In general, any graph of the vertical distribution of a quantity with respect to either time or space.
See **cross section, time section, time–height section, profile**.

vertical stability—Same as **static stability**.

vertical stretching—1. Extension, usually of a fixed volume, in the direction of the local vertical.

2. A process in which ascending vertical motion of air increases with **altitude**, or descending motion decreases with (increasing) altitude.

vertical time section—*See* **time section**.

vertical totals index—*See* **stability index**.

vertical velocity—In meteorology, the component of the **velocity** vector along the local vertical.

vertical velocity variance—The statistical **spread**, or mean-square **deviation** from the mean, of the **vertical velocity**; useful as a measure of **turbulence intensity** or **gravity wave** amplitude.
It is the vertical component of **turbulence kinetic energy** per unit mass. It determines the rate of vertical spread of **pollutants**.

vertical visibility—A subjective or instrumental evaluation of the vertical distance into a surface-based **obscuration** that an **observer** is able to see.
The height ascribed to vertical visibility is always a **ceiling** height.

vertical vorticity—The component of the **vorticity** vector along the local vertical.
References to vorticity usually imply the vertical component of that vorticity as a **scalar** quantity.

vertical wind shear—The condition produced by a change in **wind** velocity (speed and/or direction) with height.

vertical wind velocity—*See* **vertical velocity**.

vertically integrated liquid—(Abbreviated VIL.) Vertical integral of liquid **water content** obtained from **radar** observations at different **elevation** angles within a **precipitation echo**; has **dimensions** of mass per unit area.
Liquid water content M is computed from the **equivalent reflectivity factor** Z_e using the **Marshall–Palmer** drop-size distribution and is expressed as

$$M = 3.44 \times 10^{-3} Z_e^{4/7},$$

where the units of M are grams per cubic meter and Z_e has its conventional units ($mm^6\ m^{-3}$). The presence of **hail** in a **storm** produces larger VIL values than otherwise would be expected. VIL values greater than a threshold value are used to indicate the presence of large hail or other types of **severe weather**. The threshold value varies geographically, seasonally, and daily with changing environmental conditions.

very cloudy sky—A sky in which opaque **cloud cover** is between 0.9 and 1.0.

very high frequency—(Abbreviated VHF.) *See* **radio frequency band**.

very large eddy simulation—(Abbreviated VLES.) A modeling technique in which spatial **resolution** covers part, but not all, of the energy-containing subrange.
Most large-scale ocean (and **atmosphere**) **models** fall into this category.

very low frequency—(Abbreviated VLF.) *See* **radio frequency band**.

very short-range forecast—A **weather forecast** made for a time period of generally less than six hours.
Compare **nowcast**, **short-range forecast**, **long-range forecast**.

vesine—An **upvalley day wind** of the Department of Drôme in France.

VFR—The commonly used abbreviation for **visual flight rules** (formerly contact flight rules, or CFR).
In popular aviation terminology, it is used as a descriptive term for the relatively favorable weather and/or flight conditions to which visual flight rules apply. *Compare* **IFR**.

VFR between layers—A flight condition wherein an aircraft is operated under modified **visual flight rules** while in flight between two layers of clouds and/or obscuring phenomena, each of which constitutes a **ceiling**.
Compare **VFR on top**.

VFR flight—Same as **visual flight**.

VFR on top—A flight condition wherein an aircraft is operated under modified **visual flight rules** while in flight above a layer of clouds and/or an **obscuring phenomenon** sufficient to constitute a **ceiling**.
The limiting conditions for flying "VFR on top" are prescribed in Civil Air Regulations. *Compare* **VFR between layers**.

VFR terminal minimums—A set of **operational weather limits** at an airport, that is, the minimum conditions of **ceiling** and **visibility** under which **visual flight rules** may be used.

VFR weather—(Formerly called contact weather.) In aviation terminology, route or terminal weather conditions that allow operation of aircraft under **visual flight rules**.

V_g—Symbol for **deposition velocity**.

See **dry deposition.**

VHF—Abbreviation for very high frequency.
See **radio frequency band.**

VHF source—Typically, a source of **very high frequency** (VHF) radiation.
Lightning discharges are one example of VHF sources in the **atmosphere.**

video—The electrical **signal** that is obtained from a received radio signal after **detection** or demodulation; a general term pertaining to the picture signals in a television system or to the information-carrying signals that are eventually presented on the cathode ray tubes of a **radar.**

video gain—That portion of the **gain** of a **video** or **radar** receiver that occurs after **signal** detection, usually expressed in **decibels.**

video mapping—The electronic superposition of geographic or other data on a **radar** display.

video signal—1. An **analog** voltage produced by a **sensor** and conveying information, often with some information **parameter** varying periodically with time.

2. In imaging systems, a voltage representing **image** intensity as a function of position in the image.

3. In **radar** or **lidar**, the **signal** produced by the **detection** system, usually conveying the **signal intensity** as a function of **range.**

vidicon camera—A storage-type, electronically scanned photoconductive television camera tube that often has a response to **radiation** beyond the limits of the visible region; particularly used in early space applications, since no film was required.

viento zonda—Same as **zonda.**

vigil basins—Small drainage areas (up to about 25 km²) in which periodic measurements are conducted on a long-term basis.
Observations are made on both geomorphological and hydrological characteristics, including channel changes, valley-floor features, hillslopes, **reservoirs, precipitation, runoff,** and vegetation. The purpose is to document changes in the landscape and its **hydrology** over time, especially over a period of decades, and to make the data available to present and future generations of scientists. A formal Vigil Network was established on an international basis in the 1960s by U.S. Geological Survey scientists L. B. Leopold, W. W. Emmett, and R. F. Hadley.

VIL—Abbreviation for **vertically integrated liquid.**

violent storm—(Also called **storm, storm wind.**) In the **Beaufort wind scale, wind** with a speed from 56 to 63 knots (64 to 72 mph) or Beaufort Number 11 (Force 11).

virazon—(Also spelled birazon.) The very strong southwesterly **sea breeze** experienced where the coastal chains of the Andes Mountains descend steeply to the sea.
It sets in at about 10 A.M. and reaches its greatest strength at about 3 P.M. In Valparaiso, Chile, on summer afternoons it is so strong that it lifts pebbles in the streets. During the night it is followed by a light **land breeze**, the **terral** ("wind of the land").
2. A westerly **sea breeze** of Spain and Portugal.

virazones—Alternating **land** and **sea breezes** of Spain and Portugal.
See **virazon.**

virga—(Also called Fallstreifen, fallstreaks, precipitation trails.) Wisps or streaks of water or **ice** particles falling out of a **cloud** but evaporating before reaching the earth's surface as **precipitation.**
Virga is frequently seen trailing from **altocumulus** and **altostratus** clouds, but also is discernible below the bases of high-level **cumuliform** clouds from which precipitation is falling into a dry **subcloud layer**. It typically exhibits a hooked form in which the streaks descend nearly vertically just under the precipitation source but appear to be almost horizontal at their lower extremities. Such curvature of virga can be produced simply by effects of strong vertical **wind shear**, but ordinarily it results from the fact that **droplet** or **crystal** evaporation decreases the **particle** terminal **fall velocity** near the ends of the streaks. Under some conditions, virga are associated with dry **microbursts**, which are formed as a product of the **evaporation.** *See* **cloud classification.**

virtual gravity—*See* **apparent gravity.**

virtual height—The apparent height of a layer in the **ionosphere**, determined from the time required for a radio **pulse** to travel to the layer and return, assuming that the pulse propagates at the **speed of light.**

Compare **scale height**.

virtual impactor—Instrument used to sample **particles** of a given size **range** according to their inertia.

In the counterflow virtual impactor, a stream of air coming out of the instrument is used to reject particles with less than a specified inertia (size).

virtual potential temperature—The theoretical **potential temperature** of **dry air** that would have the same **density** as **moist air**.

It is used as a convenient surrogate for density in **buoyancy** calculations. The virtual potential temperature θ_v is defined by

$$\theta_v = \theta(1 + 0.61r - r_L),$$

where θ is the actual potential temperature, r is the **mixing ratio** of **water vapor**, and r_L is the mixing ratio of liquid water in the air. Temperatures must be in units of Kelvin, and mixing ratios in units of $g_{water}/g_{dry\ air}$. Because water vapor is less dense than dry air, humid air has a warmer θ_v than dry air. Liquid water **droplets**, if falling at their terminal **velocity** in air, make the air heavier and are associated with colder θ_v. For saturated or **cloudy** air, use **saturation mixing ratio** in place of r, while for unsaturated air, use $r_L = 0$. *See* **virtual temperature**.

virtual temperature—(Also called density temperature.) The virtual temperature $T_v = T(1 + r_v/\epsilon)/(1 + r_v)$, where r_v is the **mixing ratio** and ϵ is the ratio of the gas constants of air and **water vapor**, ≈ 0.622.

The virtual temperature allows the use of the dry-air **equation of state** for **moist air**, except with T replaced by T_v. Hence the virtual temperature is the **temperature** that dry **dry air** would have if its **pressure** and **density** were equal to those of a given **sample** of moist air. For typical observed values of r_v, the virtual temperature may be approximated by $T_v = (1 + 0.61\ r_v)\ T$. Some authors incorporate the density increment due to liquid or solid water into virtual temperature, in which case the definition becomes $T_v = T(1 + r_v/\epsilon)/(1 + r_v + r_l) \approx T(1 + 0.61r_v - r_l)$, where r_l is the liquid or liquid plus solid water mixing ratio.

VIS—Abbreviation for visible.

viscosity—(Also called internal friction.) The **transport** of mass motion **momentum** solely by the **random** motions of individual molecules not moving together in coherent groups.

Viscosity is a consequence of **gradients** in **velocity** fields in fluids. Sometimes described as fluid **friction** because velocity gradients in fluids are damped as a consequence of viscosity. *See* **viscous force, stress tensor, dynamic viscosity, kinematic viscosity, Newtonian friction law, Navier–Stokes equations, eddy viscosity**.

viscosity coefficient—*See* **dynamic viscosity, kinematic viscosity, eddy viscosity**.

viscous-convective subrange—The **range** of **wavenumbers** in large **Prandtl number** flows in which **viscosity** is important, resulting in **temperature** fluctuations being reduced by the strain-rate **field**, but where **thermal diffusivity** is not yet effective.

viscous dissipation—*See* **dissipation**.

viscous drag—(Or skin friction.) *See* **drag**.

viscous fluid—A fluid for which the molecular viscous effects of **diffusion** and **dissipation** can have significant effects on the flow.

The importance of **viscosity** depends on the relevant **velocity** and length scales of the flow and the viscosity of the fluid. The nondimensional measure of the relative importance of viscosity is the reciprocal of the **Reynolds number**, Re. For typical atmospheric flows, Re $> 10^7$, implying that viscous effects may be neglected relative to the leading $[O(1)]$ terms in the **Navier–Stokes equations**. However, in a **turbulent flow**, such as in the **boundary layer**, **vortex stretching** causes a continuous **nonlinear** cascade of **turbulence kinetic energy**, TKE, from large **scales** to smaller scales. The largest-scale **eddies** are responsible for the **Reynolds stresses** and conversion of mean flow **energy** into TKE. At some point in the **cascade**, the eddies will have length and velocity scales that are sufficiently reduced for the Reynolds number to be of order 1. At these scales the TKE of eddies can be converted into **internal energy** through **viscous dissipation**. These small eddies will have a much shorter timescale than the largest eddies in the flow and are thus statistically independent of the large-scale motion. For a developed turbulent flow, the rate at which the mean flow energy is converted into **turbulence** at the largest scales must be equal to the rate at which it is ultimately dissipated by viscosity by the small-scale eddies. These smallest eddies have scales that are very much larger than those of the molecular motions, and the continuum hypothesis is still valid for describing eddies of this size. Although dissipation occurs at the smallest possible eddies in the flow and the TKE is contained in the largest-scale eddies, the viscous dissipation is a leading term in the TKE budget and may not be ignored. *See* **turbulence spectrum**, **turbulence length scales**.

Tennekes, H., and J. L. Lumley, 1972: *A First Course in Turbulence*, MIT Press, 256–262.

viscous force—The force per unit volume or per unit mass arising from the action of **tangential stresses** in a moving **viscous fluid**; this force may then be introduced as a term in the **equations of motion**.

By far the most satisfactory hypothesis to date is that of Navier and Stokes, a generalization of the **Newtonian friction law**, which evaluates the **stress tensor** as directly proportional to the rate of **deformation**, the constant of proportionality being the **dynamic viscosity**. In this case, the viscous force per unit mass becomes

$$\nu \left[\nabla^2 \mathbf{V} + \frac{1}{3} \nabla (\nabla \cdot \mathbf{V}) \right],$$

where ν is the **kinematic viscosity**, $\mathbf{V}$ the **velocity** vector, and ∇ the **del operator**. The **divergence** term, $\nabla \cdot \mathbf{V}$, vanishes for an **incompressible fluid**, and the Navier–Stokes assumption is seen as leading to a simple **diffusion** of **momentum**. "It may seem a little strange that **viscosity**, which is a diffusion of momentum, can also diffuse **energy** and even **Reynolds stresses**. . . . A viscous **stress** acts both to convert mechanical **energy** into **heat** and to accelerate neighboring fluid, and this **acceleration** is an **energy transfer**. . . ." (Townsend 1956). *See* **Navier–Stokes equations**.

Townsend, A. A., 1956: *The Structure of Turbulent Shear Flow*, Cambridge University Press, p. 30.

viscous stresses—The components of the **stress tensor** remaining after the **pressure**, that is, the mean of the three normal stresses, has been subtracted out from each of the normal stresses.

See **Reynolds stresses**.

viscous subrange—The **range** of **wavenumbers** where the **dissipation** of **turbulence kinetic energy** occurs due to **viscosity**.

visentina—Strong east-northeast to east **winds** of Lake Garda in Italy.

visibility—1. The greatest distance in a given direction at which it is just possible to see and identify with the unaided eye 1) in the daytime, a prominent dark object against the sky at the **horizon**, and 2) at night, a known, preferably unfocused, moderately intense **light** source.

After visibilities have been determined around the entire horizon circle, they are resolved into a single value of **prevailing visibility** for reporting purposes. There are inherent difficulties with the conventional requirement that visibility markers be both detected and recognized. The more rigorously defined concept of the **visual range** avoids reference to recognition; thus, if the recognition requirement were dropped, the visibility could be defined as a subjective estimate of visual range. For most practical purposes, it can be defined that way now. Daytime estimates of visibility are subjective evaluations of **atmospheric attenuation** of **contrast**, while nighttime estimates represent attempts to evaluate something quite different, namely, **attenuation** of **flux density**. Thus, visibility data must be regarded as falling into two distinct classes, those obtained by day, and those by night. In U.S weather observing practice, it is the value as obtained and reported by an **observer** or by an **automatic weather station**. *See* **surface visibility**, **control-tower visibility**, **runway visual range**, **night visual range**.

2. The clarity with which an object can be seen.

visibility in clouds—As viewing the wingtip of an aircraft, gives a measure of the path droplet total cross-sectional area that is related to the inverse of the **droplet** size for a given liquid **water content**.

visibility marker—Landmark (building, church tower, house, hill, screen of trees, etc.) that is a known distance from the observing **station** and is used in determining **visibility**.

visibility meter—The general term for instruments used to make direct measurements of **visual range** in the **atmosphere** or of the physical characteristics of the atmosphere that determine the visual range.

Visibility meters may be classified according to the quantities that they measure. **Telephotometers** and **transmissometers** measure the **transmissivity** or, alternatively, the **extinction coefficient** of the atmosphere. **Nephelometers** measure the **scattering function** of the **particles** suspended in the atmosphere. A third category of visibility meters makes use of an artificial "haze" of **variable** density that is used to obscure a marker at a fixed distance from the visibility meter.

visibility object—*See* **visibility marker**.

visibility recorder—Same as **visibility meter**.

visible horizon—Same as **apparent horizon**.

visible imagery—Environmental satellite imagery sensed in the wavelengths between 0.4 and 0.75 μm in the **electromagnetic spectrum**.

Visible Infrared Spin Scan Radiometer—(Abbreviated VISSR.) The primary meteorological **sensor** providing imaging functions on the **SMS** and early **GOES** satellites.

visible radiation—**Electromagnetic radiation** capable of evoking the sensation of vision in a **normal** human observer.
> *See* **visible spectrum**

visible spectrum—The indefinite **range** of wavelengths of **visible radiation**, sometimes taken (for convenience) to lie between 400 and 700 nm.
> Radiation with wavelengths less than 400 nm is called **ultraviolet radiation**; **radiation** with wavelengths greater than 700 nm is called **infrared radiation**.

VISSR—Abbreviation for **Visible Infrared Spin Scan Radiometer**.

VISSR Atmospheric Sounder—(Abbreviated VAS.) An advanced version of the **VISSR** instrument that provided **atmospheric sounding** functions in addition to imaging functions.
> VAS was flown on **GOES**-4 through -7.

visual contact height—Same as **approach-light contact height**.

visual flight—(Also called VFR flight; formerly called contact flight.) An aircraft flight under conditions that allow navigation by visual reference to the earth's surface at a safe **altitude** and with sufficient **horizontal visibility**.
> Such a flight operates under **visual flight rules**. *See also* **VFR on top**, **VFR between layers**.

visual flight rules—(Abbreviated VFR.) A set of regulations set down by the U.S. Civil Aeronautics Board (in Civil Air Regulations) to govern the operational control of aircraft during **visual flight**.
> The abbreviation VFR is seldom used to denote the regulations themselves, but is popularly used to describe the weather and/or flight conditions to which these rules apply.

visual range—(Or daytime visual range.) The distance, under daylight conditions, at which the apparent **contrast** between a specified type of **target** and its background becomes just equal to the **threshold contrast** of an **observer**; to be distinguished from the **night visual range**.
> The visual range is a function of the atmospheric **extinction coefficient**, the **albedo** and visual angle of the target, and the observer's threshold contrast at the moment of **observation**. Only in the so-called **meteorological range** does one have a **visibility** figure dependent only upon the extinction coefficient. *See* **visual-range formula**; *compare* **visibility**.
>> Middleton, W. E. K., 1952: *Vision through the Atmosphere*, 104–122.
>> Johnson, J. C., 1954: *Physical Meteorology*, 79–90.

visual-range formula—The formula for the **daytime visual range** V,

$$V = \frac{1}{\sigma} \ln \frac{|C|}{\epsilon},$$

where there is a **contrast** C between the **horizon** target and its background viewed through an **atmosphere** of **extinction coefficient** σ by an **observer** whose momentary **threshold contrast** is ϵ.
> In practice, an important application of this formula is to the case of a black **target** (for which $C = 1$); hence

$$V = \frac{1}{\sigma} \ln \frac{1}{\epsilon}.$$

In calculations of the **meteorological range**, the threshold contrast is also assigned a constant value.

viuga—A cold north or **northeast storm** of the Russian steppes, lasting about three days.

VLES—Abbreviation for **very large eddy simulation**.

VLF—Abbreviation for very low frequency.
> *See* **radio frequency band**.

void ratio—In porous media, the volume of void space per volume of solids.

volatiles—Substances with relatively large **vapor pressures**.

Many organic substances are almost insoluble in water so that they occur primarily in a gas **phase** in contact with water, even though their vapor pressure may be very small.

volcanic aerosol—The **cloud** of **particles** injected into the **stratosphere** by explosive volcanic eruptions.

The particles consist mainly of **sulfuric acid** droplets, and their influence on **incoming solar radiation** gives rise to cooling at the earth's surface during major events. The effects can persist for years.

volume absorption coefficient—A measure of the **absorption coefficient**, with units of inverse length.

volume average—The mean of values measured within a three-dimensional volume.

Such measurements are typical of volume-scanning remote sensors such as **radar** and **lidar**. It is also used in physical and numerical simulations of the **atmosphere**. *Compare* **ensemble average, time average, ergodic condition**.

volume extinction coefficient—A measure of the depletion of **monochromatic** radiance passing through some medium in a constant direction.

The volume extinction coefficient equals the fractional depletion due to **absorption** and **scattering** when monochromatic **radiance** passes unit distance through a medium in a constant direction. Units are inverse length or km^{-1}. Values depend critically on the medium being traversed, as well as the **wavelength**. For example, in a **Rayleigh atmosphere** the volume extinction coefficient at a wavelength of 0.5 μm is less than 0.02 km^{-1}, whereas in a wet **cloud** it may exceed 100 km^{-1} at the same wavelength.

volume medium diameter—*See* **drop-size distribution**.

volume scan—In **radar**, a series of consecutive scans, either around the **horizon** or in a sector, that together **sweep** out a volume of space.

Volume scans are typically performed by conducting a series of horizontal scans, each at a progressively higher **elevation angle**. A less common method is to conduct a series of vertical scans between the horizon and the **zenith**, each at a different **azimuth** angle. Volume scans are used to develop three-dimensional views of the **reflectivity** field and, in the case of a **Doppler radar**, the **radial velocity** field associated with the targets illuminated by the radar. *See* **scanning**.

volume scattering coefficient—A measure of the **scattering coefficient**, with units of inverse length.

volume scattering function—*See* **scattering function**.

volume target—A **radar** target that fills the **pulse volume**; a **distributed target**.

voluntary observer—*See* **cooperative observer**.

Volz turbidity factor—A measure of the **turbidity** of the **atmosphere**.

The Volz turbidity factor is equal to the **aerosol optical depth** at a **wavelength** of 0.5 μm.

von Kármán's constant—(Or Kármán constant.) A constant k of the logarithmic **wind profile** in the **surface layer**.

It characterizes the dimensionless **wind shear** for statically neutral conditions,

$$\frac{\partial U}{\partial z} = \frac{u_*}{kz},$$

where z is height, u_* is **friction velocity**, and U is wind **velocity**. The Kármán constant was first determined in **hydrodynamics**. Presently a value of $k = 0.40 \pm 0.01$ is assumed. *See* **logarithmic velocity profile**.

von Kármán's law—*See* **logarithmic velocity profile**.

von Kármán vortex—A cloud **vortex** generated on the lee of an island barrier under conditions of a low-level **temperature inversion**.

The flow frequently forms a trail of counterrotating vortices, alternately placed in two parallel rows.

vortex—In its most general use, any flow possessing **vorticity**.

More often the term refers to a flow with closed streamlines or to the idealized case in which all vorticity is concentrated in a **vortex filament**.

vortex breakdown—The region of a **vortex** between a supercritical **upstream** flow incapable of supporting upstream-propagating centrifugal waves and a subcritical flow that allows downstream-propagating centrifugal waves.

The region is usually marked by abrupt swelling and often **turbulence** and reverse axial flow in the **downstream** side. Vortex breakdown is sometimes observed in **tornadoes**. *See* **hydraulic jump**.

vortex cloud street—The orientation of clouds along a **vortex filament**.

vortex filament—A line along which an infinite **vorticity** in a fluid motion is concentrated, the surrounding fluid being free of vorticity.

In an **autobarotropic** frictionless fluid, a **vortex line** always consists of the same fluid **particles**; the vortex filament is, thus, a vortex line and is the limiting case of a **vortex tube** as the cross-sectional area of the tube shrinks to zero.

vortex line—A curve tangent at every point of a **field** to the **vorticity** vector at that point.

vortex ring—A closed **vortex filament**.

vortex Rossby waves—In **tropical cyclones**, **Rossby waves** that propagate azimuthally on the radial **gradient** of relative **potential vorticity** in the cyclone circulation.

vortex sheet—A **surface of discontinuity** of **velocity** in a fluid, which may be regarded as formed by **vortex** filaments oriented normal to the **shear vector** across the surface.

Such a **velocity distribution** exhibits **Helmholtz instability**.

vortex signature—*See* **mesocyclone signature, tornadic vortex signature**.

vortex street—(Also called Kármán vortex street, vortex trail, vortex train.) Two parallel rows of alternately placed vortices along the **wake** of an obstacle in a fluid of moderate **Reynolds number**.

Fluid **drag** can be calculated from the motion of these vortices, which are stable only for a certain ratio of the width of the street to the distance between vortices along the street. *See* **vortex cloud street**.

> Batchelor, G. K., 1967: *An Introduction to Fluid Dynamics*, Cambridge University Press, pp. 261, 338, 536.

vortex stretching—The stretching (**deformation**) of a column that contains a **vortex**.

vortex thermometer—A **thermometer**, used in aircraft, that automatically corrects for **adiabatic** and frictional temperatures rises by imparting a rotary motion to the air passing the **thermal** sensing element.

By proper design of the vortex-forming chamber, the true free-air **temperature** can be obtained over a wide **range** of air speeds.

vortex trail—Same as **vortex street**.

vortex train—Same as **vortex street**.

vortex tube—The closed surface or tube consisting of the **vortex lines** passing through every point of a given closed curve.

vorticity—A **vector** measure of local rotation in a fluid flow, defined mathematically as the **curl** of the **velocity** vector,

$$\zeta = \nabla \times \mathbf{u},$$

where ζ is the vorticity, $\mathbf{u}$ the velocity, and ∇ the **del operator**.

The vorticity component normal to a small plane element is the limit of the **circulation** per unit area as the area of the element approaches zero (*see* **Stokes's theorem**). The vorticity of a **solid rotation** is twice the **angular velocity** vector. In meteorology, "the vorticity" usually refers to the vertical component of the vorticity. *See also* **relative vorticity, absolute vorticity, geostrophic vorticity, thermal vorticity, vorticity equation, curl**; *compare* **deformation, divergence**.

vorticity advection—Advection of **vorticity** by the total **wind**:

$$-\mathbf{v} \cdot \nabla_p (\zeta + f),$$

where $\mathbf{v}$ is the total **wind vector**, ∇_p is the **del operator** on an **isobaric surface**, ζ is the **relative vorticity**, and f is the planetary vorticity (the **Coriolis parameter**).

Vorticity advection is one **parameter** used in the **prediction** of vertical motion in the **atmosphere**: Positive vorticity advection corresponds to rising motion, and negative vorticity advection corresponds to sinking motion (*see* **omega equation**). Synoptically, the 500-mb level is used to evaluate vorticity advection since it is close to the **level of nondivergence** in the atmosphere where vorticity is approximately conserved.

vorticity equation—A dynamic equation for the rate of change of the **vorticity** of a **parcel**, obtained by taking the **curl** of the **vector** equation of motion.

The vorticity equation takes on slightly different forms depending on whether height or **pressure** is taken as the vertical coordinate. With height as the vertical coordinate, and with **friction** terms omitted, the vertical component of the vorticity equation is

$$\frac{D\zeta}{Dt} = -(\zeta + f)\left(\frac{\partial u}{\partial x} + \frac{\partial v}{\partial y}\right) - \frac{2\Omega\cos\phi}{a}\, v$$
$$+ \left(\frac{\partial w}{\partial y}\frac{\partial u}{\partial z} - \frac{\partial w}{\partial x}\frac{\partial v}{\partial z}\right) + \left(\frac{\partial p}{\partial x}\frac{\partial \alpha}{\partial y} - \frac{\partial p}{\partial y}\frac{\partial \alpha}{\partial x}\right),$$

where ζ is the vertical component of the **relative vorticity**; f the **Coriolis parameter**; u, v, and w the components of the **wind** velocity toward the east (x), north (y), and vertical (z); Ω the angular speed of the earth; p and α the air **pressure** and **specific volume**, respectively; a the earth's radius; and ϕ the latitude. The left-hand member of the equation represents the material rate of change of the relative vorticity of an **air parcel**. The first term on the right describes the effect of **horizontal divergence**. The second term on the right is the **Rossby parameter** times v and represents the change in vorticity resulting from latitudinal displacement. The third term on the right, often called the vertical **shear**, **twisting**, or tilting term, describes the influence of a horizontal **gradient** of **vertical velocity** in transforming vorticity about a horizontal axis to that about a vertical axis. The last term represents the generation of vorticity by pressure–volume **solenoids**. When pressure is taken as the vertical coordinate, and all differentiations are performed on an **isobaric surface**, the equation is simplified by the absence of the **solenoidal** term:

$$\frac{D\zeta_p}{Dt} = -(\zeta_p + f)\left(\frac{\partial u}{\partial x} + \frac{\partial v}{\partial y}\right) - \frac{2\Omega\cos\phi}{a}\, v + \left(\frac{\partial \omega}{\partial y}\frac{\partial u}{\partial p} - \frac{\partial \omega}{\partial x}\frac{\partial v}{\partial p}\right),$$

where ζ_p is the **isobaric vorticity** and $\omega = Dp/Dt$ is the material pressure change. If **adiabatic** processes are assumed, and **potential temperature** θ taken as vertical coordinate, all differentiations are performed on an **isentropic surface**, and the vorticity equation becomes

$$\frac{D\zeta_\theta}{Dt} = -(\zeta_\theta + f)\left(\frac{\partial u}{\partial x} + \frac{\partial v}{\partial y}\right) - \frac{2\Omega\cos\phi}{a}\, v.$$

Holton, J. R., 1992: *An Introduction to Dynamic Meteorology*, 3d edition, Academic Press, 102–113.

vorticity path—*See* **constant absolute vorticity trajectory**.

vorticity-transport hypothesis—The hypothesis that **vorticity**, and not **momentum**, is conservative in turbulent **eddy flux** due to the existence of **pressure** fluctuations.

This would apply especially if the **turbulence** were strictly two-dimensional. This hypothesis, together with that of the **mixing length**, leads to an expression for the **variation** of the **shearing stress** τ with height,

$$\frac{1}{\zeta}\frac{\partial \tau}{\partial z} = K\frac{d^2\overline{u}}{dz^2},$$

where K is the **coefficient of eddy viscosity**, and $\overline{u}$ the mean horizontal **wind**.

Voss polariscope—A modification of the **Savart polariscope** in which a **Wollaston prism** is used as an analyzer.

This instrument has a large colorless field and a high fringe-brightness level.

vriajem—Same as **friagem**.

vuthan—In Patagonia and southern South America, an intense **storm**.

VWP—Abbreviation for **VAD wind profile**.

W

wadi—In regions of the Middle East and North Africa, a **stream** bed or channel that only carries water during the **rainy season**.

In the southwest United States, the equivalent terms would be **arroyo** or wash.

wake—The region of **turbulence** immediately to the rear of a solid body in motion relative to a fluid.

Under certain conditions a series of vortices may form in the wake and extend **downstream**; such a **vortex train** in a turbulent wake is called a **vortex street**. *Compare* **lee eddies**.

wake depression—*Same as* **wake low**.

wake low—1. (Or wake depression.) In meteorology, a surface **low pressure area** or **mesolow** (or the envelope of several low pressure areas) to the rear of a **squall line**; most commonly found in squall lines with trailing **stratiform** precipitation regions, in which case the axis of the **low** is positioned near the back edge of the stratiform rain area.

2. (Or wake depression.) In fluid dynamics, a **low pressure area** on the **downstream** side of an object embedded in a flow.

wake turbulence—A disruption of airflow caused by the passage of a body through the air.

In aviation, wake turbulence has been known to cause upset of one aircraft following another. The Federal Aviation Administration has issued regulations requiring aircraft to take off, land, and fly minimum distances behind other aircraft. These distances vary by aircraft **type**.

waldsterben—German expression (literally, "forest death") for the forest dieback or decline usually associated with the effects of **acid rain**.

Walker circulation—A **direct cell** oriented along the **equator**; originally used by Bjerknes (1969) to refer to the cell induced by the contrast between the warm waters of the western Pacific and the cooler waters of the eastern Pacific.

Variability in this cell is associated with the **Southern Oscillation**. The term is now sometimes used to refer to the entire chain of east–west equatorial circulation cells that stretches around the globe.

> Bjerknes, J., 1969: *Tellus*, **18**, 820–829.

wall cloud—(Sometimes referred to as **pedestal cloud**.) A local, often abrupt lowering from a **cumulonimbus** cloud base into a low-hanging **accessory cloud**, normally a kilometer or more in diameter.

A wall cloud marks the lower portion of a very strong **updraft**, usually associated with a **supercell** or severe **multicell storm**. It typically develops near the **precipitation** region of the cumulonimbus. Wall clouds that exhibit significant rotation and vertical motions often precede **tornado** formation by a few minutes to an hour.

> Fujita, T., 1959: A detailed analysis of the Fargo tornadoes of June 20, 1957. *U.S. Wea. Bur. Res. Paper 42*, p.15.

warm-air drop—Same as **warm pool** .

warm air mass—*See* **airmass classification**.

warm anticyclone—Same as **warm high**.

warm braw—A warm, dry **foehn** wind that persists for four to eight days during the east **monsoon** in the Schouten Islands off the north coast of New Guinea.

warm cloud—Cloud that is only in the liquid **phase**; levels are not present with **temperature** below 0°C (32°F); no **ice** is present.

Any **precipitation** will originate from droplet **coalescence**. It is not to be confused with clouds extending to levels with temperature below 0°C; here, precipitation may form from the ice phase but could form by the warm cloud coalescence process.

warm conveyor belt—A narrow stream of air that transports large amounts of **heat**, moisture, and westerly **momentum**.

warm-core anticyclone—Same as **warm high**.

warm-core cyclone—Same as **warm low**.

warm-core high—Same as **warm high**.

warm-core low—Same as **warm low**.

warm-core rings—Large (roughly 300 km diameter) anticyclonically rotating **eddies** found in the slope waters to the north of the **Gulf Stream**, containing **Sargasso Sea** water in their core.

They persist for several months, being trapped in the region between the Gulf Stream and the **continental slope**, occasionally interacting with the Gulf Stream and getting destroyed in the process. A warm-core ring is formed from a large-amplitude Gulf Stream meander that pinches off to the north, trapping relatively warm Sargasso Sea water from south of the Gulf Stream within its circumferential current. *Compare* **cold-core rings**; *see* **Gulf Stream rings**.

warm cyclone—Same as **warm low**.

warm drop—Same as **warm pool**.

warm event—*See* **El Niño**.

warm fog—Fog having temperatures at all levels above 0°C (32°F).

warm front—Any nonoccluded **front**, or portion thereof, that moves in such a way that warmer air replaces colder air.

While some **occluded fronts** exhibit this characteristic, they are more properly termed warm occlusions.

warm-front-type occlusion—(Or warm occlusion.) *See* **occluded front**.

warm-front wave—A low pressure system that develops along a **warm front**.

warm high—(Or warm anticyclone; also called warm-core high, warm-core anticyclone.) At a given level in the **atmosphere**, any **high** that is warmer at its center than at its periphery; the opposite of a **cold high**.

Compare **thermal high**; *see* **cut-off high**.

warm low—(Or warm cyclone; also called warm-core low, warm-core cyclone.) At a given level in the **atmosphere**, any **low** that is warmer at its center than at its periphery; the opposite of a **cold low**.

Compare **thermal low**.

warm occluded front—(Or warm occlusion.) *See* **occluded front**

warm occlusion—(Also called warm-front-type occlusion, warm-type occlusion, warm occluded front.) *See* **occluded front**.

warm pool—(Also called warm drop, warm-air drop.) A region, or "pool," of relatively warm air surrounded by colder air; the opposite of a **cold pool**.

The common application of this term is to warm air of appreciable vertical extent isolated in high latitudes when a **cut-off high** is formed. Warm pools can be identified as **thickness** maxima on **thickness charts**.

warm rain—Rain formed from a **cloud** having temperatures at all levels above 0°C (32°F), and resulting from the droplet **coalescence** process.

warm rain process—In **cloud physics**, the process producing **precipitation** through collision between liquid **particles (cloud droplets, drizzle drops**, and **raindrops)**.

The warm rain process includes growth by **collision–coalescence** and limitations to growth by **drop breakup**. Precipitation produced by the warm rain process occurs in clouds having sufficient liquid water, **updraft**, and lifetime to sustain collision–coalescence growth to drizzle drop or raindrop sizes. Since warm base (>10°C) convective clouds of about 2 km depth typically have these features, the warm rain process is found to be active in both shallow and deep **convection** in the **Tropics** and midlatitudes. The major role of the warm rain process in thunderstorms is to **transfer** condensed water, in the form of cloud droplets, to **precipitable water**, in the the form of drizzle droplets and raindrops, by the collision–coalescence process. The warm rain process can also produce supercooled raindrops that **freeze** and become **graupel**, necessary for the rapid **glaciation** of convective tops by production of secondary **ice crystals**. This has been called the **coalescence** freezing mechanism. *See* **Hallett–Mossop process**.

warm ridge—A **ridge** characterized by relatively warm temperatures.

warm sector—That area, within the **circulation** of a **wave cyclone**, where the warm air is found.

Traditionally, it lies between the **cold front** and **warm front** of the **storm**; in the typical case, the warm sector continually diminishes in size and ultimately disappears (at the surface) as the result of **occlusion**.

warm tongue—A pronounced extension or protrusion of warm air.

warm-type occlusion—(Or warm occlusion.) *See* **occluded front**.

warm wave—(Rare.) Same as **heat wave**.

warning—Issued when a hazardous weather or hydrologic event is occurring, is imminent, or has a very high probability of occurring.

A warning is used for conditions posing a threat to life or property. *Compare* **watch**, **weather advisory**.

warning stage—The **stage**, on a fixed **river gauge**, at which it is necessary to initiate warnings for precautionary measures to be taken before **flood stage** is reached.

Wasatch winds—In the United States, strong, easterly winds, often locally referred to as canyon winds, blowing out of the mouths of the canyons of the Wasatch Mountains onto the plains of Utah.

They are produced by a strong large-scale west to east **pressure gradient** and are best developed when there is a large **high** over Wyoming or a vigorous **low** to the west or southwest in Utah or Arizona. In extreme cases they can exceed **hurricane** force. The strong winds are mainly limited to the canyons, especially in winter, but in some circumstances extend beyond them onto the plain.

Washoe zephyr—In the United States, a strong **downslope wind** on the eastern (Nevada) side of the Sierra Nevada in northern California.

This term has been applied to two different types of **wind**: 1) a **foehn** or **chinook** wind during the cold **season**; and 2) an afternoon wind down the east-facing canyons and valleys during the warm season. The latter may result from afternoon convective **mixing** or may be related to California sea-breeze effects. *See* **mountain–valley wind systems**.

washout—The removal of solid and gaseous material from the air and its **deposition** on the earth's surface due to **capture** by falling **precipitation**.

washout coefficient—A first-order **rate coefficient** that relates the rate of removal of a gas or **particle** from the **atmosphere** via **wet deposition** to its concentration.

That is, the rate of removal of the substance (in units of concentration divided by time) is directly proportional to the product of its concentration and the washout coefficient (units of inverse time).

wastewater—Water containing the waste matter (liquid or solid) of a community, manufacturing process, or industrial plant.

watch—Issued when the risk of a hazardous weather or hydrologic event has increased significantly but its occurrence, location, and/or timing is still uncertain.

It is intended to provide enough lead time so that those who need to set their plans in motion can do so. *Compare* **warning**, **weather advisory**.

water—1. A transparent, colorless, odorless, and tasteless liquid found near the surface of the earth.

In the **lithosphere, hydrosphere,** and **atmosphere** of the earth, water is found as a gas, liquid, and solid. Water falls from the clouds as **rain, hail, sleet, graupel, snow,** etc., and runs off and through soils to form creeks, streams, rivers, and lakes. In its solid form, it is referred to as **ice** or snow. Water as a liquid and as ice covers 70.8% of the surface of the earth and plays a fundamental part in the earth–atmosphere **energy balance**. Water (chemical formula H_2O) corresponds to two parts **hydrogen** and one part **oxygen** on a molecular basis; by weight, water is 11.19% hydrogen and 88.81% oxygen. Water has a **melting point** of 0°C (32°F), a **boiling point** of 100°C (212°F), and a **specific gravity** of 1.000 at 4°C (39°F), by definition.

2. Can refer to a body of water, such as a lake or a **stream**, or even a larger body of water such as a sea or part of an ocean, for example, international waters.

3. Used to describe water in specific locales; for example, hydrologists refer to soil water, **surface water**, and **groundwater**.

4. As a verb, used to describe **irrigation** corresponding to the application of water to plants, the grounds surrounding a residence, or to a garden.

water bleeding—Water forced through a **crack** in **ice** covering a body of water.

water budget—A budget of the incoming and outgoing water from a region, including **rainfall, evaporation, runoff,** and seepage; often used to estimate **evapotranspiration**.

water circulation coefficient—The ratio of a region's total **precipitation** to the amount of "external" precipitation originating as **evaporation** from the oceans as opposed to **evapotranspiration** from the land.

water cloud—Any **cloud** composed entirely of liquid water drops; to be distinguished from an **ice-crystal cloud** and from a **mixed cloud**.

Above the 0°C level some clouds may be entirely liquid within the limitations of measurement (1 crystal per 10 liters) even at low temperatures (−30°C as some lenticular clouds); on the other hand, some clouds not far below 0°C may contain considerably higher **ice** concentrations.

water content—(Also called free-water content, liquid water content.) The liquid water present within a sample of **snow** (or soil), usually expressed in percent by weight.

The water content in percent of **water equivalent** is 100 minus the **quality of snow**. *Compare* **snow density**.

water cycle—*See* **hydrologic cycle**.

water deficit—The cumulative difference between the **potential evapotranspiration** and **precipitation** during a specified period in which the precipitation is the smaller of the two.

water equivalent—The depth of water that would result from the melting of the **snowpack** or of a **snow** sample.

Thus, the water equivalent of a new **snowfall** is the same as the **amount of precipitation** represented by that snowfall. *Compare* **snow density**, **water content**.

water equivalent of snow—The amount of water that would be obtained if the **snow** sample were completely melted.

The snow sample must be obtained from a uniform volume, that is, either by melting snow in a calibrated container (e.g., an 8-in. **rain gauge**) or determined by weight (e.g., a **snow pillow**).

water equivalent of snow cover—The depth of water that would result from the melting of a **snow cover**; may be expressed either at a point or as an areal average.

water-flow pyrheliometer—An absolute **pyrheliometer**, developed by C. G. Abbot, in which the radiation-sensing element is a blackened water-calorimeter.

It consists of a cylinder, blackened on the interior, and surrounded by a special chamber through which water flows at a constant rate. The temperatures of the incoming and outgoing water, which are monitored continuously by **thermometers**, are used to compute the **intensity** of the **radiation**. This instrument was built by the Smithsonian Institution but was never widely used as a **standard** instrument. The function of the instrument as an absolute reference is currently fulfilled by the **absolute cavity radiometer**.

water hammer—A very rapidly moving **pressure wave** in a closed conduit, usually resulting from a sudden stoppage or change in the flow.

water head—*See* **head**.

water level—The height of the water surface measured above a **datum**.

water-level recorder—A device used for recording water levels in rivers, lakes, or wells, with respect to time.

water loss—Same as **evapotranspiration**.

water mass—A body of water with a common formation history, for example, **convection** caused by surface cooling, having its origin in a particular region of the ocean.

Water masses are identified by their **temperature**, **salinity**, and other properties such as nutrients or **oxygen** content. They have exclusive occupation of an oceanic region only in their formation region; elsewhere they share the ocean with other water masses with which they mix. Just as **air masses** in the **atmosphere**, water masses are physical entities with a measurable volume.

water mass formation—The **tendency** to build up the mass or density of a given **water type** (of temperature T and **salinity** S characteristics) due to air–sea fluxes or **mixing**.

water mass transformation—The conversion of seawater from one **water type** to another, by air–sea fluxes or **mixing**.

water molecule—A molecule, consisting of one central **oxygen** atom and two **hydrogen** atoms along an angle of 104°, forming an electrostatic **dipole** because of the asymmetric distribution of charge.

The dipole results in a high dielectric constant for water, whether in **vapor**, liquid, or solid form. Resonances in the intermolecular bonds result in **absorption** and **emission** of **electromagnetic radiation** at near- and thermal-infrared wavelengths, hence the importance of **water vapor** in the **greenhouse effect** in the **atmosphere** and water substance as **precipitation** in the **scattering** and absorption of **radar** waves.

water need—Quantity of water required by a specified area, to satisfy all types of water demands, for a given period of time.

water potential—The **potential energy** per unit mass of water with reference to pure water at zero **potential**.

The water potential τ is made up of several components,

$$\tau = \tau_g + \tau_m + \tau_p + \tau_o,$$

where τ_g is the component due to **gravity**, τ_m is the matrix potential that arises from the attraction of the soil matrix for water and water molecules for each other, τ_p is the **pressure** potential that is the ratio of the hydrostatic or pneumatic pressure to the **density** of water, and τ_o is the osmotic potential that is a driving force for water movement when solute movement is restrained with a semipermeable membrane

water requirement—1. Total water per unit area required by a crop for **normal** growth.

2. In plant physiology, the same as **transpiration ratio**.

water resources—Water in all states (solid, liquid, or **vapor**), in storage or in **flux** within the **hydrologic cycle**, that is necessary for a sustainable quality of life, as well as for sustaining the natural **environment**.

water sky—The dark appearance of the underside of a **cloud layer** when it is over a surface of open water.

This term is used, largely in polar regions, with references to the **sky map** in which water sky is darker than **land sky**, and much darker than **ice blink** or **snow blink**.

water smoke—Same as **steam fog**.

water snow—(Also called cooking snow.) Snow that, when melted, yields a greater-than-average amount of water; thus, any **snow** with a high water content.

water spray seeding—A process of clearing **warm fog** by enhancing drop **coalescence** growth by a high pressure and high volume water spray.

water spreading—One approach used for **artificial recharge**.

water-stage recorder—A device for obtaining a continuous record of **stage** at a point on a **stream**.

The most common recorders consist of a float-actuated pen that traces a record on a clock-driven **chart**. *See* **river gauge**.

water stress effect—The closing of the stomata by a plant in response to excessive **water loss** through **transpiration** or in response to **drought** conditions.

water structure—The arrangement of water molecules in the liquid state.

Unlike the case of **ideal gas** (**random** distribution) and ideal **crystal** (perfect order) models, there is no simple way to describe the ideal liquid water structure. It is known that the structure has short range order (similar to **ice**) but no long range order (similar to a gas), as shown by x-ray **diffraction** studies. Several competing models exist that attempt to explain the observed properties of water. Examples include the quasi-crystalline **model**, which assumes that water consists of broken-down pieces of ice; the clathrate model, which suggests that water resembles the clathrate structure of gas hydrates; and the bend-bond model, which suggests that the bonds are bent to various degrees. Other models also exist. Water has several properties of direct meteorological interest [e.g., maximum density at $+4°C$; maximum visible **refractive index** at $+1°C$; maximum thermal capacity at $+35°C$; large static dielectric constant (80) and its **frequency** variation] with which such models need to be consistent.

water-supply forecast—Statement of the expected volume of **available water** for a specified period and a specified area; may be associated with time distribution and **probability**.

water-supply sensitivity—The sensitivity of water supply systems to climatic fluctuations.

water surplus—Quantity of **available water** in excess of demand.

water table—(Also called phreatic surface, groundwater table.) Surface in a geologic medium where water pressure equals **atmospheric pressure**.

The water table separates the **saturated zone** from the **unsaturated zone**.

water-table aquifer—*See* **unconfined aquifer**.

water thermometer—*See* **seawater thermometer**.

water type—A mathematical construct representing a certain combination of **temperature** and sa-

linity and possibly nutrient and **oxygen** values, used in numerical **water mass** analysis to describe water masses.

Antarctic Bottom Water, for example, can be represented by the water type of temperature 0.3°C and salinity 34.7 **psu**, even though it may not always have exactly those properties.

water-use ratio—Same as **transpiration ratio**.

water vapor—(Also called aqueous vapor, moisture.) Water substance in **vapor** form; one of the most important of all constituents of the **atmosphere**.

Its amount varies widely in space and time due to the great variety of both "sources" of **evaporation** and "sinks" of **condensation** that provide active motivation to the **hydrologic cycle**. Approximately half of all of the atmospheric water vapor is found below 2-km **altitude**, and only a minute fraction of the total occurs above the **tropopause**. Water vapor is important not only as the raw material for **cloud** and **rain** and **snow**, but also as a vehicle for the **transport** of **energy** (**latent heat**) and as a regulator of planetary temperatures through **absorption** and **emission** of **radiation**, most significantly in the **thermal infrared** (the **greenhouse effect**). The amount of water vapor present in a given air sample may be measured in a number of different ways, involving such concepts as **absolute humidity**, **mixing ratio**, **dewpoint**, **relative humidity**, **specific humidity**, and **vapor pressure**.

water-vapor bands—Dark bands in the **solar spectrum** caused by the **absorption** of **solar radiation** by the **water vapor** contained in the earth's **atmosphere**.

water vapor feedback—The change in the radiative effect of **water vapor** in response to an external **perturbation** of the **climate**.

Water vapor, the most important greenhouse gas, absorbs only a small amount of **sunlight** but is a very efficient absorber of the earth's **thermal infrared** emission. Changes in either the amount or the vertical distribution of water vapor can therefore change the planet's ability to radiate **heat** to space. Climate models predict, without exception, that the water vapor feedback is positive. Changes in the distribution of water vapor in the middle and upper **troposphere** are inordinately important to this **feedback** process, because molecules that absorb upwelling **infrared radiation** at these altitudes emit it to space at a much colder **temperature** and therefore emit less than would be the case in their absence. The concentration of high-altitude water vapor is controlled by poorly understood dynamic and thermodynamic processes and is inadequately observed, thus contributing to uncertainty in the magnitude of the water vapor feedback.

water vapor imagery—A display of data taken in one of the **water vapor** channels, for example, 6.7 or 7.3 μm.

Atmospheric water vapor absorbs outgoing **radiation** in these regions, resulting in a decreased **temperature** being sensed by the satellite.

water vapor winds—An estimate of **wind vectors** in the middle and upper **troposphere** obtained by tracking **water vapor** targets between successive satellite images.

water year—(Or hydrologic year.) Generally, 1 October to 30 September in the Northern Hemisphere, 1 July to 30 June in the Southern Hemisphere; the **annual cycle** that is associated with the natural progression of the hydrologic seasons.

It commences with the start of the **season** of **soil moisture** recharge, includes the season of maximum **runoff** (or season of maximum groundwater **recharge**), if any, and concludes with the completion of the season of maximum **evapotranspiration** (or season of maximum soil moisture utilization).

waterfall effect—Same as **Lenard effect**.

waterlogged—Condition of land when the **water table** stands at or near the land surface, reaching into the root zone, and may be detrimental to plant growth.

watershed—1. Same as **river basin**.

In this sense, the term is most commonly applied to relatively small areas.

2. Same as **divide**.

(Apparently this use is becoming obsolete.)

waterspout—1. In general, any **tornado** over a body of water.

2. In its most common form, a nonsupercell tornado over water.

Such events consist of an intense columnar vortex (usually containing a funnel cloud) that occurs over a body of water and is connected to a cumuliform cloud. Waterspouts exhibit a five-stage, discrete life **cycle** observable from aircraft: 1) dark-spot stage; 2) spiral pattern stage; 3) spray-ring stage; 4) mature or spray-vortex stage; and 5) decay stage. Waterspouts occur most frequently in the **subtropics** during the warm **season**; more are reported in the lower Florida Keys than in any other place in the world. Funnel diameters **range** from a few up to 100 m or more; lifetimes average 5–10 minutes, but large waterspouts can persist for up to one hour.

watt—A unit of **power** equal to one **joule** per second or 10^7 ergs per second.

wave—1. Generally, any pattern with some roughly identifiable **periodicity** in time and/or space. This applies, in meteorology, to atmospheric waves in the horizontal **flow pattern** (e.g., **Rossby wave, long wave, short wave, cyclone wave, barotropic disturbance**). *See also* **inertia wave**.

2. At the surface of the ocean, a disturbance generated by **wind** action with dynamics governed by the influence of **gravity** and/or **surface tension**.
See **ocean waves**.

3. Popularly used as a synonym for "surge" or "influx," as in **tidal wave** (**storm surge**), **heat wave, cold wave**.

wave age—A measure of the time the **wind** has been acting on a **wave group**, either because the wind has been blowing for a finite length of time or because the **fetch** is limited; usually expressed as a **dimensionless number**, the **phase speed** of the peak of the **wave spectrum** divided by the **wind speed** or by the **friction velocity**.

wave amplitude—*See* **amplitude**.

wave average—Average over one or more full **wave periods**.
See **phase averaging**.

wave blocking—The blockage of ocean **surface waves** in an opposing **current**.
Blocking occurs when the **phase speed** of a **wave** equals the magnitude of the opposing current speed.

wave breaking—A complex phenomenon in which the surface of the **wave** folds or rolls over and intersects itself.
In the process it may mix (entrain) air into the water and generate **turbulence**. The causes of wave breaking are various, for example, through the wave steepening as it approaches a beach, through an interaction with other waves in **deep water**, or through the **input** of **energy** from the **wind** causing the wave to steepen and become unstable.

wave celerity—The speed of propagation of the **wave crest** or **trough**.
An alternative definition, useful for **linear** waves if many **wave** spectral components are involved, is the **frequency** multiplied by the **wavelength**, or the **angular frequency** divided by the **wavenumber**.

wave climate—The long-term **statistical** characterization of the behavior of waves in the ocean.
As an example, the seasonal variations in **significant wave height** may be characterized by calculating the monthly mean significant wave heights from several years of measurements.

wave cloud—Same as **undulatus**.
See **mountain wave cloud**.

wave crest—The highest point of a **wave motion**.

wave–current interaction—The exchange of **momentum** between an ocean **surface wave** and the near-surface current.

wave cyclone—(Also called wave depression.) A **cyclone** that forms and moves along a **front**.
The **circulation** about the cyclone center tends to produce a wavelike **deformation** of the front. The wave cyclone is the most frequent form of **extratropical cyclone** (or **low**). It was the purpose of the **wave theory of cyclones** to explain its life **cycle**.

wave depression—Same as **wave cyclone**.

wave diffraction—The change in **wave propagation** as affected by the presence of obstacles (e.g., sea walls or islands).
As the waves pass the obstacle the **wave crests** bend as the **wave** moves into the shadow of the obstacle.

wave dispersion—The spreading of waves in space due to differences in their speed and direction, in particular if speed differences arise from differences in **frequency**.
Because **dispersion** spreads the same waves over a larger area, **wave heights** decrease through dispersion. *See* **dispersion relationship**.

wave dissipation—The loss of **wave** energy, with consequent decrease in **wave height**, due to **wave**

breaking, **turbulence**, and viscous effects, and, in shallow water, due to the effects of **bottom friction**.

wave disturbance—1. (Or wave-type disturbance.) The disturbed state of a medium through which any **wave** form of **energy** is being propagated.

2. In **synoptic meteorology**, same as **wave cyclone**, but usually denoting an early stage in the **development** of a **wave cyclone**, or a poorly developed one.
See **disturbance**.

wave drag—A retarding force on the **wind** that occurs when atmospheric waves or oscillations form when air flows over mountains.

The waves **transport** momentum between the ground and critical levels of zero **wind speed** aloft. *Compare* **skin drag**, **form drag**.

wave equation—Any of a general class of **linear** and **nonlinear** scalar and **vector** partial differential equations in time and space.

An example of a wave equation is the linear scalar equation

$$\nabla^2 \phi = \frac{1}{v^2} \frac{\partial^2 \phi}{\partial t^2},$$

where v is the **phase speed**. An example of a nonlinear wave equation is the Korteweg–deVries equation

$$\frac{\partial u}{\partial t} + u \frac{\partial u}{\partial x} + \frac{\partial^3 u}{\partial x^3} = 0.$$

Electromagnetic waves are governed by a set of vector wave equations.

wave forecasting—The **prediction** of **wave** conditions in the ocean, usually based on **output** from a wave model.
See **wave modeling**.

wave frequency—1. The **frequency** of any time-harmonic (or sinusoidal) physical quantity governed by a **wave equation**.

2. In simple terms, the number of **wave crests** that pass a fixed point per second of time, equal to the inverse of the **wave period**.

However, in a **random sea**, a variety of interpretations of this are possible. A commonly used definition is the so-called zero-upcrossing **frequency**, which is the number of successive upcrossings of the **mean water level** that pass a fixed point per second of time. *See* **wave height**, **wave length**.

wave front—*See* **surface of constant phase**.

wave-front method—A method of **minimal flight** planning utilizing the physical principles that govern the propagation of a **wave front** through a medium.

Time curves are the principal reference lines.

wave generation—The action of the **wind** on the sea surface that generates ocean **surface waves** by the **transfer** of **momentum** from the **atmosphere** to the ocean.

This is an extremely complex process and still not fully understood.

wave group—A series of waves propagating together in which the **wave** direction, **wave length**, and **wave height** vary only slightly.

wave height—In simple terms, the vertical distance between a **wave crest** and the preceding or following **wave trough**.

However, in a **random sea**, a variety of interpretations are possible. A commonly used definition is the so-called zero-upcrossing wave height, which is the **range** of elevations (difference between highest crest and lowest trough) between two successive upcrossings of the **mean water level**. *See* **wave frequency**, **wave length**, **wave period**.

wave length—In simple terms, the horizontal distance between successive **wave crests** measured perpendicular to the crests.

However, in a **random sea**, a variety of interpretations of this are possible. A commonly used definition is the so-called zero-upcrossing length, which is the horizontal distance between two successive upcrossings of the **mean water level**. *See* **wave frequency**, **wave height**, **wave period**.

wave modeling—The use of mathematical and/or numerical techniques to solve the equations describing the generation, interaction, propagation, and **dissipation of waves** on the ocean surface.

The main types of wave modeling include 1) hydrodynamic modeling of fluid motions, to compute the evolution of the water surface and fluid velocities, and to compute forces on structures; 2) modeling of the propagation of waves as they approach the shore, taking account of **refraction** by varying depth; and 3) modeling of the evolution of the **spectrum** of the **wave** energy divided into components of different **frequency** and direction, in order to forecast the waves over an ocean region. Operationally, such models are used to produce wave forecasts (in much the same way as **weather forecasts** are produced) using **wind** information from **meteorological forecast** models.

wave motion—Nearly synonymous with **wave**, but more likely to be used in a restricted sense to mean a wave in which the quantity of interest is a physical displacement of matter, as, for example, waves on a string or on water.

wave of oscillation—(Or oscillatory wave.) A **wave** that results in no mean displacement of the **particles** of the fluid in the direction of motion of the wave, in contrast to a **wave of translation**.
> *Compare* **transverse wave**.

wave of translation—(Or translatory wave.) A **wave** that is accompanied by substantial net movement of the fluid in the direction of **wave motion**, although the wave propagates more rapidly than the fluid.
> Flood waves in rivers are translatory waves. *Compare* **longitudinal wave**.

wave packet—A collection of **waves**, the amplitudes of which are largest for waves with **frequency** and **wavelength** in a **range** about some central frequency and wavelength.

wave period—1. The **period** of any time-harmonic (or sinusoidal) physical quantity governed by a **wave equation**.
> 2. The time interval between successive **wave crests** or **troughs**.
> However, in a **random sea**, a variety of interpretations of this are possible. A commonly used definition is the so-called zero-upcrossing period, which is the time interval between two successive upcrossings of the **mean water level**. *See* **wave frequency, wave height, wave length**.

wave pole—(Also called wave staff.) A device for measuring sea **surface waves**.
> It consists of a weighted pole below which a disk is suspended at sufficient depth for the **wave motion** associated with deep-water waves to be negligible. The pole will then remain nearly as if anchored to the bottom, and **wave height** and **period** can be ascertained by observing or recording the length of the pole that extends above the surface. *See* **wave recorder**; *compare* **tide gauge**.

wave propagation—*See* **wave motion**.

wave recorder—An instrument for recording **ocean waves**.
> Most **wave** recorders are designed for recording **wind waves** or **swell**, that is, waves of **periods** up to about 25 seconds, but some are designed to record waves of longer periods such as **tsunamis** or **tides**. *See* **tide gauge, wave pole**.

wave set-up/set-down—An increase/decrease in the **mean water level** on a beach due to the effects of waves running up the beach and breaking.
> Set-down occurs prior to breaking, and set-up after breaking. Under some conditions the set-up can be large enough to contribute to local **flooding** and overtopping of sea defenses.

wave spectrum—The distribution of **wave** energy (for ocean **surface waves**) with **frequency** (1/period) and direction.
> The 2D **spectrum** can be expressed either as a function of frequency and direction or as a function of the 2D **wave vector**. Integrated over direction, the 1D spectrum can be expressed as a function of frequency or of **wavenumber**. For **gravity waves** the square of the **wave height** is proportional to the **potential energy** of the sea surface and the **energy spectrum** is equivalent to the spectrum of sea surface fluctuations. *See* **spectrum**.

wave speed—Same as **phase speed**.

wave staff—Same as **wave pole**.

wave steepness—*See* **steepness**.

wave system—1. **Waves** having similar direction and length in the sense that they correspond to the same peak in the **wave spectrum**.
> The **sea state** of ocean **surface waves** is often composed of a number of superimposed wave systems. *See* **wave group, wave train**.
> 2. *See* **cyclone family**.

wave theory of cyclones—(Obsolete.) A theory of **cyclone** development based upon the principles of **wave** formation on an **interface** between two fluids.

In the **atmosphere**, a **front** is taken as such an interface. *See* **Helmholtz instability**; *compare* **barrier theory, convection theory of cyclones**.

> Haurwitz, B., 1941: *Dynamic Meteorology*, 307–312.

wave theory of light—*See* **electromagnetic radiation**.

wave train—A superposition of **waves** propagating in the same direction and with almost equal **phase** speeds.

wave trough—The lowest point of a **wave motion**.

wave turbulence—**Turbulence** formed by **velocity shear** at the edges of the region of decelerated flow that forms behind an obstacle either fixed in a moving fluid (e.g., mountains) or moving relative to a fluid (e.g., aircraft).

wave-type disturbance—Same as **wave disturbance**.

wave vector—The **vector k** in the solution of a **wave equation** expressed as the product of a time-varying function and $A \exp(i\mathbf{k} \cdot \mathbf{x})$, where $\mathbf{x}$ is the **position vector** and A is constant in space and time.

The wave vector may, in general, be complex, and its real and imaginary parts need not be parallel, in which instance the **wave** is inhomogeneous.

wave velocity—Same as **phase speed**.

waveform—The pictorial representation of the shape of a **wave** showing the **amplitude** variations as a function of time; a wave as it might be displayed on a **cathode-ray tube**.

waveguide—A type of conductor used to carry **VHF** or **microwave** energy from one point to another. Most waveguides are hollow (rectangular or circular) and carefully dimensioned according to the **frequency** and **energy** to be conducted.

For low-power, short-distance applications a **coaxial cable** may be used as waveguide. Its primary use in **radar** is in providing a path for the radio energy between the **antenna** and the transmitting and receiving systems. Properly shaped open waveguides can themselves be used as radiators (horn antennas) or as feeds at the foci of reflecting antennas.

wavelength—In **radiation**, the distance between periodic spatial repetitions of an **electromagnetic wave** at a given instant of time; used extensively to classify the nature of the radiation, since most of the interactions between radiation and matter are extremely sensitive to the wavelength of the radiation.

Units are length (e.g., nm, μm, mm, cm, with conventional usage depending on which part of the **electromagnetic spectrum** is being considered).

wavelet—A member of a family of functions generated by taking translations [e.g., $w(t) \to w(t + 1)$] and scalings [e.g., $w(t) \to w(2t)$] of a function $w(t)$, called the "mother" wavelet.

The choice of $w(t)$ is limited by the condition that the square of $w(t)$ be integrable over all t. Linear combinations of wavelets are used to represent wavelike signals. The wavelet decomposition of a **signal** offers an advantage over the Fourier decomposition in that local or short-term contributions to the signal can be better represented.

wavenumber—Often 2π divided by **wavelength** but also may be simply reciprocal wavelength (used especially in **infrared spectroscopy**).

According to the first definition, wavenumber is the number of waves in a distance 2π (units are those of wavelength). *See* **amplitude**.

waves in the easterlies—Migratory wavelike disturbances in a region of persistent easterly winds (e.g., the **tropical easterlies, equatorial easterlies, polar easterlies**).

The waves move from east to west, generally more slowly than the current in which they are embedded. See **easterly wave**.

WBAN—An abbreviation of "Weather Bureau, Air Force, and Navy" used to denote observational instructions or forms that are common to the three principal meteorological agencies in the United States, or to denote certain cooperative meteorological activities or projects of the three agencies.

weak echo region—(Abbreviated WER.) A region of weak **radar echo** that is bounded on one side and above by strong echo.

It is located on the low-altitude **inflow** side of the **storm**. The WER is produced by strong **updraft** that carries **precipitation** particles to midlevels in a **convective storm** before they grow to radar-detectable sizes. In identifying a WER with **radar**, care must be taken to ensure that the strong midlevel echo is related to an updraft and not to horizontal motion of precipitation particles (e.g., the spreading **anvil**). *See also* **bounded weak echo region** (BWER).

weak-echo vault—Same as **bounded weak-echo region**.

weather—The state of the **atmosphere**, mainly with respect to its effects upon life and human activities.

As distinguished from **climate**, weather consists of the short-term (minutes to days) variations in the atmosphere. Popularly, weather is thought of in terms of **temperature, humidity, precipitation, cloudiness, visibility,** and **wind**.

2. As used in the taking of **surface weather observations**, a category of individual and combined atmospheric phenomena that must be drawn upon to describe the local atmospheric activity at the time of **observation**.

Listed weather types include **tornado, waterspout, funnel cloud, thunderstorm** and **severe storm**, liquid **precipitation (drizzle, rain, rain showers), freezing precipitation (freezing drizzle, freezing rain),** and **frozen precipitation (snow, snow pellets, snow grains, hail, ice pellets, ice crystals).** These elements, with the exception of the first three, are denoted by a letter code in the observation. With the **METAR** code, reporting weather also includes an **intensity** qualifier (light, moderate, or heavy) or proximity qualifier. The weather used in **synoptic weather observations** and **marine weather observations** is reported in two categories, "present weather" and "past weather." The "present weather" table consists of 100 possible conditions, with 10 possibilities for "past weather"; both are encoded numerically. Another method, which has the advantage of being independent of language, is the recording of **weather types** using **symbols**. There are 100 symbols that identify with the numeric codes of the synoptic observation.

3. To undergo change due to exposure to the **atmosphere**.
See also **weathering**.

weather advisory—Meteorological information issued when actual or expected weather conditions do not constitute a serious hazard but may cause inconvenience or concern.

Examples of weather advisories include small craft advisories or winter weather advisories. *Compare* **warning, watch**.

weather analysis—*See* **analysis**.

weather bureau—*See* **meteorological service, weather service**.

weather chart—*See* **synoptic chart**.

weather control—Same as **weather modification**.

weather derivatives—Business contracts, established between two firms to provide financial coverage for specific weather risks, that serve as a form of hedging against potential losses.

The weather risk is the uncertainty in cash flow and earnings due to weather variability. Risks typically covered for a firm in the energy industry are the number of heating or cooling **degree-days**, and for agribusinesses, the amount of **growing season** rainfall or the number of July days with temperatures above 32°C (90°F).

weather facsimile—(Acronym WEFAX.) A communications service provided by the **GOES, GMS,** and **Meteosat** environmental satellites.

WEFAX involves acquisition and processing of environmental satellite data on the ground and retransmission of these data at **VHF** frequencies back through the geostationary spacecraft to low-cost ground readout sites.

weather forecast—An assessment of the future state of the **atmosphere** with respect to **precipitation, clouds, winds,** and **temperature**.

Such assessments are usually made by government or private meteorologists, often using numerical simulations. Such simulations are the result of representing the atmosphere mathematically as a fluid in motion. *See also* **numerical weather prediction**.

weather forecaster—A person who predicts the weather.

weather gauge—Same as **barometer**.

weather glass—An old nautical term for **mercury barometer**.
It still enjoys some popular usage.

weather lore—Same as **weather proverb**.

weather map—Any graphical means of displaying the distribution of meteorological data over a given (usually extensive) area of the earth's surface.

weather-map type—Same as **weather type**.

weather minima—Criteria, determined by airline operators, that govern whether a particular aircraft under the control of a particular pilot may take off or land at a particular civil aerodrome.

They include **runway visual range** and **critical height** for a landing and runway visual range and **cloud ceiling** for takeoff. A variety of weather minima may be simultaneously in force at the same aerodrome.

weather modification—(Also called weather control.) In general, any effort to alter artificially the natural phenomena of the **atmosphere**.

The term usually refers to **cloud seeding** activities, but can also include constructing **windbreaks**, dissipating **fog** by the forceful addition of **heat** or water spray, or preventing **frost** formation on crops by **cloud** spray, heating, or **mixing** processes. Inadvertent weather modification refers to accidental weather effects resulting from the release of **greenhouse gases**, **aerosols**, and **dust**, or changes in **albedo** or surface properties of the earth associated with urban, industrial, or agricultural activity.

weather observation—In general, an evaluation of one or more meteorological **elements** that describe the state of the **atmosphere** either at the earth's surface or aloft.

Surface weather observations and **upper-air observations** are the major categories, with a number of subtypes, but separate from these are such categories as **radar meteorological observations**, **sferics**, and **solar-radiation observations**.

weather patrol ship—*See* **ocean station vessel**.

weather proverb—(Or weather lore.) Short, pithy, and much-used saying that expresses a well-known truth about the weather.

weather radar—Generally, any **radar** that is suitable or can be used for the **detection** of **precipitation** or **clouds**.

The general qualifications for weather radars are 1) a **wavelength** between 1 and 30 cm, 2) pulsed transmission with high peak **power** (kilowatts to megawatts), 3) relatively narrow **beamwidths**, 4) **pulse lengths** of a few microseconds or less, 5) **pulse repetition frequencies** of several hundred **hertz**, and 6) automatic **azimuth** or **elevation angle** scanning. Electronic circuits and **signal** processing permit the quantitative measurement of **radar reflectivity factor** or **signal strength** and, for **Doppler radars**, the **radial velocity**. *Compare* **wind profiler, MST radar**.

weather reconnaissance—*See* **aircraft weather reconnaissance**.

weather report—A statement of the actual values of meteorological elements observed at a specified place and time.

It is a record of an **observation**, not a forecast.

weather routing—*See* **ship routing**.

weather service—The provision of **weather forecasts**, warnings about hazardous conditions, and the collection, quality control, **verification**, archiving, and dissemination of hydrometeorological data and products.

weather ship—Same as **ocean station vessel**.

weather station—A location where **meteorological observations** such as **surface**, **upper air**, and **climatological observations** are taken.

In a general sense, the term also includes service offices that prepare **weather maps** and **charts**, issue **forecasts** and **warnings**, issue weather **briefings** to pilots, and prepare and disseminate climatological information.

weather symbol—*See* **plotting symbols**.

weather type—(Also called weather-map type.) In general, a series of generalized **synoptic situations** or patterns, usually presented in **chart** form.

Weather types are selected to represent typical **pressure** and frontal patterns and were originally devised as a method for lengthening the effective time-range of forecasts. Similar idealized patterns of **upper-air** circulation are sometimes referred to as **circulation types**.

weather warning— A hydrometeorological message issued to provide appropriate information of hazardous and potentially life-threatening weather conditions.

Examples of weather warnings include **tornado** warnings, severe **thunderstorm** warnings, **flood** and **flash flood** warnings, special marine warnings, **fire weather** red flag warnings, and winter storm warnings. *See also* **warning**.

weather warning bulletin—*See* **weather watch, weather warning**.

weather watch—A hydrometeorological message issued to provide appropriate advance notice that meteorological conditions exist in which hazardous and potentially life-threatening weather conditions may develop.

Examples include **tornado** watches, severe **thunderstorm** watches, **flood** and **flash flood** watches, **fire weather** watches, and winter storm watches. *See also* **watch**.

weather window—An interlude in a period of predominantly inclement weather during which a weather-sensitive operation may be carried out successfully.

weathered ice—Ice modified by exposure to weather; usually applied to **polar ice** where melting does not occur, such as **blue ice** areas of Antarctica.

weathering—The mechanical, chemical, or biological action of the **atmosphere, hydrometeors,** and suspended impurities on the form, color, or constitution of exposed material; to be distinguished from **erosion**.

Mechanical weathering results from the disintegrating action of high or low temperature, large changes of **temperature, frost,** or the impact of wind-borne sand or water (e.g., **frost action**, scouring, etc.). Chemical weathering is due to the chemical action of atmospheric constituents, especially acid impurities, in a moist atmosphere or in rainwater (e.g. **corrosion, oxidation,** etc.). Biological agents are mainly fungi that attack organic material (e.g., rotting, mildew, etc.).

Brooks, C. E. P., 1950: *Climate in Everyday Life,* 179–187.

Weber–Fechner law—An approximate psychophysical law relating various tactile sensations to the **intensity** of the stimulus.

The law asserts that equal increments of sensation are associated with equal increments of the **logarithm** of the stimulus. The Weber–Fechner law applies to the **detection** of **contrast** in the problem of **visual range** and to many other psychophysical problems.

Weber number—A **nondimensional number** relating to the effect of **surface tension** in a fluid system:

$$W = U \, L\rho/\sigma,$$

where U is a **characteristic velocity**, L a **characteristic length**, ρ **density**, and σ the surface tension. It is important in some problems of the action of the **wind** on a water surface.

Weddell Deep Water—The **water mass** in the center of the **cyclonic** Weddell Sea **gyre**.

It is formed by local cooling, has a **temperature** of 0.4°–0.7°C, and contributes to **Antarctic Bottom Water** formation.

Weddell Gyre Boundary—*See* **Continental Water Boundary**.

wedge—Same as **ridge**.

WEFAX—Acronym for **weather facsimile**.

Weger aspirator—An instrument of the **aspiration condenser** type that measures the concentration and **mobility** of **small ions**.

See **ion counter**.

Weibull distribution—A continuous distribution having properties of nonnegativity and positive **skewness** often used in the atmospheric sciences, especially to **model** variations in **wind speed**.

Its **probability distribution function** (pdf) is given by

$$f(x) = (\alpha/\beta)(x/\beta)^{\alpha-1}\exp[-(x/\beta)^{\alpha}],$$

where α and β are the shape and **scale** parameters, respectively, that control the form of the Weibull distribution. For $\alpha = 1$, the Weibull distribution is identical to the **gamma distribution;** for $\alpha = 3.6$, the Weibull is very similar to the **Gaussian distribution**. The scale parameter β acts to either stretch or compress the basic shape of the distribution along the x axis for a given α.

Weibull plotting position—The Weibull plotting position for the rth ranked (from largest to smallest) **datum** from a **sample** of size n is the quotient

$$\frac{r}{n+1}.$$

It is recommended for use when the form of the underlying distribution is unknown and when unbiased **exceedance probabilities** are desired. *See* **plotting position, probability paper**.

weighing rain gauge—A **recording rain gauge** in which the weight of water collected is measured as a function of time and converted to **rainfall** depth.

The weight-sensing mechanism is typically mechanical (springs). The **output** can be a pen **trace** on a strip **chart**, voltage from a **potentiometer**, punched paper tape, or other electronic means. Weighing gauges with continuous output can be used for determining **rainfall rate** and depth. *See* **Fisher & Porter rain gauge, universal rain gauge**.

weight barograph—A recording **weight barometer**.

weight barometer—A **mercury barometer** that measures **atmospheric pressure** by weighing the **mercury** in the column or the **cistern**.

weight chart—*See* **isentropic weight chart**.

weighted average—A representative value for a set of numbers $x_1, x_2, \cdots, x_n$ given by the sum $w_1 x_1 + w_2 x_2 + \cdots + w_n x_n$, where the numbers w_i are the weights.

The weights are usually assigned positive values that sum to unity. If all the weights are equal, the weighted mean reduces to the **arithmetic mean**.

weighting—The procedure of assigning different degrees of importance to values of a **variable** or to **statistics** according to some measures of reliability, areal coverage, etc.

weir—1. A low dam built across a **stream** to raise the **upstream** water level.

2. A structure built across a **stream** or channel for the purpose of measuring flow.

well-mixed layer—*See* **mixed layer**.

WER—Abbreviation for **weak echo region**.

West African tornado—*See* **tornado**.

West Antarctic Ice Sheet—That portion of the **Antarctic Ice Sheet** lying predominantly in the Western Hemisphere.

A line following the Transantarctic Mountains to the Antarctic Peninsula serves as the boundary between the East and West Antarctic Ice Sheets.

West Australian Current—The eastern **boundary current** of the subtropical **gyre** in the Indian Ocean.

It covers the eastern half of the southern Indian Ocean in a broad northward movement but does not reach to the Australian coast, where the **Leeuwin Current** flows southward instead.

west–east transport—The eastward **flux** of mass, **momentum**, **heat**, moisture or any other property of a fluid by mean motion or by **correlation** with the eastward component of motion, that is, **eddy flux**.

West Greenland Current—A current of the North Atlantic subpolar **gyre** flowing northward along the west coast of Greenland.

It is fed from the **East Greenland Current** and from the **cyclonic circulation** of the Labrador Sea and achieves a **transport** of over 30 Sv (30×10^6 m^3 s^{-1}). Some of its water turns westward at 64°N to join the **Labrador Current**; the remainder continues northward as a relatively warm current through Davis Strait and into Baffin Bay, where it can be seen as a **temperature** maximum of greater than 1°C at a depth of 500 m. This water eventually turns westward as well to feed the **Baffin Current** and through it the Labrador Current.

West Spitzbergen Current—The continuation of the **Norwegian Current**.

It carries 3–5 Sv ($3–5 \times 10^6$ m^3 s^{-1}) of water from the Atlantic into the Arctic Ocean. Another 1 Sv enters the Arctic Ocean through the Barents Sea.

West Wind Drift—*See* **Antarctic Circumpolar Current, North Atlantic Current, North Pacific Current**.

westerlies—1. (Also called circumpolar westerlies, circumpolar whirl, countertrades, middle-latitude westerlies, polar westerlies, subpolar westerlies, subtropical westerlies, temperate westerlies, zonal westerlies, westerly belt, zonal winds.) Specifically, the dominant west-to-east motion of the **atmosphere**, centered over the middle latitudes of both hemispheres.

At the earth's surface, the westerly belt (or west-wind belt, etc.) extends, on the average, from about 35° to 65° latitude. At upper levels, the westerlies extend farther equatorward and poleward. The equatorward boundary is fairly well defined by the **subtropical high pressure belt**; the poleward boundary is quite diffuse and variable. Especially in the Northern Hemisphere, even the annual average westerlies are markedly enhanced in some regions, namely, the **jet streams**. *See* **polar vortex, antitrades, tropical easterlies, zonal index.**

    2. Generally, any **winds** with components from the west.

    3. *See* **equatorial westerlies**.

westerly belt—*See* **westerlies**.

westerly trough—A **trough** in the **westerlies**.

westerly wave—An **atmospheric wave** disturbance embedded in a mean westerly flow, such as in the midlatitudes.
    Compare **easterly wave**.

westerly wind burst—A short-duration low-level westerly **wind** event along and near the **equator** in the western Pacific Ocean (and sometimes in the Indian Ocean).
    This surge may last from one day to several days and is closely linked to deep equatorial **convection** to its east. The westerly wind burst is most common during **El Niño** years from September to January and in **normal** years from October to December. It is absent in the Pacific in **La Niña** years. It is also thought to be associated with the **Madden–Julian oscillation**. The westerly winds are usually greater than 5 m s^{-1}(10 knots), and reach 15 m s^{-1}(30 knots) in well-developed systems. These intense westerly wind bursts are associated with a large cluster of deep convective clouds along the equator and are necessary precursors to the formation of **tropical cyclone twins** symmetric about the equator.

wet adiabat—Same as **moist adiabat**.

wet-adiabatic lapse rate—Same as **saturation-adiabatic lapse rate**.

wet air—A volume of air that is in a state of **saturation**.

wet-bulb depression—The difference in degrees between the **dry-bulb temperature** and the **wet-bulb temperature**.

wet-bulb potential temperature—(Also called pseudo wet-bulb potential temperature.) The **temperature** an **air parcel** would have if cooled from its initial state adiabatically to **saturation**, and thence brought to 1000 mb by a **moist-adiabatic process**.
    This temperature is conservative with respect to reversible adiabatic changes.

wet-bulb temperature—1. Isobaric wet-bulb temperature: the **temperature** an **air parcel** would have if cooled adiabatically to **saturation** at constant **pressure** by **evaporation** of water into it, all **latent heat** being supplied by the parcel.
    2. Adiabatic wet-bulb temperature (or pseudo wet-bulb temperature): the **temperature** an **air parcel** would have if cooled adiabatically to **saturation** and then compressed adiabatically to the original **pressure** in a **moist-adiabatic process**.
    This is the wet-bulb temperature as read off the **thermodynamic diagram** and is always less than the **isobaric wet-bulb temperature**, usually by a fraction of a **degree** centigrade.
    3. The **temperature** read from the **wet-bulb thermometer**.

wet-bulb thermometer—In a **psychrometer**, the **thermometer** that has the wet, muslin-covered bulb, and, therefore measures **wet-bulb temperature**.

wet climate—(Also called rain forest climate.) A **climate** with vegetation of the **rain forest** type.

wet day—(Also known as **rain day**.) A period of 24 hours in which 0.01 in. (0.2 mm) or more of **rain** is recorded.

wet deposition—The removal of atmospheric gases or **particles** through their incorporation into **hydrometeors**, which are then lost by **precipitation**.
    See **deposition**; *compare* **dry deposition**

wet-equivalent potential temperature—Same as **equivalent potential temperature**.

wet fog—**Fog** in the process of thickening with **relative humidity** near 100%, having **droplets** sufficiently large to be captured on surfaces.

wet growth—The condition in the **accretion** of **supercooled water** droplets onto **ice** in which the **temperature** of the ice remains at the **melting point** because of the release of the **heat of fusion** by the **freezing** droplets.
    The ice formed usually contains some liquid water (**spongy ice**) and its surface during growth

is mostly wet. The ice produced by wet growth is often quite clear and is called **glaze**. *Compare* **dry growth**.

wet-line correction—Depth correction to a **sounding** line measurement for that portion of the line that is below the water surface when the flow deflects the sounding line **downstream**.

wet season—*See* **rainy season**.

wet snow—Deposited **snow** that contains a great deal of liquid water.
  If **free water** entirely fills the air space in the snow it is classified as "very wet" snow.

wet spell—A period of a number of consecutive days on each of which **precipitation** exceeding a specific minimum amount has occurred.

wet year—Year in which **streamflow** records show **runoff** significantly greater than the mean annual runoff.

wetlands—Lands transitional between terrestrial and aquatic systems, where the **water table** is usually at or near the surface, or where the land is covered by shallow water.
  Wetlands must have one or more of three attributes: 1) at least periodically, the land supports predominantly **hydrophytes**; 2) the substrate is predominantly undrained hydric soil; 3) the subsoil is not soil and is saturated with water or covered by shallow water at some time during the **growing season** of each year.

wetted area—The area of the conduit wetted by water.
  In other words, this is the area through which the liquid flows.

wetted perimeter—That portion of the perimeter of a stream-channel cross section that is in contact with the water.

wetting front—Interface between soil that is unchanged from the initial state and the newly wetted zone from an **infiltration** or **irrigation** event.

whaleback cloud—An elongated **lenticular cloud**.

whip-poor-will storm—Same as **frog storm**.

whirlies—A name given by Sir Douglas Mawson (1915) to violent, snow-carrying **whirlwinds**, from a few to over one hundred meters in diameter, that occur in otherwise **calm** air on the slopes of Adélie Land, Antarctica, at about the time of the **equinoxes**.
  Over the sea, reportedly, they may lift **brash** 70–140 m into the air and form columns of water drops 1250 m high.
    Mawson, Sir Douglas, 1915: *The Home of the Blizzard*, 111–112.

whirling psychrometer—Same as **sling psychrometer**.

whirlwind—General term for a small-scale, rotating column of air.
  More specific terms are **dust whirl**, **dust devil**, **waterspout**, and **tornado**.

whistler—A type of **VLF** electromagnetic **signal** generated by some **lightning discharges**.
  Whistlers propagate along geomagnetic **field** lines and can travel back and forth several times between the Northern and Southern Hemispheres. So named from the **sound** they produce in radio receivers.

whistling meteor—Occurs when a radio **meteor** is illuminated by an unmodulated **radio wave**, the reflected **wave** being shifted in **frequency** because of the **Doppler effect**.
  The difference between the frequencies of the transmitted and reflected waves can be transformed into an audio **signal**, the frequency of which changes because of the changing motion of the meteor relative to the **receiver**. Whistling meteors get their name from the **sound** they produce in radio receivers, not because they are directly audible (although some meteors are).

white band—Ice band composed of **bubbly ice**.

White cell—An **absorption** cell devised by John White in 1942 in which **light** traverses a small volume a large and arbitrarily variable number of times.
  Such a cell, which is composed of three spherical concave mirrors with the same radius of curvature, is used to increase the optical pathlength for observing weak absorption spectra, such as those of atmospheric **trace gases**.
    White, J. U., 1942: Long optical paths of large aperture. *J. Opt. Soc. Amer.*, **32**, 285–288.

white dew—Dew that has frozen as the result of a fall in **temperature** to below **freezing** after the original formation of the **dew**.

Compare **hoarfrost**.

white frost—1. A relatively heavy coating of **hoarfrost**.

With respect to vegetation, a white frost is less damaging than a **black frost** for at least two reasons: 1) it tends to insulate the plant from further cold; and 2) it releases **latent heat** of **fusion** (albeit slight) to the **environment**.

2. Colloquial term for a deposit of fine **rime**.

white ice—Ice with a white appearance caused by the occurrence of **bubbles** within the **ice**.

The bubbles increase the **scattering** of all wavelengths of **light** in contrast to the appearance of bubble-free **blue ice**.

white noise—Random **noise** that has uniform power spectral density at every **frequency** in the **range** of interest.

white rainbow—Same as **cloudbow**.

white squall—A sudden **squall** in tropical or subtropical waters; it is so called because the usual squall cloud is absent, thus, the only warning of its approach is the whiteness of a line of broken water or whitecaps.

This may represent the **outflow** of a convective system that has recently dissipated.

whitecap—A patch of white water formed at the crest of a **wave** as it breaks, due to air being mixed into the water.

The formation of the whitecap dissipates **energy** from the wave. Whitecaps can persist after breaking has ceased, but slowly disappear as the air escapes from the water.

whiteout—(Also called milky weather.) An atmospheric optical phenomenon in which the **observer** appears to be engulfed in a uniformly white glow.

Neither shadows, **horizon**, nor clouds are discernible; sense of depth and orientation is lost; only very dark, nearby objects can be seen. Whiteout occurs over an unbroken **snow cover** and beneath a uniformly **overcast** sky, when, with the aid of the **snow blink** effect, the **light** from the sky is about equal to that from the **snow** surface. **Blowing snow** may be an additional cause. This phenomenon is experienced in the air as well as on the ground.

whole gale—1. In storm-warning terminology, a **wind** of 48 to 63 knots (55 to 72 mph).

2. In the **Beaufort wind scale**, a **wind** with a speed from 48 to 55 knots (55 to 63 mph) or Beaufort Number 10 (Force 10).

whole-gale warning—A **warning**, for marine interests, of impending winds of 48 to 63 knots (55 to 72 mph).

The storm-warning signals for this condition are 1) one square red flag with black center by day and 2) two red lanterns by night. *See* **storm warning**.

wideband radar—A **radar** system of extremely wide **bandwidth** and high range **resolution**, usually implemented by the direct **radiation** of a very short (≈ 1 ns) **video** pulse rather than the transmission of a short burst of a **carrier wave**.

Wien's displacement law—*See* **Wien's law**.

Wien's distribution law—A relation, derived on purely thermodynamic reasoning by Wien, between the **monochromatic** emittance of an ideal **blackbody** and that body's **temperature**.

$$e_\lambda / T^5 = f(\lambda T),$$

where e_λ is the monochromatic emittance (**emissive power**) of a blackbody at **wavelength** λ and **absolute** temperature T, and $f(\lambda T)$ a function that cannot be determined purely on classical thermodynamic grounds. The specification of $f(\lambda T)$ by Planck in 1901 formed the beginning of modern **quantum theory**. *Compare* **Wien's law**.

Wien's law—(Also called Wien's displacement law.) A **radiation** law that is used to relate the **wavelength** of maximum **emission** from a **blackbody** inversely to its **absolute** temperature.

It is expressed as:

$$\lambda_m = 2898/T,$$

where λ_m is the wavelength of maximum emittance in microns, and T is the blackbody temperature in Kelvin. *See also* **color temperature**.

Wild fence—A wooden enclosure about 16 ft square and 8 ft high with a **precipitation gauge** in its

center; the function of the fence is to minimize **eddies** around the gauge and thus ensure a **catch** that will be representative of the actual **rainfall** or **snowfall**.
See **rain-gauge shield**.

wild snow—(Rare.) Newly deposited **snow** that is very fluffy and unstable.
In general, it falls only during a dead **calm** at very low air temperatures and will usually have a low liquid to **snow depth** ratio. *See* **sand snow**.

williwaw—(Also spelled willywaw, willywau, willie-wa, willy-waa.) A very violent **squall** in the Straits of Magellan.
Williwaws may occur in any month but are most frequent in winter.

willy-willy—In Australia, a **dust devil**.
Also formerly used to denote a **tropical cyclone**.

Wilson cloud chamber—(Or, simply, cloud chamber.) A device that renders visible the paths of high energy subatomic **particles**.
A supersaturated **vapor** condition is created in a chamber filled with dust-free air by a sudden **adiabatic expansion** and cooling. In this **environment**, the **small ions** formed along the path of a high energy particle act as effective **condensation nuclei**. The line of **droplets** so formed can be used to mark the path. See **meteorological cloud chamber**.

wilting point—The **water content** of a soil when an indicator plant (or common agricultural crop) can no longer draw water from the soil, even if atmospheric water demand is zero (near saturated **atmosphere** and plant and atmosphere at uniform **temperature**); often estimated as a soil **water potential** at 15 bars (or atmospheres).
For many agricultural soils, this represents a loss of about half the total water-holding capacity of a soil.

wind—Air in motion relative to the surface of the earth.
Since vertical components of atmospheric motion are relatively small, especially near the surface of the earth, meteorologists use the term to denote almost exclusively the horizontal component. Vertical winds are usually identified as such. Surface winds are measured by **anemometer** and **wind vane**; winds aloft by such systems as **pilot balloon**, **radiosonde**, or aircraft navigational techniques. *See also* **circulation, general circulation, turbulence, geostrophic wind, gradient wind, local winds, Beaufort wind scale, draft, cyclone, whirlwind, squall, storm**.

wind arrow—(Also called wind shaft.) A short straight line terminating with an arrowhead centered on a **station location** on a **synoptic chart**.
It represents the direction from which the **wind** blows. The length of the line may be representative of the **wind speed**.

wind barb—*See* **barb**.

wind chill—The portion of the cooling of a human body caused by air motion.
Air motion accelerates the rate of **heat transfer** from a human body to the surrounding **atmosphere**, especially when temperatures are below about 7°C (45°F).

wind-chill factor—Same as **wind-chill index**.

wind-chill index—A means of quantifying the threat of rapid cooling during breezy or windy conditions that may result in **hypothermia** in cold conditions.
The index is used to remind the public to minimize **exposure** when outdoors and to take precautionary actions. In the late 1940s, Antarctic explorers Siple and Passel experimented with measuring the time it took to **freeze** 250 grams of water in different **temperature** and **wind** conditions. They developed empirical formulas relating these data to the rate of **heat** loss from exposed human skin. They developed the following formula which was used to determine the wind-chill index. At wind speeds of 4 mph or less, the **wind chill** temperature is the same as the actual **air temperature**:

$$T_{WC} = 0.0817(3.71V^{0.5} + 5.81 - 0.25V)(T - 91.4) + 91.4,$$

where V is **wind speed** in mph and T is temperature in °F.

wind cone—See **wind sock**.

wind corrosion—Same as **corrasion**.

wind crust—A type of **snow crust** formed by the packing action of **wind** on previously deposited **snow**.
Wind crust may break locally, but, unlike **wind slab**, does not constitute an **avalanche** hazard.

wind current—Generally, any of the quasi-permanent, large-scale **wind** systems of the **atmosphere**, for example, the **westerlies, trade winds, equatorial easterlies, polar easterlies**, etc.

wind daily run—The distance that results by integrating the **wind speed**, measured at a point, over 24 hours.

wind direction—The direction from which the **wind** is blowing.
  See also **prevailing wind direction**.

wind-direction shaft—A means of representing **wind direction** in the **plotting** of a **synoptic chart**; a straight line drawn directly **upwind** from the **station location**.
  The **wind arrow** is completed by adding the wind-speed **barbs** and **pennants** to the outer end of the shaft.

wind divide—A semipermanent feature of the atmospheric **circulation** (usually a high-pressure **ridge**) on opposite sides of which the prevailing **wind** directions differ greatly.
  A good example is the ridge that extends in winter from the **Siberian high** westward across central Europe and France, and in summer less regularly from the **Azores high** across Spain and France to central Europe. North of this the prevailing winds are from west-southwest, while south of it they are on the whole northeasterly. Köppen speaks of the **polar wind divide**, a very diffuse boundary of low pressure between the midlatitude **westerlies** and the **polar easterlies**.

wind drift—Horizontal motion in the oceanic surface layer driven by the **wind stress**.
  A rough estimate of wind drift is 3% of the **wind speed**, about 10° to the right (left) of the **wind direction** in the Northern (Southern) Hemisphere.

wind-driven current—Any **current** driven directly by the **wind stress** or indirectly by the **pressure gradients** set up by the **divergence** or **curl** of the wind stress.
  See also **surface current**.

wind-driven oceanic circulation—That part of the large-scale **circulation** of the ocean driven by and associated with the **stress** (force per unit area) applied by the **wind** at the ocean surface.
  The main principles of wind-driven ocean circulation are as follows.
  Wind stress causes **transport** of water in a thin Ekman **surface layer**
  If there is a **curl** to the **wind stress**, the **Ekman layer** transport will converge or diverge, causing **downwelling** or **upwelling**
  The downwelling or upwelling motions cause fluid columns to squash or stretch vertically, and this causes slow southward or northward flow due to conservation of **potential vorticity**.
  The southward or northward flow is determined by the Sverdrup relationship, which links the N-S transport of water with the curl of the wind stress. In order to conserve mass, the water returns to its original position via a relatively thin boundary layer adjacent to the western edge of the **basin**.

wind erosion—Same as **erosion**.

wind factor—1. In air navigation, a measure of the net effect of **wind** on the **ground speed** of an aircraft.
  It is the magnitude of the **wind vector** component parallel to the **heading**(s) of an aircraft, averaged over the entire flight; positive if a **tailwind**, negative if a **headwind**:

$$\text{Ground speed} = \text{airspeed} \pm \text{wind factor}.$$

The wind factor accounts for the retarding effect of a **crosswind**.
  2. In **oceanography**, the ratio of the **velocity** of an ocean **surface current** to the **velocity** of the **wind**.
  Its numerical value is of the order of a few hundredths and is inversely dependent upon the **depth of frictional influence** (*see* **Ekman spiral**) at that point in the **ocean**. In sufficiently shallow water, the actual depth is the controlling factor.

wind field—*See* **field**.

wind-finding—The process of determining **upper-air** winds represented by the in-flight movement of a **radiosonde balloon**, either by tracking the **radiosonde** from the ground using a **radio theodolite** or by the radiosonde itself using radio signals from aids to navigation (i.e., Loran, **GPS**, etc.).

wind flurry—A sudden and brief wind **squall**.

wind force—1. Force of the **wind** on a structure, object, etc., proportional to the square of **wind speed**.

2. Number on a progressive **scale** (**Beaufort wind scale**) corresponding to the effects produced by winds between a particular **range** of speeds

wind-induced surface heat exchange—(Abbreviated WISHE.) A hypothesis for the amplification of certain atmospheric circulations, including **tropical cyclones**, **polar lows**, and the **Madden–Julian oscillation**.

The mechanism involves a **positive feedback** between the **circulation** and **heat** fluxes from the sea surface, with stronger circulation giving rise to larger surface fluxes of heat, which are then quickly redistributed aloft by **convection**, in turn strengthening the circulation. In this theory, emphasis is placed on the surface fluxes as the principal rate-limiting process; convection serves only to redistribute heat. This can be contrasted with **conditional instability of the second kind** (CISK), in which circulations amplify through their interaction with the convection itself.

wind load—The **stress** placed upon a surface by the **wind**.
See **pressure, wind pressure**.

wind lull—A marked decrease in the **wind speed**.

wind mixing—Mixing, generally **diapycnal**, for which the **wind stress** is the primary source of **energy** to drive the turbulent **mixing** motions.

wind of 120 days—Same as **seistan**.

wind pressure—(Also called velocity pressure.) The total force exerted upon a structure by **wind**.

For a flat surface it consists of two factors, the first being the **dynamic pressure** exerted on the **windward side** of the surface (**wind load**). This is equal to $(1/2)\rho v^2$, where ρ is the air density and v is the **wind speed** normal to the surface. The second factor is the **pressure** decrease, or suction, produced on the **leeward** side of the surface, which is equal to $(1/2)c\rho v^2$, where c is a structural constant varying from -0.3 for cylindrical objects to 1.0 for long plates. The wind pressure p is the sum of these two, or

$$p = \frac{1}{2}(1 + c)\rho v^2.$$

This formula should be used with caution. The actual wind pressure depends on the shape of the object and the nature of the **environment**. For objects not presenting a flat surface to the wind, such as bridges and chimneys, complicating **aerodynamic** factors are introduced that make the above formula invalid.

wind profile—*See* **profile, logarithmic velocity profile, power-law profile**.

wind profiler—(Also called wind profiler radar, wind profiling radar.) A **radar** that is used to measure vertical profiles of the **wind**.

In general the term is applied to **Doppler radars** operating in the VHF–UHF **band** (30 MHz– 3 GHz) that determine the wind by measuring the line-of-sight **Doppler shift** of scattered signals (**Bragg scattering**) from **refractive index** fluctuations caused by **turbulence** (*see* **clear-air echo**). The turbulent **scattering** structures are assumed to be moving with the same **average velocity** as the wind. The three-dimensional **wind vector** is determined by using the **beam swinging** technique of pointing the **radar beam** in at least three different directions. Another type of wind profiler radar that operates in the VHF–UHF band uses a technique called spaced **antenna** drift (SAD). SAD radars use a single vertical-beam transmitting antenna and three or more horizontally spaced, vertical-beam receiving antennas. The horizontal wind is determined from the **cross correlation** of the received echoes and the vertical wind is determined from the Doppler shift of the echoes. In addition to measuring the wind vector, wind profiler radars can also determine several other atmospheric quantities from the **power**, mean **Doppler shift**, and Doppler spectral width of the returned **signal**. These quantities include the strength of turbulence (parameterized by the refractive index structure constant C_n^2), the **eddy** dissipation rate, atmospheric **stability, momentum flux, virtual temperature** and **heat flux** (using the **RASS** technique), and **precipitation** rates and **drop-size distributions** (from **scatter** from **hydrometeors**). **Doppler lidars** and acoustic sounders (**sodars**) can also be used as wind profilers. The scatterers for lidars are **aerosols** (for wavelengths around 10 μm) and molecules (for wavelengths less than 1 μm). Acoustic refractive index fluctuations caused by turbulence provide the scattering mechanism for sodars. *See also* **MST radar, boundary layer radar**.

wind reversal—A change in the direction of the **wind** that takes place either in time or in space, for example, the change from **easterlies** to **westerlies** or westerlies to easterlies caused by the stratospheric **quasi-biennial oscillation** and the change from easterlies to westerlies with **altitude** observed in the **troposphere** of the central equatorial Pacific.

wind ridge—A feature of hard **snow** formed by **wind**.
　　Similar to **sastrugi**, but more commonly applied to a single ridge.

wind ripple—One of a series of wavelike formations on a **snow** surface, an inch or so in height, at right angles to the direction of **wind**.

wind rose—Any one of a class of diagrams designed to show the distribution of **wind direction** experienced at a given location over a considerable period; it thus shows the **prevailing wind direction**.
　　The most common form consists of a circle from which eight or sixteen lines emanate, one for each **compass** point. The length of each line is proportional to the **frequency** of **wind** from that direction, and the frequency of **calm** conditions is entered in the center. Many variations exist; some indicate the **range** of wind speeds from each direction; some relate wind directions with other weather occurrences. *Compare* **resultant wind**, **Lambert's formula**.

wind scale—*See* **geostrophic wind scale**.

wind scoop—A saucerlike depression in the **snow** near obstructions such as trees, houses, and rocks, caused by the eddying action of the deflected **wind**.

wind sea—(Also, simply, sea.) Ocean surface **gravity waves** that are still growing or being sustained by the **wind**.
　　These waves propagate mainly in the direction of the wind, in contrast to **swell**.

wind setup—In lakes, rivers, and seas, the increase in **water level** from still conditions due to **wind**.

wind shaft—*See* **wind arrow**.

wind shear—The local **variation** of the **wind vector** or any of its components in a given direction.
　　The vertical **shear** can be expressed in terms of height $\partial V/\partial z$ or of **pressure** $\partial V/\partial p$ as the vertical coordinate. If the **wind** is **geostrophic**, the vertical shear is given by the **thermal wind equation**. The wind shear at a point is said to be **cyclonic** or **anticyclonic** according to whether the sense of rotation from the wind vector to the **shear vector** at that point is cyclonic or anticyclonic.

wind shield—1. Same as **rain-gauge shield**.
　　2. *See* **windbreak**.

wind-shift line—A line or narrow zone along which there is an abrupt change of **wind direction**. *See also* **surge line**.

wind slab—A type of **snow crust**; a patch of hard-packed **snow** that is packed as it is deposited in favored spots by the **wind** (in contrast to **wind crust**, which is packed after deposition).
　　Wind slabs can be quite rigid, but they adhere poorly to the underlying snow and hence may be readily dislodged, causing an **avalanche**.

wind sock—A tapered fabric sleeve, shaped like a truncated cone and pivoted at its larger end on a vertical standard, for the purpose of indicating **wind direction**.
　　Since the air enters the fixed end, the small end of the cone points away from the **wind**.

wind speed—Ratio of the distance covered by the air to the time taken to cover it.
　　The instantaneous speed corresponds to the case of an infinitely small time interval. The mean speed corresponds to the case of a finite time interval. It is one component of **wind** velocity, the other being **wind direction**).

wind stress—The **drag** force per unit area caused by **wind shear**.
　　For example, the wind stress on the sea surface applies a **friction** force that can drive **ocean currents**. *See* **Reynolds stresses**, **momentum flux**, **friction velocity**.

wind stress curl—The vertical component of the (mathematical) **curl** of the **surface wind stress** $\partial \tau_y/\partial x - \partial \tau_x/\partial y$ (with x and y eastward and northward coordinates, and τ_x, τ_y the corresponding components of the surface **stress**). The large-scale, long-term averaged wind stress curl contains the principal information to calculate the wind-driven **mass transport**.

wind turbine—A device for converting **wind** energy into mechanical (windmill) or electrical **energy**.

wind vane—An instrument used to indicate or measure **wind direction**.
　　It consists basically of an asymmetrical, elongated object mounted at its **center of gravity** about

a vertical axis. The end that offers the greater **resistance** to the motion of air moves to the **downwind** position. The direction of the **wind** may be determined by visual reference to an attached oriented **compass** rose. The direction may also be displayed or recorded by use of a **potentiometer** or a **digital** optical encoder.

wind vector—The two- or three-dimensional **vector** describing the instantaneous **wind** magnitude and direction at a point.

wind velocity profile—*See* **profile, logarithmic velocity profile, power-law profile.**

wind wave—A wind-generated ocean **surface wave.**
    Wind generates both **gravity waves** and **capillary waves.**

windbreak—Any device designed to obstruct **wind** flow and intended for protection against any ill effects of wind.
    Installations of this type include agricultural shelterbelts, **snow fences**, and **rain-gauge wind shields.**

windburn—Injury to plant foliage caused by strong, hot, dry **winds.**

window—1. A **band** in the **electromagnetic spectrum** that offers maximum transmission and minimal **attenuation** through a particular medium with the use of a specific sensor.
    See **atmospheric window.**
    2. Open water in a predominantly frozen river.
    It is caused by warmer water from a spring or **tributary**, or by **turbulence** over a **shoal.** See **polyn'ya.**
    3. (Obsolete.) A term for the **chaff** used for military countermeasure purposes during World War II.

windrow cloud—Popular name for **undulatus.**

winds aloft—(Also called upper winds, upper-level winds.) Generally, the **wind speeds** and **directions** at various levels in the **atmosphere** above the domain of **surface weather observations**, as determined by any of the methods of **winds-aloft observation.**

winds-aloft observation—The measurement and computation of **wind speeds** and **directions** at various levels above the surface of the earth.
    Among the methods employed are 1) the visual tracking, by **theodolite**, of ascending balloons (**pilot-balloon observation** or **rabal**); 2) the use of a **radio direction finder** to track the radio signals emitted by an ascending **radiosonde** or other type of **transponder** (**rawinsonde observation** or **rawin** observation); 3) the use of **radar** to track a balloon-borne radar target, sometimes in combination with a radiosonde (rawin observation). The tracking of high-altitude, constant-level balloons (**transosonde**) may be considered to fall within this group. Winds-aloft data are included in many aircraft observations, particularly in **aircraft weather reconnaissance** flights.

winds-aloft plotting board—A graphical aid used in the **reduction** of data from a **winds-aloft observation.**
    The board is designed so that the projections of the path of a **balloon** as a function of **azimuth** may be graphed in a convenient manner. The separation of the points that correspond to the horizontal **wind** is read from a special measuring **scale** converting distance to speed.

windstorm—*See* **storm.**

windward—The **upwind** direction from a point, for example, westward in the case of a west **wind.**

windward side—The side of a mountain, ridge, or other flow obstacle facing toward the direction of the large-scale or ridge-top **wind**; the **upwind** side; opposite of **leeward.**

WINTEM—**Aviation forecast** of winds and **temperatures** aloft at specific points.

winter—Astronomically, between the **winter solstice** and **vernal equinox** in the Northern Hemisphere, and the **summer solstice** and **autumnal equinox** in the Southern Hemisphere; the coldest **season** of the year; the "low sun" season during which the sun is over the opposite hemisphere; the "hibernal" season.
    Popularly and for most meteorological purposes, winter is taken to include December, January, and February in the Northern Hemisphere, and in the Southern Hemisphere, June, July, and August; the opposite of summer.
    2. *See* **Blackthorn winter.**

winter ice—Level **sea ice** more than eight inches thick and less than one year old; the stage that follows **young ice.**

winter monsoon—A **monsoon** resulting from the **circulation** induced by **temperature** contrasts between a cold continent and a warm ocean as occurs during the winter season.

An important example of this is the East Asian winter monsoon, which is the reversal of the East Asian **summer monsoon**.

winter solstice—For either hemisphere, the **solstice** at which the sun is above the opposite hemisphere. In northern latitudes, the time of this occurrence is approximately 22 December.

winterization—The preparation of equipment for operation in conditions of winter weather.

This applies to preparation not only for cold **temperatures**, but also for **snow**, **ice**, and strong **winds**.

wire-weight gauge—A **river gauge** in which a weight suspended on a wire is lowered to the water surface from a bridge or other overhead structure to measure the distance from a point of known **elevation** on the bridge to the water surface.

The distance is usually measured by counting the number of revolutions of a drum required to lower the weight, and a counter is provided that reads the water stage directly.

wiresonde—An **atmospheric sounding** instrument that is supported by a captive balloon and used to obtain **temperature** and **humidity** data from the ground level to a height of about 1 km.

Height is determined by means of a sensitive **altimeter**, or from the length of cable released and the angle that the cable makes with the ground. Information is telemetered to the ground through a wire cable. *See* **kytoon**.

WISHE—Abbreviation for **wind-induced surface heat exchange**.

wisperwind—A cold **night wind** blowing out of the valley of the Wisper River in Germany during **clear** weather.

withershins—(Also widdershins.) An old English term for "against the sun," **contra solem**.
Compare **deasil**.

Wolf number—Same as **relative sunspot number**.

Wolf–Wolfer number—Same as **relative sunspot number**.

Wollaston prism—A polarizing prism consisting of two calcite prisms cemented such that they deviate the two emerging **beams** (which are perpendicularly polarized) by nearly equal amounts in opposite directions.

work—A form of **energy** arising from the motion of a system against a force, existing only in the process of **energy conversion**.

Many forms of work (electrical, chemical, etc.) may be defined by analogy with mechanical work, but in meteorology the most frequently useful mathematical expression is that for the work per unit mass w done by a gaseous system in a given **reversible process** from thermodynamic state s_1 to state s_2:

$$w = \int_{s_1}^{s_2} p\,d\alpha,$$

where p is the **pressure** of the system and α its **specific volume**. The amount of work done will be a function of the particular process as well as of the initial and final states. *See* **heat**, **first law of thermodynamics**.

work function—Same as **Helmholtz function**.

working memory—The global database for an **expert system** that contains all known facts about the situation to which the expert system is currently being applied.

Workman–Reynolds effect—A mechanism for **electric charge** separation during **freezing** of slightly impure water, discovered by Workman and Reynolds (1950).

When a very dilute solution of certain salts freezes rapidly, a strong **potential** difference is established between the solid and liquid phases. For some salts, the **ice** attains negative charge, for others, positive. This mechanism was thought to play a role in the charging of thunderstorms. It is now known that it cannot account for the electrification of riming **graupel** by the shedding of charged **surface water** because the droplets **freeze** faster than the time taken for a substantial ice–water freezing potential to occur. This mechanism has been suggested as one possible mode of **thunderstorm charge separation** in those portions of a **thunderstorm** downdraft where **snow pellets** or **hail** particles sweep out **supercooled water** drops. Partial freezing and partial blow-off of a liquid film could lead to **charge separation**. The acknowledged predominance of **dry growth** of graupel in New Mexico thunderclouds led Reynolds (1953) to question the viability of this mechanism in the **atmosphere**. This should not be confused with the **Reynolds effect**.

Workman, E. J., and S. E. Reynolds, 1950: Electrical phenomena occurring during the freezing of dilute aqueous solutions and their possible relationship to thunderstorm electricity. *Phys. Rev.*, **78**, 254–259.

Reynolds, S. E., 1953: Thunderstorm-precipitation growth and electrical-charge generation. *Bull. Amer. Meteor. Soc.*, **34**, 117–123.

WSR–88D—(Abbreviation for Weather Surveillance Radar–1988 Doppler.) The **weather radar**, sometimes called **NEXRAD**, that became the operational network radar for the U.S. National Weather Service, U.S. Air Force, and Federal Aviation Administration during the early and middle 1990s.

It is a general-purpose weather radar with a **wavelength** of 10.5 cm, a **peak power** of 750 kW, selectable **pulse duration** of 1.57 or 4.0 μs, and selectable **pulse repetition frequency** from 318 to 1304 Hz. A coherent **Doppler radar**, it employs a center-fed **parabolic antenna** with a diameter of 8.5 m, producing a **beamwidth** of 0.95°. Received signals are analyzed using the method of **pulse-pair processing** to give as fundamental data the **reflectivity factor, mean Doppler velocity**, and **Doppler spread** as functions of time and location relative to the radar. The **maximum unambiguous range** is ordinarily 460 km for **reflectivity** and 115 km for Doppler information. Many kinds of computer algorithms are employed for identifying features such as vortices, downbursts, and fronts.

X

X band—*See* **radar frequency bands**.

x-ray—(Or x-radiation, Röntgen ray.) **Electromagnetic radiation** with **wavelengths** shorter than that of **ultraviolet radiation** and greater than that of **gamma radiation**.

Discovered accidentally by Röntgen in 1895. The primary mechanism for the production of x-rays is deceleration of a rapidly moving charge upon interaction with matter (**bremsstrahlung**). The x-ray spectrum from an x-ray tube consists of this **continuous spectrum** on which are superimposed narrow bands (characteristic radiation) that are a consequence of transitions between electronic **energy** levels of atoms. No sharp boundary exists between x- and ultraviolet radiation nor between x- and gamma radiation, although the latter term is usually restricted to radiation resulting from transitions between nuclear energy levels.

Boorse, H. A., and L. Motz, 1966: *The World of the Atom*, Vol. 1, 385–401.

xaloch—(Or xaloque, xoroco.) *See* **sirocco**.

XBT—Abbreviation for **expendable bathythermograph**.

xenon—(Symbol Xe.) A noble gas, **atomic number** 54, atomic weight 131.3; a heavy, unreactive, colorless **element** found in the **atmosphere** to the extent of only 0.0000087% by volume.

xerochore—A waterless **desert**.
See **biochore**.

xerophyte—A plant that grows in dry conditions where water is often limited.

Y

yagi antenna—See **Yagi-Uda antenna**.

Yagi-Uda antenna—(Often called yagi antenna.) A type of directional **antenna** comprising an elevated **linear** array of horizontal **dipoles** with different lengths and spacings, typically arranged as a driven element, **reflector** element, and several director elements.

The dipoles are organized to maximize the **radiation** in the forward direction. Radar systems requiring large antennas often use arrays of Yagi-Uda antennas (e.g., some **wind profiler radars**).

yalca—A local name for a severe **snowstorm** with a strong squally **wind** that occurs in the Andes Mountain passes of northern Peru.

yamase—A cool, onshore, easterly **wind** in the Senriku district of Japan in summer.

It originates in the Okhotsk **anticyclone** and blows over the cold sea, bringing periods of three to seven days of **clouds**, **rain**, and **fog**.

Yanai wave—(Also called mixed planetary–gravity wave.) The lowest order solution to the equation for **equatorial waves**.
See **mixed Rossby–gravity wave**.

yaw—Oscillation of a ship about the vertical axis.
See **roll, pitch, ship motion**.

year—1. The period during which the earth completes one revolution around the sun.

This has several interpretations, as follows. 1) Sidereal year: time of true revolution around the sun, that is, the time it takes the earth (as seen from the sun) to reappear at the same fixed star, equal to 365.2564 **mean solar days** or 365 days, 6 hours, 9 minutes, 10 seconds; 2) Tropical year (also called mean solar year, ordinary year): the time measured from one **vernal equinox** to the next, that is, the apparent revolution of the sun through the **zodiac**, equal to 365.2422 mean solar days or 365 days, 5 hours, 48 minutes, 46 seconds (this is not constant but decreases by about 5 seconds in one thousand years.); and 3) Calendar year: fixed by the Gregorian calendar of 365 days in ordinary years and 366 days in leap years. *See* **day**.

2. Any arbitrary 12-month period selected for a special purpose, such as the **water year**.
See also **climatological year, farmer's year, grower's year**.

Yellow Sea Warm Current—A **surface current** flowing northward along the central axis of the Yellow Sea.

The current is very shallow but strong, with speeds of 0.2 m s^{-1}. It is effectively an offshoot from the **Kuroshio** sandwiched between southward flowing coastal currents on the Chinese and Korean side. Advection of the warm Kuroshio water increases the **temperature** of the central Yellow Sea several degrees over the temperatures of the coastal regions.

yellow snow—Snow given a golden or yellow appearance by the presence of pine or cypress pollen.

yellow wind—A strong, cold, dry, west **wind** of eastern Asia, especially northern China, that sweeps across the plains in winter carrying a fine yellow **dust** from the deserts.

The deposition of this dust over thousands of years has formed the loess deposits of China.

youg—A hot **wind** during unsettled summer weather in the Mediterranean.

young ice—Newly formed, flat sea or lake **ice**, usually between 2 and 8 in. thick.
Further growth renders it **winter ice**.

Younger Dryas—The period from approximately 10 800 to 9600 BC when **climate** in the region around Greenland cooled by 5°–7°C within a few decades and recovered with similar rapidity at the end of the period; named for the expansion of the geographic range of the arctic herb *Dryas octopetala*.

Evidence is accumulating for a wider, perhaps global, extent, but the Younger Dryas was most strongly felt around the North Atlantic Ocean. Changes in the **thermohaline circulation** of the oceans are widely believed to be implicated in the onset and ending of the Younger Dryas cooling. These were in turn associated with a switch to the St. Lawrence River as the major route for meltwater reaching the North Atlantic from the retreating **continental ice** cap. Other similarly rapid oscillations in climate are recorded through much of the **last glacial** in ice cores from Greenland and in marine **sediment** cores from the North Atlantic (**Dansgaard–Oeschger events**). The period since the Younger Dryas (the Holocene) lacks such extreme excursions.

Z

Z_0—The internationally accepted symbol for the **elevation** of **mean sea level** above **chart datum**.

z-less scaling—*See* **local similarity**.

Z–R relation—A relationship between **radar reflectivity factor** Z (mm^6 m^{-3}) and **rain** rate R (mm h^{-1}).
  Empirical relationships of the form $Z = aR^b$ are often used, with $a \approx 200$ to 600 and $b \approx 1.5$ to 2.0. *See* **radar reflectivity**, **Marshall–Palmer relation**.

Z time—Same as coordinated **universal time** (UTC).
  See **zone time**.

Zanzibar Current—(Also known as the East African Coastal Current.) A western boundary current in the Indian Ocean that flows permanently northward from 10°S along the east African coast.
  During the northeast **monsoon** the Zanzibar Current flows against the **wind**, meeting the southward flowing **Somali Current** at 1°N in December, 4°S in February (when the Zanzibar Current is weakest), and at the **equator** in April. Throughout this period it continues across the equator as an **undercurrent** that feeds into the **Equatorial Countercurrent**. During the **southwest monsoon** season the current strengthens considerably, attaining speeds of 2 m s^{-1} and a **transport** of 15 Sv (15 × 10^6 m^3 s^{-1}) and feeding into the Somali Current, which flows northward during that **season**.

zastrugi—Variant spelling of **sastrugi**.

Z_{DR}—*See* **differential reflectivity**.

Zeeman effect—A shift in the **energy** levels of an isolated atom or molecule as a consequence of an external **magnetic field**.
  The energy levels of an atom (or molecule) depend on the internal forces that electrons and nuclei exert on each other as well as on any external forces (e.g., an **electromagnetic field**). An observable consequence of the Zeeman effect is the shifting and broadening of spectral lines. The Zeeman effect is the magnetic analogue of the **Stark effect**.

Zeldovich mechanism—Chemical mechanism for producing **active nitrogen** from molecular **nitrogen** in the gas **phase**.
  The pair of reactions $O + N_2 \rightarrow NO + N$ and $N + O_2 \rightarrow NO + O$ were first proposed by Russian scientist Y. B. Zeldovich in 1947. The reactions are thought to account for much of the active nitrogen formed in hot exhaust gases from combustion sources and following the rapid heating of the air during **lightning discharges**.

Zeldovich, Y. B., P. Y. Sadonikov, and D. A. Frank-Kamenetskii, 1947: Oxidation of nitrogen in combustion (M. Shelef, Transl.). *Acad. Sci. USSR*, Inst. Chem. Phys., Moscow–Leningrad.

zenith—That point, on any given observer's **celestial sphere**, that lies directly above him; the point that is elevated 90° from all points on a given observer's **astronomical horizon**.
  Diametrically opposite the zenith is the observer's **nadir**.

zenith distance—The angular distance of any celestial object from a given observer's **zenith**, measured along the **great circle** of the **celestial sphere** from the zenith to the object; the complement of the **elevation angle**.

zenithal rains—In the **Tropics** or **subtropics**, rainy seasons that recur annually or semiannually at about the time that the sun is most nearly overhead (at **zenith**).
  Within a few degrees of the **equator**, two such periods of heavy rains may occur annually, the **equinoctial rains**. Farther from the equator, these blend into a single annual summer rainy **season**.

zephyr—Any soft, **gentle breeze**.
  See **Zephyros**.

Zephyros—The ancient Greek name for the west **wind**, which is generally light and beneficial.
  On the **Tower of the Winds** at Athens it is represented by a youth wearing only a light mantle, the skirt of which is filled with flowers.

zero curtain—The layer of ground, between the **active layer** and **permafrost**, where the **temperature** remains nearly constant at 0°C.

zero layer—1. A layer that is commonly used as a reference, such as the **boundary layer** that has a height above the earth's surface of zero.
  2. A layer within which some measured quantity is zero, such as the **level of nondivergence**.

zero-order closure—An approximation to **turbulent flow** that utilizes empirical diagnostic relationships for the mean variables such as **wind speed** and **temperature**.

Because these are diagnostic equations rather than prognostic, there are no forecast equations for **turbulence** statistics of any **statistical** order including the first-order mean values. An example is the logarithmic **wind profile** in the **surface layer**, a **similarity theory** that gives the wind speed as a function of height without using the **equations of motion**. *See* **closure assumptions, first-order closure, higher-order closure, nonlocal closure.**

zero-plane displacement—1. A height scale in **turbulent flow** over tall roughness elements associated with the average level of action of **momentum transfer** between the flow and the roughness elements.

In **neutral stability** the logarithmic **wind profile** becomes a straight line only if the vertical axis is shifted by the zero-plane displacement length. Several formulas exist to relate this height scale as a function of the roughness element geometry (for example, spacing and silhouette area). Tabulated values for various surface types are published in most micrometeorological texts, for example, Oke (1987). *See* **surface roughness.**

2. (Also called displacement distance, displacement thickness.) In a viscous **homogeneous fluid**, the depth of a **boundary layer** of shearing flow over a flat plate.

Oke, T. R., 1987: *Boundary Layer Climates*, 2d ed., 435 pp.

zero-pressure balloon—A **balloon** that is not sealed.

The base of the balloon may be open or it may have a duct high on the balloon that is attached at or near the base of the balloon. The opening is used to vent excess gas when the balloon reaches **ceiling** altitude. The large plastic balloons used to carry scientific payloads to altitudes above 40 km are zero-pressure balloons.

zigzag lightning—Ordinary **lightning** of a **cloud-to-ground discharge** that appears to have a single, but very irregular, **lightning channel**.

Compare **streak lightning, forked lightning.**

zobaa—In Egypt, a lofty **whirlwind** of sand resembling a pillar, moving with great **velocity**.

See **dust whirl.**

zodiac—(Also called **zodiacal band**.) The position of the sun during the course of a year as it appears to move through successive star groups or constellations; that is, the band of the **celestial sphere**, 16° in width, through which the **ecliptic** runs centrally.

At all times this band of the heavens contains the sun, the moon, and the principal planets except Venus and Pluto. Ancient astrologers divided the zodiac into twelve equal divisions, each 30° long, and named them for the chief constellations found in each division, the twelve constellations of the zodiac. These constellations were thought by the ancient Greeks (and others) to represent characters in their mythology and were identified with them.

zodiacal band—Same as **zodiac**; also refers to the faint band of **light** seen along the **ecliptic**, which is due to **sunlight** scattering off **interplanetary dust** particles.

zodiacal counterglow—Same as **gegenschein.**

zodiacal light—A faint cone of **light** in the night sky extending upward from the **horizon** in the direction of the **ecliptic (zodiac)**.

It is seen at tropical latitudes for a few hours after **sunset** or before **sunrise**. It is mainly explained by the forward **scattering** of **sunlight** by **particles** in interplanetary space along the ecliptic.

zombie turbulence—The turbulent perturbations of **temperature** remaining in a **stable boundary layer** after all turbulent motions have died out.

Also known as the footprints of **turbulence** in the ocean boundary layer.

zonal—In meteorology, latitudinal, that is, easterly or westerly; opposed to **meridional.**

zonal circulation—Same as **zonal flow.**

zonal flow—(Also called zonal circulation.) In meteorology, the flow of air along a latitude circle; more specifically, the latitudinal (east or west) component of existing flow.

Compare **meridional flow**; *see* **zonal index.**

zonal index—A measure of strength of the **middle-latitude westerlies**, usually expressed as the horizontal **pressure** difference between 35° and 55°N latitude, or as the corresponding **geostrophic wind**.

See **circulation index, high index, low index**; *compare* **polar-easterlies index, subtropical-easterlies index, temperate-westerlies index.**

zonal kinetic energy—The **kinetic energy** of the mean **zonal wind**, obtained by averaging the component of the **wind** along a fixed latitude circle.

zonal wavenumber—The **wavenumber** in the zonal direction, that is, along a line of constant latitude.
See **wavenumber** for ambiguities in the usage of this term.

zonal westerlies—Same as **westerlies**.

zonal wind—1. The **wind**, or wind component, along the local parallel of latitude, as distinguished from the **meridional wind**.
In a horizontal **coordinate system** fixed locally with the x axis directed eastward and the y axis directed northward, the zonal wind is positive if it blows from the west and negative if from the east.

2. Same as **westerlics**.

zonal wind-speed profile—A diagram in which the speed of the **zonal flow** is one coordinate and latitude is the other.

zonality of hydrological phenomena—Variations in hydrological characteristics mainly with **altitude** or latitude.

zonally averaged models—Climate models (typically statistical–dynamical or **energy balance** models) in which the longitudinal dimension has been eliminated through latitudinal averaging, and the **meridional** transport (of **energy**, for example) by large-scale and synoptic-scale **eddies** is not explicitly resolved but is instead determined parametrically.

zonda—(Or sondo, viento zonda.) A **hot wind** in Argentina.
The name seems to apply to two winds of quite different origin. 1) The most common use is for a dry **foehn** wind descending the eastern slopes of the Andes in the central Argentine in winter, probably **polar maritime air** warmed by descent from the crest, which is some 5500 m (18 000 ft) above **sea level**. It may exceed a **velocity** of 11 m s^{-1} (25 mph). Like the foehn, it sometimes begins with a **high foehn** (zonda de altura) that overrides a layer of cold air in the lee of the mountains, eventually sweeping it away and descending to ground level as zonda de superficie (or just "zonda"), but this initial cold **air mass** is not always present. The zondas carry a lot of **dust** in the **dry season**. 2) It also describes a hot, humid north **wind** in the **pampas**, in advance of a **depression** moving eastward, and preceding the **pampero**.

zone of aeration—That portion of the **lithosphere** in which the pore spaces in rock, soil material, and soil are partially or temporarily filled with air, that is, the region above the **water table** and its accompanying **capillary fringe**.

zone of maximum precipitation—The **elevation** band on a mountain or **orographic** barrier that receives the greatest **precipitation** for a seasonal or annual average.
Typically, precipitation increases with height upward from the base of mountains, but, if the peaks are high enough, eventually a level is reached where precipitation decreases with height (**precipitation inversion**). The decrease often occurs because the cooler air at higher altitudes holds less moisture, so less **water vapor** is available to condense and precipitate. The level where the increases with height become decreases marks the zone of maximum precipitation. According to Miller (1961), the **altitude** of the zone, often between 1 and 2 km, "varies slightly from place to place, is lower in tropic than temperate zones, in humid climates than in arid, in the cold **season** than the hot, in the **wet season** than in the dry." *See also* **orographic precipitation**.
    Miller, A. A., 1961: *Climatology*, Methuen, p. 39.

zone of saturation—(Also called phreatic zone.) That portion of the **lithosphere** in which the pore spaces are filled with water, that is, the region beneath the **water table**.

zone of silence—A region surrounding or at some distance from a **sound** source, such as an explosion, in which sound cannot be detected, usually by a human without special aids, even though the source may be in **visual range**.

zone time—A system of **local time** classification, differing from coordinated **universal time** (UTC) in steps of 1 hour per 15° of longitude.
The individual time zones are categorized by the letters A, B, C, etc. (omitting J) for areas

centered on 15°E, 30°E, 45°E, etc. and by the letters N, 0, P, etc. for areas centered on 15°W, 30°W, 45°W, etc., respectively. UTC is, in this system, designated as **Z time**.

zone wind—The representative **wind** in a **standard artillery zone**.

Zürich number—Same as **relative sunspot number**.